PROBABILITY, RANDOM VARIABLES, AND RANDOM PROCESSES

PROBABILITY, RANDOM VARIABLES, AND RANDOM PROCESSES

Theory and Signal Processing Applications

JOHN J. SHYNK

Department of Electrical and Computer Engineering
University of California, Santa Barbara, CA, USA

WILEY

A JOHN WILEY & SONS, INC., PUBLICATION

Published by John Wiley & Sons, Inc., Hoboken, New Jersey.
Published simultaneously in Canada.

Library of Congress Cataloging-in-Publication Data:

Shynk, John Joseph.
Probability, random variables, and random processes : theory and signal processing applications / John J. Shynk.
p. cm.
Includes index.
ISBN 978-0-470-24209-4 (hardback)
1. Probabilities–Textbooks. 2. Stochastic processes–Textbooks. 3. Engineering–Statistical methods–Textbooks.
I. Title.
QA274.S53 2012
519.2–dc23

2012008186

Printed in the United States of America

ISBN: 9780470242094

10 9 8 7 6 5 4 3 2 1

To Tokie
In memory of A. V. and E. F.

CONTENTS

PREFACE

The goal of this book is to provide a more rigorous mathematical framework for probability, random variables, and random processes than is generally available in most undergraduate textbooks on probability and statistics for engineers. The material is designed for first-year graduate students, though much of it is accessible to seniors who have a strong mathematical background, and if some of the more theoretical sections are excluded by the instructor. The book has several features:

- Numerous detailed figures and tables are given that summarize the various techniques and show example results. These include a combination of more than 600 illustrations and plots generated using MATLAB® that are designed to reinforce the material.
- Many examples are included to illuminate subtle points that are often overlooked or not fully explained in basic textbooks. Each chapter also contains homework problems, for which there is a solution manual available to the instructor.
- Several appendices provide background material on related topics in mathematics and signals and systems so that the book is relatively self-contained. A summary of several parametric univariate distributions is included in Appendix A.
- The third part of the book describes signal processing and communications applications that a student is likely to encounter in subsequent courses in engineering. This introductory material is based on several courses offered at the University of California, Santa Barbara, and serves as a preview of applications that utilize stochastic techniques.

After the introduction in Chapter 1, which includes an overview of the material as well as a review of linear systems and frequency-domain transforms, the book is divided into three parts, each containing four chapters:

- Part I: Probability Theory; Random Variables; Multiple Random Variables; and Expectation and Moments (Chapters 2–5).
- Part II: Random Processes; Stochastic Convergence, Calculus, and Decompositions; Systems, Noise, and Spectrum Estimation; and Sufficient Statistics and Parameter Estimation (Chapters 6–9).
- Part III: Communication Systems and Information Theory; Optimal Filtering; Adaptive Filtering; and Equalization, Beamforming, and Direction Finding (Chapters 10–13). These four chapters are located at www.wiley.com/go/randomprocesses.

Chapter 2 covers basic probability with an emphasis on discrete experiments. Sample spaces, events, and fields are introduced in order to provide a framework for the abstract probability space $\{\Omega, \mathcal{F}, P\}$, which is used later for random variables and random processes. Random variables are defined in Chapter 3, which includes a description of many well-known (and not so well-known) continuous and discrete parametric families of distributions. Multiple random variables are considered in Chapter 4 where several techniques for deriving the distributions of transformations of random variables are investigated. Some multivariate distributions are also included. Chapter 5 defines the expectation of a random variable, as well as expectations of functions of random variables, moments, and the characteristic function. Conditional expectation and several of its important properties are also discussed.

By extending the characterization of multiple random variables to be indexed by time, we introduce random processes in Chapter 6. Various properties of a random process are covered, such as independence and stationarity, and we describe different types of random processes, including independent sequences, Markov chains, and martingales. Specific well-known random processes such as the Poisson and Wiener processes are developed and investigated. Chapter 7 explores other characterizations of random processes, including stochastic continuity, derivatives, integrals, and differential equations. Stochastic convergence of random sequences is also described, as well as the laws of large numbers and the central limit theorem. The power spectral density of a random process is defined in Chapter 8, which is used to describe the behavior of signals when processed (filtered) by a system. We focus on linear time-invariant systems, though some nonlinear processing is mentioned. Spectral estimation using parametric and nonparametric techniques are also discussed. Chapter 9 introduces sufficient statistics and describes several important methods for estimating the parameters of a random variable. The techniques are based on various criteria, including the mean-square error, maximum likelihood, and least squares.

The final part of the book begins with an overview of digital communications in Chapter 10, which includes an introduction to information theory. Detectors based on the maximum a posteriori and maximum likelihood criteria are derived. Optimal filtering techniques are covered in Chapter 11, focusing on the mean-square-error criterion, resulting in causal and noncausal Wiener filters. Linear prediction using a lattice filter, and Kalman filtering based on a state-space signal model are also introduced. Adaptive filtering algorithms and structures are described in Chapter 12, beginning with a discussion of steepest descent and Newton's method. Stochastic methods for studying the convergence of adaptive algorithms and their steady-state properties are described. Finally, adaptive beamforming is introduced in Chapter 13 where multiple antennas are used to collect and separate cochannel signals. Adaptive equalization is also covered, which is used to compensate for signal distortion in a transmission channel. We describe ideal and training-based methods as well as "blind" algorithms that do not require training or a pilot signal. Direction-finding algorithms for estimating the angles of arrival of the signals impinging on the antennas are also introduced.

Seven appendices provide additional background material for the topics presented in this book. They include the following:

- Summaries of univariate distributions for 22 continuous and 11 discrete random variables.
- Continuity of a function and descriptions of several functions with specific symbols used throughout the book.
- Discrete- and continuous-time frequency-domain transforms, with tables of properties and some transform pairs.
- A review of Riemann integration, a brief description of Riemann–Stieltjes and Lebesgue integrals, and a summary of useful indefinite and definite integrals.
- Identities and infinite series, mainly for discrete random variables and random sequences.
- Derivations of inequalities and bounds for expectations, such as Markov's and Chebychev's inequalities and the Cramér–Rao lower bound.
- Several matrix properties including subspaces, decompositions, and differentiation with respect to vectors.

The reader might find it useful to refer to the material in these appendices during the course of instruction, and as a self-contained review of topics from previous courses on calculus, signals and systems, and linear algebra.

The book has been designed so that it could be used for an entire academic year with each quarter covering one of the three parts mentioned earlier:

- *Quarter system.* Fall: Chapters 1–5; Winter: Chapters 6–9; Spring: Chapters 10–13.

For a semester system, it could be divided such that the material on systems, estimation, and the applications are covered during the second semester:

- *Semester system.* Fall: Chapters 1–7; Spring: Chapters 8–13.

However, the book is also easily partitioned for one quarter, two quarters, or one semester by omitting some of the advanced material in several of the chapters. For a one-quarter course (10 weeks of instruction), the instructor should be able to cover much of Chapters 1–8 by omitting, for example, stochastic calculus and spectrum estimation. For a one-semester course (15 weeks of instruction), Chapter 9 on sufficient statistics and parameter estimation might be included, again by omitting some of the earlier material. The application chapters have also been provided for interested students as a preview of material usually covered later in other engineering courses. The instructor may find it appropriate to use some of those topics as examples of systems and the signals they process when presenting the material in earlier chapters on random processes, systems, and noise.

I would like to thank S. Chandrasekaran for reviewing material in the appendices, and J. D. Gibson for providing support from the College of Engineering. I am indebted to my students in ECE 235, ECE 240A, and ECE 245, whose questions have contributed to my appreciation of the subtler aspects of several topics covered in the book. They have provided valuable feedback that led to some of the discussions and illustrative examples. I would also like to thank my colleagues in the Department of Electrical and Computer Engineering, whose interactions and insights over the years have provided me with a deeper understanding of statistical signal processing and the wide range of applications. Finally, thanks to my publisher at Wiley, George Telecki, for encouraging this project, and to Kari Capone, Dan Timek, Stephanie Loh, and Shalini Sharma for their assistance during the final stages of production.

JJS
Santa Barbara, CA
July 2012

NOTATION

Since a wide range of material is covered in this book, we present a brief overview of the notation. In many books on signal processing, the same symbol is used to denote a random process and a realization of that process, which is a deterministic waveform. In this book, we use instead the notation typical of books on probability and random processes:

- An uppercase letter denotes random variable X, random process $X(t)$, or random sequence $X[k]$, where t is continuous time and k is discrete time.
- Lowercase letter x is an outcome of X, $x(t)$ is a realization (continuous waveform) of $X(t)$, and $x[k]$ is a realization (sequence of numbers) of $X[k]$.

Typically, these letters are from the end of the Latin alphabet. One exception to the above notation is that uppercase K, M, and N usually denote (nonrandom) integers, as in the following sums of random variables:

$$\sum_{m=1}^{M} X_m, \qquad \sum_{n=1}^{N} Y_n. \tag{1}$$

If $\{K, M, N\}$ turn out to be random variables in a particular problem such as in a random sum, it will be specifically mentioned.

- A bold uppercase letter denotes random vector $\mathbf{X}$, random vector process $\mathbf{X}(t)$, or random vector sequence $\mathbf{X}[k]$.
- Bold lowercase letter $\mathbf{x}$ is a vector outcome of $\mathbf{X}$, $\mathbf{x}(t)$ is a vector realization (vector of waveforms) of $\mathbf{X}(t)$, and $\mathbf{x}[k]$ is a vector realization (vector of sequences) of $\mathbf{X}[k]$.

All vectors in this book are *column vectors*. A row vector is obtained via transpose $\mathbf{x}^T$ or complex conjugate transpose $\mathbf{x}^H$. The superscript on $\mathbf{x}^*$ denotes only complex conjugation, and does not include transpose.

- A bold uppercase letter $\mathbf{A}$ is also used for a nonrandom matrix, and a bold lowercase letter $\mathbf{a}$ for a nonrandom vector. Typically, these letters are from the beginning of the Latin alphabet.

The reader should be able to determine from the context of a discussion if $\mathbf{X}$ is a random vector or a nonrandom matrix. Two important cases are the autocorrelation matrix $\mathbf{R_{XX}}$ and autocovariance matrix $\mathbf{C_{XX}}$ which are nonrandom quantities for random vector $\mathbf{X}$.

- Calligraphic $\mathcal{E}$ is used for expectation. For example, the autocorrelation matrix above is $\mathbf{R_{XX}} \triangleq \mathcal{E}[\mathbf{XX}^T]$.

Although E is used for expectation in many books on probability, we use $\mathcal{E}$ because it is necessary in some chapters that E represent an error random variable (E is also used to represent an event in the sample space Ω).

In order to be concise in many equations throughout the book, we have written expressions like $1/2\pi j$, for example, where it is understood that all terms of $2\pi j$ are in the denominator without having to use parentheses $1/(2\pi j)$. Another example is the Gaussian probability density function (pdf):

$$f_X(x) = \frac{1}{\sqrt{2\pi\sigma^2}} \exp(-(x-\mu)^2/2\sigma^2), \tag{2}$$

where it should be clear that $2\sigma^2$ is in the denominator of the exponent (although that would not be the order of operations in most computer programming languages).

In the glossary at the end of the book, we provide a summary of the notation mentioned here, as well as the following lists of symbols and abbreviations used throughout the book: (i) general symbols and numbers, (ii) Greek symbols (often used for random variable parameters), (iii) calligraphic symbols (for special quantities and expectation $\mathcal{E}$ mentioned above), (iv) mathematical symbols, and (v) abbreviations (acronyms).

1 Overview and Background

1.1 INTRODUCTION

A physical process, either natural or synthetic (human-made), can be viewed as a *signal* with time-varying features and properties that can be modeled as a *random* process. The term signal implies that some type of *information* is represented by the physical process, and random means that future outcomes of the process are *unpredictable* to some degree. Examples of natural random signals include: ultraviolet radiation impinging on a tree, a planet revolving around a star, and a tornado moving across an open field. Synthetic examples of random signals include: a microwave signal transmitted from a cell phone to a base station, an automobile traveling from Los Angeles to San Francisco, and a baseball thrown by a pitcher to a catcher. These examples could be modeled using a function $g(x, y, z, t)$ that describes a trajectory in the three spatial dimensions $\{x, y, z\}$ as it varies over continuous time t. Obviously, these "signals" have different physical mechanisms, different levels of predictability, and contain different amounts of information. Figure 1.1 shows possible trajectories (called *realizations*) of a tornado in the x-direction as a function of time. In this single spatial dimension, we see that even the trajectories of a complicated natural process resembles our notion of a signal.

It is important to note that an observed trajectory of the tornado is one possible realization of its motion in the x-direction, which might be defined relative to a particular line of longitude. That is, before observing the actual trajectory, many realizations (even an infinite number) could occur over some region. Alternatively, if we were able to "restart" the process and allow the tornado to proceed again (repeat the "experiment") and with different atmospheric conditions, then we would expect a different trajectory, which might be only slightly different from the first realization. We denote a *random process* by $X(t)$ (uppercase letter) and a realization of the process by $x(t)$ (lowercase letter). The outcome at a specific time t_o is also random: it is a *random variable* $X(t_o)$, and a particular *outcome* at that time is $x(t_o)$. A random process is a collection of random variables that are indexed by time, as depicted in Figure 1.2 for one realization $x(t)$ and for one outcome $x(t_o)$ at $t = t_o$.

Clearly, we cannot predict the trajectory of a tornado with certainty, even if we have numerous measurements about its velocity, the ground temperature, time of day, geographic location, terrain, and so on. This is discussed further for the following simple example: a single toss of a fair coin as depicted in Figure 1.3. Excluding the "very unlikely" event that the coin lands on its edge, this experiment has, of course, only two outcomes: heads (H) and tails (T). Since we are interested only in how the coin lands (H or T) at a particular time (and not its trajectory), a single toss of the coin is modeled as a random variable and not as a random process. Given measurements such as the velocity of the coin just before impact, the angle of approach ϕ with respect to the table top, the mass of the coin, atmospheric conditions (temperature, humidity, and so on), and other relevant physical parameters, it *might* be possible to predict the outcome just before the coin lands. It would appear that an infinity of such measurements with high accuracy would be needed to reliably predict the outcome. However, it is not possible to simultaneously know the precise velocity and position of an object; taking measurements of the position changes the velocity (though very slightly) and vice versa. We can never have sufficient physical measurements about most random events in order to predict an outcome without error (unless, of course, there is trivially only one outcome).

From the previous discussion, randomness can be viewed as a *lack of complete information* about a physical process, such that we cannot predict exactly an outcome or realization before it occurs. Thus, it would be useful

Probability, Random Variables, and Random Processes: Theory and Signal Processing Applications, First Edition. John J. Shynk.

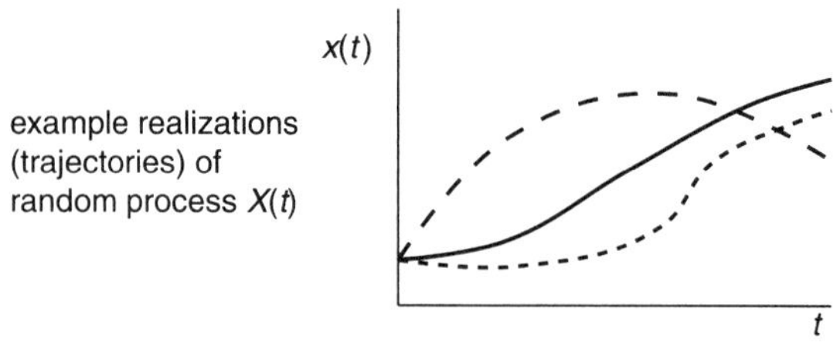

FIGURE 1.1 Three possible trajectories of a random process measured along the x-direction as a function of time t.

to develop a *model* of randomness that does not depend directly on physical attributes of a particular process, and can be applied generally to different types of signals and random experiments. For the coin-toss example using a fair coin, we expect from intuition that with an increasing number of tosses, an equal number of H's and T's will occur. This viewpoint involving *repeated experiments* is known as the *frequency interpretation* of a probability model: the frequency with which an event occurs for a large number of repeated experiments determines its probability. If the coin is tossed N times and N_H heads are observed, then we can expect for M tosses of the coin that approximately $(N_H/N) \cdot M$ heads will occur. The ratio N_H/N is the frequency of occurrence for H. Likewise, the frequency of occurrence for T is N_T/N (since $N_T + N_H = N$), and we would expect to see approximately $(N_T/N) \cdot M$ tails for M tosses of the coin.

The frequency approach to developing a probability model is consistent with our notion of the *likelihood* of observing the various outcomes in repeated experiments. It turns out that this interpretation is the intuitive basis for developing a probability model based on *three axioms*. Observe that $0 \leq N_H/N \leq 1$: the frequency of occurrence lies in the interval [0, 1]. Let us denote the probability of some event E by $P(E)$. One axiom of probability is $P(E) \geq 0$. This lower bound is appealing not only from the frequency interpretation, but also as a mathematical representation. Another axiom states that the probability of "something happening" in an experiment is 1. Combining these two axioms gives $0 \leq P(E) \leq 1$, which are the same bounds for the frequency interpretation of a random experiment.

The third axiom is more complicated: it is concerned with combinations of events, and is crucial to assigning probabilities to any event of interest in an experiment. For example, in the coin-toss experiment, consider the trivial probability $P(E = H \text{ or } T)$. Obviously, these are the only outcomes (again, excluding the possibility that the coin lands on its edge). Moreover, they are *mutually exclusive* because either H or T occurs: both cannot happen on the same toss. We find that the frequency of occurrence of either heads or tails is $(N_H + N_T)/N$; the frequencies simply add because the outcomes are "nonoverlapping" (they are disjoint). In addition, since no other outcomes are possible in this example, we must have $N_H + N_T = N$ (the total number of tosses) and $(N_H + N_T)/N = 1$ as expected. This last result also implies that the probabilities of mutually exclusive events add. This is the motivation for the third axiom. Based on these three axioms, a probability model can be developed for a random variable, which can be extended to a random *vector* and then to a random *process*.

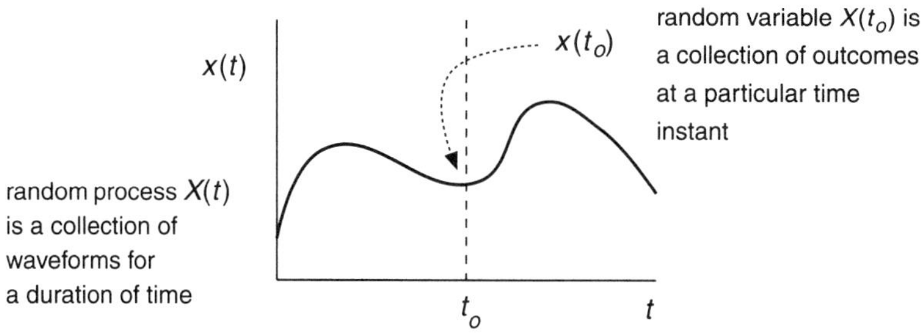

FIGURE 1.2 Realization $x(t)$ of random process $X(t)$ and outcome $x(t_o)$ of random variable $X(t_o)$ at time $t = t_o$.

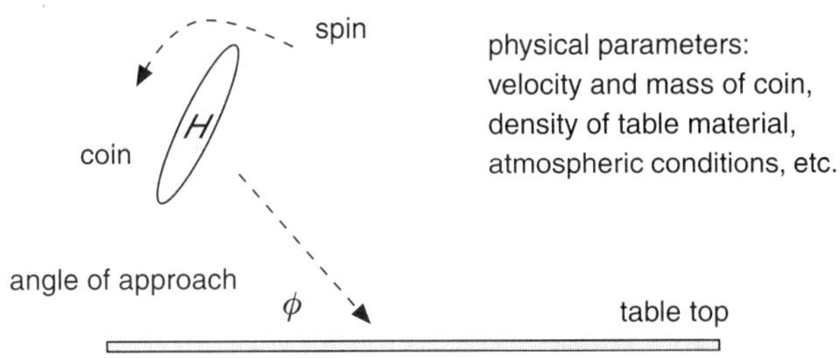

FIGURE 1.3 Physical approach to predicting whether a coin will land heads (H) or tails (T).

This book is divided into three main parts:

I. Probability, random variables, and expectation.
II. Random processes, systems, and parameter estimation.
III. Applications in signal processing and communications.

In the rest of this introductory chapter, we provide an overview of the material in the book. The notation used is summarized after the preface and in the glossary, and will be defined again in subsequent chapters. This introduction shows roughly how to proceed from a simple random experiment, like a coin toss, to more complicated random processes and *systems* in engineering that exploit or modify properties of random signals.

1.1.1 Signals, Signal Processing, and Communications

We are interested in random signals and various techniques for processing signal realizations to achieve some goal. The following broad definition is used for a signal.

Definition: Signal A *signal* is a physical quantity propagating through space and time that "contains" information about an event. It propagates from one physical location to another through some medium.

There are many signals in the modern world, such as those used in commercial radio and cable television. However, one can argue that nearly everything in the universe is a type of signal. For example, the light from an exploding star is certainly a type of signal that provides information about an extraordinary event. But a signal need not be restricted to electromagnetic radiation, which is perhaps the most familiar type of signal. A meteor passing through Earth's atmosphere could also be viewed as a signal, possibly providing information about the current location of Earth with respect to the Sun (such as the month of the year). An earthquake generates a signal that provides information about an event in the Earth's crust, and a volcanic eruption likewise imparts information caused by events far below the Earth's surface.

Figure 1.4 shows a block diagram of the three components of a *signal model*: (i) the *source* of the signal, (ii) the signal itself and the *medium* through which it propagates and, perhaps most importantly, (iii) one or more *sensors* that perceive (observe or measure) the signal. Natural signals are due to events in the physical world, such as the examples mentioned above. Other examples include the rise and fall of the tide, the movement of clouds, or a tree falling in a forest. The last example is often cited in the famous question "If a tree falls in the forest and no one is there to hear it, does it make a sound?" The falling tree and the disturbance it creates in the surrounding air can be viewed as a type of signal, and it may signify another physical event that just occurred (such as lightning striking the tree). The question above is obviously concerned with the sensor component of the signal model in Figure 1.4. Sound is the perceived vibrations in the ear drum, and thus if no one is present, then no sound will be heard. But it can be argued that the air has been disturbed by the falling tree, so a signal has been generated; moreover, additional signals (vibrations) are produced by the tree's impact with the ground.

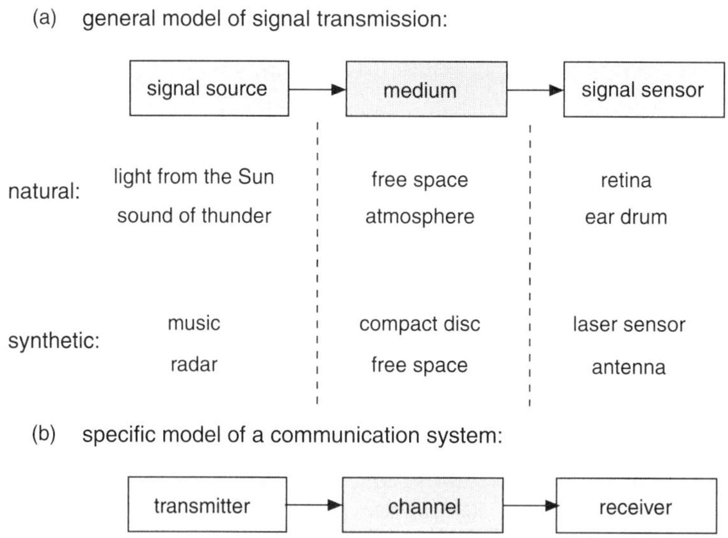

FIGURE 1.4 Signals and communications. (a) Components of a signal model. (b) Corresponding elements of a communication system.

Consider the trajectory of a jet aircraft traversing an ocean, which is moving through three-dimensional space and time. Its path contains "random" fluctuations due to turbulence that cause it to deviate from a "predictable" trajectory. If the trajectory is viewed from above (ignoring the z-axis of depth), its two-dimensional path in the x–y-plane can be traced as a function of time t. Furthermore, if the trajectory is viewed only along the x-axis, then we would see a realization of a one-dimensional signal (similar to the tornado example mentioned earlier). Thus, even an aircraft in flight can be viewed as a type of signal. If this "experiment" could be repeated, then the next realization would be different due to turbulence and course adjustments. Although random fluctuations of a signal itself might be small, we would also need to take into account the fact that perfect measurements of a signal are not possible because sensors have limited accuracy and are generally exposed to *noise*.

This leads us to the sensor component of the signal model which is of particular interest because we intend to (i) develop a probabilistic model for signals based on observations from one or more sensors, and (ii) derive techniques for modifying signals to achieve some desired goal. A probabilistic model is required because it is not possible to obtain enough measurements to predict exactly how a signal will evolve. For the coin-toss experiment, although it might be possible to predict its outcome (H or T) reasonably well from numerous physical measurements, this approach is not practical. It is better that we accept the fact that an experiment/signal is random, develop a probabilistic model for the outcomes/realizations, and then exploit that model for a particular application. Randomness at the sensor end of the signal model can be viewed as a measure of the uncertainty in a received signal because it is not possible to identify all the underlying physical mechanisms responsible for generating and modifying the signal. We cannot quantify exactly all disturbances that interfere with the signal as it propagates through some medium.

Figure 1.4 also shows a communication system model that is relevant to signals. It consists of (i) a *transmitter* for generating and sending the signal, (ii) a *channel* that includes the medium of propagation as well as any impairments encountered before reception, and (iii) a *receiver* consisting of one or more sensors for detecting the signal. The communication model explicitly assumes that some type of information is transmitted and received. Consider a synthetic signal such as that transmitted by a cell phone. Obviously, the speaker knows the information contained in the transmitted signal. Randomness can be viewed as a property of the receiver because the listener does not know what the speaker intends to say, and also due to signal distortions caused by channel impairments. The receiver has *incomplete* information. The same interpretation also holds for natural signals. For example, suppose that the intensity of solar radiation changes for a duration of time because of an explosion in the Sun's atmosphere. A satellite orbiting the Earth "views" the received signal as random (due to incomplete information), from which it might be possible with signal processing to derive information represented by the radiation (a solar event has occurred).

As shown in Chapter 10 where *information theory* is introduced, the amount of information contained in a signal is related to its degree of randomness: signals with "greater randomness" contain more information. For example, if a small planet suddenly has extreme orbital fluctuations due to an impact with a comet, that randomness provides more information than when the orbit is close to being perfectly elliptical with small variation. The same can be said for synthetic signals as depicted in Figure 1.5 (which are plotted versus the number of samples k). A sinusoidal signal with fixed amplitude a, frequency f_o, and phase ϕ is at one extreme: it is *deterministic* and thus perfectly predictable. As such, it contains essentially no useful information. A passband binary phase-shift keying (BPSK) signal is obtained by multiplying the sinusoidal *carrier* in the figure with a random sequence of ± 1s, called baseband binary pulse amplitude modulation (PAM). The BPSK waveform is also shown in Figure 1.5, which we see appears to be "more random" than the purely sinusoidal waveform.

The speech signal in Figure 1.6 exhibits even more randomness and thus contains greater information than the previous synthetic signals. The figure also shows a realization of sound caused by wind, which appears to have less structure and seems to be more random than the speech waveform. At the other extreme is "white noise" which is not predictable (and is defined precisely later); this type of signal is the "most random." We also include an example of an image that is a single *snapshot* from a video signal, and is a two-dimensional random process: it has two spatial coordinates $\{x, y\}$ and evolves in time t. Because the image in Figure 1.7 has been collected at a particular time instant of the video signal, it is modeled as a *random field*. A noisy version of the image is also shown, visually illustrating the type of effect that channel disturbances can introduce during transmission. Signal-processing techniques can be used to improve the quality of a signal by removing the additive noise to some extent.

There is a difference between information, which is related to randomness as described above, and the *meaning* of a signal. Meaning is more complex because it depends on how the signal is *interpreted* at the receiver end. For the speech example, the meaning of the signal depends on the language used in transmission, the primary language of the listener at the receiver, and many other factors such as the relationship of the speaker and listener, past experiences of the listener, the tone of voice of the speaker and so on. In this book, we are not concerned with the meaning of a signal, though we are interested in its information content as measured by its randomness. We are also interested in various types of processing of a received signal to extract and enhance this information.

Figure 1.8 illustrates the two basic types of signal processing that we will consider, though numerous variations and applications are discussed. The first is *estimation* of the parameters of some underlying model of the signal. The received signal is manipulated to derive estimators of the parameters. Perhaps the most well-known estimator is the sample mean:

$$\bar{X} = \frac{1}{N}\sum_{k=1}^{N} X[k], \tag{1.1}$$

where N samples of $X(t)$ (denoted by random sequence $X[k]$) are averaged to estimate the true mean μ_X of the signal. The second type of processing is *filtering* where the goal is to transform the signal, usually such that it has "better" properties. This is achieved by convolving the signal with filter coefficients $\{h[k]\}$:

$$Y[k] = \sum_{n=0}^{\infty} h[n]X[k-n]. \tag{1.2}$$

An example of signal processing is sound design where the signal can be manipulated so that the person speaking appears to be in a large auditorium instead of the studio where the sound was actually recorded. A combination of the two types of processing is also possible: we might want to transform the signal before parameter estimation, perhaps by filtering out some of the background noise.

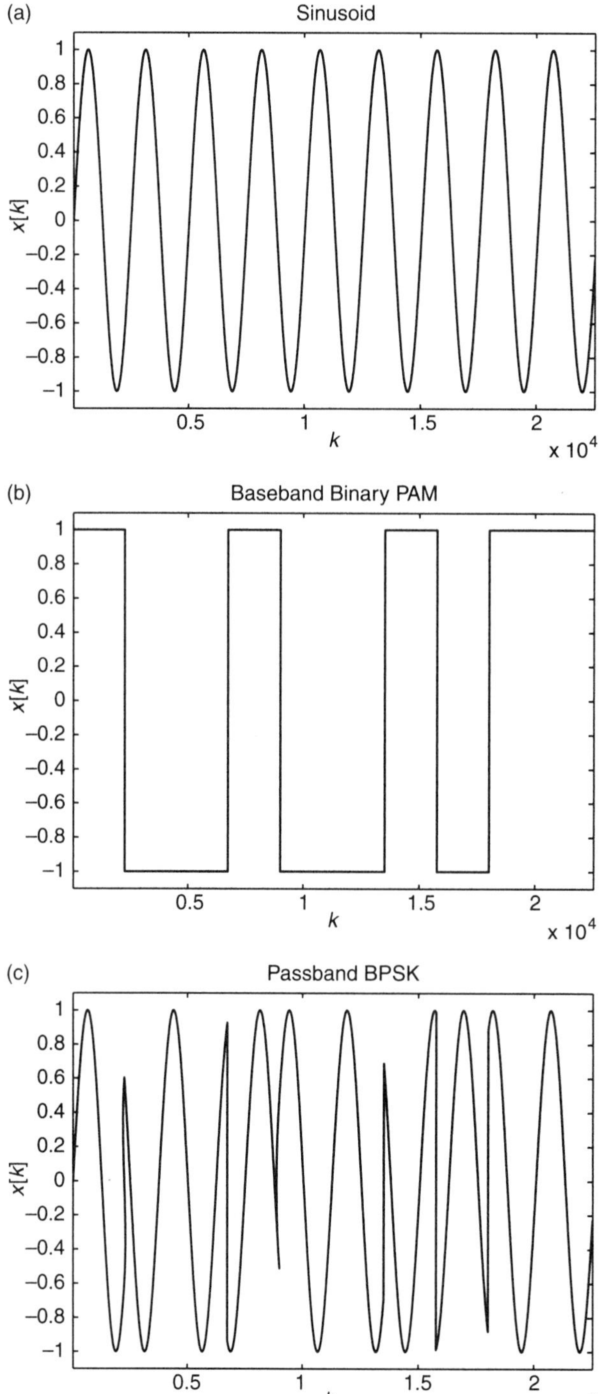

FIGURE 1.5 Examples of synthetic signals. (a) Sinusoid with fixed amplitude, phase, and frequency (nonrandom). (b) Baseband binary pulse amplitude modulation (PAM) (random). (c) Passband BPSK, obtained as the product of the signals in (a) and (b) (random).

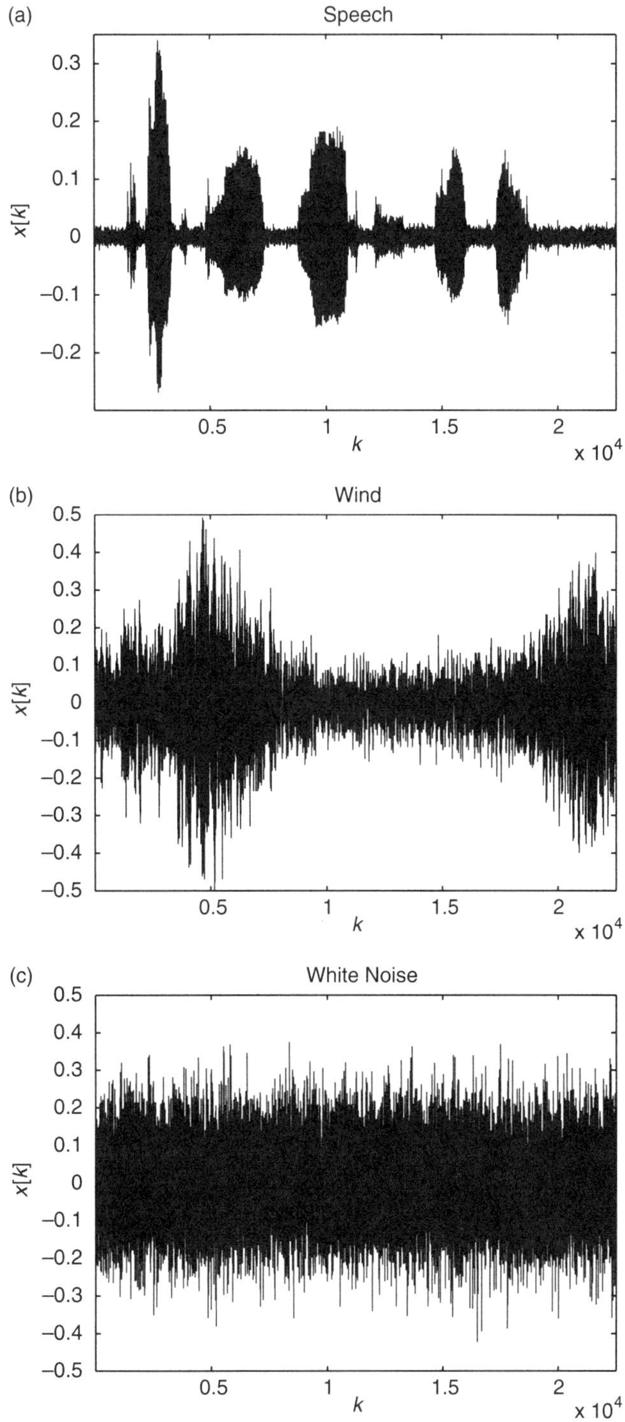

FIGURE 1.6 Examples of natural random signals. (a) Speech (Hu and Loizou, 2007). (b) Wind. (c) White Gaussian noise (though created using a pseudorandom noise generator in MATLAB).

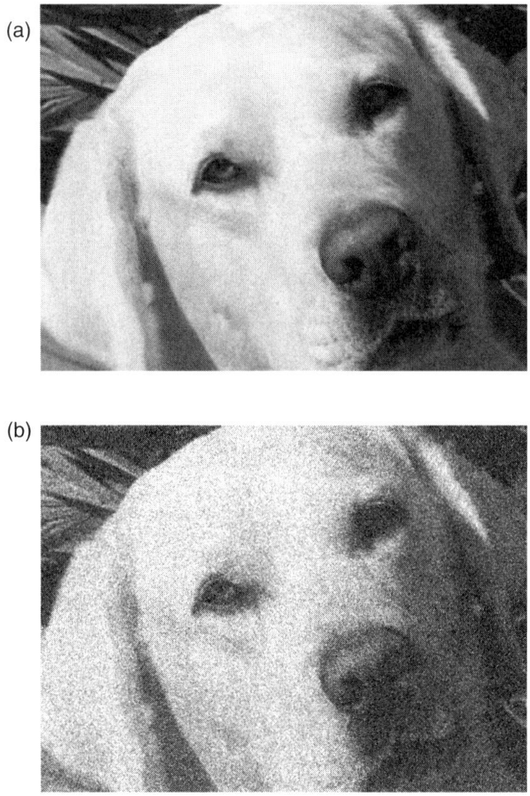

FIGURE 1.7 Random field. (a) Image from a video signal (single snapshot). (b) Image corrupted spatially with zero-mean Gaussian noise added to each pixel (picture element).

A random process can be viewed as a time-indexed collection of random variables: at each instant of time there is an underlying *distribution* of possible outcomes. Later, we give a precise description for a probability model and show how it is extended to random variables. This allows us to use various techniques from calculus and discrete mathematics to *characterize* random phenomena. These techniques are extended to random processes by including time variations. When time is included in the problem, we are usually interested in how the signal is *correlated* from one time instant to the next. Correlation is a measure of the predictability of future outcomes of a random signal, and is perhaps the most important property of a random process, which can be exploited using signal-processing techniques.

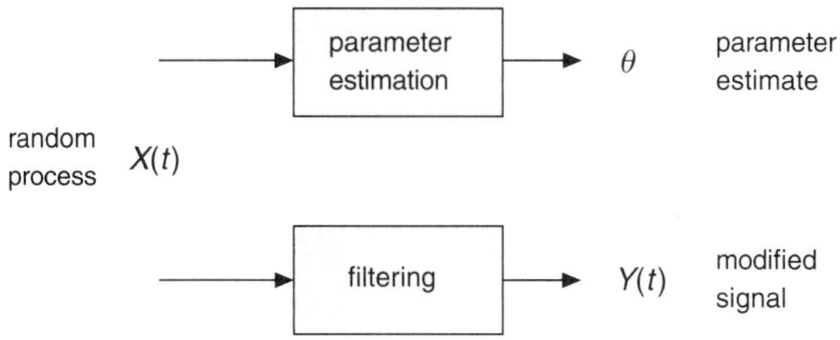

FIGURE 1.8 Two basic types of signal processing.

Finally, we discuss the spatial dimensions of random signals. As mentioned earlier, any signal can be viewed as a three-dimensional quantity (with $\{x, y, z\}$ spatial coordinates) that varies with time (continuous $t \in \mathcal{R} = (-\infty, \infty)$ or discrete $k \in \mathcal{Z} = \{\ldots, -1, 0, 1, \ldots\}$). However, in many cases it may be necessary to consider the signal only in one or two spatial dimensions. For the previous example of a planet orbiting a star, it may be convenient to view the process only on its elliptical plane, which is similar to watching a (two-dimensional) video signal. A single snapshot of the process corresponds to a single frame of the signal, which is basically equivalent to viewing a photograph. A hologram is a three-dimensional synthetic signal; viewed from any one angle (planar view) it is a two-dimensional video, and collapsing that down to a single dimension, we would see only light intensity changing with time. The vast majority of synthetic signals are one-dimensional such as those used in telecommunications. They may be propagating through three-dimensional space, but they can be represented by realizations similar to the example waveforms shown in Figure 1.5. In this book, although we focus on one-dimensional signals and present mostly synthetic examples, the techniques can be extended to any type of signal.

1.1.2 Probability, Random Variables, and Random Vectors

Next, we summarize the approach that will be taken to develop a consistent *probability space*. The reader might find it useful to return to this section when studying the early chapters of the book in order to review the overall framework:

- *Outcomes of an "experiment" are defined*. Outcomes are the observable (measurable) results or objects of a random experiment. There are basically three types depending on the number of outcomes: (i) finite, (ii) countably infinite, and (iii) uncountable. The first two cases are discrete and the third case is continuous. All outcomes of an experiment form the *sample space* Ω.
- *Events of the experiment are defined*. Event E corresponds to a feature or property that multiple outcomes share. An event coinciding with a single outcome is called an elementary event. Events are *subsets* of the sample space Ω.
- *An algebra for events called a sigma field is defined*. Since set operations on events and outcomes arise in many problems, it will be necessary to define a σ-*field*, which places a restriction on the allowable events that are assigned probabilities. The sample space and the σ-field $\mathcal{F}$ together comprise the *event space* $\{\Omega, \mathcal{F}\}$.
- *Probabilities are assigned to events according to three axioms*. These axioms allow for any probability of an event in the event space to be calculated in a consistent manner. The sample space, σ-field, and probability measure P together comprise the *probability space* $\{\Omega, \mathcal{F}, P\}$.

This framework is pictorially shown in Figure 1.9. A rectangle is used to denote the sample space Ω, and the events $\{E_n\}$ are depicted by circles. This is the so-called *Venn diagram*. Events that have some common outcomes $\{\omega_n\}$ are shown to be overlapping. A σ-field $\mathcal{F}$ is defined so that all operations of events in $\mathcal{F}$ yield an event that is also in $\mathcal{F}$. Event operations include $\cup$ (or), $\cap$ (and), ^{c} (complement), $-$ (difference), and $\oplus$ (exclusive or). A σ-field and the three axioms allow us to define a consistent probability measure P with values in the closed interval $[0, 1]$.

The previous description is suitable for relatively small sample spaces. For more complicated situations, which basically comprise most problems of interest, it is necessary that the probability space be mapped to a *random variable* defined on the real line $\mathcal{R}$. This allows us to perform complicated operations on the random variable, such as transformations from one random variable to another. We can also compute various quantities that characterize a random experiment such as moments (mean, variance, and so on). This mapping is summarized below:

- *Outcomes in the sample space Ω are mapped to numbers on the real line $\mathcal{R}$, generating random variable* X. Since engineering problems are formulated using variables and numbers, it will be necessary to map outcomes in the original sample space Ω to $\mathcal{R}$ (or integers $\mathcal{Z}$ or complex numbers $\mathcal{C}$). The original

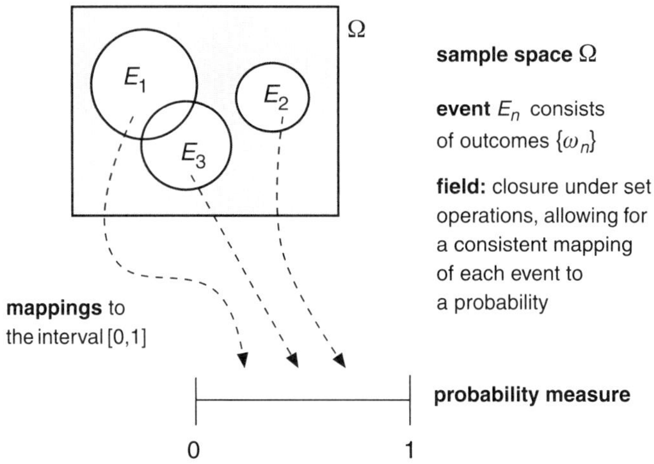

FIGURE 1.9 Pictorial representation of sample space Ω, events $\{E_n\}$ in Ω, and mappings to probabilities in the interval $[0, 1]$. A σ-field $\mathcal{F}$ is specified for Ω to describe events for which probabilities are assigned.

probability space carries over to the real line and, provided the mapping is *measurable*, the resulting random variable also has a consistent probability space.

- *Random variables can be defined directly without resorting to an underlying abstract probability space.* The probability space for random variable X is $\{\mathcal{R}, \mathcal{B}(\mathcal{R}), P_X\}$, where $\mathcal{B}(\mathcal{R})$ is the *Borel* σ-field defined by all open *intervals* on $\mathcal{R}$. With this probability framework, we can utilize standard families of random variables, such as Gaussian, binomial and so on to model various physical phenomena.

The mapping of events to intervals of random variable X is pictorially shown in Figure 1.10. The goal is to describe events in a problem of interest using the real line $\mathcal{R}$. For discrete random variables, a *probability mass function* (pmf) describes the probability of each possible outcome, which are usually denoted by some subset of the integers $\mathcal{Z}$. The probability of an event is obtained by *summing* over the integer values corresponding to the event:

$$P(a \leq X \leq b) = \sum_{x=a}^{b} p_X[x], \tag{1.3}$$

where $p_X[x] = P(X = x)$ is the pmf, and is represented by the solid circles in Figure 1.10.

For continuous random variables, a *probability density function* (pdf) describes the probability density on $\mathcal{R}$; probabilities are obtained by *integrating* the pdf over an interval of interest:

$$P(a \leq X \leq b) = \int_a^b f_X(x)\mathrm{d}x, \tag{1.4}$$

where $f_X(x)$ is the pdf and has a continuous form as shown in Figure 1.10. It is also possible to have a *mixed* random variable with discrete and continuous components. As mentioned above in the last item, once we are familiar with random variables, there is usually no need to consider the underlying abstract sample space Ω; we can operate directly on the pmf (discrete) or the pdf (continuous).

The next important generalization is the extension of mappings from Ω to a *vector* of random variables:

- *The approach used to define the probability space for a random variable is extended to define two or more related random variables comprising a random vector*. The extension to random vectors is important in most applications because the *cross-correlation* of random variables can be exploited.

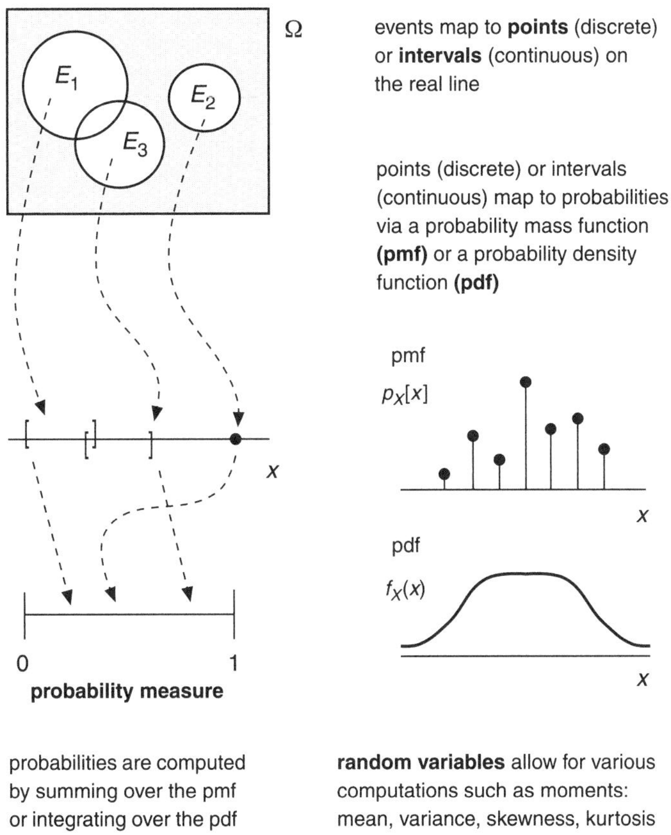

FIGURE 1.10 Pictorial representation of the mapping of events in the sample space Ω to random variable X. The probability measure of a random variable is defined by a probability mass function (pmf) (for discrete outcomes) or a probability density function (pdf) (for continuous outcomes).

This is pictorially represented in Figure 1.11 for the case of two *joint* random variables X and Y. This is a significant generalization of the probability model because it allows us to examine multiple features of outcomes (e.g., in the single toss of a die, whether the outcome is greater than three *and* if it is even). From joint probability measures, given by the pmf $p_{X,Y}[x, y]$ and pdf $f_{X,Y}(x, y)$, we can predict with varying degrees of accuracy the outcome of one random variable based on the outcome of another random variable.

Finally, we mention that when some information is known about an experiment, it can change the probability of events. With prior knowledge of some event E, the corresponding probability of another event F is called a *conditional probability* because it is conditioned on this information: the probability of F given E changes to reflect this knowledge. Conditional probabilities arise frequently in practice, either as a problem of interest or as a mechanism for computing more complicated probabilities. In the extreme case when "everything" is known about an experiment (as mentioned previously about having an infinity of physical measurements for a coin toss), we should be able to predict with probability one which outcome will occur. Thus, prior information is important in probability theory; its influence on the probabilities of events can be quantified by an extension of the three axioms of probability to include conditioning on that information.

1.1.3 Random Sequences and Random Processes

An example of a random *sequence* is several consecutive tosses of a fair coin. Each possible sequence of outcomes is called a *realization*. For example, if a fair coin is tossed 10 times, $\{H, H, H, H, H, H, H, H, H, H\}$

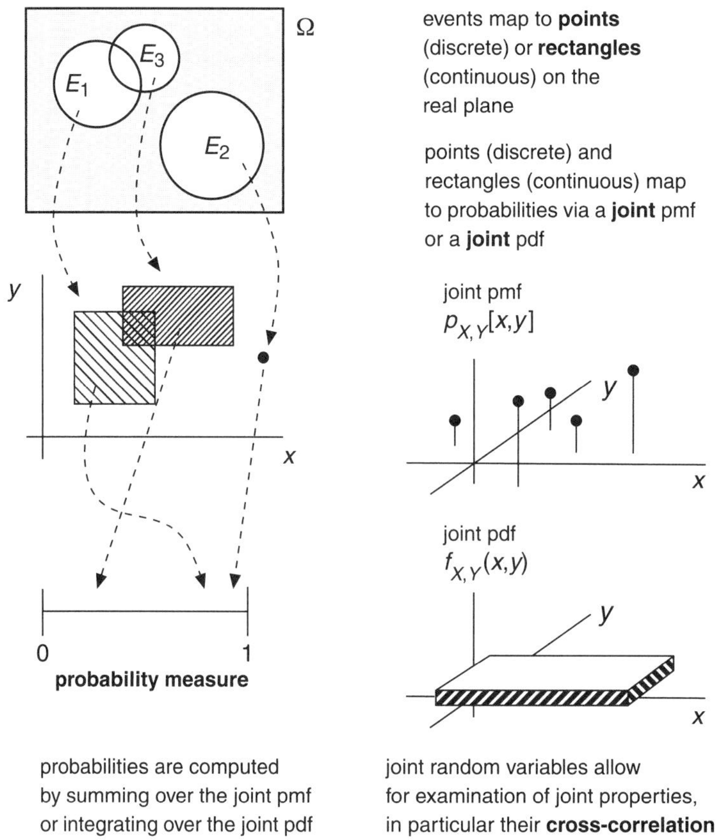

FIGURE 1.11 Pictorial representation of a mapping of events in the sample space Ω to random variables X and Y. Joint moments can be examined, and we are usually interested in the *cross-correlation* of X and Y.

is one possible realization; in fact, there are 2^{10} realizations for this experiment. We summarize the approach that will be taken to describe random sequences and random processes:

- *A collection of random variables indexed by discrete time k is defined.* A random *sequence* $X[k]$ is specified for integers $k \in \mathcal{Z}$ representing discrete time. There may be two kinds of outcomes at a time instant: either *discrete* where there is a finite or countable number of values, or *continuous* where a continuum of values (uncountable) are possible.
- *The probability measure of the collection of random variables extends to the random sequence.* Since a σ-field has been defined for joint random variables, the corresponding probability measure is well defined for any N-tuple of random variables in the sequence, even as $N \to \infty$.
- *A collection of random variables indexed by continuous time is defined.* A random *process* $X(t)$ is specified on the real line $\mathcal{R}$ representing continuous time t. As is the case for a random sequence, the outcome at any time instant might be discrete or continuous.
- *The probability measure of a collection of random variables extends to the random process.* Unlike a random sequence, the collection of random variables comprising a random process is uncountable. The extension of the probability measure to a random process requires *consistency*, which is defined in Chapter 6.

TABLE 1.1 Types of Random Processes with Examples

Random Sequence $X[k]$	Random Process $X(t)$
Discrete time, discrete outcomes (*Bernoulli sequence*)	Continuous time, discrete outcomes (*Poisson process*)
Discrete time, continuous outcomes (*Gaussian sequence*)	Continuous time, continuous outcomes (*Wiener process*)

There are four basic types of random sequences and random processes, as summarized in Table 1.1. It is also possible to have "mixed" versions where some outcomes are continuous and others are discrete, similar to a mixed random variable. In order not to be cumbersome, we use random "process" to denote either a continuous-time process or a discrete-time sequence when the discussion applies to both. We specifically refer to a random sequence if the discussion focuses only on discrete time. For a random sequence, $k \in \mathcal{Z}$ is used to index time, whereas for a random process, $t \in \mathcal{R}$ is used. A pictorial representation of the mapping from a continuous random variable to a discrete time-indexed collection of random variables (a random sequence) is illustrated in Figure 1.12. The other types of random processes are viewed in a similar manner for repeated experiments.

Example 1.1. Consider the sinusoidal signal $x(t) = \sin(\omega t)$ for $t \in \mathcal{R}$. One of the simplest types of randomization that can be introduced is a random phase $\Phi \in [0, 2\pi]$ such that we have the following random process:

$$X(t) = \sin(\omega t + \Phi). \tag{1.5}$$

The shape of the waveform is unchanged, but different outcomes of Φ modify the zero crossings. Since Φ is a continuous quantity with an infinity of outcomes (uncountable), each possible realization $x(t)$ could be plotted, yielding a family of curves called an *ensemble*, some of which are shown in Figure 1.13(a). Alternatively, the sinusoid might have a random amplitude $A \in [0, 1]$ such that

$$X(t) = A\sin(\omega t) \tag{1.6}$$

or have a random phase and a random amplitude:

$$X(t) = A\sin(\omega t + \Phi). \tag{1.7}$$

Example realizations for the last case are illustrated in Figure 1.13(b).

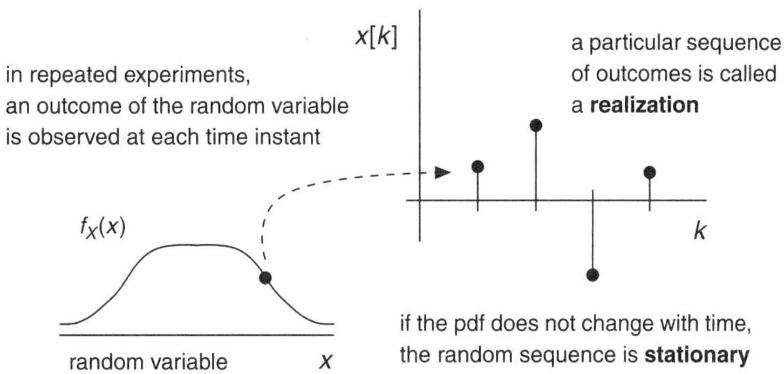

FIGURE 1.12 Pictorial representation of the connection between a random variable and a random sequence for repeated experiments (discrete time, continuous outcomes).

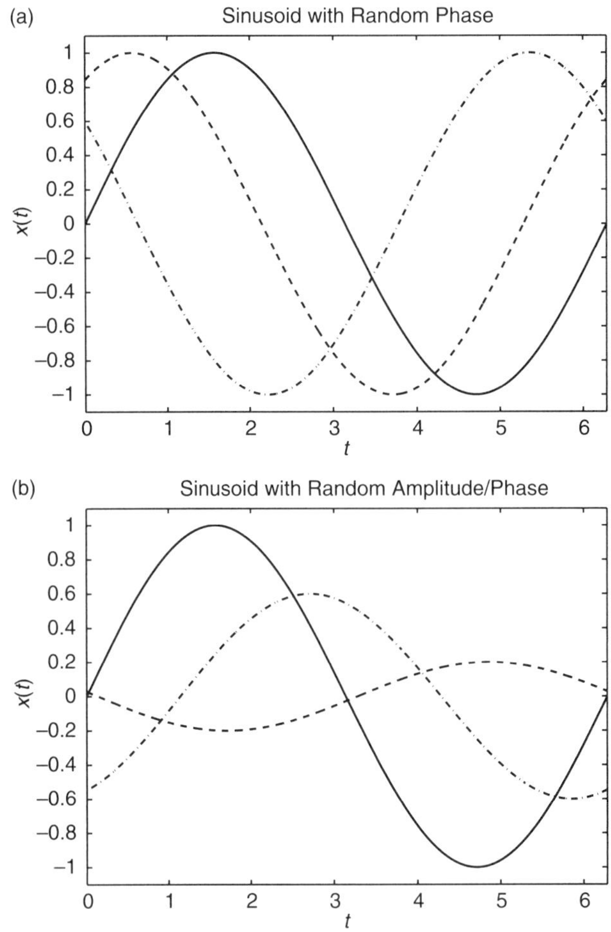

FIGURE 1.13 Sinusoidal signal realizations of $X(t) = A\sin(\omega t + \Phi)$ with $\omega = 1$. (a) Random phase Φ (with $A = 1$). (b) Random phase Φ and random amplitude A.

The quantities A and Φ are *random variables*: the probability that they take on some range of values is characterized by their pdfs. An example of the pdf $f_\Phi(\phi)$ for $\Phi \in [0, 2\pi]$ is shown in Figure 1.14; it is uniformly distributed (flat), meaning that the probability $P(\Phi \in [a, a+\epsilon])$ with $\epsilon > 0$ is the same for any a (provided that $[a, a+\epsilon]$ is a subset of $[0, 2\pi]$). Each value of random variable Φ yields a different shifted version of the waveform in Figure 1.13(a), where the outcome ϕ does not change for the duration of the sine waveform. A time-varying phase leads to a more complicated type of random signal because $\Phi[kT]$ itself is a random sequence. The argument of $\Phi[kT]$ means that the phase could change for t equal to integer multiples of some period T so that the zero-crossings of $X(t)$ are no longer regularly spaced:

$$X(t) = \sin(\omega t + \Phi[kT]). \tag{1.8}$$

The example in Figure 1.14 shows that even though the amplitude of the sinusoid is still fixed (nonrandom) at some value in [0, 1], the realization is now quite variable because of the time-varying phase. The signal looks somewhat "noisy," even though we know structurally that this behavior is due only to a time-varying phase.

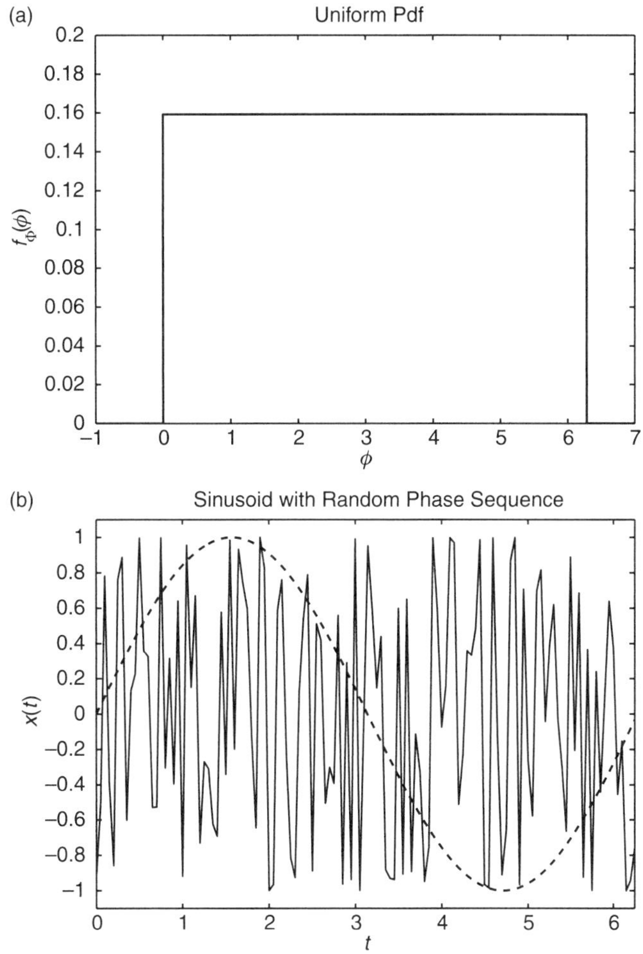

FIGURE 1.14 Sinusoid $X(t) = A\sin(\omega t + \Phi[kT])$ with random phase sequence $\Phi[kT]$ ($A = 1$, $\omega = 1$, and $T = 0.05$). (a) Pdf for the random phase: uniform on $[0, 2\pi]$. (b) Realization of the sinusoid where the phase changes independently at discrete time instants $k \in \mathcal{Z}^+ = \{0, 1, \ldots\}$. A sinusoid with zero phase shift is plotted for comparison (dashed waveform).

From the last sine wave example, we see that even with a simple type of randomness, a realization $x(t)$ can appear to be quite unpredictable. A goal in this book is to describe techniques that allow us to *characterize* the behavior of a random signal, and to show how it is processed by a *system*. Initially, one might consider examining the behavior for each realization of the ensemble, but it would quickly become clear that such an approach is not feasible (or even useful). Of course, it is true in an actual application that a specific realization of the ensemble is processed by the system, and thus we can use standard linear system techniques to determine the output for that realization. But the signal is random and we do not know beforehand which realization will occur. This is generally *not* a problem in many applications and is often desirable. For example, in a digital communication system, the receiver does not know which symbol is transmitted at any given moment; if it did, then there would be no point in transmitting that information. The receiver in such a system is designed to operate on the received signal in a manner that exploits known properties of the transmitted signal, and uses probabilistic techniques to make "good" decisions about which symbols were sent.

As mentioned earlier, random sequence $X[k]$ can be viewed as a time-indexed collection of random variables. This description applies to discrete-time instants, as in consecutive coin tosses, or to continuous time as in television and radio signals. Example realizations of a correlated Gaussian random sequence are shown in Figure 1.15(a). At any particular time instant such as $k = 500$, values of the various realizations are outcomes of random variable $X[500]$. This point is emphasized in Figure 1.15(b) which shows a plot of the cumulative distribution function (cdf) for $X[500]$; it describes the probability that the random variable lies in an open interval: $P(X[500] \leq x) \triangleq F_X(x)$. We assume that such a distribution exists for the random variable at any time instant of the random process.

A random process is characterized using probabilistic models such that even though we cannot know which realization will occur, there is an understanding of the *ensemble*. This turns out to be useful for natural random processes such as the daily temperature, as well as for synthetic random processes such as a radio signal. For example, in many applications, the average value of each realization is usually the same (often it is zero). Or, we might know that at each time instant, the value of the realization is "drawn" from a Gaussian distribution. More importantly, it is desirable to be able to specify quantitatively via *correlation* how the process at time t_1 is related to the process at a later time t_2. In many problems such as consecutive coin tosses, there is no connection between tosses; they are independent, and it is not possible to predict future outcomes from past results. Noise in an amplifier is another example of an unpredictable process, even though we may have some idea about the possible range of values. There are many processes, on the other hand, that have a high degree of predictability. For example, the temperature inside a room is relatively predictable for a short time duration; the temperature from minute to minute is usually highly correlated. Figure 1.15(c) shows an example autocorrelation function $R_{XX}[m]$ for a Gaussian random process which has a two-sided exponential form. The autocorrelation of the process at time instants 500 and 1000, for example, is the value at time lag $m = |k_2 - k_1| = 500$. As m is increased, the correlation decreases, which is typical of most random signals (though not necessarily exponentially decreasing).

1.1.4 Delta Functions

We conclude this introduction with a discussion of delta functions that will be useful throughout the book. As is often encountered in engineering courses on signals and systems, there is usually some confusion about how to represent discrete-time signals versus continuous-time signals. The same issue arises here because random processes can be defined for discrete time (sequences) as well as continuous time. There are two methods for representing a discrete-time sequence. The first method, often used in basic signal processing courses, is based on the Kronecker delta function:

$$\delta[k] \triangleq \begin{cases} 1, & k = 0 \\ 0, & \text{else}. \end{cases} \tag{1.9}$$

It is represented by a vertical line and a solid circle as illustrated in Figure 1.16(a) for a simple discrete-time realization. The process exists only for integers $k \in \mathcal{Z}$; time is not defined in between these values, and an integral cannot be computed for this realization using Riemann integration. This notation is generally adequate when there is no need to relate this process to continuous time. However, it is often convenient to have notation that allows us to describe together discrete-time sequences and continuous-time processes. Such cases can be handled using the Dirac delta function, which is illustrated for an example realization in Figure 1.16(b). The Dirac delta function is represented by an arrow, and time in between these symbols is continuous and well defined. (Strictly speaking the Dirac delta function is not a function; it is a *generalized function* as mentioned in Appendix B.) We can manipulate such waveforms using, for example, convolution and integration because

$$\int_{-\infty}^{\infty} \delta(t)\mathrm{d}t = 1. \tag{1.10}$$

The Dirac delta function is also used to represent the outcomes of a discrete random variable. For example, consider the simple communication model shown in Figure 1.17 represented by $Y = X + N$, where Y is the received random variable, X is the transmitted random variable, and N is an additive noise random variable. Suppose that X is binary, taking on the values ± 1 with equal probability (similar to the toss of a fair coin); it

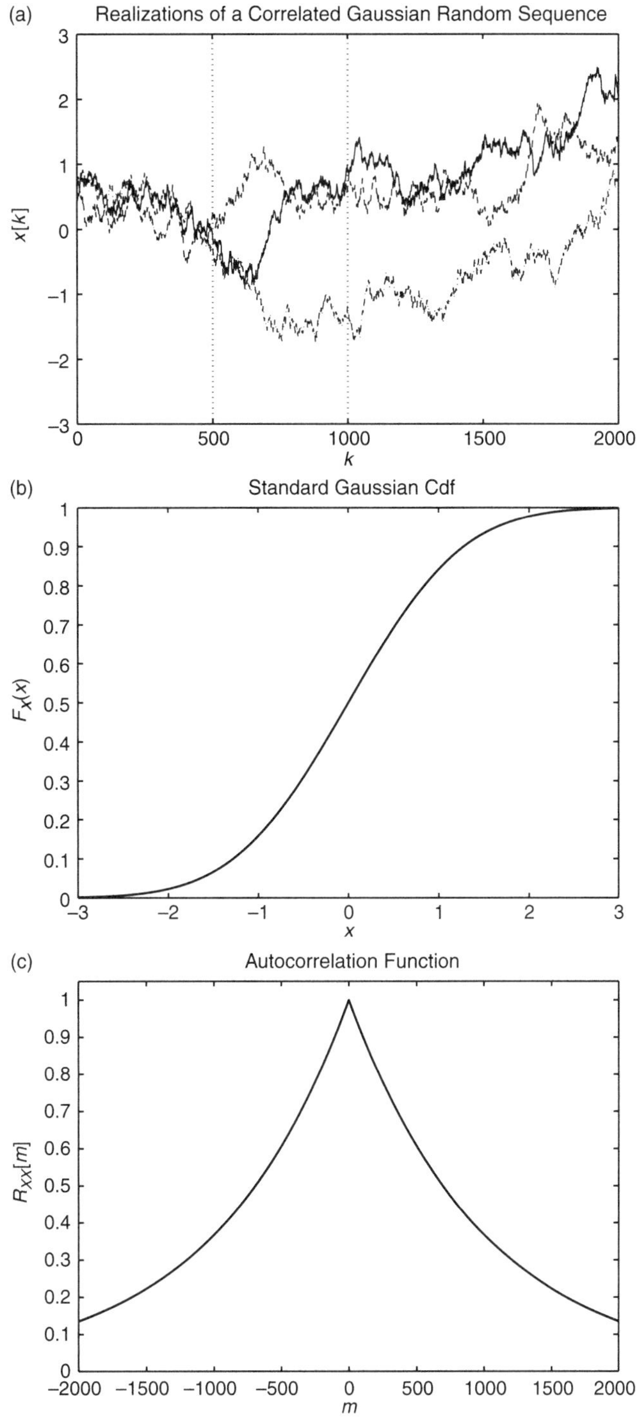

FIGURE 1.15 Correlated Gaussian sequence. (a) Example realizations. (b) Distribution function $F_X(x)$ of random variable $X(500)$ at time instant $k = 500$, which is standard Gaussian with zero mean and unit variance. (c) Autocorrelation function of the process as a function of the difference between two time instants: $m = |k_2 - k_1|$.

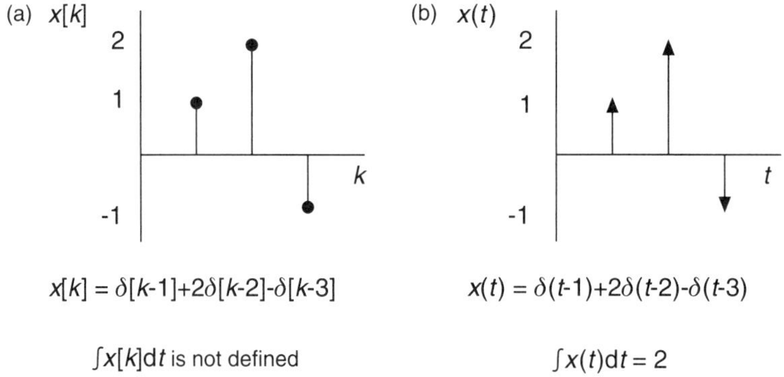

FIGURE 1.16 Discrete-time realization using (a) the Kronecker delta function $\delta[k]$ and (b) the Dirac delta function $\delta(t)$.

is known as the symmetric Bernoulli random variable. The pmf can be represented using the Kronecker delta function as follows:

$$p_X[x] = (1/2)(\delta[x-1] + \delta[x+1]). \tag{1.11}$$

Alternatively, it can be represented by a pdf based on the Dirac delta function as follows:

$$f_X(x) = (1/2)[\delta(x-1) + \delta(x+1)], \tag{1.12}$$

which is depicted in Figure 1.18. The advantage of using a pdf for discrete random variables is that it can be manipulated using methods from calculus because it is defined for the entire real line, not just integer values. As a result, in subsequent chapters we generally represent the distributions of discrete random variables using pdfs. This will also facilitate operations involving mixed random variables and mixed random processes as in Figure 1.17.

Example 1.2. Assume the noise in Figure 1.17 is Gaussian (also called normal) with zero mean so that it has the following well-known pdf ("bell-shaped" curve):

$$f_N(n) = \frac{1}{\sqrt{2\pi\sigma^2}} \exp\left(-n^2/2\sigma^2\right), \tag{1.13}$$

where σ^2 is the variance. This is known as an additive Gaussian noise (AGN) channel, which is the most widely studied channel in communications; its pdf is plotted in Figure 1.18. By using the Dirac delta function, it can

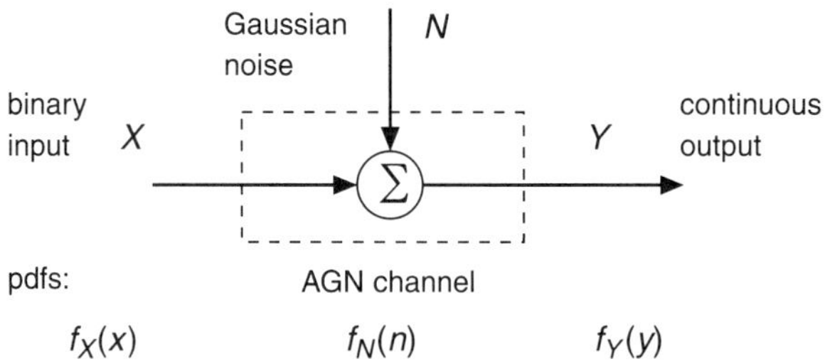

FIGURE 1.17 Additive white Gaussian noise (AWGN) channel with random binary input. X is a discrete random variable, and N and Y are continuous random variables.

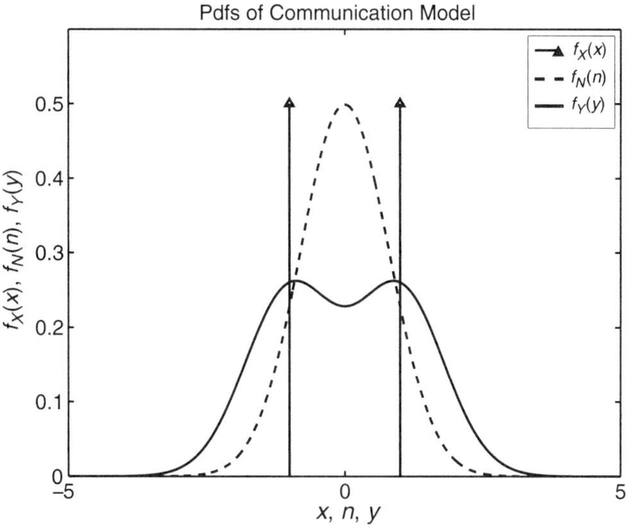

FIGURE 1.18 Pdfs for the AWGN channel. (i) Channel input X: symmetric Bernoulli random variable with pdf $f_X(x)$. (ii) Noise N: Gaussian random variable with pdf $f_N(n)$. (iii) Channel output Y: bimodal random variable with pdf $f_Y(y) = f_X(x) * f_N(n)$, where $*$ denotes convolution.

be shown that the pdf of the received random variable Y is obtained as the following convolution of $f_X(x)$ in (1.12) and $f_N(n)$ in (1.13) (see Chapter 4):

$$\begin{aligned} f_Y(y) &= \int_{-\infty}^{\infty} f_X(y-v) f_N(v) \mathrm{d}v, \\ &= \frac{1}{2\sqrt{2\pi\sigma^2}} \int_{-\infty}^{\infty} [\delta(y-v-1) + \delta(y-v+1)] \exp\left(-v^2/2\sigma^2\right) \mathrm{d}v, \\ &= \frac{1}{2\sqrt{2\pi\sigma^2}} \left[\exp\left(-(y+1)^2/2\sigma^2\right) + \exp\left(-(y-1)^2/2\sigma^2\right)\right], \end{aligned} \tag{1.14}$$

where the sifting property of the Dirac delta function has been used (see Appendix B). The pdf $f_Y(y)$ of the output is also shown in Figure 1.18. Observe in (1.14) that the Gaussian pdf has been shifted to the right (centered at 1) and to the left (centered at -1), and they are scaled by $1/2$ because the total area of $f_Y(y)$ (the sum of two shifted Gaussian pdfs) must be one. Using the Dirac delta function to represent the pdf of discrete random variable X has allowed us to easily derive $f_Y(y)$ via a continuous-time convolution.

This completes our preview of the types of methods that will be used to model random quantities. In Section 1.2, we review some basic techniques for continuous- and discrete-time signals and systems. These will be needed later when systems operating on random processes are discussed in Chapter 8, and for the signal processing and communications applications covered in the third part of the book, as well as some of the examples presented in other chapters.

1.2 DETERMINISTIC SIGNALS AND SYSTEMS

In engineering programs, students usually learn about signals and systems first in the time domain and then in the frequency domain. The focus is on *deterministic* signals such as those illustrated in Figure 1.19.

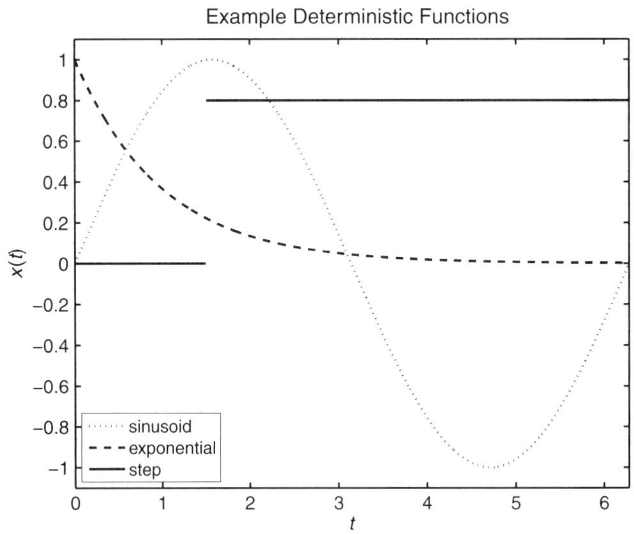

FIGURE 1.19 Examples of deterministic signals: sinusoid $x(t) = \sin(t)u(t)$, exponential $x(t) = \exp(-t)u(t)$, and step $x(t) = 0.8u(t - 1.5)$, where $u(t)$ is the unit-step function.

1.2.1 Continuous Time

Nonrandom signal $x(t)$ is a *function* of continuous time $t \in \mathcal{R}$. Consider the sinusoidal signal

$$x(t) = \sin(\omega t)u(t) \tag{1.15}$$

shown in Figure 1.19 for $t \in [0, 2\pi]$, where $u(t)$ is the unit-step (Heaviside) function (see Appendix B). Radian frequency ω (radians/second) is related to ordinary frequency f in hertz (Hz) according to $\omega = 2\pi f$. For a particular frequency ω, we can specify the value of the deterministic function $x(t)$ at any time t. Other examples of deterministic functions are shown in the figure.

The waveform can also be examined in the *frequency domain*, using either the Fourier transform

$$X(\omega) = \int_{-\infty}^{\infty} x(t)\exp(-j\omega t)\mathrm{d}t, \tag{1.16}$$

$$X(f) = \int_{-\infty}^{\infty} x(t)\exp(-j2\pi f t)\mathrm{d}t, \tag{1.17}$$

or the Laplace transform

$$X(s) = \int_{-\infty}^{\infty} x(t)\exp(-st)\mathrm{d}t, \tag{1.18}$$

where $s \triangleq \sigma + j\omega$ is a complex variable, $j = \sqrt{-1}$, and σ is neper frequency. These transforms are summarized in Appendix C, which includes a description of their properties and tables of some transform pairs. (Note that the uppercase notation used for transforms should not be confused with random variables. The reader should be able to differentiate between the two because of the argument of $X(\cdot)$ and from the context of the discussion.) Euler's formulas are given by

$$\exp(j\omega t) = \cos(\omega t) + j\sin(\omega t), \tag{1.19}$$

$$\exp(-j\omega t) = \cos(\omega t) - j\sin(\omega t), \tag{1.20}$$

and thus $\cos(\omega t)$ and $\sin(\omega t)$ are the real part and imaginary part, respectively, of $\exp(j\omega t)$. The Laplace transform $X(s)$ is a function of σ and ω, and is usually represented on the s-plane via its poles ($\times$) and zeros (o), which are values of s where $X(s)$ is infinite and zero, respectively.

The corresponding inverse transforms are

$$x(t) = \frac{1}{2\pi}\int_{-\infty}^{\infty} X(\omega)\exp(j\omega t)\mathrm{d}\omega, \tag{1.21}$$

$$= \int_{-\infty}^{\infty} X(f)\exp(2\pi f t)\mathrm{d}f, \tag{1.22}$$

$$= \oint_C X(s)\exp(st)\mathrm{d}s, \tag{1.23}$$

where the last expression is a contour integral on the complex s-plane in the region of convergence (ROC). Note that the second integral is obtained from the first integral via the transformation $\omega = 2\pi f$ and substituting $d\omega = 2\pi \mathrm{d}f$.

Example 1.3. Consider the following signal:

$$x(t) = \cos(2\pi f_o t) \tag{1.24}$$

for $t \in \mathcal{R}$, where f_o is the frequency in Hz. Its spectrum is represented mathematically using the Dirac delta function:

$$X(f) = (1/2)[\delta(f - f_o) + \delta(f + f_o)]. \tag{1.25}$$

We can also write $x(t)$ and $X(f)$ as functions of radian frequency ω as follows:

$$x(t) = \cos(\omega_o t), \tag{1.26}$$

$$X(\omega) = \pi[\delta(\omega - \omega_o) + \delta(\omega + \omega_o)]. \tag{1.27}$$

The frequency in Hz of a sinusoid is the number of times per second that one period (one "cycle") of the waveform occurs. Radian frequency is an angular quantity for which the duration of one period of a sinusoid is 2π radians. The units of ω are (radians/cycle)·(cycles/seconds) = radians/second.

These transforms can be viewed as deterministic cross-correlation functions of $x(t)$ with a complex exponential: either $\exp(-j\omega t)$, $\exp(-j2\pi f t)$, or $\exp(-st)$. For each value of ω, $X(\omega)$ indicates the strength or "amount" of that frequency contained in $x(t)$. Since $X(\omega)$ is generally complex-valued for each ω, it can be expressed in polar form as follows:

$$X(\omega) = |X(\omega)|\exp(j\phi(\omega)), \tag{1.28}$$

where

$$|X(\omega)|^2 = \mathrm{Re}^2[X(\omega)] + \mathrm{Im}^2[X(\omega)], \tag{1.29}$$

$$\phi(\omega) = \tan^{-1}\left[\frac{\mathrm{Im}(X(\omega))}{\mathrm{Re}(X(\omega))}\right] \tag{1.30}$$

are the squared magnitude and phase, respectively. The amount of the frequency component ω in $x(t)$ is given by $|X(\omega)|$.

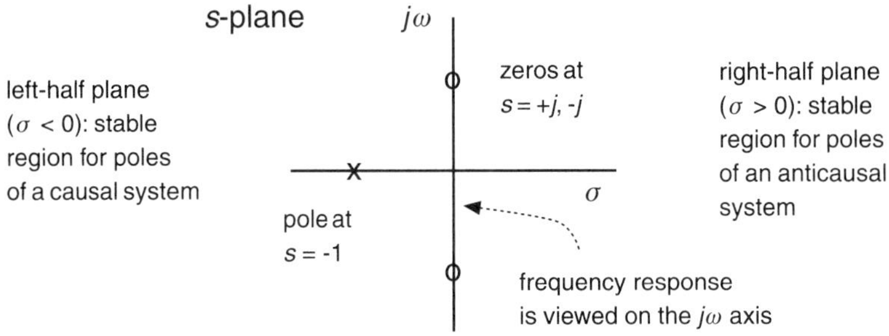

FIGURE 1.20 Location of pole and zeros on the s-plane for Laplace transform $X(s) = (s^2 + 1)/(s - 1)$ in Example 1.4.

Example 1.4. The following Laplace transform has a pole at $s = 1$ and a pair of complex zeros at $s = \pm j\omega$, as illustrated on the s-plane in Figure 1.20:

$$X(s) = \frac{s^2 + 1}{s - 1}. \tag{1.31}$$

The poles indicate various features about $x(t)$, such as boundedness ($\sigma < 0$), the rate of decay (how close σ is to zero), and if there is a sinusoidal component (when the poles are not on the real axis). The Fourier transform is obtained from $X(s)$ by evaluating it on the $j\omega$ axis, which for (1.31) is

$$X(\omega) = \frac{1 - \omega^2}{j\omega - 1} = \frac{\omega^2 - 1 + j\omega(\omega^2 - 1)}{\omega^2 + 1} = a + jb, \tag{1.32}$$

where $a = (\omega^2 - 1)/(\omega^2 + 1)$ and $b = \omega(\omega^2 - 1)/(\omega^2 + 1)$. It has the following squared magnitude and phase:

$$|X(\omega)|^2 = a^2 + b^2 = \frac{(\omega^2 - 1)^2}{\omega^2 + 1}, \tag{1.33}$$

$$\arg[X(\omega)] = \tan^{-1}(b/a) = \tan^{-1}(\omega), \tag{1.34}$$

which can be plotted separately versus ω as shown in Figure 1.21. $X(\omega)$ provides information about the frequency content of a signal; it is also called the *spectrum*. Usually, the magnitude squared is plotted in dB (decibel):

$$10 \log_{10}(|X(\omega)|^2). \tag{1.35}$$

Observe the "notches" caused by the zeros at $z = \pm j$. The phase is approximately linear in the vicinity of $\omega = 0$.

Systems covered in engineering courses are usually represented as linear time-invariant (LTI) filters with impulse-response function $h(t)$. The output of such a system with input $x(t)$ is given by the *convolution* integral:

$$y(t) = \int_0^\infty h(t - \tau)x(\tau)d\tau = \int_0^\infty x(t - \tau)h(\tau)d\tau, \tag{1.36}$$

which is symmetric in $x(t)$ and $h(t)$: $y(t) = h(t) * x(t) = x(t) * h(t)$. Linearity of the system results in an integral, and time invariance means that the impulse response is unchanged by a shift in time, giving $h(t - \tau)$ above instead of the more general form $h(t, \tau)$. The Laplace transform of $h(t)$ is the *transfer function* $H(s)$ of

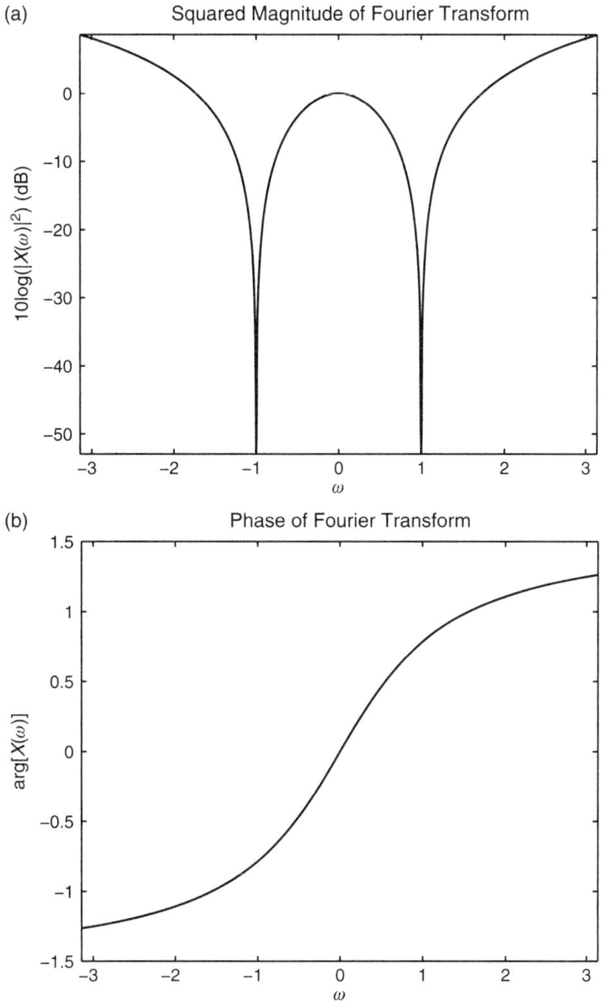

FIGURE 1.21 Fourier transform $X(\omega)$ of $X(s) = (s^2 + 1)/(s - 1)$ in Example 1.4. (a) Squared magnitude (in dB). (b) Phase.

the system, and the Fourier transform $H(\omega)$ is the corresponding *frequency response*. The output of an LTI system in the s-domain is

$$Y(s) = H(s)X(s), \tag{1.37}$$

or $Y(\omega) = H(\omega)X(\omega)$ and $Y(f) = H(f)X(f)$ in the Fourier domains.

A linear system with input $x(t)$ and output $y(t)$ can be modeled by a linear differential equation (DE) with constant coefficients $\{a_0, \ldots, a_{N-1}, b_0, \ldots, b_{M-1}\}$ as follows:

$$\begin{aligned} a_{N-1}\frac{\mathrm{d}^{N-1}y(t)}{\mathrm{d}t^{N-1}} &+ a_{N-2}\frac{\mathrm{d}^{N-2}y(t)}{\mathrm{d}t^{N-2}} + \cdots + a_1\frac{\mathrm{d}y(t)}{\mathrm{d}t} + a_0 y(t), \\ &= b_{M-1}\frac{\mathrm{d}^{M-1}x(t)}{\mathrm{d}t^{M-1}} + b_{M-2}\frac{\mathrm{d}^{M-2}x(t)}{\mathrm{d}t^{M-2}} + \cdots + b_1\frac{\mathrm{d}x(t)}{\mathrm{d}t} + b_0 x(t), \end{aligned} \tag{1.38}$$

or, more compactly, as

$$\sum_{n=0}^{N-1} a_n y^{(n)}(t) = \sum_{m=0}^{M-1} b_m x^{(m)}(t), \tag{1.39}$$

where the superscript denotes the order of the ordinary derivative. Assuming zero initial conditions, the transfer function is the following ratio of polynomials:

$$H(s) = \frac{Y(s)}{X(s)} = \frac{\sum_{m=0}^{M-1} b_m s^m}{\sum_{n=0}^{N-1} a_n s^n} \tag{1.40}$$

from which $h(t)$ can be obtained via the inverse Laplace transform (often in practice $a_0 = 1$). Such *rational functions* are important because the inverse transform can be obtained easily using a partial fraction expansion (PFE).

Although we are interested in extending similar time- and frequency-domain descriptions of deterministic signals to random signals, we generally do not consider random systems. The only exceptions are the Kalman filter, adaptive filters, and blind equalizers described later in the applications part of the book in Chapters 11–13; these filters evolve in time to satisfy some optimality criteria. Generally, we assume that a fixed system is modeled using a linear DE with constant coefficients from which the impulse-response function $h(t)$ can be derived. Of course, in many problems the actual system of interest is time varying, and a linear DE is only an approximate model of the system dynamics. Many systems, in fact, are nonlinear or may have a combination of linear and nonlinear components. Later, we consider perhaps the simplest type of nonlinear system that is *memoryless*. A memoryless system does not store past information about its input or output: the current output of a memoryless system depends only on the current input. Such a system can be represented as a function:

$$y(t) = g(x(t)), \tag{1.41}$$

where $g(\cdot)$ is time invariant and it is understood that at each instant of time, $y(t)$ is the value of that function evaluated at $x(t)$. The two types of systems described above are summarized in Figure 1.22.

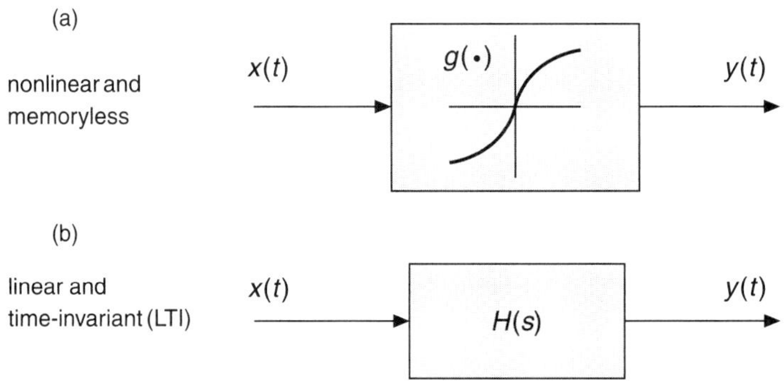

FIGURE 1.22 Types of systems considered in this book. (a) Memoryless nonlinear mapping $g(\cdot)$. (b) LTI system with impulse-response function $h(t)$ and transfer function $H(s)$.

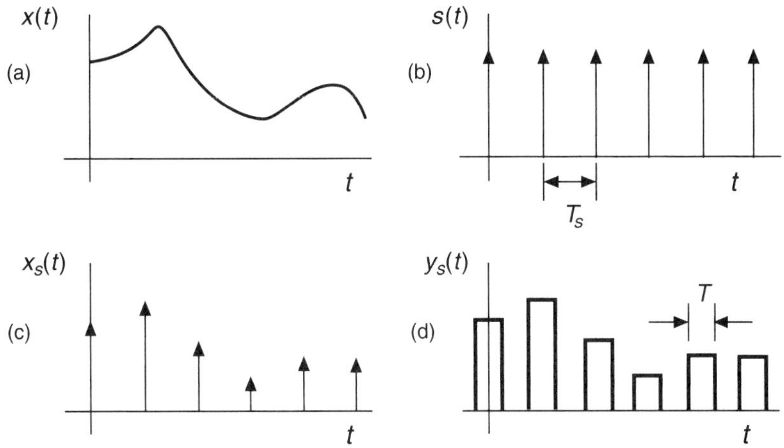

FIGURE 1.23 Illustration of the sampling process for a deterministic signal. (a) Continuous-time signal $x(t)$. (b) Sampling signal $s(t)$ based on the Dirac delta function with uniform sampling period T_s. (c) Sampled waveform $x_s(t)$. (d) Filtered waveform $y_s(t) = x_s(t) * h(t)$, where $h(t)$ is a rectangular impulse-response function with width $T < T_s$.

1.2.2 Discrete Time

Discrete-time signals can be represented using either the Kronecker delta function or the Dirac delta function. For example, if a continuous-time signal $x(t)$ is sampled to generate a discrete-time signal, it is straightforward to represent the *sampling process* as the following multiplication:

$$x_s(t) = x(t)s(t), \tag{1.42}$$

where $x_s(t)$ is the sampled waveform,

$$s(t) = \sum_{n=0}^{\infty} \delta(t - nT_s) \tag{1.43}$$

is the sampling signal, T_s is the sampling period, and $f_s = 1/T_s$ is the sampling frequency. An illustration of the sampling process is shown in Figure 1.23 for *uniform* sampling. By using the Dirac delta function, $x_s(t)$ is defined for all t, though it is nonzero only at the discrete sampling instants. We assume that the Nyquist sampling rate has been satisfied: f_s exceeds twice the highest frequency component of $x(t)$ so that aliasing does not occur (i.e., shifted replicas of the frequency response do not overlap).

From the sampling property of the Dirac delta function, the sampled waveform can be expressed as follows:

$$x_s(t) = \sum_{n=0}^{\infty} x(nT_s)\delta(t - nT_s). \tag{1.44}$$

As indicated in Figure 1.23, $x_s(t)$ is a "train" of Dirac delta functions whose areas are given by the values of $x(t)$ at the sampling instants. As mentioned before, an advantage of using the Dirac delta function is that integration is easily performed because $x_s(t)$ is defined for all t. For example, suppose we want to filter $x_s(t)$ with the continuous-time filter $h(t)$. The convolution integral yields

$$\begin{aligned} y_s(t) &= \int_0^\infty x_s(\tau)h(t-\tau)\mathrm{d}\tau, \\ &= \sum_{n=0}^{\infty} x(nT_s) \int_0^\infty \delta(\tau - nT_s)h(t-\tau)\mathrm{d}\tau, \\ &= \sum_{n=0}^{\infty} x(nT_s)h(t-nT_s), \end{aligned} \tag{1.45}$$

where the sifting property of the Dirac delta function has been used. Observe that $y_s(t)$ is also a continuous-time function, as illustrated in Figure 1.23 where $h(t)$ is a rectangular function, and assuming that its width T does not exceed the sampling period T_s. The final expression in (1.45) is a *discrete-time convolution* between the input signal and the filter. Since we are interested in $y_s(t)$ only at integer multiples of T_s in a discrete-time system, the discrete-time output can be written as follows:

$$y_s[kT_s] = \sum_{n=0}^{\infty} x[nT_s]h[kT_s - nT_s] = \sum_{n=0}^{\infty} x[nT_s]h[(k-n)T_s]. \tag{1.46}$$

In most courses on signal processing, $T_s = 1$ is assumed for mathematical convenience. The above expression then simplifies to the familiar form for a discrete-time convolution:

$$y[k] = \sum_{n=0}^{\infty} x[n]h[k-n] = \sum_{n=0}^{\infty} h[n]x[k-n], \tag{1.47}$$

which is symmetric with respect to $h[k]$ and $x[k]$, and the subscript on $y_s[k]$ has been dropped since only discrete-time quantities appear (without regard to any underlying sampling of a continuous-time waveform). Alternatively, when only discrete-time signals and systems are examined, the Kronecker delta function is used in place of the Dirac delta function, so that (1.44) becomes

$$x[k] = \sum_{n=0}^{\infty} x[n]\delta[k-n]. \tag{1.48}$$

Although this sum may initially be confusing because $x[\cdot]$ appears on both sides, note that substituting a specific value for k on the left-hand side gives the same value on the right-hand side (e.g., substituting $k = 1$ gives only the $n = 1$ term in the sum because of the delta functions). The discrete-time version of $x_s(t)$ given by $x[k]$ is illustrated in Figure 1.24. Likewise, the filter impulse-response function can be represented using the Kronecker delta function:

$$h[k] = \sum_{n=0}^{\infty} h[n]\delta[k-n], \tag{1.49}$$

and the filter output is the discrete-time convolution in (1.47). The discrete-time filter and its output $y[k]$ are also shown in Figure 1.24.

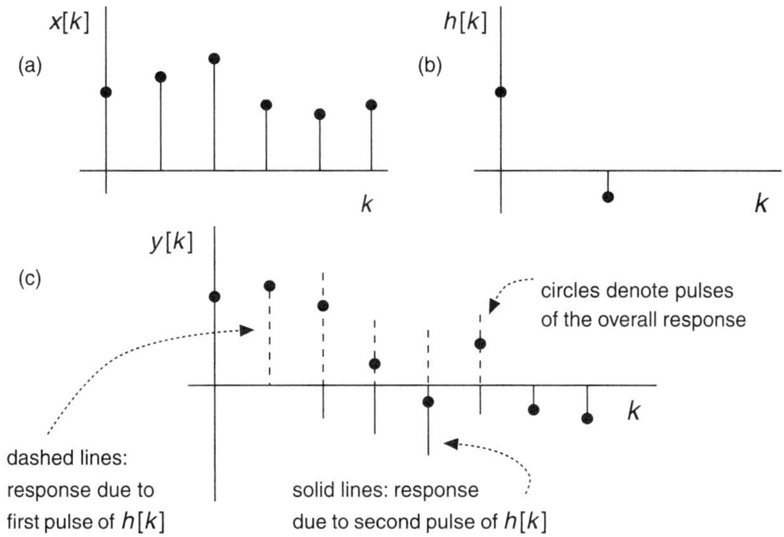

FIGURE 1.24 Discrete-time signal and filter representations. (a) Input signal $x[k]$. (b) Filter impulse-response function $h[k]$. (c) Filter output $y[k] = x[k] * h[k]$.

The Laplace transform of the sampled signal is

$$X_s(s) = \int_0^\infty x_s(t)\exp(-st)\mathrm{d}t,$$
$$= \sum_{n=0}^{\infty} x[nT_s]\int_0^\infty \delta(t - nT_s)\exp(-st)\mathrm{d}t,$$
$$= \sum_{n=0}^{\infty} x[nT_s]\exp[-nT_s s]. \tag{1.50}$$

Observe that

$$\exp(sT_s) = \exp(\sigma T_s)[\cos(\omega T_s) + j\sin(\omega T_s)] \tag{1.51}$$

is periodic in ω with period T_s. Defining $z \triangleq \exp(s)$, the z-transform is obtained from the Laplace transform above as follows:

$$X(z) = \sum_{n=0}^{\infty} x[nT_s]z^{-nT_s}. \tag{1.52}$$

The mapping $z = \exp(s)$ is a transformation from the s-plane to the z-plane as depicted in Figure 1.25. We can make the following comments about the properties of this transformation:

- The real axis on the z-plane is given by $\exp(\sigma T_s)\cos(\omega T_s)$ (the first term in (1.51)). The imaginary axis on the z-plane is given by $\exp(\sigma T_s)\sin(\omega T_s)$ (the second term in (1.51)). The unit circle corresponds to $\sigma = 0$ and is described on the z-plane by

$$\cos(\omega T_s) + j\sin(\omega T_s) = \exp(j\omega T_s). \tag{1.53}$$

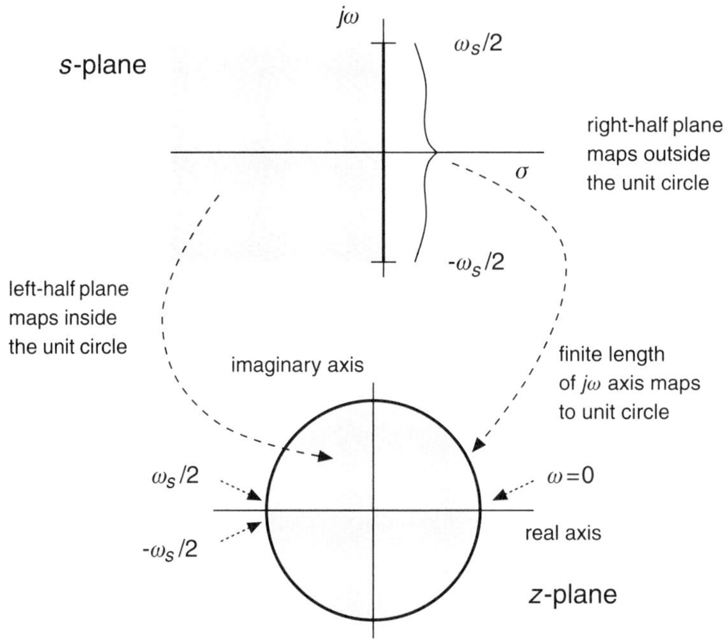

FIGURE 1.25 Mapping from the s-plane to the z-plane.

- The left half of the s-plane ($\sigma < 0$) maps inside the unit circle since $\exp(\sigma T_s) < 1$. The right half of the s-plane ($\sigma > 0$) maps outside the unit circle since $\exp(\sigma T_s) > 1$. Horizontal "strips" on the left half of the s-plane map to "pie slices" on the z-plane. When $\sigma = -\infty$, all points for ω map to the origin $z = 0$.
- The interval $[-\omega_s/2, \omega_s/2]$ on the $j\omega$ axis maps to the unit circle. All intervals on the $j\omega$ axis of the form $[-(k+1)\omega_s/2, -k\omega_s/2], [k\omega_s/2, (k+1)\omega_s/2]$ for $k \in \mathcal{Z}$ also map to the unit circle.

The frequency response of a discrete-time signal follows from (1.52) by substituting $\sigma = 0$ so that $r = 1$ and $z = \exp(j\omega)$. This yields

$$X(j\omega) = \sum_{k=0}^{\infty} x[k] \exp(-j\omega k), \tag{1.54}$$

where $T_s = 1$ has been substituted. We include j in the argument of $X(j\omega)$ so that the discrete-time Fourier transform (DTFT) is not confused with the continuous-time Fourier transform $X(\omega)$. Note in Figure 1.25 and from the comments above that $X(j\omega)$ "wraps around" on the unit circle such that $\omega = \pm\pi$ lie at the left-most point of the unit circle (for $T_s = 1$). This is easily verified from Euler's formula:

$$z = r\exp(j\omega) = r\cos(\omega) + jr\sin(\omega). \tag{1.55}$$

When $z = -1$ (the left-most point on the unit circle) and $r = 1$, the sine term must be zero and the cosine term must be -1. This occurs for all nonzero integer multiples of $\pm\pi$; the frequency response repeats when $|\omega|$ extends beyond π. This repetition is a result of the sampling process which causes the original frequency response to be replicated at integer multiples of the sampling frequency $f_s = 1/T_s$. When $T_s \neq 1$, the left-most point on the unit circle is $\pm f_s/2 = 1/2T_s$ (one-half the sampling rate). In radian frequency, the left-most point

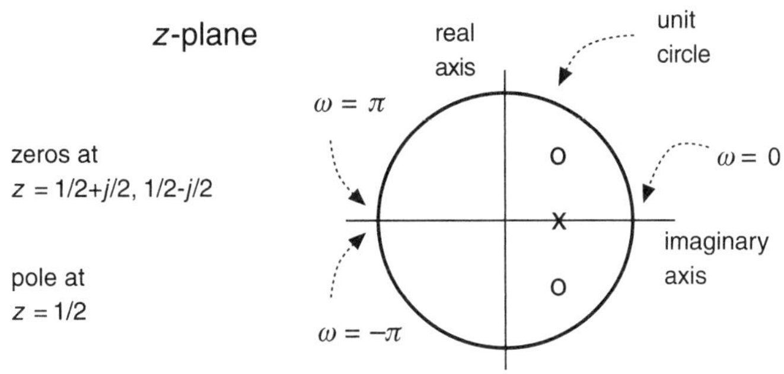

FIGURE 1.26 Location of pole and zeros on the z-plane for $X(z)$ in Example 1.5.

on the unit circle is $\pm\omega_s/2 = \pm\pi f_s$ as shown in Figure 1.25. Because the frequency response is repeated, it is usually plotted only for $|\omega| \leq \pi$ (with $T_s = 1$).

Example 1.5. Figure 1.26 shows the pole/zero locations on the z-plane for the following z-transform:

$$X(z) = \frac{z^2 - z + 1/2}{z - 1/2}. \tag{1.56}$$

The corresponding DTFT is obtained by substituting $z = \exp(j\omega)$ and evaluating $X(z)$ on the unit circle:

$$\begin{aligned} X(j\omega) &= \frac{\exp(j2\omega) - \exp(j\omega) + 1/2}{\exp(j\omega) - 1/2}, \\ &= \frac{\exp(j2\omega) - \exp(j\omega) + 1/2}{\exp(j\omega) - 1/2} \cdot \frac{\exp(-j\omega) - 1/2}{\exp(-j\omega) - 1/2}, \\ &= \frac{\exp(j\omega) - (1/2)\exp(j2\omega) + \cos(\omega) - 5/4}{5/4 - \cos(\omega)}. \end{aligned} \tag{1.57}$$

The real and imaginary parts are

$$\mathrm{Re}[X(j\omega)] = \frac{2\cos(\omega) - (1/2)\cos(2\omega) - 5/4}{5/4 - \cos(\omega)}, \tag{1.58}$$

$$\mathrm{Im}[X(j\omega)] = \frac{\sin(\omega) - (1/2)\sin(2\omega)}{5/4 - \cos(\omega)}, \tag{1.59}$$

and the corresponding magnitude and phase are plotted in Figure 1.27.

1.2.3 Discrete-Time Filters

It is straightforward to show in the z-domain that the output of the filter represented by the convolution in (1.47) is

$$Y(z) = H(z)X(z), \tag{1.60}$$

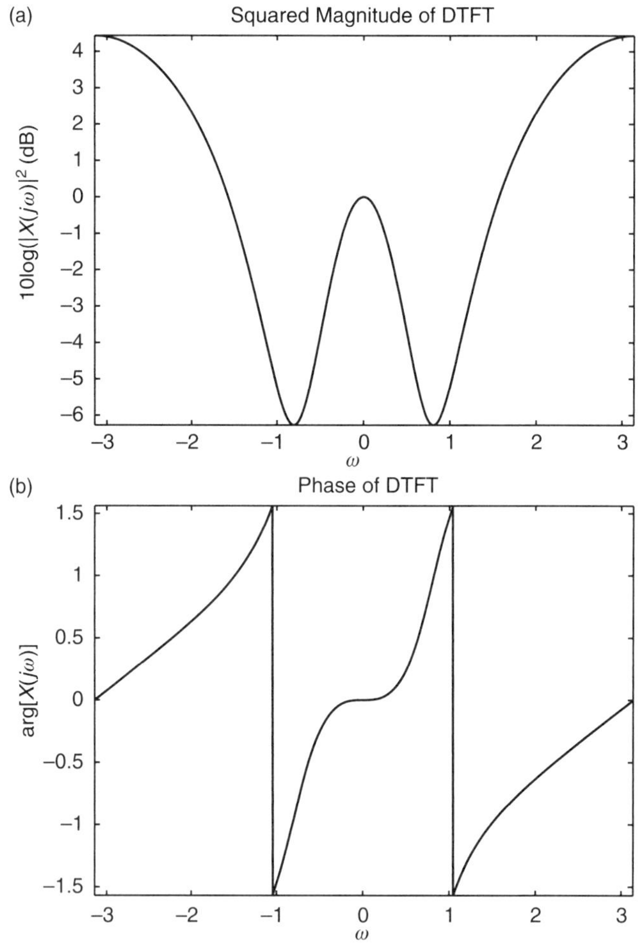

FIGURE 1.27 DTFT for $X(z)$ in Example 1.5. (a) Squared magnitude (in dB). (b) Phase.

where

$$H(z) = \sum_{k=0}^{\infty} h[k]z^{-k} \tag{1.61}$$

is the transfer function. Likewise the frequency response of the output is

$$Y(j\omega) = H(j\omega)X(j\omega). \tag{1.62}$$

If the summation limits in (1.47) and (1.61) are both finite, then the filter has a finite-impulse-response (FIR). The transfer function for this case with limits $\{0, M-1\}$ is

$$H_{\text{FIR}}(z) = \sum_{m=0}^{M-1} b_m z^{-m} \tag{1.63}$$

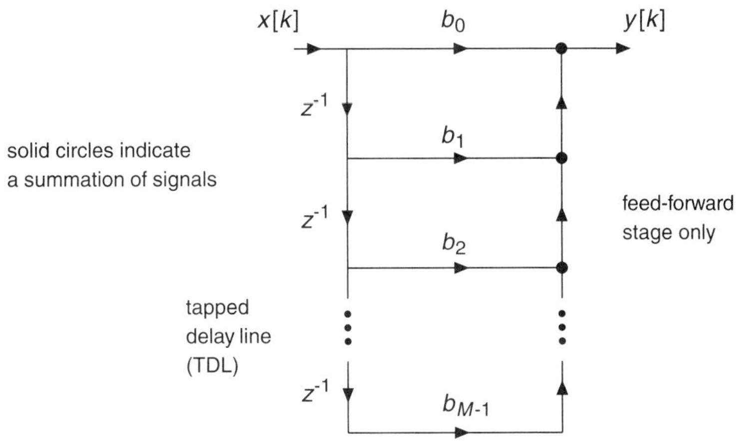

FIGURE 1.28 FIR filter implementation.

and (1.47) becomes

$$y[k] = \sum_{m=0}^{M-1} b_m x[k-m], \tag{1.64}$$

where $\{b_m\}$ are used to represent the coefficients of an FIR filter. Since the lower limit is ≥ 0, the filter is *causal*. A block diagram of an FIR filter is shown in Figure 1.28. It consists of a tapped delay line (TDL) that stores delayed values of the input $x[k]$, which in this case are $\{x[k-1], \ldots, x[k-M+1]\}$. The delay notation z^{-1} in the figure follows from $H(z)$, and can be interpreted as a delay operator such that with input $x[k]$ the output is $x[k-1]$.

If the upper limit in (1.47) is infinity, then the filter has an infinite impulse response (IIR). The transfer function for this case is

$$H_{\text{IIR}}(z) = \frac{\sum_{m=0}^{M-1} b_m z^{-m}}{\sum_{n=0}^{N-1} a_n z^{-n}}, \tag{1.65}$$

which is the discrete-time counterpart of the transfer function in (1.40) for a continuous-time system, and is derived from the following linear *difference equation* (DE) representing the filter:

$$y[k] = \sum_{m=0}^{M-1} b_m x[k-m] + \sum_{n=1}^{N-1} a_n y[k-n], \tag{1.66}$$

where $a_0 = 1$ has been assumed. Examples of IIR filter implementations are shown in Figure 1.29. There are several IIR implementations; the one shown in Figure 1.29(a) is called *direct-form I* and corresponds to a direct realization of the DE in (1.66). Since the filter is LTI, the feed-forward and feedback stages can be interchanged, leading to the second structure in Figure 1.29(b). Observe that in this configuration, the two sets of delayed signals are identical, and thus the two delay lines can be combined into a single delay line, resulting in the

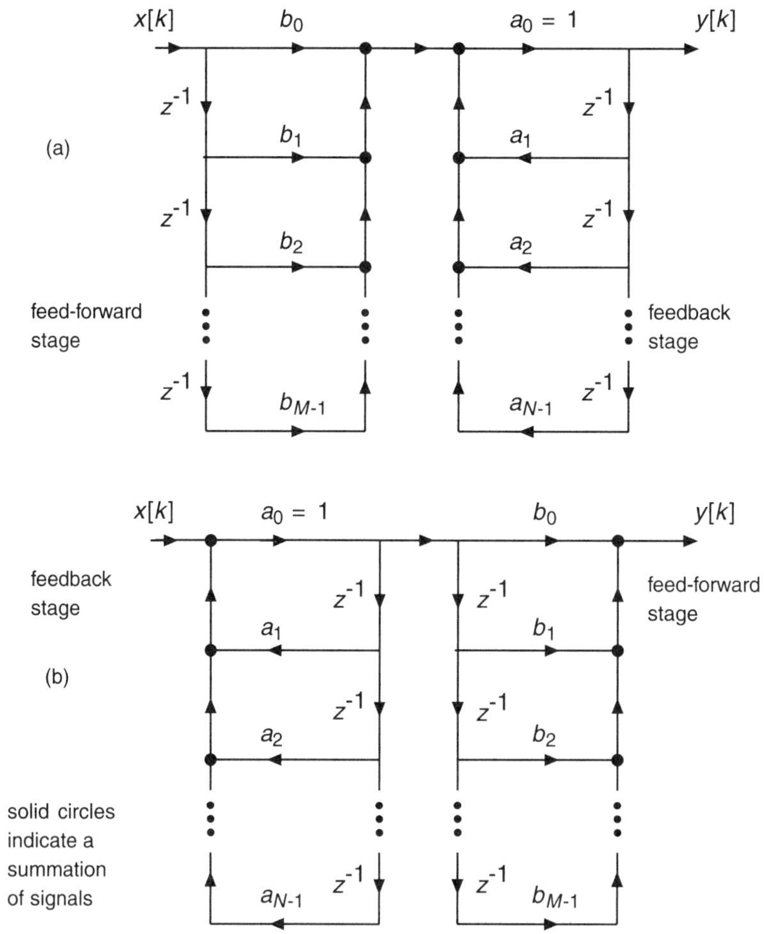

FIGURE 1.29 IIR filter implementations (for $M = N$). (a) Direct-form I. (b) Direct-form II, before combining the two descending delay lines that have identical signals, as shown in Figure 1.30(a).

direct-form II realization shown in Figure 1.30(a). This implementation has the minimum number of delays, which in general is given by $\max(M - 1, N - 1)$.

1.2.4 State-Space Realizations

Finally, we briefly summarize state-space realizations that are convenient for describing linear systems and are used for the Kalman filter introduced in Chapter 11. Such a realization is obtained by rewriting the high-order IIR DE in (1.66) as a first-order *vector* DE. This is achieved by defining internal signals of the filter to be *states* $\{x_n[k]\}$ with the *state vector*

$$\mathbf{x}[k] \triangleq [x_1[k], \ldots, x_{N-1}[k]]^T, \tag{1.67}$$

where we assume for convenience that $M = N$ as in Figure 1.30(a). Generally, the number of states would be $\max\{M - 1, N - 1\}$. Observe that the $N - 1$ states shown in direct-form II are the signals along the delay line

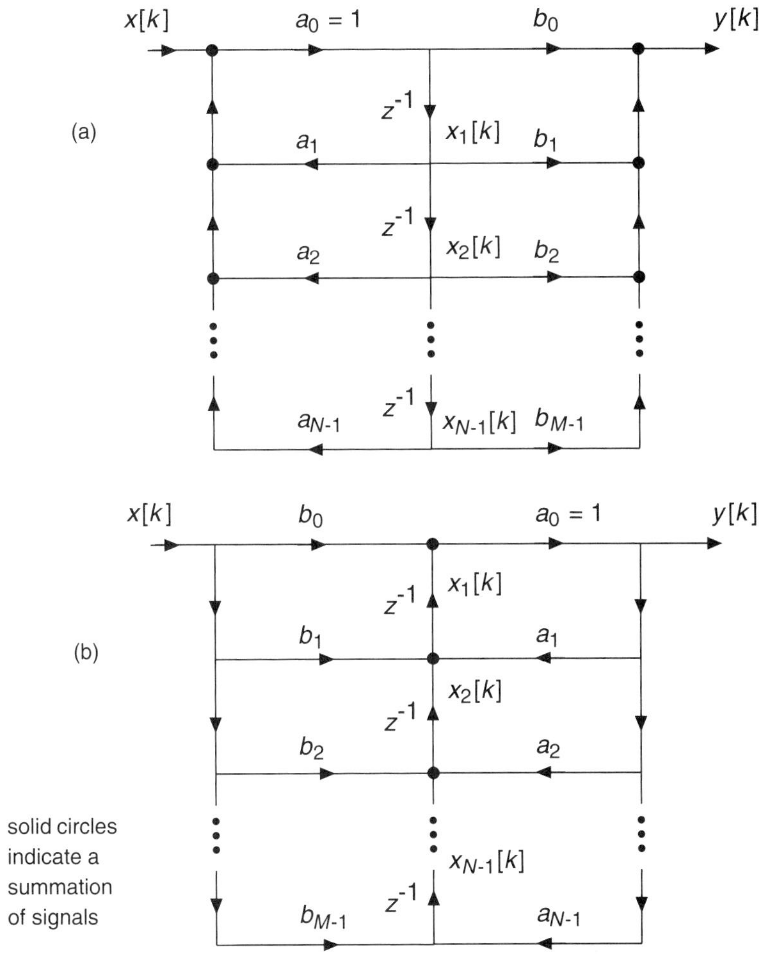

FIGURE 1.30 IIR filter implementations (for $M = N$). (a) Direct-form II (after combining the two descending delay lines in Figure 1.29(b)) with state signals defined. (b) Alternative realization with ascending delay line and different state signals defined.

so that we can write the following system of equations:

$$\begin{aligned} x_1[k+1] &= \sum_{n=1}^{N-1} a_n x_n[k] + x[k], \\ x_2[k+1] &= x_1[k], \\ &\vdots \\ x_{N-1}[k+1] &= x_{N-2}[k]. \end{aligned} \tag{1.68}$$

From the state-vector definition, this leads to the following vector recursion:

$$\mathbf{x}[k+1] = \mathbf{A}\mathbf{x}[k] + \mathbf{b}x[k], \tag{1.69}$$

where

$$\mathbf{A} \triangleq \begin{bmatrix} a_1 & a_2 & \cdots & & a_{N-1} \\ 1 & 0 & \cdots & & 0 \\ 0 & 1 & 0 & \cdots & \vdots \\ \vdots & & \ddots & \ddots & \vdots \\ 0 & \cdots & 0 & 1 & 0 \end{bmatrix}, \quad \mathbf{b} \triangleq \begin{bmatrix} 1 \\ 0 \\ \vdots \\ 0 \end{bmatrix}. \tag{1.70}$$

The DE can be rewritten in terms of the states as follows (again, with $M = N$):

$$y[k] = \sum_{m=1}^{M-1} b_m x_m[k] + b_0 \sum_{n=1}^{N-1} a_n x_n[k] + b_0 x[k], \tag{1.71}$$

$$= \sum_{n=1}^{N-1} (b_n + b_0 a_n) x_n[k] + b_0 x[k], \tag{1.72}$$

and thus the output is

$$y[k] = \mathbf{c}^T \mathbf{x}[k] + d x[k], \tag{1.73}$$

where

$$\mathbf{c} \triangleq [b_1 + b_0 a_1, \ldots, b_{N-1} + b_0 a_{N-1}]^T, \quad d = b_0. \tag{1.74}$$

Equations (1.69) and (1.73) together comprise the state-space formulation for the direct-form II filter with parameters $\{\mathbf{A}, \mathbf{b}, \mathbf{c}, d\}$.

Other state-space forms can be obtained by rearranging the direct-form II realization. For example, interchanging $x[k]$ and $y[k]$, reversing the directions of all arrows, interchanging nodes and summations, and reversing the diagram horizontally results in the "dual" realization shown in Figure 1.30(b). The two realizations in the figure have identical transfer functions (poles and zeros), but different state-space representations. Note in Figure 1.30(b) that we have again labeled the states to be at the output of the delays. The corresponding state equations are (with $M = N$)

$$\begin{aligned} x_1[k+1] &= b_1 x[k] + a_1 y[k] + x_2[k], \\ &\vdots \\ x_{N-2}[k+1] &= b_{N-2} x[k] + a_{N-2} y[k] + x_{N-1}[k], \\ x_{N-1}[k+1] &= b_{N-1} x[k] + a_{N-1} y[k]. \end{aligned} \tag{1.75}$$

Substituting $y[k] = x_1[k] + b_0 x[k]$ gives

$$\begin{aligned} x_1[k+1] &= (b_1 + b_0 a_1) x[k] + a_1 x_1[k] + x_2[k], \\ &\vdots \\ x_{N-2}[k+1] &= (b_{N-2} + b_0 a_{N-2}) x[k] + a_{N-2} x_{N-2}[k] + x_{N-1}[k], \\ x_{N-1}[k+1] &= (b_{N-1} + b_0 a_{N-1}) x[k] + a_{N-1} x_{N-1}[k], \end{aligned} \tag{1.76}$$

and thus the state-space parameters are

$$\mathbf{A} \triangleq \begin{bmatrix} a_1 & 1 & 0 & \cdots & 0 \\ a_2 & 0 & 1 & & \vdots \\ \vdots & & \ddots & \ddots & 0 \\ & \cdots & & 0 & 1 \\ a_{N-1} & 0 & \cdots & & 0 \end{bmatrix}, \quad \mathbf{b} \triangleq \begin{bmatrix} b_1 + b_0 a_1 \\ \vdots \\ b_{N-1} + b_0 a_{N-1} \end{bmatrix}, \tag{1.77}$$

$$\mathbf{c} \triangleq [1, 0, \ldots, 0]^T, \quad d = b_0. \tag{1.78}$$

Note that state transition matrix $\mathbf{A}$ for this dual realization is the transpose of that in (1.70), $\mathbf{b}$ and $\mathbf{c}$ are interchanged, and d remains the same scalar b_0.

1.3 STATISTICAL SIGNAL PROCESSING WITH MATLAB®

Although continuous-time random *processes* and discrete-time random *sequences* are covered in this book, all simulation examples generated using MATLAB®[1] are necessarily discrete-time sequences. In this section, we provide an overview of methods for generating random sequences, and of filters for modifying their properties and frequency-domain characteristics.

1.3.1 Random Number Generation

The random number generators in MATLAB provide a means for realizing a *pseudorandom* sequence $X[k]$ where the samples across time instants are independent and thus uncorrelated. There are two basic random number generators:

1. randn(N,1). Generates an $N \times 1$ vector of numbers on $\mathcal{R}$ from the standard Gaussian distribution with zero mean and unit variance.
2. rand(N,1). Generates an $N \times 1$ vector of numbers from the continuous uniform distribution on [0, 1].

Example histograms with $N = 2000$ samples for the two random number generators are shown in Figure 1.31, where n is the number of samples in a particular bin along the x-axis. By decreasing the bin width and increasing the number of samples, these histograms approach the pdfs:

$$f_X(x) = \frac{1}{\sqrt{2\pi}} \exp(-x^2/2) \text{ (standard Gaussian)}, \tag{1.79}$$

$$f_X(x) = I_{[0,1]}(x) \text{ (uniform on unit interval)}, \tag{1.80}$$

where $I(x)$ is the indicator function (see Appendix B). However, even for an infinite number of samples, the histograms and the random number generators are still only an approximation because it is not possible to generate all real numbers due to finite-precision arithmetic used in a computer.

[1] MATLAB is a registered trademark of The Mathworks, Inc., 3 Apple Hill Drive, Natick, MA.

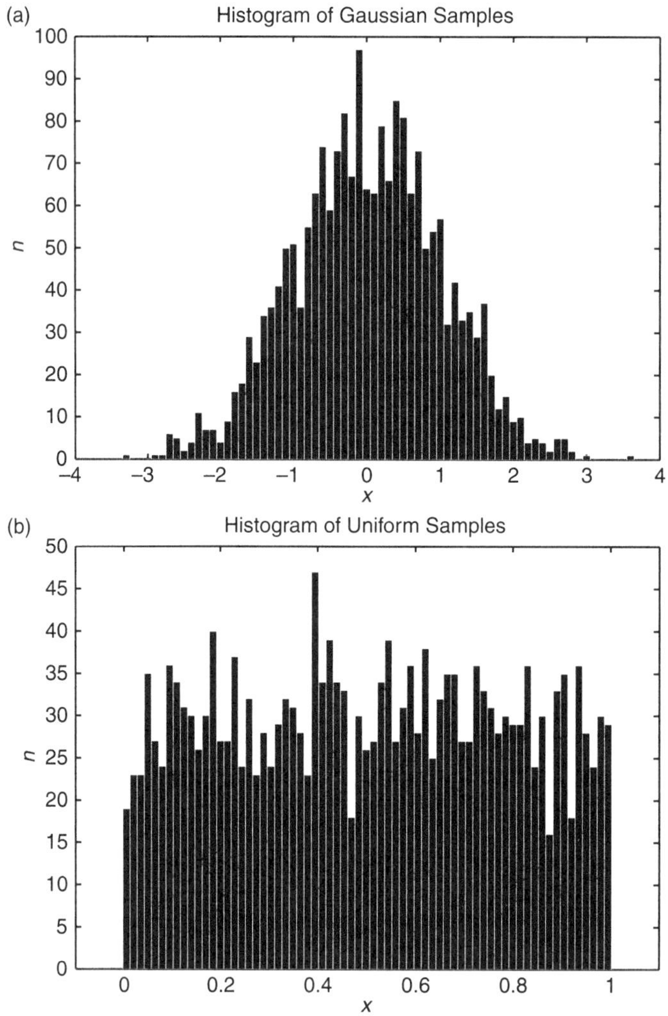

FIGURE 1.31 Histograms for two random number generators in MATLAB (2000 samples). (a) Standard Gaussian. (b) Uniform on [0, 1].

In the event that we want to change the mean and variance for the Gaussian random number generator, the random sequence is modified as follows:

$$Y[k] = \sigma_Y X[k] + \mu_Y. \tag{1.81}$$

Sequence $Y[k]$ also has a Gaussian distribution, but with mean μ_Y and variance σ_Y^2. Note that $X[k]$ is multiplied by σ_Y first, and then μ_Y is added; $\sigma_Y(X[k] + \mu_Y)$ does not produce the same result. Examples of random sequences generated by rand and randn are shown in Figure 1.32. Obviously the Gaussian realization has greater range than the uniform realization whose values are restricted to be $\in [0, 1]$. The sample at each time instant is independent of all other samples; generating *correlated* sequences using MATLAB is covered later in Chapter 5.

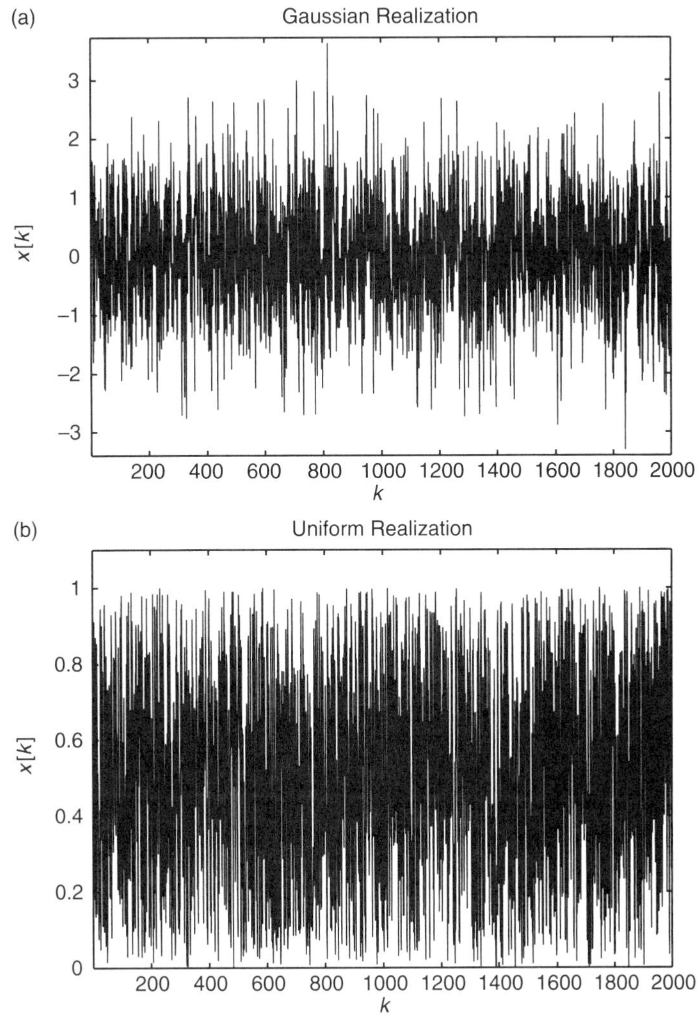

FIGURE 1.32 Example realizations of uncorrelated random sequences. (a) Standard Gaussian. (b) Uniform on [0, 1].

Random generator randn is useful because Gaussian processes arise often in practice, such as noise in a communication system. It is relatively easy to transform from one random variable to another, and this is more easily done starting with rand. If X is a uniform random variable with support [0, 1], and $F_Y(y)$ is the cdf of the desired random variable Y, then using the transformation techniques in Chapter 4, it is easily shown that Y is obtained as follows:

$$Y = F_Y^{-1}(X), \tag{1.82}$$

where $F_Y^{-1}(\cdot)$ is the inverse image of the cdf. The cdf and inverse cdf (icdf) for the standard Gaussian random variable are shown in Figure 1.33. In a computer simulation, a sample of X would be generated using rand,

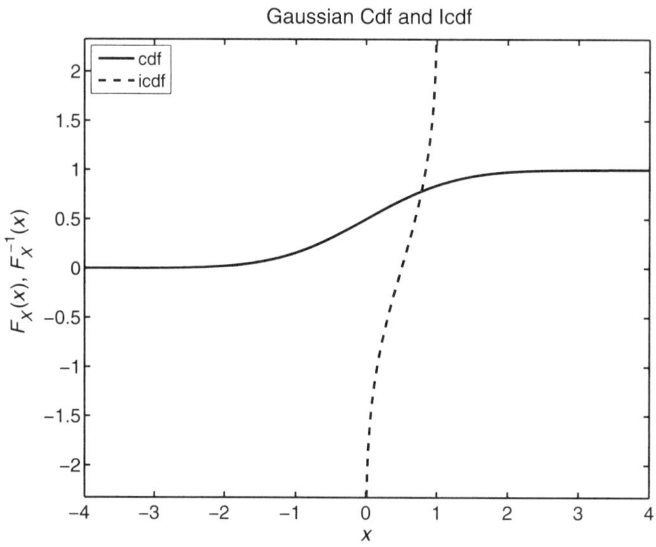

FIGURE 1.33 Gaussian cdf and icdf.

and then it would be transformed to Y via (1.82). Observe in Figure 1.33 that the range of X for the icdf is the support [0, 1] of rand. This is not a coincidence. As discussed in Chapter 3, the cdf of Y is the following probability:

$$F_Y(y) \triangleq P(Y \leq y), \tag{1.83}$$

and thus $F_Y(y) \in [0, 1]$ by the axioms of probability (this is seen in Figure 1.33). The support of $F_Y(y)$ is $\mathcal{R}$, although some random variables have finite support, so that $F_Y(y)$ is either zero or one before or after some finite value (as is the case for the uniform random variable). The range of the icdf is, in general, the entire real line as implied by the curve in Figure 1.33. The interval for rand has been chosen specifically to be uniform on [0, 1] so that it exactly matches the range of a cdf. Mapping rand samples via the icdf gives the corresponding samples on $\mathcal{R}$ for the random variable defined by the particular icdf. However, note that MATLAB has random number generators for many of the distributions covered in this book. These random number generators are implemented using random("type," "parameters") where "type" is the name of the random variable, and "parameters" is a list of the specific values of the pdf parameters (location, scale, shape, degrees of freedom, or threshold).

1.3.2 Filtering

There are two basic functions for filtering sequences in MATLAB:

- filter($\mathbf{b}, \mathbf{a}, \mathbf{x}$): generates an output vector that is the same size as the input vector $\mathbf{x}$, where $\mathbf{a} \triangleq [a_0, \ldots, a_{N-1}]^T$ and $\mathbf{b} \triangleq [b_0, \ldots, b_{M-1}]^T$ contain the filter coefficients of the linear DE in (1.66). The direct-form II configuration in Figure 1.30(a) is used.
- conv($\mathbf{h}, \mathbf{x}$): generates a $2N - 1$ output vector realizing the convolution in (1.47), but with summation range $\{0, \ldots, N - 1\}$ and for finite-length coefficient vector $\mathbf{h} = [h[0], \ldots, h[N - 1]]^T$ and input vector $\mathbf{x} = [x[k], \ldots, x[k - N + 1]]^T$.

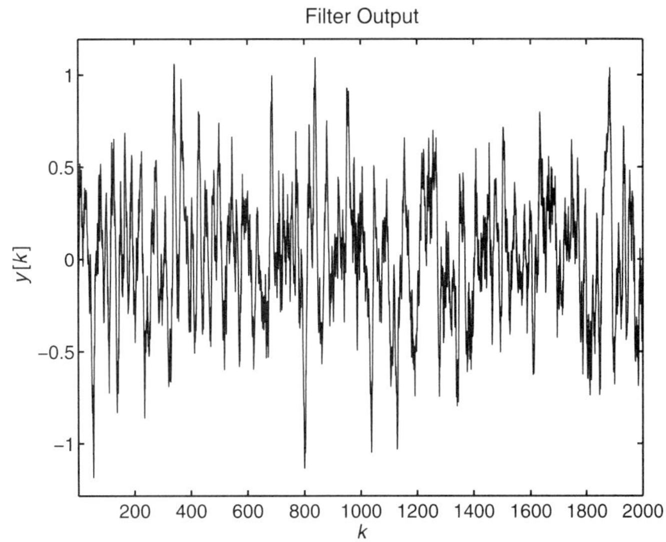

FIGURE 1.34 Example filter output realization of the Gaussian random sequence in Figure 1.32(a).

An FIR filter is realized using filter by setting $\mathbf{a} = 1$. conv also implements an FIR filter, but differs from filter because its output extends beyond the length of the input vector $\mathbf{x}$. This is achieved by assuming that values of $x[k]$ not contained in $\mathbf{x}$ are zero. As a result, filter is preferred in most filtering applications. An example of an FIR filter using filter($\mathbf{b}, \mathbf{a}, \mathbf{x}$) operating on the sequence in Figure 1.32(a) with $\mathbf{b} = (1/10)[1, \ldots, 1]^T \in \mathcal{R}^{10}$ and $\mathbf{a} = 1$ is shown in Figure 1.34. This filter has a low-pass frequency response, which we see yields an output that is less "noisy" than the input; the output sequence is correlated.

PROBLEMS

Delta Functions

1.1 (a) Find the area of $g(x)$ on $[-1, 2.5]$:

$$g(x) = 5\delta(x) + 3\delta(x-2) - 2\delta(x-3). \tag{1.84}$$

(b) Find a general expression for $h(x)$ on $\mathcal{R}^+ = [0, \infty)$:

$$h(x) = \int_{-\infty}^{x} g(u)\mathrm{d}u. \tag{1.85}$$

1.2 (a) Convolve the following functions:

$$g(x) = \sum_{n=1}^{3} 2^n \delta(x-n), \quad h(x) = \exp(-x)u(x). \tag{1.86}$$

(b) Repeat (a) for

$$g(x) = \delta(ax + b), \quad h(x) = \exp(-x)u(x). \tag{1.87}$$

1.3 (a) Find an expression for $\delta(x^2 - 1)$ in terms of two delta functions. (b) Show that $u(1/4 - x^2)$ is identical to the rectangle function.

1.4 (a) Convolve the following two functions:

$$g[k] = \sum_{n=1}^{3} 2^n \delta[k - n], \quad h[k] = \sum_{n=1}^{2} \delta[k - n]. \tag{1.88}$$

(b) Repeat (a) for

$$g[k] = \sum_{n=0}^{1} 3^n \delta[k - n], \quad h[k] = (1/3)^k u[k], \tag{1.89}$$

where $u[k]$ is the discrete unit-step function.

1.5 The Dirac delta function can be modeled using the rectangle function as follows (see Appendix B):

$$\delta(x) = \lim_{a\to 0}(1/a)\text{rect}(x/a). \tag{1.90}$$

(a) Use this definition to find an expression for $\delta(\alpha x)$ for $\alpha \neq 0$. (b) Define the derivative $\delta'(x)$ and show that

$$\int_{-\infty}^{\infty} \delta'(x)\text{d}x = 0, \quad \int_{-\infty}^{\infty} x\delta'(x)\text{d}x = -1. \tag{1.91}$$

1.6 The Dirac impulse train is defined as

$$s(x) \triangleq \sum_{n=-\infty}^{\infty} \delta(x - n). \tag{1.92}$$

(a) Find an expression for $s(\alpha x)$ for $\alpha \neq 0$. (b) Prove the following:

$$\delta(\sin(\pi x)) = (1/\pi)s(x). \tag{1.93}$$

Continuous-Time Signals and Systems

1.7 Find the impulse response $h(t)$ for the following system with input $x(t)$ and output $y(t)$:

$$\frac{\text{d}^2 y(t)}{\text{d}t^2} + 2\frac{\text{d}y(t)}{\text{d}t} + y(t) = x(t). \tag{1.94}$$

1.8 Derive the unit-step response for a system with impulse response:

$$h(t) = 2\exp(-t)u(t) + \exp(-(t - 5))u(t - 5). \tag{1.95}$$

1.9 Find Laplace transforms for the following signals and specify their ROCs:

$$x_1(t) = \exp(-5|t|), \quad x_2(t) = \text{erf}(t)u(t), \tag{1.96}$$

where $\text{erf}(t)$ is the error function.

1.10 Repeat Problem 1.9 for

$$x_1(t) = \text{rect}(t), \quad x_2(t) = 1/t, \tag{1.97}$$

1.11 Repeat Problem 1.9 for

$$x_1(t) = t\exp(-3t)u(t-1), \quad x_2(t) = \cos(2\pi t - 5)u(t-2). \tag{1.98}$$

1.12 From a PFE find the inverse Laplace transform for

$$H(s) = \frac{2s}{s^2 + 3s + 2}, \tag{1.99}$$

which has ROC $\text{Re}(s) > -1$.

1.13 Repeat Problem 1.12 for

$$H(s) = \frac{2s}{s^2 - s - 2}, \tag{1.100}$$

which has ROC $-1 < \text{Re}(s) < 2$.

1.14 Convolve the following two functions:

$$x(t) = \exp(-2t)u(t) + \exp(-t+1)u(t-1), \quad h(t) = u(t-2). \tag{1.101}$$

1.15 Find the output $y(t)$ of the following system:

$$\frac{\mathrm{d}y(t)}{\mathrm{d}t} + 4y(t) = x(t) \tag{1.102}$$

for $x(t) = \exp(-t)u(t-1)$.

1.16 The deterministic cross-correlation is computed as follows:

$$y(t) = \int_{-\infty}^{\infty} x(\tau)h(t+\tau)\mathrm{d}\tau. \tag{1.103}$$

(a) Derive an expression for $Y(\omega)$ in terms of $X(\omega)$ and $H(\omega)$. (a) Compute the cross-correlation function for the functions in Problem 1.14.

1.17 Prove Parseval's theorem for real-valued $x(t)$:

$$\int_{-\infty}^{\infty} x^2(t)\mathrm{d}t = \int_{-\infty}^{\infty} |X(f)|^2 \mathrm{d}f. \tag{1.104}$$

1.18 Find Fourier transforms for the following functions:

$$x_1(t) = 5\text{rect}(5t), \quad x_2(t) = \exp(-|t|)\cos(4\pi t). \tag{1.105}$$

1.19 Determine the frequency response $Y(\omega)$ for

$$y(t) = \frac{1}{\pi}\int_{-\infty}^{\infty} \frac{x(\tau)}{t-\tau}\,\mathrm{d}\tau. \tag{1.106}$$

Discrete-Time Signals and Systems

1.20 Find the impulse response $h[k]$ for the following system with input $x[k]$ and output $y[k]$:

$$y[k] + 2y[k-1] + 4y[k-2] = x[k] + 2x[k-1]. \tag{1.107}$$

1.21 Find the unit-step response for a system with impulse response:

$$h[k] = (1/3)^k u[k] - 2u[k-1]. \tag{1.108}$$

1.22 Determine the z-transform for the following signals and specify their ROCs:

$$x_1[k] = (k-1)^2 u[k-1], \quad x_2[k] = (1/2)^k u[k] + 4^k u[k+1]. \tag{1.109}$$

1.23 Repeat Problem 1.22 for

$$x_1[k] = 3^k u[-k+1], \quad x_2[k] = 2u[k] - 2u[k-5]. \tag{1.110}$$

1.24 Find DTFTs for the following signals:

$$x_1[k] = \mathrm{sinc}[k-1], \quad x_2[k] = k(1/2)^{k-1}u[k]. \tag{1.111}$$

1.25 Consider the sample mean

$$\bar{x}[k] = \frac{1}{k}\sum_{n=0}^{k} x[n]. \tag{1.112}$$

(a) Specify the zeros of this system for $k = 2$. (b) Rewrite $\bar{x}[k]$ in recursive form in terms of $\bar{x}[k-1]$ and sketch a digital filter implementation.

1.26 Convolve the following two functions:

$$x[k] = a^{k-1} I_{\mathcal{N}}[k], \quad h[k] = u[k] - u[k-N], \tag{1.113}$$

where $I_{\{a,b\}}[k]$ is the discrete indicator function and $\mathcal{N} = \{1, 2, \ldots\}$ are the natural numbers.

1.27 The deterministic cross-correlation function is given by

$$y[k] = \sum_{n=-\infty}^{\infty} x[n]h[k+n]. \tag{1.114}$$

Find the cross-correlation of the two functions in Problem 1.26.

1.28 Prove the derivative property of the z-transform: $kx[k] \to -z\mathrm{d}X(z)/\mathrm{d}z$.

1.29 Find the inverse z-transform for

$$H(z) = \frac{2(z+1)}{(z+1/2)(z-1/3)} \tag{1.115}$$

using Cauchy's residue theorem (see Appendix C) when the ROC is $|z| > 1/2$.

1.30 Repeat Problem 1.29 when the ROC is $1/3 < |z| < 1/2$.

1.31 Use Cauchy's residue theorem to find the inverse z-transform of the following transfer function such that $h[k]$ is a left-sided impulse response function:

$$H(z) = \frac{3}{(z+1/2)(3/2-z)}. \tag{1.116}$$

1.32 Find expressions for the magnitude and phase of the DTFT $H(j\omega)$ for

$$H(z) = \frac{5z}{(z-1/4)(z+1/5)}. \tag{1.117}$$

1.33 Repeat Problem 1.32 for the following finite-length sequence:

$$x[k] = \exp(j\omega_o k) I_{\{0,\ldots,N-1\}}[k]. \tag{1.118}$$

Sampling of Continuous-Time Signals

1.34 The continuous-time signal $x(t) = \exp(-\alpha t)u(t)$ is uniformly sampled with sampling period T_s: $x_s(t) = x(t)s(t)$, where $s(t)$ is the ideal impulse train in (1.43). (a) Find an expression for $X_s(\omega)$ showing that $X(\omega)$ is repeated at integer multiples of $2\pi/T_s$. (b) Derive the DTFT $X(j\omega)$ of the sampled signal $x[kT_s]$ and demonstrate that there is aliasing by comparing it to $X_s(\omega)$.

1.35 Instead of using the ideal impulse train, consider sampling $x(t)$ with the following rectangular pulse train:

$$p(t) = \sum_{n=-\infty}^{\infty} \text{rect}(t/T_p - nT_s), \tag{1.119}$$

where $T_p \ll T_s$. The Fourier series of $p(t)$ is

$$p(t) = \sum_{m=-\infty}^{\infty} p_m \exp(j2\pi mt/T_s) \tag{1.120}$$

with

$$p_m = \frac{1}{T_s}\int_{-T_p}^{T_p} \exp(-j2\pi mt/T_s)\mathrm{d}t. \tag{1.121}$$

Find a closed-form representation for p_m and use it to write an expression for the sampled waveform $x_p(t) = x(t)p(t)$ and its Fourier transform $X_p(\omega)$.

1.36 Consider the following continuous-time signal:

$$x(t) = 2\cos(200\pi t) + 4\cos(400\pi t) + 3\cos(600\pi t). \tag{1.122}$$

Let $x(t)$ be sampled using an ideal impulse train $s(t)$ to generate the sampled waveform $x_s(t)$. Find and sketch the Fourier transform of $x_s(t)$ for the following sampling rates. (a) $f_s = 300$ Hz. (b) $f_s = 500$ Hz. Comment about any aliasing that might occur.

1.37 The ideal reconstruction filter has frequency response:

$$H(\omega) = T_s \text{rect}(\omega/2W) = \begin{cases} T_s, & |\omega| \leq W \\ 0, & \text{else}, \end{cases} \tag{1.123}$$

where W is the one-sided bandwidth of the sampled waveform $x(t)$. The Fourier transform of the original waveform is obtained as $X(\omega) = H(\omega)X_s(\omega)$ assuming that the Nyquist rate is satisfied: $f_s = 1/T_s > 2W$. Find an expression for $x(t)$ as a convolution of the samples $\{x(kT_s)\}$ and the impulse response $h(t)$ of $H(\omega)$.

Discrete-Time Filters

1.38 A discrete-time system has the following transfer function:

$$H(z) = \frac{5(z-2)}{(z-0.3)(z^2+0.1z-0.6)}. \tag{1.124}$$

(a) Find and sketch the direct-form I and direct-form II filter structures. (b) Using a PFE and a product of first-order sections, respectively, find and sketch parallel and cascade filter structures.

1.39 Repeat Problem 1.38 for

$$H(z) = \frac{2(z^2-(5/6)z+1/6)}{(z+1/2)(z^2-1/16)}. \tag{1.125}$$

1.40 A discrete-time filter has impulse response:

$$h[k] = (0.8)^k u[k] + (-0.4)^{k-1} u[k-1]. \tag{1.126}$$

(a) Derive and sketch a direct-form II implementation for this filter. (b) Write the output $y[k]$ as a difference equation with input $x[k]$.

1.41 Consider the following system:

$$H(z) = \frac{z+2}{(z^2-(1/12)z-1/12)(z+1/2)}. \tag{1.127}$$

(a) Write a difference equation for the output $y[k]$ in terms of the input $x[k]$. (b) Derive the parameters $\{\mathbf{A}, \mathbf{b}, \mathbf{c}, d\}$ for the state-space realization defined by (1.68). (c) Repeat part (b) for the state-space realization defined by (1.75).

1.42 Repeat Problem 1.41 for the system in (1.124).

Computer Problems

1.43 Generate $N = 1000$ samples using randn and plot the resulting Gaussian sequence $x[k]$. Process $x[k]$ to generate output $y[k]$ using a filter with the following transfer function:

$$H(z) = \frac{2}{z-p}. \tag{1.128}$$

Use filter and several different values of p.

1.44 Generate $N = 1000$ samples using rand and plot the uniformly distributed random sequence $x[k]$. Convert $x[k]$ to an exponential random sequence with parameter λ using the appropriate icdf in MATLAB. Comment about the range of outcomes of the sequences for different parameter values: (a) $\lambda = 1$ and (b) $\lambda = 0.1$.

1.45 Use randn to generate 10 samples of a Gaussian random variable, map them to ± 1 via the sign function, and then insert a zero in between each bit using upsample, resulting in a sequence of 20 symbols consisting of 0s and ± 1s. Use rectpulse to map each symbol to a rectangle function of duration 100 samples, and plot the resulting PAM waveform using stairs.

FURTHER READING

The topics on signals and systems reviewed at the end of this chapter can be explored further in the following references. Dirac delta function: Bracewell (1978) and Weisstein (2003). Continuous-time linear systems: Lathi (1965), Lathi (2000), and Oppenheim, Willsky, and Nawab (1996). Digital signal processing: Hamming (1983), Oppenheim and Schafer (2009), Mitra (2010), Rabiner and Gold (1975), and Stearns and Hush (1990). State-space formulations: Kailath (1980), Mendel (1995), and Tretter (1976). Signal processing with MATLAB: Chaparro (2011), Childers (1997), Ingle and Proakis (2011), and McClellan, Schafer, and Yoder (1998).

PART I
Probability, Random Variables, and Expectation

2 Probability Theory

2.1 INTRODUCTION

In this chapter, we present a theory of probability that is based on *three axioms* and is consistent with our intuition about the probability of an event. This will require a review of some background material on set theory, which will lead to definitions of an *event* and a *field* that are needed so that probabilities can be assigned and computed. Perhaps the simplest and most widely used example of a random experiment is the toss of a single coin. Although relatively trivial, we use this experiment often to give definitions and describe properties of random events. Since there are only two outcomes in this experiment, heads (H) and tails (T), it is also a model for a digital communication system which employs a binary *alphabet* $\{0, 1\}$ (outcomes $\{-1, 1\}$ are also used when symmetry about zero is desired). Even though initially the coin-toss experiment appears to be lacking much structure, it is actually quite useful and relevant for many applications.

Another experiment that will be used to demonstrate various properties of probability is the toss of a single die, which has more complexity than the coin-toss experiment because there are multiple outcomes: $\{1, 2, 3, 4, 5, 6\}$. Although any experiment with several outcomes would be sufficient for our purposes, we use the die-toss experiment because it is familiar to most readers. This example also has relevance in communications, as a model for pulse amplitude modulation (PAM) where *symbols* are selected from a finite alphabet containing 4, 8, 16, and so on, elements for transmission across a channel.

With these two simple experiments, we can describe the important concepts needed to construct a probability space for experiments with a *finite* number of outcomes. With an understanding of these examples, the probability model can be extended to experiments with an *infinite* number of outcomes. There are three basic types of random experiments as determined by the number of outcomes: (i) a finite number of outcomes (such as a single coin toss), (ii) an infinite but *countable* number of outcomes (such as tossing a single coin ad infinitum), and (iii) a continuum (uncountable number) of outcomes (such as temperature measurements). More complicated experiments are combinations of these three types of outcomes.

It is relatively straightforward to assign probabilities to events for experiments in category (i). Extending those results to experiments in category (ii) requires some work, but is also straightforward because the outcomes are countable: there is a one-to-one mapping of outcomes to the natural numbers $\mathcal{N} = \{1, 2, \ldots\}$. Experiments in category (iii) are the most difficult to quantify because there is a continuum of outcomes: for any two values $a < b$, there exists $\epsilon > 0$ such that $a < a + \epsilon < b$. Thus, it is not immediately obvious how to assign probabilities to an uncountable number of outcomes; in fact, it turns out that the probability of a single outcome in a continuous experiment is zero. Events in continuous experiments require special consideration when specifying the probability space.

As a preview, we summarize how probabilities are generally computed for each of the three types of experiments:

- *Finite*. Probabilities are computed by *counting* the outcomes comprising an event of interest, relative to the total number of outcomes in the sample space Ω. Basic rules of *combinatorics* are used for finite problems.

Probability, Random Variables, and Random Processes: Theory and Signal Processing Applications, First Edition. John J. Shynk.

TABLE 2.1 Basic Types of Random Experiments with Examples

Outcomes	Finite	Infinite
Discrete	Tossing two coins (*finite*)	Repeated tosses of a single coin (*countably infinite*, denumerable)
Continuous	None	Temperature (*uncountable*, nondenumerable)

- *Countably infinite*. Probabilities are computed by *summing* over a discrete probability mass function that characterizes the experiment. Techniques for computing finite and *infinite sums* are used for countably infinite problems.
- *Uncountable*. Probabilities are computed by *integrating* over a continuous probability density function that characterizes the experiment. Techniques from *calculus* are used for uncountable problems.

Finite experiments can also be solved using techniques for countably infinite problems, but it is often easier to simply count outcomes as mentioned above. Countably infinite and uncountable experiments problems are not easily examined without a more rigorous characterization, as provided by the *probability mass function* (pmf) and the *probability density function* (pdf). As shown later, it is also possible to handle countably infinite experiments using methods from calculus (via the Dirac delta function), but the results essentially reduce to evaluating finite and infinite sums.

Examples for each of the three cases are summarized in Table 2.1. It is obvious that by tossing two coins there are only four outcomes: $\{HH, TT, HT, TH\}$. If the probability of at least one H is of interest, we simply count the number of such outcomes (which is three) and divide that by the total number of outcomes. Thus, the probability of at least one H is 0.75; it is denoted by $P(\text{at least one } H) = P(HH \text{ or } HT \text{ or } TH) = 0.75$. For the countably infinite example in the table, we might be interested in the probability that H first appears after the second toss. Clearly, this is more difficult to compute than the previous finite-outcome example: we need to consider all *sequences* of the form $TTHT\cdots$, $TTHH\cdots$, $TTTH\cdots$, and so on. As shown later, the pmf allows for a straightforward probability computation, which is easier than attempting to directly count all outcomes of interest. Likewise for the uncountable (continuous) case, where the pdf is used to compute probabilities via an integration over the outcomes of interest, which are described by *intervals* on $\mathcal{R}$.

Although we use the term *experiment* to describe how random outcomes are generated, in real applications the notion of experiment is artificial because events "just occur" (such as temperature fluctuations, radioactive decay, and so on). However, we use the terms experiment and *trial* not only for a *synthetic* experiment, such as tossing a die or using a computer program to generate a random number, but also for *natural* events. In the latter case, the experiment can be viewed as a model of the underlying mechanisms for such events.

A probabilistic modeling of events in the physical world is advantageous because most phenomena are too complicated to accurately represent using physical models. For example as mentioned in Chapter 1, the toss of a single coin could be modeled using many factors, such as the height from which it is tossed, the velocity and angle of the toss, and the type of surface upon which it lands, and so on. Of course, such a model for even this simple experiment is complicated and undoubtedly incomplete. By using a probabilistic model, we bypass the need for an accurate physical model, and can make predictions about the occurrence of an event with a high degree of accuracy, and which is consistent with our intuition about randomness (the *frequency* of outcomes as mentioned in Chapter 1). A probabilistic model is a powerful representation that allows us to "say something" about events that arise due to the complicated interactions of many underlying physical mechanisms.

2.2 SETS AND SAMPLE SPACES

We begin with basic definitions of sets and sample spaces, which should be a review for most readers.

Definition: Set A *set* is a collection of objects or numbers that represent those objects. These objects are called *elements* or *points* of the set.

Although it is possible to accurately describe a set using the original objects or outcomes in many experiments, it is often cumbersome, especially for a large number of elements. Instead, we represent the elements of a set using numbers—either integers, real numbers, or complex numbers—because it is more convenient mathematically, and is actually necessary when defining *random variables* in Chapter 3. For example, in experiments involving the toss of a single coin, it will be convenient to assign numbers such as $1 \equiv H$ and $0 \equiv T$ to represent the two outcomes.

Sets will be denoted by uppercase letters, usually at the beginning of the Latin alphabet, as illustrated in the following examples.

Example 2.1. Set $A = \{2, 4, 6, 8\}$ consists of four integers, and set $B = \{\ldots, -1, 0, 1, 2, \ldots\}$ consists of all integers.

The set of all integers is represented by $\mathcal{Z}$ and positive integers by $\mathcal{Z}^+$, which *includes* zero.

Example 2.2. Set $C = (1, 5)$ consists of all real numbers between 1 and 5 *excluding* 1 and 5, whereas set $D = [1, 5]$ includes the end points.

The set of all real numbers is represented by $\mathcal{R}$ and positive real numbers by $\mathcal{R}^+$, which *includes* zero. We often use the symbols $\pm\infty$ which technically are not real numbers. Although complex quantities are not used extensively in this book, they do appear in the applications part; the symbol for all complex numbers is $\mathcal{C}$.

Element a in set A is written as $a \in A$; element b that is not in set A is written as $b \notin A$. The type of set can be characterized according to the number of objects.

Definition: Cardinality The *cardinality* of a set is the number of elements in that set. The cardinality of set E is denoted by $|E|$.

A countable set can have either a finite or an infinite number of elements; the former is called *finite* (cardinality $< |\mathcal{N}|$), and the latter is *countably infinite* or *denumerable* (cardinality $= |\mathcal{N}|$). The elements of a countably infinite are in one-to-one correspondence with the natural numbers $\mathcal{N} \triangleq \{1, 2, \ldots\}$ (note that $\mathcal{N}$ does *not* include zero).

Example 2.3. $\mathcal{Z}$ and $\mathcal{Z}^+$ have the same cardinality as $\mathcal{N}$ and thus are countably infinite. The rational numbers, denoted by $\mathcal{Q}$ and having the form a/b with $a, b \in \mathcal{Z}$ $(b \neq 0)$ also have cardinality $|\mathcal{N}|$.

Example 2.4. The irrational numbers have a cardinality $> |\mathcal{N}|$ and are uncountable. Likewise, $\mathcal{R}$ and any interval on $\mathcal{R}$ have a cardinality $> |\mathcal{N}|$, and both are uncountable. "Almost all" irrational numbers are transcendental, of which π and e are perhaps the most well known of the set.

Set A in Example 2.1 is finite, whereas set B is countably infinite. If an infinite set is not countable, then it is *uncountable* or *nondenumerable* with cardinality $> |\mathcal{N}|$. Set C in Example 2.2 is uncountable. Countable sets are also called *discrete* sets, while uncountable sets are *continuous*.

These set distinctions are useful later when developing a probability space for events of an experiment. It is necessary that we precisely define events of an experiment and the corresponding *algebra* of operations which allow for a consistent probability measure. Figure 2.1 summarizes the three basic types of sets with simple examples. An example of a *mixed* set consisting of discrete and continuous components is also included. Although a mixed set is by definition uncountable (the fact that it has a discrete component does not change the fact that the continuous part is uncountable), it will be convenient to identify it as having a finite or countably infinite component when discussing random variables in Chapter 3. In the example shown in Figure 2.1, note that the union operator $\cup$ is used to describe the entire set; this set operation will be covered later.

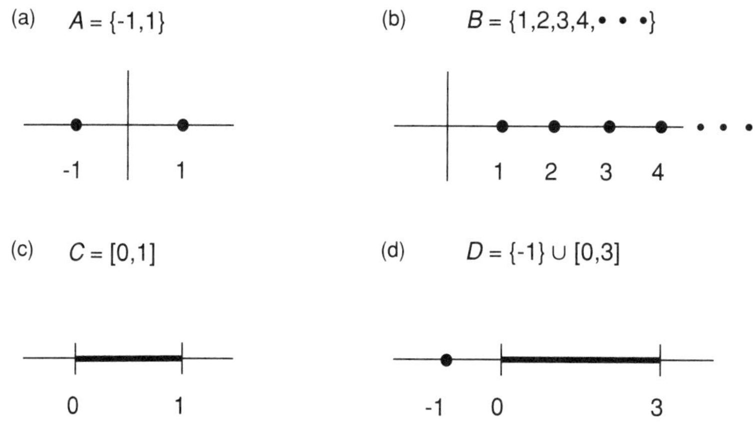

FIGURE 2.1 Examples of the three basic types of sets and a mixed set. (a) Finite. (b) Countably infinite. (c) Uncountable (an interval). (d) Mixed (an interval and one point).

Regarding set notation, we generally use braces $\{\cdot\}$ to denote a discrete set (either finite or countably infinite), such as in Example 2.1. However, it is also convenient to use braces when a *function* describes a continuous set, as demonstrated by the following example.

Example 2.5. Discrete set: $A = \{x^3 - 1 : x = 0, 2, 4\} = \{-1, 7, 63\}$. Continuous set: $B = \{x : x \geq -1\} = [-1, \infty)$. A set defined in terms of other sets: $C = \{x : x \in A \text{ or } x \in B\} = A \cup B$.

The colon (:) in this context should be read as "such that" (s.t.). The last case describes the union operation "or" ($\cup$): C is the set of all x such that x is in set A or B or *both*. Parentheses $(\cdot)$ and brackets $[\cdot]$ are also used to describe uncountable sets, such as those in Examples 2.2 and 2.5. The various types of *intervals* for uncountable sets are defined next.

Definition: Intervals and Singletons The *open interval* (a, b) is the set containing all real numbers between a and b (with $a < b$), but excluding the boundary points a and b. The *closed interval* $[a, b]$ is the set containing all real numbers between a and b, including the boundary points a and b. The *semi-open intervals* $[a, b)$ and $(a, b]$ are obtained from $[a, b]$ by excluding one of the boundary points: b is excluded in $[a, b)$ and a is excluded in $(a, b]$. The boundary points are called *singletons*, as is any individual point in $\mathcal{R}$.

We can also define intervals using the following brace notation (as was done in Example 2.5):

$$
\begin{aligned}
(a, b) &\triangleq \{x : a < x < b\}, \quad (a, b] \triangleq \{x : a < x \leq b\}, \\
[a, b) &\triangleq \{x : a \leq x < b\}, \quad [a, b] \triangleq \{x : a \leq x \leq b\}.
\end{aligned}
\tag{2.1}
$$

Since an interval is a continuous set, the following mathematical description for an open interval can be formulated. If $x \in (a, b)$, then there exists $\epsilon > 0$ such that $x + \epsilon \in (a, b)$ and $x - \epsilon \in (a, b)$. Regardless of how close x is to a or b in an *open interval*, we can always find numbers between $x \in (a, b)$ and the points a and b. Of course, this property does not apply to semi-open and closed intervals because x could be chosen as one of the boundary points.

In order to specify events and their probabilities, we define the basic components of an experiment.

Definition: Sample Space Ω The *sample space* Ω is the set of all possible outcomes in an experiment. It is also called the *universe* or the *universal set*. An element of Ω is denoted by ω and may include a subscript such as ω_n.

Example 2.6. $\Omega = \{0, 1\}$ for the two possible symbols used in a binary digital communication system. This also corresponds to the sample space for a single toss of a coin with $H \equiv 1$ and $T \equiv 0$.

Example 2.7. $\Omega = [0, \infty) = \mathcal{R}^+$ for the infinity of outcomes in an experiment using the set of nonnegative real numbers. For many experiments, $\Omega = (-\infty, \infty)$ which is the set of all real numbers $\mathcal{R}$. Note that we always use an open interval to exclude $\pm\infty$ because, as mentioned earlier, they are not real numbers.

For a group of sets in the universe Ω, it is possible to define some fundamental set relationships.

Definition: Subset $A \subset B$ A is a *subset* of B if all elements of A are in B, whereas B may have elements that are not in A. The notation $A \subseteq B$ allows for the possibility that A and B are identical sets.

Example 2.8. If $A = [0, 5]$ and $B = [0, 10)$, then $A \subset B$. The semi-open interval $(a, b]$ is a subset of the closed interval $[a, b]$. Clearly, any set is a subset of itself: $A \subset A$. If all elements of B are also in A, then the two sets are equal.

Definition: Equality $A = B$ Sets A and B are *equal* if they have exactly the same elements such that $A \subset B$ and $B \subset A$.

Since the sample space Ω contains all outcomes for an experiment, we can describe elements that exist outside of a particular set.

Definition: Complement A^c The *complement* of A is the set of elements in Ω that are not in A. Thus, $a \notin A$ implies that $a \in A^c$. Using the brace notation: $A^c = \{x : x \notin A\}$.

Example 2.9. If $\Omega = [0, 100]$ and $A = [0, 25)$, then $A^c = [25, 100]$. If $\Omega = \mathcal{R}$ and $B = (0, \infty)$, then $B^c = (-\infty, 0]$.

The complement of A is also denoted by $\bar{A}$, but we use A^c because the bar notation is used later for sample moments (such as the sample mean). Note that Ω *must* be defined in order for A^c to have meaning. We conclude this section with two important sets.

Definition: Empty Set ϕ The *empty set* is the set containing no elements. It is also sometimes called the *null* set. From the previous definitions, it is clear that $\phi = \Omega^c$.

Sometimes the empty set is denoted by { }. The empty set refers to the event "nothing occurs." As discussed later, the empty set is needed when defining an *algebra* of operations for sets.

Definition: Power Set $\mathcal{P}(\Omega)$ The *power set* contains all possible subsets of the sample space Ω. It includes the empty set ϕ as well as Ω. The power set can also be defined for any subset $E \subset \Omega$; it is denoted by $\mathcal{P}(E)$.

Example 2.10. If $\Omega = \{2, 4, 6\}$, then $\mathcal{P}(\Omega) = \{\phi, \Omega, 2, 4, 6, \{2, 4\}, \{2, 6\}, \{4, 6\}\}$ has eight elements. The power set of $E = \{2, 4\} \subset \Omega$ is $\mathcal{P}(E) = \{\phi, E, 2, 4, \{2, 4\}\}$. The power set of a countably infinite set (such as $\mathcal{N}$) is not easily visualized, but it is uncountably infinite and has the same cardinality as $\mathcal{R}$.

If E is finite, then $|\mathcal{P}(E)|$ is necessarily finite.

Proposition 2.1. *The power set of E consisting of N elements has cardinality* $|\mathcal{P}(E)| = 2^N$.

Proof. This is proved using the binomial theorem in Appendix E which states that

$$(x+y)^N = \sum_{n=0}^{N} \binom{N}{n} x^{N-n} y^n, \tag{2.2}$$

where

$$\binom{N}{n} = \frac{N!}{n!(N-n)!} \tag{2.3}$$

is a binomial coefficient (discussed later in connection with probabilities for finite sample spaces). A binomial coefficient is the number of subsets (combinations) of size n chosen from N elements. Summing over n to count all subsets and letting $x = y = 1$ in (2.2) yields the number of elements in $\mathcal{P}(E)$:

$$\sum_{n=0}^{N} \binom{N}{n} = (1+1)^N = 2^N, \tag{2.4}$$

which completes the proof. □

With these basic definitions involving sets, we can consider operations on sets.

2.3 SET OPERATIONS

Most problems of interest in probability involve operations on events. The two basic set operations are union and intersection.

Definition: Union $A \cup B$ The *union* of two sets is the set of all elements in A or B or both:

$$A \cup B = \{x : x \in A \text{ or } x \in B\}. \tag{2.5}$$

The sum notation $A + B$ is sometimes used to denote the union operation.

Definition: Intersection $A \cap B$ The *intersection* of two sets is the set of elements common to A and B:

$$A \cap B = \{x : x \in A \text{ and } x \in B\}. \tag{2.6}$$

The product notation AB is sometimes used to denote the intersection operation.

Example 2.11. If $A = \{1, 2, 3\}$ and $B = \{3, 4\}$, then $A \cup B = \{1, 2, 3, 4\}$ and $A \cap B = \{3\}$. If $E = [2, \infty)$ and $F = (-\infty, 5]$, then $E \cup F = \mathcal{R}$ and $E \cap F = [2, 5]$. If $\Omega = \mathcal{R}$, $A = (a, b)$, and $B = [a, b]$, then $AB = A$, $A \cup B = B$, $A^c B = \{a, b\}$ is finite, and $AB^c = \phi$.

The union and intersection operations can be extended to any number of sets, including an infinite collection. Examples of the notation used for N sets $\{A_n\}$ are

$$B_1 = \bigcup_{n=1}^{N} A_n, \qquad B_2 = \bigcap_{n=1}^{N} A_n. \tag{2.7}$$

For convenience, we generally use the shorthand notation AB for intersection (though this may not always be possible), but will retain the conventional notation $A \cup B$ for union because $A + B$ could be confusing later when the probabilities of events are added.

It is straightforward to show that union and intersection satisfy the following three basic laws:

- *Commutative:*

$$A \cup B = B \cup A, \qquad AB = BA. \tag{2.8}$$

- *Associative:*

$$(A \cup B) \cup C = A \cup (B \cup C), \qquad (AB)C = A(BC). \tag{2.9}$$

- *Distributive:*

$$A \cup (BC) = (A \cup B)(A \cup C), \qquad A(B \cup C) = (AB) \cup (AC). \tag{2.10}$$

These operations also satisfy De Morgan's laws which combine intersection, union, and complement as follows:

- *De Morgan's laws:*

$$(A \cup B)^c = A^c B^c, \qquad (AB)^c = A^c \cup B^c. \tag{2.11}$$

These laws are all easily verified using a Venn diagram, which is described later, and they are readily extended to any number of sets.

Next, we illustrate how intervals behave under union and intersection operations, which are not obvious for *infinite* unions and intersections arising in continuous sample spaces. For appropriate values of $\{a_n\}$, $\{b_n\}$, c, and d, the union and intersection of intervals on the real line $\mathcal{R}$ have the following properties:

- A *finite* union or intersection of *closed* intervals yields a *closed* interval:

$$\bigcup_{n=1}^{N} [a_n, b_n] = [c, d], \qquad \bigcap_{n=1}^{N} [a_n, b_n] = [c, d]. \tag{2.12}$$

- A *finite* union or intersection of *open* intervals yields an *open* interval:

$$\bigcup_{n=1}^{N} (a_n, b_n) = (c, d), \qquad \bigcap_{n=1}^{N} (a_n, b_n) = (c, d). \tag{2.13}$$

- A *countably infinite union* of *open* intervals yields an *open* interval:

$$\bigcup_{n=1}^{\infty} (a_n, b_n) = (c, d). \tag{2.14}$$

- A *countably infinite intersection* of *closed* intervals yields a *closed* interval:

$$\bigcap_{n=1}^{\infty} [a_n, b_n] = [c, d]. \tag{2.15}$$

The first two properties involving a finite number of open or closed sets are obvious. It is relatively straightforward to verify the third and fourth properties, especially after we discuss the next two properties which are not obvious. The operations in the four cases above all give the same type of interval as the original intervals.

- A *countably infinite union* of *closed* intervals can yield *any* type of interval (open, semi-open, or closed):

$$\bigcup_{n=1}^{\infty}[a_n, b_n] = (c, d) \text{ or } (c, d] \text{ or } [c, d) \text{ or } [c, d]. \tag{2.16}$$

- A *countably infinite intersection* of *open* intervals can yield *any* type of interval (open, semi-open, or closed):

$$\bigcap_{n=1}^{\infty}(a_n, b_n) = [c, d] \text{ or } (c, d] \text{ or } [c, d) \text{ or } (c, d). \tag{2.17}$$

These last two properties are interesting because the set operations can result in a type of interval that is different from the component intervals. Nothing in general can be said about the *infinite union* of *closed* intervals, nor about the *infinite intersection* of *open* intervals. The following two examples illustrate this behavior.

Example 2.12. If $(a, b_n) = (-1, 1/n)$, then $\bigcap_{n=1}^{\infty}(a, b_n) = (-1, 0]$. Observe that all the open intervals include zero, and thus their intersection—even an infinite number of them—must also include zero, so the intersection yields a semi-closed interval.

Example 2.13. If $[a_n, b] = [1/n, 1]$, then $\bigcup_{n=1}^{\infty}[a_n, b] = (0, 1]$. It is clear that since none of the sets include zero, their union can never contain zero, even an infinite number of unions. The resulting interval must be semi-open.

Variations of these examples can be devised to illustrate the other three types of intervals in (2.16) and (2.17). For convenience, Table 2.2 summarizes the possible results for set operations involving intervals. An entry in the table with multiple intervals means that the set operation *could* take on any one of those forms (though not necessarily), depending on how the initial interval in the first column is defined. We mention these results because an algebra of sets must be closed under the union and intersection operations, and thus a σ-field (defined later for $\mathcal{R}$) cannot consist solely of one type of interval.

The intersection operation is useful for defining the following important property of sets.

Definition: Mutually Exclusive Sets A and B are *mutually exclusive* if they have no elements in common such that $AB = \phi$. Mutually exclusive sets are also called *disjoint*.

This definition can be extended to any number of sets $\{A_n\}$ such that the intersection of any two sets is disjoint (called pairwise disjoint): $A_n A_m = \phi$ for all $n \neq m$. Moreover, a sample space can be divided into a collection

TABLE 2.2 Set Operations of Intervals

Type of Interval	Finite Union	Finite Intersection	Countably Infinite Union	Countably Infinite Intersection
$[a_n, b_n]$	$[c, d]$	$[c, d]$	$(c, d), (c, d], [c, d), [c, d]$	$[c, d]$
$[a_n, b_n)$	$[c, d)$	$[c, d)$	$(c, d), [c, d)$	$[c, d), [c, d]$
$(a_n, b_n]$	$(c, d]$	$(c, d]$	$(c, d), (c, d]$	$(c, d], [c, d]$
(a_n, b_n)	(c, d)	(c, d)	(c, d)	$(c, d), (c, d], [c, d), [c, d]$

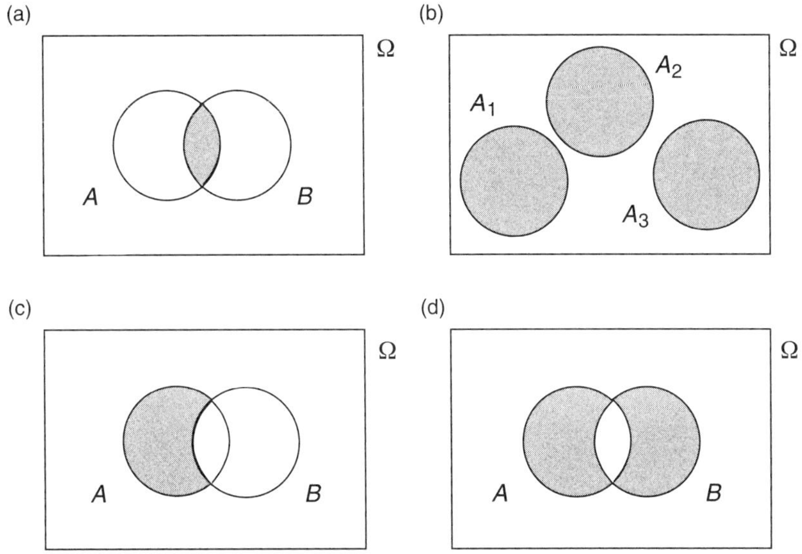

FIGURE 2.2 Set operations (denoted by the shaded regions) are visualized using Venn diagrams. (a) $A \cap B$. (b) Partition: $A_1 \cup A_2 \cup A_3 = \Omega$. (c) Difference $A - B$. (d) Exclusive or $A \oplus B$.

of disjoint sets that together contain all elements of Ω. Such a collection of sets is called a partition of the sample space.

Definition: Partition Disjoint sets $\{A_n\}$ form a *partition* of Ω if $\bigcup_n A_n = \Omega$.

Partitions of a sample space are not unique, and the number of possible partitions is finite only if Ω is finite. If the sets $\{A_n\}$ are not disjoint but we still have $\cup_n A_n = \Omega$, then $\{A_n\}$ are *collectively exhaustive*.

It is convenient at this point to utilize a *Venn diagram* to illustrate various set relationships. A Venn diagram is a simple pictorial representation of sets, often using circles to represent sets within a sample space that illustrate how they are related by their degree of overlap. Figure 2.2(a) shows a Venn diagram for $A \cap B$. Note that $A \cap B = \phi$ if A and B are disjoint, and $A \cup B = \Omega$ because in this example there are no elements outside A and B. Figure 2.2(b) shows a partition of a sample space into three disjoint sets $\{A_1, A_2, A_3\}$. The circles are nonoverlapping, and all points of the sample space are contained in the union of these three sets: $A_1 \cup A_2 \cup A_3 = \Omega$.

We conclude this section with two additional set operations that are frequently useful.

Definition: Difference $A - B$ The *difference* of sets A and B is the set of elements in A excluding those also in B. It can be written explicitly using previous set definitions as $A - B = AB^c$. The notation $A \setminus B$ is sometimes used to denote the difference of two sets.

A Venn diagram illustrating the difference of sets A and B is shown in Figure 2.2(c). Observe that if A and B are disjoint, then $A - B = A$ and $B - A = B$. The difference operator does not satisfy the commutative law: $A - B \neq B - A$ (except trivially when $B = A$ such that $A - B = \phi$). The difference of two sets is not a symmetric operation. It is easy to verify $A - B = AB^c$ using a Venn diagram.

Definition: Exclusive Or $A \oplus B$ The *exclusive or* operation yields the set of elements in A or B but *not both*: $A \oplus B = A \cup B - AB = (A - B) \cup (B - A)$. Because of this last formulation, the exclusive or operation is also called the *symmetric difference*.

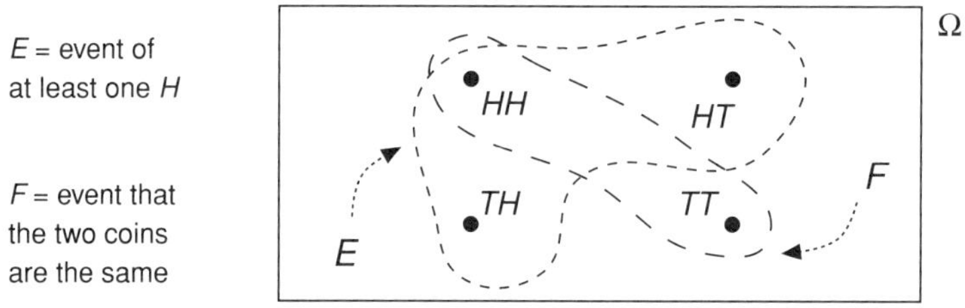

FIGURE 2.3 Sample space for two coins tossed simultaneously showing two nonelementary events.

A Venn diagram example for $A \oplus B$ is shown in Figure 2.2(d).

2.4 EVENTS AND FIELDS

The sample space contains all possible outcomes of an experiment; these outcomes are the elements (points) of Ω. However, in most applications we are not usually interested in single outcomes, but in subsets of Ω called events.

Definition: Event *Event E* of an experiment is a subset of the sample space Ω whose elements share a common feature.

Example 2.14. The experiment where two coins are tossed simultaneously has the four outcomes shown in Figure 2.3 which are the four elements of Ω. Consider the event of at least one H given by $E = \{HH, HT, TH\}$. Thus, E is not a specific outcome, but is instead a *feature* of three of the four possible outcomes. Obviously, E is a subset of Ω.

Example 2.15. For the experiment of tossing a single die, the sample space is $\Omega = \{1, 2, 3, 4, 5, 6\}$. Subset $E = \{1, 3, 5\}$ is the event that the die shows an odd value.

Example 2.16. In an experiment where two dice are tossed, the event that they show the same value is the subset $A = \{\{1, 1\}, \{2, 2\}, \{3, 3\}, \{4, 4\}, \{5, 5\}, \{6, 6\}\}$.

Example 2.17. For $\Omega = [0, 1]$, $A = [0, 1/2]$ and $B = [0, 1/4] \cup [5/8, 7/8]$ are examples of events.

Note that an element $\omega_n \in \Omega$ corresponding to a single outcome is a particular type of event.

Definition: Elementary Event An *elementary event* is a subset of Ω consisting of one element, and is the same as a single outcome.

Sometimes an elementary event is also called an atomic event, but we reserve the name *atom* for a specific component of the σ-field defined later.

Example 2.18. Subset $F = \{1\}$ in Example 2.15 is an elementary event, as is the subset $B = \{2, 2\}$ in Example 2.16.

Since our goal is to assign probabilities to events of an experiment, it is important that events be defined to satisfy *algebraic rules* that allow for consistency in the probability space. In order to be rigorous about operations involving events, we define a *field* as a collection of events that satisfy certain algebraic conditions. Such a definition allows us to include only specific subsets of Ω, which is especially important for continuous Ω.

Definition: Field $\mathcal{F}$ A *field* is the nonempty set of events of an experiment that satisfy the following conditions:

(i) If $A \in \mathcal{F}$, then $A^c \in \mathcal{F}$.

(ii) If $A_n \in \mathcal{F}$ for $n = 1, 2, \ldots, N$, where N is finite, then $\bigcup_{n=1}^{N} A_n \in \mathcal{F}$.

A field is also called an *algebra* of events. Observe that these conditions imply the following properties:

(iii) $\Omega \in \mathcal{F}$.

(iv) $\phi \in \mathcal{F}$.

(v) If $A_n \in \mathcal{F}$ for $n = 1, 2, \ldots, N$, where N is finite, then $\bigcap_{n=1}^{N} A_n \in \mathcal{F}$.

Property (v), which is obtained using one of De Morgan's laws, is sometimes included as a condition in the definition of a field, but it need not be mentioned explicitly because (i) and (ii) are sufficient.

Example 2.19. The simplest (but trivial) field is $\mathcal{F} = \{\phi, \Omega\}$.

Example 2.20. For the experiment of a single coin toss with $\Omega = \{H, T\}$, consider the nontrivial field $\mathcal{F} = \{\phi, \Omega, H, T\}$. The properties above are readily verified: (i) $H \in \mathcal{F} \Longrightarrow T \in \mathcal{F}$ (and vice versa), (ii)–(iii) $H \in \mathcal{F}, T \in \mathcal{F} \Longrightarrow H \cup T \equiv \Omega \in \mathcal{F}$, and (iv)–(v) $H \in \mathcal{F}, T \in \mathcal{F} \Longrightarrow H \cap T \equiv \phi \in \mathcal{F}$. Note that $\mathcal{F}$ is, in fact, the power set $\mathcal{P}(\Omega)$ for this experiment, and no other fields are possible, except the trivial field in Example 2.19.

In some experiments, we might first consider a set of events that do not quite comprise a field, and then include additional elements to *generate* a field. This procedure is illustrated in Example 2.21.

Example 2.21. Suppose that $\Omega = [1, \infty)$ and consider $\mathcal{F} = \{\phi, \Omega, \{[1, a)\}\}$ for every $a > 1$, that is, $\mathcal{F}$ contains all semi-open intervals of the form $[1, a)$. Obviously, $\mathcal{F}$ is not a field because $[1, a)^c = [a, \infty)$ for any $a > 1$ is not in $\mathcal{F}$. Expand the set to be $\mathcal{F} = \{\phi, \Omega, \{[1, a)\}, \{[b, \infty)\}\}$ with $b > 1$. Again, this is not a field because $[1, a) \cap [b, \infty) = [b, a)$ for any $b < a$ is not in $\mathcal{F}$. Expand the set further to be $\mathcal{F} = \{\phi, \Omega, \{[1, a)\}, \{[b, \infty)\}, \{[c, d)\}\}$ for all $a, b, c, d > 1$. We now have a field: all *finite* unions of the elements are events in $\mathcal{F}$, as are all complements of the elements. Later, we demonstrate that $\mathcal{F}$ is *not* a σ-field which is defined for an *infinity* of unions, and is described after the next set of examples.

Example 2.22. For the experiment of tossing two coins, recall that $\Omega = \{HH, TT, HT, TH\}$. The elements of the power set $\mathcal{P}(\Omega)$ are summarized in Table 2.3. Properties (i) and (ii) are easily verified for this case. From this example, it is useful to summarize some characteristics of an event. The four outcomes in the experiment

TABLE 2.3 Elements of Field $\mathcal{F} = \mathcal{P}(\Omega)$ for Example 2.22

Elementary events (the four outcomes):	$\{HH\}, \{TT\}, \{HT\}, \{TH\}$
Events defined by two outcomes:	$\{HH, TT\}, \{HH, HT\}, \{HH, TH\}, \{TT, HT\}, \{TT, TH\}, \{HT, TH\}$
Events defined by three outcomes:	$\{HH, TT, HT\}, \{HH, TT, TH\}, \{HH, HT, TH\}, \{TT, HT, TH\}$
ϕ and Ω:	$\phi = \Omega^c, \Omega = \{HH, TT, HT, TH\}$

are given by the terms in the first row of Table 2.3. These are the elementary events; all other entries in the table are not elementary events. For example, $\{HH, HT\}$ is the event that the first coin is H; $\{HH, TT\}$ is the event that the two coins have the same outcome. These two events are *not* outcomes themselves; instead, they refer to those outcomes that share a common feature. Although event $\{HH, HT, TH\}$ corresponds to those outcomes such that there is at least one H, events with three outcomes together can be more complicated to describe. Event $\{HH, TT, HT\}$ is not so easy to state; it corresponds to the situation where the coins have the same outcome *or* the first coin is H and the second coin is T. Alternatively for this event, we can say that it refers to those outcomes for which the first coin is not T *and* the second coin is not H (thus excluding TH). Since we require all unions of events in $\mathcal{F}$ to be in the field, Ω must be in $\mathcal{F}$. Likewise, $\phi \in \mathcal{F}$ so that complement and intersection are consistent for any event in $\mathcal{F}$. Observe that there are 16 elements in $\mathcal{F}$, as expected because $N = 4$; the power set for this case has cardinality $|\mathcal{P}(\Omega)| = 2^4$.

The power set $\mathcal{P}(\Omega)$ by definition is the largest field possible. Fields are not unique, as mentioned previously and as shown in Example 2.23. Generally, smaller fields are not useful for experiments with finite and even countably infinite outcomes. However, for uncountable experiments, it is necessary that we choose a σ-field that is smaller than the power set.

Example 2.23. From Example 2.19, we know that $\{\phi, \Omega\}$ is a field; it obviously satisfies properties (i)-(ii). Of course, this trivial field contains no useful information about any particular experiment. Consider the experiment of tossing a single die with outcomes $\{1, 2, 3, 4, 5, 6\}$. The power set $\mathcal{P}(\Omega)$ consists of $2^6 = 64$ elements. However, it is possible to define a smaller (nontrivial) field, though as we shall see it is not as useful as the power set. Consider the following two events: $E = \{1, 3, 5\}$ and $F = \{2, 4, 6\}$, and observe that $\mathcal{F} = \{\phi, \Omega, E, F\}$ is a field. Properties (i) and (ii) are satisfied; in particular, $E = F^c \in \mathcal{F}$ and $E \cup F = \Omega \in \mathcal{F}$. Although this is a valid field, it results in a simplification of the experiment such that we are concerned only about the outcome being odd or even; the specific number showing on the die is of no interest. In fact, by defining such a field, this "simplified" experiment is structurally identical to a single toss of a coin where, for example, $H \equiv$ even number and $T \equiv$ odd number showing on the die. Tossing a single die and using the smaller field above *simulates* tossing a single coin.

Consider again the situation where we start with some subset of Ω and then generate a field that includes the subset.

Example 2.24. For the experiment of tossing a single die, we would like to generate a field starting with the subset $\{1, 2\}$. Observe that $\{1, 2\}^c = \{3, 4, 5, 6\}$, $\{1, 2\} \cap \{3, 4, 5, 6\} = \phi$, and $\{1, 2\} \cup \{3, 4, 5, 6\} = \Omega$. Thus the field *generated* by this subset is $\mathcal{F} = \{\phi, \Omega, \{1, 2\}, \{3, 4, 5, 6\}\}$. Obviously, this field fails to capture all outcomes and events of the original experiment; in fact, it is equivalent to a binary experiment such as the toss of a single coin (as in Example 2.23). Consider starting with the subset $\{\{1, 2\}, \{3, 4\}\}$. Then it is easy to show that the generated field is $\mathcal{F} = \{\phi, \Omega, \{1, 2\}, \{3, 4\}, \{5, 6\}, \{3, 4, 5, 6\}, \{1, 2, 5, 6\}, \{1, 2, 3, 4\}\}$. In order to capture all details of the single-die experiment, we would need to choose the power set for the field with $2^6 = 64$ elements.

These last examples illustrate the difference between a "fine" field (the power set) and a "coarse" field (given by some subset of the power set). Later when conditional probability is covered, we use this distinction between fine and coarse fields to derive a useful property in probability theory when conditioning on events.

Next, we describe a property of a field $\mathcal{F}$ when Ω is finite, based on the definition of atoms.

Definition: Atoms of a Field The smallest events in $\mathcal{F}$ (excluding ϕ) are *atoms*. We denote the set of atoms by $\mathcal{A}$.

This definition means that an atom cannot be obtained from the union of other events in $\mathcal{F}$, and they are disjoint. Every other event in $\mathcal{F}$ is obtained by set operations on the elements of $\mathcal{A}$ (in order to satisfy the requirements

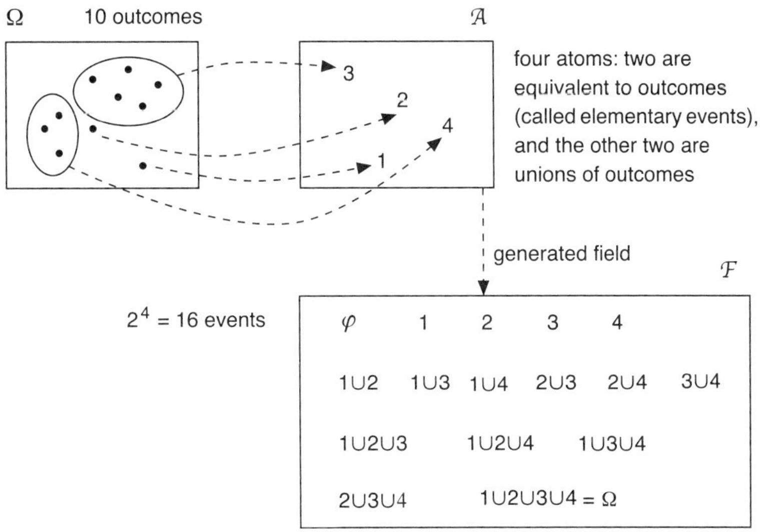

FIGURE 2.4 Example outcomes in Ω, atoms in $\mathcal{A}$, and events in $\mathcal{F}$.

of a field). A simple finite example showing the connection between outcomes in Ω, atoms in $\mathcal{A}$, and events in $\mathcal{F}$ is illustrated in Figure 2.4.

Example 2.25. For the second field in Example 2.24, the three atoms are $\mathcal{A} = \{\{1, 2\}, \{3, 4\}, \{5, 6\}\}$, which are not outcomes in Ω.

It is clear that the atoms of the power set for an experiment are the elementary events (outcomes). If individual outcomes are excluded from the field, then the atoms are nonelementary events as shown in Example 2.24.

Theorem 2.1. The number of elements in any finite field $\mathcal{F}$ is given by 2^n, where $n = |\mathcal{A}|$.

Proof. From the definition of an atom and the requirements of a field, the elements of $\mathcal{F}$ must be the power set $\mathcal{P}(\mathcal{A})$ of the atoms. We have already shown that the cardinality of the power set is 2^N for a set of size N, and thus $N = n$ completes the proof. □

This theorem is useful because we can immediately determine that $\mathcal{F}$ is *not* a field if the number of elements is not 2, 4, 8, 16, and so on. Note that Ω need not be discrete in order for the field to be finite, as demonstrated in Example 2.26.

Example 2.26. Let $\Omega = [0, \infty) = \mathcal{R}^+$ and $\mathcal{F} = \{\phi, \Omega, [0, c), [c, \infty)\}$ for some $c > 0$. Thus, although the sample space is continuous, the field has only four elements with atoms given by two subintervals: $[0, c)$ and $[c, \infty)$. Suppose instead that the atoms are $[0, c)$, $[c, d)$, and $[d, \infty)$ for some $d > c > 0$. From the theorem above, we know for this continuous sample space that the field has eight elements: $\mathcal{F} = \{\phi, \Omega, [0, c), [c, d), [d, \infty), [0, d), [c, \infty), [0, c) \cup [d, \infty)\}$. Such intervals on $\mathcal{R}$ are important for continuous sample spaces, as discussed later when we introduce Borel sets.

For problems involving a finite sample space, the definition of a field with conditions (i) and (ii) is adequate for us to specify a probability space. However, if the sample space is infinite, either countably infinite or uncountable, then it is necessary that the definition of a field be extended to include an infinite (but countable) union of events.

Definition: Sigma Field $\mathcal{F}$ A *sigma-field* (σ*-field*) is a field satisfying condition (i), with (ii) extended to a countably infinite number of subsets as follows:

(ii′) If $A_n \in \mathcal{F}$ for $n = 1, 2, \ldots$, then $\bigcup_{n=1}^{\infty} A_n \in \mathcal{F}$.

A σ-field is also called a σ*-algebra*.

Of course, this modified definition causes (v) to be extended as follows:

(v′) If $A_n \in \mathcal{F}$ for $n = 1, 2, \ldots$, then $\bigcap_{n=1}^{\infty} A_n \in \mathcal{F}$.

Example 2.27. From Example 2.21, $\mathcal{F} = \{\phi, \Omega, \{[1, a)\}, \{[b, \infty)\}, \{[c, d)\}\}$ for every $a, b, c, d > 1$ is a field for $\Omega = [1, \infty)$, but it is not a σ-field. Suppose that $[c, d)$ is written as $[c + 1/n, d)$, where n is a positive integer. Then we know from the previous discussion on the different types of intervals and Table 2.2 that

$$\bigcup_{n=1}^{\infty} [c + 1/n, d) = (c, d), \tag{2.18}$$

which is a type of interval not in $\mathcal{F}$.

Next, we consider the cardinality of infinite sets.

Definition: Beth Numbers The *beth numbers* $\{\beth_n\}$ (Forster, 1995) are used to represent the cardinality of different types of infinite sets.

In mathematics, the cardinality of the natural numbers $\mathcal{N}$ is denoted by the symbol $\beth_0$ which is called *beth null*. The beth numbers are generated by the following recursion:

$$\beth_{n+1} = 2^{\beth_n}, \tag{2.19}$$

which we see has a form similar to that in Proposition 2.1 for the cardinality of $\mathcal{P}$ for a finite set. It turns out that the cardinality of $\mathcal{R}$ is beth one $\beth_1 = 2^{\beth_0}$ which is the size of the power set for $\mathcal{N}$. Similarly, the cardinality of $\mathcal{P}(\mathcal{R})$ is beth two $\beth_2$. We are not concerned with the theory behind the cardinality of infinite sets and the nature of the different types of infinity; instead, we use beth numbers as a convenient notation to represent infinite sets that have the same cardinality. Several examples are provided in Table 2.4, some of which are defined later.

For $\Omega = \mathcal{N}$, the power set $\mathcal{P}(\mathcal{N})$ is a σ-field. In fact, the power set for *any* Ω is a σ-field, and it is clear from the previous discussions that σ-field $\mathcal{F}$ for any Ω satisfies $\{\Omega, \phi\} \subseteq \mathcal{F} \subseteq \mathcal{P}(\Omega)$.

Definition: Event Space An *event space* is the set of events in Ω that comprise the σ-field $\mathcal{F}$. The event space is written as the double $\{\Omega, \mathcal{F}\}$.

TABLE 2.4 Beth Numbers and Cardinality of Infinite Sets

Beth Number	Infinite Sets with the Same Cardinality
$\beth_0$	$\mathcal{N}, \mathcal{Q}, \mathcal{Z}$
$\beth_1$	$\mathcal{R}, \mathcal{R} - \mathcal{Q}, \mathcal{P}(\mathcal{N}), \mathcal{B}(\mathcal{R}), \mathcal{C}, [a, b] \subset \mathcal{R}$, Cantor set
$\beth_2$	$\mathcal{P}(\mathcal{R}), \mathcal{P}(\mathcal{P}(\mathcal{N}))$

The event space is designed to include events in Ω that satisfy the conditions of the specific σ-field $\mathcal{F}$. We extend the idea of generating a field described in Examples 2.21 and 2.24 to a σ-field.

Definition: Generated σ-Field For subset $\mathcal{G} \subseteq \Omega$, define the following set of σ-fields:

$$E = \{\mathcal{F} \subseteq \mathcal{P}(\Omega) : \mathcal{G} \subseteq \mathcal{F}\}. \tag{2.20}$$

The σ-field *generated* by subset $\mathcal{G}$ and denoted by $\sigma(\mathcal{G})$ is obtained as the intersection of all elements in E.

Since $\mathcal{G}$ is contained in every σ-field for Ω, $\sigma(\mathcal{G})$ is necessarily the *smallest* σ-field containing $\mathcal{G}$. If $\mathcal{G}$ itself is a σ-field, then $\sigma(\mathcal{G}) = \mathcal{G}$.

Example 2.28. Consider the finite sample space $\Omega = \{-1, 0, 1\}$. The following are all the possible σ-fields:

$$\mathcal{F}_1 = \{\phi, \Omega, -1, 0, 1, \{-1, 0\}, \{-1, 1\}, \{0, 1\}\}, \tag{2.21}$$

$$\mathcal{F}_2 = \{\phi, \Omega, -1, \{0, 1\}\}, \tag{2.22}$$

$$\mathcal{F}_3 = \{\phi, \Omega, 0, \{-1, 1\}\}, \tag{2.23}$$

$$\mathcal{F}_4 = \{\phi, \Omega, 1, \{-1, 0\}\}, \tag{2.24}$$

$$\mathcal{F}_5 = \{\phi, \Omega\}. \tag{2.25}$$

If $\mathcal{G} = \Omega$, then $\sigma(\mathcal{G}) = \mathcal{F}_1 \cap \mathcal{F}_2 \cap \mathcal{F}_3 \cap \mathcal{F}_4 \cap \mathcal{F}_5 = \mathcal{F}_5$ which is the trivial σ-field. If $\mathcal{G} = \{1\}$, then $\sigma(\mathcal{G}) = \mathcal{F}_1 \cap \mathcal{F}_4 = \mathcal{F}_4$. If $\mathcal{G} = \{0, 1\}$, then $\sigma(\mathcal{G}) = \mathcal{F}_1$, which is the power set $\mathcal{P}(\Omega)$.

We are interested in σ-fields that are *measurable* so that probabilities can be assigned in a consistent manner to the event space $\{\Omega, \mathcal{F}\}$. For finite and countably infinite experiments, there is generally no need to consider σ-fields smaller than the power set $\mathcal{P}(\Omega)$. However, for continuous sample spaces, the power set is "too large." The Vitali set described at the end of this chapter for $\Omega = [0, 1]$ is an example of a subset that is not measurable. It will be convenient from a practical sense, as well as for most applications, to consider *intervals* for continuous experiments. Also, since sample spaces are mapped to the real line $\mathcal{R}$ for the random variables in Chapter 3, we are interested in a specific type of σ-field for intervals on $\mathcal{R}$ called the Borel σ-field.

Definition: Borel σ-Field The *Borel σ-field* $\mathcal{B}(\mathcal{R})$ on the real line is the smallest σ-field generated by all open intervals in $\Omega = \mathcal{R}$. Elements of $\mathcal{B}(\mathcal{R})$ are called *Borel sets.*

The Borel σ-field is generated by starting with all open intervals in $\mathcal{R}$ of the form (a, b). Individual points (singletons) are included by observing that

$$a = \lim_{n\to\infty} (a - 1/n, a + 1/n). \tag{2.26}$$

This result allows us to include all semi-open intervals:

$$(a, b] = (a, b) \cup b, \qquad [a, b) = (a, b) \cup a. \tag{2.27}$$

Since $(-\infty, b] = (b, \infty)^c$ and $[a, \infty) = (-\infty, a)^c$, all closed intervals are also included:

$$[a, b] = [a, \infty) \cap (-\infty, b]. \tag{2.28}$$

Thus, $\mathcal{B}(\mathcal{R})$ consists of all types of intervals on $\mathcal{R}$ as well as all individual points. It is not easy to visualize subsets of $\mathcal{R}$ that lie outside of $\mathcal{B}(\mathcal{R})$; we present an example of such a subset when measure theory is discussed later. However, it turns out that these subsets do not arise in practical applications and so they are not of concern. Since we will also consider σ-fields in $\Omega = \mathcal{R}^N$ when *random vectors* are covered in Chapter 4, the above definition extends to the smallest σ-field of subsets of $\mathcal{R}^N$ denoted by $\mathcal{B}(\mathcal{R}^N)$, which is defined to be open *rectangles* on $\mathcal{R}^2$ and open hyper-rectangles on $\mathcal{R}^N$ (for $N > 2$).

Example 2.29. Examples of *Borel sets* include the following: (i) closed interval $[a, b]$, (ii) open interval (a, b), (iii) set of rational numbers $\mathcal{Q}$, (iv) set of irrational numbers $\mathcal{R} - \mathcal{Q}$, (v) $\{1, 2, 3, 4, 5, 6\}$, and (vi) the Cantor set described later.

2.5 SUMMARY OF A RANDOM EXPERIMENT

It will be useful at this point to summarize our description of a random experiment for which we want to assign a probability measure:

- The *sample space* Ω is the collection of all outcomes of an experiment. Depending on the type of experiment, it may contain a finite, countably infinite, or uncountable number of elements.
- An *event* of an experiment is a subset of Ω consisting of one or more outcomes, and which usually share some feature. Since the goal is to assign a probability measure to events that is consistent, we may not want to include all possible subsets of Ω.
- A *σ-field* $\mathcal{F}$ is a collection of subsets of Ω that satisfy the algebraic conditions in (i) and (ii′), as well as (iii), (iv), and (v′). One can envision constructing a σ-field by starting with some subset of Ω and expanding the number of subsets until the collection satisfies (i) and (ii′).
- If Ω is finite or countably infinite, then all subsets described by the power set $\mathcal{P}(\Omega)$ comprise a useful σ-field $\mathcal{F}$ for the experiment.
- If Ω is uncountable, then the Borel σ-field $\mathcal{B}(\mathcal{R})$ is a useful σ-field for the real line that consists of all intervals (open, semi-open, and closed) as well as individual points.

With these definitions of the sample space Ω and the event space $\{\Omega, \mathcal{F}\}$, we have a collection of events that satisfy the additivity property in (ii′). It is now possible to formulate a *probability space* which we denote by the triple $\{\Omega, \mathcal{F}, P\}$. We begin with a brief discussion of measure theory, of which the probability measure is a particular case.

2.6 MEASURE THEORY

Measure theory is concerned with assigning a "size" called a *measure* to subsets of a sample space Ω. Depending on the problem, this size might be a length, an area, or a volume in Euclidean space; this is known as the Lebesgue measure which is defined below. In our case, we are interested in assigning a *probability measure* to subsets of Ω that comprise the σ-field $\mathcal{F}$. For uncountable experiments, it is not possible to assign probabilities in a consistent manner to $\mathcal{P}(\mathcal{R})$. The power set of $\Omega = \mathcal{R}$ is too large (it has cardinality beth two $\beth_2$), so instead we choose $\mathcal{F}$ to be $\mathcal{B}(\mathcal{R})$, which is the Borel σ-field on the real line and has cardinality beth one $\beth_1$ (the same as $\mathcal{R}$). The Borel σ-field is useful for essentially all practical applications because we are generally interested in events described by *intervals* on the real line.

Let μ be the notation for a measure that we want to define for the event space, and denote the *measure space* by the triple $\{\Omega, \mathcal{F}, \mu\}$. (Note that μ should not be confused with the mean of a random variable covered later in Chapter 5.)

Definition: Measure A *measure* is a mapping of events $\{E_n\} \in \mathcal{F}$ to $\mathcal{R}$ that has the following properties:

- $\mu(E_n) \geq 0$ for all events $E_n \in \mathcal{F}$.
- $\mu(\phi) = 0$.

- For all events $\{E_n\}$ such that $E_n \cap E_m = \phi$ for $n \neq m$ (mutually exclusive):

$$\mu\left(\bigcup_{n=1}^{\infty} E_n\right) = \sum_{n=1}^{\infty} \mu(E_n). \tag{2.29}$$

Similar properties are shown later for the probability measure where they are known as the *three axioms of probability*.

Example 2.30. The *trivial measure* is $\mu(E_n) = 0$ for all $E_n \in \mathcal{F}$ in the event space $\{\Omega, \mathcal{F}\}$.

Example 2.31. For $\Omega = \mathcal{R}$ and $\mathcal{F} = \mathcal{B}(\mathcal{R})$, the *standard Gaussian measure* is

$$\mu(E) = \frac{1}{\sqrt{2\pi}} \int_E \exp(-x^2/2)\mathrm{d}x, \tag{2.30}$$

which turns out to be the probability $P(E)$ of event E for the standard Gaussian random variable X with zero mean and unit variance.

Example 2.32. The *Dirac measure* for event E in the event space $\{\Omega, \mathcal{F}\}$ is given by

$$\delta_\omega(E) \triangleq \begin{cases} 1, & \omega \in E \\ 0, & \text{else.} \end{cases} \tag{2.31}$$

It can be shown that $\delta_\omega(E)$ is a probability measure, and thus E is the almost sure event in Ω (see Problem 2.23).

Consider two important measures that will be useful later.

Definition: Counting Measure Let Ω be a countable sample space and $\mathcal{P}(\Omega)$ its power set. The *counting measure* for any set $E \in \mathcal{P}(\Omega)$ is

$$C(E) = \begin{cases} |E|, & E \text{ is a finite subset} \\ \infty, & \text{otherwise,} \end{cases} \tag{2.32}$$

where $|E|$ is the cardinality of E. The infinity in this definition is actually beth null $\beth_0$ mentioned above, the size of a countably infinite set.

Obviously, this measure is useful only for finite sets; it certainly provides no information about the size of uncountable events, such as intervals on $\mathcal{R}$. For this latter case, we consider the Lebesgue measure.

Definition: Lebesgue Measure on $\mathcal{R}$ The *Lebesgue measure* of interval $[a, b] \in \mathcal{B}(\mathcal{R})$ is

$$L([a, b]) \triangleq b - a, \tag{2.33}$$

which is the *length* of the interval. Since the Lebesgue measure of a single point (singleton) is defined to be zero, we also have $L((a, b)) = L((a, b]) = L([a, b)) = b - a$.

In order to extend the Lebesgue measure to N-dimensional Euclidean space, it will be convenient to define the Cartesian product.

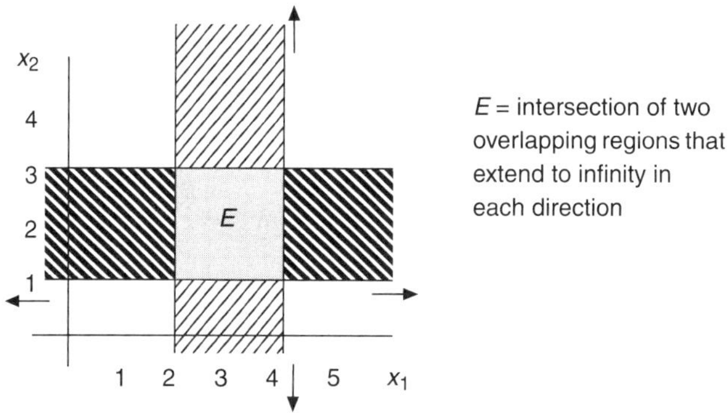

FIGURE 2.5 Cartesian product $E = [2, 4] \times [1, 3]$ for Example 2.33.

Definition: Cartesian Product The *Cartesian product* of a set of intervals $\{[a_n, b_n]\}$, $n = 1, \ldots, N$, is

$$E \triangleq \{(x_1, \ldots, x_N) : x_1 \in [a_1, b_1], \ldots, x_N \in [a_N, b_N]\}. \tag{2.34}$$

E corresponds to an ordered N-tuple, which can be represented using the following product notation:

$$E = [a_1, b_1] \times \cdots \times [a_N, b_N] = \prod_{n=1}^{N} [a_n, b_n]. \tag{2.35}$$

The Cartesian product of intervals on $\mathcal{R}$ is a rectangle on $\mathcal{R}^2$ and a hyper-rectangle on $\mathcal{R}^N$ (for $N > 2$).

Example 2.33. For intervals $[1, 3]$ and $[2, 4]$, the Cartesian product $E = \{(x_1, x_2) : x_1 \in [2, 4], x_2 \in [1, 3]\}$ is a rectangle on $\mathcal{R}^2$ described by all points on the abscissa in the interval $[2, 4]$ *intersected* with all points on the ordinate in the interval $[1, 3]$. This is depicted in Figure 2.5.

For the real line $\mathcal{R}$, we can write $\mathcal{R}^2 = \mathcal{R} \times \mathcal{R}$ (a plane), $\mathcal{R}^3 = \mathcal{R}^2 \times \mathcal{R} = \mathcal{R} \times \mathcal{R} \times \mathcal{R}$ (a hyperplane), and so on. With this geometric viewpoint, we readily see in general that the Lebesgue measure of E corresponds to the area of a rectangle, and in three dimensions it is a *volume*. For the N-dimensions, we have the following definition.

Definition: Lebesgue Measure on $\mathcal{R}^N$ The *Lebesgue measure* of the Cartesian product $E = \prod_{n=1}^{N} [a_n, b_n] \in \mathcal{B}(\mathcal{R}^N)$ is

$$L(E) \triangleq \prod_{n=1}^{N} (b_n - a_n), \tag{2.36}$$

which is the hyper-volume of the corresponding hyper-rectangle on $\mathcal{R}^N$.

Example 2.34. For the Cartesian product in Example 2.33, $L([1, 3] \times [2, 4]) = 4$ is the area of a rectangle, and $L([8, 10] \times [1, 2] \times [4, 8]) = 8$ is the volume of a rectangular cuboid.

The Cantor set described in Example 2.35 is interesting because of the way it is constructed, and because it has Lebesgue measure zero.

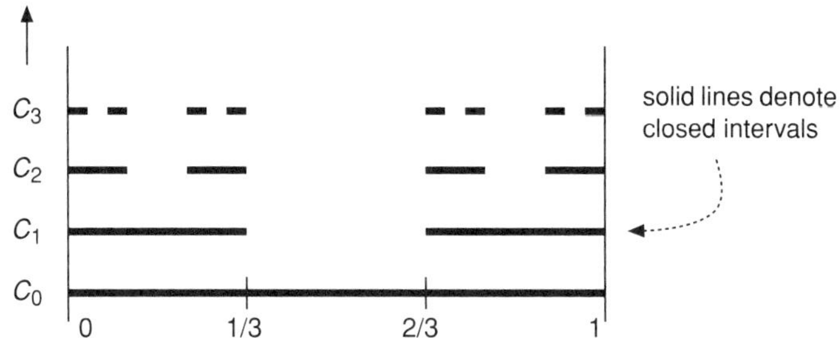

FIGURE 2.6 Sequence of C_m for $m = 1, 2, 3$, leading to the Cantor set C as $m \to \infty$.

Example 2.35 (Cantor set). The Cantor set C (Kingman and Taylor, 1966) on $[0, 1]$ is obtained by removing an infinite set of open intervals as follows:

$$C = [0, 1] - \bigcup_{m=1}^{\infty} \bigcup_{n=0}^{2^{m-1}-1} ((3n + 1)/3^m, (3n + 2)/3^m). \tag{2.37}$$

Let $C_0 = [0, 1]$ and denote successive sets C_m for $m \in \mathcal{N}$ by removing open intervals in the middle third of the remaining closed intervals. Thus

$$C_1 = [0, 1] - (1/3, 2/3) = [0, 1/3] \cup [2/3, 1] \tag{2.38}$$

such that $[0, 1]$ is divided into three regions and the middle open interval is removed. This procedure is performed again for each closed interval remaining in C_1, yielding

$$\begin{aligned} C_2 &= C_1 - (1/9, 2/9) - (7/9, 8/9), \\ &= [1, 1/9] \cup [2/9, 1/3] \cup [2/3, 7/9] \cup [8/9, 1]. \end{aligned} \tag{2.39}$$

Examples of C_m are shown in Figure 2.6. From (2.37), we see that each successive set is obtained from the previous set by scaling the intervals down by a factor of three, and then performing a union of the scaled set with a version of itself shifted by $2/3$, which is described by the general formula

$$C_m = [C_{m-1}/3] \cup [C_{m-1}/3 + 2/3]. \tag{2.40}$$

This process of removing the set of middle-third open intervals is repeated ad infinitum to generate C. Note that instead of subtracting open intervals as in (2.37), the Cantor set can also be defined in terms of intersections as follows:

$$C = \bigcap_{m=0}^{\infty} C_m. \tag{2.41}$$

The Cantor set C has the following interesting properties:

- From (2.37), the length of each open interval removed from $[0, 1]$ is

$$(3n + 2)/3^m - (3n + 1)/3^m = 1/3^m, \tag{2.42}$$

which is the Lebesgue measure of $((3n+1)/3^m, (3n+2)/3)$. For each m, 2^{m-1} open intervals are removed. Thus, by changing variables and recognizing that the sum below is geometric (see Appendix E), the total length removed is

$$\sum_{m=1}^{\infty} 2^{m-1}/3^m = \sum_{m=0}^{\infty} 2^m/3^{m-1} = 3/(1-2/3) = 1, \tag{2.43}$$

which is a surprising result because the length (Lebesgue measure) of $[0, 1]$ is also one. This occurs because at each iteration in the construction of $\{C_m\}$, the end points of the middle interval are not removed, and the fact that the Lebesgue measure of an interval is the same whether or not the end points are included:

- From the preceding result, it follows that C has Lebesgue measure zero.
- C is uncountable with the same cardinality as $[0, 1]$ (which has the same cardinality as $\mathcal{R}$, denoted by beth one $\beth_1$).
- C is a *fractal* because each iteration in its construction (removing the set of middle-third open intervals) results in a reduced-size version C_{m+1} of the previous C_m. This behavior is evident from the first few iterations in Figure 2.6.

The Cantor function based on the Cantor set is described later in Example 3.10.

The Lebesgue measure is useful in Chapter 3 where events in Ω are mapped to the real line $\mathcal{R}$ to form a *random variable*, and to $\mathcal{R}^N$ to form a *random vector*. Since the mapping of an event in Ω to a random variable is well defined (measurable), the probability measure for events in Ω carries over in a consistent manner to events given by intervals on $\mathcal{R}$ (and hyper-rectangles on $\mathcal{R}^N$).

Finally, we define the probability measure for events in the sample space Ω.

Definition: Probability Measure P The *probability measure P* for event space $\{\Omega, \mathcal{F}\}$ is a function that maps events in $\mathcal{F}$ to real numbers in $[0, 1]$, and satisfies the *three axioms of probability*. The resulting *probability space* is denoted by the triple $\{\Omega, \mathcal{F}, P\}$.

It is important to emphasize that P is defined for events in the σ-field $\mathcal{F}$, which may not include all subsets of Ω depending on the experiment. This was, of course, the purpose of constructing $\mathcal{F}$, so that a probability measure could be specified for events that is consistent with the three axioms of probability. In particular, like any measure, the probability measure must satisfy a countable additivity condition in order to "make sense," and which appeals to our intuition about how probabilities should be computed when performing set operations on events in $\mathcal{F}$.

2.7 AXIOMS OF PROBABILITY

The probability measure P of the probability space $\{\Omega, \mathcal{F}, P\}$ satisfies the following axioms:

- *Axiom 1*. $P(E) \geq 0$.
- *Axiom 2*. $P(\Omega) = 1$.
- *Axiom 3*. For $E_n \cap E_m = \phi$ and $n \neq m$, where $\{E_m\} \in \mathcal{F}$:

$$P\left(\bigcup_{m=1}^{\infty} E_m\right) = \sum_{m=1}^{\infty} P(E_m). \tag{2.44}$$

All three axioms are consistent with intuition and the frequency interpretation of probability, as described in Chapter 1. Axiom 1 obviously states that the probability measure is nonnegative. Axiom 2 states that the probability of some event in $\mathcal{F}$ occurring is 1: "something must happen." From Axioms 1 and 2, the probability of any event in $\mathcal{F}$ must lie in the range [0, 1]; events are assigned probabilities that do not exceed 1. P is a special case of the general measure μ, where $P(\Omega) = 1$ is used instead of $\mu(\phi) = 0$. This difference in Axiom 2 is the reason for $P \in [0, 1]$; all measures μ must be nonnegative but they are *not* necessarily bounded by 1 as is the probability measure.

$P(E) = 0$ does not imply that E cannot occur: there is a difference between an event with probability zero (which can occur) and the *impossible event*, which can never occur. Likewise, $P(E) = 1$ does not imply that E must occur: there is a difference between an event with probability one and the *sure event* which must happen. Events that have probability one are known as *almost sure events*. These differences are illustrated by the following example.

Example 2.36. Consider an experiment where the possible outcomes are *uniformly distributed* on $\Omega = [-1, 1]$. As discussed later for continuous sample spaces, the probability of any point in this interval is necessarily zero, for example, $P(E = 0) = 0$. However, $E = 0$ could happen. Events corresponding to numbers outside of $[-1, 1]$ are impossible events: they not only have probability zero, but they can never occur. The empty set ϕ also can never occur, because some event in $[-1, 1]$ must happen. Similarly, $P(F = [-1, 1)) = 1$ does not mean this event must happen: it is possible (though with probability zero) that event $E = 1$ happens. We are *almost sure* that $F = [-1, 1)$ will occur. On the other hand, event $F = [-1, 1] = \Omega$ is a sure event; some subset of this interval must occur, because no numbers outside this interval are possible.

Axiom 3 is the most important of the axioms in terms of ensuring a consistent probability measure; it specifies how the probabilities of several events are combined. If events are *pairwise* mutually exclusive (disjoint), then the probability of their union is the sum of their probabilities. This includes unions for an infinity of events, which is necessary for countably infinite and uncountable experiments for which a σ-field (not just a field) is required. In the next few sections, we cover some basic results in probability, as well as two important topics: *conditional probability* and *independent events*. Later, we consider probability assignments for discrete and continuous samples spaces, in preparation for the random variables discussed in Chapter 3.

2.8 BASIC PROBABILITY RESULTS

The following propositions follow from the three axioms of probability. They also illustrate how to rearrange a problem statement in terms of disjoint sets for which one can use Axiom 3 to derive the result.

Proposition 2.2. $P(\phi) = 0$.

Proof. Since $\Omega \cup \phi = \Omega$ and $\Omega \cap \phi = \phi$:

$$P(\Omega) = P(\Omega \cup \phi) = P(\Omega) + P(\phi), \tag{2.45}$$

which gives $1 = 1 + P(\phi)$ and $P(\phi) = 0$. □

Proposition 2.3. *For event $E \in \Omega$, $P(E^c) = 1 - P(E)$.*

Proof. Since $\Omega = E \cup E^c$ and $E \cap E^c = \phi$:

$$P(\Omega) = P(E^c + E) = P(E^c) + P(E) = 1, \tag{2.46}$$

which can be rewritten as $P(E^c) = 1 - P(E)$. □

Proposition 2.4. *If $E \subset F$, then $P(E) \leq P(F)$.*

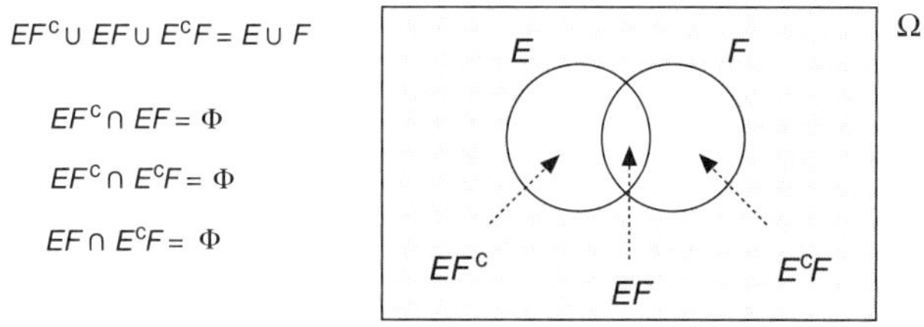

FIGURE 2.7 Partition of $E \cup F$ into three disjoint events.

Proof. Since F contains E, it can be partitioned in terms of E as follows:

$$F = E \cup E^c F. \tag{2.47}$$

Thus

$$P(F) = P(E) + P(E^c F) \geq P(E) \tag{2.48}$$

because all probabilities are nonnegative. □

Proposition 2.5, which is the most important one in this section, can be viewed as an extension of the third axiom to events that are not disjoint.

Proposition 2.5. *For events $E, F \in \Omega$:*

$$P(E \cup F) = P(E) + P(F) - P(EF). \tag{2.49}$$

Proof. Although E and F may not be disjoint, $E \cup F$ can always be written as the union of three mutually exclusive events:

$$E \cup F = EF^c \cup E^c F \cup EF \tag{2.50}$$

as illustrated in Figure 2.7. From the third axiom of probability:

$$P(E \cup F) = P(EF^c) + P(E^c F) + P(EF). \tag{2.51}$$

Observe that E and F can be expressed as the union of disjoint events: $E = EF^c \cup EF$ and $F = E^c F \cup EF$, yielding

$$P(E) = P(EF^c) + P(EF), \qquad P(F) = P(E^c F) + P(EF). \tag{2.52}$$

Substituting these expressions into (2.51) gives

$$\begin{aligned} P(E \cup F) &= P(E) - P(EF) + P(F) - P(EF) + P(EF), \\ &= P(E) + P(F) - P(EF), \end{aligned} \tag{2.53}$$

which completes the proof. □

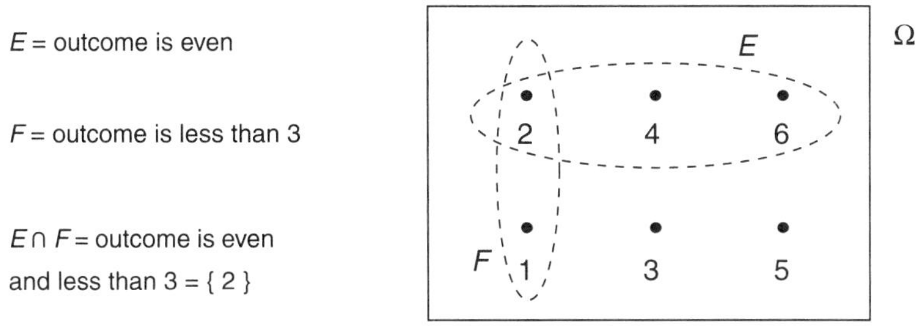

FIGURE 2.8 Venn diagram for calculating the probability of events that are not disjoint in Example 2.37.

Example 2.37. Consider the experiment of tossing a single fair die. Let event $E = \{\text{even number}\}$ and $F = \{\text{number} < 3\}$. Obviously these are not disjoint events. Since $P(E) = 1/2$, $P(F) = 1/3$, and $P(EF) = 1/6$, we have from (2.49):

$$P(E \cup F) = 1/2 + 2/3 - 1/6 = 2/3. \tag{2.54}$$

This result is easily seen from the Venn diagram in Figure 2.8.

2.9 CONDITIONAL PROBABILITY

In many applications, information is often known about some events of an experiment that can influence the probability of other events of interest. It is possible to incorporate this prior information into the calculation of probabilities. Consider two events A and B in the probability space $\{\Omega, \mathcal{F}, P\}$. If it is known that B has occurred, then the probability of A is usually different from that without this knowledge (though not necessarily). The notation $P(A|B)$ is used to denote the probability of A *given* B (also stated as given B is *true*). The following is a simple example for which it easy to compute conditional probabilities.

Example 2.38. For the experiment of tossing two fair coins, let A be the event that both coins are heads, and B the event that *at least one* coin is heads. It is easy to interpret the conditioning event (i.e., knowing this additional information) as causing the sample space to be *reduced* from four elements to three elements. Event B causes the "new" sample space to be $\{HH, HT, TH\}$ because TT is excluded, and thus we immediately have $P(A|B) = 1/3$. We can also compute $P(B|A)$, though this probability turns out to be trivial. The reduced sample space is simply $\{HH\}$, from which it is clear that B (at least one H) is always true: $P(B|A) = 1$.

$P(A|B)$ is a *conditional probability*. Although conditional probabilities were easy to calculate in Example 2.38 by examining the reduced sample space, in general it is preferable to use the following definition based on the intersection of events.

Definition: Conditional Probability The *conditional probability* of event A given event B is

$$P(A|B) \triangleq \frac{P(AB)}{P(B)} \tag{2.55}$$

assuming $P(B) \neq 0$.

We can interpret this definition using the Venn diagram for events A and B shown in Figure 2.9. By conditioning on B, the sample space is reduced from the original Ω to include only elements in B (because B has occurred). Thus, the numerator in (2.55) refers only to those elements in A that are also in B: since B is true, we are

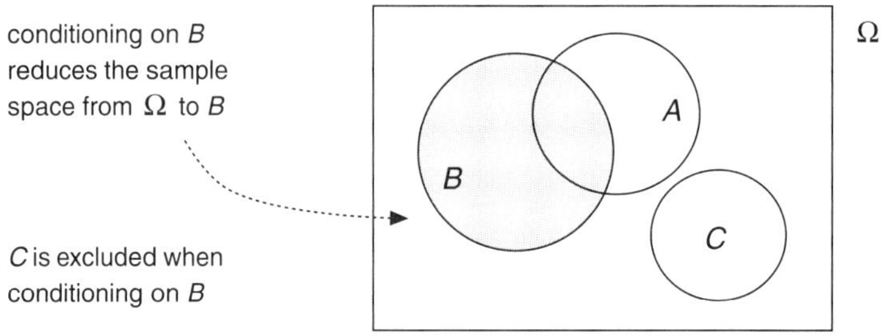

FIGURE 2.9 Conditional probability: elements in A that are not in B have zero probability in the measure $P(\cdot|B)$ (and, in fact, are impossible events).

interested in the probability of the elements in the *intersection* of A and B. Because $P(AB)$ is calculated using the original sample space based on all the original outcomes, it needs to be *normalized* to account for the fact that the sample space is now reduced. This is achieved by dividing $P(AB)$ by $P(B)$, which allows $P(\cdot|B)$ to be a consistent probability measure that satisfies the three axioms of probability, as proved in the following proposition.

Proposition 2.6. *Conditional probability $P(A|B)$ is a valid probability measure with respect to event* A.

Proof. For Axiom 1 and from the definition of conditional probability, we require that $P(AB)/P(B) \geq 0$. Since $P(AB)$ and $P(B)$ are valid probabilities in the original sample space Ω, it immediately follows that $P(A|B) \geq 0$. For Axiom 2,

$$P(\Omega|B) = \frac{P(\Omega B)}{P(B)} = \frac{P(B)}{P(B)} = 1. \tag{2.56}$$

Finally, for Axiom 3 and countably infinite disjoint events $\{A_n\}$,

$$P\left(\bigcup_{n=1}^{\infty} A_n \middle| B\right) = \frac{1}{P(B)} P\left(\bigcup_{n=1}^{\infty} A_n B\right) = \frac{1}{P(B)} \sum_{n=1}^{\infty} P(A_n B) = \sum_{n=1}^{\infty} P(A_n|B), \tag{2.57}$$

where Axiom 3 has been applied to the union of joint events $\{A_n B\}$ to complete the proof. □

We can also verify that the definition is intuitively correct using the frequency interpretation of probability for *finite* experiments where $P(B) = |B|/|\Omega|$. From the reduced sample space,

$$P(A|B) = \frac{|AB|}{|B|}. \tag{2.58}$$

This equation describes how the conditional probability was computed in Example 2.38. But note that this can be rewritten as

$$P(A|B) = \frac{|AB|/|\Omega|}{|B|/|\Omega|} = \frac{P(AB)}{P(B)}, \tag{2.59}$$

which is the same as the definition. Although we cannot apply the frequency interpretation to continuous experiments, the intuition is still the same, and the definition in (2.55) applies to any well-defined random experiment: continuous, discrete, or mixed.

It is important to note that $P(A|B)$ is a valid probability measure only with respect to event A and *not* with respect to the conditioning event B. Consider three events A, B, and C. The following expression:

$$P(A \cup B|C) = P(A|C) + P(B|C) - P(AB|C) \tag{2.60}$$

is a valid extension of the result in Proposition 2.6. The conditioning event C has been carried over to the right-hand side for each of the three terms. If the conditioning event is written in terms of a set operation such as $P(A|B \cup C)$, then in general

$$P(A|B \cup C) \neq P(A|B) + P(A|C) - P(A|BC). \tag{2.61}$$

Set operations on the conditioning event do not satisfy the axioms of probability, and in general we cannot rewrite expressions such as $P(A|B \cup C)$ in terms of probabilities conditioned on the individual events B and C.

Consider again the original probability space $\{\Omega, \mathcal{F}, P\}$. By conditioning on event B, the sample space is reduced to $B \subset \Omega$, the σ-field is reduced to the sub-σ-field $\mathcal{G} \subset \mathcal{F}$, and the probability measure is $P(\cdot|B)$ as described above. Thus, the *conditional probability space* is $\{B, \mathcal{G}, P(\cdot|B)\}$, which we illustrate by the following simple example.

Example 2.39. The sample space for the experiment of tossing two coins is $\Omega = \{HH, TT, TH, HT\}$. Let $\mathcal{F}$ be the power set with the elements summarized previously in Table 2.3. Define the conditioning event $B = \{\text{at least one } H\}$. This reduces the sample space to $B = \{HH, TH, HT\}$, and the corresponding sub-σ-field (the power set for B) is $\mathcal{G} = \{\phi, B, HH, TH, HT, \{HH, TH\}, \{HH, HT\}, \{TH, HT\}\}$.

We conclude this section with another example showing how conditioning reduces the sample space.

Example 2.40. When tossing a single die, suppose we are interested in the probability of $F = \{\text{outcome is} < 3\}$ given $E = \{\text{outcome is even}\}$. It is easily shown that $P(F|E) = 1/3$. Only outcome 2 is less than three in the reduced sample space $\{2, 4, 6\}$. For $P(E|F)$, the sample space is reduced to $\{1, 2\}$, and we immediately have $P(E|F) = 1/2$. In both cases, the conditional probabilities are the same as the corresponding probabilities without conditioning: $P(F|E) = P(F)$ and $P(E|F) = P(E)$. This is not a coincidence: it turns out that events E and F are *independent* in this experiment.

2.10 INDEPENDENCE

Independence in probability is an important property that is often assumed when modeling an experiment, which usually leads to simplified analyses. Intuitively, we can say that two events are independent if they have no influence on the outcomes of each other. It is defined as follows.

Definition: Independent Events Events A and B are *independent* if and only if

$$P(AB) = P(A)P(B). \tag{2.62}$$

Probability $P(AB)$ is known as the *joint* probability of A and B (i.e., they occur jointly). The individual probabilities $P(A)$ and $P(B)$ in the context of a joint probability are known as *marginal* probabilities. The definition of independence is clear for the following simple example.

Example 2.41. The probability of observing two heads when simultaneously tossing two fair coins is $P(HH) = P(H)P(H) = (1/2)(1/2) = 1/4$. It is obvious from a physical view of the dynamics of tossing a coin, there is no reason why the outcome of one coin should influence the other. Intuitively, and from the frequency interpretation of probability, the probabilities should split as in (2.62) for independent events. Likewise, the probability of any pair of outcomes $\{\omega_m, \omega_n\}$ when tossing two fair dice is $P(\omega_m, \omega_n) = P(\omega_m)P(\omega_n) = 1/36$, where $\omega_m \in \{1, 2, 3, 4, 5, 6\}$.

Observe from the definition of conditional probability that

$$P(A|B) = \frac{P(AB)}{P(B)} = \frac{P(A)P(B)}{P(B)} = P(A). \tag{2.63}$$

Likewise

$$P(B|A) = P(B), \tag{2.64}$$

demonstrating that independence is a symmetrical property of two events: they are *mutually independent*. Although (2.62) is the standard definition for independence, the result in (2.63) can also be taken as the definition; in fact, it is more appealing because it explicitly shows that conditioning on B has no influence on the probability of A when they are independent. In order to test whether or not two events are independent for specific numerical examples, one need only show that the left-hand side in (2.55) equals the right-hand side. That is, if both sides of the equation have the same numerical value, then the events are independent; dependent events *cannot* coincidentally have numerical equality in (2.62).

The definition of two independent events can be extended to several events.

Definition: Mutually Independent Events Events $\{A_1, \ldots, A_N\}$ are *mutually independent* if

$$P(A_m \cdots A_n) = P(A_m) \cdots P(A_n) \tag{2.65}$$

for all possible subsets of $\{A_1, \ldots, A_N\}$. This includes combinations of two events, three events, and so on, as well as all N events:

$$P(A_m A_n) = P(A_m)P(A_m), \qquad m \neq n, \tag{2.66}$$

$$P(A_m A_n A_p) = P(A_m)P(A_n)P(A_p), \qquad m \neq n \neq p, \tag{2.67}$$

$$\vdots$$

$$P(A_1 \cdots A_N) = P(A_1) \cdots P(A_N). \tag{2.68}$$

If one or more combinations above do not hold, then the events are dependent by definition. Thus, it is usually easier to demonstrate that multiple events are *dependent*. One might begin by checking all probabilities of two events, then of three events, and so on, until one of the conditions above does not hold. To demonstrate that all events are mutually independent, however, *all* combinations above must be examined.

Example 2.42. From the previous die-toss example with events $E = \{\text{even}\}$ and $F = \{\text{less than three}\}$, we know that $P(EF) = P(E)P(F)$. Consider another event $G = \{\text{divisible by three}\}$, corresponding to $G = \{3, 6\}$ for which $P(G) = 1/3$. From the Venn diagram in Figure 2.8, we find that $EG = \{6\}$ and thus $P(EG) = 1/6$. Observe that $P(FG) = 0$ (F and G are disjoint), but $P(F)P(G) = 1/9$. Events $\{E, F, G\}$ are not mutually independent.

2.11 BAYES' FORMULA

Bayes' formula is a useful equation that is used to compute conditional probabilities, or derive marginal probabilities by conditioning on events.

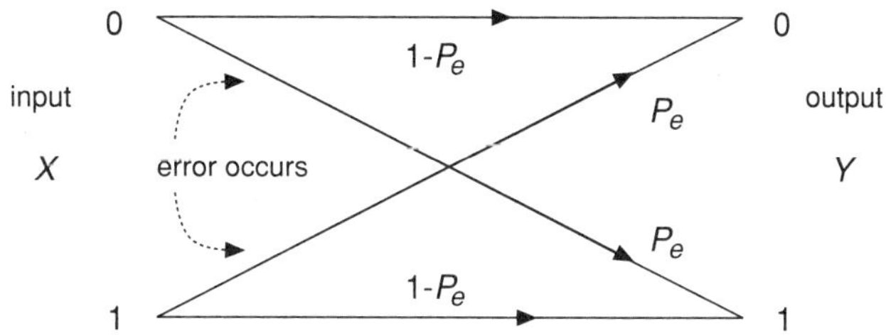

FIGURE 2.10 Binary symmetric channel (BSC) with input X and output Y. Binary $\Longrightarrow |X| = |Y| = 2$, and symmetric means $P(Y = 0|X = 0) = P(Y = 1|X = 0) \triangleq P_e$.

Theorem 2.2 (Bayes' formula). For events $\{A, B\} \in \Omega$ with $P(B) \neq 0$:

$$P(A|B) = \frac{P(B|A)P(A)}{P(B)}. \tag{2.69}$$

Likewise, for $P(A) \neq 0$:

$$P(B|A) = \frac{P(A|B)P(B)}{P(A)}. \tag{2.70}$$

Proof. From the definition of conditional probability, we can write the following two expressions:

$$P(AB) = P(B|A)P(A) = P(A|B)P(B). \tag{2.71}$$

Combining these results and solving for $P(A|B)$ or $P(B|A)$ yields Bayes' formula, which is also called *Bayes' rule*. □

Example 2.43 (Binary symmetric channel). An important example of Bayes' formula is the binary symmetric channel (BSC) shown in Figure 2.10, which is a model for bit errors that occur in a digital communication system. For a binary system, the transmitted symbol has two outcomes: we will use $\{0, 1\}$ to represent the two *binary* elementary events. This model is equivalent to the coin-toss experiment with $H \equiv 1$ and $T \equiv 0$. As shown in Chapter 3, it is convenient to represent the transmitted symbol by *random variable* X, which "randomly" takes on one of the two values. X itself is not random: it is well defined by the two possible outcomes (called the *alphabet*) and the specific probability for each value. The randomness of a random variable is due to the fact that it is not known beforehand which value will occur; X is not deterministic. Typically in a real system, $P(X = 0) = P(X = 1) = 1/2$ (analogous to a fair coin).

The received symbol can likewise be represented by random variable Y, which also has alphabet $\{0, 1\}$. We are interested in the probability of a transmission error across the channel, of which there are two types represented by conditional probabilities $P(Y = 1|X = 0)$ and $P(Y = 0|X = 1)$. The channel is *symmetric* if these two conditional probabilities are equal. The average probability of error is defined by conditioning on the two values for X:

$$P_e \triangleq P(Y = 1|X = 0)P(X = 0) + P(Y = 0|X = 1)P(X = 1), \tag{2.72}$$

which is an example of the *law of total probability* discussed later. Since the channel is symmetric, only one of the conditional probabilities needs to be examined:

$$P_e = P(Y = 1|X = 0)[P(X = 0) + P(X = 1)] = P(Y = 1|X = 0), \tag{2.73}$$

which does not require equally likely outcomes for this simplification. In a real communication system, P_e can usually be estimated by performing measurements over many trials of the experiment, by sending a stream of 0s and 1s that are known at the receiver and then counting the errors.

The receiver in a communication system requires the reverse conditional probabilities: $P(X = 0|Y = 0)$, $P(X = 0|Y = 1)$, $P(X = 1|Y = 0)$, and $P(X = 1|Y = 1)$. For example, if $Y = 0$ is received, the goal of the receiver is to decide whether $X = 0$ or $X = 1$ was transmitted. Since the input is random, the receiver cannot decide exactly which symbol was transmitted. The detection criterion, known as the maximum a posteriori (MAP) decision rule (see Chapter 10), chooses the symbol maximizing $P(X = x|Y = y)$ for $x, y \in \{0, 1\}$. Using Bayes' formula, this can be rewritten as

$$P(X = x|Y = y) = \frac{P(Y = y|X = x)P(X = x)}{P(Y = y)}, \tag{2.74}$$

where $P(Y = y|X = x)$ is generally known for the specific communication channel, as mentioned above. The prior probability $P(X = x)$ (before conditioning on Y) is also known by the receiver so that $P(Y = y)$ is readily calculated using the law of total probability as follows:

$$P(Y = y) = P(Y = y|X = 0)P(X = 0) + P(Y = y|X = 1)P(X = 1). \tag{2.75}$$

This example is continued in Section 2.12.

2.12 TOTAL PROBABILITY

The law of total probability is an extension of Bayes' formula to a *partition* of events in the sample space Ω. It is a convenient means for computing the probability of an event in terms of conditional probabilities that might be easier to compute for some problems, such as the BSC discussed in Example 2.43.

Theorem 2.3 (Total probability). Let $\{B_n\}$, $n = 1, \ldots, N$, be a partition of Ω. Then

$$P(A) = \sum_{n=1}^{N} P(A|B_n)P(B_n). \tag{2.76}$$

Proof. Since $\{B_n\}$ form a partition, $B_n \cap B_m = \phi$ and $AB_n \cap AB_m = \phi$ for $n \neq m$. We also have

$$\bigcup_{n=1}^{N} AB_n = \Omega \tag{2.77}$$

so that

$$P(A) = \sum_{n=1}^{N} P(AB_n). \tag{2.78}$$

Applying conditional probability to each term in the sum gives $P(AB_n) = P(A|B_n)P(B_n)$, which completes the proof. □

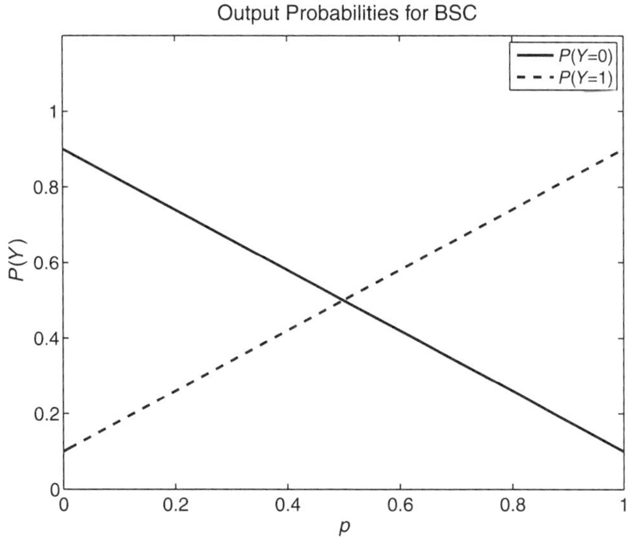

FIGURE 2.11 Output probabilities for the binary symmetric channel (BSC) in Example 2.44 as a function of $p = P(X = 1)$ for probability of error $P_e = 0.1$.

Example 2.44. Continuing with the BSC in Example 2.43 with input X and output Y, (2.75) was generated by conditioning on a partition for X; the results for each value of Y are

$$P(Y = 0) = P(Y = 0|X = 0)P(X = 0) + P(Y = 0|X = 1)P(X = 1), \quad (2.79)$$

$$P(Y = 1) = P(Y = 1|X = 0)P(X = 0) + P(Y = 1|X = 1)P(X = 1). \quad (2.80)$$

Since the prior probabilities $\{P(X = x)\}$ and the channel error probabilities $\{P(Y = y|X = x)\}$ are usually known, $\{P(Y = y)\}$ can be computed. Let $P(X = 1) = p$. Because the channel is symmetric, substituting $P(Y = 1|X = 0) = P(Y = 0|X = 1) = P_e$ and $P(Y = 1|X = 1) = P(Y = 0|X = 0) = 1 - P_e$ into (2.79) and (2.80) gives

$$P(Y = 0) = (1 - P_e)(1 - p) + P_e p, \quad (2.81)$$

$$P(Y = 1) = P_e(1 - p) + (1 - P_e)p. \quad (2.82)$$

Figure 2.11 shows plots of these two functions of p for $P_e = 0.1$ (which is a high error probability for a communication system). Clearly, $P(Y = 0) \neq P(Y = 1)$ except for $p = 1/2$ (i.e., when $P(X = 0) = P(X = 1)$).

Example 2.45 (Medical test). Another well-known application of Bayes' formula and the law of total probability is in medicine where a test has been developed to determine whether a patient has a particular disease. Consider two events: $A = \{\text{test is positive}\}$ and $B = \{\text{patient has the disease}\}$. Two types of errors can be made: $P(A|B^c)$ is known as a *false positive* and $P(A^c|B)$ is known as a *false negative*. These probabilities can be estimated from experimental trials, perhaps when the medicine is developed. Also, based on general population data, we assume that $P(B)$ is known with a high degree of accuracy. Thus, the probability that the patient has the disease given that the test is positive can be computed using Bayes' formula as follows:

$$P(B|A) = \frac{P(A|B)P(B)}{P(A)}. \quad (2.83)$$

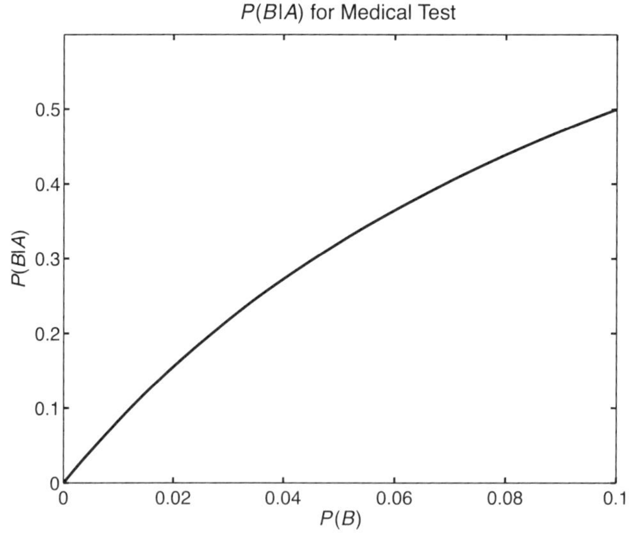

FIGURE 2.12 $P(B|A)$ in Example 2.45 as a function of $P(B)$ for $P(A^c|B) = P(A|B^c) = 0.1$.

Substituting the law of total probability gives

$$P(B|A) = \frac{P(A|B)P(B)}{P(A|B)P(B) + P(A|B^c)P(B^c)}. \tag{2.84}$$

All quantities on the right-hand side are known because $P(A|B) + P(A^c|B) = 1$ and $P(A|B^c) + P(A^c|B^c) = 1$. For example, if $P(A^c|B) = P(A|B^c) = 0.1$ and $P(B) = 0.02$, then $P(B|A) \approx 0.16$. Figure 2.12 shows a plot of $P(B|A)$ versus the probability that the general population has the disease, as it varies from 0 to 0.1.

Finally, there is a multiplicative expression for a joint probability that involves conditioning similar to that used in the law of total probability, but does not require that the events form a partition. The conditioning is performed on an increasing number of events as described in the following theorem.

Theorem 2.4. Let N events $\{B_n\}$ in Ω be such that their joint probability $P(B_1 \cdots B_N) \neq 0$. Successively conditioning on these events gives

$$P(B_1 \cdots B_N) = P(B_1)P(B_2|B_1)P(B_3|B_1B_2) \cdots P(B_N|B_1 \cdots B_{N-1}). \tag{2.85}$$

Proof. The proof is by induction. Conditioning on one event as in Bayes' formula gives

$$P(B_1B_2) = P(B_1)P(B_2|B_1). \tag{2.86}$$

Assume the following expression is true for $N - 1$ events:

$$P(B_1 \cdots B_{N-1}) = P(B_1)P(B_2|B_1)P(B_3|B_1B_2) \cdots P(B_{N-1}|B_1 \cdots B_{N-2}). \tag{2.87}$$

Then

$$P(B_1 \cdots B_N) = P(B_N | B_1 \cdots B_{N-1}) P(B_1 \cdots B_{N-1}) \tag{2.88}$$

is also a form of Bayes' formula because the intersection $B_1 \cdots B_{N-1}$ can be viewed as a single event. Substituting (2.87) into the last expression verifies the result for N events, thus completing the proof. □

2.13 DISCRETE SAMPLE SPACES

For discrete sample spaces (finite or countably infinite), it is straightforward to assign probabilities because events are expressed in terms of the individual sample points in Ω, which in turn are assigned probabilities satisfying the axioms of probability. As mentioned earlier, we generally assume $\mathcal{F} = \mathcal{P}(\Omega)$ for discrete Ω.

Example 2.46. Consider an experiment where the countable outcomes $\{x_n\}$ are the natural numbers $\mathcal{N}$. Suppose we assign probabilities to these outcomes as follows:

$$P(x_n) = (1-p)^{n-1} p, \tag{2.89}$$

where $0 < p < 1$. This probability assignment yields the *geometric random variable* described in Chapter 3. In fact, it corresponds to the experiment of successively tossing a single coin with $P(H) = p$ until the *first* H appears. Even though an infinite number of outcomes are possible, these probabilities sum to one:

$$\sum_{n=1}^{\infty} (1-p)^{n-1} p = p \sum_{n=0}^{\infty} (1-p)^n = \frac{p}{1-(1-p)} = 1, \tag{2.90}$$

where the closed-form expression for an infinite series in Appendix E has been used. This is an example of a convergent series that has been scaled by p so that it converges to 1, and is due to the fact that there is a countable number of terms. In fact, the terms of any convergent series can be assigned probabilities as long as they are scaled to sum to 1.

Next, consider the special case of a *finite* sample space Ω with *equally likely* outcomes $\{\omega_n\}$. (Obviously, a countably infinite sample space *cannot* have equally likely outcomes). The probability of any event $E \in \Omega$ for this situation is determined by *counting* the number of outcomes in E and computing the ratio

$$P(E) = \frac{|E|}{|\Omega|}, \tag{2.91}$$

where $|E|$ is the cardinality of E. For such equally likely cases, finding probabilities reduces to a problem in *combinatorics*. Computing $|\Omega|$ is usually easy, whereas finding $|E|$ often requires more work because it involves a subset of outcomes. In order to handle such problems, we need to be clear about the difference between a combination and a permutation.

Definition: Combination and Permutation A *combination* is an unordered set of distinct elements. A *permutation* is one of the many ordered sets of distinct elements.

Given N distinct elements, there is only one combination: the order of these elements does not matter, because each ordering is still the same combination. On the other hand, the various arrangements of those elements are different permutations.

Example 2.47. Consider the set $E = \{a_1, a_2, a_3\}$. There is only one combination of those elements: $\{a_2, a_1, a_3\}$ is the same combination. On the other hand, there are six permutations: in addition to the two above, they are $\{a_1, a_3, a_2\}$, $\{a_2, a_3, a_1\}$, $\{a_3, a_1, a_2\}$, and $\{a_3, a_2, a_1\}$.

It is clear from this simple example that for a set of N distinct elements, there can be only one combination, but there are $N!$ permutations. This is readily seen for the general case as follows. Since a permutation is an ordered set, meaning that the location of an element relative to the other elements must be considered, then for the first entry we can "choose" any of the N elements. For the second entry, we can choose only from the $N-1$ remaining elements (since the elements in the permutation must be distinct, we cannot choose the element already in the first entry). Similarly for the third entry, we can choose from $N-2$ elements and so on until the last entry, for which there is only one element remaining. Thus, the number of permutations of N distinct elements is

$$N(N-1)\cdots 2\cdot 1 = N!. \tag{2.92}$$

When counting the number of outcomes in event E, we must first determine whether or not their order is important. The following counting examples demonstrate some cases.

Example 2.48. Suppose we are dealt a hand of five distinct cards from a standard deck of 52 cards. There are 5! ways that we can arrange those cards, corresponding to 120 permutations. However, as we know in card games such as poker, the various permutations do not affect the strength of the hand. Whether or not we have a winning hand depends only on the combination of the cards; their order is not relevant. (This is true even for a straight such as {5, 6, 7, 8, 9}: the strength of this hand does not require that we hold the five cards in increasing or decreasing order.)

Example 2.49. Generally, when an arrangement is specified in the problem statement, permutations should be considered. Consider counting the number of social security numbers (which have nine digits) such that a digit is *not* repeated, and which does not include a zero. Then obviously the order is important: there are $9! = 362{,}880$ such social security numbers. These are all the permutations of the numbers 1 through 9 without repeated digits. This type of "experiment" is called *sampling without replacement*: once a number has been selected for a digit, it cannot be used again. Suppose now that we allow digits to be repeated. In this case, there are 9^9 such numbers (again, excluding zeros). This problem is not handled by the formula in (2.92) because the digits are no longer distinct. This scenario corresponds to *sampling with replacement*. The number of elements is readily determined by an approach similar to that used to derive (2.92): allowing digits to be repeated, there are $9\cdot 9\cdots 9 = 9^9 = 387{,}420{,}489$ numbers.

For problems involving N distinct equally likely outcomes, we are often interested in a subset of $M < N$ elements corresponding to some event. This case yields more than one combination because there are many ways to choose M elements from N without regard to order. The number of permutations for this case is easily seen to be

$$N(N-1)\cdots(N-M+1) \triangleq (N)_M, \tag{2.93}$$

where $(N)_M$ is the notation for a *falling factorial* (see Appendix B). It can be rewritten as

$$(N)_M = \frac{N(N-1)\cdots(N-M+1)}{(N-M)\cdots 2\cdot 1}[(N-M)\cdots 2\cdot 1] = \frac{N!}{(N-M)!}. \tag{2.94}$$

Since the set has M elements, there are $M!$ permutations from (2.92). The number of combinations obtained by selecting M elements from N is computed by dividing (2.94) by the number of permutations. This causes us to count all permutations of the same M elements as a single combination:

$$\frac{N!}{(N-M)!M!} = \frac{(N)_M}{M!} \triangleq \binom{N}{M}. \tag{2.95}$$

This last expression, which we state as "N choose M," is called a *binomial coefficient*. Equations (2.94) and (2.95) summarize the number of permutations and combinations, respectively, for a size-M subset chosen from N distinct elements. Some factorials and binomial coefficients are provided in Tables E.1 and E.2 in Appendix E, along with results for Stirling's formula which approximates $N!$ for large N.

Example 2.50. Suppose we are interested in the number of six-digit license plates (no letters allowed) without repeated digits (i.e., without replacement) and excluding any zeros. This problem is handled by the formula in (2.94) with $N = 9$ and $M = 6$. Clearly, order is important; the number of such license plates is

$$\frac{9!}{(9-6)!} = 504 \tag{2.96}$$

permutations. If we are concerned only with the number of license plates regardless of the order of the digits, then (2.95) is used, yielding

$$\binom{9}{6} = 84 \tag{2.97}$$

combinations. When determining the number of combinations, license plate 123456 is "equivalent" to 213456, and thus only one of them above is counted. This is handled in (2.95) by dividing by $M! = 6! = 720$ permutations of the six numbers.

These results can be generalized to an experiment where the N elements are no longer distinct, but instead can be grouped into M subsets, each of which has identical elements. The first group has N_1 indistinguishable elements, the second group has N_2 indistinguishable elements, and so on until the Mth group with N_M indistinguishable elements such that

$$\sum_{m=1}^{M} N_m = N. \tag{2.98}$$

This case reduces to the previous experiment when each group has only one element: $N_m = 1$ for $m = 1, \ldots, M$. (Note, of course, that we assume no groups have the same type of elements; otherwise, such groups should be combined into larger groups.)

Since there are no longer N distinct elements, the overall number of permutations given by $N!$ is necessarily reduced by the permutations for all groups. Since the first group has N_1 identical elements, it is clear that we must divide $N!$ by the amount $N_1!$, so that permutations in the first group are not included in the overall number. This is done for each of the groups so that the number of permutations is

$$\frac{N!}{N_1!N_2!\cdots N_M!} \triangleq \binom{N}{N_1, N_2, \ldots, N_M}. \tag{2.99}$$

The right-hand side is known as a *multinomial coefficient*, which reduces to (2.92) when all the groups have only one element.

Example 2.51. Consider an experiment where four fair coins are tossed, and let E be the event of three tails. The elements of E are $\{T, T, T, H\}$, $\{T, T, H, T\}$, $\{T, H, T, T\}$, and $\{H, T, T, T\}$, which correspond to the four possible permutations of three tails and one head. Thus $|E| = 4$ and since $|\Omega| = 2^4 = 16$, the probability is $P(E) = |E|/|\Omega| = 1/4$. In this experiment, the order of elements is important: we must take into consideration the four ways that a single H can occur. The formula in (2.92) does not apply directly because the elements are not distinct: the three tails are indistinguishable, and their order within the four elements of E is not relevant. Using (2.99), we find that $N = 4$, $N_1 = 1$, and $N_2 = 3$, which gives

$$|E| = \binom{4}{3, 1} = 4. \tag{2.100}$$

Next, we examine how the problem changes by considering event F consisting of two tails. The elements of F are $\{T, T, H, H\}$, $\{H, H, T, T\}$, $\{T, H, T, H\}$, $\{H, T, H, T\}$, $\{H, T, T, H\}$, and $\{T, H, H, T\}$; these are all the possible permutations. From (2.99),

$$|F| = \binom{4}{2, 2} = 6 \tag{2.101}$$

and the probability is $P(F) = |F|/|\Omega| = 3/8$.

Finally, we discuss further the process of *sampling* when dealing with finite sample spaces and equally likely outcomes. Consider again two previous examples: (i) drawing five cards for a hand in a game of poker and (ii) tossing a single fair die. In the first case as the cards are drawn from the deck, they are not replaced: this experiment is an example of *sampling without replacement*. The significance of this case is that when five additional cards are drawn from the same deck (for another player's hand), the sample space is changed: there are now only 47 cards. This, of course, must be considered when computing probabilities. The second example corresponds to *sampling with replacement*: each time the die is tossed, the sample space is unchanged. The same six outcomes are always possible. If the die is tossed M times, the number of permutations is

$$\underbrace{N \cdots N}_{M} = N^M. \tag{2.102}$$

Table 2.5 summarizes the formulas for determining the number of samples in a finite random experiment. The fourth case gives the number of combinations with replacement; it is derived in Problem 2.49.

Example 2.52. Suppose a fair die is tossed twice. The number of combinations with replacement from the formula in Table 2.5 is

$$\binom{6+2-1}{2} = \frac{7!}{2!5!} = 21. \tag{2.103}$$

TABLE 2.5 Number of Permutations and Combinations

Type of Sample (N Elements, Choose M)	Formula
Combinations without replacement	$\binom{N}{M} = \frac{N!}{M!(N-M)!}$
Permutations without replacement	$(N)_M = \frac{N!}{(N-M)!}$
Combinations with replacement	$\binom{N+M-1}{M} = \frac{(N+M-1)!}{M!(N-1)!}$
Permutations with replacement	N^M

Consider all possible results, arranged in groups by the outcome of the first toss as follows:

$$\begin{aligned} &(1,1),(1,2),(1,3),(1,4),(1,5),(1,6), && (2,1),(2,2),(2,3),(2,4),(2,5),(2,6), \\ &(3,1),(3,2),(3,3),(3,4),(3,5),(3,6), && (4,1),(4,2),(4,3),(4,4),(4,5),(4,6), \\ &(5,1),(5,2),(5,3),(5,4),(5,5),(5,6), && (6,1),(6,2),(6,3),(6,4),(6,5),(6,6). \end{aligned} \tag{2.104}$$

These are all the permutations with replacement, of which there are $N^M = 6^2 = 36$. Each result above is unique because order is important when counting permutations. For combinations where order is not important, $(2, 1)$ in the second group of outcomes is a repeat of $(1, 2)$ in the first set and should not be counted. Likewise, we eliminate the following outcomes in the other four sets: $\{(3, 1), (3, 2)\}$, $\{(4, 1), (4, 2), (4, 3)\}$, $\{(5, 1), (5, 2), (5, 3), (5, 4)\}$, and $\{(6, 1), (6, 2), (6, 3), (6, 4), (6, 5)\}$, respectively. Subtracting these 15 outcomes from the original 36 permutations gives 21 combinations with replacement, verifying the result in (2.103).

Although we do not actually select or draw an outcome when tossing a die, we can still view the experiment as a type of sampling from Ω. The important feature that differentiates the two types of sampling is that when counting is done to compute probabilities, $|\Omega|$ does not change when sampling with replacement. Suppose in an experiment that the sampling is done consecutively in time. When sampling with replacement, the sampled outcomes are independent because the first sample does not influence subsequent samples. This is *not* the case for sampling without replacement: the probability of a subsequent sample will depend on which elements have already been removed from the sample space Ω.

In this section, we have focused on *finite* discrete sample spaces. For countably infinite discrete sample spaces, it will be more convenient to wait until Chapter 3 on random variables where the pmf is defined. We conclude this chapter with a brief discussion of continuous sample spaces, which are also covered in greater detail in Chapter 3.

2.14 CONTINUOUS SAMPLE SPACES

For a continuous sample space with an uncountable number of outcomes, the situation is quite different because it is not possible to assign nonzero probabilities to individual sample points in such a way that they sum to 1. For example, consider an experiment where the outcomes are nonnegative real numbers $\Omega = \mathcal{R}^+$. It is clear that for any two numbers there is always an infinity of numbers in between (which is *not* the case for an experiment with a countable number of outcomes). If we attempted to assign probabilities to individual numbers, we would realize very quickly that those probabilities cannot be distributed among all the numbers in $\mathcal{R}^+$ and still satisfy the three axioms. The probability of a single number in an experiment with continuous outcomes *must* be zero. Although this result may seem to be counterintuitive at first because individual outcomes obviously do occur in a continuous experiment, some reflection shows that this must be the case. For example, the probability that $\pi = 3.14159265\cdots$ occurs is obviously "very unlikely"; it *almost surely* will not occur, which is another way of saying that $P(\pi) = 0$.

Probability assignments for continuous experiments are handled by considering *intervals* on the real line $\mathcal{R}$, *rectangles* on $\mathcal{R}^2$, and *hyper-rectangles* on $\mathcal{R}^N$. This approach not only allows for the probability assignment to satisfy the three axioms, but intervals often arise in engineering problems. For example, in an experiment involving daily temperatures, we are more likely to be interested in the probability that the temperature is between 70 and 75 degrees Fahrenheit (°F) than in a specific value like 72.354°F. By focusing on intervals for events, in effect we "convert" an uncountable experiment into one that is handled in basically the same way as a countable sample space, and is consistent with the countable additivity axiom of probability (Axiom 3).

Example 2.53. Consider the example shown in Figure 2.13 for $\Omega = [0, 1]$ where we are interested in the probabilities of three subintervals $\{E, F, G\}$. Observe that in addition to those three events, there are four other subintervals which together comprise the sample space: $\Omega = [0, 1]$ has been partitioned into seven subintervals

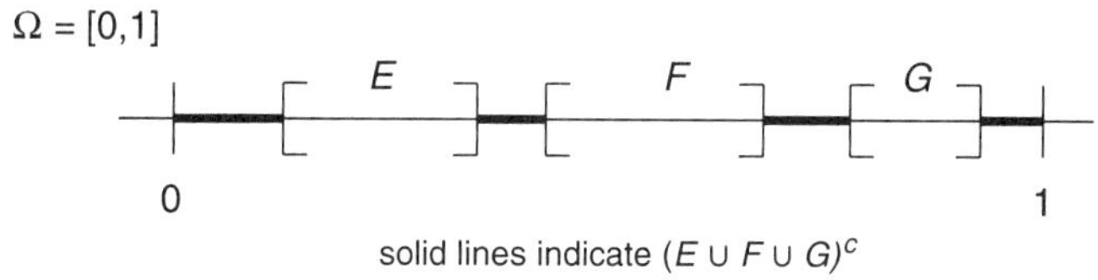

FIGURE 2.13 Nonzero probabilities are assigned to intervals for continuous sample spaces, and not to individual numbers. Using intervals, the sample space Ω is partitioned into a countable (even finite) number of events.

for which it is straightforward to assign probabilities that satisfy the axioms. Likewise, each of the subintervals can be divided further, and there would be no problem assigning probabilities to those events, provided they are not divided up into individual points. The only exception to this last remark is if we *intend* to assign nonzero probabilities to individual numbers, in which case the experiment is actually *mixed*, having both discrete and continuous components. This case is considered in Chapter 3 when mappings to different types of random variables are discussed.

For any collection of intervals on the real line, we can always find a countable number of intervals in between such that the union of all the intervals is the sample space $\Omega = \mathcal{R}$. This explains why the Borel σ-field, which contains all open, closed, and semi-open intervals on the real line (as well as all singletons) is important in the probability space $\{\mathcal{R}, \mathcal{B}(\mathcal{R}), P\}$. Not only are intervals of interest in many applications, but by specifying the Borel σ-field, we are assured that the event space is closed under all set operations (it is the smallest σ-field containing all open intervals on $\mathcal{R}$). Singletons must have probability zero, and it is possible to assign a nonzero probability to any interval such that the three axioms of probability are satisfied. This leads to the cumulative distribution function (cdf) and the pdf described in Chapter 3. The Borel σ-field is a subset of the power set $\mathcal{B}(\mathcal{R}) \subset \mathcal{P}(\mathcal{R})$, and in fact is "much smaller" with cardinality $\beth_1$ versus $\beth_2$; it has the same cardinality as the power set of a countably infinite set such as $\mathcal{Z}^+$ (see Table 2.4).

2.15 NONMEASURABLE SUBSETS OF $\mathcal{R}$

As mentioned previously, the Borel σ-field is sufficient for the probability space for practical applications, and thus we use $\mathcal{B}(\mathcal{R})$ to describe events on $\mathcal{R}$. However, $\mathcal{B}(\mathcal{R})$ is not the only subset of $\mathcal{R}$; there is in fact an infinity of subsets, of which not all are *measurable* and thus a probability measure cannot be assigned. Consider the interval $[0, 1]$ for which the Borel σ-field is a subset generated by all open intervals on $[0, 1]$. The Lebesgue measure for any such interval (a, b) is $L(a, b) = b - a$ for $b > a$, which is bounded above by 1 in this case. The Cantor set is another subset of $[0, 1]$ that is different from the Borel σ-field and is still measurable; recall that it has Lebesgue measure zero. Nonmeasurable subsets of $\mathcal{R}$ are generally difficult to describe. We illustrate one example using a construction leading to the Vitali set, which has a complicated form and would not be encountered in practical applications.

Example 2.54 (Vitali set). Consider again the closed interval $\Omega = [0, 1]$, for which we would like to construct a subset that cannot be assigned a measure (Billingsley, 1986).

For convenience in the discussion, we use the Lebesgue measure $L(a, b)$. Define two numbers x and y to be *equivalent* or *similar* if they differ by a rational number: $x \sim y$ if $x - y \in \mathcal{Q}$. The Vitali set is constructed as follows:

- For each $x \in \mathcal{R}$, define the subset $\{y : x \sim y\}$. Such a subset is called the equivalence class of x. During the construction of these classes, it is evident that each x belongs only to one such equivalence class. If x belonged to more than one class, then those classes must be the same equivalence class. All rational

TABLE 2.6 Examples of Elements in the Vitali Set V

Example Equivalence Classes	Possible Representative in [0, 1]
Rational numbers $\mathcal{Q}$	1/2
$\pi, \pi - 1/2, \pi + 30, \ldots$	$\pi - 3$
$\sqrt{2}, \sqrt{2} + 4, \sqrt{2} - 20, \ldots$	$\sqrt{2} - 1$
$e, e - 1, e - 10, \ldots$	$e - 2$

numbers $\mathcal{Q}$ are necessarily in the same equivalence class. The equivalence classes form a partition of $\mathcal{R}$ since we can account for all $x \in \mathcal{R}$ and because the classes are disjoint (they must be disjoint, or else by definition any two overlapping classes must be the same class).

- From each equivalence class, choose one "representative" value contained in [0, 1]. This is assumed to be possible by the so-called axiom of choice. It does not matter which representative value is chosen in [0, 1] for a class. For example, for the class of rational numbers $\mathcal{Q}$, we might choose 1/2 (though other choices are acceptable, such as 1/4, 2/3, and so on).
- The collection of representative numbers in [0, 1] denoted by V form the Vitali set. We have constructed this particular subset of [0, 1] for which it is possible to show that V is not measurable: unlike the Borel and Cantor sets, the Vitali set does not have a length.

Examples of elements in V are given in Table 2.6. Each equivalence class is *countably infinite* because they are defined relative to the rational numbers, which is a countable set. The cardinality of each equivalence class is the same as that of $\mathcal{Q}$ which is $\beth_0$. However, the number of equivalence classes and thus the Vitali set itself is not countable.

Assume that V is measurable such that the Lebesgue measure $L(V)$ is well defined; we show by contradiction that this assumption is not correct.

- Define the set $V(x) \triangleq \{v + x : v \in V\}$ obtained by translating the original Vitali set by $x \in \mathcal{R}$. Since the Lebesgue measure is translation invariant (see Problem 2.19), we have $L(V(x)) = L(V)$ for any x.
- Order the rational numbers in $[-1, 1]$, labeling them $\{q_n\}$ for $n \in \mathcal{N}$. Construct sets $\{V(q_n)\}$, where $V(q_n)$ is V translated by rational number q_n. Translating them by a rational number ensures that we maintain the same type of equivalence class as defined above. By using the symmetric interval $[-1, 1]$, the original interval [0, 1] is extended in both directions to have $[-1, 2]$ for the $\{V(q_n)\}$.
- The sets $\{V(q_n)\}$ are disjoint, which can be verified by contradiction as follows. Let z be an element common to $V(q_m)$ and $V(q_n)$ $(m \neq n)$, which means that for some $x_m, x_n \in V$, the common element can be written as $z = x_m + q_m = x_n + q_n$. Rearranging this expression gives $x_m - x_n = q_n - q_m \in \mathcal{Q}$ such that $x_m \sim x_n$. However, they cannot belong to the same class because by construction, V has only one element from each class, and thus the $\{V(q_n)\}$ must be disjoint.
- This is the key step: since $q_n \in [-1, 1]$, it follows that

$$[0, 1] \subseteq \bigcup_{n=1}^{\infty} V(q_n) \subseteq [-1, 2]. \tag{2.105}$$

Translating the Vitali set as described above means that the $\{V(q_n)\}$, and therefore their countable union, cannot be outside the interval $[-1, 2]$ (corresponding to the second $\subseteq$ above). This part of (2.105) is obvious. For the first $\subseteq$, the translations ensure that *all* elements of [0, 1] are included in the union of the disjoint sets $\{V(q_n)\}$. This can be seen as follows. Let x be any real number in [0, 1]. From the construction of the Vitali set, we know there exists $v \in V$ for the equivalence class containing x, and thus $x - v$ is some rational number q in $[-1, 1]$. Translating V by the rational numbers in $[-1, 1]$ ensures that *all* numbers in [0, 1] form a subset of the union in (2.105).

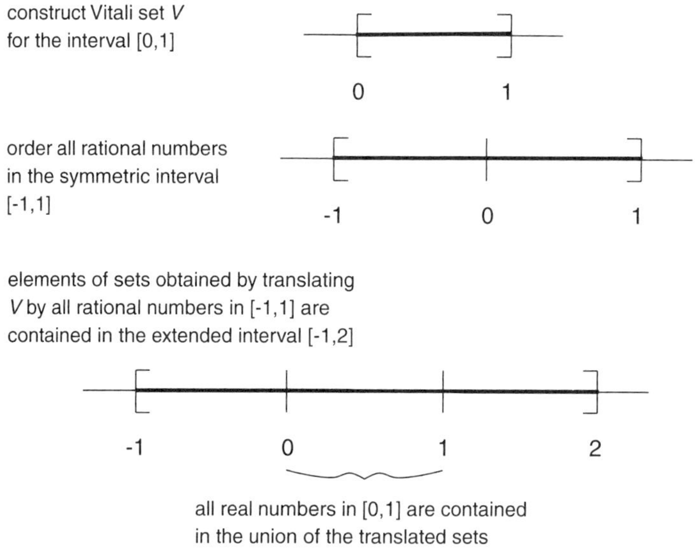

FIGURE 2.14 Intervals used in Example 2.54 describing the Vitali set.

- A summary of the various intervals is given in Figure 2.14. Since the $\{V(q_n)\}$ are disjoint, the Lebesque measure of the union is

$$L\left(\bigcup_{n=1}^{\infty} V(q_n)\right) = \sum_{n=1}^{\infty} L(V(q_n)), \tag{2.106}$$

and thus from (2.105);

$$L([0,1]) = 1 \leq L\left(\bigcup_{n=1}^{\infty} V(q_n)\right) \leq 3 = L([-1,2]). \tag{2.107}$$

It was necessary that the translations be performed on V such that all real numbers in $[0, 1]$ are contained in their union, allowing us to generate the *nonzero* lower bound $L([0, 1]) = 1$ on the measure of the countable union.

- Since the Lebesgue measure is translation invariant, we expect for the original Vitali set that

$$1 \leq \sum_{n=1}^{\infty} L(V) \leq 3. \tag{2.108}$$

But this infinite sum of a *constant* $L(V)$ cannot give a result in the range $[1, 3]$, which is a contradiction. When summing over n, either $L(V)$ is zero and the sum is zero, or $L(V)$ is nonzero and the sum is infinite.

This contradiction shows that the Vitali set does not have a length. By constructing the set in this manner, we violated the fact that the Lebesgue measure is translation invariant; as a result, the measurable assumption is incorrect.

We return to continuous sample spaces in Chapter 3 and define continuous random variables, including several well-known parametric "families" that are useful for a wide range of applications.

PROBLEMS

Set Operations

2.1 Find the power set $\mathcal{P}(\Omega)$ for (a) $\Omega = \{-1, 0, 1\}$, (b) $\Omega = \{1, 2, 3, 4\}$, and (c) $\Omega = \{[0, 1), [1, 2), [2, \infty)\}$.

2.2 Use a Venn diagram for the following. (a) Prove $(ABC)^c = A^c \cup B^c \cup C^c$. (b) Sketch $((A \cup B)C)^c$.

2.3 For $\Omega = [-10, 10]$, $A = [-1, 1]$, and $B = [-5, 0]$, find (a) AB^c, (b) $A^c \cup B$, and (c) $A \oplus B$.

2.4 For $\Omega = [-10, 10] \times [-10, 10]$, $A = [0, 3] \times [-10, 10]$, and $B = [-2, 1] \times [-10, 0]$, find (a) $A^c B$, (b) AB^c, and (c) $B - A$.

2.5 Let $\Omega = \mathcal{R}$ and define events $E_1 = \{x : x > 6\}$, $E_2 = \{x : |x - 1| < 5\}$, and $E_3 = \{x : x^2 > 4\}$. Find (a) $E_1 E_2$, (b) $E_1 \cup E_2$, (c) $E_2 E_3^c$, and (d) $E_2 \cup E_3$.

2.6 Prove De Morgan's laws without using Venn diagrams. (a) $(A \cup B)^c = A^c B^c$. (b) $(AB)^c = A^c \cup B^c$.

2.7 Prove the following without using Venn diagrams. (a) $A \cup B = (A \oplus B) \oplus (A \cap B)$. (b) $A \cap (B \oplus C) = (A \cap B) \oplus (A \cap C)$.

2.8 The Cartesian product applies to sets with a finite number of elements: if $A = \{a_1, a_2\}$ and $B = \{b_1, b_2\}$, then $A \times B = \{\{a_1, b_1\}, \{a_1, b_2\}, \{a_2, b_1\}, \{a_2, b_2\}\}$. Prove the following for general $\{A, B, C\}$. (a) $A \times (B \cup C) = (A \times B) \cup (A \times C)$. (b) $|A \times B| = |A|\,|B|$.

Fields

2.9 Consider the sample space $\Omega = \{-2, -1, 1, 2\}$. (a) For events $A = \{-1, 1\}$ and $B = \{-2, 2\}$, verify that $\{\phi, \Omega, A, B\}$ is a field. (b) For events $C = \{-2, -1\}$, $D = 1$, and $E = 2$, show that $\{\phi, \Omega, C, D, E\}$ is not a field. (c) By appending events to $\{\phi, \Omega, C, D\}$, determine if a nontrivial field can be constructed that has cardinality less than $|\mathcal{P}(\Omega)| = 16$.

2.10 Let the sample space be $\Omega = [0, 10]$. Determine if the following are fields. (a) Set of all intervals of the form $\{[0, x]\}$. (b) Set of all intervals of the form $\{[x, y]\}$.

2.11 Let $\mathcal{F}_1$ and $\mathcal{F}_2$ be σ-fields for Ω. Determine if $\mathcal{F}_1 \cap \mathcal{F}_2$ is a σ-field.

2.12 Let $\Omega = \{-2, -1, 0, 1, 2\}$. For events $A = \{-2, -1, 0\}$, $B = \{1, 2\}$, $C = \{-2, 1, 0\}$, and $D = \{-1, 2\}$, specify which of the following are fields. (a) $\{\phi, \Omega, A, B\} \cap \{\phi, \Omega, C, D\}$. (b) $\{\phi, \Omega, A, B\} \cup \{\phi, \Omega, C, D\}$.

2.13 Consider the event space $\{\Omega, \mathcal{F}\}$ and let Ω be partitioned into a finite number of sets $\{E_1, \ldots, E_N\}$. For example, a possible partition for $\Omega = \mathcal{R}$ is $E_1 = (-\infty, -1]$, $E_2 = (-1, 1]$, and $E_3 = (1, \infty)$. Demonstrate in general that $\cup E_n \in \mathcal{F}$ and $\cap E_n \in \mathcal{F}$.

2.14 Generate the smallest field possible for $\Omega = [0, 10]$ starting with $\{\phi, \Omega, 0, [0, 2)\}$, and specify the atoms of this field.

2.15 Repeat Problem 2.15 for $\Omega = \{1, 2, 3, 4\}$ starting with $\{\phi, \Omega, 1, \{3, 4\}\}$.

2.16 Prove that the natural numbers $\mathcal{N}$ are a Borel set, and likewise for the irrational numbers $\mathcal{R} - \mathcal{Q}$.

Measure

2.17 Give the counting measure C for the following sample spaces. (a) $\Omega_1 = \{1, \ldots, 10\} \cup \{1, 15, 18\}$. (b) $\Omega_2 = \{1, 2, 3\} \times \{2, 6\}$. (c) $\Omega_3 = \mathcal{Z}^+ \cap [0, 5]$.

2.18 Give the Lebesgue measure L for the following sample spaces. (a) $\Omega_1 = [0, 1] \times [2, 5]$. (b) $\Omega_2 = [1, 2] \times [10, 12] \times [2, 3] \times [-2, 0]$. (c) $\Omega_3 = \mathcal{N} \cap [0, 10]$.

2.19 Let E be an event of $\Omega = \mathcal{R}$. Prove that $L(E + a) = L(E)$ for all $a \in \Omega$, that is, the Lebesgue measure is translation invariant.

2.20 Consider the set of intervals $\{|x - n| \leq 1/3^n\}$ for $n \in \mathcal{N}$ which are defined on $\mathcal{R}^+$. For example, $[2/3, 4/3]$ is the interval for $n = 1$. Find the Lebesgue measure of this set.

2.21 Verify that the standard Gaussian measure in (2.30) is, in fact, a measure.

2.22 Let E be an event of $\{\mathcal{R}, \mathcal{B}(\mathcal{R})\}$. Show that the counting measure $C(E \cap \mathcal{Q})$ is countably additive.

2.23 Prove that the Dirac measure $\delta_\omega(E)$ is a probability measure, and explain why E is the almost sure event in Ω.

Basic Probability

2.24 Prove the following for general A and B. (a) $P(A \cup B) \geq 1 - P(A^c) - P(B^c)$. (b) $P(A) \leq P(B^c)$.

2.25 Prove the following for general E and F. (a) $P(E - F) = P(E) - P(EF)$. (b) $P(E) + P(F) + P(E^c F^c) \geq 1$.

2.26 Prove Boole's inequality:

$$P(E_1 \cup \cdots \cup E_N) \leq \sum_{n=1}^{N} P(E_n). \tag{2.109}$$

2.27 (a) Prove Bonferroni's inequality:

$$P(E_1 \cdots E_N) \geq 1 - \sum_{n=1}^{N} P(E_n^c). \tag{2.110}$$

(b) Show also

$$P(E_1 \cdots E_N) \geq \sum_{n=1}^{N} P(E_n) - N + 1. \tag{2.111}$$

2.28 Consider events $E_1 \subset E_2 \subset \cdots$ and define E to be their union. Use the axioms of probability to prove

$$\lim_{n \to \infty} P(E_n) = P(E). \tag{2.112}$$

Conditional Probability, Total Probability, and Independence

2.29 For the die-toss experiment, let $E = 5$, $F = \{\text{at least 3}\}$, and $G = \{\text{odd}\}$. Find the following probabilities. (a) $P(E|F)$. (b) $P(G|F)$. (c) $P(E|FG)$.

2.30 Determine which of the following are true. (a) If $P(E|F) = P(E|F^c)$, then E and F are independent. (b) If $P(E|F) \geq P(E)$, then $P(F|E) \geq P(F)$.

2.31 Suppose a machine has five independent components, each of which has probability 0.01 of failing. (a) Find the probability that none or all components will fail. (b) Find the probability that at least two components will fail.

2.32 For the experiment of tossing two dice, let $E = \{\text{either die is even}\}$, $F = \{\text{sum is 10}\}$, and $G = \{4\}$ on the second die. Determine if these events are mutually independent.

2.33 Let E be the event that a person has a particular disease. Consider a diagnostic medical test that has probability 0.01 of being incorrect when the patient has the illness. Let F be the event that the test is positive, and assume $P(F|E^c) = P(F^c|E) = 0.05$. Find $P(E|F^c)$ if the probability of a person having the disease (without any testing) is 0.1.

2.34 Events E and F are independent. Determine if (a) E^c and F^c are also independent and (b) likewise for E^c and F.

2.35 Prove the following for events E, F, and G. (a) $P(E|G) = P(EF|G) + P(EF^c|G)$. (b) $P(E|G) \leq P(F|G)$ if $E \subseteq F$.

2.36 Prove the following for events E, F, and G. (a) $P(E|F) = 1$ if $F \subset E$. (b) $P(EF|G) = P(E|FG)P(F|G)$.

2.37 Prove or disprove the following. (a) $P(A|B) > P(A) \Longrightarrow P(B|A) > P(B)$. (b) $P(A) > P(B) \Longrightarrow P(A|C) > P(B|C)$. (c) If $P(A) = 1/3$ and $P(A \cup B) = 1/2$, then $1/6 \leq P(B) \leq 1/2$.

2.38 Three indistinguishable boxes contain colored balls, blue (B), red (R), and yellow (Y) as follows. Box 1: 15 B, 26 R, 32 Y. Box 2: 22 B, 10 R, 18 Y. Box 3: 10 B, 23 R, 30 Y. Find the following probabilities if a single ball is selected. (a) P(B). (b) P(box 1|R). (c) P(box 2|B or Y).

Probability for Finite Sample Spaces

2.39 Four dice are tossed simultaneously. Calculate the following probabilities. (a) P(all four dice are the same). (b) P(no dice are the same). (c) P(exactly three dice are the same).

2.40 Determine the probability that at least two people in a group of N have the same birthday. Assume there are 365 equally likely days in a year (ignore leap year).

2.41 Consider a multiple-choice exam with five questions and four choices each. If a student decides each answer by guessing, find the probability of (a) no correct answers and (b) three or more correct answers.

2.42 A wallet contains 4 one-dollar bills and 6 ten-dollar bills. A bill is selected and then placed back in the wallet along with 3 bills of the same denomination. Determine the following probabilities for the total amount selected from the wallet. (a) P($11 in 2 trials). (b) P($30 in 3 trials). (c) P($22 in 4 trials).

2.43 For a three digit number $a_1a_2a_3$, the first digit has alphabet $\{0, 1\}$, and the second and third digits have alphabet $\{0, \ldots, 5\}$. Find the following probabilities. (a) $P(a_1a_2a_3 = 100)$. (b) $P(a_1a_2a_3 > 100)$. (c) $P(100 < a_1a_2a_3 < 150)$.

2.44 A circuit consists of 8 resistors, 2 inductors, and 3 capacitors. Three of the circuit elements are damaged. Find the following probabilities. (a) P(only 1 capacitor is damaged). (b) P(at least one inductor is damaged). (c) P(no resistors are damaged).

2.45 Five cards are drawn for a hand in a game of poker. Find the following probabilities. (a) P(5 hearts) (called a flush). (b) P(5 consecutive cards) (called a straight). (c) P(5 different numeric values).

2.46 A box contains the following balls that are distinguishable only by color: 4 red, 5 blue, and 8 yellow. If three balls are selected without replacement, determine the following probabilities. (a) P(3 red). (b) P(no yellow). (c) P(3 different colors).

2.47 Two dice are tossed once. Calculate the following probabilities (a) P(sum exceeds 8). (b) P(sum is between 4 and 6). (c) P(difference is ± 2).

2.48 Two dice are tossed N times. (a) Compute the probability in terms of N that the combination $\{1, 2\}$ occurs at least once. (b) Find N such that this probability exceeds 0.9.

2.49 For a sample space of size N, show that the number of combinations of M elements chosen with replacement is $\binom{N+M-1}{M}$.

Computer Problems

2.50 Use rand of MATLAB® to simulate 100 consecutive tosses of a fair die by partitioning the interval [0, 1]. Plot two different realizations of the experiment.

2.51 Use randperm to generate random permutations of $n = 5$ integers. Write a program that counts the number of trials until all 5! permutations have been randomly generated.

FURTHER READING

The various topics in this chapter can be investigated further in the following references. Set theory: Peebles (1993). Fields: Kingman and Taylor (1966), Wong and Hajek (1985), and Gray and Davisson (2004). Measure theory: Arthreya and Lahiri (2006), Billingsley (1986), and Kingman and Taylor (1966). Probability: Bertsekas and Tsitsiklis (2002), Dougherty (1990), Golberg (1984), Miller and Freund (1977), Papoulis (1990), Peebles (1993), Ross (1976), Williams (1991), and Yates and Goodman (2005).

3 Random Variables

3.1 INTRODUCTION

In Chapter 2, we introduced measure theory and defined the probability of event E in the σ-field $\mathcal{F}$ as a measure that assigns a number from $[0, 1]$ to that event. It is mathematically convenient to *map* abstract events in $\mathcal{F}$ to the real line $\mathcal{R}$. In most applications, events can be examined directly in $\mathcal{R}$ without having to consider an underlying abstract sample space Ω. Besides the fact that most problems in engineering are defined and evaluated using numbers (as opposed to abstract quantities in Ω), another reason for defining a random variable on $\mathcal{R}$ is the importance of *intervals* for continuous experiments, as discussed in Chapter 2. It is not possible to assign nonzero probabilities to individual points in a continuous experiment; instead, events with nonzero probabilities are defined by intervals on $\mathcal{R}$ such that the probability assignment satisfies three axioms. The Borel σ-field $\mathcal{B}(\mathcal{R})$ is the smallest σ-field generated by all open intervals on $\mathcal{R}$, and allows for probabilities to be assigned to all Borel sets in a consistent manner.

One goal in this chapter is to define a mapping from an abstract sample space Ω to the real line $\mathcal{R}$, yielding *random variable* X. Another goal is to describe different kinds of probability assignments to events in $\mathcal{B}(\mathcal{R})$ that are useful *models* for various applications. All the probabilities can be derived from the *cumulative distribution function* (cdf) given by $F_X(x) \triangleq P(X \leq x)$, which is the probability that X is in the semi-open interval $(-\infty, x]$. From the three axioms, we know that $0 \leq F_X(x) \leq 1$; a particular type of random variable is defined by the form or *shape* of $F_X(x)$ as x is varied on $\mathcal{R}$. Examples for discrete and continuous random variables are shown in Figure 3.1. The cdf of a discrete random variable necessarily is a "staircase" function, whereas that of a continuous random variable is continuous with a smooth appearance. (The circles in Figure 3.1(a) are defined later.) Both cdfs are *nondecreasing*. In this chapter, we will see a wide range of shapes for $F_X(x)$ that are important because a specific probability assignment is useful for modeling the "likelihood" of events in a practical application.

3.2 FUNCTIONS AND MAPPINGS

Previously, a *probability measure* was applied to the event space $\{\Omega, \mathcal{F}\}$ which satisfies three properties and provides information about the *sizes* of those events. These properties are known as the three axioms of probability. The Vitali set is an example of a subset of $\mathcal{R}$ that is not measurable. On the other hand, the Borel σ-field is measurable, which is easily demonstrated for the Lebesgue measure.

Example 3.1 (Borel set). Let $[a, b]$ for $a \leq b \in \mathcal{R}$ be a closed interval in $\mathcal{B}(\mathcal{R})$. Then (i) $L(a, b) = b - a \geq 0$ and (ii) $L(\phi) = 0$. (iii) If $[c, d]$ for $c \leq d \in \mathcal{R}$ is another interval such that $[a, b] \cap [c, d] = \phi$, then $L([a, b] \cup [c, d]) = L([a, b]) + L([c, d]) = (b - a) + (d - c)$.

Definition: Measurable Set A *measurable set* is a collection of elements in Ω for which the three properties of a measure are satisfied. $\{\Omega, \mathcal{F}\}$ is a *measurable event space* if $\mathcal{F}$ is a measurable set of events.

Probability, Random Variables, and Random Processes: Theory and Signal Processing Applications, First Edition. John J. Shynk.

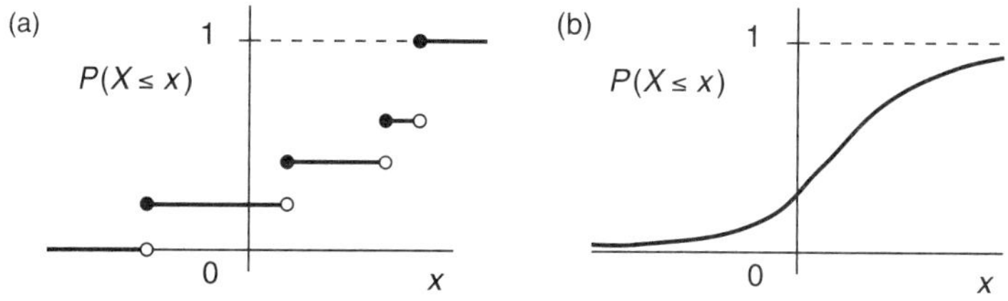

FIGURE 3.1 Examples cdfs $P(X \leq x)$. (a) Discrete random variable. (b) Continuous random variable.

Before defining a *measurable function*, consider again the abstract sample space Ω. For finite sample spaces, one can think of simple examples for Ω whose elements are not derived directly from the the real line. These include tossing a single coin or a pair of dice, and the various ways that books on a shelf can be arranged by topic. Such nonnumeric examples are more difficult to envision for countably infinite and uncountable experiments. For a continuous sample space, Ω could be the set of all sinusoidal waveforms of the form $A\sin(2\pi ft)$ for $t \in [0, T]$ with $A \in [0, 1]$. Thus, Ω contains an infinity of sinusoids, all of the same duration T and with the same frequency f, but with a continuum of amplitudes. Of course, such an experiment is essentially equivalent to the numeric sample space $\Omega = [0, 1]$. Even though we use $\{\Omega, \mathcal{F}\}$ to represent an abstract event space for the purposes of defining the measurable function called a random variable, in practice the "original" sample space usually is $\Omega = \mathcal{R}$ (i.e., there may not be an underlying abstract sample space in a particular application).

Consider the measurable event space $\{\Omega_1, \mathcal{F}_1\}$, where $\mathcal{F}_1$ is a σ-field for sample space Ω_1. Let $\{\Omega_2, \mathcal{F}_2\}$ be another measurable event space, where $\mathcal{F}_2$ is a σ-field for Ω_2. We want to describe a function of the outcomes in Ω_1 that maps them to Ω_2, and allows for a measure in the first event space to carry over to the second event space.

Definition: Measurable Function Function $g(\omega)$ with domain Ω_1 and range Ω_2 is a *measurable function* if the inverse image satisfies

$$g^{-1}(A) = \{\omega : g(\omega) \in A\} \in \mathcal{F}_1 \tag{3.1}$$

for any event $A \in \mathcal{F}_2$.

The inverse image for events in $\mathcal{F}_2$ must be contained in the original σ-field $\mathcal{F}_1$ for a measurable function. Although by definition a function always maps $\omega_1 \in \Omega_1$ to a *unique* $\omega_2 \in \Omega_2$, this is generally not the case for the inverse image. A pictorial representation of a measurable function is shown in Figure 3.2 where the inverse image of an event in $\mathcal{F}_2$ is in $\mathcal{F}_1$. Since our goal is to assign probabilities to events in $\mathcal{F}_1$, which in turn map to probabilities of the corresponding events in $\mathcal{F}_2$, it is necessary that the inverse image be *closed* with respect to $\mathcal{F}_1$ in order to satisfy the three axioms of probability.

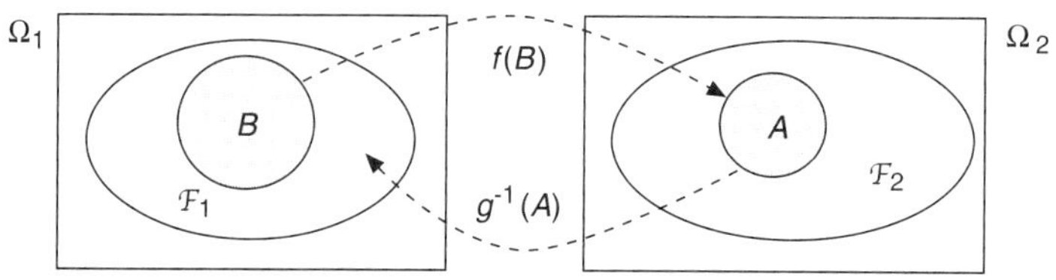

FIGURE 3.2 Pictorial representation showing a measurable function between two sample spaces where outcomes in $B \in \mathcal{F}_1$ map to elements of $A \in \mathcal{F}_2$, and the inverse image of A has elements that are not all in B, but are still in $\mathcal{F}_1$.

Example 3.2 (Measurable function). For finite samples space Ω_1, if σ-field $\mathcal{F}_1$ is the power set $\mathcal{P}(\Omega_1)$, then *any* function that maps from Ω_1 to Ω_2 is measurable. This should be obvious because the inverse image of any event in $\mathcal{F}_2$ must be in $\mathcal{F}_1$; there are no "missing" subsets in the power set by definition.

Example 3.3 (Nonmeasurable function). For the experiment of tossing a single die, consider the simplified σ-field $\mathcal{F}_1 = \{\phi, \Omega_1, \{1, 3, 5\}, \{2, 4, 6\}\}$. It was verified in Example 2.23 that this is a σ-field even though it is not the power set for this experiment. Consider the trivial mapping

$$g(\omega) = \omega, \quad \omega = 1, 2, 3, 4, 5, 6, \tag{3.2}$$

from Ω_1 to Ω_2, and let $\mathcal{F}_2$ be the power set for the elements in Ω_2 with $|\mathcal{F}_2| = 2^6 = 64$. Then, clearly the inverse images for some events in $\mathcal{F}_2$ cannot be events in $\mathcal{F}_1$. For example, the inverse image of 3 is not an event in $\mathcal{F}_1$; although 3 is an outcome of the experiment, this particular $\mathcal{F}_1$ excludes individual outcomes. By choosing $\mathcal{F}_1$ to be a subset of the power set, many potential events have been eliminated from this experiment.

It is important to note the following about a measurable function:

- Function $g(\omega)$ is defined for all *elements* (outcomes) in the sample space Ω_1, not for the events in $\mathcal{F}_1$.
- The measurability of a function depends on the *specific* $\mathcal{F}_1$ and $\mathcal{F}_2$ in the two event spaces. If either of the σ-fields are modified, then the measurability of the same function could change.

As an illustration of the last point, suppose in Example 3.3 that $\mathcal{F}_2 = \{\phi, \Omega_2\}$ which is the trivial σ-field. For this case, $g(\omega)$ is now a measurable function; in fact, *any* mapping from Ω_1 to Ω_2 would be measurable because for the two events in $\mathcal{F}_2$:

$$g^{-1}(\phi) = \{\omega : g(\omega) = \phi\} = \phi \in \mathcal{F}_1, \tag{3.3}$$

$$g^{-1}(\Omega_2) = \{\omega : g(\omega) = \Omega_2\} = \{1, 2, 3, 4, 5, 6\} = \Omega_1 \in \mathcal{F}_1. \tag{3.4}$$

Example 3.4 (Measurable function). For Example 3.3, it is clear that there is an infinity of mappings that are not measurable. A function that is measurable is

$$g(\omega) = \begin{cases} 0, & \omega = 1, 3, 5 \\ 1, & \omega = 2, 4, 6, \end{cases} \tag{3.5}$$

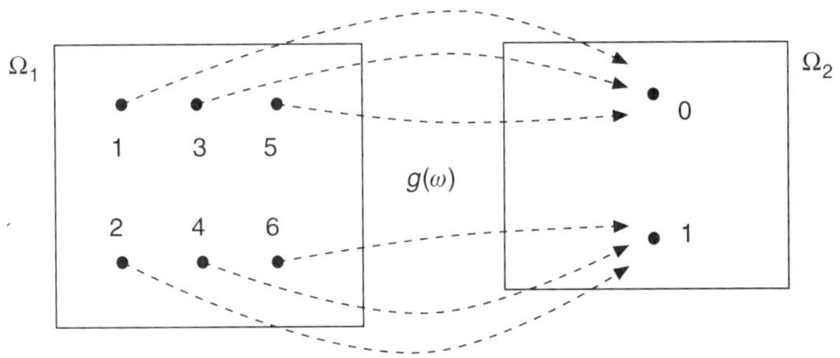

FIGURE 3.3 Mapping for Example 3.4.

which is illustrated in Figure 3.3. This function reduces the original sample space to a binary experiment with $\Omega_2 = \{0, 1\}$: we are interested in whether the die is even or odd. Assuming the power set for the σ-field such that $\mathcal{F}_2 = \{\phi, \Omega_2, 0, 1\}$, we see that this mapping is measurable:

$$g^{-1}(0) = \{\omega : g(\omega) = 0\} = \{1, 3, 5\} \in \mathcal{F}_1, \tag{3.6}$$

$$g^{-1}(1) = \{\omega : g(\omega) = 1\} = \{2, 4, 6\} \in \mathcal{F}_1. \tag{3.7}$$

From these examples with finite sample spaces, we have the following observations:

- If outcomes (elementary events) are excluded from $\mathcal{F}_1$, then it is generally easy to find functions that are *not* measurable, especially if $\mathcal{F}_2$ is the power set.
- For such cases, a function could be measurable if it condenses the number of elements in Ω_1 to a smaller number of elements in Ω_2.

This is evident in Example 3.4 where $g(\omega)$ is a many-to-one mapping. We would expect similar behavior for countably infinite and uncountable (continuous) sample spaces.

Definition: Random Variable *Random variable* X is a measurable function from Ω to the real line $\mathcal{R}$. The function for $\omega \in \Omega$ is written as $X(\omega)$.

The abstract probability space is denoted by $\{\Omega, \mathcal{F}, P\}$. The corresponding probability space for random variable X is $\{\mathcal{R}, \mathcal{B}(\mathcal{R}), P_X\}$, where $\mathcal{B}(\mathcal{R})$ is the Borel σ-field on the real line, and P_X is the probability measure for all events of $\mathcal{B}(\mathcal{R})$. Since $X(\omega)$ is measurable, any event in $\mathcal{B}(\mathcal{R})$ must have an inverse image in $\mathcal{F}$; in particular

$$X^{-1}((-\infty, x]) = \{\omega : X(\omega) \leq x\} \in \mathcal{F}. \tag{3.8}$$

(Note that in Chapter 1, $X(\omega)$ is also used for the Fourier transform of $x(t)$. Although the meaning of the notation should be clear from the context of the discussion, we will soon drop the argument of $X(\omega)$ for the mapping above and use X to denote a random variable.)

As discussed previously, a function is not measurable by itself: measurability also depends on $\mathcal{F}$, which was not specified above. Thus, if we say that $X(\omega)$ is measurable, or by calling it a random variable, then we *assume* that an appropriate $\mathcal{F}$ is used in the abstract event space, such as the power set of Ω.

Example 3.5. Let $\Omega = \{-3, -2, -1, 1, 2, 3\}$ and consider the function $X(\omega) = |\omega|$. We would like to find a σ-field $\mathcal{F}$ other than the power set $\mathcal{P}(\Omega)$ for which X is a random variable. Observe that the inverse images are $X^{-1}(1) = \{-1, 1\}$, $X^{-1}(2) = \{-2, 2\}$, and $X^{-1}(3) = \{-3, 3\}$. These are the atoms from which a σ-field for Ω can be generated (see Chapter 2). This is done by defining the σ-field to be the power set of the atoms, which yields $\mathcal{F} = \{\phi, \Omega, \{-1, 1\}, \{-2, 2\}, \{-3, 3\}, \{-3, -1, 1, 3\}, \{-3, -2, 2, 3\}, \{-2, -1, 1, 2\}\}$. This σ-field has eight elements versus $2^6 = 64$ for $\mathcal{P}(\Omega)$. $X(\omega)$ is a random variable for this event space $\{\Omega, \mathcal{F}\}$.

Example 3.6. Suppose in Example 3.5 that we expand the sample space to include all integers except zero: $\Omega = \{\ldots, -3, -2, -1, 1, 2, 3, \ldots\}$. For the function $X(\omega) = |\omega|$, the inverse images are $X^{-1}(n) = \{-n, n\}$ for $n \in \mathcal{N}$. If the σ-field $\mathcal{F}$ for Ω is generated by the atoms $\{-n, n\}$, then $X(\omega)$ is a random variable; it is a measurable function on $\mathcal{F}$. Note that the power set of these atoms is, in fact, smaller than the power set of Ω. $\{-1, 3\}$ and $\{1, 2, 3\}$ are examples of events in $\mathcal{P}(\Omega)$, but are not in the power set of the atoms $\{-n, n\}$.

Example 3.7. Consider the experiment where two coins are tossed with sample space $\Omega = \{HH, TT, HT, TH\}$ and define the mapping: $X(\omega) = \{\text{number of heads}\}$. Thus $\{TT\} \to 0$, $\{TH, HT\} \to 1$, and $\{HH\} \to 2$. If $\mathcal{F} = \mathcal{P}(\Omega)$, then $X(\omega)$ is measurable and it is a random variable. Assuming the coins are fair, we know that $P(X = 0) = 1/4 = P(X = 2)$ and $P(X = 1) = 1/2$. Suppose now that instead of the power set, $\mathcal{F} = \{\phi, \Omega, \{TT, HH\}, \{TH, HT\}\}$ which is a valid σ-field. For this case, $X(\omega)$ is *not* measurable even though it is the same mapping given above. This follows because the inverse image $X^{-1}(2) = \{HH\}$ is not in this "coarser" $\mathcal{F}$. We emphasize the following points with this example: (i) $X(\omega)$ itself is unchanged by using the coarser σ-field. (ii) Probabilities in a probability space are assigned to *events* in $\mathcal{F}$, not to the elements of Ω (unless the atoms of a countable σ-field are the outcomes of the sample space). For the coarser σ-field above, we assign probabilities only to $\{TT, HH\}$ and $\{TH, HT\}$; also, of course, $P(\{TT, HH\} \cup \{TH, HT\}) = P(\Omega) = 1$ and $P(\phi) = 0$. This demonstrates why $X(\omega)$ is not measurable: we cannot determine probabilities for $X = 0$ or $X = 2$ because they have not been assigned to the individual outcomes comprising the atom $\{TT, HH\}$. (Note that if we attempted to assign probabilities to the outcomes of Ω and not the events described by the coarser σ-field, then we are assuming the power set for $\mathcal{F}$.) One measurable function for the coarser σ-field is

$$X(\omega) = \begin{cases} 0, & \text{outcomes differ} \\ 1, & \text{outcomes are the same,} \end{cases} \tag{3.9}$$

so that $\{HT, TH\} \to 0$ and $\{HH, TT\} \to 1$. Coarse σ-fields will be explored further when expectation is covered in Chapter 5.

Although the sample space for $X(\omega)$ is the real line, not all points necessarily correspond to a mapping from elements in Ω (as is evident from the previous examples). Generally, we can say the following: if Ω is a finite (discrete) sample space, then only a finite number of points in $\mathcal{R}$ will have nonzero probability. Each point is said to have a *probability mass* which is some value in $[0, 1]$, and they must sum to 1. Points with zero probability mass can never occur: they are impossible events without corresponding events in the original sample space Ω from which they could have been mapped (otherwise they would have had nonzero probability). The probability assignment for discrete points in $\mathcal{R}$ is called a *probability mass function* (pmf).

Example 3.8. For the single coin-toss experiment, $\Omega = \{H, T\}$ and $\mathcal{F} = \{\phi, \Omega, H, T\}$. Let the mapping for random variable $X(\omega)$ be $H \to 1$ and $T \to -1$. Of course, for a fair coin $P(X = -1) = P(X = 1) = 1/2$ and the probability of any other value for X is zero. Moreover, the other numbers are impossible events. Suppose now that we allow for the event that the coin lands on its edge, denoted by E, and suppose that it maps as follows: $E \to 0$. If $P(E) \neq 0$ (though it would be a very small number), then this is a different experiment: there are actually three outcomes with nonzero probability. All other outcomes for X with zero probability are still impossible events.

The point of the last example is that for finite experiments, events with probability zero are also impossible events. Similar statements apply to countably infinite sample spaces, which when mapped to $\mathcal{R}$ yield a countable number of points that have nonzero probability mass. Intuitively, it may not seem possible that the probability mass for an infinite number of points sums to a finite value; however, the situation is analogous to an infinite series that converges (see Appendix E). Since there is an infinity of sample points, only certain pmfs sum to 1; specific examples are presented later.

The situation for continuous sample spaces is quite different. Since an uncountable number of elements in Ω map to the real line, it is not possible for *any* individual number in $\mathcal{R}$ to have a nonzero probability mass. The probability of any point on the real line for a continuous random variable is exactly zero. However, as mentioned in Chapter 2, such an event with probability zero does not mean it is an impossible event. In fact, for a continuous experiment there is an infinity of points with zero probability that could occur. Impossible outcomes correspond to those points on $\mathcal{R}$ that are excluded by the definition of the sample space. For example, choosing $\Omega = \mathcal{R}^+$ causes all negative numbers to be impossible events, whereas any nonnegative number has probability zero but *could* occur.

Since the Borel σ-field is used to describe events on the real line, nonzero probabilities are assigned to *intervals* on $\mathcal{R}$, which are events with nonzero Lebesgue measure. The probabilities assigned to $[a, b]$, $[a, b)$, $(a, b]$, and (a, b) are all identical because individual points have zero probability. As we shall see, a *probability density function* (pdf) defined for all $\mathcal{R}$ is used to describe the probability measure. The probability of a specific interval $[a, b]$ is obtained as the *area* under the pdf between a and b; technically, we say that the probability is defined with respect to Lebesgue measure.

Finally, we mention that a *mixed random variable* can be defined with continuous and discrete components. For such random variables, some points in $\mathcal{R}$ are described by a pmf, while certain intervals are described by a pdf. Of course, the sum over the pmf and the area under the pdf together must be 1 (in order to satisfy the three axioms of probability). Examples of mixed random variables are presented later.

3.3 DISTRIBUTION FUNCTION

In order to describe continuous, discrete, and mixed random variables, we first introduce the *distribution function*, from which it is possible to derive either a pmf, a pdf, or a combination of both. Once specific families of random variables have been defined, there usually is no need to consider the abstract probability space $\{\Omega, \mathcal{F}, P\}$ (which, as mentioned before, may not exist). Instead, we can work directly with the distribution function defined for the probability space $\{\mathcal{R}, \mathcal{B}(\mathcal{R}), P_X\}$.

Definition: Distribution Function The *distribution function* with domain $\mathcal{R}$ is the following probability:

$$F_X(x) \triangleq P_X((-\infty, x]) = P(X \leq x), \tag{3.10}$$

where $P_X((-\infty, x])$ is a mapping of $(\infty, x]$ in the event space $\{\mathcal{R}, \mathcal{B}(\mathcal{R})\}$ to a real number in $[0, 1]$.

This probability assignment is yet another type of mapping: from a semi-open interval on $\mathcal{R}$ to a point in the closed interval $[0, 1]$. Figure 3.4 illustrates the mapping of event E in the abstract space Ω to an interval on $\mathcal{R}$, and then to a number in $[0, 1]$ which is the probability of that interval (and of the underlying event E).

Regarding notation, an uppercase letter such as X denotes the random variable, while lowercase x is a specific value (outcome) of the random variable. The subscript on $F_X(x)$ is included to be clear that we are referring to random variable X; this will be necessary later in order to identify distribution functions for multiple random variables. However, when it is obvious from the context of the discussion, such subscripts may be dropped for notational simplicity.

The distribution function is also called the *probability distribution function* and the *cumulative probability distribution function*. In order to avoid any confusion with the pdf defined later, we refer to $F_X(x)$ simply as the *cumulative distribution function* (cdf). Observe that the cdf can be written in terms of the abstract probability space $\{\Omega, \mathcal{F}, P\}$ as follows:

$$F_X(x) = P(\{\omega : X(\omega) \leq x\}). \tag{3.11}$$

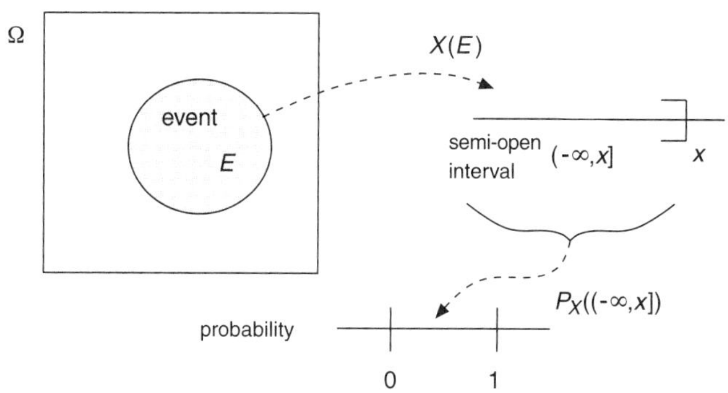

FIGURE 3.4 Mapping from $E \in \Omega$ to an interval of $\mathcal{R}$, and then to a point in the closed interval $[0, 1]$.

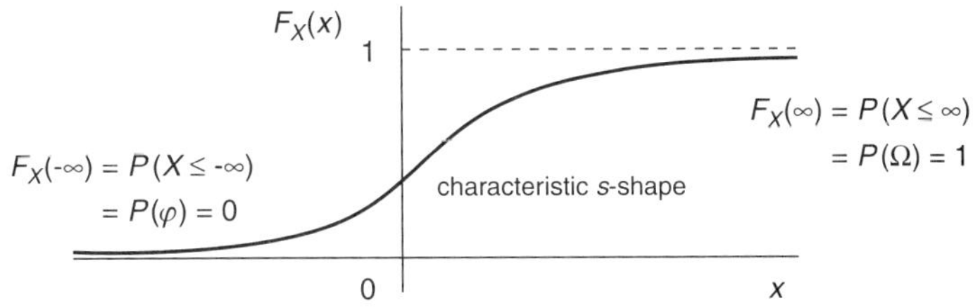

FIGURE 3.5 Example cdf $F_X(x) = P(X \leq x)$.

It is the probability assigned to all $\omega \in \Omega$ that map to $\mathcal{R}$ via the function $X(\omega)$ and do not exceed the value x.

An example cdf is shown in Figure 3.5 where we see that it has a characteristic s-shape because it is nondecreasing; this shape occurs for most "standard" pdfs because the cdf is the area under the pdf defined later. However, there are exceptions to the s-shape, for example, the cdf of the uniform pdf has straight lines, but they can be viewed as an extreme case of the s-shape.

With this definition, some basic properties of the cdf are derived:

- *End points:*

$$F_X(-\infty) = 0 \text{ and } F_X(\infty) = 1. \tag{3.12}$$

Substituting each end point and using the definition of $F_X(x)$ gives

$$F_X(-\infty) = P(X \leq -\infty) = P(\phi) = 0, \tag{3.13}$$

$$F_X(\infty) = P(X \leq \infty) = P(\Omega) = 1. \tag{3.14}$$

Note that $X \leq \infty$ covers all events in $\mathcal{F}$. On the other hand, $X \leq -\infty$ corresponds to no event in $\mathcal{F}$: it is the empty set ϕ.

- *Nondecreasing function:*

$$F_X(x_1) \leq F_X(x_2) \text{ for } x_1 \leq x_2. \tag{3.15}$$

For mutually exclusive events $A = (-\infty, x_1]$ and $B = (x_1, x_2]$:

$$\begin{aligned} F_X(x_2) &= P(X \in A \cup X \in B), \\ &= P(X \in A) + P(X \in B), \\ &= F_X(x_1) + P(x_1 < X \leq x_2) \geq F_X(x_1) \end{aligned} \tag{3.16}$$

because $P(x_1 < X \leq x_2) \geq 0$.

- *Continuous from the right:* Observe that if $F_X(x)$ has a discontinuity at x, then for $\epsilon > 0$:

$$F_X(x) = \lim_{\epsilon \to 0} F_X(x + \epsilon) = \lim_{\epsilon \to 0} P(X \leq x + \epsilon). \tag{3.17}$$

Because of the direction of the inequality, $P(X \leq x + \epsilon)$ always includes x; thus $F_X(x)$ is continuous from the right. On the other hand,

$$\lim_{\epsilon \to 0} F_X(x - \epsilon) = \lim_{\epsilon \to 0} P(X \leq x - \epsilon) \tag{3.18}$$

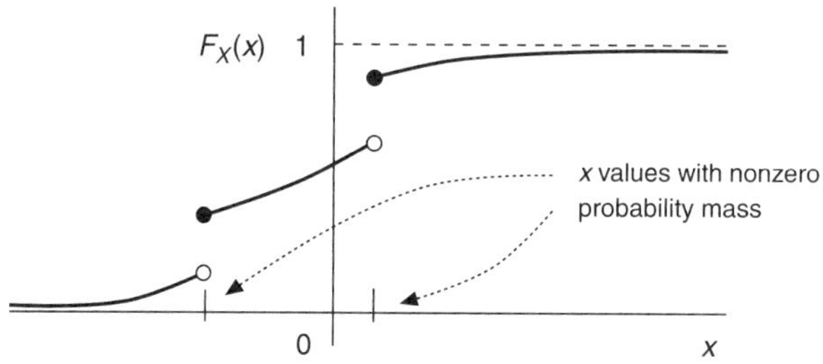

FIGURE 3.6 $F_X(x)$ with discontinuities. Solid circles indicate that the function is continuous moving from the right, whereas the open circles indicate that it is discontinuous moving from the left.

does not include x until $\epsilon = 0$ is exactly zero. Since the function has a discontinuity (a step) exactly when $\epsilon = 0$, the limit in (3.18) is not $F_X(x)$. Technically, the cdf is a right-continuous function with *left limits*, which are the function values just to the left of the discontinuities (they are also called limits from below). Definitions of continuity and limits are provided in Appendix B.

The last property is a result of the direction of the inequality in $F_X(x)$: $\leq$ and not $\geq$. Figure 3.6 shows an example of $F_X(x)$ with discontinuities. Such discontinuities occur for discrete and mixed random variables where individual points in $\mathcal{R}$ have nonzero probability mass.

Example 3.9. Let $\Omega = [-2, 1]$ and $\mathcal{F} = \mathcal{B}(\Omega)$. We would like to find the distribution function $F_X(x)$ for the mapping $X(\omega) = |\omega|$:

$$F_X(x) = P(X(\omega) \leq x) = P(|\omega| \leq x) = P(-x \leq \omega \leq \min(x, 1)) \tag{3.19}$$

for $x \in [0, 2]$. In this example, the underlying "abstract" event space is a subset of the real line with σ-field given by Borel sets in $[-2, 1]$; it has the features of a random variable, and conceivably could have been generated as a mapping from some other abstract event space. Assume the probability measure for $\{\Omega, \mathcal{F}\}$ is the Lebesgue measure $L(\cdot)$ (scaled by $1/3$), which was defined in Chapter 2. Thus

$$F_X(x) = (1/3)L(-x, \min(x, 1)) = (1/3)[\min(x, 1) + x]I_{[0,2]}(x) + I_{(2,\infty)}(x), \tag{3.20}$$

where $I_{[a,b]}(x)$ is the indicator function, and the second term has been included so that $F_X(x) = 1$ for $x > 2$. This distribution function is piecewise linear: it has slope $2/3$ on $[0, 1]$, slope $1/3$ on $(1, 2]$, and is flat beyond those intervals.

The probability space in Example 3.9 is $\{\Omega = [-2, 1], \mathcal{F} = \mathcal{B}([-2, 1]), P(\cdot) = (1/3)L(\cdot)\}$, and since P is the Lebesgue measure, the outcomes $\{\omega\}$ have a uniform distribution. Random variable $X(\omega)$ is a mapping from $[-2, 1]$ to $[0, 2]$, both of which are on the real line, and so it is actually an example of a *transformation* of a random variable. If we let Y be the random variable for Ω, then it has distribution function

$$F_Y(y) = P(\omega \leq y) = (1/3)(y + 2)I_{[-2,1]}(y) + I_{(1,\infty)}(y). \tag{3.21}$$

Different kinds of transformations for random variables are described in Chapter 4. For the rest of this chapter, we consider specific "families" of random variables without regard to an underlying abstract probability space, and use the notation X instead of $X(\omega)$. This also means by assumption that X is measurable and has the well-defined probability space $\{\mathcal{R}, \mathcal{B}(\mathcal{R}), F_X(x)\}$, where the cdf represents the probability measure P_X.

We now have a complete description of a probability space defined on the real line, as well a description of probabilities for semi-open intervals of the form $(-\infty, x]$ via the cdf. In subsequent discussions, it will be necessary to distinguish between *discrete* random variables and *continuous* random variables, as well as *mixed* random variables. Recall that a σ-field was defined for a sample space so that we can assign probabilities to events of interest in a consistent manner. The Borel σ-field is not only measurable, but is useful in practice because for a continuous random variable on $\mathcal{R}$, we are generally concerned with probabilities of intervals such as $P(a \le X \le b)$.

In order to distinguish between discrete and continuous random variables, we define what is meant by a continuous probability measure.

Definition: Absolutely Continuous Probability Measure Probability measure P on the measurable event space $\{\mathcal{R}, \mathcal{B}(\mathcal{R})\}$ is *absolutely continuous* with respect to Lebesgue measure if for some event $E \subset \mathcal{R}$ (interval or point), $L(E) = 0$ implies that $P(E) = 0$.

The phrase "with respect to Lebesgue measure" means that we are interested in defining the measure in terms of the length of an interval: if the length of the interval is zero, the probability of the interval is zero. An absolutely continuous probability measure refers to a continuous random variable. (This definition is extended to the multidimensional measurable event space $\{\mathcal{R}^N, \mathcal{B}(\mathcal{R}^N)\}$ in Chapter 4, where E is a rectangle on $\mathcal{R}^2$ and a hyper-rectangle on $\mathcal{R}^N$ for $N > 2$.)

We also define a probability measure that is not absolutely continuous.

Definition: Singular Probability Measure Probability measure P on the measurable event space $\{\mathcal{R}, \mathcal{B}(\mathcal{R})\}$ is *singular* with respect to Lebesgue measure if there exist events $\{E_n\}$ such that $L(E_n) = 0$ and $P(\cup_n E_n) = 1$.

This definition means there is a set of points on $\mathcal{R}$ (which, of course, have Lebesgue measure zero) such that the probability of their union is one. A singular probability measure refers to a discrete random variable where individual points have nonzero probability mass. (This definition can also be extended to the multidimensional measurable space $\{\mathcal{R}^N, \mathcal{B}(\mathcal{R}^N)\}$.)

Lebesgue's decomposition theorem states that a probability measure can be decomposed into an absolutely continuous component and a singular component (Billingsley, 1986), which we give without proof.

Theorem 3.1 (Lebesgue's decomposition). Probability measure P on the measurable event space $\{\mathcal{R}, \mathcal{B}(\mathcal{R})\}$ can be decomposed such that for any event E

$$P(E) = P_{ac}(E) + P_s(E), \tag{3.22}$$

where P_{ac} and P_s are absolutely continuous and singular probability measures, respectively.

This theorem allows for a *mixed random variable* that has discrete and continuous components, as described at the end of this chapter. Of course, (3.22) reduces to each case separately: either a continuous random variable with probability measure P_{ac}, or a discrete random variable with probability measure P_s.

For completeness, we mention that a singular probability measure can be decomposed further, to a part consisting of individual points (discrete) and a *singular continuous* component of uncountable events. This latter part is generally not of concern in practical problems, and thus we do not give a precise definition. However, the Cantor distribution (Richardson, 2009) described next is an example of a random variable that is not absolutely continuous; instead, it is singular continuous and does *not* have a discrete part.

Example 3.10 (Cantor function). The Cantor function $F_C(x)$ is the cdf of a random variable that is uniformly distributed on the Cantor set C, which was described in Chapter 2 in Example 2.35. We denote functions in the sequence leading to the Cantor function by $F_{C_m}(x)$ for $m \in \mathcal{N}$. Figure 3.7 shows $F_{C_m}(x)$ for three values of m, which has a staircase appearance with an increasing number of flat sections as m increases. (It is an example of the so-called "devil's staircase.") The Cantor function $F_C(x)$, obtained as $m \to \infty$, has the following properties: (i) it is a singular function that does *not* have a discrete part. (ii) It is a continuous and monotonic function (of

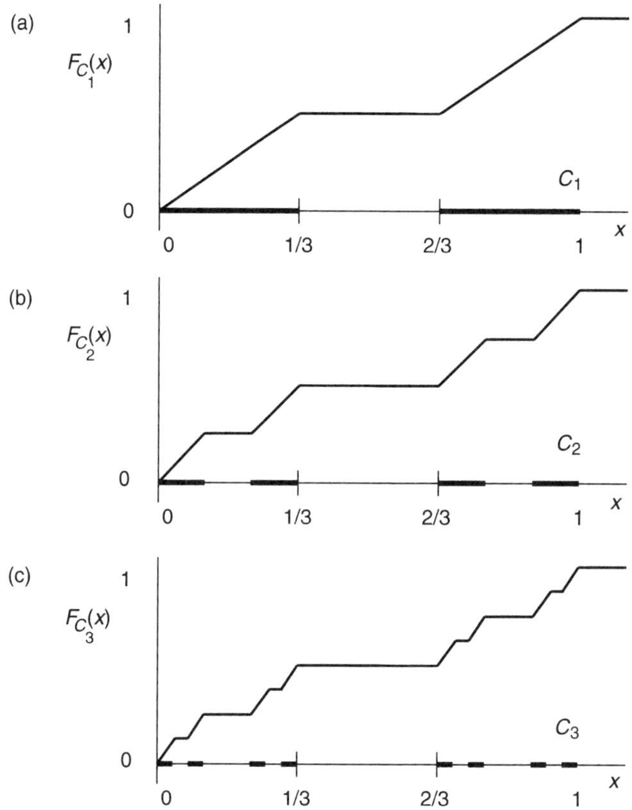

FIGURE 3.7 Cdfs for a uniform distribution on C_m for $m = 1, 2, 3$. The Cantor function $F_C(x)$ is obtained in the limit as $m \to \infty$. The $\{C_m\}$ from Figure 2.6 leading to the Cantor set are also shown (dark lines on the x-axis).

course, all cdfs are monotonic). (iii) Its derivative is zero almost everywhere because each $F_{C_m}(x)$ is constant for $x \notin C_m$ (each function is flat on the open intervals that were removed to construct C_m of the Cantor set).

The probability space $\{\mathcal{R}, \mathcal{B}(\mathcal{R}), F_X\}$ is a useful description for discrete, continuous, and mixed random variables. In Sections 3.4 and 3.5, we use this representation and the cdf $F_X(x)$ to quantify the probability mass assigned to individual points for a discrete random variable, and to compute probabilities for any interval on the real line for a continuous random variable. The pmf for a discrete random variable is relatively straightforward and is covered first. For a continuous random variable, we describe the underlying pdf which, when integrated over intervals describing an event, yields the probability of that event.

We conclude this section with a technical definition of the support of a random variable based on the cdf.

Definition: Support of a Random Variable The *support* of random variable X is given by all $x \in \mathcal{R}$ such that

$$F_X(x + \epsilon) > F_X(x - \epsilon), \tag{3.23}$$

for all $\epsilon > 0$.

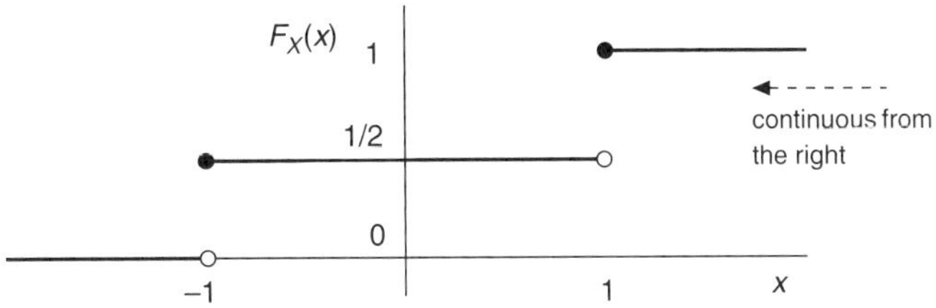

FIGURE 3.8 Distribution function for a discrete (Bernoulli) random variable.

Note that strict inequality is needed in this definition. Although $F_X(x)$ has support $\mathcal{R}$, this may not be true for the underlying pdf, and is not the case for a pmf (specified using the Kronecker delta function). Open intervals where $F_X(x)$ is flat (with zero derivative) are excluded from the support of the pdf or pmf. Such $F_X(x)$ with flat regions correspond either to a discrete random variable or a mixed random variable with discrete components. Since the cdf is right continuous, the inequality above "captures" the outcomes for a discrete random variable.

3.4 PROBABILITY MASS FUNCTION

An example of the distribution function for a discrete random variable is shown in Figure 3.8. Observe that there are two distinguishing features of this type of distribution function:

- $F_X(x)$ changes instantaneously at values of x that are outcomes of the random variable.
- $F_X(x)$ remains constant between these outcomes.

There are two useful ways of representing the outcomes and the probability assignment for a discrete random variable, using either the Kronecker delta function or the Dirac delta function $\delta(x)$. The distribution function in Figure 3.8 corresponds to the symmetric Bernoulli random variable with equally likely outcomes ± 1. This random variable would arise, for example, when tossing a single fair coin with the mapping: $H \to X = 1$ and $T \to X = -1$.

The Kronecker delta function is convenient for representing the outcomes of a discrete random variable because there are no continuity issues. Moreover, the height of a pulse is exactly the probability assigned to a specific outcome x. For Example 3.10, since $P(X = 1) = P(X = -1) = 1/2$, we can write the following pmf:

$$p_X[x] = (1/2)(\delta[x-1] + \delta[x+1]). \tag{3.24}$$

(Recall that brackets are used to distinguish the Kronecker delta function from the Dirac delta function which uses parentheses.) It is called a pmf because the function describes the amount of probability mass assigned to each outcome. The pmf for the symmetric Bernoulli example is shown in Figure 3.9(a). Observe that the locations of the delta functions show which outcomes (± 1) have nonzero probability, and the height of each line (denoted by solid circles) gives the probability ($1/2$) for each outcome. This type of representation is often used in basic probability courses because it is intuitive and clearly describes probabilities for a discrete experiment. However, since the pmf obviously is not a continuous function and the support is a subset of $\mathcal{Z}$, it cannot be integrated, which would be useful in many applications.

For continuous random variables, the cdf is differentiated to generate the underlying pdf; also, the pdf is integrated to obtain the cdf (these properties are discussed later). This cannot be done for discrete random variables using the Kronecker delta function. However, discrete problems can be handled using the Dirac delta

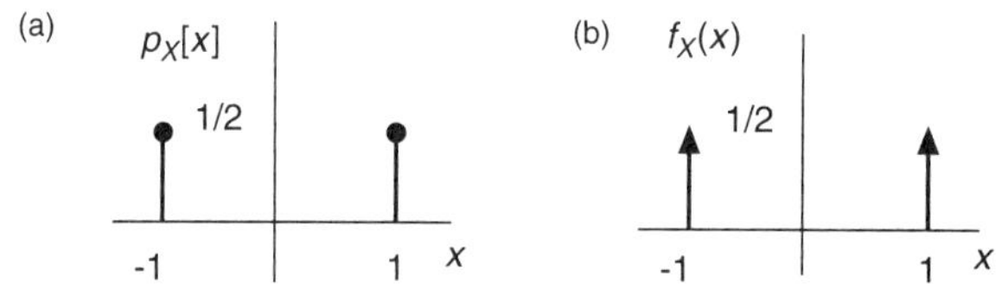

FIGURE 3.9 Symmetric Bernoulli random variable. (a) Pmf based on the Kronecker delta function. (b) Pdf based on the Dirac delta delta function.

function $\delta(x)$, which has infinite height at x, zero width, but unit *area*. Other definitions and important properties of the Dirac delta function are given in Appendix B. Thus, we can also write a pdf for the symmetric Bernoulli random variable as follows:

$$f_X(x) = (1/2)[\delta(x-1) + \delta(x+1)], \tag{3.25}$$

where it is understood that the Dirac delta function is defined by its area: $(1/2)\delta(x)$ has area $1/2$ (it is *not* the height of the Dirac delta function). This pdf is shown in Figure 3.9(b) for the symmetric Bernoulli random variable where arrows are used to distinguish it from the Kronecker delta function. The "jumps" in $F_X(x)$ are represented by the unit-step function $u(x)$ as follows:

$$F_X(x) = (1/2)[u(x+1) + u(x-1)], \tag{3.26}$$

which is shown in Figure 3.8. For convenience in some plots presented later, we will also use the illustration in Figure 3.10 where the discontinuities are emphasized by vertical lines, and where it is understood that the function is still continuous only from the right. The connection between $F_X(x)$ and $f_X(x)$ will be discussed in greater detail in Section 3.8.

We emphasize that for problems involving discrete random variables, the pdf and cdf are not continuous functions. Strictly speaking, the derivative of $F_X(x)$ does not exist for the Bernoulli random variable. However, by using the Dirac delta function, this situation can be "handled" as though the functions were continuous: wherever $F_X(x)$ has a discontinuity (a step), $f_X(x)$ has a Dirac delta function with area given by the height of the discontinuity. Likewise, $F_X(x)$ is obtained from $f_X(x)$ by integrating over all x; $F_X(x)$ remains flat until a Dirac delta function is encountered, at which point it "jumps" by an amount given by the area of the delta function (which is also the probability mass at that point).

Example 3.11. Let X be a discrete random variable with outcomes $\{1, \ldots, N\}$, and define the probability of event E to be $P(E) = C(E)/N$, where $C(E) = |E|$ is the counting measure (see Chapter 2). This probability space defines the discrete *uniform* random variable described later in this chapter. Suppose that N is even and we are interested in the event $E = \{\text{odd numbers} \cup \text{numbers} \leq N/2\}$. Thus

$$P(E) = P(\text{odd numbers}) + P(\text{numbers} \leq N/2) - P(\text{odd numbers} \leq N/2), \tag{3.27}$$

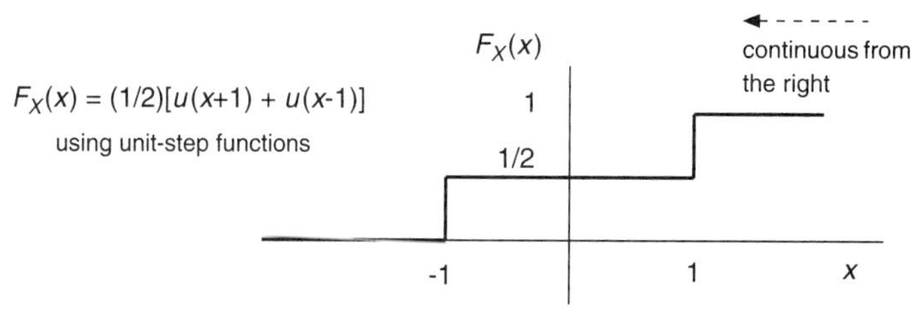

FIGURE 3.10 Distribution function for a discrete random variable with discontinuities represented by vertical lines (compare with Figure 3.8).

and from the counting measure

$$C(E) = N/2 + N/2 - N/4 = 3N/4 \Longrightarrow P(E) = 3/4. \tag{3.28}$$

Using the pmf, we can also write

$$P(E) = \sum_{x=1,3,\ldots,N-1} p_X[x] + \sum_{x=1}^{N/2} p_X[x] - \sum_{x=1,3,\ldots,N/2-1} p_X[x]. \tag{3.29}$$

Since $p_X[x] = (1/N)I_{\{1,\ldots,N\}}[x]$, it can be factored from each sum, which then gives the results in (3.28).

For a finite set of equally likely outcomes, the probability in Example 3.10 based on the counting measure is equivalent to $P(E) = |E|/|\Omega|$ discussed in Chapter 2, where we counted permutations or combinations of outcomes to find the cardinality of E. Those techniques are useful for a wide range of problems. However, when the outcomes are not equally likely and for more complicated random variables, it is necessary that probabilities be computed by summing over the pmf as in (3.29). Moreover, it is possible using the pmf to compute various functions of a random variable, such as its mean, variance, and other moments, as described in Chapter 5.

3.5 PROBABILITY DENSITY FUNCTION

The cdf is a probability measure for semi-open intervals of random variable X on the measurable space $\{\mathcal{R}, \mathcal{B}(\mathcal{R})\}$ and is given by (3.10). Since we are interested in intervals, it is instructive to consider the underlying function for $F_X(x)$, denoted by the pdf $f_X(x)$, such that when it is integrated over $(-\infty, x]$, the cdf is obtained. We can compute the probability of an event given by a union of intervals as an integral of the underlying pdf over those intervals. The following theorem, which we provide without proof, states the existence of a pdf.

Theorem 3.2 (Radon–Nikodým). If the probability measure P_X on the measurable event space $\{\mathcal{R}, \mathcal{B}(\mathcal{R})\}$ is absolutely continuous, then for random variable X there exists nonnegative function $f_X(x)$ such that

$$P(a \leq X \leq b) = \int_a^b f_X(x)\mathrm{d}x \tag{3.30}$$

for interval $[a, b] \in \mathcal{R}$.

Since the probability measure is absolutely continuous, all combinations of inequalities can be used in (3.30), such that $P(a < X < b) = P(a \leq X < b) = P(a < X \leq b) = P(a \leq X \leq b)$. Moreover, these can be expressed as

$$P(a \leq X \leq b) = F_X(b) - F_X(a), \tag{3.31}$$

which follows from

$$F_X(x) = \int_{-\infty}^{x} f_X(x)\mathrm{d}x. \tag{3.32}$$

The pdf is obtained by differentiating the cdf with respect to x:

$$f_X(x) = \frac{\mathrm{d}}{\mathrm{d}x} F_X(x), \tag{3.33}$$

which is called the Radon–Nikodým derivative of $F_X(x)$.

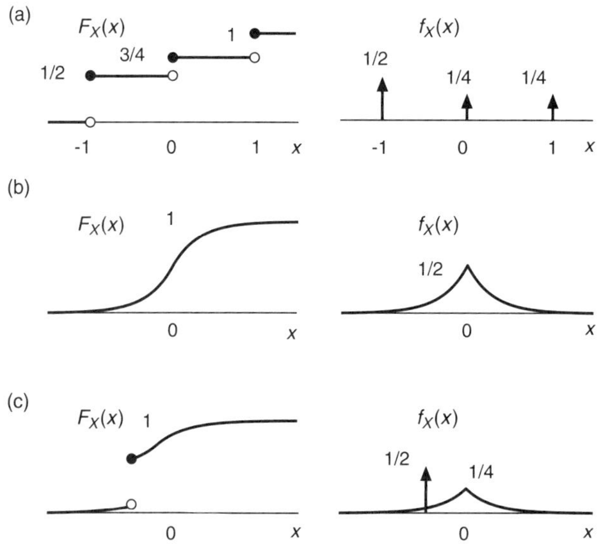

FIGURE 3.11 Example cdfs and pdfs for three types of random variables. (a) Discrete. (b) Continuous. (c) Mixed (discrete and continuous components).

In general, a density function is called a *density* because when integrated the desired quantity is obtained, which in this chapter is the probability for some random variable. The pdf itself is *not* a probability, which is a dimensionless quantity with values in [0, 1]. Suppose that random variable X corresponds to a voltage with units V. Then the pdf of X has units V^{-1}, so that when it is integrated to generate a probability, a dimensionless quantity is obtained. When a pdf is integrated over some range for X, we are assured to obtain a dimensionless probability, and if the pdf is valid, the probability will be in the range [0, 1]. (Likewise later when we consider the power spectral density $S_{XX}(f)$, its units are power (watts) per frequency (Hz). When $S_{XX}(f)$ is integrated over a range of frequencies, the corresponding power in watts is obtained.)

Strictly speaking, the Radon–Nikodým derivative is computed for absolutely continuous probability measures, which means that $F_X(x)$ needs to be an absolutely continuous function as stated in the theorem. Previously, we mentioned that $F_X(x)$ is continuous from the right. This distinction exists because $F_X(x)$ may in practice consist of absolutely continuous, singular continuous, and discrete components. The derivative of a singular continuous function is zero almost everywhere (as is the case for the Cantor function described in Example 3.10). Singular continuous functions are excluded in the Radon–Nikodým theorem because the pdf cannot be computed using (3.33). As mentioned before, cdfs with a singular continuous part are not generally encountered in practical problems, and so we are not concerned with this case.

On the other hand, a discrete component occurs frequently in many applications, and so discrete cdfs are of interest, as are mixed cdfs with absolutely continuous and discrete components. However, the discrete case is also excluded in the Radon–Nikodým theorem, and the derivative in (3.33) is not well defined. As mentioned in Section 3.4, this situation is handled by using the Dirac delta function to represent those components of $f_X(x)$ corresponding to discontinuities (steps) in $F_X(x)$. Examples of the three types of cdfs and the corresponding pdfs are shown in Figure 3.11. Note that for a mixed random variable, the area under the continuous parts plus the areas due to all the Dirac delta functions must equal 1 (as is the case for any random variable).

3.6 MIXED DISTRIBUTIONS

Starting with the mixed probability in (3.22), we would like to write an expression for a mixed cdf: one that has discontinuities with absolutely continuous parts in between the discontinuities, as depicted by the example in Figure 3.11(c). Let event E in (3.22) be $X \leq x$. Then we can write the cdf as follows:

$$P(X \leq x) = P_{ac}(X \leq x) + P_s(X \leq x). \tag{3.34}$$

This decomposition implies a conditioning on the values of $\mathcal{R}$ where the distribution has discontinuities. The subscript s means the following:

$$P_s(X \leq x) \triangleq P(X \leq x | X \in S)P(X \in S), \tag{3.35}$$

where S is the set of values at the discontinuities (the singular discrete components of the cdf). Likewise

$$P_{ac}(X \leq x) \triangleq P(X \leq x | X \in S^c)P(X \in S^c), \tag{3.36}$$

where S^c is the support of X excluding the discontinuities. (Actually, the elements of S would not be "removed" from the support of the absolutely continuous part. Since those elements have zero probability in the continuous part, they can be included in (3.36) without affecting the probability). From these definitions, (3.34) has been written in terms of conditional probabilities with conditioning based on the discontinuities. Thus

$$F_X(x) = F_{X|S^c}(x|S^c)P(X \in S^c) + F_{X|S}(x|S)P(X \in S), \tag{3.37}$$

where $F_{X|S^c}(x|S^c)P(X \in S^c)$ is the cdf of the continuous part, $F_{X|S}(x|S)$ is the cdf of the discrete part, and the probabilities $P(X \in S^c) + P(X \in S) = 1$ provide the appropriate weighting such that $F_X(x)$ is a valid cdf. The corresponding pdf is

$$f_X(x) = f_{X|S^c}(x|S^c)P(X \in S^c) + f_{X|S}(x|S)P(X \in S). \tag{3.38}$$

Observe that $P(X \in S)$ corresponds to the probability of "success" for a Bernoulli random variable; a success occurs here when X equals one of the values in S.

Example 3.12. Let $S = \{0, 1, 2\}$ with $P(X \in S) = 0.6$ and assume the following:

$$f_{X|S^c}(x|S^c) = \frac{1}{\sqrt{2\pi}} \exp(-x^2/2), \tag{3.39}$$

$$f_{X|S}(x|S) = (1/8)\delta(x) + (1/4)\delta(x-1) + (5/8)\delta(x-2). \tag{3.40}$$

The first pdf is obviously Gaussian and the second pdf is discrete with three outcomes comprising S. The weightings $P(X \in S) = 0.6$ and $P(X \in S^c) = 0.4$ ensure that $f_X(x)$ is a valid pdf, which is shown in Figure 3.12 along with its cdf.

Given a mixed cdf, it is straightforward to determine the discrete and continuous components, by first finding all the discontinuities in the function that comprise the set S. The height of each discontinuity is its probability mass, though this must be scaled so that the discrete pmf separately sums to 1. The discontinuities are removed simply by subtracting the height of each jump in the mixed cdf. The remaining function is the continuous part, and is a valid pdf with appropriate scaling. Removing the discontinuities does not mean that those outcomes are excluded from the continuous random variable (as mentioned before in connection with (3.36)); if the support of the continuous part extends across these points, then they should bc included but now have probability zero, as do all individual points of the continuous part. In Example 3.12, the support of the mixed pdf is $\mathcal{R}$. The support of the discrete part is $\{0, 1, 2\}$, whereas it is still $\mathcal{R}$ for the continuous part.

Note that (3.37) is *not* the distribution that would be obtained when two random variables are summed together. The latter type of distribution depends on a *function* of the random variables and is discussed in

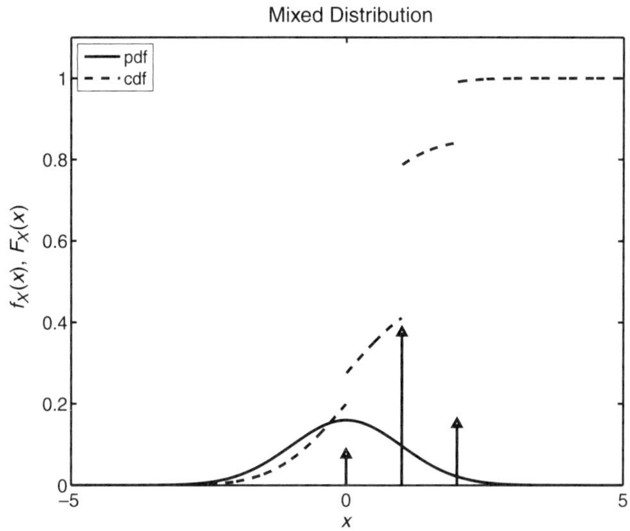

FIGURE 3.12 Mixed distribution for Example 3.12.

Chapter 3.7. Instead, a mixed random variable is actually a special case of a *mixture distribution*:

$$F_X(x) = \sum_{n=1}^{N} w_n F_{X_n}(x_n), \tag{3.41}$$

where $\{F_{X_n}(x_n)\}$ are valid cdfs (continuous or discrete), and $\{w_n\}$ are nonnegative weights such that $\sum_{n=1}^{N} w_n = 1$. The cdfs on the right-hand side of (3.41) can generally be of any form; mixture distributions are discussed in Chapter 4.

Table 3.1 summarizes event spaces used for the random variables covered later in this chapter, along with the cardinality of each sample space Ω and the corresponding σ-field $\mathcal{F}$. Beth numbers denote the cardinality of different types of infinite sets; these are described in Chapter 2 and are summarized in Table 2.4. Observe that by restricting the σ-field for continuous random variables to be $\mathcal{B}(\mathcal{R})$, it has the same cardinality as $\mathcal{R}$, whereas the power set (which is "too large") has cardinality $\mathcal{P}(\mathcal{R}) = \beth_2 = \mathcal{P}(\mathcal{P}(\mathcal{N}))$. For discrete random variables (finite and countably infinite), the sample space is shown to be a subset of $\mathcal{R}$ (N integers or $\mathcal{Z}$). The event space $\{\mathcal{R}, \mathcal{B}(\mathcal{R})\}$ still applies for the cdf $F_X(x)$ of a discrete random variable, which always has support $\mathcal{R}$. Since a pdf technically does not exist for this case, the probability of an event described by some interval is obtained via a Riemann–Stieltjes integral with integrator $F_X(x)$. However, we need not discuss that type of integration here, which is covered later in Chapter 5 on expectation and is summarized in Appendix D.

As mentioned before, a discrete random variable can be included in a continuous representation by using Dirac delta functions in the underlying pdf $f_X(x)$. The event space is $\{\mathcal{R}, \mathcal{B}(\mathcal{R})\}$, and probabilities are computed by integrating $f_X(x)$ to "capture" those delta functions within intervals specified by an event of interest. However,

TABLE 3.1 Random Variable Event Spaces $\{\Omega, \mathcal{F}\}$ and Cardinality

Type	Ω	$\lvert\Omega\rvert$	$\mathcal{F}$	$\lvert\mathcal{F}\rvert$	Example
Finite	N integers	N	$\mathcal{P}(\Omega)$	2^N	Binomial
Countably infinite	$\mathcal{Z}$	$\beth_0 = \lvert\mathcal{N}\rvert$	$\mathcal{P}(\Omega)$	$\beth_1 = \lvert\mathcal{R}\rvert$	Geometric
Continuous	$\mathcal{R}$	$\beth_1$	$\mathcal{B}(\Omega)$	$\beth_1$	Gaussian

when using a pmf to represent the distribution of a discrete random variable, the integer-based sample spaces shown in Table 3.1 are assumed. For these Ω, there is usually no need to restrict the σ-field to be smaller than the power set (which we did in some previous examples to illustrate measurability). Moreover, all elements in Ω can be assigned any distribution of probabilities provided the probability mass sums to 1.

In the rest of this chapter, we present specific *parametric families* of random variables, first continuous and then discrete. All these random variables are summarized in Appendix A, which includes various properties such as moments, the characteristic function, and entropy, all of which are covered in later chapters. Here we focus mainly on the shape and location of the pdfs and pmfs, mention some applications, and demonstrate how to compute probabilities. It will be convenient to use the indicator function to denote the support for the various random variables. It is discussed in Appendix B, but we provide the definition here for convenience:

$$I_{(a,b)}(x) \triangleq \begin{cases} 1, & a < x < b \\ 0, & \text{else.} \end{cases} \tag{3.42}$$

It is likewise defined for other types of intervals (closed, semi-closed), as well as for individual points:

$$I_{\{a,b\}}[x] \triangleq \begin{cases} 1, & x = a, b \\ 0, & \text{else.} \end{cases} \tag{3.43}$$

A pdf or pmf is multiplied by the appropriate indicator function to denote its support.

3.7 PARAMETRIC MODELS FOR RANDOM VARIABLES

In many applications where the data are random, we are interested in deriving a *distribution model* that accurately represents the observed samples. The modeling process has the following components:

- A parametric family that "best" represents the observed samples is chosen as a model for the distribution of random variable X. For example, from a histogram of observed samples, we might decide that a Gaussian distribution with pdf $f_X(x)$ is a reasonable model because the shape of the histogram is "bell-shaped."
- From observed samples of X, estimate the parameters of the distribution. For the Gaussian example, the sample mean $\bar{X}$ can be used to estimate the mean μ_X, and the sample variance S^2 for the variance σ_X^2. Several parameter-estimation techniques are discussed in Chapter 9.
- Once a parametric model has been chosen, various characteristics of the random variable can be determined, such as specific probabilities or *moments* like skewness and kurtosis. Moments and other expectations are described in Chapter 5.
- The parametric model can be used to predict future outcomes that might be used to make decisions in various applications, such as those mentioned in the next several sections for the different families of random variables.

Example 3.13. Consider the histogram of 2000 samples shown in Figure 3.13 where each bin along the x-axis has width 0.1. This histogram has the form of an exponential pdf given by

$$f_X(x) = \lambda \exp(-\lambda x) I_{[0,\infty)}(x), \tag{3.44}$$

where $\lambda > 0$ is the parameter. The mean of the exponential random variable is $\mu_X = 1/\lambda$, which can be estimated using the following sample mean:

$$\bar{x} = \frac{1}{N} \sum_{n=1}^{N} x_n, \tag{3.45}$$

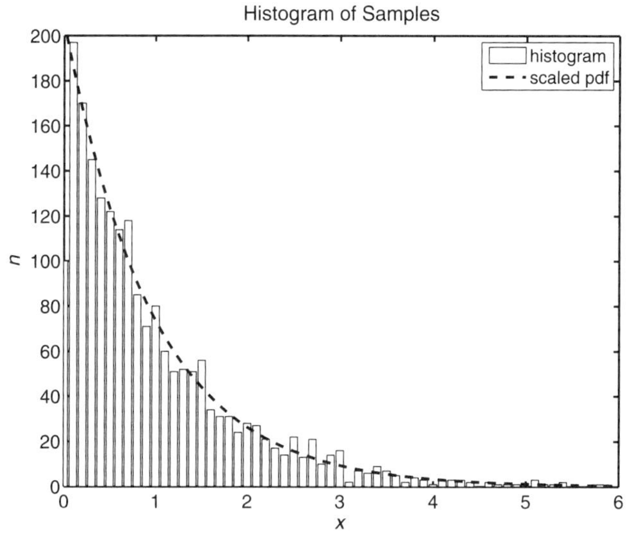

FIGURE 3.13 Histogram of 2000 samples for Example 3.13 and a pdf (scaled by 200).

where $\{x_n\}$ are the actual N outcomes. For the 2000 samples in the histogram, we find that $\bar{x} = 0.9841$ and thus the estimate of λ is 1.0162. The pdf with this value of λ is also shown in the figure (but scaled by 200 so that it aligns with the histogram). Observe that the pdf is an accurate model, which can be used to compute various probabilities such as

$$P(X > 2) = 1 - F_X(2) = \exp(-2\lambda) \approx 0.1353. \tag{3.46}$$

Of course, the pdf is accurate in this example because the data in the simulation were produced using an exponential random number generator with $\lambda = 1$.

A particular distribution is useful as a model of a random variable if its shape closely matches the histogram of observed samples. A pdf can be characterized using the following quantities that describe specific properties of $f_X(x)$:

- *Mean*. Location of a pdf:

$$\mu_X \triangleq \int_{\infty}^{\infty} x f_X(x)\mathrm{d}x. \tag{3.47}$$

- *Variance*. Width of a pdf:

$$\sigma_X^2 \triangleq \int_{-\infty}^{\infty} (x - \mu_X)^2 f_X(x)\mathrm{d}x. \tag{3.48}$$

- *Skewness*. Asymmetry of a pdf:

$$\gamma_1 \triangleq \int_{-\infty}^{\infty} [(x - \mu_X)/\sigma_X]^3 f_X(x)\mathrm{d}x = \mu_{c,3}/\sigma_X^3. \tag{3.49}$$

- *Kurtosis*. "Heavy tails" of a pdf:

$$\gamma_2 \triangleq \int_{-\infty}^{\infty} [(x - \mu_X)/\sigma_X]^4 f_X(x)\mathrm{d}x - 3 = \mu_{c,4}/\sigma_X^4 - 3, \tag{3.50}$$

where $\mu_{c,n}$ is a *central moment*. Definitions of the various moments are provided later in Chapter 5, but we assume the reader has some familiarity with the two basic moments: mean and variance. The standard deviation is the positive square-root of the variance. We introduce them here so that the shapes of the various pdfs can be discussed.

Some distributions have parameters that separately determine the mean and the variance, such as the Gaussian random variable. However, most distributions have one or more parameters that together give the mean and variance. Skewness and kurtosis are usually a complicated expression of the distribution parameters. Positive γ_1 indicates that a greater portion of the distribution lies to the left of the mean, while a negative γ_1 indicates the opposite. Three is subtracted in γ_2 so that the kurtosis of the Gaussian random variable is zero. As a result, kurtosis indicates the degree to which the distribution has heavy tails relative to those of the Gaussian distribution. For example, the Laplace distribution has $\gamma_2 = 3$, indicating that its tails are heavier; this is demonstrated later in some examples.

Various probabilities of a distribution can be described using quantiles.

Definition: Quantiles The mth *quantile* q_m of random variable X is the smallest value of x such that

$$F_X(q_m) \geq m/M \tag{3.51}$$

for $m = 1, \ldots, M - 1$.

Since m/M is not an integer for some m, the inequality is needed above for discrete random variables. For continuous random variables, we have strict equality $F_X(q_m) = m/M$. If M is even, then $q_{M/2}$ is the median m_e (defined below). $M = 4$ yields *quartiles* and $M = 100$ gives *percentiles*. Some examples of these are provided in the following sections for specific random variables.

Two measures of the location (center) of a distribution other than the mean are also discussed in the following sections.

Definition: Median The *median* of $f_X(x)$ for continuous X is obtained by finding m_e such that

$$\int_{-\infty}^{m_e} f_X(x)\mathrm{d}x = \int_{m_e}^{\infty} f_X(x)\mathrm{d}x. \tag{3.52}$$

This expression is equivalent to $F_X(m_e) = 1/2$ and it corresponds to quantile $q_{M/2}$ for M even, as discussed above. For discrete random variables, this definition needs to be modified as described in Chapter 5.

Definition: Mode The *mode* of a distribution is the value m_o where $f_X(x)$ achieves its maximum.

For many distributions, the mode can be found by differentiating the pdf with respect to x, setting the expression equal to zero, and solving for m_o. The median and the mode are also discussed further in Chapter 5.

3.8 CONTINUOUS RANDOM VARIABLES

We summarize several continuous random variables by providing expressions for $f_X(x)$ and $F_X(x)$, and describing important properties. For some continuous random variables, the cdf cannot be expressed in a closed form, including the well-known Gaussian random variable. Regarding distribution parameters, we use the following notation for consistency:

- μ for the mean (a location parameter).
- σ^2 for the variance (a scale parameter).
- σ for the standard deviation (a scale parameter).
- c for a general location parameter (other than the mean).
- α for a general scale parameter (other than the standard deviation).
- λ for a general location/scale parameter.
- r for a general shape parameter; also for degrees of freedom of Student's t distribution.
- α and β together are the location, scale, and shape parameters for the beta distribution.
- m and n for the degrees of freedom for the F-distribution.
- N for the degrees of freedom of the chi and chi-square distributions.

Location is an indication of the "center" of the pdf (it could be exactly in the middle depending on the distribution family), and scale refers to a measure of its "width." Specific definitions of these types of parameters are provided in Chapter 5. There is no definite pattern to the shape of a distribution when r is varied, as some pdfs can have a wide range of shapes. Example plots of the pdfs are included for each of the random variables; a vertical line indicates the location of the mean μ_X for a particular set of parameter values (a dashed line is used for discrete random variables).

As mentioned in Section 3.7, the random variables described here are important because they arise in various applications. That is, a specific pdf is useful because its shape is an accurate *model* for the outcomes that occur in some "experiment." For example, it is well known that many physical phenomena can be modeled quite accurately by a Gaussian random variable. Perhaps the most important example in engineering is noise, which is present to some degree in all practical systems. When discussing the various random variables, we mention some applications where the pdf is a useful model, and provide some examples of how to compute probabilities and various quantities such as the mean, median, and mode.

By the end of this chapter, the reader should recognize that all the pdfs and pmfs have basic *functional forms* in terms of the variable x, many of which are quite simple and most of which involve an exponential. The other terms in the distribution are included only so the function has unit area, yielding a valid pdf or pmf. Appendix D summarizes these functional forms in Table D.1 for the continuous random variables; the functional forms for the discrete random variables are summarized in Table E.3 of Appendix E. Note also that because of the wide range of forms for the distributions, many different functions arise in order to write expressions for the cdfs, and for other quantities of interest such as moments. For convenience, these functions are summarized in Appendix B, which includes some of their properties and plots for various parameter values.

The random variables are not presented in any special order in the following sections, though we have attempted to place together those distributions that are in some sense related to each other. For ease of reference, the summaries of the continuous and discrete distributions in Appendix A are place in alphabetical order.

3.8.1 Gaussian Random Variable (Normal)

The Gaussian random is also called the *normal* random variable. Its pdf has the familiar "bell-shaped" function:

$$f_X(x) = \frac{1}{\sqrt{2\pi\sigma^2}} \exp\left(-(x-\mu)^2/2\sigma^2\right) I_{(-\infty,\infty)}(x) \tag{3.53}$$

with mean $\mu \in \mathcal{R}$, standard deviation $\sigma > 0$, and variance σ^2. Varying parameters μ and σ over their range of values yields a *family* of Gaussian pdfs, all with the same basic shape but with different centers and widths. Figure 3.14 shows three examples of the Gaussian pdf. The *standard normal distribution* corresponds to $\mu = 0$ and $\sigma = 1$. The support of the Gaussian pdf is all $x \in \mathcal{R}$ which is emphasized by the indicator function $I_{(-\infty,\infty)}(x)$.

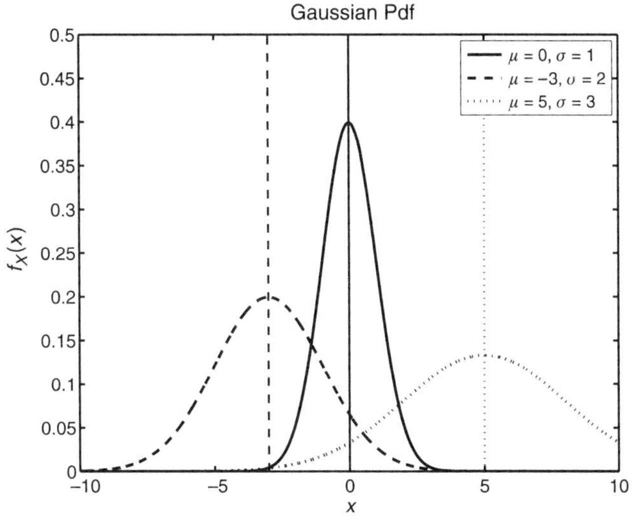

FIGURE 3.14 Examples of the Gaussian pdf. Support is $\mathcal{R}$. Vertical lines specify μ_X.

The Gaussian random variable is a useful model for many physical quantities and for a wide range of applications. It is particularly appropriate when a large number of elements contribute to a quantity of interest. Under relatively common conditions, the central limit theorem (CLT) described in Chapter 7 shows that the sum of a large number of independent random variables has a distribution that is approximately Gaussian. Applications where the Gaussian random variable is a useful model include the following:

- Noise in electrical and mechanical systems.
- Physical attributes of a large population such as height and weight.
- Exam scores of a large class of students.

The cdf of the Gaussian random variable is

$$F_X(x) = \frac{1}{\sqrt{2\pi\sigma^2}} \int_{-\infty}^{x} \exp\left(-(v-\mu)^2/2\sigma^2\right) dv, \tag{3.54}$$

which, unfortunately, does not have a closed form. Since integrals of the Gaussian pdf are needed in many problems, the Q-function is used to simplify such expressions. It is based on the standard normal pdf as follows:

$$Q(x) \triangleq \frac{1}{\sqrt{2\pi}} \int_{x}^{\infty} \exp(-v^2/2) dv. \tag{3.55}$$

Thus, when X is the standard Gaussian random variable, it is clear from (3.54) that

$$F_X(x) = 1 - Q(x). \tag{3.56}$$

Sometimes $\Phi(x)$ is used for the cdf of the standard Gaussian variable. However, since $\Phi_X(\omega)$ is the characteristic function in this book (see Chapter 5), we use the Q-function. Values of $Q(x)$ for x extending to 3.5 are provided in Appendix B. The error function $\text{erf}(x)$ and the complementary error function $\text{erfc}(x)$ are also used to simplify integral expressions of the Gaussian pdf; they are also described in Appendix B.

In order to use the Q-function for a nonstandard Gaussian random variable, a transformation must be applied to the integral. (General transformations of random variables are covered in Chapter 4.) It is relatively straightforward to show how the cdf for an arbitrary Gaussian random variable X with parameters $\{\mu, \sigma\}$ is evaluated in terms of the Q-function from the following transformation:

$$Y = (X - \mu)/\sigma, \tag{3.57}$$

where Y is the *standard* Gaussian random variable. Substituting this result into (3.56) gives the following cdf for X:

$$F_X(x) = 1 - Q\left(\frac{x - \mu}{\sigma}\right). \tag{3.58}$$

The cdf for any Gaussian random variable can be written in terms of $Q(x)$ using this formula.

Example 3.14. For a zero-mean Gaussian random variable, the probability that X is within three standard deviations is

$$P(|X| \leq 3\sigma) = F_X(3\sigma) - F_X(-3\sigma) = 1 - 2Q(3\sigma/\sigma) \approx 0.9973, \tag{3.59}$$

where $F_X(-3) = 1 - F_X(3)$ has been used, and the transformation in (3.58) allows σ to cancel in the argument.

Example 3.15. We verify that the Gaussian pdf has unit area as follows. From the transformation in (3.58), we can integrate the standard Gaussian pdf:

$$\int_{-\infty}^{\infty} f_X(x)\mathrm{d}x = \frac{1}{\sqrt{2\pi}} \int_{-\infty}^{\infty} \exp(-y^2/2)\mathrm{d}y. \tag{3.60}$$

This type of integral is usually handled by squaring the expression so that a double integral is obtained, which is transformed to polar coordinates:

$$\int_{-\infty}^{\infty}\int_{-\infty}^{\infty} \exp(-y^2/2)\exp(-z^2/2)\mathrm{d}y\mathrm{d}z = \int_{0}^{\infty}\int_{0}^{2\pi} r\exp(-r^2/2)\mathrm{d}r\,\mathrm{d}\phi, \tag{3.61}$$

where $y = r\cos(\phi)$ and $z = r\sin(\phi)$ have been used such that $r^2 = y^2 + z^2$. The Jacobian of the transformation is

$$J = \det\begin{bmatrix} \partial y/\partial r & \partial y/\partial\phi \\ \partial z/\partial r & \partial z/\partial\phi \end{bmatrix} = \det\begin{bmatrix} \cos(\phi) & -r\sin(\phi) \\ \sin(\phi) & r\cos(\phi) \end{bmatrix} = r, \tag{3.62}$$

so that $\mathrm{d}y\mathrm{d}z = |J|\,\mathrm{d}r\mathrm{d}\phi = r\,\mathrm{d}r\mathrm{d}\phi$. (The Jacobian is discussed again in Chapter 4 when general transformations of random variables are covered. It is also summarized in Appendix G.) Thus, integration by parts yields

$$\int_{0}^{\infty}\int_{0}^{2\pi} r\exp(-r^2/2)\mathrm{d}r\,\mathrm{d}\phi = 2\pi. \tag{3.63}$$

Taking the square root of (3.61) and substituting this result gives

$$\int_{-\infty}^{\infty} \exp(-y^2/2)\mathrm{d}y = \sqrt{2\pi}, \tag{3.64}$$

which completes the proof.

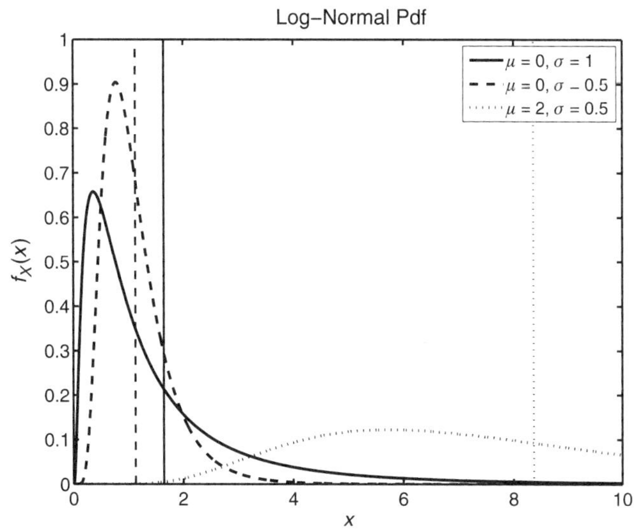

FIGURE 3.15 Examples of the log-normal pdf. Support is $\mathcal{R}^+$. Vertical lines specify μ_X.

3.8.2 Log-Normal Random Variable

The log-normal random variable is obtained via an exponential mapping of a Gaussian random variable. If X is Gaussian with parameters $\{\mu, \sigma\}$, then $Y = \exp(X)$ is log-normal distributed with the following pdf:

$$f_Y(y) = \frac{1}{y\sqrt{2\pi\sigma^2}} \exp\left(-(\ln(y) - \mu)^2/2\sigma^2\right) I_{[0,\infty)}(y), \tag{3.65}$$

where $\ln(z) \triangleq \log_e(z)$ is the natural logarithm. Because the exponential function is nonnegative, $f_Y(y)$ is nonzero only for $y \geq 0$. The name of this random variable is due to the reverse mapping: the logarithm of the log-normal random variable Y is Gaussian. Since $g(X) = \exp(X)$ is monotonically increasing, it is straightforward to derive the log-normal pdf from the Gaussian pdf using the transformation techniques in Chapter 4. From the inverse image $X = g^{-1}(Y) = \ln(Y)$, we readily obtain

$$\begin{aligned}
f_Y(y) &= f_X(g^{-1}(y)) \left|\frac{dx}{dy}\right| I_{[0,\infty)}(y), \\
&= \frac{1}{\sqrt{2\pi\sigma^2}} \exp\left(-(\ln(y) - \mu)^2/2\sigma^2\right) \left|\frac{d\ln(y)}{dy}\right| I_{[0,\infty)}(y), \\
&= \frac{1}{y\sqrt{2\pi\sigma^2}} \exp\left(-(\ln(y) - \mu)^2/2\sigma^2\right) I_{[0,\infty)}(y).
\end{aligned} \tag{3.66}$$

Examples of the pdf are shown in Figure 3.15 for various values of $\{\mu, \sigma\}$. The mean is $\mu_Y = \exp(\mu + \sigma^2/2)$ and the variance is $\sigma_Y^2 = \exp(2\mu + 2\sigma^2) - \exp(2\mu + \sigma^2)$.

The log-normal random variable is a useful model for applications where the logarithm of quantities of interest has a Gaussian distribution. Thus, based on the CLT mentioned earlier, the *product* of a large number of independent random variables tends to have a log-normal distribution. Applications where the log-normal random variable is a useful model include the following:

- Concentration of elements in the Earth's crust and their radioactivity.
- Number of words in a sentence.
- Duration of time from when a person is exposed to an infectious disease to the onset of symptoms.

Like the Gaussian random variable, it is not possible to write a closed form for the cdf of the log-normal random variable. In terms of the error function, it is

$$F_Y(y) = (1/2)\left[1 + \text{erf}\left(\frac{\ln(y) - \mu}{\sqrt{2\sigma^2}}\right)\right] I_{[0,\infty)}(y), \tag{3.67}$$

where

$$\text{erf}(x) \triangleq \frac{2}{\sqrt{\pi}} \int_0^x \exp(-v^2)dv. \tag{3.68}$$

See Appendix B for a table of values of the *complementary* error function and a discussion of some of its properties. Note that the probability in (3.67) is 0.5 when $y = \exp(\mu)$.

Example 3.16. The probability that the log-normal random variable is within 3σ when $\mu = 0$ is

$$P(Y \leq 3\sigma) = (1/2)\left[1 + \text{erf}\left(\ln(3\sigma)/\sigma\sqrt{2}\right)\right], \tag{3.69}$$

which we see, unlike the Gaussian random variable, depends on the specific value of σ. Note that σ is *not* the standard deviation of the log-normal random variable, though it is a scale parameter. We use μ and σ here (instead of the general location c and scale α parameters) to emphasize the connection between the Gaussian and log-normal random variables. Since $\ln(\sigma)$ increases with increasing σ, the probability above decreases. For example, $P(Y \leq 3\sigma) = 0.8640$ and 0.8148 for $\sigma = 1$ and 2, respectively.

Example 3.17. It is easy to verify that the mean μ_Y, median m_e, and mode m_o of the log-normal random variable are related as follows:

$$m_o < m_e < \mu_Y, \tag{3.70}$$

where $m_o = \exp(\mu - \sigma^2)$ and $m_e = \exp(\mu)$.

3.8.3 Inverse Gaussian Random Variable (Wald)

The inverse Gaussian random variable, which is also known as the Wald random variable, has the following pdf:

$$f_X(x) = \sqrt{\frac{\alpha}{2\pi x^3}} \exp(-\alpha(x-\mu)^2/2\mu^2 x) I_{[0,\infty)}(x), \tag{3.71}$$

where $\alpha > 0$ is a scale parameter and $\mu \in \mathcal{R}$ is the mean. The variance is $\sigma_X^2 = \mu^3/\alpha$. Since the cdf does not have a closed form, it can be written in terms of the Q-function similar to the Gaussian random variable:

$$F_X(x) = 1 - Q\left((x/\mu - 1)\sqrt{\alpha/x}\right) + \exp(2\alpha/\mu) Q\left((x/\mu + 1)\sqrt{\alpha/x}\right), \tag{3.72}$$

or in terms of the error function:

$$\begin{aligned} F_X(x) = {} & (1/2)\left[1 + \text{erf}\left((x/\mu - 1)\sqrt{\alpha/2x}\right)\right] \\ & + (1/2)\exp(2\alpha/\mu)\left[1 - \text{erf}\left((x/\mu + 1)\sqrt{\alpha/2x}\right)\right]. \end{aligned} \tag{3.73}$$

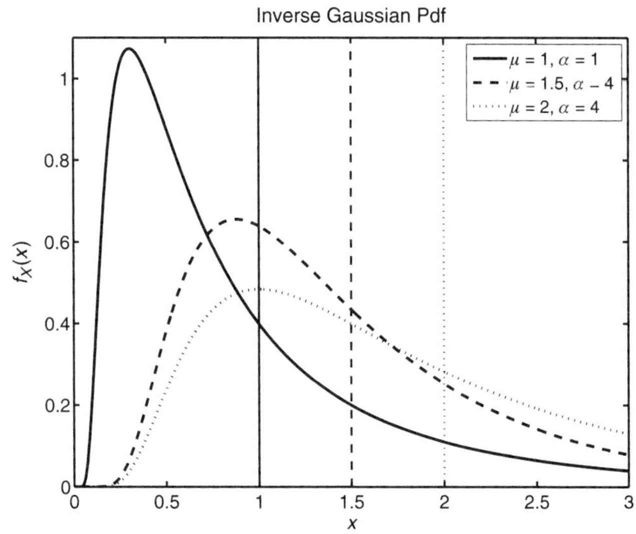

FIGURE 3.16 Examples of the inverse Gaussian pdf. Support is $\mathcal{R}^+$. Vertical lines specify μ_X.

Examples of the pdf are provided in Figure 3.16. The name of the inverse Gaussian random variable is due to the inverse relationship of its cumulant generating function (see Chapter 5) and that of the Gaussian random variable. Some applications include the following:

- Describes the time for a random walk with positive drift to reach some level (called the first passage time).
- Modeling wind speeds where low speeds are rare.
- Duration of employee strikes in various industries.

Example 3.18. The random variable given by the reciprocal $Y = 1/X$ of the Wald random variable has the following pdf:

$$f_Y(y) = \sqrt{\frac{\alpha}{2\pi y}} \exp(-\alpha y/2 - \alpha/2\mu^2 y + \alpha/\mu) I_{[0,\infty)}(y). \tag{3.74}$$

This is easily derived by starting with the cdf:

$$F_Y(y) = P(1/X \le y) = P(X \ge 1/y) = 1 - F_X(1/y), \tag{3.75}$$

where multiplying by $X > 0$ does not change the direction of the inequality. Thus

$$f_Y(y) = f_X(1/y)(1/y^2). \tag{3.76}$$

Substituting $x = 1/y$ into (3.71), multiplying the result by $1/y^2$, and rearranging the exponent gives the form in (3.74).

Example 3.19. For the one-to-one transformation $Y = \sqrt{\alpha}(X - \mu)/\sqrt{X}\mu$, it can be shown that the pdf of Y is (see Problem 4.26)

$$f_Y(y) = \frac{1}{\sqrt{2\pi}} \left(1 - y/\sqrt{4\alpha/\mu + y^2}\right) \exp(-y^2/2), \tag{3.77}$$

which has support $y \in \mathcal{R}$. This is an interesting result because it is the product of the pdf of the standard Gaussian random variable and a term that is the difference of a constant and an odd function, given by 1 and $y/\sqrt{4\alpha/\mu + y^2}$, respectively. From this result, it is easy to verify that $f_Y(y)$ has unit area because the integral of the odd part is zero. Consider next the transformation $Z = |Y|$ which has cdf

$$P(Z \leq z) = P(-z \leq Y \leq z) = \frac{1}{\sqrt{2\pi}} \int_{-z}^{z} \left(1 - y/\sqrt{4\alpha/\mu + y^2}\right) \exp(-y^2/2)\mathrm{d}y,$$

$$= \frac{1}{\sqrt{2\pi}} \int_{-z}^{z} \exp(-y^2/2)\mathrm{d}y = \frac{2}{\sqrt{2\pi}} \int_{0}^{z} \exp(-y^2/2)\mathrm{d}y, \tag{3.78}$$

where the odd term has integrated to zero. Differentiating with respect to z gives the following pdf:

$$f_Z(z) = \frac{2}{\sqrt{2\pi}} \exp(-z^2/2), \tag{3.79}$$

which corresponds to the standard *half-normal* distribution with zero mean and unit variance (described later). The support of Z is necessarily $\mathcal{R}^+$, as is obvious from $Z = |Y|$, but also so that $f_Z(z)$ has unit area. Such a pdf is similarly obtained from the transformation $Z = |U|$, where U is the standard Gaussian random variable.

3.8.4 Exponential Random Variable (One-Sided)

The exponential random variable has the following pdf:

$$f_X(x) = \lambda \exp(-\lambda x) I_{[0,\infty)}(x), \tag{3.80}$$

where the location/scale parameter $\lambda > 0$ determines the rate of decay of the function; the ratio $1/\lambda$ is referred to as the "waiting time." The cdf is

$$F_X(x) = [1 - \exp(-\lambda x)] I_{[0,\infty)}(x), \tag{3.81}$$

and the standard distribution is obtained when $\lambda = 1$. Examples of the exponential pdf are shown in Figure 3.17. The mean of the distribution is $\mu_X = 1/\lambda$ and the variance is $\sigma_X^2 = 1/\lambda^2$. Although the support above is $\mathcal{R}^+$, this can be shifted by a linear transformation of the random variable as discussed in Chapter 4. For translation parameter c, the exponential pdf is generalized to

$$f_X(x) = \lambda \exp(-\lambda(x - c)) I_{[c,\infty)}(x), \tag{3.82}$$

where the indicator function forces the pdf to be nonzero only for $x \geq c$. The exponential random variable is a useful model for numerous applications where the time until an event is of interest. Some of these include the following:

- Interarrival times of the Poisson random process (described in Chapter 6).
- Decay of radioactive particles.
- Time to failure of an electronic component.

Example 3.20. The median of the exponential random variable is derived from $F_X(0.5)$ as follows:

$$2(1 - \exp(-\lambda m_e)) = 1 \Longrightarrow m_e = \ln(2)/\lambda. \tag{3.83}$$

Since the mode of (3.80) is always zero, we have $m_o < m_e < \mu_X$.

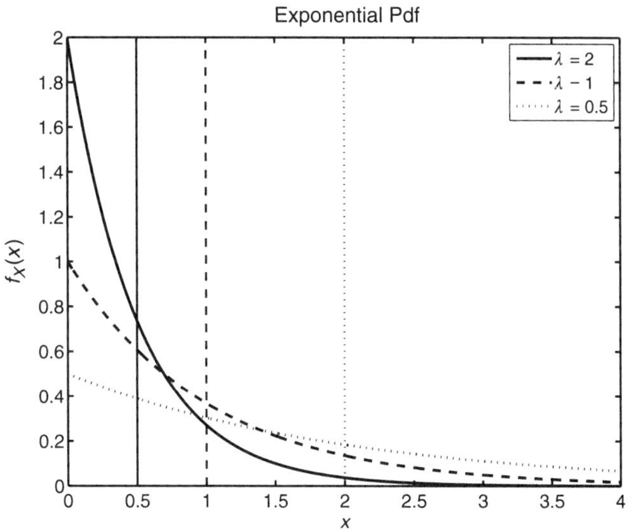

FIGURE 3.17 Examples of the exponential pdf. Support is $\mathcal{R}^+$. Vertical lines specify μ_X.

Example 3.21. The exponential random variable is a special case of the gamma random variable described later with scale parameter $\alpha = 1/\lambda$ and shape parameter $r = 1$. It is related to several other random variables, including the extreme value distribution (also described later) via the transformation:

$$Y = c - \alpha \ln(X), \tag{3.84}$$

where $c \in \mathcal{R}$ is a location parameter and $\alpha > 0$ is a scale parameter. This is shown by rearranging the cdf:

$$\begin{aligned} P(Y \le y) &= P(c - \alpha \ln(X) \le y), \\ &= P(X \ge \exp((c - y)/\alpha)) = \exp(-\lambda \exp((y - c)/\alpha))). \end{aligned} \tag{3.85}$$

The last line is the cdf of the extreme value distribution for $\lambda = 1$. Note that we used the fact that $\ln(\cdot)$ is monotonic to modify the argument above such that the direction of the inequality is unchanged.

Rearranging the cdf of the exponential random variables gives

$$P(X > x) = 1 - F_X(x) = \exp(-\lambda x) \tag{3.86}$$

from which it is straightforward to show that X has a *memoryless* property:

$$P(X > x + b|X > b) = P(X > x) \tag{3.87}$$

for any $b \ge 0$. Observe from conditional probability that

$$P(X > x + b|X > b) = \frac{P(X > x + b, X > b)}{P(X > b)} = \frac{P(X > x + b)}{P(X > b)}. \tag{3.88}$$

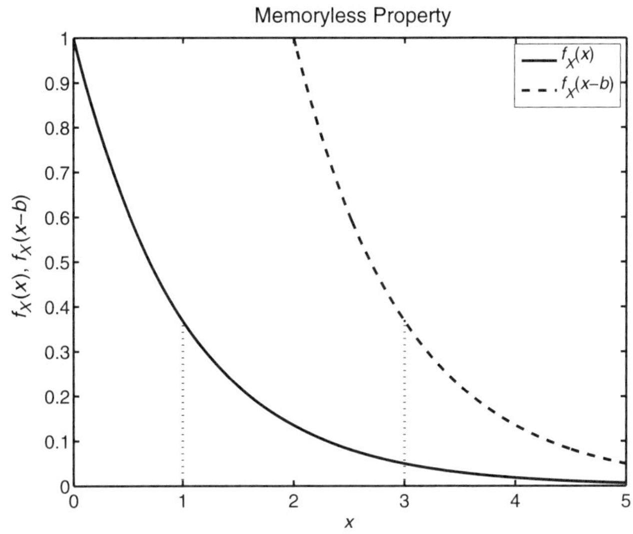

FIGURE 3.18 Memoryless property of the exponential random variable with $\lambda = 1$. The areas under the two curves to the right of the vertical dotted lines are identical. Conditioning on $X > b > 0$ is similar to a shift of $f_X(x)$ to the right by b, which in this example is $b = 2$.

Substituting the expression in (3.86) gives

$$P(X > x + b|X > b) = \frac{\exp(-\lambda(x + b))}{\exp(-\lambda b)} = \exp(-\lambda x) = P(X > x), \tag{3.89}$$

which is the desired result. Thus, given that the random variable X has already exceeded some $b \geq 0$, the probability that X is greater than $x + b$ is equivalent to evaluating $P(X > x)$, where X is the original (unshifted) exponential random variable. The memoryless property is illustrated in Figure 3.18.

Next, we demonstrate that the memoryless property actually defines the exponential pdf.

Theorem 3.3. The exponential random variable is the only continuous random variable with the memoryless property.

Proof. Consider the following conditional cdf:

$$P(X \leq x + b|X > b) = \frac{P(b < X \leq x + b)}{P(X > b)} \tag{3.90}$$

such that for the memoryless property,

$$\frac{F_X(x + b) - F_X(b)}{1 - F_X(b)} = F_X(x) - F_X(0), \tag{3.91}$$

where $F_X(x)$ is the cdf to be determined and $F_X(0) = 0$ is a boundary condition. Using techniques from calculus, we can divide each side by $(x + b) - b = x$ and let $x \to 0$ to obtain ordinary derivatives of $F_X(x)$ evaluated at $x = b$ and $x = 0$:

$$\frac{1}{1 - F_X(b)} F'_X(b) = F'_X(0) \Longrightarrow \frac{\mathrm{d}}{\mathrm{d}x} F_X(x) = [1 - F_X(x)] F'_X(0), \tag{3.92}$$

where for convenience we have replaced variable b with x in the last expression. This is a first-order differential equation for which the solution must have an exponential form: $1 - F_X(x) = a\exp(cx)$ with $a \neq 0$. Substituting this expression into (3.92) and differentiating gives

$$-ac\exp(cx) = a\exp(cx)F_X'(0) \Longrightarrow c = -F_X'(0). \tag{3.93}$$

Since the derivative of the cdf is positive for a continuous random variable, it follows that $c < 0$. Substituting $c = -\lambda$ with $\lambda \geq 0$ in the solution for $1 - F_X(x)$ gives

$$1 - F_X(x) = a\exp(-\lambda x) \Longrightarrow F_X(x) = 1 - a\exp(-\lambda x). \tag{3.94}$$

Differentiating this gives the pdf in (3.80) with $a = \lambda$, thus completing the proof. □

3.8.5 Laplace Random Variable (Double-Sided Exponential)

The Laplace random variable, which also known as the double-sided exponential random variable, has the following pdf:

$$f_X(x) = (1/2\alpha)\exp(-|x-\mu|/\alpha)I_{(-\infty,\infty)}(x), \tag{3.95}$$

where $\mu \in \mathcal{R}$ is the mean and $\alpha > 0$ is a scale parameter that determines the rate of decay of the function. This pdf is obtained from the shifted exponential pdf in (3.82) by appending its "mirror image" about $c = \mu$, letting $\alpha = 1/\lambda$, and scaling the result by $1/2$ so that the area of the two-sided pdf is one. Unlike λ of the exponential pdf, α is strictly a scale parameter that is independent of μ. The Laplace random variable X reduces to the standard exponential random variable by taking the absolute value $|X|$ (not to be confused with the cardinality of a set), and substituting $\alpha = 1/\lambda$ and $\mu = 0$. Examples of the Laplace pdf are shown in Figure 3.19. Its variance is $\sigma_X^2 = 2\alpha^2$.

The Laplace random variable is useful in applications where "long distribution tails" are an accurate model. Observe in Figure 3.20 that the Laplace and Gaussian pdfs are similarly shaped; however, the Laplace pdf has

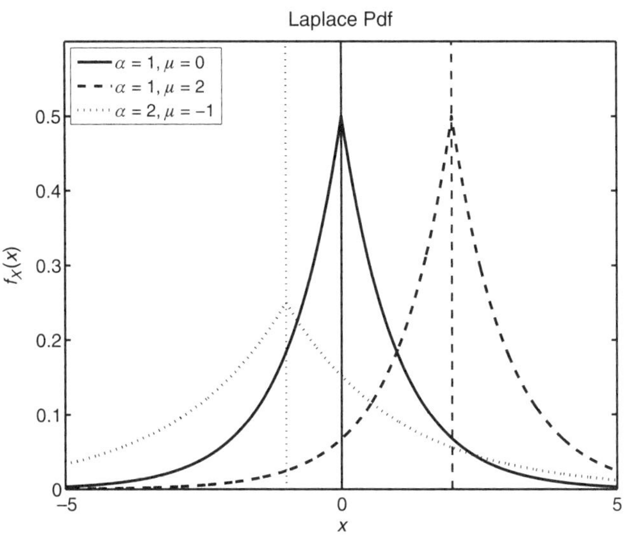

FIGURE 3.19 Examples of the Laplace pdf. Support is $\mathcal{R}$. Vertical lines specify μ_X.

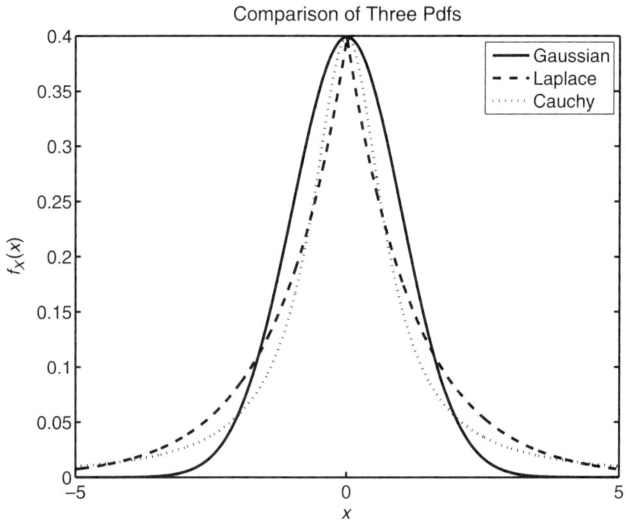

FIGURE 3.20 Gaussian ($\sigma = 1$), Laplace ($\alpha = \sqrt{\pi/2}$), and Cauchy ($\alpha = \sqrt{2/\pi}$) pdfs with identical location parameters ($\mu = c = 0$). The scale parameters were chosen so that the pdfs all have the same maximum.

a sharp peak at $x = \mu$ and the function decays more slowly, according to $\exp(-x)$ versus $\exp(-x^2)$ for the Gaussian random variable. Some applications of the Laplace random variable include the following:

- Detection of signals in impulsive noise, requiring heavier tails than Gaussian noise.
- Modeling speech signals for encoding and decoding.
- The asymmetric Laplace distribution (described in Example 3.24) has been used to model financial data such as interest rates.

The cdf of the Laplace random variable is

$$F_X(x) = \begin{cases} (1/2)\exp((x-\mu)/\alpha), & x < \mu \\ 1-(1/2)\exp(-(x-\mu)/\alpha), & x \geq \mu, \end{cases} \tag{3.96}$$

$$= (1/2)\left(1 + \text{sgn}(x-\mu)\left[1 - \exp(-|x-\mu|/\alpha)\right]\right) I_{(-\infty,\infty)}(x), \tag{3.97}$$

where

$$\text{sgn}(x) \triangleq \begin{cases} 1, & x \geq 0 \\ -1, & x < 0 \end{cases} \tag{3.98}$$

is the signum function (see Appendix B).

Example 3.22. The Laplace cdf is derived as follows. The integral is straightforward for $x \leq \mu$:

$$P(X \leq x) = (1/2\alpha)\int_{-\infty}^{x} \exp((v-\mu)/\alpha)dv,$$

$$= (1/2\alpha)\int_{-\infty}^{x-\mu} \exp(w/\alpha)dw = (1/2)\exp((x-\mu)/\alpha), \tag{3.99}$$

where we have changed variables to $w \triangleq v - \mu$. For $x > \mu$:

$$P(X \leq x) = (1/2\alpha)\int_{-\infty}^{\mu} \exp((v-\mu)/\alpha)dv + (1/2\alpha)\int_{\mu}^{x} \exp(-(v-\mu)/\alpha)dv, \tag{3.100}$$

where the integral has been split in order to remove the absolute value in the argument. Changing variables again to $w \triangleq v - \mu$ gives

$$\begin{aligned} P(X \leq x) &= (1/2\alpha)\int_{-\infty}^{0} \exp(w/\alpha)dw + (1/2\alpha)\int_{0}^{x-\mu} \exp(-w/\alpha)dw, \\ &= (1/2) - (1/2)[\exp(-(x-\mu)/\alpha) - 1], \\ &= 1 - (1/2)\exp(-(x-\mu)/\alpha), \end{aligned} \tag{3.101}$$

which together with (3.99) give the result in (3.97).

Example 3.23. The three *quartiles*, which are the quantiles with $M = 4$, divide a pdf into four regions, each with one-fourth of the area. The second quartile is the median m_e of a distribution; obviously $m_e = \mu$ for the Laplace random variable because it is a symmetric distribution. A quartile is obtained from the cdf of a continuous random variable by solving $F_X(q_m) = m/4$ with $m = 1, 2, 3$. For $m = 1$, the first expression in (3.96) is used:

$$(1/2)\exp((q_1 - \mu)/\alpha) = 1/4 \Longrightarrow q_1 = \mu - \alpha \ln(2), \tag{3.102}$$

and for $m = 3$, the second expression in (3.96) gives

$$1 - (1/2)\exp(-(q_3 - \mu)/\alpha) = 3/4 \Longrightarrow q_3 = \mu + \alpha \ln(2), \tag{3.103}$$

where $\ln(2) \approx 0.6931$.

Example 3.24. Various *asymmetric* forms of the Laplace distribution have been studied (S. Kotz, T. J. Kozubowski, and K. Podgórski); generally, they can be expressed as the following mixture of exponential forms:

$$\begin{aligned} f_X(x) = (p_1/\alpha_1)&\exp(-(x-c)/\alpha_1)I_{[c,\infty)}(x) \\ &+ (p_2/\alpha_2)\exp((x-c)/\alpha_2)I_{(-\infty,c)}(x), \end{aligned} \tag{3.104}$$

where positive $\{\alpha_1, \alpha_2\}$ provide different rates of decay of the pdf about the mode c. The two "half-pdfs" must have the same value at $x = c$ and thus $p_1/\alpha_1 = p_2/\alpha_2$. Also, in order to be a valid pdf overall, it is necessary that $p_1 + p_2 = 1$, yielding

$$p_1 = \alpha_1/(\alpha_1 + \alpha_2), \quad p_2 = \alpha_2/(\alpha_1 + \alpha_2). \tag{3.105}$$

Example pdfs for the asymmetric Laplace random variable are shown in Figure 3.21. It is straightforward to verify that the mean is (see Problem 5.29)

$$\mu_X = c + p_1\alpha_1 - p_2\alpha_2, \tag{3.106}$$

which coincides with the mode c only when $\alpha_1 = \alpha_2$ such that the distribution has a symmetric form.

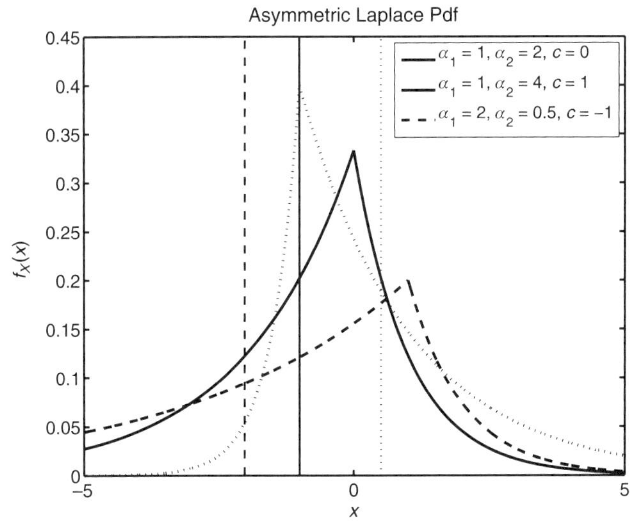

FIGURE 3.21 Examples of the asymmetric Laplace pdf in (3.104). Support is $\mathcal{R}$. Vertical lines specify μ_X.

3.8.6 Cauchy Random Variable

The Cauchy random variable has the following pdf:

$$f_X(x) = \frac{\alpha/\pi}{(x-c)^2+\alpha^2} I_{(-\infty,\infty)}(x), \tag{3.107}$$

where $\alpha > 0$ is a scale parameter and $c \in \mathcal{R}$ is a location parameter. It is also known as the Cauchy–Lorentz random variable. As shown in Chapter 5, although the Cauchy pdf obviously has a center point (given by c), the mean μ_X does *not* exist. The corresponding cdf is

$$F_X(x) = [1/2 + (1/\pi)\arctan((x-c)/\alpha)]\, I_{(-\infty,\infty)}(x), \tag{3.108}$$

and the *standard* Cauchy random variable has $c = 0$ and $\alpha = 1$. Examples of the Cauchy pdf are shown in Figure 3.22, and it is compared to the Gaussian and Laplace pdfs in Figure 3.20. Observe that the Cauchy pdf decays the slowest of the three distributions; it is heavy tailed like the Laplace distribution. Thus, it is useful in applications where "outliers" (large positive and negative values of x) are more likely to occur, compared to a model based on the Gaussian random variable.

Example 3.25. The *inflection points* of a function occur where the second derivative changes sign. For the Cauchy distribution,

$$\frac{d^2}{dx^2} f_X(x) = \frac{-2\alpha/\pi}{[(x-c)^2+\alpha^2]^2} + \frac{-6(\alpha/\pi)x}{(x-c)^2+\alpha^2}. \tag{3.109}$$

Setting this expression equal to zero and solving for x yields the two inflection points:

$$x = c \pm \alpha/\sqrt{3}. \tag{3.110}$$

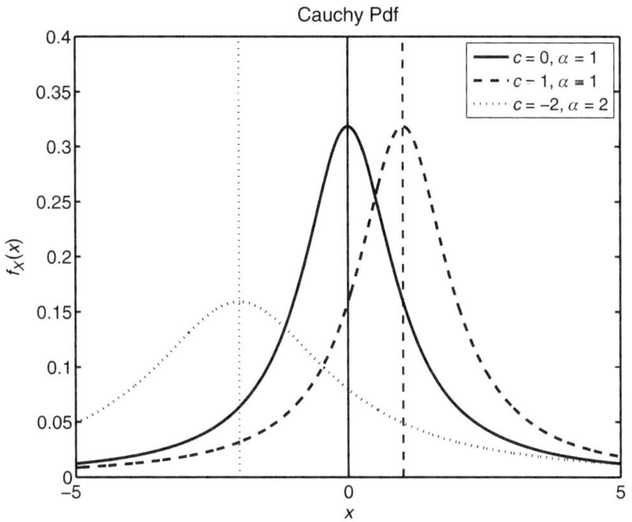

FIGURE 3.22 Examples of the Cauchy pdf. Support is $\mathcal{R}$. Vertical lines specify m_e since μ_X does not exist.

Substituting these into the cdf gives

$$F_X(c \pm \alpha/\sqrt{3}) = 1/2 + (1/\pi)\arctan(\pm 1/\sqrt{3}) = \{1/3, 2/3\}. \tag{3.111}$$

The corresponding values for the Gaussian random variable are outside this range (see Problem 3.20), supporting the fact that the Cauchy distribution has heavier tails.

Some applications of the Cauchy random variable include the following:

- The distribution of particles emanating from a point source and striking a fixed line.
- $Z = X/Y$ is the standard Cauchy random variable if X and Y are independent standard Gaussian random variables (see Chapter 4).
- Representations of earthquake geometry.

Even though the mean and other moments of the Cauchy random variable do not exist, it does have a median and a mode, both of which equal c.

Example 3.26. Without loss of generality, assume $c = 0$ so that the pdf is centered about $x = 0$. The mean of the Cauchy distribution would be computed as follows:

$$\int_{-\infty}^{\infty} x \frac{\alpha/\pi}{x^2 + \alpha^2} \mathrm{d}x = (\alpha/2\pi)\log(x^2 + \alpha^2)\Big|_{-\infty}^{\infty} = \infty - \infty, \tag{3.112}$$

which we see is undefined (it is *not* = 0). (See Appendix E for various identities involving 0 and ∞.)

Example 3.27. The quantiles for the Cauchy random variable are obtained from

$$1/2 + (1/\pi)\arctan((q_m - c)/\alpha) = m/M \Longrightarrow q_m = c + \alpha\tan((2m - M)\pi/2M) \tag{3.113}$$

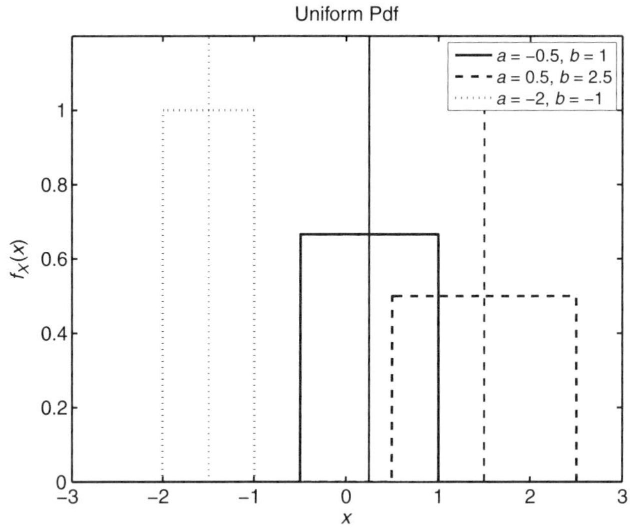

FIGURE 3.23 Examples of the continuous uniform pdf. Support is $[a, b]$. Vertical lines specify μ_X.

for $m = 1, \ldots, M - 1$. The *deciles* for $M = 10$ with $c = 0$ and $\alpha = 1$ are −3.0777, −1.3764, −0.7265, −0.3249, 0, 0.3249, 0.7265, 1.3764, and 3.0777.

3.8.7 Continuous Uniform Random Variable

The uniform random variable has a flat (rectangular) pdf over a *finite* support on $\mathcal{R}$. For the range $[a, b]$ with $a, b \in \mathcal{R}$ and $a < b$, the pdf is

$$f_X(x) = \frac{1}{b-a} I_{[a,b]}(x), \tag{3.114}$$

which can also be written in terms of the unit-step function as follows:

$$f_X(x) = \frac{1}{b-a}[u(x-a) - u(x-b)]. \tag{3.115}$$

The normalization $1/(b - a)$ is included, of course, so that $f_X(x)$ has unit area. Examples of the uniform pdf are shown in Figure 3.23. Its mean is $\mu_X = (a + b)/2$ and the variance is $\sigma_X^2 = (b - a)^2/12$, which is a well-known result for the noise model used when discrete-time signals are quantized (see Chapter 10). The standard uniform random variable is obtained when $a = 0$ and $b = 1$. Since the uniform pdf is flat, the cdf is necessarily a linearly increasing function of x:

$$\begin{aligned} F_X(x) &= \begin{cases} 0, & x < a \\ (x-a)/(b-a), & a \le x < b \\ 1, & x \ge b, \end{cases} \\ &= \frac{x-a}{b-a} I_{[a,b]}(x) + I_{(b,\infty)}(x). \end{aligned} \tag{3.116}$$

Note that the indicator function $I_{(b,\infty)}(x)$ must be included so that $F_X(x) = 1$ when x exceeds the support of $f_X(x)$. The cdf can also be expressed in terms of the ramp function as follows:

$$F_X(x) = \frac{1}{b-a}[R(x-a) - R(x-b)]I_{[a,\infty)}(x), \tag{3.117}$$

where (see Appendix B)

$$R(x) \triangleq \begin{cases} x, & x \geq 0 \\ 0, & x < 0. \end{cases} \tag{3.118}$$

A flat pdf implies that all subintervals in $[a, b]$ of the same width have the same probability: for $x_1 < x_2$, $x_3 < x_4$, and $x_1, x_2, x_3, x_4 \in [a, b]$, if $L([x_1, x_2]) = L([x_3, x_4])$, where $L(\cdot)$ is the Lebesgue measure, then

$$P(x_1 \leq X \leq x_2) = P(x_3 \leq X \leq x_4). \tag{3.119}$$

The probability of a subinterval for a uniform distribution does not depend on the location of the subinterval within the support of the random variable. The uniform random variable is useful in applications where all outcomes are "equally random." (This expression has more meaning for the discrete uniform random variable described later, where all outcomes have equal probability: the finite number of outcomes are "equally likely.") Applications include the following:

- Quantization noise in a digital system.
- Random number generators (as discussed in Chapter 1).
- It was shown in Chapter 1 that the *probability integral transform* $Y = F_X(x)$ yields the standard uniform random variable for any continuous random variable X with cdf $F_X(x)$.

Example 3.28. Observe that for $c, d \in [a, b]$:

$$\int_c^d f_X(x)\mathrm{d}x = \frac{d-c}{b-a} = \frac{L([c,d])}{L([a,b])}, \tag{3.120}$$

which confirms the equal-width subinterval property in (3.119).

Example 3.29. The standard uniform random variable X is related to the exponential random variable with parameter λ via the transformation $Y = -\ln(X)/\lambda$:

$$\begin{aligned} P(Y \leq y) &= P(-\ln(X)/\lambda \leq y) = P(X > \exp(-\lambda y)), \\ &= 1 - P(X \leq \exp(-\lambda y)) = 1 - \exp(-\lambda y), \end{aligned} \tag{3.121}$$

where $a = 0$ and $b = 1$ have been used. The last expression is the cdf of the exponential random variable.

3.8.8 Triangular Random Variable

The pdf of the triangular random variable is

$$f_X(x) = \frac{2}{b-a}\left[\frac{x-a}{c-a}I_{[a,c]}(x) + \frac{b-x}{b-c}I_{(c,b]}(x)\right][u(x-a) - u(x-b)], \tag{3.122}$$

where $a, b, c \in \mathcal{R}$ with $a < c < b$, and thus the support is $[a, b]$. Observe that $f_X(x)$ is a general triangle in that the magnitude of the slope of the increasing part (given by $1/(c-a)$) is not constrained to be the

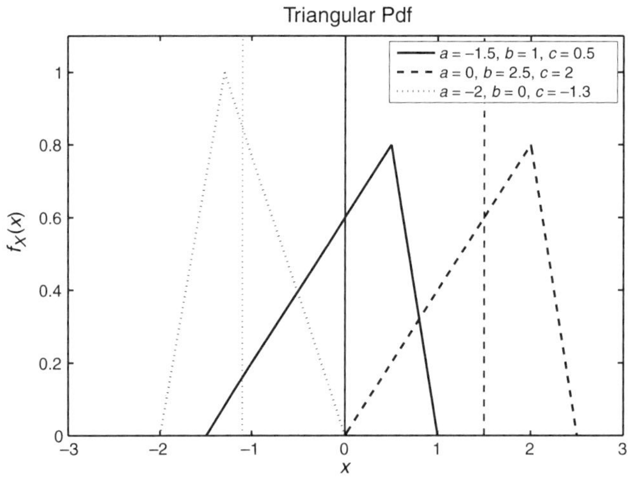

FIGURE 3.24 Examples of the triangular pdf. Support is $[a, b]$. Vertical lines specify μ_X.

same as that of the decreasing part (given by $1/(b-c)$). The mean is $\mu_X = (a+b+c)/3$ and the variance is $\sigma_X^2 = (a^2+b^2+c^2-ab-ac-bc)/18$. Example pdfs are shown in Figure 3.24. The cdf is

$$F_X(x) = \frac{1}{b-a}\left[\frac{(x-a)^2}{c-a}I_{[a,c]}(x) + \left(b-a-\frac{(b-x)^2}{b-c}\right)I_{(c,b]}(x)\right] + I_{(b,\infty)}(x), \tag{3.123}$$

where $I_{(b,\infty)}(x)$ is included above so that $F_X(x) = 1$ when x exceeds the support of $f_X(x)$. $F_X(x)$ has an inflection point at $x = c$ where it changes from being a convex function to a concave function (see the cdf example in Appendix A). The value of $F_X(x)$ at the inflection point is $(c-a)/(b-a)$, which is equal to $1/2$ only when $c = (a+b)/2$, corresponding to a symmetric triangle about c.

Applications include the following:

- Model of a distribution when the minimum, maximum, and "most likely" values are known (corresponding to estimates of a, b, and c, respectively).
- Approximation of the beta distribution (which is covered later in this chapter).
- Dithering of digital signals in order to reduce distortions that can occur during quantization and which are correlated with the original signal.

Example 3.30. The coefficient of skewness for the triangular random variable can be derived from the third central moment $\mu_{c,3}$ (defined in Chapter 5) and the variance σ_X^2 given in Appendix A:

$$\gamma_1 \triangleq \mu_{c,3}/\sigma_X^3 = (\sqrt{2}/5)\frac{(a+b-2c)(2a-b-c)(a-2b+c)}{(a^2+b^2+c^2-ab-ac-bc)^{3/2}}. \tag{3.124}$$

It is obvious from the numerator that $\gamma_1 = 0$ only when $c = (a+b)/2$; the other two terms in the numerator cannot be zero because $a < c < b$. Consider the case of $a = 0$ and $b = 1$ such that

$$\gamma_1 = (\sqrt{2}/5)\frac{(1-2c)(c+1)(2-c)}{(c^2-c+1)^{3/2}}, \tag{3.125}$$

which is plotted in Figure 3.25. The triangular distribution has negative skew when $c > 1/2$, meaning that the triangle is skewed ("leans") to the right; likewise, $\gamma_1 > 0$ for $c < 1/2$ and the triangle is skewed to the left. The figure also shows pdfs for $c = 1/10$ and $c = 3/4$ to demonstrate positive and negative skewness.

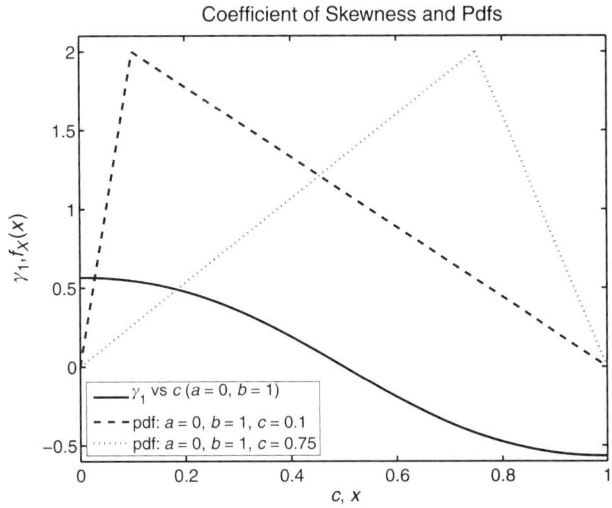

FIGURE 3.25 Coefficient of skewness for the triangular distribution in Example 3.30.

3.8.9 Rayleigh Random Variable

The Rayleigh random variable has the following pdf:

$$f_X(x) = (x/\alpha^2)\exp(-x^2/2\alpha^2)I_{[0,\infty)}(x), \tag{3.126}$$

where $\alpha > 0$ is a scale parameter. The cdf is

$$F_X(x) = [1 - \exp(-x^2/2\alpha^2]I_{[0,\infty)}(x). \tag{3.127}$$

Figure 3.26 shows examples of the Rayleigh pdf for different values of α. The mean is $\mu_X = \alpha\sqrt{\pi/2}$ and the variance is $\sigma_X^2 = (4-\pi)\alpha^2/2$. This random variable arises in applications when Gaussian random variables are summed together. Specifically, if X_1 and X_2 are independent Gaussian random variables with parameters $\{\mu = 0, \sigma\}$, then

$$Y = \sqrt{X_1^2 + X_2^2} \tag{3.128}$$

is a Rayleigh random variable with parameter $\alpha = \sigma$. This result can be derived using the transformation techniques described in Chapter 4. Applications include the following:

- Models magnitude variations of a signal (called *fading*) in a wireless communication channel, assuming there is no significant line-of-sight (LoS) propagation.
- Speckle noise in ultrasound images.
- Model for wind speed variations.

Example 3.31. The median of the Rayleigh random variable from $F_X(0.5)$ is

$$\exp(-m_e^2/2\alpha^2) = 1/2 \Longrightarrow m_e = \alpha\sqrt{2\ln(2)}. \tag{3.129}$$

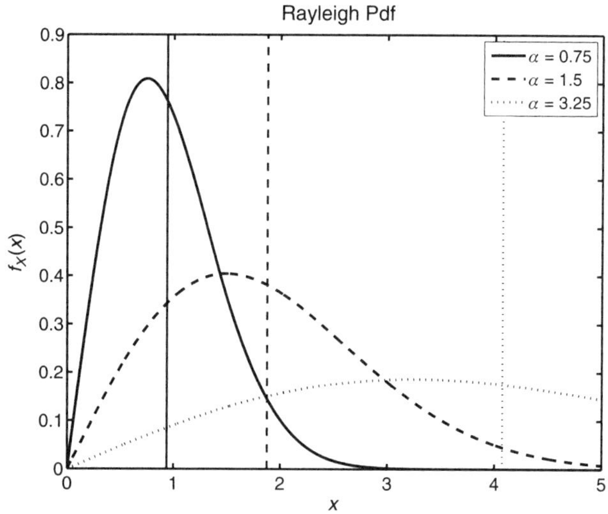

FIGURE 3.26 Examples of the Rayleigh pdf. Support is $\mathcal{R}^+$. Vertical lines specify μ_X.

Example 3.32. Since the Rayleigh pdf is unimodal, its mode is derived by differentiating $f_X(x)$ with respect to x and setting the result equal to zero:

$$(1/\alpha^2)\exp(-m_o^2/2\alpha^2) - (m_o/\alpha^2)\exp(-m_o^2/2\alpha^2)(m_o/\alpha^2) = 0. \tag{3.130}$$

Cancelling terms and solving for m_o yields

$$1 - (m_o/\alpha)^2 = 0 \Longrightarrow m_o = \alpha. \tag{3.131}$$

Thus, the mean, median, and mode are all proportional to the scale parameter α; as expected, they shift toward the right if the width of the distribution is increased.

Example 3.33. In Chapter 9, we consider several techniques for estimating the parameters of a distribution from N independent and identically distributed (iid) samples $\{X_1, \ldots, X_N\}$. In this example, we briefly describe *maximum likelihood (ML) estimation* where the parameter θ is obtained as follows:

$$\theta_{\text{ML}} = \arg\max_{\theta} \prod_{n=1}^{N} f_X(x_n), \tag{3.132}$$

and the joint distribution of $\{X_n\}$ is a product because they are iid (joint distributions are covered in Chapter 4). Since $\ln(\cdot)$ is a monotonic function, it is often easier to determine the ML estimate by maximizing the *logarithm* of the right-hand side. For the Rayleigh distribution, $\theta = \alpha$ and

$$\ln\left(\prod_{n=1}^{N} f_X(x_n)\right) = -N\ln(\alpha^2) + \ln\left(\prod_{n=1}^{N} x_n\right) - (1/2\alpha^2)\sum_{n=1}^{N} x_n^2. \tag{3.133}$$

Differentiating this expression with respect to α, setting the result equal to zero, and solving for α yields

$$\alpha_{\mathrm{ML}} = \sqrt{(1/2N)\sum_{n=1}^{N} X_n^2}, \tag{3.134}$$

which we have written in terms of the samples (instead of the outcomes $\{x_n\}$). This estimator is a random variable with various properties as described in Chapter 9.

3.8.10 Rice Random Variable

The Rice random variable can be viewed as a generalization of the Rayleigh random variable with a change in location. It has the following pdf:

$$f_X(x) = (x/\alpha^2)\exp(-(x^2+c^2)/2\alpha^2)I_0(cx/\alpha^2)I_{[0,\infty)}(x), \tag{3.135}$$

where $\alpha > 0$ is a scale parameter, $c > 0$ is a location parameter, and $I_0(x)$ is the modified Bessel function of the first kind with order zero (see Appendix B). The cdf is

$$\begin{aligned} F_X(x) &= 1 - \int_x^{\infty} (v/\alpha^2)\exp(-(v^2+c^2)/2\alpha^2)I_0(cv/\alpha^2)\mathrm{d}v, \\ &= 1 - \int_{x/\alpha}^{\infty} w\exp(-w^2/2 - c^2/2\alpha^2)I_0(cw/\alpha)\mathrm{d}w, \end{aligned} \tag{3.136}$$

where we have changed variables to $w = v/\alpha$ so that the integral is in the form of the generalized Marcum Q-function $Q_m(a, b)$. Thus

$$F_X(x) = 1 - Q_1(c/\alpha, x/\alpha)I_{[0,\infty)}(x). \tag{3.137}$$

Examples of the pdf are shown in Figure 3.27. The Rice random variable reduces to the Rayleigh random variable when $c = 0$ such that $I_0(cx/\alpha^2) = 1$. Its mean is $\mu_X = \alpha\sqrt{\pi/2}L_{1/2}(-c^2/2\alpha^2)$ and the variance is $\sigma_X^2 = c^2 + 2\alpha^2 - (\pi\alpha^2/2)L_{1/2}^2(-c^2/2\alpha^2)$, where $L_n(x)$ is a Laguerre polynomial, which can also be written in terms of the confluent hypergeometric function (see Appendix B).

Applications include the following:

- Magnitude variations of a signal in a wireless communication channel when there is significant LoS propagation (from transmission point directly to receiver).
- Noise model for magnetic resonance imaging.

Like the Rayleigh random variable, the Rice random variable can be generated from two independent Gaussian random variables with standard deviation $\sigma = \alpha$ using (3.128), except that $\{X_1, X_2\}$ have nonzero means given by $c\cos(\theta)$ and $c\sin(\theta)$ with $\theta \in [-\pi, \pi]$. Any value for the mean of X_1 can be chosen provided the mean of X_2 is related as above because the transformation in (3.128) is invariant to rotations, and due to the fact that $c^2 = c^2\cos^2\theta + c^2\sin^2\theta$. The pdf is derived in Example 3.34.

Example 3.34. Let X_1 and X_2 be independent Gaussian random variables with parameters $\{\mu_1 = c\cos(\theta), \sigma\}$ and $\{\mu_2 = c\sin(\theta), \sigma\}$. Their joint pdf (defined in Chapter 4) is simply the product of their individual (marginal) pdfs:

$$f_{X_1,X_2}(x_1, x_2) = (1/2\pi\sigma^2)\exp\left(-[x_1 - c\cos(\theta)]^2/2\sigma^2\right)\exp\left(-[x_2 - c\sin(\theta)]^2/2\sigma^2\right). \tag{3.138}$$

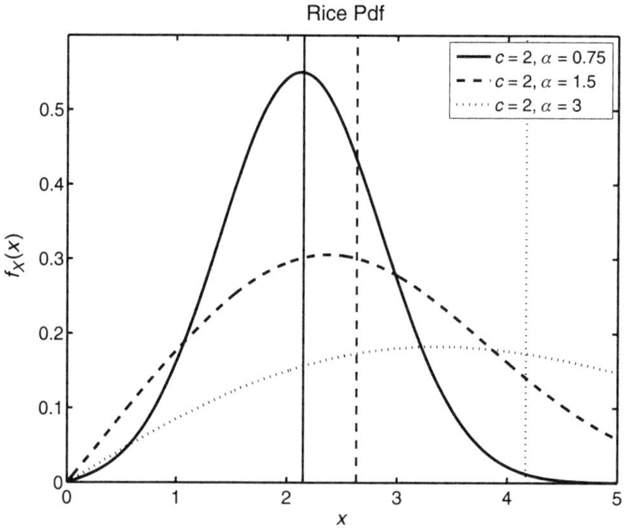

FIGURE 3.27 Examples of the Rice pdf. Support is $\mathcal{R}^+$. Vertical lines specify μ_X.

The derivation of the Rice pdf proceeds as follows. (i) Transform the random variables to polar coordinates with $R = \sqrt{X_1^2 + X_2^2}$ and $\Phi = \arctan(X_2/X_1)$. (ii) Integrate the joint pdf $f_{R,\Phi}(r, \phi)$ over ϕ resulting in $f_R(r)$, which is the Rice pdf since R is the Rice random variable by definition. For polar coordinates, we use

$$x_1 = r\cos(\phi), \quad x_2 = r\sin(\phi) \Longrightarrow r = \sqrt{x_1^2 + x_2^2}, \quad \phi = \tan^{-1}(x_2/x_1), \tag{3.139}$$

resulting in the following Jacobian matrix (see Appendix G):

$$\mathbf{J} = \begin{bmatrix} \partial r/\partial x_1 & \partial r/\partial x_2 \\ \partial\phi/\partial x_1 & \partial\phi/\partial x_2 \end{bmatrix} = \begin{bmatrix} x_1/\sqrt{x_1^2 + x_2^2} & x_2/\sqrt{x_1^2 + x_2^2} \\ -x_2/(x_1^2 + x_2^2) & x_1/(x_1^2 + x_2^2) \end{bmatrix}. \tag{3.140}$$

The Jacobian J is the determinant of $\mathbf{J}$, yielding $|J| = 1/\sqrt{x_1^2 + x_2^2} = 1/r$. Thus, the joint pdf of R and Φ is

$$f_{R,\Phi}(r, \phi) = (r/2\pi\sigma^2)\exp\left(-[r^2 + c^2 - 2rc\cos(\phi - \theta)]/2\sigma^2\right), \tag{3.141}$$

where the following identity has been used: $\cos(\phi - \theta) = \cos(\theta)\cos(\phi) + \sin(\theta)\sin(\phi)$. Integrating over ϕ gives

$$\begin{aligned} f_R(r) &= (r/2\pi\sigma^2)\exp(-[r^2 + c^2]/2\sigma^2)\int_0^{2\pi} \exp([rc/\sigma^2]\cos(\phi - \theta))\mathrm{d}\phi, \\ &= (r/\sigma^2)\exp(-[r^2 + c^2]/2\sigma^2)I_0(rc/\sigma^2), \end{aligned} \tag{3.142}$$

where we have used the fact that the integration is performed over exactly one period of the cosine to write

$$(1/2\pi)\int_0^{2\pi} \exp([rc/\sigma^2]\cos(\phi - \theta))\mathrm{d}\phi = (1/2\pi)\int_0^{2\pi} \exp([rc/\sigma^2]\cos(\phi))\mathrm{d}\phi = I_0(rc/\sigma^2), \tag{3.143}$$

which holds for any phase shift θ. The result in (3.142) is the Rice pdf in (3.135) with $\sigma = \alpha$ and $r = x$.

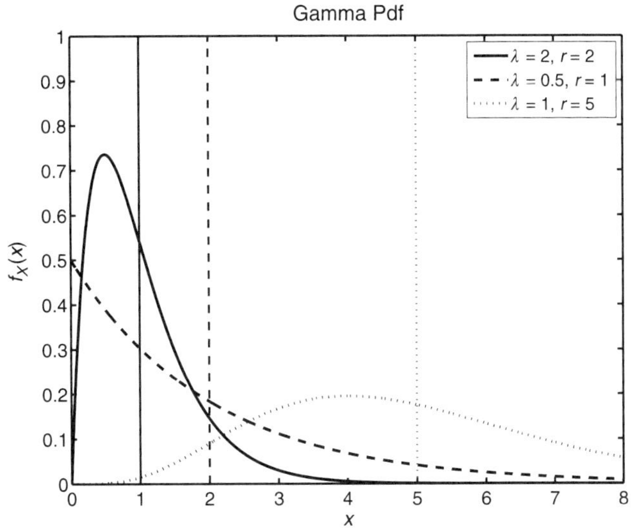

FIGURE 3.28 Examples of the gamma pdf. Support is $\mathcal{R}^+$. Vertical lines specify μ_X.

3.8.11 Gamma Random Variable (Erlang for $r \in \mathcal{N}$)

The pdf of the gamma random variable is

$$f_X(x) = \frac{1}{\Gamma(r)} x^{r-1} \lambda^r \exp(-\lambda x) I_{[0,\infty)}(x), \tag{3.144}$$

where $\lambda > 0$ is a scale parameter, $r > 0$ is a shape parameter, and $\Gamma(r)$ is the gamma function (see Appendix B). Examples of the pdf are shown in Figure 3.28. The cdf cannot be written in closed form; instead, it is convenient to express it as

$$F_X(x) = \frac{\gamma(r, \lambda x)}{\Gamma(r)} I_{[0,\infty)}(x), \tag{3.145}$$

where

$$\gamma(r, \lambda x) \triangleq \int_0^{\lambda x} v^{r-1} \exp(-v) \mathrm{d}v \tag{3.146}$$

is the *lower incomplete* gamma function. The gamma function $\Gamma(r)$ for any $r \in \mathcal{R}$ is given by $\gamma(r, \infty)$. Thus, incomplete means that the integration for $\gamma(r, \lambda x)$ is performed only for finite $\lambda x < \infty$. We may view the incomplete gamma function as serving the same role as the Q-function does for the Gaussian cdf, which likewise cannot be written in closed form.

The gamma distribution reduces to other continuous distributions discussed in this chapter for specific parameter values; these are summarized in Table 3.2. It is a generalization of the exponential random variable because it is a model of the waiting time for the rth arrival. When $r \in \mathcal{N}$, the gamma function simplifies to

$$\Gamma(r) \triangleq (r-1)! \tag{3.147}$$

and the gamma random variable has the Erlang distribution. The Nakagami distribution (Proakis, 2001), which is another model for fading in wireless communication channels (along with the Rayleigh and Rice distributions),

TABLE 3.2 Special Cases of the Gamma Random Variable

Gamma Parameters	Distribution
$r = 1$	Exponential $\{\lambda\}$
$r \in \mathcal{N}$	Erlang $\{\lambda, r\}$
$r = N, \lambda = 1/2$	Chi-square $\{N\}$

is obtained from the Gamma distribution via the transformation $Y = \sqrt{X}$ (see Problem 4.29). Its pdf is

$$f_Y(y) = \frac{2r^r}{\Gamma(r)\alpha^r} y^{2r-1} \exp(-ry^2/\alpha) I_{[0,\infty)}(y), \tag{3.148}$$

where $r \geq 1/2$ and $\alpha > 0$ are shape and scale parameters, respectively. It has mean $\mu_Y = \sqrt{r/\alpha}\Gamma(r + 1/2)/\Gamma(1/2)$, variance $\sigma_Y^2 = \alpha[1 - (1/r)\Gamma^2(r + 1/2)/\Gamma^2(r)]$, and cdf

$$F_X(y) = \gamma(r, ry^2/\alpha)/\Gamma(r). \tag{3.149}$$

Applications of the gamma distribution include the following:

- Waiting time until the rth event of a Poisson process (see Chapter 6).
- Risk analysis in the financial industry, such as a model for the amount of insurance claims.
- The gamma *process* is an independent increment process (see Chapter 6) that is based on the gamma distribution, and has been used as a model for precipitation (rainfall) amounts.

Example 3.35. Although the median is an implicit function of the parameters, the mode is relatively simple to compute. Differentiating the pdf, setting the expression equal to zero, and solving for m_o gives

$$\frac{1}{\Gamma(r)}(r-1)m_o^{r-2} \exp(-\lambda m_o) - \frac{1}{\Gamma(r)} m_o^{r-1} \exp(-\lambda m_o)\lambda = 0 \Longrightarrow m_o = (r-1)/\lambda \tag{3.150}$$

for $r > 1$. The mode is $m_o = 0$ for $r \in (0, 1]$.

Example 3.36. The standard gamma pdf is obtained when $\lambda = 1$:

$$f_X(x) = \frac{1}{\Gamma(r)} x^{r-1} \exp(-x) I_{[0,\infty)}(x), \tag{3.151}$$

which, from Example 3.35, has mode $m_o = r - 1$. Differentiating with respect to x twice gives

$$\frac{d^2}{dx^2} f_X(x) = \frac{1}{\Gamma(r)} x^{r-3} \exp(-x) \left[x^2 - 2(r-1)x + (r-1)(r-2)\right]. \tag{3.152}$$

Setting this expression equal to zero and solving for x, we find that the standard gamma pdf has two inflection points about the mode:

$$x = m_o \pm \sqrt{m_o}. \tag{3.153}$$

The pdf changes from concave to convex as x moves from m_o to the left and right of the inflection points. An example of this is shown in Figure 3.28 for $r = 5$ and $\lambda = 1$ where the mode is $m_o = 4$, and the two inflection points are located at $x = \{2, 6\}$.

Example 3.37. The mean of the gamma distribution is computed as follows:

$$\mu_X = \frac{\lambda^r}{\Gamma(r)} \int_0^\infty x^r \exp(-\lambda x) \mathrm{d}x = \frac{\lambda^{r-1}}{\Gamma(r)} \int_0^\infty \lambda x^r \exp(-\lambda x) \mathrm{d}x. \tag{3.154}$$

This integral is the rth *raw moment* of the exponential random variable which is $\Gamma(r+1)/\lambda^r$ from Appendix A. As a result, and since $\Gamma(r+1) = r\Gamma(r)$:

$$\mu_X = \Gamma(r+1)/\lambda\Gamma(r) = r/\lambda, \tag{3.155}$$

which is r times that of the exponential random variable; the same scaling by r occurs for the variance: $\sigma_X^2 = r/\lambda^2$.

3.8.12 Beta Random Variable (Arcsine for $\alpha = \beta = 1/2$, Power Function for $\beta = 1$)

The beta random variable has the following pdf:

$$f_X(x) = \frac{1}{B(\alpha, \beta)} x^{\alpha-1}(1-x)^{\beta-1} I_{[0,1]}(x), \tag{3.156}$$

where

$$B(\alpha, \beta) \triangleq \frac{\Gamma(\alpha)\Gamma(\beta)}{\Gamma(\alpha+\beta)}, \tag{3.157}$$

$\alpha, \beta > 0$ are shape parameters, and $\Gamma(a)$ is the gamma function. Along with the uniform and triangular random variables, the beta distribution is the only other continuous random variable covered in this book that has finite support. The cdf is

$$F_X(x) = \frac{B_x(\alpha, \beta)}{B(\alpha, \beta)} I_{[0,1]}(x) + I_{(1,\infty)}(x), \tag{3.158}$$

where

$$B_x(\alpha, \beta) \triangleq \int_0^x y^{\alpha-1}(1-y)^{\beta-1} \mathrm{d}y \tag{3.159}$$

is the *incomplete* beta function. The complete beta function $B(\alpha, \beta)$ above corresponds to the definite integral with $x = 1$ in $B_x(\alpha, \beta)$. The ratio in the cdf, defined as

$$I_x(\alpha, \beta) \triangleq \frac{B_x(\alpha, \beta)}{B(\alpha, \beta)}, \tag{3.160}$$

is called the *regularized* incomplete beta function (see Appendix B for a discussion of these functions). The mean is $\mu_X = \alpha/(\alpha+\beta)$ and the variance is $\sigma_X^2 = \alpha\beta/[(\alpha+\beta)^2(\alpha+\beta+1)]$.

Applications include the following:

- Modeling the underlying distribution of samples (measurements) when they are restricted to be in some finite range.
- Transmittance of solar radiation in the atmosphere.
- Model of the duration of activities required for completion of a project, as used in PERT (Project Evaluation and Review Technique) (de Neufville and Stafford, 1971).

TABLE 3.3 Shapes and Special Cases of the Beta Random Variable

Beta Parameters	Shape
$\alpha = 1, \beta > 2$ (or $\beta = 1, \alpha > 2$)	Strictly convex
$\alpha = 1, 1 < \beta < 2$ (or $\beta = 1, 1 < \alpha < 2$)	Strictly concave
$\alpha = 1, \beta = 2$ (or $\alpha = 2, \beta = 1$)	Straight line
$\alpha = \beta$	Symmetric about $1/2$
Special Cases	**Distribution**
$\alpha = \beta = 1/2$	Arcsine
$\alpha = \beta = 1$	Standard uniform [0, 1]
$\beta = 1$	Power-function $\{\alpha\}$

When the finite range of the beta random variable needs to be extended beyond $[0, 1]$ to $[a, b]$, the following more general pdf can be used:

$$f_X(x) = \frac{1}{B(\alpha, \beta)(b-a)^{\alpha+\beta-1}}(x-a)^{\alpha-1}(b-x)^{\beta-1}I_{[a,b]}(x). \tag{3.161}$$

Depending on the parameter values, the beta pdf has a wide variety of shapes. Some special cases are summarized in Table 3.3 and examples are shown in Figure 3.29. These shapes are all easily observed from the simple functional form of the pdf: $x^{\alpha-1}(1-x)^{\beta-1}$; the term $1/B(\alpha, \beta)$ is only a weighting such that $f_X(x)$ has unit area. (See Appendix D for a summary of the functional forms for continuous random variables.) If $\alpha = \beta = 1$, then $B(1, 1) = 1$ and (3.156) reduces to

$$f_X(x) = I_{[0,1]}(x), \tag{3.162}$$

which is the pdf of the continuous uniform random variable on $[0, 1]$. When $\alpha = \beta = 1/2$, the *arcsine distribution* is obtained:

$$F_X(x) = (2/\pi)\arcsin(\sqrt{x})I_{[0,1]}(x) + I_{(1,\infty)}(x) \Longrightarrow f_X(x) = 1/\pi\sqrt{x(1-x)}I_{[0,1]}(x). \tag{3.163}$$

When $\beta = 1$, a form of the *power-function distribution* is obtained:

$$f_X(x) = \alpha x^{\alpha-1}I_{[0,1]}(x), \tag{3.164}$$

where $B(\alpha, 1) = \alpha$ has been substituted. The beta distribution with parameters $\alpha = r_1$ and $\beta = r_2$ can be obtained via the following transformation:

$$X = \frac{Y_1}{Y_1 + Y_2}, \tag{3.165}$$

where Y_1 and Y_2 are gamma random variables with parameters $\{r_1, \lambda\}$ and $\{r_2, \lambda\}$, respectively. The distribution of X is obviously *not* symmetric with respect to Y_1 and Y_2.

Example 3.38. The skewness of the beta distribution is

$$\gamma_1 = \sqrt{\frac{\alpha+\beta+1}{\alpha\beta}}\frac{2(\beta-\alpha)}{(\alpha+\beta+2)} \tag{3.166}$$

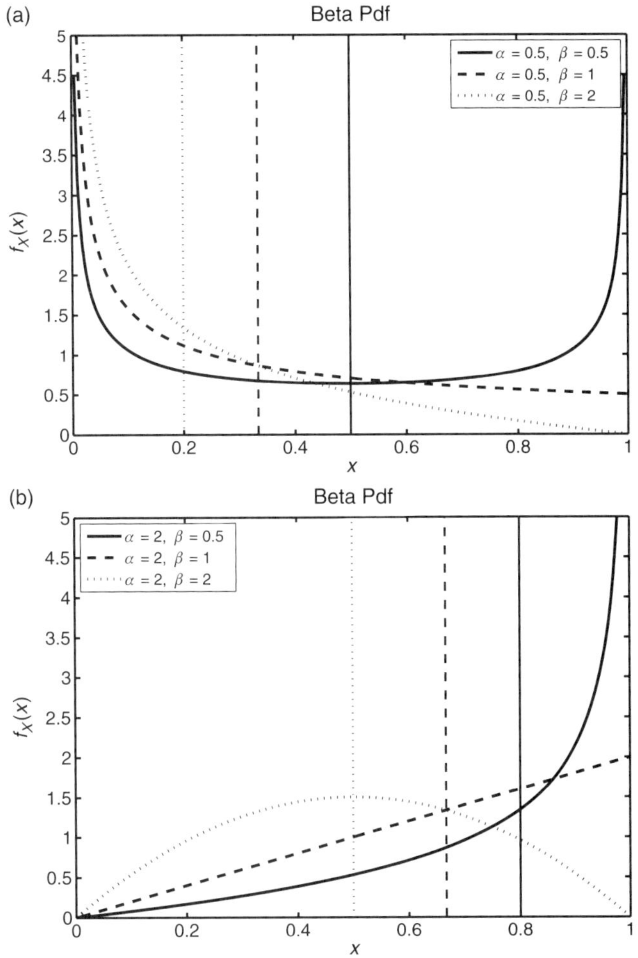

FIGURE 3.29 Examples of the Beta pdf. Support is [0, 1]. Vertical lines specify μ_X. (a) $\alpha = 0.5$. (b) $\alpha = 2$.

from which we see that $\beta - \alpha$ in the numerator determines the type of skewness. Figure 3.29(a) shows two pdfs that have positive skewness because $\beta > \alpha$, and the third pdf is symmetric because $\alpha = \beta = 1/2$. Two of the pdfs in Figure 3.29(b) have negative skewness ($\beta < \alpha$) and the other is symmetric ($\alpha = \beta = 2$).

Example 3.39. Next, we demonstrate that the exponential distribution can be obtained by the following transformation $Y = \beta X$, where X is the beta random variable with parameters $\{\alpha = 1, \beta\}$, and then letting $\beta \to \infty$. The cdf of Y is

$$P(Y \leq y) = P(\beta X \leq y) = P(X \leq y/\beta),$$

$$= \frac{B_{y/\beta}(1, \beta)}{B(1, \beta)} = \frac{1}{B(1, \beta)} \int_0^{y/\beta} (1 - v)^{\beta - 1} \mathrm{d}v, \tag{3.167}$$

which has support $[0, \beta)$ because β multiplies X. Thus

$$F_Y(y) = -\frac{1}{B(1, \beta)} (1/\beta)(1 - v)^{\beta} \Big|_0^{y/\beta} = \frac{\Gamma(1 + \beta)}{\Gamma(1)\Gamma(\beta)} (1/\beta)[1 - (1 - y/\beta)^{\beta}]. \tag{3.168}$$

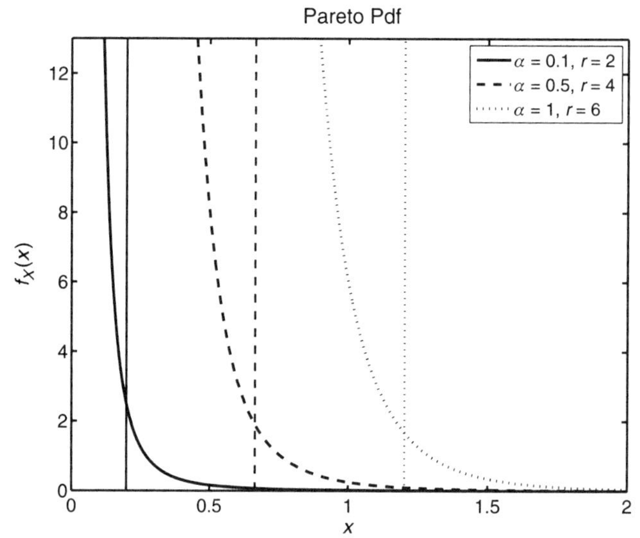

FIGURE 3.30 Examples of the Pareto pdf. Support is $[\alpha, \infty)$. Vertical lines specify μ_X.

Since $\Gamma(1) = 1$ and $\Gamma(1+\beta)/\Gamma(\beta) = \beta$, we finally have

$$\lim_{\beta\to\infty} F_Y(y) = [1 - \exp(-y)]I_{[0,\infty)}(x), \tag{3.169}$$

which is the cdf of the standard exponential random variable (with $\lambda = 1$). It can be similarly shown that if the parameters of X are $\{\alpha = r, \beta\}$, then the limiting distribution of Y is gamma with parameters $\{\lambda = 1, r\}$ (see Problem 3.22).

3.8.13 Pareto Random Variable

The Pareto random variable has the following pdf:

$$f_X(x) = (r\alpha^r/x^{r+1})I_{[\alpha,\infty)}(x), \tag{3.170}$$

where $\alpha > 0$ is a scale parameter that also defines the support via the indicator function, and $r > 0$ is a shape parameter. This distribution is characterized by a very long right tail for small values of r. The cdf is

$$F_X(x) = \left[1 - (\alpha/x)^r\right] I_{[\alpha,\infty)}(x). \tag{3.171}$$

Examples of the Pareto pdf are shown in Figure 3.30. The mean is $\mu_X = r\alpha/(r-1)$ for $r > 1$ and the variance is $\sigma_X^2 = r\alpha^2/[(r-1)^2(r-2)]$ for $\alpha > 2$ (otherwise, they are not defined). Applications include the following:

- Distributions of income and wealth.
- Error clustering in communication circuits and hard disk drives.

Example 3.40. The Pareto pdf approaches the Dirac delta function for large r:

$$\lim_{r\to\infty} f_X(x) = \delta(x - \alpha), \tag{3.172}$$

and thus the cdf approaches the unit-step function:

$$\lim_{r\to\infty} F_X(x) = u(x-\alpha). \tag{3.173}$$

First we verify that the pdf integrates to one:

$$r\alpha^r \int_\alpha^\infty x^{-r-1}\mathrm{d}x = -\alpha x^{-r}\Big|_\alpha^\infty = 1. \tag{3.174}$$

Then observe that

$$\lim_{r\to\infty} [r\alpha^r/x^{r+1}u(x-\alpha)] = (1/x)u(x-\alpha)\lim_{r\to\infty} r(\alpha/x)^r = \begin{cases} 0, & x > \alpha \\ \infty, & x = \alpha, \end{cases} \tag{3.175}$$

where $u(x)$ is the unit-step function, and by definition $\alpha/x \leq 1$. Equations (3.174) and (3.175) together define the Dirac delta function at $x = \alpha$. It is not necessary that the original function be even in order to define the Dirac delta function, as is done in Appendix B using the rectangle function.

Example 3.41 (Differential entropy). The *differential entropy* for a continuous random variable is a quantity used in information theory. It is defined as follows (see Chapter 10):

$$h(X) \triangleq -\int_{-\infty}^{\infty} f_X(x)\log(f_X(x))\,\mathrm{d}x, \tag{3.176}$$

where the logarithm can be any base, though it is usually 2 or e such that the units of $h(X)$ are bits and "nats," respectively. For the Pareto random variable:

$$\begin{aligned} h(X) &= -\int_\alpha^\infty \frac{r\alpha^r}{x^{r+1}}\ln(r\alpha^r)\mathrm{d}x + \int_\alpha^\infty \frac{r\alpha^r}{x^{r+1}}\ln(x^{r+1})\mathrm{d}x, \\ &= -\ln(r\alpha^r) + r\alpha^r(r+1)\int_\alpha^\infty \frac{\ln(x)}{x^{r+1}}\mathrm{d}x. \end{aligned} \tag{3.177}$$

The first term simplifies because the logarithm can be factored out, leaving an integral over the Pareto pdf. The indefinite integral for the second term is

$$\int \frac{\ln(x)}{x^{r+1}}\mathrm{d}x = -[r\ln(x)+1]/r^2x^r \tag{3.178}$$

so that

$$h(X) = -\ln(r\alpha^r) + r\alpha^r(r+1)[r\ln(\alpha)+1]/r^2\alpha^r = (r+1)/r + \ln(\alpha/r). \tag{3.179}$$

Differential entropies for the other continuous random variables are provided in Appendix A.

3.8.14 Weibull Random Variable

The pdf of the Weibull random variable is

$$f_X(x) = (r/\alpha)(x/\alpha)^{r-1}\exp(-(x/\alpha)^r)\,I_{[0,\infty)}(x), \tag{3.180}$$

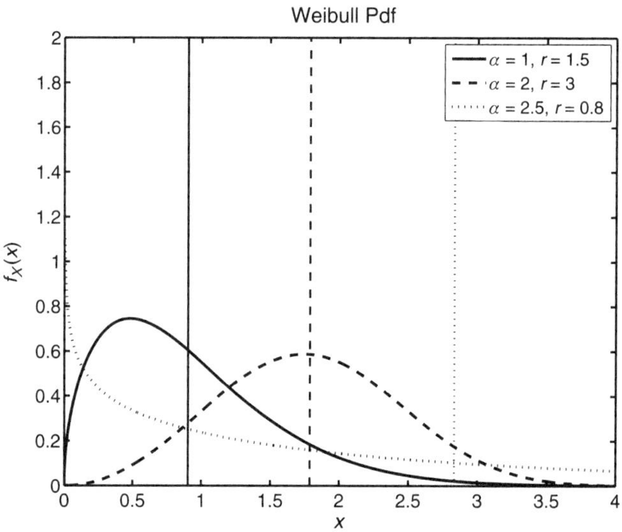

FIGURE 3.31 Examples of the Weibull pdf. Support is $\mathcal{R}^+$. Vertical lines specify μ_X.

where $\alpha > 0$ is a scale parameter and $r > 0$ is a shape parameter. A single-parameter version of the Weibull pdf is obtained when $\alpha = 1$. The Weibull random variable reduces to the exponential random variable when $r = 1$ with $\alpha = 1/\lambda$, and has the Rayleigh distribution when $r = 2$ and $\alpha \to \sqrt{2}\alpha$. The mean is $\mu_X = \alpha\Gamma(1 + 1/r)$, the variance is $\sigma_X^2 = \alpha^2[\Gamma(1 + 2/r) - \Gamma^2(1 + 1/r)]$, and the cdf is

$$F_X(x) = [1 - \exp(-(x/\alpha)^r)]I_{[0,\infty)}(x). \tag{3.181}$$

Examples of the Weibull pdf are given in Figure 3.31. Applications include the following:

- Distribution of wind speeds and a model of power from an array of wind turbines.
- Model of performance loss and the reliability of components in complex systems.
- Fading channel model in wireless communications.

Example 3.42. Observe that $F_X(\alpha) = 1 - \exp(-1) \approx 0.6321$ for any value of r. Thus, the areas of the pdfs in Figure 3.31 to the left of $x = \alpha$ are all identical, even though the pdf becomes less peaked with increasing r. Differentiating $f_X(x)$ with respect to x and setting the result equal to zero gives the mode (for $r > 1$):

$$m_o = \alpha[(r - 1)/r]^{1/r}, \tag{3.182}$$

which is 0.7071α for $r = 2$ and 0.9564α for $r = 5$. The mode is zero for $r \in (0, 1]$. The median is derived as follows:

$$1 - \exp(-(m_e/\alpha)^r) = 1/2 \Longrightarrow m_e = \alpha(\ln(2))^{1/r}, \tag{3.183}$$

which is 0.8326α for $r = 2$ and 0.9293α for $r = 5$. Comparing the expressions for m_o and m_e, we see that $m_o > m_e$ when $r > 1/(1 - \ln(2)) \approx 3.2589$. The median and mode can never exceed $x = \alpha$ because the cdf there is ≈ 0.6321 as mentioned above.

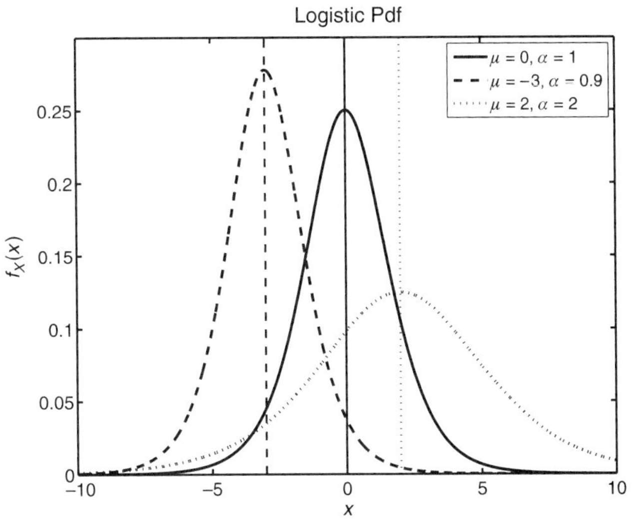

FIGURE 3.32 Examples of the logistic pdf. Support is $\mathcal{R}$. Vertical lines specify μ_X.

3.8.15 Logistic Random Variable (Sigmoid for $\{\mu = 0, \alpha = 1\}$)

The pdf of the logistic random variable is

$$f_X(x) = \frac{\exp(-(x-\mu)/\alpha)}{\alpha(1+\exp(-(x-\mu)/\alpha))^2} I_{(-\infty,\infty)}(x), \tag{3.184}$$

where $\mu \in \mathcal{R}$ is a location parameter and $\alpha > 0$ is a scale parameter. The cdf is

$$F_X(x) = [1 + \exp(-(x-\mu)/\alpha)]^{-1} I_{(-\infty,\infty)}(x). \tag{3.185}$$

Examples of the logistic pdf are shown in Figure 3.32. The *sigmoid distribution* is a special case of $F_X(x)$ with $\mu = 0$ and $\alpha = 1$. The mean is $\mu_X = \mu$ and the variance is $\sigma_X^2 = (\pi\alpha)^2/3$.

Applications include the following:

- An alternative to the Gaussian distribution when slightly heavier tails provide an improved model.
- Population growth of humans and other species.
- Model of the spread of innovations (new products and technologies) throughout societies.

Example 3.43. The pdf of the logistic random variable can be rewritten as follows:

$$\begin{aligned} f_X(x) &= \frac{\exp(-(x-\mu)/\alpha)}{\alpha[\exp(-(x-\mu)/2\alpha)]^2[\exp((x-\mu)/2\alpha) + \exp(-(x-\mu/2\alpha)]^2}, \\ &= [4\alpha\cosh^2((x-\mu)/2\alpha)]^{-1} = (1/4\alpha)\text{sech}^2((x-\mu)/2\alpha). \end{aligned} \tag{3.186}$$

The cdf of the logistic random variable can likewise be rewritten more compactly as follows (see Problem 3.18):

$$F_X(x) = (1/2)[1 + \tanh((x-\mu)/2\alpha)] I_{(-\infty,\infty)}(x). \tag{3.187}$$

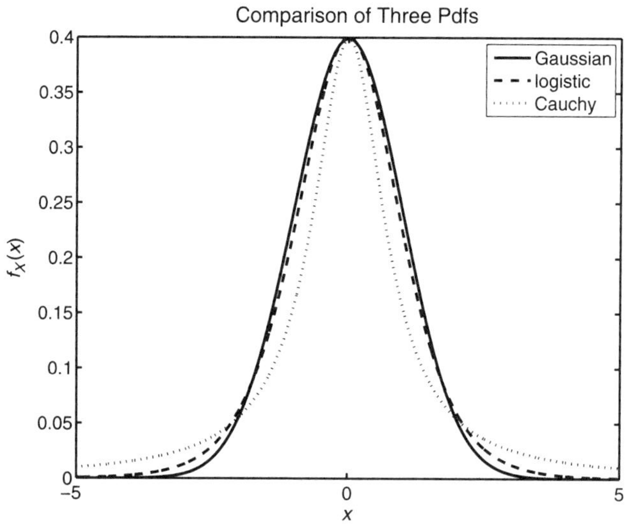

FIGURE 3.33 Gaussian ($\sigma = 1$), logistic ($\alpha = 5/8$), and Cauchy ($\alpha = \sqrt{2/\pi}$) pdfs with identical location parameters ($\mu = c = 0$). The scale parameters were chosen so that the pdfs all have the same maximum.

Observe in Figure 3.33 that the logistic pdf closely resembles the Gaussian pdf: they both have skewness $\gamma_1 = 0$, but the logistic pdf has slightly heavier tails with a kurtosis of $\gamma_2 = 6/5$ (compared to zero for the Gaussian pdf). Table 3.4 summarizes the type of tail for each of the symmetric continuous random variables covered in this book (all of which necessarily have zero skewness).

Example 3.44. Unlike most other random variables, the inverse cdf (icdf) of the logistic random variable has a simple closed form and is easy to derive. Letting $p = F_X(x)$ and rearranging the expression in (3.185) gives

$$x = F_X^{-1}(p) = \mu + \alpha \ln(p/(1-p)). \tag{3.188}$$

This result can be used to convert samples of a uniform random variable on [0, 1] to those with the logistic distribution, as described in Chapter 1. It is called an inverse probability transformation and is discussed in greater detail in Chapter 4 for other random variables. Consider random variable Y with cdf $F_Y(y)$ and mean μ_Y. An interesting inverse transformation that can be used to determine how the tails of X and Y compare is given by (Kotz, Kozubowski, and Podgórski, 2001):

$$g(y) \triangleq F_X^{-1}(F_Y(y)). \tag{3.189}$$

TABLE 3.4 Tails of Symmetric Random Variables

Random Variable	Excess Kurtosis γ_2	Type of Tail
Continuous uniform	$-6/5$	Light tailed
Triangular ($c = (a+b)/2$)	$-3/5$	Light tailed
Gaussian, logistic	$0, 6/5$	Medium tailed
Laplace, Cauchy	3, does not exist	Heavy tailed
Student's t ($r > 4$)	$6/(r-4)$	Heavy tailed

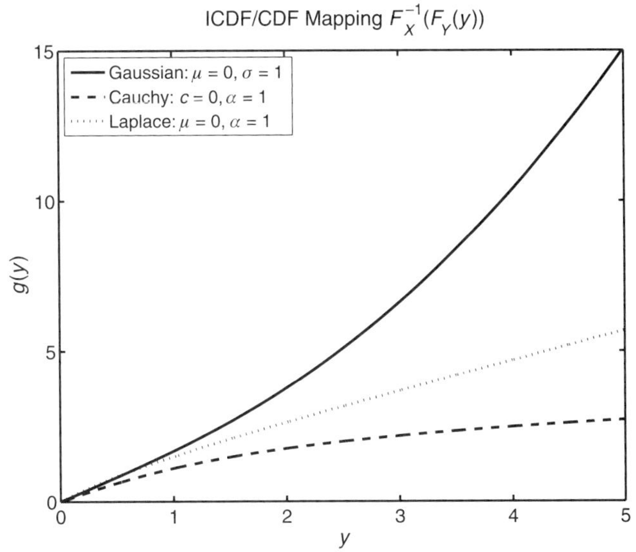

FIGURE 3.34 Transformation in Example 3.44 for determining if a distribution has lighter tails.

Specifically, if $g(y)$ is a convex function for $y \geq \mu_Y$, then Y has a lighter tail than X. Plots of $g(y)$ for logistic X and when Y is Gaussian, Laplacian, and Cauchy are shown in Figure 3.34. These results indicate that the logistic distribution is lighter tailed than the Cauchy and Laplace distributions, and heavier tailed than the Gaussian distribution, which is consistent with the summary in Table 3.4 based on kurtosis.

3.8.16 Chi Random Variable (Maxwell–Boltzmann, Half-Normal)

Let $\{X_n\}$ be N-independent Gaussian random variables with parameters $\{\mu_n, \sigma_n\}$. Then

$$X \triangleq \left[\sum_{n=1}^{N}(X_n - \mu_n)^2/\sigma_n^2\right]^{1/2} \tag{3.190}$$

is a chi random variable with the following pdf:

$$f_X(x) = \frac{x^{N-1}}{2^{N/2-1}\Gamma(N/2)}\exp(-x^2/2)I_{[0,\infty)}(x), \tag{3.191}$$

where $N \in \mathcal{N}$ is the degrees of freedom. Examples are provided in Figure 3.35. Since (3.190) is a sum of *standard* Gaussian random variables, the means and variances of the original $\{X_n\}$ do not influence the chi distribution. X has mean $\mu_X = \sqrt{2}\Gamma((N+1)/2)/\Gamma(N/2)$ and variance $\sigma_X^2 = N - 2[\Gamma((N+1)/2))/\Gamma(N/2)]^2$. The notation χ_N is often used to represent the chi random variable. Its cdf is

$$F_X(x) = [\gamma(N/2, x^2/2)/\Gamma(N/2)]I_{[0,\infty)}(x), \tag{3.192}$$

where $\gamma(a, b)$ is the lower incomplete gamma function. There are three special cases: (i) $N = 1$ gives the half-normal distribution with scale parameter $\alpha = 1$ (also called the folded-normal distribution; see Appendix A), (ii) $N = 2$ gives the Rayleigh distribution with $\alpha = 1$, and (iii) $N = 3$ gives the Maxwell–Boltzmann distribution with parameter $\alpha = 1$. The transformation X^2 has a chi-square distribution, which is more well

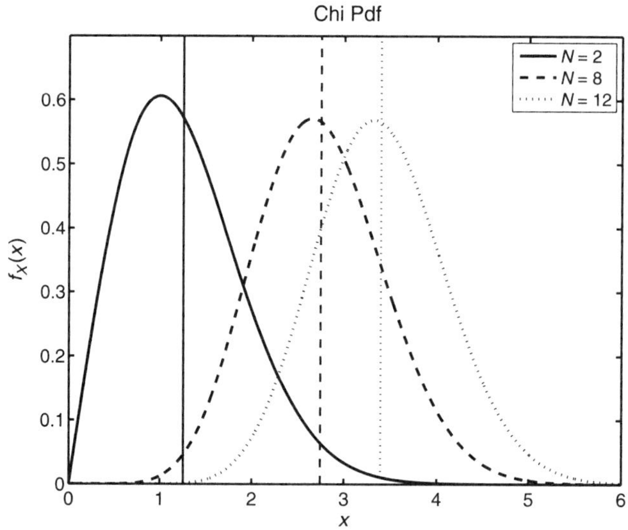

FIGURE 3.35 Examples of the chi pdf. Support is $\mathcal{R}^+$. Vertical lines specify μ_X.

known and is described in Section 3.8.17. Substituting $N = 2r$ and performing the transformation $\sqrt{\alpha/2r}X$ yields the previously mentioned Nakagami random variable where α is a scale parameter and r is a shape parameter.

Applications of the chi random variable include the following:

- Distribution of the magnitude of the velocity of a molecule in an ideal gas in a closed container (for $N = 3$, corresponding to the Maxwell–Boltzmann distribution).
- Model for normally distributed samples with missing sign information, that is, $X = |X_n - \mu_n|/\sigma_n$ (for $N = 1$, corresponding to the half-normal distribution).

Example 3.45 (Maxwell–Boltzmann). In this example, we examine the Maxwell–Boltzmann distribution in Appendix A with scale parameter α (which the chi pdf in (3.191) does not have). Substituting $N = 3$ in the chi pdf gives

$$f_X(x) = \sqrt{2/\pi}x^2 \exp(-x^2/2)I_{[0,\infty)}(x), \tag{3.193}$$

where $\Gamma(3/2) = \sqrt{\pi}/2$ has been substituted. Suppose x^2 is scaled by $1/\alpha^2$ and integrated over the support of X:

$$\sqrt{2/\pi}\int_0^\infty (x^2/\alpha^2)\exp(-x^2/2\alpha^2)\mathrm{d}x = \alpha\sqrt{2/\pi}\int_0^\infty y^2\exp(-y^2/2)\mathrm{d}y = \alpha, \tag{3.194}$$

where we have changed variables to $y^2 = x^2/\alpha^2$, and recognized that the transformed function is a valid pdf scaled by α. The pdf of the Maxwell–Boltzmann random variable with scale parameter α is

$$f_X(x) = (1/\alpha^3)\sqrt{2/\pi}x^2\exp(-x^2/2\alpha^2)I_{[0,\infty)}(x). \tag{3.195}$$

This result can be obtained more directly using the transformation techniques described in Chapter 4. A scale parameter is included in a random variable via the transformation $X = \alpha Y$ for $\alpha > 0$ and writing the cdf as follows:

$$F_X(y) = P(X \leq x) = P(Y \leq x/\alpha) = F_Y(x/\alpha). \tag{3.196}$$

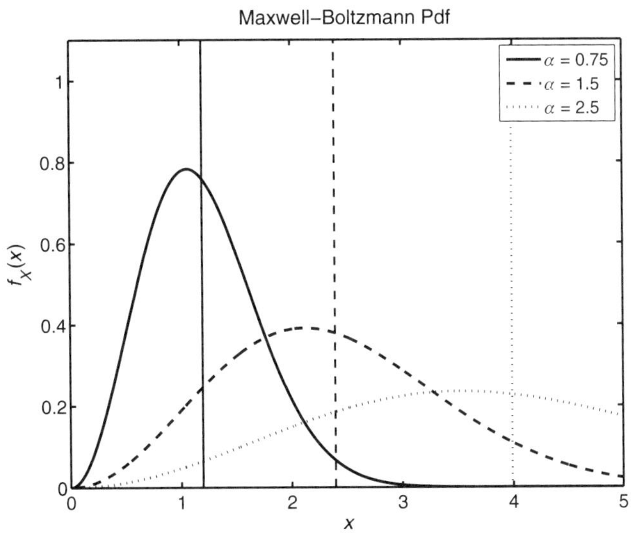

FIGURE 3.36 Examples of the Maxwell–Boltzmann pdf. Support is $\mathcal{R}^+$. Vertical lines specify μ_X.

Differentiating yields the desired result for the pdf:

$$f_X(x) = \frac{\mathrm{d}}{\mathrm{d}x} F_Y(x/\alpha) = (1/\alpha) f_Y(x/\alpha). \tag{3.197}$$

Examples of the Maxwell–Boltzmann pdf for various values of α are shown in Figure 3.36. Using a similar scaling by α on the chi random variable with $N = 3$, it is straightforward to derive the mean $\mu_X = 2\alpha\sqrt{2/\pi}$ and variance $\sigma_X^2 = \alpha^2(3\pi - 8)/\pi$ (see Problem 3.24). The cdf of the chi random variable with $N = 3$ is

$$F_X(x) = [1/\Gamma(3/2)][(1/2)\gamma(1/2, x^2/2) - (x/\sqrt{2})\exp(-x^2/2)]I_{[0,\infty)}(x), \tag{3.198}$$

where the recursion for the lower incomplete gamma function in Appendix B has been used. Thus

$$F_X(x) = (2/\sqrt{\pi})\left[(1/2)\int_0^{x^2/2} v^{-1/2}\exp(-v)\mathrm{d}v - (x/\sqrt{2})\exp(-x^2/2)\right] I_{[0,\infty)}(x). \tag{3.199}$$

In order to obtain the cdf of the Maxwell–Boltzmann random variable, we include the scale parameter α according to (3.196), yielding

$$\begin{aligned} F_X(x) &= (2/\sqrt{\pi})\left[(1/2)\int_0^{x^2/2\alpha^2} v^{-1/2}\exp(-v)\mathrm{d}v - (x/\alpha\sqrt{2})\exp(-x^2/2\alpha^2)\right], \\ &= (2/\sqrt{\pi})\int_0^{x/\alpha\sqrt{2}} \exp(-w^2)\mathrm{d}w - \sqrt{2/\pi}(x/\alpha)\exp(-x^2/2\alpha^2), \end{aligned} \tag{3.200}$$

where we have changed variables to $w^2 = v$. Since this integral is the error function, a compact form for the Maxwell–Boltzmann cdf is given by

$$F_X(x) = \left[\mathrm{erf}(x/\alpha\sqrt{2}) - \sqrt{2/\pi}(x/\alpha)\exp(-x^2/2\alpha^2)\right] I_{[0,\infty)}(x), \tag{3.201}$$

which reduces to (3.198) when $\alpha = 1$.

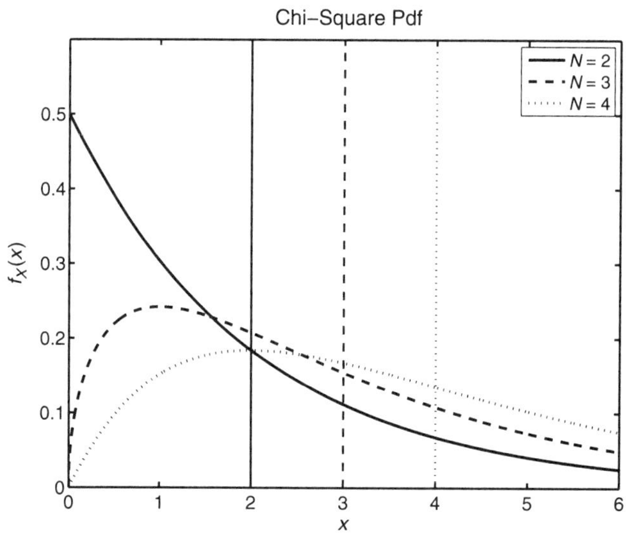

FIGURE 3.37 Examples of the chi-square pdf. Support is $\mathcal{R}^+$. Vertical lines specify μ_X.

3.8.17 Chi-Square Random Variable

Let $\{X_n\}$ be N-independent Gaussian random variables with parameters $\{\mu_n, \sigma_n\}$. Then

$$X = \sum_{n=1}^{N}(X_n - \mu_n)^2/\sigma_n^2 \tag{3.202}$$

is a chi-square random variable with the following pdf:

$$f_X(x) = \frac{x^{N/2-1}}{2^{N/2}\Gamma(N/2)}\exp(-x/2)I_{[0,\infty)}(x), \tag{3.203}$$

where $N \in \mathcal{N}$ is the degrees of freedom. Since (3.202) is a sum of standard Gaussian random variables, the means and variances of the original $\{X_n\}$ do not influence the chi-square distribution. Often the symbol χ_N^2 is used to represent the chi-square random variable, which turns out to be a special case of the gamma random variable when $N = 2r$ and $\alpha = 1/2$. The chi-square random variable reduces to the exponential random variable with parameter $\lambda = 1/2$ when $N = 2$. The cdf does not have a closed form, but can be expressed as follows:

$$F_X(x) = \frac{1}{2^{N/2}\Gamma(N/2)}\int_0^x z^{N/2-1}\exp(-z/2)dz = \frac{\gamma(N/2, x/2)}{\Gamma(N/2)}I_{(-\infty,\infty)}(x), \tag{3.204}$$

where $\gamma(a, b)$ is the lower incomplete gamma function. Examples of the pdf are provided in Figure 3.37. The mean is $\mu_X = N$ and the variance is $\sigma_X^2 = 2N$. The chi-square distribution is related to other distributions via the transformations summarized in Table 3.5. Of course, the last two results in the table follow from the first line because they are special cases of the chi random variable.

Applications include the following:

- Finding confidence intervals for the variance from samples of a Gaussian random variable.
- Hypothesis testing to decide if the variance estimate obtained from Gaussian samples equals a particular value.

TABLE 3.5 Transformations to the Chi-Square Random Variable

Random Variable X	Transformation	Distribution
Chi $\{N\}$	X^2	χ_N^2
Standard Gaussian	X^2	χ_1^2
Uniform [0, 1]	$-\ln(X^2)$	χ_2^2
Rayleigh $\{\alpha = 1\}$	X^2	χ_2^2
Maxwell–Boltzmann $\{\alpha = 1\}$	X^2	χ_3^2

Example 3.46. The uniform random variable transformation in Table 3.5 can be derived by starting with the cdf of $Y = -\ln(X^2)$:

$$F_Y(y) = P(Y \le y) = P(\ln(X^2) \ge -y) = 1 - P(X < \exp(-y/2)), \tag{3.205}$$

where we have used the fact that the logarithm is a monotonic function. Since X is uniform on [0, 1], its cdf simplifies to $xI_{[0,1]}(x)$ so that

$$F_Y(y) = 1 - \exp(-y/2) \Longrightarrow f_Y(y) = (1/2)\exp(-y/2)I_{[0,\infty)}(x), \tag{3.206}$$

which is the pdf of χ_2^2. Note that the logarithm expands the finite support of X to be $\mathcal{R}^+$ for Y.

Example 3.47. For $N > 2$, the derivative of the chi-square pdf is

$$\frac{\mathrm{d}}{\mathrm{d}x} f_X(x) = \frac{(N/2-1)x^{N/2-2}}{2^{N/2}\Gamma(N/2)} \exp(-x/2) - (1/2)\frac{x^{N/2-1}}{2^{N/2}\Gamma(N/2)} \exp(-x/2). \tag{3.207}$$

Setting this expression equal to zero and cancelling terms gives the mode $m_o = N - 2$. For $N = 2$:

$$\frac{\mathrm{d}}{\mathrm{d}x} f_X(x) = -(1/2)\frac{1}{2^{N/2}\Gamma(N/2)} \exp(-x/2) = 0 \Longrightarrow m_o = 0, \tag{3.208}$$

and for $N = 1$, (3.207) simplifies to the following expression after setting it equal to zero and cancelling terms: $x^{-3/2} + x^{-1/2} = 0$. Because the two remaining terms no longer have opposite signs as they did for $N > 2$, the mode is again $m_o = 0$.

The chi-square distribution arises in an estimation problem as follows. Let $\{X_1, \ldots, X_N\}$ be samples independently drawn from a Gaussian pdf with parameters $\{\mu, \sigma\}$. Consider the following sample mean:

$$\bar{X} \triangleq \frac{1}{N}\sum_{n=1}^{N} X_n, \tag{3.209}$$

which is commonly used to estimate the mean of the underlying distribution. It is easily shown that $\bar{X}$ is Gaussian with mean μ and variance σ^2/N (see Chapter 5). The sample variance is given by

$$\mathcal{S}^2 \triangleq \frac{1}{N-1}\sum_{n=1}^{N}(X_n - \bar{X})^2, \tag{3.210}$$

which is a function of the sample mean (and not the actual mean μ_X). It can be shown that

$$Y \triangleq (N-1)S^2/\sigma^2 \tag{3.211}$$

is a chi-square random variable with $N-1$ degrees of freedom (see Problem 4.35).

Example 3.48 (Confidence intervals). Let Y be a random variable that is used to estimate a parameter of some distribution. A confidence interval with *confidence level* $1-\alpha_c$ is defined to be the following probability:

$$P(y_l \le Y \le y_u) = 1-\alpha_c, \tag{3.212}$$

where $\{y_l, y_u\}$ are the lower and upper confidence limits. The distribution of Y is required, from which it is possible to determine the confidence limits for a particular α_c. Consider the sample variance in (3.210) which has distribution χ^2_{N-1}. The confidence interval is

$$P(y_l \le Y \le y_u) = P(y_l/(N-1)S^2 \le \sigma^2 \le y_u/(N-1)S^2) = 1-\alpha_c. \tag{3.213}$$

For a specific value of α_c, the confidence limits $\{y_l, y_u\}$ are determined from the icdf of the chi-square distribution. By computing S^2 from the samples $\{X_n\}$, we can say with probability $1-\alpha_c$ (i.e., "with $100(1-\alpha_c)\%$ confidence") that σ lies in the following range:

$$y_l/(N-1)S^2 \le \sigma^2 \le y_u/(N-1)S^2 \Longrightarrow \sqrt{y_l/(N-1)S^2} \le \sigma \le \sqrt{y_u/(N-1)S^2}. \tag{3.214}$$

For example, suppose the following $N = 10$ samples have been collected from the standard Gaussian random variable: 0.8369, −0.7223, −0.7215, −0.2012, −0.0205, 0.2789, 1.0583, 0.6217, −1.7506, and 0.6973. The sample mean is $\bar{x} = 0.0077$ and the sample variance is $s^2 = 0.7687$. For a 95% confidence level, the following values are obtained from the chi-square icdf with $N-1 = 9$ degrees of freedom: $y_l = F_Y^{-1}(\alpha_c/2) = F_Y^{-1}(0.025) = 2.7004$ and $y_u = F_Y^{-1}(1-\alpha_c/2) = F_Y^{-1}(0.975) = 19.0228$. Substituting these into (3.214) gives the following 95% confidence intervals:

$$0.3903 \le \sigma^2 \le 2.7497 \Longrightarrow 0.6248 \le \sigma \le 1.6582. \tag{3.215}$$

Recall that the actual parameter value is $\sigma = 1$ for the standard Gaussian pdf.

Example 3.49 (Hypothesis testing). In hypothesis testing, we are interested in choosing between two or more hypotheses about the value of a parameter from samples of a random variable. For the unknown variance σ^2 of a distribution, the hypotheses $\{H_n\}$ might be

$$H_0 : \sigma^2 = \sigma_o^2, \quad H_1 : \sigma^2 > \sigma_o^2, \tag{3.216}$$

where σ_o^2 is some specific value. Assume again that we are interested in the unknown variance of the standard Gaussian random variable and use the sample variance S^2 as an estimate. For the chi-square random variable in (3.211) and *significance level* α_s, the null hypothesis H_0 would be rejected if

$$(N-1)S^2/\sigma_o^2 > F_Y^{-1}(1-\alpha_s), \tag{3.217}$$

where $F_Y^{-1}(\cdot)$ is the icdf of the chi-square random variable with $N-1$ degrees of freedom. Let $\sigma_o^2 = 1.1$ and $\alpha_s = 0.05$. Using the samples from Example 3.48, which yielded $s^2 = 0.7687$, we find that

$$(N-1)s^2/\sigma_o^2 = 9(0.7687)/1.1 = 6.2894. \tag{3.218}$$

Since this does not exceed $F_Y^{-1}(1-0.05) = 16.9190$ for $N-1=9$ degrees of freedom, H_0 is not rejected: we decide that the variance is $\sigma^2 = 1.1$. In fact, it would be necessary that the sample variance be $s^2 > 2.0679$ in order to reject H_0. Consider instead the hypotheses:

$$H_0 : \sigma^2 = \sigma_o^2, \quad H_1 : \sigma^2 < \sigma_o^2. \tag{3.219}$$

For this case, H_0 is rejected if

$$(N-1)S^2/\sigma_o^2 < F_Y^{-1}(\alpha_s). \tag{3.220}$$

Again, the null hypothesis is not rejected because (3.218) exceeds $F_Y^{-1}(0.05) = 3.3251$. A sample variance of $s^2 < 0.4064$ would be needed to reject H_0.

3.8.18 *F*-Distribution

Let $\{X_1, X_2\}$ be independent chi-square random variables with degrees of freedom m and n, respectively. Then

$$X = \frac{X_1/m}{X_2/n} \tag{3.221}$$

has the F-distribution with pdf

$$f_X(x) = \frac{\Gamma((m+n)/2)(m/n)^{m/2}x^{(m-2)/2}}{\Gamma(m/2)\Gamma(n/2)(1+(m/n)x)^{(m+n)/2}} I_{[0,\infty)}(x), \tag{3.222}$$

which is not symmetric with respect to m and n. An interesting property is that $1/X$ also has an F distribution, but with the degrees of freedom interchanged, which follows from the ratio above. Examples of the pdf are provided in Figure 3.38. The cdf is

$$F_X(x) = B_{mx/(mx+n)}(m/2, n/2) I_{[0,\infty)}(x), \tag{3.223}$$

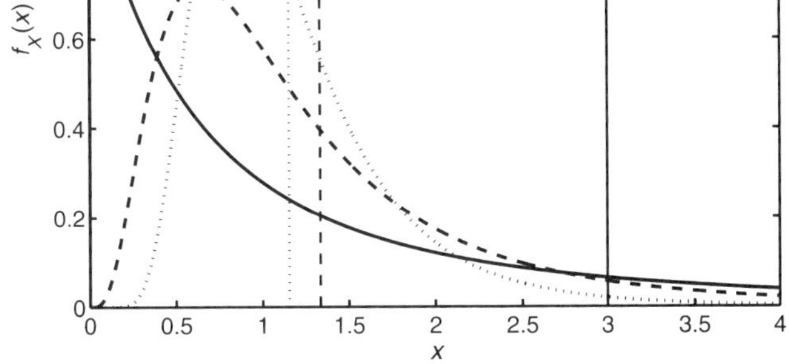

FIGURE 3.38 Examples of the F-distribution pdf. Support is $\mathcal{R}^+$. Vertical lines specify μ_X.

where $B_y(a, b)$ is the incomplete beta function (see Appendix B). The mean is $\mu_X = n/(n-2)$ for $n > 2$ and the variance is $\sigma_X^2 = 2n^2(m+n-2)/[m(n-2)^2(n-4)]$ for $n > 4$.

Example 3.50. Similar to the chi-square random variable in Example 3.47, the mode of the F distribution is $m_o = 0$ for $m = \{1, 2\}$, where m is the degrees of freedom of the chi-square random variable in the numerator of the ratio in (3.221). For $m \geq 2$, it can be shown that (see Problem 3.16)

$$m_o = (m/n)(m-2)/(n+2). \tag{3.224}$$

The F distribution arises in the following situation. Suppose m samples are collected from a Gaussian distribution with parameters $\{\mu_X, \sigma\}$, yielding $\{X_1, \ldots, X_m\}$. Similarly, let $\{Z_1, \ldots, Z_n\}$ be samples from another Gaussian distribution with the *same* variance σ^2 but with different mean μ_Z. Let $\bar{X}$ and $\bar{Z}$ be sample means and assume that all the samples are mutually independent. Then the random variable

$$Y \triangleq \left(\sum_{l=1}^{m}(X_l - \bar{X})^2/(m-1)\right)\left(\sum_{k=1}^{n}(Z_l - \bar{Z})^2/(n-1)\right)^{-1} = S_X^2/S_Z^2 \tag{3.225}$$

has an F distribution with $\{m-1, n-1\}$ degrees of freedom. Since each term in the ratio is a sample variance, Y is a measure of *similarity* of the two variances for two different sets of samples.

Applications of the F-distribution include the following:

- Hypothesis testing to determine from samples of two Gaussian random variables if they have the same variance.
- Evaluating the accuracy of a least-squares (LS) model of samples of the affine form $Y = aX + b$, where $\{a, b\}$ are estimated parameters (LS estimation is covered in Chapter 9).

Example 3.51. Since it is assumed that the variances of the underlying distributions in (3.225) are the same, random variable Y based on samples $\{X_n\}$ and $\{Z_n\}$ can be used to test the null hypothesis $H_0 : \sigma_X^2 = \sigma_Z^2$. If the other hypothesis is $H_1 : \sigma_X^2 > \sigma_Z^2$, then H_0 is rejected if

$$Y = S_X^2/S_Z^2 > F_Y^{-1}(1-\alpha_s), \tag{3.226}$$

where $F_Y^{-1}(\cdot)$ is the icdf of the F-distribution with parameters $\{m-1, n-1\}$, and α_s is the level of significance. Let $\sigma_X^2 = 2$, $\sigma_Z^2 = 1$, and $m = n = 10$. Assume we have the following samples for X: -0.3757, -2.2156, -0.5581, -0.2034, -3.3011, -1.9199, -2.5678, 1.5675, -0.2010, 1.5737, and for Z: 0.5593, 0.4784, -0.6794, 0.2850, -1.3329, -0.7240, -0.6636, 0.1984, -1.7949, -1.3877. The sample means are $\bar{x} = -0.8201$ and $\bar{z} = -0.5061$, and the ratio in (3.225) is $y = s_X^2/s_Z^2 = 2.7586/0.7167 = 3.8490$. Since Y exceeds $F_Y^{-1}(0.95) = 3.1789$, the null hypothesis is rejected. If the other hypothesis is instead $H_1 : \sigma_X^2 < \sigma_Z^2$, then H_0 is rejected if

$$1/Y = S_Z^2/S_X^2 > F_{1/Y}^{-1}(1-\alpha_s), \tag{3.227}$$

where the icdf $F_{1/Y}^{-1}(\cdot)$ is obtained from $F_Y^{-1}(\cdot)$ by interchanging $m-1$ and $n-1$. For the numerical values above, $1/y = 0.2598$ which does not exceed $F_{1/Y}^{-1}(0.95) = 3.1789$, so that H_0 would not be rejected. Note that $F_{1/Y}^{-1}(\cdot) = F_Y^{-1}(\cdot)$ in this example because $m = n$.

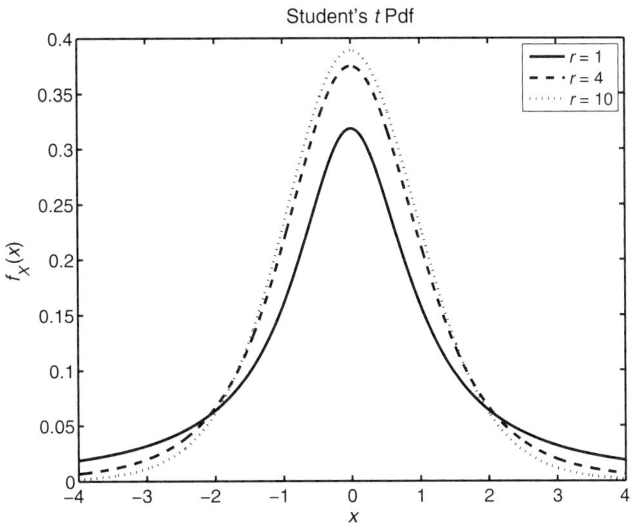

FIGURE 3.39 Examples of Student's t pdf. Support is $\mathcal{R}$. $\mu_X = 0$ for $r > 1$ and does not exist for $r = 1$.

3.8.19 Student's *t* Distribution

Let X_1 be the standard Gaussian random variable and X_2 a chi-square random variable with $N = r$ degrees of freedom. For independent X_1 and X_2, the following ratio:

$$X = X_1/\sqrt{X_2/r} \tag{3.228}$$

is Student's t random variable with pdf

$$f_X(x) = \frac{1}{\sqrt{r\pi}} \frac{\Gamma((r+1)/2)}{\Gamma(r/2)} \frac{1}{(1+x^2/r)^{(r+1)/2}} I_{(-\infty,\infty)}(x), \tag{3.229}$$

and $r \in \mathcal{N}$ is the degrees of freedom. There are two special cases: (i) when $r = 1$, X reduces to the Cauchy random variable with parameters $\{c = 0, \alpha = 1\}$, and (ii) when $r \to \infty$, the standard Gaussian random variable is obtained (see Problem 4.2). Example pdfs are shown in Figure 3.39. Its cdf is

$$F_X(x) = B_{(x+\sqrt{x^2+r})/2\sqrt{x^2+r}}(r/2, r/2) I_{(-\infty,\infty)}(x), \tag{3.230}$$

where $B_y(a, b)$ is the incomplete beta function mentioned earlier. The mean is $\mu_X = 0$ for $r > 1$ and the variance is $\sigma_X^2 = r/(r-2)$ for $r > 2$.

Student's t distribution arises as follows. Let $\{X_1, \ldots, X_r\}$ be samples drawn from a Gaussian distribution with parameters $\{\mu, \sigma\}$. The sample mean $\bar{X}$ is also Gaussian, and the transformation $Y_1 = \sqrt{r}(\bar{X} - \mu)/\sigma$ yields the standard Gaussian random variable. From a previous section, we know that $Y_2 = (r-1)S^2/\sigma^2$ has a chi-square distribution with degrees of freedom $r - 1$. It turns out that Y_1 and Y_2 are independent, and the following ratio gives Student's t distribution with parameter $r - 1$:

$$Y = \frac{Y_1}{\sqrt{Y_2/(r-1)}} = \frac{\bar{X} - \mu}{\sqrt{S^2/r}}. \tag{3.231}$$

Applications include the following:

- Finding confidence intervals for the mean from samples of a Gaussian random variable.
- An alternative to the Gaussian distribution when heavier tails provide more accuracy (see Table 3.4).

Example 3.52. Similar to the approach used in Example 3.48 to derive confidence intervals for the variance based on the chi-square random variable, we can rewrite (3.231) as follows:

$$P(y_l\sqrt{S^2/r} \leq \bar{X} - \mu \leq y_u\sqrt{S^2/r}) = P(\bar{X} - y_u\sqrt{S^2/r} \leq \mu \leq \bar{X} - y_l\sqrt{S^2/r}), \tag{3.232}$$

where $\{y_l, y_u\}$ are the lower and upper confidence limits, and the minus sign on μ caused them to be interchanged in the last expression. Setting this result equal to $1 - \alpha_c$ and substituting $\bar{X}$ and S^2 yields the $100(1 - \alpha_c)\%$ confidence interval, with $y_l = F_Y^{-1}(\alpha_c/2)$ and $y_u = F_Y^{-1}(1 - \alpha_c/2)$, where $F_Y^{-1}(\cdot)$ is the icdf of Student's t distribution with $r - 1$ degrees of freedom. Since $F_Y(y)$ is symmetric about the origin, $y_l = -y_u$ can be substituted (which is not the case for the confidence limits in Example 3.48 because the chi-square distribution used there is not symmetric). Suppose we have the following $N = 10$ samples from the standard Gaussian distribution: $-0.8674, -0.4742, 0.2224, 1.8713, 0.1100, -0.4113, 0.5112, -1.1991, -0.0964$, and 0.4458, which give sample mean $\bar{x} = 0.0112$ and sample variance $s^2 = 0.7531$. For $\alpha_c = 0.05$, we have $y_l = F_Y^{-1}(0.975) = -2.2622$ and $y_u = F_Y^{-1}(0.025) = 2.2622$, which give the following 95% confidence interval:

$$-0.6021 \leq \mu \leq 0.6246. \tag{3.233}$$

Recall that the actual mean is $\mu = 0$ because X is standard Gaussian.

3.8.20 Extreme Value Distribution (Type I: Gumbel)

Of the three types of extreme-value distributions, we describe only type I, which is also known as the Gumbel distribution. It has the following pdf:

$$f_X(x) = (1/\alpha)\exp(\pm(x - c)/\alpha)\exp(-\exp(\pm(x - c)/\alpha))I_{(-\infty,\infty)}(x), \tag{3.234}$$

where $c \in \mathcal{R}$ is a location parameter and $\alpha > 0$ is a scale parameter. There are two versions of the pdf depending on whether the sign in each exponent is + (the minimum case) or each is − (the maximum case). We consider only the maximum case which has cdf

$$F_X(x) = \exp(-\exp(-(x - c)/\alpha))I_{(-\infty,\infty)}(x). \tag{3.235}$$

Example pdfs are shown in Figure 3.40. Like the logistic distribution, it is straightforward to derive the icdf: for $p = F_X(x)$, it is given by

$$x = F_X^{-1}(p) = c - \alpha\ln(-\ln(p)). \tag{3.236}$$

Applications of the Gumbel distribution include the following:

- Distribution of the maximum of N samples of an exponential-type random variable for large N, called the Nth-order statistic (see Chapter 4).
- Model for extreme events, such as maximum earthquake motions, breaking strength of metals, and high winds.

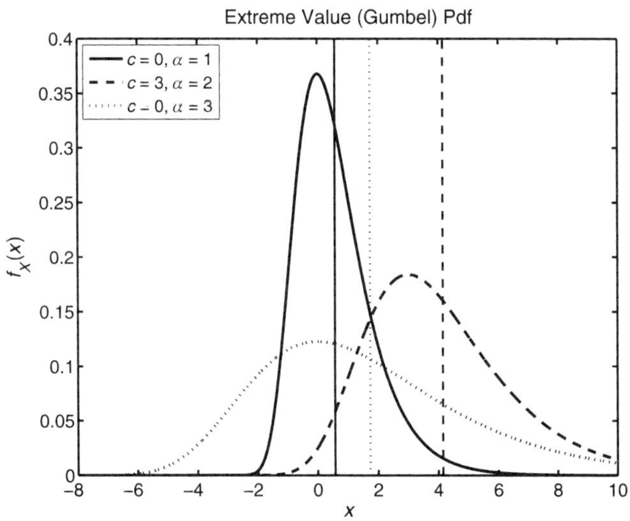

FIGURE 3.40 Examples of the extreme value (Gumbel) pdf. Support is $\mathcal{R}$. Vertical lines specify μ_X.

Example 3.53. From the cdf, we find that the median is

$$\exp(-\exp(-(m_e - c)/\alpha)) = 1/2 \Longrightarrow -\exp(-(m_e - c)/\alpha) = \ln(1/2),$$
$$\Longrightarrow -(m_e - c)/\alpha = \ln(\ln(2)) \Longrightarrow m_e = c - \alpha\ln(\ln(2)). \tag{3.237}$$

Thus, $m_e \approx c + (0.3665)\alpha$ which always exceeds the mode $m_o = c$. It can be shown that the mean is $\mu_X = c + \alpha\gamma$, where $\gamma \approx 0.5772$ is the Euler–Mascheroni constant, and thus $\mu_X \approx c + (0.5772)\alpha$ which always exceeds the median. These results imply that the pdf leans to the left and has positive skewness: $\gamma_1 = (12\sqrt{6}/\pi^3)\zeta(3) \approx 1.1396$, where $\zeta(v)$ is the zeta function, and $\zeta(3) \approx 1.2021$ is known as Apéry's constant. The variance of the Gumbel distribution is $\sigma_X^2 = (\pi\alpha)^2/6$.

Example 3.54. The Gumbel distribution is obtained from the transformation $X = c - \alpha\ln(Y)$, where Y is an exponential random variable with parameter $\lambda = 1$:

$$P(X \leq x) = P([c - x]/\alpha \leq \ln(Y)),$$
$$= 1 - P(Y \leq \exp([c - x]/\alpha)). \tag{3.238}$$

Substituting the cdf of the exponential random variable gives the desired result:

$$F_X(x) = 1 - [1 - \exp(-\exp(-[x - c]/\alpha))], \tag{3.239}$$

which reduces to (3.235).

3.9 DISCRETE RANDOM VARIABLES

As discussed previously, the outcomes of a discrete random variable can be expressed using either the Kronecker delta function or the Dirac delta function, resulting in a pmf or pdf, respectively. Let the outcomes for discrete random variable X be $\mathcal{Z}^+$ with nonzero probabilities $P(X = x)$. The pmf is written as

$$p_X[x] = \sum_{n=0}^{\infty} p_X[n]\delta[x - n], \tag{3.240}$$

and the pdf is

$$f_X(x) = \sum_{n=0}^{\infty} p_X[n]\delta(x-n). \tag{3.241}$$

Although the sums above are infinite, only the term for $n = x$ is nonzero because of the delta functions. The indicator function is not needed to specify the support for X (as was used for continuous random variables) because it is evident from the sums above. This is also the case for the cdf where the unit-step function $u(x)$ is used, and which is obtained by integrating the pdf:

$$F_X(x) = \int_{-\infty}^{x} f_X(v)\mathrm{d}v = \sum_{n=0}^{\infty} p_X[n]u(x-n). \tag{3.242}$$

In order to have a consistent representation for all random variables (including those with mixed distributions), the pdf representation is generally used in this section. This allows us to use integration later when computing various quantities, such as the mean described in Chapter 5:

$$\mu_X = \int_{-\infty}^{\infty} x f_X(x)\mathrm{d}x = \sum_{n=0}^{\infty} p_X[n] \int_{-\infty}^{\infty} x\delta(x-n)\mathrm{d}x = \sum_{n=0}^{\infty} n p_X[n], \tag{3.243}$$

where the sifting property of the Dirac delta function has been used to obtain the last expression. The summaries in Appendix A include expressions for the pdf and the pmf. The following notation for the various parameters is used in this section:

- p for probability of "success."
- $q \triangleq 1 - p$ for probability of "failure."
- N for number of trials/outcomes (binomial, hypergeometric, uniform) or sum of random variables (negative binomial).
- M for number of success objects (hypergeometric).
- α for general scale parameter (Poisson and zeta).

3.9.1 Bernoulli Random Variable

The Bernoull random variable has two outcomes labeled $\{0, 1\}$ with the following pdf:

$$f_X(x) = q\delta(x) + p\delta(x-1), \tag{3.244}$$

where $p = P(X = 1) \in [0, 1]$ is the probability of a success (such as heads in a coin toss) and $q \triangleq 1 - p$ is the probability of a failure. The *symmetric* Bernoulli random variable has outcomes $\{-1, 1\}$ (see Appendix A). The cdf readily follows, by replacing the Dirac delta functions with unit-step functions:

$$F_X(x) = qu(x) + pu(x-1). \tag{3.245}$$

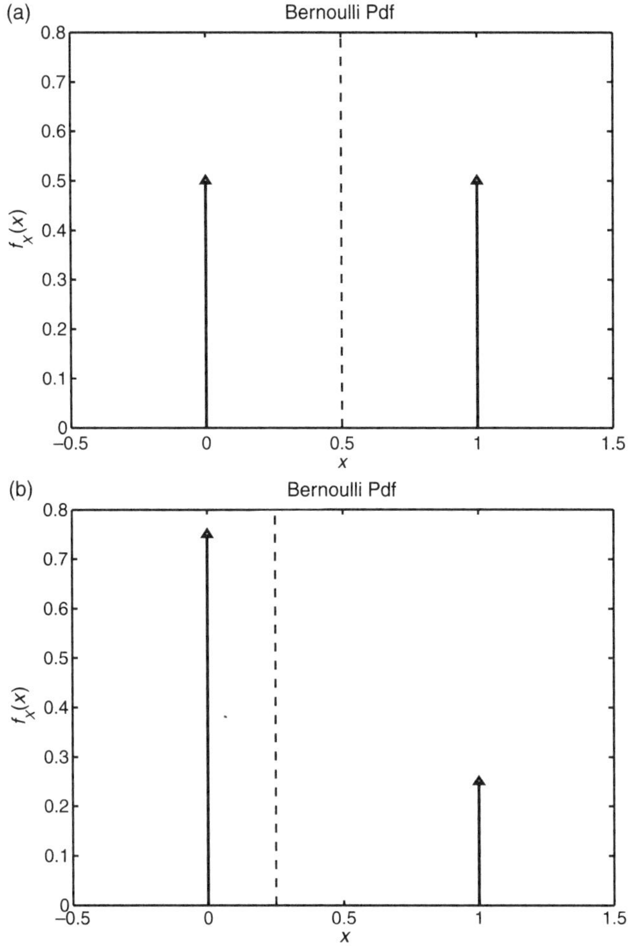

FIGURE 3.41 Examples of the Bernoulli pdf. (a) $p = 0.5$. (b) $p = 0.25$. Support is $\{0, 1\}$. Dashed lines specify μ_X.

Examples of the pdf are shown in Figure 3.41. The mean is $\mu_X = p$ and the variance is $\sigma_X^2 = pq$. Applications include the following:

- Many experiments involving two outcomes such as $\{H, T\}$ in a single coin toss, or complementary events such as even or odd outcomes when tossing a single die.
- A binary digit (bit) in a digital communication system.

Example 3.55. Consider the abstract probability space $\{\Omega, \mathcal{F}, P\}$, where Ω may be uncountable. Let the *indicator random variable* be a mapping of some event $E \in \mathcal{F}$ to $\mathcal{R}$ as follows:

$$X(\omega) = I_E(\omega). \tag{3.246}$$

Since X has only two outcomes, it is a Bernoulli random variable with

$$p = P(X = 1) = P(E), \quad q = P(X = 0) = P(E^c). \tag{3.247}$$

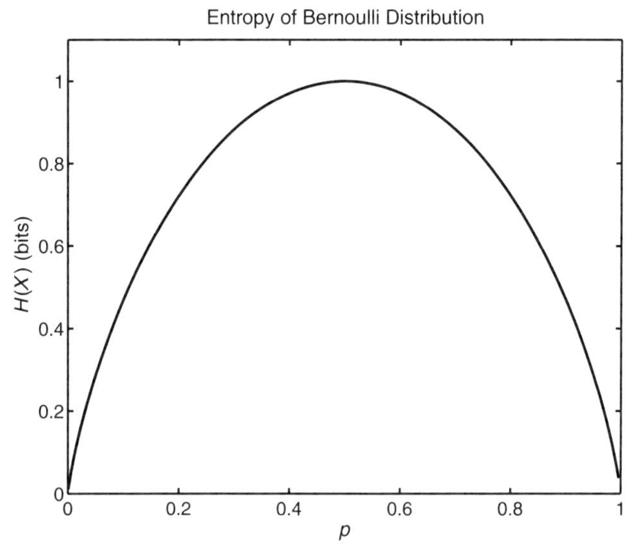

FIGURE 3.42 Entropy $H(x)$ (bits) of the Bernoulli random variable.

Suppose Ω represents the outcomes of a single toss of a die. If, for example, E is the event that the number on the die exceeds four, then X is Bernoulli with $p = 1/3$ and $q = 2/3$.

Example 3.56 (Entropy). The *entropy* of a discrete random variable is a measure of the amount of *information* or uncertainty associated with the distribution of outcomes. It is defined as follows:

$$H(X) = -\sum_{n=0}^{N} p_X[n] \log (p_X[n]), \tag{3.248}$$

where the logarithm is any base, though it is usually 2 or e such that the units of $H(X)$ are bits and "nats," respectively. It is described in greater detail in Chapter 10. Here, we briefly consider the entropy of the Bernoulli random variable:

$$H(X) = p \log(p) + q \log(q), \tag{3.249}$$

which is plotted in Figure 3.42. Observe that $H(X)$ is greatest when $p = 1/2$ (greatest uncertainty) and smallest when $p = \{0, 1\}$ (no uncertainty). The information content is zero in the latter case because we know with probability one that the outcome is either zero or one, respectively.

3.9.2 Binomial Random Variable

The binomial random variable is generated by repeating a Bernoulli experiment N times (all independent) and counting the number of successes. The pdf is

$$f_X(x) = \sum_{n=0}^{N} \binom{N}{n} p^n q^{N-n} \delta(x - n), \tag{3.250}$$

where $p \in [0, 1]$ and $q \triangleq 1 - p$. The binomial coefficient

$$\binom{N}{n} \triangleq \frac{N!}{n!(N-n)!} \tag{3.251}$$

is the number of permutations possible for n successes of N Bernoulli trials. The description above for the binomial distribution corresponds to sampling a Bernoulli random variable with replacement, as discussed for finite sample spaces Ω in Chapter 2. The binomial probabilities for sample size N can be computed using the following recursive formula (see Problem 3.37):

$$P(X = x) = [(N - x + 1)p/qx]\, P(X = x - 1). \tag{3.252}$$

The cdf is

$$F_X(x) = \sum_{n=0}^{N} \binom{N}{n} p^n q^{N-n} u(x - n), \tag{3.253}$$

where for a particular value of $x < N$, the unit-step function is nonzero only for $n \le x$; terms in the sum for $n > x$ are excluded (equal to zero). We can also write the cdf without using the unit-step function as follows:

$$F_X(x) = \sum_{n=0}^{\lfloor x \rfloor} \binom{N}{n} p^n q^{N-n}, \tag{3.254}$$

where $\lfloor x \rfloor$ is the floor function that gives the largest integer not exceeding x. The sum is well defined using this notation, and the cdf is constant for noninteger values of x as required for a discrete random variable. Example pdfs of the binomial random variable are provided in Figure 3.43. The mean is $\mu_X = Np$ and the variance is $\sigma_X^2 = Npq$. Applications include the following:

- Repeated Bernoulli trials (such as tossing a single coin N times, or simultaneously tossing N coins), where we are interested in the number of successes out of N.
- Predicting the number of defective products in a large sample size N.

Example 3.57. Consider $N = 3$ tosses of a fair coin with pdf

$$f_X(x) = \sum_{n=0}^{3} \binom{3}{n} (1/2)^n (1/2)^{3-n} \delta(x - n) = (1/8) \sum_{n=0}^{3} \binom{3}{n} \delta(x - n). \tag{3.255}$$

Suppose we want to calculate $P(X = 3)$. Because (3.250) is a pdf, this probability can be calculated by integrating over a small interval centered at $x = 3$ as follows:

$$P(X = 3) = \int_{3-\epsilon}^{3+\epsilon} f_X(x) \mathrm{d}x = (1/8)\binom{3}{3} = 1/8, \tag{3.256}$$

where small $\epsilon \in (0, 1)$ allows us to "capture" the delta function at $x = 3$. Consider another example:

$$P(X > 1) = \int_{1+\epsilon}^{\infty} f_X(x) \mathrm{d}x = (1/8)\left[\binom{3}{2} + \binom{3}{3}\right] = (1/8)(3 + 1) = 1/2, \tag{3.257}$$

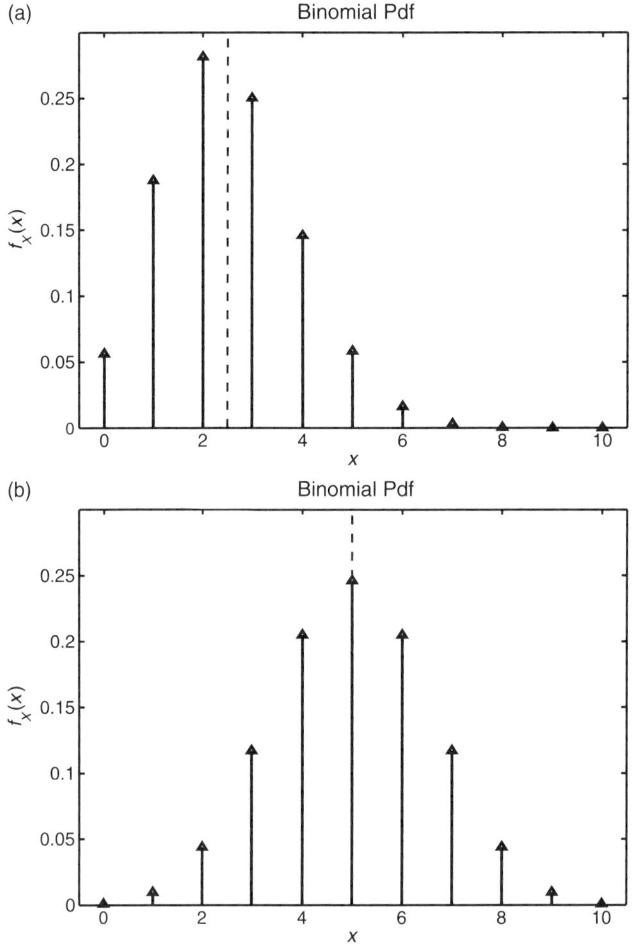

FIGURE 3.43 Examples of the binomial pdf. (a) $N = 10$ and $p = 0.25$. (b) $N = 10$ and $p = 0.5$. Support is $\{0, \ldots, N\}$. Dashed lines specify μ_X.

where the lower limit has been chosen to exclude outcomes $\{0, 1\}$. This result could also have been computed using the cdf as follows:

$$P(X > 1) = 1 - P(X \leq 1) = 1 - (1/8)\sum_{n=0}^{3}\binom{3}{n}u(1-n). \tag{3.258}$$

For the four terms in the sum, observe that $u(0) = u(1) = 1$, whereas $u(-2) = u(-1) = 0$ such that

$$P(X > 1) = 1 - (1/8)\left[\binom{3}{0} + \binom{3}{1}\right] = 1/2. \tag{3.259}$$

This example illustrates how to compute probabilities for discrete random variables using the pdf and cdf, based on the Dirac delta function and the unit-step function, respectively. However, the reader may find it conceptually easier to work with the pmf and simply add the various probabilities. Of course, summing over the pmf is equivalent to integrating over the pdf because the Dirac delta functions are located at the same outcomes of X as the Kronecker delta functions of the pmf.

3.9.3 Geometric Random Variable (with Support $\mathcal{Z}^+$ or $\mathcal{N}$)

The geometric random variable is often defined as the number of Bernoulli trials needed to achieve the first success *after x failures*. It is sometimes called the "waiting-time" random variable where time is defined by the number of failures (it is not actual time). Since the experiments are independent, we can write

$$P(X = x) = \underbrace{q \cdots q}_{x \text{ terms}} p = pq^x I_{\mathcal{Z}^+}[x], \tag{3.260}$$

where $p \in [0, 1]$ and $q \triangleq 1 - p$. The mean is $\mu_X = q/p$ and the variance is $\sigma_X^2 = q/p^2$. Another form of the geometric random variable is obtained by observing $x - 1$ failures before the first success, which changes the exponent of q to $x - 1$:

$$P(X = x) = pq^{x-1} I_{\mathcal{N}}[x]. \tag{3.261}$$

This distribution can also be interpreted as the probability that the first success requires x trials. It is important to note that the support for this version does not include zero (hence the subscript $\mathcal{N}$ on the indicator function), whereas zero is included in (3.260). Thus, we refer to (3.261) as the distribution for the *shifted* geometric random variable (it is also summarized in Appendix A). The first form of the geometric random variable in (3.261) has the following pdf and cdf:

$$f_X(x) = \sum_{n=0}^{\infty} pq^n \delta(x - n), \quad F_X(x) = \sum_{n=0}^{\infty} pq^n u(x - n). \tag{3.262}$$

Examples of the geometric pdf are given in Figure 3.44. Applications include the following:

- Repeated Bernoulli trials (such as repeatedly tossing a single coin) where we are interested in the first success after x failures.
- Markov chains (see Chapter 6) and queueing theory.

Example 3.58. Consider a sequence of bits with outcomes $\{0, 1\}$ that are equally likely. Then

$$P(X > x) = 1 - F_X(x) = 1 - p \sum_{n=0}^{\infty} q^n u(x - n) = 1 - p \sum_{n=0}^{x} q^n \tag{3.263}$$

because $u(x - n) = 1$ for $n \in \{0, \ldots, x\}$ and is zero for $n > x$. As mentioned in Example 3.57, this result is also obtained by adding up the corresponding terms in the pmf. Using the closed-form result in Appendix E for the series above gives

$$P(X > x) = 1 - p(1 - q^{x+1})/(1 - q) = q^{x+1}. \tag{3.264}$$

Unlike the other discrete random variables, there is a simple closed form for the cdf:

$$F_X(x) = 1 - P(X > x) = (1 - q^{x+1}) I_{\mathcal{Z}^+}[x], \tag{3.265}$$

where it is understood that the cdf is flat between integer values of x.

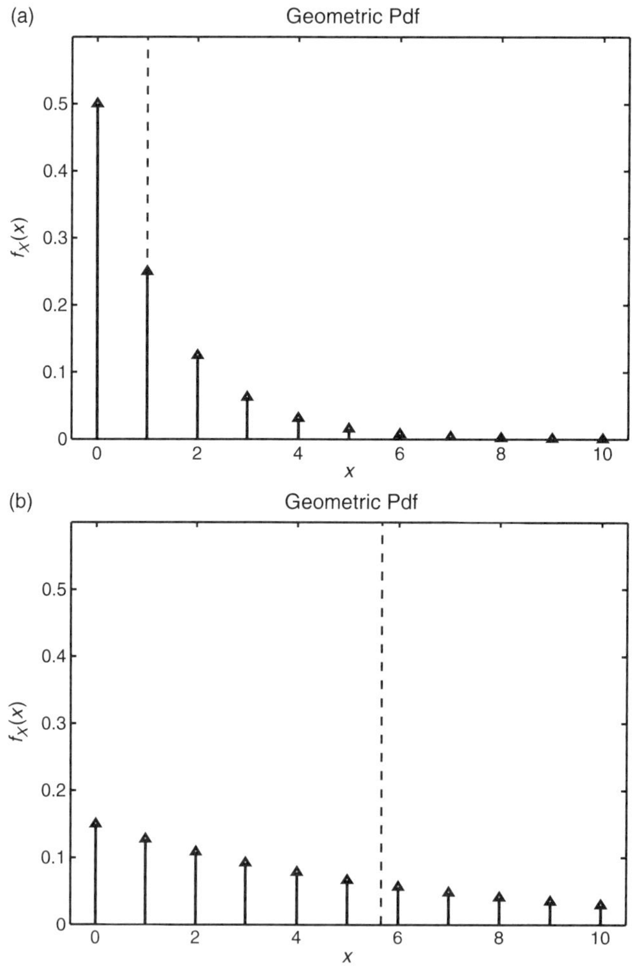

FIGURE 3.44 Examples of the geometric pdf. (a) $p = 0.5$. (b) $p = 0.15$. Support is $\mathcal{Z}^+$. Dashed lines specify μ_X.

Theorem 3.4. The geometric random variable is the only discrete random variable with the *memoryless* property.

Proof. From conditional probability:

$$P(X > x + b|X > b) = P(X > x + b)/P(X > b) = [1 - F_X(x + b)]/[1 - F_X(b)]. \tag{3.266}$$

Substituting the cdf for the geometric random variable gives

$$P(X > x + b|X > b) = q^{x+b+1}/q^{b+1} = q^x \neq 1 - F_X(x), \tag{3.267}$$

which shows that the form of the pdf in (3.260) does *not* have the memoryless property in (3.87). Consider instead the *shifted* geometric form in (3.261) which has the following cdf:

$$F_X(x) = (1 - q^x)I_{\mathcal{N}}[x]. \tag{3.268}$$

Substituting this expression into the conditional probability above gives

$$P(X > x + b|X > b) = q^{x+b}/q^{b} = q^{x} = 1 - F_X(x) \tag{3.269}$$

demonstrating that the shifted geometric pdf with support $\mathcal{N}$ has the memoryless property (the zero outcome must be excluded). It is clear that the only function satisfying

$$P(X > x + b)/P(X > b) = P(X > x) \tag{3.270}$$

for integer x is of the form a^x such that $a^{x+b}/a^b = a^x$. Thus

$$P(X > x) = a^x \Longrightarrow P(X \leq x) = 1 - a^x = F_X(x) \tag{3.271}$$

from which we see there is zero probability mass at $x = 0$, and $a \in [0, 1]$ in order for $1 - a^x$ to be a valid distribution. Consider the following closed-form expression for a finite sum (see Appendix E):

$$\sum_{n=0}^{x-1} a^n = \frac{1 - a^x}{1 - a} = \sum_{m=1}^{x} a^{m-1}, \tag{3.272}$$

where $m = n + 1$ has been substituted so that the sum starts at 1 (to match the smallest value of the support $\mathcal{N}$). Rearranging this expression gives

$$(1 - a)\sum_{m=1}^{x} a^{m-1} = 1 - a^x, \tag{3.273}$$

which is the cdf in (3.271). The corresponding pmf for $x \in \mathcal{N}$ is

$$p_X[x] = (1 - a)a^{x-1} I_{\mathcal{N}}[x]. \tag{3.274}$$

Substituting $a = q$ and $1 - a = p$ completes the proof. □

Example 3.59. The median of the geometric random variable is derived using the cdf as follows:

$$F_X(m_e) \geq 1/2 \Longrightarrow (1 - q^{m_e+1}) \geq 1/2. \tag{3.275}$$

Rearranging the expression and taking the logarithm of both sides gives

$$(m_e + 1)\log_2(q) \leq \log_2(1/2) = -1 \Longrightarrow m_e \geq -1/\log_2(q) - 1, \tag{3.276}$$

where dividing by $\log_2(q)$ has reversed the direction of the inequality. Choosing m_e to be the smallest integer that satisfies the inequality yields the ceiling function (see Appendix B):

$$m_e = \lceil -1/\log_2(q) \rceil - 1. \tag{3.277}$$

Table 3.6 summarizes how the binomial and geometric random variables are related to the Bernoulli random variable. It also includes the negative binomial random variable, of which the geometric distribution is a special case when $N = 1$.

TABLE 3.6 Bernoulli Trials and Related Random Variables

Random Variable X	Distribution
x successes in N trials	Binomial $\{N, p\}$
x failures before first success	Geometric $\{p\}$ (support $\mathcal{Z}^+$)
x trials before first success	Shifted geometric $\{p\}$ (support $\mathcal{N}$)
x failures before Nth success	Negative binomial $\{N, p\}$

3.9.4 Negative Binomial Random Variable (Pascal)

The negative binomial random variable is generated by summing together N geometric random variables, all with the same parameter $p \in [0, 1]$: it is the probability of x failures before N successes. One possible sequence of Bernoulli outcomes is that all N successes occur immediately after the first failure; this has probability $p^N q$, where $q \triangleq 1 - p$. The number of permutations of $x = 1$ failures in the first $N + 1$ trials is $\binom{N}{1} = N$:

$$\underbrace{\overbrace{\{0, 1, \ldots, 1\}}^{N+1 \text{ terms}}, \{1, 0, 1, \ldots, 1\}, \ldots, \{1, \ldots, 1, 0, 1, 1\}, \{1, \ldots, 1, 0, 1\}}_{N \text{ permutations}}. \tag{3.278}$$

Generalizing this result to x failures and considering all permutations gives

$$P(X = x) = \binom{x + N - 1}{x} p^N q^x I_{\mathcal{Z}^+}[x]. \tag{3.279}$$

Note that there is only one permutation for $x = 0$ failures. There are always N successes by definition; the binomial coefficient provides the number of permutations of x failures that can occur among the successes (of course, the last outcome must be a success for any N). Thus, the pdf of the negative binomial random variable is

$$f_X(x) = \sum_{n=0}^{\infty} \binom{n + N - 1}{n} p^N q^n \delta(x - n), \tag{3.280}$$

and the cdf is

$$F_X(x) = \sum_{n=0}^{\infty} \binom{n + N - 1}{n} p^N q^n u(x - n). \tag{3.281}$$

Examples of the pdf are shown in Figure 3.45. The mean is $\mu_X = Nq/p$ and the variance is $\sigma_X^2 = Nq/p^2$. Obviously the negative binomial random variable reduces to the geometric random variable when $N = 1$. It is possible to generalize the negative binomial pdf such that $N \in \mathcal{R}^+$, in which case it is known as the Pólya distribution. For integer-valued N as derived above, the negative binomial distribution is also known as the Pascal distribution. (The reader should be aware that there are different versions of the negative binomial pdf. For example, one version is identical to the pdf given here, but with p and q interchanged.) Applications include the following:

- Alternative model for the Poisson distribution. (See Example 3.62 for a derivation of the negative binomial distribution starting with the Poisson random variable.)
- Model for accident proneness.

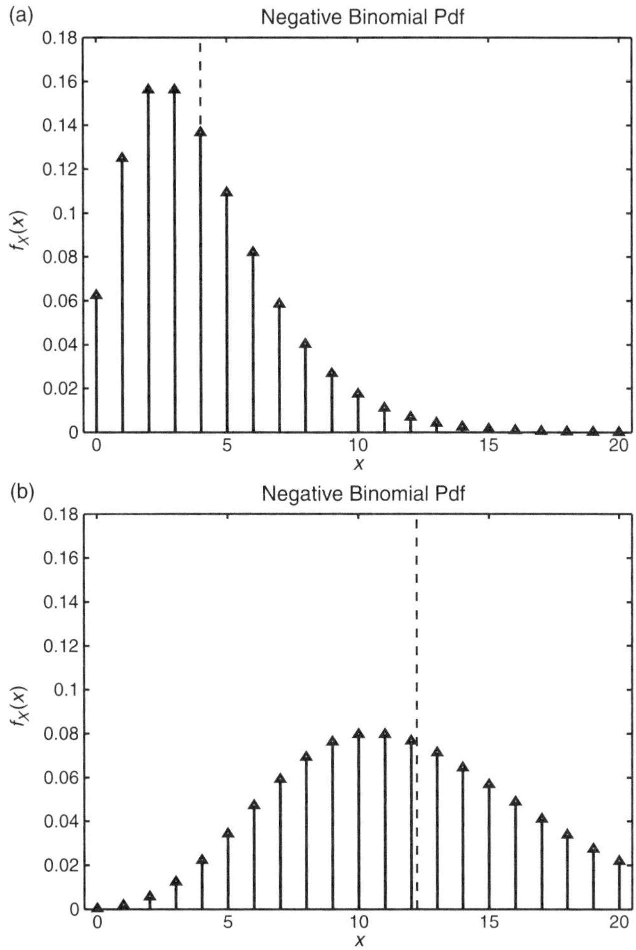

FIGURE 3.45 Examples of the negative binomial pdf. (a) $N = 4$ and $p = 0.5$. (b) $N = 10$ and $p = 0.45$. Support is $\mathcal{Z}^+$. Dashed lines specify μ_X.

Example 3.60. Let $N = 2$ and assume $p = q = 1/2$. Then for small $\epsilon \in (0, 1)$,

$$\begin{aligned} P(2 \leq X \leq 4) &= \int_{2-\epsilon}^{4+\epsilon} f_X(x)\mathrm{d}x = p^2\left[\binom{3}{2}q^2 + \binom{4}{3}q^3 + \binom{5}{4}q^4\right], \\ &= p^2[2q^2 + 4q^3 + 5q^4] = (1/2)[1/2 + 3/8 + 5/16]. \end{aligned} \tag{3.282}$$

If we were to toss a fair coin, then with probability $19/32 \approx 0.5938$ there will be $x = 2$, 3, or 4 tails before there are $N = 2$ heads.

Example 3.61 (Krishnamoorthy, 2006). Let X have a negative binomial distribution with parameters $\{N, p\}$. Then it can be shown for integer k that

$$P(X \leq k) = P(Y \geq N), \tag{3.283}$$

where Y is binomial with parameters $\{N+k, p\}$. In order to rewrite the probability on the left-hand side, we recognize that event $E = \{\leq k \text{ failures before the } N\text{th success}\}$ is equivalent to event $F = \{\leq k+N \text{ experiments with } N \text{ successes}\}$. For the binomial random variable, the probability of n successes in $k+N$ experiments is

$$P(Y=n) = \binom{k+N}{n} p^n q^{k+N-n}. \tag{3.284}$$

Consider event $G = \{\geq N \text{ successes in } k+N \text{ experiments}\}$. The key step is to recognize that although F and G are different events, they have the same probability. Thus

$$P(G) = \sum_{n=N}^{k+N} \binom{k+N}{n} p^n q^{k+N-n}, \tag{3.285}$$

which is the same as $P(Y \geq N)$.

3.9.5 Poisson Random Variable

The Poisson random variable arises in many applications where counting is involved. It has the following pdf:

$$f_X(x) = \sum_{n=0}^{\infty} \exp(-\alpha)\frac{\alpha^n}{n!}\delta(x-n), \tag{3.286}$$

where $\alpha > 0$ is both a location and shape parameter. The cdf is

$$F_X(x) = \sum_{n=0}^{\infty} \exp(-\alpha)\frac{\alpha^n}{n!}u(x-n). \tag{3.287}$$

Examples of the pdf are shown in Figure 3.46. The mean and variance are the same: $\mu_X = \sigma_X^2 = \alpha$. Applications include the following:

- Numerous counting problems such as radioactive decay over some time duration, the number of typographical errors on a book page, and the number of phone calls received at a telephone company central office for some time interval.
- An approximation to the binomial distribution (see Chapter 4).

In Chapter 6, we discuss the Poisson counting process where time t is incorporated into the Poisson pdf, such that α becomes $\alpha = \lambda t$ with rate parameter λ.

Example 3.62. The Poisson distribution can be used to derive the negative binomial distribution when α has a gamma pdf with parameter q/p. Because α itself is now random, conditioning can be used to write the Poisson pdf as follows (see Chapter 4 for a discussion of conditional pdfs):

$$f_{X|\alpha}(x|\alpha) = \frac{\alpha^x}{x!}\exp(-\alpha)I_{[0,\infty)}(x). \tag{3.288}$$

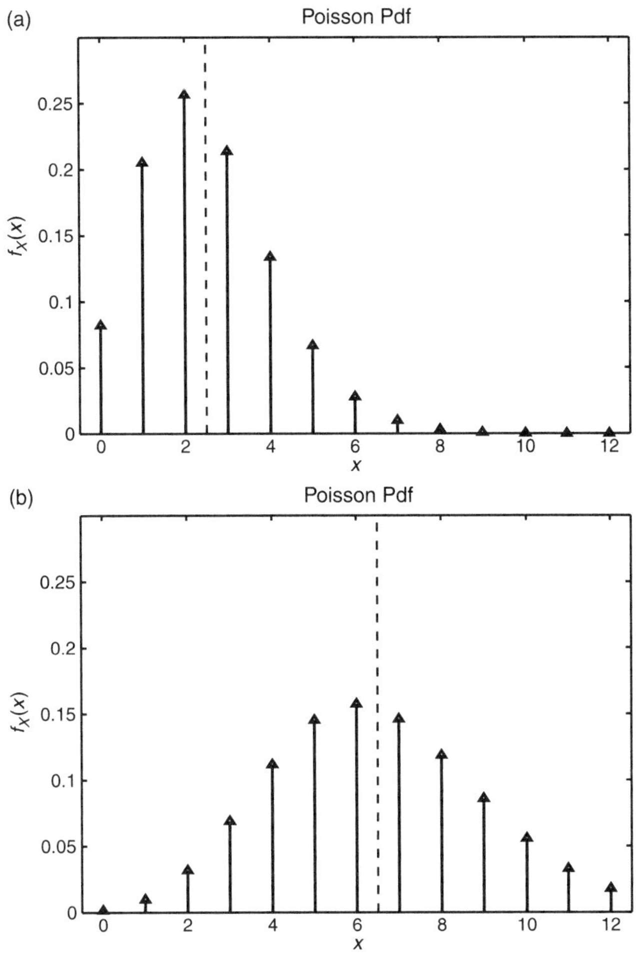

FIGURE 3.46 Examples of the Poisson pdf. (a) $\alpha = 2.5$. (b) $\alpha = 6.5$. Support is $\mathcal{Z}^+$. Dashed lines specify μ_X.

(Since uppercase α is identical to the Latin A, we use lowercase α in the subscript without confusion; it is still a random variable.) Conditioning on α fixes it at a particular value so that it is no longer random. Multiplying this expression by the pdf of α and integrating over its support gives the pdf for X as follows:

$$\int_0^\infty f_{X|\alpha}(x|\alpha) f_\alpha(\alpha) \mathrm{d}\alpha = \int_0^\infty f_{X,\alpha}(x, \alpha) \mathrm{d}\alpha = f_X(x). \tag{3.289}$$

This expression is a continuous mixture of distributions $\{f_{X|\alpha}(x|\alpha)\}$ that are parameterized by α (mixtures are also covered in Chapter 4). When $f_{X|\alpha}(x|\alpha)$ is Poisson with parameter α and $f_\alpha(\alpha)$ is a gamma pdf with parameters $\{r = m, \lambda = p/q\}$, it can be shown that $f_X(x)$ is a negative binomial pdf with parameters $\{p, N = m\}$:

$$\begin{aligned} f_X(x) &= \int_0^\infty \sum_{n=0}^\infty \frac{\alpha^n}{n!} \exp(-\alpha)\delta(x-n) \frac{1}{\Gamma(m)} \alpha^{m-1} (p/q)^m \exp(-\alpha p/q) \mathrm{d}\alpha, \\ &= \frac{(p/q)^m}{\Gamma(m)} \sum_{n=0}^\infty \frac{1}{n!} \delta(x-n) \int_0^\infty \alpha^{n+m-1} \exp(-\alpha/q) \mathrm{d}\alpha, \end{aligned} \tag{3.290}$$

where $1 + p/q = 1/q$ has been used in the exponential. If $v = \alpha/q$ and $d\alpha = q\, dv$ are substituted, the integral has the form of a gamma function:

$$\int_0^\infty \alpha^{n+m-1} \exp(-\alpha/q) d\alpha = q^{n+m} \int_0^\infty v^{n+m-1} \exp(-v) dv, \\ = q^{n+m} \Gamma(n+m). \tag{3.291}$$

Thus,

$$f_X(x) = \frac{p^m}{\Gamma(m)} \sum_{n=0}^{\infty} \frac{q^n}{n!} \Gamma(n+m) \delta(x-n) = \sum_{n=0}^{\infty} \frac{\Gamma(n+m)}{\Gamma(m) n!} p^m q^n \delta(x-n). \tag{3.292}$$

For integers $m, x \in \mathcal{Z}^+$, we have from Appendix B:

$$\frac{\Gamma(n+m)}{\Gamma(m)} = \frac{(n+m-1)!}{(m-1)!} \tag{3.293}$$

so that the coefficient in (3.292) is

$$\frac{\Gamma(n+m)}{\Gamma(m) n!} = \frac{n+m-1}{(m-1)! n!} = \binom{n+m-1}{n}, \tag{3.294}$$

which is the form used by the negative binomial pdf in (3.280).

Example 3.63. Similar to (3.283) which relates the negative binomial and binomial distributions, we can write

$$P(X \leq k) = P(Y > \alpha), \tag{3.295}$$

where X is Poisson with parameter α and Y has a gamma distribution with parameters $\{\lambda = 1, r = k+1\}$. From Appendix B, the lower incomplete gamma function can be expressed as the following series:

$$\gamma(a, b) = \exp(-b) \sum_{n=0}^{\infty} \frac{(a-1)! x^{a+b}}{(a+n)!}. \tag{3.296}$$

From the cdf of the gamma random variable Y with the parameters above, we have

$$P(Y > \alpha) = 1 - F_Y(\alpha) = 1 - \frac{\gamma(k+1, \alpha)}{\Gamma(k+1)}. \tag{3.297}$$

Substituting $\Gamma(k+1) = k!$ and (3.296) gives

$$P(Y > \alpha) = 1 - \exp(-\alpha) \frac{1}{k!} \sum_{n=0}^{\infty} \frac{k! \alpha^{k+1+n}}{(k+1+n)!}. \tag{3.298}$$

Cancelling $k!$ and changing variables in the sum to $m = k+1+n$ yields the desired result:

$$P(Y > \alpha) = 1 - \exp(-\alpha) \sum_{m=k+1}^{\infty} \frac{\alpha^m}{m!} = \exp(-\alpha) \sum_{m=0}^{k} \frac{\alpha^m}{m!} = P(X \leq k). \tag{3.299}$$

3.9.6 Hypergeometric Random Variable

The hypergeometric random variable is similar to the binomial random variable but with the following two differences: (i) there are $N \in \mathcal{N}$ objects to choose from and (ii) sampling is done *without replacement*: once an object has been chosen, it is removed before the next object is selected. Of the total number of objects, there are two types: one of size M ("success") and the other of size $N - M$ ("failure"). The objects are not necessarily denoted as success or failure; in practice we need only distinguish between two different kinds of objects, one of which is the object of interest corresponding to a success. If n items are chosen, there are $\binom{N}{n}$ combinations without regard to the type of objects. We need to keep track of the number of successes m in a sample of size n. Consider the following scenario:

$$\underbrace{1, \ldots, 1}_{M \text{ terms}} \; \underbrace{0, \ldots, 0}_{N-M \text{ terms}}, \tag{3.300}$$

where 1 = success and 0 = failure. Since there are M success objects, there are $\binom{M}{m}$ combinations; likewise, there are $\binom{N-M}{n-m}$ combinations of the failure objects. Thus, the probability of m successes is the following ratio:

$$P(X = m) = \frac{\binom{M}{m}\binom{N-M}{n-m}}{\binom{N}{n}}. \tag{3.301}$$

This result was obtained using combinatorics: we counted the number of different ways of having m successes and $n - m$ failures, and then divided the product of those by the number of ways of selecting n objects. (This approach is same as that used in Chapter 2 for finite Ω with equally likely outcomes, where the cardinality of some event E is used to calculate the ratio $P(E) = |E|/|\Omega|$.) The pdf of the hypergeometric random variable is

$$f_X(x) = \sum_{m=\max(0,n+M-N)}^{\min(M,n)} \frac{\binom{M}{m}\binom{N-M}{n-m}}{\binom{N}{n}} \delta(x - m). \tag{3.302}$$

Examples are provided in Figure 3.47. These limits are required so that the lower number of each binomial coefficient does not exceed the upper number. If it is understood that the binomial coefficient is defined to be zero in such cases, then the limits of the sum can be simplified to $\{0, N\}$. The cdf is

$$F_X(x) = \sum_{m=\max(0,n+M-N)}^{\min(M,n)} \frac{\binom{M}{m}\binom{N-M}{n-m}}{\binom{N}{n}} u(x - m). \tag{3.303}$$

Hypergeometric X has mean $\mu_X = nM/N$ and variance is $\sigma_X^2 = nM(N - M)(N - n)/N^2(N - 1)$. Applications include the following:

- Quality control of products where sampling is used to determine the probabilities of defective items.
- Estimating the size of an animal population, by "tagging" a sample (size M), reintroducing them into the population, and then sampling again (size n) to estimate N.

Example 3.64. Consider the case of $n = 1$ such that the pdf reduces to

$$\begin{aligned} f_X(x) &= \sum_{m=0}^{1} \frac{\binom{M}{m}\binom{N-M}{1-m}}{\binom{N}{1}} \delta(x - m), \\ &= [(N - M)/N]\delta(x) + (M/N)\delta(x - 1). \end{aligned} \tag{3.304}$$

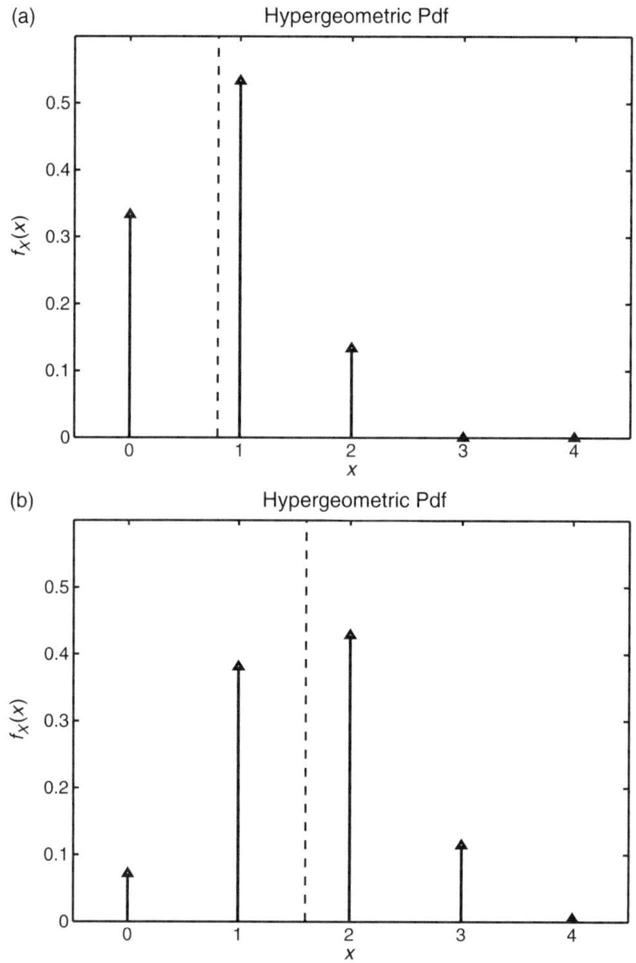

FIGURE 3.47 Examples of the hypergeometric pdf. (a) $N = 10$, $M = 4$, and $n = 2$. (b) $N = 10$, $M = n = 4$. Support is $\{\max(0, n + M - N), \ldots, \min(M, n)\}$. Dashed lines specify μ_X.

If we define $p \triangleq M/N$, then for $n = 1$ the hypergeometric distribution reduces to the Bernoulli distribution. It is demonstrated in Chapter 4 that the binomial distribution with p defined this way approximates the hypergeometric distribution for any n provided that both M and N are large.

Example 3.65. Let the number of items in a shipment be $N = 100$ and suppose $M = 15$ are defective. For a sample of $n = 5$ items, the probability that x samples are defective is given by the following expression:

$$P(X = x) = \frac{\binom{15}{x}\binom{100-15}{5-x}}{\binom{100}{5}}. \tag{3.305}$$

The probabilities for $x = 0, 1, 2, 3, 4$, and 5 are 0.4357, 0.4034, 0.1377, 0.0216, 0.0015, and 3.99×10^{-5}, respectively.

3.9.7 Discrete Uniform Random Variable

The discrete uniform random variable is the finite counterpart of the continuous uniform random variable described previously. It has the following pdf:

$$f_X(x) = \frac{1}{N}\sum_{n=a}^{b}\delta(x-n), \tag{3.306}$$

where $a, b \in \mathcal{Z}$, $a < b$, and $N \triangleq b - a + 1$ is the number of outcomes. The cdf is

$$F_X(x) = \frac{1}{N}\sum_{n=a}^{b}u(x-n). \tag{3.307}$$

Example pdfs are shown in Figure 3.48. The standard discrete uniform pdf is obtained when $a = 1$ and $b = N$. Applications include the following:

- Experiments with equally likely outcomes such as tossing a single die.
- Pulse amplitude modulation (PAM) in a digital communication system (see Chapter 10).

Example 3.66. The mean of X is

$$\mu_X = \frac{1}{N}\int_{-\infty}^{\infty}\sum_{n=a}^{b}x\delta(x-n)\mathrm{d}x = \frac{1}{N}\sum_{n=a}^{b}n. \tag{3.308}$$

The sum can be written in two ways: first by counting up relative to a and then by counting down relative to b:

$$\begin{aligned}\mu_X &= \frac{1}{N}[a+(a+1)+\cdots+(a+N-1)] = \frac{1}{N}\left(aN+\sum_{k=0}^{N-1}k\right),\\ &= \frac{1}{N}[(b-N+1)+\cdots+(b-1)+b] = \frac{1}{N}\left(bN-\sum_{k=0}^{N-1}k\right).\end{aligned} \tag{3.309}$$

Since the last two expressions in each line are equal, they can be added together and divided by two to give the final result: $\mu_X = (a+b)/2$. The following finite-sum formula in Appendix E can also be used to derive the mean:

$$\sum_{n=1}^{N}n = N(N+1)/2 \Longrightarrow \sum_{n=0}^{N-1}n = N(N-1)/2. \tag{3.310}$$

Changing variables to $m = n - a$ yields

$$\frac{1}{N}\sum_{m=0}^{b-a}(m+a) = \frac{1}{N}\left[a(b-a+1)+\sum_{m=0}^{b-a}m\right], \tag{3.311}$$

and thus

$$(1/N)\left[aN+(b-a+1)(b-a)/2\right] = (a+b)/2. \tag{3.312}$$

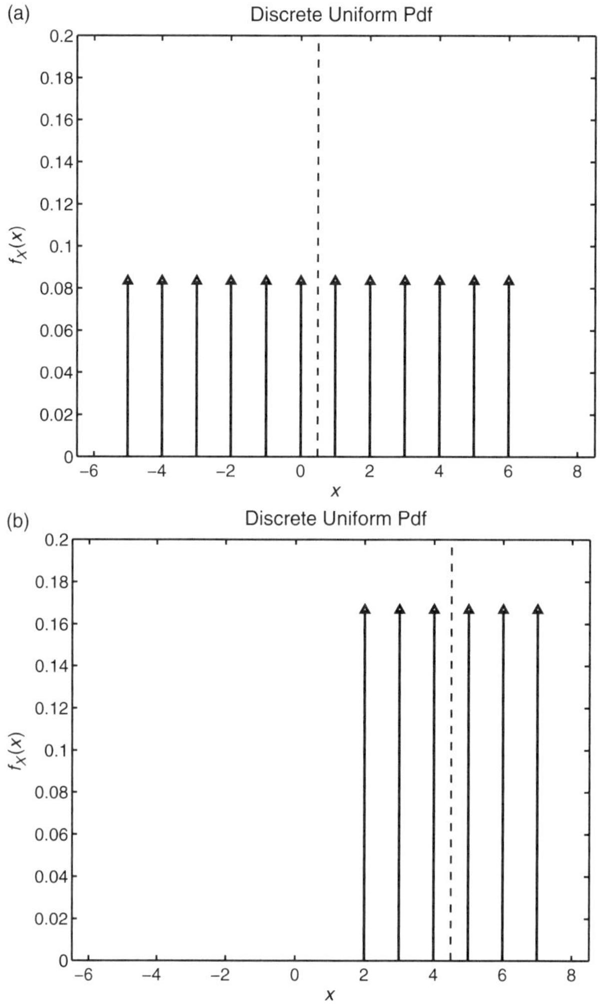

FIGURE 3.48 Examples of the discrete uniform pdf. (a) $a = -5$ and $b = 6$. (b) $a = 2$ and $b = 7$. Support is $\{a, \ldots, b\}$. Dashed lines specify μ_X.

Using similar techniques, it can be shown that the variance is (see Problem 5.36)

$$\sigma_X^2 = (N^2 - 1)/12. \tag{3.313}$$

Note that the mean is identical to that of the continuous uniform random variable, and the variances are similar, though $\{a, b\}$ and N are restricted to be integers above.

3.9.8 Logarithmic Random Variable (Log-Series)

The logarithmic random variable has the following pdf:

$$f_X(x) = \frac{-1}{\ln(q)} \sum_{n=1}^{\infty} (p^n/n)\delta(x - n) \tag{3.314}$$

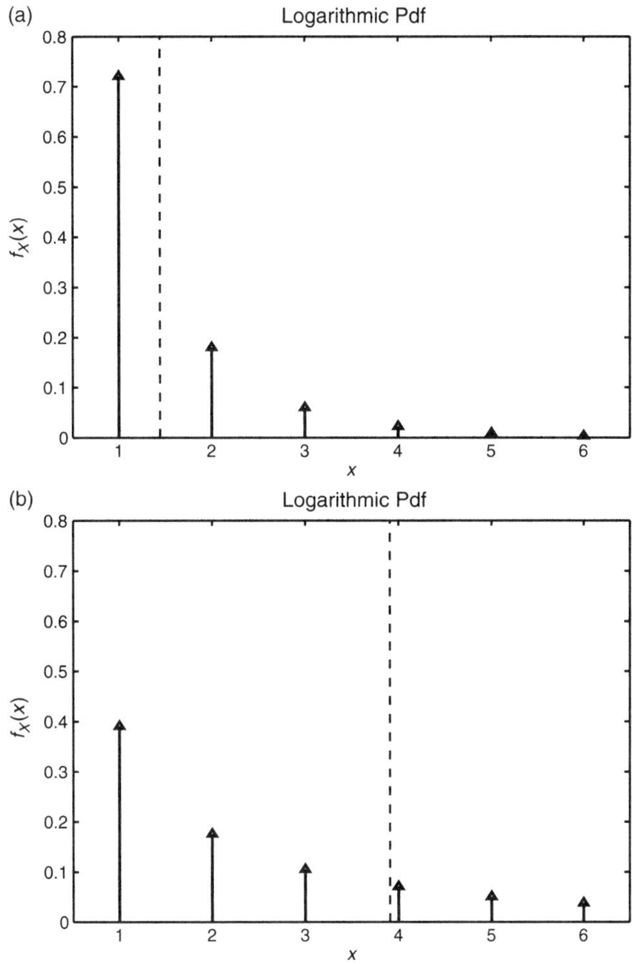

FIGURE 3.49 Examples of the logarithmic pdf. (a) $p = 0.5$. (b) $p = 0.9$. Support is $\mathcal{N}$. Dashed lines specify μ_X.

with $p \in [0, 1]$ and $q \triangleq 1 - p$. The minus sign is needed because the normalization $\ln(q)$ is negative, which is included so that $f_X(x)$ is a valid pdf. The logarithmic distribution, which is also known as the *log-series distribution*, follows from the following Maclaurin series:

$$\ln(1 - p) = -\sum_{n=1}^{\infty} p^n/n. \tag{3.315}$$

Dividing both sides by $\ln(1 - p) = \ln(q)$, we see from this series that the expression in (3.314) sums to 1. Examples of the pdf are shown in Figure 3.49. The cdf is

$$F_X(x) = \frac{-1}{\ln(q)} \sum_{n=1}^{\infty} (p^n/n) u(x - n). \tag{3.316}$$

Applications of the logarithmic distribution include the following:

- Model for the abundance of some species in a region.
- Distribution of the amount of a product purchased by a consumer over some duration of time.

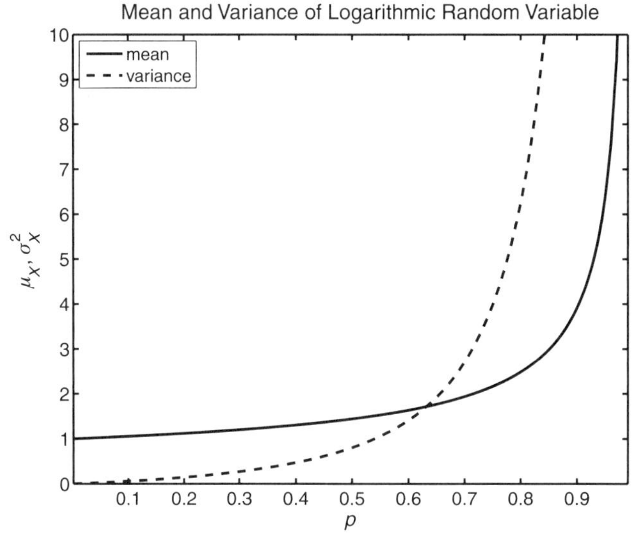

FIGURE 3.50 Mean and variance of the logarithmic random variable as p is varied.

Example 3.67. Consider the following ratio of logarithmic pmfs:

$$\frac{p_X[x+1]}{p_X[x]} = \frac{x}{x+1}p, \tag{3.317}$$

which demonstrates that the logarithmic pmf is strictly decreasing. (See Appendix E for a summary of the functional forms for discrete random variables.) As a result, its mode is necessarily $m_o = 1$. The mean is

$$\begin{aligned}\mu_X &= \frac{-1}{\ln(q)}\int_{-\infty}^{\infty}\sum_{n=1}^{\infty} x(p^n/n)\delta(x-n)\mathrm{d}x = \frac{-1}{\ln(q)}\sum_{n=1}^{\infty} p^n,\\ &= \frac{-1}{\ln(q)}[1/(1-p)-1] = \frac{-1}{\ln(q)}(p/q).\end{aligned} \tag{3.318}$$

Observe in Figure 3.50 that μ_X increases dramatically as $p \to 1$, indicating that the tail of the pdf is quite long when p is very close to 1, unlike the plots in Figure 3.49 where p is relatively small. This long tail is due to $\ln(q)$ approaching zero in the denominator of the pdf as $q \to 0$. The variance is $\sigma_X^2 = -p[p+\ln(q)]/q^2\ln^2(q)$, which is also shown in Figure 3.50 (see Problem 5.36).

3.9.9 Zeta Random Variable (Zipf)

The zeta random variable is based on the Riemann zeta function:

$$\zeta(\alpha) \triangleq \sum_{n=1}^{\infty}\frac{1}{n^\alpha}, \tag{3.319}$$

where $\alpha > 1$ is real-valued (the zeta function is also defined for complex α, in which case we require $\mathrm{Re}(\alpha) > 1$). Note that $\zeta(\alpha)$ is a power series as discussed in Appendix B. The zeta pdf is

$$f_X(x) = \frac{1}{\zeta(\alpha)}\sum_{n=1}^{\infty}\frac{1}{n^\alpha}\delta(x-n), \tag{3.320}$$

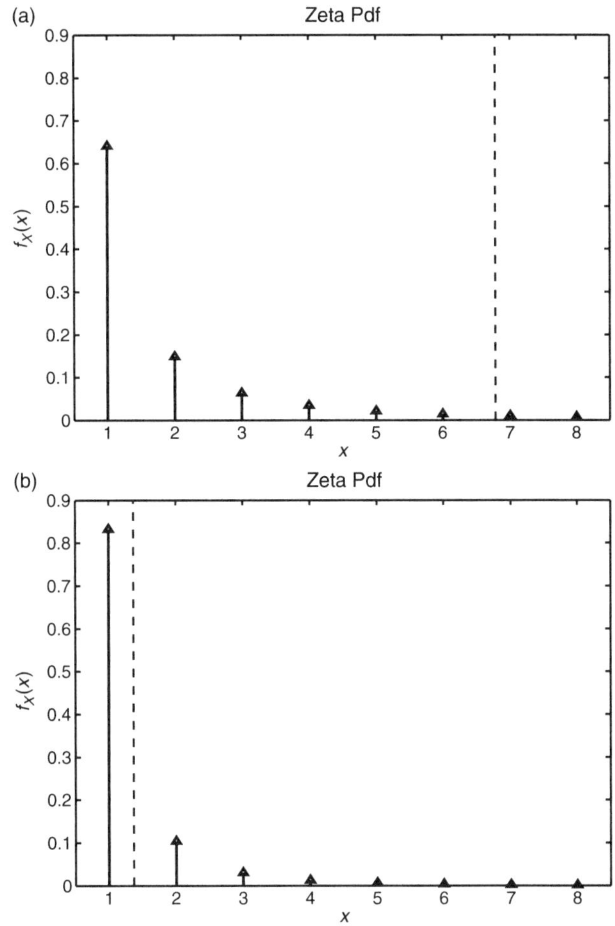

FIGURE 3.51 Examples of the zeta pdf. (a) $\alpha = 2.1$. (b) $\alpha = 3$. Support is $\mathcal{N}$. Dashed lines specify μ_X.

where $\zeta(\alpha)$ is a normalization factor required so that the pdf integrates to one. The cdf is

$$F_X(x) = \frac{1}{\zeta(\alpha)} \sum_{n=1}^{\infty} \frac{1}{n^{\alpha}} u(x-n). \tag{3.321}$$

Examples of the pdf are shown in Figure 3.51. Observe that $P(X=1) = F_X(1) = 1/\zeta(\alpha)$ increases quickly with increasing α so that the pmf tends to be dominated by the outcome $X = 1$. The zeta distribution can be viewed as as special case (as $N \to \infty$) of the *Zipf distribution* (also called Zipf's law) which has pdf

$$f_Y(y) = \frac{1}{H_N(\alpha)} \sum_{n=1}^{N} \frac{1}{n^{\alpha}} \delta(y-n), \tag{3.322}$$

where

$$H_n(\alpha) \triangleq \sum_{m=1}^{n} \frac{1}{m^{\alpha}} \tag{3.323}$$

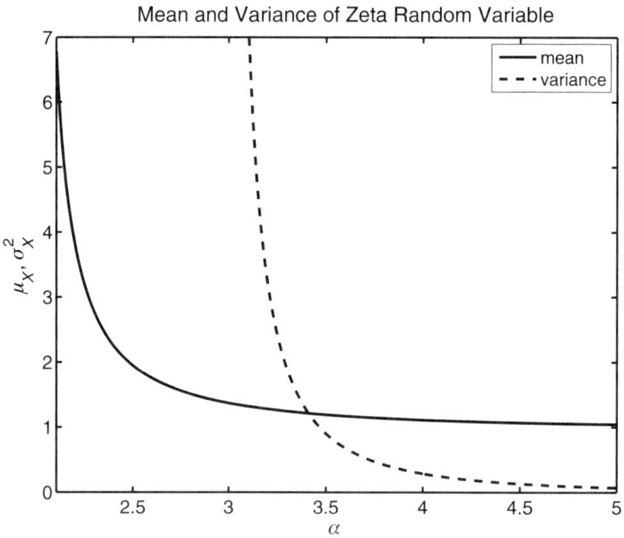

FIGURE 3.52 Mean and variance of the zeta random variable as α is varied.

is the nth generalized harmonic number (see Appendix B). Applications of these distributions include the following:

- Distribution of the frequency of words in a large book.
- Model where most outcomes rarely occur.

Example 3.68. Like the logarithmic distribution, the zeta pmf is strictly decreasing:

$$\frac{p_X[x+1]}{p_X[x]} = \left(\frac{x}{x+1}\right)^{\alpha}, \tag{3.324}$$

and so the mode is $m_o = 1$. The mean is quite easy to calculate:

$$\mu_X = \frac{1}{\zeta(\alpha)} \sum_{n=1}^{\infty} \frac{1}{n^{\alpha}} \int_{-\infty}^{\infty} x\delta(x-n)\mathrm{d}x = \frac{1}{\zeta(\alpha)} \sum_{n=1}^{\infty} \frac{n}{n^{\alpha}} = \frac{\zeta(\alpha-1)}{\zeta(\alpha)}. \tag{3.325}$$

The infinite sum converges only for $\alpha - 1 > 1 \Longrightarrow \alpha > 2$. Observe in Figure 3.52 that the mean is quite high for α close to 2, indicating that significant probability mass is in the tail (i.e., a very long tail). The mean approaches one with increasing α because the tail probabilities become very small and nearly all the probability mass is located at $X = 1$, as mentioned above. For example, when $\alpha = 5$, we have $P(X = 2) = 0.0301$, $P(X = 3) = 0.0040$, and $P(X = 4) = 0.0009$. The pmf would need to be plotted on a logarithmic scale in order to view the probability mass for larger values of α. The variance is $\sigma_X^2 = [\zeta(\alpha)\zeta(\alpha-2) - \zeta^2(\alpha-1)]/\zeta^2(\alpha)$ for $\alpha > 3$, which is also shown in Figure 3.52.

PROBLEMS

Random Variables and Distributions

3.1 Let $\Omega = \{-3, -2, -1, 0, 1, 2, 3\}$. Show the mapping and specify the outcomes for each of the following random variables. (a) $X_1(\omega) = \text{sgn}(\omega)$. (b) $X_2(\omega) = \omega^2$. (c) $X_3(\omega) = I_{\{-3,-1,1,3\}}[\omega] + 2I_{\{-2,0,2\}}[\omega]$. If the outcomes of Ω are equally likely, give the cdf $F_X(x)$ for each of the three random variables.

3.2 Repeat Problem 3.1 for $\Omega = \{1, \ldots, 10\}$ and the following random variables. (a) $X_1(\omega) = u(\omega - 4.5)$. (b) $X_2(\omega) = (10 - \omega/2)I_{\{2,4,6,8,10\}}[\omega]$. (c) $X_3(\omega) = \text{sgn}(\omega - 2)$.

3.3 Repeat Problem 3.1 for $\Omega = [-1, 1]$ and the following random variables. (a) $X_1(\omega) = \exp(\omega)$. (b) $X_2(\omega) = \cos(\omega\pi/2)$. (c) $X_3(\omega) = \text{sgn}(\omega)$. Assume the outcomes of Ω are uniformly distributed in order to find $F_X(x)$.

3.4 Let Ω be the sample space for the toss of a single die and define $X(\omega) = I_{\{1,2\}}(\omega) + 2I_{\{3,4,5\}}(\omega) - I_{\{6\}}(\omega)$. Find the smallest σ-field for Ω such that $X(\omega)$ is a random variable, and specify the probability of each outcome.

3.5 Describe the smallest σ-field for Ω such that $X(\omega) = \lfloor \omega \rfloor$ is a random variable for the following sample spaces. (a) $\Omega_1 = \mathcal{R}^+$. (b) $\Omega_2 = [0, 4)$.

3.6 Determine if each of the following are valid distribution functions for random variable X. (a) $\sum_{n=1}^{\infty} p^n u(x - n)$ for $x \in \mathcal{Z}$ and $0 < p < 1$. (b) $x[u(x) - u(x - 2)]$ for $x \in \mathcal{R}$. (c) $(1/2)[u(x) + \exp(x - 1)u(x - 1)]$ for $x \in \mathcal{R}$.

3.7 (a) Find and sketch the cdf for the following pdf:

$$f_X(x) = (1/4)[u(x-1) - u(x+1)] + (1/6)\delta(x-1) + (1/3)\delta(x-2). \tag{3.326}$$

(b) Repeat (a) for

$$f_X(x) = \begin{cases} 0, & x < -1 \\ x + 1, & -1 \leq x < 0 \\ (1/2)\exp(-x), & 0 \leq x. \end{cases} \tag{3.327}$$

3.8 For $\Omega = [-1, 2]$ and $X(\omega) = \omega^2 - 1$, find expressions for $F_X(x)$ and $f_X(x)$ assuming that the probability measure on Ω is the Lebesgue measure scaled by $1/3$.

Continuous Random Variables

3.9 For the exponential random variable X with parameter $\lambda = 2$, compute the following probabilities. (a) $P(1 < X < 3)$. (b) $P(X > 5)$. (c) $P(X > 5|X > 3)$. (d) $P(X > 2|X > 3)$.

3.10 Let X be Gaussian with parameters $\{\mu, \sigma\}$. (a) Write $P(x_1 \leq X \leq x_2)$ in terms of the cdf for the standard Gaussian random variable, which we denote by $F_o(x)$. For $\mu = 0$ and $\sigma = 2$, compute the following probabilities. (b) $P(-2 < X < 1)$. (c) $P(|X| > 2)$.

3.11 The distribution of grades on an exam is approximately normal with mean 60 and standard deviation 5. Find the following probabilities. (a) $P(X > 90)$. (b) $P(40 < X < 65)$. (c) $P(|X - \mu| \leq 2\sigma)$.

3.12 Let Y have a log-normal distribution with the parameter values in Problem 3.10. Compute the following probabilities. (a) $P(1 < Y < 3)$. (b) $P(Y > 2)$. Compare the last result with $P(X > 2)$ for X Gaussian with the same parameter values.

3.13 Prove that the beta distribution is symmetric about $x = 1/2$ when the parameters are equal: $\alpha = \beta$.

3.14 Derive the cdf of the Rayleigh random variable with parameter α, and use it to verify the mean of the half-normal distribution.

3.15 For Gaussian random variable X with parameters $\{\mu = 2, \sigma = 1\}$, write the following probabilities in terms of the Q-function. (a) $P(|X| > \sigma)$. (b) $P(-3 \leq X - \mu \leq 1)$. Rewrite your answers in terms of the error function.

3.16 Derive the expression in (3.224) for the mode of the F-distribution.

3.17 For X Rayleigh distributed with parameter $\alpha = 2$, find (a) $P(1 \leq X \leq 3)$, (b) $P(X \leq 4 \text{ or } X > 3)$, and (c) $P(X > 4 \text{ or } X \leq 3)$.

3.18 Verify the expression in (3.187) for the cdf $F_X(x)$ of the logistic random variable.

3.19 Derive an expression for the median m_e of the *asymmetric* Laplace distribution in (3.104) in terms of its parameters $\{\alpha_1, \alpha_2, c\}$.

3.20 Find the inflection points of the Gaussian random variable of the form $x = a \pm b$, and compare them to those of the Cauchy random variable in Example 3.25.

3.21 Derive an expression for the median m_e of the triangular distribution in (3.122) in terms of its parameters $\{a, b, c\}$.

3.22 If X has a beta distribution with parameters $\{\alpha = r, \beta\}$, show that the limiting distribution of $Y = \beta X$ as $\beta \to \infty$ is gamma with parameters $\{\lambda = 1, r\}$.

3.23 Let X be uniform on $[0, 10]$. Suppose we randomly choose a value x_o to divide this interval into two subintervals. Find the pdf of random variable Y corresponding to the *longer* subinterval.

3.24 Using a transformation of variables on the chi distribution, show that the mean of the Maxwell–Boltzmann distribution with scale parameter α is $\mu_X = 2\alpha\sqrt{2/\pi}$.

3.25 A three-parameter version of the Weibull random variable is obtained by the transformation $X + c$ where $c \geq 0$. Find expressions for the median, mode, and any inflection points for this shifted distribution.

Discrete Random Variables

3.26 For a Poisson random variable X with parameter $\alpha = 2$, compute the following probabilities. (a) $P(X > 3)$. (b) $P(2 < X \leq 5)$. (c) $P(X = 3|2 < X \leq 5)$.

3.27 A fair coin is tossed five times. Find the following probabilities. (a) P(three heads). (b) P(either zero heads or zero tails). (c) P(number of heads exceeds number of tails). (d) P(three heads|greater than two heads).

3.28 (a) A fair coin is tossed until the first H occurs. Compute the probability that three tosses are required. (b) A fair coin is tossed until the second H occurs. Compute the probability that five tosses are required. (c) A coin with $P(H) = 2/3$ is tossed until the third T. Find the probability of five heads.

3.29 A coin is tossed until either H occurs or until the Nth toss. Let $P(H) = p$ and define X to be a random variable for the number of tosses. Find an expression for the pmf $p_X[x]$.

3.30 Find the mode of the binomial pdf in terms of its parameters $\{N, p\}$.

3.31 A fair coin is sequentially tossed four times. Define random variable X to be the number of tails preceding the first head. Find the pdf $f_X(x)$.

3.32 For the binary symmetric channel (BSC) with input X, output Y, and alphabet $\{0, 1\}$, the probability of a channel error is $P(Y = 0|X = 1) = P(Y = 1|X = 0) = 0.1$. In order to improve transmission accuracy, each bit is repeated twice: 000 is sent for a 0 and 111 is sent for a 1. Find the probability of a symbol error when majority rule is used to determine whether a 1 or a 0 was transmitted.

3.33 Two people take turns tossing a single fair coin until the second H occurs. Intuitively, the person tossing first has a higher probability of winning. Determine if this is correct by calculating the probability of winning for each person.

3.34 For the binomial random variable X with parameters $\{p = 1/4, N = 5\}$, find (a) $P(X \leq 2)$ and (b) $P(X \leq 2 \text{ or } X > 3)$. (c) Determine x such that $P(X > x) < 0.01$.

3.35 Suppose a professor has a collection of fifteen different books on probability which are numbered for convenience. Each day a graduate student randomly borrows two books and returns them at the end of the day. Find the following probabilities and specify the type of discrete distribution. (a) Book four is chosen twice in the first five days. (b) Book one is not chosen until the third day.

3.36 If we are interested in $P(X > k)$ for the geometric random variable, we can write for $0 < \epsilon < 1$

$$\int_{k+\epsilon}^{\infty} f_X(x)\mathrm{d}x = \sum_{n=k+1}^{\infty} pq^n, \tag{3.328}$$

where $n = k$ is excluded in the sum because $X > k$. Find a closed-form expression for this sum.

3.37 Derive the recursive expression in (3.252) for the binomial distribution.

3.38 Suppose a discrete random variable has the following property: $P(X = x + 1) = aP(X = x)$ for $x \in \mathcal{Z}^+$. Find an expression for $p_X[x]$ and show that $1/2 < P(X \text{ is even}) < 1$.

3.39 Demonstrate that the pdf of the Poisson random variable with parameter α is nonincreasing for non-integer $x \in [\lfloor \alpha \rfloor, \infty)$ and integer $x \in [\alpha - 1, \infty)$.

Mixed Random Variables

3.40 Find the pdfs of the continuous and discrete parts for the following mixed distribution:

$$F_X(x) = 1/4 + (1/6)u(x+1) + (1/3)u(x) + (1/4)\text{sgn}(x - \mu)[1 - \exp(-\alpha|x - \mu|)]. \tag{3.329}$$

Verify that $F_X(x)$ is a valid cdf.

3.41 Consider the following pdf:

$$f_X(x) = (1/8)\delta(x-1) + (1/8)\delta(x-2) + (1/4)[u(x-1.5) - u(x-2.5)] + (1/2)\exp(-x)u(x). \tag{3.330}$$

(a) Find $P(X > 1)$ and $P(X \leq 2)$. (b) Write an expression for the cdf $F_X(x)$.

Computer Problems

3.42 Simulate the experiment in Problem 3.23 using rand. Generate a histogram for Y consisting of 1000 samples and observe how closely it matches the theoretical pdf derived in that problem.

3.43 Use MATLAB® to generate the $g(y)$ curves in Figure 3.34 of Example 3.44. Include the corresponding results for the uniform random variable on $[-1, 1]$ and the triangular random variable with $c = (a + b)/2$, both of which are known to have light tails.

3.44 Using Student's t distribution, find the 90% and 95% confidence intervals from $N = 100$ samples of a Gaussian random variable generated by MATLAB with parameters $\{\mu = 2, \sigma = 1\}$. Find the corresponding confidence intervals for the variance using the chi-square distribution.

FURTHER READING

Several references for the parametric continuous and discrete variables described in this chapter are provided in Appendix A. Probability and random variables can be explored further in the following references: Davenport, Jr. (1970), Gray and Davisson (2004), Larson and Shubert (1979), Leon–Garcia (2008), Mood, Graybill, and Boes (1974), Olofsson (2005), Papoulis (1965), Parzen (1960), Peebles (1993), Robinson (1985), Shiryayev (1984), Stark and Woods (2002), and Viniotis (1998).

4 Multiple Random Variables

4.1 INTRODUCTION

A goal of this chapter is to extend the definition of a random variable to two or more *joint* random variables and to investigate how they are related. We first describe approximations of some random variables covered in Chapter 3. These not only provide insight about how these random variables are similar, but the approximations can also be used to perform computations that would otherwise be more difficult using the probability density function (pdf) of the original random variable. We then describe joint distributions that are used to derive probabilities of events for several random variables, and can be represented by a *random vector*. This leads to *conditional* distributions where probabilities are computed based on a conditioning event or a conditioning random variable. An example of the conditional pdf $f_{Y|X}(y|x)$ for a specific value x is shown in Figure 4.1(a). Conditioning is important because it describes the extent to which two random variables are correlated and thus their "mutual information."

We also consider *independence*, which is a property used in Chapters 6 and 7 to characterize random *processes*. Techniques for deriving the pdf and cumulative distribution function (cdf) of a random variable that is a *transformation* of another random variable are presented along with several examples. For example, a specific nonlinear transformation of an exponential random variable leads to the logistic random variable shown in Figure 4.1(b). The transformations examined in this chapter include an important set of random variables known as the *order statistics*. We conclude with a discussion of *mixtures*, which are not transformations of random variables, but instead are weighted combinations of their pdfs.

4.2 RANDOM VARIABLE APPROXIMATIONS

The approximations described in this section are not special cases of a random variable for specific finite values of the pdf parameters, but are *limiting* cases when one or more of the parameters becomes very large. Table 4.1 summarizes some related discrete random variables that are discussed first.

4.2.1 Binomial Approximation of Hypergeometric

Consider the hypergeometric probability mass function (pmf) with parameters $N \in \mathcal{N}$, $M \in \{0, \ldots, N\}$, and $n \in \{1, \ldots, N\}$:

$$P(X = x) = \frac{\binom{M}{x}\binom{N-M}{n-x}}{\binom{N}{n}} = \frac{\binom{Np}{x}\binom{N-Np}{n-x}}{\binom{N}{n}}, \tag{4.1}$$

Probability, Random Variables, and Random Processes: Theory and Signal Processing Applications, First Edition. John J. Shynk.

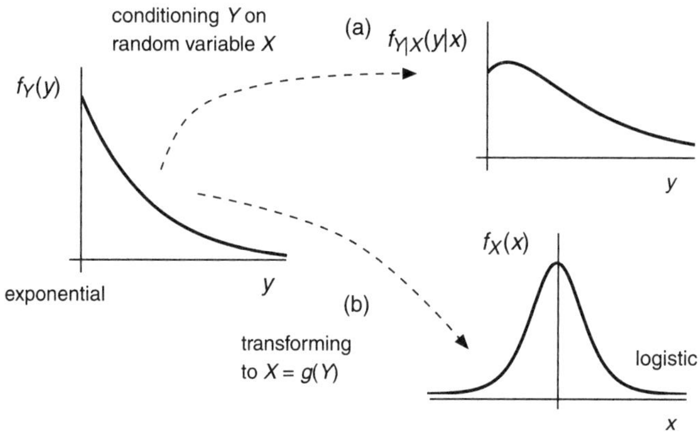

FIGURE 4.1 Modifying a random variable. (a) Conditioning on a related random variable X for a specific value x to give $f_{Y|X}(y|x)$. (b) Transforming to a new random variable X with pdf $f_X(x)$.

where $M = Np$ has been substituted. Let $M \to \infty$ and $N \to \infty$ such that their ratio remains a constant $p \triangleq M/N$. Expanding the binomial coefficients, combining terms to obtain $\binom{n}{x}$, and then canceling all common terms in the numerator and denominator yields

$$\frac{(Np)!}{x!(Np-x)!}\frac{(N-Np)!}{(n-x)!(N-Np-n+x)!}\frac{n!(N-n)!}{N!}$$
$$= \binom{n}{x}\prod_{k=0}^{x-1}(Np-k)\prod_{m=0}^{n-x-1}(N-Np-m)\frac{1}{\prod_{k=0}^{n-1}(N-k)}. \tag{4.2}$$

There are $x + (n - x) = n$ terms in the numerator (excluding the $\binom{n}{x}$ factor in front), which is the same number of terms in the denominator. Thus, we can factor the denominator into two parts and group them with the products in the numerator as follows (and noting that $x \le n$):

$$\binom{n}{x}\prod_{k=0}^{x-1}\frac{(Np-k)}{(N-k)}\prod_{m=0}^{n-x-1}\frac{(N-Np-m)}{(N-x-m)} \xrightarrow{N\to\infty} \binom{n}{x}\prod_{k=0}^{x-1}p\prod_{m=0}^{n-x-1}q,$$
$$= \binom{n}{x}p^x q^{n-x}, \tag{4.3}$$

TABLE 4.1 Asymptotic Approximations of Random Variables

Distribution	Conditions	Approximation
Hypergeometric $\{N, M, n\}$	$N, M \to \infty$, $M/N = p$	Binomial $\{n, p\}$
Binomial $\{n, p\}$	$n \to \infty$, $p \to 0$, $np = \alpha$	Poisson $\{\alpha\}$
Negative binomial $\{m, p\}$	$m \to \infty$	Poisson $\{m(1-p)\}$
Binomial $\{N, p\}$	N large	Gaussian $\{\mu = Np, \sigma^2 = Npq\}$
Hypergeometric $\{N, M, n\}$	$N, M \to \infty$, M/N constant	Gaussian $\{\mu = nM/N,$ $\sigma^2 = n(M/N)(1-M/N)\}$
Poisson $\{N\alpha\}$	N large	Gaussian $\{\mu = N\alpha, \sigma^2 = N\alpha\}$
Chi-square $\{N\}$	N large	Gaussian $\{\mu = N, \sigma^2 = 2N\}$
Student's t $\{r\}$	r large	Gaussian $\{\mu = 0, \sigma^2 = 1\}$
Gamma $\{\lambda, r\}$	r large	Gaussian $\{\mu = r/\lambda, \sigma^2 = r/\lambda^2\}$

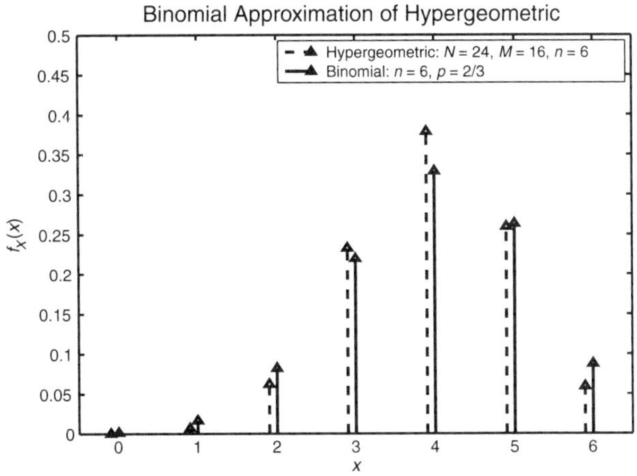

FIGURE 4.2 Example of a binomial approximation of the hypergeometric random variable. (The hypergeometric pdf has been shifted slightly to the left for ease of comparison.)

where the following have been used:

$$\lim_{N\to\infty} \frac{Np-k}{N-k} = \lim_{N\to\infty} \frac{Np}{N} = p, \tag{4.4}$$

$$\lim_{N\to\infty} \frac{N-Np-k}{N-x-k} = \lim_{N\to\infty} \frac{N-Np}{N} = 1-p \triangleq q. \tag{4.5}$$

The last expression in (4.3) is the pmf of the binomial random variable with parameters $\{n, p\}$. An example of this approximation is shown in Figure 4.2 for $p = M/N = 16/24$ and $n = 6$.

4.2.2 Poisson Approximation of Binomial

The Poisson random variable approximates the binomial random variable with parameters $\{n, p\}$ when the probability of success $p \in [0, 1]$ is small and the number of trials $n \in \mathcal{Z}^+$ is large. The key condition is that $np \triangleq \alpha > 0$ remains constant as $n \to \infty$ and $p \to 0$, where α is the parameter of the Poisson random variable. The binomial pmf can be rearranged as follows:

$$\begin{aligned}\lim_{n\to\infty} \binom{n}{x} p^x q^{n-x} &= \lim_{n\to\infty} \binom{n}{x} (\alpha/n)^x (1-\alpha/n)^{n-x}, \\ &= \lim_{n\to\infty} \frac{n!}{(n-x)!n^x} (\alpha^x/x!)(1-\alpha/n)^n (1-\alpha/n)^{-x},\end{aligned} \tag{4.6}$$

where the constraint above on the parameters is enforced by substituting $p = \alpha/n$ and $q \triangleq 1-p$. Examining the three terms involving n, we find that

$$\lim_{n\to\infty} (1-\alpha/n)^{-x} = 1, \tag{4.7}$$

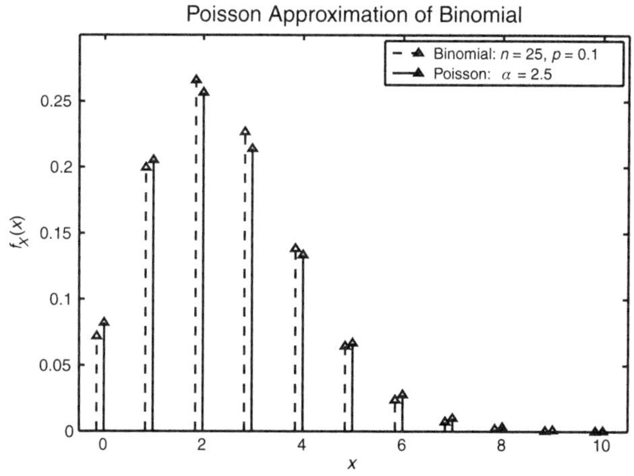

FIGURE 4.3 Poisson approximation of the binomial random variable. (The binomial pdf has been shifted slightly to the left for ease of comparison.)

$$\lim_{n\to\infty} \frac{n!}{(n-x)!n^x} = \lim_{n\to\infty} \frac{n(n-1)\cdots(n-x+1)}{n^x} = 1, \tag{4.8}$$

$$\lim_{n\to\infty} (1-\alpha/n)^n = \exp(-\alpha). \tag{4.9}$$

The first result is obvious, and the second result follows because the largest exponent of n in the numerator is x, which is the same as that in the denominator. The third result is obtained from the binomial theorem (see Appendix B):

$$\begin{aligned}(1-\alpha/n)^n &= 1 + n(-\alpha/n) + n(n-1)(-\alpha/n)^2/2! + n(n-1)(n-2)(-\alpha/n)^3/3! \\ &\quad + \cdots + (-\alpha/n)^n, \\ &= 1 - \alpha + [(n-1)/n]\alpha^2/2! - [(n-1)(n-2)/n^2]\alpha^3/3! \\ &\quad + \cdots + (-\alpha/n)^n. \end{aligned} \tag{4.10}$$

As $n \to \infty$, all ratios involving n approach one and $(1/n)^n \to 0$ such that

$$\lim_{n\to\infty} (1-\alpha/n)^n = \sum_{m=0}^{\infty} (-\alpha)^m/m!, \tag{4.11}$$

which is a series expansion for $\exp(-\alpha)$. Combining these results gives the Poisson distribution with parameter α:

$$\lim_{n\to\infty} \binom{n}{x} p^x q^{n-x} = \exp(-\alpha)\frac{\alpha^x}{x!} = P(X = x). \tag{4.12}$$

An example of this approximation is shown in Figure 4.3 where $p = 0.1$ and $n = 25$ for the binomial pdf, and thus $\alpha = np = 2.5$ for the Poisson pdf.

A Poisson approximation of the negative binomial random variable given in Table 4.1 is considered in Problem 4.3.

4.2.3 Gaussian Approximations

The previous discrete random variables can also be approximated by a Gaussian random variable as summarized in Table 4.1. The approximations are all derived by using the central limit theorem (CLT) discussed later in Chapter 7, which we mention below. Gaussian approximations of the chi-square random variable and Student's t distribution shown in the table are examined in the problems at the end of this chapter.

Theorem 4.1 (Central limit theorem). Let $\{X_1, \ldots, X_N\}$ be independent and identically distributed (iid) random variables with finite mean μ_X and finite variance σ_X^2. Then

$$\lim_{N\to\infty} P\left(\frac{1}{\sigma_X\sqrt{N}}\sum_{n=1}^{N}(X_n - \mu_X) \leq z\right) = \frac{1}{\sqrt{2\pi}}\int_{-\infty}^{z} \exp(-x^2/2)\mathrm{d}x, \tag{4.13}$$

which is the standard Gaussian cdf with parameters $\{\mu = 0, \sigma = 1\}$.

Proof. See Chapter 7. □

Observe that subtracting μ_X and dividing by $\sqrt{N\sigma_X^2}$ causes the sum to have zero mean and unit variance. Basically, if a "large" number of normalized iid random variables are summed together, the distribution of the sum is approximately standard Gaussian. In order to derive a Gaussian approximation for a discrete random variable using the CLT, we must choose parameters that satisfy the large number requirement of the theorem. (Independent random variables are defined later in this chapter, and are analogous to independent events described in Chapter 2.)

4.2.4 Gaussian Approximation of Binomial

The binomial random variable is obtained as the sum of N iid Bernoulli random variables $\{X_k\}$ with probability of success p. The mean and variance of the binomial random variable are Np and Npq, respectively. From the CLT,

$$Y = \frac{1}{N\sqrt{pq}}\sum_{n=1}^{N}(X_n - Np) \tag{4.14}$$

has a distribution that is approximately standard Gaussian. We can view the sum above as a transformation from N random variables to a single random variable (transformations of random variables are covered later in this chapter). Alternatively, for large N the pdf of $Z = \sum_{n=1}^{N} X_n$ is given approximately by

$$f_Z(z) \approx \frac{1}{\sqrt{2\pi Npq}}\exp(-(z - Np)^2/2Npq), \tag{4.15}$$

which is Gaussian with mean Np and variance Npq. Note that the mean and variance are *not* controlled by different parameters: for a given value of p (and thus q), a specific Gaussian approximation is obtained for large N. An example of this approximation is shown in Figure 4.4 for $\mu = Np = 2.5$ and $\sigma^2 = Npq = 2.25$.

4.2.5 Gaussian Approximation of Poisson

This approximation is obtained using the same approach as above by recognizing that the sum of N iid Poisson random variables with parameter α is also Poisson but with parameter $N\alpha$ (see Problem 4.25). If each term in

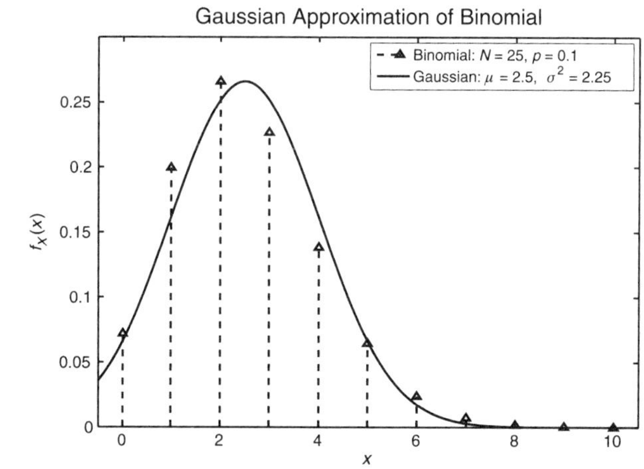

FIGURE 4.4 Example Gaussian approximation of the binomial random variable.

the sum has mean α and variance α, then

$$Y = \frac{1}{\sqrt{N\alpha}} \sum_{n=1}^{N} (X_n - \alpha)^2 \tag{4.16}$$

is approximately standard Gaussian. Alternatively, for large N the pdf of $Z = \sum_{n=1}^{N} X_n$ is given approximately by

$$f_Z(z) \approx \frac{1}{\sqrt{2\pi N\alpha}} \exp(-(z - N\alpha)^2/2N\alpha), \tag{4.17}$$

which has mean $N\alpha$ and variance $N\alpha$. As in the Gaussian approximation for the binomial random variable, the mean and variance are not controlled independently by different parameters, and in fact they are identical. An example of this approximation is shown in Figure 4.5 for $\mu = \sigma^2 = N\alpha = 5$.

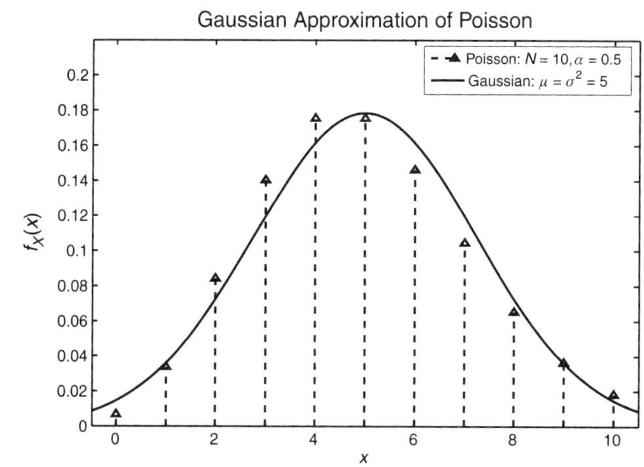

FIGURE 4.5 Example Gaussian approximation of the Poisson random variable.

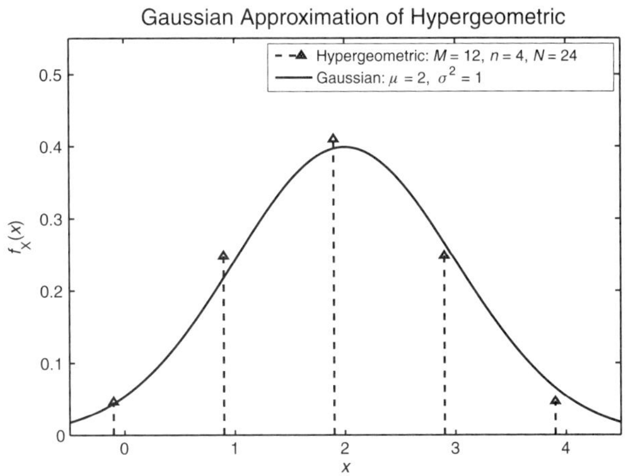

FIGURE 4.6 Example Gaussian approximation of the hypergeometric random variable.

4.2.6 Gaussian Approximation of Hypergeometric

Finally, a Gaussian approximation of the hypergeometric random variable is readily obtained because the Gaussian pdf can be used to approximate the binomial random variable, which in turn approximates the hypergeometric random variable as described previously. Thus, using the same condition $\lim_{N\to\infty} M/N = p$, where p is a constant, the Gaussian approximation has mean nM/N and variance $n(M/N)(1 - M/N)$ (where $q \triangleq 1 - p = 1 - M/N$ has been used). Again, the mean and variance are not independent parameters. An example of this distribution is shown in Figure 4.6 for $N = 24$, $M = 12$, and $n = 4$.

The last three Gaussian approximations in Table 4.1 are examined in the problems at the end of this chapter. Note that these approximations are not based on a sum of the random variables in the first column of the table; instead, all three approach a Gaussian distribution when the degrees of freedom parameter is large. However, recall from Chapter 3 that these particular random variables can be derived as a sum of other random variables, with the number in the sum given by the degrees of freedom. Thus, the CLT also applies to these random variables for large N (chi-square) and large r (Student's t and gamma).

4.3 JOINT AND MARGINAL DISTRIBUTIONS

We are interested in a *joint* probabilistic description for multiple random variables so that their relationship can be quantified. This is important in many applications such as estimation (see Chapters 9 and 11) where properties of a random variable or a random process can be estimated or predicted from observations of another random quantity. The joint pdf of two random variables X and Y is derived in a manner similar to that used for a single random variable in Chapter 3. We defer the description of mappings from an abstract probability space $\{\Omega, \mathcal{F}, P\}$ to multiple random variables until the section on *random vectors*. Here, we simply give the joint pdf as a definition.

Definition: Joint Probability Density Function The *joint pdf* of random variables X and Y is denoted by $f_{X,Y}(x, y)$ from which probabilities are computed as follows:

$$P(x_1 \leq X \leq x_2, y_1 \leq Y \leq y_2) = \int_{x_1}^{x_2} \int_{y_1}^{y_2} f_{X,Y}(x, y) \mathrm{d}y \mathrm{d}x. \tag{4.18}$$

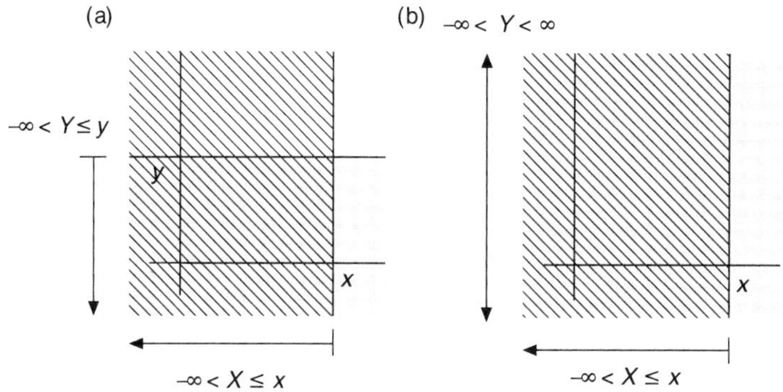

FIGURE 4.7 Joint probability distributions. (a) $P(X \leq x, Y \leq y)$ is computed by integrating the joint pdf over $(-\infty, y] \times (-\infty, x]$. (b) $P(X \leq x)$ is obtained by integrating the joint pdf over $(-\infty, \infty] \times (-\infty, x]$. (The edges of the rectangles extending to the ends of the x-y axes continue to infinity.)

This probability measure is defined on the measurable space $\{\mathcal{R}^2, \mathcal{B}(\mathcal{R}^2)\}$ and is absolutely continuous with respect to Lebesgue measure on $\mathcal{R}^2$.

The probability above is obtained by integrating the joint pdf over the Cartesian product $[x_1, x_2] \times [y_1, y_2]$, which is a rectangle on $\mathcal{R}^2$. The resulting intersection is illustrated in Figure 4.7(a) for $(-\infty, x] \times (-\infty, y]$.

It is possible to generate separate pdfs for X and Y from $f_{X,Y}(x, y)$ which, in this context, are called *marginal pdfs*. Observe that

$$P(X \leq x) = P(X \leq x, -\infty \leq Y \leq \infty), \tag{4.19}$$

which follows because $\{X \leq x\} \cap \{-\infty \leq Y \leq \infty\} = \{X \leq x\}$ as depicted in Figure 4.7(b). The sure event $-\infty < Y < \infty$ does not influence the probability measure of X. From the joint pdf, (4.19) is given by

$$F_X(x) = \int_{-\infty}^{\infty} \int_{-\infty}^{x} f_{X,Y}(x, y) \mathrm{d}x \mathrm{d}y \tag{4.20}$$

which after differentiating with respect to x and using Leibniz's integral rule (see Appendix D) gives the marginal pdf for X:

$$f_X(x) = \int_{-\infty}^{\infty} f_{X,Y}(x, y) \mathrm{d}y. \tag{4.21}$$

Likewise, the marginal pdf for Y is

$$f_Y(y) = \int_{-\infty}^{\infty} f_{X,Y}(x, y) \mathrm{d}x. \tag{4.22}$$

Summarizing, we simply integrate the joint pdf over the *entire support* of one random variable to obtain the marginal pdf of the other random variable. The marginal cdfs are more easily obtained as follows:

$$F_X(x) = F_{X,Y}(x, \infty), \qquad F_Y(y) = F_{X,Y}(\infty, y). \tag{4.23}$$

Example 4.1. Figure 4.8 shows an example of the joint pdf for Gaussian random variables X and Y. The parameters are $\mu_X = 1$, $\mu_Y = 0$, $\sigma_X^2 = 2$, and $\sigma_Y^2 = 1$, and the correlation coefficient is $\rho = 0.9$ (see Chapter

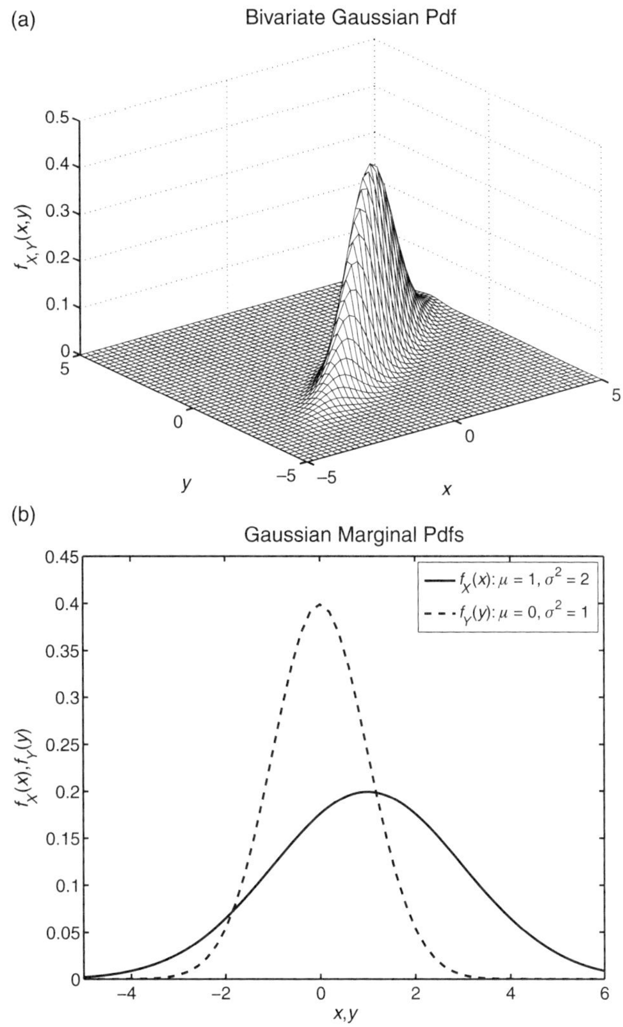

FIGURE 4.8 Gaussian random variables. (a) Bivariate pdf with $\rho = 0.9$. (b) Marginal pdfs.

5). This *bivariate* Gaussian pdf and the marginal pdfs for X and Y shown in the figure are discussed later in this chapter.

Example 4.2. For the following joint pdf:

$$f_{X,Y}(x, y) = 6\exp(-2x - 3y)I_{[0,\infty)}(x)I_{[0,\infty)}(y), \tag{4.24}$$

the marginal pdfs are

$$\begin{aligned} f_X(x) &= 6\int_0^\infty \exp(-2x - 3y)I_{[0,\infty)}(x)\mathrm{d}y, \\ &= -2\exp(-2x)I_{[0,\infty)}(x)\exp(-3y)\big|_0^\infty = 2\exp(-2x)I_{[0,\infty)}(x), \end{aligned} \tag{4.25}$$

$$f_Y(y) = 3\exp(-3x)I_{[0,\infty)}(y). \tag{4.26}$$

Since the joint pdf is the product of the marginal pdfs, it turns out that X and Y are *independent* random variables.

4.4 INDEPENDENT RANDOM VARIABLES

Independent random variables arise in many applications, and it is often assumed as an approximation in order to derive results that otherwise would not be possible in some problems.

Definition: Independent Random Variables Random variables X and Y are *independent* if and only if

$$F_{X,Y}(x, y) = F_X(x)F_Y(y), \tag{4.27}$$

where $F_X(x)$ and $F_Y(y)$ are the marginal cdfs.

Since a cdf is the probability of an event, this definition is consistent not only with intuition but also from the definition based on probabilities in Chapter 2: $P(AB) = P(A)P(B)$ for independent events A and B. It is straightforward to show from (4.27) that the joint pdf also splits into a product:

$$f_{X,Y}(x, y) = f_X(x)f_Y(y) \tag{4.28}$$

for independent random variables X and Y (we could also state this as the definition of independence).

Example 4.3. Let X be the standard Gaussian random variable and Y be exponential with parameter λ. If they are independent, then we can immediately write their joint (bivariate) pdf as

$$f_{X,Y}(x, y) = \frac{\lambda}{\sqrt{2\pi}} \exp(-x^2/2 - \lambda y) I_{[0,\infty)}(y) I_{(-\infty,\infty)}(x). \tag{4.29}$$

This can be extended to additional independent random variables. If Z is uniform on $[-1, 1]$, then the joint pdf of all three random variables is

$$f_{X,Y,Z}(x, y, z) = \frac{\lambda}{2\sqrt{2\pi}} \exp(-x^2/2 - \lambda y) I_{[0,\infty)}(y) I_{(-\infty,\infty)}(x) I_{[-1,1]}(z), \tag{4.30}$$

which is the previous expression scaled by $1/2$, and it includes an indicator function giving the support of Z.

Example 4.4. Let random variable X have the standard Gaussian pdf and Y be uniform on $[-2, 2]$. If these random variables are independent, their joint pdf is

$$f_{X,Y}(x, y) = \frac{1}{4\sqrt{2\pi}} \exp(-x^2/2) I_{(-\infty,\infty)}(x) I_{[-2,2]}(y). \tag{4.31}$$

The indicator functions clearly define the support so that it can be plotted in three dimensions, as shown in Figure 4.9. From a functional viewpoint, we see that for independent random variables, the basic shape of the pdf for X remains unchanged over the support of Y: it is bell-shaped. Likewise the basic shape of Y is unchanged over the support of X: it is flat (uniform). This is the defining property of independent random variables: regardless of the outcome of one random variable, the shape of the marginal pdf of the other random variable remains unchanged (though it is scaled so that the overall bivariate pdf has unit volume).

Next, we define the conditional distribution of two random variables X and Y, which is a useful probability measure when they are *not* independent, and is an indication of the degree to which the two variables are related.

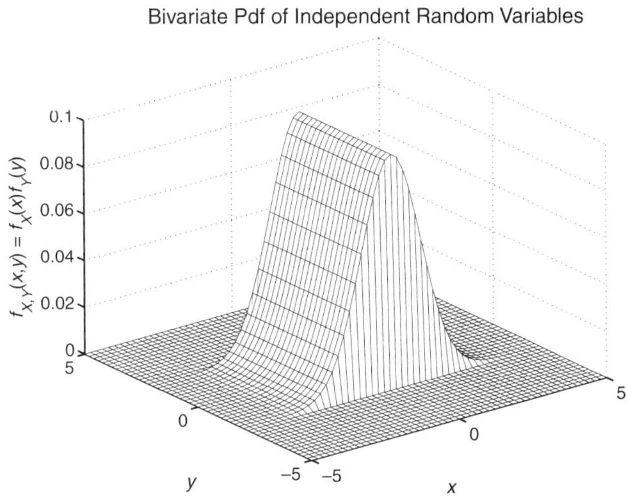

FIGURE 4.9 Bivariate pdf of independent Gaussian and uniform random variables.

4.5 CONDITIONAL DISTRIBUTION

The conditional distribution of random variable X is a probability measure on the measurable space $\{\mathcal{R}, \mathcal{B}(\mathcal{R})\}$ *conditioned* on an event or another random variable. This conditioning, in effect, provides information about the experiment and modifies the sample space. The notion of conditioning is the same as that discussed in Chapter 2 in connection with the probabilities of events; it is extended here to the cdf and pdf of a random variable.

Definition: Conditional Distribution Function The *cdf* of random variable X conditioned on event E is

$$F_{X|E}(x|E) \triangleq \frac{P(X \leq x, E)}{P(E)}, \tag{4.32}$$

where it is assumed that $P(E) \neq 0$.

Generally, there is no reason to condition on an event with zero probability. Observe that $F_{X|E}(x|E)$ is a valid cdf (probability measure) with respect to random variable X: it has the properties described in Chapter 3. It is *not* a probability measure with respect to the conditioning event E. The conditional cdf is depicted in Figure 4.10 where event A in the abstract sample space Ω corresponds to $\{X \leq x\}$ for random variable X. Since we are interested in random variables, event E might represent $\{Y \leq y\}$ for some other random variable Y. However, other intervals might be of interest when conditioning, such as $\{0 \leq Y \leq y\}$, $\{Y > y\}$, and so on. The conditional cdf for event $E = \{Y \leq y\}$ is

$$F_{X|E}(x|E) = \frac{P(X \leq x, Y \leq y)}{P(Y \leq y)} = \frac{F_{X,Y}(x, y)}{F_Y(y)}. \tag{4.33}$$

In the abstract sample space, conditioning on E reduces Ω to E so that we examine only a subset of A contained in E, depicted by the intersection of events in Figure 4.10. When these events are mapped to the x–y-plane, the intersection $A \cap E$ maps to the region $\{X \leq x\} \cap \{Y \leq y\}$. The joint cdf in (4.33) is defined only in this region, and it is normalized by $F_Y(y)$ so that the conditional cdf is a valid probability measure.

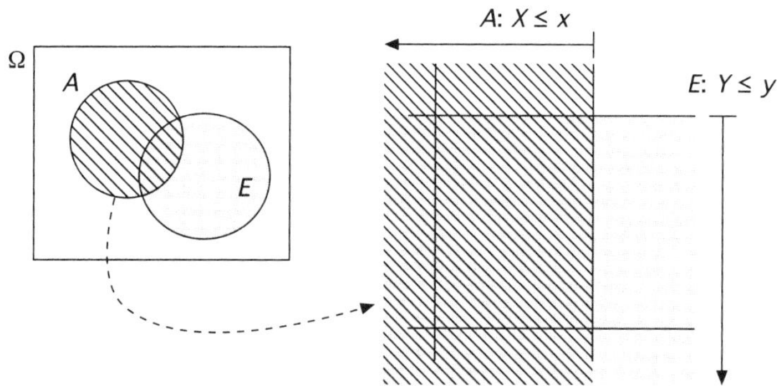

FIGURE 4.10 Example of conditioning event E in the abstract sample space Ω and its mapping to random variable Y.

If $F_{X|E}(x|E)$ is absolutely continuous with respect to x, then from the Radon–Nikodým theorem in Chapter 3, we can write the following conditional pdf:

$$f_{X|E}(x|E) = \frac{\mathrm{d}}{\mathrm{d}x} F_{X|E}(x|E). \tag{4.34}$$

Substituting (4.33) for $E = \{Y \leq y\}$ gives

$$f_{X|E}(x|E) = \frac{1}{F_Y(y)} \frac{\partial}{\partial x} F_{X,Y}(x, y). \tag{4.35}$$

Note that the derivative does not give the joint pdf $f_{X,Y}(x, y)$ because we have not differentiated with respect to the conditioning random variable y. Next, the definition of a conditional distribution is extended to conditioning with respect to *random variable* Y.

We begin with a probability defined on an interval, which is an event with nonzero probability, and let the length of the interval approach zero. For small $\epsilon > 0$:

$$F_{X|Y}(x|y \leq Y \leq y+\epsilon) = \frac{P(X \leq x, y \leq Y \leq y+\epsilon)}{P(y \leq Y \leq y+\epsilon)} = \frac{\int_{-\infty}^{x} \int_{y}^{y+\epsilon} f_{X,Y}(x, y)\mathrm{d}y\mathrm{d}x}{P(y \leq Y \leq y+\epsilon)}. \tag{4.36}$$

Substituting

$$P(y \leq Y \leq y+\epsilon) \approx \epsilon f_Y(y), \tag{4.37}$$

which is the area of a rectangle approximating the integral of $f_Y(y)$ over an interval of width ϵ, the following approximation is obtained:

$$F_{X|Y}(x|y \leq Y \leq y+\epsilon) \approx \frac{1}{\epsilon f_Y(y)} \int_{-\infty}^{x} \int_{y}^{y+\epsilon} f_{X,Y}(x, y)\mathrm{d}y\mathrm{d}x. \tag{4.38}$$

Similarly, we can substitute

$$\int_{y}^{y+\epsilon} f_{X,Y}(x, y)\mathrm{d}y \approx \epsilon f_{X,Y}(x, y), \tag{4.39}$$

and letting $\epsilon \to 0$ yields

$$F_{X|Y}(x|y) = \lim_{\epsilon\to 0} F_{X|Y}(x|y \le Y \le y+\epsilon),$$

$$= \lim_{\epsilon\to 0} \frac{\int_{-\infty}^{x} \epsilon f_{X,Y}(x,y)\mathrm{d}y}{\epsilon f_Y(y)} = \frac{\int_{-\infty}^{x} f_{X,Y}(x,y)\mathrm{d}y}{f_Y(y)}, \tag{4.40}$$

where ϵ has cancelled in the last expression. Differentiating this result with respect to x gives the conditional pdf:

$$f_{X|Y}(x|y) = \frac{\partial}{\partial x} F_{X|Y}(x|y) = \frac{f_{X,Y}(x,y)}{f_Y(y)}. \tag{4.41}$$

Note the difference between this expression where we have conditioned on random variable $Y = y$, and the result in (4.35) where the conditioning is on event E which is an interval of Y.

Example 4.5. Consider the bivariate Gaussian distribution where the random variables both have zero mean and unit variance such that

$$f_{X,Y}(x,y) = \frac{1}{2\pi\sqrt{1-\rho^2}} \exp\left(-\frac{1}{2(1-\rho^2)}\left(x^2 - 2\rho xy + y^2\right)\right), \tag{4.42}$$

where ρ is the correlation coefficient. The marginal pdf for Y is

$$f_Y(y) = \int_{-\infty}^{\infty} f_{X,Y}(x,y)\mathrm{d}x = \frac{1}{\sqrt{2\pi}} \exp\left(-y^2/2\right). \tag{4.43}$$

Dividing these two expressions gives the conditional pdf

$$f_{X|Y}(x|y) = \frac{1}{\sqrt{2\pi(1-\rho^2)}} \exp\left(-\frac{1}{2(1-\rho^2)}\left(x^2 - 2\rho xy + y^2\right) + y^2/2\right),$$

$$= \frac{1}{\sqrt{2\pi(1-\rho^2)}} \exp\left(-\frac{1}{2(1-\rho^2)}(x-\rho y)^2\right), \tag{4.44}$$

which is also Gaussian. Comparing with the standard Gaussian marginal pdf of X, we see that the conditioning results in a shift of the mean from zero to $\mu_{X|Y} = \rho y$, and the variance changes from 1 to $\sigma^2_{X|Y} = 1 - \rho^2$. Since X and Y are *correlated*, the conditioning provides "information" about X. Specifically, observing $Y = y$ causes X to conditionally have mean ρy. For example, suppose $\rho = 0.9$ and $y = 5$ is observed. The corresponding marginal and conditional Gaussian pdfs are depicted in Figure 4.11. With this conditioning, and because Y is highly correlated with X (due to the large positive correlation coefficient), we conclude that an outcome of X (conditionally) is more likely to be near $5\rho = 0.45$. This is reinforced by the smaller variance $1 - \rho^2 = 0.19$ compared with 1 in the marginal pdf. On the other hand, if $\rho = 0$ such that X and Y are uncorrelated (and thus independent because they are jointly Gaussian), the conditional pdf of X is the same as its marginal pdf: Y provides no information about X.

This type of conditional information can be exploited in many applications to make decisions or predictions. Some of these are discussed later in the application chapters. We conclude this section with a brief introduction to a topic in information theory (see Chapter 10).

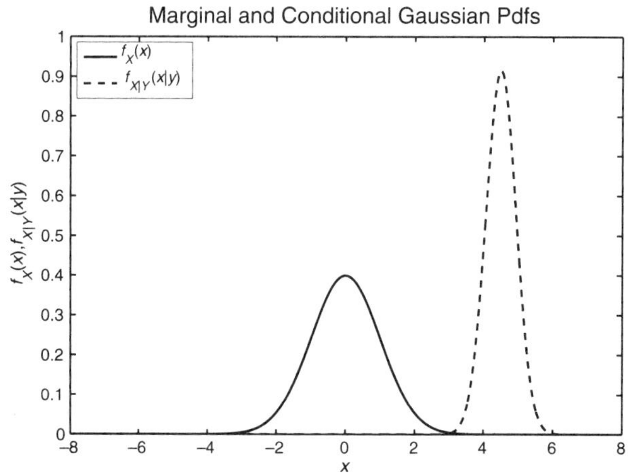

FIGURE 4.11 Marginal and conditional Gaussian pdfs in Example 4.5, with parameters $\{\mu = 0, \sigma = 1\}$ and $\{\mu_{X|Y} = 5, \sigma_{X|Y} = 0.4359\}$. Conditioning leads to a pdf with smaller variance (dashed curve).

Example 4.6. The *average mutual information* of random variables X and Y is

$$I(X;Y) \triangleq \int_{-\infty}^{\infty}\int_{-\infty}^{\infty} f_{X,Y}(x,y)\log(f_{X|Y}(x|y)/f_X(x))\mathrm{d}x\mathrm{d}y, \tag{4.45}$$

where the logarithm is usually base two or e. $I(X;Y)$ is a measure of the amount of information that X "contains" about Y and vice versa. Observe that if X and Y are uncorrelated, then $f_{X|Y}(x|y) = f_X(x)$ and

$$I(X;Y) = \int_{-\infty}^{\infty}\int_{-\infty}^{\infty} f_{X,Y}(x,y)\log(f_X(x)/f_X(x))\mathrm{d}x\mathrm{d}y = 0. \tag{4.46}$$

When two random variables are uncorrelated, they provide no information about each other. For jointly Gaussian random variables with the joint distribution in Example 4.5, it is straightforward to show that (see Problem 4.14)

$$I(X;Y) = -(1/2)\ln(1-\rho^2). \tag{4.47}$$

This expression is plotted in Figure 4.12, demonstrating that the mutual information increases dramatically with increasing correlation (positive or negative ρ). Chapter 5 explores correlation further, showing how the shape of the joint distribution is influenced by the degree of correlation.

4.6 RANDOM VECTORS

The definition of a random variable can be extended to a random *vector* by considering multiple mappings (functions) of outcomes in the event space $\{\Omega, \mathcal{F}\}$ to N-dimensional Euclidean space $\mathcal{R}^N = \mathcal{R} \times \cdots \times \mathcal{R}$. Let the mappings be denoted by the column vector of functions $\mathbf{X}(\omega) \triangleq [X_1(\omega), \ldots, X_N(\omega)]^T$, where $X_n(\omega)$ maps $\omega \in \Omega$ to the nth coordinate of $\mathcal{R}^N$. Thus, instead of open intervals in the case of a single (one-dimensional) mapping to $\mathcal{R}$, the vector mapping is a generalization such that events map to open hyper-rectangles in $\mathcal{R}^N$. The notation can be simplified as follows:

$$\mathbf{X}(\omega) \rightarrow \mathbf{X} \triangleq [X_1, \ldots, X_N]^T, \tag{4.48}$$

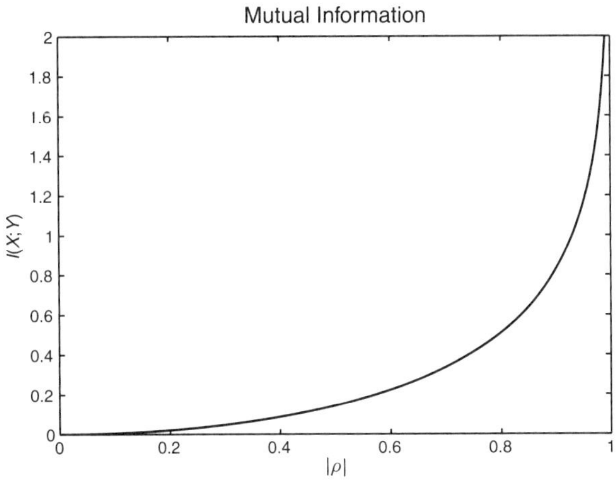

FIGURE 4.12 Mutual information for the standard Gaussian random variables in Example 4.6 with correlation coefficient ρ.

where $\mathbf{X} \in \mathcal{R}^N$ is the random vector. An example mapping for $N = 2$ is shown in Figure 4.13 where event $A \in \Omega$ corresponds to a semi-infinite rectangle $(-\infty, a] \times (-\infty, b] \in \mathcal{R}^2$. Let B be the hyper-rectangle in $\mathcal{B}(\mathcal{R}^N)$ given by the Cartesian product $\prod_{n=1}^{N}[a_n, b_n]$ with $b_n > a_n$. Recall that although the definition of $\mathcal{B}(\mathcal{R})$ is based on open intervals on $\mathcal{R}$, it also contains all semi-open and closed intervals, as well as all individual points. This is likewise the case for random vectors with $N > 1$: $\mathcal{B}(\mathcal{R}^N)$ includes all open, semi-open, and closed hyper-rectangles, as well as all points of $\mathcal{R}^N$. The vector of mappings $\mathbf{X}(\omega)$ is measurable if (i) each of its components is a measurable function (as in the scalar case) and (ii) the intersection of the individual inverse images $\{X_n^{-1}(B)\}$ is in $\mathcal{F}$:

$$\begin{aligned} \mathbf{X}^{-1}(B) &= \{\omega : X_n(\omega) \in [a_n, b_n], n = 1, \ldots, N\}, \\ &= \bigcap_{n=1}^{N} X_n^{-1}([a_n, b_n]) \triangleq A \in \mathcal{F}. \end{aligned} \tag{4.49}$$

This last condition is important: the intersection defines the hyper-rectangle representing $B \in \mathcal{R}^N$, which must have an inverse image $A \in \mathcal{F}$ in the abstract sample space Ω.

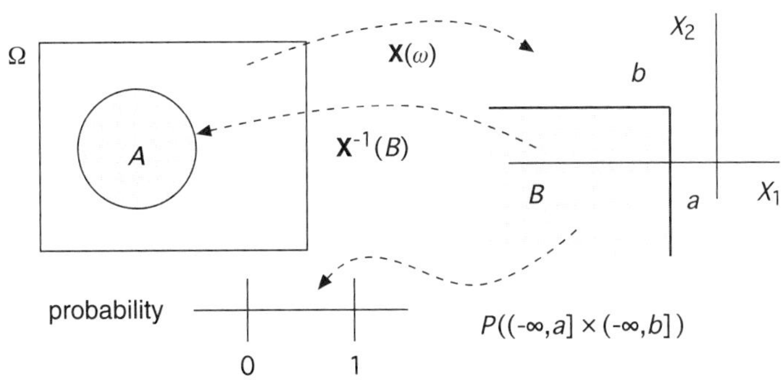

FIGURE 4.13 Mapping from Ω to $\mathcal{R}^2$, inverse image A of event B, and probability mapping to a point in the closed interval [0, 1].

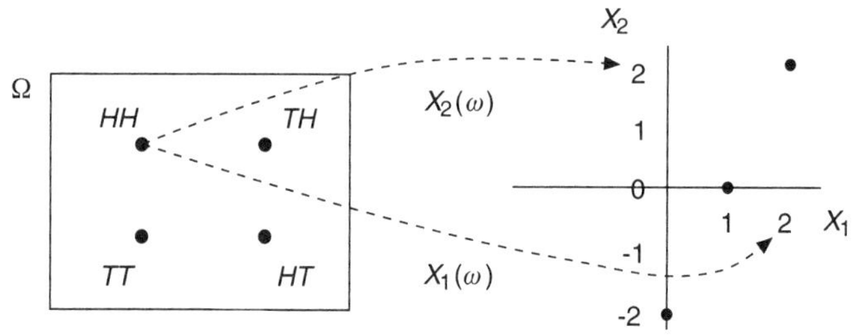

FIGURE 4.14 Random vector for Example 4.7 showing the two mappings for HH, corresponding to outcome $\mathbf{x} = [2, 2]^T$ in $\mathcal{R}^2$.

Example 4.7. Consider the experiment of tossing two fair coins and define the following random variables: $X_1(\omega)$ = number of heads and $X_2(\omega)$ = number of heads minus the number of tails. Obviously, there are $2^2 = 4$ outcomes in the original sample space Ω which are equally likely, and the power set $\mathcal{P}(\Omega) = \mathcal{F}$ has $2^4 = 16$ elements. Random variable X_1 has outcomes $\{0, 1, 2\}$ and X_2 has outcomes $\{-2, 0, 2\}$, as shown in Figure 4.14. Although the Cartesian product $\{0, 1, 2\} \times \{-2, 0, 2\} \in \mathcal{R}^2$ has nine points, only three have nonzero probability: $\mathbf{x} = \{[2, 2]^T, [1, 0]^T, [0, -2]^T\}$, as indicated by the solid circles in the figure. The various pdfs are

$$f_{X_1}(x_1) = (1/4)\delta(x_1) + (1/2)\delta(x_1 - 1) + (1/4)\delta(x_1 - 2), \tag{4.50}$$

$$f_{X_2}(x_2) = (1/2)\delta(x_2) + (1/4)\delta(x_2 - 2) + (1/4)\delta(x_2 + 2), \tag{4.51}$$

$$\begin{aligned} f_{X_1,X_2}(x_1, x_2) &= (1/4)\delta(x_1 - 2, x_2 - 2) + (1/2)\delta(x_1 - 1, x_2) \\ &\quad + (1/4)\delta(x_1, x_2 + 2), \end{aligned} \tag{4.52}$$

where $\delta(x_1, x_2) = \delta(x_1)\delta(x_2)$ (see Appendix B). The marginal pdfs are obtained by inspection, and they are easily verified by integrating over the joint pdf. Note that X_1 and X_2 are not independent: $f_{X_1,X_2}(x_1, x_2) \neq f_{X_1}(x_1)f_{X_2}(x_2)$. Next, let B be the event $\{X_1 < 2, X_2 < 0\}$ which corresponds to a semi-infinite rectangle containing only outcome $[0, -2]^T$. The inverse image is

$$\begin{aligned} \mathbf{X}^{-1}(B) &= X_1^{-1}((-\infty, 2)) \cap X_2^{-1}((-\infty, 0)), \\ &= \{TT, TH, HT\} \cap TT = TT \in \mathcal{F}, \end{aligned} \tag{4.53}$$

which is a specific outcome in Ω. Consider another event C given by $X_1 \geq 1$ with inverse image:

$$\mathbf{X}^{-1}(C) = X_1^{-1}([1, \infty)) = \{TH, HT, HH\} \in \mathcal{F}, \tag{4.54}$$

which we see is an event (not a single outcome) in the power set $\mathcal{P}(\Omega)$.

About the notation, a bold uppercase letter $\mathbf{X}$ is used to represent a random *vector* whose outcome is denoted by the bold lowercase letter $\mathbf{x}$. However, we also use a bold uppercase letter to represent a (nonrandom) *matrix*, such as $\mathbf{R_{XX}}$ for the autocorrelation matrix of $\mathbf{X}$ discussed later. The reader should be able to distinguish between random vector $\mathbf{X}$ and a nonrandom matrix $\mathbf{R_{XX}}$ from the context of the discussion.

In the next few sections, we describe some multivariate random variables. It turns out that there are not many widely used multivariate pdfs for *dependent* random variables. In fact, it is possible to define different versions of a multivariate pdf such that they all have marginal pdfs of the same distribution type. We provide an example of this later for exponential random variables. On the other hand, when the random variables are independent, the marginal pdfs are simply multiplied together to obtain the multivariate joint pdf. An example of this is given next for the bivariate uniform pdf.

4.6.1 Bivariate Uniform Distribution

Let $\{X_1, X_2\}$ be jointly uniform with the following joint pdf:

$$\begin{aligned} f_{X_1,X_2}(x_1, x_2) &= \begin{cases} c, & a_1 \leq x_1 \leq b_1, a_2 \leq x_2 \leq b_2 \\ 0, & \text{else,} \end{cases} \\ &= cI_{[a_1,b_1]}(x_1)I_{[a_2,b_2]}(x_2). \end{aligned} \tag{4.55}$$

Since a valid pdf must integrate to one:

$$\int_{a_1}^{b_1}\int_{a_2}^{b_2} f_{X_1,X_2}(x_1, x_2)\mathrm{d}x_1\mathrm{d}x_2 = c(b_1 - a_1)(b_2 - a_2) = cL([b_1 - a_1] \times [b_2 - a_2]), \tag{4.56}$$

so that the constant is $c = 1/[(b_1 - a_1)(b_2 - a_2)]$. The Lebesgue measure $L(\cdot)$ gives the area of the rectangle defined by the Cartesian product $[b_1 - a_1] \times [b_2 - a_2]$. From the multivariate pdf, the marginal pdf of X_1 is computed as follows:

$$f_{X_1}(x_1) = \int_{-\infty}^{\infty} f_{X_1,X_2}(x_1, x_2)\mathrm{d}x_2 = \frac{1}{(b_2 - a_2)(b_1 - a_1)}\int_{-\infty}^{\infty} I_{[a_1,b_1]}(x_1)I_{[a_2,b_2]}(x_2)\mathrm{d}x_2. \tag{4.57}$$

The indicator function for X_2 determines the limits of integration, such that

$$f_{X_1}(x_1) = \frac{1}{b_1 - a_1}I_{[a_1,b_1]}(x_1) \tag{4.58}$$

and likewise for X_2:

$$f_{X_2}(x_2) = \frac{1}{b_2 - a_2}I_{[a_2,b_2]}(x_2). \tag{4.59}$$

Since $f_{X_1,X_2}(x_1, x_2) = f_{X_1}(x_1)f_{X_2}(x_2)$, the random variables are independent. An example of the bivariate uniform pdf as well as the marginal pdfs are shown in Figure 4.15. Note that the marginal pdfs are not simply "slices" along the two axes of the bivariate pdf; they are scaled to individually have unit area.

4.6.2 Multivariate Gaussian Distribution

The multivariate Gaussian pdf for random vector $\mathbf{X} \in \mathcal{R}^N$ is

$$f_{\mathbf{X}}(\mathbf{x}) = \frac{1}{(2\pi)^{N/2}|\mathbf{C_{XX}}|^{1/2}} \exp\left(-(1/2)(\mathbf{x} - \boldsymbol{\mu}_{\mathbf{X}})^T \mathbf{C}_{\mathbf{XX}}^{-1}(\mathbf{x} - \boldsymbol{\mu}_{\mathbf{X}})\right), \tag{4.60}$$

where $\boldsymbol{\mu} \triangleq [\mu_1, \ldots, \mu_N]^T$ and μ_n is the mean of X_n. $\mathbf{C_{XX}} = \mathcal{E}[(\mathbf{X} - \boldsymbol{\mu}_{\mathbf{X}})(\mathbf{X} - \boldsymbol{\mu}_{\mathbf{X}})^T]$ is a covariance matrix where $\mathcal{E}[\cdot]$ is expectation (described in Chapter 5), $\mathbf{C}_{\mathbf{XX}}^{-1}$ is its inverse (assuming it is nonsingular), and $|\mathbf{C_{XX}}|$

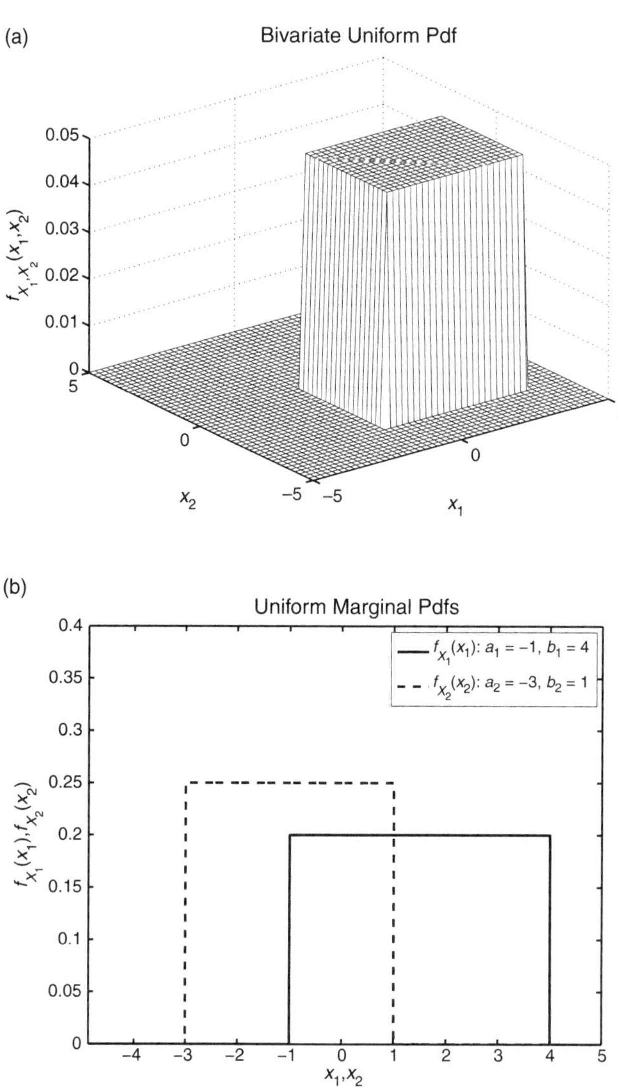

FIGURE 4.15 Uniform random variables. (a) Bivariate pdf (independent random variables). (b) Marginal pdfs with parameters.

is the determinant. An example of the pdf for $N = 2$ was shown in Figure 4.8 (for random variables X and Y). The corresponding bivariate Gaussian pdf is often written in terms of the scalar quantities as follows:

$$f_{X_1,X_2}(x_1, x_2) = \frac{1}{2\pi\sigma_{X_1}\sigma_{X_2}\sqrt{1-\rho^2}} \exp\left(-\frac{1}{2(1-\rho^2)}\left(\frac{(x_1-\mu_{X_1})^2}{\sigma_{X_1}^2} - \frac{2\rho(x_1-\mu_{X_1})(x_2-\mu_{X_2})}{\sigma_{X_1}\sigma_{X_2}} + \frac{(x_2-\mu_{X_2})^2}{\sigma_{X_2}^2}\right)\right), \tag{4.61}$$

where

$$\rho \triangleq \mathcal{E}[(X_1 - \mu_{X_1})(X_2 - \mu_{X_2})]/\sigma_{X_1}\sigma_{X_2} \tag{4.62}$$

is the correlation coefficient. These parameters are related to $\mathbf{C_{XX}}$ in (4.60) as follows:

$$\mathbf{C_{XX}} = \begin{bmatrix} \sigma_{X_1}^2 & \rho\sigma_{X_1}\sigma_{X_2} \\ \rho\sigma_{X_1}\sigma_{X_2} & \sigma_{X_2}^2 \end{bmatrix}. \tag{4.63}$$

The marginal pdfs are obtained by integrating $f_{X_1X_2}(x_1, x_2)$ as was done for the bivariate uniform pdf in Section 4.5. In order to simplify the derivations so that the technique is more easily seen, consider the case of $\mu_{X_1} = \mu_{X_2} = 0$. For notational convenience, let $\sigma_1^2 \equiv \sigma_{X_1}^2$ and $\sigma_2^2 \equiv \sigma_{X_2}^2$. Then

$$\begin{aligned} f_{X_1}(x_1) &= \frac{1}{2\pi\sigma_1\sigma_2\sqrt{1-\rho^2}} \int_{-\infty}^{\infty} \exp\left(-\frac{1}{2(1-\rho^2)}\left(\frac{x_1^2}{\sigma_1^2} - \frac{2\rho x_1 x_2}{\sigma_1\sigma_2} + \frac{x_2^2}{\sigma_2^2}\right)\right) \mathrm{d}x_2, \\ &= \frac{1}{2\pi\sigma_1\sigma_2\sqrt{1-\rho^2}} \exp\left(-\frac{x_1^2}{2\sigma_1^2(1-\rho^2)}\right) \\ &\quad \cdot \int_{-\infty}^{\infty} \exp\left(-\frac{1}{2(1-\rho^2)}\left(-\frac{2\rho x_1 x_2}{\sigma_1\sigma_2} + \frac{x_2^2}{\sigma_2^2}\right)\right) \mathrm{d}x_2, \end{aligned} \tag{4.64}$$

where the term not containing x_2 has been factored out. Completing the square yields

$$\begin{aligned} f_{X_1}(x_1) = &\frac{1}{2\pi\sigma_1\sigma_2\sqrt{1-\rho^2}} \exp\left(-\frac{x_1^2}{2\sigma_1^2(1-\rho^2)}\right) \\ &\cdot \int_{-\infty}^{\infty} \exp\left(-\frac{1}{2(1-\rho^2)}\left[\left(\frac{x_2 - \rho x_1\sigma_2/\sigma_1}{\sigma_2}\right)^2 - \frac{\rho^2 x_1^2}{\sigma_1^2}\right]\right) \mathrm{d}x_2. \end{aligned} \tag{4.65}$$

Factoring the right-most term in the exponent gives

$$\begin{aligned} f_{X_1}(x_1) = &\frac{1}{\sqrt{2\pi\sigma_1^2}} \exp\left(-\frac{x_1^2 - \rho^2 x_1^2}{2\sigma_1^2(1-\rho^2)}\right) \int_{-\infty}^{\infty} \frac{1}{\sqrt{2\pi\sigma_2^2(1-\rho^2)}} \\ &\cdot \exp\left(-\frac{1}{2(1-\rho^2)}\left(\frac{x_2 - \rho x_1\sigma_2/\sigma_1}{\sigma_2}\right)^2\right) \mathrm{d}x_2. \end{aligned} \tag{4.66}$$

The integral is 1 because the integrand is a Gaussian pdf with mean $\rho x_1\sigma_2/\sigma_1$ and variance $(1-\rho^2)\sigma_2^2$. Since the exponent of the term outside the integral simplifies when canceling $(1-\rho^2)$, we finally have the marginal pdf:

$$f_{X_1}(x_1) = \frac{1}{\sqrt{2\pi\sigma_1^2}} \exp\left(-x_1^2/2\sigma_1^2\right). \tag{4.67}$$

A similar result is obtained for $f_{X_2}(x_2)$ by integrating (4.65) with respect to x_1 (of course, this case is obtained immediately by recognizing that the bivariate Gaussian pdf is symmetric with respect to x_1 and x_2). Examples of the Gaussian bivariate pdf are provided in Figure 4.16.

For the general case when $N > 1$, any lower order joint Gaussian pdf is obtained by keeping only those parameters associated with the random variables of interest. For example, if the original joint pdf has three variables $\{X_1, X_2, X_3\}$, and we are interested in the reduced-order joint pdf for $\{X_2, X_3\}$, then only the following changes are needed:

$$\mathbf{X} = [X_1, X_2, X_3]^T \to [X_2, X_3]^T, \tag{4.68}$$

$$\boldsymbol{\mu}_{\mathbf{X}} = [\mu_{X_1}, \mu_{X_2}, \mu_{X_3}]^T \to [\mu_{X_2}, \mu_{X_3}]^T, \tag{4.69}$$

$$\mathbf{C_{XX}} = \begin{bmatrix} C_{11} & C_{12} & C_{13} \\ C_{21} & C_{22} & C_{23} \\ C_{31} & C_{32} & C_{33} \end{bmatrix} \to \begin{bmatrix} C_{22} & C_{23} \\ C_{32} & C_{33} \end{bmatrix}. \tag{4.70}$$

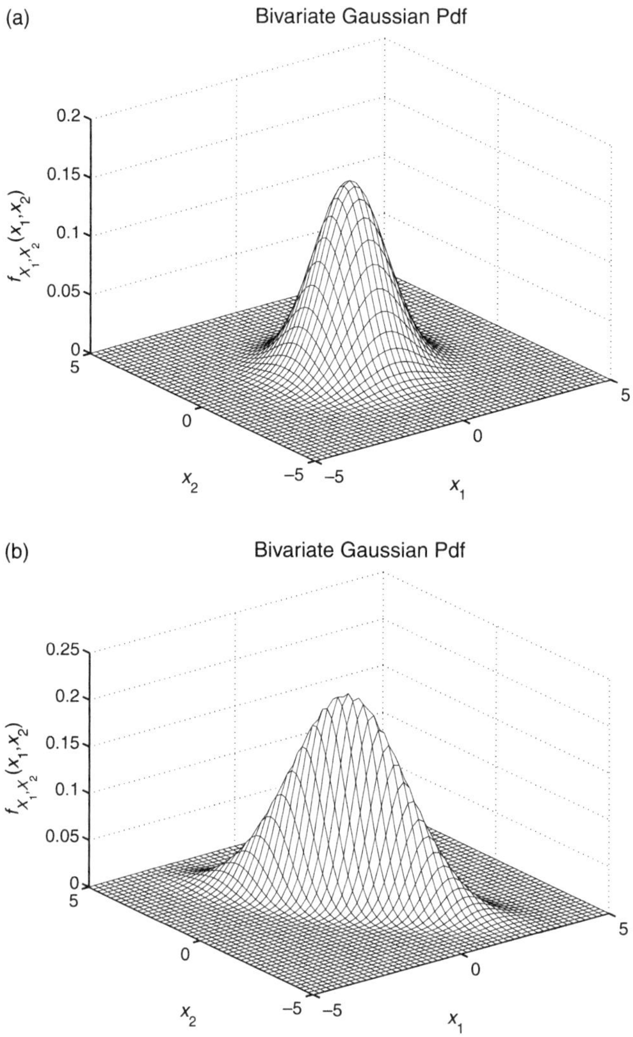

FIGURE 4.16 Gaussian bivariate pdfs with $\mu_{X_1} = \mu_{X_2} = 0$ and $\sigma_{X_1}^2 = 1$. (a) $\sigma_{X_2}^2 = 1$ and $\rho = 0$ (independent random variables). (b) $\sigma_{X_2}^2 = 2$ and $\rho = -0.8$.

The exponent on $(2\pi)^{N/2}$ is reduced from $N = 3$ to 2 for this example in order to provide the proper normalization so that the joint pdf of the lower order random vector has unit volume.

4.6.3 Multivariate Student's *t* Distribution

The multivariate pdf of Student's *t* distribution with *N* components is

$$f_{\mathbf{X}}(\mathbf{x}) = \frac{\Gamma((r+N)/2)}{\Gamma(r/2)(r\pi)^{N/2}|\tilde{\mathbf{C}}_{\mathbf{XX}}|^{1/2}}(1 + (\mathbf{x} - \boldsymbol{\mu}_{\mathbf{X}})^T \tilde{\mathbf{C}}_{\mathbf{XX}}^{-1}(\mathbf{x} - \boldsymbol{\mu}_{\mathbf{X}})/r)^{-(r+N)/2}, \tag{4.71}$$

where $r > 0$ is the degrees of freedom. $\tilde{\mathbf{C}}_{\mathbf{XX}}$ is a covariance matrix with 1s along the main diagonal, and where the other elements are the same as $\mathbf{C}_{\mathbf{XX}}$ of the multivariate Gaussian pdf. $\tilde{\mathbf{C}}_{\mathbf{XX}}$ and $\mathbf{C}_{\mathbf{XX}}$ are identical only when all the random variables have unit variance. For $N = 2$, the bivariate pdf is

$$f_{X_1,X_2}(x_1, x_2) = \frac{\Gamma((r+2)/2)}{\Gamma(r/2)k\pi\sqrt{1-\rho^2}}(1 + [1/r(1-\rho^2)][(x_1 - \mu_{X_1})^2 \\ -2\rho(x_1 - \mu_{X_1})(x_2 - \mu_{X_2}) + (x_2 - \mu_{X_2})^2])^{-(r+2)/2}, \tag{4.72}$$

which has been written in this expanded form by substituting

$$\tilde{\mathbf{C}}_{\mathbf{XX}} = \begin{bmatrix} 1 & \rho \\ \rho & 1 \end{bmatrix}. \tag{4.73}$$

Examples of Student's t bivariate pdf with $\boldsymbol{\mu}_{\mathbf{X}} = \mathbf{0}$ and $r = 1$ degree of freedom are shown in Figure 4.17 for comparison with the Gaussian bivariate pdfs in Figure 4.16. Observe that for this low value of r, Student's t pdfs are more peaked than the corresponding Gaussian pdfs. (Recall from Chapter 3 that as $r \to \infty$, Student's t distribution approaches the standard Gaussian pdf.)

4.6.4 Multinomial Distribution

The multinomial distribution is a generalization of the binomial distribution to M different kinds of outcomes. There are M random variables $\{X_1, \ldots, X_M\}$ and the probabilities of the different types of outcomes are $\{p_1, \ldots, p_M\} \in [0, 1]$ such that $\sum_{m=1}^{M} p_m = 1$. For N trials with $\sum_{m=1}^{M} X_m = N$, the joint pmf is

$$\begin{aligned} p_{X_1,\ldots,X_M}[x_1, \ldots, x_M] &= N! \prod_{m=1}^{M} \frac{p_m^{x_m}}{x_m!} I_{\{N\}}\left[\sum_{m=1}^{M} x_m\right], \\ &= \binom{N}{x_1, \ldots, x_M} \prod_{m=1}^{M} p_m^{x_m} I_{\{N\}}\left[\sum_{m=1}^{M} x_m\right], \end{aligned} \tag{4.74}$$

where

$$\binom{N}{x_1, \ldots, x_M} \triangleq \frac{N!}{\prod_{m=1}^{M} x_m!} = \frac{\Gamma(N+1)}{\prod_{m=1}^{M} \Gamma(x_m + 1)} \tag{4.75}$$

is a multinomial coefficient and $\Gamma(\cdot)$ is the gamma function. The mean and variance of the mth component are $\mu_{X_m} = Np_m$ and $\sigma_{X_m}^2 = Np_m q_m$ with $q_m \triangleq 1 - p_m$. The random variables are always negatively correlated:

$$\rho_{X_m X_n} = -\sqrt{p_m p_n / q_m q_n}. \tag{4.76}$$

The multivariate cdf is

$$F_{X_1,\ldots,X_M}(x_1, \ldots, x_M) = N! \sum_{n_1=0}^{x_1} \cdots \sum_{n_{M-1}=0}^{x_{M-1}} \frac{1}{n_1! \cdots (N - \sum_{m=1}^{M-1} n_m)!} \prod_{m=1}^{M} p_m^{n_m}, \tag{4.77}$$

where only $M - 1$ sums are needed because x_m is determined by the indicator function in (4.74), that is, $x_M = 1 - \sum_{m=1}^{M-1} x_m$. Examples of the pmf for $M = 3$ are shown in Figure 4.18 (showing only x_1 and x_2).

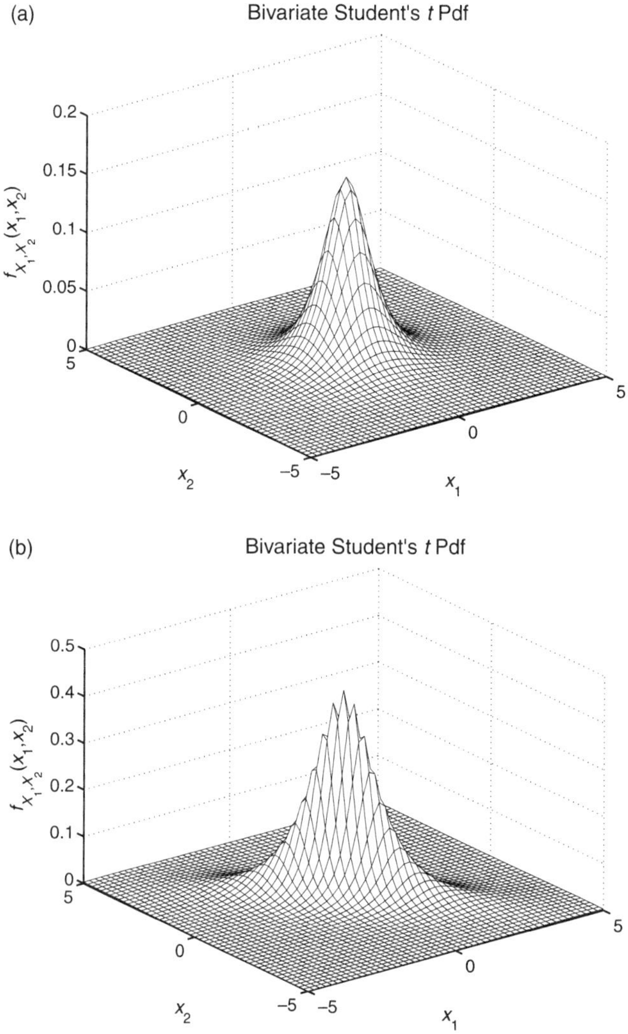

FIGURE 4.17 Bivariate Student's t pdfs with $\mu_{X_1} = \mu_{X_2} = 0$ and $r = 1$. (a) $\rho = 0$ (independent random variables). (b) $\rho = -0.8$.

4.6.5 Multivariate Hypergeometric Distribution

The multivariate hypergeometric distribution is an extension of the hypergeometric pmf from two types of objects to $L \in \mathcal{N}$ types of objects. There are still $N \in \mathcal{N}$ objects overall, consisting of M_1 objects of type 1, M_2 objects of type 2, and so on up to M_L objects of type L such that $\sum_{m=1}^{L} M_m = N$. (Recall for the hypergeometric distribution that there are M objects of one type and $N - M$ objects of another type.) The multivariate hypergeometric distribution models sampling *without* replacement of size $n \leq N$, whereas the multinomial distribution assumes sampling with replacement of size N; thus, they have the same relationship as the hypergeometric and binomial random variables. The multivariate pmf for sample size n is

$$p_{X_1,\ldots,X_L}[x_1,\ldots,x_L] = \frac{\prod_{m=1}^{L}\binom{M_m}{x_m}}{\binom{N}{n}} I_{\{n\}}\left[\sum_{m=1}^{L} x_m\right] I_{\{0,\ldots,N\}}[n] \prod_{m=1}^{L} I_{\{0,\ldots,M_m\}}[x_m], \tag{4.78}$$

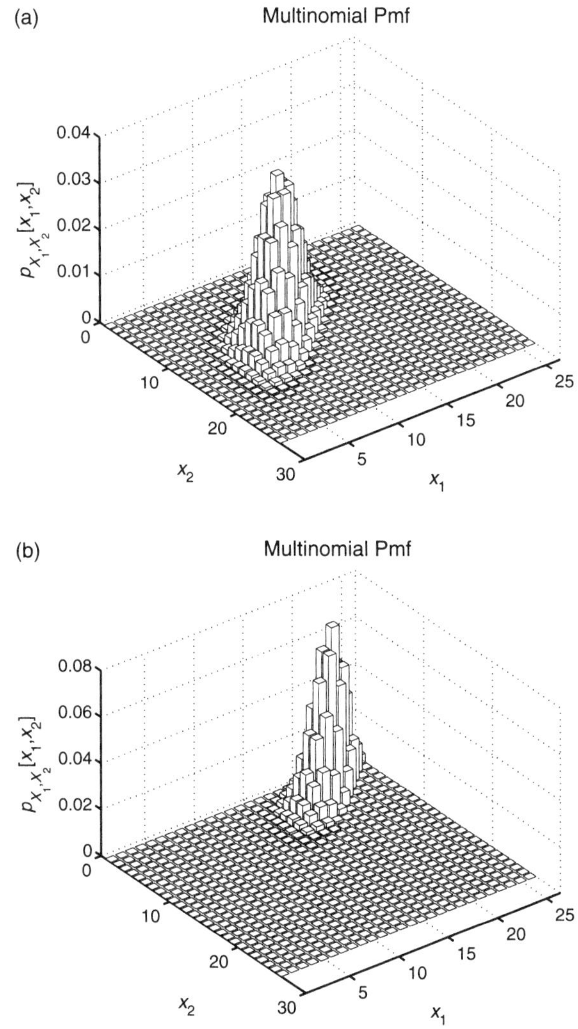

FIGURE 4.18 Multinomial pmfs for $M = 3$ (showing only x_1 and x_2). (a) $N = 25$ and $\{p_1, p_2, p_3\} = \{0.4, 0.4, 0.2\}$. (b) $N = 25$ and $\{p_1, p_2, p_3\} = \{0.8, 0.1, 0.1\}$.

where the indicator functions specify the ranges of all values. The mean and variance of the mth random variable is $\mu_{X_m} = nM_m/N$ and $\sigma^2_{X_m} = nM_m(N - M_m)(N - n)/N^2(N - 1)$. Like the multinomial distribution, the random variables are always negatively correlated:

$$\rho_{X_k X_l} = -\sqrt{M_k M_l/(N - M_k)(N - M_l)}. \tag{4.79}$$

Examples of the pmf for $L = 3$ are shown in Figure 4.19 (showing only x_1 and x_2). Note that an outcome x_n cannot exceed M_n, and the distribution is relatively narrow in each direction because n is close to N (see the mean and variance expressions above).

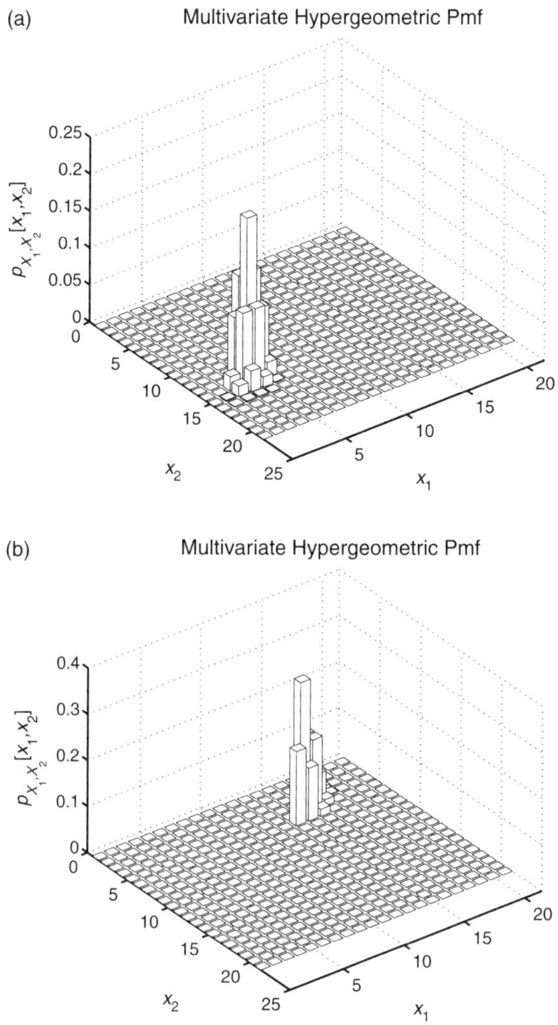

FIGURE 4.19 Multivariate hypergeometric pmfs for $L = 3$ (showing only x_1 and x_2). (a) $N = 25$, $n = 20$, and $\{M_1, M_2, M_3\} = \{5, 15, 5\}$. (b) $N = 25$, $n = 20$, and $\{M_1, M_2, M_3\} = \{18, 5, 2\}$.

4.6.6 Bivariate Exponential Distributions

As mentioned previously, it is possible to define different multivariate distributions for a random vector that is a generalization of the univariate case. For example, the following is one possible bivariate pdf with an exponential form (Gumbel, 1960):

$$f_{X_1,X_2}(x_1, x_2) = [(1 + \alpha x_1)(1 + \alpha x_2) - \alpha] \cdot \exp(-(x_1 + x_2 + \alpha x_1 x_2)) I_{[0,\infty)}(x_1) I_{[0,\infty)}(x_2) \tag{4.80}$$

with parameter $\alpha \in [0, 1]$. The bivariate cdf is

$$F_{X_1,X_2}(x_1, x_2) = [1 + \exp(-x_1 - x_2 - \alpha x_1 x_2) - \exp(-x_1) - \exp(x_2)] I_{[0,\infty)}(x_1) I_{[0,\infty)}(x_2) \tag{4.81}$$

from which we see the marginal cdfs are (by letting $x_1 \to \infty$ and $x_2 \to \infty$):

$$F_{X_1}(x_1) = 1 - \exp(-x_1), \qquad F_{X_2}(x_2) = 1 - \exp(-x_2). \tag{4.82}$$

These are standard exponential distributions with parameter $\lambda = 1$. Note that the random variables are not independent unless $\alpha = 0$. It can be shown that the correlation coefficient is

$$\rho = -1 - (1/\alpha)\exp(1/\alpha)\int_1^{\infty}(1/v)\exp(-v/\alpha)\mathrm{d}v \tag{4.83}$$

such that $\rho \in [-0.4037, 0]$ because $\alpha \in [0, 1]$.

Another bivariate exponential distribution has the following pdf:

$$\begin{aligned} f_{X_1,X_2}(x_1, x_2) = {} & [1 + \alpha(2\exp(-x_1) - 1)(2\exp(-x_2) - 1)] \\ & \cdot \exp(-x_1 - x_2)I_{[0,\infty)}(x_1)I_{[0,\infty)}(x_2), \end{aligned} \tag{4.84}$$

and cdf:

$$\begin{aligned} F_{X_1,X_2}(x_1, x_2) = {} & (1 - \exp(-x_1))(1 - \exp(-x_2)) \\ & \cdot[1 + \alpha\exp(-x_1 - x_2)]I_{[0,\infty)}(x_1)I_{[0,\infty)}(x_2), \end{aligned} \tag{4.85}$$

where $\alpha \in [-1, 1]$. As in (4.82), the marginal pdfs are standard exponential with parameter $\lambda = 1$. The correlation coefficient is (see Problem 5.35)

$$\rho = \alpha/4, \tag{4.86}$$

such that $\rho \in [-1/4, 1/4]$. The random variables are independent only when $\alpha = 0$. It is interesting that although (4.80) and (4.84) have the same marginal pdfs, the correlation coefficients are quite different. These bivariate pdfs are shown in Figure 4.20 for $\alpha = 1/2$. Other versions of the bivariate exponential pdf are possible (see Problem 4.11), as is the case for other families of univariate random variables extended to multivariate distributions.

4.7 GENERATING DEPENDENT RANDOM VARIABLES

One can easily generate a multivariate pdf $f_{\mathbf{X}}(\mathbf{x})$ for N *independent* marginal pdfs as the product $\prod_{n=1}^{N} f_{X_n}(x_n)$, and this can be done even if the marginal pdfs are from different distribution families as illustrated in previous examples. Unfortunately, there are not many widely used multivariate distributions for *dependent* random variables, as in the case of the bivariate Gaussian and Student's t distributions where the correlation coefficient ρ appears explicitly. Dependency can be introduced for other random variables by implementing the following procedure, which we describe for the bivariate case ($N = 2$). This method does not yield a closed-form expression for the desired joint pdf, but instead is a means of generating *samples* for a computer simulation:

- Let $\mathbf{X} = [X_1, X_2]^T$ be a random vector with a bivariate standard Gaussian pdf and a prespecified value for the correlation coefficient $\rho \neq 0$. Note that for the standard bivariate pdf, the mean is zero and $\mathbf{C_{XX}} = \tilde{\mathbf{C}}_{\mathbf{XX}}$. This matrix was defined in (4.73) in connection with the bivariate Student's t distribution; it differs from $\mathbf{C_{XX}}$ by having 1s along the main diagonal. Random variables $\{X_1, X_2\}$ individually have marginal distributions that are standard Gaussian.
- Apply the probability transformation $Y_n = F_X(X_n)$, $n = 1, 2$, to each random variable where $F_X(\cdot)$ is the cdf of the standard Gaussian random variable. The $\{Y_n\}$ are now uniform on [0, 1] (see Theorem 4.2) and they are dependent due to the original dependence of $\{X_n\}$.
- Now that the random variables are uniform, each can be transformed to a new random variable by using the inverse of the corresponding cdf (see Theorem 4.3 and the discussion in Chapter 1). Thus, we form $Z_n = F_Z^{-1}(Y_n)$, $n = 1, 2$, where $F_Z^{-1}(\cdot)$ is the inverse cdf (icdf) of the desired random variable. The transformations to the dependent $\{Z_n\}$ need not be the same.

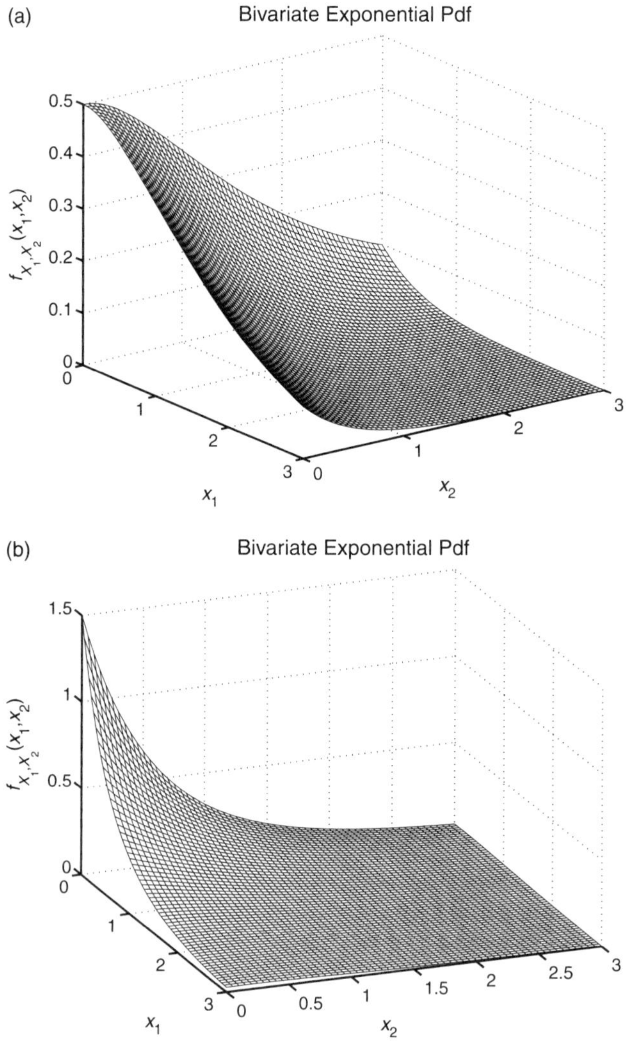

FIGURE 4.20 Bivariate exponential pdfs with $\alpha = 1$. (a) Distribution in (4.80). (b) Distribution in (4.84).

Summarizing the approach: for $\{X_1, X_2\}$ distributed according to the standard bivariate Gaussian distribution with correlation coefficient ρ, the transformed random variables $\{Z_1, Z_2\}$ are obtained as follows:

$$Z_n = F_{Z_n}^{-1}(F_X(X_n)), \qquad n = 1, 2, \tag{4.87}$$

where subscript n has been included on $F_{Z_n}(\cdot)$ to emphasize that the mappings to each Z_n can differ. Although this procedure is capable of generating dependent random variables with an arbitrary bivariate pdf, the resulting dependency between Z_1 and Z_2 is *not* the same as the *linear* dependency given by ρ of the bivariate Gaussian random variables $\{X_1, X_2\}$ (see Chapter 5 for a discussion of this linear relationship).

Theorem 4.2 (Probability transformation). Let X be a continuous random variable with strictly increasing cdf $F_X(X)$. Random variable $Y = F_X(X)$ is uniformly distributed on [0, 1].

Proof. Note that the argument of $F_X(X)$ is random variable X (not an outcome x). As a result, $Y = F_X(X)$ is a mapping such that we can write

$$P(Y \leq y) = P(F_X(X) \leq y) = P(X \leq F_X^{-1}(y)), \tag{4.88}$$

where the inverse image $F_X^{-1}(y)$ is well-defined because $F_X(\cdot)$ is strictly increasing (one-to-one). The last expression is the cdf of X such that

$$P(Y \leq y) = F_X(F_X^{-1}(y)) = yI_{[0,1]}(y), \tag{4.89}$$

where the indicator function has been included to emphasize that the range of $F_X(\cdot)$ is [0, 1], which is the support of Y. Finally, the pdf of Y is

$$f_Y(y) = \frac{\mathrm{d}}{\mathrm{d}y} F_Y(y) = I_{[0,1]}(y), \tag{4.90}$$

which is the uniform distribution on [0, 1]. □

The technique used in this proof for computing the pdf due to a transformation (mapping) is called the "cdf approach." It is explored later in this chapter when general transformations of random variables are covered.

Theorem 4.3 (Inverse probability transformation). Let random variable Y be uniformly distributed on [0, 1], and let $F_Z(z)$ be a strictly increasing cdf. Random variable $Z = F_Z^{-1}(Y)$ has distribution $F_Z(z)$.

Proof. As in Theorem 4.2, we can write

$$P(Z \leq z) = P(F_Z^{-1}(Y) \leq z) = P(Y \leq F_Z(z)) = F_Y(F_Z(z)). \tag{4.91}$$

Since $F_Y(y) = yI_{[0,1]}(y)$ is a linear function of its argument, the expression simplifies:

$$P(Z \leq z) = F_Z(z)I_{[0,1]}(F_Z(z)) = F_Z(z). \tag{4.92}$$

The last line follows because $F_Z(z)$ has range [0, 1] by definition so that the indicator function can be dropped. □

The condition that the cdfs in the previous theorems be strictly increasing yields relatively simple transformation results. More general cdfs can be considered, but more work is required to describe those results as shown in Section 4.8.

Example 4.8. Figure 4.21(a) shows a histogram of 100,000 samples from a bivariate Gaussian distribution with mean vector $\boldsymbol{\mu}_\mathbf{X} = \mathbf{0}$ and covariance matrix

$$\mathbf{C_{XX}} = \begin{bmatrix} 1 & -0.75 \\ -0.75 & 1 \end{bmatrix}, \tag{4.93}$$

which has correlation coefficient $\rho = -0.75$. The data were generated using mvnrand from MATLAB® which is the multidimensional version of randn. The negative correlation coefficient causes the distribution to be elongated along the $x_2 = -x_1$ axis. Figure 4.21(b) shows the resulting bivariate uniform distribution when $\{X_1, X_2\}$ are individually transformed using the probability transformation technique above. (Note that because the variables are correlated, the distribution is not uniformly flat on its support.) Random variables $\{Y_1, Y_2\}$ also have a negative correlation. Figure 4.21(c) shows the bivariate distribution when each of the uniform random

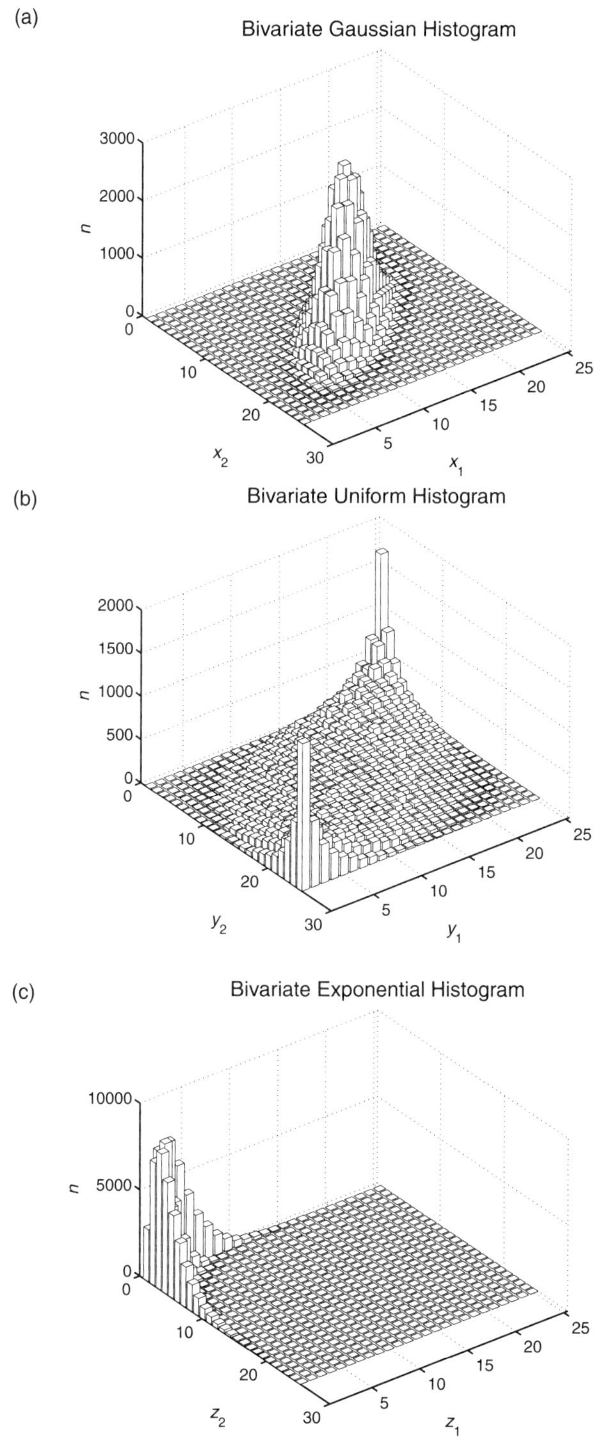

FIGURE 4.21 Histograms of bivariate random variables. (a) Bivariate Gaussian with $\rho = -0.75$. (b) Probability transformation to bivariate uniform. (c) Inverse probability transformation to bivariate exponential.

variables is inverse probability transformed using the icdf for an exponential random variable with parameter $\lambda = 0.5$. The icdf for probability p is

$$z = -(1/\lambda)\ln(1-p), \tag{4.94}$$

yielding exponential random variables $\{Z_1, Z_2\}$ which are also negatively correlated. The estimated correlation coefficients for random vectors **X**, **Y**, and **Z** computed from the samples using corrcoef are $\hat{\rho} = -0.7471$, -0.7325, and -0.5162, respectively.

4.8 RANDOM VARIABLE TRANSFORMATIONS

The families of random variables discussed in Chapter 3 are important for many applications. Often, their pdfs are transformed, either because of some physical mechanism (like a communication channel) or by intentional manipulation to generate variables that have more useful properties. We consider the following three basic types of transformations:

- One random variable is a function of another random variable $Y = g(X)$, where the mapping $g(\cdot)$ is known and nonrandom.
- Different functions $\{g_n(\cdot)\}$ map several random variables to another set of random variables.
- A time-indexed set of random variables representing a *random process* are filtered by a system to generate a modified random process.

Usually, we are interested in how the pdfs of the random variables are changed by such transformations. The third case is covered in Chapter 8 when systems operating on random processes are discussed, and where we are also interested in how the correlation is affected by filtering. In this section, we focus on the first case involving a single random variable:

$$Y = g(X), \tag{4.95}$$

where $g(\cdot)$ is a *memoryless* mapping. Some special mappings of multiple random variables for the second case are covered later in this chapter.

It will be convenient to consider discrete and continuous random variables separately. Transformations of discrete random variables are straightforward because the pmf is directly related to the probabilities of events. For continuous random variables, it will be necessary to use techniques from calculus to perform the transformation. In addition to the type of random variable, the form of the function is important. Specifically, we need to know whether or not the function is monotonic.

Definition: Monotonic Function Function $y = g(x)$ is *monotonic increasing* if $g(x) \leq g(y)$ for $x \leq y$. The function is *monotonic decreasing* if $g(x) \geq g(y)$ for $x \leq y$. With strict inequalities $g(x) < g(y)$ and $g(x) > g(y)$, the function is *strictly monotonic*.

A monotonic function can be a many-to-one mapping due to regions where the function is flat (the derivative is zero). Strictly monotonic functions are also known as one-to-one mappings.

4.8.1 Transformations of Discrete Random Variables

The following transformation results apply to finite or countably infinite random variables:

- *One-to-one (strictly monotonic)* $g(X)$. Observe that the probability of a mapping can be rewritten as follows:

$$P(Y = y) = P(g(X) = y) = P(X = g^{-1}(y)), \tag{4.96}$$

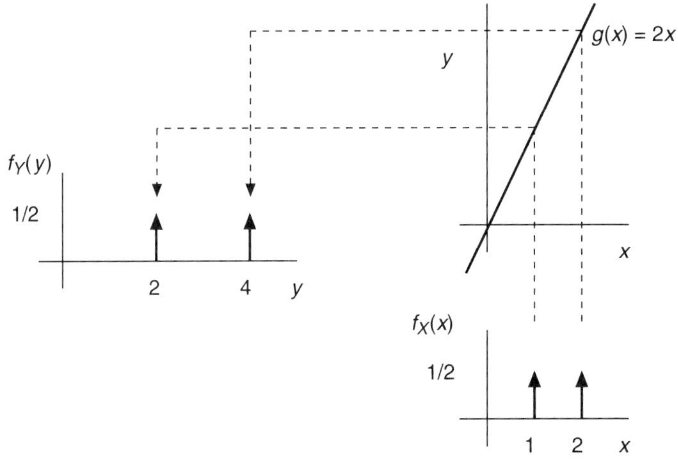

FIGURE 4.22 Probability mapping for a one-to-one transformation of a discrete random variable.

where $g^{-1}(y)$ is the inverse image. A simple example of a one-to-one mapping is illustrated in Figure 4.22 (using pdfs and the Dirac delta function for discrete random variables). The probability mass of one value of X carries over to the corresponding Y value. Because the function is one-to-one, Y necessarily has the same number of outcomes as X and the relative probabilities do not change. This type of mapping is essentially a *reassignment* of the probabilities to different outcomes.

Although a function by definition maps its argument to a single point, this is generally not true for the inverse image. For a monotonic function with flat regions, the function is a many-to-one mapping.

- *Monotonic* $g(X)$. The function is partitioned along the x-axis into flat regions and one-to-one regions. The strictly monotonic regions are handled separately as described above where the probability mass for each x carries over to the corresponding value of y. In flat regions, the *cumulative* probability mass of all values of x that map to y is the probability for that particular y. This result is true even for nonmonotonic functions, but we must be careful to consider all distinct regions of x that map to the same y, as described next.
- *Arbitrary* $g(X)$. Suppose that N different values of x map to the same y. Since X is discrete and the x values are distinct (mutually exclusive), we can always write the probability in terms of a union of the values of X:

$$P(Y = y) = P\left(\bigcup_{n=1}^{N}\{X = x_n\}\right) = \sum_{n=1}^{N} P(X = x_n) = \sum_{n=1}^{N} P(X = g_n^{-1}(y)),$$

where the subscript on $g_n^{-1}(y)$ identifies the different values of x that map to the same y. N need not be finite: the result could include a countably infinite number of values of x that map to the same y. It is only necessary that we keep track of all x values so that the corresponding cumulative probability mass is assigned to the correct value y. A simple nonmonotonic example for a finite random variable is illustrated in Figure 4.23.

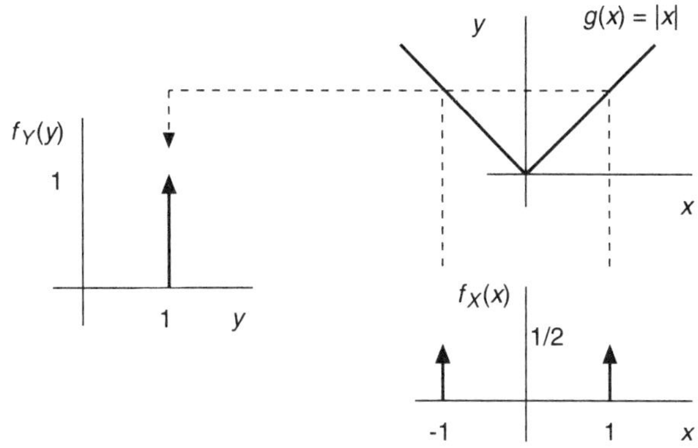

FIGURE 4.23 Probability mapping for a many-to-one transformation of a discrete random variable.

Example 4.9. Let X be a geometric random variable with parameter p and suppose the mapping is $g(X) = u(X-3)$ (a shifted unit-step function). As a result, $X = \{0, 1, 2\}$ map to $Y = 0$ and $X \geq 3$ map to $Y = 1$. Clearly, Y is a Bernoulli random variable with

$$P(Y=0) = p(p^2 - 3p + 3), \quad P(Y=1) = 1 - p(p^2 - 3p + 3), \tag{4.97}$$

where the last term is the probability of success.

4.8.2 Transformations of Continuous Random Variables

For continuous random variables, we first consider one-to-one mappings and then generalize the result to arbitrary functions. Regions where the derivative is zero must be handled in a different manner. Four different techniques for performing the transformation in (4.95) are described: they are known as (i) the pdf or *direct* method, (ii) the cdf or *indirect* method, (iii) the delta function method, and (iv) the indicator function method.

- *Indirect method for one-to-one* $g(X)$. Although this case is similar to that for discrete random variables, we cannot start with $P(Y = y)$ because it is zero for a continuous random variable. Instead, we examine the cdf $F_Y(y)$ and consider strictly increasing functions and strictly decreasing functions separately. First, assume that $g(X)$ is strictly increasing such that we can write

$$\begin{aligned} F_Y(y) &= P(Y \leq y) = P(g(X) \leq y), \\ &= P(X \leq g^{-1}(y)) = F_X(g^{-1}(y)), \end{aligned} \tag{4.98}$$

where the inverse image $g^{-1}(y)$ is unique for each y because the function is one-to-one. The corresponding pdf is obtained by differentiating this result, yielding

$$\begin{aligned} f_Y(y) &= \frac{\mathrm{d}}{\mathrm{d}y} F_Y(y) = \frac{\mathrm{d}}{\mathrm{d}y} F_X(g^{-1}(y)), \\ &= \frac{\mathrm{d}}{\mathrm{d}x} F_X(g^{-1}(y)) \frac{\mathrm{d}x}{\mathrm{d}y} = f_X(g^{-1}(y)) \frac{\mathrm{d}x}{\mathrm{d}y}, \end{aligned} \tag{4.99}$$

where the chain rule has been used and $x = g^{-1}(y)$ has been substituted.

For a strictly decreasing function, observe that the next-to-last expression in (4.98) is instead

$$F_Y(y) = P(X \geq g^{-1}(y)) = 1 - F_X(g^{-1}(y)), \tag{4.100}$$

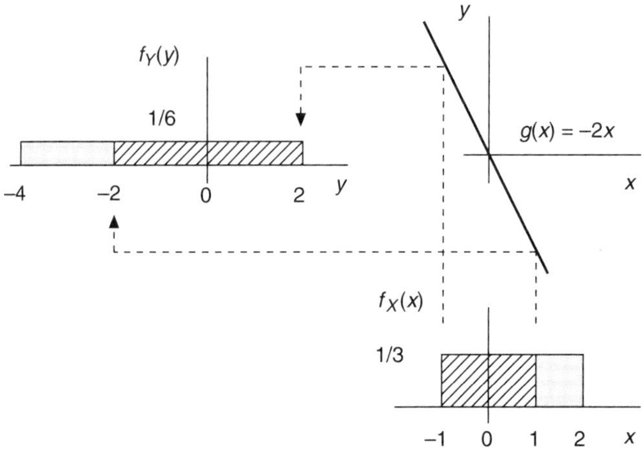

FIGURE 4.24 Mapping for a continuous random variable when the function is strictly decreasing. Observe from the shaded areas that $P(Y \leq -2) = P(X \geq 1)$ because the mapping "reverses" the pdf.

where the direction of the inequality is reversed. This can be seen from the example in Figure 4.24 where the distributions are uniform. Differentiating this expression gives

$$f_Y(y) = -f_X(g^{-1}(y))\frac{\mathrm{d}x}{\mathrm{d}y}, \tag{4.101}$$

which is identical to (4.99) except for the minus sign. However, since the derivative $\mathrm{d}x/\mathrm{d}y$ is negative for a strictly decreasing function, (4.101) can be replaced with

$$f_Y(y) = f_X(g^{-1}(y))\left|\frac{\mathrm{d}x}{\mathrm{d}y}\right|, \tag{4.102}$$

which also describes (4.99) and thus holds for any one-to-one mapping. Finally, since it is usually easier to calculate the derivative of $y = g(x)$, the following equivalent result is generally used:

$$f_Y(y) = \left.\frac{f_X(x)}{|g'(x)|}\right|_{x=g^{-1}(y)} \tag{4.103}$$

where $g'(x)$ is the ordinary derivative of the function with respect to x.

The previous indirect approach can also be used for an arbitrary function, but it it easier to use the following direct method.

- *Direct method for arbitrary* $g(X)$. Assume there are no regions where the derivative of the function is zero. The resulting transformation has a form similar to (4.103), which cannot be directly evaluated when $g'(x) = 0$. The case where the function has flat regions is described later.

Consider the nonmonotonic function shown in Figure 4.25 for which we see that each value of y is obtained from two different values of x (except $x = 0$). Although we consider this simple case for a uniform random variable, it will be evident that the final result is easily generalized to any number of intervals that map from x to y (including countably infinite). Start with the probability of y in the small interval $[y, y + \mathrm{d}y]$:

$$\begin{aligned} P(y \in [y, y + \mathrm{d}y]) &= P(x \in [x_1, x_1 + \mathrm{d}x_1] \cup x \in [x_2, x_2 + \mathrm{d}x_2]), \\ &= P(x \in [x_1, x_1 + \mathrm{d}x_1]) + P(x \in [x_2, x_2 + \mathrm{d}x_2]), \end{aligned} \tag{4.104}$$

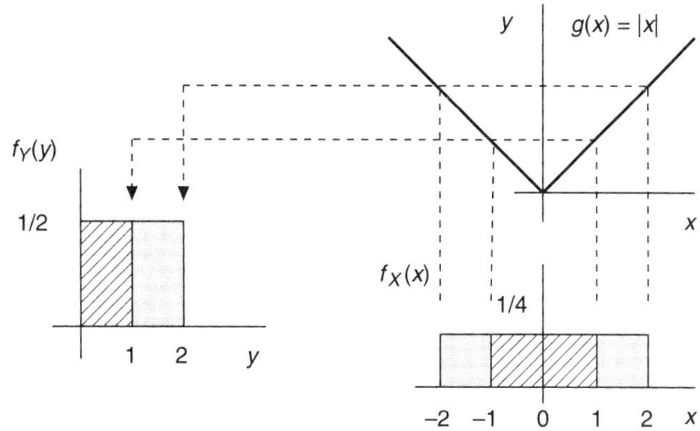

FIGURE 4.25 Mapping for a continuous random variable when the function is not monotonic. The two shaded areas for X indicate intervals that map to the same interval of Y (for a many-to-one function).

where the second equality follows because the intervals have been chosen to be disjoint. These probabilities can be approximated by the areas of rectangles as follows:

$$f_Y(y)|dy| \approx f_X(x_1)|dx_1| + f_X(x_2)|dx_2|, \tag{4.105}$$

where the intervals are sufficiently small so that $|dx_1| \approx |dx_2| \triangleq |dx|$. As is done in differential calculus, the values $\{x_1, x_2, y\}$ could be the center points of the various intervals, but the choice is not important because the size of each interval will tend to zero in the final result. The absolute value $|dx_n|$ guarantees that the areas are positive in the event that dx_1 or dx_2 in (4.104) are negative (the pdfs are, of course, always nonnegative). Finally, dividing this result by dy, substituting the inverse image $x = g^{-1}(y)$, and generalizing the approach to N different regions of the function gives

$$f_Y(y) = \sum_{n=1}^{N} \frac{f_X(x_n)}{|g'(x_n)|}\Bigg|_{x_n = g_n^{-1}(y)} \tag{4.106}$$

where the subscript on $g_n(x)$ denotes the nth region of the function corresponding to x_n.

The number of regions of the function $g(x)$ is easily determined as illustrated by the example in Figure 4.26. Horizontal lines are drawn through the function to determine where they intersect $y = g(x)$ for various values of y. In this example, there are three distinct regions: (i) (x_4, ∞) maps one-to-one to (y_2, ∞), (ii) $(-\infty, x_1]$ maps one-to-one to $(-\infty, y_1]$, and (iii) three subregions of $(x_1, x_4]$ map to the same interval $(y_1, y_2]$. These subregions are labeled in the figure as $\{(x_1, x_2], (x_2, x_3], (x_3, x_4]\}$. (Note that it is not important which end is closed on the intervals since $g(X)$ is assumed to be continuous.) Because the shape of the function in each subregion differs (which is not necessarily the case), a subscript is included on $g_n(\cdot)$ in (4.106). The horizontal lines allow one to easily see regions where the function changes from increasing to decreasing, and vice versa, indicating where many-to-one mappings are located.

Sketching horizontal lines through the function is equivalent to finding the roots of $g(x) - y = 0$ for various values of y. For example, consider $y = g(x) = x^2 - 1$ for which the roots of $x^2 - y - 1 = 0$ are $x = \pm\sqrt{y+1}$. Two values of x map to the same value of y (except for $y = 0$) so that $N = 2$ in (4.106). Thus, in general, we can replace (4.106) with the following simplified form:

$$f_Y(y) = \sum_{n=1}^{N} \frac{f_X(x_n)}{|g'(x_n)|} \tag{4.107}$$

with the understanding that the $\{x_n\}$ are the *roots* of $g(x) - y = 0$ for $n = 1, \ldots, N$.

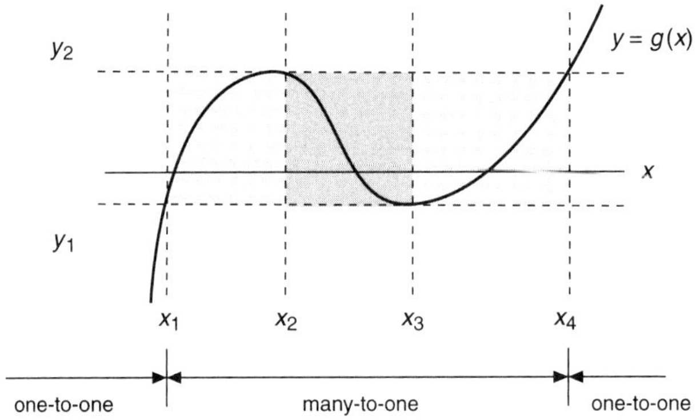

FIGURE 4.26 Regions of a nonmonotonic function for the mapping from X to Y are determined by horizontal lines intersecting $g(x)$.

Example 4.10. Let X be the Laplace random variable with parameters $\{\mu = 0, \alpha\}$ and consider the mapping $Y = X^2$ for which $N = 2$. Since $g'(x) = 2x$, we have

$$f_Y(y) = \frac{1}{2\alpha} \sum_{n=1}^{2} \frac{1}{|2x_n|} \exp(-|x_n|/\alpha), \tag{4.108}$$

which has support $\mathcal{R}^+$. The roots of $g(x) - y = 0$ are $x_n = \pm\sqrt{y}$ so that

$$\begin{aligned} f_Y(y) &= \frac{1}{4\alpha\sqrt{y}}[\exp(-|\sqrt{y}|/\alpha) + \exp(-| - \sqrt{y}|/\alpha)], \\ &= \frac{1}{2\alpha\sqrt{y}} \exp(-\sqrt{y}/\alpha) I_{[0,\infty)}(y), \end{aligned} \tag{4.109}$$

where the absolute values have been ignored in the final result. An example pdf is shown in Figure 4.27 for $\alpha = 1$, where the pdf (dashed line) increases to infinity at the origin because of $\sqrt{y}$ in the denominator. This is due to the fact that $g'(x)$ approaches zero at $x = 0$. (The chi-square random variable with one degree of freedom, which is obtained as the square of a Gaussian random variable, exhibits similar asymptotic behavior at the origin.)

For an arbitrary function, it is usually more difficult to apply the indirect (cdf) method because it can be cumbersome to keep track of all the different regions of the x-axis that map to the same values of y. However, it is useful for relatively simple functions such as the one in Example 4.11.

Example 4.11. For the transformation $Y = X^2$, we examine the following probability:

$$P(Y \leq y) = P(X^2 \leq y) = P(-\sqrt{y} \leq X \leq \sqrt{y}), \tag{4.110}$$

which can be written in terms of the cdf for X:

$$F_Y(y) = F_X(\sqrt{y}) - F_X(-\sqrt{y}). \tag{4.111}$$

Differentiating with respect to y yields the general result

$$f_Y(y) = \frac{1}{2\sqrt{y}}[f_X(\sqrt{y}) + f_X(-\sqrt{y})]. \tag{4.112}$$

Substituting the pdf for the Laplace random variable gives (4.109) in Example 4.10.

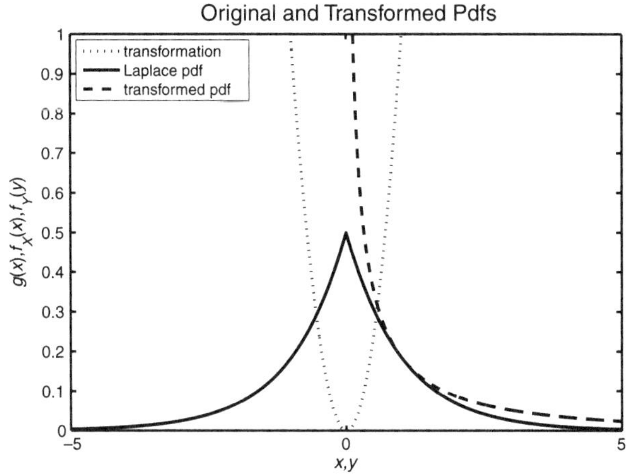

FIGURE 4.27 Pdfs for Example 4.10 with $\alpha = 1$ for the Laplace pdf and the transformation $Y = X^2$.

- *Functions with flat intervals.* Next, we describe how to handle transformations when the function has one or more intervals where the derivative is zero. An example of a function with one such region is shown in Figure 4.28. Observe that the function can be partitioned into multiple components: (i) strictly increasing or strictly decreasing regions and (ii) flat regions where $g'(x) = 0$. Those regions corresponding to the first case are handled by the formula in (4.106). Consider the flat region (lightly shaded) in Figure 4.28 which occurs on $[x_2, x_3]$, and observe that

$$P(Y = y) = P(X \in [x_2, x_3]) = \int_{x_2}^{x_3} f_X(x) \mathrm{d}x \triangleq P_Y(y), \tag{4.113}$$

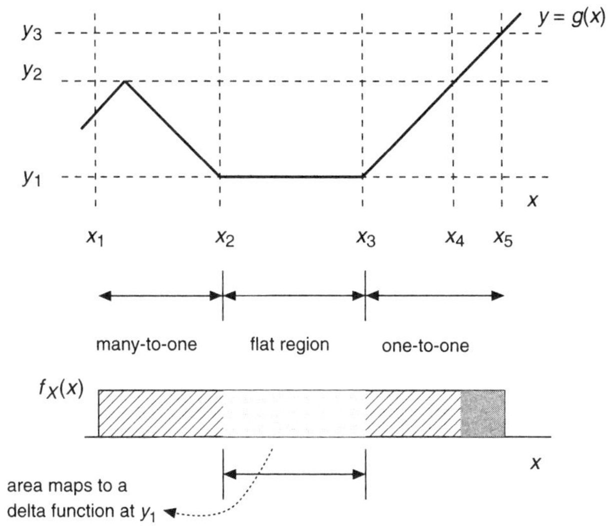

FIGURE 4.28 Example of a nonmonotonic function with one flat region. The area of $f_X(x)$ for $x \in [x_2, x_3]$ maps to a delta function at $y = y_1$.

which is not zero because $x_3 > x_2$. The area under $f_X(x)$ is the probability $P(Y = y)$, for which we use the shorthand notation $P_Y(y)$ (it is *not* the pmf $p_Y[y]$). We may combine this result with (4.106) to generate a *mixed* pdf for Y, assuming there are M flat regions:

$$f_Y(y) = \sum_{n=1}^{N} \left. \frac{f_X(x_n)}{|g'(x_n)|} \right|_{x_n = g_n^{-1}(y)} + \sum_{m=1}^{M} P_Y(y_m)\delta(y - y_m), \tag{4.114}$$

where $\delta(y)$ is the Dirac delta function. Note that for the example in Figure 4.28, a horizontal line just below y_2 intersects $g(x)$ three times for $x \in [x_1, x_2] \cup [x_3, x_4]$; these intervals map to $[y_1, y_2]$. Interval $[x_4, x_5]$ maps one-to-one to $[y_2, y_3]$. Since the uniform pdf is zero beyond $[x_1, x_5]$, no other regions need to be considered for this transformation.

In the extreme case where the function consists only of flat regions (as in a staircase function), the mapping converts a continuous random variable to a purely discrete random variable. This mapping is described later in this chapter where the indicator function is used to denote the various flat regions.

Example 4.12. Let X be an exponential random variable with parameter λ which is mapped as follows:

$$g(X) = \begin{cases} 0, & X < 0 \\ 2X, & 0 \le X \le 1 \\ 2, & X > 1, \end{cases} \tag{4.115}$$

yielding a mixed random variable. Using the indirect method for the linear region, the cdf of $Y = g(X)$ for $X \in [0, 1]$ is

$$F_Y(y) = P(2X \le y) = P(X \le y/2) = F_X(y/2). \tag{4.116}$$

Differentiating this expression with respect to y gives

$$\frac{\mathrm{d}}{\mathrm{d}y} F_X(y/2) = (1/2) f_X(y/2) = (\lambda/2)\exp(-\lambda y/2) I_{[0,2]}(y), \tag{4.117}$$

where the indicator function has been included to specify the support of the continuous part of $f_Y(y)$. The discrete part is obtained as follows:

$$P(Y = 2) = P(X > 1) = \lambda \int_1^{\infty} \exp(-\lambda x)\mathrm{d}x = \exp(-\lambda), \tag{4.118}$$

and thus the overall pdf of Y is

$$f_Y(y) = (\lambda/2)\exp(-\lambda y/2) I_{[0,2]}(y) + \exp(-\lambda)\delta(y - 2). \tag{4.119}$$

It is straightforward to show that this pdf has unit area. The transformation and the two pdfs are illustrated in Figure 4.29.

- *Delta function method for arbitrary* $g(X)$. It is possible to reformulate the transformation procedure for continuous functions using the following property of the Dirac delta function when it is a *composition* with smooth function $g(x)$ (see Appendix B):

$$\delta(g(x)) = \sum_{n=1}^{N} \frac{\delta(x - x_n)}{|g'(x_n)|}, \tag{4.120}$$

where $\{x_n\}$ are the roots of $g(x) = 0$. We demonstrate this property with the following example.

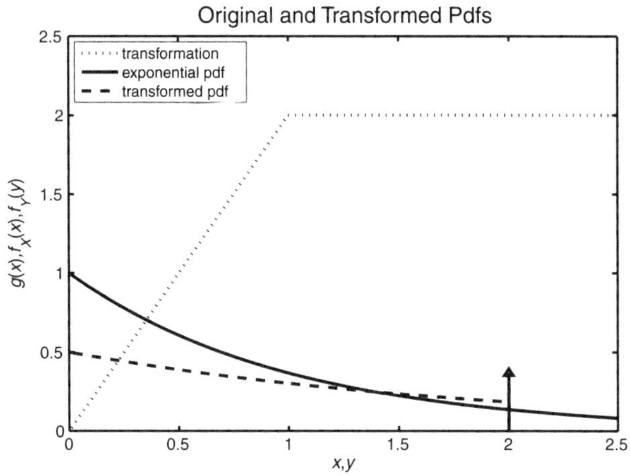

FIGURE 4.29 Pdfs for the random variable transformation in Example 4.12 where X is exponential with $\lambda = 1$.

Example 4.13. Let the transformation be

$$g(x) = x^2 - 4x + 3 = (x - 1)(x - 3) \tag{4.121}$$

for which $g'(x) = 2x - 4$. The roots of $g(x) = 0$ are $x_1 = 1$ and $x_2 = 3$ so that $g'(1) = -2$ and $g'(3) = 2$. The expression in (4.120) gives

$$\delta(x^2 - 3x + 2) = (1/2)[\delta(x - 3) - \delta(x - 1)]. \tag{4.122}$$

We are interested in the roots of the argument $g(x) = 0$ of the delta functions in (4.120) because $\delta(x)$ is nonzero only when $x = 0$. In Example 4.13, $g(x)$ has two roots such that delta functions are located at $x =$ 1 and 3. The scaling for each term is due to the following result from calculus when changing variables for monotonic function $g(x)$:

$$\int_a^b h(g(x))|g'(x)|\mathrm{d}x = \int_{g(a)}^{g(b)} h(x)\mathrm{d}x. \tag{4.123}$$

Recall that the Dirac delta function is characterized by how it behaves when integrated. Using the result above with $h(x) = \delta(x)$, $a = -\infty$, and $b = \infty$ yields

$$\int_{-\infty}^{\infty} \delta(g(x))|g'(x)|\mathrm{d}x = \int_{-\infty}^{\infty} \delta(x)\mathrm{d}x, \tag{4.124}$$

and rearranging the integrands gives

$$\delta(g(x)) = \frac{1}{|g'(x)|}\delta(x). \tag{4.125}$$

Equation (4.120) is a generalization of this result for nonmonotonic functions where $g(x) = 0$ has N roots.

Comparing (4.120) and (4.106), which is used to determine the pdf of $Y = g(X)$, we recognize their similarity. In fact, from the sifting property of the Dirac delta function:

$$\int_{-\infty}^{\infty} f_X(x)\delta(g(x))\mathrm{d}x = \int_{-\infty}^{\infty} f_X(x)\sum_{n=1}^{N}\frac{\delta(x-x_n)}{|g'(x_n)|}\mathrm{d}x = \sum_{n=1}^{N}\frac{f_X(x_n)}{|g'(x_n)|}, \tag{4.126}$$

where $\{x_n\}$ are the roots of $g(x) = 0$. In order to obtain $f_Y(y)$ from the transformation, $\{x_n\}$ must be the roots of $g(x) - y = 0$ as discussed after (4.107). Thus, we can replace $\delta(g(x))$ with $\delta(g(x) - y)$ in expression (4.126), giving the final result:

$$f_Y(y) = \int_{-\infty}^{\infty} f_X(x)\delta(g(x)-y)\mathrm{d}x. \tag{4.127}$$

Since the roots of $g(x) - y = 0$ are determined by the value of y, the right-hand side is a function of y, as is the case in the original expression in (4.107).

The advantage of this approach is that (4.120) is used instead of (4.106) with the following modification:

$$\delta(g(x)-y) = \sum_{n=1}^{N}\frac{\delta(x-x_n)}{|g'(x_n)|}, \tag{4.128}$$

where $\{x_n\}$ are the roots of $g(x) - y = 0$ (*not* of $g(x) = 0$). The sum of delta functions is substituted into (4.127), and the sifting property produces $f_Y(y)$, which is the same result as (4.106). The usefulness of the delta-function method will be more apparent later when transformations of two or more random variables are discussed.

Example 4.14. For the transformation,

$$Y = X^2 - 4X + 2, \tag{4.129}$$

the roots of $g(x) - y = x^2 - 4x + 2 - y = 0$ are $x_{1,2} = 2 \pm \sqrt{2+y}$, from which we find that the support of Y is $[-2, \infty)$. Since

$$g'(x_1) = 2\sqrt{2+y}, \qquad g'(x_2) = -2\sqrt{2+y}, \tag{4.130}$$

the pdf of Y is

$$f_Y(y) = \frac{1}{2\sqrt{2+y}}\left[f_X(2+\sqrt{2+y}) + f_X(2-\sqrt{2+y})\right]. \tag{4.131}$$

Figure 4.30 shows $f_Y(y)$ when X is the standard Gaussian random variable.

- *Indicator function method*. In many applications, transformations from continuous signals to strictly discrete amplitude signals is done intentionally. For example in digital communications, the signum function is used as a decision device to map a continuous waveform to two values: ± 1. Similarly, the unit-step function can be used to map a continuous waveform to $\{0, 1\}$. In Chapter 5, we describe how to approximate a continuous random variable by a sequence of discrete random variables generated by a set of staircase functions.

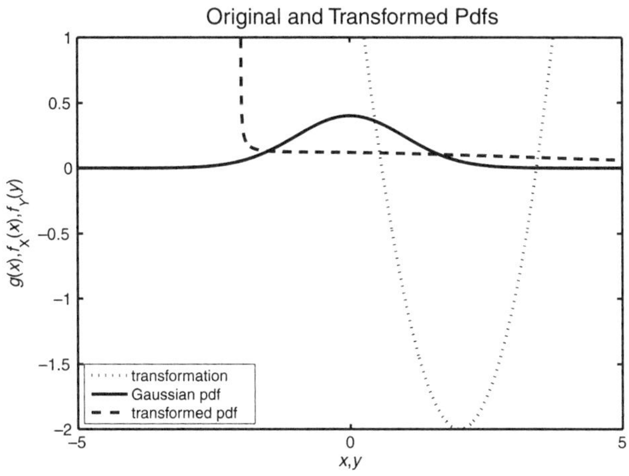

FIGURE 4.30 Pdfs for the random variable transformation in Example 4.14 where X is standard Gaussian.

It will be convenient to use the indicator function $I_{[a,b]}(x)$ with $a < b$ for transformations, by replacing its argument with random variable X:

$$Y = I_{[a,b]}(X). \tag{4.132}$$

This mapping from a continuous random variable X (with arbitrary pdf) yields a Bernoulli random variable Y with outcomes $\{0, 1\}$. In order to find $f_Y(y)$, we need only determine the area under the pdf of X that maps to 0; the remaining area necessarily maps to 1. The signum function is another means for transforming a continuous random variable to a discrete one. Figure 4.31 illustrates the following mapping:

$$Y = \text{sgn}(X), \tag{4.133}$$

where X has a uniform pdf.

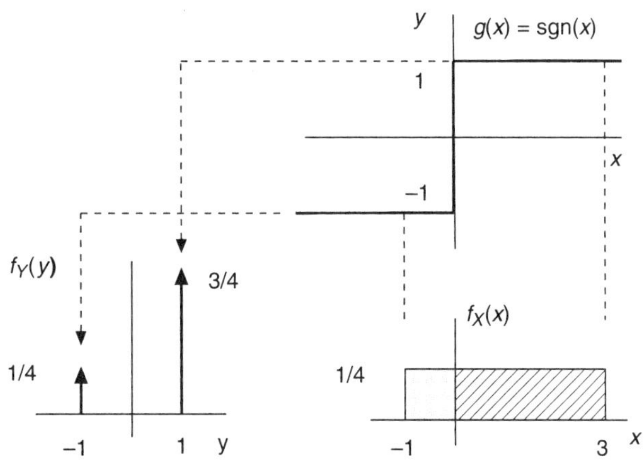

FIGURE 4.31 Mapping of a continuous random variable to a discrete random variable. Areas under $f_X(x)$ for positive and negative intervals map to delta functions of $f_Y(y)$.

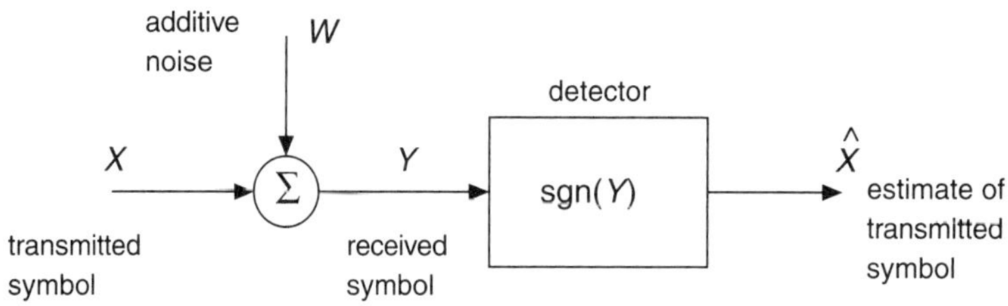

FIGURE 4.32 Simplified block diagram of a binary digital communication system.

Returning to the indicator function and assuming outcomes $\{0, 1\}$ for Y, $f_Y(y)$ is derived as follows:

$$\begin{aligned} P(Y=1) = P\big(I_{[a,b]}(X)=1\big) &= P(a \le X \le b), \\ &= F_X(b) - F_X(a) = p, \quad (4.134) \\ P(Y=0) = P\big(I_{[a,b]}(X)=0\big) &= 1 - P(a \le X \le b), \\ &= 1 - [F_X(b) - F_X(a)] = 1 - p \triangleq q, \quad (4.135) \end{aligned}$$

where p is the probability of a success for the Bernoulli pdf. This result is quite general and can be used to represent a range of transformations from continuous to discrete random variables. For example, the indicator function and the expressions above can be used to realize the unit-step function by letting $[a, b] = [0, \infty)$. For the signum function, the interval is also $[a, b] = [0, \infty)$, but outcome 0 in (4.135) is replaced with -1. For any random variable X with a symmetric pdf about the origin, the resulting symmetric Bernoulli random variable Y necessarily has equally likely outcomes: $p = q = 1/2$.

Example 4.15. As mentioned above, the signum function is used in digital communications to *detect* at discrete time instants whether 1 or -1 was transmitted. A simplified block diagram of such a system with additive noise is shown in Figure 4.32. Assume the symbols with alphabet ± 1 are transmitted randomly every T seconds, and correspond to a symmetric Bernoulli random variable at each discrete time instant. The additive noise W is assumed to be a Gaussian random variable with parameters $\{\mu = 0, \sigma\}$. Although X is a discrete random variable, the received symbol Y is continuous and can take on any value in $\mathcal{R}$ because of the additive noise. It is shown in a subsequent section that when X and W are independent, the pdf of $Y = X + W$ is generated by the following convolution:

$$f_Y(y) = \int_{-\infty}^{\infty} f_X(v) f_W(y - v)\mathrm{d}v. \quad (4.136)$$

Since $f_X(x)$ is a discrete random variable whose pdf is described by Dirac delta functions, the convolution is easy to compute:

$$\begin{aligned} f_Y(y) &= \int_{-\infty}^{\infty} (1/2)\,[\delta(v-1) + \delta(v+1)]\, \frac{1}{\sqrt{2\pi\sigma^2}} \exp\left(-(y-v)^2/2\sigma^2\right)\mathrm{d}v, \\ &= (1/2)\left[\frac{1}{\sqrt{2\pi\sigma^2}} \exp\left(-(y-1)^2/2\sigma^2\right) + \frac{1}{\sqrt{2\pi\sigma^2}} \exp\left(-(y+1)^2/2\sigma^2\right)\right], \\ &= (1/2)\left[f_{Y|X}(y|\,-1) + f_{Y|X}(y|\,+1)\right], \quad (4.137) \end{aligned}$$

where the sifting property of the Dirac delta function has been used. This result is the sum of two conditional Gaussian pdfs, one centered at 1 and the other at -1. The overall pdf of Y and the two conditional pdfs are

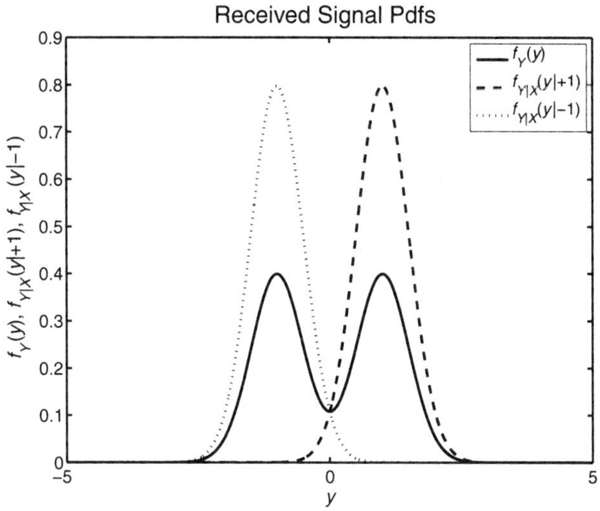

FIGURE 4.33 Bimodal pdf of the received random variable Y for the binary digital communication system in Example 4.15, and the conditional pdfs $f_{Y|X}(y|-1)$ and $f_{Y|X}(y|+1)$.

shown in Figure 4.33. By factoring out the common exponential term in (4.137), this result can be rewritten as a product involving the hyperbolic cosine function:

$$\begin{aligned} f_Y(y) &= \frac{1}{\sqrt{8\pi\sigma^2}} \exp\left(-(y^2+1)/2\sigma^2\right)\left[\exp\left(-y/\sigma^2\right)+\exp\left(y/\sigma^2\right)\right], \\ &= \frac{1}{\sqrt{2\pi\sigma^2}} \exp\left(-(y^2+1)/2\sigma^2\right)\cosh(y/\sigma^2), \end{aligned} \tag{4.138}$$

which is *bimodal* with modes located at $x = \pm 1$. The signum function detector "decides" which symbol was transmitted as follows:

$$\hat{X} = \text{sgn}(Y), \tag{4.139}$$

where $\hat{X}$ denotes an estimate of X. (We arbitrarily choose $\hat{X} = 1$ when $Y = 0$.) From the symmetry of the problem, we know that $P(\hat{X} = 1) = P(\hat{X} = -1) = 1/2$ if the symbols are equally likely. This does not mean that the correct value of X will be chosen. If $X = -1$ is transmitted and the value of the noise exceeds 1, then the signum function detector will incorrectly decide that 1 was transmitted. The probability of an error, given by the conditional probabilities $P(\hat{X} = 1|X = -1) = P(\hat{X} = -1|X = 1)$, is a measure of performance for the receiver. It is shown in Chapter 10 that these probabilities depend on the variance of W: a larger noise variance results in a greater probability of error.

The transformation method using the indicator function can be generalized as follows:

$$Y = \sum_{m=1}^{M} y_m I_{[a_m, b_m]}(X), \tag{4.140}$$

where interval $[a_m, b_m]$ denotes the range of values of X such that $g(X)$ maps to y_m. The corresponding probability is

$$P(Y = y_m) = P(a_m \leq X \leq b_m) = \int_{-\infty}^{\infty} f_X(x) I_{[a_m, b_m]}(x) \mathrm{d}x \triangleq P_Y(y_m), \tag{4.141}$$

which is the same notation used for the previous transformations to mixed random variables. Thus, the pdf of Y is obtained from (4.114) by retaining only the discrete part:

$$f_Y(y) = \sum_{m=1}^{M} P_Y(y_m)\delta(y - y_m), \tag{4.142}$$

where it is understood that the intervals $\{[a_m, b_m]\}$ corresponding to the $\{P_Y(y_m)\}$ are nonoverlapping, and they may form a partition of the x-axis.

Example 4.16. Let X be uniformly distributed on $[0, 3]$ and consider the mapping:

$$Y = g(X) = 2I_{[0,1/2]}(X) + I_{[1,2]}(X) - I_{[5/2,3]}(X), \tag{4.143}$$

which is not a partition of $[0, 3]$. This means that some intervals of X map to zero so that the outcomes of Y are $\{-1, 0, 1, 2\}$. Since X is uniformly distributed, the Lebesgue measure can be used to write the probabilities:

$$P(0 \le X \le 1/2) = (1/3)L[0, 1/2] = 1/6, \tag{4.144}$$

$$P(1 \le X \le 2) = (1/3)L[1, 2] = 1/3, \tag{4.145}$$

$$P(5/2 \le X \le 3) = (1/3)L[5/2, 3] = 1/6, \tag{4.146}$$

where $1/3$ is the height of $f_X(x)$. The probability that X lies in the two internal intervals $\{[1/2, 1], [2, 5/2]\}$ is $1/3$, obtained using the Lebesgue measure, or more simply by subtracting from 1 the three probabilities above. The pdf of discrete random variable Y is

$$f_Y(y) = (1/6)[2\delta(y - 1) + \delta(y - 2) - \delta(y + 1) + 2\delta(y)], \tag{4.147}$$

where the arguments of the delta functions, which specify the outcomes of Y, should not be confused with their areas which give the probabilities of the discrete outcomes.

Table 4.2 summarizes the four basic transformation techniques used to find $f_Y(y)$ for the mapping $Y = g(X)$ when X is a continuous random variable.

4.9 IMPORTANT FUNCTIONS OF TWO RANDOM VARIABLES

In this section, we consider transformations from two independent random variables to a single random variable. Except for the sum of random variables described first, we provide derivations only for the case when both random variables are continuous; the results for discrete random variables, or when one is continuous and the other is discrete (as in Example 4.15) can be determined from of the continuous results.

4.9.1 Sum: $Z = X + Y$

The cdf (indirect) method will be used to derive the pdf of Z when X and Y are independent continuous random variables. Starting with the joint cdf $F_{X,Y}(x, y)$, the cdf of Z is written as the following double integral:

$$F_Z(z) = P(X + Y \le z) = \iint_{x+y\le z} f_{X,Y}(x, y)\mathrm{d}x\mathrm{d}y,$$

$$= \int_{-\infty}^{\infty}\int_{-\infty}^{z-y} f_{X,Y}(x, y)\mathrm{d}x\mathrm{d}y, \tag{4.148}$$

TABLE 4.2 Transformation Methods for $Y = g(X)$ and Continuous Random Variable X

Cdf (Indirect) Method
Start with the cdf of Y: $F_Y(y)$ (assume monotonic $g(X)$)
Write $F_Y(y)$ in terms of $F_X(\cdot)$: $P(Y \leq y) = P(X \leq g^{-1}(y)) = F_X(g^{-1}(y))$
Differentiate with respect to y: $f_Y(y) = \frac{\mathrm{d}}{\mathrm{d}y} F_X(g^{-1}(y)) = f_X(g^{-1}(y)) \frac{\mathrm{d}x}{\mathrm{d}y}$

Pdf (Direct) Method
Find all roots of $g(x) = 0$ and write them in terms of y: $x_n = g^{-1}(y)$
Compute $g'(x)$ and evaluate the ratio $f_X(x)/g'(x)$ for each x_n
Sum over all roots: $f_Y(y) = \sum_{n=1}^{N} f_X(x_n)/|g'(x_n)|$
Substitute $x_n = g^{-1}(y)$

Delta Function Method
Find all roots of $g(x) - y = 0$
Rewrite $\delta(g(x) - y)$ using $\delta(g(x) - y) = \sum_{n=1}^{N} \delta(x - x_n)/|g'(x_n)|$
Integrate and apply the sifting property: $f_Y(y) = \int_{-\infty}^{\infty} f_X(x)\delta(g(x) - y)\mathrm{d}x$

Indicator Function Method
Express $g(X)$ as a sum of weighted indicator functions
Determine probabilities $P_Y(y_m) = P(a_m \leq X \leq b_m)$
Weight and sum the delta functions: $f_Y(y) = \sum_{m=1}^{M} P_Y(y_m)\delta(y - y_m)$

where in the last expression we have rewritten $x + y \leq z$ as $x \leq z - y$ and used it to limit the range of the inner integral. Differentiating with respect to z gives

$$f_Z(z) = \int_{-\infty}^{\infty} f_{X,Y}(z - y, y)\mathrm{d}y. \tag{4.149}$$

This is a general result for dependent X and Y. When they are independent, the joint pdf splits, giving the final result

$$f_Z(z) = \int_{-\infty}^{\infty} f_X(z - y) f_Y(y)\mathrm{d}y, \tag{4.150}$$

which we recognize is a *convolution* of the two marginal pdfs: $f_Z(z) = f_X(x) * f_Y(y)$.

Suppose now that X is continuous and Y is discrete, for which (4.150) becomes

$$f_Z(z) = \int_{-\infty}^{\infty} f_X(z - y) \sum_{n=-\infty}^{\infty} p_Y[n]\delta(y - n)\mathrm{d}y, \tag{4.151}$$

where the pdf of Y based on the Dirac delta function has been substituted. From the sifting property, the convolution becomes

$$f_Z(z) = \sum_{n=-\infty}^{\infty} f_X(z - n) p_Y[n]. \tag{4.152}$$

This expression is a weighted sum of translated versions of the continuous pdf for X. The weightings are given by the pmf of Y which ensure that $f_Z(z)$ has unit area; it is an example of a *mixture* covered later in this chapter.

Finally, consider when both X and Y are discrete, which is handled by starting with an expression similar to (4.148), but with equality:

$$P(Z = z) = P(X + Y = z) = \sum_{x+y=z} p_{X,Y}[x, y]. \tag{4.153}$$

There is necessarily a single sum because as x is varied, the equality $x + y = z$ forces y to be $z - x$. Assuming independence and replacing the y argument of the joint pmf with $z - x$ yields the desired result, which is a discrete convolution:

$$p_Z[z] = \sum_{x=-\infty}^{\infty} p_X[x]p_Y[z - x]. \tag{4.154}$$

Example 4.17. Let X and Y be iid exponential random variables with parameter λ. The pdf of $Z = X + Y$ is

$$f_Z(z) = \lambda^2 \int_{-\infty}^{\infty} \exp(-\lambda(z - y)) \exp(-\lambda y) I_{[0,\infty)}(z - y) I_{[0,\infty)}(y) \mathrm{d}y, \tag{4.155}$$

where indicator functions have been used to keep track of the support of each function, which is especially important when the arguments of the integrand are changed. The limits of integration are $[0, \infty) \cap (-\infty, z] = [0, z]$, yielding

$$f_Z(z) = \lambda^2 \exp(-\lambda z) \int_0^z \mathrm{d}y = \lambda^2 z \exp(-\lambda z) I_{[0,\infty)}(z), \tag{4.156}$$

which is the Erlang (gamma) distribution with parameters $\{\lambda, r = 2\}$.

Example 4.18. Let X_1 and X_2 be iid binomial random variables with parameters $\{N, p\}$. The pmf of $Y = X_1 + X_2$ is derived using the formula in (4.154):

$$\begin{aligned} p_Y[y] &= \sum_{x_1=-\infty}^{\infty} \binom{N}{x_1} p^{x_1} q^{N-x_1} \binom{N}{y - x_1} p^{y-x_1} q^{N-y+x_1} I_{\{0,\dots,N\}}[x_1] I_{\{0,\dots,N\}}[y - x_1], \\ &= p^y q^{2N-y} \sum_{x_1=0}^{y} \binom{N}{x_1}\binom{N}{y - x_1} I_{\{0,\dots,2N\}}[y], \end{aligned} \tag{4.157}$$

where the discrete indicator function specifies the proper limits for the sum. It can be shown (see Problem 4.30) that a summation of the form in the last expression simplifies to $\binom{2N}{y}$. The result is a binomial pmf with parameters $\{2N, p\}$:

$$p_Y[y] = \binom{2N}{y} p^y q^{2N-y} I_{\{0,\dots,2N\}}[y]. \tag{4.158}$$

4.9.2 Difference: $Z = X - Y$

Using a similar approach as in Section 4.9.1, the cdf of the difference is

$$\begin{aligned} F_Z(z) = P(X - Y \le z) &= \iint_{x-y\le z} f_{X,Y}(x, y)\mathrm{d}x\mathrm{d}y, \\ &= \int_{-\infty}^{\infty} \int_{-\infty}^{z+y} f_{X,Y}(x, y)\mathrm{d}x\mathrm{d}y. \end{aligned} \tag{4.159}$$

Differentiating with respect to z gives

$$f_Z(z) = \int_{-\infty}^{\infty} f_{X,Y}(z+y, y)\mathrm{d}y = \int_{-\infty}^{\infty} f_X(z+y)f_Y(y)\mathrm{d}y, \quad (4.160)$$

where independence of the random variables gives the last expression. This integral is the deterministic *cross-correlation* between the two pdfs: $f_Z(z) = f_Y(y) \star f_X(x)$ (see the definition in Appendix C). Unlike convolution, it is not symmetric, which we would expect because $Z = X - Y$ is not symmetric with respect to X and Y:

$$f_Z(z) = \int_{-\infty}^{\infty} f_X(x)f_Y(x-z)\mathrm{d}x. \quad (4.161)$$

The corresponding results for discrete random variables are

$$p_Z[z] = \sum_{y=-\infty}^{\infty} p_X[z+y]p_Y[y] = \sum_{x=-\infty}^{\infty} p_X[x]p_Y[x-z]. \quad (4.162)$$

Example 4.19. Let X and Y be iid Bernoulli random variables with parameter p, such that the outcomes of $Z = X - Y$ are $\{-1, 0, 1\}$. Its pmf is

$$\begin{aligned} p_Z[z] &= \sum_{y=-\infty}^{\infty} p^{z+y}q^{1-y-z}p^y q^{1-y} I_{\{0,1\}}[z+y]I_{\{0,1\}}[y], \\ &= q^2(p/q)^z \sum_{y=0}^{1}(p/q)^{2y} I_{\{0,1\}}[z+y], \end{aligned} \quad (4.163)$$

where the second indicator function gives the summation limits $\{0, 1\}$. Depending on z, the first indicator function restricts the values of x as follows: $z = -1 \Rightarrow y = 1$, $z = 0 \Rightarrow y = \{0, 1\}$, and $z = 1 \Rightarrow y = 0$. Thus

$$\begin{aligned} p_Z[z] &= q^2[(p/q)^{-1}(p/q)^2\delta[z+1] + (1+(p/q)^2)\delta[z] + (p/q)\delta[z-1]], \\ &= pq\delta[z+1] + (q^2+p^2)\delta[z] + pq\delta[z-1], \end{aligned} \quad (4.164)$$

which is a symmetric (even) pmf and does not correspond to any of the standard parametric distributions.

4.9.3 Product: $Z = XY$

Consider next the case where the independent random variables X and Y are multiplied. Again using the cdf method:

$$\begin{aligned} F_Z(z) = P(XY \le z) &= \iint_{xy \le z} f_{X,Y}(x, y)\mathrm{d}x\mathrm{d}y, \\ &= \int_{-\infty}^{0}\int_{z/y}^{\infty} f_{X,Y}(x, y)\mathrm{d}x\mathrm{d}y + \int_{0}^{\infty}\int_{-\infty}^{z/y} f_{X,Y}(x, y)\mathrm{d}x\mathrm{d}y. \end{aligned} \quad (4.165)$$

Because y can be negative, which would change the direction of the inequality when dividing, the outer integral must be split over two regions: $y \in (-\infty, 0]$ and $y \in [0, \infty)$. The corresponding inequalities that determine the limits of the inner integral are $x \geq z/y$ and $x \leq z/y$, respectively. Differentiating both terms with respect to z and using the chain rule gives

$$\begin{aligned} f_Z(z) &= -\int_{-\infty}^{0} (1/y) f_{X,Y}(z/y, y) dy + \int_{0}^{\infty} (1/y) f_{X,Y}(z/y, y) dy, \\ &= \int_{-\infty}^{\infty} \frac{1}{|y|} f_{X,Y}(z/y, y) dy. \end{aligned} \tag{4.166}$$

When X and Y are independent, the final result is

$$f_Z(z) = \int_{-\infty}^{\infty} \frac{1}{|y|} f_X(z/y) f_Y(y) dy. \tag{4.167}$$

For discrete random variables, the expression is

$$p_Z[z] = P(XY = z) = \sum_{x} \sum_{y=z/x} p_X[x] p_Y[y], \tag{4.168}$$

where it is understood that z/x is an integer; one must be careful to account for all terms in the product for the discrete case.

Example 4.20. Let X be a geometric random variable with parameter p, and Y be Bernoulli, also with parameter p. The support of the product $Z = XY$ is $\mathcal{Z}^+$ which is the same as that of X. Because Y is Bernoulli, we expect that Z will have greater probability mass at $Z = 0$ (compared with the geometric random variable) because $Y = 0$ occurs with probability $q \triangleq 1 - p$. The pdf is derived by considering $Z = 0$ and $Z \neq 0$ separately:

$$\begin{aligned} P(Z = 0) &= P(X = 0 \text{ or } Y = 0), \\ &= P(X = 0) + P(Y = 0) - P(X = 0)P(Y = 0), \\ &= p + q - pq = 1 - pq, \end{aligned} \tag{4.169}$$

where independence has been used in the second line. For $Z > 0$:

$$\begin{aligned} P(Z = z) &= P(X = z \text{ and } Y = 1), \\ &= P(X = z)P(Y = 1) = p^2 q^z. \end{aligned} \tag{4.170}$$

Combining both results yields the following pmf:

$$p_Z[z] = (1 - pq)\delta[z] + p^2 \sum_{n=1}^{\infty} q^n \delta[z - n]. \tag{4.171}$$

Note that $p^2\delta[z]$ can added and subtracted so that the sum includes $n = 0$:

$$p_Z[z] = q\delta[z] + p \sum_{n=0}^{\infty} pq^n \delta[z - n], \tag{4.172}$$

which is a mixture of a geometric pdf weighted by p and the pmf $\delta[z]$ of a "random" variable with the single outcome zero, weighted by q.

Next, we derive this result using (4.168) by summing over y and substituting the values of y in the second sum:

$$p_Z[z] = \sum_y \sum_{x=z/y} pq^x p^y q^{1-y} = \sum_{x:x\cdot 0=z} pq^x q + \sum_{x=z} pq^x p. \tag{4.173}$$

The first sum contributes only for $z = 0$, and since $y = 0$ in the product, the sum is over all values of x:

$$\sum_{x:x\cdot 0=0} pq^x q = pq \sum_{x=0}^{\infty} q^x = pq/(1-q) = q. \tag{4.174}$$

The second sum reduces to a single term for each value of z:

$$\sum_{x:x\cdot 1=z} pq^x p = p^2 \sum_{x=z} q^x = p^2 q^z I_{Z^+}[z]. \tag{4.175}$$

Combining both results gives the pmf in (4.172).

Example 4.21. Suppose that X has an exponential distribution with parameter λ, and Y is Bernoulli with parameter p. Intuitively, we expect the product $Z = XY$ to be a mixed random variable with nonzero probability mass at $Z = 0$ given by

$$P(Z = 0) = P(XY = 0) = P(Y = 0) = q. \tag{4.176}$$

Consider starting with the cdf and conditioning as follows:

$$\begin{aligned} P(Z \le z) &= P(Z \le z|Y = 0)P(Y = 0) + P(Z \le z|Y = 1)P(Y = 1), \\ &= qu(z) + pP(Z \le z|Y = 1). \end{aligned} \tag{4.177}$$

The first term is a step function because $P(Z \le z|Y = 1) = 1$ for $z \ge 0$. For the second term:

$$\begin{aligned} P(Z \le z|Y = 1) &= P(XY \le z|Y = 1), \\ &= P(X \le z) = \lambda \exp(-\lambda z)u(z). \end{aligned} \tag{4.178}$$

Thus, the pdf is mixed as follows:

$$f_Z(z) = q\delta(z) + p\lambda \exp(-\lambda z) I_{[0,\infty)}(z). \tag{4.179}$$

Using instead the expression in (4.167) gives

$$\begin{aligned} f_Z(z) &= \lambda \int_0^{\infty} (1/y)[p\delta(y-1) + q\delta(y)] \exp(-\lambda z/y) \mathrm{d}y, \\ &= p\lambda \exp(-\lambda z) I_{[0,\infty)}(z) + q \int_0^{\infty} (\lambda/y)\delta(y) \exp(-\lambda z/y) \mathrm{d}y, \end{aligned} \tag{4.180}$$

where the integral limits are determined by the support of Y, which is also the support of Z. The last integral is zero when $z > 0$ because the exponential decays faster to zero than $1/y$ approaches infinity as $y \to 0$ (note that the argument of the exponential also includes $1/y$). It can be shown that the last integral is the Dirac delta function $\delta(z)$ (see Problem 4.31), and the mixed pdf in (4.179) is obtained again.

4.9.4 Quotient (Ratio): $Z = X/Y$

For the ratio of two random variables, we can write the cdf as follows:

$$F_Z(z) = P(X/Y \leq z) = \iint_{x/y \leq z} f_{X,Y}(x, y)\mathrm{d}x\mathrm{d}y,$$

$$= \int_{-\infty}^{0} \int_{zy}^{\infty} f_{X,Y}(x, y)\mathrm{d}x\mathrm{d}y + \int_{0}^{\infty} \int_{-\infty}^{zy} f_{X,Y}(x, y)\mathrm{d}x\mathrm{d}y. \tag{4.181}$$

Differentiating with respect to z and assuming independence of the random variables gives

$$f_Z(z) = -\int_{-\infty}^{0} y f_{X,Y}(zy, y)\mathrm{d}y + \int_{0}^{\infty} y f_{X,Y}(zy, y)\mathrm{d}y,$$

$$= \int_{-\infty}^{\infty} |y| f_X(zy) f_Y(y)\mathrm{d}y. \tag{4.182}$$

This is the so-called *ratio distribution* (Marsaglia, 2006), which is also called the quotient distribution. For discrete random variables, the pmf is obtained as follows:

$$p_Z[z] = \sum_{y} \sum_{x=zy} p_Y[y] p_X[x], \tag{4.183}$$

where one must be careful that Y has probability mass only for nonzero values because of the division X/Y.

Example 4.22. Suppose that X and Y are iid Gaussian random variables with parameters $\{\mu = 0, \sigma\}$. The pdf of $Z = X/Y$ is

$$f_Z(z) = \frac{1}{2\pi\sigma^2} \int_{-\infty}^{\infty} |y| \exp(-(zy)^2/2\sigma^2) \exp(-y^2/2\sigma^2)\mathrm{d}y,$$

$$= \frac{1}{\pi\sigma^2} \int_{0}^{\infty} y \exp(-y^2(z^2+1)/2\sigma^2)\mathrm{d}y. \tag{4.184}$$

Substituting $v \triangleq y\sqrt{z^2+1}$ and $\mathrm{d}v = \mathrm{d}y\sqrt{z^2+1}$ gives

$$f_Z(z) = \frac{1}{\pi\sigma^2(z^2+1)} \int_{0}^{\infty} v \exp(-v^2/2\sigma^2)\mathrm{d}v. \tag{4.185}$$

Finally, let $w \triangleq v^2/2\sigma^2$ such that $\mathrm{d}w = v\mathrm{d}v/\sigma^2$ and

$$f_Z(z) = \frac{1}{\pi(z^2+1)} \int_{0}^{\infty} \exp(-w)\mathrm{d}w = \frac{1}{\pi(z^2+1)}, \tag{4.186}$$

which is the standard Cauchy pdf.

Example 4.23. Suppose in Example 4.22 that X and Y still have zero mean, but different variances σ_X^2 and σ_Y^2. The ratio pdf is

$$f_Z(z) = \frac{1}{2\pi\sigma_X\sigma_Y}\int_{-\infty}^{\infty}|y|\exp(-(zy)^2/2\sigma_X^2)\exp(-y^2/2\sigma_Y^2)\mathrm{d}y,$$

$$= \frac{1}{\pi\sigma_X\sigma_Y}\int_0^{\infty} y\exp(-y^2[z^2/2\sigma_X^2 + 1/2\sigma_Y^2])\mathrm{d}y,$$

$$= \frac{1}{\pi\sigma_X\sigma_Y}\int_0^{\infty} y\exp(-y^2/2\alpha^2)\mathrm{d}y, \tag{4.187}$$

where we have substituted

$$z^2/2\sigma_X^2 + 1/2\sigma_Y^2 = 1/2\alpha^2. \tag{4.188}$$

We recognize from Appendix A that the integral is proportional to the mean of the half-normal distribution:

$$\int_0^{\infty} y\exp(-y^2/2\alpha^2)\mathrm{d}y = a\alpha\sqrt{2/\pi} = \alpha^2 \tag{4.189}$$

with proportionality constant $a = \sqrt{\pi\alpha^2/2}$. Substituting this result gives the final result:

$$f_Z(z) = \frac{\alpha^2}{\pi\sigma_X\sigma_Y} = \frac{\sigma_X\sigma_Y}{\pi(\sigma_X^2 + z\sigma_Y^2)} = \frac{\sigma_X/\sigma_Y}{\pi[(\sigma_X/\sigma_Y)^2 + z^2]}, \tag{4.190}$$

which is another Cauchy pdf.

Example 4.24. In the previous examples, the two random variables in the ratio have identical support. Consider next the ratio of two discrete random variables with different support. Let X be geometric with parameter p, and Y be symmetric Bernoulli with outcomes $\{-1, 1\}$ so that we are not concerned with dividing by zero. It is clear that the support of $Z = X/Y$ is $\mathcal{Z}$. Consider the following probabilities:

$$P(Z = z) = \begin{cases} P(X = 0), & z = 0 \\ P(X = z, Y = 1), & z > 0 \\ P(X = -z, Y = -1), & z < 0. \end{cases} \tag{4.191}$$

Since X and Y are independent, we have the following pmf:

$$p_Z[z] = p\delta[z] + p^2q^zI_{\mathcal{N}}[z] + pq^{-z+1}I_{\{\ldots,-1\}}[z]. \tag{4.192}$$

Consider next using the formula in (4.183):

$$p_Z[z] = \sum_{y=-1,1}\sum_{x=zy,x\in\mathcal{Z}^+} pq^x p^{(1+y)/2}q^{(1-y)/2}, \tag{4.193}$$

where the exponents of p and q differ from the standard Bernoulli distribution because the outcomes here are ± 1 (see Appendix A). Summing over y gives two terms:

$$p_Z[z] = \sum_{x=z,x\in\mathcal{Z}^+} p^2q^x + \sum_{x=-z,x\in\mathcal{Z}^+} pq^{x+1} \tag{4.194}$$

TABLE 4.3 Transformations of Gaussian Random Variables

Transformation	Conditions on iid $\{X_n\}$	Random Variable Y {parameters}
$Y = \exp(X)$	None	Log-normal $\{\mu, \sigma\}$
$Y = \|X\|$	$\mu = 0$	Half-normal $\{\alpha = \sigma\}$
$Y = X_1/X_2$	$\mu = 0, \sigma = 1$	Cauchy $\{c = 0, \alpha = 1\}$
$Y = X_1 \pm X_2$	None	Gaussian $\{0 \text{ or } 2\mu, 2\sigma\}$
$Y = \sqrt{X_1^2 + X_2^2}$	$\mu = 0$	Rayleigh $\{\alpha = \sigma\}$
$Y = \sum_{n=1}^{N} X_n^2$	$\mu = 0, \sigma = 1$	Chi-square $\{N\}$
$Y = \tan^{-1}(X_1/X_2)$	$\mu = 0, \sigma = 1$	Uniform $[0, 2\pi]$

from which we find that both contribute to the pmf together only for $z = 0$. Note that $x = -z$ and $x \in \mathcal{Z}^+$ in the second term is equivalent to $z \in \mathcal{Z}^-$ (which includes zero), and thus

$$p_Z[z] = p^2 q^z I_{\mathcal{Z}^+}[z] + pq^{-z+1} I_{\mathcal{Z}^-}[z]. \tag{4.195}$$

This is the same as (4.192), which is seen by separating the $z = 0$ term from each component and combining them to give $p^2 + pq = p$ for $\delta[z]$.

4.10 TRANSFORMATIONS OF RANDOM VARIABLE FAMILIES

In this section, we summarize transformation results for some of the parametric distributions discussed in Chapter 3. Derivations are provided only for a few of the transformations using the previous techniques; others are considered in the problems at the end of this chapter.

4.10.1 Gaussian Transformations

Let $\{X_1, \ldots, X_N\}$ be iid Gaussian random variables with parameters $\{\mu, \sigma\}$. The results for several transformations are provided in Table 4.3.

Example 4.25 ($Y = \exp(X)$). Since the exponential function is monotonic, the cdf method yields

$$\begin{aligned} F_Y(y) &= P(\exp(X) \le y) = P(X \le \ln(y)), \\ &= \frac{1}{\sqrt{2\pi\sigma^2}} \int_{-\infty}^{\ln(y)} \exp(-(x-\mu)^2/2\sigma^2) \mathrm{d}x. \end{aligned} \tag{4.196}$$

Differentiating gives

$$\begin{aligned} f_Y(y) &= \frac{1}{\sqrt{2\pi\sigma^2}} \exp(-(\ln(y)-\mu)^2/2\sigma^2) \frac{\mathrm{d}\ln(y)}{\mathrm{d}y}, \\ &= \frac{1}{y\sqrt{2\pi\sigma^2}} \exp(-(\ln(y)-\mu)^2/2\sigma^2) I_{[0,\infty)}(y), \end{aligned} \tag{4.197}$$

which is the log-normal pdf with parameters $\{\mu, \sigma\}$. This result is also easily obtained using the pdf (direct) method.

Example 4.26 ($Y = g(X_1/X_2)$). This pdf was derived previously in Example 4.22 using the general formula in (4.182). Here, we use the delta function transformation method which gives (with $\mu = 0$ and $\sigma = 1$)

$$
\begin{aligned}
f_Y(y) &= \frac{1}{2\pi}\int_{-\infty}^{\infty}\int_{-\infty}^{\infty} \exp(-(x_1^2 + x_2^2)/2)\delta(x_1/x_2 - y)\mathrm{d}x_1\mathrm{d}x_2, \\
&= \frac{1}{2\pi}\int_{-\infty}^{\infty} |x_2| \exp(-(x_2^2 y^2 + x_2^2)/2)\mathrm{d}x_2, \\
&= \frac{1}{\pi}\int_{0}^{\infty} x_2 \exp(-x_2^2(y^2 + 1)/2)\mathrm{d}x_2,
\end{aligned}
\tag{4.198}
$$

where

$$
\delta(x_1/x_2 - y) = |x_2|\delta(x_1 - x_2 y), \tag{4.199}
$$

and the sifting property has been used when integrating over x_1. Recognizing that the integral is similar to that in (4.189), we obtain the following result after the appropriate change of variables:

$$
f_Y(y) = \frac{1}{\pi(1 + y^2)}, \tag{4.200}
$$

which again is the standard Cauchy distribution.

4.10.2 Exponential Transformations

Let $\{X_1, \ldots, X_N\}$ be iid exponential random variables with parameter λ. The results for several transformations are provided in Table 4.4.

Example 4.27 ($Y = X_1 - X_2$). Using the delta function method:

$$
\begin{aligned}
f_Y(y) &= \int_{-\infty}^{\infty}\int_{-\infty}^{\infty} \lambda^2 \exp(-\lambda(x_1 + x_2))\delta(x_1 - x_2 - y)u(x_1)u(x_2)\mathrm{d}x_1\mathrm{d}x_2, \\
&= \lambda^2 \int_{-\infty}^{\infty} \exp(-\lambda(2x_2 + y))u(x_2 + y)u(x_2)\mathrm{d}x_2.
\end{aligned}
\tag{4.201}
$$

TABLE 4.4 Transformations of Exponential Random Variables

Transformation	Conditions on iid $\{X_n\}$	Random Variable Y {parameters}
$Y = aX$	$a > 0$	Exponential $\{\lambda/a\}$
$Y = \exp(-\lambda X)$	None	Uniform (0, 1]
$Y = \sqrt{2\alpha^2 X/\lambda}$	None	Rayleigh $\{\alpha\}$
$Y = r\exp(X)$	$r \geq 1$	Pareto $\{r, \alpha = \lambda\}$
$Y = \mu - \alpha\ln(\exp(-X)/(1 - \exp(-X)))$	$\lambda = 1$	Logistic $\{\mu, \alpha\}$
$Y = X^{1/r}$	$\lambda \rightarrow 1/\lambda^r$	Weibull $\{r, \alpha = 1/\lambda\}$
$Y = X_1 - X_2$	None	Laplace $\{\mu = 0, \alpha = 1/\lambda\}$
$Y = \min(X_1, X_2)$	None	Exponential $\{2\lambda\}$
$Y = \sum_{n=1}^{N} X_n$	None	Gamma $\{r = N, \alpha = \lambda\}$

Both unit-step functions affect only the lower limit as follows:

$$\begin{aligned} f_Y(y) &= \lambda^2 \exp(-\lambda y) \int_{\min(0,-y)}^{\infty} \exp(-2\lambda x_2) \mathrm{d}x_2, \\ &= (\lambda/2) \exp(-\lambda y) \exp(-2\lambda \min(0, -y)). \end{aligned} \tag{4.202}$$

By examining the sign of y:

$$f_Y(y) = \begin{cases} (\lambda/2) \exp(-\lambda y), & y \geq 0 \\ (\lambda/2) \exp(-\lambda y + 2\lambda y), & y < 0, \end{cases} \tag{4.203}$$

we obtain

$$f_Y(y) = (\lambda/2) \exp(-\lambda |y|), \tag{4.204}$$

which is the Laplace pdf with zero mean and scale parameter $\alpha = 1/\lambda$.

Example 4.28 ($Y = \exp(-\lambda X)$). Using the pdf (direct) method:

$$\begin{aligned} f_Y(y) &= \left. \frac{f_X(x_n)}{|g'(x_n)|} \right|_{x_n = g^{-1}(y)}, \\ &= \left. \frac{\lambda \exp(-\lambda x) u(x)}{|\lambda \exp(-\lambda x)|} \right|_{x_n = g^{-1}(y)} = u(x)|_{x_n = g^{-1}(y)}, \end{aligned} \tag{4.205}$$

where $g'(x) = -\lambda \exp(-\lambda x)$ has been substituted. Since $g(x)$ is monotonic, there is only one $x_n = x$, and the range is $y \in (0, 1]$ because $x \in [0, \infty)$. Substituting $x = -\ln(y)/\lambda$ gives

$$f_Y(y) = u(-\ln(y)/\lambda) = I_{(0,1]}(y), \tag{4.206}$$

which is the pdf of a uniform random variable on (0, 1]. The last expression follows because $u(-\ln(y)/\lambda) = 1$ when $-\ln(y)/\lambda > 0 \Rightarrow \ln(y) < 0$, which means $y \in (0, 1]$.

4.10.3 Chi-Square Transformations

Let $\{X_1, X_2\}$ be independent chi-square random variables with degrees of freedom $\{m, n\}$. The results for several transformations are provided in Table 4.5.

TABLE 4.5 Transformations of Chi-Square Random Variables

Transformation	Conditions on Independent $\{X_p\}$	Random Variable Y {parameters}
$Y = aX$	$a > 0$	Gamma $\{r = n/2, \lambda = 1/2a\}$
$Y = \frac{X_1/m}{X_2/n}$	None	F $\{m, n\}$
$Y = X_1/\sqrt{X_2/r}$	X_1 standard Gaussian	Student's t $\{r\}$
$Y = X_1 + X_2$	None	Chi-square $\{N = 2n\}$
$Y = X_1/(X_1 + X_2)$	None	Beta $\{\alpha = n/2, \beta = n/2\}$

Example 4.29 $(Y = (X_1/m)/(X_2/n))$. The delta function method yields

$$f_Y(y) = \frac{1}{2^{(m+n)/2}\Gamma(m/2)\Gamma(n/2)} \int_{-\infty}^{\infty}\int_{-\infty}^{\infty} x_1^{m/2-1} x_2^{n/2-1}$$
$$\cdot \exp(-(x_1 + x_2)/2)\delta([(x_1/m)/(x_2/n)] - y)u(x_1)u(x_2)\mathrm{d}x_1\mathrm{d}x_2. \tag{4.207}$$

Substituting

$$\delta(nx_1/mx_2 - y) = (m/n)|x_2|\delta(x_1 - (m/n)x_2 y) \tag{4.208}$$

gives

$$f_Y(y) = \frac{1}{2^{(m+n)/2}\Gamma(m/2)\Gamma(n/2)} \int_{-\infty}^{\infty} (m/n)|x_2|((m/n)x_2 y)^{m/2-1} x_2^{n/2-1}$$
$$\cdot \exp(-((m/n)x_2 y + x_2)/2)u((m/n)x_2 y)u(x_2)\mathrm{d}x_2,$$
$$= \frac{(m/n)^{m/2}}{2^{(m+n)/2}\Gamma(m/2)\Gamma(n/2)} y^{m/2-1} \int_0^{\infty} x_2^{(m+n)/2-1}$$
$$\cdot \exp(-((m/n)y + 1)x_2/2)\mathrm{d}x_2. \tag{4.209}$$

If the integral is premultiplied by $[(m/n)y + 1]/2$, then the combination is a noncentral moment of the exponential pdf (see Appendix A):

$$\frac{(m/n)y + 1}{2} \int_0^{\infty} x_2^{(m+n)/2-1} \exp(-((m/n)y + 1)x_2/2)\mathrm{d}x_2 = \frac{\Gamma((m+n)/2 - 1)}{([(m/n)y + 1]/2)^{(m+n)/2-1}} \tag{4.210}$$

such that

$$f_Y(y) = \frac{(m/n)^{m/2}}{2^{(m+n)/2}\Gamma(m/2)\Gamma(n/2)} \frac{\Gamma((m+n)/2)}{([(m/n)y + 1]/2)^{(m+n)/2-1}} \frac{y^{m/2-1}}{((m/n)y + 1)/2}. \tag{4.211}$$

Cancelling terms and substituting the beta function

$$B(m/2, n/2) \triangleq \frac{\Gamma(m/2 + n/2)}{\Gamma(m/2)\Gamma(n/2)} \tag{4.212}$$

yields

$$f_Y(y) = \frac{(m/n)^{m/2} y^{m/2-1}}{B(m/2, n/2)((m/n)y + 1)^{(m+n)/2}}, \tag{4.213}$$

which is the F-distribution with parameters $\{m, n\}$.

4.11 TRANSFORMATIONS OF RANDOM VECTORS

The previous techniques for transforming one random variable to another random variable can be extended to multiple random variables. We will not derive all the results in detail, but instead present the main ideas behind transformations of continuous random variables, and then summarize the results for various types of transformations, including those for discrete random variables.

Consider random vector $\mathbf{X} \triangleq [X_1, \ldots, X_N]^T$ and its transformation to random vector $\mathbf{Y} = [Y_1, \ldots, Y_N]^T$ according to $Y_n = g_n(\mathbf{X})$, where $\{g_n(\cdot)\}$ are known functions (possibly nonlinear). These functions could be strictly increasing, strictly decreasing, nonmonotonic, may contain flat regions, or may have some combination of these features over various intervals. The functions might also map to a subset of the elements of $\mathbf{X}$ to $\mathbf{Y}$. We first consider the case where the N functions are either strictly increasing or strictly decreasing. For this case, the inverse images $\{g_n^{-1}(\mathbf{Y})\}$ are well defined.

- *Continuous random vector*. The transformation between continuous random vectors is a straightforward generalization of the result in (4.106), obtained using the direct method for mapping one random variable to another. The joint pdf for $\mathbf{Y}$ is

$$f_{\mathbf{Y}}(\mathbf{y}) = |J(\mathbf{x})|^{-1} \, f_{\mathbf{X}}(\mathbf{x})\big|_{x_n = g_n^{-1}(\mathbf{y})} \tag{4.214}$$

for $n = 1, \ldots, N$, where $J(\mathbf{x})$ is the Jacobian determinant computed as follows:

$$J(\mathbf{x}) \triangleq \det[\mathbf{J}(\mathbf{x})] = \det \begin{bmatrix} \partial y_1/\partial x_1 & \cdots & \partial y_1/\partial x_N \\ \vdots & \cdots & \vdots \\ \partial y_N/\partial x_1 & \cdots & \partial y_N/\partial x_N \end{bmatrix}, \tag{4.215}$$

and $\mathbf{J}(\mathbf{x})$ is the Jacobian matrix (see Appendix G). The argument of $\mathbf{J}(\mathbf{x})$ denotes the independent variables, such that the elements of the matrix are functions of $\mathbf{x}$. Note that the absolute value of $J(\mathbf{x})$ is used in (4.214), and should not be confused with the determinant in (4.215), which could be positive or negative. As was done for the transformation of a single random variable, this result can be generalized to include nonmonotonic functions by considering for each function $g_n(\mathbf{X})$ different intervals for the components of $\mathbf{X}$ that map to the same interval of y_n. Also, regions where any of the partial derivatives in the Jacobian are zero will cause $f_{\mathbf{Y}}(\mathbf{y})$ to have a discrete component, resulting in a mixed pdf.

For some mappings, it is easier to compute the Jacobian for the inverse transformations as follows:

$$J(\mathbf{y}) \triangleq \det \begin{bmatrix} \partial x_1/\partial y_1 & \cdots & \partial x_1/\partial y_N \\ \vdots & \cdots & \vdots \\ \partial x_N/\partial y_1 & \cdots & \partial x_N/\partial y_N \end{bmatrix} \tag{4.216}$$

so that the following is used instead of (4.214):

$$f_{\mathbf{Y}}(\mathbf{y}) = |J(\mathbf{y})| \, f_{\mathbf{X}}(\mathbf{x})\big|_{x_n = g_n^{-1}(\mathbf{y})}. \tag{4.217}$$

The Jacobian in this expression is already a function of $\mathbf{y}$ so that the substitution $x_n = g_n^{-1}(\mathbf{y})$ applies only to $f_{\mathbf{X}}(\mathbf{x})$.

Example 4.30. Let X_1 and X_2 be independent standard Gaussian random variables, and consider the following transformations:

$$Y_1 = \sqrt{X_1^2 + X_2^2}, \quad Y_2 = \tan^{-1}(X_1/X_2). \tag{4.218}$$

It is clear that the support of Y_1 is $\mathcal{R}^+$ and that of Y_2 is $[0, 2\pi]$. In order to find the joint distribution $f_{\mathbf{Y}}(\mathbf{y})$, we need to find (i) the Jacobian determinant and (ii) the inverse transformations. For (i):

$$J(x_1, x_2) = \det \begin{bmatrix} x_1/\sqrt{x_1^2 + x_2^2} & x_2/\sqrt{x_1^2 + x_2^2} \\ x_2/\left(x_1^2 + x_2^2\right) & -x_1/\left(x_1^2 + x_2^2\right) \end{bmatrix}, \\ = -1/\sqrt{x_1^2 + x_2^2}. \tag{4.219}$$

For (ii), we rearrange the transformations as follows:

$$Y_1^2 = X_1^2 + X_2^2, \quad X_1^2 = X_2^2 \tan^2(Y_2). \tag{4.220}$$

Solving these for X_1 and X_2 yields

$$X_1 = Y_1 \sin(Y_2), \quad X_2 = Y_1 \cos(Y_2). \tag{4.221}$$

Substituting these expressions into (4.214) gives

$$f_{\mathbf{Y}}(\mathbf{y}) = \left(x_1^2 + x_2^2\right)^{-1/2} (1/2\pi) \exp\left(-(1/2)\left[x_1^2 + x_2^2\right]\right)\Big|_{x_n = g_n^{-1}(\mathbf{y})}, \\ = |y_1|(1/2\pi)\exp(-(1/2)(y_1^2 \sin^2[y_2) + y_1^2 \cos^2(y_2)]), \\ = (1/2\pi) I_{[0,2\pi]}(y_2)|y_1| \exp(-y_1^2/2) I_{\mathcal{R}^+}(y_1), \tag{4.222}$$

from which we see that Y_1 and Y_2 are independent Rayleigh and uniform random variables, respectively. For completeness, note that the other Jacobian is

$$J(y_1, y_2) = \det \begin{bmatrix} \sin(y_2) & y_1 \cos(y_2) \\ \cos(y_2) & -y_1 \sin(y_2) \end{bmatrix} = -y_1 \tag{4.223}$$

whose magnitude is the leading term in the second line of (4.222).

- *Discrete random vector*. The transformation between discrete random vectors requires that we find all possible outcomes of the transformation, and then determine how the probability mass for each component of $\mathbf{X}$ carries over to that for each component of $\mathbf{Y}$. This approach is illustrated by the following example.

Example 4.31. Let $\mathbf{X}$ have $N = 2$ independent components that are Poisson random variables with parameters α_1 and α_2. Consider the following transformations:

$$Y_1 = X_1 + X_2, \quad Y_2 = X_1^2 \tag{4.224}$$

such that Y_1 has the same support as X_1 and X_2, but Y_2 excludes integers that do not have integer-valued square roots. For random variable Y_2,

$$p_{Y_2}[y_2] = P(X_1^2 = y_2) = P(X_1 = \sqrt{y_2}) = \frac{\alpha_1^{\sqrt{y_2}}}{\sqrt{y_2}!} \exp(-\alpha_1), \tag{4.225}$$

where $y_2 \in \{0, 1, 4, 9, 16, \ldots\}$. This is not a standard parametric distribution. For random variable Y_1,

$$p_{Y_1}[y_1] = \exp(-(\alpha_1 + \alpha_2)) \sum_{x=-\infty}^{\infty} \frac{\alpha_1^x}{x!} \frac{\alpha_2^{y_1-x}}{(y_1 - x)!} I_{\mathcal{Z}^+}[x] I_{\mathcal{Z}^+}[y_1 - x],$$

$$= \exp(-(\alpha_1 + \alpha_2)) \sum_{x=0}^{y_1} \frac{\alpha_1^x \alpha_2^{y_1-x}}{x!(y_1 - x)!} I_{\mathcal{Z}^+}[y_1], \tag{4.226}$$

where the second indicator function in the first equation causes the upper limit of the sum to be y_1 because $y_1 - x \in \mathcal{Z}^+$ means that x cannot exceed y_1 (otherwise, $y_1 - x$ is negative). The last result can be rewritten using the binomial formula if we multiply and divide by $y_1!$ as follows:

$$p_{Y_1}[y_1] = (1/y!) \exp(-(\alpha_1 + \alpha_2)) \sum_{x=0}^{y_1} \frac{\alpha_1^x \alpha_2^{y_1-x} y_1!}{x!(y_1 - x)!} I_{\mathcal{Z}^+}[y_1],$$

$$= (1/y_1!) \exp(-(\alpha_1 + \alpha_2)) \sum_{x=0}^{y_1} \binom{y_1}{x} \alpha_1^x \alpha_2^{y_1-x} I_{\mathcal{Z}^+}[y_1],$$

$$= (1/y!) \exp(-(\alpha_1 + \alpha_2))(\alpha_1 + \alpha_2)^{y_1} I_{\mathcal{Z}^+}[y_1], \tag{4.227}$$

which is Poisson with parameter $\alpha_1 + \alpha_2$.

4.12 SAMPLE MEAN $\bar{X}$ AND SAMPLE VARIANCE S^2

One of the more important functions of several random variables is the *sample mean* which is also known as the *arithmetic mean*. (The geometric mean is $(X_1 \cdots X_N)^{1/N}$.) Consider an experiment with repeated trials, such as consecutive coin tosses, where N outcomes are observed corresponding to N random variables $\{X_1, \ldots, X_N\}$. This type of repeated experiment can be viewed as a discrete-time random sequence which is covered in Chapter 6. For our purposes in this section, we only need to consider N random variables regardless of how they are generated, and which may or may not be mutually dependent. The sample mean is generally used to provide a measure of the average value of the resulting outcomes: it is an *estimate* of the mean.

Definition: Sample Mean The *sample mean* of random variables $\{X_1, \ldots, X_N\}$ is

$$\bar{X} \triangleq \frac{1}{N} \sum_{n=1}^{N} X_n, \tag{4.228}$$

which itself is a random variable.

In subsequent chapters, we discuss some interesting properties of $\bar{X}$. Here, we mention that although the sample mean is an unbiased estimator of μ_X for iid $\{X_n\}$, it is not robust because of its sensitivity to *outliers*. These are large outcomes (positive or negative) that are unlikely to occur, but can happen, causing the sample mean to be skewed above or below the actual mean. One way to improve the estimate is to increase N so that the effect of one outlier is minimized; alternatively, we might use other estimators of the average (or "center" of the pdf) such as the median and mode discussed in Chapter 5.

Definition: Sample Variance The *sample variance* of random variables $\{X_1, \ldots, X_N\}$ is

$$S^2 \triangleq \frac{1}{N-1}\sum_{n=1}^{N}(X_n - \bar{X})^2, \tag{4.229}$$

which itself is a random variable.

The factor $1/(N-1)$ is included (instead of $1/N$) so that S^2 is an unbiased estimator of σ_X^2 for iid $\{X_n\}$.

Example 4.32. Let $\{X_n\}$ be uniformly distributed on $[-1, 1]$. When $N = 2$, the pdf of $\bar{X}$ is obtained as a convolution of two identical rectangular pdfs, resulting in a triangular pdf for the sum as shown in Figure 4.34(a). Increasing N causes the rectangular pdf to be repeatedly convolved with previous convolution results. Successive pdfs obtained with increasing N are also illustrated in Figure 4.34(a). Observe that the pdf of the sample mean begins to resemble a Gaussian distribution. This behavior is characterized by the CLT, which was

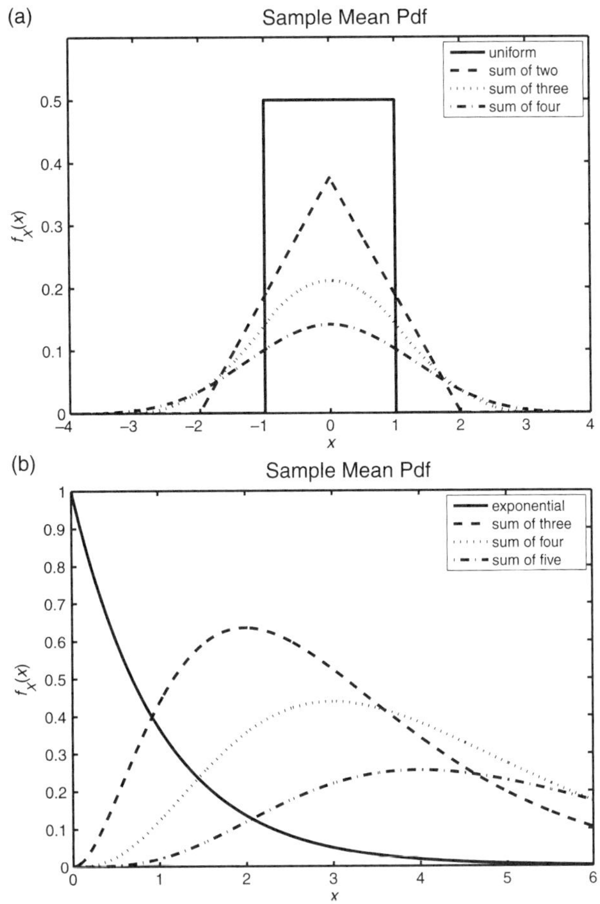

FIGURE 4.34 Sample mean pdfs for increasing N in Example 4.32. (a) $\{X_n\}$ are iid uniform on $[-1, 1]$. (b) $\{X_n\}$ are iid exponential ($\lambda = 1$). The pdfs have been scaled for visual clarity.

discussed earlier (and is seen again in Chapter 7); it gives conditions when the sum of independent random variables approach a Gaussian pdf as $N \to \infty$. Another example of the pdf for the sample mean is shown in Figure 4.34(b) when the distribution of the samples is exponential with parameter $\lambda = 1$. Note that even though the exponential pdf is not symmetric (like the uniform and Gaussian pdfs), the pdf of the sample mean quickly approaches the shape of a Gaussian pdf.

4.13 MINIMUM, MAXIMUM, AND ORDER STATISTICS

Consider N random variables $\{X_1, \ldots, X_N\}$ that are iid with pdf $f_X(x)$. The goal in this section is to derive pdfs for the minimum $X_{(1)} \triangleq \min(X_1, \ldots, X_N)$ (called the first-order statistic), the maximum $X_{(N)} \triangleq \max(X_1, \ldots, X_N)$ (called the Nth-order statistic), as well as the overall *order statistics*.

Definition: Order Statistics The *order statistics* are the random variables $\{X_1, \ldots, X_N\}$ arranged in increasing order: $\{X_{(1)}, \ldots, X_{(N)}\}$.

There may be some confusion about ordering random variables according to size because they are random and each can assume a range of values. Of course, once we have an actual sample (collection of outcomes) given by $\{x_1, \ldots, x_N\}$, they can be ordered according to size. However, ordered random variables $\{X_{(1)}, \ldots, X_{(N)}\}$ can be described in a probabilistic sense, as we show now using the cdf transformation method.

Theorem 4.4 (Order statistics). The pdf of the nth-order statistic $X_{(n)}$ is

$$f_{X_{(n)}}(x) = \frac{N!}{(n-1)!(N-n)!} F_X^{n-1}(x)[1 - F_X(x)]^{N-n} f_X(x), \tag{4.230}$$

where $f_X(x)$ and $F_X(x)$ are the pdf and cdf, respectively, of iid $\{X_1, \ldots, X_N\}$, and $n \in \{1, \ldots, N\}$.

Proof. The cdf of the nth-order statistic is

$$\begin{aligned} F_{X_{(n)}}(x) &= P(X_{(n)} \le x) = P(\text{at least } n \text{ of the } X_m \le x), \\ &= \sum_{m=n}^{N} \binom{N}{m} P(X \le x)^m (1 - P(X \le x))^{N-m}, \\ &= \sum_{m=n}^{N} \binom{N}{m} F_X^m(x)(1 - F_X(x))^{N-m}. \end{aligned} \tag{4.231}$$

The key step is the first line: since we are interested in ordered samples, $X_{(n)} \le x$ requires that *no fewer* than n of the random variables be $\le x$. More than n random variables are allowed in the expression because we are guaranteed that the nth-order statistic is $\le x$. The expression in the sum is the number of permutations where n or more of the N $\{X_m\}$ are $\le x$. This result corresponds to a binomial pdf with probability of success $p = F_X(x) = P(X \le x)$ and probability of failure $q = 1 - F_X(x) = P(X > x)$. (Note that $0 \le F_X(x) \le 1$ as required for p in the binomial pdf.) As long as n or more of the random variables are $\le x$, the probability will be nonzero. Differentiating with respect to x and using the chain rule yields the pdf:

$$\begin{aligned} f_{X_{(n)}}(x) = \sum_{m=n}^{N} \binom{N}{m} f_X(x) \big[& m F_X^{m-1}(x)(1 - F_X(x))^{N-m} \\ & - (N-m) F_X^m(x)(1 - F_X(x))^{N-m-1} \big]. \end{aligned} \tag{4.232}$$

Although not obvious, this expression can be simplified. Observe that the subtracted quantities differ only by one term, and many other terms cancel. Factoring out $Nf_X(x)$ gives

$$f_{X_{(n)}}(x) = Nf_X(x)\left[\sum_{m=n}^{N}\binom{N-1}{m-1}F_X^{m-1}(x)(1-F_X(x))^{N-m}\right.$$
$$\left. - \sum_{m=n}^{N}\binom{N-1}{m}F_X^{m}(x)(1-F_X(x))^{N-m-1}\right], \tag{4.233}$$

where we have used

$$m\binom{N}{m} = N\binom{N-1}{m-1}, \quad (N-m)\binom{N}{m} = N\binom{N-1}{m}. \tag{4.234}$$

The Nth term in the second sum is zero because the corresponding binomial coefficient is zero $\binom{N-1}{N} \triangleq 0$; this causes the upper limit to be $N-1$ instead of N. Let $l = m-1$ in the first sum so that

$$f_{X_{(n)}}(x) = Nf_X(x)\left[\sum_{l=n-1}^{N-1}\binom{N-1}{l}F_X^{l}(x)(1-F_X(x))^{N-1-l}\right.$$
$$\left. - \sum_{m=n}^{N-1}\binom{N-1}{m}F_X^{m}(x)(1-F_X(x))^{N-m-1}\right]. \tag{4.235}$$

As a result of the subtraction, only the $(n-1)$st term in the first sum remains:

$$f_{X_{(n)}}(x) = Nf_X(x)\binom{N-1}{n-1}F_X^{n-1}(x)(1-F_X(x))^{N-n}, \tag{4.236}$$

which is the same as (4.230). □

Theorem 4.5 (First-order statistic: minimum). The pdf of $X_{(1)} \triangleq \min(X_1, \ldots, X_N)$ is

$$f_{X_{(1)}}(x) = N[1-F_X(x)]^{N-1}f_X(x), \tag{4.237}$$

and the cdf is

$$F_{X_{(1)}}(x) = 1-[1-F_X(x)]^N \tag{4.238}$$

for iid $\{X_1, \ldots, X_N\}$.

Proof. This pdf is a special case of Theorem 4.4 and is easily proved by substituting $n = 1$. The cdf is derived directly as follows:

$$\begin{aligned} F_{X_{(1)}}(x) &= 1 - P(X_1 > x, \ldots, X_N > x), \\ &= 1-[1-F_1(x)]\cdots[1-F_N(x)] = 1-[1-F_X(x)]^N, \end{aligned} \tag{4.239}$$

which can also be derived from (4.231) by substituting $n = 1$ and recognizing that it is a sum over a binomial-type distribution excluding the $m = 0$ term. □

Theorem 4.6 (*N*th-order statistic: maximum). The pdf of $X_{(N)} \triangleq \max(X_1, \ldots, X_N)$ is

$$f_{X_{(N)}}(x) = NF_X^{N-1}(x)f_X(x), \tag{4.240}$$

and the cdf is

$$F_{X_{(N)}}(x) = F_X^N(x) \tag{4.241}$$

for iid $\{X_1, \ldots, X_N\}$.

Proof. This pdf is also a special case of Theorem 4.4 and is proved by substituting $n = N$. The cdf is derived as follows:

$$F_{X_{(N)}}(x) = P(X_1 \leq x, \ldots, X_N \leq x) = F_1(x) \cdots F_N(x) = F_X^N(x), \tag{4.242}$$

which can also be obtained from (4.231) by substituting $n = N$. □

Example 4.33. Let $\{X_1, \ldots, X_N\}$ be iid samples from a uniform distribution on [0, 1] which has the following pdf and cdf:

$$f_X(x) = I_{[0,1]}(x), \quad F_X(x) = xI_{[0,1]}(x). \tag{4.243}$$

The pdf of the *n*th-order statistic from Theorem 4.4 is

$$f_{X_{(n)}}(x) = \frac{N!}{(n-1)!(N-n)!} x^{n-1}(1-x)^{N-n} I_{[0,1]}(x), \tag{4.244}$$

which gives the following pdfs for the minimum and maximum:

$$f_{X_{(1)}}(x) = N(1-x)^{N-1} I_{[0,1]}(x), \quad f_{X_{(N)}}(x) = Nx^{N-1} I_{[0,1]}(x). \tag{4.245}$$

Examples of these for the uniform [0, 1] distribution are shown in Figure 4.35. Observe that the pdf of the first-order statistic is located near zero and that it becomes more concentrated there as N is increased. A similar effect is observed for the *N*th-order statistic except that it is located near 1. Increasing N does not, of course, affect the underlying pdf which is still uniform on [0, 1]; it changes the exponent in (4.242) so that with more samples it is more probable that the first sample will be close to zero, and likewise for the *N*th sample being close to 1. In fact, the pdfs of the first and *N*th-order statistics approach Dirac delta functions as $N \to \infty$:

$$\lim_{N\to\infty} f_{X_{(N)}}(x) = f_X(x) \lim_{N\to\infty} NF_X^{N-1}(x) = I_{[0,1]}(x) \lim_{N\to\infty} Nx^{N-1}. \tag{4.246}$$

Since

$$\int_{-\infty}^{\infty} Nx^{N-1} I_{[0,1]}(x) \mathrm{d}x = 1 \tag{4.247}$$

for any $N > 0$, the above limit is basically another definition of the Dirac delta function (see Appendix B): as $N \to \infty$, the height of the function at $x = 1$ approaches infinity, its width approaches zero, and its area is fixed

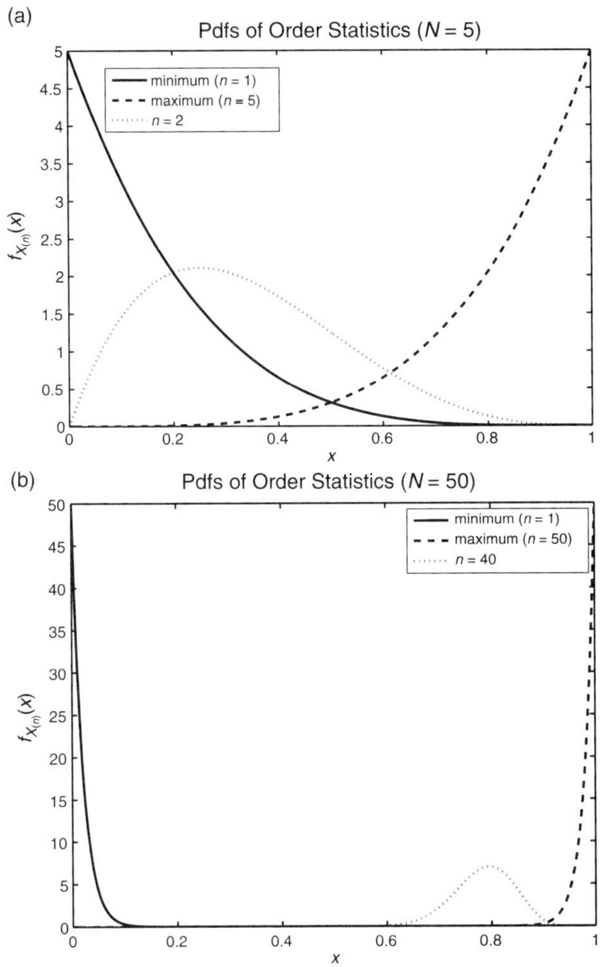

FIGURE 4.35 Order statistic pdfs for the uniform [0, 1] random variable in Example 4.33. (a) $N = 5$ samples. (b) $N = 50$ samples.

at 1. The same behavior occurs for the first order statistic at $x = 0$. We have also included in the figure the pdfs for the $n = 2$ and 40 order statistics, with $N = 5$ and 50, respectively.

Example 4.34. Consider the order statistics for the standard Gaussian random variable. The pdfs of the minimum and maximum are

$$f_{X_{(1)}}(x) = \frac{N Q^{N-1}(x)}{\sqrt{2\pi}} \exp(-x^2/2), \tag{4.248}$$

$$f_{X_{(N)}}(x) = \frac{N[1 - Q(x)]^{N-1}}{\sqrt{2\pi}} \exp(-x^2/2), \tag{4.249}$$

where $Q(x)$ is the Q-function. Examples of these pdfs are shown in Figure 4.36. Since this pdf does not have finite support, the samples do not "bunch up" near a maximum or minimum value as they do for the uniform

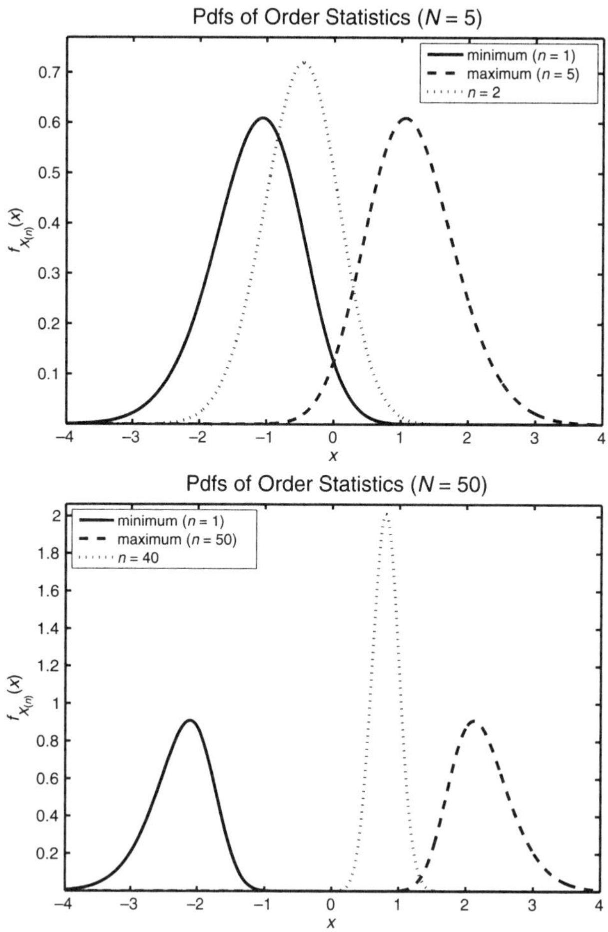

FIGURE 4.36 Order statistic pdfs for the standard Gaussian random variable in Example 4.34. (a) $N = 5$ samples. (b) $N = 50$ samples.

distribution. Instead, as N is increased, the pdfs move increasingly to the left and to the right, which makes sense because with more samples it is more likely that a very small and a very large sample will occur.

4.14 MIXTURES

Let $\{f_{X_n}(x_n)\}$ be the pdfs of random variables $\{X_n\}$ for $n = 1, \ldots, N$.

Definition: Probability Mixture Random variable Y is a *probability mixture* if it can be expressed as the following weighted sum:

$$f_Y(y) = \sum_{n=1}^{N} a_n f_{X_n}(y), \tag{4.250}$$

where $a_n \in [0, 1]$ and $\sum_{n=1}^{N} a_n = 1$. It is also called a *finite mixture*.

A mixture is *not* a sum of the random variables like the sample mean. The sum in (4.250) is a *convex combination* because the coefficients are (i) nonnegative and (ii) they sum to 1. These two properties ensure that $f_Y(y)$ is a valid pdf: $f_Y(y) \geq 0$ and

$$\int_{-\infty}^{\infty} f_Y(y)\mathrm{d}y = \int_{-\infty}^{\infty} \sum_{n=1}^{N} a_n f_{X_n}(y)\mathrm{d}y,$$

$$= \sum_{n=1}^{N} a_n \int_{-\infty}^{\infty} f_{X_n}(y)\mathrm{d}y = \sum_{n=1}^{N} a_n = 1. \tag{4.251}$$

Note that different pdfs can be used in (4.250). If the pdfs turn out to be from the same family, differing only by the values of the distribution parameters, then the probability mixture is known as a parametric mixture.

Definition: Parametric Mixture (Discrete) Random variable Y is a *parametric mixture* if it can be expressed as the weighted sum

$$f_Y(y) = \sum_{n=1}^{N} a_n f_X(y;\theta_n), \tag{4.252}$$

where $a_n \in [0, 1]$, $\sum_{n=1}^{N} a_n = 1$, and θ_n is a parameter of random variable X.

The sum is performed only over different values of the parameter; there is only one type ("family") of pdf in the sum. The corresponding mixture can also be defined when the parameter is continuous.

Definition: Parametric Mixture (Continuous) Random variable Y is a *continuous mixture* if it can be expressed as the following integral:

$$f_Y(y) = \int_{-\infty}^{\infty} a(\theta) f_X(y;\theta)\mathrm{d}\theta, \tag{4.253}$$

where $a(\theta) \geq 0$ and $\int_{-\infty}^{\infty} a(\theta)\mathrm{d}\theta = 1$. It is also called an *infinite mixture*.

Although $a(\theta)$ satisfies the properties required of a pdf, it need not be a pdf. If it turns out that $a(\theta)$ is a valid pdf, then θ can be viewed as a random variable which we might explicitly indicate by using $f_\Theta(\theta)$.

Example 4.35. The negative binomial random variable Y is an example of a continuous mixture (even though Y itself is discrete). Let X be a Poisson random variable with parameter α, and assume α is a gamma random variable with parameters $\{\lambda = p/q, r\}$ ($q \triangleq 1 - p$). Then from (4.253):

$$f_Y(y) = \int_{-\infty}^{\infty} f_X(y;\alpha) f_\alpha(\alpha)\mathrm{d}\alpha,$$

$$= \int_{-\infty}^{\infty} \sum_{n=0}^{\infty} \exp(-\alpha)\frac{\alpha^n}{n!}\delta(y-n)$$

$$\cdot \frac{1}{\Gamma(r)}\alpha^{r-1}(p/q)^r \exp(-(p/q)\alpha)u(\alpha)\mathrm{d}\alpha. \tag{4.254}$$

It was shown in Example 3.62 that this expression can be rearranged into the form of the negative binomial pdf:

$$f_Y(y) = \sum_{n=0}^{\infty} \binom{n+r-1}{n} p^r q^n \delta(y-n), \tag{4.255}$$

which has parameters $\{p, N = r\}$.

PROBLEMS

Approximations of Random Variables

4.1 Show that the chi-square random variable with parameter N is Gaussian with parameters $\{\mu = N, \sigma = \sqrt{2N}\}$ as $N \to \infty$.

4.2 Demonstrate that Student's t distribution with parameter r is approximately standard Gaussian for large r.

4.3 Show that the negative binomial distribution with parameters $\{N, p\}$ is Poisson with parameter $\alpha = N(1-p)$ as $N \to \infty$.

4.4 Show that the gamma distribution with parameters $\{\lambda, r\}$ is Gaussian with parameters $\{\mu = r/\lambda, \sigma = \sqrt{r}/\lambda\}$ as $r \to \infty$.

Joint, Marginal, and Conditional Distributions

4.5 (a) Find the marginal pdfs for the following joint pdf:

$$f_{X,Y}(x, y) = y \exp(-x - y) I_{[0,\infty)}(x) I_{[0,\infty)}(y). \tag{4.256}$$

(b) Derive the conditional pdfs $f_{Y|X}(y|x)$ and $f_{X|Y}(x|y)$.

4.6 Repeat Problem 4.5 for (a)

$$f_{X,Y}(x, y) = (1/2) x y I_{[0,2]}(x) I_{[0,x]}(y) \tag{4.257}$$

and (b)

$$f_{X,Y}(x, y) = (1/2)(3x + y) I_{[0,1]}(x) I_{[0,1]}(y). \tag{4.258}$$

4.7 Two dice are tossed sequentially. Let $X = \{\text{sum is even}\}$ and $Y = \{\text{first die is even}\}$. Determine if X and Y are independent.

4.8 Let X be uniform on $[0, a]$ with $a > 0$, and define the event $B = \{X > b\}$ such that $0 < b < a$. Derive an expression for $f_{X|B}(x|B)$.

4.9 For the bivariate exponential pdf in (4.80), find (a) the conditional pdf $f_{X_1|X_2}(x_1|x_2)$ and (b) an expression for $P(X_1 > X_2 | X_2 = 2)$.

4.10 Derive the cdf in (4.85) for one version of bivariate exponential random variables.

4.11 Another version of the bivariate exponential distribution has the following cdf:

$$F_{X_1,X_2}(x_1, x_2) = \left[1 - \exp(-x_1) - \exp(-x_2) + \exp\left(-\sqrt{x_1^2 + x_2^2}\right)\right] I_{[0,\infty)}(x_1) I_{[0,\infty)}(x_2). \tag{4.259}$$

Find expressions for (a) the bivariate pdf and (b) the marginal pdfs.

4.12 Let X_1 and X_2 be independent Poisson random variables with parameters α_1 and α_2. Based on the result in Problem 4.25, $Y = X_1 + X_2$ is Poisson with parameter $\alpha_1 + \alpha_2$. Show that the conditional pmf of X_1 is binomial when conditioned on $Y = y$ (a fixed number).

4.13 Consider the following joint pmf:

$$p_{X,Y}[x, y] = c\alpha^{x+y} I_{\{x,x+1,\cdots\}}[y] I_{\{0,1,\ldots\}}[x], \tag{4.260}$$

where $\alpha \in (0, 1)$. (a) Find c such that this is a valid pmf. (b) Derive the corresponding marginal pmfs and (c) the conditional pmf $p_{Y|X}[y|x]$.

4.14 Derive the expression for mutual information in Example 4.6.

Transformations of Random Variables

4.15 For an exponential random variable with parameter λ, show that the transformation $Y = \sqrt{2\sigma^2 X/\lambda}$ is Rayleigh with parameter $\alpha = \sigma/\lambda$.

4.16 Let X be uniform on $[-1, 2]$. Find the pdf for the following transformations. (a) $Y = \exp(-X)$. (b) $Z = X^2$.

4.17 Let X be a continuous uniform random variable on $[-a, a]$. Find the pdf for $Y = g(X)$, where

$$g(x) = \begin{cases} 0, & x < 0 \\ 3x, & 0 \le x \le b \\ 3b, & x > b. \end{cases} \tag{4.261}$$

4.18 (a) Find the pdf of $Y = \text{sgn}(X)$ for X Gaussian with parameters $\{\mu = 2, \sigma = 1\}$. (b) Repeat part (a) for $Z = X^3$ and standard Gaussian X.

4.19 Let X and Y be independent uniform random variables on $[0, 1]$. Show that for $V = X/Y$:

$$f_V(v) = (1/2)I_{[0,1)}(v) + (1/2v^2)I_{[1,\infty)}(v). \tag{4.262}$$

4.20 For independent discrete random variables X and Y, derive the expression in (4.168) for the pmf of $Z = XY$, and comment about the result for $Z = 0$.

4.21 Let X be a binomial random variable with parameters $\{N = 4, p\}$. Find the pdf of $Z = X + Y$, when Y is continuous uniform on $[0, 2]$ and is independent of X. Sketch $f_Z(z)$ for $p = 1/2$.

4.22 Demonstrate for exponential X with parameter λ that $Y = \exp(X)$ is Pareto with parameters $\{\alpha = \lambda, r = 1\}$.

4.23 Assume that X_1 and X_2 are independent geometric random variables with support $\mathcal{N}$. Show that

$$P(X_1 = x|X_1 + X_2 = y) = [1/(y-1)]I_{\{1,\ldots,y-1\}}[x], \tag{4.263}$$

which is a discrete uniform distribution.

4.24 Show that $Y = X_1/(X_1 + X_2)$ is a beta random variable with parameters $\{\alpha = N/2, \beta = N/2\}$, when X_1 and X_2 are iid chi-square random variables with parameter N.

4.25 Demonstrate that the sum of M independent Poisson random variables with parameter α is also Poisson, but with parameter $M\alpha$.

4.26 Derive the pdf in (3.77) due to a transformation of the inverse Gaussian random variable.

4.27 Derive the logistic distribution for the transformation of an exponential random variable shown in Table 4.4. (This is the example shown at the beginning of the chapter in Figure 4.1.)

4.28 (a) For X_1 and X_2 iid exponential random variables with parameter λ, demonstrate that $Y = X_1/X_2$ has the following pdf:

$$f_Y(y) = \frac{1}{(1+y)^2} I_{[0,\infty)}(y). \tag{4.264}$$

(b) Use this result to show that $Z = \mu - \alpha \ln(X_1/X_2)$ has the same logistic distribution as in Problem 4.27.

4.29 Show that the Nakagami random variable with pdf in (3.148) is obtained via the transformation $Y = \sqrt{X}$, where X has a gamma distribution with parameters $\{\lambda = \alpha/r, r\}$.

4.30 Verify the following sum of binomial coefficients:

$$\binom{2N}{m} = \sum_{n=0}^{m} \binom{N}{n}\binom{N}{m-n}. \tag{4.265}$$

4.31 Show that

$$\delta(z) = \int_0^\infty (\lambda/y)\delta(y)\exp(-\lambda z/y)dy. \tag{4.266}$$

4.32 Let X be a random variable with equally likely outcomes $\{-2, -1, 0, 1, 2\}$. Define random vector $\mathbf{Y} \triangleq [Y_1, Y_2]^T$, where

$$Y_1 \triangleq X + 2, \quad Y_2 \triangleq X^2. \tag{4.267}$$

Find an expression for the joint pdf $f_{\mathbf{Y}}(\mathbf{y})$, and determine if Y_1 and Y_2 are independent.

4.33 Repeat Problem 4.32 where X is the standard Gaussian random variable.

4.34 Let X_1 and X_2 be iid exponential random variables with parameter λ. Find the joint pdf of $\mathbf{Y} = [Y_1, Y_2]^T$ for

$$Y_1 = X_1 - X_2, \quad Y_2 = X_1 + X_2. \tag{4.268}$$

4.35 Let $\{X_1, \ldots, X_N\}$ be iid Gaussian random variables with parameters $\{\mu, \sigma\}$. Show that $(N-1)S^2/\sigma^2$ in (3.211) has a chi-square distribution with $N-1$ degrees of freedom where S^2 is the sample variance, which is based on the sample mean $\bar{X}$.

Order Statistics

4.36 Let $\{X_1, \ldots, X_N\}$ be samples from the Laplace distribution with parameters $\{\mu = 0, \alpha\}$. Find pdfs for the minimum $X_{(1)}$ and the maximum $X_{(N)}$.

4.37 Repeat Problem 4.36 for the Poisson random variable with parameter α.

4.38 Repeat Problem 4.36 for the Cauchy random variable with parameters $\{c = 0, \alpha\}$.

4.39 For X_1 and X_2 exponential random variables with parameters λ_1 and λ_2, find the cdf and pdf of the minimum $X_{(1)}$.

4.40 Let $\{X_1, \ldots, X_N\}$ be samples from a uniform distribution with support $[\theta - 1/2, \theta + 1/2]$, where θ is a fixed parameter. Find an expression for the pdf of the order statistics $\{X_{(n)}\}$.

4.41 Let $\{X_1, \ldots, X_N\}$ be iid uniform samples on $[0, \theta]$. Show that the joint pdf of $\{X_{(m)}, X_{(n)}\}$ for $1 \le m < n \le N$ is

$$f_{X_{(m)},X_{(n)}}(x_m, x_n) = \binom{N}{m-1, n-m-1, N-n}(1/\theta^N)x_m^{m-1}(x_n - x_m)^{n-m-1}$$
$$\cdot (\theta - x_n)^{N-n} I_{[0,x_n]}(x_m) I_{[x_m,\theta]}(x_n), \tag{4.269}$$

which uses a multinomial coefficient.

Mixtures

4.42 Let X be a Gaussian random variable with parameters $\{\mu, \sigma\}$. Show that the following parametric mixture is also Gaussian:

$$f_Y(y) = \int_{-\infty}^{\infty} a(\mu) f_X(y; \mu) \mathrm{d}\mu, \tag{4.270}$$

where $a(\mu)$ is the standard Gaussian pdf.

4.43 Consider the following mixture:

$$f_Y(y) = p f_{X_1}(y) + q f_{X_2}(y), \tag{4.271}$$

where X_1 is uniform on $(0, b]$, X_2 is uniform on $(b, 1]$, $p \in [0, 1]$, and $q \triangleq 1 - p$. Find an expression for the cdf $F_Y(y)$ and determine the median m_e.

Computer Problems

4.44 Generate dependent beta random variables using cdf and icdf from MATLAB and the technique described by (4.87). Let the correlation coefficient be $\rho = 0.8$, and use hist3 to plot histograms of (i) the bivariate Gaussian distribution, (ii) the inverse transformation to the bivariate uniform distribution, and (iii) the bivariate beta distribution. Let $\alpha_1 = \alpha_2 = 1$ and $\beta_1 = \beta_2 = 4$. Use corrcoef to compute the correlation coefficient from samples of the bivariate beta distribution.

4.45 Repeat Problem 4.44 to obtain dependent Laplace random variables. Let the correlation coefficient be $\rho = -0.6$, and the parameters of the Laplace distribution be $\mu_1 = \mu_2 = 0$, $\alpha_1 = 1$, and $\alpha_2 = 2$.

FURTHER READING

Additional material on multiple random variables and transformations can be found in most of the references given at the end of Chapters 2 and 3. References for specific multivariate distributions are provided in Appendix A. Topics near the end of this chapter can be investigated further in the following references. Mixtures: Johnson, Kemp, and Kotz (2005) and Johnson, Kotz, and Balakrishnan (1995). Order statistics: Lehmann (1959), Mood, Graybill, and Boes (1974), and Lehmann (1983).

5 Expectation and Moments

5.1 INTRODUCTION

The expectation of a random variable is one measure of the location and center of its probability density function (pdf). It is the "expected value" in the sense that with repeated trials, the expectation would be observed *on average*, which is also called the *mean*. Other measures of the center of a distribution, such as the median and the mode, are discussed in this chapter, as well as expectations of functions of a random variable called *moments*. Four important moments that provide much information about the location and shape of a random variable are summarized in Figure 5.1. Basically, all expectations are integrals of a function *with respect to a probability measure*. Thus, it can be written as a Lebesgue integral on the abstract probability space $\{\Omega, \mathcal{F}, P\}$, or as a Riemann–Stieltjes integral on $\{\mathcal{R}, \mathcal{B}(\mathcal{R}), F_X(x)\}$ for random variable X. In previous chapters, we used the Riemann integral to (i) calculate probabilities from pdfs and (ii) derive pdfs for various functions of random variables. Generally, we have assumed that the pdf is continuous or can be represented using Dirac delta functions for discrete and mixed random variables.

5.2 EXPECTATION AND INTEGRATION

Consider first the abstract probability space $\{\Omega, \mathcal{F}, P\}$, and let X be a random variable that maps outcomes in Ω to $\mathcal{R}$.

Definition: Expectation The *expectation* of random variable X is

$$\mathcal{E}[X] \triangleq \int_{\Omega} X(\omega)\mathrm{d}P(\omega), \tag{5.1}$$

which is the Lebesgue integral of function $X(\omega)$ and P is the probability measure for the event space $\{\Omega, \mathcal{F}\}$.

The Lebesgue integral is briefly described in Appendix D. All outcomes that map from Ω to $\mathcal{R}$ are weighted by the probability measure P above to give the expectation. We can also write $\mathcal{E}[X]$ entirely in terms of the random variable with probability space $\{\mathcal{R}, \mathcal{B}(\mathcal{R}), F_X(x)\}$ as follows:

$$\mathcal{E}[X] = \int_{-\infty}^{\infty} x\mathrm{d}F_X(x), \tag{5.2}$$

where $F_X(x)$ is the cumulative distribution function (cdf) of X. This expression is the Lebesgue–Stieltjes integral of function $g(x) = x$; it is also briefly described in Appendix D. In the next few sections, we explore how these integrals are computed, first for discrete random variables and then for continuous random variables. Later, the integrals are generalized to compute expectations of functions of a random variable, leading to *moments* (such as variance and skewness) and the *characteristic function*.

Probability, Random Variables, and Random Processes: Theory and Signal Processing Applications, First Edition. John J. Shynk.

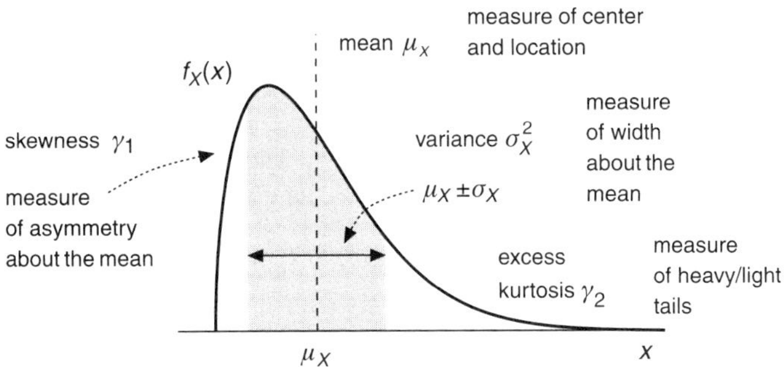

FIGURE 5.1 Four important moments of random variable X.

5.3 INDICATOR RANDOM VARIABLE

For the abstract probability space $\{\Omega, \mathcal{F}, P\}$, a basic type of random variable is the indicator random variable.

Definition: Indicator Random Variable The *indicator random variable* $X(\omega)$ is a mapping of event $E \in \Omega$ to the real line $\mathcal{R}$ as follows:

$$X(\omega) = I_E(\omega) = \begin{cases} 1, & \omega \in E \\ 0, & \text{else.} \end{cases} \tag{5.3}$$

The mapping for the indicator random variable is shown in Figure 5.2. Observe that E^c necessarily maps to zero. The indicator random variable is essentially the same as the indicator function, except that its argument ω is random as specified by the probability space. Its expectation turns out to be the probability of event E:

$$\mathcal{E}[X] = \int_\Omega I_E(\omega)\mathrm{d}P(\omega) = P(E), \tag{5.4}$$

and also

$$\int_\Omega I_{E^c}(\omega)\mathrm{d}P(\omega) = P(E^c) = 1 - P(E). \tag{5.5}$$

These probabilities are shown using Dirac delta functions in Figure 5.2. It is clear that since there are only two outcomes (E and E^c) for the indicator random variable, the probability mass function (pmf) of X is Bernoulli with probability of "success" $p = P(E)$:

$$P(X = 1) = P(E) = p, \qquad P(X = 0) = P(E^c) = 1 - p \triangleq q, \tag{5.6}$$

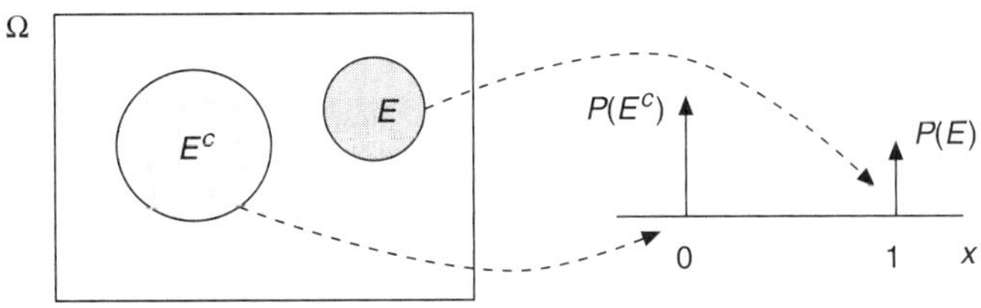

FIGURE 5.2 Mapping of the indicator function for event E to the indicator random variable X on the real line $\mathcal{R}$.

where we have used the notationally simpler X to represent $X(\omega)$; recall that $X = x$ actually means the set $\{\omega : X(\omega) = x\}$. The toss of a single die where event $E = \{\text{even number}\}$ is an example of an indicator random variable with $P(E) = 1/2$.

Example 5.1. A more complicated example is given by the Dirichlet function where $\Omega = \mathcal{R}$ and $E = \mathcal{Q}$ is the set of all rational numbers. For the indicator random variable, the rational numbers map to 1, and the irrational numbers (given by E^c) map to 0. The probability of event E is

$$P(E) = \int_{\Omega} I_{\mathcal{Q}}(\omega)\mathrm{d}P(\omega) = 0, \tag{5.7}$$

and thus $P(E^c) = 1$: when randomly selecting a real number, we will "almost surely" choose an irrational number. Although both E and E^c have an infinity of elements, the cardinality of the irrational numbers is the same as that of $\mathcal{R}$ (denoted by $\beth_1$), whereas the cardinality of $\mathcal{Q}$ is smaller, and is the same as that of the integers $\mathcal{Z}$ (given by $\beth_0$). This is an example where the Riemann integral cannot be used to compute a probability because no useful partition of $\mathcal{R}$ is possible. On the other hand, it is straightforward to calculate $P(E)$ using the Lebesgue integral because the probability weighting is applied to *sets* (see Appendix D for more discussion about the different types of integration).

Note that the expression in (5.3) happens to be identical to the Dirac measure covered in Chapter 2:

$$I_E(\omega) = \delta_\omega(E). \tag{5.8}$$

However, here we are using $I_E(\omega)$ as a particular type of random variable with probability measure $P(\omega)$. Later, we say more about the Dirac measure as a probability measure in the context of expectation.

5.4 SIMPLE RANDOM VARIABLE

A simple random variable is a generalization of the indicator random variable where instead of two events, N *mutually exclusive* events in $\mathcal{F}$ that form a partition of Ω are mapped to N values in $\mathcal{R}$.

Definition: Simple Random Variable *Simple random variable* X has the form

$$X(\omega) = \sum_{n=1}^{N} x_n I_{E_n}(\omega), \tag{5.9}$$

where x_n is the value in $\mathcal{R}$ assigned to event E_n, and the $\{E_n\}$ form a partition of Ω.

A simple random variable is essentially the same as a simple function (see Appendix D), except that its argument ω is random as determined by the probability space $\{\Omega, \mathcal{F}, P\}$. It follows from this definition that

$$P(X = x_n) = P(E_n) = \int_{\Omega} I_{E_n}(\omega)\mathrm{d}P(\omega). \tag{5.10}$$

Clearly, any discrete random variable with a finite number of outcomes is a simple random variable because it is readily represented by (5.9). The expectation is

$$\mathcal{E}[X] = \int_{\Omega} \sum_{n=1}^{N} x_n I_{E_n}(\omega)\mathrm{d}P(\omega) = \sum_{n=1}^{N} x_n P(X = x_n) = \sum_{n=1}^{N} x_n p_X[x_n], \tag{5.11}$$

where (5.10) has been used, and $p_X[x_n]$ is the pmf of X. The expectation of a finite random variable is the sum of all possible outcomes weighted by their probabilities. This result is consistent with the frequency interpretation of probability discussed in Chapter 2, as well as our intuition about the "average outcome" for an experiment.

Example 5.2. Let $\Omega = \{1, 2, 3, 4, 5, 6, 7, 8, 9\}$ and $\mathcal{F} = \mathcal{P}(\Omega)$. Define $E_1 = \{1, 2\}$, $E_2 = \{3, 4, 5, 6\}$, and $E_3 = \{7, 8, 9\}$, which together form a partition of Ω. The following is a simple random variable:

$$X(\omega) = \begin{cases} -1, & \omega \in E_1 \\ 0, & \omega \in E_2 \\ 1, & \omega \in E_3, \end{cases} \tag{5.12}$$

such that $x_1 = -1$, $x_2 = 0$, and $x_3 = 1$. If the outcomes in Ω are equally likely, then $P(X = x_1) = 2/9$, $P(X = x_2) = 4/9$, and $P(X = x_3) = 3/9$, yielding $\mathcal{E}[X] = (-1)(2/9) + (0)(4/9) + (1)(3/9) = 1/9$.

5.5 EXPECTATION FOR DISCRETE SAMPLE SPACES

For discrete random variables with N outcomes, the expectation $\mathcal{E}[X]$ is given by (5.11). The expectation for a discrete random variable with a *countably infinite* number of outcomes is derived by considering a sequence of simple random variables as follows. Assume initially that X is nonnegative. It is straightforward to show that a sequence of simple random variables exists that tends to $X(\omega)$ in the limit. For example, assume that $\Omega = \mathcal{Z}^+$ and let $X(\omega) = \omega$ be the identity mapping so that the outcomes of $X(\omega)$ are nonnegative integers. Define the following sequence of simple random variables:

$$X_n(\omega) \triangleq \begin{cases} \omega, & X(\omega) \leq n \\ n, & \text{else} \end{cases} \tag{5.13}$$

for $n \in \mathcal{Z}^+$. The outcomes of this sequence of simple random variables are 0, $\{0, 1\}$, $\{0, 1, 2\}$, and so on. Note that by construction, this sequence is nondecreasing for every $\omega \in \Omega$: $X_0(\omega) \leq X_1(\omega) \leq \cdots X_n(\omega) \leq \cdots$. For example, when $\omega = 2$ the outcomes are $X_0(\omega) = 0$, $X_1(\omega) = 0$, and $X_n(\omega) = 2$ for $n \geq 2$. From the expectation in (5.11) for a finite sum, we can write

$$\mathcal{E}[X_n] = \sum_{m=0}^{n} mP(X = m) + nP(X > n). \tag{5.14}$$

Since $\lim_{n\to\infty} X_n(\omega) = X(\omega)$ for every $\omega \in \Omega$, the expectation is

$$\begin{aligned} \mathcal{E}[X] &= \mathcal{E}\left[\lim_{n\to\infty} X_n\right] = \lim_{n\to\infty} \mathcal{E}[X_n], \\ &= \lim_{n\to\infty} \sum_{m=0}^{n} mP(X = m) + \lim_{n\to\infty} nP(X > n) = \sum_{m=0}^{\infty} x_m p_X[x_m], \end{aligned} \tag{5.15}$$

where in the last expression $\{x_m\}$ represent the nonnegative integer-valued outcomes. Also, we have assumed that $\lim_{n\to\infty} P(X > n) = 0$ faster than $n \to \infty$, which is required for the expectation to exist. This means that the pmf does not have "heavy tails." If those two terms approach 0 and ∞ at the same rate, then the result $0 \cdot \infty$ is not defined (see Appendix E). Or, if $n \to \infty$ faster than $P(X > n) \to 0$, then the expectation is infinite. (The

Cauchy random variable is an example of a heavy-tailed continuous random variable whose mean is not defined and second moment is infinite. It is the only parametric random variable discussed in this book for which this is the case.) The result in (5.15) is a straightforward extension of the expectation of a finite random variable to one that is infinite but countable. For random variables that take on negative values, the expectation of the negative and positive parts can be determined separately, treating them both as nonnegative random variables, and then the overall expectation is obtained by subtracting the two partial results: $\mathcal{E}[X] = \mathcal{E}_+[X] - \mathcal{E}_-[X]$. However, this is not necessary for discrete random variables on $\mathcal{Z}$ for which we can write

$$\mathcal{E}[X] = \sum_{n=-\infty}^{\infty} x_n p_X[x_n]. \tag{5.16}$$

Example 5.3. The expectation of the Poisson random variable is

$$\mathcal{E}[X] = \sum_{n=0}^{\infty} n \exp(-\alpha) \frac{\alpha^n}{n!}. \tag{5.17}$$

When infinite sums are encountered, it is convenient to manipulate the expression such that terms can be factored, leaving a sum over a pmf that equals 1. The functional forms for discrete pmfs are given in Table E.3 in Appendix E. Also, it might be possible to rearrange an expression so that it is a series expansion of a known function. Observe that the first term in the sum is zero such that

$$\mathcal{E}[X] = \exp(-\alpha) \sum_{n=1}^{\infty} n \frac{\alpha^n}{n!} = \exp(-\alpha) \sum_{n=1}^{\infty} \frac{\alpha^n}{(n-1)!}. \tag{5.18}$$

Changing variables to $m = n - 1$ yields

$$\mathcal{E}[X] = \exp(-\alpha) \sum_{m=0}^{\infty} \frac{\alpha^{m+1}}{m!} = \exp(-\alpha)\alpha \sum_{m=0}^{\infty} \frac{\alpha^m}{m!}. \tag{5.19}$$

Since the last sum is the series expansion for $\exp(\alpha)$ (see Appendix E), cancelling terms gives $\mathcal{E}[X] = \alpha$.

Example 5.4. The expectation of the geometric random variable is

$$\mathcal{E}[X] = \sum_{n=0}^{\infty} npq^n, \tag{5.20}$$

which can be evaluated by noting that n appears as a multiplier and in the exponent. Differentiation can be used to find such expectations:

$$\mathcal{E}[X] = pq \frac{\mathrm{d}}{\mathrm{d}q} \left(\sum_{n=0}^{\infty} q^n \right) = pq \frac{\mathrm{d}}{\mathrm{d}q} \left(\frac{1}{1-q} \right) = pq \frac{1}{(1-q)^2}. \tag{5.21}$$

Since $q \triangleq 1 - p$, we finally have $\mathcal{E}[X] = q/p$.

We return to expectation written as an integral over $\mathcal{R}$. Recall that the cdf for a discrete random variable X can be written as a weighted sum of unit-step functions:

$$F_X(x) = \sum_{n=-\infty}^{\infty} p_X[x_n]u(x - x_n), \tag{5.22}$$

where $\{x_n\}$ are the outcomes of X and $p_X[x_n]$ is the pmf. Substituting

$$\mathrm{d}F_X(x) = \sum_{n=-\infty}^{\infty} p_X[x_n]\mathrm{d}u(x - x_n) \tag{5.23}$$

into (5.2) gives

$$\mathcal{E}[X] = \sum_{n=-\infty}^{\infty} p_X[x_n] \int_{-\infty}^{\infty} x\mathrm{d}u(x - x_n). \tag{5.24}$$

Although the unit-step function $u(x)$ technically does not have a derivative at $x = 0$, we can write an integral in terms of the Dirac delta function as follows:

$$\int_{-\infty}^{\infty} x\mathrm{d}u(x - x_n) = \int_{-\infty}^{\infty} x\delta(x - x_n)\mathrm{d}x, \tag{5.25}$$

where in the last expression, the Dirac delta function is treated as an ordinary function even though it is a generalized function. Note also that we can write the Lebesgue integral

$$\int_{-\infty}^{\infty} x\mathrm{d}u(x - x_n) = \int_{-\infty}^{\infty} x\mathrm{d}\delta_{x_n}(x), \tag{5.26}$$

where $\delta_{x_n}(x)$ is the previously mentioned Dirac measure, which in this case $= 1$ only when $x = x_n$.

Since $\delta(x - x_n)$ in (5.25) is under an integral, the meaning is clear:

$$\int_{-\infty}^{\infty} x\delta(x - x_n)\mathrm{d}x = \int_{-\infty}^{\infty} x\mathrm{d}\delta_{x_n}(x) = x_n \tag{5.27}$$

such that (5.24) leads to the result in (5.16). The expressions in (5.24) and (5.25) imply that the pdf for a discrete random variable can be written as

$$f_X(x) = \sum_{n=-\infty}^{\infty} p_X[x_n]\delta(x - x_n), \tag{5.28}$$

which is the form introduced in Chapter 3. Although $f_X(x)$ for a discrete random variable is not a continuous function, if the Dirac delta function is interpreted as a measure, then we understand how to evaluate *integrals* involving such $f_X(x)$. Moreover, for function $g(x)$, we have

$$\int_{-\infty}^{\infty} g(x)\delta(x - x_n)\mathrm{d}x = \int_{-\infty}^{\infty} g(x)\mathrm{d}\delta_{x_n}(x) = g(x_n). \tag{5.29}$$

This is the form used later when evaluating expectations of functions of a discrete random variable.

Example 5.5. Consider again the geometric random variable whose pdf can be written as follows using the Dirac delta function:

$$f_X(x) = \sum_{n=0}^{\infty} pq^n \delta(x - n). \tag{5.30}$$

Its expectation is

$$\mathcal{E}[X] = \int_{-\infty}^{\infty} x f_X(x) \mathrm{d}x = \sum_{n=0}^{\infty} pq^n \int_{-\infty}^{\infty} x\delta(x - n) \mathrm{d}x = \sum_{n=0}^{\infty} npq^n, \tag{5.31}$$

which is the expression in (5.20), and can be evaluated using the differentiation approach in (5.21).

5.6 EXPECTATION FOR CONTINUOUS SAMPLE SPACES

The expectation of a continuous random variable can also be derived using a sequence of simple random variables. This is done by approximating the continuous random variable with a series of staircase mappings (as is done for a continuous function and the Lebesgue integral in Appendix D). Assume that $X(\omega)$ in (5.1) is nonnegative and define the following simple random variable:

$$X_n(\omega) \triangleq \begin{cases} m/2^n, & m/2^n \leq X(\omega) < (m+1)/2^n \\ 2^n, & X(\omega) \geq 2^n \end{cases} \tag{5.32}$$

for $0 \leq m \leq 2^{2n} - 1$. This is a countable random variable that approximates the original uncountable random variable X. Note that by construction, the $\{X_n(\omega)\}$ are nondecreasing for every $\omega \in \Omega$: $X_0(\omega) \leq X_1(\omega) \leq \cdots X_n(\omega) \leq \cdots$. The first simple random variable $X_1(\omega)$ is shown in Figure 5.3. The representation above is similar to quantization used in signal processing where continuous waveforms are converted to discrete signals (see Chapter 10). It is also used to describe the Lebesgue integral, which is summarized in Appendix D. Finally, note that a similar approach was used for countably infinite discrete random variables in (5.13), with the identity mapping $X(\omega) = \omega$ and $\Omega = \mathcal{Z}^+$. The simple random variable in (5.32) assumes an arbitrary nonnegative mapping $X(\omega)$. Noninteger steps $\{m/2^n\}$ are used because they must approach zero with increasing n in order to obtain a continuous random variable in the limit, as described in Theorem 5.1.

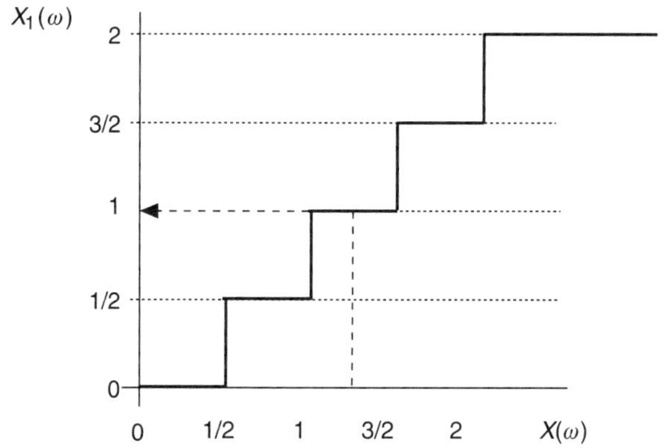

FIGURE 5.3 Approximation of $X(\omega)$ by the simple random variable $X_1(\omega)$.

Theorem 5.1. A continuous nonnegative random variable is the limit of a sequence of simple random variables.

Proof. Using the notation in (5.9) for a simple random variable, we can rewrite the nth approximation for $X(\omega)$ as follows:

$$X_n(\omega) = \sum_{m=0}^{2^{2n}-1} (m/2^n) I_{\{m/2^n \leq X(\omega) < (m+1)/2^n\}}(\omega) + 2^n I_{\{X(\omega) \geq 2^n\}}(\omega). \tag{5.33}$$

Observe that

$$\sup_{m \geq n} |X(\omega) - X_m(\omega)| \leq 2^{-n}, \tag{5.34}$$

where 2^{-n} is the width of each interval (excluding the last step in the staircase which extends to ∞). The staircase approximations are nondecreasing for every $\omega \in \Omega$, with increasingly finer steps as n increases, and the range of the steps extends to cover $\mathcal{R}^+$. As $n \to \infty$, the approximations converge to the original random variable: $X_n(\omega) \to X(\omega)$ as the right-hand side above tends to zero. (Convergence of a sequence of random variables indexed by n is discussed further in Chapter 7.) □

From this representation, the expectation for a continuous (nonnegative) random variable is

$$\mathcal{E}[X] = \lim_{n\to\infty} \sum_{m=0}^{2^{2n}-1} (m/2^n) P(m/2^n \leq X(\omega) < (m+1)/2^n) + \lim_{n\to\infty} 2^n P(X(\omega) \geq 2^n), \tag{5.35}$$

$$= \lim_{n\to\infty} \sum_{m=0}^{2^{2n}-1} x_{m,n} P(X \in [m/2^n, (m+1)/2^n)), \tag{5.36}$$

where $x_{m,n} \triangleq m/2^n$, and we have assumed that $2^n \to \infty$ slower than $P(X(\omega) \geq 2^n) \to 0$ in order to drop the last term. Note again that $x_{m,n}$ is not integer valued as is the case for x_n in (5.16) for discrete random variables. This is the Lebesgue integral over nonnegative x with probability measure P_X:

$$\mathcal{E}[X] = \int_0^\infty x \mathrm{d}P_X(x). \tag{5.37}$$

The expectation for an arbitrary continuous random variable is obtained by subtracting the expectation for negative outcomes from that of positive outcomes: $\mathcal{E}[X] = \mathcal{E}_+[X] - \mathcal{E}_-[X]$.

Example 5.6. Let X be an exponential random variable with cdf:

$$F_X(x) = [1 - \exp(-\lambda x)] I_{[0,\infty)}(x). \tag{5.38}$$

The expectation of simple random variable $X_1(\omega)$ is

$$\begin{aligned} \mathcal{E}[X_1] = {} & (0)P(0 \leq X < 1/2) + (1/2)P(1/2 \leq X < 1) \\ & (1)P(1 \leq X < 3/2) + (3/2)P(3/2 < X < 2) + (2)P(X \geq 2). \end{aligned} \tag{5.39}$$

From the cdf, we can write

$$F_X(a) - F_X(b) = \exp(-\lambda b) - \exp(-\lambda a) \tag{5.40}$$

so that

$$\mathcal{E}[X_1] = (1/2)[\exp(-\lambda/2) - \exp(-\lambda)] + [\exp(-\lambda) - \exp(-3\lambda/2)] \\ + (3/2)[\exp(-3\lambda/2) - \exp(-2\lambda)] + 2\exp(-2\lambda). \quad (5.41)$$

For $\lambda = 1$, we find from this expression that $\mathcal{E}[X_1] = 0.6664$. As n is increased and the expectations are computed, it is easy to see that $\mathcal{E}[X_n] \to \mathcal{E}[X] = 1$. Note, in particular, that

$$P(X \geq 2^n) = 1 - F_X(2^n) = \exp(-2^n\lambda), \quad (5.42)$$

and thus

$$\lim_{n\to\infty} 2^n \exp(-2^n\lambda) = 0, \quad (5.43)$$

that is, the probability in the last term of (5.35) tends to zero faster than $2^n \to \infty$.

In the mapping of random variable $X(\omega)$ from $\{\Omega, \mathcal{R}, P\}$ to $\{\mathcal{R}, \mathcal{B}(\mathcal{R}), F_X(x)\}$, the expectation can be computed over $\mathcal{R}$ using the Riemann–Stieltjes integral:

$$\mathcal{E}[X] = \int_{-\infty}^{\infty} x\mathrm{d}F_X(x), \quad (5.44)$$

where $F_X(x)$ is the cdf of X. If $F_X(x)$ is differentiable everywhere such that $\mathrm{d}F_X(x) = f_X(x)\,\mathrm{d}x$ where $f_X(x)$ is the pdf, then

$$\mathcal{E}[X] = \int_{-\infty}^{\infty} xf_X(x)\mathrm{d}x. \quad (5.45)$$

This expression, which is the more well-known definition of expectation, is the continuous counterpart of the expectation for a discrete random variable in (5.15); the summation has been "replaced" by an integration. For mixed random variables on $\mathcal{R}$ or the abstract sample space Ω, the Riemann–Stieltjes and Lebesgue integrals, respectively, are used to compute expectations. Examples of these are provided in Appendix D.

Example 5.7. The expectation of a Gaussian random variable is easily computed because it is symmetric about parameter μ. However, note that this symmetry does not always guarantee a well-defined expectation as illustrated later for the Cauchy random variable. When a function is symmetric about some μ, we can write $f_X(x+\mu) = f_X(-x+\mu)$. Multiplying $f_X(x)$ by an odd function about μ and integrating yields zero. Let the odd function be $g(x) = \mu - x$. Then

$$\int_{-\infty}^{\infty} g(x)f_X(x)\mathrm{d}x = \int_{-\infty}^{\infty} (x-\mu)f_X(x)\mathrm{d}x = 0 \quad (5.46)$$

provided the integral of $g(x)f_X(x)$ is finite at each limit. Rearranging this expression gives the final result:

$$\mathcal{E}[X] = \int_{-\infty}^{\infty} xf_X(x)\mathrm{d}x = \mu\int_{-\infty}^{\infty} f_X(x)\mathrm{d}x = \mu. \quad (5.47)$$

Example 5.8. For the exponential random variable, integration by parts (see Appendix D) is often used to evaluate the expectation (as well as other moments described later):

$$\int_a^b u \mathrm{d}v = uv|_a^b - \int_a^b v \mathrm{d}u. \tag{5.48}$$

Let $u \equiv \lambda x$ and $\mathrm{d}v \equiv \exp(-\lambda x)$ such that

$$\begin{aligned}\mathcal{E}[X] &= \int_0^\infty x\lambda \exp(-\lambda x)\mathrm{d}x, \\ &= \lambda x(-1/\lambda)\exp(-\lambda x)|_0^\infty - \int_0^\infty \lambda(-1/\lambda)\exp(-\lambda x)\mathrm{d}x, \\ &= 0 + \int_0^\infty \exp(-\lambda x)\mathrm{d}x = 1/\lambda.\end{aligned} \tag{5.49}$$

Example 5.9. For the Cauchy random variable with location parameter $c = 0$, the expectation is

$$\begin{aligned}\mathcal{E}[X] &= (\alpha/\pi)\int_{-\infty}^\infty \frac{x}{x^2+\alpha^2}\mathrm{d}x, \\ &= (\alpha/2\pi)\ln(x^2+\alpha^2)|_{-\infty}^\infty = \infty - \infty.\end{aligned} \tag{5.50}$$

Although x is an odd function and $f_X(x)$ is even, $\mathcal{E}[X]$ does not exist because the integral evaluated at each limit is not finite: $\infty - \infty$ is not defined (see Appendix B).

5.7 SUMMARY OF EXPECTATION

In this section, we summarize the main results for the expectation of a random variable. For the abstract probability space $\{\Omega, \mathcal{F}, P\}$, the expectation of random variable X is

$$\mathcal{E}[X] = \int_\Omega X(\omega)\mathrm{d}P(\omega) \qquad \text{(Lebesgue integral)}, \tag{5.51}$$

where $P(\omega)$ is the probability measure. In the probability space $\{\mathcal{R}, \mathcal{B}(\mathcal{R}), F_X(x)\}$, the expectation of random variable X is the following integral:

$$\mathcal{E}[X] = \int_{-\infty}^\infty x\mathrm{d}F_X(x) \qquad \text{(Riemann–Stieltjes integral)}, \tag{5.52}$$

where x is the integrand and the cdf $F_X(x)$ is the *integrator*. If $F_X(x)$ is differentiable everywhere, then this last expression can be expressed as an ordinary Riemann integral by substituting $\mathrm{d}F_X(x) = f_X(x)\mathrm{d}x$:

$$\mathcal{E}[X] = \int_{-\infty}^\infty xf_X(x)\mathrm{d}x \qquad \text{(Riemann integral)}, \tag{5.53}$$

where $f_X(x)$ is the pdf (and $xf_X(x)$ is the integrand). This result can be used for a discrete random variable (as well as a mixed random variable) if Dirac delta functions are used to represent the pdf. For a discrete random variable with pmf $p_X[x]$, the expectation is also evaluated as follows:

$$\mathcal{E}[X] = \int_{-\infty}^{\infty} x \sum_{n=-\infty}^{\infty} p_X[x_n]\delta(x - x_n)\mathrm{d}x = \sum_{n=-\infty}^{\infty} x_n p_X[x_n] \qquad \text{(infinite sum)}, \tag{5.54}$$

where the sifting property of the Dirac delta function has been used. For the rest of this book, we will generally use the expressions in (5.53) and (5.54). These results can be extended to expectations of *functions* of random variable X, as well as to functions of a random vector (which are considered later in this chapter).

Finally, note that sometimes it can be confusing to identify in problems with multiple random variables which distribution is used in an expectation. For this reason, and to be more precise, we can write the expectation of X as $\mathcal{E}_X[x]$, where the subscript indicates that the sum or integral is over the pmf or pdf, respectively, of X. This subscript notation will be especially useful for the conditional expectations described later. When it is obvious which distribution is involved, the shorthand notation $\mathcal{E}[X]$ is used so that the expressions are not cumbersome.

5.8 FUNCTIONAL VIEW OF THE MEAN

The expectation $\mathcal{E}[X]$ is also called the *mean* of random variable X: it is one measure of the center and location of the pdf. We would like to provide more insight about the mean by viewing it simply as the area of the product of two functions. Observe that the integrand in (5.53) is the product of the pdf $f_X(x)$ (which, of course, is always nonnegative) and x which is an *odd* function about the origin. Let $f_X(x)$ be the standard Gaussian pdf with parameters $\{\mu = 0, \sigma = 1\}$. Figure 5.4(a) shows plots of x and $f_X(x)$ superimposed, as well as their product for $x \in [-5\sigma, 5\sigma]$. Since $g(x) = x$ is an odd function, it inverts $f_X(x)$ in the product $xf_X(x)$ for $x < 0$. When the result is integrated, positive and negative areas of $xf_X(x)$ cancel each other; they cancel exactly when $\mu = 0$ because $f_X(x)$ is an even function about the origin. We also show the cumulative area under the *product:*

$$A(x) \triangleq \int_{-\infty}^{x} v f_X(v)\mathrm{d}v \tag{5.55}$$

as x is varied over the support of $f_X(x)$. ($A(x)$ can be viewed as a weighted type of "cdf" with weighting x applied to $f_X(x)$.) Obviously, $\mathcal{E}[X] = \lim_{x\to\infty} A(x)$.

Figure 5.4(b) shows the result when the Gaussian pdf is shifted to the right with $\mu = 1$. In this case, the positive and negative areas of $xf_X(x)$ do not cancel exactly: the overall area is positive, reflecting that the pdf is shifted. Thus, we see that by multiplying $f_X(x)$ with a function that crosses the x-axis at the origin, the product $xf_X(x)$ is weighted to reflect where $f_X(x)$ is located (centered) relative to the origin. Similar results are shown in Figure 5.5 for the exponential random variable, which is zero for $x < 0$, and thus the cumulative area $A(x)$ is never negative.

An important feature to observe from this functional view is that when $f_X(x)$ is multiplied by $g(x) = x$ and integrated, more emphasis is placed on large values of x. This property explains why the Cauchy pdf, which is known to have "heavy tails," does not have a mean (it is undefined). By viewing the product $xf_X(x)$ simply as a function and integrating, the area on each side of the center of the pdf may not be finite, which is exactly the case for the Cauchy pdf. The heavy tails of $f_X(x)$ allow large values of x (via the function $g(x) = x$) to dominate: $f_X(x)$ does not decay sufficiently fast with increasing $|x|$. By way of analogy, this behavior is similar to an infinite series that does not converge. Thus, even though the Cauchy pdf does have a central point (given by the mode and median), its mean does not exist because $\infty - \infty$ is not defined as mentioned earlier. This result demonstrates that the product of even and odd functions does not necessarily integrate to zero.

From this functional view, we can conclude that other odd functions provide some measure of the center of a distribution. Examples of these include all high-order moments with odd exponents, such as $\mathcal{E}[X^3]$, $\mathcal{E}[X^5]$, and so on. Later, we use a similar functional description to explain why the variance (second-order central moment) is a measure of the *width* of a distribution.

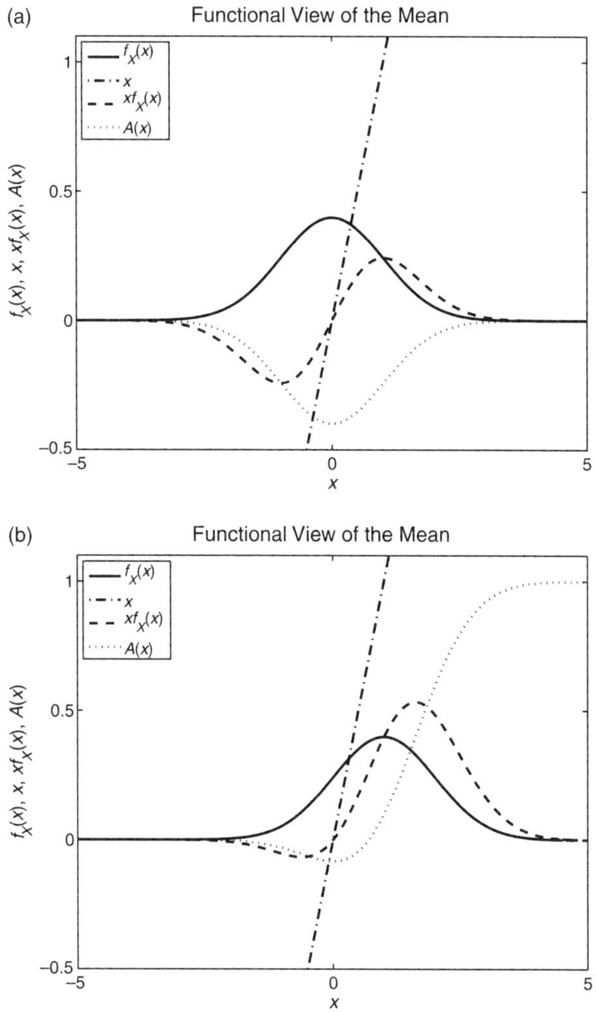

FIGURE 5.4 Functional view of the mean, showing $f_X(x)$, x, $xf_X(x)$, and $A(x)$ for a Gaussian pdf. (a) $\mu = 0$ and $\sigma = 1$. (b) $\mu = 1$ and $\sigma = 1$.

5.9 PROPERTIES OF EXPECTATION

The following properties of expectation apply to discrete, continuous, and mixed random variables:

- *Indicator function.* The expectation of the indicator function is a probability:

$$\mathcal{E}\left[I_{[a,b]}(x)\right] = P(a \leq X \leq b). \tag{5.56}$$

This is easily seen as follows:

$$\int_{-\infty}^{\infty} I_{[a,b]}(x) f_X(x)\mathrm{d}x = \int_a^b f_X(x)\mathrm{d}x = F_X(b) - F_X(a), \tag{5.57}$$

where $F_X(x)$ is the cdf of $f_X(x)$.

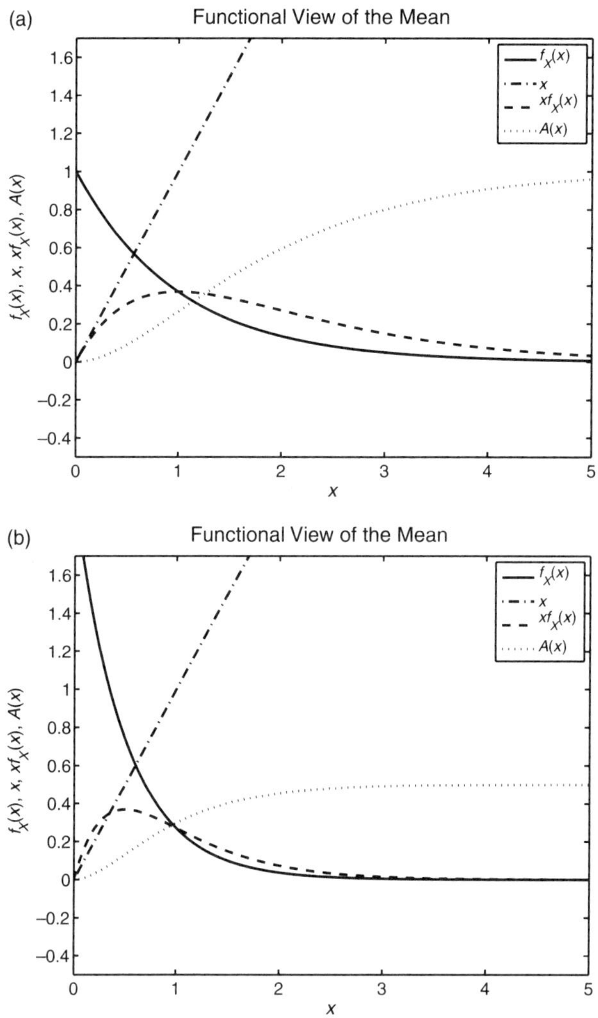

FIGURE 5.5 Functional view of the mean, showing $f_X(x), x, xf_X(x)$, and $A(x)$ for an exponential pdf. (a) $\lambda = 2$. (b) $\lambda = 1$

- *Linearity.* Expectation is a linear operator:

$$\mathcal{E}[aX + bY] = a\mathcal{E}[X] + b\mathcal{E}[Y], \tag{5.58}$$

  where $\{a, b\}$ are constant coefficients.
- *Nonnegative.* If X is nonnegative such that $P(X < 0) = 0$, then $\mathcal{E}[X] \geq 0$ because $f_X(x) \geq 0$.
- *Symmetry.* If there exists a value μ about which the pdf is symmetric, which is the case for a Gaussian random variable, then μ is the expected value (the mean). However, this is not true for some distributions such as the Cauchy random variable mentioned earlier, where the integral is not finite when evaluated at each limit.
- *Independence.* If $\{X_1, \ldots, X_N\}$ are independent random variables, then

$$\mathcal{E}[X_1 \cdots X_N] = \prod_{n=1}^{N} \mathcal{E}[X_n]. \tag{5.59}$$

This follows because the joint pdf splits into the product of marginal pdfs:

$$\mathcal{E}[X_1 \cdots X_N] = \int_{-\infty}^{\infty} \cdots \int_{-\infty}^{\infty} x_1 \cdots x_N f_{X_1,\ldots,X_N}(x_1, \ldots, x_N) \mathrm{d}x_1 \cdots \mathrm{d}x_N,$$

$$= \prod_{n=1}^{N} \int_{-\infty}^{\infty} x_n f_{X_n}(x_n) \mathrm{d}x_n. \tag{5.60}$$

- *Inequality.* If $X \leq Y$ almost surely, then

$$\mathcal{E}[X] \leq \mathcal{E}[Y]. \tag{5.61}$$

A simple example is $Y = |X|$. Almost surely means that no outcomes of X exceed those of Y, except possibly for individual points which collectively have probability zero (see Chapter 2). If y_o is the smallest value in the support of $f_Y(y)$, then

$$\mathcal{E}[X] = \int_{-\infty}^{\infty} x f_X(x) \mathrm{d}x \leq \int_{-\infty}^{\infty} y_o f_X(x) \mathrm{d}x = y_o \leq \mathcal{E}[Y]. \tag{5.62}$$

The last property leads to the following theorem.

Theorem 5.2. For general random variable X,

$$\mathcal{E}[X] \leq |\mathcal{E}[X]| \leq \mathcal{E}[|X|]. \tag{5.63}$$

Proof. The first inequality is obvious. For the second inequality,

$$|\mathcal{E}[X]| \leq \int_{-\infty}^{\infty} |x f_X(x)| \mathrm{d}x = \int_{-\infty}^{\infty} |x| f_X(x) \mathrm{d}x = \mathcal{E}[|X|] \tag{5.64}$$

because $f_X(x)$ is nonnegative for all x. This result is also obtained using the inequality property in (5.61) with $Y = |X|$ and noting that $X \leq |X|$ and $-X \leq |X|$. □

Theorem 5.3. For nonnegative continuous random variable X,

$$\mathcal{E}[X] = \int_0^{\infty} P(X \geq x) \mathrm{d}x. \tag{5.65}$$

For nonnegative discrete random variable X,

$$\mathcal{E}[X] = \sum_{n=1}^{\infty} P(X \geq n), \tag{5.66}$$

where the sum starts at $n = 1$ (not $n = 0$).

Proof. We provide a proof only for the continuous case. Substituting the pdf gives

$$\int_0^{\infty} P(X \geq x) \mathrm{d}x = \int_0^{\infty} \left[\int_x^{\infty} f_X(v) \mathrm{d}v \right] \mathrm{d}x. \tag{5.67}$$

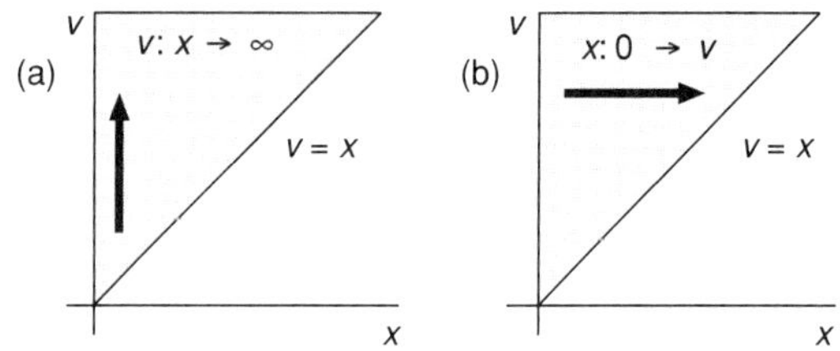

FIGURE 5.6 Region of integration for (5.67). (a) Integrating vertically first with respect to v. (b) Integrating horizontally first with respect to x.

Figure 5.6(a) shows the (shaded) region of integration in the x–v-plane. Integrating first with respect to x requires that the limits be changed so that it is performed horizontally as shown in Figure 5.6(b). This gives an inner integral over a constant:

$$\begin{aligned}\int_0^\infty P(X \geq x)\mathrm{d}x &= \int_0^\infty \left[\int_0^v f_X(v)\mathrm{d}x\right]\mathrm{d}v, \\ &= \int_0^\infty f_X(v)\left[\int_0^v \mathrm{d}x\right]\mathrm{d}v, \\ &= \int_0^\infty v f_X(v)\mathrm{d}v = \mathcal{E}[X]. \end{aligned} \tag{5.68}$$

The proof for the discrete case requires that the order of the two sums be interchanged (see Problem 5.7). □

- *Markov's inequality.* The expression in (5.67) is related to Markov's inequality (see Appendix F):

$$P(X \geq x_o) \leq \mathcal{E}[X]/x_o, \tag{5.69}$$

where $x_o > 0$ is a constant.

Example 5.10. Consider the exponential random variable X with cdf

$$F_X(x) = [1 - \exp(-\lambda x)]I_{\mathcal{R}^+}(x). \tag{5.70}$$

For a continuous random variable, $P(X > x) = P(X \geq x)$ (equality may or may not be included: they both equal $1 - F_X(x)$). Thus

$$P(X \geq x) = \exp(-\lambda x), \tag{5.71}$$

and since X is nonnegative, we have from (5.65):

$$\mathcal{E}[X] = \int_0^\infty \exp(-\lambda x)\mathrm{d}x = (-1/\lambda)\exp(-\lambda x)|_0^\infty = 1/\lambda. \tag{5.72}$$

Directly computing $\mathcal{E}[X]$ requires integration by parts as was done in Example 5.8. The expression in (5.67) is simpler to use because, in effect, $P(X \geq x)$ includes one of the integrals via the cdf.

Example 5.11. For the geometric random variable X with cdf

$$F_X(x) = (1 - q^{x+1})I_{Z^+}[x], \tag{5.73}$$

the expectation is

$$\begin{aligned}\mathcal{E}[X] &= \sum_{n=1}^{\infty} P(X \geq n) = \sum_{n=1}^{\infty}[1 - P(X < n)], \\ &= \sum_{n=1}^{\infty}[1 - F_X(n) + pq^n].\end{aligned} \tag{5.74}$$

Substituting the cdf and using the result for the infinite series in Appendix E gives

$$\mathcal{E}[X] = (p+q)\sum_{n=0}^{\infty} q^n - 1 = 1/(1-q) - 1 = q/p, \tag{5.75}$$

which is the mean in Example 5.4 obtained by directly computing $\mathcal{E}[X]$. Observe that we needed to include the probability mass at n (the last line of (5.74)) because of the equality in $P(X \geq n)$; subtracting the cdf, which also includes equality, requires that pq^n be added to the expression. The term at $n = 0$ (equal to 1) also needed to be subtracted so that the sum begins at zero, and allows us to use a closed-form expression for the infinite sum.

5.10 EXPECTATION OF A FUNCTION

The expectation of function $Y = g(X)$ of random variable X can be computed in two ways: (i) deriving the pdf $f_Y(y)$ and then calculating $\mathcal{E}_Y[Y]$ or (ii) directly calculating $\mathcal{E}_X[g(X)]$ based on $f_X(x)$. We demonstrate the equivalence of both approaches in the following theorem.

Theorem 5.4. The expectation of function $Y = g(X)$ of random variable X is

$$\mathcal{E}_Y[Y] = \mathcal{E}_X[g(X)]. \tag{5.76}$$

Proof. (Discrete random variable). The proof is a straightforward application of the mapping process from X to Y:

$$\mathcal{E}_Y[Y] = \sum_{y=-\infty}^{\infty} y p_Y[y] = \sum_{y=-\infty}^{\infty} y \sum_{x:g(x)=y} p_X[x]. \tag{5.77}$$

The inner sum accounts for all values of x (and the corresponding probability mass) that map to value y. The double sum is the same as a single sum over all possible values of $g(x)$, weighted by $p_X[x]$:

$$\sum_{y=-\infty}^{\infty} y \sum_{x:g(x)=y} p_X[x] = \sum_{x=-\infty}^{\infty} g(x) p_X[x] = \mathcal{E}_X[g(X)]. \tag{5.78}$$

(Continuous random variable). Assume that $g(X)$ is monotonically increasing where each x maps to a unique y. Thus

$$\mathcal{E}_Y[Y] = \int_{-\infty}^{\infty} y f_Y(y)\mathrm{d}y = \int_{-\infty}^{\infty} y f_X(g^{-1}(y)) \left| \frac{\mathrm{d}}{\mathrm{d}y} g^{-1}(y) \right| \mathrm{d}y. \tag{5.79}$$

Substituting $y = g(x)$ and $x = g^{-1}(y)$ gives

$$\mathcal{E}_Y[Y] = \int_{-\infty}^{\infty} g(x) f_X(x) \left| \frac{\mathrm{d}x}{\mathrm{d}y} \right| \mathrm{d}y = \int_{-\infty}^{\infty} g(x) f_X(x)\mathrm{d}x = \mathcal{E}_X[g(X)], \tag{5.80}$$

where $\mathrm{d}x/\mathrm{d}y > 0$ for a monotonically increasing function. Methods for more general functions are the same as those used in Chapter 4 for transformations of random variables. □

Example 5.12. Let X be a Gaussian random variable with parameters $\{\mu, \sigma\}$. From Chapter 4, we know that $Y = g(X) = \exp(X)$ is log-normal with the same parameters. Computing the expectation over X gives

$$\begin{aligned} \mathcal{E}_X[g(X)] &= \int_{-\infty}^{\infty} \exp(x) \frac{1}{\sqrt{2\pi\sigma^2}} \exp\left(-(x-\mu)^2/2\sigma^2\right) \mathrm{d}x, \\ &= \frac{1}{\sqrt{2\pi\sigma^2}} \int_{-\infty}^{\infty} \exp\left(x - (x-\mu)^2/2\sigma^2\right) \mathrm{d}x. \end{aligned} \tag{5.81}$$

Completing the square in the exponent yields

$$\mathcal{E}_X[g(X)] = \frac{1}{\sqrt{2\pi\sigma^2}} \int_{-\infty}^{\infty} \exp\left(-([x - (\mu+\sigma^2)]^2 - (\mu+\sigma^2)^2 + \mu^2)/2\sigma^2\right) \mathrm{d}x, \tag{5.82}$$

and factoring out the constant gives

$$\begin{aligned} \mathcal{E}_X[g(X)] &= \exp\left(\mu + \sigma^2/2\right) \frac{1}{\sqrt{2\pi\sigma^2}} \int_{-\infty}^{\infty} \exp\left(-(x - (\mu+\sigma^2))^2/2\sigma^2\right) \mathrm{d}x, \\ &= \exp\left(\mu + \sigma^2/2\right), \end{aligned} \tag{5.83}$$

where the last integral $= 1$ because the integrand is a Gaussian pdf with parameters $\{\mu + \sigma^2, \sigma\}$. This result can be verified by directly calculating the mean of the log-normal pdf as follows:

$$\begin{aligned} \mathcal{E}_Y[Y] &= \int_{0}^{\infty} \frac{1}{y\sqrt{2\pi\sigma^2}} y \exp\left(-(\ln(y)-\mu)^2/2\sigma^2\right) \mathrm{d}y, \\ &= \frac{1}{\sqrt{2\pi\sigma^2}} \int_{-\infty}^{\infty} \exp(x) \exp\left(-(x-\mu)^2/2\sigma^2\right) \mathrm{d}x = \mathcal{E}_X[g(X)], \end{aligned} \tag{5.84}$$

where the transformation $x = \ln(y)$ has been used with $\mathrm{d}y = \exp(x)\mathrm{d}x$, which changes the limits to $\{-\infty, \infty\}$.

5.11 CHARACTERISTIC FUNCTION

The characteristic function is the expectation of a particular *complex-valued* function of a random variable. It is useful for systematically determining moments of a random variable, as well as for computing functions of variables such as the sum of two independent random variables covered in Chapter 4.

Definition: Characteristic Function The *characteristic function (CF)* of random variable X is

$$\Phi_X(\omega) \triangleq \mathcal{E}[\exp(j\omega X)], \tag{5.85}$$

where $j = \sqrt{-1}$ and $\omega \in [0, 2\pi]$.

Substituting $g(X) = \exp(j\omega X)$ into (5.80) gives

$$\Phi_X(\omega) = \int_{-\infty}^{\infty} f_X(x)\exp(j\omega x)\mathrm{d}x, \tag{5.86}$$

which we see is similar to the Fourier transform of the pdf (the exponent is positive instead of negative as in the usual definition of the Fourier transform). Like the Fourier transform, the following conditions are sufficient for $\Phi_X(\omega)$ to exist on the support of $f_X(x)$:

- Finite number of discontinuities.
- Bounded variation (see Appendix B).
- Absolutely integrable:

$$\int_{-\infty}^{\infty} |f_X(x)|\mathrm{d}x < \infty. \tag{5.87}$$

Note that all the distributions summarized in Appendix A have a CF (even Cauchy), though some are complicated and are not given there.

Let $\{X_1, \ldots, X_N\}$ be independent random variables with arbitrary pdfs. The CF of the sum of these random variables can be written as a product:

$$\Phi_{X_1+\cdots+X_N}(\omega) = \mathcal{E}\left[\exp\left(j\omega \sum_{n=1}^{N} X_n\right)\right] = \mathcal{E}\left[\prod_{n=1}^{N} \exp(j\omega X_n)\right]. \tag{5.88}$$

The initial expectation is performed over the joint pdf of the $\{X_n\}$. However, since the random variables are independent, the joint pdf splits into a product of N marginal pdfs, yielding

$$\Phi_{X_1+\cdots+X_N}(\omega) = \prod_{n=1}^{N} \mathcal{E}[\exp(j\omega X_n)] = \prod_{n=1}^{N} \Phi_{X_n}(\omega), \tag{5.89}$$

where the nth expectation in the product is over the pdf of X_n.

The following theorem describes the moment-generating property of the CF (high-order moments are covered later in this chapter).

Theorem 5.5. Noncentral (raw) moments of random variable X can be generated using the CF as follows:

$$\mathcal{E}[X^n] = (-j)^n \frac{\mathrm{d}^n}{\mathrm{d}\omega^n} \Phi_X(\omega)\bigg|_{\omega=0}. \tag{5.90}$$

Proof. (Power series approach). Expanding the exponential using a MacLauren series for the exponential (see Appendix E) yields

$$\mathcal{E}\left[\exp(j\omega X)\right] = \mathcal{E}\left[1 + j\omega X + \frac{(j\omega X)^2}{2!} + \cdots + \frac{(j\omega X)^k}{k!} + \frac{(j\omega X)^{k+1}}{(k+1)!} + \cdots\right]. \tag{5.91}$$

Using the identity

$$(-j)^n \frac{\mathrm{d}^n \Phi_X(\omega)}{\mathrm{d}\omega^n} = \frac{\mathrm{d}^n \Phi_X(\omega)}{\mathrm{d}(j\omega)^n}, \tag{5.92}$$

and differentiating n times removes the leading $n-1$ terms in the series expansion:

$$\frac{\mathrm{d}^n}{\mathrm{d}(j\omega)^n}\Phi_X(\omega) = \mathcal{E}\left[X^n + j\omega X^{n+1} + (1/2)(j\omega)^2 X^{n+2} + \cdots\right]. \tag{5.93}$$

Substituting $\omega = 0$ removes all terms beyond the first term which no longer depends on ω because of the successive differentiations:

$$\left.\frac{\mathrm{d}^n}{\mathrm{d}(j\omega)^n}\Phi_X(\omega)\right|_{\omega=0} = \mathcal{E}[X^n + 0 + \cdots] = \mathcal{E}[X^n]. \tag{5.94}$$

(Definition approach). Alternatively from the definition of the CF:

$$\begin{aligned}\left.\frac{\mathrm{d}^n}{\mathrm{d}(j\omega)^n}\mathcal{E}\left[\exp(j\omega X)\right]\right|_{\omega=0} &= \left.\int_{-\infty}^{\infty}\frac{\mathrm{d}^n}{\mathrm{d}(j\omega)^n}\exp(j\omega x)\, f_X(x)\mathrm{d}x\right|_{\omega=0}, \\ &= \left.\int_{-\infty}^{\infty} x^n \exp(j\omega x)\, f_X(x)\mathrm{d}x\right|_{\omega=0}. \end{aligned} \tag{5.95}$$

Evaluating this expression at $\omega = 0$ removes the exponential from the integrand:

$$\left.\frac{\mathrm{d}^n}{\mathrm{d}(j\omega)^n}\mathcal{E}\left[\exp(j\omega X)\right]\right|_{\omega=0} = \int_{-\infty}^{\infty} x^n f_X(x)\mathrm{d}x = \mathcal{E}[X^n], \tag{5.96}$$

which completes the proof. □

This theorem also demonstrates that the moments of a random variable together describe a pdf: collectively the moments contain the same "information" as the pdf.

Example 5.13. The CF of the exponential random variable is

$$\begin{aligned}\Phi_X(\omega) &= \int_0^{\infty} \lambda \exp(-\lambda x)\exp(j\omega x)\mathrm{d}x, \\ &= \lambda \int_0^{\infty} \exp(-x(\lambda - j\omega))\mathrm{d}x = \frac{\lambda}{\lambda - j\omega}. \end{aligned} \tag{5.97}$$

The first moment (mean) is computed as follows:

$$\mathcal{E}[X] = \left.\frac{\mathrm{d}}{\mathrm{d}(j\omega)}\left(\frac{\lambda}{\lambda - j\omega}\right)\right|_{\omega=0} = \left.\frac{\lambda}{(\lambda - j\omega)^2}\right|_{\omega=0} = 1/\lambda. \tag{5.98}$$

Example 5.14. The CF of the binomial random variable is

$$\Phi_X(j\omega) = \int_{-\infty}^{\infty} \sum_{n=0}^{N} \binom{N}{n} p^n q^{N-n} \exp(j\omega x)\delta(x-n)\mathrm{d}x,$$

$$= q^N \sum_{n=0}^{N} \binom{N}{n} \left[(p/q)\exp(j\omega)\right]^n, \tag{5.99}$$

where $q \triangleq 1-p$, and the Dirac delta function has produced $\exp(j\omega x)$ at $x=n$. Using the binomial formula

$$(x+y)^N = \sum_{n=0}^{N} \binom{N}{n} x^{N-n} y^n, \tag{5.100}$$

we can write

$$\Phi_X(j\omega) = q^N[1+(p/q)\exp(j\omega)]^N = [q+p\exp(j\omega)]^N. \tag{5.101}$$

The mean is then obtained as follows:

$$\mathcal{E}[X] = \frac{\mathrm{d}}{\mathrm{d}(j\omega)}[q+p\exp(j\omega)]^N \bigg|_{\omega=0},$$

$$= N[q+p\exp(j\omega)]^{N-1} p\exp(j\omega)\big|_{\omega=0} = Np, \tag{5.102}$$

where $q+p=1$ in the brackets after substituting $\omega = 0$.

Note that j is included in the argument of $\Phi_X(j\omega)$ for a *discrete* pmf because it is analogous to the discrete-time Fourier transform (DTFT) of a discrete-time sequence. It is written in terms of the pmf as follows:

$$\Phi_X(j\omega) = \sum_{n=-\infty}^{\infty} p_X[n]\exp(j\omega n). \tag{5.103}$$

Example 5.15. The CF of the Cauchy pdf is

$$\Phi_X(\omega) = \int_{-\infty}^{\infty} \frac{\alpha/\pi}{(x-c)^2+\alpha^2} \exp(j\omega x)\mathrm{d}x,$$

$$= \exp(j\omega c)\frac{1}{2\pi}\int_{-\infty}^{\infty} \frac{2\alpha}{v^2+\alpha^2}\exp(j\omega v)\mathrm{d}v, \tag{5.104}$$

where we have changed variables in the second expression to $v = x - c$. From a table of Fourier transform pairs (see Appendix C), we recognize that this integral (including the factor $1/2\pi$) is $\exp(-\alpha|\omega|)$. Thus, it is interesting that the Cauchy random variable has a CF:

$$\Phi_X(\omega) = \exp(j\omega c - \alpha|\omega|) \tag{5.105}$$

although its moments do not exist or are infinite. Consider the mean:

$$\mathcal{E}[X] = \left.\frac{\mathrm{d}}{\mathrm{d}(j\omega)}\exp(j\omega c - \alpha|\omega|)\right|_{\omega=0},$$

$$= (j\omega c - \alpha|\omega|)\exp(j\omega c - \alpha|\omega|)\left.\left(c - \alpha\frac{\mathrm{d}|j\omega|}{\mathrm{d}(j\omega)}\right)\right|_{\omega=0}, \quad (5.106)$$

where $|\omega| = |j\omega|$ has been substituted. Since the derivative of the absolute function is not defined at $\omega = 0$, the mean $\mathcal{E}[X]$ does not exist.

We conclude this section with the definition of the multivariate CF for a random vector.

Definition: Multivariate Characteristic Function The *multivariate characteristic function* is the following joint expectation:

$$\Phi_{\mathbf{X}}(\boldsymbol{\omega}) \triangleq \mathcal{E}_{\mathbf{X}}[\exp(j\boldsymbol{\omega}^T\mathbf{X})], \quad (5.107)$$

where $\mathbf{X} = [X_1, \ldots, X_N]^T$, $\boldsymbol{\omega} = [\omega_1, \ldots, \omega_N]^T$, and the expectation is over the joint pdf of $\mathbf{X}$.

Example 5.16. The CF of the multivariate Gaussian random variable with mean vector $\boldsymbol{\mu}_{\mathbf{X}}$ and covariance matrix $\mathbf{C}_{\mathbf{XX}}$ is

$$\Phi_{\mathbf{X}}(\boldsymbol{\omega}) = \frac{1}{(2\pi)^{N/2}|\mathbf{C}_{\mathbf{XX}}|^{1/2}}\int_{-\infty}^{\infty}\cdots\int_{-\infty}^{\infty}\exp(-(1/2)(\mathbf{x}-\boldsymbol{\mu}_{\mathbf{X}})^T\mathbf{C}_{\mathbf{XX}}^{-1}(\mathbf{x}-\boldsymbol{\mu}_{\mathbf{X}}))$$

$$\cdot\ \exp(j\boldsymbol{\omega}^T\mathbf{x})\mathrm{d}x_1\cdots\mathrm{d}x_N. \quad (5.108)$$

It can be shown that $\Phi_{\mathbf{X}}(\boldsymbol{\omega})$ has a form resembling $\Phi_X(\omega)$ in Appendix A for a single Gaussian random variable (see Problem 5.42). It is given by

$$\Phi_{\mathbf{X}}(\boldsymbol{\omega}) = \exp\left(j\boldsymbol{\omega}^T\boldsymbol{\mu}_{\mathbf{X}} - (1/2)\boldsymbol{\omega}^T\mathbf{C}_{\mathbf{XX}}\boldsymbol{\omega}\right). \quad (5.109)$$

Example 5.17. Consider the sum $Z = X + Y$, where X and Y are independent random variables with CF given by

$$\Phi_Z(\omega) = \mathcal{E}_{XY}[\exp(j\omega(X+Y))],$$

$$= \int_{-\infty}^{\infty}\int_{-\infty}^{\infty}\exp(j\omega x)\exp(j\omega y)f_{X,Y}(x,y)\mathrm{d}x\mathrm{d}y. \quad (5.110)$$

Since X and Y are independent, $f_{X,Y}(x,y) = f_X(x)f_Y(y)$ and

$$\Phi_Z(\omega) = \mathcal{E}_X[\exp(j\omega X)]\mathcal{E}_Y[\exp(j\omega Y)] = \Phi_X(\omega)\Phi_Y(\omega). \quad (5.111)$$

From Fourier transform properties, we know that a product in the frequency domain corresponds to a convolution in the time domain. Since the inverse transforms are pdfs, we have the convolution $f_Z(z) = f_X(x) * f_Y(y)$, which is the same result derived in Chapter 4 using transformation techniques. Problem 5.18 considers the difference of random variables $Z = X - Y$.

5.12 CONDITIONAL EXPECTATION

The conditional probability of events A and B in the abstract probability space $\{\Omega, \mathcal{F}, P\}$ was introduced in Chapter 2. There we discussed how conditioning on event B reduces the sample space for event A such that we can write

$$P(A|B) = \frac{P(AB)}{P(B)} \tag{5.112}$$

assuming $P(B) \neq 0$. In Chapter 3, we considered conditional pdfs which have a form analogous to (5.112):

$$f_{X|Y}(x|y) = \frac{f_{X,Y}(x, y)}{f_Y(y)}, \qquad f_{Y|X}(y|x) = \frac{f_{X,Y}(x, y)}{f_X(x)}, \tag{5.113}$$

and likewise for conditional pmfs. The conditional expectation, which is also called the *conditional mean*, follows from these expressions.

Definition: Conditional Expectation The *conditional expectation* of random variable X given that random variable Y takes on the value y is

$$\mathcal{E}[X|Y = y] \triangleq \int_{-\infty}^{\infty} x f_{X|Y}(x|y)\mathrm{d}x. \tag{5.114}$$

Observe that $\mathcal{E}[X|Y = y]$ is a nonrandom function of y, such that we can write $\mathcal{E}[X|Y = y] = g(y)$. The conditional expectation given *random variable* Y is itself a random variable because it is a function of Y: $\mathcal{E}[X|Y] = g(Y)$. One can obtain $\mathcal{E}[X|Y]$ by first evaluating $\mathcal{E}[X|Y = y] = g(y)$ as a function of y, and then replacing y with Y. The result $Z \triangleq g(Y)$ is a random variable whose distribution $f_Z(z)$ is determined by applying the appropriate transformation to $f_Y(y)$. Consider a simple example that illustrates the difference between the two types of conditional expectation.

Example 5.18. Let $Y = X + V$, where X is the symmetric Bernoulli random variable taking on the values ± 1 with equal probability, and V is Gaussian with parameters $\{\mu = 0, \sigma\}$. The pdf of Y, as well as the conditional pdf of Y given $X = \pm 1$, were discussed in Example 4.15. Using the conditional pdf for $X = 1$ gives

$$\begin{aligned} \mathcal{E}[Y|X = 1] &= \int_{-\infty}^{\infty} y f_{Y|X}(y|1)\mathrm{d}y, \\ &= \frac{1}{\sqrt{2\pi\sigma^2}} \int_{-\infty}^{\infty} y \exp\left(-(y-1)^2/2\sigma^2\right) \mathrm{d}y = 1. \end{aligned} \tag{5.115}$$

This result is obvious because the conditional pdf of Y given that $X = 1$ is Gaussian with mean $= 1$. Likewise, the conditional pdf of Y given $X = -1$ is Gaussian with mean -1. Next condition on random variable X. From the previous results, it is evident that $\mathcal{E}[Y|X = x] = x$, and thus

$$\mathcal{E}[Y|X] = X. \tag{5.116}$$

Suppose now V has mean μ_V. Then it is straightforward to show that $\mathcal{E}[Y|X = 1] = \mu_V + 1$, $\mathcal{E}[Y|X = -1] = \mu_V - 1$, yielding

$$\mathcal{E}[Y|X] = \mu_V + X. \tag{5.117}$$

Example 5.19. Consider the following *random* sum:

$$Y = \sum_{m=1}^{N} X_m, \tag{5.118}$$

where $\{X_m\}$ are independent and identically distributed (iid) Bernoulli random variables, each with parameter p. Assume that N is a positive discrete random variable with unknown distribution. If $N = n$ is fixed, then the pdf of Y is binomial with parameters $\{n, p\}$. Thus

$$\mathcal{E}[Y|N = n] = np \tag{5.119}$$

is a number and

$$\mathcal{E}[Y|N] = Np \tag{5.120}$$

is a random variable.

Example 5.20. Suppose we want to estimate random variable Y using a function of a second random variable X. This application is related to *estimation theory* which is covered in Chapter 9. Let the estimate of Y be denoted by $\hat{Y}$ and assume it is given by $\hat{Y} = g(X)$. Consider the performance measure (cost function) known as the mean-square error (MSE) whose expectation over the joint pdf of X and Y is

$$\begin{aligned}\mathcal{E}[(Y - g(X))^2] &= \int_{-\infty}^{\infty}\int_{-\infty}^{\infty} [y - g(x)]^2 f_{X,Y}(x, y)\mathrm{d}x\mathrm{d}y,\\ &= \int_{-\infty}^{\infty}\int_{-\infty}^{\infty} [y - g(x)]^2 f_{Y|X}(y|x) f_X(x)\mathrm{d}x\mathrm{d}y.\end{aligned} \tag{5.121}$$

Differentiating this expression with respect to $g(x)$ gives

$$\frac{\partial}{\partial g(x)}\mathcal{E}[(Y - g(X))^2] = -2\int_{-\infty}^{\infty}\int_{-\infty}^{\infty} [y - g(x)] f_{Y|X}(y|x) f_X(x)\mathrm{d}x\mathrm{d}y, \tag{5.122}$$

and setting the result equal to zero yields

$$\begin{aligned}\int_{-\infty}^{\infty} f_X(x)\int_{-\infty}^{\infty} y f_{Y|X}(y|x)\mathrm{d}y\mathrm{d}x &= \int_{-\infty}^{\infty} f_X(x)g(x)\int_{-\infty}^{\infty} f_{Y|X}(y|x)\mathrm{d}y\mathrm{d}x,\\ &= \int_{-\infty}^{\infty} f_X(x)g(x)\mathrm{d}x,\end{aligned} \tag{5.123}$$

where the second integral with respect to y on the right-hand side $= 1$. This equality holds for all x and, moreover, since $f_X(x) \geq 0$ we have

$$g(x) = \int_{-\infty}^{\infty} y f_{Y|X}(y|x)\mathrm{d}y = \mathcal{E}[Y|X = x]. \tag{5.124}$$

Finally, since X is a random variable, we can write

$$\hat{Y} = g(X) = \mathcal{E}[Y|X], \tag{5.125}$$

which also is a random variable. This minimum MSE (MMSE) estimator of Y given X is also known as the *conditional mean estimator*.

5.13 PROPERTIES OF CONDITIONAL EXPECTATION

Next, we present several properties of conditional expectation, many of which are straightforward extensions of those for ordinary expectation.

- *Linearity*. For constants $a, b \in \mathcal{R}$:

$$\mathcal{E}[aX_1 + bX_2|Y] = a\mathcal{E}[X_1|Y] + b\mathcal{E}[X_2|Y]. \tag{5.126}$$

  This result is obvious from the definition of expectation.
- *Independence*. For independent X and Y,

$$\mathcal{E}[Y|X] = \mathcal{E}[Y]. \tag{5.127}$$

  Since $f_{Y|X}(y|x) = f_Y(y)$ for every x, the conditional expectation simplifies to

$$\mathcal{E}[Y|X = x] = \int_{-\infty}^{\infty} y f_{Y|X}(y|x)\mathrm{d}y = \int_{-\infty}^{\infty} y f_Y(y)\mathrm{d}y = \mathcal{E}[Y]. \tag{5.128}$$

- *Function of conditioning variable*. For $g(Y)$ a function of Y, we have the following three results:

$$\mathcal{E}[g(Y)|Y] = g(Y), \tag{5.129}$$

$$\mathcal{E}[Xg(Y)|Y] = g(Y)\mathcal{E}[X|Y], \tag{5.130}$$

$$\mathcal{E}[X|Y, g(Y)] = \mathcal{E}[X|Y]. \tag{5.131}$$

  The first expression is intuitively obvious, but we derive it starting with the following conditional cdf:

$$\begin{aligned} P(Y \le y|Y = y_o) &= \begin{cases} 1, & y \ge y_o \\ 0, & y < y_o, \end{cases} \\ &= u(y - y_o), \end{aligned} \tag{5.132}$$

  where $u(y)$ is the unit-step function. This result is valid regardless of the marginal pdf $f_Y(y)$. Differentiating with respect to y gives a shifted Dirac delta function $\delta(y - y_o)$ and

$$\mathcal{E}[g(Y)|Y = y_o] = \int_{-\infty}^{\infty} g(y)\delta(y - y_o)\mathrm{d}y = g(y_o). \tag{5.133}$$

  Since this holds for all values of Y, we obtain the result in (5.129). Likewise for (5.130),

$$\begin{aligned} P(X \le x, Y \le y|Y = y_o) &= \begin{cases} P(X \le x|Y = y_o), & y \ge y_o \\ 0, & y < y_o, \end{cases} \\ &= P(X \le x|Y = y_o)u(y - y_o), \end{aligned} \tag{5.134}$$

and the derivative with respect to x and y is $f_{X|Y}(x|y_o)\delta(y-y_o)$, yielding

$$\begin{aligned}\mathcal{E}[Xg(Y)|Y=y_o] &= \int_{-\infty}^{\infty}\int_{-\infty}^{\infty} xg(y)f_{X|Y}(x|y_o)\delta(y-y_o)\mathrm{d}x\mathrm{d}y,\\ &= \int_{-\infty}^{\infty} xf_{X|Y}(x|y_o)\mathrm{d}x\int_{-\infty}^{\infty} g(y)\delta(y-y_o)\mathrm{d}y,\\ &= g(y_o)\mathcal{E}[X|Y=y_o].\end{aligned} \tag{5.135}$$

This holds for all values of Y, giving the result in (5.130), which is also known as the *factorization* property of conditional expectation. For the property in (5.131), $g(Y)$ provides no new information beyond what is given by Y; in fact, it contains less information than Y if $g(\cdot)$ is not one-to-one. As a result, the cdf is the same as in (5.134):

$$P(X\le x|Y=y_o, g(Y)=g(y_o)) = P(X\le x|Y=y_o)u(y-y_o) \tag{5.136}$$

such that

$$\begin{aligned}\mathcal{E}[X|Y=y_o, g(Y)=g(y_o)] &= \int_{-\infty}^{\infty}\int_{-\infty}^{\infty} xf_{X|Y}(x|y_o)\delta(y-y_o)\mathrm{d}x\mathrm{d}y,\\ &= \mathcal{E}[X|Y=y_o]\end{aligned} \tag{5.137}$$

from which (5.131) follows.

- *Nested expectations with two variables.* For random variables X and Y,

$$\mathcal{E}[\mathcal{E}[X|Y]] = \mathcal{E}[X]. \tag{5.138}$$

The outer expectation is over Y:

$$\begin{aligned}\mathcal{E}[\mathcal{E}[X|Y]] &= \int_{-\infty}^{\infty} f_Y(y)\mathcal{E}[X|Y=y]\mathrm{d}y = \int_{-\infty}^{\infty} f_Y(y)\int_{-\infty}^{\infty} xf_{X|Y}(x|y)\mathrm{d}x\mathrm{d}y,\\ &= \int_{-\infty}^{\infty}\int_{-\infty}^{\infty} xf_{X,Y}(x,y)\mathrm{d}x\mathrm{d}y = \int_{-\infty}^{\infty} xf_X(x)\mathrm{d}x = \mathcal{E}[X],\end{aligned} \tag{5.139}$$

where (5.113) has been used. This proof illustrates that $\mathcal{E}[X|Y]$, which is a random variable, can be written as an integral only for specific values of $Y=y$, as is done in the first line.

Example 5.21. For the conditions in Example 5.19, we have from (5.120)

$$\mathcal{E}[Y] = \mathcal{E}_N[\mathcal{E}[Y|N]] = \mathcal{E}[Np] = \mathcal{E}[N]p, \tag{5.140}$$

where $p=\mathcal{E}[X_m]$. By using nested expectations, it is quite simple to find the expectation $\mathcal{E}[Y]$ which is a nontrivial function of independent random variables. It was not necessary to compute the joint pdf of $\{X_m\}$ and N because we are interested only in the *expectation* of random variable Y.

- *Orthogonality.* For random variables X and Y and arbitrary function $g(\cdot)$:

$$\mathcal{E}[(X-\mathcal{E}[X|Y])g(Y)] = 0. \tag{5.141}$$

In this formulation, $\mathcal{E}[X|Y]$ can be viewed as an estimator of X given observation Y (as mentioned previously), and thus $E \triangleq X-\mathcal{E}[X|Y]$ is an "error" random variable.

The *orthogonality principle* above shows that this error and any function of Y are orthogonal: $\mathcal{E}[Eg(Y)] = 0$. Orthogonality of random variables is discussed again later in this chapter. Rearranging the expression gives

$$\mathcal{E}[Xg(Y)] = \mathcal{E}[\mathcal{E}[Xg(Y)|Y]], \tag{5.142}$$

where we have used the (reverse) factorization property in (5.130) on the right-hand side. Consider the inner expectation for $Y = y$:

$$\mathcal{E}[Xg(Y)|Y = y] = \int_{-\infty}^{\infty} xg(y) f_{X|Y}(x|y)\mathrm{d}x. \tag{5.143}$$

The outer expectation on the right-hand side of (5.142) gives

$$\int_{-\infty}^{\infty} f_Y(y) \int_{-\infty}^{\infty} xg(y) f_{X|Y}(x|y)\mathrm{d}x\mathrm{d}y = \int_{-\infty}^{\infty}\int_{-\infty}^{\infty} xg(y) f_{X,Y}(x, y)\mathrm{d}x\mathrm{d}y, \tag{5.144}$$

which is the left-hand side of (5.142). Observe from (5.144) that the inner expectation on the right-hand side of (5.142) is over $X|Y = y$, and the outer expectation is over Y, whereas the expectation on the left-hand side is over the joint pdf of X and Y. Thus, one needs to be careful how to interpret and evaluate expectations such as the outer expectation in (5.141), which involves a combination of terms.

Consider the following expectation:

$$\mathcal{E}[g(Y)a(X, Y) - g(Y)b(Y)] = 0, \tag{5.145}$$

which is similar to (5.141) except that function $a(X, Y)$ replaces X and function $b(Y)$ replaces $\mathcal{E}[X|Y]$. It turns out that if the condition in (5.145) holds for *any* function $g(Y)$, then $b(Y)$ must be the conditional expectation of the first term multiplying $g(Y)$, that is, $b(Y) = \mathcal{E}[a(X, Y)|Y]$. This is a useful result because it can be used to show the equivalence of quantities under an expectation by rearranging (5.145) as follows (Larson and Shubert, 1979):

$$\mathcal{E}[g(Y)a(X, Y)] = \mathcal{E}[g(Y)b(Y)]. \tag{5.146}$$

In general, even though this equality of the expectations holds for any function $g(Y)$, we *cannot* conclude that $a(X, Y)$ and $b(Y)$ are equal; we can only say that $b(Y) = \mathcal{E}[a(X, Y)|Y]$ is the conditional expectation. However, if $a(X, Y) \to a(Y)$ is a function only of Y, then $b(Y) = \mathcal{E}[a(Y)|Y] = a(Y)$ due to the property in (5.129).

Example 5.22. We illustrate how to use this last result by proving again the factorization property. From (5.141) using $h(Y)$ instead of $g(Y)$, we can write the following expectation for the left-hand side of (5.130):

$$\mathcal{E}[Xg(Y)h(Y)] = \mathcal{E}[\mathcal{E}[Xg(Y)|Y]h(Y)], \tag{5.147}$$

which holds for any $h(Y)$. Next, define $c(Y) \triangleq g(Y)h(Y)$ which is also a function of Y. Using (5.141) gives

$$\mathcal{E}[Xc(Y)] = \mathcal{E}[\mathcal{E}[X|Y]c(Y)] = \mathcal{E}[\mathcal{E}[X|Y]g(Y)h(Y)] \tag{5.148}$$

for any $c(Y) = g(Y)h(Y)$, which means for any $h(Y)$ because $g(Y)$ is given. Since the left-hand sides of (5.147) and (5.148) are the same, we have from the right-hand sides:

$$\mathcal{E}[\mathcal{E}[Xg(Y)|Y]h(Y)] = \mathcal{E}[\mathcal{E}[X|Y]g(Y)h(Y)], \tag{5.149}$$

which holds for any $h(Y)$. Thus, from the above discussion, we know that the two quantities under the expectations multiplying $h(Y)$ are equal: $\mathcal{E}[Xg(Y)|Y] = \mathcal{E}[X|Y]g(Y)$, which is the factorization property.

The orthogonality principle is an important property used in MSE estimation theory; it is covered in greater detail in Chapters 11 and 12.

- *Nested expectations with three variables.* For random variables X, Y, and Z,

$$\mathcal{E}_{Y|Z}[\mathcal{E}_{X|Y,Z}[X|Y, Z]|Z] = \mathcal{E}_{X|Z}[X|Z]. \tag{5.150}$$

As mentioned before, we need to write the conditioning for a specific value z:

$$\begin{aligned}\mathcal{E}_{Y|Z}[\mathcal{E}_{X|Y,Z}[X|Y, Z]|Z = z] &= \int_{-\infty}^{\infty} \mathcal{E}[X|Y = y, Z = z] f_{Y|Z}(y|z)\mathrm{d}y, \\ &= \int_{-\infty}^{\infty}\int_{-\infty}^{\infty} x f_{X|Y,Z}(x|y, z) f_{Y|Z}(y|z)\mathrm{d}y\mathrm{d}x. \end{aligned}\tag{5.151}$$

Rewriting each conditional pdf gives

$$\mathcal{E}_{Y|Z}[\mathcal{E}_{X|Y,Z}[X|Y, Z]|Z = z] = \frac{1}{f_Z(z)}\int_{-\infty}^{\infty}\int_{-\infty}^{\infty} x \frac{f_{X,Y,Z}(x, y, z)}{f_{Y,Z}(y, z)} f_{Y,Z}(y, z)\mathrm{d}y\mathrm{d}x. \tag{5.152}$$

Two pdfs cancel each other, and Y is integrated out of the remaining joint pdf, yielding

$$\begin{aligned}\mathcal{E}_{Y|Z}[\mathcal{E}_{X|Y,Z}[X|Y, Z]|Z = z] &= \frac{1}{f_Z(z)}\int_{-\infty}^{\infty} x f_{X,Z}(x, z)\mathrm{d}x, \\ &= \int_{-\infty}^{\infty} x f_{X|Z}(x|z)\mathrm{d}x = \mathcal{E}_{X|Z}[X|Z = z], \end{aligned}\tag{5.153}$$

which holds for all z, thus giving the right-hand side of (5.150).

The previous result applies in general for three random variables, and the nesting can be used to solve a complicated expectation by appropriate conditioning. After Example 5.23 which provides some background, we prove a theorem that also involves three random variables, but requires that they have successively *coarser* σ-fields to achieve a specific interesting property.

Example 5.23. Consider the discrete sample space Ω and define two σ-fields such that

$$\mathcal{F}_2 \subset \mathcal{F}_1, \tag{5.154}$$

which means that $\mathcal{F}_2$ is coarser than $\mathcal{F}_1$; we also know that the cardinality is $|\mathcal{F}_2| < |\mathcal{F}_1|$. Let the *atoms* of $\mathcal{F}_1$ and $\mathcal{F}_2$ be denoted by $\{E_n\}$ and $\{F_m\}$, respectively (see Chapter 2 for a discussion of atoms). An illustration of the atoms for the experiment of tossing a single die is shown in Figure 5.7, where $\mathcal{F}_1$ is the power set $\mathcal{P}(\Omega)$ with $|\mathcal{F}_1| = 64$ and $\mathcal{F}_2 = \{\phi, \Omega, \text{even}, \text{odd}\}$ with $|\mathcal{F}_2| = 4$. It is clear that atoms $\{F_m\}$ are not as "fine" as atoms $\{E_n\}$, and thus (5.154) holds for this example. In Theorem 5.6, we extend (5.154) to three σ-fields in order to demonstrate the *smoothing property* of conditional expectation.

Consider discrete random variable X with outcomes $x \in \mathcal{Z}$, which are the atoms of the σ-field $\mathcal{F}_X = \mathcal{P}(\mathcal{Z})$ (the power set of those outcomes). If the σ-field $\mathcal{F}_Y$ of another discrete random variable Y is a subset of $\mathcal{F}_X$, then that means the atoms (outcomes y) of $\mathcal{F}_Y$ are subsets of the atoms (outcomes x) of $\mathcal{F}_X$. This relationship is easily visualized if Y is generated as a many-to-one mapping from X: a particular y may be obtained from many x, but only one y has an inverse image to a specific x. This is evident for the example in Figure 5.7 where $x = \{1, 2, 3, 4, 5, 6\}$, $y = \{\text{even}, \text{odd}\}$, $\{2, 4, 6\} \rightarrow$ even, and $\{1, 3, 5\} \rightarrow$ odd. Since we are discussing random variables, we could use $\{0, 1\}$ instead of {even, odd}.

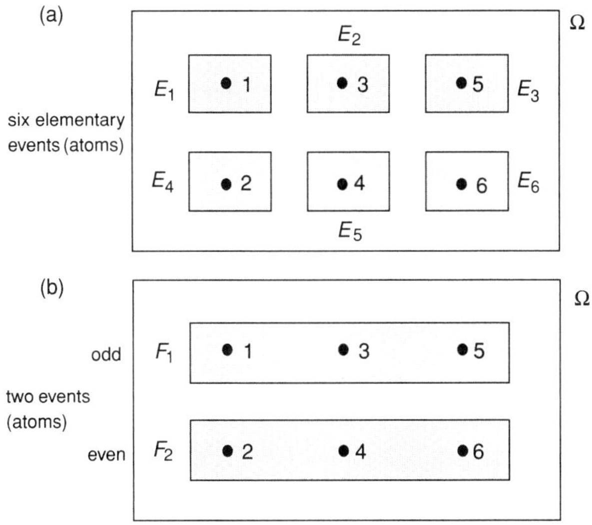

FIGURE 5.7 Example atoms for two σ-fields for the experiment of tossing a single die in Example 5.23. (a) Events $\{E_n\}$ are the six outcomes with fine σ-field given by the power set $\mathcal{P}(\Omega)$. (b) Events $\{F_m\}$ are the even and odd atoms with coarse σ-field given by $\{\phi, \Omega, \text{even}, \text{odd}\}$.

Theorem 5.6 (Smoothing). Let X, Y, and Z be discrete random variables on $\mathcal{Z}$ with three different σ-fields such that

$$\mathcal{F}_Z \subset \mathcal{F}_Y \subset \mathcal{F}_X. \tag{5.155}$$

Conditioning successively on Y and Z is the same as conditioning only on Z:

$$\mathcal{E}_{Z|Y}[\mathcal{E}_{X|Y}[X|Y]|Z] = \mathcal{E}_{X|Z}[X|Z]. \tag{5.156}$$

For the reverse conditioning order,

$$\mathcal{E}_{Y|Z}[\mathcal{E}_{X|Z}[X|Z]|Y] = \mathcal{E}_{X|Z}[X|Z]. \tag{5.157}$$

Observe that the conditional expectations in (5.156) and (5.157) are not symmetric; by interchanging the order of the conditionings, the result is still a conditional expectation with respect to the random variable with the *coarser* σ-field. It is possible to generate these random variables such that (5.155) holds by using functions $Y = g(X)$ and $Z = h(Y)$, which are many-to-one mappings. As a result, X necessarily has the least coarse σ-field of the three random variables; it is not possible to write X as a function of Y and Z because of the many-to-one mappings.

Proof. (Discrete random variables). Let the atoms of Y and Z be $\{y_n\}$ for $n = 1, \ldots, |\mathcal{F}_Y|$, and $\{z_m\}$ for $m = 1, \ldots, |\mathcal{F}_Z|$. Because $\mathcal{F}_Z$ is coarser than $\mathcal{F}_Y$, the key step in the proof is to recognize that

$$\begin{aligned} p_{Y,Z}[y_n, z_m] &= \begin{cases} p_Y[y_n], & y_n = z_m \\ 0, & \text{else,} \end{cases} \\ &= p_Y[y_n] I_{\{z_m\}}[y_n], \end{aligned} \tag{5.158}$$

where $I_{\{z_m\}}[y_n]$ is the discrete indicator function. This probability is nonzero only if outcome y_n from the finer σ-field is identical to a particular z_m in the coarser σ-field, because y_n is associated with only one outcome of Z. Thus

$$p_{Y|Z}[y_n|z_m] = \frac{p_{Y,Z}[y_n, z_m]}{p_Z[z_m]} = \frac{p_Y[y_n]}{p_Z[z_m]} I_{\{z_m\}}[y_n]. \tag{5.159}$$

The outer expectation in (5.156) is evaluated over all outcomes $\{y_n\}$ such that

$$\begin{aligned} \mathcal{E}_{Y|Z}[\mathcal{E}_{X|Y}[X|y_n]|z_m] &= \sum_{n=1}^{|\mathcal{F}_Y|} \mathcal{E}_{X|Y}[X|y_n] p_{Y|Z}[y_n|z_m], \\ &= \sum_{n:y_n=z_m} \frac{p_Y[y_n]}{p_Z[z_m]} \mathcal{E}_{X|Y}[X|y_n]. \end{aligned} \tag{5.160}$$

Assuming $x \in \mathcal{Z}$, the inner expectation is

$$\begin{aligned} \mathcal{E}_{Y|Z}[\mathcal{E}_{X|Y}[X|y_n]|z_m] &= \sum_{n:y_n=z_m} \frac{1}{p_Z[z_m]} \sum_{x=-\infty}^{\infty} x p_{X|Y}[x|y_n] p_Y[y_n], \\ &= \frac{1}{p_Z[z_m]} \sum_{x=-\infty}^{\infty} x \sum_{n:y_n=z_m} p_{X,Y}[x, y_n]. \end{aligned} \tag{5.161}$$

Again, since $\mathcal{F}_Z \subset \mathcal{F}_Y$ such that one or more y_n are equal to z_n, we have

$$\sum_{n:y_n=z_m} p_{X,Y}[x, y_n] = p_{X,Z}[x, z_m], \tag{5.162}$$

which gives

$$\begin{aligned} \mathcal{E}_{Y|Z}[\mathcal{E}_{X|Y}[X|Y = y_n]|Z = z_m] &= \frac{1}{p_Z[z_m]} \sum_{x=-\infty}^{\infty} x p_{X,Z}[x, z_m], \\ &= \sum_{x=-\infty}^{\infty} x p_{X|Z}[x|z_m] = \mathcal{E}_{X|Z}[X|Z = z_m]. \end{aligned} \tag{5.163}$$

Because the expectations hold for all outcomes $\{y_n\}$ and $\{z_m\}$, they can be replaced with random variables Y and Z, thus completing the proof.

For the conditional expectation in (5.157), observe from (5.158) that

$$p_{Z|Y}[z_m|y_n] = \frac{p_{Z,Y}[z_m, y_n]}{p_Y[y_n]} = \frac{p_Y[y_n]}{p_Y[y_n]} I_{\{z_m\}}[y_n] = I_{\{z_m\}}[y_n] \tag{5.164}$$

such that

$$\begin{aligned} \mathcal{E}_{Z|Y}[\mathcal{E}_{X|Z}[X|z_m]|y_n] &= \sum_{m=1}^{|\mathcal{F}_Z|} \mathcal{E}_{X|Z}[X|z_m] p_{Z|Y}[z_m|y_n], \\ &= \sum_{m:y_n=z_m} \mathcal{E}_{X|Z}[X|z_m] = \mathcal{E}_{X|Z}[X|z_m]. \end{aligned} \tag{5.165}$$

The last expression is due to the fact that only one term in the sum is nonzero because only one atom z_m (of the coarser $\mathcal{F}_Z$) equals y_n. The final result is obtained by replacing y_n and z_m with Y and Z. □

The expression in (5.156) is called the smoothing property of conditional expectation because averaging over finer atoms that comprise atoms of the coarse conditioning does not change the original conditional expectation. We can say the coarser conditioning has a "smoothing" effect on the finer conditioning.

Example 5.24. In this example, we demonstrate how to create coarser σ-fields by mapping random variable X to other random variables. Consider tossing a fair die for which X is a random variable with outcomes $\Omega = \{1, 2, 3, 4, 5, 6\}$, each with probability $1/6$. The largest σ-field is the power set $\mathcal{P}(\Omega)$ consisting of 2^6 elements. Define a mapping from X to Y as follows:

$$Y = \begin{cases} 0, & X = 1, 2 \\ -1, & X = 3, 4 \\ 1, & X = 5, 6. \end{cases} \tag{5.166}$$

The σ-field for Y has $2^3 = 8$ elements and obviously it is coarser: $\mathcal{F}_Y \subset \mathcal{F}_X$. Note that random variable Y can be viewed in two ways: (i) having the same sample space as X but with coarser σ-field $\mathcal{F}_Y$ or (ii) having the reduced sample space $\Omega_Y = \{0, -1, 1\}$ with σ-field given by the power set for Ω_Y. We are interested in the first interpretation where Y is defined to have a coarser σ-field.

Next, define a mapping from Y to Z as follows:

$$Z = |Y| = \begin{cases} 0, & Y = 0 \\ 1, & Y = -1, 1, \end{cases} \tag{5.167}$$

for which $\mathcal{F}_Z \subset \mathcal{F}_Y$ has only $2^2 = 4$ elements: $\{\phi, \Omega, 0, 1\}$. The many-to-one mappings from X to Y and then Y to Z are illustrated in Figure 5.8. Since the mappings are measurable, the probability assignment for X carries over to Y and then to Z, resulting in the following pdfs:

$$f_X(x) = (1/6)\sum_{n=1}^{6} \delta(x - n), \tag{5.168}$$

$$f_Y(y) = (1/3)[\delta(y + 1) + \delta(y) + \delta(y - 1)], \tag{5.169}$$

$$f_Z(z) = (1/3)\delta(z) + (2/3)\delta(z - 1). \tag{5.170}$$

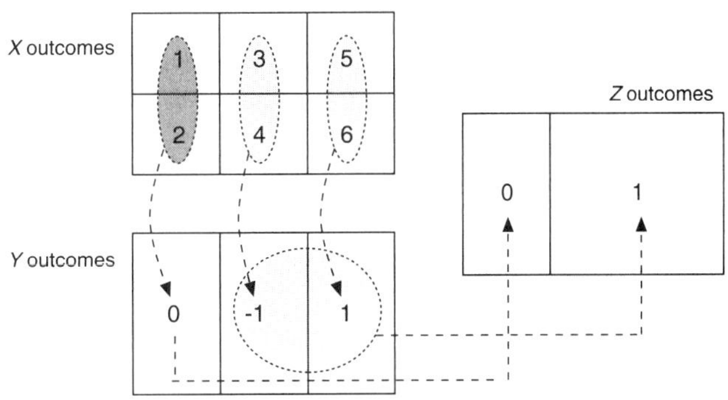

FIGURE 5.8 Mappings $X \to Y$ and $Y \to Z$ corresponding to increasingly coarser σ-fields in Example 5.24.

TABLE 5.1 $p_{X|Z}[x|z]$ **for Example 5.24**

	$X = 1$	2	3	4	5	6
$Z = 0$	1/2	1/2	0	0	0	0
$Z = 1$	0	0	1/4	1/4	1/4	1/4

In order to verify the smoothing property of conditional expectation, we first examine $\mathcal{E}[X|Z]$ for which we need the conditional pmf $p_{X|Z}[x|z]$ summarized in Table 5.1. Thus

$$\mathcal{E}[X|Z = 0] = \sum_{x=1}^{6} x p_{X|Z}[x|0] = (1/2)(1 + 2) = 3/2, \tag{5.171}$$

$$\mathcal{E}[X|Z = 1] = \sum_{x=1}^{6} x p_{X|Z}[x|1] = (1/4)(3 + 4 + 5 + 6) = 9/2. \tag{5.172}$$

Since $\mathcal{E}[X|Z]$ is a function of random variable Z, we can use the results above to write

$$\mathcal{E}[X|Z] = 3/2 + 3Z. \tag{5.173}$$

$\mathcal{E}[\mathcal{E}[X|Y]|Z]$ is examined next to show that conditioning on random variable Y (which has a finer σ-field) gives the same result as $\mathcal{E}[X|Z]$. For the inner expectation, the conditional pmf $p_{X|Y}[x|y]$ summarized in Table 5.2 yields

$$\mathcal{E}[X|Y = 0] = \sum_{x=1}^{6} x p_{X|Y}[x|0] = (1/2)(1 + 2) = 3/2, \tag{5.174}$$

$$\mathcal{E}[X|Y = -1] = \sum_{x=1}^{6} x p_{X|Y}[x| - 1] = (1/2)(3 + 4) = 7/2, \tag{5.175}$$

$$\mathcal{E}[X|Y = 1] = \sum_{x=1}^{6} x p_{X|Y}[x|1] = (1/2)(5 + 6) = 11/2 \tag{5.176}$$

for which the general expression is

$$\mathcal{E}[X|Y] = 7/2 + 2\,\mathrm{sgn}(Y + 1)\mathrm{sgn}(Y - 1/2), \tag{5.177}$$

TABLE 5.2 $p_{X|Y}[x|y]$ **for Example 5.24**

	$X = 1$	2	3	4	5	6
$Y = 0$	1/2	1/2	0	0	0	0
$Y = -1$	0	0	1/2	1/2	0	0
$Y = 1$	0	0	0	0	1/2	1/2

TABLE 5.3 $p_{Y|Z}[y|z]$ **for Example 5.24**

	$Y = 0$	-1	1
$Z = 0$	1	0	0
$Z = 1$	0	1/2	1/2

where sgn(·) is the signum function. Completing the derivation by evaluating the outer expectation, $p_{Y|Z}[y|z]$ in Table 5.3 gives

$$\begin{aligned}\mathcal{E}[\mathcal{E}[X|Y]|Z=0] &= \sum_{y=-1,0,1} (7/2 + 2\,\text{sgn}(y+1)\text{sgn}(y-1/2))p_{Y|Z}[y|0], \\ &= 3/2, \end{aligned} \tag{5.178}$$

$$\begin{aligned}\mathcal{E}[\mathcal{E}[X|Y]|Z=1] &= \sum_{y=-1,0,1} (7/2 + 2\,\text{sgn}(y+1)\text{sgn}(y-1/2))p_{Y|Z}[y|1], \\ &= (1/2)(7/2 + 11/2) = 9/2, \end{aligned} \tag{5.179}$$

which are the same results as before.

For convenience, we summarize the various properties of conditional expectation in Table 5.4. The third property (Independence 2) is proved in Problem 11.14, and is used to derive the Kalman filter in that chapter. Comparing these results, we can make the following additional observations:

- The first nested expectation is actually the orthogonality principle rearranged where we have brought $g(Y)$ inside the inner expectation using the factorization property. Comparing this expression with the next line, we see that it is a generalized form of that nested expectation with $g(Y)X$ instead of X.
- The first smoothing property can be viewed as a special case of the nested expectation result in the previous line when the σ-fields of the random variables are successively coarser.
- Technically, the last line in the table is not a smoothing property because the inner expectation is already conditioned on the random variable with the coarser σ-field. Instead, it has the "appearance" of the independence result in the second line of the table, such that the outer expectation has no effect.

Finally, we present a proof of the smoothing property based on the orthogonality property in Table 5.4 (Larson and Shubert, 1979). It is simpler than the one given earlier, and is more general because it applies to

TABLE 5.4 Properties of Conditional Expectation

Property	Conditional Expectation	Conditions
Linearity	$\mathcal{E}[aX_1 + bX_2|Y] = a\mathcal{E}[X_1|Y] + b\mathcal{E}[X_2|Y]$	a, b constants
Independence 1	$\mathcal{E}[Y|X] = \mathcal{E}[Y]$	X, Y independent
Independence 2	$\mathcal{E}[X|Y, Z] = \mathcal{E}[X|Y] + \mathcal{E}[X|Z] - \mathcal{E}[X]$	Y, Z independent, X, Y, Z jointly Gaussian
Function of Y	$\mathcal{E}[g(Y)|Y] = g(Y)$	None
Function of Y	$\mathcal{E}[X|Y, g(Y)] = \mathcal{E}[X|Y]$	None
Factorization	$\mathcal{E}[Xg(Y)|Y] = g(Y)\mathcal{E}[X|Y]$	None
Orthogonality	$\mathcal{E}[(X - \mathcal{E}[X|Y])g(Y)] = 0$	None
Nested 1	$\mathcal{E}[\mathcal{E}[Xg(Y)|Y]] = \mathcal{E}[Xg(Y)]$	None
Nested 2	$\mathcal{E}[\mathcal{E}[X|Y]] = \mathcal{E}[X]$	None
Nested 3	$\mathcal{E}[\mathcal{E}[X|Y, Z]|Z] = \mathcal{E}[X|Z]$	None
Smoothing	$\mathcal{E}[\mathcal{E}[X|Y]|Z] = \mathcal{E}[X|Z]$	$\mathcal{F}_Z \subset \mathcal{F}_Y \subset \mathcal{F}_X$
"Smoothing"	$\mathcal{E}[\mathcal{E}[X|Z]|Y] = \mathcal{E}[X|Z]$	$\mathcal{F}_Z \subset \mathcal{F}_Y \subset \mathcal{F}_X$

continuous and discrete random variables, but perhaps does not directly provide the same intuition about coarse and fine σ-fields.

Proof. (Arbitrary random variables). The condition $\mathcal{F}_Z \subset \mathcal{F}_Y \subset \mathcal{F}_X$ is achieved because the random variables are assumed to be functions of each other. First, rewrite the orthogonality property in terms of X and Z:

$$\mathcal{E}[Xg(Z)] = \mathcal{E}[\mathcal{E}[X|Z]g(Z)], \tag{5.180}$$

and replace X on both sides with $\mathcal{E}[X|Y]$ such that

$$\mathcal{E}[\mathcal{E}[X|Y]g(Z)] = \mathcal{E}[\mathcal{E}[\mathcal{E}[X|Y]g(Z)|Z]], \tag{5.181}$$

where $g(Z)$ on the right-hand side has been brought inside the second expectation using the factorization property. Note that if X were a function of Y given by $X = c(Y)$, then this expression would simplify to (5.180) because $\mathcal{E}[X = c(Y)|Y] = c(Y) = X$ by the fourth line in Table 5.4. Thus, X is not a function of Y, and hence has a finer σ-field. The next step in the derivation is to write $g(Z) = g(h(Y))$ for some function $h(\cdot)$ so that Z has a coarser σ-field. The left-hand side above becomes

$$\mathcal{E}[\mathcal{E}[X|Y]g(h(Y))] = \mathcal{E}[\mathcal{E}[Xg(h(Y))|Y] = \mathcal{E}[Xg(h(Y))], \tag{5.182}$$

where the factorization and orthogonality properties have been used. This last expectation is, of course, equivalent to the left-hand side in (5.180), which means the right-hand sides of (5.180) and (5.181) are equal:

$$\mathcal{E}[g(Z)\mathcal{E}[X|Z]] = \mathcal{E}[g(Z)\mathcal{E}[\mathcal{E}[X|Y]|Z]]. \tag{5.183}$$

Since this holds for any $g(Z)$, and the quantities are functions only of Z (similar to the result in Example 5.22), we have $\mathcal{E}[X|Z] = \mathcal{E}[\mathcal{E}[X|Y]|Z]$, which is the smoothing property.

For the final line in the table,

$$\mathcal{E}[\mathcal{E}[X|Z]|Y] = \mathcal{E}[g(Z)|Y] = \mathcal{E}[g(h(Y))|Y] = g(h(Y)), \tag{5.184}$$

where $\mathcal{E}[X|Z] = g(Z)$ is a function of Z (not the same $g(\cdot)$ used in the proof above), and again we assume that $Z = h(Y)$ is a function of Y so that it has a coarser σ-field. The last result follows from the fourth line in the table. Since $g(h(Y)) = \mathcal{E}[X|Z]$ by definition, this completes the proof. □

5.14 LOCATION PARAMETERS: MEAN, MEDIAN, AND MODE

A parameter that characterizes the center and location of a pdf is useful as a simple single descriptor of a random variable, as was used in Chapter 3 when we described the properties of several distributions. There are three such measures that generally give different values, although they coincide for some distributions.

Definition: Mean The *mean* of random variable X is its expectation:

$$\mu_X \triangleq \mathcal{E}[X]. \tag{5.185}$$

The mean of a random variable can be viewed as that value obtained by averaging the outcomes over a very large number of experiments. It can also be viewed functionally as the area of $f_X(x)$ weighted by the odd function $g(x) = x$ which places increasing emphasis on larger values (positive and negative) of x.

Definition: Median The *median* m_e of continuous random variable X is the midpoint of the cdf:

$$F_X(m_e) = 1/2 \Longrightarrow \int_{-\infty}^{m_e} f_X(x)\mathrm{d}x = \int_{m_e}^{\infty} f_X(x)\mathrm{d}x = 1/2, \tag{5.186}$$

which is equivalent to $P(X \leq m_e) = P(X \geq m_e) = 1/2$. For a discrete random variable, this last expression is modified to two conditions:

$$P(X \leq m_e) \geq 1/2 \quad \text{and} \quad P(X \geq m_e) \geq 1/2, \tag{5.187}$$

where the directions of the inequalities relative to $1/2$ are the *same*. Substituting the pdfs yields

$$\int_{-\infty}^{m_e} \sum_{n=-\infty}^{\infty} P(X = n)\delta(x - n)\mathrm{d}x \geq 1/2 \Longrightarrow \sum_{n=-\infty}^{m_e} p_X[n] \geq 1/2, \tag{5.188}$$

$$\int_{m_e}^{\infty} \sum_{n=-\infty}^{\infty} P(X = n)\delta(x - n)\mathrm{d}x \geq 1/2 \Longrightarrow \sum_{n=m_e}^{\infty} p_X[n] \geq 1/2. \tag{5.189}$$

Since there is probability mass only for integer values of n, it is clear that in order that both sums include m_e and both exceed $1/2$, the median must be an integer (corresponding to an outcome of X). The following example illustrates this definition.

Example 5.25. Consider the Poisson distribution for which we cannot write an explicit expression for the median m_e (unlike many of the random variables in Appendix A). Suppose $\alpha = 2.5$ and observe that

$$P(X \leq 2) = \exp(-2.5)[1 + 2.5/1! + (2.5)^2/2!] \approx 0.5438 > 1/2, \tag{5.190}$$

and thus

$$P(X > 2) \approx 1 - 0.5438 = 0.4562. \tag{5.191}$$

For the second inequality in (5.187),

$$P(X \geq 2) \approx 0.4562 + \exp(-2.5)(2.5)^2/2! \approx 0.7127 > 1/2 \tag{5.192}$$

from which we conclude that $m_e = 2$ for this value of α.

Although the median is sometimes chosen as the midpoint between two outcomes for a discrete random variable, which would be $X = 2.5$ in Example 5.25, the definition in (5.187) forces the median to be an integer outcome of the random variable. The median is often important because it is not influenced by "heavy tails," as is the mean (discussed previously for the Cauchy distribution). For this reason, the median is sometimes considered in statistical studies as a measure of the center because it is not affected by rare extreme events called *outliers*.

Example 5.26. For the discrete uniform random variable with $N = b - a + 1$ outcomes $\{a, \ldots, b\}$, the two expressions in (5.189) defining the median are

$$\sum_{n=a}^{m_e} (1/N) \geq 1/2 \Longrightarrow m_e - a + 1 \geq N/2 \Longrightarrow m_e \geq (a + b - 1)/2, \tag{5.193}$$

$$\sum_{n=m_e}^{b} (1/N) \geq 1/2 \Longrightarrow b - m_e + 1 \geq N/2 \Longrightarrow m_e \leq (a + b + 1)/2, \tag{5.194}$$

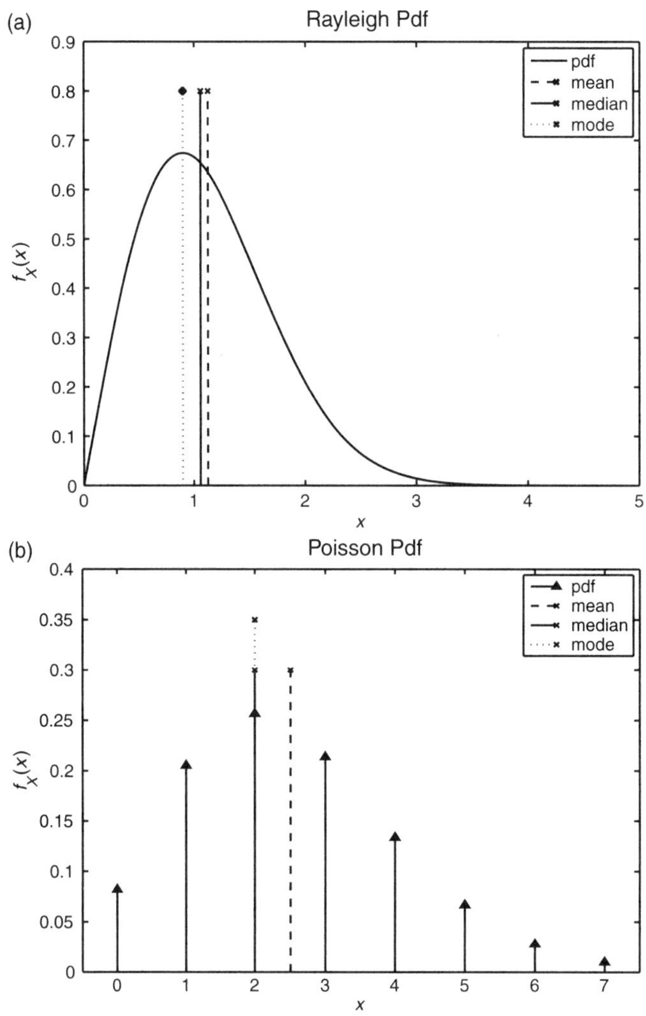

FIGURE 5.9 Examples showing the mean, median, and mode. (a) Rayleigh distribution ($\sigma = 0.9$): $\mu_X = \sigma/\sqrt{\pi/2}$, $m_e = \sigma\sqrt{\ln(4)}$, $m_o = \sigma$. (b) Poisson distribution ($\alpha = 2.5$): $\mu_X = \alpha$, $m_e =$ implicit from (5.187), $m_o = \lfloor \alpha \rfloor$.

which simplify because the sums are performed over a constant. If $a + b$ is odd (N is even), two different values of m_e are obtained, and thus from our definition the median does not exist. If $a + b$ is even (N is odd), then m_e is unique with value given by the floor function $m_e = \lfloor (a + b + 1)/2 \rfloor$. Sometimes $m_e = (a + b)/2$ is used, but this yields a non-integer value which is not an outcome of X when $a + b$ is odd.

Definition: Mode The *mode* m_o of random variable X is the value x corresponding to the global maximum of the pdf.

Examples of the mean, median, and mode for continuous and discrete distributions are shown in Figure 5.9. For pdfs that have several maxima, the corresponding values of x are called *modes*, and in such cases the pdf is *multimodal*. An example of a *bimodal* pdf is shown in Figure 5.10.

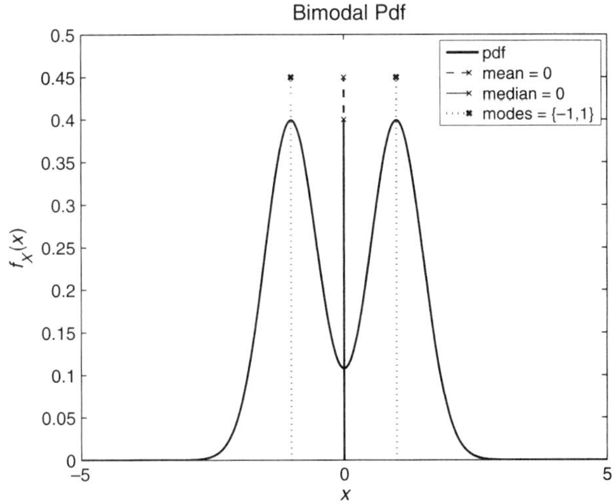

FIGURE 5.10 Example of a bimodal distribution, obtained as the sum of two standard Gaussian pdfs (each scaled by 0.5), called a mixture.

Proposition 5.1. *The mean of a random variable is the value* a *that minimizes the MSE* $\mathcal{E}[(X-a)^2]$.

Proof. Expand the expectation as follows:

$$\mathcal{E}[(X-a)^2] = \mathcal{E}[X^2] - 2a\mathcal{E}[X] + a^2. \tag{5.195}$$

Differentiating with respect to a gives

$$\frac{\partial}{\partial a}\mathcal{E}[(X-a)^2] = -2\mathcal{E}[X] + 2a. \tag{5.196}$$

Setting the result equal to zero and solving for a completes the proof:

$$a = \mathcal{E}[X] = \mu_X. \tag{5.197}$$

The MMSE is the variance σ_X^2. □

Proposition 5.2. *The median of a random variable is the value* a *that minimizes the mean absolute error (MAE)* $\mathcal{E}[|X-a|]$.

Proof. Expand the expectation as follows:

$$\begin{aligned}\mathcal{E}[|X-a|] &= \int_{-\infty}^{\infty} |x-a| f_X(x)\mathrm{d}x, \\ &= \int_{a}^{\infty} (x-a) f_X(x)\mathrm{d}x - \int_{\infty}^{a} (x-a) f_X(x)\mathrm{d}x.\end{aligned} \tag{5.198}$$

Differentiating with respect to a via the chain rule and setting the result equal to zero gives

$$-\int_{a}^{\infty} f_X(x)\mathrm{d}x + \int_{-\infty}^{a} f_X(x)\mathrm{d}x = 0, \tag{5.199}$$

which is the same as (5.186), yielding $a = m_e$. □

Example 5.27. For Gaussian random variable X with parameters $\{\mu, \sigma\}$, the mean, median, and mode are all equal to μ. For the exponential random variable X with parameter λ, the mean, median, and mode are $1/\lambda$, $\ln(2)/\lambda$, and zero, respectively. The mean was derived in (5.49) and the mode is obviously zero. The median is obtained from

$$\int_0^{m_e} f_X(x)\mathrm{d}x = \lambda \int_0^{m_e} \exp(-\lambda x)\mathrm{d}x = 1 - \exp(-\lambda m_e).$$

Setting this expression equal to 1/2 gives

$$\exp(-\lambda m_e) = 1/2 \Longrightarrow m_e = \ln(2)/\lambda. \tag{5.200}$$

5.15 VARIANCE, COVARIANCE, AND CORRELATION

Measures of *dispersion* or spread of a random variable are considered next.

Definition: Variance The *variance* of random variable X is

$$\sigma_X^2 \triangleq \mathcal{E}[(X - \mu_X)^2], \tag{5.201}$$

where $\mu_X = \mathcal{E}[X]$ is the mean.

The variance is *centered* in the sense that the mean of the random variable is subtracted before computing the expectation; it is one of the *central moments* defined in a subsequent section. The variance can be expressed in terms of the mean and the mean-square as follows:

$$\sigma_X^2 = \mathcal{E}[X^2 - 2X\mu_X + \mu_X^2] = \mathcal{E}[X^2] - \mu_X^2. \tag{5.202}$$

We summarize several properties of the variance:

- *Linear transformation*. For $aX + b$:

$$\mathrm{var}[aX + b] = a^2\sigma_X^2, \tag{5.203}$$

  where $\{a, b\} \in \mathcal{R}$ are constants. The variance changes when the random variable is *scaled* (slope a of the transformation), but is invariant to a translation (intercept b of the transformation). This is expected because a translation changes the mean and the variance is a centered expectation.
- *Variance of a sum*. The variance of the sum of two *dependent* random variables X and Y is

$$\begin{aligned} \mathrm{var}[X + Y] &= \mathcal{E}[(X + Y - (\mu_X + \mu_Y))^2], \\ &= \mathcal{E}[(X - \mu_X)^2 + (Y - \mu_Y)^2 + 2(X - \mu_X)(Y - \mu_Y)], \\ &= \sigma_X^2 + \sigma_Y^2 + 2\mathrm{cov}[X, Y], \end{aligned} \tag{5.204}$$

  where $\mathrm{cov}[X, Y] \triangleq \mathcal{E}[(X - \mu_X)(Y - \mu_Y)]$ is the covariance C_{XY} of X and Y (and is covered later in this section). If X and Y are independent random variables, then

$$\mathrm{var}[X + Y] = \sigma_X^2 + \sigma_Y^2. \tag{5.205}$$

  Actually, this result requires only that the random variables be *uncorrelated* with $C_{XY} = 0$.

- *Variance of a linear combination*. The previous result is extended to a weighted sum of several random variables as follows:

$$\text{var}\left[\sum_{n=1}^{N} a_n X_n\right] = \sum_{n=1}^{N} a_n^2 \sigma_{X_n}^2 + \sum_{m=1}^{N} \sum_{n=1,n\neq m}^{N} a_m a_n C_{X_m X_n} \tag{5.206}$$

  for constants $\{a_n\} \in \mathcal{R}$.
- *Variance of a difference*. The variance of the difference of two *dependent* random variables is

$$\text{var}[X - Y] = \sigma_X^2 + \sigma_Y^2 - 2C_{XY}. \tag{5.207}$$

  If X and Y are uncorrelated with $C_{XY} = 0$, then $\text{var}[X - Y] = \sigma_X^2 + \sigma_Y^2$ is the sum of their variances; it must be a sum (not a difference) because it is not possible for the variance to be negative.

Theorem 5.7 (Law of total variance). For dependent random variables:

$$\text{var}[X] = \mathcal{E}[\text{var}[X|Y]] + \text{var}[\mathcal{E}[X|Y]]. \tag{5.208}$$

Proof. Conditioning on Y and using a form similar to (5.202) gives

$$\text{var}[X] = \mathcal{E}[\mathcal{E}[X^2|Y]] - (\mathcal{E}[\mathcal{E}[X|Y]])^2, \tag{5.209}$$

where the outer expectations are over Y. Using (5.202) again to rewrite the first expectation gives

$$\begin{aligned} \text{var}[X] &= \mathcal{E}[\text{var}[X|Y] + \left(\mathcal{E}[X|Y]\right)^2] - \left(\mathcal{E}[(\mathcal{E}[X|Y]]\right)^2, \\ &= \mathcal{E}[\text{var}[X|Y]] + \mathcal{E}[(\mathcal{E}[X|Y])^2] - (\mathcal{E}[\mathcal{E}[X|Y]])^2. \end{aligned} \tag{5.210}$$

The last two terms combine to give another variance:

$$\mathcal{E}[(\mathcal{E}[X|Y])^2] - (\mathcal{E}[\mathcal{E}[X|Y]])^2 = \text{var}[\mathcal{E}[X|Y]]. \tag{5.211}$$

Substituting this result into (5.210) completes the proof. □

This law has a form similar to that of the law of total probability in Chapter 2.

Definition: Standard Deviation The *standard deviation* σ_X of random variable X is the positive square root of the variance σ_X^2.

The standard deviation is often used as a measure of dispersion for a random variable because, unlike the variance, it has the same units as the mean (and the random variable itself). Like the variance, σ_X is changed when the random variable is scaled, and is unaffected by a translation:

$$\text{stdev}[aX + b] = |a|\sigma_X \tag{5.212}$$

for constants $a, b \in \mathcal{R}$.

Example 5.28. Consider the standard Gaussian random variable with parameters $\{\mu = 0, \sigma = 1\}$. Figure 5.11 shows values of the random variable that are within one, two, and three standard deviations from the mean. One can show that $P(|X| \leq a\sigma)$ is 0.683, 0.955, 0.997, 0.9999, and 0.999999 for $a = 1, 2, 3, 4$, and 5, respectively.

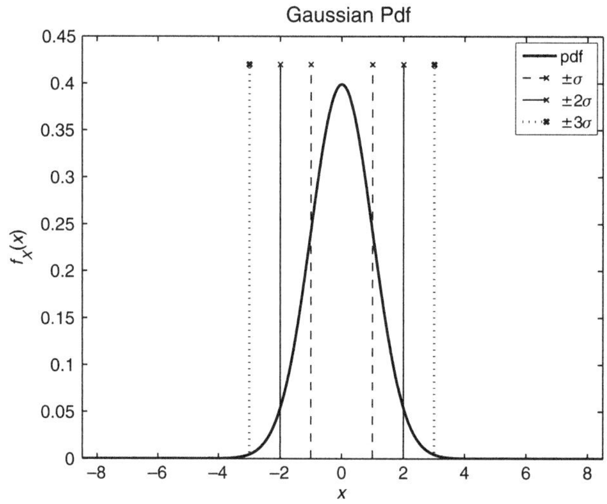

FIGURE 5.11 Standard Gaussian pdf showing one, two, and three standard deviations from the mean. The corresponding probabilities within these intervals (extending on both sides of the mean) are 0.683, 0.955, and 0.997, respectively.

Definition: Coefficient of Variation The *coefficient of variation* of random variable X is

$$c_X \triangleq \frac{\sigma_X}{\mu_X} \tag{5.213}$$

assuming $\mu_X \neq 0$.

It is a measure of the *variability* of the pdf relative to the mean that is independent of the units of X because the standard deviation (and not the variance) is used in the ratio.

The covariance can be viewed as an extension of the variance of one random variable to that involving two random variables.

Definition: Covariance and Cross-Correlation The *covariance* of random variables X and Y is

$$C_{XY} \triangleq \mathcal{E}[(X - \mu_X)(Y - \mu_Y)]. \tag{5.214}$$

It can be rewritten as

$$C_{XY} = \mathcal{E}[XY] - \mu_X\mathcal{E}[Y] - \mu_Y\mathcal{E}[X] + \mu_X\mu_Y = R_{XY} - \mu_X\mu_Y, \tag{5.215}$$

where

$$R_{XY} \triangleq \mathcal{E}[XY] \tag{5.216}$$

is the *cross-correlation* of X and Y.

Observe that $C_{XY} = R_{XY}$ when at least one of the random variables has zero mean. The cross-correlation is a measure of the *linear* relationship between X and Y, as discussed later when the *correlation coefficient* is defined. Properties of the covariance include the following:

- *Uncorrelated.* Random variables X and Y are uncorrelated when $C_{XY} = 0$ such that $R_{XY} = \mathcal{E}[X]\mathcal{E}[Y]$.
- *Symmetric.* $C_{XY} = C_{YX}$.
- *Scaling and translation.* $\text{cov}[aX + b, cY + d] = acC_{XY}$ for constants $a, b, c, d \in \mathcal{R}$.

The proofs of these properties are left as exercises.

Example 5.29. Let X_1 and X_2 be dependent random variables with *identical* means, and define $Y_1 \triangleq X_1 + X_2$ and $Y_2 \triangleq X_1 - X_2$. The covariance of Y_1 and Y_2 is

$$\begin{aligned} C_{Y_1Y_2} &= \mathcal{E}[(X_1 + X_2)(X_1 - X_2)] - \mu_{Y_1}\mu_{Y_2}, \\ &= \mathcal{E}[X_1^2] + \mathcal{E}[X_2^2] = R_{Y_1Y_2} \end{aligned} \tag{5.217}$$

because $\mu_{Y_2} = \mathcal{E}[X_1 - X_2] = 0$.

Note that when $Y = X$, $C_{XY} \to C_{XX} = \sigma_X^2$, and $R_{XY} \to R_{XX} = \mathcal{E}[X^2]$. Although the symbols C_{XX} and R_{XX} are generally not standard notation in this random variable context, they are used for the random *processes* covered in Chapter 6. There they depend on time instants $\{t_1, t_2\}$ of the process: $C_{XX}(t_1, t_2) \triangleq \mathcal{E}[[X(t_1) - \mu_X(t_1)][X(t_2) - \mu_X(t_2)]]$ and $R_{XX}(t_1, t_2) \triangleq \mathcal{E}[X(t_1)X(t_2)]$, where they are known as autocovariance and autocorrelation *functions*, respectively.

5.16 FUNCTIONAL VIEW OF THE VARIANCE

Based on a functional approach similar to that used to interpret the mean, we demonstrate how the variance is a measure of the width of a pdf. Since the variance is a central moment where the mean has been subtracted such that the pdf is shifted to be centered about zero, we assume the mean of X is zero and instead examine

$$\mathcal{E}[X^2] = \int_{-\infty}^{\infty} x^2 f_X(x)\mathrm{d}x. \tag{5.218}$$

Let $f_X(x)$ be standard Gaussian with parameters $\{\mu = 0, \sigma = 1\}$. Figure 5.12(a) shows a plot of the product $x^2 f_X(x)$ (dashed line). Since x^2 and $f_X(x)$ are nonnegative, (5.218) must also be nonnegative. If the pdf is wider (larger σ), then we can expect that $x^2 f_X(x)$ will have greater area because x^2 is a rapidly increasing nonnegative function. This is demonstrated in Figure 5.12(b) where we see that the area under the dashed line has increased for $\sigma^2 = 2$. Unlike x in the product $xf_X(x)$ for the mean, there is no zero crossing for x^2: it is an even function about the origin that equally weights $f_X(x)$ on each side of $x = 0$ and with positive sign. We have also plotted the cumulative area

$$B(x) = \int_{-\infty}^{x} v^2 f_X(v)\mathrm{d}v \tag{5.219}$$

as x is varied over the support of $f_X(x)$; of course, $\mathcal{E}[X^2] = \lim_{x\to\infty} B(x)$. This function resembles a cdf, but for the weighted pdf $x^2 f_X(x)$.

From this result, we can conclude that other even functions about the mean are also measures of the width of a pdf. Examples of these include high-order central moments with even exponents such as $\mathcal{E}[(X - \mu_X)^4]$, $\mathcal{E}[(X - \mu_X)^6]$, and so on, as well as $\mathcal{E}[|X - \mu_X|]$. For the last case with $\mu_X = 0$, note that

$$\mathcal{E}[|X|] = \int_{-\infty}^{\infty} |x| f_X(x)\mathrm{d}x = \int_{0}^{\infty} xf_X(x)\mathrm{d}x - \int_{-\infty}^{0} xf_X(x)\mathrm{d}x. \tag{5.220}$$

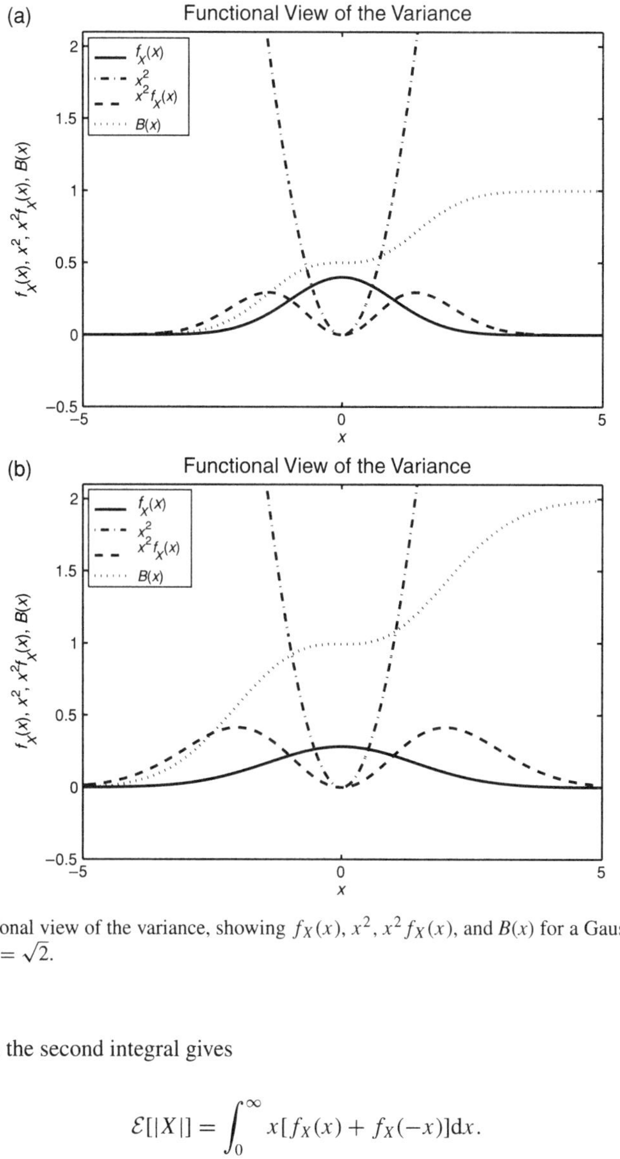

FIGURE 5.12 Functional view of the variance, showing $f_X(x)$, x^2, $x^2 f_X(x)$, and $B(x)$ for a Gaussian pdf. (a) $\mu = 0$ and $\sigma = 1$. (b) $\mu = 0$ and $\sigma = \sqrt{2}$.

Changing variables in the second integral gives

$$\mathcal{E}[|X|] = \int_0^\infty x[f_X(x) + f_X(-x)]\mathrm{d}x. \tag{5.221}$$

Since $x \geq 0$ and $f_X(x)$ is nonnegative, we see that if $f_X(x)$ were to become wider, the area under $x[f_X(x) + f_X(-x)]$ is always positive and increases, though not as fast as $x^2 f_X(x)$.

5.17 EXPECTATION AND THE INDICATOR FUNCTION

Sometimes it is convenient to write various probabilistic quantities in terms of an expectation using the indicator function. For random variable X with pdf $f_X(x)$ and cdf $F_X(x)$, observe that

$$P(a \leq X \leq b) = \int_{-\infty}^{\infty} I_{[a,b]}(x) f_X(x)\mathrm{d}x = \mathcal{E}\left[I_{[a,b]}(X)\right], \tag{5.222}$$

where in this context, $I_{[a,b]}(X)$ is a particular nonlinear transformation of random variable X. Likewise, we can write the cdf as follows:

$$F_X(x) = P(X \leq x) = \mathcal{E}\left[I_{(-\infty,x]}(X)\right]. \tag{5.223}$$

We might also be interested in various expectations of the indicator function itself. For example, observe that

$$\begin{aligned} \text{var}\left[I_{[a,b]}(X)\right] &= \mathcal{E}\left[\left(I_{[a,b]}(X) - \mathcal{E}[I_{[a,b]}(X)]\right)^2\right], \\ &= \mathcal{E}\left[I^2_{[a,b]}(X)\right] - P^2(a \leq X \leq b), \end{aligned} \tag{5.224}$$

where (5.222) has been substituted for the mean of the indicator function. Since squaring $I_{[a,b]}(X)$ gives the same indicator function, we finally have

$$\begin{aligned} \text{var}\left[I_{[a,b]}(X)\right] &= P(a \leq X \leq b) - P^2(a \leq X \leq b), \\ &= P(a \leq X \leq b)[(1 - P(a \leq X \leq b)]. \end{aligned} \tag{5.225}$$

As with any function of random variable X, this variance can be computed without first finding the pdf of the mapping. A special case of the result above is

$$\text{var}\left[I_{(-\infty,x]}(X)\right] = F_X(x)[1 - F_X(x)]. \tag{5.226}$$

For random variables X and Y, the following covariance can be computed using a set operation:

$$\begin{aligned} \text{cov}\left[I_{[a,b]}(X)I_{[c,d]}(Y)\right] &= \mathcal{E}\left[I_{[a,b]}(X)I_{[c,d]}(Y)\right] - \mathcal{E}\left[I_{[a,b]}(X)\right]\mathcal{E}\left[I_{[c,d]}(Y)\right], \\ &= P(\{a \leq X \leq b\} \cap \{c \leq Y \leq d\}) - P(a \leq X \leq b)P(c \leq Y \leq d). \end{aligned} \tag{5.227}$$

Since the indicator function takes on only two values $\{0, 1\}$, its expectation is nonnegative and cannot exceed 1. This is clear from (5.222) where $0 \leq P(a \leq X \leq b) \leq 1$. Likewise, the variance of the indicator function is bounded: from (5.225) we have

$$0 \leq P(a \leq X \leq b)[1 - P(a \leq X \leq b)] \leq 1 - P(a \leq X \leq b) \leq 1. \tag{5.228}$$

The covariance in (5.227) is similarly bounded.

Example 5.30. Let X be standard Gaussian with parameters $\{\mu = 0, \sigma = 1\}$. Then

$$\mathcal{E}\left[I_{[0,\infty)}(X)\right] = 1/2 \tag{5.229}$$

and

$$\text{var}\left[I_{[0,\infty)}(X)\right] = P(0 \leq X < \infty)[1 - P(0 \leq X < \infty)] = 1/4. \tag{5.230}$$

5.18 CORRELATION COEFFICIENTS

Correlation is an important measure of the degree to which two or more random variables have some relationship. For example, if X and Y are correlated, then outcomes of X might be used to predict outcomes of Y. Correlation also means that if X is conditioned on Y (or vice versa), then the conditional pdf of X given Y differs from the marginal pdf. We observed this previously in Chapter 4 for the bivariate Gaussian pdf in Example 4.5. In this section, we discuss *normalized* measures of correlation that can also be used to generate random variables with a specified degree of correlation.

Definition: Pearson's Correlation *Pearson's correlation coefficient* for random variables X and Y is

$$\rho_{XY} \triangleq \frac{C_{XY}}{\sigma_X \sigma_Y}. \tag{5.231}$$

Generally, when a correlation coefficient is mentioned, it refers to Pearson's ρ_{XY} (and not the other coefficients described below).

Due to the normalization, ρ_{XY} is dimensionless and bounded, which allows for more meaningful interpretations of the degree of correlation between two random variables.

Theorem 5.8. The correlation coefficient is bounded as follows:

$$|\rho_{XY}| \leq 1. \tag{5.232}$$

Proof. From the Cauchy–Schwarz inequality in Appendix F:

$$|\mathcal{E}[(X-\mu_X)(Y-\mu_Y)]| \leq \sqrt{\mathcal{E}[X-\mu_X]\mathcal{E}[Y-\mu_Y]} = \sigma_X \sigma_Y. \tag{5.233}$$

Dividing both sides by $\sigma_X \sigma_Y$ completes the proof. □

Suppose that U and V are zero-mean uncorrelated random variables, each with unit variance. In order to derive a pair of random variables $\{X, Y\}$ with a specific correlation coefficient ρ_{XY}, we can perform the following transformation:

$$X = U, \qquad Y = \rho_{XY} U + \sqrt{1-\rho_{XY}^2}\, V. \tag{5.234}$$

Then $\sigma_X^2 = 1$, $\sigma_Y^2 = \rho_{XY}^2 + 1 - \rho_{XY}^2 = 1$, and

$$\mathcal{E}[XY] = \rho_{XY}\mathcal{E}[U^2] + \sqrt{1-\rho_{XY}}\,\mathcal{E}[UV] = \rho_{XY} \tag{5.235}$$

so that X and Y are zero-mean, unit-variance random variables with correlation coefficient ρ_{XY}. From this transformation, a scatter plot can be used to show how correlated random variables behave when ρ_{XY} is varied. This is done by randomly generating pairs of numbers $\{x, y\}$, transforming them as in (5.234), and plotting many such pairs. Several examples for various ρ_{XY} are shown in Figure 5.13 where the original $\{U, V\}$ are independent standard Gaussian random variables. Observe that the "constellations" range from circular ($\rho_{XY} = 0$) to linear ($\rho_{XY} = \pm 1$). The corresponding pdfs for four of these cases are shown in Figure 5.14 where we find that contours of constant values of the pdf have the same shapes as the scatter plots. They range from circular ($\rho_{XY} = 0$), to elliptical ($0 < |\rho_{XY}| < 1$), and then to linear ($|\rho_{XY}| = 1$). (The linear case is not shown because three-dimensional plots are not easily realized when the pdf collapses to a vertical plane.)

However, note that the correlation coefficient does *not* describe all types of dependencies between two random variables, only a *linear dependency* as seen in Figure 5.13. This point is illustrated by the following simple example.

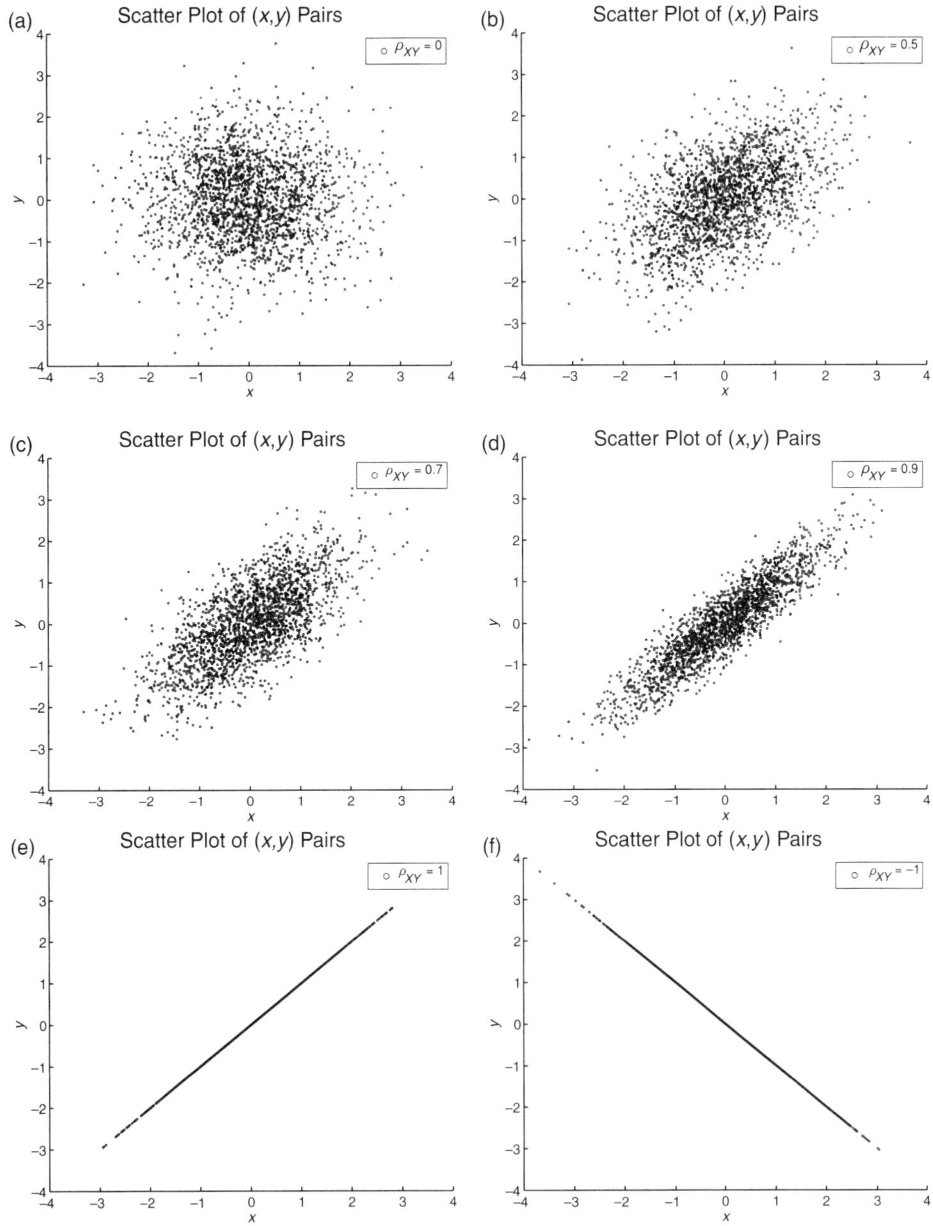

FIGURE 5.13 Scatter plots of correlated random variables $\{X, Y\}$ for several values of the correlation coefficient. (a) $\rho_{XY} = 0$. (b) $\rho_{XY} = 0.5$. (c) $\rho_{XY} = 0.7$. (d) $\rho_{XY} = 0.9$. (e) $\rho_{XY} = 1$. (f) $\rho_{XY} = -1$.

Example 5.31. Let X be a symmetric Bernoulli random variable with outcomes $\{-1, 1\}$ and probability of success $p = 1/2$. The transformation $Y = |X|$ has pdf $f_Y(y) = \delta(x - 1)$ so that

$$f_{X,Y}(x, y) = \begin{cases} 1/2, & x = -1, y = 1 \\ 1/2, & x = 1, y = 1 \\ 0, & \text{else}. \end{cases} \tag{5.236}$$

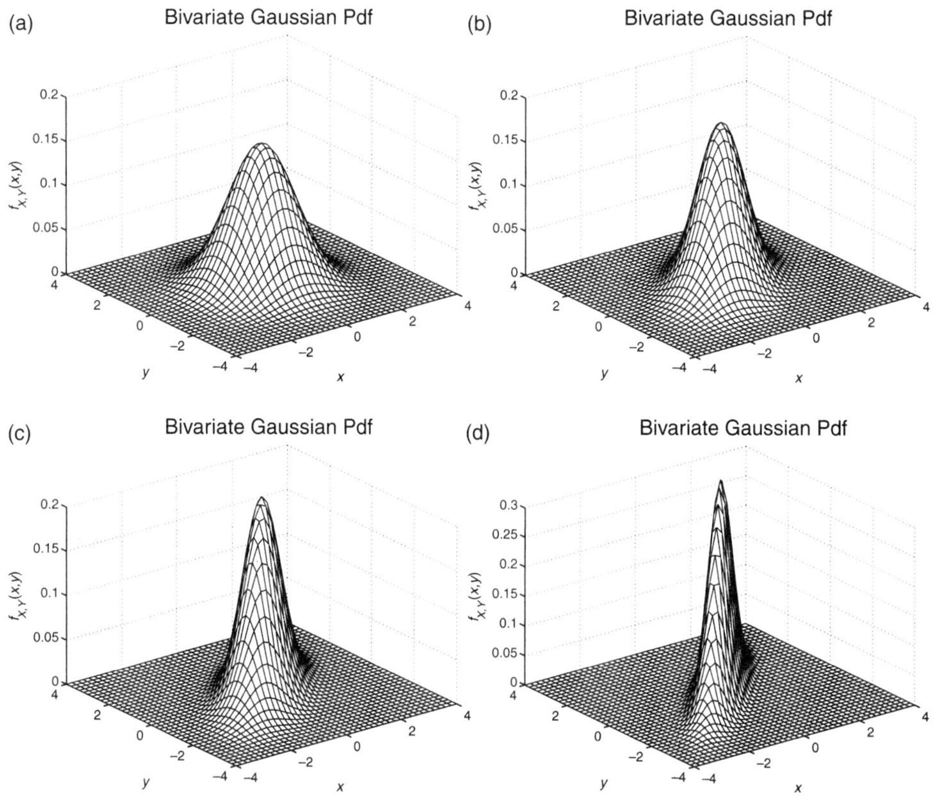

FIGURE 5.14 Bivariate Gaussian pdfs of correlated random variables $\{X, Y\}$ for several values of the correlation coefficient. (a) $\rho_{XY} = 0$. (b) $\rho_{XY} = 0.5$. (c) $\rho_{XY} = 0.7$. (d) $\rho_{XY} = 0.9$.

Obviously X and Y are dependent, but they are uncorrelated

$$\mathcal{E}[XY] = (1/2)(-1)(1) + (1/2)(1)(1) = 0, \tag{5.237}$$

so that $\rho_{XY} = 0$.

The following theorem describes a linear relationship between random variables.

Theorem 5.9. $|\rho_{XY}| = 1$ if and only if $Y = aX + b$, where $\{a, b\} \in \mathcal{R}$ are constants.

Proof. Only if: Define $U \triangleq X/\sigma_X$ and $V \triangleq Y/\sigma_Y$, and consider the variance of their difference:

$$\text{var}[U - Y] = \sigma_U^2 + \sigma_V^2 - 2C_{UV} = 2(1 - \rho_{UV}). \tag{5.238}$$

This follows from

$$\sigma_U^2 = \mathcal{E}[(U - \mu_U)^2] = \mathcal{E}[(X - \mu_X)^2]/\sigma_X^2 = 1, \tag{5.239}$$

and likewise $\sigma_V^2 = 1$. When $\rho_{UV} = \rho_{XY} = 1$, we see that var$[U - V] = 0$ which means $U - V$ takes on only one value, given by some $c \in \mathcal{R}$; its pdf is $f_{U-V}(z) = \delta(z - c)$. Thus, $U - V$ equals c with probability one and

$$Y = \sigma_Y(U - c) = (\sigma_Y/\sigma_X)X - c\sigma_Y, \tag{5.240}$$

which, of course, is a linear (affine) function with slope $a = \sigma_Y/\sigma_X$ and ordinate $b = -c\sigma_Y$. A similar result can be shown for $\rho_{UV} = \rho_{XY} = -1$ using var$[U + V]$. (The if part of the theorem is straightforward, and is left as an exercise.) □

Definition: Pearson's Sample Correlation *Pearson's sample correlation coefficient* for random variables X and Y based on N pairs of samples $\{X_n, Y_n\}$ is

$$\bar{\rho}_{XY} \triangleq \frac{\sum_{n=1}^{N}(X_n - \bar{X})(Y_n - \bar{Y})}{\sqrt{\sum_{n=1}^{N}(X_n - \bar{X})^2 \sum_{n=1}^{N}(Y_n - \bar{Y})^2}}, \tag{5.241}$$

where $\bar{X}$ is the sample mean defined in (4.228) and likewise for $\bar{Y}$.

The samples $\{X_n, Y_n\}$ are random variables; actual outcomes $\{x_n, y_n\}$ would be substituted to obtain a specific value of the sample correlation coefficient. This expression can be rewritten as follows (see Problem 5.33):

$$\bar{\rho}_{XY} = \frac{1}{N-1}\sum_{n=1}^{N}\frac{(X_n - \bar{X})(Y_n - \bar{Y})}{\sqrt{S_X^2 S_Y^2}}, \tag{5.242}$$

where

$$S_X^2 \triangleq \frac{1}{N-1}\sum_{n=1}^{N}(X_n - \bar{X})^2 \tag{5.243}$$

is the sample variance for X and likewise for S_Y^2. The square roots $S_X \triangleq \sqrt{S_X^2}$ and $S_Y \triangleq \sqrt{S_Y^2}$ are sample standard deviations.

It is straightforward to show that $|\bar{\rho}_{XY}| \leq 1$ for any set of outcomes by defining the following sample-mean-shifted vectors: $\tilde{\mathbf{X}} = [X_1 - \bar{X}, \ldots, X_N - \bar{X}]^T$ and $\tilde{\mathbf{Y}} = [Y_1 - \bar{Y}, \ldots, Y_N - \bar{Y}]^T$ such that

$$\bar{\rho}_{XY} = \frac{\tilde{\mathbf{X}}^T\tilde{\mathbf{Y}}}{\sqrt{\tilde{\mathbf{X}}^T\tilde{\mathbf{X}}\tilde{\mathbf{Y}}^T\tilde{\mathbf{Y}}}}. \tag{5.244}$$

The bounds above follow directly from the Cauchy–Schwarz inequality for vectors (see Appendix G).

Definition: Spearman's Rank Correlation *Spearman's rank correlation coefficient* is

$$\tau_S \triangleq \frac{\sum_{n=1}^{N}(r_{X,n} - \bar{r}_X)(r_{Y,n} - \bar{r}_Y)}{\sqrt{\sum_{n=1}^{N}(r_{Y,n} - \bar{r}_X)^2 \sum_{n=1}^{N}(r_{Y,n} - \bar{r}_Y)^2}}, \tag{5.245}$$

where $\{r_{X,n}\}$ and $\{r_{Y,n}\}$ are the ranks (orderings) of samples $\{X_n\}$ and $\{Y_n\}$, respectively. The corresponding sample means of the ranks are $\bar{r}_X$ and $\bar{r}_Y$.

It is important to note that the actual values of the samples $\{X_n\}$ and $\{Y_n\}$ are *not* used to compute τ_S, only their rankings. Spearman's rank correlation is also bounded as $|\tau_S| \leq 1$, since the Cauchy–Schwarz inequality applies to the ratio of vectors described by (5.245), even though they contain rankings.

Before describing the last measure of correlation based on samples of random variables, we give the following definitions regarding how samples are ranked.

Definition: Concordant and Discordant Let $\{X_n, Y_n\}$ be samples of random variables X and Y. The pairs $\{X_n, Y_n\}$ and $\{X_m, Y_m\}$ $(n \neq m)$ are *concordant* if either

$$\{X_n < X_m \text{ and } Y_n < Y_m\} \quad \text{or} \quad \{X_n > X_m \text{ and } Y_n > Y_m\}. \tag{5.246}$$

They are *disconcordant* if either

$$\{X_n < X_m \text{ and } Y_n > Y_m\} \quad \text{or} \quad \{X_n > X_m \text{ and } Y_n < Y_m\}. \tag{5.247}$$

They are neither concordant nor disconcordant if

$$\{X_n = X_m\} \quad \text{or} \quad \{Y_n = Y_m\}. \tag{5.248}$$

In the last condition involving equal values called a *tie*, only one of the equalities is needed to eliminate the pairs as concordant or discordant. With N samples, the number of combinations of pairs (choosing 2 from N) is

$$\binom{N}{2} = \frac{N!}{2!(N-2)!} = N(N-1)/2. \tag{5.249}$$

Let N_c be the number of concordant pairs, N_d the number of discordant pairs, and define the difference $\Delta_{cd} \triangleq N_c - N_d$. We would also like to consider the possibility of tied values. Suppose there are N_X groups of tied values for $\{X_n\}$ and let $N_{X,m}$ be the number of values in the mth group. Obviously, $N_{X,m} \geq 2$: at least two values are needed to have a tie. We want to account for when more than two samples have the same value. Likewise, let $N_{Y,m}$ be the number of values in the mth group of ties for $\{Y_n\}$ where there are N_Y such groups. The next definition of correlation takes into account whether the samples are concordant or discordant.

Definition: Kendall's Rank Correlation *Kendall's rank correlation coefficient* is

$$\tau_K \triangleq \frac{\Delta_{cd}}{\left(\left[\frac{N(N-1)}{2} - \sum_{m=1}^{N_X} N_{X,m}(N_{X,m}-1)/2\right]\left[\frac{N(N-1)}{2} - \sum_{m=1}^{N_Y} N_{Y,m}(N_{Y,m}-1)/2\right]\right)^{1/2}}. \tag{5.250}$$

When there are no ties, Kendall's rank correlation reduces to

$$\tau_K = \frac{\Delta_{cd}}{N(N-1)/2} \tag{5.251}$$

whose denominator is the first term in the denominator of (5.250). We have written (5.250) in a form that is similar to Pearson's correlation coefficient, where the numerator is a *rank* correlation and the denominator provides normalization such that $|\tau_K| \leq 1$. The subtractions in the denominator are needed because Δ_{cd} does not account for ties; we must subtract the number of tied values from the number of combinations $N(N-1)/2$. For *continuous* random variables, we can generally use the simplified form in (5.251) because the probability of a tie is zero.

5.19 ORTHOGONALITY

Orthogonality is a property of two random variables that is useful for applications such as parameter estimation (Chapter 9) and signal estimation (Chapter 11).

Definition: Orthogonal Random variables X and Y are *orthogonal* if $\mathcal{E}[XY] = 0$.

This definition follows from a generalized form of orthogonal functions: two deterministic functions $g_1(x)$ and $g_2(x)$ are orthogonal if

$$\int_{-\infty}^{\infty} g_1(x)g_2(x)w(x)\mathrm{d}x = 0, \tag{5.252}$$

where $w(x)$ is a positive weighting function. In our case, the weighting function is the joint pdf of X and Y, and the integration is performed over two variables:

$$\int_{-\infty}^{\infty}\int_{-\infty}^{\infty} xyf_{X,Y}(x, y)\mathrm{d}x\mathrm{d}y = 0. \tag{5.253}$$

The connections between independence, uncorrelated, and orthogonal for two random variables are described in the following theorem.

Theorem 5.10. For random variables X and Y:

- Independence implies uncorrelated:

$$f_{X,Y}(x, y) = f_X(x)f_Y(y) \Longrightarrow \mathcal{E}[XY] = \mathcal{E}[X]\mathcal{E}[Y]. \tag{5.254}$$

- Uncorrelated and orthogonal are the same when at least one of the random variables has zero mean:

$$\mathcal{E}[XY] = \mathcal{E}[X]\mathcal{E}[Y] \text{ and } \mu_X = 0 \text{ or } \mu_Y = 0 \text{ (or both)} \Longrightarrow \mathcal{E}[XY] = 0. \tag{5.255}$$

Proof. The proofs follow immediately from the statements. □

Although uncorrelated and orthogonal are the same for zero-mean random variables, they are very different features. These two properties are often confused because in many applications, the random variables have zero mean, either because that is the nature of the problem or transformations were used to generate zero-mean random variables. From (5.253), we see that orthogonality depends on the structure of the pdf relative to the product xy: whether or not the function $xyf_{X,Y}(x, y)$ has an equal number of positive and negative regions over its support. From this observation, we find that when X and Y are both positive or both negative, they cannot be orthogonal.

On the other hand, since correlation is a measure of the linear dependence of two random variables, uncorrelated just means that no such dependence exists. This linear dependence is evident in Figure 5.13 when the correlation coefficient ρ is close to ± 1. In general, uncorrelated random variables are not independent. However, for some distributions they are also independent as demonstrated in the following examples.

Example 5.32. Let X and Y be jointly Gaussian random variables with the following bivariate pdf:

$$\begin{aligned} f_{X,Y}(x, y) = {} & \frac{1}{2\pi\sigma_X\sigma_Y\sqrt{1-\rho_{XY}^2}} \exp\left(-\left((x-\mu_X)^2/\sigma_X^2\right.\right. \\ & \left.\left. - 2\rho(x-\mu_X)(y-\mu_Y)/\sigma_X\sigma_Y + (y-\mu_Y)^2/\sigma_Y^2\right)/2(1-\rho_{XY}^2)\right). \end{aligned} \tag{5.256}$$

If X and Y are uncorrelated ($\rho_{XY} = 0$), then $f_{X,Y}(x, y)$ can be expressed as the product of two marginal pdfs:

$$f_{X,Y}(x, y) = \frac{1}{\sqrt{2\pi}\sigma_X} \exp\left(-(x - \mu_X)^2/2\sigma_X^2\right) \frac{1}{\sqrt{2\pi}\sigma_Y} \exp\left(-(y - \mu_Y)^2/2\sigma_Y^2\right),$$
$$= f_X(x) f_Y(y). \tag{5.257}$$

Although uncorrelated Gaussian random variables are *usually* independent, there are exceptions, but typically for artificially constructed scenarios as illustrated in Example 5.33.

Example 5.33. Let X be a Gaussian random variable with parameters $\{\mu = 0, \sigma\}$. Suppose we form $Y = ZX$, where Z is a symmetric Bernoulli random variable that is independent of X, with outcomes $\{-1, 1\}$ and $p = P(Z = 1) = 1 - q$. The cdf of Y is

$$F_Y(y) = P(Y \leq y|Z = 1)p + P(Y \leq y|Z = -1)q,$$
$$= P(X \leq y|Z = 1)p + P(X \geq -y|Z = -1)q. \tag{5.258}$$

The conditioning can be dropped in both terms because X and Z are independent:

$$F_Y(y) = P(X \leq y)p + P(X \geq -y)q = P(X \leq y) = F_X(y), \tag{5.259}$$

where we have used the fact that $f_X(x)$ is an even function. Thus, Y has exactly the same distribution as X. Obviously, X and Y are not independent: for $X = x$, Y must be $+x$ or $-x$. On the other hand, they are uncorrelated when $p = q = 1/2$:

$$C_{XY} = \mathcal{E}[XY] = \mathcal{E}[X^2 Z],$$
$$= \mathcal{E}[X^2]\mathcal{E}[Z] = \sigma^2(p - q) = 0. \tag{5.260}$$

Although X and Y are individually Gaussian, they are *not jointly* Gaussian. This is the reason why they can be uncorrelated but not independent.

Example 5.34. Next, we want to derive the joint pdf of random variables X and Y in Example 5.33. Starting with the joint cdf and assuming $p = 1/2$:

$$F_{X,Y}(x, y) = P(X \leq x, XZ \leq y|Z = 1)P(Z = 1) + P(X \leq x, XZ \leq y|Z = -1)P(Z = -1),$$
$$= (1/2)P(X \leq x, X \leq y) + (1/2)P(X \leq x, X \geq -y),$$
$$= (1/2)F_X(\min(x, y)) + (1/2)[F_X(x) - F_X(-y)]u(x + y), \tag{5.261}$$

where the unit-step function ensures that the second term is nonzero only for $x \geq -y$. The second term can also be written as

$$F_X(x) - F_X(-y) = \int_{-y}^{x} f_X(x)\mathrm{d}x, \tag{5.262}$$

where $f_X(x)$ is the standard Gaussian pdf. This cdf is quite different from the product $F_X(x)F_Y(y)$ for independent random variables. Figure 5.15 shows individual plots of the first and second terms above, as well as the overall cdf. Observe the "ridge" in the first plot at $x = y$ caused by $\min(x, y)$, and the abrupt flat region at $x = -y$ in the second plot due to $u(x + y)$. The function $F_X(x) - F_X(-y)$ would, of course, be negative when $x < -y$, but the unit-step function truncates it so that $F_X(x) - F_X(-y)$ is always nonnegative, thus creating

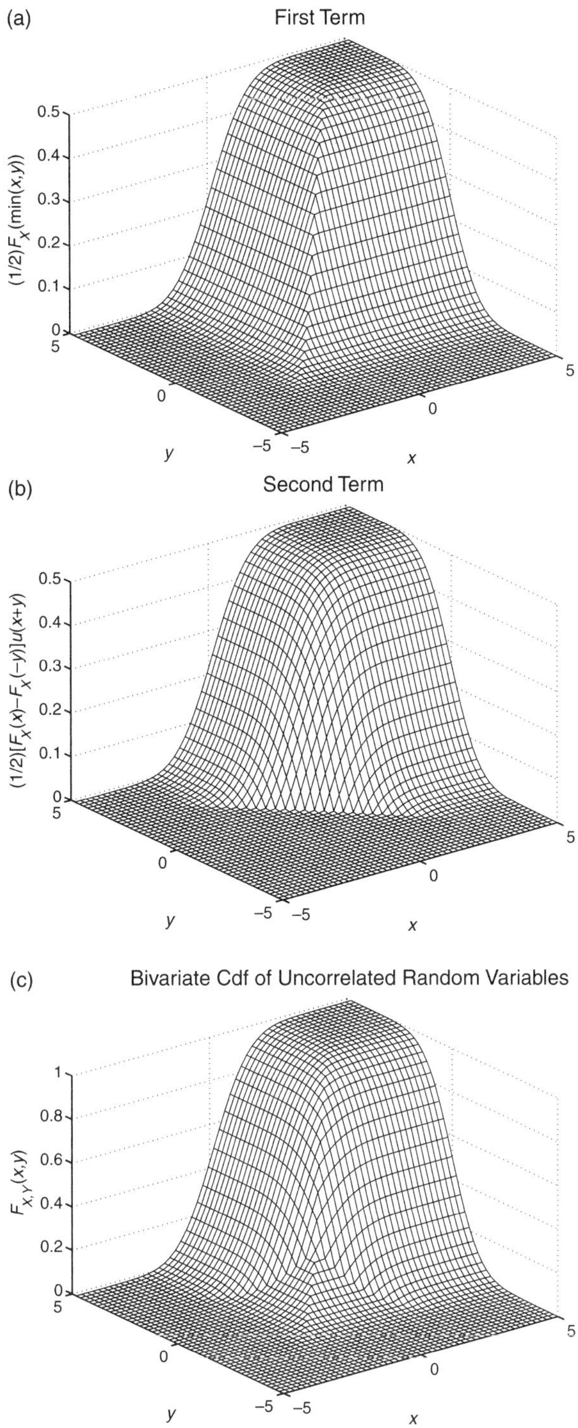

FIGURE 5.15 Bivariate cdf of uncorrelated random variables in Example 5.34. (a) First term: $(1/2)F_X(\min(x, y))$. (b) Second term: $(1/2)[F_X(x) - F_X(-y)]u(x + y)$. (c) Sum of (a) and (b).

the "floor" at zero in the plot. The second term in (5.261) can be rearranged as

$$F_X(x) - F_X(-y) = F_X(x) + F_X(y) - 1 \tag{5.263}$$

demonstrating that it is a symmetric function with respect to x and y. The pdf for (5.261) is derived using

$$\min(x, y) = (1/2)(x + y - |x - y|) \tag{5.264}$$

such that

$$\frac{\partial}{\partial x}\min(x, y) = (1/2)[1 - \operatorname{sgn}(x - y)], \qquad \frac{\partial^2}{\partial x \partial y}\min(x, y) = \delta(x - y), \tag{5.265}$$

where $d\operatorname{sgn}(y)/dy = 2\delta(y)$. The overall pdf includes delta functions along the $x = y$ line due to the first term of the cdf (see Problem 5.37).

5.20 CORRELATION AND COVARIANCE MATRICES

Consider random vector $\mathbf{X} = [X_1, \ldots, X_N]^T$ consisting of N random variables $\{X_n\}$.

Definition: Autocorrelation Matrix The *autocorrelation matrix* of random vector $\mathbf{X} \in \mathcal{R}^N$ is

$$\mathbf{R_{XX}} \triangleq \mathcal{E}[\mathbf{X}\mathbf{X}^T] = \begin{bmatrix} \mathcal{E}[X_1^2] & \mathcal{E}[X_1X_2] & \cdots & \mathcal{E}[X_1X_N] \\ \mathcal{E}[X_2X_1] & \mathcal{E}[X_2^2] & & \\ \vdots & \vdots & \ddots & \\ \mathcal{E}[X_NX_1] & \cdots & & \mathcal{E}[X_N^2] \end{bmatrix}. \tag{5.266}$$

The (m, n)th component is $\mathcal{E}[X_mX_n]$, where m is the row and n is the column.

Let $\mathbf{Y} = [Y_1, \ldots, Y_M]^T$ consist of M random variables $\{Y_m\}$.

Definition: Cross-Correlation Matrix The *cross-correlation matrix* of random vectors $\mathbf{X} \in \mathcal{R}^N$ and $\mathbf{Y} \in \mathcal{R}^M$ is

$$\mathbf{R_{XY}} \triangleq \mathcal{E}[\mathbf{X}\mathbf{Y}^T]. \tag{5.267}$$

Assume that the random variables of a random vector are mutually correlated to some degree. The elements of a random vector can be transformed via multiplication by a matrix to a new set of random variables that are uncorrelated. Consider the transformation $\mathbf{Y} = \mathbf{AX}$, where $\mathbf{A} \in \mathcal{R}^{N\times N}$ so that

$$\mathbf{R_{YY}} = \mathcal{E}[\mathbf{A}\mathbf{X}\mathbf{X}^T\mathbf{A}^T] = \mathbf{A}\mathbf{R_{XX}}\mathbf{A}^T. \tag{5.268}$$

The elements of $\mathbf{Y}$ are uncorrelated if $\mathbf{R_{YY}}$ is a diagonal matrix. This is achieved via the eigendecomposition (see Appendix G)

$$\mathbf{R_{XX}} = \mathbf{Q}\boldsymbol{\Lambda}\mathbf{Q}^T, \tag{5.269}$$

where the diagonal matrix $\boldsymbol{\Lambda}$ contains the eigenvalues of $\mathbf{R_{XX}}$. The columns of $\mathbf{Q}$ are the corresponding eigenvectors such that $\mathbf{Q}^T\mathbf{Q} = \mathbf{I}$, where $\mathbf{I} \in \mathcal{R}^{N\times N}$ is an identity matrix, and thus $\mathbf{Q}^{-1} = \mathbf{Q}^T$. By choosing

$\mathbf{A} = \mathbf{Q}^{-1}$, we have the desired diagonal result: $\mathbf{R_{YY}} = \mathbf{\Lambda}$. If

$$\mathbf{A} = \mathbf{Q}^{-1}\mathbf{\Lambda}^{-1/2}, \tag{5.270}$$

then random variables are transformed such that $\mathbf{R_{YY}} = \mathbf{I}$.

Definition: Covariance Matrix The *covariance matrix* of random vectors $\mathbf{X} \in \mathcal{R}^N$ and $\mathbf{Y} \in \mathcal{R}^M$ is

$$\begin{aligned}\mathbf{C_{XY}} &\triangleq \mathcal{E}[(\mathbf{X} - \boldsymbol{\mu}_\mathbf{X})(\mathbf{Y} - \boldsymbol{\mu}_\mathbf{Y})^T], \\ &= \begin{bmatrix} \mathcal{E}[(X_1 - \mu_{X_1})(Y_1 - \mu_{Y_1})] & \cdots & \mathcal{E}[(X_1 - \mu_{X_1})(Y_M - \mu_{Y_M})] \\ \vdots & \ddots & \vdots \\ \mathcal{E}[(X_N - \mu_{X_N})(Y_1 - \mu_{Y_1})] & \cdots & \mathcal{E}[(X_N - \mu_{X_N})(Y - \mu_{Y_M})] \end{bmatrix},\end{aligned} \tag{5.271}$$

where $\boldsymbol{\mu}_\mathbf{X} \triangleq \mathcal{E}[\mathbf{X}]$ and $\boldsymbol{\mu}_\mathbf{Y} \triangleq \mathcal{E}[\mathbf{Y}]$.

This expression is an extension of scalar covariance to several variables. A covariance matrix can be expressed in terms of a cross-correlation matrix as follows:

$$\mathbf{C_{XY}} = \mathbf{R_{XY}} - \boldsymbol{\mu}_\mathbf{X}\boldsymbol{\mu}_\mathbf{Y}^T. \tag{5.272}$$

Definition: Autocovariance Matrix The *autocovariance matrix* of random vector $\mathbf{X} \in \mathcal{R}^N$ is obtained from (5.271) when $\mathbf{Y} = \mathbf{X}$:

$$\mathbf{C_{XX}} \triangleq \mathcal{E}[(\mathbf{X} - \boldsymbol{\mu}_\mathbf{X})(\mathbf{X} - \boldsymbol{\mu}_\mathbf{X})^T], \tag{5.273}$$

where each diagonal element is a variance. Similar to (5.272):

$$\mathbf{C_{XX}} = \mathbf{R_{XX}} - \boldsymbol{\mu}_\mathbf{X}\boldsymbol{\mu}_\mathbf{X}^T. \tag{5.274}$$

A covariance matrix has the following properties:

- *Asymmetric.* $\mathbf{C_{XY}} = \mathbf{C}_\mathbf{YX}^T$. This follows from $\mathbf{C}_\mathbf{XY}^T = \mathcal{E}[((\mathbf{X} - \boldsymbol{\mu}_\mathbf{X})(\mathbf{Y} - \boldsymbol{\mu}_\mathbf{Y})^T)^T] = \mathcal{E}[(\mathbf{Y} - \boldsymbol{\mu}_\mathbf{Y})(\mathbf{X} - \boldsymbol{\mu}_\mathbf{X})^T] = \mathbf{C_{YX}} \neq \mathbf{C_{XY}}$.
- *Matrix scaling and vector translation.*

$$\text{cov}[\mathbf{AX} + \mathbf{a}, \mathbf{BY} + \mathbf{b}] = \mathbf{AC_{XY}B}^T \tag{5.275}$$

  for constant $\{\mathbf{A}, \mathbf{B}, \mathbf{a}, \mathbf{b}\}$ and random $\{\mathbf{X}, \mathbf{Y}\}$, all with appropriate dimensions. Since the mean values have changed to $\mathcal{E}[\mathbf{AX} + \mathbf{a}] = \mathbf{A}\boldsymbol{\mu}_\mathbf{X} + \mathbf{a}$ and $\mathcal{E}[\mathbf{BY} + \mathbf{b}] = \mathbf{B}\boldsymbol{\mu}_\mathbf{Y} + \mathbf{b}$, subtracting them from $\mathbf{AX} + \mathbf{a}$ and $\mathbf{BY} + \mathbf{b}$ gives

$$\mathcal{E}[(\mathbf{AX} - \mathbf{A}\boldsymbol{\mu}_\mathbf{X})(\mathbf{BY} - \mathbf{B}\boldsymbol{\mu}_\mathbf{Y})^T] = \mathbf{A}\mathcal{E}[(\mathbf{X} - \boldsymbol{\mu}_\mathbf{X})(\mathbf{Y} - \boldsymbol{\mu}_\mathbf{Y})^T]\mathbf{B}^T = \mathbf{AC_{XY}B}^T. \tag{5.276}$$

- *Addition of random vectors.*

$$\text{cov}[\mathbf{X} + \mathbf{Z}, \mathbf{Y}] = \mathbf{C_{XY}} + \mathbf{C_{ZY}} \tag{5.277}$$

for random $\{\mathbf{X}, \mathbf{Y}, \mathbf{Z}\}$ with appropriate dimensions. This follows from

$$\begin{aligned}\mathcal{E}[(\mathbf{X}+\mathbf{Z}-(\boldsymbol{\mu}_\mathbf{X}+\boldsymbol{\mu}_\mathbf{Z}))(\mathbf{Y}-\boldsymbol{\mu}_\mathbf{Y})^T] &= \mathcal{E}[(\mathbf{X}-\boldsymbol{\mu}_\mathbf{X})(\mathbf{Y}-\boldsymbol{\mu}_\mathbf{Y})^T]+\mathcal{E}[(\mathbf{Z}-\boldsymbol{\mu}_\mathbf{Z})(\mathbf{Y}-\boldsymbol{\mu}_\mathbf{Y})^T], \\ &= \mathbf{C}_{\mathbf{XY}}+\mathbf{C}_{\mathbf{ZY}}. \end{aligned} \tag{5.278}$$

These properties also hold for the cross-correlation matrix $\mathbf{R}_{\mathbf{XY}}$. The autocovariance matrix has the following additional properties:

- *Symmetric.* $\mathbf{C}_{\mathbf{XX}} = \mathbf{C}_{\mathbf{XX}}^T$.
- *Positive semidefinite (PSD).* For any vector $\mathbf{v} \neq \mathbf{0}$:

$$\mathbf{v}^T\mathbf{C}_{\mathbf{XX}}\mathbf{v} = \mathcal{E}[\mathbf{v}^T(\mathbf{X}-\boldsymbol{\mu}_\mathbf{X})(\mathbf{X}-\boldsymbol{\mu}_\mathbf{X})^T\mathbf{v}] = \mathcal{E}[(\mathbf{v}^T(\mathbf{X}-\boldsymbol{\mu}_\mathbf{X}))^2] \geq 0, \tag{5.279}$$

which means that all eigenvalues of $\mathbf{C}_{\mathbf{XX}}$ are nonnegative. If the matrix is nonsingular (has inverse $\mathbf{C}_{\mathbf{XX}}^{-1}$), then the eigenvalues are strictly positive and $\mathbf{C}_{\mathbf{XX}}$ is positive definite (PD) (strict inequality $>$ in (5.279) for any vector $\mathbf{v} \neq \mathbf{0}$).

These properties also hold for the autocorrelation matrix $\mathbf{R}_{\mathbf{XX}}$.

5.21 HIGHER ORDER MOMENTS AND CUMULANTS

In this section, we extend the definitions of mean and variance, which are first- and second-order moments, to higher order moments.

Definition: Moment The *n*th *moment* of random variable X is

$$\mu_n \triangleq \mathcal{E}[X^n]. \tag{5.280}$$

This type of expectation where the mean is not subtracted is also referred to as a *raw moment* (noncentral).

Example 5.35. For the Poisson random variable with parameter α, the second moment can be derived by first finding a closed form for the following expression:

$$\mathcal{E}[X(X-1)] = \exp(-\alpha)\sum_{n=0}^{\infty} n(n-1)\frac{\alpha^n}{n!} = \exp(-\alpha)\sum_{n=2}^{\infty}\frac{\alpha^n}{(n-2)!}, \tag{5.281}$$

which allows us to cancel terms in the denominator and change the lower limit to $n = 2$. We can factor α^2 and change variables in the sum to $m = n - 2$, yielding

$$\mathcal{E}[X(X-1)] = \alpha^2\sum_{m=0}^{\infty}\exp(-\alpha)\frac{\alpha^m}{m!} = \alpha^2, \tag{5.282}$$

where the last sum is over a valid pmf and is equal to 1. Substituting $\mathcal{E}[X] = \alpha$ yields $\mu_2 = \alpha + \alpha^2$ and $\sigma_X^2 = \mu_2 - (\mathcal{E}[X])^2 = \alpha + \alpha^2 - \alpha^2 = \alpha$.

Example 5.36. The second moment of the Cauchy distribution with parameters $\{c = 0, \alpha\}$ is

$$\mu_2 = (\alpha/\pi)\int_{-\infty}^{\infty}\frac{x^2}{x^2+\alpha^2}\mathrm{d}x = (\alpha/\pi)\int_{-\infty}^{\infty}\left(1-\frac{\alpha^2}{x^2+\alpha^2}\right)\mathrm{d}x, \tag{5.283}$$

which is rewritten this way to remove x^2 from the numerator. Thus

$$\mu_2 = (\alpha/\pi)\int_{-\infty}^{\infty} \mathrm{d}x - (\alpha^3/\pi)\int_{-\infty}^{\infty} \frac{1}{x^2+\alpha^2}\mathrm{d}x = \infty. \tag{5.284}$$

When integrating products of functions, one generally can use integration by parts. However, note for this example that a simple manipulation (the reverse of combining terms over a common denominator) gives two readily integrated terms. Since the second integrand is a Cauchy distribution scaled by α^2, it integrates to α^2. The first integral, of course, is infinity. So, even though the mean of the Cauchy distribution is not defined, the second (noncentral) moment is defined, but it happens to be infinity; since it is not useful, in Appendix A we say that it does not exist. This result does not imply that the Cauchy distribution has infinite width (recall that the variance is a measure of width); only that because of the heavy tails, the integral defining the second moment is not finite (recall the functional interpretation of the variance given earlier).

Example 5.37. Let $\{X_n\}$ be N iid random variables with mean μ_X and variance σ_X^2, and consider the second moment of the sample mean:

$$\begin{aligned}\mathcal{E}[\bar{X}^2] &= \frac{1}{N^2}\sum_{n=1}^{N}\sum_{m=1}^{N}\mathcal{E}[X_n X_m], \\ &= \frac{1}{N^2}\sum_{n=1}^{N}\sum_{m=1,m\neq n}^{N}\mathcal{E}[X_n]\mathcal{E}[X_m] + \frac{1}{N^2}\sum_{n=1}^{N}\mathcal{E}[X_n^2],\end{aligned} \tag{5.285}$$

where the sum of N^2 terms has been split into two sums: one for $m = n$ (N terms) and the other for $m \neq n$ ($(N^2 - N)$ terms). Independence has been used in the first sum to split the expectations into products. Thus

$$\mathcal{E}[\bar{X}^2] = (1/N^2)(N^2 - N)\mu_X^2 + (1/N^2)N(\sigma_X^2 + \mu_X^2) = \mu_X^2 + \sigma_X^2/N. \tag{5.286}$$

The expectation of the sample mean is

$$\mathcal{E}[\bar{X}] = \frac{1}{N}\sum_{n=1}^{N}\mathcal{E}[X_n] = \mu_X, \tag{5.287}$$

which demonstrates that the sample mean is an *unbiased estimator* of μ_X. Its variance is

$$\sigma_{\bar{X}}^2 = \mathcal{E}[\bar{X}^2] - \mathcal{E}^2[\bar{X}] = \sigma_X^2/N, \tag{5.288}$$

which shows that the quality of $\bar{X}$ as an estimator improves (i.e., decreasing variance) with an increasing number of samples N. An example of this result for Gaussian $\{X_n\}$ is shown later in Figure 9.8.

Definition: Central Moment The nth *central moment* of random variable X with mean μ_X is

$$\mu_{c,n} \triangleq \mathcal{E}[(X - \mu_X)^n]. \tag{5.289}$$

The variance is the second central moment: $\sigma_X^2 \equiv \mu_{c,2}$. The central moment of a random variable whose pdf is symmetric about the mean is zero for any odd order n. This is readily seen by examining the definition:

$$\mu_{c,n} = \int_{-\infty}^{\infty}(x - \mu_X)^n f_X(x)\mathrm{d}x. \tag{5.290}$$

Observe that $(x - \mu_X)^n$ is an even function for n even, and is an odd function for n odd. Since the product of an even function ($f_X(x)$ in this case) and an odd function is odd, the area above is necessarily zero for n odd (unless, of course, it does not exist as in the Cauchy distribution).

Theorem 5.11. The nth central moment of a Gaussian random variable X with parameters $\{\mu, \sigma\}$ is

$$\mu_{c,n} = \begin{cases} \sigma^n n!/2^{n/2}(n/2)!, & n \text{ even} \\ 0, & n \text{ odd.} \end{cases} \tag{5.291}$$

Proof. (Papoulis, 1965). The case of odd n was discussed above for a general distribution using a symmetry argument. Let $Y \triangleq X - \mu$ and rearrange the expression for the area of the Gaussian pdf as follows (see the pdf integrals in Appendix D):

$$\int_{-\infty}^{\infty} \exp\left(-y^2/2\sigma^2\right) dy = \sqrt{2\pi\sigma^2}. \tag{5.292}$$

Substitute $a \triangleq 1/2\sigma^2$:

$$\int_{-\infty}^{\infty} \exp\left(-ay^2\right) dy = \sqrt{\pi/a}, \tag{5.293}$$

and repeatedly differentiate both sides of this equation with respect to a. The results for $n = 2$, 4, and 6 differentiations are

$$\int_{-\infty}^{\infty} (-y^2) \exp\left(-ay^2\right) dy = a^{-3/2}(-1/2)\sqrt{\pi}, \tag{5.294}$$

$$\int_{-\infty}^{\infty} y^4 \exp\left(-ay^2\right) dy = a^{-5/2}(1/2)(3/2)\sqrt{\pi}, \tag{5.295}$$

$$\int_{-\infty}^{\infty} (-y^6) \exp\left(-ay^2\right) dy = a^{-7/2}(1/2)(3/2)(-5/2)\sqrt{\pi}. \tag{5.296}$$

Since the minus sign always cancels on both sides, and by observing the pattern with increasing n, we find that in general:

$$\int_{-\infty}^{\infty} y^{2n} \exp\left(-ay^2\right) dy = a^{-(2n+1)/2}(1/2)^n \sqrt{\pi} \prod_{m=1}^{n}(2m - 1). \tag{5.297}$$

The product is a type of factorial that includes only odd positive integers; it is known as a double factorial, and can be rewritten as

$$\prod_{m=1}^{n}(2m - 1) = (2n)!/2^n n!. \tag{5.298}$$

Substituting a gives

$$\begin{aligned} \int_{-\infty}^{\infty} y^{2n} \exp\left(-y^2/2\sigma^2\right) dy &= (2\sigma^2)^{(2n+1)/2}(1/2)^n \sqrt{\pi}(2n)!/2^n n!, \\ &= \sigma^{2n}\sqrt{2\pi\sigma^2}(2n)!/2^n n!. \end{aligned} \tag{5.299}$$

Since n is accompanied by 2 throughout the expression, we can define $m \triangleq 2n$ for m even. Dividing by $\sqrt{2\pi\sigma^2}$ completes the proof. □

Thus, all central Gaussian moments are proportional to the variance for n even:

$$\mu_{c,n} = [n!/2^{n/2}(n/2)!]\mu_{c,2}^{n/2} \qquad \text{(Gaussian)}. \tag{5.300}$$

The fourth central moment has the simple expression: $\mu_{c,4} = 3\mu_{c,2}^2 = 3\sigma^4$. The expressions for the raw moments of a Gaussian random variable are more complicated, but they reduce to (5.291) when $\mu = 0$.

Definition: Standardized Moment The nth *standardized moment* of random variable X is the nth central moment divided by σ_X^n:

$$\tilde{\mu}_{c,n} \triangleq \mu_{c,n}/\sigma_X^n. \tag{5.301}$$

The exponent of the denominator ensures that the standardized moment is dimensionless. There are two trivial cases, $\tilde{\mu}_{c,1} = 0$ and $\tilde{\mu}_{c,2} = 1$, and two important cases, *skewness* $\tilde{\mu}_{c,3}$ and *kurtosis* $\tilde{\mu}_{c,4}$. These are discussed after cumulants are defined.

Definition: Cumulant The nth *cumulant* of random variable X is

$$\kappa_n \triangleq c_X^{(n)}(0), \tag{5.302}$$

where $c_X^{(n)}(t)$ is the nth ordinary derivative of the cumulant generating function $c_X(t)$.

The cumulant generating function is defined later in this chapter. The mean and variance of X are the first two cumulants: $\mu_X \equiv \kappa_1 = c_X^{(1)}(0)$ and $\sigma_X^2 \equiv \kappa_2 = c_X^{(2)}(0)$. Although these two quantities are raw and central moments, respectively, higher order cumulants are generally more complicated and do not correspond exactly to either type of moment.

Skewness is a measure of the *asymmetry* of a distribution: whether more mass of a pmf or area of a pdf is concentrated to the left (positive skew) or right (negative skew) of "center." For unimodal distributions, the following relationships generally hold for the mean, mode m_o, and median m_e of X, but not always. Negative skew: $\mu_X < m_e < m_o$. Positive skew: $m_o < m_e < \mu_X$. For some distributions, it is possible for the mean and median to be interchanged in these expressions.

Definition: Skewness The *skewness* of random variable X is

$$\gamma_1 = \tilde{\mu}_{c,3} \triangleq \mu_{c,3}/\sigma_X^3. \tag{5.303}$$

It is also defined in terms of cumulants as follows:

$$\gamma_1 \triangleq \kappa_3/\kappa_2^{3/2}, \tag{5.304}$$

where the symbol γ_1 is often used.

For symmetric pdfs where the mean, median, and mode are all the same, such as a Gaussian distribution, the skewness is zero. Note from the distribution summaries in Appendix A that most of the random variables covered in this book have positive skewness; only a few have zero or negative skewness.

Example 5.38. Figure 5.16 shows three pdfs to illustrate the three types of skewness. The skewness of the log-normal pdf is

$$\gamma_1 = (\exp(\sigma^2) + 2)\sqrt{\exp(\sigma^2) - 1}, \tag{5.305}$$

which is positive for any $\sigma^2 > 0$. The binomial pdf has skewness

$$\gamma_1 = (1 - 2p)/\sqrt{Npq}, \tag{5.306}$$

which is negative when $p > 1/2$. The Laplace distribution obviously has zero skewness for any $\alpha > 0$.

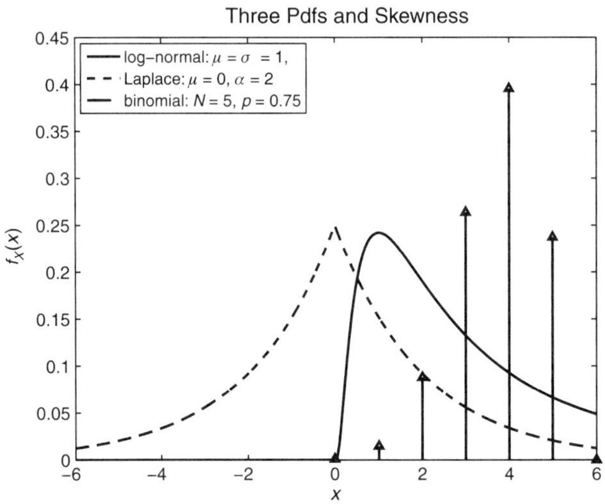

FIGURE 5.16 Examples of pdfs showing the three types of skewness: (i) positive $\gamma_1 \approx 6.18$ (log-normal), (ii) $\gamma_1 = 0$ (Laplace), and (iii) negative $\gamma_1 \approx -1.19$ (binomial).

Another standardized moment that is useful for characterizing a distribution is the kurtosis.

Definition: Kurtosis The *kurtosis* of random variable X is

$$\tilde{\mu}_{c,4} \triangleq \mu_{c,4}/\sigma_X^4, \tag{5.307}$$

where $\mu_{c,4}$ is the fourth central moment. A more common form known as the *excess kurtosis* is defined in terms of cumulants as follows:

$$\gamma_2 \triangleq \kappa_4/\kappa_2^2 = \tilde{\mu}_{c,4} - 3, \tag{5.308}$$

where the symbol γ_2 is often used.

The excess kurtosis is usually the preferred quantity because it is zero for a Gaussian random variable, and thus "excess" means relative to Gaussian. The kurtosis is considered to be a measure of how peaked or flat the pdf of a random variable is relative to that of a Gaussian pdf. However, from the functional view presented later, it is apparent that kurtosis is more important as a measure of *heavy tails* in a distribution (again, relative to the Gaussian distribution).

Example 5.39. Figure 5.17 compares the kurtosis of three pdfs. The excess kurtosis of the Rayleigh pdf is

$$\gamma_2 = \frac{24\pi - 16 - 6\pi^2}{(4-\pi)^2} \approx 0.2451, \tag{5.309}$$

indicating that the "heaviness" of its tails are not too different from that of the Gaussian pdf ($\gamma_2 = 0$). On the other hand, $\gamma_2 = 6$ for the exponential pdf. Table 3.4 summarizes the types of tails for several symmetric distributions covered in Chapter 3.

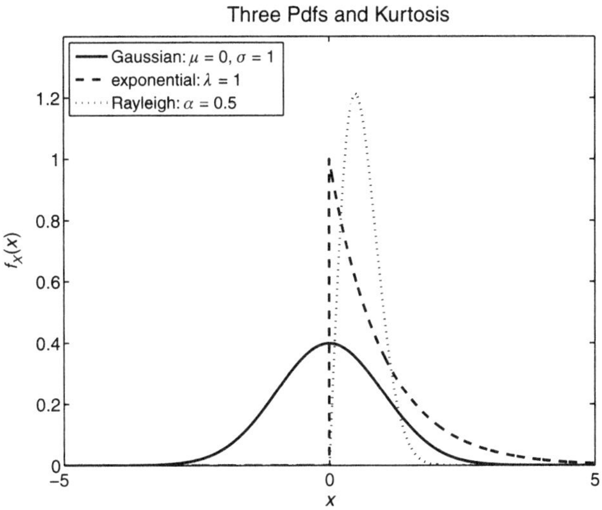

FIGURE 5.17 Examples of three pdfs illustrating different levels of kurtosis: (i) standard Gaussian ($\gamma_2 = 0$), (ii) exponential with $\lambda = 1$ ($\gamma_2 = 6$), and (iii) Rayleigh with $\alpha = 0.5$ ($\gamma_2 \approx 0.2451$).

As mentioned in Chapter 3 and at the beginning of this chapter, four moments are typically used to characterize the location and shape of a distribution: (i) mean μ_X (first raw moment), (ii) variance σ_X^2 (second central moment), (iii) skewness γ_1 (third standardized central moment), and (iv) excess kurtosis γ_2 (translated fourth standardized central moment). Higher order moments generally are not used to describe a distribution, though they have been considered in various signal processing applications. For notational convenience and for comparison purposes, we have included a summary of the definitions of moments, cumulants, and generating functions (presented later) in Table 5.5. As has been done earlier in this book, sometimes we do not include subscript X on the various quantities to keep the notation simple.

TABLE 5.5 Summary of Moments, Cumulants, and Generating Functions

Symbol	Definition
$\mu_n = \mathcal{E}[X^n] = m_X^{(n)}(0)$	nth raw moment
$\mu_X = \mu_1 = M_X^{(n)}(0)$	Mean (first raw moment)
$\mu_{c,n} = \mathcal{E}[(X - \mu_X)^n]$	nth central moment
$\sigma_X^2 = \mu_{c,2}$	Variance (second central moment)
$c_X = \sigma_X / \mu_X$	Coefficient of variation
$\tilde{\mu}_{c,n} = \mu_{c,n} / \sigma_X^n$	nth standardized central moment
$\kappa_n = c_X^{(n)}(0)$	nth cumulant
$\gamma_1 = \tilde{\mu}_{c,3} = \mu_{c,3}/\sigma_X^3 = \kappa_3/\kappa_2^{3/2}$	Skewness (standardized third central moment)
$\tilde{\mu}_{c,4} = \mu_{c,4}/\sigma_X^4$	Kurtosis (standardized fourth central moment)
$\gamma_2 = \kappa_4/\kappa_2^2 = \tilde{\mu}_{c,4} - 3$	Excess kurtosis
Generating Functions	**Type**
$m_X(t) = \mathcal{E}[\exp(tX)] = \sum_{n=0}^{\infty} t^n \mu_n / n!$	Moment generating function
$M_X(t) = \mathcal{E}[\exp(t(X - \mu_X))] = \sum_{n=0}^{\infty} t^n \mu_{c,n} / n!$	Central moment generating function
$c_X(t) = \ln(m_X(t)) = \sum_{n=1}^{\infty} t^n \kappa_n / n!$	Cumulant generating function
$\Phi_X(\omega) = \mathcal{E}[\exp(j\omega X)] = \sum_{n=1}^{\infty} (j\omega)^n \mu_n / n!$	Characteristic function
$G_X(t) = \mathcal{E}[t^X] = \sum_{x=0}^{\infty} p_X[x] t^x$	Probability generating function

5.22 FUNCTIONAL VIEW OF SKEWNESS

We can use a functional approach to show that skewness is a measure of the asymmetry of a distribution. Recall for a pdf that is symmetric about the mean, all odd central moments are zero; this occurs because $(X - \mu)^n$ is an odd function when n is odd, and thus its expectation is zero. When a pdf is less symmetric about its mean, odd central moments increase in value (except for the first central moment $\mathcal{E}[X - \mu]$ which is always zero). Consider the third-central moment:

$$\mu_{c,3} = \mathcal{E}[(X - \mu_X)^3] = \int_{-\infty}^{\infty} (x - \mu_X)^3 f_X(x)\mathrm{d}x. \tag{5.310}$$

Figure 5.18 shows examples of $(x - \mu_X)^3 f_X(x)$ for exponential and Rayleigh pdfs, both of which have unit variance so that $\mu_{c,3}$ is the same as skewness. We have also plotted the cumulative area

$$C(x) = \int_{-\infty}^{x} (v - \mu_X)^3 f_X(v)\mathrm{d}v \tag{5.311}$$

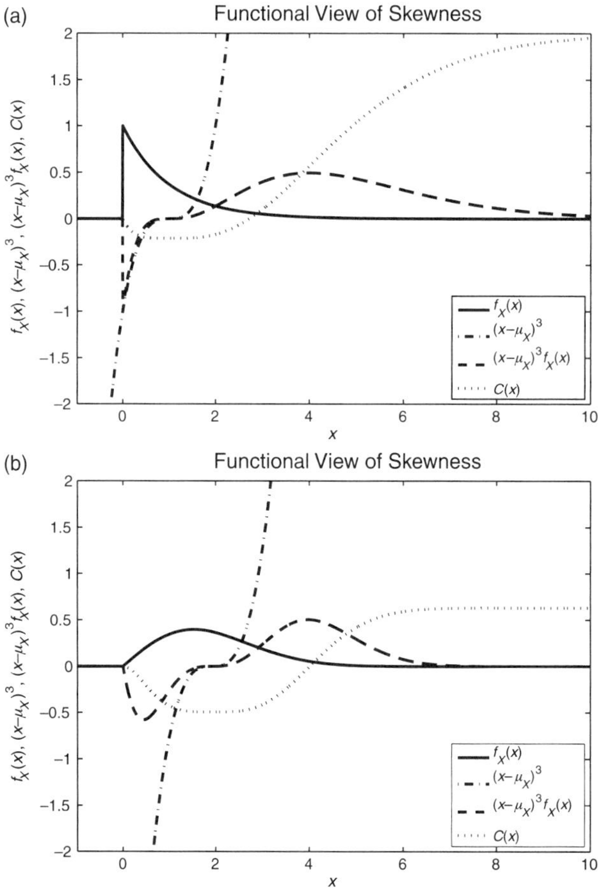

FIGURE 5.18 Functional view of skewness, showing $f_X(x)$, $(x - \mu_X)^3$, $(x - \mu_X)^3 f_X(x)$, and $C(x)$. (a) Exponential with $\lambda = 1$ ($\mu_X = 1$). (b) Rayleigh with $\alpha = \sqrt{2/(4 - \pi)}$ ($\mu_X = 1.9131$).

as x is varied over the support of $f_X(x)$, and where $\mathcal{E}[(X - \mu_X)^3] = \lim_{x\to\infty} C(x)$. Observe that the area under the dashed curve eventually becomes positive with increasing x because the pdf "leans" to the left, and thus tapers off more slowly to the right while $(x - \mu_X)^3$ increases dramatically. Both of these distribution have positive skewness. Other high-order odd central moments can be used to measure such asymmetry, but the third-order central moment is commonly used.

Example 5.40. In this example, we demonstrate how the exponential random variable with mean λ and variance $1/\lambda^2$ has constant skewness for any value of λ. From the definition:

$$\gamma_1 = \frac{\mathcal{E}[(X - 1/\lambda)^3]}{1/\lambda^3} = \mathcal{E}[(\lambda X - 1)^3]. \tag{5.312}$$

Note that $Y = \lambda X$ has the standard exponential distribution (with mean 1) so that the final expectation above is independent of λ. We leave it as an exercise for the reader to show that $\mathcal{E}[(Y - 1)^3] = 2$. From the summaries in Appendix A, we see that the exponential distribution is one of the few pdfs described in this book whose skewness is a nonzero integer, independent of its parameter.

5.23 FUNCTIONAL VIEW OF KURTOSIS

Finally, we provide a functional view of kurtosis demonstrating that it depends on the degree to which the pdf has heavy tails. The fourth-central moment is

$$\mu_{c,4} = \mathcal{E}[(X - \mu_X)^4] = \int_{-\infty}^{\infty} (x - \mu_X)^4 f_X(x)\mathrm{d}x. \tag{5.313}$$

Figure 5.19 shows an example of $(x - \mu_X)^4 f_X(x)$ for Laplace and Gaussian pdfs. We have also plotted the cumulative area

$$D(x) = \int_{-\infty}^{x} (v - \mu_X)^4 f_X(v)\mathrm{d}v \tag{5.314}$$

as x is varied over the support of $f_X(x)$, and where $\mathcal{E}[(X - \mu_X)^4] = \lim_{x\to\infty} D(x)$. In both cases, the variance has been set to one so that $\mu_{c,4}$ is the same as the kurtosis (subtracting three gives the excess kurtosis). Observe that the dashed line for the Laplace pdf has greater area and thus greater kurtosis than the Gaussian pdf. This occurs because the tails decrease more slowly, according to $\exp(-x)$ versus $\exp(-x^2)$, while $(x - \mu_X)^4$ increases rapidly as x is varied above and below the mean.

Example 5.41. From the summaries in Appendix A, we find that several distributions have an excess kurtosis that is independent of the pdf parameters. This result is easy to show for the Gaussian random variable which has fourth-central moment: $\mu_{c,4} = 3\sigma^2$, yielding $\gamma_2 = \mu_{c,4}/\sigma^2 - 3 = 0$. For the exponential distribution, we find a result similar to skewness in Example 5.40:

$$\frac{\mathcal{E}[(X - 1/\lambda)^4)}{1/\lambda^4} = \mathcal{E}[(\lambda X - 1)^4] = 9 \tag{5.315}$$

since λX is independent of λ. Subtracting three gives $\gamma_2 = 6$. Finally, consider the Laplace distribution with parameters $\{\mu, \alpha\}$ and variance $2\alpha^2$:

$$\frac{\mathcal{E}[(X - \mu)^4]}{(2\alpha^2)^2} = \frac{\mathcal{E}[Y^4]}{4\alpha^4} = \frac{1}{8\alpha^5}\int_{-\infty}^{\infty} y^4 \exp(-|y|/2\alpha)\mathrm{d}y \tag{5.316}$$

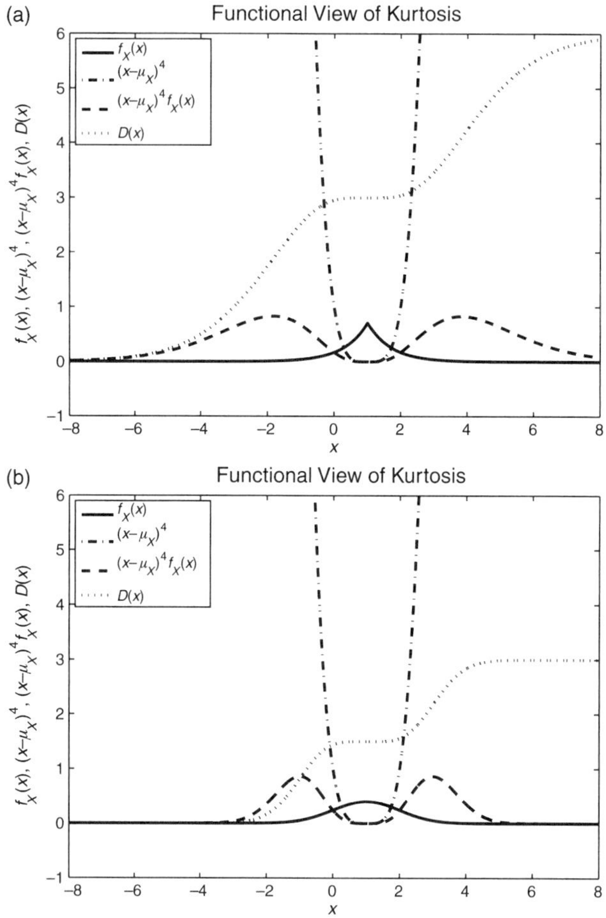

FIGURE 5.19 Functional view of kurtosis, showing $f_X(x)$, $(x-\mu_X)^4$, $(x-\mu_X)^4 f_X(x)$, and $D(x)$. (a) Laplace with $\{\mu = 1, \alpha = \sqrt{2}\}$. (b) Gaussian with $\{\mu = 1, \sigma = 1\}$.

where Y is Laplacian with parameters $\{\mu = 0, \alpha\}$. Changing variables to $z = y/\alpha$ gives:

$$\gamma_2 = (1/8)\int_{-\infty}^{\infty} z^4 \exp(-|z|/2)\mathrm{d}z - 3 = 3. \tag{5.317}$$

Obviously for those random variables with a kurtosis that is independent of the scale parameter, there is a connection between the fourth central moment $\mu_{c,4}$ and the normalization in the denominator given by the variance squared. This connection was explicitly shown for the Gaussian random variable in Theorem 5.11.

5.24 GENERATING FUNCTIONS

In this section, we consider expectations similar to the CF that can also be used to generate moments as well as cumulants.

Definition: Moment Generating Function The *moment generating function (MGF)* of random variable X is

$$m_X(t) \triangleq \mathcal{E}[\exp(tX)] = \sum_{n=0}^{\infty} \frac{t^n \mathcal{E}[X^n]}{n!} = \sum_{n=0}^{\infty} \frac{t^n \mu_n}{n!} \tag{5.318}$$

for $t \in \mathcal{R}$ and with $\mu_0 = 1$. The nth moment is generated as follows:

$$\mu_n = m_X^{(n)}(0), \tag{5.319}$$

where $m_X^{(n)}(t)$ is the nth ordinary derivative of $m_X(t)$ with respect to t.

The MGF does not always exist whereas the CF usually does. Recall that the CF includes $j = \sqrt{-1}$ in the exponent of the exponential $\Phi_X(\omega) = \mathcal{E}[\exp(j\omega X)]$, and is similar to the Fourier transform. The moment-generating property of $m_X(t)$ is easily verified by expanding the last expression in (5.318):

$$m_X(t) = 1 + t\mu_1 + \frac{t^2\mu_2}{2!} + \frac{t^3\mu_3}{3!} + \cdots. \tag{5.320}$$

For example, differentiating twice gives

$$m_X^{(2)}(t) = 0 + 0 + \mu_2 + t\mu_3 + \cdots. \tag{5.321}$$

Obviously, terms subsequent to $t\mu_3$ can be nonzero for $t \neq 0$ because of the higher powers of t. Setting $t = 0$ in this expression removes those terms, leaving only the second moment because that term is not a function of t:

$$m_X^{(2)}(t)\Big|_{t=0} = \mu_2. \tag{5.322}$$

Example 5.42. Let X be the standard Gaussian random variable. The transformation $Y = \sigma X + \mu$ yields another Gaussian random variable with parameters $\{\mu, \sigma\}$. The MGF of Y is

$$\begin{aligned} m_X(t) &= \mathcal{E}[\exp(t(\sigma X + \mu))], \\ &= \exp(t\mu) \int_{-\infty}^{\infty} \exp(t\sigma x) \frac{1}{\sqrt{2\pi}} \exp\left(-x^2/2\right) \mathrm{d}x, \\ &= \exp(t\mu) \frac{1}{\sqrt{2\pi}} \int_{-\infty}^{\infty} \exp\left((2t\sigma x - x^2)/2\right) \mathrm{d}x. \end{aligned} \tag{5.323}$$

Completing the square gives $2t\sigma x - x^2 = (x - t\sigma)^2 - t^2\sigma^2$. Substituting this result into the expression above and recognizing that the integral is the area under a Gaussian pdf with parameters $\{\mu = 0, \sigma\}$ yields

$$\begin{aligned} m_X(t) &= \exp(t\mu) \frac{1}{\sqrt{2\pi}} \int_{-\infty}^{\infty} \exp\left(t^2\sigma^2/2\right) \exp\left(-(x - t\sigma)^2/2\right) \mathrm{d}x, \\ &= \exp\left(t\mu + t^2\sigma^2/2\right). \end{aligned} \tag{5.324}$$

Observe that

$$m_X^{(1)}(0) = (\mu + t\sigma^2)\exp(t\mu + t^2\sigma^2/2)\big|_{t=0} = \mu = \mu_1 \tag{5.325}$$

and

$$\begin{aligned} m_X^{(2)}(0) &= \sigma^2 \exp(t\mu + t^2\sigma^2/2)\big|_{t=0} + (\mu + t\sigma^2)^2 \exp(t\mu + t^2\sigma^2/2)\big|_{t=0}, \\ &= \sigma^2 + \mu^2 = \mu_2. \end{aligned} \tag{5.326}$$

Examples of the MGF and its derivatives are shown in Figure 5.20. The moments are obtained from the plots by locating where the derivative curves cross $t = 0$. We can conclude from the first plot for the standard Gaussian variable (with zero mean) that the odd moments are zero because every odd-order derivative is necessarily an odd function. This follows from (5.291) for the Gaussian central moments, which coincide with the raw moment results in the first plot because $\mu = 0$.

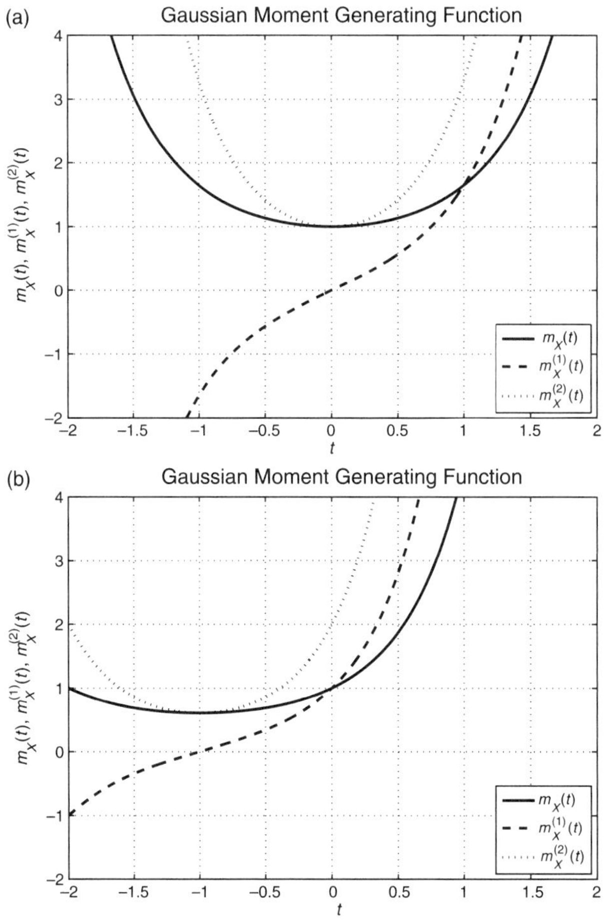

FIGURE 5.20 Gaussian moment generating function (MGF) and its derivatives for Example 5.42. (a) $\mu = 0$ and $\sigma = 1$. (b) $\mu = 1$ and $\sigma = 1$

Example 5.43. Let X be a binomial random variable with parameters $\{N, p\}$ (and $q \triangleq 1 - p$). Its MGF is

$$m_X(t) = \sum_{n=0}^{N} \binom{N}{n} p^n q^{N-n} \exp(tn),$$
$$= \sum_{n=0}^{N} \binom{N}{n} [p \exp(t)]^n q^{N-n} = [p \exp(t) + q]^N, \tag{5.327}$$

where the binomial formula in Appendix E has been used. Observe that

$$m_X^{(1)}(0) = N[p\exp(t) + q]^{N-1} p \exp(t)\big|_{t=0} = Np = \mu_1 \tag{5.328}$$

and

$$m_X^{(2)}(0) = N(N-1)[p\exp(t)+q]^{N-2} p^2 \exp(t)\big|_{t=0} + N\,[p\exp(t)+q]^{N-1} p\exp(t)\big|_{t=0},$$
$$= p^2 N(N-1) + Np = N^2p^2 + Npq = \mu_2. \tag{5.329}$$

The following extension of the MGF is perhaps not as well known.

Definition: Central Moment Generating Function The *central moment generating function (CMGF)* of random variable X is

$$M_X(t) \triangleq \mathcal{E}[\exp(t(X - \mu_X))] = \sum_{n=0}^{\infty} \frac{t^n \mathcal{E}[(X-\mu_X)^n]}{n!} = \sum_{n=0}^{\infty} \frac{t^n \mu_{c,n}}{n!} \tag{5.330}$$

for $t \in \mathcal{R}$ and with $\mu_{c,0} = 1$. The nth central moment is generated as follows:

$$\mu_{c,n} = M_X^{(n)}(0). \tag{5.331}$$

The CMGF can also be written in terms of the MGF:

$$M_X(t) = \exp(-\mu_X t)\mathcal{E}[\exp(tX)] = \exp(-\mu_X t) m_X(t). \tag{5.332}$$

The following generating function is used to derive cumulants.

Definition: Cumulant Generating Function The *cumulant generating function (CGF)* of random variable X is the logarithm of the MGF:

$$c_X(t) \triangleq \ln(\mathcal{E}[\exp(tX)]) = \ln(m_X(t)) = \sum_{n=1}^{\infty} \frac{t^n \kappa_n}{n!}. \tag{5.333}$$

The mth cumulant is obtained by differentiating $c_X(t)$ m times with respect to t and setting $t = 0$ in the result:

$$\kappa_n = c_X^{(m)}(0). \tag{5.334}$$

Note that the lower limit of the sum is 1 (because of the logarithm), whereas it is zero for the previous generating functions.

Example 5.44. The logarithm of the MGF for the Gaussian random variable in Example 5.42 is

$$c_X(t) = t\mu + t^2\sigma^2/2. \tag{5.335}$$

The cumulants are easily obtained as follows:

$$\kappa_1 = c_X^{(1)}(0) = \mu + t\sigma^2\big|_{t=0} = \mu, \tag{5.336}$$

$$\kappa_2 = c_X^{(2)}(0) = \sigma^2, \tag{5.337}$$

$$\kappa_n = 0, \quad n \geq 3. \tag{5.338}$$

As mentioned earlier, the first and second cumulants for any distribution are the mean and variance, but higher order cumulants are not necessarily zero as is the case for a Gaussian random variable.

Example 5.45. Consider the exponential random variable with MGF:

$$m_X(t) = \lambda \int_0^\infty \exp(tx)\exp(-\lambda x)\mathrm{d}x = \frac{\lambda}{\lambda - t}, \tag{5.339}$$

which has CGF:

$$c_X(t) = \ln(\lambda) - \ln(\lambda - t). \tag{5.340}$$

Since

$$c_X^{(n)}(t) = (n-1)!/(\lambda - t)^n, \tag{5.341}$$

the cumulants have the following simple form:

$$\kappa_n = (n-1)!/\lambda^n. \tag{5.342}$$

Finally, we present a generating function that applies only to discrete random variables.

Definition: Probability Generating Function The *probability generating function (PGF)* of discrete random variable X is

$$G_X(t) \triangleq \mathcal{E}[t^X] = \sum_{x=0}^{\infty} p_X[x]t^x \tag{5.343}$$

for $t \in \mathcal{R}$. Probabilities are obtained by differentiating $G_X(t)$ with respect to t and setting $t = 0$ as follows:

$$P(X = m) = G_X^{(m)}(0)/m!. \tag{5.344}$$

The final result includes $m!$ in the denominator because the first nonzero term after differentiating m times is $m!p_X[m]$. The following properties are easily verified:

- *Sum of probabilities.* $G_X(1) = 1$ (i.e., $p_X[x]$ is a valid pmf).
- *Characteristic function (CF).* $\Phi_X(j\omega) = G_X(\exp(j\omega))$.
- *Expectation.* $\mathcal{E}[X] = G_X^{(1)}(1)$
- *Moment generating function (MGF).* $m_X(t) = G_X(\exp(t))$.

- *Factorial moment generating function (FMGF).* $\mathcal{E}[(X)_n] = G_X^{(n)}(1)$ where $(X)_n$ is a falling factorial (see Appendix B).
- *Variance.* $\sigma_X^2 = G_X^{(2)}(1) + G_X^{(1)}(1)[1 - G_X^{(1)}(1)]$.
- *Independent X and Y.* $G_{X+Y}(t) = G_X(t)G_Y(t)$.

Note that $G_X(t)$ used for the FMGF also applies for continuous random variables, assuming the expectation exists for $t = 1$.

Example 5.46. For the geometric random variable:

$$G_X(t) = \sum_{x=0}^{\infty} pq^x t^x = \frac{p}{1-qt} \tag{5.345}$$

for which we verify that $G(1) = 1$ and

$$\mathcal{E}[X] = \left.\frac{pq}{(1-qt)^2}\right|_{t=1} = q/p, \tag{5.346}$$

$$P(X=1) = \left.\frac{pq}{(1-qt)^2}\right|_{t=0} = pq. \tag{5.347}$$

5.25 FOURTH-ORDER GAUSSIAN MOMENT

A fourth-order moment for Gaussian random variables is provided in the following theorem, which is useful later in Chapter 12 on adaptive filtering.

Theorem 5.12. Let $\{X_1, X_2, X_3, X_4\}$ be zero-mean jointly Gaussian random variables. Their joint moment is

$$\mathcal{E}[X_1X_2X_3X_4] = \mathcal{E}[X_1X_2]\mathcal{E}[X_3X_4] + \mathcal{E}[X_1X_3]\mathcal{E}[X_2X_4] + \mathcal{E}[X_1X_4]\mathcal{E}[X_2X_3]. \tag{5.348}$$

Proof. Define the Gaussian random vector $\mathbf{X} \triangleq [X_1, X_2, X_3, X_4]^T$ whose CF is given by (5.109) after substituting $\mathbf{R_{XX}}$ and $\boldsymbol{\mu} = \mathbf{0}$:

$$\Phi_{\mathbf{X}}(\boldsymbol{\omega}) = \exp\left(-\boldsymbol{\omega}^T \mathbf{R}_{\mathbf{XX}}^{-1}\boldsymbol{\omega}\right), \tag{5.349}$$

where $\boldsymbol{\omega} = [\omega_1, \omega_2, \omega_3, \omega_4]^T$. Because $\mathbf{X}$ has zero mean, this function is real valued (it is not a function of j as are most CFs). Typically such exponentials can be analyzed using a Maclaurin series:

$$\Phi_{\mathbf{X}}(\boldsymbol{\omega}) = 1 + (1/2)\boldsymbol{\omega}^T\mathbf{R_{XX}}\boldsymbol{\omega} + (1/8)(\boldsymbol{\omega}^T\mathbf{R_{XX}}\boldsymbol{\omega})^2 + \cdots. \tag{5.350}$$

where $\cdots$ represent high-order terms. The fourth moment is computed as follows:

$$\mathcal{E}[X_1X_2X_3X_4] = \left.\frac{\partial^4 \Phi_{\mathbf{X}}(\boldsymbol{\omega})}{\partial\omega_1\partial\omega_2\partial\omega_3\partial\omega_4}\right|_{\boldsymbol{\omega}=\mathbf{0}}, \tag{5.351}$$

which after differentiation and substituting $\boldsymbol{\omega} = \mathbf{0}$, only the third term in (5.350) remains:

$$\mathcal{E}[X_1X_2X_3X_4] = (1/8)\left.\frac{\partial^4(\boldsymbol{\omega}^T\mathbf{R_{XX}}\boldsymbol{\omega})^2}{\partial\omega_1\partial\omega_2\partial\omega_3\partial\omega_4}\right|_{\boldsymbol{\omega}=\mathbf{0}}. \tag{5.352}$$

The autocorrelation matrix $\mathbf{R_{XX}}$ has 16 elements of the form $\mathcal{E}[X_m X_n]$ for $m, n = 1, 2, 3, 4$ so that $(\boldsymbol{\omega}^T \mathbf{R_{XX}} \boldsymbol{\omega})^2$ has $16^2 = 256$ terms given by

$$(\boldsymbol{\omega}^T \mathbf{R_{XX}} \boldsymbol{\omega})^2 = \sum_{k=1}^{4} \sum_{l=1}^{4} \sum_{m=1}^{4} \sum_{n=1}^{4} \omega_k \omega_l \omega_m \omega_n \mathcal{E}[X_k X_l] \mathcal{E}[X_m X_n]. \tag{5.353}$$

However, because $\boldsymbol{\omega} = \mathbf{0}$ is substituted after differentiation, only those terms where the subscripts of $\{\omega_k, \omega_l, \omega_m, \omega_n\}$ are all different remain because the other terms have one or more ω with an exponent exceeding 1. There are $4! = 24$ such nonzero terms, but only three are unique with each occurring eight times. Thus, the factor $1/8$ in (5.352) cancels and we finally have

$$\mathcal{E}[X_1 X_2 X_3 X_4] = \mathcal{E}[X_1 X_2] \mathcal{E}[X_3 X_4] + \mathcal{E}[X_1 X_3] \mathcal{E}[X_2 X_4] + \mathcal{E}[X_1 X_4] \mathcal{E}[X_2 X_3]. \tag{5.354}$$

□

It is convenient that the fourth-order moment reduces to a sum of second-order moments for zero-mean Gaussian random variables. In fact, it can be shown that all even-order moments for jointly Gaussian random variables with zero mean reduce to a function of their raw second moments, and that the odd-order moments are all zero (similar to that shown previously for a single Gaussian random variable). When the $\{X_n\}$ have the same variance σ^2 (or when they are samples of the same random variable X), (5.348) reduces to

$$\mathcal{E}[X_1 X_2 X_3 X_4] = 3\sigma^2, \tag{5.355}$$

which is the result in (5.291) for $n = 2$ assuming zero mean.

5.26 EXPECTATIONS OF NONLINEAR TRANSFORMATIONS

Generally, it is difficult to obtain a closed-form expression for the expectation of nonlinear functions of several random variables. However, if the random variables are jointly Gaussian with nonzero cross-correlation, then it is possible to compute useful expectations when the nonlinear functions are *memoryless*. This is convenient because memoryless nonlinearities often arise in engineering problems, and Gaussian noise is inevitably present. The following theorem provides an interesting result relating the derivative of a cross-correlation function to derivatives of the nonlinear functions. Thus, if differentiating a nonlinear function leads to a simpler expression, the expectation is derived more easily. Although the theorem has been proved in general for N jointly Gaussian random variables, we provide the result only for the case of $N = 2$, and without proof (Price, 1958; McMahon, 1964; Pawula, 1967).

Theorem 5.13 (Price and an extension). Let X_1 and X_2 be jointly Gaussian random variables with correlation coefficient ρ, and suppose they are transformed via memoryless nonlinear functions: $g_1(X_1)$ and $g_2(X_2)$. Define the following cross-correlation of the transformed random variables:

$$R_g \triangleq \mathcal{E}[g_1(X_1) g_2(X_2)]. \tag{5.356}$$

Then

$$\frac{\partial^n R_g}{\partial \rho^n} = \mathcal{E}\left[\frac{\partial^n g_1(X_1)}{\partial X_1^n} \frac{\partial^n g_2(X_2)}{\partial X_2^n}\right]. \tag{5.357}$$

For the more general function $h(X_1, X_2)$ and cross-correlation

$$R_h \triangleq \mathcal{E}[h(X_1, X_2)] \tag{5.358}$$

differentiating yields

$$\frac{\partial^n R_h}{\partial \rho^n} = \mathcal{E}\left[\frac{\partial^{2n} h(X_1, X_2)}{\partial X_1^n \partial X_2^n}\right]. \tag{5.359}$$

Example 5.47. Consider the "hard-limiting" nonlinearity given by the signum function:

$$g_1(x) = g_2(x) = \text{sgn}(x). \tag{5.360}$$

The first derivative is the Dirac delta function:

$$\frac{\partial}{\partial x} g_m(x) = 2\delta(x). \tag{5.361}$$

Suppose the zero-mean random variables $\{X_1, X_2\}$ have unit variance. Inserting (5.361) into (5.357) gives

$$\begin{aligned}\frac{\partial R_g}{\partial \rho} &= 4\mathcal{E}[\delta(X_1)\delta(X_2)], \\ &= \frac{4}{2\pi\sqrt{1-\rho^2}} \int_{-\infty}^{\infty}\int_{-\infty}^{\infty} \delta(x_1)\delta(x_2)\exp\left(-[x_1^2 - 2\rho x_1 x_2 + x_2^2]/2(1-\rho^2)\right) \mathrm{d}x_1 \mathrm{d}x_2 \\ &= \frac{2}{\pi\sqrt{1-\rho^2}},\end{aligned} \tag{5.362}$$

where the bivariate Gaussian pdf has been substituted. Observe that the delta functions have greatly simplified the double integral because the sifting property causes the exponential function to be evaluated only at $x_1 = x_2 = 0$. This example illustrates the usefulness of the theorem: differentiating some nonlinear functions (especially those containing unit-step and signum functions) can result in much simpler forms to evaluate under the expectation. Finally, R_g is obtained by integrating (5.362) with respect to ρ, which requires a boundary condition to determine the lower limit of integration. When $\rho = 0$ and $\{X_1, X_2\}$ are uncorrelated, $R_g = 0$ because the expectation in (5.356) splits, and each term is zero because the nonlinear function is odd. Thus

$$\int_0^{R_g} \mathrm{d}R_g = \int_0^{\rho} \frac{2}{\pi\sqrt{1-\rho^2}} \mathrm{d}\rho \tag{5.363}$$

and

$$R_g = \mathcal{E}[\text{sgn}(X_1)\text{sgn}(X_2)] = (2/\pi)\sin^{-1}(\rho), \tag{5.364}$$

which is the desired expectation. This result is plotted in Figure 5.21 versus the correlation coefficient ρ, where $|R_g| \leq 1$ as expected because of the signum functions.

Example 5.48. For the same conditions on the random variables as in Example 5.47, consider the more general nonlinear function

$$h(x_1, x_2) = |x_1 + x_2| - |x_1 - x_2|, \tag{5.365}$$

which corresponds to a "rectifier." Since this function cannot be split into the product of two functions, the theorem extension in (5.359) is used. Differentiating yields

$$\frac{\partial h(x_1, x_2)}{\partial x_1} = \text{sgn}(x_1 + x_2) - \text{sgn}(x_1 - x_2) \tag{5.366}$$

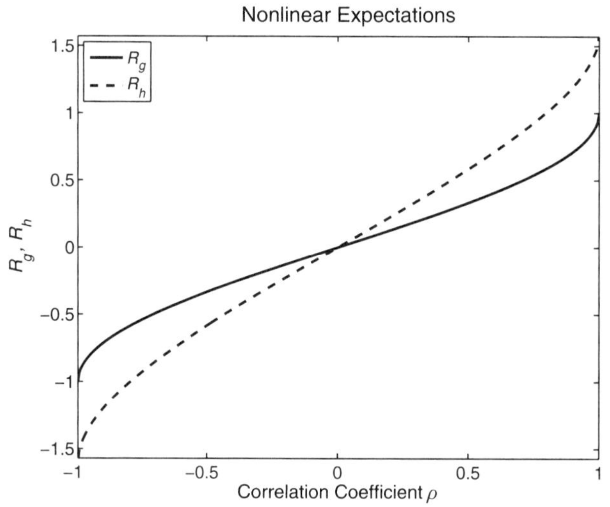

FIGURE 5.21 Nonlinear expectations in (5.364) and (5.372) of Examples 5.47 and 5.48 versus the correlation coefficient ρ.

and

$$\frac{\partial^2 h(x_1, x_2)}{\partial x_1 \partial x_2} = 2[\delta(x_1 + x_2) - \delta(x_1 - x_2)]. \tag{5.367}$$

Again, the delta functions greatly simplify the expectation:

$$\begin{aligned}\mathcal{E}\left[\frac{\partial^2 h(X_1, X_2)}{\partial X_1 \partial X_2}\right] &= \frac{1}{\pi\sqrt{1-\rho^2}} \int_{-\infty}^{\infty} \int_{-\infty}^{\infty} [\delta(x_1 + x_2) - \delta(x_1 - x_2)] \\ &\quad \cdot \exp\left(-(x_1^2 - 2\rho x_1 x_2 + x_2^2)/2(1-\rho^2)\right) \mathrm{d}x_1 \mathrm{d}x_2, \\ &= \frac{1}{\pi\sqrt{1-\rho^2}} \int_{-\infty}^{\infty} \left[\exp\left(-x_1^2/(1+\rho)\right) + \exp\left(-x_1^2/(1-\rho)\right)\right] \mathrm{d}x_1. \end{aligned} \tag{5.368}$$

The integrand consists of two Gaussian pdfs, both with zero mean but different variances: $(1-\rho)/2$ and $(1+\rho)/2$. Using these for σ^2 in the Gaussian integral

$$\int_{-\infty}^{\infty} \exp(-x^2/2\sigma^2) \mathrm{d}x = \sqrt{2\pi\sigma^2}, \tag{5.369}$$

we obtain

$$\mathcal{E}\left[\frac{\partial^2 h(X_1, X_2)}{\partial X_1 \partial X_2}\right] = \frac{\sqrt{\pi(1+\rho)} + \sqrt{\pi(1-\rho)}}{\pi\sqrt{1-\rho^2}} = \frac{1}{\sqrt{\pi}}\left[\frac{1}{\sqrt{1-\rho}} + \frac{1}{\sqrt{1+\rho}}\right]. \tag{5.370}$$

Finally, with boundary condition $R_h = 0$ for $\rho = 0$, this expression is integrated with respect to ρ as follows:

$$\int_0^{R_h} \mathrm{d}R_h = \frac{1}{\sqrt{\pi}} \int_0^{\rho} \left[\frac{1}{\sqrt{1-\rho}} + \frac{1}{\sqrt{1+\rho}}\right] \mathrm{d}\rho, \tag{5.371}$$

which gives the desired expectation:

$$R_h = \mathcal{E}[|X_1 + X_2| - |X_1 - X_2|] = \frac{2}{\sqrt{\pi}}\left[\sqrt{1+\rho} - \sqrt{1-\rho}\right]. \tag{5.372}$$

This result is also plotted in Figure 5.21 versus the correlation coefficient ρ. It is interesting that $|R_h| \approx 1.5958$ when the random variables are perfectly correlated with $|\rho| = 1$, which is less than $|R_h| = 2$ when $X_1 = \pm X_2$. This example demonstrates that perfectly correlated and equality (with probability one) are different properties of random variables.

PROBLEMS

Simple Random Variable

5.1 Let X be a Laplace random variable with parameters $\{\mu = 0, \alpha\}$. Define the mapping

$$Y \triangleq \begin{cases} 1, & |X| \le 1 \\ 1/2, & |X| > 1 \end{cases}. \tag{5.373}$$

Write an expression for simple random variable Y and give its pmf.

5.2 Demonstrate that the binomial random variable is a simple random variable.

5.3 Partition the real line $\Omega = \mathcal{R}$ into the following three intervals: $(-\infty, 0]$, $(0, 2]$, and $(2, \infty)$, and define

$$Y \triangleq \begin{cases} -1, & X \le 0 \\ 0, & 0 < X \le 2 \\ 1, & X > 2, \end{cases} \tag{5.374}$$

where X is the standard Gaussian random variable. Write an expression for the simple random variable Y and give its pmf.

Expectation

5.4 (a) Show that if X is a nonnegative discrete random variable with $\mathcal{E}[X] = 0$, then $P(X = 0) = 1$. (b) Repeat part (a) with $\mathcal{E}[X] = 1$ for $P(X = 1) = 1$.

5.5 Derive $\mathcal{E}[X]$ for the binomial random variable with parameters $\{N, p\}$ (without using the CF or MGF).

5.6 Repeat Problem 5.5 for the hypergeometric random variable with parameters $\{N, M, n\}$.

5.7 For nonnegative discrete random variable X, prove $\mathcal{E}[X] = \sum_{n=1}^{\infty} P(X \ge n)$.

5.8 For continuous random variable X with finite expectation, prove $\lim_{x\to\infty} |x| f_X(x) = 0$.

5.9 Derive $\mathcal{E}[X]$ for the log-normal random variable with parameters $\{\mu, \sigma\}$ (without using the CF or MGF).

5.10 Repeat Problem 5.9 for the beta random variable with parameters $\{\alpha, \beta\}$.

Expectation of a Function

5.11 Let X be Gaussian with parameters $\{\mu = 0, \sigma\}$. (a) Find $\mathcal{E}[Y]$ and the variance σ_Y^2 for $Y = u(X)$, where $u(\cdot)$ is the unit-step function. (b) Repeat part (a) for $Z = u(X - 1)$.

5.12 Using X from Problem 5.11, find $\mathcal{E}[Y]$ and the variance σ_Y^2 for $Y = |X|$ (Y has the half-normal distribution).

5.13 Find the expectation of $Y = \text{sgn}(X - 5)$, where X is a geometric random variable with parameter p and support $\mathcal{Z}^+$.

5.14 For X uniformly distributed on $[-1, 1]$, find the expectation of $Y = \exp(|X - 1/2|)$.

5.15 Write $\mathcal{E}[u(X - c)]$ in terms of the cdf $F_X(x)$, where $u(\cdot)$ is the unit-step function. Use (a) the definition of expectation and then (b) conditional expectation. (c) Repeat this problem for $\mathcal{E}[\text{sgn}(X - c)]$.

5.16 Let X be exponential with parameter $\lambda > 1$. Find (a) $\mathcal{E}[\exp(X)]$ and (b) $\mathcal{E}[(1 - X)I_{[0,1]}(X)]$ where $I(\cdot)$ is the indicator function.

5.17 Let $\{X_1, \ldots, X_N\}$ be iid exponential random variables with parameter λ. Find $\mathcal{E}[X_{(1)}]$ and $\mathcal{E}[X_{(N)}]$, where $X_{(1)}$ and $X_{(N)}$ are the minimum and maximum (order statistics).

5.18 Using CFs, show that the pdf of $Z = X - Y$ is the deterministic cross-correlation $f_Z(z) = f_Y(y) \star f_X(x)$, when X and Y are independent (see (4.160)).

Conditional Expectation

5.19 Let X and Y have the triangular bivariate pdf $f_{X,Y}(x, y) = [u(y) - u(y - 1)][u(x + y - 1) - u(x - y + 1)]$. Find $\mathcal{E}[XY]$ and $\mathcal{E}[Y|X]$.

5.20 The mean of the random sum $Y = \sum_{n=1}^{N} X_n$ is $\mathcal{E}[Y] = \mathcal{E}[X]\mathcal{E}[N]$, when $\{X_n\}$ are iid and independent of N. Find the variance of Y by conditioning on N.

5.21 Find $\mathcal{E}[X|X > 2]$ for the binomial random variable X with parameters $\{N, p\}$.

5.22 Let $Y = \prod_{n=1}^{N} X_n$, where $\{X_n\}$ are iid uniform random variables on $[0, 1]$, and N is Poisson with parameter α. Find $\mathcal{E}[Y]$ by conditioning on N assuming that $\{X_n\}$ and N are independent.

5.23 For the Laplace random variable with parameters $\{\mu = 0, \alpha\}$, find (a) $\mathcal{E}[X||X| < 1]$ and (b) $\mathcal{E}[|X||X < 1]$.

5.24 Let X and Y be iid uniform random variables on $[-1, 1]$. Find (a) $\mathcal{E}[XY|X > Y]$ and (b) $\mathcal{E}[X - Y|X > Y]$.

5.25 Repeat Problem 5.24 for discrete uniform random variables on $\{1, \ldots, N\}$.

Location Parameters

5.26 Find the mean, mode, and median of the Rayleigh distribution with parameter α.

5.27 Derive the median and the mode for the extreme value (Gumbel) distribution with parameters $\{c, \alpha\}$.

5.28 Repeat Problem 5.27 for the log-normal distribution with parameters $\{\mu, \sigma\}$. (The mean was derived in Problem 5.9.)

5.29 Derive the mean for the asymmetric Laplace distribution in (3.106), and find an expression for the median m_e.

Variance, Covariance, and Correlation

5.30 (a) Derive the mean and variance of the Erlang random variable with parameters $\{\alpha, r\}$ using the definition of expectation. (b) Repeat part (a) using results for the exponential random variable.

5.31 Using the definition of covariance, find C_{XY} for the following joint pdfs where $\{a, b\}$ are constants. (a) $f_{X,Y}(x, y) = axy[u(x) - u(x - 1)][u(y) - u(y - 1)]$. (b) $f_{X,Y}(x, y) = ax\exp(-b(x + y))u(x)u(y)$.

5.32 Find $\mathcal{E}[X^2]$ and σ_X^2 for the geometric random variable (a) with support $\mathcal{Z}^+$ and then (b) with support $\mathcal{N}$.

5.33 Derive the expression for Pearson's sample correlation coefficient in (5.242).

5.34 Consider the following estimate of Y based on random variable X: $\hat{Y} = a_1 X + a_2 X^2$. (a) Find expressions for the optimal coefficients $\{a_1, a_2\}$ that minimize $\mathcal{E}[(Y - \hat{Y})^2]$. (b) Evaluate these expressions when X and Y are jointly Gaussian with zero mean, unit variance, and correlation coefficient ρ_{XY}.

5.35 Derive the correlation coefficient in (4.86) for the bivariate exponential distribution.

5.36 Derive the variance for (a) the discrete uniform random variable on $\{1, \ldots, N\}$ and (b) the logarithmic random variable with parameter p.

5.37 Derive the joint pdf $f_{X,Y}(x, y)$ for the cdf in (5.261).

Generating Functions

5.38 (a) Find the MGF $m_X(t)$ for a Poisson random variable with parameter α. (b) Find the CF $\Phi_X(j\omega)$ for a binomial random variable with parameters $\{N, p\}$.

5.39 Derive the CF $\Phi_X(\omega)$ for (a) the Laplace random variable with parameters $\{\mu, \alpha\}$ and (b) the Erlang random variable with parameters $\{\lambda, r\}$.

5.40 Prove the following upper bound for the CF:

$$|\Phi_X(\omega)| \leq \Phi_X(0) = 1. \tag{5.375}$$

5.41 Use the CF to find $\mathcal{E}[X^2]$ when (a) X is Poisson with parameter α and (b) Y is Laplacian with parameters $\{\mu = 0, \alpha\}$.

5.42 Derive the CF in (5.109) for the multivariate Gaussian random variable.

Moments and Cumulants

5.43 For a geometric random variable with parameter p, find an expression for $\mathcal{E}[X(X-1)\cdots(X-n)]$.

5.44 Derive the nth central moment $\mu_{c,n}$ of the continuous uniform random variable on $[a, b]$.

5.45 Find all cumulants for the Poisson random variable with parameter α.

5.46 Derive an expression for the third-order cumulant κ_3 in terms of raw moments $\mu_n = \mathcal{E}[X^n]$ for $n = 1, 2, 3$.

Expectations of Nonlinear Functions

5.47 Find an expression for $\mathcal{E}[u(X_1)u(X_2 - 1)]$, where $\{X_1, X_2\}$ are jointly Gaussian, zero-mean random variables with correlation coefficient ρ.

5.48 Repeat Problem 5.48 for $\mathcal{E}[X_1 u(X_1) X_2 u(X_2 - 1)]$.

Computer Problems

5.49 Derive the optimal coefficients $\{a_1, a_2\}$ for estimating Y that minimize $\mathcal{E}[(Y - \hat{Y})^2]$, where $\hat{Y} = a_1 + a_2 X$. Use mvnrnd to generate 1000 samples of $\{X, Y\}$ for $\mu_X = 1, \mu_Y = 2, \sigma_X^2 = \sigma_Y^2 = 1$, and $\rho_{XY} = 0.8$. Generate a scatter plot for the pairs of samples, and plot the line based on the optimal coefficients for the parameters given above. Generate a second figure for the same scenario except with $\rho_{XY} = -0.8$.

5.50 Use mvnrnd to generate 1000 samples of $\{X, Y\}$ for $\mu_X = 0, \mu_Y = 1, \sigma_X^2 = 1, \sigma_Y^2 = 2$, and $\rho_{XY} = 0.9$. Generate a scatter plot for the pairs of samples and use MATLAB® to compute (i) Pearson's sample correlation coefficient $\bar{\rho}_{XY}$, (ii) Spearman's rank correlation coefficient τ_S, and (iii) Kendall's rank correlation coefficient τ_K.

5.51 For the random sum in (5.118), let N be geometric with parameter $p = 0.1$, and $\{X_n\}$ be iid exponential with parameter $\lambda = 1$. We know from conditioning that when N and $\{X_n\}$ are independent, $\mathcal{E}[Y] = \mathcal{E}[\mathcal{E}[Y|N]] = \mathcal{E}[X_n]\mathcal{E}[N] = (1/\lambda)q/p = 9$. Use MATLAB to generate an outcome n for random variable N, and then average n exponential samples to estimate $\mathcal{E}[Y|N]$. Repeat these steps for 1000 samples of N to compute the sample mean $\bar{Y}$ for estimating $\mathcal{E}[Y]$.

FURTHER READING

Additional material on expectation can be found in many of the references provided in Chapters 2–4. Several references for integration are given at the end of Appendix D.

PART II
Random Processes, Systems, and Parameter Estimation

6 Random Processes

6.1 INTRODUCTION

A random process is a collection of random variables that are *indexed by time*. Let $\mathcal{T}$ be the set of time instants over which the random process is defined. This set could be continuous time such as $\mathcal{T} = \mathcal{R}$ or $\mathcal{R}^+$, or it might be discrete time with $\mathcal{T} = \mathcal{Z}$ or $\mathcal{Z}^+$. A discrete-time random process is also called a *random sequence* or *time series*, and the time instants are usually equally spaced, though it is not necessary. In this chapter, we often use the term random process in a general sense, to include continuous-time processes as well as discrete-time sequences. When it is necessary to restrict our discussion specifically to discrete time, we will use random sequence. A random process is also known as a *stochastic process* (stochastic from Greek *stokhastikos* means "capable of guessing").

Figure 6.1 shows an example realization of a Gaussian random process. We are interested in exploring various *characterizations* of such a random process in order to have some understanding of its properties. This function of time was generated by filtering an independent sequence of zero-mean Gaussian random variables. As a result, the random variables associated with several time instants are jointly Gaussian. We also know that the filter has caused the output to be a *correlated* sequence, such that there is now some degree of predictability of future outcomes. In this chapter, we describe several other features of a random process, including: stationarity, independent increments, martingales, and the Markov property. Several specific random processes are considered, and we focus on their correlation properties. In Chapter 7, we cover additional properties of a random process such as stochastic convergence, continuity, derivatives, and integrals. A study of stochastic differential equations will prepare us for the systems described in Chapter 8 that *filter* random processes to modify their properties, as was done in the Gaussian example above.

6.2 CHARACTERIZATIONS OF A RANDOM PROCESS

A random process can be defined in a manner similar to that used for a random variable. Assume there exists a sample space $\mathcal{S}$ containing all possible outcomes of duration $\mathcal{T}$ of the random process. We denote an outcome by $\zeta \in \mathcal{S}$ and use this notation for a random process for two reasons: (i) to differentiate it from $\omega \in \Omega$ used previously for the outcomes of an abstract sample space and (ii) to avoid confusion when we consider sinusoidal signals and Fourier transforms based on radian frequency ω. Each outcome of $\mathcal{S}$ is a deterministic function of time and is called a *realization* of the random process. We assume there is an underlying probability space $\{\mathcal{S}, \mathcal{F}, P\}$ that defines the random process. For outcome $\zeta \in \mathcal{S}$, $x(t, \zeta)$ represents the realization. The *ensemble* of realizations, which together comprise all possible outcomes of the random process, is denoted by the uppercase function $X(t, \zeta)$. Examples of two continuous-time realizations are illustrated in Figure 6.2. Observe that $x(t, \zeta_n)$ is the entire waveform for all $t \in \mathcal{T}$ corresponding to outcome ζ_n.

Example 6.1. Consider a random process where the ensemble of realizations for $t \in \mathcal{T} = \mathcal{R}$ are cosine waveforms with different amplitudes. For a specific $\zeta \in \mathcal{S}$, a realization of the process is

$$x(t, \zeta) = \zeta^a \cos(2\pi f_c t), \tag{6.1}$$

Probability, Random Variables, and Random Processes: Theory and Signal Processing Applications, First Edition. John J. Shynk.

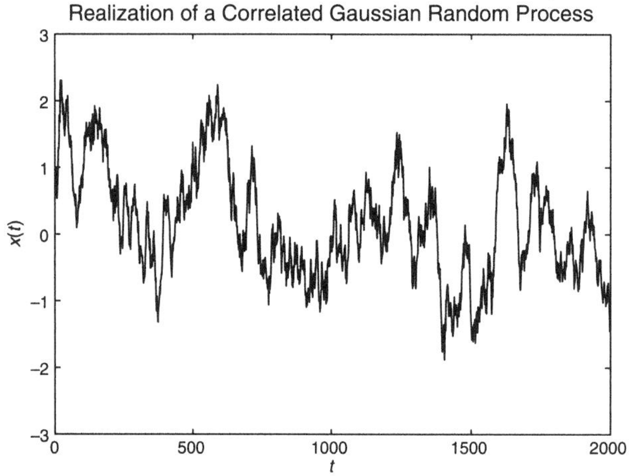

FIGURE 6.1 Realization of a Gaussian random process.

where ζ^a is the amplitude corresponding to realization ζ. The range of values for ζ^a might be uniform on [0, 1], for example. This random process has some structure (not much randomness) because only the amplitude is random for the ensemble of waveforms. For such structured cases, we can describe the sample space more simply; in this case $\mathcal{S} = [0, 1]$ *represents* all possible realizations of the random process with the understanding that the outcomes are given by (6.1). In fact, this random process is essentially equivalent to a uniform random variable on [0, 1]. Suppose now that the realizations also have different phase shifts:

$$x(t, \zeta) = \zeta^a \cos(2\pi f_c t + \zeta^p), \tag{6.2}$$

where $\{\zeta^a, \zeta^p\}$ are the amplitude and phase of realization ζ. For this structured realization, the sample space could be described by $\mathcal{S} = [0, 1] \times [-\pi, \pi]$, where [0, 1] and $[-\pi, \pi]$ are the possible amplitude and phase values, respectively. It is understood that an outcome $\zeta \in \mathcal{S}$ is a waveform with a specific amplitude and phase. This random process is essentially equivalent to a random vector with two independent components that are uniform on [0, 1] and $[-\pi, \pi]$.

In general, each $\zeta \in \mathcal{S}$ corresponds to a specific realization (waveform) as depicted in Figure 6.2. However, if the waveforms have some structure as in Example 6.1, we can describe $\mathcal{S}$ more simply in terms of random *parameters* of the waveform (such as amplitude or phase) that differ over the ensemble of realizations. These simple cases are sometimes equivalent to random variables or random vectors. We describe random processes with more complicated structures later in this chapter.

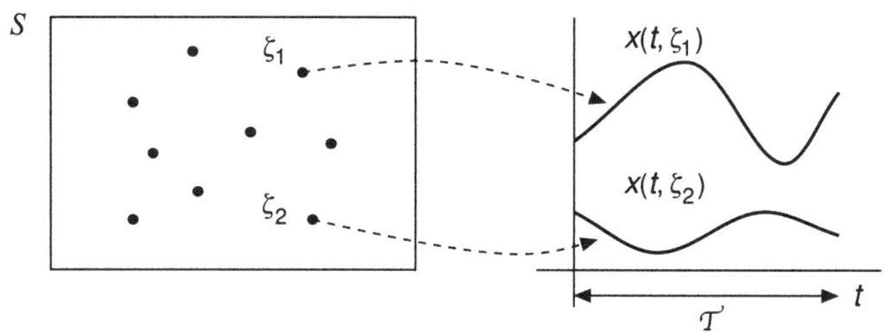

FIGURE 6.2 Realizations of a random process for the entire time interval $\mathcal{T}$ viewed as outcomes of $\mathcal{S}$.

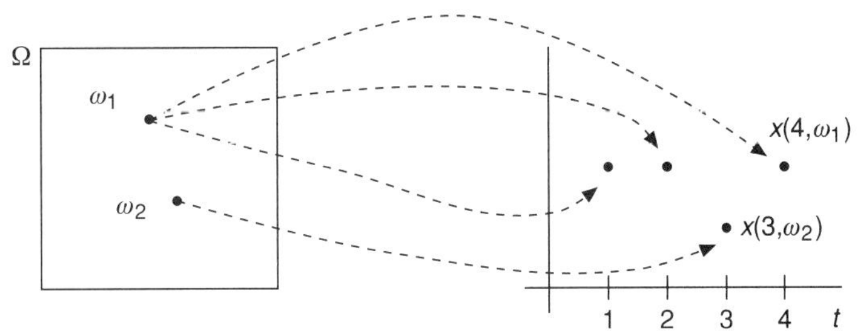

FIGURE 6.3 Values of a random sequence at different times viewed as outcomes of random variables.

A random process can also be viewed as the outcomes of experiments (random variables) sequentially in time. At each time instant (continuous or discrete) there is an underlying probability space $\{\Omega, \mathcal{F}, P\}$ which may or may not be the same over the entire set $\mathcal{T}$ (the σ-field $\mathcal{F}$ and probability assignment P differ from those used to describe the random process above with sample space $\mathcal{S}$). For example, repeated tosses of a fair coin yield a random sequence with an underlying probability space that remains fixed. For each $t \in \mathcal{T}$, an outcome is "drawn" from the probability distribution at that time. Thus, as we move forward in time, a realization of the process is constructed from the outcomes of individual experiments. An example of this mechanism for a discrete-time sequence and Ω with two outcomes is illustrated in Figure 6.3. Observe that $x(3, \omega_2)$ is the value of outcome ω_2 at time instant $t = 3$. It is important to note that the elements of $\mathcal{S}$ in Figure 6.2 refer to the *realizations* of the random process, whereas the elements of Ω in Figure 6.3 are the *individual outcomes* of the experiment *at a specific time instant*: $X(3, \omega)$ is the random variable at time instant 3 whose particular outcome is $x(3, \omega_2)$.

This second description based on Ω may represent how the realizations are actually generated, such as repeated coin tosses mentioned above where a sequence of heads and tails is generated. This series of outcomes on $\mathcal{T}$ is a realization of a Bernoulli random sequence (which is discussed in detail later). If we restart the experiment, then a different realization of the process would almost surely be obtained when the outcomes of the sequential coin tosses are observed. Even though the realizations are actually generated (constructed) in this way, we can still use the first description for a random process, where each sequence of heads and tails corresponds to a unique ζ in the underlying sample space $\mathcal{S}$ for the random process.

Consider another example based on temperature variation during a single day. Again, we could use either description for the realizations of this process. At each instant of time, we can imagine that the temperature is the outcome of some random variable (perhaps a truncated Gaussian distribution). Alternatively, we can assume that all possible realizations of temperature are described by the continuous sample space $\mathcal{S}$, and one realization occurring during the day is a specific ζ. Unlike the coin-toss example, it is not obvious how the temperature process is actually generated; it depends on many physical factors of which there are too many to consider. Nevertheless, both descriptions are useful representations of a random process which we can examine using probabilistic techniques.

The previous examples illustrate an important property of a random process that will be of particular interest: whether or not the outcomes are *correlated* at different time instants. Obviously for the coin toss experiment, the outcomes are independent: future outcomes do not depend on the results of previous coin tosses. This may not be the case for other types of random processes which might be highly correlated. The temperature example is an "experiment" where the outcomes are correlated: we would expect that the temperature at a future time will be relatively close to temperatures observed in the recent past. Correlation is perhaps the most important property of a random process for many applications; we investigate it further in Chapter 8, and it is used later in the application chapters.

A summary of the notation for a random process is provided in Figure 6.4. Although the dependence on ζ is explicitly shown in the figure (and in the figure for Example 6.2), it will be suppressed in subsequent discussions to simplify the notation. Thus, $X(t)$ denotes a random process, $x(t)$ is a single realization (the observed waveform) on $\mathcal{T}$, $X(t_o)$ is a random variable at time t_o, and $x(t_o)$ is a single outcome of the random variable at time t_o (which is also the value of the realization $x(t)$ at that time).

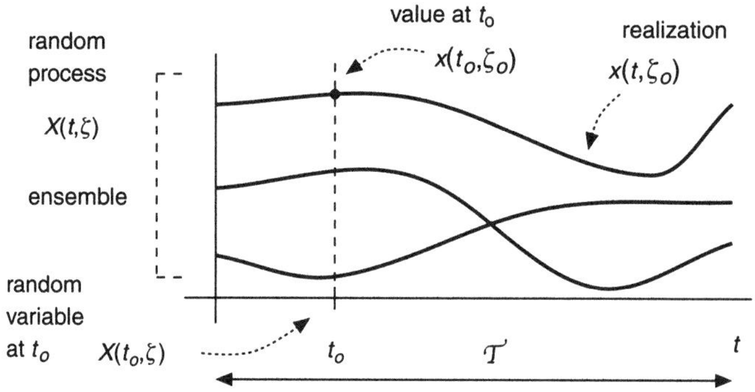

FIGURE 6.4 Summary of notation for a random process. (i) $X(t, \zeta)$: random process consisting of an *ensemble* of realizations $\zeta \in \mathcal{S}$. (ii) $x(t, \zeta_o)$: *realization* of $X(t, \zeta)$ (a deterministic function of time) corresponding to $\zeta_o \in \mathcal{S}$. (iii) $X(t_o, \zeta_o)$: random *variable* with outcomes across all realizations at t_o. (iv) $x(t_o, \zeta_o)$: value at t_o of realization $\zeta_o \in \mathcal{S}$, corresponding to an outcome of the random variable at time t_o.

Example 6.2. Before giving a general probabilistic description for a random process, consider another relatively simple random process defined as follows:

$$Y(t) = X[u(t) - u(t - T)], \tag{6.3}$$

where X and T are independent random variables with cumulative distribution functions (cdfs) $F_X(x)$ and $F_T(t)$. Because of the unit-step function $u(t)$, we have

$$Y(t) = \begin{cases} X, & T > t \\ 0, & T \le t \end{cases} \tag{6.4}$$

for $t \in \mathcal{T} = \mathcal{R}^+$. This type of process could arise in a system where Y is connected to X by some communication link and T is the time when the link is broken. Thus, at a particular time instant t, $Y(t)$ is zero if the outcome of T is $\le t$, meaning the link is broken before (or at) time t. Example realizations are shown in Figure 6.5 where we see there are two effects of the randomness: (i) an outcome of X gives the nonzero amplitude of $Y(t)$ and (ii) an outcome of T determines when $Y(t)$ drops to zero. One way to characterize this process is to derive its cdf as a function of time, by conditioning on T as follows:

$$P(Y(t) \le y) = P(Y(t) \le y|T \le t)P(T \le t) + P(Y(t) \le y|T > t)P(T > t). \tag{6.5}$$

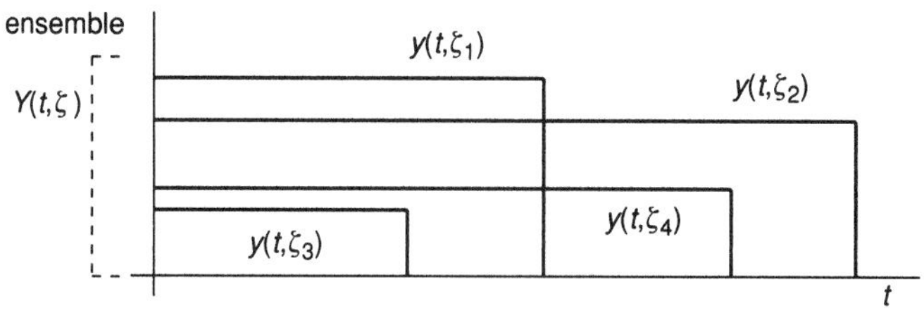

FIGURE 6.5 Example realizations for the random process in Example 6.2. Each ζ_n denotes a specific pair of outcomes for X and T.

From the description above:

$$P(Y(t) \leq y|T \leq t) = u(y) \tag{6.6}$$

because $Y(t) = 0$ for this conditioning, and thus with probability one it is less than any positive y. For the other conditional probability:

$$P(Y(t) \leq y|T > t) = P(X \leq y) = F_X(y) \tag{6.7}$$

because the link is not broken and $Y(t) = X$. Combining these results gives the following *time-varying* cdf:

$$\begin{aligned} F_{Y(t)}(y) &= u(y)F_T(t) + F_X(y)[1 - F_T(t)], \\ &= F_T(t)[u(y) - F_X(y)] + F_X(y), \end{aligned} \tag{6.8}$$

where the time dependence (causing $Y(t)$ to drop to zero) is due to the first term in the last equation. $Y(t)$ is an example of a *nonstationary* process (defined later) because the distribution changes with time. The corresponding probability density function (pdf) is

$$f_{Y(t)}(y) = F_T(t)\delta(y) + f_X(y)[1 - F_T(t)], \tag{6.9}$$

where $f_X(y)$ is the pdf of X and $\delta(y)$ is the Dirac delta function, which demonstrates that $Y(t)$ at each finite t is a mixed random variable. It is also a probability *mixture* with weightings $\{F_T(t), 1 - F_T(t)\}$ that obviously sum to 1.

Example 6.3. Consider a random process whose realizations are defined as follows:

$$X(t) = A\exp(-\lambda t) \tag{6.10}$$

for $t \in \mathcal{R}^+$ and with $\lambda > 0$. Let A be a uniform random variable on $[0, 1]$ so that the ensemble of realizations of $X(t)$ are decaying exponentials with different initial amplitudes. It has the following raw moments:

$$\mathcal{E}[X^n(t)] = \mathcal{E}[A^n]\exp(-\lambda nt) = \exp(-\lambda nt)/(n + 1). \tag{6.11}$$

The cdf at time t is

$$\begin{aligned} P(X(t) \leq x) &= P(A\exp(-\lambda t) \leq x) = P(A \leq x\exp(\lambda t)), \\ &= x\exp(\lambda t)I_{[0,1]}(x\exp(\lambda t)) + I_{(1,\infty)}(x\exp(\lambda t)), \\ &= x\exp(\lambda t)I_{[0,\exp(-\lambda t)]}(x) + I_{(\exp(-\lambda t),\infty)}(x), \end{aligned} \tag{6.12}$$

where the cdf of the uniform random variable has been substituted to obtain the last expression. Differentiating this result with respect to x gives the pdf:

$$f_{X(t)}(x) = \exp(\lambda t)I_{[0,\exp(-\lambda t)]}(x), \tag{6.13}$$

which we see is uniform on $[0, \exp(-\lambda t)]$. (Note that the indicator functions are used only to explicitly show the support; they are not differentiated with respect to x.) As t increases, the range of outcomes of the pdf, given by the subscript of the indicator function, becomes increasingly compressed toward zero.

6.3 CONSISTENCY AND EXTENSION

In the last two examples, the cdf of a random process was derived for specific time instants t. It is not always straightforward to characterize individual realizations of a random process as in those examples. Generally, in order to make the probabilistic description of a random process more precise, it will be necessary to define the *joint probability* distribution of the random variables across all subsets of time interval $\mathcal{T}$. We can view a random process as an extension of a random vector with each component of the vector indexed by an element of $\mathcal{T}$, which can have an infinity of elements and may or may not be countable. From this random vector description, a joint distribution is specified from which joint properties of the random variables are determined, such as their joint moments, independence, correlation, orthogonality, and so on.

Consider the subset of discrete time instants $\{t_1, \ldots, t_N\} \in \mathcal{T}$ which need not be in increasing order, and there may be gaps in time from the original elements in $\mathcal{T}$. Although we normally view time as moving forward and usually without gaps (as when the time instants are uniformly spaced), we are free to choose any subset of $\mathcal{T}$ and with the elements in any order because we are considering a collection (set) of random variables. As was done for a random vector defined in Chapter 4, the joint cdf for this collection of random variables is specified as follows:

$$F_{X(t_1),\ldots,X(t_N)}(x_1, \ldots, x_N) = P(X(t_1) \le x_1, \ldots, X(t_N) \le x_N), \tag{6.14}$$

where the lower-case arguments $\{x_k\}$ are the outcomes at time instants $\{t_k\}$. This cdf has essentially the same form as that for a random vector, except here we use time arguments of the random process to distinguish among the various components instead of subscripts used previously for the elements of a random vector.

There is fundamentally no difference between the cdf of a random vector and that for a finite subset of elements of a random process. In order for the probability model of a random process to "make sense," the cdf for *any* subset of $\mathcal{T}$ must be consistent. Generally speaking: (i) the probability measure must be well defined for any permutation of elements indexed by a subset of $\mathcal{T}$ and (ii) the cdf of this subset of random variables should be readily obtained from that of a *larger* subset. To be more specific, we define the notion of consistency for the probability measure of a subset of random variables.

Definition: Consistency of a Probability Measure Probability measure P is *consistent* if for finite-dimensional distributions of random variables the following two conditions hold:

- For subset $\{t_1, \ldots, t_N\} \in \mathcal{T}$ and events $\{E_1, \ldots, E_N\}$ for the random variables, let $\{\pi(1), \ldots, \pi(N)\}$ be a permutation of the subscripts. Then

$$P(X(t_{\pi(1)}) \in E_{\pi(1)}, \ldots, X(t_{\pi(N)}) \in E_{\pi(N)}) = P(X(t_1) \in E_1, \ldots, X(t_N) \in E_N). \tag{6.15}$$

  If the random variables and events in the probability measure are permuted, the same probability is obtained as with the original ordering of the random variables and associated events. This condition ensures that the probability of any finite-dimensional subset of random variables does not depend on a particular order: the ordering does not change the overall composite event given by $\{E_1, \ldots, E_N\}$ in the argument of $P(\cdot)$.
- For the larger subset $\{t_1, \ldots, t_N, t_{N+1} \ldots, t_{N+M}\} \in \mathcal{T}$,

$$\begin{aligned} P(X(t_1) \in E_1, \ldots, X(t_N) \in E_N) = P(X(t_1) \in E_1, \ldots, X(t_N) \in E_N, \\ X(t_{N+1}) \in \Omega_{N+1}, \ldots, X(t_{N+M}) \in \Omega_{N+M}), \end{aligned} \tag{6.16}$$

  where Ω_k denotes the sample space of the random variable at time instant t_k. Since $X(t_{N+m}) \in \Omega_m$ for $m = 1, \ldots, M$, are sure events, the composite events in both arguments of (6.16) are effectively the same and so the probability is unchanged. This condition ensures there is a consistent method for deriving probabilities of subsets of events from a higher order joint probability, and gives the ability to obtain marginal probabilities from joint probabilities.

These two conditions of consistency for a probability measure are intuitively appealing. We now state the Kolmogorov extension theorem (without proof) which specifies the existence of a random process associated with a set of finite-dimensional distributions. It is also called the Kolmogorov existence theorem.

Theorem 6.1 (Kolmogorov extension). Let the probability measure $P(X(t_1), \ldots, X(t_N))$ be consistent for any collection of sample times $\{t_1, \ldots, t_N\}$ with $t_n \in \mathcal{T}$ and any size $N \in \mathcal{N}$. Then there exists a random process $X(t)$ with $t \in \mathcal{T}$ from which any finite-dimensional distribution can be generated.

Basically, this theorem extends a finite collection of random variables $\{X(t_1), \ldots, X(t_N)\}$ to the random *process* $X(t)$ by (i) defining a set of times $\mathcal{T}$ such that $t \in \mathcal{T}$ and (ii) ensuring that all random variables indexed by any finite subset of $\mathcal{T}$ are consistent. The set of elements in $\mathcal{T}$ may be finite, countably infinite, or uncountable (continuous). The theorem relates to the second description of a random process where the process is obtained by sequentially selecting samples from random variables indexed by time. If the conditions of the theorem are satisfied, then the random process exists and is described by an underlying probability space $\{\mathcal{S}, \mathcal{F}, P\}$ with *realizations* of the process as outcomes. The two different characterizations of the same random process are connected as follows. If random variable $X(t_n)$ has sample space Ω_n for $n \in \mathcal{N}$, then the sample space of the random process is $\mathcal{S} = \Omega_1 \times \Omega_2 \times \cdots$. Realization ζ is a particular waveform of random variable outcomes extended over the sample space $\mathcal{S}$. If $\Omega_n = \Omega$ is time invariant for all n (called stationary), then the sample space "simplifies" to $\mathcal{S} = \Omega \times \Omega \times \cdots = \Omega^{|\mathcal{T}|}$, where the possible outcomes at each time instant are the same and $|\mathcal{T}|$ is the cardinality of $\mathcal{T}$. The probability measure must be consistent as given in the previous definition. It cannot, in general, be written in terms of the probability measure at individual discrete-time instants, unless the random variables are independent. The probability measure incorporates any dependence that may exist across time instants, which obviously cannot be extrapolated from a single time instant. The Kolmogorov extension theorem extends the discrete-time description to continuous time and allows for variation in the probability space of the random variables across time instants, provided the probability measure is consistent.

Example 6.4. Consider sequential coin tosses where $\Omega = \{H, T\}$ and assume for simplicity that $\mathcal{T} = \{t_1, t_2\}$. The sample space of the random sequence is $\mathcal{S} = \{HH, HT, TH, TT\}$, where the time index is implied by the ordered outcomes. Alternatively, we can write $\mathcal{S} = \Omega \times \Omega = \Omega^2$ using the sample space for the coin toss. Suppose instead that we toss a coin and then a die. For the first time instant, $\Omega_1 = \{H, T\}$ whereas $\Omega_2 = \{1, 2, 3, 4, 5, 6\}$ for the second time instant. Thus, it is not possible to write $\mathcal{S}$ as $\Omega^{|\mathcal{T}|}$ because Ω changes; this is an example of a nonstationary random process where the distribution of outcomes varies with time. Instead, we use $\mathcal{S} = \Omega_1 \times \Omega_2 = \{\{H, 1\}, \{T, 1\}, \ldots, \{H, 6\}, \{T, 6\}\}$ where again the time index is implied by the ordered outcomes.

6.4 TYPES OF RANDOM PROCESSES

Next, the four basic types of random processes are summarized, depending on whether $\mathcal{T}$ and the random variables are continuous or discrete. For clarity and when necessary, we distinguish between a continuous-time process and a discrete-time sequence using the following notation:

$$\textit{Random sequence:} \quad X[k],\ k \in \mathcal{T} \subseteq \mathcal{Z}, \tag{6.17}$$

$$\textit{Random process:} \quad X(t),\ t \in \mathcal{T} \subseteq \mathcal{R}, \tag{6.18}$$

where brackets denote a *sequence* (time series), which is typically done for discrete-time signals (deterministic or random). Simple examples for each of the four types of realizations are shown in Figure 6.6 where in the discrete-time cases, the time instants are assumed to be uniformly spaced (which is also implied by using $\mathcal{Z}$ above).

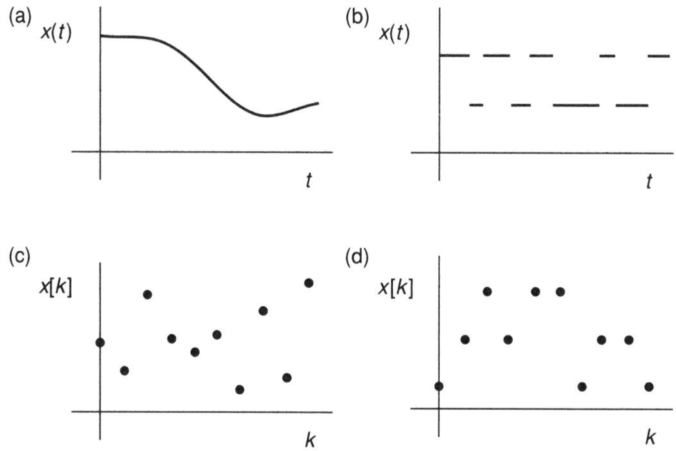

FIGURE 6.6 Example realizations of random processes. (a) Continuous time, continuous amplitude. (b) Continuous time, discrete amplitude (two values). (c) Discrete time, continuous amplitude. (d) Discrete time, discrete amplitude (three values).

In addition to the four cases in the figure, it is possible that the underlying probability space is *mixed* where the random variables have continuous and discrete components. A simple mixed process is given in Example 6.2 where $Y(t)$ has nonzero probability mass at $y = 0$. Other issues and properties of a random process are evident from the previous descriptions. For example, the random variables may change with time (e.g., from Gaussian to exponential), or the same random variable may exist for all $\mathcal{T}$, but one or more of its parameters could change (e.g., Gaussian with time-varying mean). In this chapter, we cover important examples of random processes, discuss their properties and some applications for which they are relevant *models*, and examine the correlation of random variables at different time instants.

We begin with some general properties or *characterizations* of a random process that arise because the process is a collection of random variables indexed by time.

6.5 STATIONARITY

There are several different types of stationarity for a random process. Generally, stationarity refers to the degree to which the probabilistic model of the time-indexed random variables is "constant" with time.

Definition: First-Order Stationary Random process $X(t)$ is *first-order stationary* if the pdf does not depend on time:

$$f_{X(t)}(x) = f_X(x), \quad \text{for all } t \in \mathcal{T}. \tag{6.19}$$

As a result, all moments (raw and central), cumulants, and so on, are constant. For this type of stationarity, we cannot say anything about moments across two or more time instants, such as correlation and covariance.

Definition: *N*th-Order Stationary Random process $X(t)$ is *Nth-order stationary* if the joint probability distribution of the process at N time instants does not change by any time shift τ of all $\{t_1, \ldots, t_N\}$:

$$f_{X(t_1),\ldots,X(t_N)}(x_1, \ldots, x_N) = f_{X(t_1-\tau),\ldots,X(t_N-\tau)}(x_1, \ldots, x_N). \tag{6.20}$$

This type of stationarity is obviously stronger than first-order stationarity; it implies that moments across time instants (up to $N-1$ time differences) are unchanged due to a time shift. We can view Nth-order stationarity as though there is a "sliding window" (of width N) within which the joint pdf does not change.

Definition: Strictly Stationary Random process $X(t)$ is *strictly stationary* if it is Nth-order stationary for all N as $N \to \infty$.

This is the strongest form of stationarity: all statistics of the random process are constant for all time differences and all time shifts. A strictly stationary process implies (6.20) which in turn implies (6.19). For all three cases, the marginal distributions are identical: the type of distribution (Gaussian, Poisson, and so on) is the same and the parameters of the pdf do not change. In general, (6.19) does not imply (6.20). An important exception occurs when the random process is independent such that all joint distributions can be expressed as products of the marginal distributions. This corresponds to an independent and identically distributed (iid) random process, which is discussed in Section 6.6.

In many applications, all three of these definitions of stationarity are often stronger than is needed to characterize a random process. A strictly stationary random process is probably too strong a requirement and is unnecessary for most practical scenarios. A random process that is Nth-order stationary is also quite strong because all moments across N time instants are assumed to be constant, which may be unrealistic in practice. For many problems, we are usually interested only in the first- and second-order moments (especially correlation) and whether or not those are time invariant. This will become evident later when we discuss correlation in detail and its frequency-domain counterpart (via the Fourier transform) known as the power spectral density (PSD).

We repeat the definition of correlation given in Chapter 5 for random variables, applied here to a random process.

Definition: Autocorrelation Function The *autocorrelation function* of random process $X(t)$ is

$$R_{XX}(t_1, t_2) \triangleq \mathcal{E}[X(t_1)X(t_2)], \tag{6.21}$$

which, in general, is a function of the specific values $\{t_1, t_2\}$. The definition for random sequence $X[k]$ is

$$R_{XX}[k_1, k_2] \triangleq \mathcal{E}[X[k_1]X[k_2]]. \tag{6.22}$$

Unlike correlation for random variables, $R_{XX}(t_1, t_2)$ is a *function*: it depends on the particular time instants.

Definition: Uncorrelated Process Random process $X(t)$ is *uncorrelated* if

$$R_{XX}(t_1, t_2) = \begin{cases} \mathcal{E}[X^2(t_1)], & t_1 = t_2 \\ \mathcal{E}[X(t_1)]\mathcal{E}[X(t_2)], & t_1 \neq t_2. \end{cases} \tag{6.23}$$

Likewise for random sequence $X[k]$:

$$R_{XX}[k_1, k_2] = \begin{cases} \mathcal{E}[X^2[k_1]], & k_1 = k_2 \\ \mathcal{E}[X[k_1]]\mathcal{E}[X[k_2]], & k_1 \neq k_2. \end{cases} \tag{6.24}$$

If the process has zero mean, the autocorrelation function is zero, except when its arguments are equal, in which case it is the variance at that particular time.

Definition: Autocovariance Function The *autocovariance function* of random process $X(t)$ is

$$C_{XX}(t_1, t_2) \triangleq \mathcal{E}[(X(t_1) - \mathcal{E}[X(t_1)])(X(t_2) - \mathcal{E}[X(t_2)])]. \tag{6.25}$$

The definition for random sequence $X[k]$ is

$$C_{XX}[k_1, k_2] \triangleq \mathcal{E}[(X[k_1] - \mathcal{E}[X[k_1]])(X[k_2] - \mathcal{E}[X[k_2]])]. \tag{6.26}$$

If the process is uncorrelated, the autocovariance function is zero except when its arguments are equal, which is the variance (even for nonzero mean). The following type of stationarity is often assumed in many engineering applications where systems operate on random signals.

Definition: Wide-Sense Stationary Random process $X(t)$ is *wide-sense stationary* if the mean is constant and the autocovariance function depends only on the time difference and not on specific time instants:

$$\mathcal{E}[X(t)] = \mu_X, \quad \text{for all } t \in \mathcal{T} \tag{6.27}$$

and

$$C_{XX}(t_1, t_2) = C_{XX}(t_2 - t_1), \quad \text{for all } t_1, t_2 \in \mathcal{T}. \tag{6.28}$$

The time difference $t_2 - t_1$ is also called the time *lag*, and when a process is wide-sense stationary, $\tau \triangleq t_2 - t_1$ is usually substituted such that

$$C_{XX}(t_2 - t_1) = C_{XX}(\tau). \tag{6.29}$$

For a random sequence,

$$C_{XX}[k_2 - k_1] = C_{XX}[m], \tag{6.30}$$

where $m \triangleq k_2 - k_1$. The corresponding autocorrelation functions are $R_{XX}(\tau)$ and $R_{XX}[m]$.

For uncorrelated processes, $C_{XX}(\tau) = \sigma_X^2\delta(\tau)$ and $C_{XX}[m] = \sigma_X^2\delta[m]$. These are discussed further in Chapter 8 where *white noise* is defined.

Example 6.5. The autocorrelation function for the random process in Example 6.3 is

$$R_{XX}(t, t+\tau) = \mathcal{E}[A^2]\exp(-\lambda(2t+\tau)) = (1/3)\exp(-\lambda(2t+\tau)), \tag{6.31}$$

which demonstrates that $X(t)$ is *not* wide-sense stationary. The mean also varies with time: $\mu_X(t) = (1/2)\exp(-\lambda t)$.

Wide-sense stationary processes are examined extensively in Chapter 8, and they are generally assumed for the various signals in the application chapters.

From symmetry in the definition, the autocorrelation function is even: $R_{XX}(\tau) = R_{XX}(-\tau)$, and clearly it is nonnegative for lag zero: $R_{XX}(0) = \mathcal{E}[X^2(t)] \geq 0$. The following theorem describes an additional property of the autocorrelation function.

Theorem 6.2. The autocorrelation function is maximum for zero lag:

$$|R_{XX}(\tau)| \leq R_{XX}(0). \tag{6.32}$$

Proof. From the Cauchy–Schwarz inequality in Appendix F,

$$|R_{XX}(\tau)| = |\mathcal{E}[X(t)X(t+\tau)]| \leq \sqrt{\mathcal{E}[X^2(t)]\mathcal{E}[X^2(t+\tau)]}. \tag{6.33}$$

Since the process is wide-sense stationary, $\mathcal{E}[X^2(t)] = \mathcal{E}[X^2(t+\tau)] = R_{XX}(0)$, which completes the proof. □

The previous definitions of correlation and covariance can be extended to two random processes.

Definition: Cross-Correlation and Covariance The *cross-correlation function* of random processes $X(t)$ and $Y(t)$ is

$$R_{XY}(t_1, t_2) \triangleq \mathcal{E}[X(t_1)Y(t_2)], \tag{6.34}$$

and the *covariance function* is

$$C_{XY}(t_1, t_2) \triangleq \mathcal{E}[(X(t_1) - \mathcal{E}[X(t_1)])(Y(t_2) - \mathcal{E}[Y(t_2)]]. \tag{6.35}$$

For random sequences $X[k]$ and $X[k]$,

$$R_{XY}[k_1, k_2] \triangleq \mathcal{E}[X[k_1]Y[k_2]], \tag{6.36}$$

$$C_{XY}[k_1, k_2] \triangleq \mathcal{E}[(X[k_1] - \mathcal{E}[X[k_1]])(Y[k_2] - \mathcal{E}[Y[k_2]])]. \tag{6.37}$$

When the processes are wide-sense stationary, these expressions become $R_{XY}(\tau), C_{XY}(\tau), R_{XY}[m]$, and $C_{XY}[m]$, where $\tau \triangleq t_2 - t_1$ and $m \triangleq k_2 - k_1$.

Note that the wide-sense stationary results in the previous definition are *not* symmetric: for example, $R_{XY}(\tau) \neq R_{XY}(-\tau)$; instead $R_{XY}(\tau) = R_{YX}(-\tau)$ and likewise for the other quantities.

For completeness, we also include the definition of cyclostationarity; it is used only briefly later in Chapter 7.

Definition: Cyclostationary Random process $X(t)$ is *wide-sense cyclostationary* if the mean is constant and the autocovariance function is periodic with period T_o:

$$C_{XX}(\tau) = C_{XX}(\tau + T_o), \quad \text{for any } \tau. \tag{6.38}$$

6.6 INDEPENDENT AND IDENTICALLY DISTRIBUTED

Since a random process is viewed as a collection of random variables indexed by time, we can extend the idea of independent random variables to a process. Consider the following example of an independent random sequence.

Example 6.6. At each instant of time $k \in \mathcal{T} = \mathcal{Z}^+$, an outcome $x[k]$ of the random sequence $X[k]$ is drawn from a Bernoulli random variable with fixed parameter p. This is equivalent to a series of coin tosses with $P(H) = p$. At each time instant, the coin is tossed, the outcome is observed, and the result is "appended" to prior outcomes to construct a realization (a discrete-time function). The pdf of the random variable remains unchanged over k, which means the random sequence is first-order stationary. Because each toss is independent, the random sequence is also strictly stationary. Assume the coin has been tossed N times for which $X[k]$ has 2^N possible realizations. An example is shown in Figure 6.7, where $H \equiv 1$ and $T \equiv 0$. The set of 2^N realizations is referred to as the *ensemble* of the random sequence: each realization is an outcome of the underlying probability space. For this discrete and finite example, we can easily express the sample space of the random sequence in terms of $\Omega = \{H, T\}$ for the Bernoulli random variable as the following Cartesian product: $\mathcal{S} = \Omega \times \cdots \times \Omega = \Omega^N$. Its elements $\{\zeta\}$ are the 2^N *permutations*: $\{H, \ldots, H, H\}, \{H, \ldots, H, T\}, \ldots, \{T, \ldots, T, T\}$. Since the sequence is indexed by time (a direction is implied), we must consider all permutations, not just combinations; realization $\{H, \ldots, H, H, T\}$ is different from $\{H, \ldots, H, T, H\}$ even though each set has only one T. Since the random variables are independent and stationary (p does not change), each $\zeta \in \mathcal{S}$ with m heads has probability $p^m q^{N-m}$ with $q \triangleq 1 - p$.

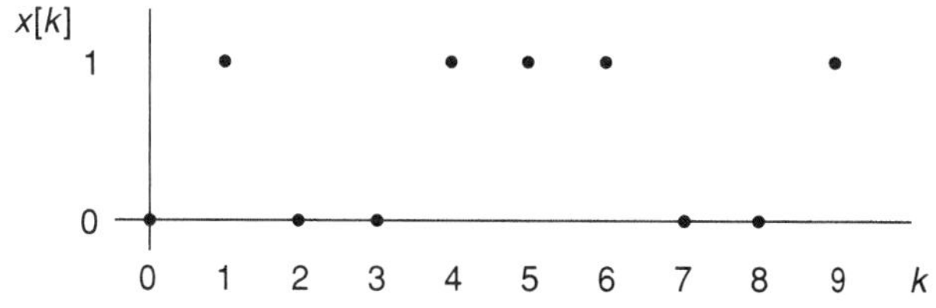

FIGURE 6.7 Example realization $x[k]$ of a Bernoulli random sequence for $\mathcal{T} = \{0, \ldots, 9\}$.

It is readily shown for this example that the probability measure is consistent. Obviously $p^m q^{N-m}$ does not depend on the order of outcomes of the random variables: any permutation of m heads and $N - m$ tails has the same probability. Consider a size-M subset of N random variables; the probability of this subset is computed as follows. If the subset contains m heads and $M - m$ tails, then

$$P(\underbrace{H, H, \ldots, H}_{m \text{ terms}}, \underbrace{T, \ldots, T}_{M-m \text{ terms}}) = P(\underbrace{H, H, \ldots, H}_{m \text{ terms}}, \underbrace{T, \ldots, T}_{M-m \text{ terms}}, \underbrace{\Omega, \ldots, \Omega}_{N-M \text{ terms}}),$$
$$= p^m q^{M-m} 1^{N-M} = p^m q^{M-m}, \tag{6.39}$$

where we have used independence and $P(\Omega) = 1$. The probability of the smaller subset is calculated from the larger subset by including $N - M$ sure events Ω.

From this example, we find that the second part of the definition of consistency can be stated for general distributions as follows. Consider the joint cdf of random process $X(t)$ at N time instants given by $F_{X(t_1),\ldots,X(t_N)}(x_1, \ldots, x_N)$. The corresponding cdf for a subset of $M < N$ time instants is obtained as follows:

$$F_{X(t_1),\ldots,X(t_M)}(x_1, \ldots, x_M) = F_{X(t_1),\ldots,X(t_M),X(t_{M+1}),\ldots,X(t_N)}(x_1, \ldots, x_M, \infty, \ldots, \infty), \tag{6.40}$$

where, without loss of generality, $t_M < t_N$ has been assumed. The corresponding result for the joint pdf is

$$f_{X(t_1),\ldots,X(t_M)}(x_1, \ldots, x_M) =$$
$$\underbrace{\int_{-\infty}^{\infty} \cdots \int_{-\infty}^{\infty}}_{N-M \text{ integrals}} f_{X(t_1),\ldots,X(t_{M+1}),\ldots,X(t_N)}(x_1, \ldots, x_{M+1}, \ldots, x_N) \mathrm{d}x_{M+1} \cdots \mathrm{d}x_N. \tag{6.41}$$

We have seen this type of expression in Chapter 4 involving random vectors. The difference here for a random process is that these properties must hold for *all subsets* of time instants in $\mathcal{T}$ in order to have consistency. Of course, these properties need not be verified: we simply specify the type of random process for a particular application, with the constraint that its probability measure be consistent.

The Bernoulli random sequence is known to be iid, which is perhaps the most fundamental kind of random process.

Definition: IID Random Process Random process $X(t)$ is iid if the random variables at all time instants $t \in \mathcal{T}$ are mutually independent and have the same distribution $F_{X(t)}(x) = F_X(x)$.

As shown in Example 6.5 for the Bernoulli random sequence, an iid random sequence of duration N has a sample space of the form $\mathcal{S} = \Omega^N$.

Example 6.7. Successive tosses of a fair die for $k \in \mathcal{Z}^+$ is an iid random sequence. This is similar to a Bernoulli random sequence, except that the random variables have a discrete uniform distribution with six outcomes $\Omega = \{1, 2, 3, 4, 5, 6\}$.

Example 6.8. The random number generator available with most signal processing software (such as MATLAB®) is an example of an algorithm that constructs iid random sequences (though technically it is a *pseudorandom* number (PRN) generator because eventually the numbers repeat, though usually that point is not reached in simulations). An ideal uniform random number generator provides a sequence of iid numbers drawn from [0, 1]. The distribution of this sequence can be transformed to a different distribution using the techniques described in Chapter 4.

Example 6.9. A Gaussian random sequence $X[k]$ is iid provided any finite set of random variables indexed by $k \in \mathcal{T} = \mathcal{Z}^+$ are mutually independent, and each has the same Gaussian distribution with fixed parameters $\{\mu, \sigma\}$. In general, a Gaussian random process requires that the set of random variables be jointly Gaussian; for the iid case, however, it is sufficient that each random variable be Gaussian because the joint pdf is the product of the marginal pdfs, which are all identical.

Although iid random processes are used in many applications such as digital communications, the fact that the samples at different times are independent can be overly restrictive. There is no "predictability" in the process: previous outcomes provide no information about what to expect for future outcomes. Some degree of predictability is important in many applications which we discuss in subsequent chapters.

6.7 INDEPENDENT INCREMENTS

In this section, we consider an important type of random process that can be viewed as a generalization of an iid process, and of which some well-known random processes are special cases. Let $X(t)$ be a random process for $t \in \mathcal{T}$. Since $X(t)$ at any time instant is a random variable, the difference (increment) $X(t_1) - X(t_2)$ for fixed t_1 and t_2 is also a random variable. This result follows from the techniques in Chapter 4 used for transformations of random variables: the pdf of $X(t_1) - X(t_2)$ can be computed from those of $X(t_1)$ and $X(t_2)$.

Definition: Independent Increment Process Random process $X(t)$ has *independent increments* if for any $t_1 < t_2 < t_3 < t_4$:

$$f_{X(t_2)-X(t_1),X(t_4)-X(t_3)}(x_2 - x_1, x_4 - x_3) = f_{X(t_2)-X(t_1)}(x_2 - x_1) f_{X(t_4)-X(t_3)}(x_4 - x_3). \tag{6.42}$$

This definition is a straightforward extension of independent random variables applied to *disjoint increments* of a random process. However, note that the increments can be contiguous, overlapping at individual time instants, such as t_2 and t_3 for the following:

$$\begin{aligned} f_{X(t_2)-X(t_1),X(t_3)-X(t_2),X(t_4)-X(t_3)}(x_2 - x_1, x_3 - x_2, x_4 - x_3) &= f_{X(t_2)-X(t_1)}(x_2 - x_1) \\ &\cdot f_{X(t_3)-X(t_2)}(x_3 - x_2) f_{X(t_4)-X(t_3)}(x_4 - x_3). \end{aligned} \tag{6.43}$$

The definition also applies to the situation where t_1 is specified relative to the origin (starting point of the process) for which $X(t_0) = x_0 = 0$ is fixed (nonrandom). Thus

$$f_{X(t_1)-X(t_0),X(t_2)-X(t_1)}(x_1 - x_0, x_2 - x_1) = f_{X(t_1)}(x_1) f_{X(t_2)-X(t_1)}(x_2 - x_1). \tag{6.44}$$

Specific examples of independent increment processes are provided later. First, we extend this type of process to include stationarity.

Definition: ISI Random Process The distribution of increments $X(t_N - \tau) - X(t_M - \tau)$ of an *independent and stationary increment* (isi) process is unchanged for any time shift τ and for any $t_N > t_M$.

This stationarity applies only to the *increments* of a random process, not the process itself as described in a previous section. As a result, increments of the *same duration* are iid. This property follows from the definition of strictly stationary, and it is not necessary to consider all joint distributions of increments (as generally required for strict stationarity) because the increments are independent. An independent random process that is first-order stationary is also strictly stationary because joint moments split into products of moments (since the joint pdf is a product of the marginal pdfs). The same applies to increments of an isi process.

Next, we consider a representation for some distributions that is also useful for a random sequence.

Definition: Infinitely Divisible Distribution $F_X(x)$ is an *infinitely divisible distribution* if there exist iid random variables $\{X_{n,1}, \ldots, X_{n,n}\}$ such that

$$Y_n = \sum_{m=1}^{n} X_{n,m}, \quad \text{for every } n \in \mathcal{N} \tag{6.45}$$

have the same distribution as X, and thus are equivalent random variables.

The $\{X_{n,m}\}$ have two subscripts to emphasize that $\{Y_n\}$ are obtained as sums of *different* random variables. Otherwise, for a single subscript, we could write $Y_n = Y_{n-1} + X_n$ such that Y_n and Y_{n-1} do not have the same distribution, as required in the definition. Since the random variables are iid, the distribution of their sum is obtained as $n-1$ convolutions of the same pdf. As a result, if X has finite support, it is not infinitely divisible; examples include the beta distribution and the continuous and discrete uniform random variables. Using characteristic functions (CFs), the convolutions correspond to products:

$$\Phi_X(\omega) = \Phi_{Y_n}(\omega) = \prod_{m=1}^{n} \Phi_{X_{n,m}}(\omega) = \Phi_{X_n}^{n}(\omega). \tag{6.46}$$

The last line follows because the random variables for every n are iid. Thus, if the CF of X can be written as the power of another CF for every $n \in \mathcal{N}$, then X is infinitely divisible. Several distributions in Appendix A with this property are summarized in Table 6.1.

Example 6.10. Consider the gamma random variable with parameters $\{\lambda, r\}$ and the CF given in the table:

$$\Phi_X(\omega) = \left(\frac{\lambda}{\lambda - j\omega}\right)^r = \left[\left(\frac{\lambda}{\lambda - j\omega}\right)^{r/n}\right]^n, \tag{6.47}$$

TABLE 6.1 Examples of Infinitely Divisible Distributions

Random Variable X {parameters}	CF: $\Phi_X(\omega)$ or $\Phi_X(j\omega)$
Cauchy $\{c, \alpha\}$	$\exp(j\omega c - \alpha\lvert\omega\rvert)$
Gamma $\{\lambda, r\}$	$[\lambda/(\lambda - j\omega)]^r$
Geometric $\{p\}$	$p[1 - q\exp(j\omega)]^{-1}$
Gaussian $\{\mu, \sigma\}$	$\exp(j\omega\mu - \sigma^2\omega^2/2)$
Negative binomial $\{p, N\}$	$p^N/[1 - q\exp(j\omega)]^N$
Poisson $\{\alpha\}$	$\exp(\alpha(\exp(j\omega) - 1))$
Student's t $\{r\}$	$K_{r/2}(\sqrt{r}\lvert\omega\rvert)\sqrt{(\sqrt{r}\lvert\omega\rvert)^r} \,/\, \Gamma(r/2)2^{r/2-1}$

where the $\{X_{n,1}, \ldots, X_{n,n}\}$ are iid gamma random variables with parameters $\{\lambda, r/n\}$. This example illustrates that a different set of random variables is used as n is varied, and thus the need for two subscripts on $X_{n,m}$. Similarly, for the Gaussian random variable:

$$\begin{aligned}\Phi_X(\omega) &= \exp(j\omega\mu - \sigma^2\omega^2/2) = \left(\exp([j\omega\mu - \sigma^2\omega^2/2]/n)\right)^n, \\ &= \left[\exp(j\omega(\mu/n) - (\sigma^2/n)\omega^2/2)\right]^n, \end{aligned} \tag{6.48}$$

where the $\{X_{n,1}, \ldots, X_{n,n}\}$ are iid Gaussian with parameters $\{\mu/n, \sigma^2/n\}$.

It is not always straightforward to demonstrate infinite divisibility for other random variables, such as Student's t distribution in the table. Also note that the $\{X_{n,1}, \ldots, X_{n,n}\}$ may not have the same type of distribution as X; this can be verified for the geometric random variable. The logarithmic random variable is infinitely divisible if the support is shifted to $\mathcal{Z}^+$ (instead of $\mathcal{N}$) (Johnson, Kemp, and Kotz, 2005).

For an isi random process, any increment can be divided into a sum of smaller increments. In particular, an interval of length $t_2 - t_1$ can be divided into N contiguous subintervals, each of length $(t_2 - t_1)/N$. These smaller increments are also independent as well as stationary, such that an *isi random process has infinite divisibility*. This discussion of infinite divisibility for random variables leads to the following property of an isi random sequence.

Theorem 6.3. Random sequence $X[k]$ is an isi sequence if and only if it can be expressed as the sum of iid random variables.

Proof. Only if: Let $X[k]$ for $k \in \mathcal{Z}^+$ be an isi sequence, and define $Y[k] \triangleq X[k] - X[k-1]$ which is necessarily an iid sequence. Observe that

$$X[k] = \sum_{n=1}^{k}(X[n] - X[n-1]) = \sum_{n=1}^{k} Y[n], \tag{6.49}$$

where $X[0] = 0$ has been assumed. The first equality follows because all terms except $X[k]$ cancel. Thus $X[k]$ has been written as the sum of iid random variables. *If:* Let $X[k]$ be the sum of iid random variables $\{Y[n]\}$ as in the second sum above. Observe that for any $k_2 > k_1$:

$$X[k_2] - X[k_1] = \sum_{n=1}^{k_2} Y[n] - \sum_{n=1}^{k_1} Y[n] = \sum_{n=k_1+1}^{k_2} Y[n]. \tag{6.50}$$

This shows that all disjoint increments are independent because they do not contain any common terms of $\{Y[n]\}$. Also, the distribution of any increment is unchanged by a shift in time because $\{Y[n]\}$ are identically distributed, which completes the proof. □

Note that different size increments generally have different distributions; independent and stationary increments are *not* necessarily identically distributed.

6.8 MARTINGALES

In this section, we present an interesting property of a random sequence that will be useful in Chapter 7 where it is decomposed into two distinct components. Without loss of generality, assume the process starts at time $t = 0$.

Definition: Martingale Sequence Random sequence $X[k]$ with $\mathcal{E}[|X[k]|] < \infty$ is a *martingale* if

$$\mathcal{E}[X[k]|X[0], \ldots, X[k-1]] = X[k-1], \quad \text{for all } k \geq 1. \tag{6.51}$$

It is important to note that there is no expectation on the right-hand side: this particular conditional mean is the most recent *random variable* $X[k-1]$ of the sequence. Rewriting this expression, observe that

$$\mathcal{E}[X[k] - X[k-1]|X[0], \ldots, X[k-1]] = 0, \tag{6.52}$$

where $X[k-1]$ has been brought inside the expectation because it is part of the conditioning. This expression shows that on average the *change* in a martingale sequence from one time instant to the next is zero.

Example 6.11. Let $Y[k]$ be a Bernoulli random sequence where a coin with $p = P(H)$ is tossed repeatedly, and define $X[k]$ be the net amount of a gambler's winnings at time k. When the coin shows H, the gambler wins a fixed amount $c > 0$; otherwise the gambler loses the same amount c. Thus

$$X[k+1] = \begin{cases} X[k] + c, & Y[k] = H \\ X[k] - c, & Y[k] = T. \end{cases} \tag{6.53}$$

Conditioning on all previous net winnings gives

$$\begin{aligned} \mathcal{E}[X[k+1]|X[0], \ldots, X[k]] &= p(X[k] + c) + q(X[k] - c), \\ &= X[k] + (p - q)c, \end{aligned} \tag{6.54}$$

where $q \triangleq 1 - p$. As a result, $X[k]$ is a martingale only when the game is fair with $p = q = 1/2$.

For Examples 6.12 and 6.13, the definition of a martingale sequence is generalized as follows.

Definition: Generalized Martingale Sequence Random sequence $Y[k]$ with $\mathcal{E}[|Y[k]|] < \infty$ is a *martingale with respect to sequence* $X[k]$ if

$$\mathcal{E}[Y[k]|X[0], \ldots, X[k-1]] = Y[k-1], \quad \text{for all } k \geq 1. \tag{6.55}$$

The right-hand side is the previous sample of $Y[k]$, and not of the conditioning sequence.

Example 6.12. A useful quantity in parameter estimation is the likelihood function (see Chapter 9). Assume that we are interested in estimating the scalar parameter θ of pdf $f_X(x)$ for an iid random sequence. For example, the parameter might be μ of a Gaussian random variable or the probability of heads p for a coin toss. Consider two hypotheses regarding the value of the parameter: H_0: $\theta = \theta_0$ and H_1: $\theta = \theta_1$. It is possible to have multiple hypotheses, but for simplicity we consider only two hypotheses in this example. The likelihood ratio is defined as

$$\Lambda(x) \triangleq \frac{f_X(x; \theta_0)}{f_X(x; \theta_1)}, \tag{6.56}$$

where the value of the parameter is explicitly shown in the argument of each pdf. Consider k samples $\{X[1], \ldots, X[k]\}$ of a random sequence corresponding to independent random variables. The likelihood ratio for these k samples is generalized to (Larson and Shubert, 1979)

$$\Lambda[k] \triangleq \prod_{m=1}^{k} \frac{f_X(X[m];\theta_0)}{f_X(X[m];\theta_1)}, \tag{6.57}$$

which itself can be viewed as a random sequence. The randomness in $\Lambda[k]$ is due to the fact that the pdfs have been written as functions of the random variables $\{X[m]\}$ (and not their outcomes). In this type of estimation problem, the form of the pdf is known and only the parameter is estimated. For this example, there are two values for θ, both of which are known; we want to decide which value of θ is "most likely" to be correct given the observed samples. Consider the following conditional expectation for $\Lambda[k]$:

$$\begin{aligned} \mathcal{E}[\Lambda[k]|X[1], \ldots, X[k-1]] = \mathcal{E}\Bigg[& \frac{f_X(X[k];\theta_0)}{f_X(X[k];\theta_1)} \\ & \cdot \prod_{m=1}^{k-1} \frac{f_X(X[m];\theta_0)}{f_X(X[m];\theta_1)} \Bigg| X[1], \ldots, X[k-1] \Bigg], \end{aligned} \tag{6.58}$$

where we have substituted $\Lambda[k]$ and factored the ratio of pdfs at time instant k. Since the random variables are independent, the expectation splits into a product:

$$\begin{aligned} \mathcal{E}[\Lambda[k]|X[1], \ldots, X[k-1]] = & \mathcal{E}\left[\prod_{m=1}^{k-1} \frac{f_X(X[m];\theta_0)}{f_X(X[m];\theta_1)} \Bigg| X[1], \ldots, X[k-1] \right] \\ & \cdot \mathcal{E}\left[\frac{f_X(X[k];\theta_0)}{f_X(X[k];\theta_1)} \right]. \end{aligned} \tag{6.59}$$

The conditioning in the second expectation is not needed because $X[k]$ is not part of the conditioning and the random variables are independent. Since the first expectation is $\Lambda[k-1]$:

$$\mathcal{E}[\Lambda[k]|X[1], \ldots, X[k-1]] = \Lambda[k-1]\mathcal{E}\left[\frac{f_X(X[k];\theta_0)}{f_X(X[k];\theta_1)} \right], \tag{6.60}$$

demonstrating that $\Lambda[k]$ is a martingale with respect to $X[k]$ only if the expectation on the right-hand side equals 1. This occurs if the *actual* parameter is θ_1 such that

$$\mathcal{E}\left[\frac{f_X(X[k];\theta_0)}{f_X(X[k];\theta_1)} \right] = \int_{-\infty}^{\infty} \frac{f_X(x_k;\theta_0)}{f_X(x_k;\theta_1)} f_X(x_k;\theta_1)\mathrm{d}x_k = \int_{-\infty}^{\infty} f_X(x_k;\theta_0)\mathrm{d}x_k = 1. \tag{6.61}$$

If $\Lambda[k]$ is defined with θ_0 and θ_1 interchanged, then θ_0 would be used in the result above: $\Lambda[k]$ is a martingale if its *denominator* depends on the actual parameter.

Example 6.13 (de Moivre's martingale). When $p \neq q$ in Example 6.11, it is still possible to find a martingale by defining the sequence

$$Y[k] = (q/p)^{X[k]}, \tag{6.62}$$

where $X[k]$ is the original Bernoulli sequence. Thus

$$\begin{aligned} \mathcal{E}[Y[k+1]|X[1], \ldots, X[k]] &= \mathcal{E}\left[(q/p)^{X[k+1]} |X[1], \ldots, X[k] \right], \\ &= p(q/p)^{X[k]+c} + q(q/p)^{X[k]-c}, \end{aligned} \tag{6.63}$$

where the conditioning in (6.53) has been used. Rearranging this expression gives

$$\begin{aligned}\mathcal{E}[Y[k+1]|X[1],\ldots,X[k]] &= (q/p)^{X[k]}[p(q/p)^c + q(q/p)^{-c}],\\ &= Y[k][p(q/p)^c + q(p/q)^c] \qquad (6.64)\end{aligned}$$

from which we see $Y[k]$ is a martingale with respect to $X[k]$ provided that $c = 1$: the gambler must win or lose one unit at each time instant k.

The following theorem is another generalization of a martingale sequence where the most recent random sample in the conditioning is not necessarily at the previous time instant.

Theorem 6.4. For martingale sequence $X[k]$ and $n > k$,

$$\mathcal{E}[X[n]|X[0],\ldots,X[k-1]] = X[k-1], \quad \text{for } 1 < k \leq n. \qquad (6.65)$$

Proof. Include an inner expectation conditioned on $\{X[0],\ldots,X[n-1]\}$ such that

$$\begin{aligned}\mathcal{E}[X[n]|X[0],\ldots,X[k-1]] &= \mathcal{E}[\mathcal{E}[X[n]|X[0],\ldots,X[n-1]]|X[0],\ldots,X[k-1]],\\ &= \mathcal{E}[X[n-1]|X[0],\ldots,X[k-1]]. \qquad (6.66)\end{aligned}$$

The second equality follows because $X[k]$ is assumed to be a martingale. Another inner expectation conditioned on $\{X[0],\ldots,X[n-2]\}$ can be included such that the last expression becomes

$$\begin{aligned}\mathcal{E}[X[n-1]|X[0],\ldots,X[k-1]] &= \mathcal{E}[\mathcal{E}[X[n-1]|X[0],\ldots,X[n-2]]|X[0],\ldots,X[k-1]],\\ &= \mathcal{E}[X[n-2]|X[0],\ldots,X[k-1]]. \qquad (6.67)\end{aligned}$$

Continuing this process for $X[n-3]$, $X[n-4]$, and so on, until $X[k-1]$ is reached, the left-hand side of (6.66) becomes

$$\mathcal{E}[X[n]|X[0],\ldots,X[k-1]] = \mathcal{E}[X[k-1]|X[0],\ldots,X[k-1]] = X[k-1], \qquad (6.68)$$

which completes the proof. □

This theorem can be extended to a random sequence $Y[k]$ that is a martingale with respect to another sequence $X[k]$:

$$\mathcal{E}[Y[n]|X[0],\ldots,X[k-1]] = Y[k-1], \quad \text{for } 1 < k \leq n. \qquad (6.69)$$

For completeness, we include the following definitions.

Definition: Submartingale and Supermartingale Random sequence $X[k]$ is a *submartingale* if

$$\mathcal{E}[X[k]|X[0],\ldots,X[k-1]] \geq X[k-1], \quad \text{for all } k \geq 1, \qquad (6.70)$$

and a *supermartingale* if

$$\mathcal{E}[X[k]|X[0],\ldots,X[k-1]] \leq X[k-1], \quad \text{for all } k \geq 1. \qquad (6.71)$$

Because equality is included in these definitions, a martingale is necessarily a submartingale and a supermartingale. Note that the Bernoulli sequence in Example 6.11 is a submartingale for $p > q$ and a supermartingale for $p < q$ (since $c > 0$).

Finally, we provide definitions of martingales for random processes where $\mathcal{T}$ is continuous time.

Definition: Martingale Process Random process $X(t)$ with $\mathcal{E}[|X(t)|] < \infty$ is a *martingale* if

$$\mathcal{E}[X(t_2)|X(t), t \leq t_1] = X(t_1), \quad \text{for } t_1 < t_2. \tag{6.72}$$

Random process $Y(t)$ with $\mathcal{E}[|Y(t)|] < \infty$ is a *martingale with respect to process* $X(t)$ if

$$\mathcal{E}[Y(t_2)|X(t), t \leq t_1] = Y(t_1), \quad \text{for } t_1 < t_2. \tag{6.73}$$

For a continuous-time martingale process, the expectation reduces to the random variable at the most recent time that is *included in the conditioning*. The following theorem is important for independent increment processes, of which there are several well-known examples (some are discussed later).

Theorem 6.5. An independent-increment process $X(t)$ with zero mean is a martingale.

Proof. From the linearity of expectation,

$$\mathcal{E}[X(t_2)|X(t), t \leq t_1] = \mathcal{E}[X(t_2) - X(t_1)|X(t), t \leq t_1] + \mathcal{E}[X(t_1)|X(t), t \leq t_1]. \tag{6.74}$$

Since the process has independent increments, the conditioning in the first expectation on the right-hand side can be dropped:

$$\mathcal{E}[X(t_2)|X(t), t \leq t_1] = \mathcal{E}[X(t_2) - X(t_1)] + \mathcal{E}[X(t_1)|X(t), t \leq t_1]. \tag{6.75}$$

Applying the zero-mean assumption gives

$$\mathcal{E}[X(t_2)|X(t), t \leq t_1] = 0 + \mathcal{E}[X(t_1)|X(t), t \leq t_1] = X(t_1), \tag{6.76}$$

where we have used the fact that the conditioning on the right-hand side includes $X(t_1)$, thus completing the proof. □

Note that the zero-mean assumption is required.

Example 6.14. Since the Wiener process $W(t)$ described later is an independent increment process, it is also a martingale. We demonstrate this property by directly evaluating the conditional expectation. It is shown later in Chapter 9 on parameter estimation that for jointly Gaussian random variables X and Y, the conditional expectation is

$$\mathcal{E}[X|Y] = \mu_X + (C_{XY}/\sigma_Y^2)(Y - \mu_Y). \tag{6.77}$$

For a Wiener process, the various quantities are $X = X(t_2)$, $Y = X(t_1)$, $\mu_X = \mu_Y = 0$, $C_{XY}(t_1, t_2) = \min(t_1, t_2) = t_1$, and $\sigma_Y^2 = t_1$. Thus, we immediately have

$$\mathcal{E}[X(t_2)|X(t_1)] = X(t_1), \quad \text{for } t_1 < t_2. \tag{6.78}$$

6.9 MARKOV SEQUENCE

The next property of a random sequence involves conditioning across several time instants.

Definition: Markov Sequence *Markov sequence* $X[k]$ is a time-indexed set of random variables that satisfies the following Markov property:

$$P(X[k] = x_k|X[1] = x_1, \ldots, X[k-1] = x_{k-1}) = P(X[k] = x_k|X[k-1] = x_{k-1}). \tag{6.79}$$

A Markov sequence is also called a *Markov chain*.

For a Markov sequence, the probability of an outcome at a particular time instant given the outcomes at all previous time instants requires conditioning only on the most recent outcome; the complete history of outcomes is not needed. The property in (6.79) is sometimes referred to as *one-step* Markov; it can be generalized to m steps as follows.

Definition: Order-*m* Markov Sequence An *order-m Markov sequence* $X[k]$ has the following property for $k > m$:

$$\begin{aligned} P(X[k] = x_k|X[1] = x_1, \ldots, X[k-m-1] = x_{k-m-1}, \\ X[k-m] = x_{k-m}, \ldots, X[k-1] = x_{k-1}), \\ = P(X[k] = x_k|X[k-m] = x_{k-m}, \ldots, X[k-1] = x_{k-1}). \end{aligned} \tag{6.80}$$

Although a Markov sequence can be described for a countably infinite set of outcomes, most Markov applications involve sequences with a finite number of outcomes. Such random sequences can be represented by a finite number of states.

Definition: States of a Markov Chain The *states* of a Markov chain are the possible outcomes of the random variable $X[k]$ at any time instant k.

For notational convenience, we represent the one-step conditional probability as follows:

$$p_{mn} \triangleq P(X[k] = n|X[k-1] = m). \tag{6.81}$$

The process is assumed to be stationary such that $\{p_{mn}\}$ do not depend on the specific time k. States are usually numbered sequentially using subsets of $\mathcal{Z}^+$, and thus can be viewed as a mapping from the actual outcomes to numbers that are more convenient to manipulate.

Assuming a finite number of states, a Markov chain can be represented by a *state transition matrix*. Let the number of states be M and label them $1, \ldots, M$, to represent the outcomes of $X[k]$. The column vector $\mathbf{p}[k] \triangleq [p_1[k], \ldots, p_M[k]]^T \in [0, 1]^M$ contains the probabilities of the sequence being in each of the M possible states at time instant k. (The state probabilities $p_m[k]$ should not be confused with the probability mass function (pmf) $p_X[x]$.) The state transition matrix $\mathbf{P} \in [0, 1]^{M \times M}$ has elements

$$\mathbf{P} \triangleq \begin{bmatrix} p_{11} & \cdots & p_{1M} \\ \vdots & \cdots & \vdots \\ p_{M1} & \cdots & p_{MM} \end{bmatrix} \tag{6.82}$$

with $\{p_{mn}\}$ given by (6.81). From the definition of conditional probability, consider the nth element of $\mathbf{p}[k]$:

$$\begin{aligned} p_n[k] &= \sum_{m=1}^{M} P(X[k] = n, X[k-1] = m), \\ &= \sum_{m=1}^{M} P(X[k] = n | X[k-1] = m) P(X[k-1] = m), \end{aligned} \tag{6.83}$$

where the sum is performed over all outcomes (states) of the random sequence at time $k-1$. Using the notation for the elements of $\mathbf{P}$, we have

$$p_n[k] = \sum_{m=1}^{M} p_{mn} p_m[k-1], \tag{6.84}$$

which is the law of total probability. For the entire state vector, the transition from time instant $k-1$ to k is the following *recursion*:

$$\mathbf{p}^T[k] = \mathbf{p}^T[k-1]\mathbf{P}, \tag{6.85}$$

which we could write as column vectors: $\mathbf{p}[k] = \mathbf{P}^T \mathbf{p}[k-1]$. However, the row-vector form above will be used in order to be consistent with most books on Markov chains. This form is preferable because it is more natural to view p_{mn} as moving from state m to n (left-to-right on the subscript), which requires that the elements of $\mathbf{p}[k-1]$ be weighted by the nth column of $\mathbf{P}$ to give $p_n[k]$.

Obviously, the elements of $\mathbf{P}$ must be nonnegative because they are probabilities. If an element p_{mn} is zero, the transition from state m to state n cannot happen. However, the reverse transition from state n to m is possible provided $p_{nm} \neq 0$; $\mathbf{P}$ is generally *not* a symmetric matrix. From time instant $k-1$ to k, something must happen: either we move to a different state or remain in the same state. This means that each *row* of $\mathbf{P}$ sums to 1. Summarizing, the two defining properties of $\mathbf{P}$ are

$$p_{mn} \geq 0, \quad m, n = 1, \ldots, M, \tag{6.86}$$

$$\sum_{n=1}^{M} p_{mn} = 1, \quad m = 1, \ldots, M. \tag{6.87}$$

A state transition matrix is also known as a *stochastic matrix*, and since its rows sum to 1, $\mathbf{P}$ is specifically called a *right stochastic matrix*.

Example 6.15. An example *state-transition diagram* for a Markov chain with three states is shown in Figure 6.8. Suppose it has the following transition matrix:

$$\mathbf{P} = \begin{bmatrix} 0.1 & 0.3 & 0.6 \\ 0 & 0.5 & 0.5 \\ 0.9 & 0.1 & 0 \end{bmatrix}. \tag{6.88}$$

When the sequence enters state 3, it must transition to a different state in the next time instant because $p_{33} = 0$. (For zero transition probabilities, an arrow is generally not shown in the state transition diagram). Note that all rows above sum to 1 whereas the columns do not (though, of course, depending on the problem one or more columns could sum to 1).

We summarize some important properties of a Markov chain.

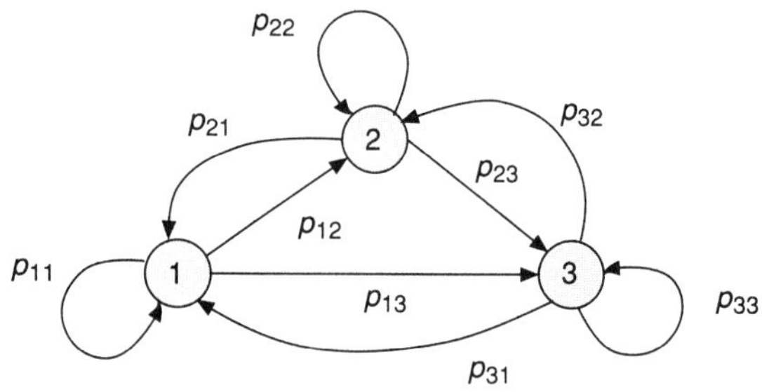

FIGURE 6.8 Example of a Markov chain with three states.

Definition: Irreducible A Markov chain is *irreducible* if any state can be reached from any other state at some point in time.

This means there exists $k > 0$ such that

$$P(X[k] = n|X[0] = m) > 0. \tag{6.89}$$

Irreducible does not imply that all the transition probabilities in $\mathbf{P}$ are nonzero. Even though two elements of $\mathbf{P}$ are zero in Example 6.15, we see that it is possible to reach any state (one can easily verify this from the state transition diagram). In particular, even though the sequence cannot transition from state 1 to state 2 because $p_{12} = 0$, it is possible to reach state 2 by first transitioning to state 3.

Example 6.16. A Markov chain with the following state transition matrix is *not* irreducible:

$$\mathbf{P} = \begin{bmatrix} 0.5 & 0.5 & 0 & 0 \\ 0.2 & 0.8 & 0 & 0 \\ 0 & 0 & 0.1 & 0.9 \\ 0 & 0 & 0.3 & 0.7 \end{bmatrix}. \tag{6.90}$$

It is not possible to reach states 3 and 4 at any time from states 1 and 2, and vice versa.

It turns out that $\mathbf{P}$ in Example 6.16 has two communicating classes.

Definition: Communicating Class A *communicating class* in a Markov chain is a subset of states such that those states can be reached from each other at some point in time, and which no states communicate with states outside the class.

Thus, an irreducible Markov chain has only one communicating class. For Example 6.16, we can view the two communicating classes as two separate irreducible Markov chains.

It is often convenient to examine types of state transitions other than one step. Consider the r-step transition from state m to state n with the following conditional probability:

$$P(X[k] = n|X[k-r] = m) \triangleq p_{mn}^{(r)}. \tag{6.91}$$

In Theorem 6.6, we demonstrate how to obtain this transition probability from lower order transition probabilities.

Theorem 6.6 (Chapman–Kolmogorov). For Markov chain $X[k]$, an r-step transition probability is computed from the $(r-1)$-step transition probabilities as follows:

$$p_{mn}^{(r)} = \sum_{q=1}^{M} p_{mq}^{(r-1)} p_{qn}, \tag{6.92}$$

where $p_{mq}^{(1)} \equiv p_{mq}$.

Proof. From the law of total probability in (6.83),

$$P(X[r] = n) = \sum_{q=1}^{M} P(X[r] = n | X[r-1] = q) P(X[r-1] = q). \tag{6.93}$$

Conditioning on the initial state $X[0] = m$ gives

$$\begin{aligned} P(X[r] = n | X[0] = m) = \sum_{q=1}^{M} & P(X[r] = n | X[0] = m, X[r-1] = q) \\ & \cdot P(X[r-1] = q | X[0] = m), \end{aligned} \tag{6.94}$$

and using the previous notation,

$$p_{mn}^{(r)} = \sum_{q=1}^{M} P(X[r] = n | X[0] = m, X[r-1] = q) p_{mq}^{(r-1)}. \tag{6.95}$$

Since $X[k]$ is a Markov chain, we can substitute

$$P(X[r] = n | X[0] = m, X[r-1] = q) = P(X[r] = n | X[r-1] = q) = p_{qn}, \tag{6.96}$$

which completes the proof. □

Defining the r-step state transition matrix $\mathbf{P}^{(r)}$ with elements $\{p_{mn}^{(r)}\}$, the sum in (6.92) gives the following recursion:

$$\mathbf{P}^{(r)} = \mathbf{P}\mathbf{P}^{(r-1)}. \tag{6.97}$$

Repeatedly applying the Chapman–Kolmogorov equation yields

$$\mathbf{P}^{(r)} = \mathbf{P}\mathbf{P}\mathbf{P}^{(r-2)} = \mathbf{P}^{r}, \tag{6.98}$$

where $\mathbf{P}$ is the original one-step transition matrix.

Definition: Transient, Recurrent, and Absorbing States A state is *transient* if after entering that state, there is a nonzero probability that the state will not be reached again. A *recurrent state* will be encountered again with probability one. It is not possible to leave an *absorbing state* once it is reached.

Obviously for absorbing state m, $p_{mm} = 1$ and $p_{mn} = 0$ for $m \neq n$. A mathematical description of transient and recurrent states can be given as follows. Define the conditional probability:

$$\tilde{p}_n^{(k)} \triangleq P(X[k] = n, X[m] \neq n, m = 1, \ldots, k-1 | X[0] = n), \tag{6.99}$$

which is the probability that the sequence *returns* to the initial state n only at time instant k. Then

$$\sum_{k=1}^{\infty} \tilde{p}_n^{(k)} \triangleq \tilde{p}_n \tag{6.100}$$

is the probability that the sequence returns to state n at *some* point in time. Thus, $\tilde{p}_n = 1$ for a recurrent state whereas $\tilde{p}_n < 1$ for a transient state. It is not necessary that $\tilde{p}_n = 0$ for a transient state, which may or may not be reached again. This description, which is not easy to evaluate, leads to the following theorem, which we state without proof.

Theorem 6.7. State n is transient if and only if

$$\sum_{r=1}^{\infty} p_{nn}^{(r)} < \infty, \tag{6.101}$$

and is recurrent if and only if

$$\sum_{r=1}^{\infty} p_{nn}^{(r)} = \infty. \tag{6.102}$$

As a result, we can examine only the diagonal elements of the transition matrix **P**, which are the probabilities that a Markov chain stays in the same state for the next transition.

Example 6.17. Consider the following transition matrix:

$$\mathbf{P} = \begin{bmatrix} 0 & 0 & 0 & 1 \\ 1/3 & 1/3 & 1/3 & 0 \\ 0 & 0 & 1 & 0 \\ 1 & 0 & 0 & 0 \end{bmatrix}. \tag{6.103}$$

This simple example is easily characterized using the state transition diagram shown in Figure 6.9. State 3 is obviously an absorbing state, and states 1 and 4 are recurrent states. State 2 is a transient state because the probability that it will not be reached again is $2/3$.

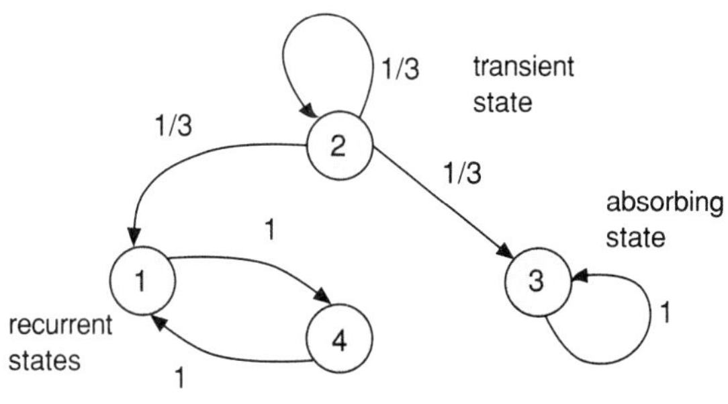

FIGURE 6.9 State transition diagram for Example 6.17.

Using the result in (6.98), it is possible to compute the state probabilities at any time in the future provided the initial state $\mathbf{p}[0]$ is known. The recursive form in (6.85) corresponds to a linear system with "output" $\mathbf{p}[k]$. It is, in fact, a type of state-space model but with zero input and written in row-vector form (see Chapter 1). It is well known that the behavior of such a linear system is controlled by the eigenvalues of $\mathbf{P}$, which are identical to the poles of the system. These are obtained by solving a characteristic equation.

Definition: Characteristic Equation The *characteristic equation* of a Markov chain with transition matrix $\mathbf{P}$ is

$$\det(\mathbf{P} - \lambda \mathbf{I}) = 0, \tag{6.104}$$

where $\mathbf{I} \in \mathcal{R}^{M \times M}$ is an identity matrix, λ is a scalar, and $\det(\mathbf{A})$ is the determinant of square matrix $\mathbf{A}$.

The characteristic equation is an Mth-order polynomial in λ that has at most M unique solutions corresponding to the *eigenvalues* of $\mathbf{P}$.

Example 6.18. Consider the following second-order state-transition matrix:

$$\mathbf{P} = \begin{bmatrix} 0.2 & 0.8 \\ 0.5 & 0.5 \end{bmatrix}. \tag{6.105}$$

The characteristic equation is

$$\det \begin{bmatrix} 0.2 - \lambda & 0.8 \\ 0.5 & 0.5 - \lambda \end{bmatrix} = \lambda^2 - 0.7\lambda - 0.3, \tag{6.106}$$

$$= (\lambda - 1)(\lambda + 0.3) = 0, \tag{6.107}$$

which yields eigenvalues $\{-0.3, 1\}$.

The following is a useful property of $\mathbf{P}$ because its rows sum to 1.

Theorem 6.8. The eigenvalues of state transition matrix $\mathbf{P}$ are bounded as $|\lambda| \leq 1$. At least one eigenvalue is equal to 1.

Proof. Start with the eigenvalue equation:

$$\mathbf{Pq} = \lambda \mathbf{q} \tag{6.108}$$

for nonzero *right* eigenvector $\mathbf{q} \in \mathcal{R}^M$. Choose the mth row of $\mathbf{Pq}$ such that $|q_m| \geq |q_n|$ for $1 \leq n \leq M$:

$$\sum_{n=1}^{M} p_{mn} q_n = \lambda q_m. \tag{6.109}$$

Since the elements of $\mathbf{P}$ are nonnegative, we have the following upper bound:

$$|\lambda q_m| \leq \sum_{n=1}^{M} p_{mn} |q_n|. \tag{6.110}$$

Because q_m has the largest magnitude of all elements in $\mathbf{q}$, every $|q_n|$ in the sum can be replaced with $|q_m|$ to give another upper bound, and $|q_m|$ can be factored out:

$$\sum_{n=1}^{M} p_{mn}|q_n| \leq |q_m| \sum_{n=1}^{M} p_{mn} = |q_m|, \tag{6.111}$$

where we have used the fact that the rows of $\mathbf{P}$ sum to 1 (i.e., summing over the column index n). Combining these results gives

$$|\lambda q_m| = |\lambda||q_m| \leq |q_m| \Longrightarrow |\lambda| \leq 1. \tag{6.112}$$

Note that this result holds for any of the eigenvalues of $\mathbf{P}$; we did not choose a specific eigenvalue λ and right eigenvector $\mathbf{q}$ above. Finally, since the rows of $\mathbf{P}$ sum to 1, we can write

$$\mathbf{P1} = \mathbf{1}, \tag{6.113}$$

where $\mathbf{1} \triangleq [1, \ldots, 1]^T \in \mathcal{R}^M$. This is an eigenvalue equation with right eigenvector $\mathbf{1}$ and eigenvalue $\lambda = 1$; thus, at least one of the eigenvalues of $\mathbf{P}$ is equal to 1, which completes the proof. □

Assume that $\mathbf{P}$ has M linearly independent *left* eigenvectors, and consider the following decomposition of $\mathbf{P}$:

$$\mathbf{V}^T\mathbf{P} = \mathbf{\Lambda}\mathbf{V}^T \Longrightarrow \mathbf{P} = \mathbf{V}^{-T}\mathbf{\Lambda}\mathbf{V}^T, \tag{6.114}$$

where the rows of $\mathbf{V}^T$ are the left eigenvectors of $\mathbf{P}$, and $\mathbf{\Lambda}$ is a diagonal matrix containing the eigenvalues $\{\lambda_m\}$ (see Appendix G). Expanding this as a sum of outer products yields

$$\mathbf{P} = \sum_{m=1}^{M} \lambda_m \mathbf{s}_m \mathbf{v}_m^T, \tag{6.115}$$

where $\{\mathbf{v}_m^T\}$ are the rows of $\mathbf{V}^T$ (columns of $\mathbf{V}$), and $\{\mathbf{s}_m\}$ are the columns of $\mathbf{V}^{-T}$ (rows of $\mathbf{V}^{-1}$). Note that the left and right eigenvectors are not the same unless $\mathbf{P}$ happens to be symmetric, which is generally not the case for a Markov chain. The left eigenvectors are not unique: $a\mathbf{v}_n^T$ for $a \neq 0$ is also an eigenvector associated with λ_n. Generally for convenience, the eigenvectors are normalized to have unit length: $\mathbf{v}_m^T\mathbf{v}_m = 1$. Note that $\mathbf{P}^k$ can also be decomposed as above: $\mathbf{P}^k$ and $\mathbf{P}$ have the same eigenvectors, and the only difference is that $\mathbf{\Lambda}^k$ appears in place of $\mathbf{\Lambda}$. Using this eigendecomposition, the state probabilities at time instant k can be expressed in a nonrecursive form as follows:

$$\mathbf{p}^T[k] = \sum_{m=1}^{M} \lambda_m^k \mathbf{p}^T[0]\mathbf{s}_m\mathbf{v}_m^T = \sum_{m=1}^{M} c_m \lambda_m^k \mathbf{v}_m^T, \tag{6.116}$$

where $c_m \triangleq \mathbf{p}^T[0]\mathbf{s}_m$ is a scalar. This expression has the form of a geometric progression from which it is evident that in order for $\mathbf{p}[k]$ to be finite with increasing k, $|\lambda_m|$ must not exceed 1 for any m. Although the left eigenvectors and eigenvalues of $\mathbf{P}$ are required in (6.116), it is not necessary that we find $\{\mathbf{s}_m\}$ because only the $\{c_m\}$ are needed, and these are derived from the initial conditions as demonstrated in the example below. Finally, we mention that it is possible to rewrite the results above in terms of right eigenvectors by defining the state transition matrix to be $\mathbf{P}^T$. However, as mentioned earlier, most books on Markov chains use the definition of $\mathbf{P}$ with $P(X[k] = n|X[k-1] = m]) = p_{mn}$.

The steady-state probabilities of the states, given by $\lim_{k\to\infty} \mathbf{p}[k] = \mathbf{p}[\infty]$, are obtained from all terms in (6.116) with $\lambda = 1$ since the other terms with $|\lambda| < 1$ decay to zero. (Example 6.20 examines $\mathbf{P}$ where two eigenvalues have unit magnitude.) Assuming the nth eigenvalue is 1, the steady-state solution is

$$\mathbf{p}[\infty] = \lim_{k\to\infty} \sum_{m=1,m\neq n}^{M} c_m \lambda_m^k \mathbf{v}_m + c_n \mathbf{v}_n = c_n \mathbf{v}_n, \tag{6.117}$$

which is proportional to the eigenvector associated with $\lambda_n = 1$. (Although we have written this expression as a column vector, $\mathbf{v}_m^T$ is still a left eigenvector of $\mathbf{P}$.)

Example 6.19. For $\mathbf{P}$ in Example 6.18, the eigenvectors are obtained by solving the following two matrix equations:

$$\mathbf{v}_1^T(\mathbf{P} + 0.3\mathbf{I}) = \mathbf{v}_1^T \begin{bmatrix} 0.5 & 0.8 \\ 0.5 & 0.8 \end{bmatrix} = \mathbf{0}, \tag{6.118}$$

$$\mathbf{v}_2^T(\mathbf{P} - \mathbf{I})\mathbf{q}_2 = \mathbf{v}_2^T \begin{bmatrix} -0.8 & 0.8 \\ 0.5 & -0.5 \end{bmatrix} = \mathbf{0}, \tag{6.119}$$

resulting in

$$\mathbf{v}_1^T = [1, -1], \quad \mathbf{v}_2^T = [1, 1.6], \tag{6.120}$$

and

$$\mathbf{p}[k] = c_1(-0.3)^k \begin{bmatrix} 1 \\ -1 \end{bmatrix} + c_2 \begin{bmatrix} 1 \\ 1.6 \end{bmatrix}. \tag{6.121}$$

Assume the initial state probability vector is $\mathbf{p}[0] = [0.6, 0.4]^T$. From the steady-state result in (6.117), we know the terms must sum to one so that

$$c_2 + (1.6)c_2 = 1 \Longrightarrow c_2 = 1/2.6 \approx 0.384. \tag{6.122}$$

From the initial state:

$$\mathbf{p}[0] = c_1 \begin{bmatrix} 1 \\ -1 \end{bmatrix} + c_2 \begin{bmatrix} 1 \\ 1.6 \end{bmatrix} = \begin{bmatrix} 0.6 \\ 0.4 \end{bmatrix}. \tag{6.123}$$

The first line gives $c_1 + c_2 = 0.6 \Longrightarrow c_1 = 0.6 - 1/2.6 \approx 0.215$. We could also have obtained c_1 and c_2 by solving the two linear equations in (6.123). However, it is usually easier to first find one coefficient from the steady-state term of the decomposition as done above (because at least one eigenvalue = 1). This is especially useful when $M = 3$ because after one coefficient is known, it is easy to find the other two by solving only two of the three linear equations.

In general, $\mathbf{P}$ is a nonnegative matrix.

Definition: Positive and Nonnegative Matrices Matrix $\mathbf{\Lambda}$ is *positive* if all elements are strictly positive. It is *nonnegative* if all elements are nonnegative.

A positive matrix is *not* the same as a positive-definite matrix where all *eigenvalues* are strictly positive. As we saw in Example 6.18 for positive $\mathbf{P}$, the eigenvalues can be negative. If $\mathbf{P}$ happens to be positive, then the following theorem describes some useful properties, which we provide without proof.

Theorem 6.9 (Perron–Frobenius). For positive matrix $\mathbf{A} \in (0, \infty)^{M \times M}$:

- There exists a real eigenvalue $\lambda_o > 0$ such that all other eigenvalues satisfy $|\lambda| < \lambda_o$.
- The multiplicity of λ_o is one.
- There exist left and right eigenvectors $\mathbf{u}$ and $\mathbf{v}$ associated with λ_o: $\mathbf{v}^T\mathbf{A} = \lambda_o \mathbf{v}^T$ and $\mathbf{Aq} = \lambda_o \mathbf{q}$ such that $\mathbf{v}$ and $\mathbf{q}$ have only positive elements.
- There are no other positive eigenvectors.
- λ_o is bounded by the elements of $\mathbf{A}$ as follows:

$$\min_m \sum_{n=1}^{M} a_{mn} \leq \lambda_o \leq \max_m \sum_{n=1}^{M} a_{mn}. \tag{6.124}$$

The last property confirms that $\lambda_o = 1$ for state-transition matrix $\mathbf{P}$ of a Markov chain. If $\mathbf{P}$ is only nonnegative, then not all results of this theorem apply; instead $|\lambda| \leq \lambda_o$ (possible equality), and the associated left and right eigenvectors are nonnegative.

Example 6.20. Consider the following nonnegative transition matrix:

$$\mathbf{P} = \begin{bmatrix} 0 & 1 \\ 1 & 0 \end{bmatrix} \tag{6.125}$$

for which $\det(\mathbf{P} - \lambda\mathbf{I}) = \lambda^2 - 1 = 0$ gives eigenvalues ± 1. For this simple example where $\mathbf{P}$ is nonnegative, the first result of the Perron–Frobenius theorem does not hold with strict inequality. The eigenvectors associated with $\lambda = \{-1, 1\}$ are $(1/\sqrt{2})[-1, 1]^T$ and $(1/\sqrt{2})[1, 1]^T$, respectively. The expression in (6.116) for this example is

$$\mathbf{p}[k] = \frac{1}{\sqrt{2}} \left[c_1(-1)^k \begin{bmatrix} -1 \\ 1 \end{bmatrix} + c_2 \begin{bmatrix} 1 \\ 1 \end{bmatrix} \right], \tag{6.126}$$

which does *not* have a steady state because of $\lambda = -1$. This sequence continually alternates between states 1 and 2.

Example 6.21. Consider the positive matrix

$$\mathbf{P} = \begin{bmatrix} 1/3 & 1/3 & 1/3 \\ 1/4 & 1/2 & 1/4 \\ 2/3 & 1/6 & 1/6 \end{bmatrix} \tag{6.127}$$

for which (6.116) is

$$\mathbf{p}[k] = c_1 \begin{bmatrix} 0.669 \\ 0.595 \\ 0.446 \end{bmatrix} + c_2(-0.204)^k \begin{bmatrix} 0.798 \\ -0.247 \\ -0.550 \end{bmatrix} + c_3(0.204)^k \begin{bmatrix} -0.603 \\ 0.778 \\ -0.175 \end{bmatrix}. \tag{6.128}$$

The eigenvector associated with $\lambda = 1$ is the only positive eigenvector; all parts of the Perron–Frobenius theorem hold for this $\mathbf{P}$. The steady-state result is

$$\mathbf{p}[\infty] = c_1 \begin{bmatrix} 0.669 \\ 0.595 \\ 0.446 \end{bmatrix}, \tag{6.129}$$

with $c_1 = 0.585$ so that the elements of the vector sum to 1.

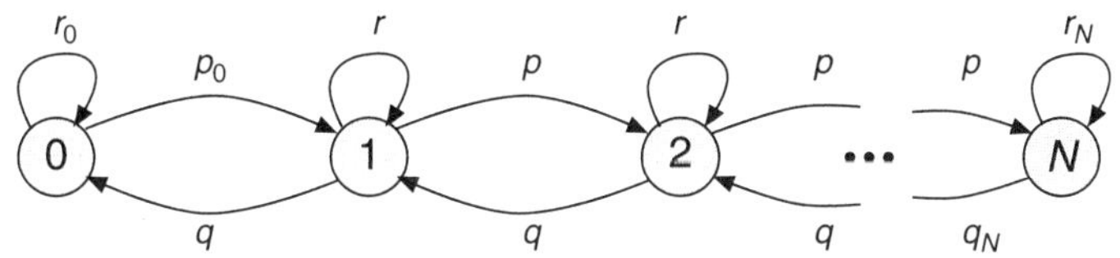

FIGURE 6.10 State transition diagram for the one-dimensional random walk in Example 6.22.

Example 6.22. In this example, we consider a Markov chain known as a *one-dimensional random walk* as depicted in Figure 6.10. There are $N+1$ states, and transitions can occur only to adjacent states or the sequence can remain in the current state. The transition matrix $\mathbf{P} \in \mathcal{R}^{(N+1)\times(N+1)}$ has the following form:

$$\mathbf{P} = \begin{bmatrix} r_0 & p_0 & 0 & \cdots & & 0 \\ q & r & p & & & \vdots \\ 0 & q & \ddots & \ddots & & \vdots \\ \vdots & & \ddots & & p & 0 \\ \vdots & & & & r & p \\ 0 & \cdots & \cdots & 0 & q_N & r_N \end{bmatrix}, \tag{6.130}$$

which is *tridiagonal*. Note that $p+q+r=1$ for the internal states, whereas $r_0+p_0=1$ and $q_N+r_N=1$ for states 0 and N, respectively. Such a transition matrix would arise in a gambling contest involving two players A and B, where the maximum amount of winnings is N. Let the diagram in Figure 6.10 represent the state $X[k]$ of A's winnings at time instant k; obviously, B's winnings are $N-X[k]$. This contest can be simulated by tossing a single biased coin with $P(H)=p$ which is the probability that A wins the toss; the probability that B wins the toss is $q \triangleq 1-p$. For the coin-tossing contest, $r=0$ (because someone must win at each toss), and the game ends when a player's accumulated winnings are N. This means that states 0 and N are absorbing states: $r_0=r_N=1$. Suppose initially that $X[0]=n$: player A is in state n so that player B's winnings are $N-n$. The probability that player A loses the game is given by $P(X[K]=0|X[0]=n)$, where $K \triangleq \min\{k \geq 0 : X[k]=0 \text{ or } X[k]=N)\}$ is a random variable for the duration of the game. It can be shown that (see Problem 6.26)

$$P(X[K]=0|X[0]=n) = \begin{cases} (p^{N-n}q^n - q^N)/(p^N - q^N), & p \neq q \\ (N-n)/N, & p=q=1/2. \end{cases} \tag{6.131}$$

It is interesting that the probability of A losing increases significantly when p is only slightly less than $1/2$, which is due to the nonlinear form in (6.131). Examples are shown in Figure 6.11 for $N=20$. These plots are *not* pmfs: each vertical line is the specific conditional probability above.

Example 6.23. The Bernoulli sequence described later is a useful model for sequences of bits in a digital communication system. Suppose we are interested in counting specific bit patterns such as the first occurrence of 1, 1 of the Bernoulli sequence $X[k]$ with outcomes $\{0, 1\}$. In order to detect this pattern, we can construct a *finite-state machine* (FSM) (the term used in digital logic design) which is equivalent to a discrete-time Markov chain. Let $Y[k]$ be the state at time instant k. Three states are needed in the chain as follows. Let state 0 be the initial state. If 1 is received, $Y[k]$ moves to state 1; otherwise it stays in state 0. If 1 is received while in state 1, $Y[k]$ moves to state 2; otherwise it moves to state 0. The state transition diagram is shown in Figure 6.12 where the transition labels are the received bits. Assuming the bits are equally likely, the transition probabilities are all $1/2$. When state 3 is reached, bit sequence 1, 1 is detected. Note that dashed transitions have been

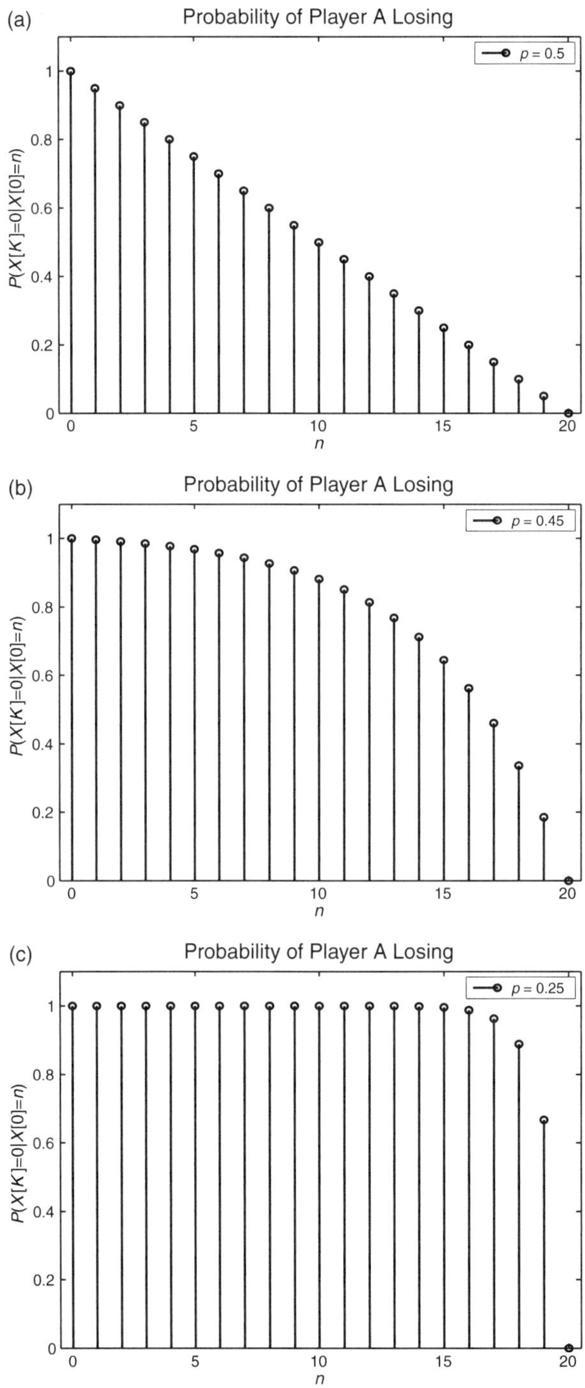

FIGURE 6.11 Probability of player A losing in Example 6.22 with n initial winnings and the total amount is $N = 20$. (a) $p = 0.5$. (b) $p = 0.45$. (c) $p = 0.25$.

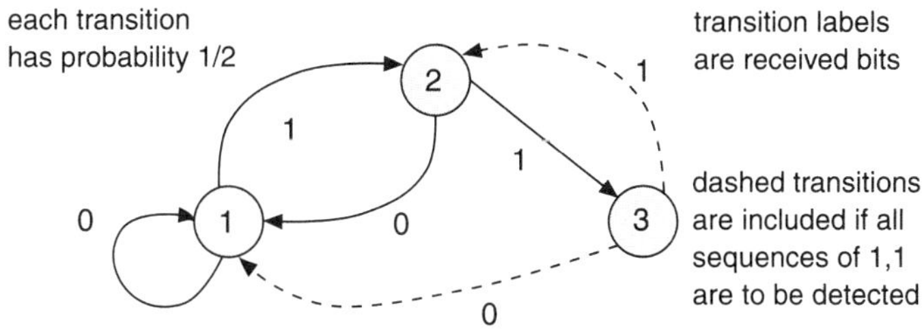

FIGURE 6.12 State transition diagram for the finite-state machine in Example 6.23.

included to show the state diagram if we are interested in all nonoverlapping 1, 1 bit sequences (not just the first occurrence). In the state diagram with solid transition lines, the transition matrix is

$$\mathbf{P} = \begin{bmatrix} 1/2 & 1/2 & 0 \\ 1/2 & 0 & 1/2 \\ 0 & 0 & 1 \end{bmatrix}, \tag{6.132}$$

where $p_{33} = 1$ has been included only to indicate that state 3 is an absorbing state. Of course, once state 3 is reached, future bits are no longer examined in this example.

Suppose we want to know the average number of bits N required to reach state 3. This can be determined by a technique known as a *first-step analysis* which conditions on all the state transitions. Consider the following conditional expectations:

$$n_1 \triangleq \mathcal{E}[N|Y[0] = 1], \quad n_2 \triangleq \mathcal{E}[N|Y[0] = 2], \quad n_3 \triangleq \mathcal{E}[N|Y[0] = 3], \tag{6.133}$$

where n_m is the average number of bits required to reach state 3 given that $Y[0]$ is in state m. Obviously, $n_3 = 0$; the other two cases are derived using the Markov property of the state diagram. Observe first that

$$\mathcal{E}[N|Y[0] = 1] = 1 + \mathcal{E}[N-1|Y[0] = 1], \tag{6.134}$$

$$\mathcal{E}[N|Y[0] = 2] = 1 + \mathcal{E}[N-1|Y[0] = 2], \tag{6.135}$$

where the leading 1 in each expression is just the count difference between N and $N - 1$. From the law of total probability:

$$n_1 = 1 + \sum_{m=1}^{3} \mathcal{E}[N-1|Y[1] = m]P(Y[1] = m|Y[0] = 1), \tag{6.136}$$

$$n_2 = 1 + \sum_{m=1}^{3} \mathcal{E}[N-1|Y[1] = m]P(Y[1] = m|Y[0] = 2), \tag{6.137}$$

where we have conditioned on the "first step" (i.e., the state transition from $Y[0]$ to $Y[1]$). Although we want to find n_1, which is the number of bits required to reach state 3 when starting in state 1, a first-step analysis requires that the equations above be solved simultaneously. The key point is it to recognize that because of stationarity:

$$\mathcal{E}[N-1|Y[1] = m] = \mathcal{E}[N|Y[0] = m], \tag{6.138}$$

and moving from $Y[0]$ to $Y[1]$ requires one bit. Substituting the probabilities from the state transition matrix (and $n_3 = 0$) gives

$$n_1 = 1 + (1/2)n_1 + (1/2)n_2, \quad n_2 = 1 + (1/2)n_1, \tag{6.139}$$

and thus the average number of bits is $n_1 = 6$. The average number of bits to reach state 3 when starting in state 2 is also obtained: $n_2 = 4$.

6.10 MARKOV PROCESS

Consider the continuous-time random process $X(t)$ at several time instants $t_1 < \cdots < t_{N-1} < t_N$ such that we can write the joint pdf $f_{X(t_1),\ldots,X(t_N)}(x(t_1),\ldots,x(t_N))$. In general, there may be some dependence across time instants of the process, or the random variables may be independent, in which case the joint pdf splits into the product of the N marginal pdfs $\{f_{X(t_k)}(x(t_k))\}$. We are interested in the Markov property for continuous-time random processes.

Definition: Markov Process Random process $X(t)$ is a *Markov process* if the conditional pdf simplifies as

$$f_{X(t_N)|X(t_1),\ldots,X(t_{N-1})}(x(t_N)|x(t_1),\ldots,x(t_{N-1})) = f_{X(t_N)|X(t_{N-1})}(x(t_N)|x(t_{N-1})). \tag{6.140}$$

The Markov property can also be expressed using a conditional cdf:

$$\begin{aligned} P(X(t_N) \le x(t_N)|X(t_1) = x(t_1),\ldots,X(t_{N-1}) = x(t_{N-1})),\\ = P(X(t_N) \le x(t_N)|X(t_{N-1}) = x(t_{N-1})). \end{aligned} \tag{6.141}$$

As in the discrete-time case, the conditioning reduces to include only the random variable at the most recent time instant defined by the conditioning. The history of a particular realization before time t_{N-1} is not relevant.

The term Markov process is often used to describe any process with the property in the definition above, for continuous or discrete time. If the process has a finite or countable number of states, then it is generally called a Markov chain. The Markov chains covered in Section 6.9 were examined only for discrete time; it is also possible to consider continuous-time Markov chains (an example is the Poisson process described later). If a Markov process has continuous realizations, then it is known as a *diffusion process* (an example is the Wiener process described later). The various types of Markov processes are summarized in Table 6.2. In this section, we briefly describe some properties of continuous-time Markov processes with continuous realizations.

The Chapman–Kolmogorov equation also applies to continuous-time processes.

Theorem 6.10 (Chapman–Kolmogorov). For Markov random process $X(t)$,

$$\begin{aligned} f_{X(t_3)|X(t_1)}(x(t_3)|x(t_1)) = \int_{-\infty}^{\infty} f_{X(t_3)|X(t_2)}(x(t_3)|x(t_2))\\ \cdot f_{X(t_2)|X(t_1)}(x(t_2)|x(t_1))\mathrm{d}x(t_2) \end{aligned} \tag{6.142}$$

for $t_1 < t_2 < t_3$.

TABLE 6.2 Categories of Markov Processes with Examples

	Countable Number of States	Continuous Realizations
Discrete time k	Markov chain (finite state machine)	None
Continuous time t	Continuous-time Markov chain (Poisson process)	Diffusion process (Wiener process)

Proof. Recall that a marginal pdf is obtained by integrating a joint pdf as follows:

$$\begin{aligned} f_{X(t_3)}(x(t_3)) &= \int_{-\infty}^{\infty} f_{X(t_2),X(t_3)}(x(t_2), x(t_3))\mathrm{d}x(t_2), \\ &= \int_{-\infty}^{\infty} f_{X(t_3)|X(t_2)}(x(t_3)|x(t_2)) f_{X(t_2)}(x(t_2))\mathrm{d}x(t_2), \end{aligned} \tag{6.143}$$

where a conditional probability has been substituted in the second line. Conditioning on $X(t_1)$ gives

$$\begin{aligned} f_{X(t_3)|X(t_1)}(x(t_3)|x(t_1)) = \int_{-\infty}^{\infty} & f_{X(t_3)|X(t_1),X(t_2)}(x(t_3)|x(t_1), x(t_2)) \\ & \cdot f_{X(t_2)|X(t_1)}(x(t_2)|x(t_1))\mathrm{d}x(t_2). \end{aligned} \tag{6.144}$$

Since $X(t)$ is a Markov process, we can substitute

$$f_{X(t_3)|X(t_1),X(t_2)}(x(t_3)|x(t_1), x(t_2)) = f_{X(t_3)|X(t_2)}(x(t_3)|x(t_2)), \tag{6.145}$$

which completes the proof. □

Theorem 6.11. If $X(t)$ is an independent-increment random process, then it is also a Markov process.

Proof. Consider the conditional pdf $f_{X(t_3)|X(t_1),X(t_2)}(x(t_3)|x(t_1), x(t_2))$ of $X(t)$ at three time instants $t_1 < t_2 < t_3$. For the increment $X(t_3) - X(t_2)$, we can write

$$f_{X(t_3)|X(t_1),X(t_2)}(x(t_3)|x(t_1), x(t_2)) = f_{X(t_3)-X(t_2)|X(t_1),X(t_2)}(x(t_3) - x(t_2)|x(t_1), x(t_2)). \tag{6.146}$$

The right-hand side is obtained using the transformation techniques in Chapter 4 for the linear transformation $X(t_3) \to X(t_3) - X(t_2)$. This is easily done above because of the conditioning on $X(t_2)$ so that in effect a constant is subtracted from $X(t_3)$. Likewise, since the conditioning also includes $X(t_1)$, another linear transformation $X(t_2) \to X(t_2) - X(t_1)$ yields

$$\begin{aligned} & f_{X(t_3)-X(t_2)|X(t_1),X(t_2)}(x(t_3) - x(t_2)|x(t_1), x(t_2)), \\ & = f_{X(t_3)-X(t_2)|X(t_2)-X(t_1),X(t_2)}(x(t_3) - x(t_2)|x(t_2) - x(t_1), x(t_2)). \end{aligned} \tag{6.147}$$

Due to the conditioning, the form of the original conditional pdf is unchanged: only translations have occurred. Since $X(t)$ is an independent-increment process, $X(t_3) - X(t_2)$ and $X(t_2) - X(t_1)$ are independent so that $X(t_2) - X(t_1)$ can be dropped from the conditioning:

$$\begin{aligned} & f_{X(t_3)-X(t_2)|X(t_2)-X(t_1),X(t_2)}(x(t_3) - x(t_2)|x(t_2) - x(t_1), x(t_2)), \\ & \qquad = f_{X(t_3)-X(t_2)|X(t_2)}(x(t_3) - x(t_2)|x(t_2)). \end{aligned} \tag{6.148}$$

Finally, since $X(t_2)$ is the conditioning variable, $X(t_3) - X(t_2)$ can be replaced by $X(t_3)$, resulting in yet another translation so that

$$f_{X(t_3)-X(t_2)|X(t_2)}(x(t_3) - x(t_2)|x(t_2)) = f_{X(t_3)|X(t_2)}(x(t_3)|x(t_2)), \tag{6.149}$$

which completes the proof. □

6.11 RANDOM SEQUENCES

Now that we have described some basic properties of random processes, we consider important cases with specific distributions and properties. Discrete-time random sequences are covered in this section, and continuous-time random processes are described in Section 6.12.

6.11.1 Bernoulli Sequence

We begin with a formal definition of the Bernoulli random sequence that was discussed in Example 6.6 as a model for repeated coin tosses.

Definition: Bernoulli Sequence *Bernoulli sequence* $X[k]$ is a set of iid random variables $\{X[0], X[1], \ldots\}$ with outcomes $\{0, 1\}$, where $P(X[k] = 1) = p$ and $P(X[k] = 0) = q \triangleq 1 - p$.

This sequence is strictly stationary. Each random variable can be viewed as a *trial* of which there are two outcomes. Outcome 1 is referred to as a success while 0 is called a failure. The expectation of a Bernoulli sequence is $\mathcal{E}[X[k]] = p$, the second moment is $\mathcal{E}[X^2[k]] = p$, and the variance is $\text{var}[X[k]] = p - p^2 = pq$. These, of course, are the same results for the Bernoulli random variable because the pdf does not change with time in the definition above.

We can also examine related events that are modeled by random variables discussed in Chapter 3. Let N_1 be the *number of successes in N trials*. Then N_1 is a binomial random variable with pmf

$$P(N_1 = n) = \binom{N}{n} p^n q^{N-n} I_{\{0,\ldots,N\}}[n]. \tag{6.150}$$

Let N_2 be the *number of trials required to have* $r \in \mathcal{Z}^+$ *successes*. Then N_2 is the negative binomial random variable with pmf

$$P(N_2 = n) = \binom{n + r - 1}{r - 1} p^r q^n I_{\mathcal{Z}^+}[n]. \tag{6.151}$$

A special case for N_2 is obtained when $r = 1$, resulting in the geometric random variable with support $\mathcal{Z}^+$.

If the outcomes are changed to $\{-1, 1\}$, then $X[k]$ is a *symmetric* Bernoulli sequence. An example realization of this sequence with $p = q = 1/2$ is shown in Figure 6.13.

6.11.2 Bernoulli Scheme

A Bernoulli scheme is a generalized version of the Bernoulli sequence that has more than two outcomes at each instant of time.

Definition: Bernoulli Scheme *Bernoulli scheme* $X[k]$ is a set of iid random variables $\{X[0], X[1], \ldots\}$ with discrete alphabet $\{1, \ldots, N\}$ such that $P(X[k] = n) \triangleq p_n$ and $\sum_{n=1}^{N} p_n = 1$.

Bernoulli schemes are strictly stationary because the random variables are iid. Consecutive tosses of a die with $P(X[k] = n) = 1/6$ for $n = 1, \ldots, 6$, is an example of a Bernoulli scheme. For $N > 2$, the pdf of the random variables technically is *not* Bernoulli; also, it is not necessarily uniform as is the case for the fair die example. It is called a Bernoulli scheme only because it is a natural extension of a Bernoulli sequence to multiple outcomes. An example realization of a Bernoulli scheme with four equally likely outcomes is shown in Figure 6.14. A multilevel Bernoulli scheme is used for the transmitted symbols in digital communications, as described in Chapter 10, where it is known as pulse amplitude modulation (PAM). However, for such systems, an outcome extends for some fixed duration of continuous time as determined by the *pulse shape*, resulting in a random process $X(t)$. An example realization using rectangular pulses is shown later in Figure 6.29.

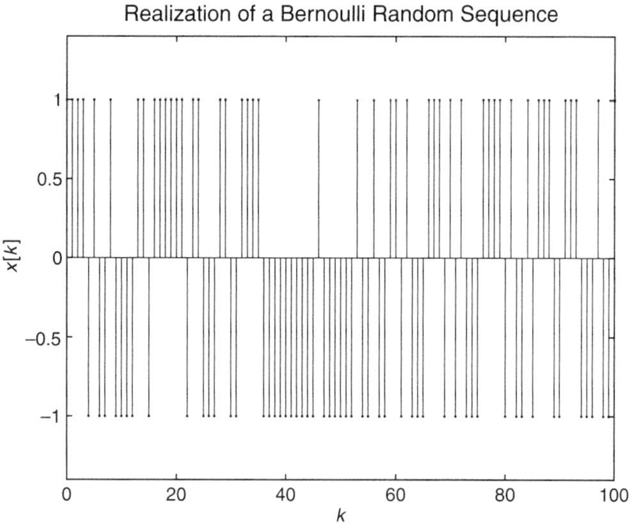

FIGURE 6.13 Example realization $x[k]$ of a symmetric Bernoulli random sequence with $p = 0.5$. (Vertical lines are included for ease of viewing.)

6.11.3 Independent Sequences

The Bernoulli sequence and the Bernoulli scheme are examples of iid random sequences. Basically, any of the random variables covered in Chapter 3 can be indexed by time to generate an iid random sequence. This includes discrete random variables such as geometric and Poisson, as well as continuous random variables such as Gaussian and exponential. Figure 6.15 shows an example realization of an iid exponential random sequence with parameter $\lambda = 1$ where at each discrete time instant k, a sample from $f_X(x) = \exp(-x)u(x)$ is selected.

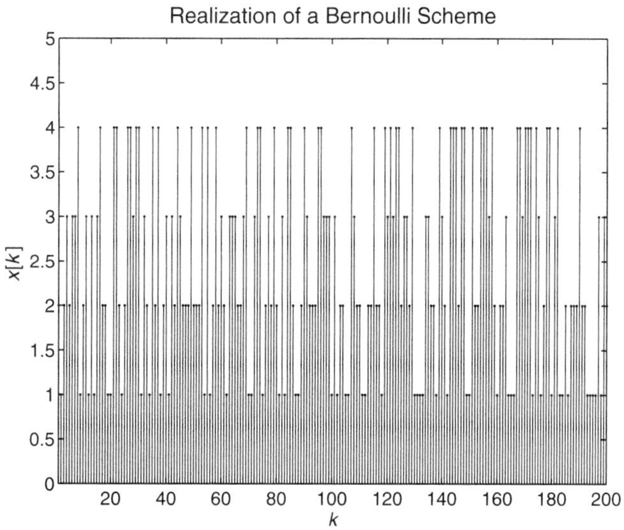

FIGURE 6.14 Example realization $x[k]$ of a Bernoulli scheme with a uniform distribution and outcomes $\{1, 2, 3, 4\}$. (Vertical lines are included for ease of viewing.)

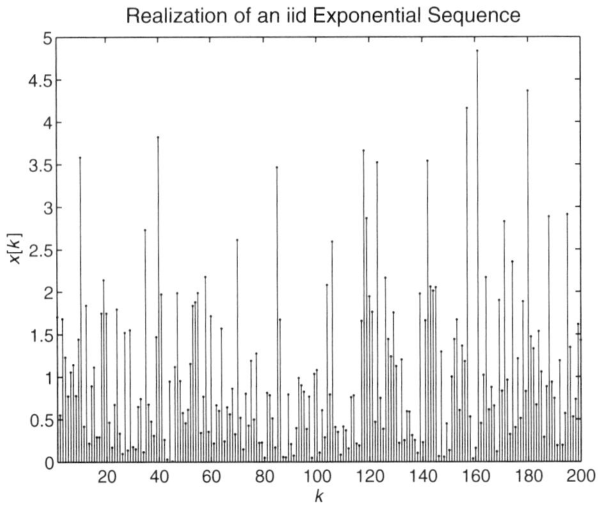

FIGURE 6.15 Example realization $x[k]$ of an iid exponential sequence with $\lambda = 1$. (Vertical lines are included for ease of viewing.)

Consider removing the identically distributed condition such that $\{X[k]\}$ are independent but do not necessarily have the same distribution for each k. A simple example of this is to allow the parameter of the previous exponential sequence to vary. For example, suppose that $\lambda[k] = 2 + 4/(k + 1)$ for $k \in \mathcal{Z}^+$. Each random variable is still exponential, but the parameter varies from a maximum of six to a minimum of two. In Chapter 7, we define different types of convergence of a sequence as $k \to \infty$. This simple example is a random sequence that *converges in distribution* to an exponential random variable with parameter $\lambda = 2$.

An independent random sequence is perhaps the most basic type of sequence from which numerous other types of random sequences can be generated. For example, the Bernoulli random walk and the binomial counting sequence described next are generated by *summing* the outcomes of a Bernoulli random sequence. Such a summation corresponds to a type of *filter* operating on the Bernoulli input to generate an output sequence that is no longer independent. This type of model depicted in Figure 6.16 is widely used to describe random sequences, as discussed in Chapter 8. The filter could be linear or nonlinear, and it may be time varying or time invariant. The realization shown in Figure 6.1 and described in the introduction was generated by processing an independent Gaussian sequence with a linear time-invariant filter. In Chapter 7, we describe differential equations with constant coefficients that are used to model linear systems. Such a system can be represented by an impulse response function, and in the frequency domain by a transfer function.

6.11.4 Bernoulli Random Walk

We return to the Bernoulli sequence and describe a "random walk" that is not independent nor identically distributed.

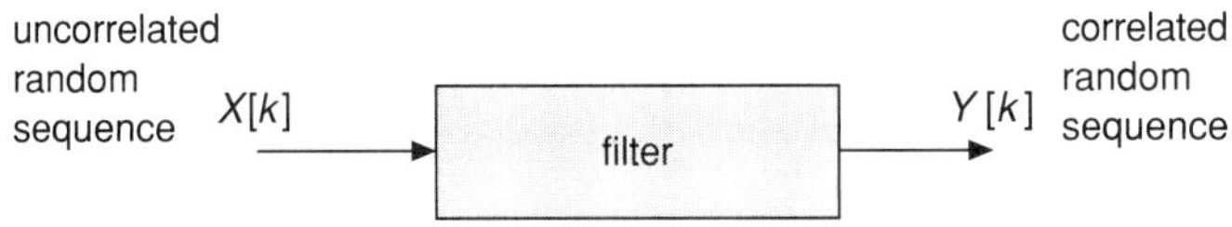

FIGURE 6.16 Filter operating on an iid sequence.

Definition: Bernoulli Random Walk Let $W[k]$ be a symmetric Bernoulli sequence with outcomes $\{-1, 1\}$ such that $P(W[k] = 1) = p = P(W[k] = -1) = q = 1/2$. The random sequence

$$X[k] = \sum_{m=1}^{k} W[m] \tag{6.152}$$

with $X[0] = W[0] = 0$ is a *Bernoulli random walk*.

Obviously, the sum sequence $X[k]$ is not Bernoulli: it can take on any integer $\mathcal{Z}$ with increasing k. The Bernoulli random walk derives its name from the pdf of the underlying random variables comprising the sum in (6.152). It is called a random walk because the sequence takes unit steps randomly in positive and negative directions. Note that (6.152) can be rewritten as

$$X[k] = X[k-1] + W[k], \tag{6.153}$$

which has the form of an infinite-impulse-response (IIR) filter with a single pole at $z = 1$. The impulse response is given by the discrete unit-step function:

$$u[k] \triangleq \begin{cases} 1, & k \in \mathcal{Z}^+ \\ 0, & \text{else.} \end{cases} \tag{6.154}$$

The Bernoulli random walk can be also be viewed as the output of a discrete-time filter with input given by an independent sequence of Kronecker delta functions with random amplitudes $\{-1, 1\}$ (i.e., a symmetric Bernoulli sequence). An example realization is shown in Figure 6.17. The first and second moments of the underlying Bernoulli sequence are $\mathcal{E}[W[k]] = 0$ and $\mathcal{E}[W^2[k]] = 1$. As a result, $X[k]$ evolves in time by taking on values centered about zero because $\mathcal{E}[X[k]] = 0$, but with linearly increasing variance:

$$\text{var}[X[k]] = \mathcal{E}[X^2[k]] = \sum_{m=1}^{k} \mathcal{E}[W^2[m]] = k. \tag{6.155}$$

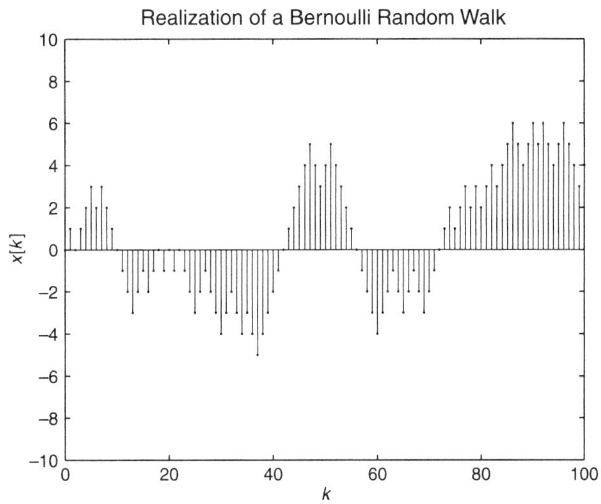

FIGURE 6.17 Example realization of a Bernoulli random walk with $p = 0.5$.

Observe also that

$$X[k+N] - X[k] = \sum_{m=k+1}^{k+N} W[m] \tag{6.156}$$

is independent of $X[k]$. This independence holds for any set of nonoverlapping *intervals* of $X[k]$; thus, the Bernoulli random walk is an example of an isi sequence. However, note that $X[k]$ itself is *not* stationary because its variance changes with time.

Although the Bernoulli random walk is technically defined with $p = q = 1/2$, it can be generalized such that $p \neq q$. For this case, the walk would not be centered about zero; instead, the mean is time varying:

$$\mathcal{E}[X[k]] = \sum_{m=1}^{k} \mathcal{E}[W[m]] = (p-q)k, \tag{6.157}$$

which is either increasing $(p > q)$ or decreasing $(p < q)$ because $\mathcal{E}[W[m]] = p - q$ when the outcomes are $\{-1, 1\}$. The second moment is

$$\mathcal{E}[X^2[k]] = k\mathcal{E}[W^2[k]] + (k^2 - k)(\mathcal{E}[W[k]])^2 = k + (p-q)^2(k^2 - k). \tag{6.158}$$

Using the fact that $1 - (p-q)^2 = 4pq$, the variance is an increasing function of k:

$$\text{var}[X(k)] = 4pqk. \tag{6.159}$$

The autocorrelation and autocovariance functions are derived in Problem 6.13.

6.11.5 Binomial Counting Sequence

Consider again the Bernoulli random sequence with outcomes $\{0, 1\}$.

Definition: Binomial Counting Sequence Let $W[k]$ be a Bernoulli random sequence with $P(W[k] = 1) = p$ and $P(W[k] = 0) = q$. The random sequence

$$X[k] = \sum_{m=1}^{k} W[m] \tag{6.160}$$

with $X[0] = W[0] = 0$ is a *binomial counting sequence*.

Unlike the previous Bernoulli random walks, the binomial counting sequence cannot decrease. This, of course, is due to outcome 0 instead of -1. From the sum in (6.160) and Theorem 6.3, we find that the binomial counting sequence is another example of an isi sequence. For a fixed value of k, random variable $X[k]$ has a binomial pmf:

$$P(X[k] = n) = \binom{k}{n} p^n q^{k-n} I_{\{0,\ldots,k\}}[n]. \tag{6.161}$$

The binomial counting sequence can also be represented as the output of the IIR filter in (6.153) with impulse response given by the discrete unit-step function. The major difference here is that the Kronecker delta functions of the input sequence are not uniformly spaced in time because the Bernoulli outcomes are $\{0, 1\}$. Thus, the amplitude of a realization could remain constant for several time instants because of a string of Bernoulli zeros. A similar behavior occurs for the Poisson counting process described later, which is the continuous-time

TABLE 6.3 Random Sequences Based on Bernoulli $W[k]$

Random Sequence $X[k] = \sum_{m=1}^{k} W[m]$	Outcomes of $W[m]$	Parameter
Bernoulli random walk	$\{-1, 1\}$	$p = 1/2$
Generalized Bernoulli random walk	$\{-1, 1\}$	$0 < p < 1$
Binomial counting sequence	$\{0, 1\}$	$0 < p < 1$

counterpart of the binomial counting process. A summary of the two random walks and the counting sequence based on the Bernoulli sequence is provided in Table 6.3.

Next, we investigate the *interarrival time* between successive Bernoulli 1s in the binomial counting sequence. Since the counting process is isi, any interval in $\mathcal{T}$ can be examined; thus, it is convenient to examine the occurrence of the first 1 which we denote by random variable Y. We make this choice because event $\{Y > t\}$ is equivalent to event $\{X(t) = 0\}$. For the first occurrence of a 1 to be after time instant k, we can write

$$P(Y > k) = P(X[k] = 0) = \binom{k}{0} q^k = q^k, \tag{6.162}$$

where (6.161) has been used. Thus

$$P(Y \leq k) = (1 - q^k) I_{\mathcal{N}}[k], \tag{6.163}$$

which corresponds to the cdf of the geometric random variable with support $\mathcal{N}$, and has the following pmf:

$$p_Y[k] = p(1 - q)^{k-1} I_{\mathcal{N}}[k]. \tag{6.164}$$

(Note that this is the *shifted* geometric distribution in Appendix A.)

The mean of the binomial counting sequence is $\mathcal{E}[X[k]] = pk$, the second moment is

$$\mathcal{E}[X^2[k]] = k\mathcal{E}[W^2[k]] + (k^2 - k)(\mathcal{E}[W[k]])^2 = pk + p^2(k^2 - k), \tag{6.165}$$

and the variance is $\text{var}(X[k]) = pk - p^2 k = pqk$. An example realization is shown in Figure 6.18. Its autocorrelation function is

$$R_{XX}[k_1, k_2] = \mathcal{E}\left[\sum_{m=1}^{k_1} W[m] \sum_{n=1}^{k_2} W[n]\right]. \tag{6.166}$$

In order to continue, assume that $k_2 > k_1$; the reverse case is easily found later due to symmetry of $R_{XX}[k_1, k_2]$. Since $\{W[k]\}$ are independent, the expectation of each cross-term is a product of expectations:

$$\begin{aligned} R_{XX}[k_1, k_2] &= \sum_{m=1}^{k_1} \left(\mathcal{E}[W^2[m]] + \sum_{n=1, n\neq m}^{k_2} \mathcal{E}[W[m]]\mathcal{E}[W[n]] \right), \\ &= \sum_{m=1}^{k_1} [p + p^2(k_2 - 1)] = pk_1[1 + p(k_2 - 1)]. \end{aligned} \tag{6.167}$$

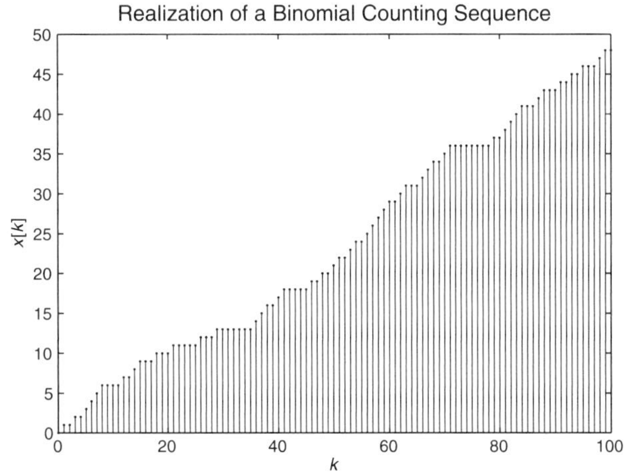

FIGURE 6.18 Example realization of a binomial counting sequence with outcomes {0, 1} and $p = 0.5$.

The autocovariance function is obtained by subtracting $\mathcal{E}[X[k_1]]\mathcal{E}[X[k_2]] = p^2k_1k_2$:

$$C_{XX}[k_1, k_2] = pk_1[1 + p(k_2 - 1)] - p^2k_1k_2 = pqk_1. \tag{6.168}$$

Thus, for general $\{k_1, k_2\}$:

$$R_{XX}[k_1, k_2] = pq\min(k_1, k_2) + p^2k_1k_2, \tag{6.169}$$

$$C_{XX}[k_1, k_2] = pq\min(k_1, k_2), \tag{6.170}$$

from which we see that $X[k]$ is *not* wide-sense stationary. Since the variance at time instant k is $\text{var}[X[k]] = pqk$, the correlation coefficient is

$$\rho = \frac{pq\min(k_1, k_2)}{\sqrt{(pqk_1)(pqk_2)}} = \frac{\min(k_1, k_2)}{\sqrt{k_1k_2}}. \tag{6.171}$$

These quantities are summarized in Table 6.4 along with the results for the Bernoulli walk and the random processes discussed in Section 6.12. It is interesting that the correlation coefficient for some well-known random processes have the same form and are independent of their parameters.

TABLE 6.4 Correlation and Covariance of Random Processes

Process	Autocorrelation Function	Autocovariance Function	Correlation Coefficient
Bernoulli walk	$4pq\min(k_1, k_2) + (p-q)^2k_1k_2$	$4pq\min(k_1, k_2)$	$\min(k_1, k_2)/\sqrt{k_1k_2}$
Binomial counting	$pq\min(k_1, k_2) + p^2k_1k_2$	$pq\min(k_1, k_2)$	$\min(k_1, k_2)/\sqrt{k_1k_2}$
Poisson counting	$\lambda\min(t_1, t_2) + \lambda^2t_1t_2$	$\lambda\min(t_1, t_2)$	$\min(t_1, t_2)/\sqrt{t_1t_2}$
"Semirandom" telegraph	$\exp(-2\lambda\|t_2 - t_1\|)$	$\exp(-2\lambda\|t_2 - t_1\|) - \exp(-2\lambda(t_1 + t_2))$	see (6.206)
"Randomized" telegraph	$\exp(-2\lambda\|t_2 - t_1\|)$	$\exp(-2\lambda\|t_2 - t_1\|)$	$\exp(-2\lambda\|t_2 - t_1\|)$
Wiener	$\alpha\epsilon^2\min(t_1, t_2)$	$\alpha\epsilon^2\min(t_1, t_2)$	$\min(t_1, t_2)/\sqrt{t_1t_2}$

6.12 RANDOM PROCESSES

Next, we consider some random processes that evolve continuously over time. As was the case with random sequences, the time-indexed random variables can have discrete or continuous outcomes. We begin with a sequence of independent random variables having a Poisson distribution, and then allow the intervals between time instants to approach zero in order to generate a continuous-time random process that can have a discrete change in amplitude at any time instant. (Note that example plots of the various random processes are actually discrete-time realizations because, obviously, MATLAB can only plot a finite number of outcomes for a process.)

6.12.1 Poisson Counting Process

The outcomes of the Poisson random variable are discrete (countably infinite), given by all nonnegative integers $\mathcal{Z}^+$. Define the time interval $\Delta t \triangleq t/k$, where $t \in \mathcal{T} = \mathcal{R}^+$ is continuous time and $k \in \mathcal{N}$. Consider the random process defined as follows:

$$X(t) \triangleq X[k\Delta t], \tag{6.172}$$

where $X[k]$ (with $\Delta t = 1$) is the binomial counting sequence in (6.160). The argument on the right-hand side has been generalized such that a Bernoulli success can occur for an arbitrary interval duration (not only $\Delta t = 1$). We assume the following conditions such that as $\Delta t \to 0$, $X(t)$ is the Poisson counting process defined for all $t \in \mathcal{T}$:

- $X[k\Delta t]$ is a binomial random variable with a probability of success p that is proportional to the time duration:

$$p = P(X[k\Delta t] = 1) = \lambda\Delta t, \tag{6.173}$$

  where λ is a *rate parameter*.
- Interval Δt is sufficiently small so that there is at most one Bernoulli success in any interval. The probability of two or more successes in any interval is assumed to be zero.

For the binomial counting process and these conditions, the probability of n successes is

$$P(X[k\Delta t] = n) = \binom{k}{n}(\lambda\Delta t)^n(1 - \lambda\Delta t)^{k-n}, \quad n = 0, \ldots, k \tag{6.174}$$

We considered this binomial distribution in Chapter 4 where it was approximated by a Poisson distribution. Specifically, if $k \to \infty$ such that the interval $\Delta t \to 0$, and thus $p \to 0$ with the constraint that $kp = \lambda t < \infty$ is constant, then (6.174) is approximated as follows:

$$P(X(t) = n) \approx \frac{(\lambda t)^n}{n!}\exp(-\lambda t), \quad n \in \mathcal{Z}^+, \tag{6.175}$$

which is the Poisson distribution with *time-dependent* parameter $\alpha = \lambda t$.

Definition: Poisson Counting Process The *Poisson counting process* $X(t)$ is a discrete-outcome, continuous-time random process with $X(0) = 0$ such that the probability of an outcome at time t is given by the Poisson distribution in (6.175) with parameter $\alpha = \lambda t$.

Like the binomial counting sequence, a realization of the Poisson process has only positive unit-amplitude steps. However, unlike the binomial counting sequence where outcomes occur only at discrete time instants, the amplitude of the Poisson counting process can change at *any* time $t \in \mathcal{T}$. An example realization is

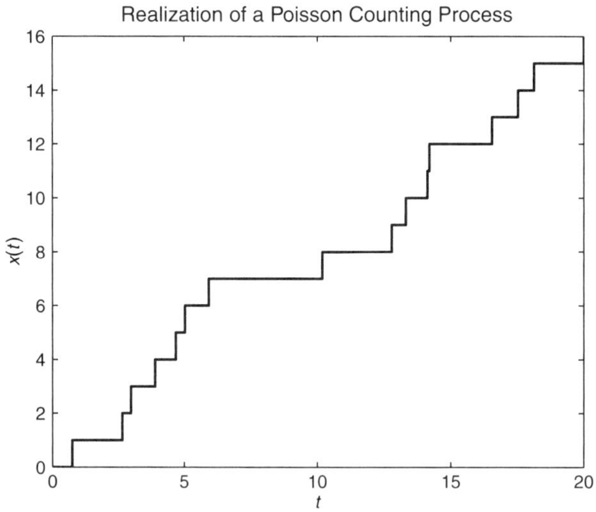

FIGURE 6.19 Example realization of the Poisson counting process with rate parameter $\lambda = 1$.

shown in Figure 6.19. The mean and variance of the Poisson process are readily obtained from (6.175): $\mathcal{E}[X(t)] = \text{var}[X(t)] = \lambda t$. Since these moments depend on time, the process is nonstationary.

Example 6.24. The Poisson process $X(t)$ is an accurate model for many physical processes such as radioactive decay and defects along a telephone line. Suppose that customers enter a store at a rate of $\lambda = 5$ per hour. Various probabilities are easily determined, such as (a) the probability that 8 customers arrive over a three-hour period, (b) the probability that 3 customers arrive in an hour given that none arrived the previous hour, and (c) the probability that 1 customer arrives in an hour and 2 customers arrive the next hour. The corresponding probability descriptions are (a) $P(X(3) = 8)$, (b) $P(X(2) - X(1) = 3 | X(1) - X(0) = 0)$, and (c) $P(X(1) = 1, X(2) = 3)$, which are all defined relative to $X(0) = 0$. Part (a) is straightforward:

$$P(X(3) = 8) = \frac{(5 \cdot 3)^8}{8!} \exp(-5 \cdot 3) \approx 0.0194. \tag{6.176}$$

In part (b), the conditioning can be dropped because the increments are independent:

$$P(X(2) - X(1) = 3) = \frac{(5 \cdot 1)^3}{3!} \exp(-5 \cdot 1) \approx 0.0056. \tag{6.177}$$

Part (c) is handled by rewriting the expression in terms of independent increments: $P(X(1) = 1, X(2) = 3) = P(X(1) = 1, X(2) - X(1) = 2) = P(X(1) = 1)P(X(2) - X(1) = 2)$. Thus

$$P(X(1) = 1, X(2) = 3) = \frac{(5 \cdot 1)^1}{1!} \exp(-5 \cdot 1) \frac{(5 \cdot 1)^2}{2!} \exp(-5 \cdot 1) \approx 0.0028. \tag{6.178}$$

The Poisson process is a particular type of Lévy random process.

Definition: Lévy Random Process The *Lévy random process* $X(t)$ is any continuous-time random process with the following properties: (i) isi, (ii) zero initial outcome $X(0) = 0$ (almost surely), and (iii) right continuous (almost surely) with left limits.

Recall for right-continuous function $x(t)$ that $\lim_{\epsilon\to 0} x(t+\epsilon) = x(t)$ for $\epsilon > 0$ and $t \in \mathcal{T}$. It has a left limit at t if $\lim_{\epsilon\to 0} x(t-\epsilon)$ exists, which is not necessarily $x(t)$ from the right. The cdf of a discrete random variable is an example of a right-continuous function with left limits.

Other examples of a Lévy random process are the gamma and Wiener processes described later. Since the Poisson process is isi, events in nonoverlapping intervals are independent and their distributions depend only on the durations of the intervals, not on their location in $\mathcal{T}$. This implies for duration $\tau > 0$ that

$$P(X(t) - X(t-\tau) = n) = \frac{(\lambda\tau)^n}{n!}\exp(-\lambda\tau). \tag{6.179}$$

The expression in (6.175) is a special case of (6.179) for the interval $[0, t]$, by letting $t - \tau = 0$ such that $\tau = t$ and $X(t-\tau) \to X(0) = 0$.

We can also determine the distribution of the *interarrival* time between two occurrences (steps) of the Poisson random process. Because the Poisson process is an isi process, any interval in $\mathcal{T}$ can be examined such as the time of the first occurrence, which we denote by random variable Y (as we did previously for the binomial counting process). If the time of the first occurrence is $> t$, then there are zero occurrences during $[0, t]$ (and vice versa):

$$P(Y > t) = P(X(t) = 0) = \exp(-\lambda t) \tag{6.180}$$

for $t \geq 0$. From this result, the cdf of Y is $F_Y(t) = P(Y \leq t) = 1 - P(Y > t) = [1 - \exp(-\lambda t)]I_{[0,\infty)}(t)$, which we recognize is the cdf of an *exponential* random variable with parameter λ. Thus, the pdf of Y is

$$f_Y(t) = \lambda\exp(-\lambda t)I_{[0,\infty)}(t), \tag{6.181}$$

which has mean $\mathcal{E}[Y] = 1/\lambda$ and variance $\text{var}[Y] = 1/\lambda^2$. Note that the plot in Figure 6.19 was actually generated by using samples of the exponential distribution for the interarrival times, in order to provide values of t where the realization changes (jumps).

This result can be extended to specify the distribution for the time until the nth occurrence denoted by random variable Y_n. Since the intervals between occurrences are independent, it is convenient to write Y_n as a sum of those intervals as follows:

$$Y_n = \sum_{k=1}^{n}(Y_k - Y_{k-1}). \tag{6.182}$$

Because Y_n is a sum of independent exponential random variables, we conclude from Example 4.17 in Chapter 4 that Y_n is an Erlang (gamma) random variable with parameters $\{\lambda, r = n\}$:

$$f_{Y_n}(t) = \frac{\lambda^n t^{n-1}}{(n-1)!}\exp(-\lambda t)\, I_{[0,\infty)}(t), \tag{6.183}$$

which has mean $\mathcal{E}[Y_n] = n/\lambda$ and variance $\text{var}[Y_n] = n/\lambda^2$. As expected, the mean and variance increase with the number of occurrences n.

Next, we show that the distribution of the first arrival time Y in $[0, t]$ *conditioned* on a single arrival during that interval is uniform. Since $X(t)$ is a Poisson random variable for the number of arrivals in $[0, t]$, we have from conditional probability:

$$\begin{aligned} P(Y \leq y|X(t) = 1) &= P(Y \leq y, X(t) = 1)/P(X(t) = 1), \\ &= P(X(y) = 1, X(t) = 1)/P(X(t) = 1), \end{aligned} \tag{6.184}$$

where $Y \leq y \Longrightarrow X(y) = 1$ has been used. Since

$$P(X(y) = 1, X(t) = 1) = P(X(y) = 1, X(t) - X(y) = 0), \tag{6.185}$$

and the Poisson process has independent increments, (6.184) becomes

$$\begin{aligned} P(Y \leq y|X(t) = 1) &= P(X(y) = 1)P(X(t) - X(y) = 0)/P(X(t) = 1), \\ &= \frac{\lambda y \exp(-\lambda y)\exp(-\lambda(t-y))}{\lambda t \exp(-\lambda t)} = (y/t)I_{[0,t]}(y), \end{aligned} \tag{6.186}$$

where the indicator function has been included to emphasize that $y \in [0, t]$. This result is the cdf of a uniform random variable. It turns out that the conditional distribution of arrival times conditioned on $X(t) = n$ arrivals in $[0, t]$ is the distribution of the order statistics for n iid samples of a random variable that is uniform on $[0, t]$ (see Problem 6.37).

Observe that the Poisson counting process can be modeled as a filtered train of Dirac delta functions. The impulse response of the filter is the continuous unit-step function $u(t)$, and the duration between impulses has an exponential distribution. Filtering a signal with a unit-step function is equivalent to an integrator, which is the continuous-time counterpart of the summation filters of the Bernoulli walk and the binomial counting process discussed earlier. This is not surprising because in the derivation of the Poisson counting process, we started with the binomial counting process and let the time interval $\Delta t \to 0$. The input of the filter for the binomial counting process consists of Kronecker delta functions where the duration from one to the next has the geometric distribution in (6.164). In effect, the discrete unit-step function of the filter model used there is replaced by the continuous unit-step function for the Poisson counting process when $\Delta t \to 0$. The reason for the nonuniform spacing between the Dirac delta functions is due to strings of zeros that occur for the binomial counting process from which it was derived.

In order to calculate the autocorrelation function for $X(t)$, the joint probability distribution of the Poisson process at two time instants $\{t_1, t_2\}$ is required. This can be obtained using conditional probability for $t_2 > t_1$ as follows:

$$P(X(t_1) = n_1, X(t_2) = n_2) = P(X(t_2) = n_2|X(t_1) = n_1)P(X(t_1) = n_1), \tag{6.187}$$

where $n_2 > n_1$. Due to the memoryless property of the exponential distribution (see Chapter 3), the conditional probability has an Erlang distribution with parameters $\{\lambda(t_2 - t_1), r = n_2 - n_1\}$. Given n_1 occurrences by time t_1, the distribution depends only on the number of additional occurrences $n_2 - n_1$ during time interval $(t_1, t_2]$. Thus

$$\begin{aligned} P(X(t_1) = n_1, X(t_2) = n_2) &= \frac{[\lambda(t_2 - t_1)]^{n_2 - n_1}}{(n_2 - n_1)!} \exp(-\lambda(t_2 - t_1)) \\ &\quad \cdot \frac{(\lambda t_1)^{n_1}}{n_1!} \exp(-\lambda t_1), \\ &= \frac{[\lambda(t_2 - t_1)]^{n_2 - n_1}(\lambda t_1)^{n_1}}{(n_2 - n_1)! n_1!} \exp(-\lambda t_2), \end{aligned} \tag{6.188}$$

such that the exponential term is a function only of t_2. This simplifies to

$$P(X(t_1) = n_1, X(t_2) = n_2) = \frac{1}{n_2!}[\lambda(t_2 - t_1)]^{n_2 - n_1}(\lambda t_1)^{n_1}\binom{n_2}{n_1}\exp(-\lambda t_2) \tag{6.189}$$

For $n_2 > n_1$ and $t_2 > t_1$, the expectation of the autocorrelation function is a double summation because (6.189) is a discrete joint pdf (even though the process is defined for continuous time):

$$\begin{aligned} R_{XX}(t_1, t_2) &= \sum_{n_1=0}^{n_2}\sum_{n_2=0}^{\infty} n_1 n_2 \frac{[\lambda(t_2 - t_1)]^{n_2 - n_1}(\lambda t_1)^{n_1}}{n_2!}\binom{n_2}{n_1}\exp(-\lambda t_2), \\ &= \sum_{n_2=0}^{\infty} \frac{n_2}{n_2!}\exp(-\lambda t_2)(\lambda t_2)^{n_2} \\ &\quad \cdot \sum_{n_1=0}^{n_2} n_1 \binom{n_2}{n_1}(t_1/t_2)^{n_1}[1 - t_1/t_2]^{n_2 - n_1}. \end{aligned} \tag{6.190}$$

Since the second sum gives the mean Np of a binomial distribution with parameters $\{p = t_1/t_2, N = n_2\}$, we obtain

$$R_{XX}(t_1, t_2) = \sum_{n_2=0}^{\infty} \frac{n_2}{n_2!} \exp(-\lambda t_2)(\lambda t_2)^{n_2} n_2(t_1/t_2),$$

$$= (t_1/t_2) \sum_{n_2=0}^{\infty} n_2^2 \frac{(\lambda t_2)^{n_2}}{n_2!} \exp(-\lambda t_2), \tag{6.191}$$

which is the second moment $\alpha(1+\alpha)$ of a Poisson distribution with parameter $\alpha = \lambda t_2$, yielding

$$R_{XX}(t_1, t_2) = (t_1/t_2)\lambda t_2(1+\lambda t_2) = \lambda t_1(1+\lambda t_2) \tag{6.192}$$

for $t_2 > t_1$. Finally, for arbitrary $\{t_1, t_2\}$:

$$R_{XX}(t_1, t_2) = \lambda \min(t_1, t_2) + \lambda^2 t_1 t_2, \tag{6.193}$$

which does not depend on the lag $\tau = t_2 - t_1$, and thus $X(t)$ is not wide-sense stationary. A plot of $R_{XX}(t_1, t_2)$ is provided in Figure 6.20(a) where a saddle point is seen at the origin. Since $\mathcal{E}[X(t)] = \lambda t$, the autocovariance function is

$$C_{XX}(t_1, t_2) = R_{XX}(t_1, t_2) - \mathcal{E}[X(t_1)]\mathcal{E}[X(t_2)] = \lambda \min(t_1, t_2). \tag{6.194}$$

These results for the Poisson process are also summarized in Table 6.4, where we see the correlation coefficient has a form similar to that of the binomial counting process (except that it is defined for continuous time). The correlation coefficient is plotted in Figure 6.20(b).

The type of Poisson process where λ is constant is called *homogeneous* (or uniform). The rate parameter can be generalized to be a function of time $\lambda(t)$, corresponding to situations where the rate of events varies with time (such as at different times of the day). The distribution for this *nonhomogeneous* (or nonuniform) Poisson process is

$$P(X(t) = n) = \frac{1}{n!}\left(\int_0^t \lambda(v)\mathrm{d}v\right)^n \exp\left(-\int_0^t \lambda(v)\mathrm{d}v\right). \tag{6.195}$$

The mean and variance of $X(t)$ again depend on t: $\mathcal{E}[X(t)] = \mathrm{var}[X(t)] = \int_0^t \lambda(v)\mathrm{d}v$, which reduces to λt for the homogeneous case. The increments of $X(t)$ are independent, but they are not stationary as is the case for the homogeneous process. Similar to (6.179), we can write the probability of outcomes for any duration τ by changing the limits of integration:

$$P(X(t) - X(t-\tau) = n) = \frac{1}{n!}\left(\int_{t-\tau}^t \lambda(v)\mathrm{d}v\right)^n \exp\left(-\int_{t-\tau}^t \lambda(v)\mathrm{d}v\right). \tag{6.196}$$

The autocorrelation function for the nonhomogeneous case is obtained from (6.193) by replacing $\min(t_1, t_2)$ with $\int_0^{\min(t_1,t_2)} \lambda(v)\mathrm{d}v$.

Finally, we generalize the Poisson counting process such that interarrival times are described by a *gamma* distribution (instead of exponential):

$$f_Y(y) = \frac{y^{r-1}\lambda^r}{\Gamma(r)} \exp(-\lambda y) I_{[0,\infty)}(y) \tag{6.197}$$

with parameters $\{\lambda, r > 0\}$.

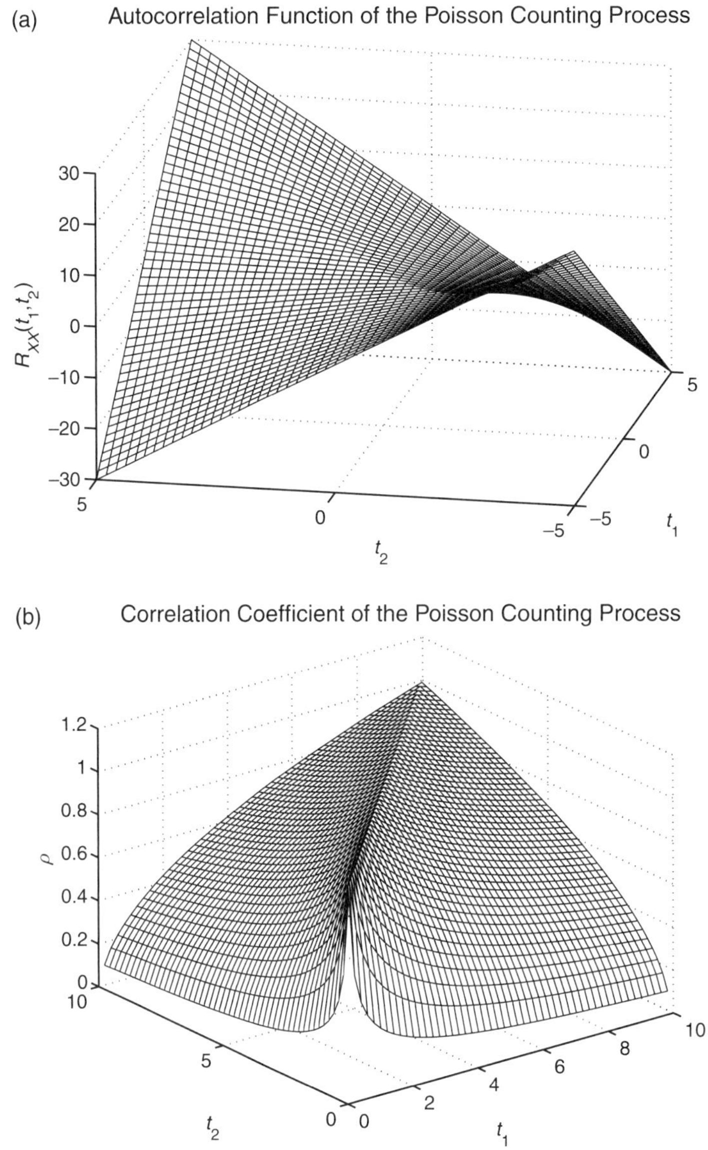

FIGURE 6.20 Correlation of the Poisson counting process with $\lambda = 1$. (a) Autocorrelation function $R_{XX}(t_1, t_2)$. (b) Correlation coefficient ρ. This is also a plot of the correlation coefficient for the Wiener process, as well as the Bernoulli walk and the binomial counting process (but for integer values of $\{t_1, t_2\}$).

Definition: Gamma Random Process The *gamma random process* is a Lévy random process with gamma interarrival times.

Example realizations are shown in Figure 6.21 where we see that the counting rate is reduced with an increase in r because the mean of the gamma distribution increases with r. The Poisson realization in Figure 6.19 lies in between those shown here, as expected because the exponential interarrival distribution corresponds to $r = 1$.

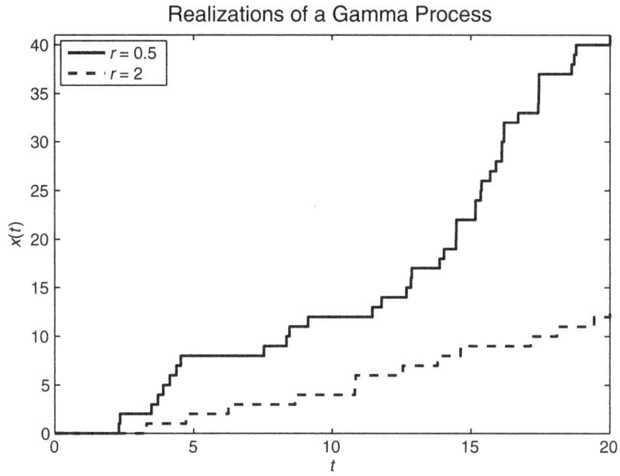

FIGURE 6.21 Example realizations of a gamma process with rate parameter $\lambda = 1$ and different values of r.

6.12.2 Random Telegraph Signal

Definition: Random Telegraph Signal The *random telegraph signal* is derived from the Poisson counting process $X(t)$ as follows:

$$Y(t) = \begin{cases} 1, & X(t) \text{ even} \\ -1, & X(t) \text{ odd}, \end{cases} \tag{6.198}$$

which can be written simply as $Y(t) = (-1)^{X(t)}$, where $X(t)$ has the distribution in (6.175) on $\mathcal{T} = [0, t]$.

Rather than counting the number of steps, $Y(t)$ indicates only whether the count is *even or odd*. Since $X(0) = 0$, the initial condition for the random telegraph signal is $Y(0) = 1$. $Y(t)$ is a continuous-time process that maintains its current value (1 or -1) until the underlying poisson process $X(t)$ changes.

Although it would be appear that the outcomes are equally likely, it is true only in the limit as $t \to \infty$. For finite t, the probability that the Poisson process $X(t)$ has an even number of steps is

$$P(X(t) \text{ even}) = \sum_{m=0,\text{even}}^{\infty} \exp(-\lambda t)(\lambda t)^m / m!, \tag{6.199}$$

where the Poisson distribution of counts in (6.175) has been used. By factoring $\exp(-\lambda t)$, we recognize that the infinite series is the hyperbolic cosine (see Appendix E):

$$P(X(t) \text{ even}) = \exp(-\lambda t) \cosh(\lambda t) \tag{6.200}$$

and likewise for $X(t)$ odd:

$$P(X(t) \text{ odd}) = \exp(-\lambda t) \sinh(\lambda t). \tag{6.201}$$

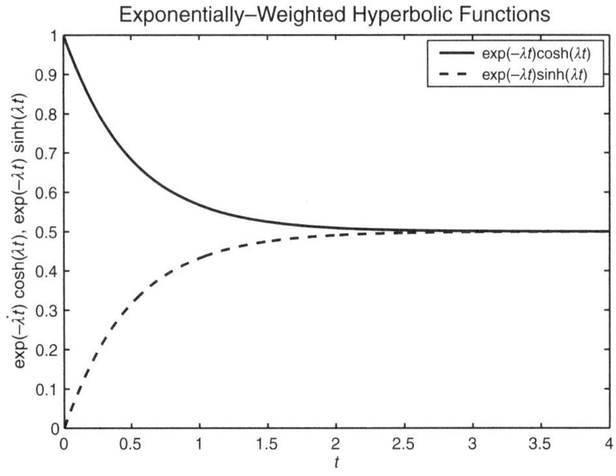

FIGURE 6.22 Pdf components of the random telegraph signal with $\lambda = 1$.

Combining these results gives the following pdf for $Y(t)$:

$$f_{Y(t)}(y) = \exp(-\lambda t)\left[\cosh(\lambda t)\delta(y-1) + \sinh(\lambda t)\delta(y+1)\right], \tag{6.202}$$

where $\delta(y)$ is the Dirac delta function. Note that $f_{Y(t)}(y)$ is a valid pdf:

$$\int_{-\infty}^{\infty} f_{Y(t)}(y)\mathrm{d}y = \exp(-\lambda t)\left[\cosh(\lambda t) + \sinh(\lambda t)\right] = 1. \tag{6.203}$$

The plots shown in Figure 6.22 for $\lambda = 1$ demonstrate that with increasing t, the components of (6.202) weighting $\delta(y-1)$ and $\delta(y+1)$ individually decay, but always sum to 1 at any time t . The two components are equal (to $1/2$) only in the limit as $t \to \infty$; however, for practical purposes, they are nearly equal after about three to four time constants where the time constant of an exponential function is $\tau \approx 1/\lambda$.

From a filtering perspective, it is not possible to generate a random telegraph signal using a time-invariant filter operating on a train of Dirac delta functions. Recall that this was possible for the Poisson counting process, where the filter has an impulse response given by the unit-step function. The filter in this case would be rectangular with a time-varying duration, as determined by the intervals between the delta functions.

The mean of the random telegraph process is

$$\mathcal{E}[Y(t)] = \exp(-\lambda t)\left[(+1)\cosh(\lambda t) + (-1)\sinh(\lambda t)\right] = \exp(-2\lambda t), \tag{6.204}$$

and since the second moment is $\mathcal{E}[Y^2(t)] = 1$, the variance is $\mathrm{var}[Y(t)] = 1 - \exp(-4\lambda t)$, which varies with time. Asymptotically as $t \to \infty$, the random telegraph signal has zero mean and unit variance, which is the same as the symmetric Bernoulli sequence. Example realizations of the random telegraph signal with $\lambda = 1$ are shown in Figure 6.23. The *randomized* version of the random telegraph signal is given by $Z(t) = WY(t)$, where W is a symmetric Bernoulli random variable with equally likely outcomes $\{-1, 1\}$ that is independent of $Y(t)$. (The original version of the random telegraph signal is also called "semirandom.") With this additional randomness such that the initial condition is $Y(0) = W$, we no longer know which value, 1 or -1, of the process corresponds to an even number of jumps.

It can be shown that its autocovariance function is (see Problem 6.31)

$$C_{ZZ}(t_1, t_2) = \exp(-2\lambda|t_2 - t_1|) = C_{ZZ}(\tau), \tag{6.205}$$

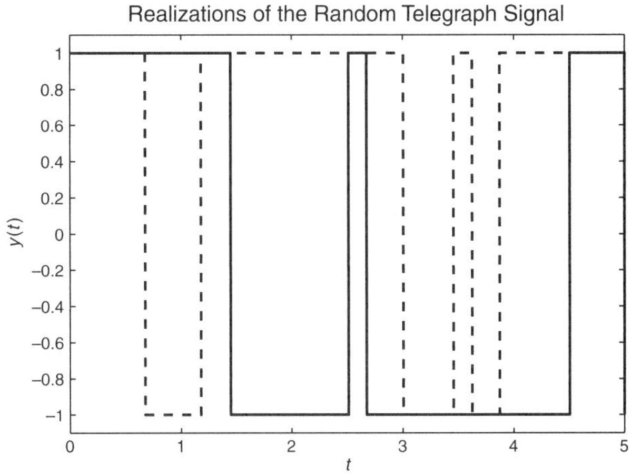

FIGURE 6.23 Example realizations of the random telegraph signal with $\lambda = 1$.

which is wide-sense stationary. $C_{ZZ}(t_1, t_2)$ is also the correlation coefficient ρ because the variance of $Z(t)$ is 1 for all t; it is plotted in Figure 6.24. The correlation coefficient for the "semirandom" version with $Y(0) = 1$ is given by (see Problem 6.32)

$$\rho = \frac{\exp(-2\lambda|t_2 - t_1|) - \exp(-2\lambda(t_1 + t_2))}{\sqrt{1 - \exp(-4\lambda t_1) - \exp(-4\lambda t_2) + \exp(-4\lambda(t_1 + t_2))}}, \tag{6.206}$$

which cannot be simplified. However, note that for large $\{t_1, t_2\}$ the correlation coefficients for both versions of the random telegraph signal are approximately the same. Correlation results for the two random telegraph signals are also summarized in Table 6.4.

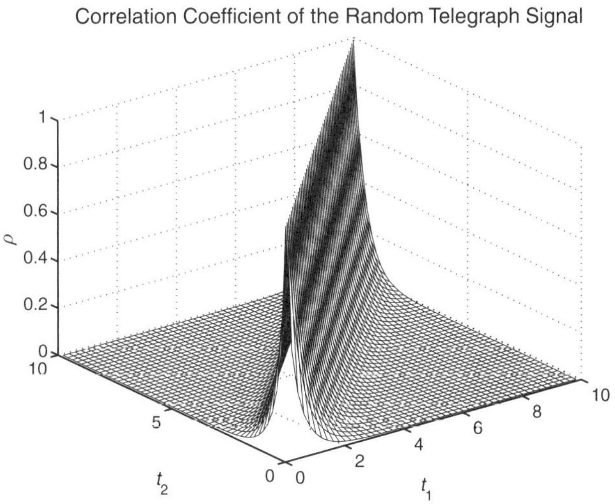

FIGURE 6.24 Correlation coefficient for the "randomized" telegraph signal ($\lambda = 1$). The correlation coefficient for the "semirandom" telegraph signal (with $Y(0) = 1$) is similar to this plot except for small values of $\{t_1, t_2\}$.

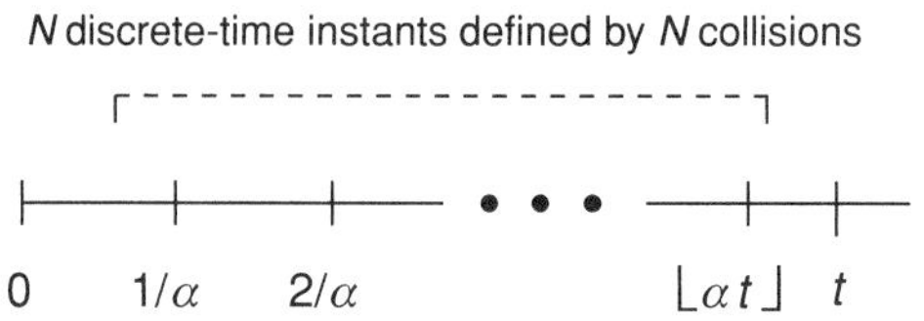

FIGURE 6.25 Discrete-time instants used in the derivation of the Wiener process $X(t)$.

6.12.3 Wiener Process

The Wiener process is a model of the physical phenomenon known as Brownian motion where a particle in a fluid moves erratically because of frequent random collisions with molecules of the fluid. The model can be derived by starting with the symmetric Bernoulli sequence $W[k]$, but with outcomes $\{-\epsilon, \epsilon\}$ for small $\epsilon > 0$. Assume that particle collisions occur equally spaced in time with rate α, which means they happen at integer multiples of $1/\alpha$. The total number of collisions by time t is $N = \lfloor \alpha t \rfloor$, where $\lfloor a \rfloor$ is the floor function retaining the largest integer $\leq a$. The various quantities are depicted in Figure 6.25. Define the random process

$$X(t) = \sum_{m=1}^{\lfloor \alpha t \rfloor} W[m], \tag{6.207}$$

which is similar to a Bernoulli random walk with $p = 1/2$ (see Table 6.3); the differences here are (i) the discrete-time intervals have duration $1/\alpha$ instead of 1, (ii) the amplitudes are $\pm\epsilon$ instead of ± 1, and (iii) the upper limit in the sum is $\lfloor \alpha t \rfloor$ instead of k. Because of this upper limit, $X(t)$ maintains its current value until $t \to t + 1/\alpha$ when another term would be included in the sum due to another collision. This description is the first step toward extending the symmetric Bernoulli sequence to be a function of *continuous time t*. As was done for the Poisson counting process, the Wiener process is obtained by letting $1/\alpha \to 0$, resulting in a continuous-time process. Additionally, the amplitude $\epsilon \to 0$ so that the random variable at each time instant is continuous with a well-defined pdf. Since $\mathcal{E}[W[m]] = 0$ and $\mathcal{E}[W^2[m]] = \epsilon^2$, it is follows that $\mathcal{E}[X(t)] = 0$ and $\mathcal{E}[X^2(t)] = \epsilon^2 \lfloor \alpha t \rfloor$.

Suppose now that the current time t coincides exactly with an integer multiple of $1/\alpha$, and assume $f_{X(t)}(x)$ exists. Consider the pdf at the next time instant $t + 1/\alpha$ and apply the law of total probability such that

$$\begin{aligned} f_{X(t+1/\alpha)}(x) = {} & f_{X(t+1/\alpha)}(x|W[m] = \epsilon)P(W[m] = \epsilon) \\ & + f_{X(t+1/\alpha)}(x|W[m] = -\epsilon)P(W[m] = -\epsilon), \end{aligned} \tag{6.208}$$

where by assumption $m = \lfloor \alpha t \rfloor$ corresponds exactly to time instant t. This partitioning describes the two events that can happen in the process as it moves from time t to time $t + 1/\alpha$. In order to rewrite (6.208) in terms of the pdf at time t, we recognize that $X(t + 1/\alpha) = X(t) + \epsilon$ and $X(t + 1/\alpha) = X(t) - \epsilon$ are linear transformations of random variable $X(t)$. From the transformation results in Chapter 4:

$$f_{X(t+1/\alpha)}(x|W[m] = \epsilon) = f_{X(t)}(x - \epsilon), \tag{6.209}$$

$$f_{X(t+1/\alpha)}(x|W[m] = -\epsilon) = f_{X(t)}(x + \epsilon). \tag{6.210}$$

Substituting these expressions into (6.208) gives

$$f_{X(t+1/\alpha)}(x) = (1/2)[f_{X(t)}(x - \epsilon) + f_{X(t)}(x + \epsilon)], \tag{6.211}$$

where $P(W[m] = \epsilon) = P(W[m] = -\epsilon) = 1/2$ has been used.

Since t and ϵ are assumed to be continuous variables, Taylor series expansions can be used to represent the functions. The left-hand side of (6.211) is expanded about t as follows:

$$f_{X(t+1/\alpha)}(x) = f_{X(t)}(x) + (1/\alpha)\frac{\partial}{\partial t} f_{X(t)}(x) + \cdots. \tag{6.212}$$

Likewise, the terms on the right-hand side of (6.211) are expanded about x:

$$f_{X(t)}(x+\epsilon) = f_{X(t)}(x) + \epsilon\frac{\partial}{\partial x} f_{X(t)}(x) + \epsilon^2\frac{\partial^2}{\partial x^2} f_{X(t)}(x) + \cdots, \tag{6.213}$$

$$f_{X(t)}(x-\epsilon) = f_{X(t)}(x) - \epsilon\frac{\partial}{\partial x} f_{X(t)}(x) + \epsilon^2\frac{\partial^2}{\partial x^2} f_{X(t)}(x) - \cdots. \tag{6.214}$$

These equations simultaneously represent $f_{X(t+1/\alpha)}(x)$ as expansions along the time t and amplitude x axes. Since the goal is to let $\epsilon \to 0$ (creating a continuous-amplitude function) and $\alpha \to \infty$ (allowing the process to change at any time), we ignore all high-order terms because they depend either on $1/\alpha^m$ for $m \geq 2$ or ϵ^m for $m \geq 3$. Note that we must retain terms with ϵ^2 because (i) the two terms with ϵ cancel each other and (ii) the product $\alpha\epsilon^2$ is assumed to be a constant.

Substituting (6.212)–(6.214) into (6.211) yields

$$(1/\alpha)\frac{\partial}{\partial t} f_{X(t)}(x) = \epsilon^2\frac{\partial^2}{\partial x^2} f_{X(t)}(x) + \cdots, \tag{6.215}$$

where $\cdots$ indicates all high-order terms from both expansions. Finally, letting $\alpha \to \infty$ and $\epsilon \to 0$ such that $\alpha\epsilon^2$ is a constant, and dropping all high-order terms, gives a second-order partial differential equation (DE):

$$\frac{\partial}{\partial t} f_{X(t)}(x) = \alpha\epsilon^2\frac{\partial^2}{\partial x^2} f_{X(t)}(x). \tag{6.216}$$

The solution of this equation is the pdf of the Wiener process (see Problem 6.33).

Definition: Wiener Process The *Wiener process* is a continuous-time random process with the following Gaussian pdf:

$$f_{X(t)}(x) = \frac{1}{\sqrt{2\pi\alpha\epsilon^2 t}} \exp\left(-x^2/2\alpha\epsilon^2 t\right), \tag{6.217}$$

which has zero mean and variance $\sigma_X^2(t) = \alpha\epsilon^2 t$. The initial condition is $X(0) = 0$.

The Wiener process is a continuous-time, continuous-amplitude version of the Bernoulli random walk where its amplitude can change at any time instant $t \in \mathcal{T} = \mathcal{R}^+$, and the amplitude can take on any value in $\mathcal{R}$. Thus, like the Poisson counting process, the Wiener process can be viewed as the output of a filter that is an integrator. The difference here is that the model assumes the Bernoulli sequence has outcomes $\pm\epsilon$ (instead of $\{0, 1\}$ for the Poisson counting process). As demonstrated in Chapter 8, a Wiener process can be generated by integrating a white-noise process with flat PSD given by $\alpha\epsilon^2$. Although the mean of the Wiener process is always zero, its variance $\alpha\epsilon^2 t$ grows linearly with time, such that $X(t)$ is a nonstationary random process. Example realizations of the Wiener process are shown in Figure 6.26.

The Wiener process can be defined in more general terms, assuming that it is an isi process (and thus is a type of Lévy process).

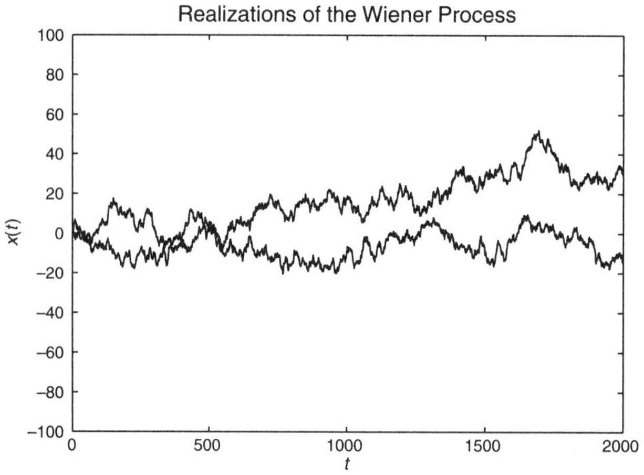

FIGURE 6.26 Example realizations of the Wiener process in (6.217) with $\alpha\epsilon^2 = 1$.

Definition: Generalized Wiener Process The *generalized Wiener process* $X(t)$ is an isi process with $X(0) = 0$ such that

$$f_{X(t_2)-X(t_1)}(x) = \frac{1}{\sqrt{2\pi\alpha\epsilon^2(t_2 - t_1)}} \exp\left(-x^2/2\alpha\epsilon^2(t_2 - t_1)\right) \tag{6.218}$$

for $t_2 > t_1$.

The variance of $X(t_2) - X(t_1)$ is $\alpha\epsilon^2(t_2 - t_1)$, which reduces to $\alpha\epsilon^2 t$ in (6.217) when $t_2 = t$ and $t_1 = 0$. The autocovariance function of $X(t_2)$ and $X(t_1)$ is

$$\begin{aligned} C_{XX}(t_1, t_2) &= \mathcal{E}[X(t_2)X(t_1)] = R_{XX}(t_1, t_2), \\ &= \mathcal{E}[(X(t_2) - X(t_1) + X(t_1))X(t_1)], \\ &= \mathcal{E}[(X(t_2) - X(t_1))X(t_1)] + \mathcal{E}[X^2(t_1)]. \end{aligned} \tag{6.219}$$

Since the increments are independent, the first expectation in the last line splits into a product of expectations such that

$$\begin{aligned} C_{XX}(t_1, t_2) &= \mathcal{E}[X(t_2) - X(t_1)]\mathcal{E}[X(t_1)] + \mathcal{E}[X^2(t_1)], \\ &= \mathcal{E}[X^2(t_1)] = \alpha\epsilon^2 t_1, \end{aligned} \tag{6.220}$$

where we have used the fact that the process has zero mean to remove the first term. This result assumed $t_2 > t_1$; for arbitrary $\{t_1, t_2\}$:

$$C_{XX}(t_1, t_2) = \alpha\epsilon^2 \min(t_1, t_2) = R_{XX}(t_1, t_2), \tag{6.221}$$

which is also the autocorrelation function because the process has zero mean; it is plotted in Figure 6.27. Note that the autocovariance function of the Poisson counting process has the same shape as this plot, as do those of the Bernoulli walk and Poisson counting process (but at integer values of $\{t_1, t_2\}$). The correlation results for the Wiener process are also summarized in Table 6.4.

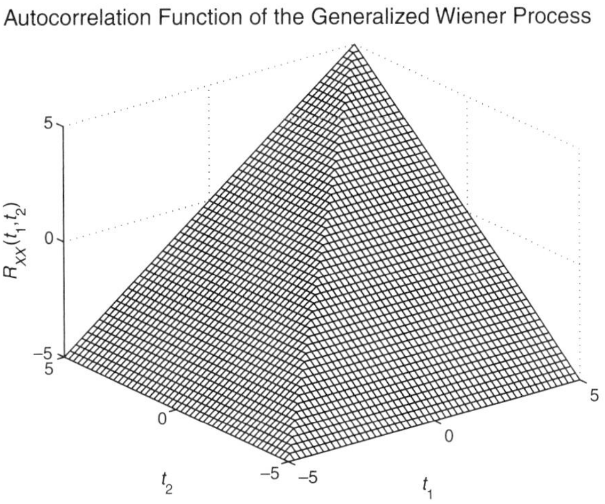

FIGURE 6.27 Autocorrelation function of the generalized Wiener process with $\alpha\epsilon^2 = 1$.

6.12.4 Gaussian Process

Consider a random process with outcomes given by the Gaussian distribution.

Definition: Gaussian Process *Gaussian process* $X(t)$ is a set of random variables indexed by continuous time such that for any N time instants $\{t_1, \ldots, t_N\}$, their joint pdf is Gaussian:

$$f_{X(t_1),\ldots,X(t_N)}(x_1, \ldots, x_N) = \frac{1}{\sqrt{(2\pi)^N |\mathbf{C_{XX}}|}} \exp\left(-(\mathbf{x} - \boldsymbol{\mu})^T \mathbf{C}_{\mathbf{XX}}^{-1}(\mathbf{x} - \boldsymbol{\mu})\right), \tag{6.222}$$

where $\mathbf{X} \triangleq [X(t_1), \ldots, X(t_N)]^T$, $\mathbf{x} \triangleq [x_1, \ldots, x_N]^T$, $\boldsymbol{\mu} = \mathcal{E}[\mathbf{X}]$, $\mathbf{C_{XX}} = \mathcal{E}[(\mathbf{X} - \boldsymbol{\mu})(\mathbf{X} - \boldsymbol{\mu})^T]$ is the autocovariance matrix, and $|\mathbf{C_{XX}}|$ is the determinant of $\mathbf{C_{XX}}$.

An example realization of an iid Gaussian process is illustrated in Figure 6.28(a) for $\boldsymbol{\mu} = \mathbf{0}$ and $\mathbf{C_{XX}} = \mathbf{I}$. Since all the previous processes are based on a sum or integral of a Bernoulli sequence, the autocorrelation function is determined by how the process is constructed. The Gaussian process, on the other hand, is not defined in such a manner: the covariance matrix is an independent *parameter* of the process, given by $\mathbf{C_{XX}}$ in (6.222). Like a Gaussian random vector, a Gaussian random process is specified completely by its mean vector and its covariance matrix (which could be functions of time).

Suppose we want to generate a realization of a *correlated* Gaussian process. Let $\mathbf{Y}$ be a vector of N uncorrelated Gaussian random variables with zero mean and unit variance: $\mathcal{E}[\mathbf{Y}] = \mathbf{0}$ and $\mathcal{E}[\mathbf{YY}^T] = \mathbf{I}$. Let $\mathbf{C_{XX}}$ be the desired covariance matrix (which is also the autocorrelation matrix because the mean is zero). Since $\mathbf{C_{XX}}$ is symmetric and positive definite (PD), it can be decomposed using a Cholesky factorization as follows (see Appendix G):

$$\mathbf{C_{XX}} = \mathbf{BB}^T, \tag{6.223}$$

where $\mathbf{B}$ is a lower triangular matrix with positive diagonal elements. By forming the transformation:

$$\mathbf{X} = \mathbf{BY}, \tag{6.224}$$

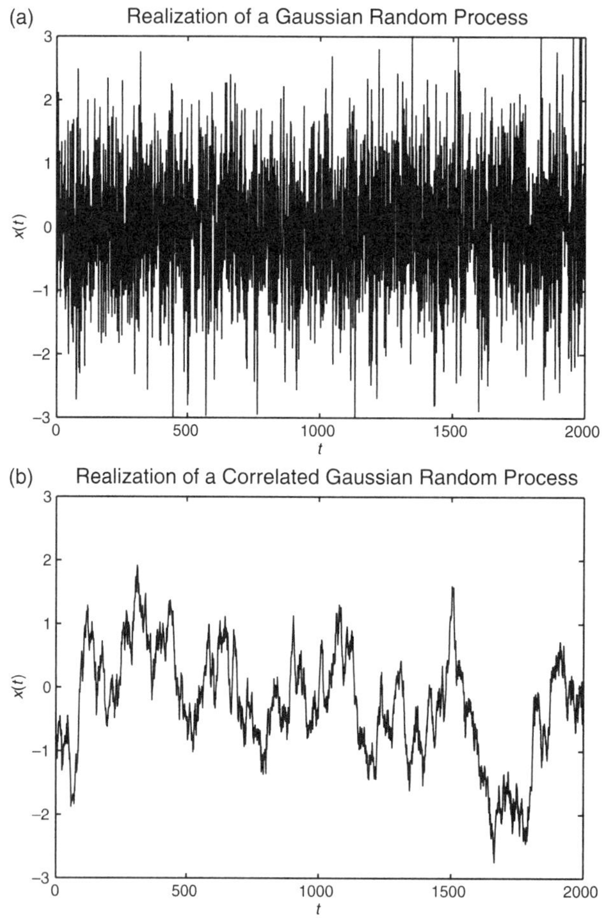

FIGURE 6.28 Example realizations of a Gaussian process. (a) Independent and identically distributed (iid) with parameters $\{\mu = 0, \sigma = 1\}$. (b) Wide-sense stationary with $\mu = 0$ and $\mathcal{E}[X(t_1)X(t_2)] = R_{XX}(t_2 - t_1) = R_{XX}(\tau) = \exp(-|\tau|/100)$.

we obtain the desired autocorrelation matrix:

$$\mathcal{E}[\mathbf{X}\mathbf{X}^T] = \mathbf{B}\mathcal{E}[\mathbf{Y}\mathbf{Y}^T]\mathbf{B}^T = \mathbf{C}_{\mathbf{XX}} = \mathbf{R}_{\mathbf{XX}} \text{ (for zero mean).} \tag{6.225}$$

An example realization of a correlated Gaussian process is shown in Figure 6.28(b) with $R_{XX}(\tau) = \exp(-|\tau|/100)$. Observe that this realization appears to be more "predictable" than the iid process in Figure 6.28(a); the realization in Figure 6.1 presented at the beginning of this chapter is another realization of this correlated Gaussian random process.

6.12.5 Pulse Amplitude Modulation

A random process known as PAM has the following representation:

$$X(t) = \sum_{n=-\infty}^{\infty} A_n \text{rect}((t - nT_o - T)/T_o - 1/2), \tag{6.226}$$

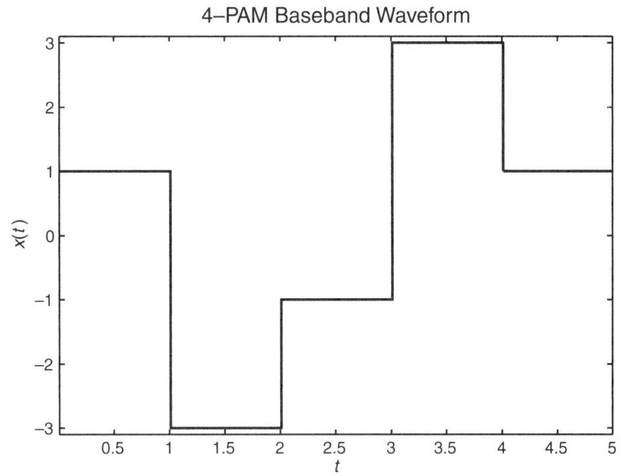

FIGURE 6.29 Example realization $x(t)$ of four-level PAM with $T_o = 1$ and $T = 0$.

where T_o is the duration of one symbol, $\{A_n\}$ are iid random variables, and T is a uniform random variable on $[0, T_o]$ that models the delay of an entire realization. An example realization is shown in Figure 6.29 where the outcomes of A_n are $\{-3, -1, 1, 3\}$, $T_o = 1$, and $T = 0$. Because of the rectangle function, PAM with outcomes ± 1 resembles the random telegraph signal, except that it has more structure: the waveform can change only at integer multiples of the symbol period T_o, and a change occurs only when $A_n \neq A_{n+1}$. Some properties of this random signal are considered in Problem 6.36. A special case where the outcomes of the $\{A_n\}$ are ± 1 is sometimes called a *random binary signal*.

6.12.6 Random Sine Signals

Finally, we return to sinusoidal waveforms that have random components as first discussed in Chapter 1. Such signals are important in communications where sine waves are *modulated* by random processes in order to transmit information between different locations using high-frequency *carriers*.

Consider the following random process:

$$X(t) = A \sin(2\pi f_c t + \Phi), \tag{6.227}$$

where A is the amplitude, f_c is the carrier frequency, and Φ is a phase shift, for which we describe various degrees of randomness. Initially, assume that A and Φ are independent random variables that do not change with time. For example, Φ might be uniformly distributed on $[0, 2\pi]$, and A might be Gaussian with zero mean and unit variance. A realization of $X(t)$ is obtained when A and Φ take on particular values from their distributions. Three such realizations for $f_c = 1$ Hz are shown in Figure 6.30(a), each of which has a different amplitude and phase.

The random sine wave can be generalized by allowing the amplitude and phase to be random processes:

$$X(t) = A(t) \sin(2\pi f_c t + \Phi(t)). \tag{6.228}$$

For the time-varying case, the process of multiplying a sine wave by $A(t)$ is called amplitude modulation (AM); similarly, including a time-varying phase $\Phi(t)$ yields phase modulation (PM). (It is also possible to vary the frequency f_c, resulting in frequency modulation (FM).) Figures 6.30(b) and 6.30(c) show examples of AM and PM for $A(t)$ and $\Phi(t)$ implemented as Bernoulli schemes with amplitudes $\{1, 2, 3, 4\}$ and phases $\{2\pi/5, 4\pi/5, 6\pi/5, 8\pi/5\}$, corresponding to four *symbols* each. The carrier frequency is $f_c = 10$ Hz, and the

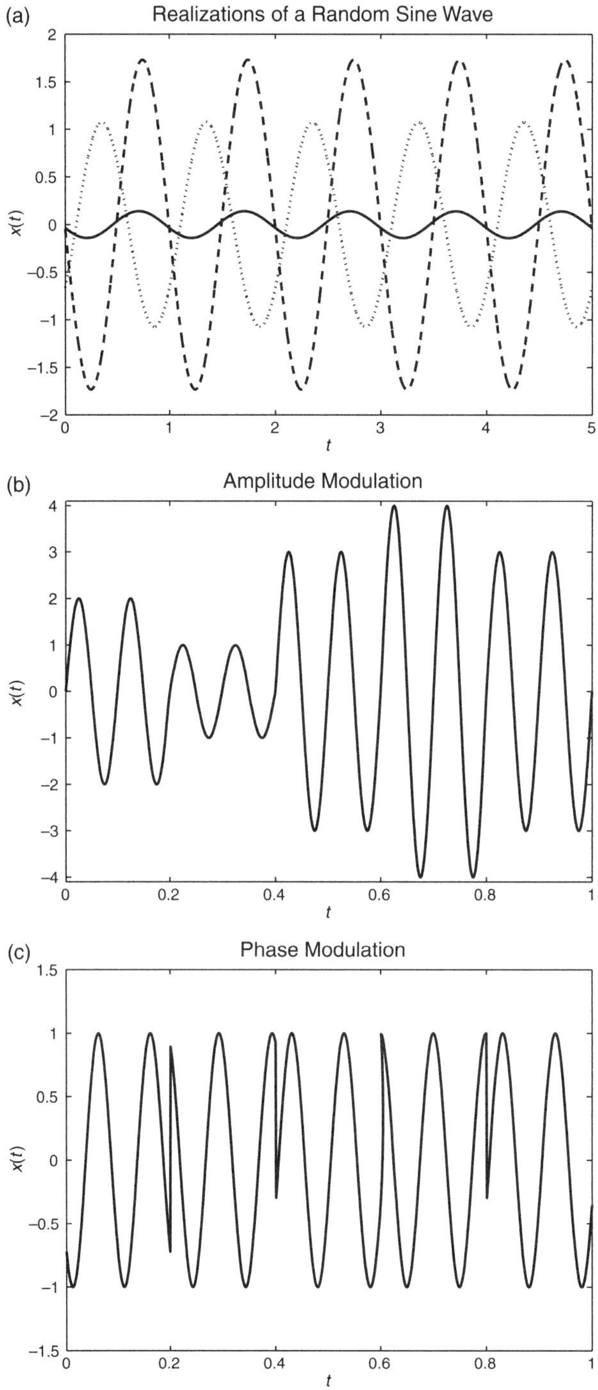

FIGURE 6.30 Random sine waveforms. (a) Realizations where the amplitude and phase are random variables. (b) Amplitude shift keying (ASK). (c) Phase-shift keying (PSK).

amplitude and phase can change only every 0.2 s (the symbol period). Because the modulating waveforms $A(t)$ and $\Phi(t)$ take on only a finite number of values, they are examples of *digital modulation*, and are called amplitude shift keying (ASK) and phase shift keying (PSK), respectively. Note that the PAM signal discussed in Section 6.11 is a baseband version of ASK because it does not include the sine carrier. As is the case for PAM, digital modulation of a sinusoidal waveform has much structure because (i) the number of amplitudes and phases is finite and (ii) they can change only at integer multiples of a fixed symbol period. This type of structure is exploited in the receiver to make *decisions* about the transmitted symbols, as discussed in Chapter 10 on digital communications.

PROBLEMS

Basic Random Processes

6.1 Suppose the sample space representing a random process is $\mathcal{S} = \{1, 2, 3, 4\}$ and let $X(t, \zeta) = \exp(-\zeta t)u(t)$ for $\mathcal{T} = \mathcal{R}^+$. For equally likely ζ, specify the probability space $\{\Omega, \mathcal{F}, P\}$ for the random variables at $t = 1$ and $t = 2$, and sketch all realizations.

6.2 Let the sample space representing a random process be $\mathcal{S} = [0, 1]$ such that $X(t, \zeta) = \zeta(t-2)u(t-3) + \exp(-\zeta t)u(t)$ for $\mathcal{T} = [0, 5]$. For uniformly distributed ζ, give the pdfs for the random variables at $t = 2$ and $t = 4$.

6.3 Consider the following random process which increases linearly with time:

$$Y(t) = at + X, \tag{6.229}$$

where a is a constant and X is uniformly distributed on $[-1, 1]$. Find the cdf and pdf of $Y(t)$ for $\mathcal{T} = \mathcal{R}^+$.

6.4 Let $X(t) = A\exp(-t)u(t)$, where A has a discrete uniform distribution with outcomes $\{-1, 0, 1, 2\}$. Find the pdf of $X(t)$ for $\mathcal{T} \in \mathcal{R}^+$.

6.5 Consider the following integral random process:

$$Y(t) = \int_0^t X(\tau)\mathrm{d}\tau. \tag{6.230}$$

Let $X(t)$ be the same as in Problem 6.4 except that A is a standard Gaussian random variable. Find the pdf and cdf of $Y(t)$.

Correlation and Stationarity

6.6 Find the mean, variance, autocorrelation function, and autocovariance function for $Y(t)$ in Problem 6.3.

6.7 Let $Y(t) = X(t) + V$ for $\mathcal{T} = \mathcal{R}^+$, where $X(t)$ is wide-sense stationary with mean μ_X and autocorrelation function $R_{XX}(\tau)$. Find expressions for $R_{YY}(\tau)$ and $C_{YY}(\tau)$ when (a) $V = v$ is a constant and (b) V is a random variable independent of $X(t)$ with mean μ_V and variance σ_V^2.

6.8 Define the following random process for $\mathcal{T} = [0, 1]$:

$$Y(t) = X_1 t - X_2, \tag{6.231}$$

where $\{X_1, X_2\}$ are independent random variables uniformly distributed on $[-1, 1]$. Find the cdf, mean $\mu_Y(t)$, and autocorrelation function $R_{YY}(t_1, t_2)$, and decide if the process is wide-sense stationary.

6.9 Define the following random process for $\mathcal{T} = \mathcal{R}$:

$$Y(t) = X_1 \cos(\omega_o t) + X_2 \sin(\omega_o t), \tag{6.232}$$

where ω_o is fixed and $\{X_1, X_2\}$ are independent symmetric Bernoulli random variables with outcomes $\{-1, 1\}$ such that $P(X_1 = 1) = P(X_2 = 1) = p$. Determine if $Y(t)$ is wide-sense stationary for any value of p.

6.10 For $\mathcal{T} = \mathcal{R}$, define

$$X(t) = \cos(\omega_o t + \Phi), \tag{6.233}$$

where Φ is uniform on $[-\pi, \pi]$. (a) Find an expression for $R_{XX}(t_1, t_2)$ and determine if $X(t)$ is wide-sense stationary. (b) Repeat part (a) for sine instead of cosine.

6.11 Consider the following random sequences:

$$Y_1[k] = X[k] + 2X[k-1], \tag{6.234}$$

$$Y_2[k] = X[k] + (1/2)Y_2[k-1], \tag{6.235}$$

where $X[k]$ is a zero-mean iid sequence for $\mathcal{T} = \mathcal{Z}^+$. (a) Determine if $Y_1[k]$ and $Y_2[k]$ are wide-sense stationary and (b) find their cross-correlation function $\mathcal{E}[Y_1[k_1]Y_2[k_2]]$.

6.12 Suppose $X(t)$ is a wide-sense stationary Gaussian process with nonzero mean and autocorrelation function $R_{XX}(\tau)$. (a) Show that $f_{X(t)}(x) = f_X(x)$ for all $t \in \mathcal{T}$, that is, it is also first-order stationary. (b) Determine if $X(t)$ is strictly stationary.

6.13 Derive the autocorrelation and autocovariance functions in Table 6.4 for the generalized Bernoulli random walk.

6.14 For random sequence $X[k]$ defined on $\mathcal{T} = \mathcal{N}$, assume we alternately toss a coin for even k and a die for odd k. Label the coin outcomes $\{0, 1\}$ and assume the usual six outcomes for the die. (a) Derive an expression for $R_{XX}[m_1, m_2]$. (b) Find $P(X[k] = X[k-1])$ for all k.

6.15 Consider random sequence $X[k]$ of iid standard Gaussian random variables defined on $\mathcal{T} = \mathcal{Z}^+$. Let $Y[k] = X[k] + \alpha X[k-1]$ for $\mathcal{T} = \mathcal{N}$ and $\alpha \in \mathcal{R}$. (a) Find an expression for $R_{YY}[m]$. (b) Determine if $Y[k]$ is an independent increment sequence. (c) Repeat parts (a) and (b) for $Z[k] = X[k] + \alpha X[k-2]$.

Martingales

6.16 Consider the following product sequence:

$$Y[k] = \alpha \prod_{n=1}^{k} X[n], \tag{6.236}$$

where random sequence $X[k]$ is iid with outcomes in $\mathcal{R}^+$. Find α such that $Y[k]$ is a martingale.

6.17 Consider the sum

$$Y[k] = \sum_{n=1}^{k} X[n], \tag{6.237}$$

where $X[k]$ is an iid sequence with nonzero mean μ_X. Determine if $Z[k] = Y[k] - k\mu_X$ is a martingale.

6.18 Determine if the following sequence is a martingale:

$$Y[k] = \left(\sum_{n=1}^{k} X[n]\right)^2 - k\sigma_X^2, \tag{6.238}$$

where $X[k]$ is a zero-mean iid sequence with variance σ_X^2.

Markov Chains

6.19 A Markov chain has three states $\{1, 2, 3\}$ with transition matrix

$$\mathbf{P} = \begin{bmatrix} 0.1 & 0.3 & 0.6 \\ 0 & 0.5 & 0.5 \\ 0.2 & 0.2 & 0.6 \end{bmatrix}. \tag{6.239}$$

Find (a) $P(X[2] = 3|X[0] = 1)$, (b) $P(X[2] = 3)$, and (c) $P(X[1] = 2|X[2] = 3)$. Assume the initial states are equally likely.

6.20 Derive the more general form of the Chapman–Kolmogorov equation given by

$$p_{mn}^{(r)} = \sum_{q=1}^{M} p_{mq}^{(s)} p_{qn}^{(r-s)} \tag{6.240}$$

for $1 < s < r$.

6.21 A Markov chain has three states $\{1, 2, 3\}$ with transition matrix

$$\mathbf{P} = \begin{bmatrix} 0.2 & 0.4 & 0.4 \\ 0.1 & 0.2 & 0.7 \\ 0.8 & 0.1 & 0.1 \end{bmatrix}. \tag{6.241}$$

For initial state $\mathbf{p}[0] = [0.2, 0.3, 0.5]^T$, find an expression for $\mathbf{p}[k]$ and determine the steady-state probabilities.

6.22 Repeat Problem 6.21 for a Markov chain with four states $\{1, 2, 3, 4\}$ and transition matrix

$$\mathbf{P} = \begin{bmatrix} 0.1 & 0.9 & 0 & 0 \\ 0.5 & 0.5 & 0 & 0 \\ 0 & 0 & 0.4 & 0.6 \\ 0 & 0 & 0.3 & 0.7 \end{bmatrix}. \tag{6.242}$$

Let $\mathbf{p}[0] = [0.1, 0.2, 0.5, 0.2]^T$.

6.23 Consider the following transition matrix for Markov chain $X[k]$ with outcomes $\{-1, 1\}$:

$$\mathbf{P} = \begin{bmatrix} 0.1 & 0.9 \\ 0.5 & 0.5 \end{bmatrix}. \tag{6.243}$$

Find probabilities for sequences $-1, -1, -1, -1, 1, -1$ and $1, 1, 1, 1, 1, -1$.

6.24 Suppose we want to extend the Bernoulli bit counter in Example 6.23 to detect the first occurrence of the sequence $\{1, 0, 1\}$. (a) Derive a state transition diagram for this scenario and (b) use a first-step analysis to find the expected number of bits received before detecting the $\{1, 0, 1\}$ sequence.

6.25 Consider the following difference equation:

$$Y[k] = \alpha Y[k-1] + X[k] \tag{6.244}$$

with $Y[0] = 0$. Let $X[k]$ be an iid sequence for $k \geq 0$, and assume $|\alpha| < 1$. Determine if $X[k]$ and $Y[k]$ are Markov chains.

6.26 Verify the probability in (6.131).

Random Processes

6.27 Let $X(t)$ be a Poisson process with parameter $\lambda = 3$. Find the following quantities. (a) $P(X(2) > 2)$. (b) $P(X(1) = 1, X(2) = 2)$. (c) The average waiting time until $X(t) = 5$.

6.28 For the Poisson counting process $X(t)$, show that $P(X(t_o) = n|X(t) = N)$ for $t_o < t$ and $n < N$ has a binomial distribution with parameters $\{N, p = t_o/t\}$

6.29 Let $X(t)$ be a Poisson process with parameter $\lambda = 2$. If $X(4) = 3$ arrivals, find the following. (a) Two of the arrivals occur after $t = 2$. (b) One arrival occurs before $t = 1$.

6.30 Find the mean, variance, skewness, and excess kurtosis of the randomized telegraph signal $Z(t) = WY(t)$.

6.31 Derive the autocovariance function in (6.205) for the randomized telegraph signal $Z(t)$.

6.32 Derive the correlation coefficient in (6.206) for the semirandom telegraph signal $X(t)$.

6.33 Verify that the pdf of the Wiener process satisfies the second-order DE in (6.216).

6.34 Let $X(t)$ be the standard Wiener process with variance $\alpha\epsilon^2 t$. Find the distribution of the process $Y(t) = |X(t)|$.

6.35 For the Wiener process, we know $f_{X(t_1),X(t_2)}(x_1, x_2) = f_{X(t_1)}(x_1) f_{X(t_2)-X(t_1)}(x_2 - x_1)$ because it has independent increments. Write an expression for the joint pdf of $X(1)$ and $X(2)$.

6.36 Let $\{0, 1\}$ be equally likely amplitudes for the PAM process in (6.226). Assuming that $T_o = 0$, find expressions for $\mu_X(t)$ and $R_{XX}(t_1, t_2)$. In order to find the autocorrelation function, you should condition the expectation on whether or not $\{t_1, t_2\}$ are located within the same symbol.

6.37 Derive the conditional distribution of arrival times for the Poisson counting process $X(t)$ when the conditioning is $X(t) = n$.

Computer Problems

6.38 Let the outcomes of random sequence $X[k]$ be $\{-1, 0, 1\}$ with $P(0) = 1/2$ and $P(-1) = P(1) = 1/4$. Generate realizations of $X[k]$ using rand by partitioning $[0, 1]$ into three intervals to give the three possible outcomes.

6.39 Generate realizations of a gamma process $X(t)$ with parameters $\lambda = 2$ and $r = \{1/2, 1, 3/2\}$. This is done by using the gamma distribution in MATLAB to determine interarrival times.

6.40 Use randn and a Cholesky decomposition to generate realizations of a correlated Gaussian sequence $X[k]$ with zero mean and autocorrelation matrix:

$$\mathbf{R}_{XX} = \begin{bmatrix} 21/16 & 5/8 & 1/4 \\ 5/8 & 21/16 & 5/8 \\ 1/4 & 5/8 & 21/16 \end{bmatrix}. \tag{6.245}$$

6.41 Simulate the Markov chain in Problem 6.21 using `rand` partitioned into subintervals in order to generate the conditional probabilities of **P**. By counting state occurrences for a sufficiently long interval of time, estimate the steady-state probabilities and determine how they compare to the theoretical values found in Problem 6.21.

FURTHER READING

The following references provide more information about the various characterizations of random processes: Davenport (1970), Doob (1953), Larson and Shubert (1979), Gray and Davisson (2004), Hajek (2009), Leon–Garcia (2008), Papoulis (1965), Parzen (1962), and Viniotis (1998). Some specific topics can be explored further in the following references. Markov processes: Hoel, Port, and Stone (1972), Karlin and Taylor (1975), and Taylor and Karlin (1984). Martingales: Athreya and Lahiri (2006), Billingsley (1985), and Larson and Shubert (1979).

7 Stochastic Convergence, Calculus, and Decompositions

7.1 INTRODUCTION

In this chapter, we consider additional properties of random processes that further characterize their behavior. These include convergence of a random sequence $X[k]$ as $k \to \infty$, and its counterpart for a random process $X(t)$: continuity at $t \in \mathcal{T}$, where $\mathcal{T}$ is a continuous set of time instants. Derivatives and integrals of a random process are also defined, as well as differential equations (DEs). These are useful because in Chapter 8, we consider signals and linear systems, which are modeled by DEs with constant coefficients. With this background, we can examine how a linear system alters probabilistic features of a random process, especially its correlation. Finally, we describe how a random signal is decomposed into additive components that separately provide greater insight about the original process.

Consider the realization of a Poisson process shown in Figure 7.1. In this chapter, we are interested in different types of continuity for a *process* that may or may not apply to each realization. Even though the *realization* shown in the figure is not continuous, there are stochastic definitions of continuity where the Poisson process is considered to be continuous. Various definitions of continuity apply to a single time instant t or for all $t \in \mathcal{T}$. Similar stochastic definitions are used for the derivative at $t \in \mathcal{T}$ and the integral over some interval $[t_o, t] \subset \mathcal{T}$. As shown later, it is more useful to have such stochastic definitions that apply to the *ensemble* of a random process rather than to individual realizations. The same approach applies to the convergence of a random sequence, which we describe first.

7.2 STOCHASTIC CONVERGENCE

Convergence of a random sequence to a constant or a random variable can be useful in many applications such as parameter estimation (see Chapter 9). There are several kinds of convergence with varying degrees of strictness, but they all provide information about the behavior of a random sequence as $k \to \infty$. They also lead to the *laws of large numbers* which we discuss in this chapter.

In Chapter 2, we described almost sure and sure events for the abstract probability space $\{\Omega, \mathcal{F}, P\}$. Event E is *almost sure* if $P(E) = 1$ or, alternatively, if $P(E^c) = 0$. Almost sure does *not* imply that E will occur. Event E is *sure* if it must happen and no other event can occur. Likewise, there is a distinction between an event E with probability zero $P(E) = 0$, but which could occur, and an event that can never occur: the *impossible* event.

Example 7.1. Consider a Bernoulli random sequence $X[k]$ for $k \in \mathcal{Z}^+$ with outcomes $\{0, 1\}$ and $P(X[k] = 0) = P(X[k] = 1) = 0.5$. The probability that an infinite sequence contains all 1s is obviously zero though such a sequence *could* occur. The event that $X[k] = 0$ for *some* k as $k \to \infty$ is almost sure; it may not happen and thus is not sure.

Probability, Random Variables, and Random Processes: Theory and Signal Processing Applications, First Edition. John J. Shynk.

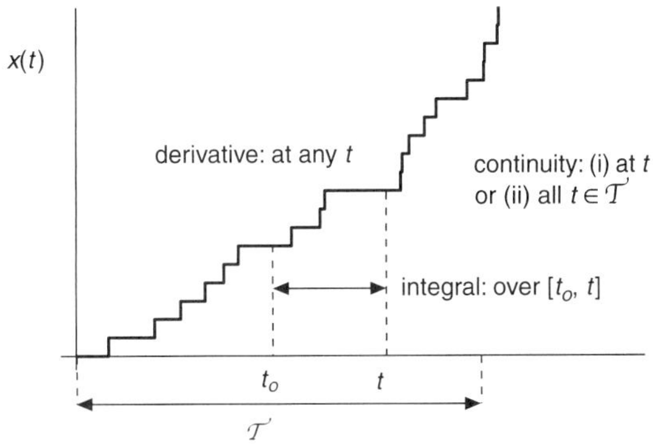

FIGURE 7.1 Realization of a Poisson random process. Properties of the process include: continuity at time instant t or for the entire duration $\mathcal{T}$, the derivative at t, and the integral over the interval $[t_o, t] \subset \mathcal{T}$.

From Example 7.1, we see that the notion of an almost sure event arises when an infinite number of observations are involved. We apply these definitions to the convergence of a random sequence $X[k]$ (which also generally apply to a sequence of random variables $\{X_1, X_2, \ldots\}$).

Definition: Almost Sure Convergence Random sequence $X[k]$ *converges almost surely* if

$$P\left(\lim_{k\to\infty} X[k] = X\right) = 1, \tag{7.1}$$

where X is a random variable or a constant. This type of convergence is also called *convergence with probability one*. The following notation is often used: $X[k] \xrightarrow{a.s.} X$. In terms of the underlying sample space $\mathcal{S}$ for a sequence of random variables (see Chapter 6), we can state more precisely that

$$P\left(\zeta \in \mathcal{S} : \lim_{k\to\infty} X[k, \zeta] = X(\zeta)\right) = 1, \tag{7.2}$$

where $X(\zeta)$ is a random variable or a constant.

If a realization does not converge, then either it diverges to $\pm\infty$ or it oscillates and never settles down.

Definition: Convergent Realization Realization $x[k, \zeta]$ *converges* to $x(\zeta)$ if for every $\epsilon > 0$, there exists a value of k after which

$$|x[k, \zeta] - x(\zeta)| < \epsilon, \tag{7.3}$$

If $x(\zeta)$ is the same value for all $\zeta \in \mathcal{S}$, then sequence $X[k]$ converges to a constant. This is illustrated in Figure 7.2(a) for three realizations, all of which converge to the same number (at the center of a 2ϵ interval on the vertical axis). Otherwise, it is a random variable whose distribution is determined by $\zeta \in \mathcal{S}$: a specific ζ gives a specific value $x(\zeta)$. This is illustrated in Figure 7.2(b) where two realizations converge to one value and a third realization converges to a different outcome of the random variable.

Definition: Sure Convergence Random sequence $X[k]$ *converges surely* if

$$\lim_{k\to\infty} X[k, \zeta] = X(\zeta) \text{ for all } \zeta \in \mathcal{S}, \tag{7.4}$$

where $X(\zeta)$ is a random variable or a constant.

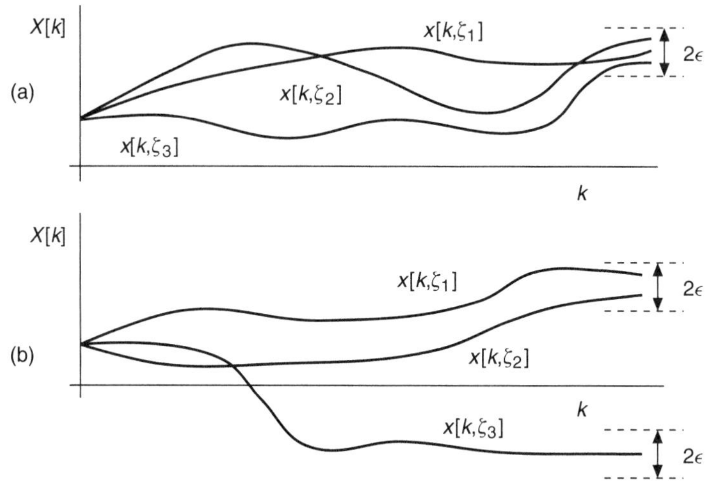

FIGURE 7.2 Illustration of converging realizations. (a) Convergence to a constant. (b) Convergence to a random variable. (Continuous representations of sequences are shown for ease of viewing.)

For sure convergence, probability is not part of the definition; all realizations of the random sequence converge, even those with zero probability. Sure convergence is obviously stronger than almost sure convergence, which "ignores" nonconvergent realizations that have *zero* probability. However, sure convergence is not usually considered because in practice we are not concerned with realizations that have zero probability.

Example 7.2. Let the realizations of a random sequence be described as follows:

$$X[k, \zeta] = \zeta^k u[k], \tag{7.5}$$

where $u[k]$ is the discrete unit-step function, and the outcomes of the sample space $\mathcal{S}$ are uniformly distributed on $(0, 1]$. (Recall from the last chapter that the elements of $\mathcal{S}$ are actually the individual realizations. However, for some random processes with structure such as the one in the example, it is usually possible to describe those realizations by the outcomes of a random variable or a random vector.) If $\zeta = 0.5$, then the corresponding realization is $X[k, 0.5] = (0.5)^k u[k]$ which converges to zero. Obviously, $X[k, \zeta]$ converges to zero for all $\zeta \in (0, 1)$; only $X[k, 1]$ "converges" to 1 (it is actually a constant for all k). The random variable $X(\zeta)$ at convergence is Bernoulli with outcomes $\{0, 1\}$, where $P(X(\zeta) = 1) = p = 0$ and $P(X(\zeta) = 0) = q \triangleq 1 - p = 1$. Thus, $X(\zeta)$ is an example of a random variable that is almost surely a constant (equal to zero). Since all converged values are finite, $X[k]$ converges surely. Suppose instead that

$$X[k, \zeta] = k\zeta^k u[k]. \tag{7.6}$$

This random sequence does not converge surely because it diverges when $\zeta = 1$. However, it converges almost surely because $P(\zeta = 1) = 0$, and $\zeta^k \to 0$ faster than $k \to \infty$ in the product $k\zeta^k$ for $\zeta \in (0, 1)$.

Example 7.3. Let the realizations of a random sequence be given by

$$X[k, \zeta] = \exp(\zeta k)u[k], \tag{7.7}$$

where the three outcomes $\mathcal{S} = \{-1, 0, 1\}$ are equally likely. It is clear that $X[k]$ does not converge almost surely because it diverges when $\zeta = 1$ which has *nonzero* probability. Likewise, if the outcomes of $\mathcal{S}$ are uniform on $[-1, 1]$, $X[k]$ does not converge almost surely. Realizations corresponding to $\zeta < 0$ converge to zero with $P(\lim_{k\to\infty} X[k] = 0) = 1/2$, the realization for $\zeta = 0$ "converges" to 1, and all the other realizations (with a combined probability of 1/2) diverge.

Almost sure convergence implies that each realization of the random sequence with nonzero probability converges to a constant. This constant may not be the same for all realizations, in which case the sequence still converges but to a random variable. Since almost sure convergence applies to individual realizations, it is a strong condition compared with other types of convergence. For example, convergence in probability is a weaker type of convergence.

Definition: Convergence in Probability Random sequence $X[k]$ *converges in probability* if for every $\epsilon > 0$

$$\lim_{k\to\infty} P(|X[k] - X| > \epsilon) = 0, \tag{7.8}$$

where X is a random variable or a constant. The following notation is often used: $X[k] \xrightarrow{p} X$.

Comparing with (7.1), we see that the limit here is taken *outside* the probability expression. The counterpart to (7.2) for convergence in probability is

$$\lim_{k\to\infty} P(\zeta \in \mathcal{S} : |X[k, \zeta] - X(\zeta)| > \epsilon) = 0. \tag{7.9}$$

The probability in (7.8) is actually a sequence itself that evolves with k. This definition of convergence is not concerned with individual realizations as are sure and almost sure convergence. Individual realizations are not restricted to stay within the range 2ϵ after some k. Some realizations can exit the range as long as others enter in order to satisfy the limit of the *sequence of probabilities* above. This is depicted in Figure 7.3.

Example 7.4. Consider the following random sequence:

$$Y[k] = \frac{1}{k}\sum_{m=1}^{k} X[m], \tag{7.10}$$

where $\{X[m]\}$ are iid Gaussian random variables with parameters $\{\mu = 0, \sigma\}$. It is clear that $Y[k]$ is also Gaussian with zero mean, but with decreasing variance σ^2/k. This variance is easily computed because of the independence assumption (see Chapter 5). In the following derivation, we show that $Y[k]$ converges in probability to zero. From the definition,

$$P(|Y[k] - 0| > \epsilon) = P(Y[k] > \epsilon) + P(Y[k] \leq -\epsilon) = 2P(Y[k] > \epsilon), \tag{7.11}$$

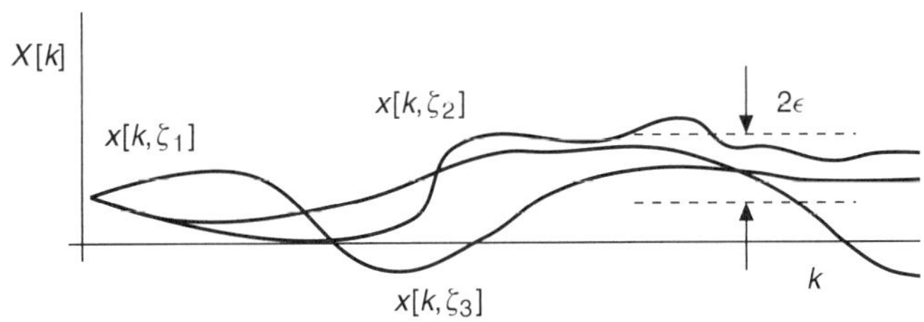

FIGURE 7.3 Illustration of convergence in probability. (Continuous realizations of the sequence are shown for ease of viewing.)

which simplifies because the probability density function (pdf) of $Y(k)$ is even. From Chapter 3, we know that

$$\begin{aligned} P(Y[k] > \epsilon) &= \frac{1}{\sqrt{2\pi(\sigma^2/k)}} \int_{\epsilon}^{\infty} \exp\left(-v^2/2(\sigma^2/k)\right)\mathrm{d}v, \\ &= \frac{1}{\sqrt{2\pi}} \int_{\epsilon/(\sigma/\sqrt{k})}^{\infty} \exp\left(-v^2/2\right)\mathrm{d}v = Q(\sqrt{k}\epsilon/\sigma), \end{aligned} \tag{7.12}$$

which is the Q-function described in Appendix B. This probability is a deterministic sequence, as mentioned above, and is not concerned with individual realizations. For convergence in probability, we require only that (7.12) converge to zero for every $\epsilon > 0$. From the properties of the Q-function, this is the case:

$$\lim_{k\to\infty} Q(\sqrt{k}\epsilon/\sigma) = 0 \tag{7.13}$$

so that $Y[k]$ converges in probability to a constant.

Another type of convergence that is not concerned with individual realizations is convergence in mean.

Definition: Convergence in Mean Random sequence $X[k]$ *converges in mean* if

$$\lim_{k\to\infty} \mathcal{E}[|X[k] - X|^m] = 0, \tag{7.14}$$

where X is a random variable or a constant and $m \in \mathcal{N}$. The following notation is often used: $X[k] \xrightarrow{L^m} X$, where L^m is the L^m-norm (Euclidean norm when $m = 2$).

It can be shown that L^m convergence implies L^n convergence when $m > n$ (see Problem 7.4). The most widely used form of L^m convergence is *mean-square (MS) convergence* with $m = 2$; the following notation is also used for this case: $X[k] \xrightarrow{m.s.} X$. The expectation above is a deterministic sequence, as was the probability in (7.8). It is not concerned with individual realizations, only that the ensemble $|X[k] - X|^m$ approach zero. It turns out that this type of convergence is stronger than convergence in probability as shown later.

Example 7.5. For the random sequence in Example 7.4, we need to show for MS convergence that

$$\lim_{k\to\infty} \mathcal{E}[Y^2[k]] = 0. \tag{7.15}$$

This readily follows because $Y[k]$ has zero mean and we already know that the variance of $Y[k]$ is σ^2/k.

Consider random sequence $X[k]$ where for a specific time instant k, the cumulative distribution function (cdf) is $F_{X[k]}(x) = P(X[k] \le x)$. We define one more type of convergence, which turns out to be weaker than all the previous definitions of convergence.

Definition: Convergence in Distribution Random sequence $X[k]$ *converges in distribution* if

$$\lim_{k\to\infty} F_{X[k]}(x) = F_X(x) \tag{7.16}$$

for all x assuming the cdf $F_X(x)$ exists. The following notation is often used: $X[k] \xrightarrow{d} X$.

TABLE 7.1 Summary of Convergence for Random Sequence $X[k]$

Type of Convergence	Definition
Sure: $X[k] \xrightarrow{s} X$	$\lim_{k\to\infty} X[k,\zeta] = X(\zeta)$ for all $\zeta \in S$
Almost sure: $X[k] \xrightarrow{a.s.} X$	$P(\lim_{k\to\infty} X[k] = X) = 1$
In probability: $X[k] \xrightarrow{p} X$	$\lim_{k\to\infty} P(\lvert X[k] - X\rvert > \epsilon) = 0$ for every $\epsilon > 0$
In mean: $X[k] \xrightarrow{L^m} X$	$\lim_{k\to\infty} \mathcal{E}[\lvert X[k] - X\rvert^m] = 0$ for $m \in \mathcal{N}$
In distribution: $X[k] \xrightarrow{d} X$	$\lim_{k\to\infty} F_{X[k]}(x) = F_X(x)$

Example 7.6. Consider the random process in Example 6.2 whose distribution we rewrite here assuming that T is a discrete random variable such that $t \to k \in \mathcal{Z}^+$:

$$F_{Y[k]}(y) = u(y)F_T[k] + F_X(y)(1 - F_T[k]). \tag{7.17}$$

It is clear that $Y[k]$ converges in distribution:

$$\lim_{k\to\infty} F_{Y[k]}(y) = u(y) \tag{7.18}$$

because $F_T[k] \to 1$. The corresponding pdf is $\delta(y)$, meaning that $Y[k]$ actually converges in distribution to zero.

Table 7.1 summarizes the five types of convergence and Figure 7.4 illustrates how they are related.

Theorem 7.1. Convergence in mean implies convergence in probability.

Proof. From Bienaymé's inequality in Appendix F with $X[k] - X$ substituted for X and $\alpha = 0$, we can write

$$P(|X[k] - X| \geq \epsilon) \leq (1/\epsilon^2)\mathcal{E}[|X[k] - X|^m] \tag{7.19}$$

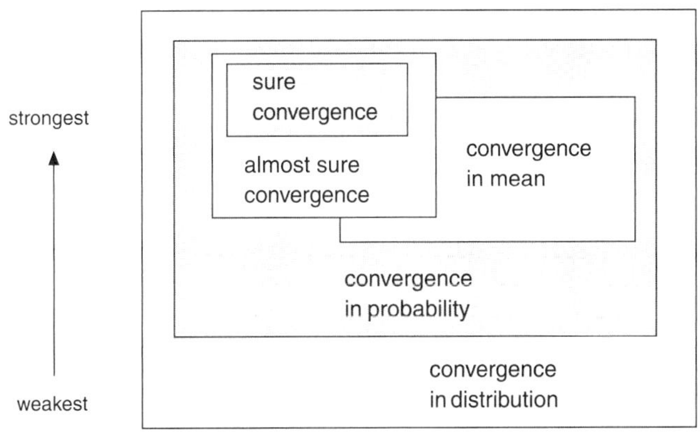

FIGURE 7.4 Relationships among the five types of convergence for a random sequence: s $\subseteq$ a.s. $\subseteq$ p $\subseteq$ d and $L^m \subseteq$ p $\subseteq$ d. Generally, we cannot conclude how sure convergence and almost sure convergence are related to convergence in mean.

for every $\epsilon > 0$. Letting $k \to \infty$ gives

$$\lim_{k\to\infty} P(|X[k] - X| \geq \epsilon) \leq (1/\epsilon^2) \lim_{k\to\infty} \mathcal{E}[|X[k] - X|^m]. \tag{7.20}$$

Assuming convergence in the mean such that the right-hand side is zero completes the proof. □

Theorem 7.2. MS convergence implies that the mean and variance also converge:

$$X[k] \xrightarrow{m.s.} X \Rightarrow \lim_{k\to\infty} \mathcal{E}[X[k]] = \mathcal{E}[X] \text{ and } \lim_{k\to\infty} \text{var}[X[k]] = \text{var}[X]. \tag{7.21}$$

Proof. Mean: Since expectation is linear, $|\mathcal{E}[X[k]] - \mathcal{E}[X]| = |\mathcal{E}[X[k] - X]|$, and from the Cauchy–Schwarz inequality (see Appendix F with $g(X) = X[k] - X$ and $h(X) = 1$):

$$|\mathcal{E}[X[k] - X]| \leq \sqrt{\mathcal{E}[(X[k] - X)^2]}. \tag{7.22}$$

If the argument of the last expression tends to zero (i.e., MS convergence), then $\mathcal{E}[X[k]]$ converges to $\mathcal{E}[X]$. *Variance*: From the triangle inequality (see Appendix F):

$$\sqrt{\mathcal{E}[X^2[k]]} = \sqrt{\mathcal{E}[(X[k] - X + X)^2]} \leq \sqrt{\mathcal{E}[(X[k] - X)^2]} + \sqrt{\mathcal{E}[X^2]}. \tag{7.23}$$

Moving the last term to the left-hand side gives

$$\sqrt{\mathcal{E}[X^2[k]]} - \sqrt{\mathcal{E}[X^2]} \leq \sqrt{\mathcal{E}[(X[k] - X)^2]}. \tag{7.24}$$

If the right-hand side tends to zero, then the second-order moment $\mathcal{E}[X^2[k]]$ converges to $\mathcal{E}[X^2]$, and $\mathcal{E}[X^2[k]] - \mathcal{E}^2[X[k]] = \text{var}[X[k]]$ converges to $\text{var}[X]$. □

Since random variable X at convergence may not be known, the following theorem can be used to verify MS convergence.

Theorem 7.3 (Cauchy criterion). Random sequence $X[k]$ converges in MS if and only if

$$\lim_{k_1,k_2\to\infty} \mathcal{E}[(X[k_1] - X[k_2])^2] = 0. \tag{7.25}$$

Proof. We prove the only if part of the theorem. Rewriting (7.25) yields

$$\begin{aligned}\mathcal{E}[(X[k_1] - X + X - X[k_2])^2] &= \mathcal{E}[(X[k_1] - X)^2] + 2\mathcal{E}[(X[k_1] - X)(X - X[k_2])] \\ &\quad + \mathcal{E}[(X - X[k_2])^2].\end{aligned} \tag{7.26}$$

By the Cauchy–Schwarz inequality, the middle term on the right-hand side is bounded as

$$\begin{aligned}\mathcal{E}[(X[k_1] - X)(X - X[k_2])] &\leq |\mathcal{E}[(X[k_1] - X)]\mathcal{E}[(X - X[k_2])]|, \\ &\leq \sqrt{\mathcal{E}[(X[k_1] - X)^2]\mathcal{E}[(X - X[k_2])^2]}.\end{aligned} \tag{7.27}$$

Thus, if $\mathcal{E}[(X[k_1] - X)^2] \to 0$ and $\mathcal{E}[(X - X[k_2])^2] \to 0$, the limit in (7.25) holds. □

The Cauchy criterion in the above MS sense is an extension of the deterministic case in real analysis where a sequence of numbers $\{x_k\}$ converges if and only if

$$\lim_{k_1,k_2\to\infty} |x_{k_1} - x_{k_2}| = 0 \tag{7.28}$$

for subsequences $\{x_{k_1}\}$ and $\{x_{k_2}\}$. Cauchy's criterion can also be used for the other forms of convergence which we summarize here:

$$\textit{In probability}: \quad \lim_{k_1,k_2\to\infty} P(|X[k_1,\zeta] - X[k_2,\zeta]| > \epsilon) = 0 \text{ for all } \epsilon > 0, \tag{7.29}$$

$$\textit{Sure}: \quad \lim_{k_1,k_2\to\infty} |X[k_1,\zeta] - X[k_2,\zeta]| = 0 \text{ for all } \zeta \in \mathcal{S}, \tag{7.30}$$

$$\textit{Almost sure}: \quad \lim_{k_1,k_2\to\infty} |X[k_1,\zeta] - X[k_2,\zeta]| = 0 \text{ for all } \zeta \in E \subset \mathcal{S}, \tag{7.31}$$

where $P(E) = 1$. The last expression can also be written as

$$P\left(\zeta \in \mathcal{S} : \lim_{k_1\to\infty} X[k_1,\zeta] = \lim_{k_2\to\infty} X[k_2,\zeta]\right) = 1 \tag{7.32}$$

The following theorem is useful for martingale sequences, which we state without proof.

Theorem 7.4 (Martingale convergence). If $X[k]$ is a martingale sequence with $\mathcal{E}[X^2[k]] < \infty$, then it converges almost surely to random variable X or a constant as $k \to \infty$.

Next, we present two additional properties for MS convergence that emphasize the techniques used in Theorems 7.2 and 7.3. The goal in the proofs is to rewrite the expectation of a sequence of interest to be bounded by an expectation of other sequences which we know converge in the MS. A bound can usually be obtained from the Cauchy–Schwarz or triangle inequalities, as used in the theorems above and demonstrated next (Stark and Woods, 2002).

- *Product with a random variable.* Let $X[k]$ converge in MS to X and let Y be a random variable. Assume they have finite second moments: $\mathcal{E}[X^2[k]] < \infty$ and $\mathcal{E}[Y^2] < \infty$. Then

$$\lim_{k\to\infty} \mathcal{E}[X[k]Y] = \mathcal{E}[XY]. \tag{7.33}$$

  Subtracting XY and using the Cauchy–Schwarz inequality gives

$$|\mathcal{E}[X[k]Y - XY]| \leq |\mathcal{E}[X[k] - X]||\mathcal{E}[Y]| \leq \sqrt{\mathcal{E}[(X[k] - X)^2]\mathcal{E}[Y^2]}. \tag{7.34}$$

  Since $X[k] \xrightarrow{m.s.} X$ and $\mathcal{E}[Y^2]$ is finite, the right-hand side tends to zero and $X[k]Y \xrightarrow{m.s.} XY$.
- *Product of random sequences.* Let $X[k]$ be as above and assume $Y[k]$ converges in the MS to Y. Then, generally we cannot say that $X[k]Y[k]$ converges to XY in the MS. From the triangle inequality:

$$\begin{aligned}\sqrt{\mathcal{E}[(X[k]Y[k] - XY)^2]} &= \sqrt{\mathcal{E}[(X[k](Y[k] - Y) + (X[k] - X)Y)^2]} \\ &\leq \sqrt{\mathcal{E}[X^2[k](Y[k] - Y)^2]} + \sqrt{\mathcal{E}[(X[k] - X)^2 Y^2]},\end{aligned} \tag{7.35}$$

where $X[k]Y$ has been added and subtracted in the first line. Because of the products under the expectations, MS convergence of each sequence individually cannot be used. Note that if $X[k]$ and $Y[k]$ are uncorrelated, then those expectations split:

$$\sqrt{\mathcal{E}[(X[k]Y[k]-XY)^2]} \leq \sqrt{\mathcal{E}[X^2[k]]}\sqrt{\mathcal{E}[(Y[k]-Y)^2]} + \sqrt{\mathcal{E}[(X[k]-X)^2]}\sqrt{\mathcal{E}[Y^2]}, \tag{7.36}$$

which is a trivial result. Returning to the general case, we have from the Cauchy–Schwarz inequality:

$$\mathcal{E}[X^2[k](Y[k]-Y)^2] \leq \sqrt{\mathcal{E}[X^4[k]]}\sqrt{\mathcal{E}[(Y[k]-Y)^4]}, \tag{7.37}$$

$$\mathcal{E}[(X[k]-X)^2Y^2] \leq \sqrt{\mathcal{E}[(X[k]-X)^4]}\sqrt{\mathcal{E}[Y^4]}. \tag{7.38}$$

In order to use the MS assumptions of $X[k]$ and $Y[k]$, we need second-order moments on the right-hand side (not fourth order). However, if $X[k]$ and $Y[k]$ are Gaussian, then we know from Chapter 5 that a fourth-order moment is obtained from a second-order moment:

$$\mathcal{E}[X^4[k]] = 3\mathcal{E}^2[X^2[k]], \quad \mathcal{E}[(X[k]-X)^4] = 3\mathcal{E}^2[(X[k]-X)^2], \tag{7.39}$$

and likewise for $Y[k]$ and Y. Since $X[k] \xrightarrow{m.s.} X$, $Y[k] \xrightarrow{m.s.} Y$, and their second moments are finite, the right-hand side of (7.35) tends to zero in the limit for Gaussian sequences. If $X[k] \xrightarrow{L^4} X$ and $Y[k] \xrightarrow{L^4} Y$ (which is a stronger condition than MS convergence), then $X[k]Y[k] \xrightarrow{L^2} XY$ without the Gaussian assumption.

The following additional properties are examined in Problem 7.6.

- *Sum of random sequences.* If $X[k] \xrightarrow{m.s.} X$ and $Y[k] \xrightarrow{m.s.} Y$, then $X[k]+Y[k] \xrightarrow{m.s.} X+Y$.
- *Convergence of cross-correlation.* For $X[k]$ and $Y[k]$ MS convergent as in the preceding line (Hajek, 2009):

$$\lim_{k\to\infty} \mathcal{E}[X[k]Y[k]] = \mathcal{E}[XY]. \tag{7.40}$$

This result is similar to convergence of the mean and variance in Theorem 7.2. Note that (7.40) is not the same as MS convergence of $X[k]Y[k]$ examined in (7.35).

7.3 LAWS OF LARGE NUMBERS

The laws of large numbers describe how a specific function of a random sequence behaves when the time argument approaches infinity. They are special cases of the previous convergence definitions regarding convergence of the *sample mean* to a random variable or a constant. Consider the iid random sequence $X[k]$ with mean μ_X and finite variance $\sigma_X^2 < \infty$. Define the following sample mean which itself is a random sequence:

$$\bar{X}[k] \triangleq \frac{1}{k}\sum_{m=1}^{k} X[m], \tag{7.41}$$

and is an estimator of the ensemble mean μ_X (estimators and their properties are discussed in Chapter 9).

Theorem 7.5 (Weak law of large numbers). The sample mean estimator $\bar{X}[k]$ converges in probability to μ_X:

$$\lim_{k\to\infty} P(|\bar{X}[k] - \mu_X| \geq \epsilon) = 0 \tag{7.42}$$

for every $\epsilon > 0$.

Proof. Since $X[k]$ is assumed to be an iid sequence, the expectation of the sample mean is

$$\mathcal{E}[\bar{X}[k]] = \frac{1}{k}\sum_{m=1}^{k}\mathcal{E}[X[m]] = \frac{1}{k}k\mu_X = \mu_X \tag{7.43}$$

so that it is an *unbiased* estimator, and its variance is

$$\mathcal{E}[(\bar{X}[k] - \mu_X)^2] = \mathcal{E}\left[\left(\frac{1}{k}\sum_{m=1}^{k}(X[m] - \mu_X)\right)^2\right] = \frac{1}{k^2}\sum_{m=1}^{k}\sigma_X^2 = \sigma_X^2/k. \tag{7.44}$$

Chebyshev's inequality in Appendix F yields

$$P(|\bar{X}[k] - \mu_X| \geq \epsilon) \leq \frac{\sigma_X^2}{k\epsilon^2}, \tag{7.45}$$

where the variance above has been substituted. Thus, for every $\epsilon > 0$:

$$\lim_{k\to\infty} P(|\bar{X}[k] - \mu_X| \geq \epsilon) \leq \lim_{k\to\infty}\frac{\sigma_X^2}{k\epsilon^2} = 0, \tag{7.46}$$

which completes the proof. □

As discussed previously, this probability is a deterministic sequence. It is a "weak law" because it is based on convergence in probability. The strong law of large numbers refers to *almost sure* convergence of the sample mean, which is a stronger condition because it is concerned with individual realizations. Kronecker's lemma is a useful result for proving the strong law of large numbers which we provide without proof.

Lemma 7.1 (Kronecker). Assume that nonrandom sequence $x[k]$ for $k \in \mathcal{N}$ converges to c as follows:

$$\sum_{k=1}^{\infty} x[k]/a_k = c < \infty, \tag{7.47}$$

where $0 = a_0 < a_1 \leq a_2 \leq \cdots$ is an increasing set of coefficients. Then

$$\lim_{k\to\infty}(1/a_k)\sum_{m=1}^{k} x[m] = 0. \tag{7.48}$$

Theorem 7.6 (Strong law of large numbers). The sample mean estimator $\bar{X}[k]$ converges almost surely to μ_X:

$$P\left(\lim_{k\to\infty}\bar{X}[k] = \mu_X\right) = 1. \tag{7.49}$$

Proof. (Larson and Shubert, 1979). Define the zero-mean sequence

$$Y[k] \triangleq \sum_{m=1}^{k}(X[m]-\mu_X)/m, \tag{7.50}$$

which has variance

$$\text{var}[Y[k]] = \sum_{m=1}^{k}(1/m^2)\mathcal{E}[(X[m]-\mu_X)^2] = \sigma_X^2\sum_{m=1}^{k}(1/m^2), \tag{7.51}$$

where we have used the fact that $X[k]$ is an iid sequence. This variance is finite because the sum converges (see Appendix E). Observe that $Y[k]$ is a martingale

$$\begin{aligned}\mathcal{E}[Y[k]|Y[1],\ldots,Y[k-1]] &= \mathcal{E}\left[\sum_{m=1}^{k}(1/m)(X[m]-\mu_X)\Big|Y[1],\ldots,Y[k-1]\right],\\ &= \mathcal{E}\left[(1/k)(X[k]-\mu_X)+Y[k-1]\,|Y[1],\ldots,Y[k-1]\right],\\ &= (1/k)\mathcal{E}[X[k]-\mu_X]+Y[k-1] = Y[k-1].\end{aligned} \tag{7.52}$$

By the previous martingale convergence theorem (see Theorem 7.4), $Y[k]$ converges almost surely to random variable Y (or possibly a constant) as $k\to\infty$. Since a realization of $Y[k]$ has the following form:

$$y[k] = \sum_{m=1}^{k}(x[m]-\mu_X)/m, \tag{7.53}$$

we can apply Kronecker's lemma, yielding

$$\lim_{k\to\infty}\frac{1}{k}\sum_{m=1}^{k}(x[m]-\mu_X) = 0 \tag{7.54}$$

with $a_k = k$ in the lemma. Thus, almost all realizations of the form $\bar{x}[k] = (1/k)\sum_{m=1}^{k}x[m]$ converge to μ_X, which completes the proof. □

Example realizations of the sample mean for iid standard Gaussian sequence $X[k]$ are shown in Figure 7.5, demonstrating that they approach zero.

7.4 CENTRAL LIMIT THEOREM

The central limit theorem (CLT) describes an important property of a sum of independent random variables.

Theorem 7.7 (Central limit). Let $X[k]$ be an iid random sequence with mean μ_X and variance σ_X^2, and define the following function of the sample mean:

$$Y[k] \triangleq \frac{\bar{X}[k]-\mu_X}{\sigma_X/\sqrt{k}}. \tag{7.55}$$

Then $\lim_{k\to\infty} Y[k]$ has the standard Gaussian distribution.

This theorem refers to a specific convergence in distribution.

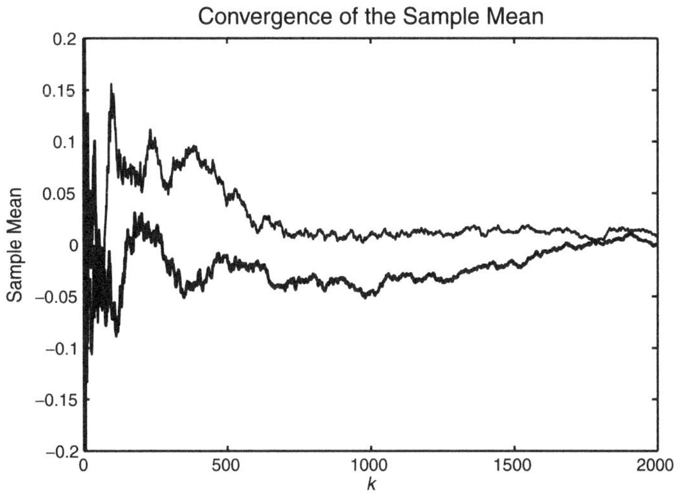

FIGURE 7.5 Example realizations of the sample mean $\bar{X}[k]$ for an iid standard Gaussian random sequence.

Proof. The characteristic function of the sample mean is

$$\Phi_{\bar{X}[k]}(\omega) = \mathcal{E}[\exp(j\omega\bar{X}[k])] = \Phi_{\sum_{n=1}^{k} X[n]}(\omega/k),$$

$$= \prod_{n=1}^{k} \Phi_X(\omega/k) = \Phi_X^k(\omega/k), \tag{7.56}$$

where $1/k$ scaling $\sum_{n=1}^{k} X[n]$ has been grouped with ω, and the iid assumption gives the last result. It is straightforward to show that for the transformation in (7.55) (see Problem 7.13):

$$\Phi_{Y[k]}(\omega) = \left[\exp(-j\omega\mu_X/\sqrt{k}\sigma_X)\Phi_X(\omega/\sqrt{k}\sigma_X)\right]^k. \tag{7.57}$$

For notational convenience, define $\omega_k \triangleq \omega/\sqrt{k}\sigma_X$ and write the following series expansion for an exponential:

$$\Phi_X(\omega_k) = \mathcal{E}[\exp(j\omega_k X)] = \mathcal{E}\left[\sum_{n=0}^{\infty} \frac{(j\omega_k X)^n}{n!}\right] = \sum_{n=0}^{\infty} \frac{(j\omega_k)^n \mathcal{E}[X^n]}{n!},$$

$$= 1 + j\omega_k\mathcal{E}[X] - \omega_k^2\mathcal{E}[X^2]/2 + \sum_{n=3}^{\infty} \frac{(j\omega_k)^n\mathcal{E}[X^n]}{n!}, \tag{7.58}$$

where the first three terms in the sum have been shown separately. Likewise for the exponential term in (7.57):

$$\exp(-j\omega_k\mu_X) = \sum_{n=0}^{\infty} \frac{(-j\omega_k\mu_X)^n}{n!},$$

$$= 1 - j\omega_k\mu_X - \omega_k^2\mu_X^2/2 + \sum_{n=3}^{\infty} \frac{(-j\omega_k\mu_X)^n}{n!}. \tag{7.59}$$

Substituting $\mathcal{E}[X^2] = \sigma_X^2 + \mu_X^2$ into (7.58) and multiplying the two series above, we find that most product terms for $n \in \{0, 1, 2\}$ cancel such that

$$\Phi_{Y[k]}(\omega) = [1 - \omega_k^2\sigma_X^2/2 + \cdots]^k, \tag{7.60}$$

where $\cdots$ denotes all high-order terms with exponent of ω_k^n exceeding $n = 2$. In the limit as $k \to \infty$, these terms are ignored because of $\sqrt{k}$ in the denominator of ω_k, yielding

$$\lim_{k\to\infty} [1 - \omega^2/2k + \cdots]^k = \exp(-\omega^2/2), \tag{7.61}$$

where the exponential limit result in (4.9) has been used. This last expression is the characteristic function of the standard Gaussian random variable, thus completing the proof. □

Since $\bar{X}$ has mean μ_X and variance σ_X^2/k, the expression above is just a transformation of $\bar{X}[k]$ to a zero mean random variable with unit variance. The mean is obviously zero and the variance is

$$\frac{\mathcal{E}[(\bar{X}[k] - \mu_X)^2]}{\sigma_X^2/k} = \frac{\sigma_X^2/k}{\sigma_X^2/k} = 1. \tag{7.62}$$

The important feature, of course, is that the limiting distribution is Gaussian, regardless of the original distribution provided the mean exists and it has finite variance. Even without the proof above, we know from Chapter 4 that the pdf of a sum of independent random variables is a convolution of the individual pdfs. There we showed some examples for continuous random variables, demonstrating that with repeated convolutions the pdf of the sum becomes progressively smoother and begins to have a Gaussian appearance. This is the case even for discrete random variables as demonstrated in Example 7.7 for the discrete uniform random variable. The Gaussian approximations of discrete random variables described in Chapter 4 are special cases of the CLT.

Recall from Chapter 6 that random variable X is *infinitely divisible* if it can be written as the sum of k iid random variables for every $k \in \mathcal{N}$, and all of which have the *same* distribution as X. Examples include the Poisson, gamma, and Student's t distributions. The sample mean used in the CLT differs from the sum for an infinitely divisible random variable because it is *shifted* by μ_X and *scaled* by $\sqrt{k}/\sigma_X$. This normalization causes the terms to have zero mean and unit variance, and is the reason why $Y[k]$ converges in distribution to a standard Gaussian random variable. An infinitely divisible random variable does not assume this normalization so that it is possible for the sum and its terms to have the same distribution.

Example 7.7. Suppose that $X[k]$ has a discrete uniform distribution with outcomes $\{1, 2\}$ which has mean $\mu_X = (1+2)/2 = 1.5$ and variance $\sigma_X^2 = (2^2 - 1)/12 = 1/4$. Then (7.55) is

$$Y[k] = \frac{(1/k)\sum_{m=1}^k X[m] - 1.5}{\sqrt{1/4k}} = \frac{1}{\sqrt{k/4}}\left[\sum_{m=1}^k X[m] - 1.5k\right], \tag{7.63}$$

which has zero mean and unit variance for any $k \geq 1$. Even though $X[k]$ originally has only positive integer outcomes, the expression above allows for negative and noninteger outcomes. For the sum alone, the probability mass function (pmf) is obtained from successive discrete convolutions and is nonzero only for integer values. The shifting and scaling above yields real numbers as is evident in Table 7.2 for the first four

TABLE 7.2 Illustration of the CLT for Example 7.7

k	Sum Outcomes	Shifted/Scaled Outcomes	Probability Mass
2	$2, 3, 4$	$-1.414, 0, 1.414$	$1/4, 2/4, 1/4$
3	$3, 4, 5, 6$	$-1.732, -0.577, 0.577, 1.732$	$1/8, 3/8, 3/8, 1/8$
4	$4, 5, 6, 7, 8$	$-2, -1, 0, 1, 2$	$1/16, 4/16, 6/16, 4/16, 1/16$
5	$5, 6, 7, 8, 9, 10$	$-2.236, -1.342, -0.447, 0.447, 1.342, 2.236$	$1/32, 5/32, 10/32, 10/32, 5/32, 1/32$

values of k. Obviously, k must be large before the tails of the pdf for the sum start to approximate those of a Gaussian pdf.

Example 7.8. Let $X[k]$ be an iid exponential sequence with parameter λ which has mean $\mu_X = 1/\lambda$ and variance $\sigma_X^2 = 1/\lambda^2$. For this example,

$$Y[k] = \frac{(1/k)\sum_{m=1}^{k} X[m] - 1/\lambda}{(1/\lambda)/\sqrt{k}} = (\lambda/\sqrt{k})\sum_{m=1}^{k} X[m] - \sqrt{k}. \tag{7.64}$$

Since the random variables in the sum are iid,

$$\Phi_{\sum_{m=1}^{k} X[m]}(j\omega) = \prod_{m=1}^{k} \Phi_{X[m]}(\omega) = [\lambda/(\lambda - j\omega)]^k, \tag{7.65}$$

which is the characteristic function of an Erlang (gamma) distribution with parameters $\{\lambda, r = k\}$. Since the sum is scaled by $\lambda/\sqrt{k}$ (i.e., divided by its standard deviation), the parameters change to $\{\sqrt{k}, r = k\}$. The mean and variance of this scaled distribution are $k/\sqrt{k} = \sqrt{k}$ and $k/(\sqrt{k})^2 = 1$, respectively, so that $Y[k]$ in (7.64) has zero mean and unit variance. Examples of the pdf for $Y[k]$ are shown in Figure 7.6 for different values of k, which have been shifted to the right by $\sqrt{k}$ in order to better view their shapes. The pdf of a Gaussian random variable with mean $\mu = \sqrt{20} \approx 4.4721$ and unit variance is also shown for comparison.

Example 7.9. In order to use the CLT, we should be able to recognize from the problem statement that the random variable of interest can be modeled as the sum of iid random variables. For example, consider scores for a comprehensive exam in a school district where it is known that the mean is $\mu_X = 70$ and the standard deviation is $\sigma_X = 10$. Then, for a class size of 10, various probabilities involving the sample mean can be approximated using the expression in (7.55). For the bounds in $P(70 \leq \bar{X}[10] \leq 75)$, we calculate

$$y_1 = \frac{70 - 70}{10/\sqrt{10}} = 0, \quad y_2 = \frac{75 - 70}{10/\sqrt{10}} \approx 1.5811, \tag{7.66}$$

from which

$$\begin{aligned} P(y_1 \leq Y[10] \leq y_2) &= F_Y(1.5811) - F_Y(0), \\ &= Q(0) - Q(1.5811) \approx 0.4431, \end{aligned} \tag{7.67}$$

where $F_Y(x)$ is the standard Gaussian cdf and $Q(\cdot)$ is the Q-function (note that $Q(x) = 1 - F_Y(x)$).

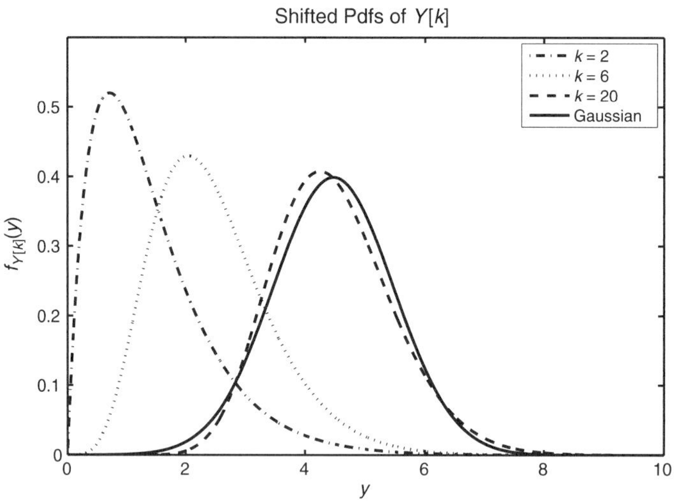

FIGURE 7.6 Pdfs of $Y[k]$ in Example 7.8 (shifted to the right by $\sqrt{k}$ for ease of viewing).

7.5 STOCHASTIC CONTINUITY

In the next few sections, we examine properties of continuous-time random processes that are covered in courses on calculus regarding functions: continuity, derivatives, and integrals. Since the process is random, we might consider examining the continuity of each realization and possibly compute derivatives or integrals depending on the application. Of course, it is not feasible to perform this for all individual realizations, and even if this were possible, it is not clear that such a characterization would provide us with useful information about the collective behavior of the *ensemble*. As was done in previous chapters on random variables and random processes, it is preferable that we develop probabilistic formulations for these calculus considerations. DEs representing linear systems with random inputs are also discussed, which will be seen again in Chapter 8. We begin with the definition of continuity for a nonrandom function.

Definition: Continuous Function Function $x(t)$ is *continuous* at t_o if for $\delta > 0$ there exists $\epsilon > 0$ such that

$$|t - t_o| < \epsilon \Rightarrow |x(t) - x(t_o)| < \delta. \tag{7.68}$$

In general, δ depends on ϵ and t_o which we can write as the function $\delta(\epsilon, t_o)$. This definition is consistent with our intuition about continuity: small variations in the argument of a function do not cause abrupt changes in the value of the function. This is depicted in Figure 7.7 for an arbitrary continuous function. Other types of continuity for a nonrandom function are discussed in Appendix B where examples are provided.

As was done for convergence of a random sequence, we consider various stochastic definitions of continuity for a random process. These are (i) sure continuous, (ii) sample continuous, (iii) continuous with probability one, (iv) continuous in probability, (v) continuous in distribution, and (vi) MS continuous. The definition of MS continuous turns out to be the most useful one because it utilizes the autocorrelation function $R_{XX}(t_1, t_2)$. When derivatives and integrals of a random process are considered, only MS definitions are given. The reason for our lengthy discussion of continuity for deterministic functions in Appendix B is that MS continuity of a random process reduces to continuity of the autocorrelation function.

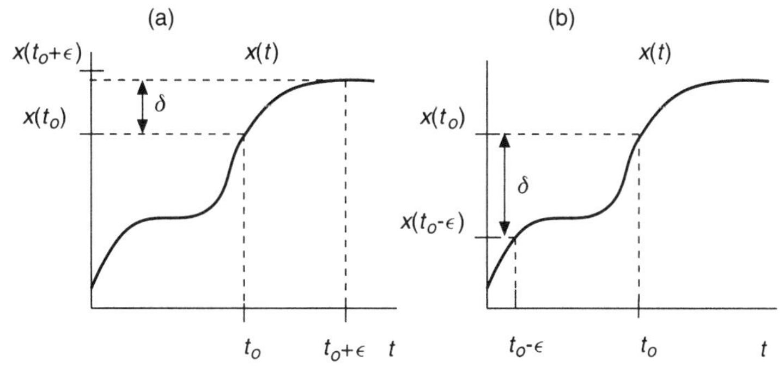

FIGURE 7.7 Example of a continuous function at t_o: $\delta \to 0$ as $\epsilon \to 0$. $x(t)$ is (a) continuous from the right and (b) continuous from the left.

Definition: Sure Continuous Random process $X(t)$ is *sure continuous* on $\mathcal{T}$ if

$$\lim_{\epsilon \to 0} |X(t+\epsilon, \zeta) - X(t, \zeta)| = 0 \text{ for all } t \in \mathcal{T} \text{and every } \zeta \in S. \tag{7.69}$$

Since ζ represents a particular realization of $X(t)$, sure continuous is basically the definition of continuity for a deterministic function applied to each realization in $\mathcal{S}$. Sure continuous means that *every realization is continuous* on $\mathcal{T}$. This is the strongest definition of continuity for a random process, although in practice it is not very useful. For example, if one realization happens to be discontinuous at some time t_o, the random process is not sure continuous even though this single realization is not representative of the ensemble of the process, and it may even have probability zero. Using a probabilistic formulation leads to the next definition of continuity.

Definition: Sample Continuous Random process $X(t)$ is *sample continuous* on $\mathcal{T}$ if

$$P\left(\lim_{\epsilon \to 0} |X(t+\epsilon) - X(t)| \neq 0 \text{ for all } t \in \mathcal{T}\right) = 0, \tag{7.70}$$

which can be rewritten as

$$P(X(t) \text{ is continuous at } t \text{ for all } t \in \mathcal{T}) = 1. \tag{7.71}$$

An alternative expression similar to (7.69) is

$$P\left(\zeta \in \mathcal{S} : \lim_{\epsilon \to 0} |X(t+\epsilon, \zeta) - X(t, \zeta)| \neq 0 \text{ for all } t \in \mathcal{T}\right) = 1. \tag{7.72}$$

All discontinuous realizations in the interval $\mathcal{T}$ which collectively have probability zero are ignored in this definition; the random process is still considered to be continuous. A random process that is sample continuous is also called *almost sure continuous*. However, it is important to note that this definition refers to the realizations for the *entire interval* $\mathcal{T}$. The next definition of continuity refers to the ensemble of realizations at a specific $t \in \mathcal{T}$.

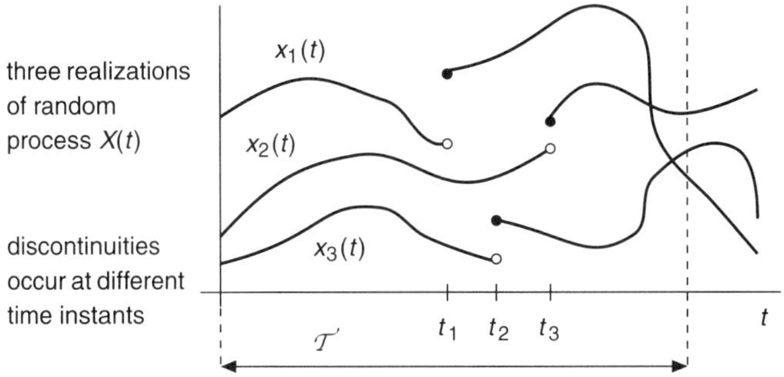

FIGURE 7.8 Illustrative example of realizations with discontinuities at different times $t \in \mathcal{T}$: they are not continuous from the left. The process is continuous with probability one, but is not sample continuous, nor sure continuous.

Definition: Continuous with Probability One Random process $X(t)$ is *continuous with probability one* at $t \in \mathcal{T}$ if

$$P\left(\lim_{\epsilon\to 0} |X(t+\epsilon) - X(t)| = 0\right) = 1, \tag{7.73}$$

which can be rewritten as

$$P(X(t) \text{ is continuous at } t) = 1. \tag{7.74}$$

Another expression similar to (7.69) is

$$P\left(\zeta \in \mathcal{S} : \lim_{\epsilon\to 0} |X(t+\epsilon,\zeta) - X(t,\zeta)| = 0\right) = 1. \tag{7.75}$$

This definition of continuity means that $\lim_{\epsilon\to 0} |X(t+\epsilon) - X(t)| = 0$ occurs *almost sure* at the *specific* $t \in \mathcal{T}$. Figure 7.8 shows example realizations of a process that is continuous with probability one, where the realizations have discontinuities at *different* times of $\mathcal{T}$. If for any $t \in \mathcal{T}$, a countable number of realizations with overall probability zero have a discontinuity, $X(t)$ is continuous with probability one. On the other hand, $X(t)$ is *not* sample continuous: every realization has a discontinuity at some $t \in \mathcal{T}$. The following example further illustrates this difference.

Example 7.10. Consider the digital communication signal (random process) known as binary pulse amplitude modulation (PAM) which can be expressed as

$$X(t) = \sum_{k=0}^{\infty} B[k][u(t-kT_o) - u(t-(k+1)T_o - T_d)], \tag{7.76}$$

where $B[k]$ is an iid symmetric Bernoulli sequence with outcomes $\{-1, 1\}$, $u(t)$ is the unit-step function such that $u(t) - u(t-T_d)$ is a rectangular "pulse," $T_d < T_o$ is the duration of each pulse, and T_o is the period between pulses. Each realization is a waveform of pulses with amplitudes ± 1 and discontinuities only at integer multiples of T_o. Two example realizations are shown in Figure 7.9(a) for $T_d = T_o/2$ (which is called 50% duty cycle because the pulse returns to zero for half the pulse period). Observe that a discontinuity occurs when adjacent random variables, $B[k]$ and $B[k-1]$, take on different values. It is clear that no realization is continuous and

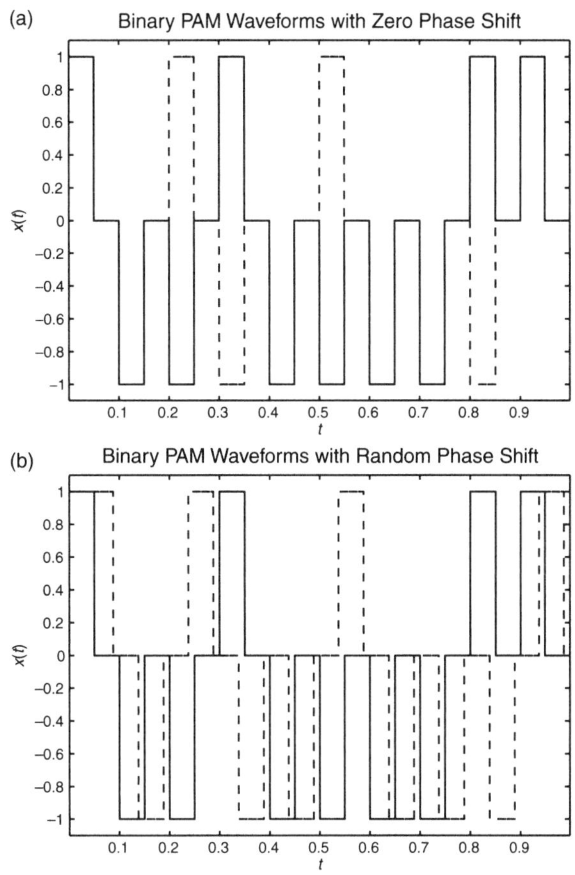

FIGURE 7.9 Example realizations of binary PAM. (a) Zero phase shift. (b) Random phase shift. (When there is no phase shift and the pulses of the two waveforms are identical, the solid line partially obscures the dashed line.)

thus $X(t)$ is not a sample continuous process. Also, since the discontinuities occur at the same time instants $T_o, T_o + T_d, 2T_o, 2T_o + T_d$, and so on, $X(t)$ is not continuous with probability one. However, suppose now that each realization has a random phase shift Φ such that the random process is

$$X(t) = \sum_{k=0}^{\infty} B[k][u(t - kT_o - \Phi) - u(t - (k+1)T_o - T_d - \Phi)]. \tag{7.77}$$

Assume that Φ is uniform on $[0, T_o - T_d)$ so that the pulses in any realization do not extend into an adjacent time interval. The probability of a discontinuity occurring at a particular t is zero because Φ is a *continuous* random variable. Two example realizations with random phase shifts are shown in Figure 7.9(b). Even though all realizations are still not sample continuous, $X(t)$ with a random phase shift is continuous with probability one, which is defined across the ensemble for each $t \in \mathcal{T}$ and not for the duration of the interval.

The remaining definitions of continuity do not consider individual realizations, and instead focus on the distribution of the process or its moments.

Definition: Continuous in Probability Random process $X(t)$ is *continuous in probability* at $t \in \mathcal{T}$ if for every $\delta > 0$

$$\lim_{\epsilon \to 0} P(|X(t+\epsilon) - X(t)| \geq \delta) = 0. \tag{7.78}$$

An expression similar to (7.75) is given by

$$\lim_{\epsilon \to 0} P\left(\zeta \in \mathcal{S} : |X(t+\epsilon, \zeta) - X(t, \zeta)| \geq \delta\right) = 0. \tag{7.79}$$

Compared to continuity with probability one, the limit here is taken outside the probability and therefore is a weaker definition. In order to evaluate (7.78), we require either the joint distribution $f_{X(t+\epsilon),X(t)}(x(t+\epsilon), x(t))$ or the pdf of the difference $f_{X(t+\epsilon)-X(t)}(x(t+\epsilon) - x(t))$.

Example 7.11. Consider the generalized Wiener process which has the following pdf (see Chapter 6):

$$f_{X(t_2)-X(t_1)}(x) = \frac{1}{\sqrt{2\pi(t_2 - t_1)}} \exp\left(-x^2/2(t_2 - t_1)\right). \tag{7.80}$$

Substituting $t_2 = t + \epsilon$ and $t_1 = t$, (7.78) becomes

$$P(|X(t+\epsilon) - X(t)| \geq \delta) = 2\int_{\delta}^{\infty} \frac{1}{\sqrt{2\pi\epsilon}} \exp\left(-x^2/2\epsilon\right)\mathrm{d}x = 2Q(\delta/\sqrt{\epsilon}), \tag{7.81}$$

where we have used the symmetry of the pdf, and $Q(\cdot)$ is the Q-function. The generalized Wiener process is continuous in probability because $\lim_{\epsilon \to 0} Q(\delta/\sqrt{\epsilon}) = 0$ for every $\delta > 0$.

The weakest type of continuity is based on the cdf.

Definition: Continuous in Distribution Random process $X(t)$ is *continuous in distribution* at $t \in \mathcal{T}$ if

$$\lim_{\epsilon \to 0} F_{X(t+\epsilon)}(x) = F_{X(t)}(x) \tag{7.82}$$

for every $x \in \mathcal{R}$.

This limit refers only to the continuity of the cdf itself. If the process is iid, then it is obviously continuous in distribution because the cdf is independent of t.

Example 7.12. Consider the standard Wiener process with cdf

$$F_{X(t+\epsilon)}(x) = \int_{x}^{\infty} \frac{1}{\sqrt{2\pi(t+\epsilon)}} \exp\left(-z^2/2(t+\epsilon)\right)\mathrm{d}z = Q(x/\sqrt{t+\epsilon}). \tag{7.83}$$

Since the Q-function is continuous,

$$\lim_{\epsilon \to 0} Q(x/\sqrt{t+\epsilon}) = Q(x/\sqrt{t}), \tag{7.84}$$

demonstrating that the Wiener process is continuous in distribution. (This result also follows from Example 7.11 because continuous in probability implies continuous in distribution.)

Example 7.13. Consider again random process $Y(t)$ in Example 6.2, of which we considered a discrete-time version earlier in Example 7.6. For convenience, we repeat the cdf:

$$F_{Y(t)}(y) = u(y)F_T(t) + F_X(y)[1 - F_T(t)]. \tag{7.85}$$

It is clear that this random process is continuous in distribution provided $F_T(t)$ is a continuous function. Moreover, we can conclude from the realizations in Figure 6.5 that $Y(t)$ is continuous with probability one (again, assuming that T is a continuous random variable), but is not sample continuous nor sure continuous because of the unit-step functions that define the process: $Y(t) = X[u(t) - u(t - T)]$.

Finally, we discuss the most widely used definition of continuity which is based on the autocorrelation function of the random process. In general, we can define *continuity in the mth mean* for $m = 1, 2, \ldots$ However, the most important case is $m = 2$ corresponding to MS continuity.

Definition: Mean-Square Continuous Random process $X(t)$ is *MS continuous* at $t \in \mathcal{T}$ if $\mathcal{E}[X^2(t)] < \infty$ and

$$\lim_{\epsilon \to 0} \mathcal{E}[(X(t+\epsilon) - X(t))^2] = 0. \tag{7.86}$$

Example 7.14. For the generalized Wiener process, which is an independent-increment process, we have

$$\mathcal{E}[(X(t+\epsilon) - X(t))^2] = \int_{-\infty}^{\infty} \frac{1}{\sqrt{2\pi\epsilon}} x^2 \exp\left(-x^2/2\epsilon\right) \mathrm{d}x = \epsilon \tag{7.87}$$

because the increment process is Gaussian with zero mean and variance ϵ. Since this expression tends to zero as $\epsilon \to 0$, the generalized Wiener process is MS continuous.

Theorem 7.8. If random process $X(t)$ is MS continuous, then it is also continuous in the mean.

Proof. Since the squared mean of a random variable cannot exceed its MS, the same is true for a random process at two different time instants:

$$\mathcal{E}[(X(t+\epsilon) - X(t))^2] \geq \mathcal{E}^2[X(t+\epsilon) - X(t)]. \tag{7.88}$$

Thus

$$\begin{aligned}
\lim_{\epsilon \to 0} \mathcal{E}[(X(t+\epsilon) - X(t))^2] = 0 &\Rightarrow \lim_{\epsilon \to 0} \mathcal{E}^2[X(t+\epsilon) - X(t)] = 0, \\
&\Rightarrow \lim_{\epsilon \to 0} \mathcal{E}[X(t+\epsilon) - X(t)] = 0, \\
&\Rightarrow \lim_{\epsilon \to 0} \mathcal{E}[X(t+\epsilon)] = \mathcal{E}[X(t)],
\end{aligned} \tag{7.89}$$

which completes the proof. □

Continuity in the mean generally does not imply MS continuity. Also, if the mean $\mu_X(t)$ is not a continuous function, then $X(t)$ is not MS continuous, due to the following lower bound:

$$\mathcal{E}[(X(t+\epsilon) - X(t))^2] \geq (\mathcal{E}[X(t+\epsilon)] - \mathcal{E}[X(t)])^2. \tag{7.90}$$

If the right-hand side does not approach zero for $t \in \mathcal{T}$ as $\epsilon \to 0$, then the MS continuity condition in (7.86) is not achieved.

Observe that the second moment in (7.86) can be rewritten in terms of the autocorrelation function as follows:

$$\mathcal{E}[(X(t+\epsilon)-X(t))^2] = R_{XX}(t+\epsilon, t+\epsilon) + R_{XX}(t,t) - R_{XX}(t+\epsilon, t) - R_{XX}(t, t+\epsilon). \tag{7.91}$$

From this expression, we find that MS continuity is achieved if the autocorrelation function is continuous with respect to *both* arguments. We need only examine the continuity of $R_{XX}(t_1, t_2)$ when approaching the line $t_1 = t_2 \in \mathcal{T}$, and not for every $\{t_1, t_2\}$.

Theorem 7.9. Random process $X(t)$ with $\mathcal{E}[X^2(t)] < \infty$ is MS continuous at $t \in \mathcal{T}$ if and only if $R_{XX}(t_1, t_2)$ is continuous at $t_1 = t_2 = t$.

Proof. (Larson and Shubert, 1979). *If:* From (7.91), if $R_{XX}(t_1, t_2)$ is continuous with respect to both arguments at t, then the right-hand side $\to 0$. From the definition of continuity for a function of two variables, this means that for every $\delta > 0$ there exists $\epsilon > 0$ such that

$$\sqrt{(t_1-t)^2 + (t_2-t)^2} < \epsilon \Rightarrow |R_{XX}(t_1, t_2) - R_{XX}(t,t)| < \delta. \tag{7.92}$$

Only if: Consider the difference

$$R_{XX}(t_1, t_2) - R_{XX}(t,t) = R_{XX}(t_1, t_2) - R_{XX}(t, t_1) + R_{XX}(t, t_1) - R_{XX}(t,t), \tag{7.93}$$

where $R_{XX}(t, t_1)$ has been added and subtracted. The triangle inequality for absolute value yields

$$|R_{XX}(t_1, t_2) - R_{XX}(t,t)| \leq |R_{XX}(t_1, t_2) - R_{XX}(t, t_1)| + |R_{XX}(t, t_1) - R_{XX}(t,t)|. \tag{7.94}$$

From the definition of the autocorrelation function,

$$R_{XX}(t_1, t_2) - R_{XX}(t_1, t) = \mathcal{E}[X(t_1)(X(t_2) - X(t))], \tag{7.95}$$

$$R_{XX}(t, t_1) - R_{XX}(t, t) = \mathcal{E}[X(t)(X(t_1) - X(t))], \tag{7.96}$$

and using the Cauchy–Schwarz inequality:

$$|R_{XX}(t_1, t_2) - R_{XX}(t_1, t)| \leq \sqrt{\mathcal{E}[X^2(t_1)]\mathcal{E}[(X(t_2) - X(t))^2]}, \tag{7.97}$$

$$|R_{XX}(t, t_1) - R_{XX}(t, t)| \leq \sqrt{\mathcal{E}[X^2(t)]\mathcal{E}[(X(t_1) - X(t))^2]}. \tag{7.98}$$

Substituting $t_1 = t + \epsilon_1$ and $t_2 = t + \epsilon_2$, and using the definition of MS continuity in (7.86), we find that as $\epsilon_1, \epsilon_2 \to 0$, the right-hand side of (7.94) tends to zero, thus completing the proof. □

For MS continuity, we are interested only in the continuity of $R_{XX}(t_1, t_2)$ at $t_1 = t_2 = t$. We need to evaluate continuity starting from any $\{t_1, t_2\} \in \mathcal{T}$, by examining $R_{XX}(t_1, t_2)$ with $t_1 = t + \epsilon_1$ and $t_2 = t + \epsilon_2$ and letting ϵ_1 and ϵ_2 separately approach zero. This ensures that continuity is not evaluated only along the $t_1 = t_2$ line, so that we avoid the mistake of examining only the behavior of $R_{XX}(t+\epsilon, t+\epsilon)$ as $\epsilon \to 0$ (although that is usually an easier condition to check first to show that a random process is *not* MS continuous). Finally, we mention that if $X(t)$ is MS continuous for every $t \in \mathcal{T}$, then $R_{XX}(t_1, t_2)$ is continuous for all $t_1, t_2 \in \mathcal{T}$ (see Problem 7.18).

Example 7.15. The Poisson counting process has the following autocorrelation function:

$$R_{XX}(t_1, t_2) = \lambda \min(t_1, t_2) + \lambda^2 t_1 t_2, \tag{7.99}$$

where $\lambda > 0$ is the rate parameter. From Theorem 7.9:

$$\begin{aligned}
&\lim_{\epsilon_1, \epsilon_2 \to 0} |R_{XX}(t+\epsilon_1, t+\epsilon_2) - R_{XX}(t,t)|, \\
&\quad = \lim_{\epsilon_1, \epsilon_2 \to 0} |\lambda \min(t+\epsilon_1, t+\epsilon_2) + \lambda^2 (t+\epsilon_1)(t+\epsilon_2) - \lambda t - \lambda^2 t^2|, \\
&\quad = \lim_{\epsilon_1, \epsilon_2 \to 0} |\lambda \min(\epsilon_1, \epsilon_2) + \lambda^2 (t\epsilon_1 + t\epsilon_2 + \epsilon_1\epsilon_2)| = 0.
\end{aligned} \tag{7.100}$$

The Poisson process is MS continuous even though the realizations are not sample continuous as illustrated in Figure 7.10. A plot of the autocorrelation function was shown in Figure 6.20, from which we see that $R_{XX}(t_1, t_2)$ is continuous at $t_1 = t_2 \in \mathcal{T}$.

Example 7.16. Realizations of the "semirandom" telegraph signal $Y(t)$ are generated from the Poisson counting process $X(t)$ as follows (see Chapter 6): $Y(t) = (-1)^{X(t)}$. With each jump in the Poisson process, the random telegraph signal changes sign, keeping track of when the number of increments is even or odd. If $Y(t)$ is multiplied by a symmetric Bernoulli random variable with equally likely outcomes $\{-1, 1\}$, and which is independent of $Y(t)$, then the "randomized" form of the telegraph signal is obtained. This introduces an ambiguity (greater randomness) because we no longer know which value, -1 or 1, corresponds to an even number of increments. Example realizations are shown in Figure 7.11. Both forms of the random telegraph signal have the following autocorrelation function (see Problem 6.3 and Table 6.4):

$$R_{YY}(t_1, t_2) = \exp(-2\lambda |t_2 - t_1|), \tag{7.101}$$

which is shown in Figure 7.12(a). This process is MS continuous:

$$\lim_{\epsilon_1, \epsilon_2 \to 0} |R_{YY}(t+\epsilon_1, t+\epsilon_2) - R_{YY}(t,t)| = \lim_{\epsilon_1, \epsilon_2 \to 0} |\exp(-2\lambda|\epsilon_1 - \epsilon_2|) - 1| = 0. \tag{7.102}$$

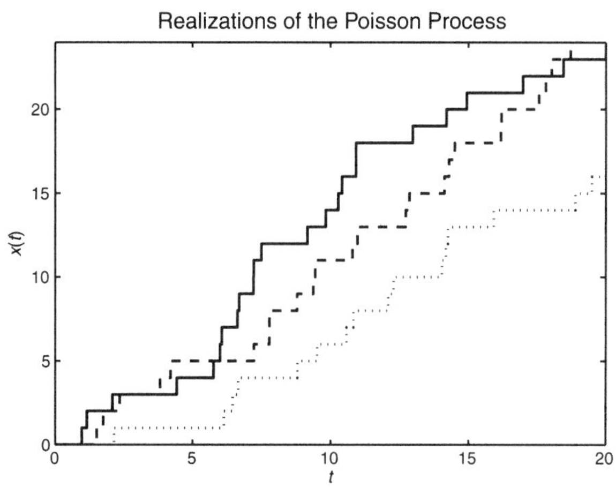

FIGURE 7.10 Example realizations of the Poisson counting process showing that individually they are not sample continuous, although the process overall is MS continuous.

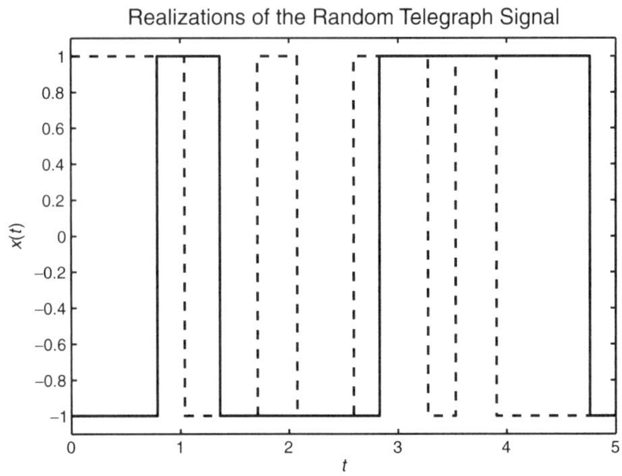

FIGURE 7.11 Realizations of the random telegraph signal, showing that individually they are not sample continuous, although the process is MS continuous.

The randomized telegraph signal is actually wide-sense stationary because $\tau = t_2 - t_1$ can be substituted; the autocorrelation function $R_{YY}(\tau)$ depends only on one variable, and its mean is zero. Although the autocorrelation function of the semirandom telegraph signal also depends only on the lag τ, it is not wide-sense stationary because its mean varies with time. For $R_{XX}(\tau)$, MS continuity reduces to continuity only at one point: $\tau = 0$.

Theorem 7.10. If random process $X(t)$ is wide-sense stationary, then it is MS continuous if and only if $R_{XX}(\tau)$ is continuous at $\tau = 0$.

Proof. This is a special case of Theorem 7.9 such that (7.91) becomes

$$\mathcal{E}[(X(t+\epsilon) - X(t))^2] \leq R_{XX}(0) - R_{XX}(\epsilon) + R_{XX}(0) - R_{XX}(\epsilon). \tag{7.103}$$

The proof is the basically the same (and is simpler because $R_{XX}(\tau)$ is a function of only one variable). □

For the wide-sense stationary case, we need to examine the continuity of $R_{XX}(\tau)$ at $\tau = 0$, approaching from two directions $\tau \pm \epsilon$ (unlike $R_{XX}(t_1, t_2)$ whose continuity at $t_1 = t_2 = t$ is examined from all directions on the t_1–t_2-plane).

Example 7.17. For the random telegraph signal in Example 7.16, the autocorrelation function can be written as

$$R_{XX}(\tau) = \exp(-2\lambda|\tau|), \tag{7.104}$$

which is continuous at the origin:

$$\lim_{\tau\to 0} |R_{XX}(\tau) - R_{XX}(0)| = \lim_{\tau\to 0} |\exp(-2\lambda|\tau|) - 1| = 0. \tag{7.105}$$

Figure 7.12(b) shows $R_{XX}(\tau)$ which we see corresponds to a "slice" through $R_{XX}(t_1, t_2)$ along any line perpendicular to the $t_1 = t_2$ line in Figure 7.12(a). The autocorrelation function of a wide-sense stationary process necessarily has this type of symmetry about $\tau = t_2 - t_1$.

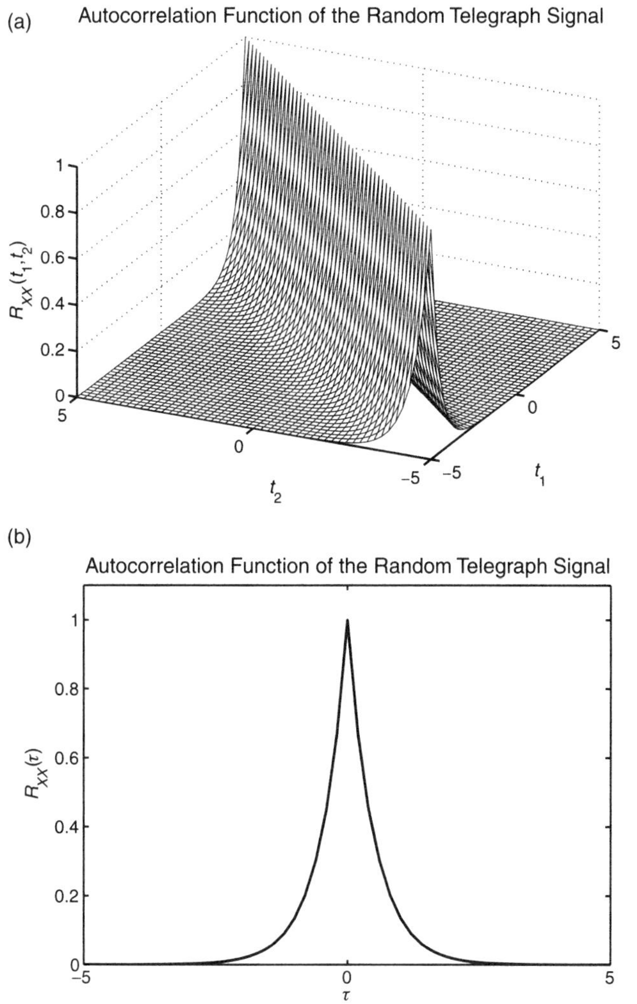

FIGURE 7.12 Autocorrelation function of the random telegraph signal illustrating that it is continuous at $t_1 = t_2$. (a) $R_{XX}(t_1, t_2)$ on the t_1–t_2-plane. (b) $R_{XX}(\tau)$ versus τ. Since $X(t)$ is wide-sense stationary, $R_{XX}(\tau)$ is examined for continuity only at $\tau = 0$.

Example 7.18. Consider a white-noise random process with autocorrelation function

$$R_{XX}(\tau) = \sigma_X^2 \delta(\tau). \tag{7.106}$$

(White noise is discussed in detail in Chapter 8.) Because the Dirac delta function is infinite at the origin, all such processes are *not* MS continuous, even if a process is iid so that its distribution is time invariant and thus continuous in distribution (which is a weaker type of continuity).

Table 7.3 summarizes the six types of continuity, and Figure 7.13 illustrates how they are related. We conclude this section with a proof of one of those relationships.

TABLE 7.3 Summary of Continuity for Random Process $X(t)$

Type of Continuity	Definition
Sure (for all $t \in \mathcal{T}$)	$\lim_{\epsilon\to 0} \lvert X(t+\epsilon,\zeta) - X(t,\zeta)\rvert = 0$ for all $\zeta \in S$
Sample continuous (for all $t \in \mathcal{T}$)	$P\left(\lim_{\epsilon\to 0} \lvert X(t+\epsilon) - X(t)\rvert \neq 0\right) = 0$
With probability one (at $t \in \mathcal{T}$)	$P\left(\lim_{\epsilon\to 0} \lvert X(t+\epsilon) - X(t)\rvert = 0\right) = 1$
In probability (at $t \in \mathcal{T}$)	$\lim_{\epsilon\to 0} P\left(\lvert X(t+\epsilon) - X(t)\rvert \geq \delta\right) = 0$ for every $\delta > 0$
Mean-square (at $t \in \mathcal{T}$)	$\lim_{\epsilon\to 0} \mathcal{E}[\lvert X(t+\epsilon) - X(t)\rvert^2] = 0$
In distribution (at $t \in \mathcal{T}$)	$\lim_{\epsilon\to 0} F_{X(t+\epsilon)}(x) = F_{X(t)}(x)$
Mean-Square Continuity	**Condition**
Nonstationary (at $t \in \mathcal{T}$)	$R_{XX}(t_1, t_2)$ continuous at $t_1 = t_2 = t$
Wide-sense stationary (at $t \in \mathcal{T}$)	$R_{XX}(\tau)$ continuous at $\tau = 0$

Theorem 7.11. If random process $X(t)$ is MS continuous, then it is also continuous in probability.

Proof. This result follows from Chebyshev's inequality (see Appendix F):

$$\lim_{\epsilon\to 0} P(\lvert X(t+\epsilon) - X(t)\rvert \geq \delta) \leq \lim_{\epsilon\to 0} \frac{\mathcal{E}[(X(t+\epsilon) - X(t))^2]}{\delta^2}. \tag{7.107}$$

For MS continuity, the right-hand side is zero, thus completing the proof. □

7.6 DERIVATIVES AND INTEGRALS

Derivatives and integrals arise in many applications and are used to model systems and solve a wide range of engineering problems. Often, we are interested in filtering a random process and would like to know how its properties are changed when it is convolved with the filter impulse response function. Moreover, lumped linear systems are represented by differential and integro-differential equations, and it will be useful to know how to interpret such equations for random processes.

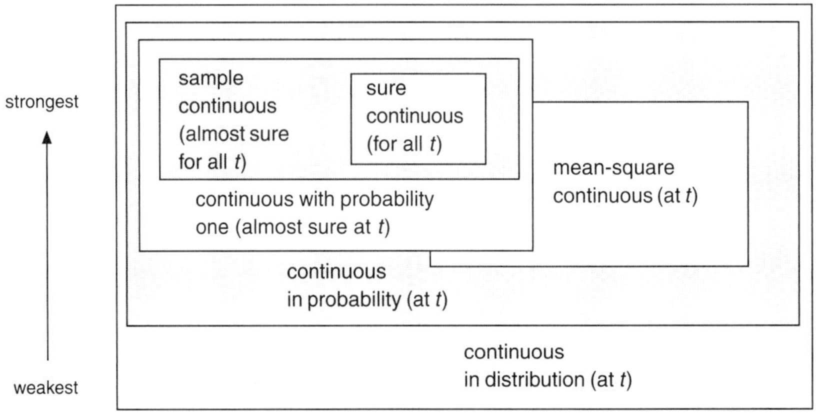

FIGURE 7.13 Relationships among the six types of continuity for a random process. Sure $\subseteq$ sample continuous $\subseteq$ w.p.1 $\subseteq$ p $\subseteq$ d and $L^2 \subseteq$ p $\subseteq$ d. Generally, we cannot conclude how MS continuity is related to continuity with probability one, sample continuous, and sure continuous.

We consider only MS definitions for derivatives and integrals of a random process, which again lead to an evaluation of the autocorrelation function $R_{XX}(t_1, t_2)$. The derivative of a nonrandom function is defined as follows.

Definition: Derivatives Function $x(t)$ has a *derivative* at $t \in \mathcal{T}$ if the following limit exists:

$$x'(t) = \frac{\mathrm{d}x}{\mathrm{d}t} \triangleq \lim_{\epsilon \to 0} \frac{x(t+\epsilon) - x(t)}{\epsilon}. \tag{7.108}$$

It has a second derivative at $t \in \mathcal{T}$ if the following limit exists:

$$x''(t) = \frac{\mathrm{d}^2 x}{\mathrm{d}t^2} \triangleq \lim_{\epsilon \to 0} \frac{x(t+\epsilon) - 2x(t) + x(t-\epsilon)}{\epsilon^2}. \tag{7.109}$$

For higher order ordinary derivatives, we also use the notation $x^{(m)}(t)$, so that $x'(t) = x^{(1)}(t)$ and $x''(t) = x^{(2)}(t)$. Combining this definition with the approach used for MS continuity leads to the MS derivative.

Definition: Mean-Square Derivative Random process $X(t)$ has a *MS derivative* at $t \in \mathcal{T}$ if

$$\lim_{\epsilon \to 0} \mathcal{E}\left[\left(\frac{X(t+\epsilon) - X(t)}{\epsilon} - X'(t)\right)^2\right] = 0, \tag{7.110}$$

where $X'(t)$ represents the derivative.

The first term is the finite approximation in (7.108) except that $X(t)$ is a random process. The second term $X'(t)$ represents the derivative process generated by the finite approximation in the limit. If $X'(t)$ is known beforehand, then we can evaluate this expression because the expectation is computed before the limit is taken. However, the derivative process is usually not known, so instead the Cauchy criterion is used to determine if a random process has an MS derivative:

$$\lim_{\epsilon_1 \to 0, \epsilon_2 \to 0} \mathcal{E}\left[\left(\frac{X(t+\epsilon_1) - X(t)}{\epsilon_1} - \frac{X(t+\epsilon_2) - X(t)}{\epsilon_2}\right)^2\right] = 0. \tag{7.111}$$

Consider the following outer term in the expectation:

$$\begin{aligned} \mathcal{E}\left[\left(\frac{X(t+\epsilon_1) - X(t)}{\epsilon_1}\right)^2\right] &= (1/\epsilon_1^2)[R_{XX}(t+\epsilon_1, t+\epsilon_1) + R_{XX}(t,t) \\ &\quad - R_{XX}(t, t+\epsilon_1) - R_{XX}(t+\epsilon_1, t)], \end{aligned} \tag{7.112}$$

and likewise for the outer term involving ϵ_2. Similarly, for the two cross-terms in (7.111):

$$\begin{aligned} \mathcal{E}\left[\frac{X(t+\epsilon_1) - X(t)}{\epsilon_1}\frac{X(t+\epsilon_2) - X(t)}{\epsilon_2}\right] &= (1/\epsilon_1\epsilon_2)[R_{XX}(t+\epsilon_1, t+\epsilon_2) + R_{XX}(t,t) \\ &\quad - R_{XX}(t, t+\epsilon_2) - R_{XX}(t+\epsilon_1, t)]. \end{aligned} \tag{7.113}$$

Although these two expression are similar, note that the cross-terms in (7.113) depend on ϵ_1 and ϵ_2. This means that in the limit as ϵ_1 and ϵ_2 approach zero, the existence of the derivative of $R_{XX}(t_1, t_2)$ is evaluated from every direction on the t_1–t_2-plane. This is not the case for the outer terms in (7.112) because each depends only on ϵ_1 or ϵ_2 (not both).

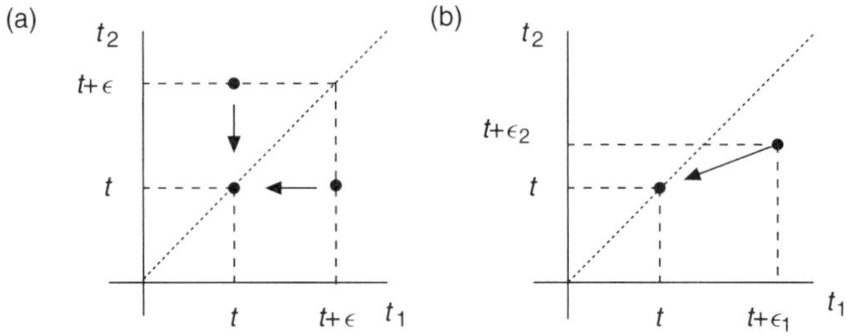

FIGURE 7.14 MS derivative based on $R_{XX}(t_1, t_2)$. (a) Depiction of the partial derivatives computed sequentially in (7.115). (b) Existence of the MS derivative using (7.114).

Theorem 7.12. Random process $X(t)$ with autocorrelation function $R_{XX}(t_1, t_2)$ has an MS derivative at $t \in \mathcal{T}$ if and only if

$$\lim_{\epsilon_1 \to 0, \epsilon_2 \to 0} (1/\epsilon_1\epsilon_2)[R_{XX}(t+\epsilon_1, t+\epsilon_2) + R_{XX}(t,t) - R_{XX}(t, t+\epsilon_2) - R_{XX}(t+\epsilon_1, t)] \tag{7.114}$$

exists at $t_1 = t_2 = t$.

We provide a proof for the if part of the theorem.

Proof. If the limit in (7.114) exists, then so does the limit of (7.112) because (7.114) approaches $\{t, t\}$ from every direction as mentioned above. Since all four terms (two outer terms and two cross-terms) have the same limit, combining them gives the result in (7.111). □

Note that the condition in (7.114) is *not* the same as the existence of $\partial^2 R_{XX}(t_1, t_2)/\partial t_1 \partial t_2$ at $t_1 = t_2 = t$ because even though the latter may be finite at t, the notation does not "take into account" all approaches to $\{t, t\}$ on the t_1–t_2-plane. It is calculated in the following two ways, as depicted in Figure 7.14(a):

$$\frac{\partial^2}{\partial t_1 \partial t_2} R_{XX}(t_1, t_2) = \frac{\partial}{\partial t_1}\left(\frac{\partial}{\partial t_2} R_{XX}(t_1, t_2)\right) = \frac{\partial}{\partial t_2}\left(\frac{\partial}{\partial t_1} R_{XX}(t_1, t_2)\right). \tag{7.115}$$

In the first case, the partial derivative is computed parallel to the t_2 axis (t_1 is fixed), and then along the t_1 axis (with t_2 fixed). The order of differentiation is reversed in the second case. From the figure, we see that existence of the derivative is examined only along two perpendicular lines intersecting at $\{t, t\}$. On the other hand, the condition in (7.114) is clear because it takes into account all approaches to $\{t, t\}$ on the t_1–t_2-plane, as shown by the example in Figure 7.14(b).

The following sufficient condition is often used for the MS derivative, which we state without proof. Note that it requires continuity of the partial derivatives.

Theorem 7.13. Random process $X(t)$ with autocorrelation function $R_{XX}(t_1, t_2)$ has an MS derivative at $t \in \mathcal{T}$ if the partial derivatives in (7.115) exist and are continuous for every $t_1 = t_2 = t \in \mathcal{T}$.

Example 7.19. Consider the Poisson process with the autocorrelation function in (7.99) which is plotted in Figure 6.20. It is convenient to use the identity (see Appendix E)

$$\min(t_1, t_2) = (1/2)(t_1 + t_2 - |t_1 - t_2|) \tag{7.116}$$

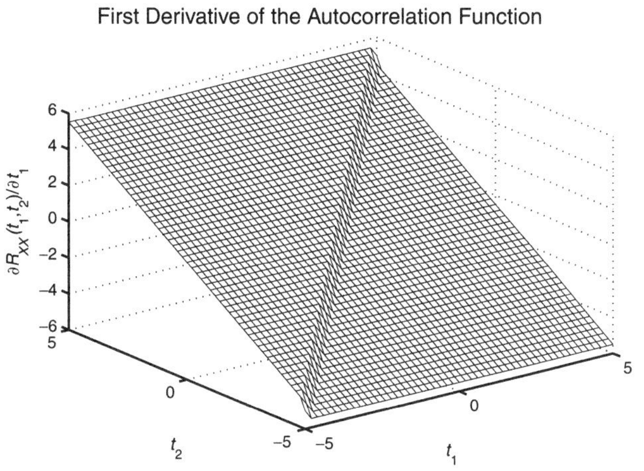

FIGURE 7.15 Derivative of the autocorrelation function for the Poisson counting process with $\lambda = 1$: $\partial R_{XX}(t_1, t_2)/\partial t_1$ showing a step.

so that

$$R_{XX}(t_1, t_2) = (\lambda/2)(t_1 + t_2 - |t_1 - t_2|) + \lambda^2 t_1 t_2. \tag{7.117}$$

Differentiating gives

$$\frac{\partial}{\partial t_1} R_{XX}(t_1, t_2) = (\lambda/2)\mathrm{sgn}(t_2 - t_1) + \lambda^2 t_2, \tag{7.118}$$

$$\frac{\partial^2}{\partial t_1 \partial t_2} R_{XX}(t_1, t_2) = \lambda\delta(t_2 - t_1) + \lambda^2, \tag{7.119}$$

which we see does not exist at $t_1 = t_2 = t$ because of the Dirac delta function; the first derivative in (7.118) is plotted in Figure 7.15. The Poisson counting process does not have an MS derivative.

For a wide-sense stationary process, the autocorrelation function $R_{XX}(\tau)$ is examined at $\tau = 0$.

Theorem 7.14. Wide-sense stationary random process $X(t)$ has an MS derivative for every $t \in \mathcal{T}$ if and only if $\mathrm{d}^2 R_{XX}(\tau)/\mathrm{d}\tau^2$ exists at $\tau = 0$.

We prove the if part of the theorem.

Proof. In order to show (7.111), the outer term in (7.112) for wide-sense stationary $R_{XX}(\tau)$ becomes

$$\mathcal{E}\left[\left(\frac{X(t+\epsilon_1) - X(t)}{\epsilon_1}\right)^2\right] = (1/\epsilon_1^2)[2R_{XX}(0) - 2R_{XX}(\epsilon_1)]. \tag{7.120}$$

Using the definition for the second ordinary derivative in (7.109) and the fact that $R_{XX}(\tau)$ is an even function, we have the following limit at $\tau = 0$:

$$\lim_{\epsilon_1 \to 0} (1/\epsilon_1^2)[2R_{XX}(0) - 2R_{XX}(\epsilon_1)] = -\frac{\mathrm{d}^2}{\mathrm{d}\tau^2} R_{XX}(\tau)\Big|_{\tau=0} = -R''_{XX}(0).$$

Similarly for the cross-term in (7.113):

$$\mathcal{E}\left[\left(\frac{X(t+\epsilon_1)-X(t)}{\epsilon_1}\right)\left(\frac{X(t+\epsilon_2)-X(t)}{\epsilon_2}\right)\right] = (1/\epsilon_1\epsilon_2)[R_{XX}(\epsilon_2-\epsilon_1)+R_{XX}(0) \\ -R_{XX}(\epsilon_1)-R_{XX}(\epsilon_2)]. \tag{7.121}$$

Using (7.108) twice at $\tau = 0$ and $\tau = \epsilon_2$ gives

$$\lim_{\epsilon_1\to 0}(1/\epsilon_1\epsilon_2)[R_{XX}(\epsilon_2-\epsilon_1)-R_{XX}(\epsilon_2)+R_{XX}(0)-R_{XX}(-\epsilon_1)] \\ = (1/\epsilon_2)[-R'_{XX}(\epsilon_2)+R'_{XX}(0)], \tag{7.122}$$

where

$$R'_{XX}(0) = \left.\frac{\mathrm{d}}{\mathrm{d}\tau}R_{XX}(\tau)\right|_{\tau=0}. \tag{7.123}$$

Thus

$$\lim_{\epsilon_2\to 0} -(1/\epsilon_2)[R'_{XX}(\epsilon_2)-R'_{XX}(0)] = -R''_{XX}(0). \tag{7.124}$$

Combining all four results gives

$$\lim_{\epsilon_1\to 0,\epsilon_2\to 0}\mathcal{E}\left[\left(\frac{X(t+\epsilon_1)-X(t)}{\epsilon_1}-\frac{X(t+\epsilon_2)-X(t)}{\epsilon_2}\right)^2\right] \\ = 2R''_{XX}(0)-2R''_{XX}(0)=0, \tag{7.125}$$

which completes the proof. □

Unlike the general nonstationary case, there is no issue about the second derivative in Theorem 7.14, which gives a necessary and sufficient condition. This follows because $R_{XX}(\tau)$ is a function of only one variable τ so that the derivative is examined from only two directions: $\tau \pm \epsilon$. Finally, note that since $R_{XX}(\tau)$ is an even function, its derivative is an odd function:

$$\frac{\mathrm{d}}{\mathrm{d}\tau}R_{XX}(\tau) = -\frac{\mathrm{d}}{\mathrm{d}\tau}R_{XX}(-\tau), \tag{7.126}$$

and thus, if it exists, $R'_{XX}(0) = 0$. Of course, this does not mean that the second derivative is also zero, because $R''_{XX}(\tau)$ must be computed before $\tau = 0$ is substituted.

Next, since $X'(t)$ is a random variable, we provide expressions for its mean at t:

$$\mathcal{E}[X'(t)] = \mathcal{E}\left[\lim_{\epsilon\to 0}\frac{X(t+\epsilon)-X(t)}{\epsilon}\right], \\ = \lim_{\epsilon\to 0}(1/\epsilon)\mathcal{E}[X(t+\epsilon)-X(t)] = \frac{\mathrm{d}}{\mathrm{d}t}\mathcal{E}[X(t)], \tag{7.127}$$

and autocorrelation function:

$$\begin{aligned} R_{X'X'}(t_1, t_2) &= \lim_{\epsilon_1, \epsilon_2 \to 0} \mathcal{E}\left[\frac{(X(t_1+\epsilon_1) - X(t_1))(X(t_2+\epsilon_2) - X(t_2))}{\epsilon_1 \epsilon_2}\right], \\ &= \lim_{\epsilon_1, \epsilon_2 \to 0} (1/\epsilon_1\epsilon_2)[R_{XX}(t_1+\epsilon_1, t_2+\epsilon_2) + R_{XX}(t_1, t_2) \\ &\quad - R_{XX}(t_1+\epsilon_1, t_2) - R_{XX}(t_1, t_2+\epsilon_2)], \\ &= \frac{\partial^2}{\partial t_1 \partial t_2} R_{XX}(t_1, t_2). \end{aligned} \tag{7.128}$$

Note the similarity between the last expression and the sufficient condition in Theorem 7.13 used to determine if the MS derivative exists. Above we see that the autocorrelation function of the derivative process is the second derivative of the autocorrelation function of the original process $X(t)$, and that $R_{X'X'}(t_1, t_2)$ is defined for all values of $\{t_1, t_2\}$. The theorem states that $R_{X'X'}(t_1, t_2)$ exist and be continuous for all $t_1 = t_2 = t \in \mathcal{T}$, which means that the power of the derivative process must be finite:

$$R_{X'X'}(t, t) = \mathcal{E}[(X'(t))^2] < \infty. \tag{7.129}$$

For a wide-sense stationary process with $\tau \triangleq t_2 - t_1$ (see Problem 7.19):

$$R_{X'X}(\tau) = \frac{\mathrm{d}}{\mathrm{d}\tau} R_{XX}(\tau), \quad R_{XX'}(\tau) = -\frac{\mathrm{d}}{\mathrm{d}\tau} R_{XX}(\tau) \tag{7.130}$$

and

$$R_{X'X'}(\tau) = -\frac{\mathrm{d}}{\mathrm{d}\tau} R_{X'X}(\tau) = \frac{\mathrm{d}}{\mathrm{d}\tau} R_{XX'}(\tau) = -\frac{\mathrm{d}^2}{\mathrm{d}\tau^2} R_{XX}(\tau). \tag{7.131}$$

When the autocorrelation function has discontinuities at $t_1 = t_2 = t$ (or $\tau = 0$ in the wide-sense stationary case), which means the MS derivative does not exist, the Dirac delta function can be used to represent the derivative of $R_{XX}(t_1, t_2)$.

Example 7.20. For the random telegraph signal:

$$\frac{\mathrm{d}}{\mathrm{d}\tau} R_{XX}(\tau) = \frac{\mathrm{d}}{\mathrm{d}\tau} \exp(-2\lambda|\tau|) = -2\lambda \exp(-2\lambda|\tau|)\mathrm{sgn}(\tau) \tag{7.132}$$

and

$$\frac{\mathrm{d}^2}{\mathrm{d}\tau^2} R_{XX}(\tau) = 4\lambda^2 \exp(-2\lambda|\tau|) - 4\lambda \exp(-2\lambda|\tau|)\delta(\tau), \tag{7.133}$$

where $\mathrm{sgn}^2(\tau) = 1$ has been used and the derivative of the signum function is $2\delta(\tau)$. The random telegraph signal does not have an MS derivative which is evident from the plots in Figure 7.12.

Definition: Mean-Square Integral The *MS integral* of random process $X(t)$, denoted by $Y(t) \triangleq \int_{t_o}^{t} X(t)\mathrm{d}t$ for $[t_o, t] \in \mathcal{T}$, is defined by

$$\lim_{N \to \infty} \mathcal{E}\left[\left(\int_{t_o}^{t} X(t)\mathrm{d}t - \sum_{n=0}^{N-1} X(t_o + n\Delta_N)\Delta_N\right)^2\right] = 0, \tag{7.134}$$

where $\Delta_N \triangleq (t - t_o)/N$ is the size of the intervals which form a partition on $[t_o, t]$.

The sum in this definition is a staircase approximation as is used to describe the Riemann sum for a nonrandom function (see Appendix D). The integral is approximated by the total area of N nonoverlapping rectangles which becomes more accurate as N is increased: the intervals become smaller in the partition and the rectangles more closely follow the function. Observe that $Y(t)$ at time t is a random variable which can be described in terms of its mean and autocorrelation function, as was done for the MS derivative.

For notational convenience, define the sum in (7.134) as follows:

$$Y_N \triangleq \sum_{n=0}^{N-1} X(t_o + n\Delta_N)\Delta_N \tag{7.135}$$

so that (7.134) becomes

$$\lim_{N\to\infty} \mathcal{E}[(Y(t) - Y_N)^2] = 0. \tag{7.136}$$

Since the MS integral $Y(t)$ may not be known, the Cauchy criterion can be used to determine if the MS integral exists:

$$\lim_{M,N\to\infty} \mathcal{E}[(Y_M - Y_N)^2] = 0, \tag{7.137}$$

which can be written in terms of the autocorrelation function of $X(t)$.

Theorem 7.15. Random process $X(t)$ with autocorrelation function $R_{XX}(t_1, t_2)$ has an MS integral on $[t_o, t] \in \mathcal{T}$ if and only if $\int_{t_o}^{t}\int_{t_o}^{t} R_{XX}(t_1, t_2)\mathrm{d}t_1\mathrm{d}t_2$ exists.

Proof. Consider the cross-terms from the Cauchy criterion in (7.137):

$$\begin{aligned}
\mathcal{E}[Y_M Y_N] &= \mathcal{E}\left[\sum_{m=1}^{M} X(t_m)\Delta t_m \sum_{n=1}^{N} X(t_n)\Delta t_n\right], \\
&= \sum_{m=1}^{M}\sum_{n=1}^{N} R_{XX}(t_m, t_n)\Delta t_m \Delta t_n.
\end{aligned} \tag{7.138}$$

The right-hand side is a staircase approximation of the integral of the deterministic function $R_{XX}(t_1, t_2)$ such that in the limit:

$$\lim_{M,N\to\infty} \mathcal{E}[Y_M Y_N] = \int_{t_o}^{t}\int_{t_o}^{t} R_{XX}(t_1, t_2)\mathrm{d}t_1\mathrm{d}t_2. \tag{7.139}$$

Consider one of the outer terms in the product:

$$\begin{aligned}
\lim_{M\to\infty} \mathcal{E}[Y_M^2] &= \lim_{M\to\infty} \mathcal{E}\left[\sum_{m=1}^{M} X(t_m)\Delta t_m \sum_{n=1}^{M} X(t_n)\Delta t_n\right], \\
&= \lim_{M\to\infty} \sum_{m=0}^{M}\sum_{n=0}^{M} R_{XX}(t_m, t_n)\Delta t_m \Delta t_n, \\
&= \int_{t_o}^{t}\int_{t_o}^{t} R_{XX}(t_1, t_2)\mathrm{d}t_1\mathrm{d}t_2,
\end{aligned} \tag{7.140}$$

and likewise for $\lim_{N\to\infty} \mathcal{E}[Y_N^2]$. Combining these results according to (7.137) yields zero, which completes the proof. □

Note that since (7.140) is a double integral of a nonrandom function, the MS integral always exists when the limits $[t_o, t]$ are finite.

Theorem 7.16. If $X(t)$ is a wide-sense stationary process such that $R_{XX}(t_1, t_2) = R_{XX}(\tau)$, the MS integral of $X(t)$ exists on $[t_o, t]$ if and only if the following exists:

$$2\int_0^{t-t_o} (t - t_o - \tau)R_{XX}(\tau)\mathrm{d}\tau. \tag{7.141}$$

Proof. Since the autocorrelation function of a wide-sense stationary process depends only on a single variable (the time lag), we perform the following transformation of variables. Let $\tau \triangleq t_2 - t_1$ and $\alpha \triangleq t_1 + t_2$, yielding the following Jacobian of the transformation (see Appendix G):

$$J = \det(\mathbf{J}) = \det\begin{bmatrix} \partial\tau/\partial t_1 & \partial\tau/\partial t_2 \\ \partial\alpha/\partial t_1 & \partial\alpha/\partial t_2 \end{bmatrix} = \det\begin{bmatrix} -1 & 1 \\ 1 & 1 \end{bmatrix} = -2. \tag{7.142}$$

Thus, $\mathrm{d}t_1\mathrm{d}t_2 = (1/|J|)\mathrm{d}\tau\mathrm{d}\alpha = (1/2)\mathrm{d}\tau\mathrm{d}\alpha$ such that (7.140) becomes

$$\begin{aligned}\int_{t_o}^{t}\int_{t_o}^{t} R_{XX}(t_1, t_2)\mathrm{d}t_1\mathrm{d}t_2 &= (1/2)\int_{t_o}^{t}\int_{t_o}^{t} R_{XX}(t_2 - t_1)\mathrm{d}t_1\mathrm{d}t_2, \\ &= \int_{-(t-t_o)}^{t-t_o}\int_{-(t-t_o-|\tau|)}^{t-t_o-|\tau|} R_{XX}(\tau)\mathrm{d}\alpha\mathrm{d}\tau.\end{aligned} \tag{7.143}$$

Since the inner integral is performed over a constant, the expression reduces to

$$\begin{aligned}\int_{t_o}^{t}\int_{t_o}^{t} R_{XX}(t_1, t_2)\mathrm{d}t_1\mathrm{d}t_2 &= \int_{-(t-t_o)}^{t-t_o} (t - t_o - |\tau|)R_{XX}(\tau)\mathrm{d}\tau, \\ &= 2\int_0^{t-t_o} (t - t_o - \tau)R_{XX}(\tau)\mathrm{d}\tau,\end{aligned} \tag{7.144}$$

where in the last expression we have used the fact that $R_{XX}(\tau)$ is an even function. □

When $X(t)$ is wide-sense stationary, $R_{XX}(t_1, t_2)$ takes on the same value for all $\{t_1, t_2\}$ with identical $\tau = t_2 - t_1$. As illustrated in Figure 7.16, $R_{XX}(t_1, t_2)$ has the same value along lines that are rotated 45° counterclockwise with respect to the t_1-t_2 axes. $R_{XX}(\tau)$ corresponds to any "slice" through $R_{XX}(t_1, t_2)$ that is perpendicular to those lines of constant correlation as shown in the figure. Observe in the derivation that $(t - t_o - |\tau|)$ is a triangular function on $[-(t - t_o), t - t_o]$. The original integration is performed over the square defined by $[t_o, t] \times [t_o, t]$. Since $R_{XX}(t_1, t_2)$ is constant along lines perpendicular to $t_1 = t_2$, the integration can instead be performed over $\tau \in [-(t - t_o), t - t_o]$. However, in order to obtain the same result, $R_{XX}(\tau)$ for each τ needs to be weighted by an amount given by the length of the rotated line inside the square. As seen in the figure, these lines form a triangle with a maximum at $\tau = 0$ and is zero outside $\tau \in [-(t - t_o), t - t_o]$. The triangular weighting is only an "artifact" of the change of variables; it is not due to any other inherent property of random process $X(t)$, except for the fact that $R_{XX}(\tau)$ depends only on one variable.

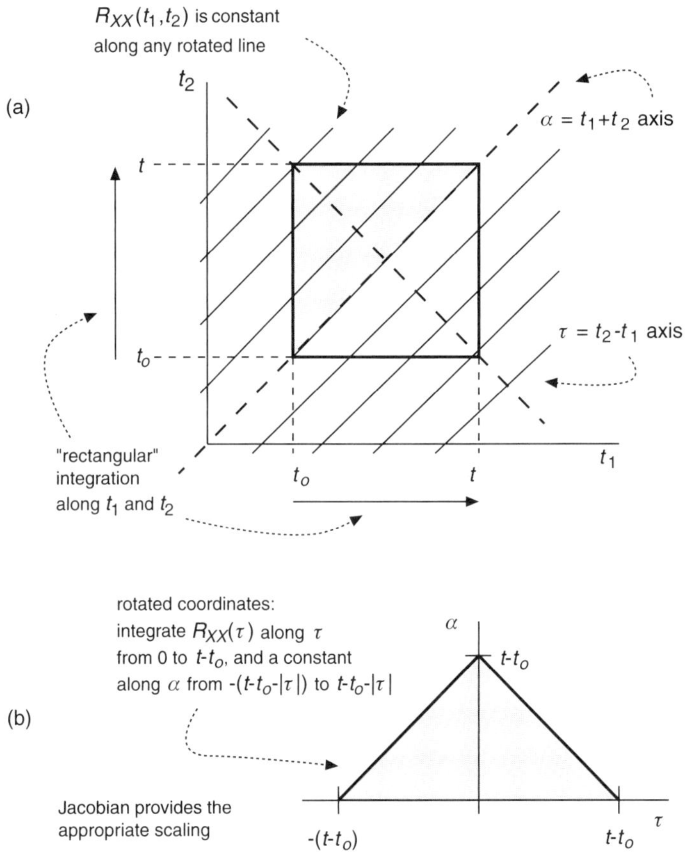

FIGURE 7.16 Regions of integration for $R_{XX}(t_1, t_2)$. (a) Original axes showing lines on the t_1–t_2-plane where $R_{XX}(t_1, t_2) = R_{XX}(\tau)$ is constant. (b) Rotated axes with integration limits indicated by the shaded triangular region.

Finally, note that the MS integral is a random variable with the following mean:

$$\mathcal{E}\left[\int_{t_o}^{t} X(t)\mathrm{d}t\right] = \int_{t_o}^{t} \mathcal{E}[X(t)]\mathrm{d}t = \int_{t_o}^{t} \mu_X(t)\mathrm{d}t, \tag{7.145}$$

second moment:

$$\begin{aligned}\mathcal{E}\left[\int_{t_o}^{t}\int_{t_o}^{t} X(t_1)X(t_2)\mathrm{d}t_1\mathrm{d}t_2\right] &= \int_{t_o}^{t}\int_{t_o}^{t} \mathcal{E}[X(t_1)X(t_2)]\mathrm{d}t_1\mathrm{d}t_2,\\ &= \int_{t_o}^{t}\int_{t_o}^{t} R_{XX}(t_1, t_2)\mathrm{d}t_1\mathrm{d}t_2,\end{aligned} \tag{7.146}$$

and variance:

$$\begin{aligned}\mathrm{var}\left[\int_{t_o}^{t} X(t)\mathrm{d}t\right] &= \mathcal{E}\left[\int_{t_o}^{t}\int_{t_o}^{t} (X(t_1) - \mu_X(t_1))(X(t_2) - \mu_X(t_2))\mathrm{d}t_1\mathrm{d}t_2\right],\\ &= \int_{t_o}^{t}\int_{t_o}^{t} C_{XX}(t_1, t_2)\mathrm{d}t_1\mathrm{d}t_2,\end{aligned} \tag{7.147}$$

where $C_{XX}(t_1, t_2)$ is the autocovariance function of $X(t)$. When $X(t)$ is wide-sense stationary, these equations reduce to

$$\mathcal{E}\left[\int_{t_o}^{t} X(t)\mathrm{d}t\right] = (t - t_o)\mu_X, \tag{7.148}$$

$$\mathcal{E}\left[\int_{t_o}^{t}\int_{t_o}^{t} X(t_1)X(t_2)\mathrm{d}t_1\mathrm{d}t_2\right] = 2\int_0^{t-t_o} (t - t_o - \tau)R_{XX}(\tau)\mathrm{d}\tau, \tag{7.149}$$

$$\mathrm{var}\left[\int_{t_o}^{t} X(t)\mathrm{d}t\right] = 2\int_0^{t-t_o} (t - t_o - \tau)C_{XX}(\tau)\mathrm{d}\tau, \tag{7.150}$$

which are functions of the interval $t - t_o$. The second moment of the integral process is the same expression used to verify the existence of the MS integral, that is, (7.146) must exist.

Example 7.21. For the randomized telegraph signal, $C_{XX}(\tau) = \exp(-2\lambda|\tau|)$ such that the variance above is

$$\begin{aligned}\mathrm{var}\left[\int_{t_o}^{t} X(\tau)\mathrm{d}\tau\right] &= 2\int_0^{t-t_o} (t - t_o - \tau)\exp(-2\lambda|\tau|)\mathrm{d}\tau, \\ &= 2\exp(-2\lambda(t - t_o))\int_0^{t-t_o} v\exp(2\lambda v)\mathrm{d}v,\end{aligned} \tag{7.151}$$

where we have dropped the absolute value in the argument of the exponential and changed variables to $v \triangleq t - t_o - \tau$. The integral is evaluated using integration by parts:

$$\begin{aligned}\int_0^{t-t_o} v\exp(2\lambda v)\mathrm{d}v &= [(v/2\lambda)\exp(2\lambda v) - (1/2\lambda)^2\exp(2\lambda v)]\Big|_0^{t-t_o}, \\ &= \exp(2\lambda(t - t_o))[(t - t_o)/2\lambda - 1/4\lambda^2] + 1/4\lambda^2.\end{aligned} \tag{7.152}$$

Substituting this result into the expression above gives

$$\mathrm{var}\left[\int_{t_o}^{t} X(\tau)\mathrm{d}\tau\right] = (t - t_o)/\lambda + (1/2\lambda^2)[\exp(-2\lambda(t - t_o)) - 1], \tag{7.153}$$

which is also the second moment because $R_{XX}(\tau) = C_{XX}(\tau)$ for this zero-mean random process.

Summarizing the stochastic calculus results, the MS definitions of continuity and the derivative of random process $X(t)$ at $t \in \mathcal{T}$ require that the autocorrelation function $R_{XX}(t_1, t_2)$ be continuous and differentiable at $t_1 = t_2 = t$, approaching from any direction on the t_1–t_2-plane. Similarly, the MS integral of random process $X(t)$ on $[t_o, t]$ requires that the double integral of $R_{XX}(t_1, t_2)$ on $[t_o, t] \times [t_o, t]$ exist. As mentioned in Appendix D which provides some background on integration, differentiable functions are a subset of continuous functions, and continuous functions are a subset of integrable functions. These relationships also apply for the previous MS definitions because $R_{XX}(t_1, t_2)$ is a nonrandom function. The following theorem states these results.

Theorem 7.17. For random process $X(t)$ with $\mathcal{E}[X^2(t)] < \infty$: (i) if the MS derivative exists at $t \in \mathcal{T}$, then $X(t)$ is MS continuous at t, and (ii) if $X(t)$ is MS continuous on $[t_o, t] \in \mathcal{T}$, then its MS integral on that interval exists.

Proof. (i) Rearrange the expression used for the definition of MS continuity as follows:

$$\mathcal{E}[(X(t+\epsilon) - X(t))^2] = \epsilon^2 \mathcal{E}\left[\left(\frac{X(t+\epsilon) - X(t)}{\epsilon}\right)^2\right]. \tag{7.154}$$

Since $X(t)$ has an MS derivative at t, the limit of the expectation on the right-hand exists (is finite). And because $\lim_{\epsilon \to 0} \epsilon^2 = 0$, we can write

$$\lim_{\epsilon \to 0} \epsilon^2 \lim_{\epsilon \to 0} \mathcal{E}\left[\left(\frac{X(t+\epsilon) - X(t)}{\epsilon}\right)^2\right] = \lim_{\epsilon \to 0} \epsilon^2 \mathcal{E}\left[\left(\frac{X(t+\epsilon) - X(t)}{\epsilon}\right)^2\right] = 0 \tag{7.155}$$

demonstrating that $X(t)$ is MS continuous. (ii) From the result in Problem 7.18, $R_{XX}(t_1, t_2)$ is continuous for all $t_o \leq t_1, t_2 \leq t$, and so the MS integral of $X(t)$ exists. □

7.7 DIFFERENTIAL EQUATIONS

Many systems in engineering are modeled by linear DEs with constant coefficients to describe how various quantities (signals, voltages, currents, and so on) change with time. For convenience in our discussion, we provide a simple example for a second-order RLC circuit.

Example 7.22. A parallel RLC circuit (with resistor R, inductor L, and capacitor C) has an independent current source i_s that is switched on at time $t = 0$. The current through the inductor can be modeled by the following second-order DE:

$$a_2 \frac{d^2}{dt^2} i(t) + a_1 \frac{d}{dt} i(t) + a_0 i(t) = i_s u(t), \tag{7.156}$$

where the coefficients $\{a_0, a_1, a_2\}$ are functions of $\{R, L, C\}$ and $u(t)$ is the unit-step function. A circuit diagram is provided in Figure 7.17 for which it can be shown

$$a_2 = LC, \quad a_1 = L/R, \quad a_0 = 1. \tag{7.157}$$

Depending on the coefficient values, the solution of (7.156) can take on one of three forms:

$$\textit{Overdamped}: \quad i(t) = [c_1 \exp(-\alpha_1 t) + c_2 \exp(-\alpha_2 t) + d]u(t), \tag{7.158}$$

$$\textit{Underdamped}: \quad i(t) = [(c_1 \cos(\omega_d t) + c_2 \sin(\omega_d t)) \exp(-\alpha t) + d]u(t), \tag{7.159}$$

$$\textit{Critically damped}: \quad i(t) = [c_1 \exp(-\alpha_1 t) + c_2 t \exp(-\alpha_2 t) + d]u(t), \tag{7.160}$$

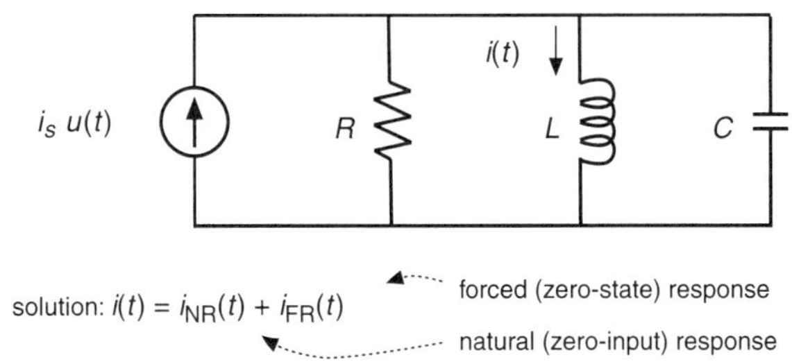

FIGURE 7.17 Parallel RLC circuit. Current source i_s is switched on at $t = 0$ as indicated by the unit step-function $u(t)$.

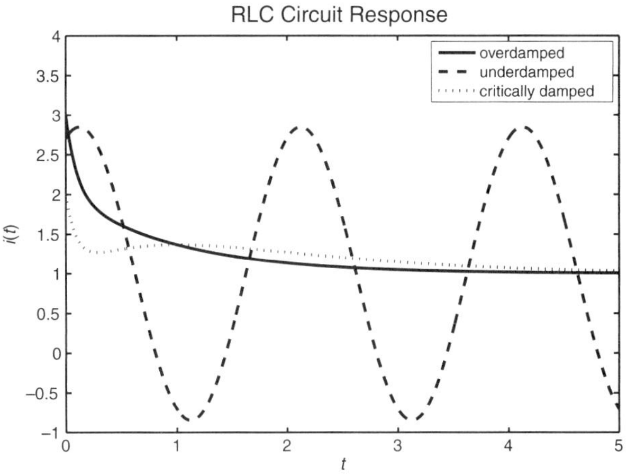

FIGURE 7.18 Examples of the three types of responses for a parallel RLC circuit.

where $\{\alpha, \alpha_1, \alpha_2\}$ and ω_d are functions of $\{a_0, a_1, a_2\}$, $\{c_1, c_2\}$ are constants determined by the initial conditions $\{i(0), i'(0)\}$ of the circuit, and d is the steady-state current (as $t \to \infty$). Note that the $\{c_1, c_2\}$ are not usually the same set of values across the three different cases. In general for such a linear system, superposition is used to rewrite any of the expressions above as follows:

$$i(t) = i_{\mathrm{NR}}(t) + i_{\mathrm{FR}}(t), \tag{7.161}$$

where $i_{\mathrm{NR}}(t)$ is the *natural response* (NR) and $i_{\mathrm{FR}}(t)$ is the *forced response* (FR); they are also called the *zero-input response* and the *zero-state response*, respectively. The natural response is derived by setting the right-hand side of (7.156) equal to zero, resulting in a *homogeneous* DE. The forced response ignores the initial conditions due to the energy-storage devices $\{L, C\}$, and is based on the impulse-response function of the system as described below. Since this parallel circuit has a current source and the inductor eventually functions as a short circuit, we find that $d = i_s$ in all three cases: it is the steady-state current when the exponentials decay to zero. Examples of the three types of responses are shown in Figure 7.18.

We would like to extend this description of a linear system to one for random processes. Consider the following nth-order linear *stochastic DE*:

$$a_n \frac{\mathrm{d}^n}{\mathrm{d}t^n} X(t) + a_{n-1} \frac{\mathrm{d}^{n-1}}{\mathrm{d}t^{n-1}} X(t) + \cdots + a_1 \frac{\mathrm{d}}{\mathrm{d}t} X(t) + a_0 X(t) = Y(t), \tag{7.162}$$

where $\{a_k\}$ are constant coefficients. Since $X(t)$ and $Y(t)$ are random processes, it is not clear how such a system can be solved since there is an ensemble of realizations for $Y(t)$ and $X(t)$. However, based on the previous MS definitions of the derivative, we can determine how the *mean* and *correlation functions* evolve over time assuming that equality holds in the MS sense for each $t \in \mathcal{T}$. Generally, n initial conditions are needed for an nth-order DE: these are provided by the *random variables* $\{X(t_o), X^{(1)}(t_o), \ldots, X^{(n-1)}(t_o)\}$, where the superscripts indicates the order of the differentiation. Generalizing the MS derivative, the nth-order derivative $\mathrm{d}^n X(t)/\,\mathrm{d}t^n \triangleq X^{(n)}(t)$ exists in the MS sense at t if the following partial derivative exists and is continuous at $\{t_1, t_2\} = \{t, t\}$:

$$\left.\frac{\partial^{2n}}{\partial t_1^n \partial t_2^n} R_{XX}(t_1, t_2)\right|_{t_1=t_2=t}. \tag{7.163}$$

Since the DE is linear with constant coefficients, $Y(t)$ can be viewed as the input of a linear system with output $X(t)$. Such a system is expressed as

$$X(t) = X_{\text{NR}}(t) + \int_{t_o}^{t} h(t-\tau)Y(\tau)\mathrm{d}\tau, \tag{7.164}$$

where $h(t)$ is the nonrandom impulse response and $X_{\text{NR}}(t)$ is the natural response of the system that is due to the nonzero initial conditions. The second term on the right-hand side is the forced response which we denote by $X_{\text{FR}}(t)$. The convolution integral is assumed to exist in the MS sense.

We emphasize that the system itself is not random: the coefficients $\{a_m\}$ are assumed to be constant. As a result, since the system is linear and time invariant (LTI), the impulse response does not change if the input is shifted in time. The natural response $X_{\text{NR}}(t)$ can be expressed in terms of the initial conditions $\{X(t_o), X^{(1)}(t_o), \ldots, X^{(n-1)}(t_o)\}$ as follows:

$$X_{\text{NR}}(t) = \sum_{m=0}^{n-1} g_m(t)X^{(m)}(t_o)u(t), \tag{7.165}$$

where $g_m(t)$ is the deterministic response of the system when $X^{(m)}(t_o) = 1$ and the other initial conditions are zero. This type of decomposition is possible because the system is linear and superposition applies. In Chapter 8, we extend these results to provide a *frequency-domain* description of a random process filtered by an LTI system to determine how the autocorrelation function of the output $X(t)$ is related to that of the input $Y(t)$ using *power spectral densities*.

Example 7.23. In the previous circuit example with deterministic signals, the initial conditions at $t_o = 0$ for the overdamped case of the homogeneous DE are related to the coefficients as follows:

$$i(0) = c_1 + c_2, \quad i'(0) = -\alpha_1 c_1 - \alpha_2 c_2 \tag{7.166}$$

so that

$$c_1 = [\alpha_2 i(0) + i'(0)]/(\alpha_2 - \alpha_1), \tag{7.167}$$

$$c_2 = [-\alpha_1 i(0) - i'(0)]/(\alpha_2 - \alpha_1), \tag{7.168}$$

where $d = 0$ in the solution for the homogeneous DE. (Technically, the initial conditions are defined at $t = 0^+$ to include any discontinuities at $t = 0$, such as the current source modeled by the unit-step function.) Observe that if $i(0) = 1$ and $i'(0) = 0$, then $c_1 = \alpha_2/(\alpha_2 - \alpha_1)$ and $c_2 = -\alpha_1/(\alpha_2 - \alpha_1)$; likewise if $i(0) = 0$ and $i'(0) = 1$, then $c_1 = 1/(\alpha_2 - \alpha_1)$ and $c_2 = -1/(\alpha_2 - \alpha_1)$. From these results, we obtain the functions

$$g_0(t) = [\alpha_2 \exp(-\alpha_1 t) - \alpha_1 \exp(-\alpha_2 t)]/(\alpha_2 - \alpha_1), \tag{7.169}$$

$$g_1(t) = [\exp(-\alpha_1 t) - \exp(-\alpha_2 t)]/(\alpha_2 - \alpha_1), \tag{7.170}$$

which are weighted by $i(0)$ and $i'(0)$, respectively, in (7.165) to give the natural response:

$$\begin{aligned} i_{\text{NR}}(t) = {} & [1/(\alpha_2 - \alpha_1)]\big([i(0)\alpha_2 + i'(0)]\exp(-\alpha_1 t) \\ & -[i(0)\alpha_1 + i'(0)]\exp(-\alpha_2 t)\big)\, u(t). \end{aligned} \tag{7.171}$$

In order to complete the expression for the natural response, we need $i(0)$ and $i'(0)$ for this particular circuit. The capacitor and inductor are described by the following voltage/current equations:

$$i_C(t) = C\frac{\mathrm{d}v_C(t)}{\mathrm{d}t}, \quad v_L(t) = L\frac{\mathrm{d}i_L(t)}{\mathrm{d}t}, \tag{7.172}$$

which imply that $v_C(t)$ and $i_L(t)$ cannot change instantaneously. If all voltages and currents are zero for $t < 0$, then $i_L(0) = 0$ and $v_C(0) = 0 \Rightarrow v_L(0) = 0$ from which $i(0) = 0$ and $i'(0) = 0$. Because the initial conditions are zero in this example, the natural response is $i_{\mathrm{NR}}(t) = 0$.

Often when systems are analyzed, the initial conditions are assumed to be zero, and only the forced response in terms of the impulse-response function $h(t)$ is considered. The transfer function $H(s)$ is derived from the Laplace transform of the DE:

$$(a_2s^2 + a_1s + a_0)I(s) = I_s/s \Rightarrow H(s) = \frac{1}{a_2s^2 + a_1s + a_0}. \tag{7.173}$$

Rearranging this expression gives

$$H(s) = \frac{1/a_2}{s^2 + (a_1/a_2)s + a_0/a_2} = \frac{1/a_2}{(s+\alpha_1)(s+\alpha_2)}, \tag{7.174}$$

where $\{\alpha_1, \alpha_2\}$ are the poles of $H(s)$. Since $a_0 = 1$ in this example, $a_2 = 1/\alpha_1\alpha_2$ and $a_1 = (\alpha_1 + \alpha_2)/\alpha_1\alpha_2$. From a partial fraction expansion (PFE), we obtain the inverse Laplace transform:

$$h(t) = \frac{\alpha_1\alpha_2}{\alpha_2 - \alpha_1}[\exp(-\alpha_1 t) - \exp(-\alpha_2 t)]u(t). \tag{7.175}$$

Since the input is a step function, the forced response simplifies to

$$i_{\mathrm{FR}}(t) = I_s\int_0^t h(t-\tau)\mathrm{d}\tau. \tag{7.176}$$

Substituting the impulse response gives

$$\begin{aligned} i_{\mathrm{FR}}(t) &= \frac{\alpha_1\alpha_2 I_s}{\alpha_2 - \alpha_1}\left[\int_0^t \exp(-\alpha_1(t-\tau))\mathrm{d}\tau - \int_0^t \exp(-\alpha_2(t-\tau))\mathrm{d}\tau\right]u(t), \\ &= \frac{\alpha_2 I_s}{\alpha_2 - \alpha_1}[1 - \exp(-\alpha_1 t)]u(t) - \frac{\alpha_1 I_s}{\alpha_2 - \alpha_1}[1 - \exp(-\alpha_2 t)]u(t), \\ &= \left[\frac{\alpha_1 I_s}{\alpha_2 - \alpha_1}\exp(-\alpha_2 t) - \frac{\alpha_2 I_s}{\alpha_2 - \alpha_1}\exp(-\alpha_1 t) + I_s\right]u(t), \end{aligned} \tag{7.177}$$

which is the overall response $i(t)$ since $i_{\mathrm{NR}}(t) = 0$. This expression has the same form as (7.158) so that

$$c_1 = -\frac{\alpha_2 I_s}{\alpha_2 - \alpha_1}, \quad c_2 = \frac{\alpha_1 I_s}{\alpha_2 - \alpha_1}, \quad d = i_s. \tag{7.178}$$

It is clear that in the limit $\lim_{t\to\infty} i_{\mathrm{FR}}(t) = i_s$ as expected.

If the input is a random process in Example 7.23, then the same $\{g_m(t)\}$ would be obtained because they are deterministic. However, the natural response would be random because the initial conditions in (7.165) are random variables. As mentioned earlier, the derivatives of the linear DE and the convolution integral are all defined in the MS sense.

Next, we characterize the mean and autocorrelation function of random process $X(t)$ described by the DE. We will find that the resulting equations are similar to those of the deterministic signals in the previous circuit example. The mean function of the mth term of the DE in (7.162) is

$$\mu_X^{(m)}(t) \triangleq \mathcal{E}\left[\frac{\mathrm{d}^m}{\mathrm{d}t^m}X(t)\right] = \frac{\mathrm{d}^m}{\mathrm{d}t^m}\mathcal{E}[X(t)]. \tag{7.179}$$

Applying this result to the entire DE gives

$$\sum_{m=0}^{n} a_m \mu_X^{(m)}(t) = \mu_Y(t). \tag{7.180}$$

In order to solve this equation, the following initial conditions are needed: $\{\mu_X^{(n-1)}(0), \ldots, \mu_X^{(1)}(0), \mu_X(0)\}$. Interchanging derivatives and expectations, the following cross-correlation function is obtained:

$$\begin{aligned} R_{XY}(t_1, t_2) &= \mathcal{E}[X(t_1)Y(t_2)], \\ &= \mathcal{E}\left[X(t_1)\sum_{m=0}^{n} a_m \frac{\partial^m}{\partial t_2^m}X(t_2)\right] = \sum_{m=0}^{n} a_m \frac{\partial}{\partial t_2}R_{XX}(t_1, t_2) \end{aligned} \tag{7.181}$$

for $t_1, t_2 \geq 0$. This is another linear DE but in terms of correlation functions instead of random processes. Comparing with (7.162), we see that for any fixed t_1, which can be viewed as a parameter, the DE above is a function of t_2. Likewise, we can write

$$R_{YY}(t_1, t_2) = \sum_{m=0}^{n} a_m \frac{\partial^m}{\partial t_1^m} R_{XY}(t_1, t_2), \tag{7.182}$$

which is a function of $t_1 \geq 0$ and is parameterized by $t_2 \geq 0$. The solution $R_{XY}(t_1, t_2)$ of (7.182) can be substituted into (7.181) from which it is possible to solve for $R_{XX}(t_1, t_2)$ given the initial conditions. In Chapter 8, we demonstrate that it is easier to evaluate the autocorrelation functions of the input and output of a linear system using Fourier and Laplace transforms. A preview is given in Example 7.24.

Example 7.24. Consider the following second-order DE:

$$a_2 \frac{\mathrm{d}^2}{\mathrm{d}t^2}X(t) + a_1 \frac{\mathrm{d}}{\mathrm{d}t}X(t) + a_0 X(t) = Y(t) \tag{7.183}$$

for $t > 0$ where $Y(t)$ is a zero-mean process with autocorrelation function $R_{YY}(\tau) = \delta(\tau)$. Assume $\{a_o, a_1, a_2\}$ are chosen such that the system is overdamped. Observe from (7.179) that

$$a_2 \frac{\mathrm{d}^2}{\mathrm{d}t^2}\mu_X(t) + a_1 \frac{\mathrm{d}}{\mathrm{d}t}\mu_X(t) + a_0 \mu_X(t) = \mu_Y(t) = 0 \tag{7.184}$$

is an unforced DE. From (7.158) (with $d = 0$), the solution has the following form:

$$\mu_X(t) = [c_1 \exp(-\alpha_1 t) + c_2 \exp(-\alpha_2 t)]u(t), \tag{7.185}$$

where $\{c_1, c_2\}$ are determined by the initial conditions $\{\mu'_X(0), \mu_X(0)\}$, and $\{\alpha_1, \alpha_2\}$ are the real-valued roots of the characteristic equation:

$$a_2 s^2 + a_1 s + a_0 = 0. \tag{7.186}$$

Applying the unilateral Laplace transform to (7.184) and including the initial conditions gives

$$a_2 s^2 \mu_X(s) - a_2 s \mu_X(0) - a_2 \mu'_X(0) + a_1 s \mu_X(s) - a_1 \mu_X(0) + a_0 \mu_X(s) = 0 \tag{7.187}$$

so that a PFE yields

$$\mu_X(s) = [a_2 \mu'_X(0) + (a_2 s + a_1)\mu_X(0)]/(a_2 s^2 + a_1 s + a_0) = \frac{c_1}{s+\alpha_1} + \frac{c_2}{s+\alpha_2}. \tag{7.188}$$

The inverse Laplace transform gives

$$\mu_X(t) = c_1 \exp(-\alpha_1 t) u(t) + c_2 \exp(-\alpha_2 t) u(t). \tag{7.189}$$

Of course, if the initial conditions are zero, then $\mu_X(t) = 0$ for $t \geq 0$. For nonzero initial conditions, $\mu_X(t) \to 0$ assuming that $\alpha_1, \alpha_2 > 0$. Thus, even though the input $Y(t)$ is a wide-sense stationary process, the output $X(t)$ is not wide-sense stationary because of the time-varying mean. However, since the mean does converge, $X(t)$ is *asymptotically* wide-sense stationary if $R_{XX}(t_1, t_2) \to R_{XX}(\tau)$. This nonstationarity arises because the current source I_s is switched on at time $t = 0$ and energy is stored in $\{L, C\}$ (corresponding to nonzero initial conditions).

For the cross-correlation function, (7.182) yields

$$a_2 \frac{\partial^2}{\partial t_1^2} R_{XY}(t_1, t_2) + a_1 \frac{\partial}{\partial t_1} R_{XY}(t_1, t_2) + a_0 R_{XY}(t_1, t_2) = R_{YY}(t_1, t_2) = \delta(t_1 - t_2) \tag{7.190}$$

for $t_1, t_2 > 0$. As mentioned before, the solution of this DE is a function of t_1 while t_2 is viewed as a parameter. Since the right-hand side is the Dirac delta function, the solution is the impulse-response function $h(t_1 - t_2)$ of the system represented by the DE; it is a function of t_1 and is shifted to the right by t_2 because the delta function is likewise shifted. Applying the Laplace transform (operating only on t_1) gives

$$\begin{aligned} a_2 s^2 R_{XY}(s, t_2) - a_2 s R_{XY}(0, t_2) - a_2 R'_{XY}(0, t_2) + a_1 R_{XY}(s, t_2) \\ - a_1 R_{XY}(0, t_2) + a_0 R_{XY}(s, t_2) = \exp(-s t_2), \end{aligned} \tag{7.191}$$

where the shift property of the Laplace transform has been used to obtain the result on the right-hand side (see Appendix C). Note that since t_2 is assumed to be fixed in this expression, we use the following ordinary derivative notation to specify an initial condition at $t_1 = 0$:

$$R'_{XY}(0, t_2) \triangleq \left. \frac{\partial}{\partial t_1} R_{XY}(t_1, t_2) \right|_{t_1 = 0}. \tag{7.192}$$

Thus

$$R_{XY}(s, t_2) = [\exp(-s t_2) + a_2 R'_{XY}(0, t_2) + (a_2 s + a_1) R_{XY}(0, t_2)]/(a_2 s^2 + a_1 s + a_0). \tag{7.193}$$

Assuming zero initial conditions $R'_{XY}(0, t_2) = R_{XY}(0, t_2) = 0$, the Laplace transform of the forced response is

$$R_{XY}(s, t_2) = \exp(-s t_2) \left(\frac{c_3}{s+\alpha_1} + \frac{c_4}{s+\alpha_2} \right), \tag{7.194}$$

where $\{c_3, c_4\}$ differ from those obtained for the mean $\mu_X(t)$. The inverse Laplace transform gives

$$R_{XY}(t_1, t_2) = \left[c_3 \exp(-\alpha_1 (t_1 - t_2)) + c_4 \exp(-\alpha_2 (t_1 - t_2)) \right] u(t_1 - t_2), \tag{7.195}$$

which is a function of the time difference $t_1 - t_2$. However, note that this is a cross-correlation function and does not imply that $X(t)$ is wide-sense stationary; we need to derive an expression for $R_{XX}(t_1, t_2)$ in order to determine stationarity. For fixed t_2, $R_{XY}(t_1, t_2)$ is a right-sided function of t_1 that is nonzero only for $t_1 > t_2$, and decays exponentially to zero on $\mathcal{R}^+$, as shown in Figure 7.19(a). Alternatively for fixed t_1, the function can be viewed as an exponentially increasing function on $[0, t_1)$. This latter viewpoint is taken when evaluating the DE in (7.181) which becomes

$$a_2 \frac{\partial^2}{\partial t_2^2} R_{XX}(t_1, t_2) + a_1 \frac{\partial}{\partial t_2} R_{XX}(t_1, t_2) + a_0 R_{XX}(t_1, t_2) = R_{XY}(t_1, t_2) \tag{7.196}$$

for this second-order example, where $R_{XY}(t_1, t_2)$ is the solution in (7.195). Since t_1 is viewed as a parameter, this equation describes how $R_{XX}(t_1, t_2)$ varies with t_2 for each value of t_1. Assuming zero initial conditions, the Laplace transform yields the following forced response:

$$R_{XX}(t_1, s) = R_{XY}(t_1, s)/[a_2 s^2 + a_1 s + a_0] = R_{XY}(t_1, s)\left(\frac{c_5}{s + \alpha_1} + \frac{c_6}{s + \alpha_2}\right), \tag{7.197}$$

where $\{c_5, c_6\}$ differ from all previous coefficients, though the poles $\{\alpha_1, \alpha_2\}$ are the same as before (they characterize the DE and do not depend on the initial conditions). The input $R_{XY}(t_1, s)$ on the right-hand side is the Laplace transform of $R_{XY}(t_1, t_2)$ in (7.195) with respect to t_2. Instead of substituting the transform, we perform a convolution of (7.195) with the impulse-response function of the system in (7.197) given by

$$h(t_2) = [c_5 \exp(-\alpha_1 t_2) + c_6 \exp(-\alpha_2 t_2)]u(t_2). \tag{7.198}$$

The convolution is

$$\begin{aligned} R_{XX}(t_1, t_2) &= \int_0^\infty R_{XY}(t_1, \tau) h(t_2 - \tau) d\tau, \\ &= \int_0^{\min(t_1, t_2)} c_3 c_5 \exp(-\alpha_1 (t_1 - \tau) \exp(-\alpha_1 (t_2 - \tau)) d\tau, \\ &+ \int_0^{\min(t_1, t_2)} c_3 c_6 \exp(-\alpha_1 (t_1 - \tau) \exp(-\alpha_2 (t_2 - \tau)) d\tau, \\ &+ \int_0^{\min(t_1, t_2)} c_4 c_5 \exp(-\alpha_2 (t_1 - \tau) \exp(-\alpha_1 (t_2 - \tau)) d\tau, \\ &+ \int_0^{\min(t_1, t_2)} c_4 c_6 \exp(-\alpha_2 (t_1 - \tau) \exp(-\alpha_2 (t_2 - \tau)) d\tau, \end{aligned} \tag{7.199}$$

where the unit-step functions have determined the upper limit for each integral. The first term yields

$$\begin{aligned} & c_3 c_5 \exp(-\alpha_1 (t_1 + t_2)) \int_0^{\min(t_1, t_2)} \exp(2\alpha_1 \tau) d\tau \\ &= \frac{c_3 c_5}{2\alpha_1} \left[\exp(-\alpha_1 |t_1 - t_2|) - \exp(-\alpha_1 (t_1 + t_2))\right], \end{aligned} \tag{7.200}$$

where $\min(t_1, t_2) = (1/2)(t_1 + t_2 - |t_1 - t_2|)$ has been used. The fourth term is obtained from the result above via the substitutions $c_3 c_5 \to c_4 c_6$ and $\alpha_1 \to \alpha_2$. The second and third terms are somewhat more complicated, but have the same basic form as above with two terms: one is a function of $|t_1 - t_2|$ and the other is a weighted

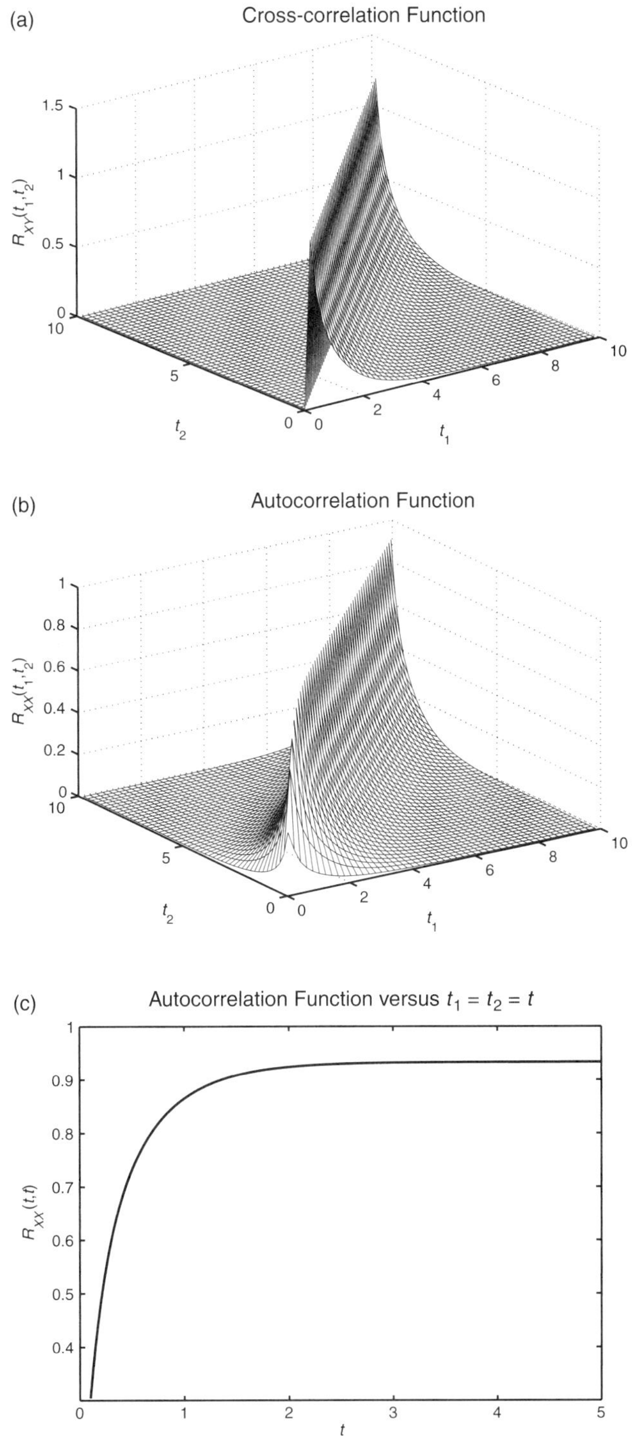

FIGURE 7.19 Correlation functions for Example 7.24. (a) Cross-correlation function $R_{XY}(t_1, t_2)$. (b) Autocorrelation function $R_{XX}(t_1, t_2)$. (c) $R_{XX}(t_1, t_2)$ versus $t_1 = t_2 = t$.

sum of t_1 and t_2 (see Problem 7.22). Thus, we find that although $X(t)$ is nonstationary, it is asymptotically wide-sense stationary as $t_1, t_2 \to \infty$:

$$R_{XX}(t_1, t_2) \to \frac{c_3c_5}{2\alpha_1}\exp(-\alpha_1|t_1 - t_2|) + \frac{c_4c_6}{2\alpha_2}\exp(-\alpha_2|t_1 - t_2|) + \frac{c_3c_6 + c_4c_5}{\alpha_1 + \alpha_2}\exp(-(\alpha_1 + \alpha_2)|t_1 - t_2|/2), \tag{7.201}$$

where only exponentials with $|t_1 - t_2|$ are retained (i.e., those with $t_1 + t_2$ tend to zero). The nonstationarity is evident from the example in Figures 7.19(b) and 7.19(c) where we see that $R_{XX}(t_1, t_2)$ varies when $t_1 = t_2 = t$ for t_1 and t_2 near zero.

Example 7.24 is somewhat complicated because the correlations are functions of two variables $\{t_1, t_2\}$. In Chapter 8, we focus on wide-sense stationary random processes for which the correlations are functions of the single variable τ. This simplifies the analysis above such that the properties of a linear system modeled by a DE are more easily investigated in the frequency domain via the power spectral density (PSD). The analysis is similar to that for deterministic signals, except that here the "signals" are the mean and correlation functions of a random process.

Finally, we mention that the DE in (7.162) can be generalized to include derivatives of the input as follows:

$$\sum_{m=0}^{n_a} a_m \frac{\mathrm{d}^m}{\mathrm{d}t^m} X(t) = \sum_{m=0}^{n_b} b_m \frac{\mathrm{d}^m}{\mathrm{d}t^m} Y(t). \tag{7.202}$$

All the previous results can be extended to this case. As is done for a DE describing deterministic signals, the Laplace transform is used to provide the equivalent representation in the s-domain. For the general DE in (7.202), the transfer function is a ratio of two polynomials in s with constant coefficients: it is a *rational function* of s.

7.8 DIFFERENCE EQUATIONS

Similar results can be derived for a random sequence $Y[k]$ that is processed by a discrete-time system represented by the following linear difference equation:

$$\sum_{m=0}^{n_a} a_m X[k-m] = \sum_{m=0}^{n_b} b_m Y[k-m] \tag{7.203}$$

yielding the output $X[k]$. Taking expectations gives the following equation for the mean:

$$\sum_{m=0}^{n_a} a_m \mu_X[k-m] = \sum_{m=0}^{n_b} b_m \mu_Y[k-m], \tag{7.204}$$

and it is straightforward to generate similar equations for the various correlation functions:

$$\sum_{m=0}^{n_a} a_m R_{XX}[k_1 - m, k_2] = \sum_{m=0}^{n_b} b_m R_{YX}[k_1 - m, k_2], \tag{7.205}$$

$$\sum_{m=0}^{n_a} a_m R_{XY}[k_1 - m, k_2] = \sum_{m=0}^{n_b} b_m R_{YY}[k_1 - m, k_2], \tag{7.206}$$

where the random sequences are not necessarily wide-sense stationary. The first equation is derived by replacing k with k_1 in (7.203), which is then multiplied by $X[k_2]$ and the expectation is taken. The second equation is obtained in the same way except that (7.203) is multiplied by $Y[k_2]$. These difference equations can be analyzed using methods similar to those of the previous example for DEs except that the z-transform is used instead of the Laplace transform. We defer that discussion to Chapter 8 where we focus on wide-sense stationary signals, though a first-order example is considered in Problem 7.24.

7.9 INNOVATIONS AND MEAN-SQUARE PREDICTABILITY

In Sections 7.10 and 7.11, we describe models of a random sequence that are based on the *predictability* of future outcomes given past outcomes. In many engineering problems, we are interested in a random sequence that can be modeled or *decomposed* as follows:

$$Y[k] = X[k] + W[k] \tag{7.207}$$

for $k \in \mathcal{T}$, where $X[k]$ is a random signal of interest and $W[k]$ is a "noise" process. Usually, $W[k]$ is assumed to be an uncorrelated Gaussian sequence with zero mean, in which case it is known as *additive white Gaussian noise* (AWGN). In this model, we only have access to $Y[k]$ which is sometimes called the measurement sequence, and the goal is to estimate or predict $X[k]$. In a communication system, $X[k]$ is the transmitted signal, $W[k]$ is receiver noise (caused by the transmission medium, receiver amplifiers, and so on), and $Y[k]$ is the received signal. In subsequent chapters, we describe various techniques and algorithms for estimating $X[k]$. Here, we focus on rewriting $Y[k]$ in terms of a predicable part and an unpredictable part known as the *innovations*, which is the "new information" in $X[k]$ that cannot be estimated from past outcomes. Although the innovations can be defined for a continuous-time random process, we describe it only for a discrete-time random sequence. If the time-indexed random variables of a random sequence are not independent, then the sequence has a predictable part.

Definition: Predictable Part of a Random Sequence The conditional mean

$$X_{\text{MS}}[k] \triangleq \mathcal{E}[X[k]|X[0], \ldots, X[k-1]] \tag{7.208}$$

is defined to be the *MS predictable part* of $X[k]$.

Random variable $X_{\text{MS}}[k]$ can be viewed as an estimate of $X[k]$ at time instant k based on previous samples $\{X[0], \ldots, X[k-1]\}$. If the process is independent, the estimate of $X[k]$ is just its mean $\mathcal{E}[X[k]]$, which might be known about the random sequence or can usually be estimated. If the process is not independent, then (7.208) is a deterministic *function* of the conditioning variables: for specific values $\{x_0, \ldots, x_{k-1}\}$ of $\{X[0], \ldots, X[k-1]\}$, we can compute a value of $X_{\text{MS}}[k]$ in (7.208).

Recall from Chapter 5 that the conditional mean is the minimum MS error (MMSE) estimator, where in this case the MSE is

$$\mathcal{E}[(X[k] - g(X[0], \ldots, X[k-1]))^2], \tag{7.209}$$

and $g(\cdot)$ is the deterministic function mentioned above used to estimate the current sample from all previous samples. For specific conditioning values, the conditional mean is computed as follows:

$$\begin{aligned}
&\mathcal{E}[X[k]|X[0] = x_0, \ldots, X[k-1] = x_{k-1}], \\
&\quad = \int_{-\infty}^{\infty} x_k f_{X[k]|X[0],\ldots,X[k-1]}(x_k|x_0, \ldots, x_{k-1}) \mathrm{d}x_k = g(x_0, \ldots, x_{k-1}).
\end{aligned} \tag{7.210}$$

In general, the joint pdf associated with k time instants of the random sequence is needed. If the sequence happens to be Markov, the conditional mean simplifies to $\mathcal{E}[X[k]|X[0], \ldots, X[k-1]] = \mathcal{E}[X[k]|X[k-1]]$ and

$$\begin{aligned}\mathcal{E}[X[k]|X[k-1]=x_{k-1}] &= \int_{-\infty}^{\infty} x_k f_{X[k]|X[k-1]}(x_k|x_{k-1})\mathrm{d}x_k, \\ &= \frac{1}{f_{X[k-1]}(x_{k-1})}\int_{-\infty}^{\infty} x_k f_{X[k],X[k-1]}(x_k, x_{k-1})\mathrm{d}x_k \end{aligned} \tag{7.211}$$

so that only the bivariate pdf of the random sequence is needed at time instants $k-1$ and k. This is intuitive: for a Markov process, only information from the most recent sample $X[k-1]$ is needed, which means the prediction requires only the joint pdf of $X[k-1]$ and $X[k]$. Alternatively, if the sequence is not Markov, we can still use the expression in (7.211), but it would be called a *one-step predictor*: it is an estimate of the current sample based only on using the most recent sample, and not all previous samples as in (7.208).

Example 7.25. Let $X[k]$ be a Gaussian sequence with constant parameters $\{\mu_X, \sigma_X\}$, and assume it is Markov so that we need examine only its bivariate pdf:

$$\begin{aligned} f_{X[k],X[k-1]}(x_k, x_{k-1}) &= \frac{1}{2\pi\sigma_X^2\sqrt{1-\rho^2}}\exp\left(-\frac{1}{2\sigma_X^2(1-\rho^2)}\right. \\ &\cdot\left.\left[(x_k-\mu_X)^2 - 2\rho(x_k-\mu_X)(x_{k-1}-\mu_X) + (x_{k-1}-\mu_X)^2\right]\right), \end{aligned} \tag{7.212}$$

where $\rho \triangleq \mathcal{E}[(X[k]-\mu_X)(X[k-1]-\mu_X)]/\sigma_X^2$ is its correlation coefficient. Dividing this joint pdf by the marginal pdf at time $k-1$ gives the following conditional pdf:

$$\begin{aligned} f_{X[k]|X[k-1]}(x_k|x_{k-1}) &= \frac{f_{X[k],X[k-1]}(x_k, x_{k-1})}{f_{X[k-1]}(x_{k-1})} = \frac{1}{\sqrt{2\pi}\sigma_X}\exp\left(-\frac{1}{2\sigma_X^2(1-\rho^2)}\right. \\ &\cdot\left.\left[(x_k-\mu_X)^2 - 2\rho(x_k-\mu_X)(x_{k-1}-\mu_X) + (x_{k-1}-\mu_X)^2\right] + \frac{1}{2\sigma_X^2}(x_{k-1}-\mu_X)^2\right), \\ &= \frac{1}{\sqrt{2\pi\sigma_X^2(1-\rho^2)}}\exp\left(-\frac{1}{2\sigma_X^2(1-\rho^2)}\left[x_k-(1+\rho)\mu_X-\rho x_{k-1}\right]^2\right), \end{aligned} \tag{7.213}$$

where the last result has been obtained by completing the square. Since the *conditional* pdf is Gaussian with mean $\rho x_{k-1} + (1+\rho)\mu_X$, the conditional mean is

$$X_{\mathrm{MS}}[k] = \rho X[k-1] + (1+\rho)\mu_X, \tag{7.214}$$

which is the predictable part of the *Gauss–Markov sequence*. Observe that $X_{\mathrm{MS}}[k]$ is a *linear function* of the most recent sample $X[k-1]$; it is a characteristic of a Gaussian sequence that the conditional mean estimator is linear. (Actually, the result above is *affine* because it includes the additive constant term with μ_X.) If the process is uncorrelated with $\rho = 0$, the predictor is a constant given by the mean of the sequence.

Returning to the general decomposition problem where the predictor is based on all past samples $\{X[0], \ldots, X[k-1]\}$, if the sequence is independent then

$$X_{\mathrm{MS}}[k] = \mathcal{E}[X[k]] = \mu_X. \tag{7.215}$$

The conditional mean estimate of $X[k]$ for this case is its mean as mentioned above: $\{X[0], \ldots, X[k-1]\}$ provide no information about $X[k]$. An uncorrelated Gaussian sequence (with $\rho = 0$) is also an independent

TABLE 7.4 Expectation Properties of Random Sequence $X[k]$

Property	Expectation Result
Uncorrelated	$\mathcal{E}[X[k-1]X[k]] = \mathcal{E}[X[k-1]]\mathcal{E}[X[k]]$
Orthogonal	$\mathcal{E}[X[k-1]X[k]] = 0$
Mean-square unpredictable	$\mathcal{E}[X[k]\|X[k-1]] = \mathcal{E}[X[k]]$
First-order Markov	$\mathcal{E}[X[k]\|X[k-1], X[k-2]] = \mathcal{E}[X[k]\|X[k-1]]$
Martingale	$\mathcal{E}[X[k]\|X[k-1]] = X[k-1]$

sequence. However, for a general random sequence, independence is not required to obtain the result in (7.215); instead, we can state the following condition.

Definition: Mean-Square Unpredictable Random sequence $X[k]$ is *MS unpredictable* if

$$\mathcal{E}[X[k]|X[0], \ldots, X[k-1]] = \mathcal{E}[X[k]]. \tag{7.216}$$

MS unpredictability is a weaker condition than independence, but is stronger than uncorrelated. This is readily shown by examining the joint expectation $\mathcal{E}[X[0], \ldots, X[k-1], X[k]]$. If $X[k]$ and $\{X[0], \ldots, X[k-1]\}$ are uncorrelated (or independent), then the expectation splits:

$$\mathcal{E}[X[0], \ldots, X[k-1], X[k]] = \mathcal{E}[X[0], \ldots, X[k-1]]\mathcal{E}[X[k]]. \tag{7.217}$$

This result is also obtained using (7.216) as follows:

$$\begin{aligned}\mathcal{E}[X[0], \ldots, X[k-1], X[k]] &= \mathcal{E}[X[0], \ldots, X[k-1], \mathcal{E}[X[k]|X[0], \ldots, X[k-1]]],\\ &= \mathcal{E}[X[0], \ldots, X[k-1], \mathcal{E}[X[k]]],\\ &= \mathcal{E}[X[0], \ldots, X[k-1]]\mathcal{E}[X[k]].\end{aligned} \tag{7.218}$$

The first equality is always true (nested expectations) and the second equality is due to the condition in (7.216). In the last expression, $\mathcal{E}[X(k)]$ is factored out because the inner expectation is a constant: it does not depend on any of the quantities of the outer expectation. From the last result, (7.216) implies that $X[k]$ and $\{X[0], \ldots, X[k-1]\}$ are uncorrelated; in general, uncorrelatedness does not imply (7.216). Independence always implies (7.216), but the reverse is generally not true. For convenience, we summarize several properties of expectation for random sequence $X[k]$ in Table 7.4.

Given that $X_{\text{MS}}[k]$ is the predictable part of $X[k]$ in the conditional-mean sense, this implies there is a nonpredictable part called the innovations.

Definition: Innovations The *innovations* $V[k]$ of random sequence $X[k]$ is

$$V[k] \triangleq X[k] - X_{\text{MS}}[k]. \tag{7.219}$$

It is the *new information* in $X[k]$ that is not MS predictable from previous samples of the random sequence.

Observe that the expected value of $V[k]$ conditioned on previous samples is zero for $k > 0$:

$$\begin{aligned}\mathcal{E}[V[k]|X[0], \ldots, X[k-1]] &= \mathcal{E}[X[k]|X[0], \ldots, X[k-1]]\\ &\quad - \mathcal{E}[X_{\text{MS}}[k]|X[0], \ldots, X[k-1]],\\ &= X_{\text{MS}}[k] - X_{\text{MS}}[k] = 0.\end{aligned} \tag{7.220}$$

Since $X_{\mathrm{MS}}[k]$ is already conditioned on previous samples, a second expectation with the same conditioning gives $X_{\mathrm{MS}}[k]$ again because it is basically a constant under the second conditional expectation. The mean of $V[k]$ for $k > 0$ is also zero:

$$\begin{aligned}\mathcal{E}[V[k]] &= \mathcal{E}[X[k]] - \mathcal{E}[X_{\mathrm{MS}}[k]], \\ &= \mathcal{E}[X[k]] - \mathcal{E}[\mathcal{E}[X[k]|X[0], \ldots, X[k-1]]], \\ &= \mathcal{E}[X[k]] - \mathcal{E}[X[k]] = 0,\end{aligned} \tag{7.221}$$

where again nested expectations have been used. Thus, from (7.220) and (7.221):

$$\mathcal{E}[V[k]|X[0], \ldots, X[k-1]] = \mathcal{E}[V[k]] = 0, \tag{7.222}$$

so that $V[k]$ is the MS *unpredictable part* of $X[k]$. Moreover, it can be shown that $V[k]$ itself is MS unpredictable.

Theorem 7.18. The innovations $V[k]$ of random sequence $X[k]$ is MS unpredictable:

$$\mathcal{E}[V[k]|V[0], \ldots, V[k-1]] = \mathcal{E}[V[k]]. \tag{7.223}$$

Proof. This result is verified by proving the following:

$$\mathcal{E}[V[k]|V[0], \ldots, V[k-1]] = \mathcal{E}[V[k]|X[0], \ldots, X[k-1]] \tag{7.224}$$

because we have already shown that the right-hand side is $\mathcal{E}[V[k]]$. From the definition of the innovations, observe that

$$\begin{aligned}V[k-1] &= X[k-1] - \mathcal{E}[X[k-1]|X[0], \ldots, X[k-2]], \\ &\vdots \\ V[1] &= X[1] - \mathcal{E}[X[1]|X[0]], \\ V[0] &= X[0].\end{aligned} \tag{7.225}$$

These expressions follow because of the manner in which the innovations sequence is defined. Since the innovations can be expressed as a function of $\{X[0], \ldots, X[k-1]\}$, the smoothing property of conditional expectation (see Chapter 5) can be used to write

$$\begin{aligned}\mathcal{E}[V[k]|X[0], \ldots, X[k-1]] &= \mathcal{E}[\mathcal{E}[V[k]|X[0], \ldots, X[k-1]]|V[0], \ldots, V[k-1]], \\ &= \mathcal{E}[V[k]|V[0], \ldots, V[k-1]].\end{aligned} \tag{7.226}$$

The first equality holds because it is a nested expectation. The second equality follows because the $\{V[0], \ldots, V[k-1]\}$ are obtained from $\{X[0], \ldots, X[k-1]\}$, which means they have a coarser σ-field. Thus, the conditioning on $\{X[0], \ldots, X[k-1]\}$ can be ignored, which completes the proof. □

Theorem 7.19. The innovations $V[k]$ and the MS predictable part $X_{\mathrm{MS}}[k]$ are orthogonal:

$$\mathcal{E}[X_{\mathrm{MS}}[k]V[k]] = 0. \tag{7.227}$$

Proof. Substituting the definition of $V[k]$ gives

$$\mathcal{E}[X_{\mathrm{MS}}[k](X[k] - X_{\mathrm{MS}}[k])] = \mathcal{E}[X_{\mathrm{MS}}[k]X[k]] - \mathcal{E}[X_{\mathrm{MS}}^2[k]]. \tag{7.228}$$

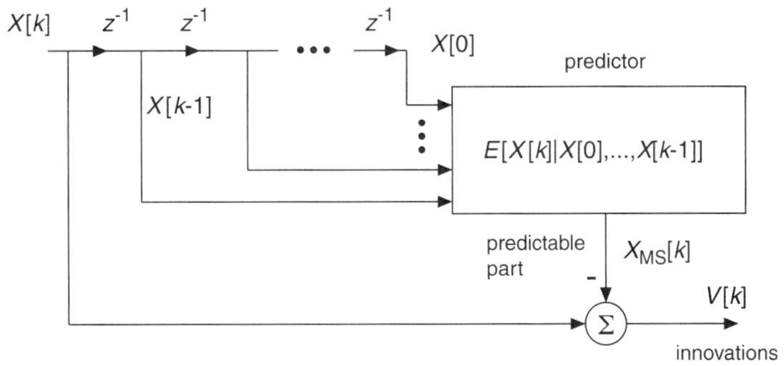

FIGURE 7.20 System configuration for generating the predictable part $X_{\text{MS}}[k]$ and the innovations $V[k]$ of random sequence $X[k]$. System $\mathcal{E}[X[k]|X[k-1], \ldots, X[0]]$ is a function of $\{X[k-1], \ldots, X[0]\}$ (and does not include $X[k]$). The system may be nonlinear, but it is time-invariant because the sequence is assumed to be wide-sense stationary.

Observe that

$$\begin{aligned}\mathcal{E}[X_{\text{MS}}^2[k]] &= \mathcal{E}[X_{\text{MS}}[k]\mathcal{E}[X[k]|X[0], \ldots, X[k-1]]], \\ &= \mathcal{E}[\mathcal{E}[X_{\text{MS}}[k]X[k]|X[0], \ldots, X[k-1]]] = \mathcal{E}[X_{\text{MS}}[k]X[k]],\end{aligned} \tag{7.229}$$

which gives zero in (7.227). The second equality follows because $X_{\text{MS}}[k]$ is a function of $\{X[0], \ldots, X[k-1]\}$ and can be brought inside the expectation where it is a constant with respect to that conditioning (called the *factorization* property in Chapter 5). The last expression is due to the property of nested expectations. □

From the previous discussion, we see that a random sequence can be decomposed into the sum of a predictable part $X_{\text{MS}}[k]$ (in the MS sense) and the innovations $V[k]$ (the MS unpredictable part) as follows:

$$X[k] = X_{\text{MS}}[k] + V[k]. \tag{7.230}$$

In this sense, a random sequence consists of an underlying "signal" and an additive "noise" sequence. This model implies that we can express either component of the original sequence as the output of a system (which is not necessarily linear). Figure 7.20 shows a block diagram of how the predictable part and the innovations are generated from $X[k]$.

Example 7.26. For the Gauss–Markov sequence in Example 7.25, the innovations sequence is

$$V[k] = X[k] - \rho X[k-1] - (1+\rho)\mu_X, \tag{7.231}$$

which is an intuitive result: the innovations sequence is obtained by subtracting from $X[k]$ the part that is predictable from $X[k-1]$. The "strength" of that prediction is determined by the size of the correlation coefficient ρ. An expression involving the mean is also subtracted because the mean is known and "predictable." In particular, note that if $\rho = 0$, then the process has no predictable part and the innovations sequence is equal to the original sequence with its mean removed. Figure 7.21 shows discrete-time linear filter implementations for generating $V[k]$ from $X[k]$ and vice versa (where, for convenience, $\mu_X = 0$ has been assumed). Observe that the filter generating $V[k]$ has a finite-impulse response (FIR), whereas that generating $X[k]$ has an infinite impulse response (IIR). The IIR filter has a pole at ρ and is stable because $|\rho| \leq 1$ (technically, the filter is only marginally stable if $|\rho| = 1$).

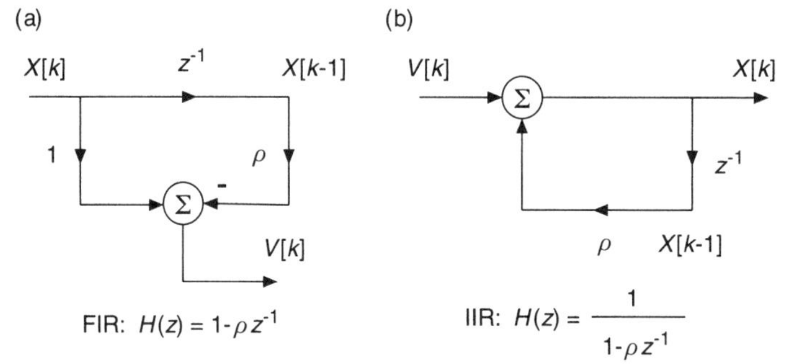

FIGURE 7.21 Filters for generating the sequences in Example 7.26 with $\mu_X = 0$. (a) Innovations sequence $V[k]$. (b) Original sequence $X[k]$.

Example realizations for the Gauss–Markov process and its innovations are shown in Figure 7.22. The sequence was generated as follows:

$$X[k] = \rho X[k-1] + W[k], \tag{7.232}$$

where $W[k]$ is an uncorrelated Gaussian sequence with zero mean and unit variance. This expression can be written in nonrecursive form as follows:

$$X[k] = \sum_{n=0}^{k} \rho^n W[k-n]. \tag{7.233}$$

Thus $\mu_X = 0$ and

$$\begin{aligned}\sigma_X^2 &= \sum_{n=0}^{k} \operatorname{var}[\rho^n W[k-n]] = \sum_{n=0}^{k} \rho^{2n} \operatorname{var}[W[k-n]], \\ &= \sum_{n=0}^{k} \rho^{2n} = [1-\rho^{2(k+1)}]/(1-\rho^2).\end{aligned} \tag{7.234}$$

Since the sequence starts at $k = 0$, the variance is time varying so that $X[k]$ is not stationary. However, it is asymptotically stationary as $k \to \infty$ with asymptotic variance $\sigma_X^2 = 1/(1-\rho^2)$. For practical purposes in the simulation, we can assume that the output is stationary after about three to five time constants of the system, which for this single-pole filter is approximately $\tau \approx 1/(1-\rho)$. The variance above is $\sigma_X^2 \approx 50.25$ (often, it is convenient to scale $W[k]$ so that $X[k]$ has unit variance). Observe that the one-step predicted sequence in the figure is essentially indistinguishable from the original sequence, which is to be expected because $X_{\text{MS}}[k] = 0.99X[k-1]$ is just a scaled and shifted version of $X[k]$. The innovations sequence, on the other hand, has a noisy appearance due to the MS unpredictability of $V[k]$.

7.10 DOOB–MEYER DECOMPOSITION

Next, we describe another decomposition of random sequence $X[k]$ that turns out to be a reformulation of the previous innovations decomposition. This decomposition is also additive as in (7.230), but the component sequences have properties differing from those of $X_{\text{MS}}[k]$ and $V[k]$, although they are related to each other

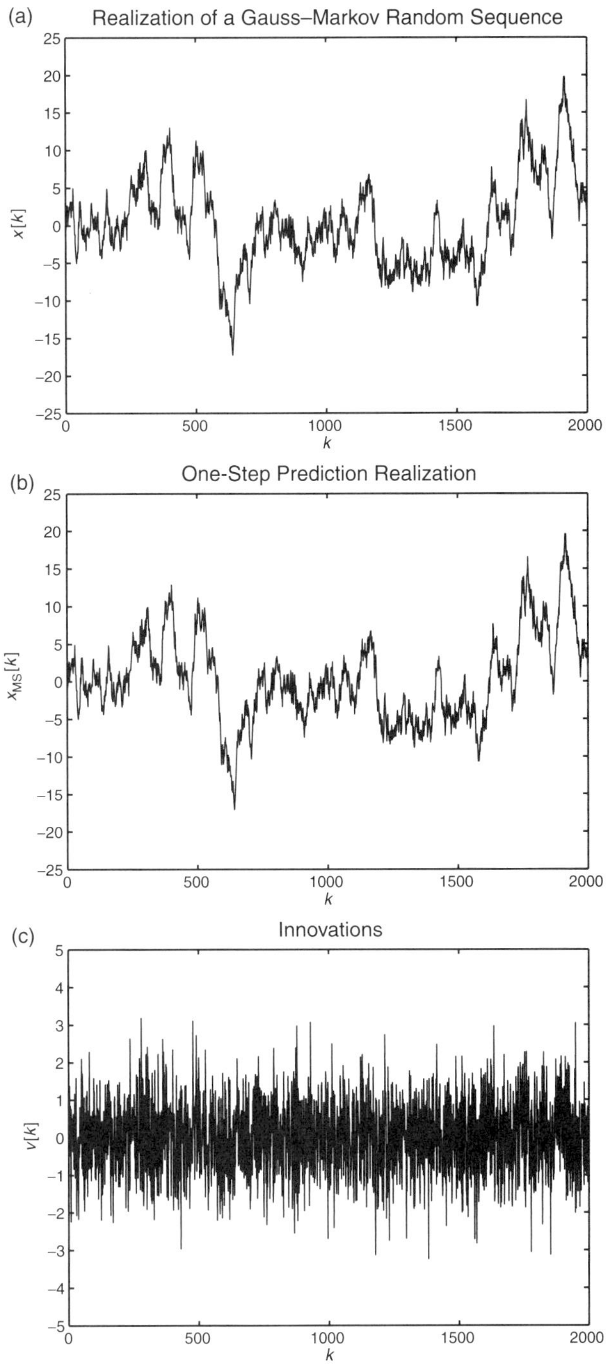

FIGURE 7.22 Realizations for Example 7.26 with $\rho = 0.99$ and $\mu_X = 0$. (a) Gauss-Markov sequence $X[k]$. (b) MS predictable part $X_{\text{MS}}[k]$. (c) Innovations $V[k]$. (Note the change in the amplitude scale for the innovations.)

as demonstrated later. By examining the *difference* $X[k] - X[k-1]$, we find that one component of the decomposed $X[k]$ is a *martingale* sequence.

Theorem 7.20 (Doob–Meyer). Random sequence $X[k]$ with $\mathcal{E}[|X[k]|] < \infty$ for $k \in \mathcal{Z}^+$ can be expressed uniquely as the sum of two sequences as follows:

$$X[k] = X_{\mathrm{DM}}[k] + E[k], \tag{7.235}$$

where $X_{\mathrm{DM}}[k]$ is an MS predictable sequence and $E[k]$ is a martingale.

Note that the Doob-Meyer (DM) predictable part $X_{\mathrm{DM}}[k]$ differs from $X_{\mathrm{MS}}[k]$ of the innovations-based decomposition. (Although we refer to (7.235) as the DM decomposition, it is more precisely known as the Doob decomposition for discrete-time sequences. Doob-Meyer is the corresponding decomposition for continuous-time processes.)

Proof. (Larson and Shubert, 1979). We begin by using the following notation to define a sequence of *increments*:

$$\Delta X[k] \triangleq X[k] - X[k-1] \tag{7.236}$$

The DM predictable part of the increment sequence given $\{X[0], \ldots, X[k-1]\}$ is the following conditional mean:

$$\Delta X_{\mathrm{DM}}[k] \triangleq \mathcal{E}[\Delta X[k]|X[0], \ldots, X[k-1]] \tag{7.237}$$

from which an error sequence is defined:

$$\Delta E[k] \triangleq \Delta X[k] - \Delta X_{\mathrm{DM}}[k]. \tag{7.238}$$

Note the similarity between $\{X_{\mathrm{MS}}[k], V[k]\}$ of the innovations decomposition and $\{\Delta X_{\mathrm{DM}}[k], \Delta E[k]\}$ of the DM decomposition. Because the DM predictable part $\Delta X_{\mathrm{DM}}[k]$ is defined for $\Delta X[k]$ (instead of $X[k]$), $\Delta E[k]$ is not the same as the innovations (later we show that it is a sum of all the innovations).

Observe that

$$X[k] = \sum_{m=0}^{k-1} \Delta X[m] \tag{7.239}$$

because all terms in the increments cancel except for $X[k]$ when $m = k-1$, and $X[-1] = 0$ has been assumed. Similarly, the following sequences based on increments can be defined:

$$X_{\mathrm{DM}}[k] \triangleq \sum_{m=0}^{k-1} \Delta X_{\mathrm{DM}}[m], \quad E[k] \triangleq \sum_{m=0}^{k-1} \Delta E[m] \tag{7.240}$$

so that

$$X[k] = X_{\mathrm{DM}}[k] + E[k], \tag{7.241}$$

which is the desired decomposition. We emphasize that $X_{\mathrm{DM}}[k]$ is the predictable part of $X[k]$ defined by the conditional mean in (7.237). It is assumed that $X_{\mathrm{DM}}[0] = 0$ because $X[k] = 0$ for $k < 0$, and thus there are no prior samples that can be used to predict $X[0]$. As a result, $E[0] = X[0]$ and $E[k] = 0$ for $k < 0$.

The expression in (7.241) is the DM decomposition: $X_{\text{DM}}[k]$ is the MS predictable part of the sequence (defined in terms of $\Delta X(k)$), and $E[k]$ is a martingale sequence which we now demonstrate. Observe from (7.238) that

$$\begin{aligned}\mathcal{E}[\Delta E[k]|X[0], \ldots, X[k-1]] &= \mathcal{E}[\Delta X[k] - \Delta X_{\text{DM}}[k]|X[0], \ldots, X[k-1]], \\ &= \Delta X_{\text{DM}}[k] - \Delta X_{\text{DM}}[k] = 0,\end{aligned} \tag{7.242}$$

where we have used the fact that by definition $\Delta X_{\text{DM}}[k]$ is already conditioned on $\{X[0], \ldots, X[k-1]\}$ and is not changed by this expectation. Since $\{E[0], \ldots, E[k-1]\}$ are functions of $\{X[0], \ldots, X[k-1]\}$, they are based on a courser σ-field, and thus from the smoothing property of expectation used previously:

$$\begin{aligned}\mathcal{E}[\Delta E[k]|X[0], \ldots, X[k-1]] &= \mathcal{E}[\mathcal{E}[\Delta E[k]|X[0], \ldots, X[k-1]]|E[0], \ldots, E[k-1]]], \\ &= \mathcal{E}[\Delta E[k]|E[0], \ldots, E[k-1]].\end{aligned} \tag{7.243}$$

Substituting the terms in the increment yields

$$\begin{aligned}\mathcal{E}[\Delta E[k]|E[0], \ldots, E[k-1]] &= \mathcal{E}[E[k] - E[k-1]|E[0], \ldots, E[k-1]], \\ &= \mathcal{E}[E[k]|E[0], \ldots, E[k-1]] - E[k-1].\end{aligned} \tag{7.244}$$

Since the left-hand side is zero from (7.242) and (7.243), we finally have

$$\mathcal{E}[E[k]|E[0], \ldots, E[k-1]] = E[k-1] \tag{7.245}$$

demonstrating that $E[k]$ is a martingale. □

Next, we derive how the innovations and DM decompositions are related. From the definitions of $\Delta X_{\text{DM}}[k]$ and $\Delta E[k]$, we find that

$$\begin{aligned}\Delta X_{\text{DM}}[k] &= \mathcal{E}[X[k]|X[0], \ldots, X[k-1]] - \mathcal{E}[X[k-1]|X[0], \ldots, X[k-1]], \\ &= X_{\text{MS}}[k] - X[k-1],\end{aligned} \tag{7.246}$$

where $X_{\text{MS}}[k]$ is the MS predictable part in the innovations decomposition. Combining this result with (7.238) gives

$$\begin{aligned}\Delta E[k] &= X[k] - X[k-1] - (X_{\text{MS}}[k] - X[k-1]), \\ &= X[k] - X_{\text{MS}}[k] = V[k],\end{aligned} \tag{7.247}$$

which is the innovations. As a result, the martingale component of the DM decomposition is the sum of all the innovations samples:

$$E[k] = \sum_{m=0}^{k} \Delta E[m] = \sum_{m=0}^{k} V[m]. \tag{7.248}$$

The various components for the two types of sequence decompositions are summarized in Figure 7.23.

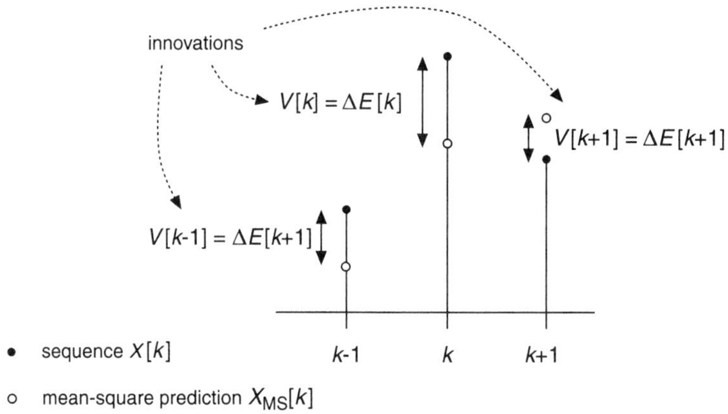

FIGURE 7.23 Illustration showing the MS predicted samples $X_{\mathrm{MS}}[k]$, the innovations $V[k]$, and the martingale component $E[k]$ for the DM decomposition of $X[k]$.

Since $E[k]$ is the sum of all the innovations, we can show that it is a martingale using a more direct approach than in the previous proof. Substituting the sum of innovations:

$$\mathcal{E}[E[k]|E[0],\ldots,E[k-1]] = \sum_{m=0}^{k} \mathcal{E}[V[m]|E[0],\ldots,E[k-1]], \tag{7.249}$$

and separating the kth term gives

$$\begin{aligned}\mathcal{E}[E[k]|E[0],\ldots,E[k-1]] &= \sum_{m=0}^{k-1} \mathcal{E}[V[m]|E[0],\ldots,E[k-1]] \\ &\quad + \mathcal{E}[V[k]|E[0],\ldots,E[k-1]] = E[k-1].\end{aligned} \tag{7.250}$$

The first term on the right-hand side is the desired result $E[k-1]$ because it is included in the conditioning; the second term is zero because $\{E[0],\ldots,E[k-1]\}$ depend only on the innovations samples $\{V[0],\ldots,V[k-1]\}$ up to time $k-1$, which are MS unpredictable and have zero mean.

Finally, using the previous results we show how to transition from the DM decomposition to the innovations decomposition:

$$\begin{aligned}X[k] &= X_{\mathrm{DM}}[k] + E[k], \\ &= \sum_{m=0}^{k} \Delta X_{\mathrm{DM}}[m] + \sum_{m=0}^{k} V[m], \\ &= \sum_{m=0}^{k} X_{\mathrm{MS}}[m] - \sum_{m=0}^{k-1} X[m] + \sum_{m=0}^{k} V[m].\end{aligned} \tag{7.251}$$

Separating the $m = k$ terms from the first and third sums gives

$$X[k] = X_{\mathrm{MS}}[k] + V[k] + \sum_{m=0}^{k-1}(X_{\mathrm{MS}}[m] - X[m] + V[m]), \tag{7.252}$$

which is the desired result because $V[m] \triangleq X[m] - X_{\mathrm{MS}}[m]$ so that the last sum is zero.

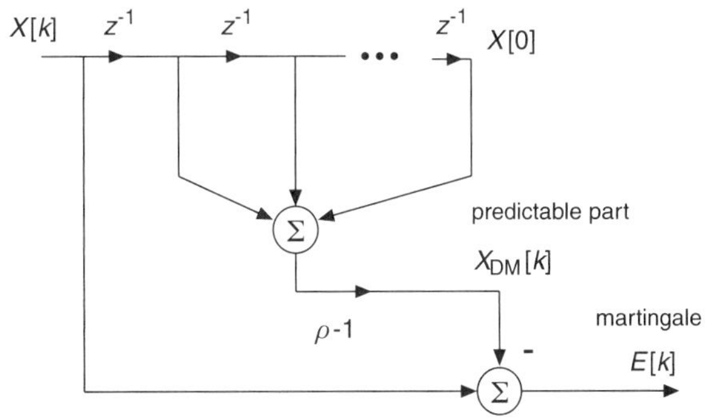

FIGURE 7.24 Filters for generating the sequences in Example 7.27 with $\mu_X = 0$. (a) Predictable part $X_{\mathrm{DM}}[k]$. (b) Martingale sequence $E[k]$.

Example 7.27. Continuing with Example 7.26, we evaluate the DM decomposition for the Gauss–Markov sequence when $\mu_X = 0$. The predictable part is

$$\begin{aligned} X_{\mathrm{DM}}[k] &= \sum_{m=0}^{k} X_{\mathrm{MS}}[m] - \sum_{m=0}^{k-1} X[m], \\ &= \sum_{m=1}^{k} \rho X[m-1] - \sum_{m=0}^{k-1} X[m] = \sum_{m=0}^{k-1} (\rho - 1) X[m], \end{aligned} \tag{7.253}$$

and the martingale sequence is

$$E[k] = X[k] - \sum_{m=0}^{k-1} (\rho - 1) X[m]. \tag{7.254}$$

Alternatively, by summing over the innovations in (7.231) (with $\mu_X = 0$):

$$\begin{aligned} E[k] &= \sum_{m=0}^{k} (X[m] - \rho X[m-1]), \\ &= X[k] + \sum_{m=0}^{k-1} X[m] - \rho \sum_{m=1}^{k} X[m-1]. \end{aligned} \tag{7.255}$$

Changing variables in the last sum gives the result in (7.254). Figure 7.24 provides a block diagram of the DM decomposition showing how the DM predictable part and the martingale component are generated. Example realizations of $X_{\mathrm{DM}}[k]$ and $E[k]$ are given in Figure 7.25 for comparison with the original sequence and its innovations in Figure 7.20.

7.11 KARHUNEN–LOÈVE EXPANSION

In the final section of this chapter, we return to continuous-time random processes and describe a decomposition that is a weighted sum of orthonormal functions known as the Karhunen–Loève expansion (KLE).

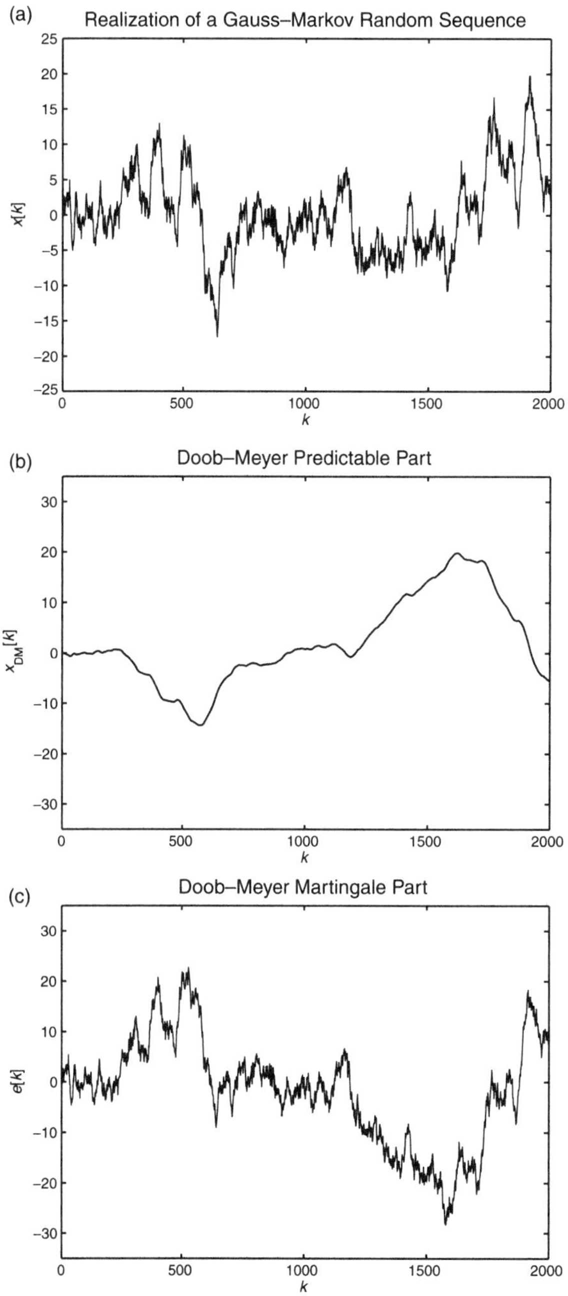

FIGURE 7.25 Realizations for Example 7.27 with $\rho = 0.99$ and $\mu_X = 0$. (a) Gauss–Markov sequence $X[k]$. (b) MS predictable part $X_{\text{DM}}[k]$. (c) Martingale $E[k]$.

Theorem 7.21 (Karhunen-Loève). Let $X(t)$ be a zero-mean random process with autocorrelation function $R_{XX}(t_1, t_2)$ defined on the finite interval $\mathcal{T} = [0, T]$. It can be decomposed as follows:

$$X(t) = \sum_{n=1}^{\infty} X_n \phi_n(t), \tag{7.256}$$

where $\{X_n\}$ are random variables generated as follows:

$$X_n = \int_0^T X(t)\phi_n(t)\mathrm{d}t, \tag{7.257}$$

and $\{\phi_n(t)\}$ are orthonormal eigenfunctions that satisfy

$$\int_0^T R_{XX}(t_1, t_2)\phi_n(t_2)\mathrm{d}t_2 = \lambda_n \phi_n(t_1), \quad t_1 \in \mathcal{T} \tag{7.258}$$

with eigenvalues $\{\lambda_n\}$.

The sum in (7.256) and the integral in (7.257) are defined in the MS sense, whereas the last integral is over the product of deterministic functions. The autocorrelation function in (7.258) is also called the *kernel* of the integral equation.

Proof. Let the sum in (7.256) be finite with N elements so that it is a sequence. For MS convergence, we require that

$$\lim_{N\to\infty} \mathcal{E}\left[\left(X(t) - \sum_{n=1}^{N} X_n\phi_n(t)\right)^2\right] = 0. \tag{7.259}$$

Expanding the expectation yields

$$\begin{aligned} \mathcal{E}\left[\left(X(t) - \sum_{n=1}^{N} X_n\phi_n(t)\right)^2\right] &= \mathcal{E}[X^2(t)] - 2\sum_{n=1}^{N}\mathcal{E}[X(t)X_n]\phi_n(t) \\ &+ \sum_{n=1}^{N}\sum_{m=1}^{N}\mathcal{E}[X_nX_m]\phi_n(t)\phi_m(t). \end{aligned} \tag{7.260}$$

Substituting (7.257) into the second term on the right-hand side gives

$$\begin{aligned} \mathcal{E}[X(t)X_n] &= \int_0^T \mathcal{E}[X(t)X(t_2)]\phi_n(t_2)\mathrm{d}t_2, \\ &= \int_0^T R_{XX}(t, t_2)\phi_n(t_2)\mathrm{d}t_2 = \lambda_n\phi_n(t), \end{aligned} \tag{7.261}$$

where (7.258) has been used. Likewise for the last term in (7.260):

$$
\begin{aligned}
\mathcal{E}[X_n X_m] &= \int_0^T \int_0^T \mathcal{E}[X(t_1)X(t_2)]\phi_n(t_1)\phi_m(t_2)\mathrm{d}t_1\mathrm{d}t_2, \\
&= \int_0^T \int_0^T R_{XX}(t_1, t_2)\phi_n(t_1)\phi_m(t_2)\mathrm{d}t_1\mathrm{d}t_2, \\
&= \int_0^T \lambda_m \phi_m(t_1)\phi_n(t_1)\mathrm{d}t_1 = \lambda_m \delta[m-n],
\end{aligned} \tag{7.262}
$$

where we have used the fact that the eigenfunctions are orthonormal:

$$
\int_0^T \phi_n(t)\phi_m(t)\mathrm{d}t = \delta[m-n], \tag{7.263}
$$

which is the Kronecker delta function. This demonstrates that the random variables in the KLE are orthogonal. Combining all the results above, (7.260) becomes

$$
\begin{aligned}
\mathcal{E}\left[\left(X(t) - \sum_{n=1}^{N} X_n\phi_n(t)\right)^2\right] &= R_{XX}(t,t) - 2\sum_{n=1}^{N}\lambda_n\phi_n^2(t) \\
&\quad + \sum_{n=1}^{N}\sum_{m=1}^{N}\lambda_m\delta[m-n]\phi_n(t)\phi_m(t), \\
&= R_{XX}(t,t) - \sum_{n=1}^{N}\lambda_n\phi_n^2(t),
\end{aligned} \tag{7.264}
$$

where the delta function has reduced the double sum to a single sum, which is then canceled by the middle term. Letting $N \to \infty$:

$$
R_{XX}(t,t) - \lim_{N\to\infty}\sum_{n=1}^{N}\lambda_n\phi_n^2(t) = 0, \tag{7.265}
$$

such that (7.259) is obtained. The final result is due to Mercer's theorem below (with $t = t_1 = t_2$). □

Note that $R_{XX}(t,t) \neq R_{XX}(0)$ unless $X(t)$ is wide-sense stationary: the two time arguments must be shown explicitly for nonstationary $X(t)$.

Theorem 7.22 (Mercer). The autocorrelation function of random process $X(t)$ can be decomposed as follows:

$$
R_{XX}(t_1, t_2) = \sum_{n=1}^{\infty}\lambda_n\phi_n(t_1)\phi_n(t_2), \tag{7.266}
$$

where $\{\lambda_n\}$ and $\{\phi_n(t)\}$ are the previously defined eigenvalues and eigenfunctions, respectively, of the KLE.

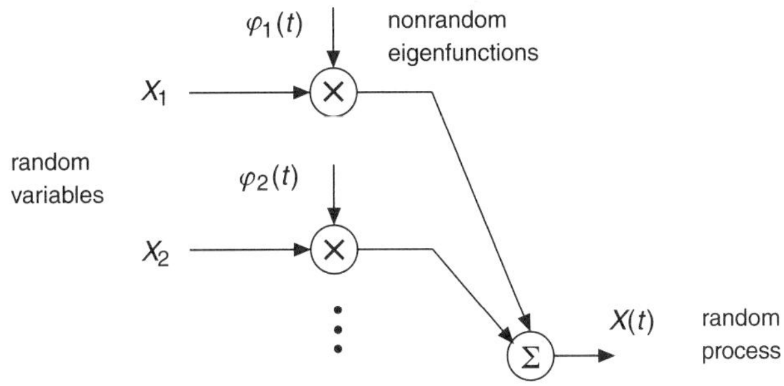

FIGURE 7.26 KLE of random process $X(t)$ in terms of eigenfunctions $\{\phi_n(t)\}$ weighted by random variables $\{X_n\}$ for $n \in \mathcal{N}$.

Proof. Substituting (7.256) gives

$$\mathcal{E}[X(t_1)X(t_2)] = \sum_{n=1}^{\infty}\sum_{m=1}^{\infty} \mathcal{E}[X_n X_m]\phi_n(t_1)\phi_m(t_2),$$
$$= \sum_{n=1}^{\infty}\sum_{m=1}^{\infty} \lambda_m \delta[m-n]\phi_n(t_1)\phi_m(t_2). \tag{7.267}$$

Using the delta function result in (7.262) eliminates the sum over m and gives the expression in (7.266). □

Observe that the KLE is similar to a matrix eigendecomposition (see Appendix G). Instead of a matrix, the autocorrelation function in (7.258) determines the type of eigenfunctions. A diagram of the decomposition is shown in Figure 7.26. In Chapter 10, we find that this architecture is similar to a matched filter receiver.

Example 7.28. Consider the Wiener process with autocorrelation function $R_{XX}(t_1, t_2) = c\min(t_1, t_2)$, where $c = \alpha\epsilon^2$ (see Table 6.4). Assuming that $X(t)$ is defined on $\mathcal{T} = [0, T]$, (7.258) gives

$$c\int_0^T \min(t_1, t_2)\phi_n(t_2)\mathrm{d}t_2 = c\int_0^{t_1} t_2\phi_n(t_2)\mathrm{d}t_2 + c\int_{t_1}^T t_1\phi_n(t_2)\mathrm{d}t_2,$$
$$= \lambda_n\phi_n(t_1) \tag{7.268}$$

for some $t_1 \in \mathcal{T}$. As is done to solve integro-differential equations (such as those used to describe second-order RLC circuits), this expression is differentiated to obtain an explicit expression for the function in the integrand. From the chain rule of differentiation and Leibniz's integral rule (see Appendix E):

$$\frac{\partial}{\partial t_1}\int_0^{t_1} t_2\phi_n(t_2)\mathrm{d}t_2 = t_1\phi_n(t_1), \tag{7.269}$$

$$\frac{\partial}{\partial t_1}\int_{t_1}^T t_1\phi_n(t_2)\mathrm{d}t_2 = -t_1\phi_n(t_1) + \int_{t_1}^T \phi_n(t_2)\mathrm{d}t_2. \tag{7.270}$$

Differentiating (7.268) and substituting these expressions yields a simplified integral equation that defines the eigenfunctions:

$$c\int_{t_1}^{T} \phi_n(t_2)\mathrm{d}t_2 = \lambda_n \frac{\partial}{\partial t_1}\phi_n(t_1). \tag{7.271}$$

This is easily differentiated a second time with respect to t_1 to give a second-order DE:

$$\frac{\mathrm{d}^2}{\mathrm{d}t_1^2}\phi_n(t_1) = -(c/\lambda_n)\phi_n(t_1). \tag{7.272}$$

The unilateral Laplace transform is

$$s^2\Phi_n(s) - s\phi_n(0) - \phi_n'(0) = -(c/\lambda_n)\Phi_n(s). \tag{7.273}$$

Assuming zero initial conditions, the characteristic equation is

$$s^2 + c/\lambda_n = 0. \tag{7.274}$$

If the autocorrelation matrix is positive definite (PD), then $\lambda_n > 0$ (see Problem 7.28) so that the roots of the characteristic equation are $s = j\sqrt{c/\lambda_n}$ (recall that $c > 0$). Thus, the general form of the solution is

$$\phi_n(t) = a_1 \sin(t\sqrt{c/\lambda_n}) + a_2 \cos(t\sqrt{c/\lambda_n}). \tag{7.275}$$

From $\phi_n(0) = 0$, we have $a_2 = 0$. The solution for λ_n is obtained by the following condition at $t = T$:

$$\phi_n'(T) = 0 \Rightarrow a_1\sqrt{c/\lambda_n}\cos(T\sqrt{c/\lambda_n}) = 0, \tag{7.276}$$

which means that $T\sqrt{c/\lambda}$ is an odd integer multiple of $\pi/2$:

$$T\sqrt{c/\lambda_n} = (2n-1)\pi/2 \Rightarrow \lambda_n = 4cT^2/(2n-1)^2\pi^2 = cT^2/(n-1/2)^2\pi^2 \tag{7.277}$$

for $n \in \mathcal{N}$. Substituting this result into (7.275) gives

$$\phi_n(t) = a_1 \sin\left((n-1/2)\pi t/T\right). \tag{7.278}$$

In order that the eigenfunctions have unit energy, it is straightforward to show $a_1 = \sqrt{2/T}$. The first three eigenfunctions are plotted in Figure 7.27(a). Note that the same waveforms are obtained on $[0, T]$ for any $T > 0$, though the amplitude decreases with increasing T so that the functions maintain unit energy. In Figure 7.27(b), we show approximations of $R_{XX}(t,t) = \mathcal{E}[X^2(t)]$ using Mercer's formula with $N = 1$ and 6 terms in the summation. For $c = 1$, note that $R_{XX}(t,t) = \min(t,t) = t$ is linear, for which the approximation is quite accurate for only a few terms in the sum. The derivatives of the curves are zero at $t = 0$ and T because of the boundary conditions for each of the eigenfunctions in the sum.

Generally, it is difficult to derive expressions for the eigenfunctions $\{\phi_n(t)\}$, and they may need to be solved numerically. In Example 7.29, we demonstrate this difficulty for a relatively simple random process.

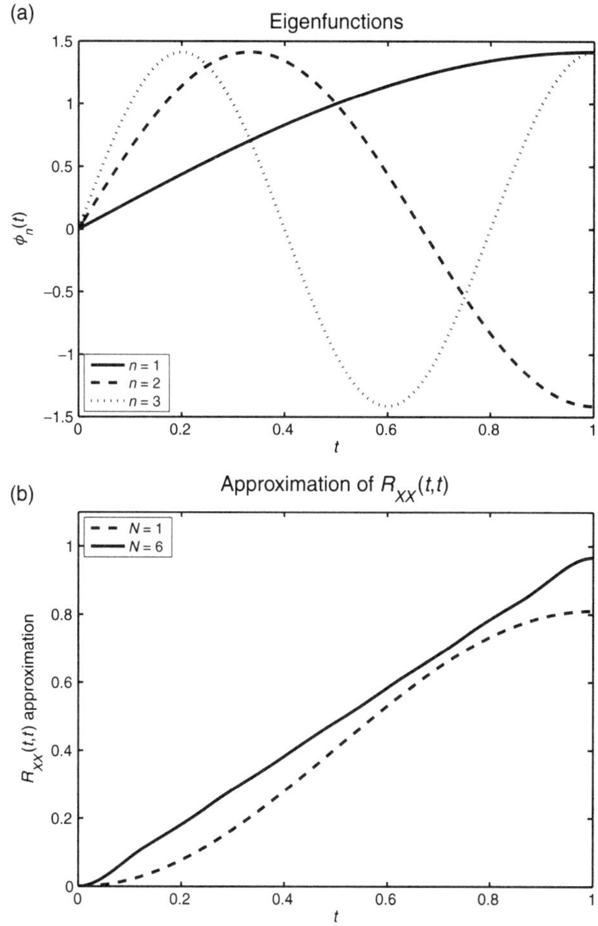

FIGURE 7.27 KLE for Example 7.28 with $T = 1$. (a) Example eigenfunctions for $R_{XX}(t_1, t_2)$. (b) Approximation of $R_{XX}(t, t)$ using a finite sum of terms in Mercer's formula.

Example 7.29. Consider the following random process defined on $[0, T]$:

$$X(t) = \sum_{m=1}^{M} A_m g_m(t), \tag{7.279}$$

where $\{g_m(t)\}$ are deterministic functions. The only randomness in $X(t)$ are the random variables $\{A_m\}$, which we assume have zero mean and finite variances $\{\sigma_m^2\}$. When the random variables are independent, the autocorrelation function is

$$R_{XX}(t_1, t_2) = \sum_{m=1}^{M} \sigma_m^2 g_m(t_1) g_m(t_2). \tag{7.280}$$

From the integral equation,

$$\sum_{m=1}^{M} \sigma_m^2 \int_0^T g_m(t_1) g_m(t_2) \phi_n(t_1) \mathrm{d}t_2 = \lambda_n \phi_n(t_1) \tag{7.281}$$

from which we find that the eigenfunctions depend on the specific form of $\{g_m(t)\}$. The key step in the derivation is to factor $g_m(t_1)$ from the integral and use the fact that the other terms on the left-hand side are independent of t_1 to write

$$\sum_{m=1}^{M} \sigma_m^2 g_m(t_1) \int_0^T g_m(t_2) \phi_n(t_1) \mathrm{d}t_2 = \sum_{m=1}^{M} b_{m,n} g_m(t_1), \tag{7.282}$$

where

$$b_{m,n} \triangleq \sigma_m^2 \int_0^T g_m(t_2) \phi_n(t) \mathrm{d}t_2. \tag{7.283}$$

Although the eigenfunctions are needed to give specific values for $\{b_{m,n}\}$, we see that by equating (7.282) to the right-hand side of (7.281), the eigenfunctions are *linear combinations* of the $\{g_m(t)\}$:

$$\phi_n(t) = \sum_{m=1}^{M} c_{m,n} g_m(t), \tag{7.284}$$

where $c_{m,n} \triangleq b_{m,n}/\lambda_n$. In order to proceed, additional information is needed about $\{g_m(t)\}$. For example, we might assume they have unit energy $\int_0^T g_m^2(t)\mathrm{d}t = 1$, by appropriately scaling the variances $\{\sigma_m^2\}$. More importantly, we require the deterministic cross-correlations $\int_0^T g_m(t) g_n(t)\mathrm{d}t$, which directly influence the realizations of $X(t)$ and properties of the autocorrelation matrix in (7.280). In general for this example, it is not possible to derive closed-form expressions for the $\{c_{m,n}\}$; in fact, for $M = 2$ the solutions are not unique (Ludeman, 2003). In Chapter 10 on digital communications, we present an expansion similar to the KLE called *Gram–Schmidt orthogonalization*, which is used for deterministic signals similar to the $\{g_n(t)\}$ functions in this example. The orthogonalization leads to a convenient representation of the transmitted waveforms from which it is possible to derive the optimal detector for signals in a noisy channel.

In the final example, we show how a Fourier series expansion of a random process differs from the KLE.

Example 7.30. Suppose that random process $X(t)$ is wide-sense stationary and MS *periodic* such that

$$\mathcal{E}[(X(t) - X(t-T))^2] = 2R_{XX}(0) - 2R_{XX}(T) = 0. \tag{7.285}$$

For this case, $X(t)$ is cyclostationary (see Chapter 6), and can be expanded as a Fourier series

$$X(t) = \sum_{n=-\infty}^{\infty} X_n \exp(j2\pi nt/T), \tag{7.286}$$

where $\{X_n\}$ are random variables:

$$X_n = \frac{1}{T} \int_0^T X(t) \exp(-j2\pi nt/T) \mathrm{d}t. \tag{7.287}$$

The sum and integral above are defined in the MS sense. Observe that this expansion resembles the KLE with the following differences: (i) $\{\phi_n(t)\}$ are replaced by complex exponentials and (ii) the sum extends for all $n \in \mathcal{Z}$. The KLE assumes that $X(t)$ exists only on $[0, T]$; if we imagine that the process is repeated with period T such that (7.285) holds, then a Fourier series representation can be derived. Since $R_{XX}(\tau)$ is periodic with period T (see Problem 7.33), it can also be expanded using a Fourier series:

$$R_{XX}(\tau) = \sum_{n=-\infty}^{\infty} r_n \exp(j2\pi n\tau/T) \tag{7.288}$$

with Fourier series coefficients:

$$r_n = \frac{1}{T}\int_0^T R_{XX}(\tau)\exp(-j2\pi n\tau/T)\mathrm{d}\tau. \tag{7.289}$$

This integral equation is analogous to (7.258) for the KLE. Using a derivation similar to that for the KLE, it can be shown that a Fourier series for $X(t)$ exists in the MS sense (see Problem 7.34).

PROBLEMS

Stochastic Convergence

7.1 Determine which of the following random sequences converge (i) surely or (ii) almost surely. (a) $X_1[k, \zeta] = \zeta(k-1)/k$ with $\mathcal{S} = [0, 1]$. (b) $X_2[k, \zeta] = (1/\zeta)(k-1)/k$ with $\mathcal{S} = [0, \infty)$. (c) $X_3[k, \zeta] = \sin(2\pi\zeta k)$ with $\mathcal{S} = [-1, 1]$.

7.2 Determine which of the following random sequences converge in MS. (a) $X_1[k, \zeta] = \zeta/k$ with $\mathcal{S} = [0, 1]$. (b) $X_2[k, \zeta] = \zeta(k-1)/k$ with $\mathcal{S} = [0, 1]$. (c) $X_3[k, \zeta] = \exp(\zeta k)$ with $\mathcal{S} = [-1, 1]$.

7.3 Suppose that $X[k]$ converges in probability to the constant a, and $Y[k]$ converges in probability to the constant b. Show from the definition that $X[k] + Y[k]$ converges in probability to $a + b$.

7.4 Prove that L^m convergence implies L^n convergence for $m > n$.

7.5 Let $Y[k] = p^k \prod_{n=1}^{k} X[k]$, where $X[k]$ is an iid Bernoulli sequence with parameter p. Determine if this sequence converges almost surely for the following outcomes for $X[k]$. (a) $\{-1, 1\}$. (b) $\{0, 1\}$.

7.6 For $X[k] \xrightarrow{m.s.} X$ and $Y[k] \xrightarrow{m.s.} Y$, prove the following. (a) $X[k] + Y[k] \xrightarrow{m.s.} X + Y$. (b) $\lim_{k\to\infty} \mathcal{E}[X[k]Y[k]] = \mathcal{E}[XY]$.

Laws of Large Numbers

7.7 Verify the weak law of large numbers for an iid Bernoulli sequence with outcomes $\{0, 1\}$ and parameter p.

7.8 Let $X[k]$ be an iid random sequence for which the weak law of large numbers holds. Determine if the weak law also applies to $Y[k] = g(X[k])$, where $g(\cdot)$ is a memoryless function.

7.9 Suppose the conditions for the weak law of large numbers are relaxed such that $X[k]$ is not an independent sequence. For the autocovariance function $C_{XX}[m] = \exp(-|m|)$ for $m \in \mathcal{Z}$, determine if the property in (7.42) is still valid.

Central Limit Theorem

7.10 Let $X[k]$ be an iid sequence of exponential random variables with parameter $\lambda = 1$ and define the sum $Y[k] \triangleq (1/k)\sum_{n=1}^{k} X[n]$. Calculate the following probabilities using (i) the exact distribution of $Y[k]$ and (ii) a Gaussian approximation based on the CLT. (a) $P(Y[5] > 2)$. (b) $P(1 < Y[10] \leq 2)$.

7.11 Repeat Problem 7.10 for a Bernoulli random variable with parameter $p = 3/4$ and the following probabilities. (a) $P(Y[3] < 2)$. (b) $P(1 \leq Y[4] \leq 2)$.

7.12 For the binary symmetric channel (BSC) (see Chapter 2), let the probability of a bit error be $p_e = 0.01$. If 500 bits are transmitted across this channel, use the CLT to compute the probability that there will be greater than 5 errors.

7.13 Derive the transformation result in (7.57) used in the proof of the CLT.

7.14 Consider a Poisson random variable Y with parameter $\alpha = 10$. Recognizing that this distribution is obtained by summing together ten iid Poisson random variables, we can apply the CLT and the result in (4.17). Find $P(Y > 8)$ using the Gaussian approximation.

Continuity, Derivatives, and Integrals

7.15 Let $X(t) = u(t) - u(t - T)$, where $u(t)$ is the unit-step function and T is a uniform random variable on $[1, 2]$. Determine if $X(t)$ is MS continuous.

7.16 Suppose that $X(t) = \zeta \cos(2\pi f_o t)$ for $\mathcal{T} = \mathcal{R}$, where f_o is a constant frequency and $\zeta \in \mathcal{S} = [-1, 1]$ is uniformly distributed. Determine if $X(t)$ is (a) continuous with probability one and (b) continuous in MS.

7.17 Prove that the MS derivative $X'(t)$ is uncorrelated with wide-sense stationary process $X(t)$.

7.18 Prove that if $X(t)$ is MS continuous for every $t \in \mathcal{T}$, then $R_{XX}(t_1, t_2)$ is a continuous function for all $t_1, t_2 \in \mathcal{T}$.

7.19 Derive the correlation results in (7.130) and (7.131) for the MS derivative when $X(t)$ is wide-sense stationary.

7.20 Determine if the following random process has an MS derivative and MS integral: $Y(t)$ with $R_{YY}(\tau) = 1 - \tau^2$ for $|\tau| \in [0, 1]$.

7.21 Determine if $X(t)$ in Problem 7.15 has an MS derivative and MS integral.

7.22 Complete the derivation for $R_{XX}(t_1, t_2)$ in Example 7.24.

7.23 Similar to the approach in Example 7.22, derive expressions for $R_{XY}(t_1, t_2)$ and $R_{XX}(t_1, t_2)$ for the following first-order DE:

$$a_1 \frac{\mathrm{d}}{\mathrm{d}t} X(t) + a_0 X(t) = Y(t) \tag{7.290}$$

for $t > 0$ with $R_{YY}(\tau) = \delta(\tau)$.

Random Sequences and Decompositions

7.24 Consider the following first-order difference equation:

$$X[k] + 0.5X[k-1] = Y[k]. \tag{7.291}$$

Let $R_{YY}[m] = \delta[k]$ such that $Y[k]$ is an uncorrelated sequence with zero mean. Assume zero initial conditions for the difference equation, and find expressions for (a) $R_{XY}[k_1, k_2]$ and (b) $R_{XX}[k_1, k_2]$ using z-transforms.

7.25 Suppose that a zero-mean Markov sequence $X[k]$ has the following conditional pdf:

$$f_{X[k]|X[k-1]}(x[k]|x[k-1]) = \frac{1}{\sqrt{2\pi\sigma^2}} \exp(-(x[k] - ax^2[k-1])^2/2\sigma^2), \tag{7.292}$$

where $a > 0$. Specify (a) the MS predictable part $X_{\text{MS}}[k] = \mathcal{E}[X[k]|X[0], \ldots, X[k-1]]$ and (b) the DM predictable part $\Delta X_{\text{DM}}[k] = \mathcal{E}[X[k] - X[k-1]|X[0], \ldots, X[k-1]]$.

7.26 For the binomial counting sequence $X[k]$ defined in (6.160), find the DM predictable part $\Delta X_{\text{DM}}[k]$ and give the components of the decomposition $X[k] = X_{\text{DM}}[k] + E[k]$.

Karhunen–Loève Expansion

7.27 Let the autocorrelation function of $X(t)$ be $R_{XX}(t_1, t_2) = \sigma^2\delta(t_2 - t_1)$. Show that any set of orthonormal eigenfunctions can be used in the KLE for $X(t)$.

7.28 The kernal of the integral equation in (7.258) is PD if

$$\int_0^\infty \int_0^\infty g(t_1) R_{XX}(t_1, t_2) g(t_2) \mathrm{d}t_1 \mathrm{d}t_2 > 0 \tag{7.293}$$

for any $g(t) \neq 0$ on $[0, T]$. Use this result to demonstrate that the eigenvalues $\{\lambda_n\}$ of the KLE are positive.

7.29 Use Mercer's theorem to find expressions for the following integrals in terms of the eigenvalues:

$$\int_0^T R_{XX}(t, t)\mathrm{d}t, \qquad \int_0^T \int_0^T R_{XX}^2(t_1, t_2)\mathrm{d}t_1 \mathrm{d}t_2. \tag{7.294}$$

7.30 For the following autocorrelation function,

$$R_{XX}(t_1, t_2) = 2 + \sin(2\pi t_1/T)\sin(2\pi t_2/T), \tag{7.295}$$

use Mercer's theorem to find the eigenvalues $\{\lambda_n\}$ and eigenfunctions $\{\phi_n(t)\}$.

7.31 Determine the KLE for wide-sense stationary random process $X(t)$ defined on $[0, T]$ with autocorrelation function

$$R_{XX}(\tau) = T - |\tau|. \tag{7.296}$$

7.32 Repeat Problem 7.31 for the following random process:

$$X(t) = a\cos(2\pi f_c t + \Phi), \tag{7.297}$$

where $\{a, f_c\}$ are nonrandom and Φ is uniform on $[-\pi, \pi]$.

7.33 Let $X(t)$ be a wide-sense stationary random process with zero mean that is MS periodic with period T. Show that the autocorrelation function $R_{XX}(\tau)$ is periodic with the same period T.

7.34 (a) Demonstrate that the Fourier series expansion for an MS periodic random process exists in the MS sense. The approach used for the KLE can also be used here, except that $\{X_n\}$ are complex-valued because of the complex exponentials. (b) Show that $\mathcal{E}[X_n X_m^*] = r_n\delta[m - n]$.

Computer Problems

7.35 Let $X[k]$ be an iid random sequence with a beta distribution and parameters $\alpha = \beta = 1/2$ (corresponding to the arcsine distribution). Use MATLAB® to compute and plot the pdf of $Y[k]$ in (7.55), demonstrating that it approaches a Gaussian pdf with increasing k. Since the sample mean does not have a closed-form pdf in this case, use convolve and include the shifting and scaling in (7.55) for each k.

7.36 Use MATLAB to generate plots similar to those in Figure 7.25 for the DM decomposition in Example 7.27. Let $\mu_X = 0$ and plot realizations of the predictable part $X_{\text{DM}}[k]$ and the martingale $E[k]$ for $\rho = \pm 0.9$.

7.37 Using finite sums for Mercer's formula, plot approximations of the autocorrelation function $R_{XX}(t_1, t_2)$ for the Wiener process. Use meshgrid and mesh to plot the functions in three dimensions, showing they approach the actual result in Figure 6.27. Note that the approximations shown in Figure 7.27 for Example 7.28 correspond to the autocorrelation function only along the $t_1 = t_2 = t$ line.

FURTHER READING

The material in this chapter can be explored further in several of the references at the end of Chapter 3. Some specific topics can be found in the following references. Stochastic calculus: Gardner (1990), Larson and Shubert (1979), Parzen (1962), and Stark and Woods (2002). Doob–Meyer decomposition: Doob (1953) and Larson and Shubert (1979). Karhunen–Loève expansion: Helstrom (1995), Ludeman (2003), Schonhoff and Giordana (2006), and Van Trees (1968).

8 Systems, Noise, and Spectrum Estimation

8.1 INTRODUCTION

In this chapter, we consider systems that filter a random process to achieve a particular goal, such as removing additive noise or compensating for channel distortion in communications. Generally, we are interested in linear time-invariant (LTI) systems that are described by an impulse-response function $h(t)$ and the corresponding transfer function $H(\omega)$. Such systems can be represented by a linear differential equation (DE) with constant coefficients, so they include a combination of integration and differentiation of the signals. In Chapter 7, we considered mean-square (MS) definitions of derivatives and integrals of a random process, which led to an evaluation of the autocorrelation function $R_{XX}(t_1, t_2)$. Table 8.1 summarizes the results for MS continuity, derivatives, and integrals; the wide-sense stationary case, where the autocorrelation function $R_{XX}(\tau)$ depends only on the time lag $\tau \triangleq t_2 - t_1$, is also included. In this chapter, we focus on wide-sense stationary processes such that

$$\mu_X(t) = \mu_X, \quad R_{XX}(t, t+\tau) = R_{XX}(\tau) \tag{8.1}$$

for all $t \in \mathcal{T}$. Techniques are developed for investigating how a random process is modified when it is processed by an LTI system, in terms of its time response as well as its frequency response. Discrete-time systems are also considered where the convolution of an LTI system is a sum rather than an integral, though integration can be used if the impulse response function is modeled using Dirac delta functions.

Perhaps the most important result in this chapter is the power spectral density (PSD) $S_{XX}(\omega)$, which is the Fourier transform of $R_{XX}(\tau)$. When $X(t)$ is represented only by the quantities in (8.1), we are assuming a *partial characterization* of the random process. Often in many applications, we are not interested in the particular distribution of the random process, nor in other quantities such as high-order moments. The description in (8.1) is quite useful for a wide range of random processes and applications. Recall in Chapter 1 that deterministic signals and LTI systems are often examined in the frequency domain. For deterministic $x(t)$, the output of an LTI system with impulse-response function $h(t)$ is the convolution $y(t) = h(t) * x(t)$, and the result in the frequency domain is $Y(\omega) = H(\omega)X(\omega)$, where $H(\omega)$ is the transfer function. For a wide-sense stationary random process, we show in this chapter that the corresponding frequency-domain expression is $S_{YY}(\omega) = |H(\omega)|^2 S_{XX}(\omega)$, as illustrated in Figure 8.1. This similar result occurs because it turns out that the *eigenfunctions* of $R_{XX}(\tau)$ for $\tau \in \mathcal{R}$ are complex exponentials, just as they are for an LTI system. This is demonstrated later after we revisit correlation in Section 8.2 and discuss *ergodicity*, which is another important property of a random process.

8.2 CORRELATION REVISITED

In Chapter 7, we investigated the DE:

$$a_n \frac{\mathrm{d}^n}{\mathrm{d}t^n} X(t) + a_{n-1} \frac{\mathrm{d}^{n-1}}{\mathrm{d}t^{n-1}} X(t) + \cdots + a_1 \frac{\mathrm{d}}{\mathrm{d}t} X(t) + a_0 X(t) = Y(t), \tag{8.2}$$

Probability, Random Variables, and Random Processes: Theory and Signal Processing Applications, First Edition. John J. Shynk.

TABLE 8.1 Mean-Square Results for a Random Process

Continuity:	$R_{XX}(t_1, t_2)$ is continuous at $t_1 = t_2 = t$ or $R_{XX}(\tau)$ is continuous at $\tau = 0$
Derivative:	$\partial^2 R_{XX}(t_1, t_2)/\partial t_1 \partial t_2$ exists at $t_1 = t_2 = t$ or $\partial^2 R_{XX}(\tau)/\partial \tau^2$ exists at $\tau = 0$
Integral:	$\int_{t_o}^{t} \int_{t_o}^{t} R_{XX}(t_1, t_2) \mathrm{d}t_1 \mathrm{d}t_2$ exists or $2 \int_0^{t-t_o} (t - t_o - \tau) R_{XX}(\tau) \mathrm{d}\tau$ exists

where $\{a_1, \ldots, a_n\}$ are fixed coefficients, random process $Y(t)$ is the input of the linear system represented by the DE, and $X(t)$ is the output (note that the roles of $X(t)$ and $Y(t)$ are reversed in Chapter 7 and (8.2)) . Since $Y(t)$ is random, it is generally not possible to write an expression for each realization of $X(t)$. We first considered evaluating the mean:

$$a_n \frac{\mathrm{d}^n}{\mathrm{d}t^n} \mu_X(t) + a_{n-1} \frac{\mathrm{d}^{n-1}}{\mathrm{d}t^{n-1}} \mu_X(t) + \cdots + a_1 \frac{\mathrm{d}}{\mathrm{d}t} \mu_X(t) + a_0 X(t) = \mu_Y(t) \tag{8.3}$$

for $t > 0$, where it is assumed that the derivatives and expectations can be interchanged. Solving this equation for $\mu_X(t)$ provides an expression for the average behavior of the output $X(t)$. However, if $\mu_Y(t)$ is zero for all t and if the initial conditions are zero, which is often the case in many applications, then $\mu_X(t)$ will be zero for $t \in \mathcal{T}$. As a result, the DE for the mean is of limited use for providing information about the ensemble behavior of $X(t)$.

We also investigated the autocorrelation function, obtained by replacing t with t_2, multiplying each term of the DE by $X(t_1)$, and taking expectations:

$$a_n \frac{\partial^n}{\partial t_2^n} R_{XX}(t_1, t_2) + a_{n-1} \frac{\partial^{n-1}}{\partial t_2^{n-1}} R_{XX}(t_1, t_2) + \cdots + a_1 \frac{\partial}{\partial t_2} R_{XX}(t_1, t_2) + a_0 R_{XX}(t_1, t_2) = R_{XY}(t_1, t_2) \tag{8.4}$$

for $t_1, t_2 > 0$. This allowed us to solve for $R_{XX}(t_1, t_2)$ with input $R_{XY}(t_1, t_2)$, which in turn was provided by a second DE obtained by multiplying (8.2) with $Y(t_2)$ and again taking expectations:

$$a_2 \frac{\partial^2}{\partial t_1^2} R_{XY}(t_1, t_2) + a_1 \frac{\partial}{\partial t_1} R_{XY}(t_1, t_2) + a_0 R_{XY}(t_1, t_2) = R_{YY}(t_1, t_2) \tag{8.5}$$

for $t_1, t_2 > 0$. From this expression, we can investigate the second-order behavior of a random process that is described by a linear DE. In particular, if the processes have zero mean, the variance is examined by evaluating the autocorrelation function at $t_1 = t_2 = t$. Similar results can be obtained for random sequences generated by linear difference equations.

The autocorrelation function of wide-sense stationary random process $X(t)$ is defined as an expectation in terms of its joint probability density function (pdf) as follows:

$$\begin{aligned} R_{XX}(\tau) &\triangleq \mathcal{E}[X(t)X(t+\tau)], \\ &= \int_{-\infty}^{\infty} \int_{-\infty}^{\infty} x_1 x_2 f_{X(t),X(t+\tau)}(x_1, x_2) \mathrm{d}x_1 \mathrm{d}x_2. \end{aligned} \tag{8.6}$$

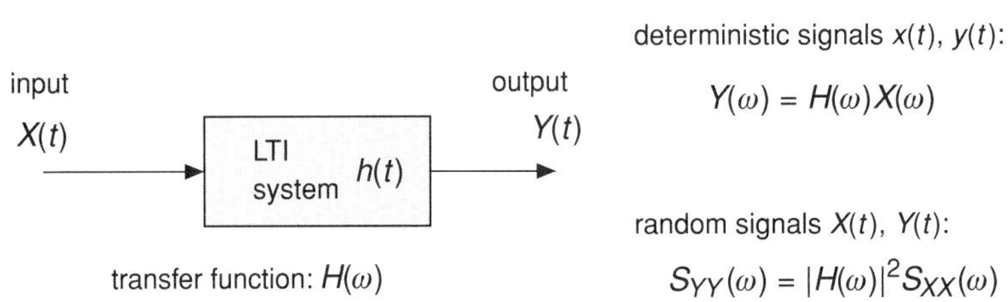

FIGURE 8.1 Input/output frequency-domain characterization of LTI systems for deterministic signals and wide-sense stationary random signals.

Consider the following estimate of the autocorrelation function:

$$\bar{R}_{XX}(\tau) \triangleq \frac{1}{2T} \int_{-T}^{T} X(t)X(t+\tau)\mathrm{d}t, \tag{8.7}$$

where the integral is defined in the MS sense. The corresponding expression for the mean is

$$\bar{\mu}_X \triangleq \frac{1}{2T} \int_{-T}^{T} X(t)\mathrm{d}t. \tag{8.8}$$

Since $X(t)$ is a random process, the estimates $\bar{R}_{XX}(\tau)$ and $\bar{\mu}_X$ are random variables, so we can evaluate their expectations:

$$\mathcal{E}[\bar{\mu}_X] = \frac{1}{2T} \int_{-T}^{T} \mathcal{E}[X(t)]\mathrm{d}t = \mu_X, \tag{8.9}$$

$$\mathcal{E}[\bar{R}_{XX}(\tau)] = \frac{1}{2T} \int_{-T}^{T} \mathcal{E}[X(t)X(t+\tau)]\mathrm{d}t = R_{XX}(\tau), \tag{8.10}$$

which show that these estimates are *unbiased*: on average they are the same as the actual quantities (refer to Chapter 9 for properties of estimators).

Suppose we compute time averages over a single realization $x(t)$ such that

$$\bar{\bar{R}}_{XX}(\tau) = \frac{1}{2T} \int_{-T}^{T} x(t)x(t+\tau)\mathrm{d}t, \tag{8.11}$$

$$\bar{\bar{\mu}}_X = \frac{1}{2T} \int_{-T}^{T} x(t)\mathrm{d}t, \tag{8.12}$$

where the double-bar notation is used for these estimates. Observe that $\bar{\bar{R}}_{XX}(\tau)$ is a deterministic autocorrelation function because the averaging is performed over a specific realization of $X(t)$, and likewise for the deterministic mean $\bar{\bar{\mu}}$. Comparing the deterministic autocorrelation function in (8.11) and the ensemble definition in (8.6), we see they have similar forms. The integral of the ensemble autocorrelation function is performed over the values of all realizations at time instants t and $t+\tau$, weighted by the joint pdf, whereas the integral of the deterministic autocorrelation function is performed over a single realization (without any weighting). These are depicted in Figure 8.2 for the autocorrelation function, and in Figure 8.3 for the mean.

The time-averaged autocorrelation function is a measure of similarity between a realization and a time-shifted version of itself. This is illustrated in Figure 8.4 for an artificial rectangular realization and different lags τ. $\bar{\bar{R}}_{XX}(\tau)$ achieves its maximum at $\tau = 0$ when $x(t)$ and $x(t+\tau)$ are most similar (identical), and it steadily decreases as τ increases, indicating that functionally the waveforms are less similar. We have the same interpretation for the ensemble autocorrelation function, except that the averaging is performed as a joint moment of two random variables: $X(t)$ and $X(t+\tau)$. This is done instead of performing time averages over all possible realizations of the random process; recall, that our goal in Chapter 7 was to use probabilistic descriptions of a random process for the ensemble so that we need not examine every realization.

Since a random process is a function of time, this measure of similarity provided by the autocorrelation function implies a degree of predictability in the process such that past values contain "information" about future values. Most signal processing algorithms operating on a random process exploit correlation information to achieve some goal, such as estimating a transmitted signal that is corrupted by noise. We explore correlation further for random signals and the systems that process those signals. Before proceeding with filters, we discuss the concept of *ergodicity* where for some random processes

$$\lim_{T\to\infty} \bar{\mu}_X = \mu_X, \quad \lim_{T\to\infty} \bar{R}_{XX}(\tau) = R_{XX}(\tau). \tag{8.13}$$

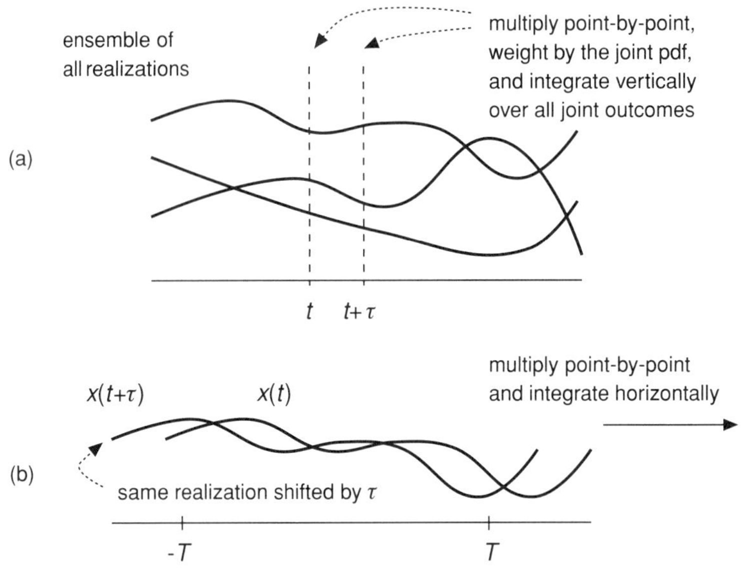

FIGURE 8.2 Computing the autocorrelation function. (a) $R_{XX}(\tau)$ is obtained by integrating over all possible outcomes at two time instants t and $t+\tau$, weighted by the joint pdf. (b) $\bar{\bar{R}}_{XX}(\tau)$ is obtained by averaging the product of a realization with itself shifted by τ over $[-T, T]$.

In the limit, a time average of *any single realization* of $X(t)$ is exactly the same as the expectation over the ensemble of all realizations.

8.3 ERGODICITY

Consider first ergodicity in the mean. A time average $\bar{\mu}_X$ for the mean of random process $X(t)$ is given in (8.8). We know that if $X(t)$ has a constant mean, then $\mathcal{E}[\bar{\mu}_X] = \mu_X$: the expected value of the time average, even for finite T, is the mean of the random process so that $\bar{\mu}_X$ is an unbiased estimator of μ_X. Ergodicity in the mean is a stronger property such that the time average itself converges to the mean as $T \to \infty$.

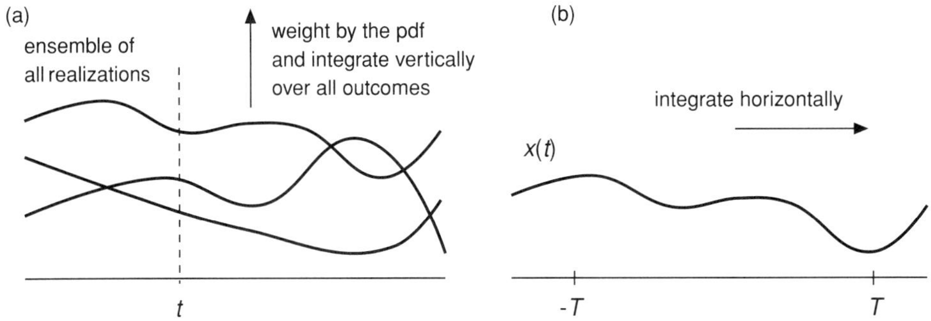

FIGURE 8.3 Computing the mean. (a) μ_X is obtained by integrating over all possible outcomes at time t weighted by the pdf. (b) $\bar{\mu}_X$ is obtained by averaging a realization over $[-T, T]$.

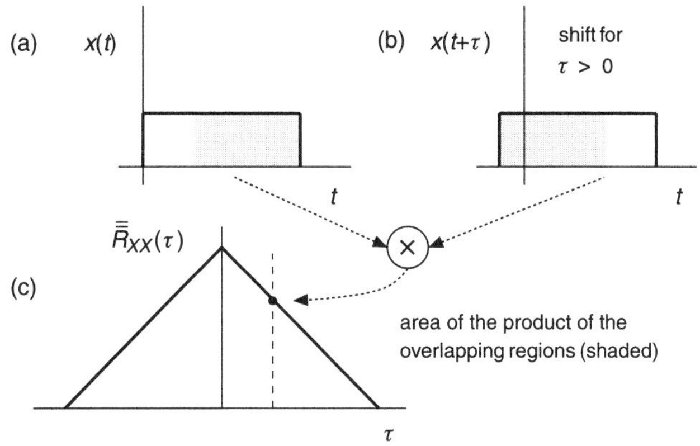

FIGURE 8.4 Deterministic autocorrelation function. (a) Realization $x(t)$ of random process $X(t)$. (b) Time-shifted version of the realization. (c) Resulting time-average autocorrelation function $\bar{\bar{R}}_{XX}(\tau)$.

Definition: Ergodic in Mean (Process) Random process $X(t)$ with constant mean μ_X is *ergodic in the mean* if

$$\lim_{T\to\infty}\frac{1}{2T}\int_{-T}^{T}X(t)\mathrm{d}t=\mu_X. \tag{8.14}$$

When the time average approaches its expectation as implied above, its *variance* approaches zero. This is a very useful property: the mean of a random process can be accurately estimated by averaging any *realization* over a sufficiently long time interval. Ergodicity in the mean is verified as follows.

Theorem 8.1 (Ergodic in Mean). Random process $X(t)$ with autocovariance function $C_{XX}(\tau)$ is ergodic in the mean if and only if

$$\lim_{T\to\infty}\frac{1}{2T}\int_{-2T}^{2T}\left(1-\frac{|\tau|}{2T}\right)C_{XX}(\tau)\mathrm{d}\tau=0, \tag{8.15}$$

where

$$C_{XX}(\tau)=R_{XX}(\tau)-\mu_X^2. \tag{8.16}$$

Proof. Consider the variance of the integral in (8.14):

$$\sigma_\mu^2=E\left[\left(\frac{1}{2T}\int_{-T}^{T}X(t)\mathrm{d}t-\mu_X\right)^2\right]=\frac{1}{4T^2}\int_{-T}^{T}\int_{-T}^{T}C_{XX}(t_1,t_2)\mathrm{d}t_1\mathrm{d}t_2. \tag{8.17}$$

Since the integrals exist in the MS sense, the theorem states that $\sigma_\mu^2\to 0$. As the variance of the time average approaches zero, its pdf becomes a Dirac delta function at μ_X. Thus, we need only show that the double integral in (8.17) is equivalent to the single integral in (8.15). A similar equivalence was shown in Chapter 7 for the MS integral involving the autocorrelation function $R_{XX}(\tau)$, so those details will not be repeated. The result in Figure 7.16 for $R_{XX}(\tau)$ also applies here for $C_{XX}(\tau)$ with limits of integration changed from $[t_o, t]$ to $[-T, T]$.

Since $C_{XX}(\tau)$ mentioned in the theorem depends only on the time lag τ, $X(t)$ is wide-sense stationary: $C_{XX}(t_1,t_2)=C_{XX}(t_2-t_1)=C_{XX}(\tau)$ with $\tau\triangleq t_2-t_1$. The transformation of variables $\tau\triangleq t_2-t_1$ and $\alpha\triangleq t_1+t_2$ causes a rotation of the axes as illustrated in Figure 7.16. The range of integration over τ is obviously

$[-2T, 2T]$, but those for α depend on τ. It is evident from Figure 7.16 that in order for the double integral to stay within the square region, the range for α must be $[-(2t - |\tau|), 2T - |\tau|]$. The Jacobian matrix is (see Appendix G)

$$\mathbf{J} = \begin{bmatrix} \partial\tau/\partial t_1 & \partial\tau/\partial t_2 \\ \partial\alpha/\partial t_1 & \partial\alpha/\partial t_2 \end{bmatrix} = \begin{bmatrix} -1 & 1 \\ 1 & 1 \end{bmatrix} \tag{8.18}$$

so that

$$\begin{aligned} \sigma_\mu^2 &= \frac{1}{4T^2|J|} \int_{-2T}^{2T} \int_{-(2T-|\tau|)}^{2T-|\tau|} C_{XX}(\tau)\mathrm{d}\alpha\mathrm{d}\tau, \\ &= \frac{1}{8T^2} \int_{-2T}^{2T} 2(2T - |\tau|)C_{XX}(\tau)\mathrm{d}\tau, \\ &= \frac{1}{2T} \int_{-2T}^{2T} \left(1 - \frac{|\tau|}{2T}\right) C_{XX}(\tau)\mathrm{d}\tau, \end{aligned} \tag{8.19}$$

where $J = \det(\mathbf{J}) = 2$ is the Jacobian of the transformation. Letting $\sigma_\mu^2 \to 0$ as $T \to \infty$ completes the proof. □

The triangular weighting centered at $\tau = 0$ with unit height and width $4T$ is due to the fact that the autocovariance is a function of a single variable, even though the original integration is performed over two variables. When the axes are rotated to align with variable τ, integration along the other variable α results in the triangular weighting. The weighting ensures that the integration with respect to the transformed axes is the same as that for the original axes, and exploits the fact that $C_{XX}(t_1, t_2)$ is constant along lines perpendicular to the $t_1 = t_2$ line on the t_1–t_2-plane, as was shown previously in Figure 7.16 for $R_{XX}(t_1, t_2)$ (again, the limits of integration in that figure should be changed to $[-T, T]$ for $C_{XX}(t_1, t_2)$ used here).

Since the expression in (8.15) is not always easily evaluated, we provide the following sufficient condition for ergodicity in the mean.

Theorem 8.2. Random process $X(t)$ with autocovariance function $C_{XX}(\tau)$ is ergodic in the mean if

$$\int_{-\infty}^{\infty} |C_{XX}(\tau)|\,\mathrm{d}\tau < \infty. \tag{8.20}$$

Proof. Taking the absolute value of the right-hand-side of (8.19) gives

$$\begin{aligned} \sigma_\mu^2 &= \frac{1}{2T} \left| \int_{-2T}^{2T} \left(1 - \frac{|\tau|}{2T}\right) C_{XX}(\tau)\mathrm{d}\tau \right|, \\ &\le \frac{1}{2T} \int_{-2T}^{2T} \left| \left(1 - \frac{|\tau|}{2T}\right) C_{XX}(\tau) \right| \mathrm{d}\tau \le \frac{1}{2T} \int_{-2T}^{2T} |C_{XX}(\tau)|\,\mathrm{d}\tau. \end{aligned} \tag{8.21}$$

If (8.20) is true, then the right-hand side approaches zero because $T \to \infty$ in the denominator. □

Definition: Ergodic in Mean (Sequence) Random sequence $X[k]$ with constant mean μ_X is *ergodic in the mean* if

$$\lim_{T\to\infty} \frac{1}{2T+1} \sum_{k=-T}^{T} X[k] = \mu_X. \tag{8.22}$$

The scale factor $1/(2T+1)$ is required for the sum to be an unbiased estimate of μ_X because there are $2T+1$ terms.

The corresponding conditions in terms of the autocovariance function $C_{XX}[m]$ are (see Problem 8.1)

$$\lim_{T\to\infty} \frac{1}{2T+1} \sum_{m=-2T}^{2T} \left(1 - \frac{|m|}{2T+1}\right) C_{XX}[m] = 0 \tag{8.23}$$

and

$$\sum_{m=-\infty}^{\infty} |C_{XX}[m]| < \infty. \tag{8.24}$$

Example 8.1. Suppose that $X(t)$ is an uncorrelated process with $R_{XX}(\tau) = \mathcal{E}[X(t)X(t+\tau)] = \mathcal{E}[X(t)]\mathcal{E}[X(t+\tau)] = \mu_X^2$ for $\tau \neq 0$ and has finite variance σ_X^2 so that

$$\begin{aligned} C_{XX}(\tau) &= \begin{cases} \mathcal{E}[X^2(t)] - \mu_X^2, & \tau = 0 \\ 0, & \text{else}, \end{cases} \\ &= \sigma_X^2 \delta(\tau). \end{aligned} \tag{8.25}$$

Observe that

$$\sigma_\mu^2 = \frac{1}{2T} \int_{-2T}^{2T} \left(1 - \frac{|\tau|}{2T}\right) \sigma_X^2 \delta(\tau) \mathrm{d}\tau = (1/2T)\sigma_X^2, \tag{8.26}$$

where the sifting property of the Dirac delta function has been used. This result obviously tends to zero as $T \to \infty$. The sufficient condition is also satisfied:

$$\int_{-\infty}^{\infty} \sigma_X^2 \delta(\tau) \mathrm{d}\tau = \sigma_X^2 < \infty. \tag{8.27}$$

Example 8.2. Consider a random process with zero mean and autocorrelation function $R_{XX}(\tau) = \exp(-|\tau|) = C_{XX}(\tau)$. The sufficient condition for ergodicity in the mean yields

$$\int_{-\infty}^{\infty} |\exp(-|\tau|)| \mathrm{d}\tau = 2 \int_{0}^{\infty} \exp(-\tau) \mathrm{d}\tau = 2 < \infty. \tag{8.28}$$

This example illustrates that it is generally easier to check the sufficient condition: if that condition is not satisfied, then the necessary and sufficient condition in (8.15) should be checked. For completeness, we have

$$\begin{aligned} \frac{1}{2T} \int_{-2T}^{2T} \left(1 - \frac{|\tau|}{2T}\right) \exp(-|\tau|) \mathrm{d}\tau &= \frac{1}{T} \int_{0}^{2T} \left[\exp(-\tau) - (\tau/2T)\exp(-\tau)\right] \mathrm{d}\tau, \\ &= -(1/T)\exp(-\tau)\Big|_0^{2T} + (1/2T^2)(\tau+1)\exp(-\tau)\Big|_0^{2T}, \\ &= (1/T)[1 - \exp(-2T)] \\ &\quad + (1/2T^2)[(2T+1)\exp(-2T) - 1], \end{aligned} \tag{8.29}$$

which approaches zero as $T \to \infty$.

Ergodicity can be extended to other expectations; in particular, consider ergodicity in *correlation*. The ability to estimate the mean and autocorrelation function of a random process from a single realization is useful in many applications, and especially for a linear system as described in subsequent sections.

Definition: Ergodic in Correlation Random process $X(t)$ with autocorrelation function $R_{XX}(\tau)$ is *ergodic in correlation* if

$$\lim_{T\to\infty} \frac{1}{2T} \int_{-T}^{T} X(t)X(t+\tau)\mathrm{d}t = R_{XX}(\tau). \tag{8.30}$$

The time average over a realization is an unbiased estimate of $R_{XX}(\tau)$ even for finite T:

$$\mathcal{E}\left[\frac{1}{2T}\int_{-T}^{T} X(t)X(t+\tau)\mathrm{d}t\right] = R_{XX}(\tau). \tag{8.31}$$

Proceeding as was done for ergodicity in the mean, we can evaluate the variance of the autocorrelation function and show that it converges to zero as $T \to \infty$. In order to simplify the notation, define the following product random process:

$$Y_\tau(t) \triangleq X(t)X(t+\tau), \tag{8.32}$$

which has mean $\mu_{Y_\tau} = R_{XX}(\tau)$ and autocovariance function

$$C_{Y_\tau Y_\tau}(\alpha) = R_{Y_\tau Y_\tau}(\alpha) - R_{XX}^2(\tau). \tag{8.33}$$

A different process $Y_\tau(t)$ is defined for each value of τ; its mean is independent of t and it is wide-sense stationary because its autocovariance function depends on a single variable α.

Theorem 8.3 (Ergodic in correlation). Random process $X(t)$ is ergodic in correlation if and only if

$$\lim_{T\to\infty} \int_{-T}^{T} \left(1 - \frac{|\alpha|}{2T}\right) C_{Y_\tau Y_\tau}(\alpha)\mathrm{d}\alpha = 0. \tag{8.34}$$

Proof. The variance of $Y_\tau(t)$ is

$$\sigma_{Y_\tau}^2 = \mathcal{E}\left[\left|\frac{1}{2T}\int_{-T}^{T} Y_\tau(t)\mathrm{d}t - \mu_{Y_\tau}\right|^2\right], \tag{8.35}$$

which has the same form as in Theorem 8.1. Thus, we can immediately write

$$\sigma_{Y_\tau}^2 = \frac{1}{4T^2}\int_{-T}^{T}\int_{-T}^{T} C_{Y_\tau Y_\tau}(t_1, t_2)\mathrm{d}t_1\mathrm{d}t_2, \tag{8.36}$$

where $C_{Y_\tau Y_\tau}(t_1, t_2) \triangleq \mathcal{E}[Y_\tau(t_1)Y_\tau(t_2)] - \mu_{Y_\tau}^2 = R_{Y_\tau Y_\tau}(t_1, t_2) - R_{XX}^2(\tau)$. A transformation of variables yields

$$\sigma_{Y_\tau}^2 = \frac{1}{2T}\int_{-2T}^{2T} \left(1 - \frac{|\alpha|}{2T}\right) C_{Y_\tau Y_\tau}(\alpha)\mathrm{d}\alpha. \tag{8.37}$$

Letting $\sigma_{Y_\tau}^2 \to 0$ as $T \to \infty$ completes the proof. □

Since (8.37) has the same form as that obtained for ergodicity in the mean, replacing $C_{XX}(\tau)$ with $C_{Y_\tau Y_\tau}(\alpha)$, it is clear that the following condition is sufficient for ergodicity in correlation:

$$\int_{-\infty}^{\infty} C_{Y_\tau Y_\tau}(\alpha)\mathrm{d}\alpha < \infty. \tag{8.38}$$

Observe that $R_{Y_\tau Y_\tau}(\alpha)$ in (8.33) is a fourth-order moment in terms of the original random process:

$$\begin{aligned} R_{Y_\tau Y_\tau}(\alpha) &= \mathcal{E}[Y_\tau(t)Y_\tau(t+\alpha)], \\ &= \mathcal{E}[X(t)X(t+\tau)X(t+\alpha)X(t+\tau+\alpha)]. \end{aligned} \tag{8.39}$$

Example 8.3. Generally, it is not easy to write an expression for the fourth-order moment $R_{Y_\tau Y_\tau}(\alpha)$. However, if $X(t)$ is a zero-mean Gaussian process, then all even moments can be expressed in terms of the second moment. Specifically, if $\{X(t_n)\}$ for $n = 1, 2, 3, 4$, are jointly Gaussian, then

$$\begin{aligned} \mathcal{E}[X(t_1)X(t_2)X(t_3)X(t_4)] &= \mathcal{E}[X(t_1)X(t_2)]\mathcal{E}[X(t_3)X(t_4)] \\ &\quad + \mathcal{E}[X(t_1)X(t_3)]\mathcal{E}[X(t_2)X(t_4)] \\ &\quad + \mathcal{E}[X(t_1)X(t_4)]\mathcal{E}[X(t_2)X(t_3)]. \end{aligned} \tag{8.40}$$

Using this result for $R_{Y_\tau Y_\tau}(\alpha)$ yields

$$\begin{aligned} R_{Y_\tau Y_\tau}(\alpha) &= \mathcal{E}[X(t)X(t+\tau)X(t+\alpha)X(t+\tau+\alpha)] \\ &= R_{XX}^2(\tau) + R_{XX}^2(\alpha) + R_{XX}(\tau+\alpha)R_{XX}(\alpha-\tau) \end{aligned} \tag{8.41}$$

and

$$C_{Y_\tau Y_\tau}(\alpha) = R_{XX}^2(\alpha) + R_{XX}(\tau+\alpha)R_{XX}(\alpha-\tau). \tag{8.42}$$

Problem 8.6 investigates if the zero-mean Gaussian process $X(t)$ with $R_{XX}(\tau)$ in Example 8.2 is ergodic in correlation.

We can also describe ergodicity of $X(t)$ in *distribution*. As was done for the mean and autocorrelation function, the goal is to replace an expectation with a time average. Recall in Chapter 5 that the cumulative distribution function (cdf) of a random variable can be expressed as an expectation using the indicator function. Extending this to a random process yields

$$\mathcal{E}\left[I_{(-\infty,x]}(X(t))\right] = F_{X(t)}(x) \tag{8.43}$$

for which we give the following definition.

Definition: Ergodic in Distribution Random process $X(t)$ is *ergodic in distribution* if for any x

$$\lim_{T\to\infty} \frac{1}{2T}\int_{-T}^{T} I_{(-\infty,x]}(X(t))\mathrm{d}t = F_{X(t)}(x). \tag{8.44}$$

It is instructive to elaborate on the kinds of realizations that would be obtained when using the indicator function and over which we would average to estimate the distribution. Suppose the argument $X(t)$ of the indicator function is an iid standard Gaussian process. An example realization of $X(t)$ is shown in Figure 8.5(a). If $x = \infty$ in (8.43), the corresponding realization for $I_{[0,\infty)}(X(t))$ is obviously a constant 1; likewise, if $x = -\infty$, the realization is a constant 0. For all other values of x, a realization of $I_{(-\infty,x]}(X(t))$ would fluctuate between 0

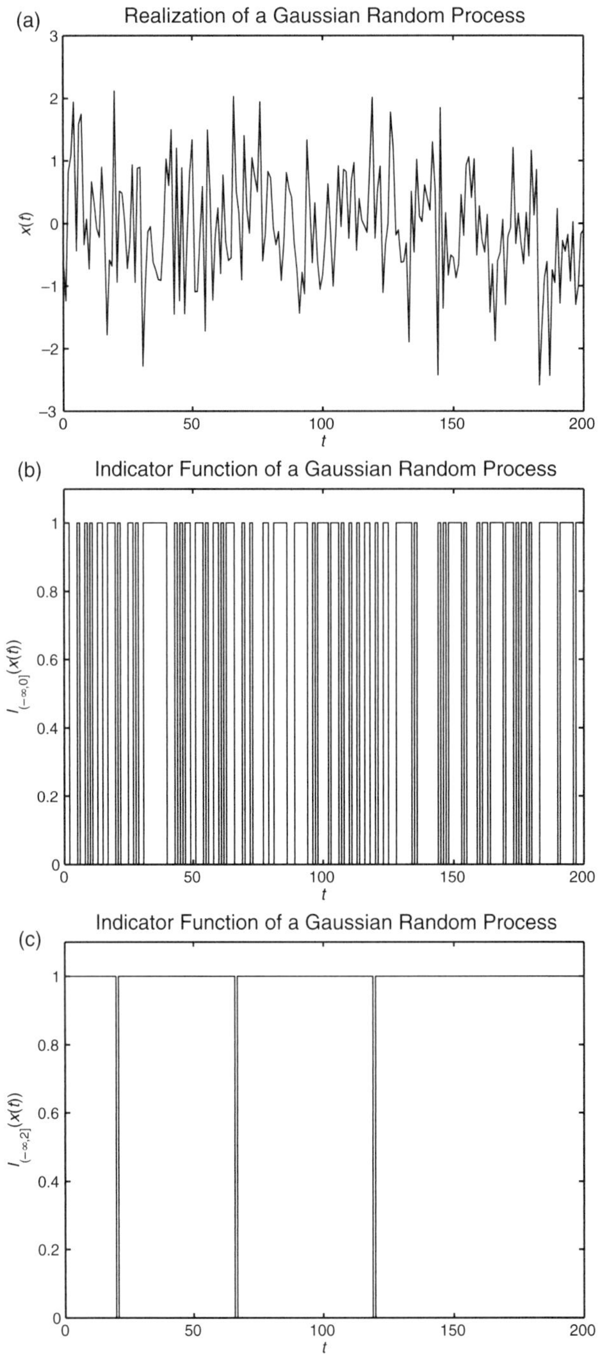

FIGURE 8.5 Random processes and the indicator function. (a) Example realization of a Gaussian random process $X(t)$. (b) Realization of $I_{(-\infty,0]}(X(t))$. (c) Realization of $I_{(-\infty,2]}(X(t))$.

and 1. Such realizations would have periods of sudden change, but there would also be relatively flat regions. When $x = 0$ for this example, we should find that $I_{(-\infty,0]}(X(t))$ spends roughly equal amounts of time at the two values $\{0, 1\}$; an example of such a realization is shown in Figure 8.5(b). Thus, when averaging over that realization, we would expect to obtain 0.5, which is the same as $F_{X(t)}(0)$. For values of $x > 0$, the time average over a given realization should be > 0.5, and for $x < 0$ it should be < 0.5. An example realization of $I_{(-\infty,2]}(X(t))$ for $x = 2$ is shown in Figure 8.5(c). This is the basic idea of how the cdf is estimated as an expectation of a binary process via the indicator function that "keeps track" of when $X(t)$ exceeds x.

Based on the previous results for ergodicity in the mean and correlation, we can readily provide a necessary and sufficient condition for ergodicity in distribution. In order to simplify the notation, define the random process

$$Z_x(t) \triangleq I_{(-\infty,x]}(X(t)), \tag{8.45}$$

where the subscript x denotes the specific value used in the indicator function. Its expectation is

$$\begin{aligned}\mathcal{E}[Z_x(t)] &= \int_{-\infty}^{\infty} I_{(-\infty,x]}(x(t)) f_{X(t)}(x(t))\mathrm{d}x(t),\\ &= \int_{-\infty}^{x} f_{X(t)}(x(t))\mathrm{d}x(t) = F_{X(t)}(x)\end{aligned} \tag{8.46}$$

showing that it is an unbiased estimator of the cdf. The autocorrelation function is

$$\begin{aligned}R_{Z_xZ_x}(\tau) &= \mathcal{E}\left[I_{(-\infty,x]}(X(t))I_{(-\infty,x]}(X(t+\tau))\right],\\ &= \int_{-\infty}^{\infty}\int_{-\infty}^{\infty} I_{(-\infty,x]}(x(t))I_{(-\infty,x]}(x(t+\tau))\\ &\quad \cdot f_{X(t),X(t+\tau)}(x(t), x(t+\tau))\mathrm{d}x(t)\mathrm{d}x(t+\tau),\\ &= \int_{-\infty}^{x}\int_{-\infty}^{x} f_{X(t),X(t+\tau)}(x(t), x(t+\tau))\mathrm{d}x(t)\mathrm{d}x(t+\tau),\\ &= P(X(t) \le x, X(t+\tau) \le x) = F_{X(t),X(t+\tau)}(x, x),\end{aligned} \tag{8.47}$$

which is a joint cdf for random process $X(t)$ at two different time instants and with the same argument x. Since $X(t)$ is assumed to be wide-sense stationary, $Z_x(t)$ is likewise wide-sense stationary so that $R_{Z_xZ_x}(\tau)$ depends only on the time lag τ. The corresponding autocovariance function is

$$C_{Z_xZ_x}(\tau) = R_{Z_xZ_x}(\tau) - F^2_{X(t)}(x). \tag{8.48}$$

Theorem 8.4 (Ergodic in distribution). Random process $X(t)$ is ergodic in distribution $F_{X(t)}(x)$ for any x if and only if

$$\lim_{T\to\infty} \int_{-2T}^{2T} \left(1 - \frac{\tau}{2T}\right) C_{Z_xZ_x}(\tau)\mathrm{d}\tau = 0. \tag{8.49}$$

Proof. The proof is similar to that used for ergodicity in the mean, except that $C_{Z_xZ_x}(\tau)$ is substituted for $C_{XX}(\tau)$. □

A sufficient condition for ergodicity in distribution is

$$\int_{-\infty}^{\infty} |C_{Z_xZ_x}(\tau)|\mathrm{d}\tau < \infty. \tag{8.50}$$

Example 8.4. If random process $X(t)$ is iid, then it is ergodic in distribution (as well as ergodic in mean). This follows because the autocorrelation function splits into a product $R_{Z_xZ_x}(\tau) = F_{X(t)}(x)F_{X(t+\tau)}(x)$ so that $C_{Z_xZ_x}(\tau) = 0$ for any $\tau \neq 0$ because $F_{X(t+\tau)}(x) = F_{X(t)}(x)$. For $\tau = 0$:

$$\begin{aligned} R_{Z_xZ_x}(0) &= \mathcal{E}\left[I^2_{(-\infty,x]}(X(t))\right] = \mathcal{E}\left[I_{(-\infty,x]}(X(t))\right], \\ &= \int_{-\infty}^{x} f_{X(t)}(x(t))\mathrm{d}x(t) = P(X(t) \leq x) = F_{X(t)}(x), \end{aligned} \tag{8.51}$$

and

$$C_{Z_xZ_x}(\tau) = [F_{X(t)}(x) - F^2_{X(t)}(x)]\delta(\tau) = F_{X(t)}(x)[1 - F_{X(t)}(x)]\delta(\tau). \tag{8.52}$$

$C_{Z_xZ_x}(0)$ is the variance of $Z_x(t)$ which in this case is $P(X(t) \leq x)P(X(t) > x)$. As a result, $Z_z(t)$ for a specific t can be viewed as a Bernoulli random variable with probability of success $F_{X(t)}(x)$. Substituting this into (8.49) gives

$$\frac{1}{2T}\int_{-2T}^{2T}\left(1 - \frac{|\tau|}{2T}\right)F_{X(t)}(x)[1 - F_{X(t)}(x)]\delta(\tau)\mathrm{d}\tau = (1/2T)F_{X(t)}(x)[1 - F_{X(t)}(x)]. \tag{8.53}$$

Since $F_{X(t)}(x)$ is bounded, the last result tends to zero as $T \to \infty$.

8.4 EIGENFUNCTIONS OF $R_{XX}(\tau)$

Consider the integral equation for the Karhunen-Loève expansion (KLE) in (7.258), but with the finite limits $[0, T]$ extended to $\pm\infty$:

$$\int_{-\infty}^{\infty} R_{XX}(t_1 - t_2)\phi_n(t_2)\mathrm{d}t_2 = \lambda_n\phi_n(t_1), \quad t_1 \in \mathcal{R}, \tag{8.54}$$

where $\{\phi_n(t)\}$ are eigenfunctions and $\{\lambda_n\}$ are eigenvalues. Since $X(t)$ is wide-sense stationary, the argument of the autocorrelation function depends on the difference of the two time instants. As a result, this integral has the form of a *convolution*, corresponding to the LTI system in Figure 8.1, given by

$$y(t) = \int_{-\infty}^{\infty} h(t - \tau)x(\tau)\mathrm{d}\tau. \tag{8.55}$$

For an LTI system, complex exponentials are eigenfunctions:

$$\begin{aligned} \int_{-\infty}^{\infty} h(t - \tau)\exp(j\omega_n\tau)\mathrm{d}\tau &= \int_{-\infty}^{\infty} h(v)\exp(j\omega_n(t - v))\mathrm{d}v, \\ &= H(\omega_n)\exp(j\omega_n t), \end{aligned} \tag{8.56}$$

where we have changed variables to $v = t - \tau$, and

$$H(\omega_n) = \int_{-\infty}^{\infty} h(v)\exp(-j\omega_n v)\mathrm{d}v \tag{8.57}$$

is the frequency response of $h(t)$ evaluated at the particular frequency ω_n. This is a fundamental result covered in basic courses on signals and systems: the output has the same frequency as the input, and only its magnitude and phase are possibly different, as determined by $H(\omega_n)$. Since the autocorrelation function is just a specific type of function in the convolution of (8.54), the complex exponentials are also eigenfunctions with the PSD $S_{XX}(\omega)$ used in place of $H(\omega)$. Substituting

$$\phi_n(t) = \exp(j\omega_n t), \quad \lambda_n = S_{XX}(\omega_n) \tag{8.58}$$

into (8.54) and rearranging the result gives

$$\int_{-\infty}^{\infty} R_{XX}(t_1 - t_2)\exp(-j\omega_n(t_1 - t_2))\mathrm{d}t_2 = S_{XX}(\omega_n). \tag{8.59}$$

Changing variables to $\tau \triangleq t_1 - t_2$ and letting $\omega_n \to \omega$, we find that $S_{XX}(\omega)$ is the Fourier transform of $R_{XX}(\tau)$, as mentioned in the introduction:

$$\int_{-\infty}^{\infty} R_{XX}(\tau)\exp(-j\omega\tau)\mathrm{d}\tau = S_{XX}(\omega). \tag{8.60}$$

Unlike the KLE for $X(t)$ on $\mathcal{T} = [0, T]$ where a countable number of eigenfunctions are obtained, the number of eigenfunctions above for $\mathcal{T} = \mathcal{R}$ is uncountable. Also, recall from Chapter 7 that when $R_{XX}(\tau)$ is periodic with period T, a Fourier series representation of the process can be derived. The result in this section can be viewed as an "extension" of the Fourier series result to a Fourier transform for the aperiodic function $R_{XX}(\tau)$. However, it is important to note that although (8.54) is a convolution, $R_{XX}(\tau)$ is *not* being used as a filter for $X(t)$. In Section 8.5, we derive the PSD using a different approach and show how it is related to the *power* of a random process. Various properties of the PSD are described, and later in the chapter we prove the result in Figure 8.1 that shows how $S_{YY}(\omega)$ is related to $S_{XX}(\omega)$ when $X(t)$ is filtered by $H(\omega)$.

8.5 POWER SPECTRAL DENSITY

We begin with a brief review of power and energy for nonrandom signals.

Definition: Energy The *energy* of deterministic waveform $x(t)$ is defined as

$$E \triangleq \int_{-\infty}^{\infty} x^2(t)\mathrm{d}t. \tag{8.61}$$

If $x(t)$ is periodic on $\mathcal{R}$, then E is infinite. For example, for $x(t) = \cos(\omega t)$:

$$E = \int_{-\infty}^{\infty} \cos^2(\omega t)\mathrm{d}t = \int_{-\infty}^{\infty} (1/2)[1 + \cos(2\omega t)]\mathrm{d}t = \infty.$$

For such signals, we examine instead the average power.

Definition: Average Power The *average power* of deterministic $x(t)$ is defined as

$$P \triangleq \lim_{T\to\infty} \frac{1}{2T}\int_{-T}^{T} x^2(t)\mathrm{d}t. \tag{8.62}$$

For the cosine function, the average power is finite:

$$P = \lim_{T\to\infty} \frac{1}{2T}\int_{-T}^{T} (1/2)[1 + \cos(2\omega t)]\mathrm{d}t = 1/2. \tag{8.63}$$

Unlike an infinite-length periodic signal, the energy E of a finite-duration nonrandom signal is finite and thus has zero average power P.

Definition: Energy and Power Signals An *energy signal* has finite energy and zero power. A *power signal* has infinite energy and finite power. They are characterized concisely as $0 < E < \infty$ for an energy signal and $0 < P < \infty$ for a power signal.

The rectangle function rect(t) is an example of an energy signal whereas the sine function is a power signal.

For random process $X(t)$, we are interested in realizations $\{x(t)\}$ defined on $\mathcal{R}$, and so it is a power signal. In order to properly define the average power of a random process, we need to introduce an expectation over the ensemble of realizations. First, consider the average power of the specific realization $x(t)$ on $[-T, T]$ as follows:

$$P_T \triangleq \frac{1}{2T} \int_{-T}^{T} x^2(t) \mathrm{d}t. \tag{8.64}$$

Using Parseval's theorem (see Appendix C), the corresponding result in the frequency domain is

$$P_T = \frac{1}{2\pi} \int_{-\infty}^{\infty} (1/2T)|X(\omega)|^2 \mathrm{d}\omega, \tag{8.65}$$

where $X(\omega)$ is the Fourier transform of $x(t)$ on $[-T, T]$. The argument ω distinguishes this uppercase symbol from the time-domain random process $X(t)$. However, note that $X(\omega)$ can be viewed as a *realization* of a random process *in the frequency domain*, obtained by performing a Fourier transform of realization $x(t)$ of $X(t)$.

Since $x(t)$ is any realization, it can be replaced in (8.64) with $X(t)$ and the meaning of the integration is clear. This is similar to that done for conditional expectation in Chapter 5 when we conditioned on a random variable instead of a specific outcome. Recall there that initially we considered expectations of the form $\mathcal{E}[X|Y = y]$ which is a function of outcome y. This was extended to conditioning that is random: $\mathcal{E}[X|Y]$ is interpreted to be a function of random variable Y. Likewise, (8.64) is a function of a realization that we want to generalize to be a function of random process $X(t)$. As a result, realization $X(\omega)$ in (8.65) can be replaced by the corresponding random process in the frequency domain so that in both expressions above, P_T becomes a random variable.

We define the average power of a random process by taking the expectation of P_T and letting $T \to \infty$:

$$P_{XX} \triangleq \lim_{T\to\infty} \mathcal{E}[P_T] = \frac{1}{2\pi} \int_{-\infty}^{\infty} \lim_{T\to\infty} (1/2T)\mathcal{E}[|X(\omega)|^2] \mathrm{d}\omega, \tag{8.66}$$

where it is assumed that the limit and the integration can be interchanged. Defining the following limit:

$$S_{XX}(\omega) \triangleq \lim_{T\to\infty} (1/2T)\mathcal{E}[|X(\omega)|^2], \tag{8.67}$$

we have

$$P_{XX} = \frac{1}{2\pi} \int_{-\infty}^{\infty} S_{XX}(\omega) \mathrm{d}\omega. \tag{8.68}$$

This last expression shows that $S_{XX}(\omega)$ is a *density function* for power: when it is integrated over some range of ω, it provides the average power over those frequencies. The average power of the random process over the entire range $\omega \in \mathcal{R}$ is P_{XX}. The PSD $S_{XX}(\omega)$ is the same quantity discussed in Section 8.4 where we considered eigenfunctions of $R_{XX}(\tau)$; it has units of watts/radians.

The final step is the derivation of a useful expression for $S_{XX}(\omega)$. Substituting the Fourier transform for each term of $|X(\omega)|^2 = X^*(\omega)X(\omega)$ into (8.67) gives (the superscript * denotes complex conjugation)

$$\begin{aligned} S_{XX}(\omega) &= \lim_{T\to\infty} \frac{1}{2T}\mathcal{E}\left[\int_{-T}^{T} X(t_1)\exp(-j\omega t_1)\mathrm{d}t_1 \int_{-T}^{T} X(t_2)\exp(j\omega t_2)\mathrm{d}t_2\right], \\ &= \lim_{T\to\infty} \frac{1}{2T}\int_{-T}^{T}\int_{-T}^{T} R_{XX}(t_1, t_2)\exp(-j\omega(t_1 - t_2))\mathrm{d}t_1\mathrm{d}t_2, \end{aligned} \tag{8.69}$$

where we have used different variables of integration and assumed that random process $X(t)$ is real valued. The autocorrelation function $R_{XX}(t_1, t_2) \triangleq \mathcal{E}[X(t_1)X(t_2)]$ is defined for $t_1, t_2 \in [-T, T]$ before taking the limit. Assume also that $X(t)$ is wide-sense stationary such that $R_{XX}(t_1, t_2) = R_{XX}(t_2 - t_1)$ and let $\tau \triangleq t_1 - t_2$. Using the same transformation of variables as was done for ergodicity, the PSD becomes

$$\begin{aligned} S_{XX}(\omega) &= \lim_{T\to\infty} \frac{1}{2T}\int_{-2T}^{2T} (2T - |\tau|)R_{XX}(\tau)\exp(-j\omega\tau)\mathrm{d}\tau, \\ &= \lim_{T\to\infty} \int_{-2T}^{2T} R_{XX}(\tau)\exp(-j\omega\tau)\mathrm{d}\tau \\ &\quad - \lim_{T\to\infty} \int_{-2T}^{2T} (1/2T)|\tau|R_{XX}(\tau)\exp(-j\omega\tau)\mathrm{d}\tau. \end{aligned} \tag{8.70}$$

If the following boundedness condition is assumed:

$$\int_{-\infty}^{\infty} |\tau|R_{XX}(\tau)\mathrm{d}\tau < \infty, \tag{8.71}$$

the second integral above becomes zero in the limit, yielding

$$S_{XX}(\omega) = \int_{-\infty}^{\infty} R_{XX}(\tau)\exp(-j\omega\tau)\mathrm{d}\tau, \tag{8.72}$$

which is identical to the result in (8.60) that was derived using eigenfunctions. The condition in (8.71) requires that the autocorrelation function decay to zero sufficiently fast, which means that the random process cannot be "too correlated." We summarize these results in the following definition.

Definition: Power Spectral Density The *power spectral density* (*PSD*) of wide-sense stationary process $X(t)$ with autocorrelation function $R_{XX}(\tau)$ is

$$S_{XX}(\omega) = \int_{-\infty}^{\infty} R_{XX}(\tau)\exp(-j\omega\tau)\mathrm{d}\tau, \tag{8.73}$$

where ω is radian frequency. In terms of ordinary frequency f, where $\omega = 2\pi f$, the PSD is

$$S_{XX}(f) = \int_{-\infty}^{\infty} R_{XX}(\tau)\exp(-j2\pi f\tau)\mathrm{d}\tau. \tag{8.74}$$

The autocorrelation function is recovered from the inverse Fourier transform of the PSD as follows:

$$R_{XX}(\tau) = \frac{1}{2\pi}\int_{-\pi}^{\pi} S_{XX}(\omega)\exp(j\omega\tau)\mathrm{d}\omega \tag{8.75}$$

or by substituting $\omega = 2\pi f$:

$$R_{XX}(\tau) = \int_{-\infty}^{\infty} S_{XX}(f)\exp(j2\pi f\tau)\mathrm{d}f. \tag{8.76}$$

For completeness, we include other forms of the PSD depending on the type of random signal and the type of transform.

- *Random process, Laplace transform:*

$$S_{XX}(s) = \int_{-\infty}^{\infty} R_{XX}(\tau)\exp(-s\tau)\mathrm{d}\tau, \tag{8.77}$$

$$R_{XX}(\tau) = \frac{1}{2\pi j}\int_{c-j\infty}^{c+j\infty} S_{XX}(s)\exp(s\tau)\mathrm{d}s. \tag{8.78}$$

- *Random sequence, discrete-time Fourier transform (DTFT):*

$$S_{XX}(j\omega) = \sum_{m=-\infty}^{\infty} R_{XX}[m]\exp(-j\omega m), \tag{8.79}$$

$$R_{XX}[m] = \frac{1}{2\pi}\int_{-\pi}^{\pi} S_{XX}(j\omega)\exp(j\omega m)\mathrm{d}\omega. \tag{8.80}$$

- *Random sequence, z-transform:*

$$S_{XX}(z) = \sum_{m=-\infty}^{\infty} R_{XX}[m]z^{-m}, \tag{8.81}$$

$$R_{XX}[m] = \frac{1}{2\pi j}\oint_{C} S_{XX}(z)z^{m-1}\mathrm{d}z. \tag{8.82}$$

Table 8.2 provides some example autocorrelation functions and their PSDs. Appendix G discusses the regions of convergence (ROCs) for the Laplace and z-transforms. Figure 8.6 provides an example of an exponential autocorrelation function and its PSD. When we are interested in the frequency properties (spectrum) of a random process, the PSD in terms of ω or f is typically used, whereas the PSD described by the Laplace and z-transforms tend to be more useful when studying systems with random inputs.

TABLE 8.2 Example PSDs

$R_{XX}(\tau)$	$S_{XX}(\omega)$	$S_{XX}(s)$
White noise: $\delta(\tau)$	1	1
Exponential: $\exp(-\alpha\|\tau\|)$	$2\alpha/(\alpha^2+\omega^2)$	$2\alpha/(\alpha^2-s^2)$
Rectangular: $\mathrm{rect}(\alpha\tau)$	$(1/\|\alpha\|)\mathrm{sinc}(\omega/2\pi\alpha)$	$(2\alpha/\|\alpha\|)\sinh(s/2\alpha)/s$
Triangular: $1-\alpha\|\tau\|/T$	$(1/\|\alpha\|)\mathrm{sinc}^2(\omega/2\pi\alpha)$	$(4\alpha^2/\|\alpha\|)\sinh^2(s/2\alpha)/s^2$
$R_{XX}[m]$	$S_{XX}(j\omega)$	$S_{XX}(z)$
White noise: $\delta[m]$	1	1
Exponential: $\alpha^{\|m\|}$	$(1-\alpha^2)/[1+\alpha^2-2\alpha\cos(\omega)]$	$(1-\alpha^2)/[(1-\alpha z^{-1})(1-\alpha z)]$

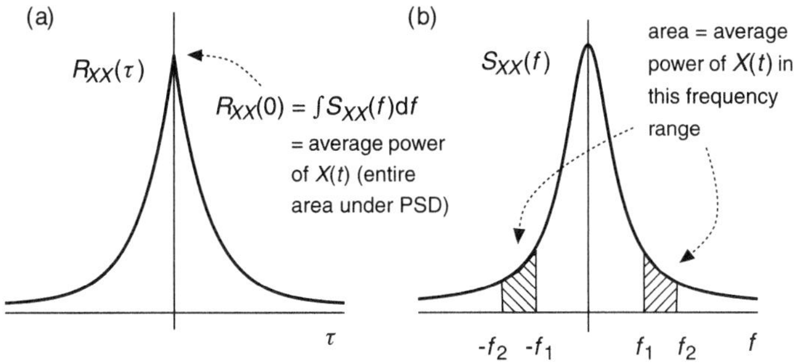

FIGURE 8.6 Autocorrelation function and PSD. (a) Exponential $R_{XX}(\tau)$. (b) $S_{XX}(f)$ (watts/Hz). Average power in a particular frequency range is obtained by integrating the PSD over that range.

Properties of the PSD include the following:

- *Average power.* The second moment of $X(t)$ is

$$\mathcal{E}[X^2(t)] = R_{XX}(0) = \frac{1}{2\pi}\int_{-\infty}^{\infty} S_{XX}(\omega)\mathrm{d}\omega, \tag{8.83}$$

  which is obtained by letting $\tau = 0$ in (8.76).
- *Real-valued.* $S_{XX}(\omega)$ is a real-valued function of ω. Since $R_{XX}(\tau)$ is an even function of τ, we can write

$$\begin{aligned} S_{XX}(\omega) &= \int_0^{\infty} R_{XX}(\tau)\big[\exp(-j\omega\tau) + \exp(j\omega\tau)\big]\mathrm{d}\tau, \\ &= 2\int_0^{\infty} R_{XX}(\tau)\cos(\omega\tau)\mathrm{d}\tau, \end{aligned} \tag{8.84}$$

  which does not depend on j.
- *Even.* $S_{XX}(\omega)$ is an even function. This follows from (8.84) because $\cos(\omega\tau)$ is an even function.
- *Nonnegative.* $S_{XX}(\omega) \geq 0$ because of the original definition in (8.67).
- *White noise.* $S_{XX}(\omega) = 1$ when $X(t)$ is a white-noise process with autocorrelation function $R_{XX}(\tau) = \delta(\tau)$. This result follows from the sifting property of the Dirac delta function:

$$S_{XX}(\omega) = \int_{-\infty}^{\infty} \delta(\tau)\exp(-j\omega\tau)\mathrm{d}\tau = 1. \tag{8.85}$$

  White noise is discussed later in this chapter.
- *Derivative process.* For $Y(t) = X'(t)$:

$$S_{YY}(\omega) = \omega^2 S_{XX}(\omega). \tag{8.86}$$

  This follows from $R_{YY}(\tau) = -\mathrm{d}^2R_{XX}(\tau)/\mathrm{d}\tau^2$ derived in Chapter 7 and the derivative property of the Fourier transform.
- *Spectral factorization theorem.* When $S_{XX}(s)$ is a rational function of s, it can be factored as follows:

$$S_{XX}(s) = S_{XX}^{+}(s)S_{XX}^{-}(s), \tag{8.87}$$

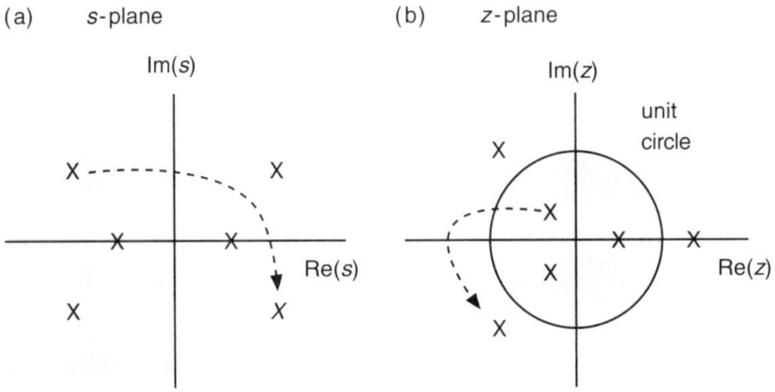

FIGURE 8.7 Example pole locations for the PSD. (a) Poles for $S_{XX}(s) = S_{XX}^{+}(s)S_{XX}^{-}(s)$. (b) Poles for $S_{XX}(z) = S_{XX}^{+}(z)S_{XX}^{-}(z)$. The poles in the shaded regions correspond to the causal part of the PSD.

where the poles and zeros of $S_{XX}^{+}(s)$ are on the left-half of the s-plane and those of $S_{XX}^{-}(s)$ are on the right-half. It is shown in Section 8.6 that $S_{XX}(s)$ can be generated as the output of a causal filter with transfer function $H(s)$ when the input has a flat PSD. It is given by the product

$$S_{XX}(s) = H(s)H(-s). \tag{8.88}$$

Since $H(s)$ is stable, all its poles are on the left-half of the s-plane and correspond to those of $S_{XX}^{+}(s)$. Likewise, the poles of the anticausal $H(-s)$ are on the right-half of the s-plane and correspond to those of $S_{XX}^{-}(s)$. (We assume that $H(s)$ is strictly stable with no poles on the imaginary axis.)

- *Pole symmetry.* Due to the spectral factorization property, the poles of $S_{XX}(s)$ have the following symmetry:

$$\text{pole at } s = -\sigma_1 + j\omega_1 \Longrightarrow \text{pole at } s = \sigma_1 - j\omega_1. \tag{8.89}$$

Examples are shown in Figure 8.7(a) where a double symmetry is evident. The symmetry about the real axis is due to the fact that the autocorrelation function is even, yielding complex conjugate pairs. The symmetry about the imaginary axis follows because $S_{XX}(s)$ is an even function of ω. The dashed line in the figure shows which two complex poles are "related" in (8.89) due to the factorization.

- *Zero symmetry.* Of course, there is no restriction on the zeros of $H(s)$ in order that it be stable. However, it will be convenient later in Chapter 11 that only the zeros on the left-half of the s-plane be included in $S_{XX}^{+}(s)$, so that the zeros have the same symmetry as the poles. The reason for this particular factorization is that the Wiener filter covered there requires that $S_{XX}^{+}(s)$ be inverted such that the zeros become poles, and these should be on the left-half of the s-plane for stability. (For this reason, we also assume there are no zeros on the imaginary axis.) The condition when a transfer function has all its zeros on the left-half of the s-plane is known as *minimum phase*. The group delay is less when $S_{XX}^{+}(s)$ excludes any zeros from the right-half of the s-plane.

The corresponding properties for a random sequence with PSD $S_{XX}(j\omega)$ are left as an exercise for the reader. However, we give the spectral factorization property for $S_{XX}(z)$:

$$S_{XX}(z) = S_{XX}^{+}(z)S_{XX}^{-}(z), \tag{8.90}$$

where the poles and zeros of $S^+_{XX}(z)$ are inside the unit circle on the z-plane, and those of $S^-_{XX}(s)$ are outside the unit circle. The result corresponding to (8.88) for transfer function $H(z)$ is

$$S_{XX}(z) = H(z)H(z^{-1}), \tag{8.91}$$

and the pole symmetry is

$$\begin{aligned} &\text{pole at } z = r_1 \exp(j\omega_1) = r_1[\cos(\omega) + j\sin(\omega)], \\ \Longrightarrow\ &\text{pole at } z = (1/r_1)\exp(-j\omega_1) = (1/r_1)[\cos(\omega) - j\sin(\omega)], \end{aligned} \tag{8.92}$$

which also applies to the zeros. Figure 8.7(b) shows some examples, where the dashed line indicates which two complex poles are "related" in (8.92) due to the factorization.

Although in subsequent sections the various forms of the PSD are considered, that is, $S_{XX}(\omega)$, $S_{XX}(f)$, $S_{XX}(s)$, $S_{XX}(j\omega)$, and $S_{XX}(z)$, we usually state a result for only one form, and the other results can be inferred.

8.6 POWER SPECTRAL DISTRIBUTION

In Chapter 3 on random variables, we defined the cdf $F_X(x)$ and pdf $f_X(x)$ for random variable X, which are related as follows:

$$f_X(x) = \frac{\mathrm{d}}{\mathrm{d}x} F_X(x) \tag{8.93}$$

provided $F_X(x)$ is differentiable. In Chapter 5 on expectation and moments, we defined the expectation of function $g(X)$ of random variable X as follows:

$$\mathcal{E}[g(X)] = \int_{-\infty}^{\infty} g(x)\mathrm{d}F_X(x), \tag{8.94}$$

which is a Riemann–Stieltjes integral with integrator $F_X(x)$. The cdf $F_X(x) = P(X \le x)$ denotes a particular probability whereas $f_X(x)$ is a density function that is integrated in order to obtain a probability:

$$P(X \le x) = \int_{-\infty}^{x} f_X(x)\mathrm{d}x. \tag{8.95}$$

The PSD is also a density function that is integrated to obtain the average power of wide-sense stationary process $X(t)$. For example, the overall average power is

$$\int_{-\infty}^{\infty} S_{XX}(f)\mathrm{d}f = R_{XX}(0), \tag{8.96}$$

and the average power in some frequency range $f \in [-B, B]$ is

$$P_{XX} = \int_{-B}^{B} S_{XX}(f)\mathrm{d}f. \tag{8.97}$$

Similar to the cdf, we can describe a power spectral *distribution* function.

Definition: Power Spectral Distribution The *power spectral distribution* $S_{XX}^D(f)$ of wide-sense stationary random process $X(t)$ is defined by

$$R_{XX}(\tau) = \int_{-\infty}^{\infty} \exp(j2\pi f\tau)\mathrm{d}S_{XX}^D(f), \tag{8.98}$$

where $R_{XX}(\tau)$ is the autocorrelation function.

In order to be consistent with the notation $f_X(x)$ and $F_X(x)$, we could use $s_X(f)$ and $S_X(f)$ for the PSD and power spectral distribution function, respectively. However, uppercase $S_{XX}(f)$ is commonly used in engineering for the PSD, and so we use a superscript to denote the power spectral distribution $S_{XX}^D(f)$.

It is interesting to note that the expression in (8.98) has a form similar to that of the characteristic function with $\omega = 2\pi f$ (see Chapter 5):

$$\Phi_X(\omega) = \int_{-\infty}^{\infty} \exp(j\omega x)\mathrm{d}F_X(x) \tag{8.99}$$

though (8.98) is an inverse Fourier transform and $\Phi_X(\omega)$ is a Fourier transform (with $-\omega$ in place of ω). Since $S_{XX}^D(f)$ is a *distribution*, it has the usual properties:

- *Limits.*

$$\lim_{f\to-\infty} S_{XX}^D(f) = 0, \qquad \lim_{f\to\infty} S_{XX}^D(f)/R_{XX}(0) = 1. \tag{8.100}$$

- *Nondecreasing.* For $f_1 \leq f_2$,

$$S_{XX}^D(f_1) \leq S_{XX}^D(f_2). \tag{8.101}$$

- *Continuous from the right.*

The only caveat above is that $S_{XX}^D(f)$ must be normalized by $R_{XX}(0) = \int_{-\infty}^{\infty} \mathrm{d}S_{XX}^D(f)$ in order to have an upper limit of 1 in (8.100), because the overall average power of the random process is not restricted to be 1. Although the two density functions $f_X(x)$ and $S_{XX}(f)$ are real-valued and nonnegative, an important difference between them is that the PSD must be an even function. This means that the power spectral distribution function has an additional property:

- *Normalized center.* $S_{XX}^D(0)/R_{XX}(0) = 1/2$.

If $S_{XX}^D(f)$ is continuous, then $\mathrm{d}S_{XX}^D(f) = S_{XX}(f)\mathrm{d}f$. If $S_{XX}^D(f)$ has step discontinuities, then Dirac delta functions can be used to represent $S_{XX}(f)$ (similar to that used by $f_X(x)$ for discrete and mixed random variables).

Example 8.5. From Table 8.2, we find that $S_{XX}(f) = 2\beta\alpha/(4\pi^2 f^2 + \alpha^2)$ for $R_{XX}(\tau) = \beta\exp(-\alpha|\tau|)$. The corresponding power spectral distribution function is

$$\begin{aligned} S_{XX}^D(f) &= 2\beta\alpha \int_{-\infty}^{f} \frac{1}{4\pi^2 f^2 + \alpha^2}\mathrm{d}f, \\ &= (\beta/\pi)\tan^{-1}(2\pi f/\alpha)\Big|_{-\infty}^{f} = (\beta/\pi)\tan^{-1}(2\pi f/\alpha) + \beta/2. \end{aligned} \tag{8.102}$$

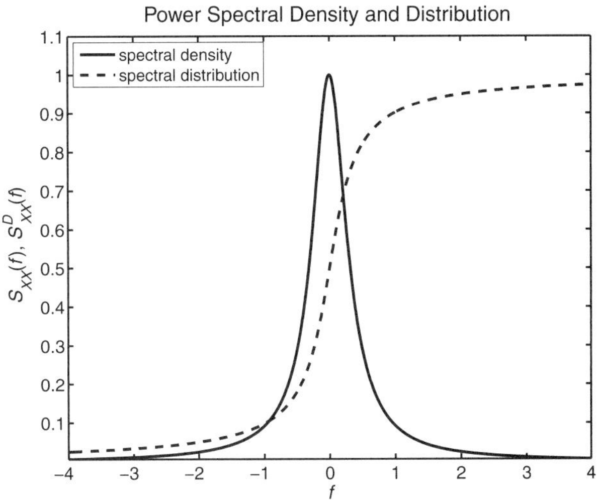

FIGURE 8.8 PSD and power spectral distribution for Example 8.5 with $\alpha = 2$ and $\beta = 1$.

Observe that $S^D_{XX}(\infty) = \beta$ which is the power of the random process in this case: $R_{XX}(0) = \beta$. Plots of $S_{XX}(f)$ and $S^D_{XX}(f)$ for $\beta = 1$ and $\alpha = 2$ are shown in Figure 8.8.

Example 8.6. Suppose the spectral distribution is piecewise linear as follows:

$$S^D_{XX}(f) = \begin{cases} 0, & f < -f_1 \\ f/f_1 + 1, & |f| \le f_1 \\ 2, & f > f_1, \end{cases} \tag{8.103}$$

where $f_1 > 0$. The corresponding PSD is

$$S_{XX}(f) = \begin{cases} 0, & |f| > f_1 \\ 1/f_1, & |f| \le f_1, \end{cases} \tag{8.104}$$

which is a rectangle with area 2: $S_{XX}(f) = (1/f_1)\text{rect}(f/2f_1)$, and thus the autocorrelation function is

$$R_{XX}(\tau) = 2\text{sinc}(2f_1\tau). \tag{8.105}$$

The overall average power is $R_{XX}(0) = 2$ which is the same as $S^D_{XX}(\infty)$.

8.7 CROSS-POWER SPECTRAL DENSITY

The cross-PSD for random processes $X(t)$ and $Y(t)$ can be derived using the same approach used for $S_{XX}(\omega)$. From Parseval's theorem, the power can be written as follows:

$$P_{XY} = \frac{1}{2\pi}\int_{-\infty}^{\infty} S_{XY}(\omega)\mathrm{d}\omega, \tag{8.106}$$

where

$$S_{XY}(\omega) \triangleq \lim_{T\to\infty} (1/2T)\mathcal{E}[X(\omega)Y^*(\omega)]. \tag{8.107}$$

Substituting Fourier transforms gives

$$\begin{aligned} S_{XY}(\omega) &= \lim_{T\to\infty} \frac{1}{2T}\mathcal{E}\left[\int_{-T}^{T} X(t_1)\exp(-j\omega t_1)\mathrm{d}t_1 \int_{-T}^{T} Y(t_2)\exp(j\omega t_2)\mathrm{d}t_2\right], \\ &= \lim_{T\to\infty} \frac{1}{2T}\int_{-T}^{T}\int_{-T}^{T} R_{XY}(t_1, t_2)\exp(-j\omega(t_1 - t_2))\mathrm{d}t_1\mathrm{d}t_2. \end{aligned} \tag{8.108}$$

It is clear from the derivation for $S_{XX}(\omega)$ that the steps are essentially the same here, leading to the following result which we state as a definition.

Definition: Cross-Power Spectral Density The *cross-PSD* of wide-sense stationary processes $X(t)$ and $Y(t)$ with cross-correlation function $R_{XY}(\tau) \triangleq \mathcal{E}[X(t)Y(t+\tau)]$ is

$$S_{XY}(\omega) = \int_{-\infty}^{\infty} R_{XY}(\tau)\exp(-j\omega\tau)\mathrm{d}\tau, \tag{8.109}$$

where ω is radian frequency.

(Note that some books define the cross-correlation function to be $\mathcal{E}[X(t+\tau)Y(t)]$.) The other forms of the PSD with arguments f and s, as well as those for random sequences with arguments $j\omega$ and z, can also be defined for the cross-PSD.

The cross-PSD of random processes $\{X(t), Y(t)\}$ has the following properties. Similar properties hold for discrete-time sequences $\{X[k], Y[k]\}$ with cross-PSD $S_{XY}(j\omega)$.

- *Asymmetric with respect to subscripts.* $S_{YX}(\omega) = S_{XY}(-\omega) = S^*_{XY}(\omega)$. From the cross-correlation property $R_{YX}(\tau) = R_{XY}(-\tau)$:

$$S_{YX}(\omega) = \int_{-\infty}^{\infty} R_{YX}(\tau)\exp(-j\omega\tau)\mathrm{d}\tau = \int_{-\infty}^{\infty} R_{XY}(-\tau)\exp(-j\omega\tau)\mathrm{d}\tau. \tag{8.110}$$

  Changing variables to $\alpha = -\tau$ gives

$$S_{YX}(\omega) = \int_{-\infty}^{\infty} R_{XY}(\alpha)\exp(-j(-\omega)\alpha)\mathrm{d}\alpha = S_{XY}(-\omega), \tag{8.111}$$

  which is the same as $S^*_{XY}(\omega)$ because $\{X(t), Y(t)\}$ are real-valued.
- *Even part.* $\mathrm{Re}[S_{XY}(\omega)]$ is an even function. From the definitions of $S_{XY}(\omega)$ and $S_{YX}(\omega)$:

$$\mathrm{Re}[S_{XY}(\omega)] = \int_{-\infty}^{\infty} R_{XY}(\tau)\cos(\omega\tau)\mathrm{d}\tau = \mathrm{Re}[S_{YX}(\omega)], \tag{8.112}$$

  where $\cos(-\omega\tau) = \cos(\omega\tau)$ has been used. From the first property, we can write $\mathrm{Re}[S_{XY}(\omega)] = \mathrm{Re}[S_{XY}(-\omega)]$, which is the definition of an even function.
- *Odd part.* $\mathrm{Im}[S_{XY}(\omega)]$ is an odd function. Using an approach similar to that in the previous proof:

$$\mathrm{Im}[S_{XY}(\omega)] = -\int_{-\infty}^{\infty} R_{XY}(\tau)\sin(\omega\tau)\mathrm{d}\tau = -\mathrm{Im}[S_{YX}(\omega)], \tag{8.113}$$

where $\sin(-\omega\tau) = -\sin(\omega\tau)$ has been used. Again from the first property, $\text{Im}[S_{XY}(\omega)] = -\text{Im}[S_{XY}(-\omega)]$, which is the definition of an odd function.

- *Orthogonality.* $S_{XY}(\omega) = 0$ if $X(t)$ and $Y(t)$ are orthogonal with $R_{XY}(\tau) = 0$. Since $R_{XY}(\tau) = 0$ for all τ, we immediately have $S_{XY}(\omega) = 0$.
- *Uncorrelated.* $S_{XY}(\omega) = 2\pi\mu_X\mu_Y\delta(\omega)$ if $X(t)$ and $Y(t)$ are uncorrelated with $R_{XY}(\tau) = \mathcal{E}[X(t)]\mathcal{E}[Y(t+\tau)]$. Since the processes are uncorrelated, $R_{XY}(\tau) = \mathcal{E}[X(t)Y(t+\tau)] = \mu_X\mu_Y$ is a constant. Thus

$$S_{XY}(\tau) = \mu_X\mu_Y \int_{-\infty}^{\infty} \exp(-j\omega\tau)\mathrm{d}\tau = 2\pi\mu_X\mu_Y\delta(\omega) \tag{8.114}$$

because the Fourier transform of a constant is the Dirac delta function. The corresponding result for ordinary frequency is

$$S_{XY}(f) = \mu_X\mu_Y\delta(f). \tag{8.115}$$

If either process has zero mean, which is often the case in practice, the cross-PSD is zero.

We can also define a distribution for the cross-PSD.

Definition: Cross-Power Spectral Distribution The *cross-power spectral distribution* $S_{XY}^D(f)$ is

$$R_{XY}(\tau) = \int_{-\infty}^{\infty} \exp(j2\pi f\tau)\mathrm{d}S_{XY}^D(f), \tag{8.116}$$

where $R_{XY}(\tau)$ is the cross-correlation function of random processes $\{X(t), Y(t)\}$.

The cross-power spectral distribution has the same three properties as $S_{XX}^D(f)$, and if it is differentiable:

$$\mathrm{d}S_{XY}^D(f) = S_{XY}(f)\mathrm{d}f. \tag{8.117}$$

Example 8.7. In Section 8.8, we consider various systems with random inputs; here we describe a simple example of a moving average (MA) filter (also covered later in this chapter). Suppose that random sequence $X[k]$ with autocorrelation function $R_{XX}[m] = \delta[m]$ is processed by a filter with impulse response $h[k] = \delta[k] + \delta[k-1]$ to generate the output sequence $Y[k] = X[k] + X[k-1]$. The cross-correlation function is

$$\begin{aligned} R_{XY}[m] &= \mathcal{E}[X[k]X[k+m]] + \mathcal{E}[X[k]X[k+m-1]], \\ &= \delta[m] + \delta[m-1] \end{aligned} \tag{8.118}$$

with cross-PSD

$$\begin{aligned} S_{XY}(j\omega) &= \sum_{m=-\infty}^{\infty} (\delta[m] + \delta[m-1])\exp(-j\omega m), \\ &= 1 + \exp(-j\omega) = 2\exp(-j\omega/2)\cos(\omega/2). \end{aligned} \tag{8.119}$$

Observe that

$$\text{Re}[S_{XY}(j\omega)] = 1 + \cos(\omega), \quad \text{Im}[S_{XY}(j\omega)] = \sin(\omega), \tag{8.120}$$

which are even and odd functions, respectively. Alternatively, we can write

$$\begin{aligned} R_{YX}[m] &= \mathcal{E}[X[k]X[k+m]] + \mathcal{E}[X[k-1]X[k+m]], \\ &= \delta[m] + \delta[m+1], \end{aligned} \tag{8.121}$$

so that

$$S_{YX}(j\omega) = 1 + \exp(j\omega) = 2\exp(j\omega/2)\cos(\omega/2), \tag{8.122}$$

which verifies $S_{YX}(j\omega) = S_{XY}(-j\omega) = S^*_{XY}(\omega)$.

In many engineering courses, only unilateral transforms are studied because the signals and systems are assumed to be causal. However, since the autocorrelation function is an even function, the PSD is defined using *bilateral* transforms. Appendix C contains several tables of Fourier transforms, DTFTs, Laplace transforms, and z-transforms. In Section 8.8, we investigate how the PSD of random process $X(t)$ is altered when filtered by a causal LTI system. Thus, a combination of unilateral and bilateral transforms are used to investigate such filtered signals. Of course, we can always use bilateral transforms and include a unit-step function in the integrand to obtain the unilateral result, so one must be careful to use the correct transform pair from a table of bilateral transforms (and whether or not the time-domain function is multiplied by $u(t)$).

8.8 SYSTEMS WITH RANDOM INPUTS

Although many systems have multiple inputs and multiple outputs, in this chapter we focus on single-input single-output (SISO) systems as depicted in Figure 8.9(a). Let the input be random process $X(t)$ which has known properties such as its first and second moments, possibly its pdf, and whether or not it is uncorrelated. Typically, the goal in the analysis or design of a system is to modify $X(t)$ in order to generate an output $Y(t)$ that has some desired characteristics. For example, $X(t)$ might contain a communication signal embedded in additive noise, and the goal of the system is to filter $X(t)$ such that $Y(t)$ has reduced noise, and without adversely affecting the underlying signal of interest. Figure 8.9(b) shows the configuration for an *equalizer* which is designed to compensate for the distorting effects of the channel $H(z)$ through which the transmitted sequence $X[k]$ propagates. Figure 8.9(c) shows a *system identification* configuration where $G(z)$ is a model of $H(z)$, obtained from measurements of the input and output sequences $X[k]$ and $Y[k]$. Of course, in an actual implementation, the various filters operate on realizations of the random sequences. As was done in previous chapters, a probabilistic representation will be used so that we can describe how the ensemble of realizations are processed.

Generally, systems can be categorized as being (i) linear or nonlinear and (ii) time-invariant or time varying. A linear system satisfies superposition such that if deterministic signals $x_1(t)$ and $x_2(t)$ generate outputs $y_1(t)$ and $y_2(t)$, respectively, then $c_1y_1(t) + c_2y_2(t)$ is the output for input $c_1x_1(t) + c_2x_2(t)$, where $\{c_n\}$ are constant coefficients. Although most physical systems are nonlinear and can be difficult to analyze and control, fortunately many are approximated reasonably well by linear models, though possibly only over a finite dynamic range and for a limited time duration. In this chapter, we focus on LTI systems because they are modeled by an impulse-response function whose output is derived by a convolution (a review of such systems for continuous and discrete time is provided in Chapter 1).

However, first, we briefly consider nonlinear systems that are memoryless where the output at any time instant depends only on the input at the same time.

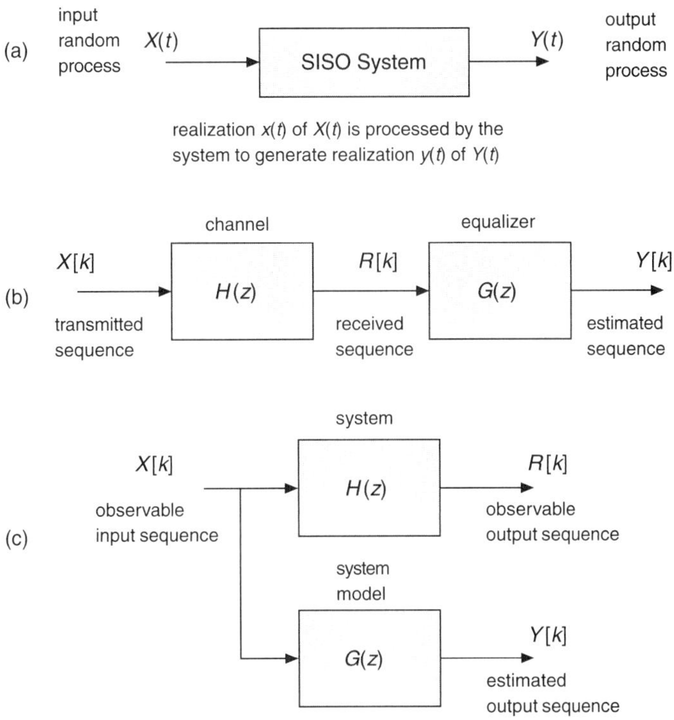

FIGURE 8.9 LTI systems. (a) SISO system. (b) Channel equalization. (c) System identification.

8.8.1 Nonlinear Systems

Memoryless nonlinear systems can be viewed as a *transformation* or mapping of the input random variable at each time instant such that

$$Y(t) = g(X(t)), \tag{8.123}$$

where $g(\cdot)$ is a deterministic function. Common examples in engineering include $X^2(t)$ and the signum function $\text{sgn}(X(t))$. Since the system has no memory, whereby it does not store past input values, the transformation techniques used for random variables in Chapter 4 can be applied here. If the function changes with time, represented by the notation $g(X(t), t)$, we need only consider the current transformation to determine the properties of the output at that time. Of course, if the transformation changes, then we can expect that $Y(t)$ is a nonstationary process even if $X(t)$ is stationary.

Example 8.8. Consider the following nonlinear filter for random sequence $X[k]$:

$$Y[k] = \text{sgn}(X[k]), \tag{8.124}$$

and assume that

$$X[k] = S[k] + W[k], \tag{8.125}$$

where $S[k]$ is an iid symmetric Bernoulli sequence with outcomes ± 1, and $W[k]$ is an additive white Gaussian noise (AWGN) sequence with variance σ_W^2. It is clear that $Y[k]$ is also a Bernoulli sequence with the same outcomes ± 1. Assuming that $S[k]$ and $W[k]$ are independent, we know from Chapter 4 that $X[k]$ has the following pdf:

$$\begin{aligned} f_{X[k]}(x) &= f_{S[k]}(s) * f_{W[k]}(w), \\ &= \frac{1}{2\sqrt{2\pi\sigma_W^2}} \int_{-\infty}^{\infty} [\delta(x-1-v) + \delta(x+1-v)] \exp(-v^2/2\sigma_W^2) \mathrm{d}v, \\ &= \frac{1}{2\sqrt{2\pi\sigma_W^2}} \left[\exp(-(x-1)^2/2\sigma_W^2) + \exp(-(x+1)^2/2\sigma_W^2)\right]. \end{aligned} \tag{8.126}$$

An example of this bimodal pdf was shown previously in Figure 4.33. The pdf of the sequence at the output of the filter is

$$f_{Y[k]}(y) = (1/2)[\delta(y-1) + \delta(y+1)]. \tag{8.127}$$

This type of nonlinear filter is an example of a *detector* in communications: either 1 or -1 is transmitted which is corrupted by Gaussian noise, leading to the pdf shown in Figure 4.33. It turns out that $\mathrm{sgn}(X[k])$ is the optimum decision device for estimating $S[k]$ from $X[k]$ when the symbols ± 1 are equally likely. (Signal detection is discussed further in Chapter 10.) Various moments of $Y[k]$ can be computed directly from the pdf of $X[k]$:

$$\mathcal{E}[Y[k]] = \int_{-\infty}^{\infty} \mathrm{sgn}(x) f_{X[k]}(x) \mathrm{d}x = 0 \tag{8.128}$$

because $\mathrm{sgn}(\cdot)$ is an odd function and the pdf of $X[k]$ is even. Also since $Y^2[k] = 1$,

$$\mathcal{E}[Y^2[k]] = \int_{-\infty}^{\infty} y^2[k] f_{X[k]}(x) \mathrm{d}x = 1. \tag{8.129}$$

These results are also readily obtained using $f_{Y[k]}(y)$.

If the outcomes of $S[k]$ are not equally likely such that $P(S[k]=1) = p$ and $P(S[k]=-1) = 1-p \triangleq q$, then the pdf of $X[k]$ is

$$f_{X[k]}(x) = \frac{1}{\sqrt{2\pi\sigma_W^2}} \left[p \exp(-(x-1)^2/2\sigma_W^2) + q \exp(-(x+1)^2/2\sigma_W^2)\right], \tag{8.130}$$

which is shown in Figure 8.10 for $p = 3/4$. The pdf of the signum function output is

$$f_{Y[k]}(y) = (0.7386)\delta(y-1) + (0.2614)\delta(y+1), \tag{8.131}$$

where because $p \neq q$:

$$\int_{0}^{\infty} f_{X[k]}(x) \mathrm{d}x = 0.7386, \quad \int_{-\infty}^{0} f_{X[k]}(x) \mathrm{d}x = 0.2614. \tag{8.132}$$

The output has the following moments:

$$\mathcal{E}[Y[k]] = 0.4772, \quad \mathcal{E}[Y^2[k]] = 1. \tag{8.133}$$

For this scenario, the signum function is not the optimum detector because the pdf of $X[k]$ is no longer an even function. If $p > 1/2$, then $S[k] = 1$ is more likely to have been transmitted, which needs to be taken into account in the receiver. This means that the zero-crossing of the signum function detector should be shifted to the left of the origin by an amount that depends on the value of p.

Additional examples of nonlinear systems are covered in Chapter 12.

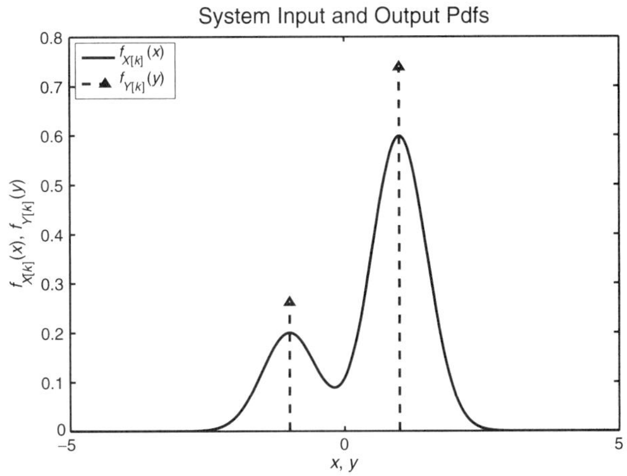

FIGURE 8.10 Pdf of the symmetric Bernoulli sequence in Example 8.8 with outcomes ± 1 and $p = 3/4$ in AWGN with unit variance.

8.8.2 Linear Systems

For a linear system that is not necessarily time-invariant, random process $Y(t)$ at the filter output is obtained as the following integral:

$$Y(t) = \int_{-\infty}^{t} h(t, \tau)X(\tau)\mathrm{d}\tau, \tag{8.134}$$

where $X(t)$ is the input random process, and $h(t, \tau)$ is the deterministic impulse-response function of the system. The principle of superposition for a linear system leads to this integral for a time-varying system: $h(t, \tau)$ depends independently on t and τ. For an LTI system, $h(t, \tau) = h(t - \tau)$ depends only on the time difference, not on t and τ individually. The impulse-response function is obtained by letting the input be the Dirac delta function $\delta(t)$ and using its sifting property:

$$y(t) = \int_{-\infty}^{t} h(t - \tau)\delta(\tau)\mathrm{d}\tau = h(t). \tag{8.135}$$

The delta function produces the value of $h(t - \tau)$ at the location of $\delta(\tau)$ which is at $\tau = 0$. For a time-varying system, the impulse response is

$$y(t) = \int_{-\infty}^{t} h(t, \tau)\delta(\tau)\mathrm{d}\tau = h(t, 0). \tag{8.136}$$

Suppose now that the impulse function is shifted by 1. Then, for the two cases:

$$y(t) = \int_{-\infty}^{t} h(t, \tau)\delta(\tau - 1)\mathrm{d}\tau = h(t, 1) \quad \text{(linear)}, \tag{8.137}$$

$$y(t) = \int_{-\infty}^{t} h(t - \tau)\delta(\tau - 1)\mathrm{d}\tau = h(t - 1) \quad \text{(LTI)}. \tag{8.138}$$

The output of the LTI system is also shifted whereas the output of the time-varying system is more complicated. In order for a system to be *causal*, the following conditions are required:

$$h(t, \tau) = 0, \quad \text{for } t < \tau \quad \text{(linear)}, \tag{8.139}$$

$$h(t - \tau) = 0, \quad \text{for } t < \tau \Longrightarrow h(t) = 0 \text{ for } t < 0 \quad \text{(LTI)}. \tag{8.140}$$

Example 8.9. Consider a linear system with impulse response

$$h(t, \tau) = \exp(-t\tau)u(\tau)u(t - \tau), \tag{8.141}$$

where the unit-step function $u(t - \tau)$ ensures $t > \tau$ for a causal system. Obviously, this is a time-varying system because for each value of τ the system response changes. For an LTI system, we have

$$h(t - \tau) = \exp(-(t - \tau))u(\tau)u(t - \tau) \tag{8.142}$$

whose shape is unchanged for different values of τ. Let the system input be $x(t) = \exp(-t)u(t)$. Then for the time-varying system:

$$\begin{aligned} y(t) &= \int_0^t \exp(-t\tau)\exp(-\tau)\mathrm{d}\tau, \\ &= \int_0^t \exp(-\tau(t+1))\mathrm{d}\tau = \frac{1}{t+1}[1 - \exp(-t^2 - t)]u(t). \end{aligned} \tag{8.143}$$

The corresponding result for the LTI system is

$$\begin{aligned} y(t) &= \int_0^t \exp(-(t - \tau))\exp(-\tau)\mathrm{d}\tau, \\ &= \int_0^t \exp(-t)\mathrm{d}\tau = t\exp(-t)u(t). \end{aligned} \tag{8.144}$$

For shifted input $x(t) = \exp(-(t - 1))u(t - 1)$, the output of the time-varying system is

$$\begin{aligned} y(t) &= \int_1^t \exp(-t\tau)\exp(-(\tau - 1))\mathrm{d}\tau, \\ &= \frac{\exp(1)}{t+1}[\exp(-t - 1) - \exp(-t^2 - t)]u(t - 1), \\ &= \frac{1}{t+1}[\exp(-t) - \exp(-t^2 - t + 1)]u(t - 1), \end{aligned} \tag{8.145}$$

and for the LTI system:

$$\begin{aligned} y(t) &= \int_1^t \exp(-(t - \tau))\exp(-(\tau - 1))\mathrm{d}\tau, \\ &= \int_1^t \exp(-(t - 1))\mathrm{d}\tau = (t - 1)\exp(-(t - 1))u(t - 1). \end{aligned} \tag{8.146}$$

The results for both systems are shown in Figure 8.11, where we see that a shifted input yields a more complicated output for the time-varying system, rather than the simple time shift for the LTI system.

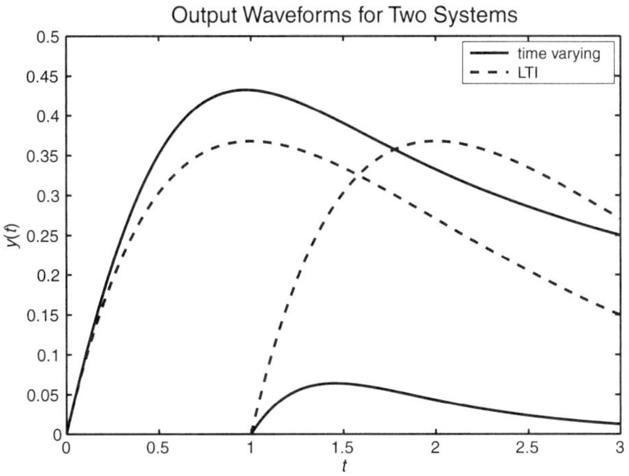

FIGURE 8.11 Output examples for the time-varying and LTI systems in Example 8.9. For the time-varying system, a time shift of the input does not result in a simple time shift of the output.

Next, we are interested in signal descriptions of an LTI system with a random input. The cross-correlation function between the system input and output is

$$R_{XY}(\tau) = \mathcal{E}[X(t)Y(t+\tau)], \\ = \mathcal{E}\left[X(t)\int_{-\infty}^{\infty} h(t+\tau-\alpha)X(\alpha)\mathrm{d}\alpha\right], \tag{8.147}$$

where the integral is defined in the MS sense. Interchanging the integral and expectation gives

$$R_{XY}(\tau) = \int_{-\infty}^{\infty} h(t+\tau-\alpha)\mathcal{E}[X(t)X(\alpha)]\mathrm{d}\alpha, \\ = \int_{-\infty}^{\infty} h(t+\tau-\alpha)R_{XX}(\alpha-t)\mathrm{d}\alpha, \tag{8.148}$$

where it is assumed that $X(t)$ is wide-sense stationary. Defining $v \triangleq \alpha - t$ and rewriting this expression yields

$$R_{XY}(\tau) = \int_{-\infty}^{\infty} h(\tau - v)R_{XX}(v)\mathrm{d}v, \tag{8.149}$$

which we recognize is a convolution between $R_{XX}(\tau)$ and $h(\tau)$:

$$R_{XY}(\tau) = R_{XX}(\tau) * h(\tau). \tag{8.150}$$

Likewise, we can show that

$$R_{YX}(\tau) = R_{XX}(\tau) * h(-\tau) \tag{8.151}$$

and (see Problem 8.17)

$$R_{YY}(\tau) = R_{XX}(\tau) * h(\tau) * h(-\tau). \tag{8.152}$$

The PSD of $R_{XY}(\tau)$ is

$$\begin{aligned} S_{XY}(f) &= \int_{-\infty}^{\infty} R_{XY}(\tau)\exp(-j2\pi f\tau)\mathrm{d}\tau, \\ &= \int_{-\infty}^{\infty}\int_{-\infty}^{\infty} h(\tau-\alpha)R_{XX}(\alpha)\exp(-j2\pi f\tau)\mathrm{d}\alpha\mathrm{d}\tau. \end{aligned} \tag{8.153}$$

Substituting $t = \tau - \alpha$ gives

$$\begin{aligned} S_{XY}(f) &= \int_{-\infty}^{\infty}\int_{-\infty}^{\infty} h(t)R_{XX}(\alpha)\exp(-j2\pi f(t+\alpha))\mathrm{d}\alpha\mathrm{d}t, \\ &= \int_{-\infty}^{\infty} h(t)\exp(-j2\pi ft)\mathrm{d}t \int_{-\infty}^{\infty} R_{XX}(\alpha)\exp(-j2\pi f\alpha)\mathrm{d}\alpha, \\ &= H(f)S_{XX}(f), \end{aligned} \tag{8.154}$$

where we have used the fact that $S_{XX}(f)$ is an even function. Similarly,

$$S_{YX}(f) = H(-f)S_{XX}(f) \tag{8.155}$$

and (see Problem 8.18)

$$S_{YY}(f) = H(f)H(-f)S_{XX}(f). \tag{8.156}$$

For a real-valued impulse response function, $H(-f) = H^*(f)$ so that

$$S_{YY}(f) = |H(f)|^2 S_{XX}(f). \tag{8.157}$$

Table 8.3 summarizes the results for radian frequency ω, as well as those using $H(s)$. The corresponding results for random sequences with $H(j\omega)$ and $H(z)$ are also included.

Example 8.10. In this example, we demonstrate the effect of filtering the uncorrelated Gaussian sequence shown in Figure 8.12(a) which has zero mean and unit variance. The filter impulse response is

$$h[k] = \frac{1}{N}\sum_{m=0}^{N-1}\delta[k-m] \tag{8.158}$$

TABLE 8.3 LTI Systems and PSDs

Continuous-Time Process	Discrete-Time Sequence
$S_{YY}(\omega) = \|H(\omega)\|^2 S_{XX}(\omega)$	$S_{YY}(j\omega) = \|H(j\omega)\|^2 S_{XX}(j\omega)$
$S_{YY}(s) = H(s)H(-s)S_{XX}(s)$	$S_{YY}(z) = H(z)H(z^{-1})S_{XX}(z)$
$S_{XY}(\omega) = H(\omega)S_{XX}(\omega)$	$S_{XY}(j\omega) = H(j\omega)S_{XX}(j\omega)$
$S_{XY}(s) = H(s)S_{XX}(s)$	$S_{XY}(z) = H(z)S_{XX}(z)$
$S_{YX}(\omega) = H(-\omega)S_{XX}(\omega)$	$S_{YX}(j\omega) = H(-j\omega)S_{XX}(j\omega)$
$S_{YX}(s) = H(-s)S_{XX}(s)$	$S_{YX}(z) = H(z^{-1})S_{XX}(z)$

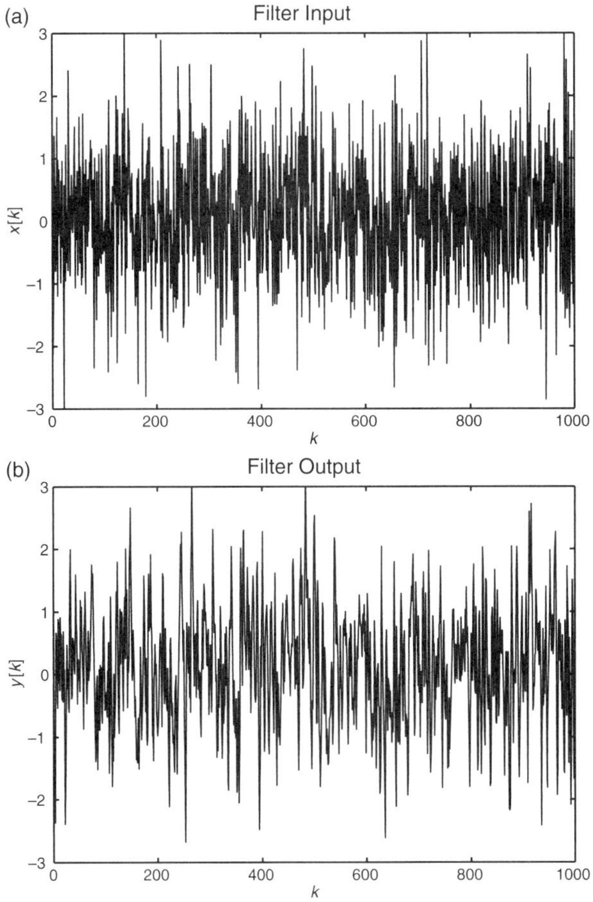

FIGURE 8.12 Realizations for the LTI system in Example 8.10. (a) Uncorrelated Gaussian input $X[k]$. (b) Correlated filter output $Y[k]$.

with $N = 3$. This MA filter has a finite impulse response (FIR) (see Chapter 1) with transfer function

$$\begin{aligned} H(j\omega) &= (1/3)\sum_{k=0}^{2}\exp(-j\omega k), \\ &= (1/3)\exp(-j\omega)\left(\exp(j\omega) + 1 + \exp(-j\omega)\right), \\ &= (1/3)\exp(-j\omega)[1 + 2\cos(\omega)]. \end{aligned} \tag{8.159}$$

Since $S_{XX}(j\omega) = 1$, the PSD of the output is

$$S_{YY}(j\omega) = |H(j\omega)|^2 = (1/9)[1 + 2\cos(\omega)]^2, \tag{8.160}$$

and the autocorrelation function is

$$R_{YY}[m] = (1/9)(3\delta[m] + 2\delta[m-1] + 2\delta[m+1] + \delta[m-2] + \delta[m+2]), \tag{8.161}$$

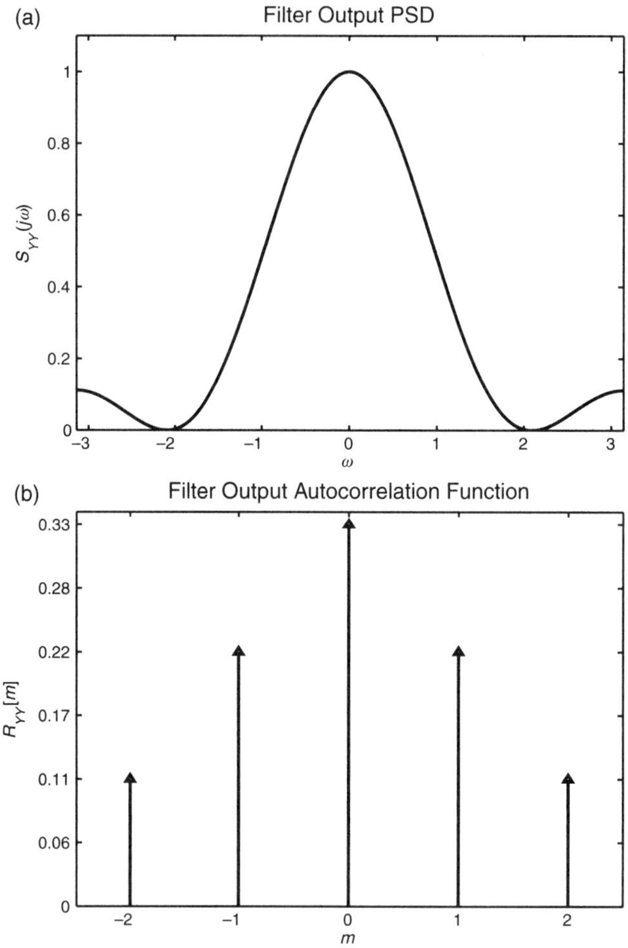

FIGURE 8.13 Filter results in Example 8.10 for an uncorrelated Gaussian sequence. (a) Output PSD. (b) Output autocorrelation function.

which is also easily obtained by convolving $h[k]$ and $h[-k]$. $S_{YY}(j\omega)$ and $R_{YY}[m]$ are shown in Figure 8.13, and an example realization of the filter output is shown in Figure 8.12(b) where it is evident that the output is correlated.

8.9 PASSBAND SIGNALS

So far, we have focused on random signals whose PSDs are located at the origin which are referred to as *baseband* signals. In most communication systems, one is interested in signals whose PSDs are located about some center frequency f_c, where typically $f_c \gg 0$. Such signals are called *passband signals*. Figure 8.14 provides examples of $S_{XX}(f)$ for baseband and passband signals. Observe that the one-sided bandwidth (for $f \geq 0$) of the baseband signal is B whereas it is $2B$ for the passband signal. The reason for this is that a passband signal is often generated from a baseband signal via *modulation* of a sinusoidal signal with carrier frequency f_c. Typically, for a passband signal the bandwidth is $B \ll f_c$, for which case it is called a *narrowband passband* signal.

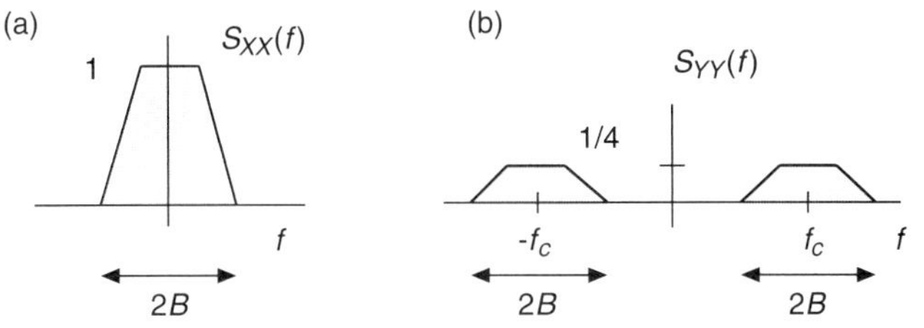

FIGURE 8.14 PSDs. (a) Baseband random process $X(t)$. (b) Passband random process $Y(t)$ obtained by modulating $X(t)$.

Let $X(t)$ be a random process with baseband PSD $S_{XX}(f)$ and one-sided bandwidth B. Suppose $X(t)$ is multiplied by a sinusoidal waveform with known carrier frequency f_c:

$$Y(t) = X(t)\cos(2\pi f_c t). \tag{8.162}$$

In order to determine $S_{YY}(f)$, the autocorrelation function is derived:

$$R_{YY}(\tau) = \mathcal{E}[X(t)\cos(2\pi f_c t)X(t+\tau)\cos(2\pi f_c(t+\tau))]. \tag{8.163}$$

A random phase Φ uniformly distributed on $[-\pi, \pi]$ is usually included in the argument of the cosine. This is done in practice to model the fact that the cosine is not necessarily perfectly aligned with the starting time of the process. Assuming the random phase and the random process are uncorrelated gives

$$\begin{aligned} R_{YY}(\tau) &= \mathcal{E}[X(t)X(t+\tau)]\mathcal{E}[\cos(2\pi f_c t + \Phi)\cos(2\pi f_c(t+\tau) + \Phi)], \\ &= (1/2)R_{XX}(\tau)\mathcal{E}[\cos(2\pi f_c \tau) + \cos(2\pi f_c(2t+\tau) + 2\Phi)], \\ &= (1/2)R_{XX}(\tau)\cos(2\pi f_c \tau), \end{aligned} \tag{8.164}$$

where we have used the fact that the cosine averaged over the random phase is zero. From the modulation property of the Fourier transform (see Appendix C):

$$S_{YY}(f) = (1/4)[S_{XX}(f - f_c) + S_{XX}(f + f_c)]. \tag{8.165}$$

The narrowband passband process in (8.162) can be generalized to include the random phase *process* $\Phi(t)$ as follows:

$$Y(t) = X(t)\cos(2\pi f_c t + \Phi(t)), \tag{8.166}$$

assuming that $X(t)$ and $\Phi(t)$ are uncorrelated. Using trigonometric identities, this expression can be converted to the following *quadrature* form:

$$Y(t) = X_I(t)\cos(2\pi f_c t) - X_Q(t)\sin(2\pi f_c t), \tag{8.167}$$

where

$$X_I(t) \triangleq X(t)\cos(\Phi(t)), \quad X_Q(t) \triangleq X(t)\sin(\Phi(t)) \tag{8.168}$$

are the *in-phase and quadrature components*, respectively, of the passband process. When there is no phase process, $Y(t)$ has only an in-phase component: $X_I(t) = X(t)$ and $X_Q(t) = 0$.

It turns out that the autocorrelation functions (and PSDs) of $X_I(t)$ and $X_Q(t)$ in (8.168) are not readily obtained because we have made no assumption about the correlation properties of $\Phi(t)$. However, assuming that $Y(t)$ is wide-sense stationary, then some properties of $X_I(t)$ and $X_Q(t)$ can be derived. First, it is clear that

$$\mathcal{E}[Y(t)] = \mathcal{E}[X_I(t)] = \mathcal{E}[X_Q(t)] = 0. \tag{8.169}$$

If this were not the case, then from (8.167) we would find that $\mathcal{E}[Y(t)]$ is time varying, which contradicts our assumption that $Y(t)$ is wide-sense stationary. Next, the autocorrelation function $\mathcal{E}[Y(t)Y(t+\tau)] = R_{YY}(\tau)$ is derived. In order that this depend only on τ, observe that

$$\begin{aligned}\mathcal{E}[Y(t)Y(t+\tau)] &= \mathcal{E}[X_I(t)X_I(t+\tau)\cos(2\pi f_c t)\cos(2\pi f_c(t+\tau))] \\ &\quad + \mathcal{E}[X_Q(t)X_Q(t+\tau)\sin(2\pi f_c t)\sin(2\pi f_c(t+\tau))] \\ &\quad - \mathcal{E}[X_I(t)X_Q(t+\tau)\cos(2\pi f_c t)\sin(2\pi f_c(t+\tau))] \\ &\quad - \mathcal{E}[X_Q(t)X_I(t+\tau)\sin(2\pi f_c t)\cos(2\pi f_c(t+\tau))].\end{aligned} \tag{8.170}$$

Rewriting this expression using trigonometric identities gives

$$\begin{aligned}R_{YY}(\tau) &= (1/2)\big[(R_{X_IX_I}(\tau) + R_{X_QX_Q}(\tau))\cos(2\pi f_c\tau) \\ &\quad + (R_{X_IX_I}(\tau) - R_{X_QX_Q}(\tau))\cos(2\pi f_c(2t+\tau)) \\ &\quad - (R_{X_QX_I}(\tau) - R_{X_IX_Q}(\tau))\sin(2\pi f_c\tau) \\ &\quad - (R_{X_QX_I}(\tau) + R_{X_IX_Q}(\tau))\sin(2\pi f_c(2t+\tau))\big].\end{aligned} \tag{8.171}$$

Since $Y(t)$ is wide-sense stationary, the terms with t must be zero, leading to the following conditions on the correlation functions:

$$R_{X_IX_I}(\tau) = R_{X_QX_Q}(\tau), \tag{8.172}$$

$$R_{X_IX_Q}(\tau) = -R_{X_QX_I}(\tau) = -R_{X_IX_Q}(-\tau), \tag{8.173}$$

where the last equality holds for any cross-correlation function. For a narrowband passband process, we see that the cross-correlation between the in-phase and quadrature components must be an odd function (which is not true in general). As a result,

$$R_{X_IX_Q}(0) = 0. \tag{8.174}$$

With these conditions, $R_{YY}(\tau)$ also has a quadrature form:

$$R_{YY}(\tau) = R_{X_IX_I}(\tau)\cos(2\pi f_c\tau) - R_{X_QX_I}(\tau)\sin(2\pi f_c\tau). \tag{8.175}$$

Observe that $R_{X_IX_I}(\tau)$ and $R_{X_QX_Q}(\tau)$ do not equal $R_{YY}(\tau)$ except when $\tau = 0$:

$$R_{X_IX_I}(0) = R_{X_QX_Q}(0) = R_{YY}(0). \tag{8.176}$$

Thus, all three processes have the same variance (recall they also all have zero mean). We can also make statements about the corresponding PSDs. From the modulation property of the Fourier transform:

$$\begin{aligned}S_{YY}(f) &= (1/2)[S_{X_IX_I}(f-f_c) + S_{X_IX_I}(f+f_c)] \\ &\quad - (1/2j)[S_{X_QX_I}(f-f_c) - S_{X_QX_I}(f+f_c)].\end{aligned} \tag{8.177}$$

Since the autocorrelation function is a quadrature signal, the PSD necessarily has an imaginary component. From (8.176), all three PSDs have the same average power:

$$\int_{-\infty}^{\infty} S_{X_I X_I}(f)\mathrm{d}f = \int_{-\infty}^{\infty} S_{X_Q X_Q}(f)\mathrm{d}f = \int_{-\infty}^{\infty} S_{YY}(f)\mathrm{d}f. \tag{8.178}$$

Passband signals are considered again in Chapter 10 when communication systems are introduced.

8.10 WHITE NOISE

Noise appears to some degree in nearly all real signals and systems. Of particular interest is noise that has a flat PSD.

Definition: White-Noise Process Wide-sense stationary random process $X(t)$ with flat PSD

$$S_{XX}(f) = N_o/2 \tag{8.179}$$

for $f \in \mathcal{R}$ is called *white noise*. It has zero mean, its autocorrelation function is

$$R_{XX}(\tau) = (N_o/2)\delta(\tau), \tag{8.180}$$

and thus $C_{XX}(\tau) = R_{XX}(\tau)$.

The units of N_o are watts/Hz. It is called white noise because (i) the waveforms have a noisy appearance (and sound like noise in the audible frequency range) and (ii) it includes all frequencies equally (a flat spectrum), similar to white light in the visible range of the electromagnetic spectrum (which is flat across wavelength). This definition does not state anything about the distribution of white noise, which can be of any type, though a Gaussian pdf is often assumed.

Theorem 8.5. A white-noise random process with the autocorrelation function in (8.180) must have zero mean.

Proof. This result is a consequence of the defining property in (8.179). Let $Y(t) = X(t) + \mu_Y$, where $X(t)$ has zero mean and the autocorrelation function in (8.180). Assume that μ_Y is nonzero. The autocorrelation function of $Y(t)$ is

$$R_{YY}(\tau) = \mathcal{E}[(X(t) + \mu_Y)(X(t+\tau) + \mu_Y)] = (N_o/2)\delta(\tau) + \mu_Y^2, \tag{8.181}$$

which has PSD

$$S_{YY}(f) = N_o/2 + \mu_Y^2\delta(f). \tag{8.182}$$

Since there is a delta function at the origin, $Y(t)$ cannot be white noise as defined by (8.179), unless its mean is zero. □

Recall in Chapter 6 that $X(t)$ is an uncorrelated process if

$$R_{XX}(\tau) = \begin{cases} \mathcal{E}[X^2(t)], & \tau = 0 \\ \mathcal{E}[X(t)]\mathcal{E}[X(t+\tau)], & \tau \neq 0, \end{cases} \tag{8.183}$$

which has the form $R_{XX}(\tau) = \alpha\delta(\tau) + \mu_X^2$, similar to that in (8.181). Thus, white-noise with a flat PSD implies an uncorrelated process, but the reverse holds only for zero mean.

A white-noise process with a *Gaussian* distribution is obtained as the derivative of the Wiener process: $X(t) = Y'(t)$, where $Y(t)$ has the following autocorrelation function:

$$R_{YY}(t_1, t_2) = c \min(t_1, t_2) \tag{8.184}$$

with $c = \alpha\epsilon^2$ (see Chapter 6). Similar to the derivation leading to (7.119) for the Poisson process, we have the following for the derivative for the autocorrelation function of the Wiener process:

$$R_{XX}(t_1, t_2) = \frac{\partial^2}{\partial t_1 \partial t_2} R_{YY}(t_1, t_2) = c\delta(t_2 - t_1), \tag{8.185}$$

which is the same as (8.180). Thus, although the MS derivative does not technically exist because of the delta functions at all $t_1 = t_2$, we find that the derivative of the Wiener process is white noise.

A white-noise process is not physically realizable because it has infinite average power (equal to the variance):

$$\sigma_X^2 = C_{XX}(0) = R_{XX}(0) = \int_{-\infty}^{\infty} S_{XX}(f)\mathrm{d}f = \infty. \tag{8.186}$$

However, in practice $X(t)$ is usually bandlimited, typically due to various filters in a system so that the PSD is flat only in some frequency range of interest and is zero elsewhere:

$$S_{XX}(f) = \begin{cases} N_o/2, & |f| \leq B \\ 0, & \text{else}, \end{cases} \tag{8.187}$$

where B is the one-sided bandwidth. A process with this type of PSD is technically not white, although it is called *bandlimited white noise* when B is much larger than the bandwidth of the signal of interest. The PSD in (8.187) corresponds to a baseband signal because it is nonzero about $f = 0$. A passband signal has a PSD of the form:

$$S_{XX}(f) = \begin{cases} N_o/2, & |f| \in [f_c - B/2, f_c + B/2] \\ 0, & \text{else}, \end{cases} \tag{8.188}$$

where f_c denotes the center frequency. An example of the spectrum of a passband signal is shown in Figure 8.15 for an amplitude modulated (AM) signal:

$$X(t) = M(t)\cos(2\pi f_c t) + N(t), \tag{8.189}$$

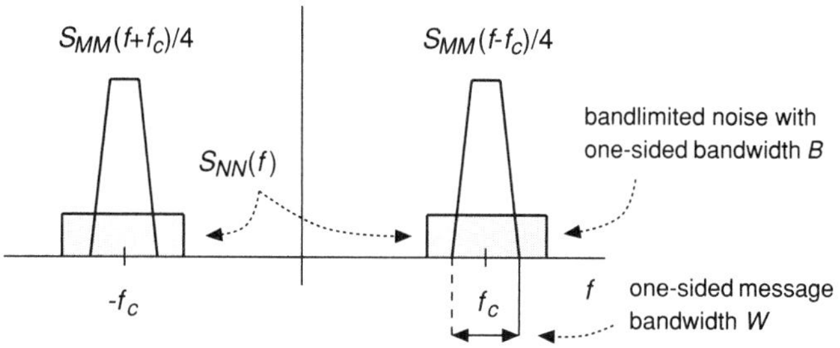

FIGURE 8.15 Power spectrum components of an AM signal in bandlimited white noise. $N(t)$ has the appearance of white noise because its bandwidth B is greater than the message bandwidth W. (Note that $S_{NN}(f)$ is shown separately from the signal spectrum $[S_{MM}(f + f_c) + S_{MM}(f - f_c)]/4$. The two spectra should be added together to give the overall $S_{XX}(f)$.)

where $M(t)$ is the random message signal, f_c is the passband center frequency, and $N(t)$ is bandlimited white noise. It is relatively straightforward to show the following using the previous results for passband signals:

$$S_{XX}(f) = (1/4)[S_{MM}(f - f_c) + S_{MM}(f + f_c)] + S_{NN}(f) \tag{8.190}$$

assuming $M(t)$ and $N(t)$ are uncorrelated.

For bandlimited white noise, the variance is finite:

$$\sigma_X^2 = \int_{-B}^{B} S_{XX}(f)\mathrm{d}f = N_o B. \tag{8.191}$$

Since the one-sided bandwidth B is specified as above, the factor of $1/2$ is included with N_o for the PSD in (8.187) because it is canceled by the factor of 2 in the overall (two-sided) bandwidth $2B$ (in order to simplify the expression). For the AM signal, assume that $M(t)$ has zero mean. If the power of the random message is

$$P_M = R_{MM}(0) = \int_{-W}^{W} S_{MM}(f)\mathrm{d}f, \tag{8.192}$$

then the signal-to-noise ratio (SNR) of the received signal is

$$\text{SNR} = \frac{P_M}{N_o B}. \tag{8.193}$$

The SNR is an important measure of the quality of the received signal: it determines the degree to which the message $M(t)$ can be reliably recovered. Observe that the SNR decreases when B is increased because greater noise is present. One goal of front-end filters in a receiver is to remove as much noise as possible without distorting the message signal; the bandwidth B of such filters is chosen to be as close as possible to the message bandwidth W while maintaining a relatively flat response over the frequency range of the message.

The autocorrelation function for bandlimited white noise is

$$\begin{aligned} R_{XX}(\tau) &= \int_{-B}^{B} (N_o/2)\exp(j2\pi f\tau)\mathrm{d}f, \\ &= (N_o/2)[\exp(j2\pi B\tau) - \exp(-j2\pi B\tau)]/(j2\pi\tau) = N_o B\,\mathrm{sinc}(2B\tau), \end{aligned} \tag{8.194}$$

where $\mathrm{sinc}(\tau) \triangleq \sin(\pi\tau)/\pi\tau$ is the sinc function. Some examples are shown in Figure 8.16 for different noise bandwidths where we see that the width of the main lobe is inversely proportional to B. It is shown in Appendix B that

$$\lim_{B\to\infty} B\,\mathrm{sinc}(2\tau) = \delta(\tau), \tag{8.195}$$

and thus

$$\lim_{B\to\infty} N_o B\,\mathrm{sinc}(2B\tau) = (N_o/2)\delta(\tau), \tag{8.196}$$

which is the autocorrelation function for the infinite bandwidth PSD.

Next, we describe how a white-noise *process* and a white-noise *sequence* are related and how they differ. Recall in Chapter 6 that random sequence $X[k]$ is uncorrelated if

$$\begin{aligned} R_{XX}[m] &= \begin{cases} \mathcal{E}[X^2[k]], & m = 0 \\ \mathcal{E}[X[m]]\mathcal{E}[X[k+m]], & m \neq 0, \end{cases} \\ &= (\sigma_X^2 + \mu_X^2)\delta[m] + \mu_X^2, \end{aligned} \tag{8.197}$$

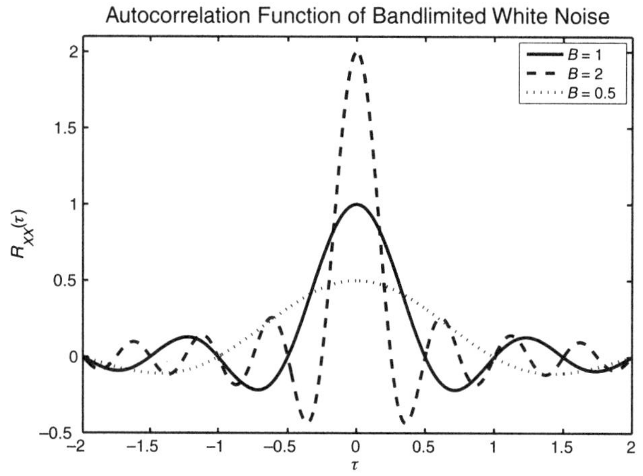

FIGURE 8.16 Autocorrelation function of bandlimited white noise for different one-sided bandwidths B.

where σ_X^2 is a *finite* variance. It is demonstrated in Problem 8.13 that if $X[k]$ is a white-noise sequence (with a flat PSD), it must have zero mean, as was shown previously for a continuous-time random process. Assume the mean is zero such that $R_{XX}[m] = \sigma_X^2 \delta[m]$. The PSD is obtained from the DTFT:

$$S_{XX}(j\omega) = \sum_{m=-\infty}^{\infty} R_{XX}[m] \exp(-jm\omega) = \sigma_X^2, \quad |\omega| \leq \pi, \tag{8.198}$$

which is defined on the unit circle of the z-plane. The average power (variance) is the area under the PSD given by

$$R_{XX}[0] = \frac{1}{2\pi} \int_{-\pi}^{\pi} S_{XX}(j\omega) \mathrm{d}\omega = \sigma_X^2, \tag{8.199}$$

where the $1/2\pi$ factor is due to the inverse DTFT (see the area property of the DTFT in Appendix C). The variance is *not* infinite because the PSD is integrated over a finite interval for a random sequence. We summarize the previous results in the following definition.

Definition: White-Noise Sequence Wide-sense stationary random sequence $X[k]$ with flat PSD

$$S_{XX}(j\omega) = \sigma_X^2 \tag{8.200}$$

for $\omega \in [-\pi, \pi]$ is called *white noise*. It has zero mean, its autocorrelation function is

$$R_{XX}[m] = \sigma_X^2 \delta[m], \tag{8.201}$$

where $\sigma_X^2 = C_{XX}[0]$ is the variance of the random sequence, and $C_{XX}[m] = R_{XX}[m]$.

Assume that $X[k]$ is generated by sampling the continuous-time process $X(t)$. With this sampling framework (which is discussed in Chapter 1 for deterministic signals), the frequency bounds on the unit circle are $[-\omega_s/2, \omega_s/2]$, where $\omega_s = 2\pi f_s$ and f_s is the sampling frequency. The reason that $[-\pi, \pi]$ is generally used in digital signal models is due to the fact that $f_s = 1/T_s = 1$ is often chosen for mathematical convenience so that $\pm\omega_s/2 = \pm\pi f_s \rightarrow \pm\pi$. This simplification is useful when analyzing signals only in the discrete-time domain. However, when it is necessary to implement a system in real (continuous) time, one must take into account the sampling rate used to generate the sequence, and it is necessary to use $[-\omega_s/2, \omega_s/2]$ for the frequency-domain interval.

It was shown in Chapter 1 that the z-transform is obtained from the Laplace transform via $z = \exp(s)$, which maps the imaginary axis on the s-plane to the unit circle on the z-plane. As a result, the PSD on $[-\omega_s/2, \omega_s/2]$ repeats for integer multiples of ω_s such that

$$S_{X_sX_s}(\omega) = \sigma_X^2 \sum_{k=-\infty}^{\infty} \text{rect}(\omega/\omega_s - k) = \sigma_X^2, \tag{8.202}$$

where the nonoverlapping adjacent rectangle functions together give a flat response for all $\omega \in \mathcal{R}$ as illustrated in Figure 8.17. The PSD can also be expressed in terms of ordinary frequency f and the sampling rate f_s as follows:

$$S_{X_sX_s}(f) = \sigma_X^2 \sum_{k=-\infty}^{\infty} \text{rect}(f/f_s - k) = \sigma_X^2. \tag{8.203}$$

It is important to note that $S_{X_sX_s}(f)$ is the PSD of the sampled process $X_s(t)$, not of the original continuous process $X(t)$. In order to sample $X(t)$ and then be able to reconstruct it from those samples, $S_{XX}(f)$ must be *bandlimited*. Thus, the spectrum of $X(t)$ is actually

$$S_{XX}(f) = \sigma_X^2 \text{rect}(f/f_s). \tag{8.204}$$

When $X(t)$ is sampled with rate f_s, the PSD is repeated at integer multiples of f_s as mentioned above to give the flat spectrum in Figure 8.17(b). This repeated spectrum is flat only because the spectrum of $X(t)$ is a rectangle function that is nonzero on $[-f_s/2, f_s/2]$. (Of course, such ideal functions are not possible in

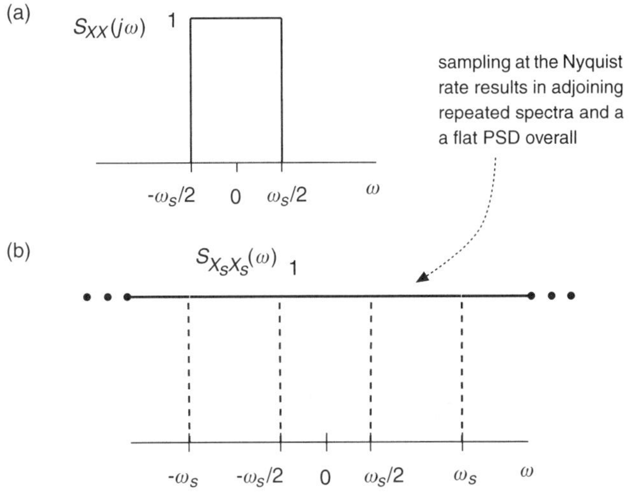

FIGURE 8.17 PSD for a white-noise random sequence $X[k]$. (a) $S_{XX}(j\omega)$ on the unit circle. (b) $S_{X_sX_s}(\omega)$ of the sampled process $X_s(t)$.

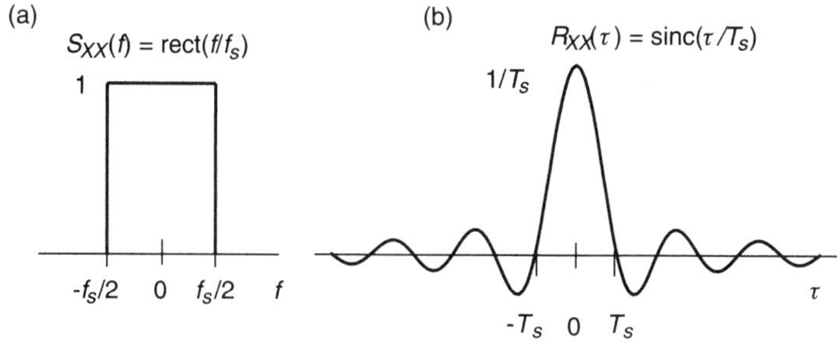

FIGURE 8.18 Sampling random process $X(t)$. (a) PSD of $X(t)$ corresponding to white-noise sequence $X[k]$. (b) Autocorrelation function of $X(t)$.

actual implementations; they are used here to illustrate the connection between $X(t)$ and $X[k]$.) As a result, the autocorrelation function of the continuous-time process $X(t)$ is *not* a Dirac delta function; it is instead

$$R_{XX}(\tau) = \sigma_X^2 T_s \text{sinc}(\tau/T_s) \tag{8.205}$$

as shown in Figure 8.18. A white-noise sequence *cannot* be obtained by sampling a white-noise process (which is not physically realizable as mentioned earlier).

There appears to be an inconsistency between a white-noise process and a white-noise sequence, but this is due to the fact that a white-noise sequence is defined for discrete-time instants, and requires only that samples spaced at integer multiples of $T_s = 1/f_s$ be uncorrelated. This leads to the Kronecker delta function for the autocorrelation function. A white-noise process is defined for continuous time and requires that the process at time instants t and $t + \tau$ for *every* $\tau \neq 0$ (including infinitesimally small) be uncorrelated. This leads to the Dirac delta function for the autocorrelation function.

Although we cannot realize a white-noise process, it is useful for theoretical models of continuous-time systems as discussed in Chapter 1. We can view a random sequence as though it was generated by sampling the nonwhite random process with rectangular PSD in Figure 8.18. This sequence has samples only at integer multiples of T_s, which are precisely where the autocorrelation function of the nonwhite random process $X(t)$ is zero as illustrated in Figure 8.18. Thus, the samples of a random sequence are *uncorrelated* provided the rectangle extends for the full interval $[-f_s/2, f_s/2]$. If the rectangular PSD of process $X(t)$ does not extend for the entire interval, the corresponding random sequence obtained via sampling is necessarily correlated. For example, if the rectangle is nonzero only in the range $[-f_s/4, f_s/4]$, then we know from Fourier transform properties that the autocorrelation function will be wider with its first zero crossing at $2T_s$. As a result, $R_{XX}(\tau)$ is nonzero at $\tau = T_s$, and since $X[k]$ is generated with sample rate $f_s = 1/T_s$, it cannot be an uncorrelated sequence.

The previous discussion illustrates theoretically how a white-noise sequence is generated by sampling a bandlimited random process. In practice, however, realizations of a white-noise sequence with finite variance are easily generated using a pseudorandom number generator such as rand or randn in MATLAB®.

8.11 BANDWIDTH

In Section 8.10, we described a bandlimited white-noise process where its PSD has been modified by a filter such that it is nonzero only in the frequency range $f \in [-B, B]$. This ideal spectrum can be represented by the rectangle function:

$$S_{XX}(f) = (N_o/2)\text{rect}(f/2B) \tag{8.206}$$

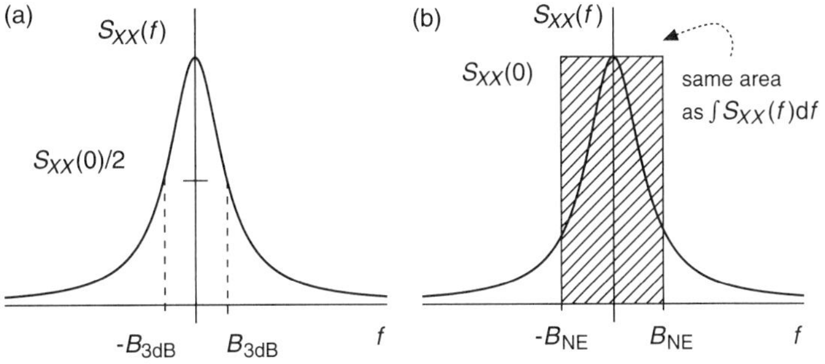

FIGURE 8.19 PSD for an exponential autocorrelation function. (a) 3-dB bandwidth. (b) Noise-equivalent bandwidth.

whose autocorrelation is the sinc function in (8.194). Obviously not all PSDs have such a well-defined bandwidth. For example, Figure 8.19 shows the PSD of the exponential autocorrelation function $R_{XX}(\tau) = \exp(-\alpha|\tau|)$ given by

$$S_{XX}(f) = (1/\pi)\frac{\alpha}{\alpha^2 + f^2}, \tag{8.207}$$

where $\alpha > 0$. For this type of PSD, we can define an *effective bandwidth*. This approach is often taken in filter design where the one-sided bandwidth B (positive frequencies) is defined such that the transfer function satisfies

$$|H(B_{3\text{dB}})|^2 = (1/2)|H(0)|^2. \tag{8.208}$$

If $|H(f)|^2$ is expressed in dB, then the cutoff frequencies $\pm B$ have a magnitude squared that is 3 dB below the maximum at $f = 0$. We can apply a similar definition to the PSD.

Definition: Half-Power Bandwidth The *half-power bandwidth* $B_{3\text{dB}}$ of a PSD is the range of positive frequencies such that

$$S_{XX}(f) \geq S_{XX}(B_{3\text{dB}}) = (1/2)S_{XX}(0). \tag{8.209}$$

It is also called the *3-dB bandwidth*.

For the example in (8.207), the cut-off frequencies are

$$B_{3\text{dB}} = \pm\alpha \tag{8.210}$$

as shown in Figure 8.19(a) where the two-sided 3-dB bandwidth is 2α. This definition can also be applied to PSDs centered about some carrier frequency f_c such as the previous AM signal example in Figure 8.15. For that case, $S_{XX}(0)$ would be replaced with $S_{XX}(f_c)$ in the expression for $B_{3\text{dB}}$.

The following is another measure of bandwidth.

Definition: Noise-Equivalent Bandwidth The *noise-equivalent bandwidth* B_{NE} of random process $X(t)$ is

$$B_{\text{NE}} \triangleq (1/2)\frac{\int_{-\infty}^{\infty} S_{XX}(f)\text{d}f}{S_{XX}(0)}, \tag{8.211}$$

which is the bandwidth of bandlimited white noise that has the same power as $X(t)$.

This definition is illustrated in Fig. 8.19(b). Basically, we find the width of a rectangle with height $S_{XX}(0)$ such that the rectangle and $S_{XX}(f)$ have the same area:

$$\int_{-B_{\mathrm{NE}}}^{B_{\mathrm{NE}}} S_{XX}(0)\mathrm{d}f = 2B_{\mathrm{NE}}S_{XX}(0) = \int_{-\infty}^{\infty} S_{XX}(f)\mathrm{d}f. \tag{8.212}$$

Rearranging this expression gives the formula in the definition. For the exponential autocorrelation function:

$$(1/\pi)\int_{-\infty}^{\infty} \frac{\alpha}{\alpha^2+f^2}\mathrm{d}f = (1/\pi)\tan^{-1}(f/\alpha)\Big|_{-\infty}^{\infty} = (1/\pi)(\pi/2+\pi/2) = 1, \tag{8.213}$$

which turns out to be independent of α so that

$$B_{\mathrm{NE}} = \alpha\pi/2. \tag{8.214}$$

If the autocorrelation function is already known, then B_{NE} can be obtained without integrating the PSD by using

$$B_{\mathrm{NE}} = (1/2)\frac{R_{XX}(0)}{S_{XX}(0)}. \tag{8.215}$$

Observe for the exponential autocorrelation function that the noise-equivalent bandwidth is greater than the 3-dB bandwidth: $\alpha\pi/2 > \alpha$. This result is true in general:

$$B_{\mathrm{NE}} \geq B_{\mathrm{3dB}} \tag{8.216}$$

as we now demonstrate. Since $S_{XX}(f)$ is an even function, only $f \geq 0$ needs to be considered. Also, because both types of bandwidth are defined relative to the maximum of $S_{XX}(f)$, let $S_{XX}(0) = 1$ without loss of generality. Then

$$\begin{aligned} B_{\mathrm{NE}} &= \int_0^{\infty} S_{XX}(f)\mathrm{d}f, \\ &= \int_0^{B_{\mathrm{3dB}}} S_{XX}(f)\mathrm{d}f + \int_{B_{\mathrm{3dB}}}^{\infty} S_{XX}(f)\mathrm{d}f \geq \int_0^{B_{\mathrm{3dB}}} S_{XX}(f)\mathrm{d}f, \end{aligned} \tag{8.217}$$

where we have used the fact that $S_{XX}(f) \geq 0$. If $S_{XX}(f)$ is rectangular, then $B_{\mathrm{3dB}} = B_{\mathrm{NE}}$.

A third measure of bandwidth is related to the standard deviation σ_X of a pdf where the following normalized PSD has the properties of a pdf:

$$\frac{S_{XX}(f)}{\int_{-\infty}^{\infty} S_{XX}(f)\mathrm{d}f} = \frac{S_{XX}(f)}{R_{XX}(0)}, \tag{8.218}$$

which is nonnegative and has unit area. Since $S_{XX}(f)$ is an even function, the following definition gives a type of standard deviation for the normalized PSD.

Definition: Root-Mean-Square Bandwidth The *root-mean-square (RMS) bandwidth* of random process $X(t)$ is

$$B_{\mathrm{RMS}} \triangleq \sqrt{\int_{-\infty}^{\infty} f^2[S_{XX}(f)/R_{XX}(0)]\mathrm{d}f}. \tag{8.219}$$

For the previous exponential autocorrelation function and the PSD in (8.213):

$$\frac{1}{R_{XX}(0)}\int_{-\infty}^{\infty} f^2 S_{XX}(f)\mathrm{d}f = (\alpha/\pi)\int_{-\infty}^{\infty}\frac{f^2}{\alpha^2+f^2}\mathrm{d}f. \tag{8.220}$$

Unfortunately, this is equivalent to the second moment of a Cauchy distribution which we know does not exist; the RMS bandwidth is not useful for this particular PSD. Another example is considered in Problem 8.24.

8.12 SPECTRUM ESTIMATION

Observations of a random process can be used to estimate its underlying mechanism from which it may be possible to predict future outcomes. Familiar examples for which no satisfactory long-term models have been devised include stock market returns and weather forecasting. In the rest of this chapter, we consider signal models that describe how the power spectrum of a random process is distributed across a range of frequencies. There are basically two approaches to *spectrum estimation*: (i) *parametric* and (ii) *nonparametric*. The parametric approach assumes that a random process has an underlying model represented by filter with a white-noise input. Nonparametric PSD estimators, on the other hand, make no assumption about the underlying structure of the random process. The autocorrelation function is estimated and its Fourier transform is computed to derive the PSD. The basic estimator for this approach is known as the *periodogram*.

8.12.1 Periodogram

For wide-sense stationary random sequence $X[k]$ with finite variance, the autocorrelation function is

$$R_{XX}[m] = \mathcal{E}[X[k]X[k+m]], \tag{8.221}$$

where $m \in \mathcal{Z}$ is the time lag. Consider the following estimate of $R_{XX}[m]$:

$$\hat{R}_{XX}[m] = \frac{1}{N}\sum_{k=0}^{N-|m|-1} x[k]x[k+m], \tag{8.222}$$

where $x[k]$ is a realization of the random sequence and we have used $|m|$ in the sum because $R_{XX}[m]$ is an even function. This is a time average over a single realization, and thus we assume that the random sequence is ergodic in correlation so that the estimate has meaning. Since $|m| \leq N-1$, it is desirable that N be sufficiently large so that $\hat{R}_{XX}[m]$ "captures" important details of the correlation because increasingly fewer samples are used for larger values of $|m|$ in the estimate. This property also causes the estimator to be biased for $m \neq 0$:

$$\mathcal{E}[\hat{R}_{XX}[m]] = \frac{1}{N}\sum_{k=0}^{N-|m|-1} R_{XX}[m] = \frac{N-|m|}{N}R_{XX}[m] \tag{8.223}$$

although it is *asymptotically unbiased* as $N \to \infty$. If an unbiased estimator is desired, the estimator could be scaled as follows:

$$\frac{N}{N-|m|}\hat{R}_{XX}[m], \tag{8.224}$$

but it can be shown that it has greater variance than the biased estimator in (8.222).

The PSD $S_{XX}(j\omega)$ is the DTFT of the autocorrelation function $R_{XX}[m]$. The periodogram is likewise a Fourier transform, but of the biased autocorrelation estimator.

Definition: Periodogram The *periodogram* of random sequence $X[k]$ is the DTFT of a biased estimator of its autocorrelation function:

$$P(j\omega) \triangleq \sum_{m=-(N-1)}^{N-1} \hat{R}_{XX}[m]\exp(-j\omega m), \tag{8.225}$$

where $\hat{R}_{XX}[m]$ is given in (8.222). If the sum extends over $\{-(L-1), \ldots, L-1\}$ for $L < N$, but $\hat{R}_{XX}[m]$ is still based on the total number of samples N, then $P(j\omega)$ is known as a *correlogram.*

It is possible to write $P(j\omega)$ in terms of the DTFT of realization $x[k]$ as follows:

$$X(j\omega) = \sum_{k=0}^{N-1} x[k]\exp(-j\omega k). \tag{8.226}$$

Excluding the scale factor $1/N$, observe that (8.222) is a convolution of $x[k]$ with $x[-k]$. Since the DTFT of a reversed (real) sequence is $X(-j\omega)$, (8.225) becomes (see Problem 8.25)

$$P(j\omega) = \frac{1}{N}|X(j\omega)|^2. \tag{8.227}$$

Example 8.11. We present a computer simulation of the following random sequence:

$$X[k] = \cos(2\pi f_1 k + \Phi_1) + 2\cos(2\pi f_2 k + \Phi_2) + \cos(2\pi f_3 k + \Phi_3) + V[k], \tag{8.228}$$

where $\{f_1, f_2, f_3\} = \{100, 200, 300\}$ Hz. The phases $\{\Phi_n\}$ are independent random variables that are uniform on $[-\pi, \pi]$, and $V[k]$ is a white-noise sequence with variance $\sigma_V^2 = 0.01$. Figure 8.20 shows a realization of the random sequence and the corresponding periodogram using a rectangle "window" on the realization. Windowing of the data is achieved by multiplying a realization sample-by-sample with some function, usually of the same duration; windowing is discussed in detail later. Observe that the periodogram clearly shows the three sinusoids.

Next, we derive the mean and variance of the periodogram in order to evaluate the quality of the estimator. Taking the expectation of (8.225) gives

$$\begin{aligned}\mathcal{E}[P(j\omega)] &= \sum_{m=-(N-1)}^{N-1} \mathcal{E}[\hat{R}_{XX}[m]]\exp(-j\omega m), \\ &= \sum_{m=-(N-1)}^{N-1} \frac{N-|m|}{N} R_{XX}[m]\exp(-j\omega m),\end{aligned} \tag{8.229}$$

which we see is the DTFT of the actual autocorrelation function multiplied by the triangular function (window) $1 - |m|/N$. Because of this weighting and the fact that the limits of the DTFT are finite so that only a small segment of a realization is used in the transform, the periodogram is biased. This effect is evident in Example 8.11 where the ideal spectrum has a flat component due to the noise, but this is not the case in the simulation shown in Figure 8.20(b). The triangular weighting is known as a *Bartlett window*, and the product in (8.229) is a convolution in the frequency domain:

$$\mathcal{E}[P(j\omega)] = \frac{1}{2\pi}\int_{-\pi}^{\pi} S_{XX}(j\alpha)W_B(j(\omega-\alpha))\mathrm{d}\alpha, \tag{8.230}$$

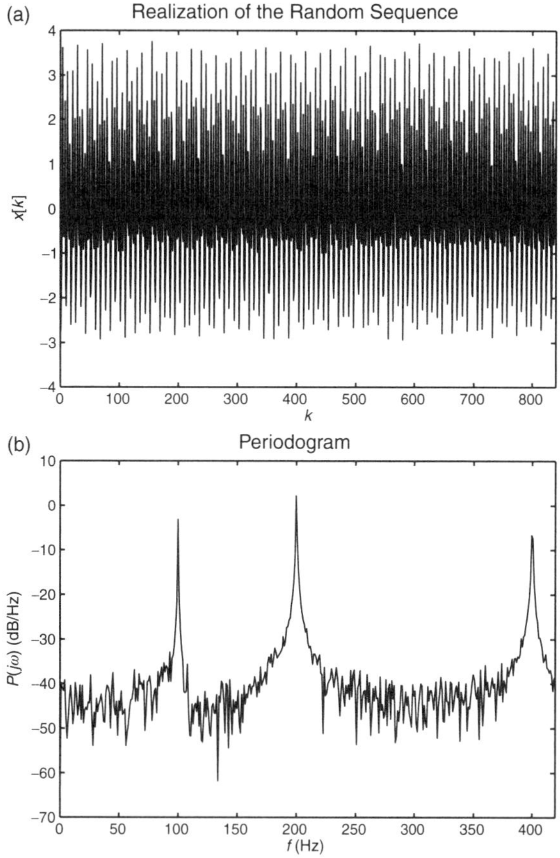

FIGURE 8.20 Simulation results for Example 8.11. (a) Realization of three sinusoids in white noise. (b) Periodogram using a rectangular window on the samples.

where $S_{XX}(j\omega)$ is the PSD of the random sequence and

$$W_B(j\omega) = \sum_{m=-(N-1)}^{N-1} w_B[m]\exp(-j\omega m) = (1/N)\frac{\sin^2(\omega N/2)}{\sin^2(\omega/2)} \tag{8.231}$$

is the DTFT of the Bartlett window:

$$w_B[m] = \begin{cases} 1 - |m|/N, & |m| \leq N-1 \\ 0, & \text{else.} \end{cases} \tag{8.232}$$

An example of the Bartlett window and its DTFT are shown in Figure 8.21 for $N = 15$ (where (8.231) is scaled by $1/N$ so that the main lobe has a maximum of 0 dB).

In order to calculate the variance of the periodogram, we first examine its autocorrelation function, which is the following expectation evaluated at two different frequencies:

$$\begin{aligned} \mathcal{E}[P(j\omega_1)P(j\omega_2)] = \frac{1}{N^2}\sum_{k_1=0}^{N-1}\sum_{k_2=0}^{N-1}\sum_{k_3=0}^{N-1}\sum_{k_4=0}^{N-1} \mathcal{E}[X[k_1]X[k_2]X[k_3]X[k_4]] \\ \cdot \exp(-j\omega_1(k_1-k_2) - j\omega_2(k_3-k_4)). \end{aligned} \tag{8.233}$$

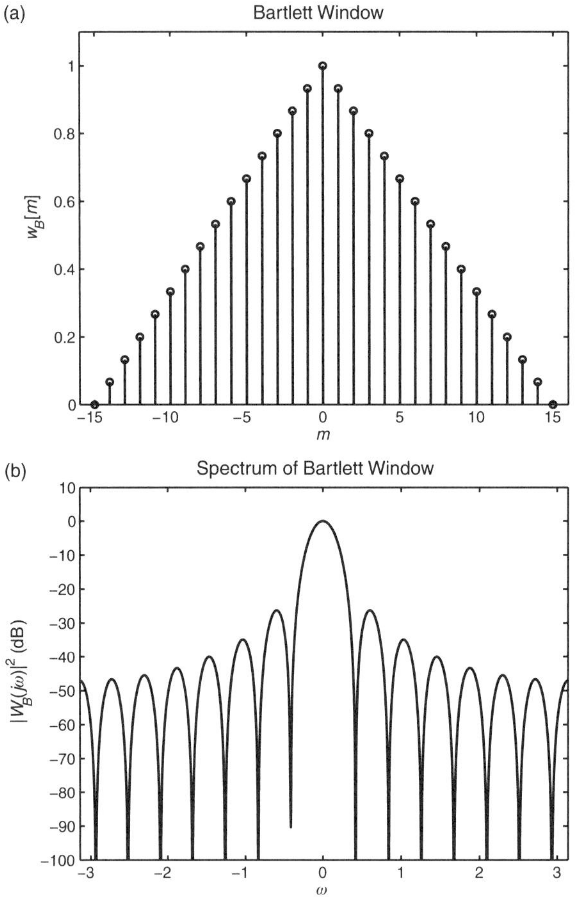

FIGURE 8.21 Bartlett window for $N = 15$. (a) Time domain. (b) Spectrum (scaled by $1/N$).

Since this expression is difficult to evaluate for a general random sequence, we assume that $X[k]$ is Gaussian white noise with finite variance σ_X^2. This allows us to replace the fourth-order moment with σ_X^4 (see the derivation in Chapter 5) when pairs of arguments are equal: (i) $k_1 = k_2$ and $k_3 = k_4$, (ii) $k_1 = k_3$ and $k_2 = k_4$, and (iii) $k_1 = k_4$ and $k_2 = k_3$; otherwise, the expectation is zero. For case (i), the exponential reduces to 1 and the four sums become

$$\sum_{k_1=0}^{N-1} \sum_{k_3=0}^{N-1} \sigma_X^4 = \sigma_X^4 N^2. \tag{8.234}$$

For case (ii),

$$\sum_{k_1=0}^{N-1} \sum_{k_2=0}^{N-1} \sigma_X^4 \exp(-j\omega_1(k_1 - k_2) - j\omega_2(k_1 - k_2)) = \sigma_X^4 \sum_{k_1=0}^{N-1} \sum_{k_2=0}^{N-1} \exp(-j(\omega_1 + \omega_2)(k_1 - k_2)), \tag{8.235}$$

and similarly for case (iii),

$$\sum_{k_1=0}^{N-1} \sum_{k_2=0}^{N-1} \sigma_X^4 \exp(-j\omega_1(k_1 - k_2) - j\omega_2(k_2 - k_1)) = \sigma_X^4 \sum_{k_1=0}^{N-1} \sum_{k_2=0}^{N-1} \exp(-j(\omega_1 - \omega_2)(k_1 - k_2)). \tag{8.236}$$

Combining these three results gives

$$\mathcal{E}[P(j\omega_1)P(j\omega_2)] = \frac{\sigma_X^4}{N^2}\left[N^2 + \sum_{k_1=0}^{N-1}\sum_{k_2=0}^{N-1}\exp(-j(\omega_1+\omega_2)(k_1-k_2)) + \sum_{k_1=0}^{N-1}\sum_{k_2=0}^{N-1}\exp(-j(\omega_1-\omega_2)(k_1-k_2))\right]. \tag{8.237}$$

Consider the first double sum which we see is actually the product of two sums:

$$\sum_{k_1=0}^{N-1}\sum_{k_2=0}^{N-1}\exp(-j(\omega_1+\omega_2)(k_1-k_2)) = \sum_{k_1=0}^{N-1}\exp(-j(\omega_1+\omega_2)k_1)\sum_{k_2=0}^{N-1}\exp(j(\omega_1+\omega_2)k_2). \tag{8.238}$$

Each sum is the DTFT of a rectangle sequence that is nonzero on $\{0, \ldots, N-1\}$. From the DTFT properties in Appendix C,

$$\sum_{k_1=0}^{N-1}\exp(-j(\omega_1+\omega_2)k_1) = \exp(-j(\omega_1+\omega_2)(N-1)/2)\frac{\sin((\omega_1+\omega_2)N/2)}{\sin((\omega_1+\omega_2)/2)}. \tag{8.239}$$

Since the two sums in (8.238) form a complex conjugate pair, the phase component in (8.239) cancels. A similar result is obtained for the other double sum in (8.237), such that the autocorrelation function becomes

$$\mathcal{E}[P(j\omega_1)P(j\omega_2)] = \sigma_X^4\left[1 + \frac{\sin^2((\omega_1+\omega_2)N/2)}{N^2\sin^2((\omega_1+\omega_2)/2)} + \frac{\sin^2((\omega_1-\omega_2)N/2)}{N^2\sin^2((\omega_1-\omega_2)/2)}\right]. \tag{8.240}$$

The autocovariance function is

$$\text{cov}[P(j\omega_1), P(j\omega_2)] = \mathcal{E}[P(j\omega_1)P(j\omega_2)] - \mathcal{E}[P(j\omega_1)]\mathcal{E}[P(j\omega_2)], \tag{8.241}$$

and from (8.229) with $R_{XX}[m] = \sigma_X^2\delta[m]$ (because the sequence is white), we have

$$\sum_{m=-(N-1)}^{N-1}\frac{N-|m|}{N}\sigma_X^2\delta[m]\exp(-j\omega m) = \sigma_X^2. \tag{8.242}$$

Thus

$$\text{cov}[P(j\omega_1), P(j\omega_2)] = \sigma_X^4\left[\frac{\sin^2((\omega_1+\omega_2)N/2)}{N^2\sin^2((\omega_1+\omega_2)/2)} + \frac{\sin^2((\omega_1-\omega_2)N/2)}{N^2\sin^2((\omega_1-\omega_2)/2)}\right]. \tag{8.243}$$

The variance is obtained when $\omega_1 = \omega_2 = \omega$:

$$\text{var}[P(j\omega)] = \sigma_X^4\left[\frac{\sin^2(\omega N)}{N^2\sin^2(\omega)} + 1\right], \tag{8.244}$$

which, using l'Hôpital's rule, we see is bounded as $\sigma_X^4 \le \text{var}[P(j\omega)] \le 2\sigma_X^4$.

Figure 8.22(a) shows the autocovariance function of the periodogram for $N = 5$ and $\sigma_X^2 = 1$. Observe that the function reaches a maximum at the origin and at the four corner points $\omega_1 = \pm\omega_2 = \pm\pi$. The covariance is relatively small when $\omega_1 \neq \pm\omega_2$, and it is exactly zero when $\omega_1 = 2\pi m/N$ and $\omega_2 = 2\pi n/N$ for $m, n \in \mathcal{Z}$ and $m \neq n$. The variance plotted in Figure 8.22(b) corresponds to a "slice" through the function at $\omega = \omega_1 = \omega_2$.

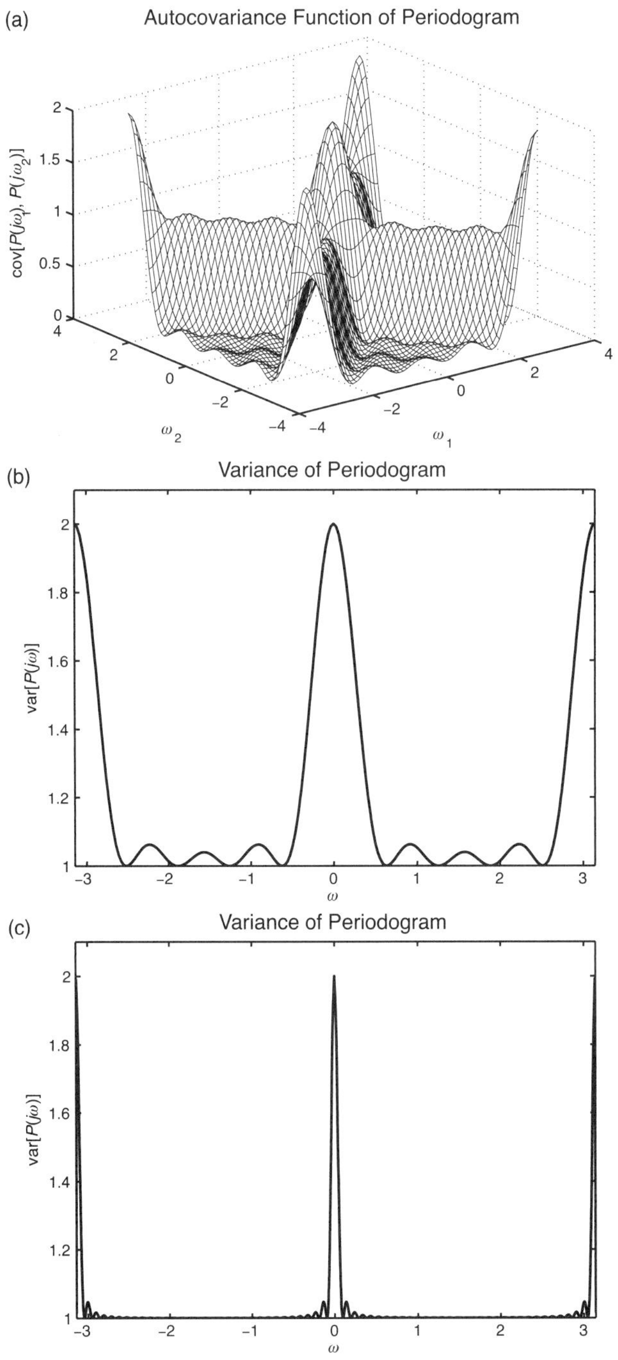

FIGURE 8.22 Autocovariance function and variance of the periodogram for a white Gaussian sequence with variance $\sigma_X^2 = 1$. (a) $\text{cov}[P(j\omega_1), P(j\omega_2)]$ for $N = 5$. (b) $\text{var}[P(j\omega)]$ for $N = 5$. (c) $\text{var}[P(j\omega)]$ for $N = 33$.

When N is increased, the covariance decreases for $\omega_1 \neq \pm\omega_2$. It still has value 2 at the origin and the four points $\omega_1 = \pm\omega_2 = \pm\pi$, and it decreases to 1 when $\omega_1 = \pm\omega_2$. This is demonstrated by the variance in Figure 8.22(c) for $N = 33$. The variance does not decrease to zero with increasing N (generally, we prefer that the variance of an estimator continue to decrease as discussed later in Chapter 9).

In the following sections, we consider two variations of the standard periodogram for estimating the PSD of a random sequence: (i) smoothing the periodogram and (ii) windowing the realization.

8.12.2 Smoothed Periodogram

One variation of the periodogram performs windowing on $\hat{R}_{XX}[m]$ such that it is multiplied by a function as follows:

$$P_S(j\omega) = \sum_{m=-(N-1)}^{N-1} w[m]\hat{R}_{XX}[m]\exp(-j\omega m), \tag{8.245}$$

where the subscript S denotes smoothed. The weighting $w[m]$ performs the following: (i) *truncates* $\hat{R}_{XX}[m]$ to the range $m \in \{-(M-1), \ldots, M-1\}$, where typically $M \ll N$ so that only the more accurate autocorrelation estimates with small time lags are included, and (ii) shapes the autocorrelation estimate which in turn *smooths* the PSD estimate. Note that $w[m]$ should be an even function (as is $R_{XX}[m]$) so that $P_S(j\omega)$ retains the same properties as a PSD. The smoothing property is evident because the DTFT of the product on the right-hand side above is replaced by a convolution in the frequency domain:

$$P_S(j\omega) = \frac{1}{2\pi}\int_{-\pi}^{\pi} W(j\alpha)P(j(\omega-\alpha))\mathrm{d}\alpha, \tag{8.246}$$

where

$$W(j\omega) = \sum_{m=-(M-1)}^{M-1} w[m]\exp(-j\omega m) \tag{8.247}$$

is the DTFT of the window and $P(j\omega)$ is the standard periodogram. Since $M \ll N$, corresponding to a narrow window, its transform $W(j\omega)$ is relatively wide, so that convolving it with the periodogram causes $P_S(j\omega)$ to be smoother than $P(j\omega)$.

For the standard periodogram $P(j\omega)$, we found that $\mathcal{E}[P(j\omega)]$ is the DTFT of a weighted form of $R_{XX}[m]$ where the weighting is a Bartlett window with overall length $2N-1$. This triangular weighting comes about because $\hat{R}_{XX}[m]$ is a biased estimator of $R_{XX}[m]$, which is used because it has lower variance than the unbiased estimator. The windowing described here for smoothing the periodogram differs in that it is intentionally applied to the autocorrelation function estimate $\hat{R}_{XX}[m]$. Moreover, $M \ll N$ so that values of $\hat{R}_{XX}[m]$ with large time lags are removed; as mentioned before, those autocorrelation estimates are less accurate because fewer terms are used in the sum due to the smaller upper limit $N - |m| - 1$.

The mean of $P_S(j\omega)$ is

$$\mathcal{E}[P_S(j\omega)] = \sum_{m=-(N-1)}^{N-1} w[m]w_B[m]R_{XX}[m]\exp(-j\omega m), \tag{8.248}$$

where we have included the Bartlett window $w_B[m]$ that appears in (8.229). Thus, the actual autocorrelation function $R_{XX}[m]$ is weighted by the composite window $w[m]w_B[m]$. However, since $M \ll N$, the Bartlett window is relatively flat in the range $\{-(M-1), \ldots, M-1\}$ such that $w[m]$ dominates the product of the windows: $w[m]w_B[m] \approx w[m]$. Ignoring the taper of $w_B[m]$ in the range $\{-(M-1), \ldots, M-1\}$ has little effect on the mean. In the frequency domain, we have the approximation

$$\mathcal{E}[P_S(j\omega)] \approx \frac{1}{2\pi}\int_{-\pi}^{\pi} W(j\alpha)S_{XX}(j(\omega-\alpha))\mathrm{d}\alpha, \tag{8.249}$$

where $W(j\alpha)$ is the DTFT of the windowing explicitly applied to the autocorrelation estimate.

The autocovariance function of the smoothed periodogram is

$$\begin{aligned}\operatorname{cov}[P_S(j\omega_1), P_S(j\omega_2)] = \frac{1}{4\pi^2}\int_{-\pi}^{\pi}\int_{-\pi}^{\pi} & W(j\alpha_1)W(j\alpha_2)\\ & \cdot\operatorname{cov}[P(j(\omega_1-\alpha_1)), P(j(\omega_2-\alpha_2))]\mathrm{d}\alpha_1\mathrm{d}\alpha_2.\end{aligned} \tag{8.250}$$

As in the analysis of the standard periodogram, we assume that $X[k]$ is Gaussian white noise with variance σ_X^2 so that the covariance in the integrand can be derived from (8.243). In order to obtain a simple expression, assume N is large so that the DTFT of the Bartlett window in (8.231) is approximated by a train of Dirac delta functions (Oppenheim and Schafer, 1975; Tretter, 1976):

$$W_B(j\omega) = \frac{\sin^2(\omega N/2)}{N\sin^2(\omega/2)} \approx \sum_{n=-\infty}^{\infty} 2\pi\delta(\omega - 2\pi n). \tag{8.251}$$

It is evident from l'Hôpital's rule that $W_B(j\omega)$ has peaks of height N at integer multiples of 2π. As N is increased, those peaks approach infinity and the width of each "pulse" of $W_B(j\omega)$ approaches zero. Since $w_B[m]$ is an even function, the area property of a PSD can be used to write:

$$\frac{1}{2\pi}\int_{-\pi}^{\pi}\frac{\sin^2(\omega N/2)}{N\sin^2(\omega/2)}\mathrm{d}\omega = w_B[0] = 1. \tag{8.252}$$

From the example in Figure 8.23, we see that with increasing N the DTFT of the Bartlett window for $\omega \in [-\pi, \pi]$ approaches the Dirac delta function with area 2π, and these repeat at frequencies that are integer multiples

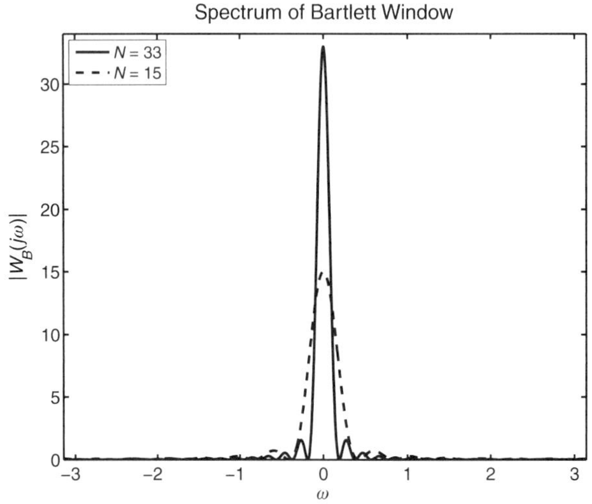

FIGURE 8.23 Bartlett window spectra (scaled by $1/N$, linear scale).

of 2π. It is this narrowing of the Bartlett DTFT pulse that leads to the impulse-like variance at the origin in Figure 8.22(c).

Substituting the approximation of (8.251) into (8.243) gives two delta functions:

$$\sigma_X^4 \left[\frac{\sin^2((\omega_1+\omega_2)N/2)}{N^2 \sin^2((\omega_1+\omega_2)/2)} + \frac{\sin^2((\omega_1-\omega_2)N/2)}{N^2 \sin^2((\omega_1-\omega_2)/2)} \right] \approx \frac{2\pi\sigma_X^4}{N} [\delta(\omega_1+\omega_2) + \delta(\omega_1-\omega_2)] \tag{8.253}$$

in the frequency range $\omega \in [-\pi, \pi]$. Substituting this into (8.250) with the appropriate arguments allows us to reduce the double integral to a single integral:

$$\begin{aligned} \text{cov}[P_S(j\omega_1), P_S(j\omega_2)] \approx \frac{\sigma_X^4}{2\pi N} \int_{-\pi}^{\pi} & [W(j(\omega_1+\omega_2-\alpha_2)) \\ & + W(j(\omega_1-\omega_2+\alpha_2))] W(j\alpha_2) d\alpha_2. \end{aligned} \tag{8.254}$$

The variance is obtained when $\omega_1 = \omega_2 = \omega$:

$$\text{var}[P_S(j\omega)] \approx \frac{\sigma_X^4}{2\pi N} \int_{-\pi}^{\pi} [W(j(2\omega-\alpha_2))W(j\alpha_2) + W^2(j\alpha_2)] d\alpha_2. \tag{8.255}$$

Assuming that M is sufficiently large (but still $\ll N$) so that $W(j\omega)$ is narrow, the integral of the first term is small compared to the second term so that it can be ignored. This yields the approximation

$$\text{var}[P_S(j\omega)] \approx \frac{\sigma_X^4}{2\pi N} \int_{-\pi}^{\pi} W^2(j\alpha_2) d\alpha_2. \tag{8.256}$$

In order to compare this variance result to that for $P(j\omega)$ in (8.244), we need a measure of the size of the integral above. From Parseval's theorem,

$$\frac{1}{2\pi} \int_{-\pi}^{\pi} W^2(j\alpha_2) d\alpha_2 = \sum_{-(M-1)}^{M-1} w^2[m] \leq 2M-1, \tag{8.257}$$

where the window has been normalized so that $0 \leq w[m] \leq 1$ for all m in order to obtain the upper bound. This yields

$$\text{var}[P_S(j\omega)] \approx (1/N)(2M-1)\sigma_X^4. \tag{8.258}$$

Comparing with the lower bound σ_X^4 for the variance of $P(j\omega)$ in (8.244), we find that var$[P_S(j\omega)]$ is reduced by choosing M of the window to be much smaller than the number of samples N used in the autocorrelation estimate. Again, by truncating the estimated autocorrelation function, those terms with large time lags and thus greater variance are excluded. However, for small values of M, the DTFT of the window is relatively wide, which means that the periodogram will not provide good *resolution* of individual PSD components, especially if they are close to each other in frequency.

Example 8.12. Consider using the following *rectangular window* with $M < N$:

$$w_R[m] = \begin{cases} 1, & |m| \leq M-1 \\ 0, & \text{else,} \end{cases} \tag{8.259}$$

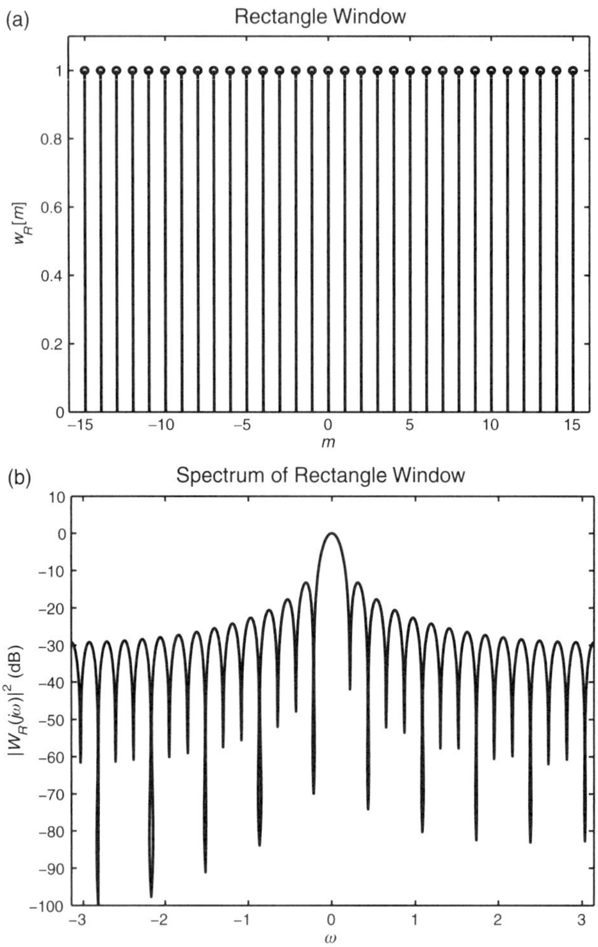

FIGURE 8.24 Rectangle window for $M = 16$. (a) Time domain. (b) Spectrum (scaled by $1/(2M - 1)$).

which has DTFT

$$W_R(j\omega) = \frac{\sin((2M - 1)\omega/2)}{\sin(\omega/2)}. \tag{8.260}$$

An example is shown in Figure 8.24, which we see has a narrower main lobe than the Bartlett window in Figure 8.21, but the sidelobes are higher. It is well known in discrete-time signal processing that there is generally a tradeoff between the width of the main lobe (in the passband) and the height of the sidelobes (in the stopband). However, the rectangular window should not be used for the smoothed periodogram because $W_R(j\omega)$ has negative values and the PSD estimate should be nonnegative. The Bartlett window can be used because its DTFT is nonnegative, as is evident from (8.231). Figure 8.25 shows an example of the smoothed periodogram for the signal in Example 8.11 with three sinusoids and $M = N/2$. Observe that even though the spectrum appears to be less noisy (reduced variance) than that in Figure 8.20, the resolution is less. Convolving the standard periodogram with the spectrum of the Bartlett window causes the peaks to widen such that closely spaced sinusoids might be difficult to separate. This example illustrates that even though the variance of an estimator can be reduced, the overall quality may not be improved when considering other features of an estimator such as its ability to resolve and separate signals.

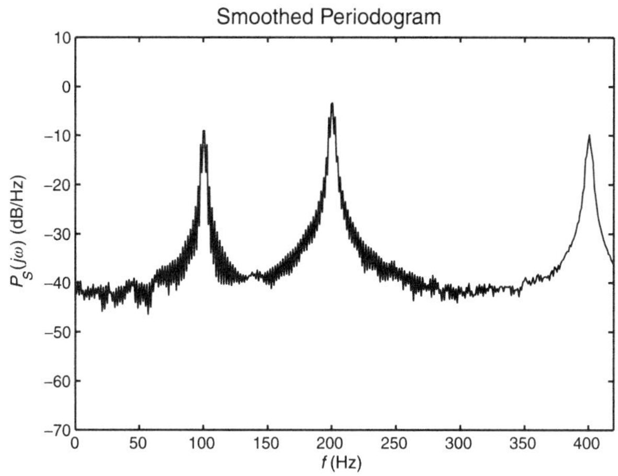

FIGURE 8.25 Smoothed periodogram for Example 8.12.

8.12.3 Modified Periodogram

Another variation of the standard periodogram windows the realization itself (instead of the autocorrelation estimate), resulting in the so-called modified periodogram.

Definition: Modified Periodogram The *modified periodogram* of random sequence $X[k]$ is

$$P_M(j\omega) \triangleq \frac{1}{\sum_{k=0}^{N-1} w^2[k]} \left| \sum_{k=0}^{N-1} w[k]x[k]\exp(-j\omega k) \right|^2, \tag{8.261}$$

where $w[k]$ is a window applied to realization $x[k]$.

If the window is rectangular as depicted in Figure 8.24, the modified periodogram reduces to the standard periodogram when the window is of duration $2N - 1$. Observe that the frequency response of the rectangle window has a relatively narrow main lobe, but the sidelobes are high compared with those of the Hamming window shown in Figure 8.26. The *Hamming window* is the following raised-cosine function:

$$w_H[k] = \begin{cases} 0.54 + 0.46\cos(2\pi k/(2N-1)), & |k| \le N-1 \\ 0, & \text{else.} \end{cases} \tag{8.262}$$

Although the DTFT of the Hamming window does not have a simple form, it can be written in terms of $W_R(j\omega)$. The first term above with 0.54 corresponds to a rectangle function, as does the magnitude of the cosine term if it is rewritten using Euler's formula:

$$0.46\cos(2\pi m/(2N-1)) = 0.23\exp(j2\pi m/(2N-1)) + 0.23\exp(-j2\pi m/(2N-1)). \tag{8.263}$$

Thus, the spectrum is the sum of three terms:

$$\begin{aligned} W_H(j\omega) = {} & 0.54W_R(j\omega) + 0.23W_R(j(\omega - 2\pi/(2N-1))) \\ & + 0.23W_R(j(\omega + 2\pi/(2N-1))). \end{aligned} \tag{8.264}$$

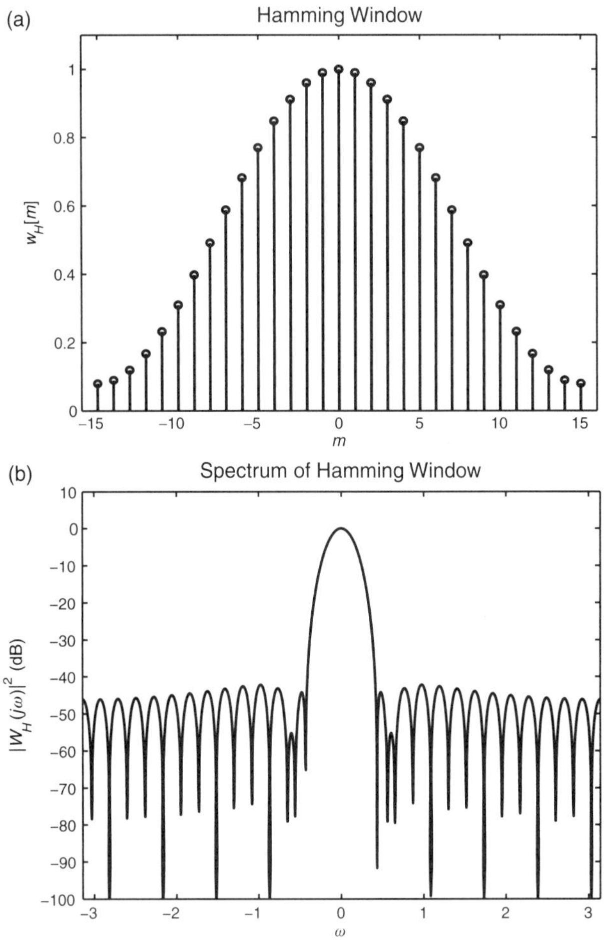

FIGURE 8.26 Hamming window for $N = 16$. (a) Time domain. (b) Spectrum (scaled by $1/0.54(2N - 1)$).

Multiplying a sequence by a window results in a convolution in the frequency domain between the DTFT of the signal and the frequency response of the window. This explains why using only a finite number of samples when computing the periodogram, which is equivalent to multiplying an infinite-length sequence by a finite-length rectangular window, results in a "noisy" PSD estimate. Let $X(j\omega)$ be the DTFT of realization $x[k]$ assuming it has infinite length. The DTFT of the windowed sequence can be written as the following convolution:

$$X_W(j\omega) = \int_{-\pi}^{\pi} X(j\alpha)W(j(\omega - \alpha))\mathrm{d}\alpha, \tag{8.265}$$

where $W(j\omega)$ is the DTFT of the window. The corresponding periodogram is

$$P_M(j\omega) = \frac{1}{N}|X_W(j\omega)|^2. \tag{8.266}$$

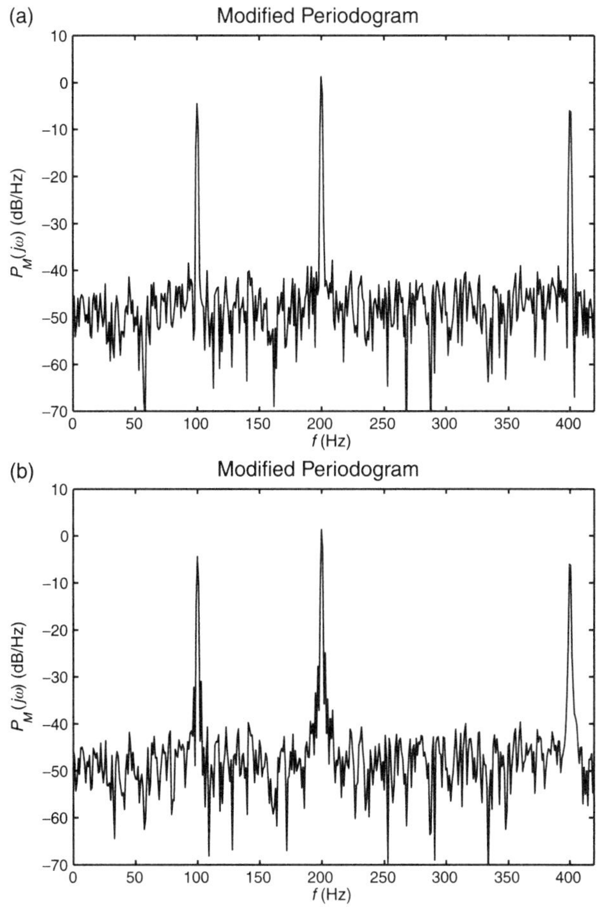

FIGURE 8.27 Modified periodogram for Example 8.13 using (a) the Hamming window and (b) the Bartlett window.

Example 8.13. For the signal in Example 8.11, improved sidelobe suppression is achieved by using the Hamming window as demonstrated in Figure 8.27. The corresponding result using the Bartlett window is also shown, which gives similar performance.

It is important to note that the modified periodogram differs from the smoothed periodogram in that the former windows a realization of the random sequence. The corresponding convolution in the frequency domain reduces the effect of the additive noise in the sequence itself so that a more accurate autocorrelation estimate is obtained. The smoothed periodogram applies a window to the autocorrelation estimate as opposed to the realization. Generally in practice, it is preferable to process (window or filter) the original samples before computing an autocorrelation estimate, rather than trying to process (smooth) the autocorrelation function.

Finally, we briefly mention two averaging methods known as the *methods of Bartlett and Welch* that operate on the standard periodogram rather than on a realization of the random sequence. Bartlett's method of averaging first partitions the N samples of realization $x[k]$ into K nonoverlapping sections, then computes a periodogram for each section using a rectangular window, and finally averages the K periodograms to obtain the overall PSD estimate. Welch's method is an extension of Bartlett's method that (i) windows the realization $x[k]$ as in the modified periodogram and (ii) allows the Bartlett sections to overlap. These approaches typically provide less noisy estimates than the modified periodogram, but the resolution is not always as good.

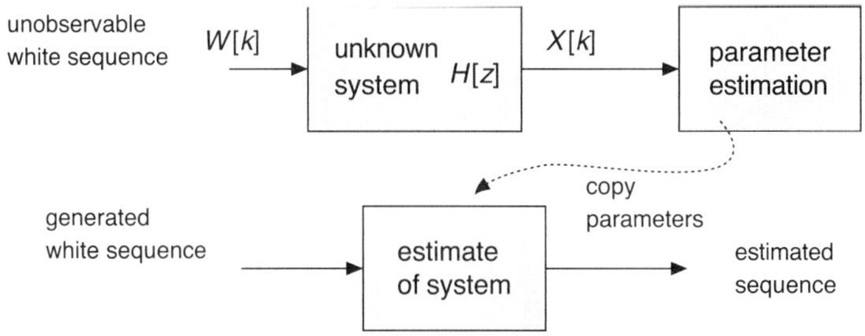

FIGURE 8.28 System identification framework for parametric spectrum analysis.

8.13 PARAMETRIC MODELS

The parametric approach to spectrum estimation is a type of *system identification* as illustrated in Figure 8.28. The model assumes that random sequence $X[k]$ is generated as the output of an unknown filter with transfer function $H(z)$, whose input is the white-noise sequence $W[k]$. The basic types of systems are known as autoregressive (AR), MA, and ARMA models, and are described in the next few sections. For a particular realization $x[k]$, the goal is to estimate the parameters of $H(z)$ such that when the estimated system is driven by white noise, its output closely approximates $x[k]$. Some issues with this approach include: (i) the type of model, (ii) the order of the model (number of parameters), and (iii) the method of calculating the parameter values. Since the parametric estimate is derived only from the output $X[k]$, the modeling approach is sometimes called "blind" because the input $W[k]$ is not observable. The more general system identification problem assumes that both the input and the output of the model are available; techniques for estimating filter parameters for that case are mentioned briefly at the end of this chapter.

Given the previous model framework, correlation is an important feature of $X[k]$ that is exploited to estimate the model parameters. Since the input is white noise with variance σ_W^2, we know that the autocorrelation function of the output is the inverse z-transform of the following PSD:

$$S_{YY}(z) = \sigma_W^2 H(z)H(z^{-1}). \tag{8.267}$$

Obviously the model parameters determine the degree of correlation since the input has autocorrelation function $R_{WW}[m] = \sigma_W^2 \delta[m]$, and thus the choice of model is an important consideration. In applications involving natural signals, a sequence is actually more complicated than can represented by the transfer function $H(z)$. Such models are only approximations for how the random sequence might have been generated. AR, MA, and ARMA models are considered because they are well-known LTI filters. It is straightforward to derive an estimate for the AR model because the problem formulation is linear, whereas it is nonlinear for the MA and ARMA models.

8.13.1 Autoregressive Model

An AR model for random sequence $X[k]$ is described by the following *recursive* difference equation:

$$X[k] = \sum_{n=1}^{N} a_n X[k-n] + W[k], \tag{8.268}$$

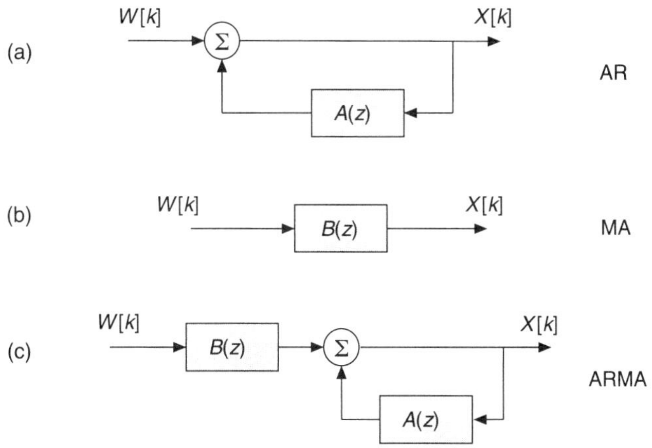

FIGURE 8.29 Random sequence models. (a) AR. (b) MA. (c) ARMA.

where N is the order of the model. Parameter a_0 can be included on the left-hand side to weight $X[k]$, but often it is chosen to be 1. This model is an infinite-impulse-response (IIR) filter operating on the input sequence $W[k]$ as shown in Figure 8.29(a). It is compactly written in vector form as follows:

$$X[k] = \mathbf{a}^T \mathbf{X}[k-1] + W[k], \tag{8.269}$$

where $\mathbf{a} \triangleq [a_1, \ldots, a_N]^T$ and $\mathbf{X}[k-1] \triangleq [X[k-1], \ldots, X[k-N]]^T$. The transfer function of the system is

$$H(z) = \frac{1}{1 - \sum_{n=1}^{N} a_n z^{-n}} \triangleq \frac{1}{1 - A(z)}. \tag{8.270}$$

For an AR model, $H(z)$ has only poles; the more general case with poles and zeros is the ARMA model discussed later. For a stable and causal system, the poles of $H(z)$ are all inside the unit circle. Since $W[k]$ is a white sequence, the PSD of $X[k]$ is

$$S_{XX}(z) = H(z)H(z^{-1}) = \frac{\sigma_W^2}{[1 - A(z)][1 - A(z^{-1})]}. \tag{8.271}$$

Consider the following first-order AR model with $N = 1$:

$$X[k] = aX[k-1] + W[k], \tag{8.272}$$

which is a filter with a single pole at $z = a$ and requires $|a| < 1$ for the system to be stable. By substituting for $X[k-1]$, then $X[k-2]$, and so on, the recursive expression above can be rewritten in the following nonrecursive form:

$$\begin{aligned} X[k] &= a(aX[k-2] + W[k-1]) + W[k], \\ &= a^2(aX[k-3] + W[k-2]) + aW[k-1] + W[k], \\ &\vdots \\ &= \sum_{n=0}^{\infty} a^n W[k-n]. \end{aligned} \tag{8.273}$$

Since $W[k]$ has zero mean, $\mathcal{E}[X[k]] = 0$ for all k. The autocorrelation function is

$$\mathcal{E}[X[k]X[k+m]] = \sum_{n=0}^{\infty}\sum_{p=0}^{\infty} a^{n+p}\mathcal{E}[W[k-p]W[k+m-n]],$$

$$= \sum_{n=0}^{\infty}\sum_{p=0}^{\infty} a^{n+p}\sigma_W^2\delta[m-n+p]. \tag{8.274}$$

Changing variables to $l = m + p$ and assuming for now that m is positive yields

$$R_{XX}[m] = \sigma_W^2 \sum_{n=0}^{\infty}\sum_{l=m}^{\infty} a^{n-m+l}\delta[l-n],$$

$$= a^{-m}\sigma_W^2 \sum_{n=m}^{\infty} a^{2n} = \frac{a^m\sigma_W^2}{1-a^2}. \tag{8.275}$$

Because the autocorrelation function is an even function, we have for $m \in \mathcal{Z}$:

$$R_{XX}[m] = \frac{a^{|m|}\sigma_W^2}{1-a^2}, \tag{8.276}$$

which gives the following variance:

$$\sigma_X^2 = R_{XX}[0] = \frac{\sigma_W^2}{1-a^2}. \tag{8.277}$$

Figure 8.30 shows examples of the AR(1) autocorrelation function for positive and negative values of a (with $\sigma_W^2 = 1$). The correlation increases when the pole of the AR model is close to the unit circle ($|a| \to 1$), and when the pole is negative the correlation is alternately positive (for m even) and negative (for m odd).

The autocorrelation function can also be derived from the PSD:

$$S_{XX}(z) = \frac{\sigma_W^2}{(1-az^{-1})(1-az)},$$

$$= \frac{\sigma_W^2}{1-a^2}\left(\frac{1}{1-az^{-1}} + \frac{1}{1-az}\right), \tag{8.278}$$

which has been written as a partial fraction expansion (PFE). The first term corresponds to a right-sided function, and the second term gives a left-sided function. The inverse z-transform is

$$R_{XX}[m] = \frac{\sigma_W^2}{1-a^2}\left(a^{-m}u[m] + a^m u[-m+1]\right) = \frac{a^{-|m|}\sigma_W^2}{1-a^2}, \tag{8.279}$$

which is the result in (8.276). Realizations of AR(1) sequences with the correlations in Figure 8.30 are shown in Figure 8.31. Observe that the output signals have more "structure" than the input signal, and the output with $a = 0.9$ appears to be more predictable than when $a = -0.9$.

For the general AR sequence in (8.268), it is not possible to obtain a simple explicit expression for the autocorrelation function. Instead, we can write a recursive expression that is similar to the random sequence description itself. Consider

$$\mathcal{E}[X[k-m]X[k]] = \sum_{n=1}^{N} a_n\mathcal{E}[X[k-m]X[k-n]] + \mathcal{E}[X[k-m]W[k]] \tag{8.280}$$

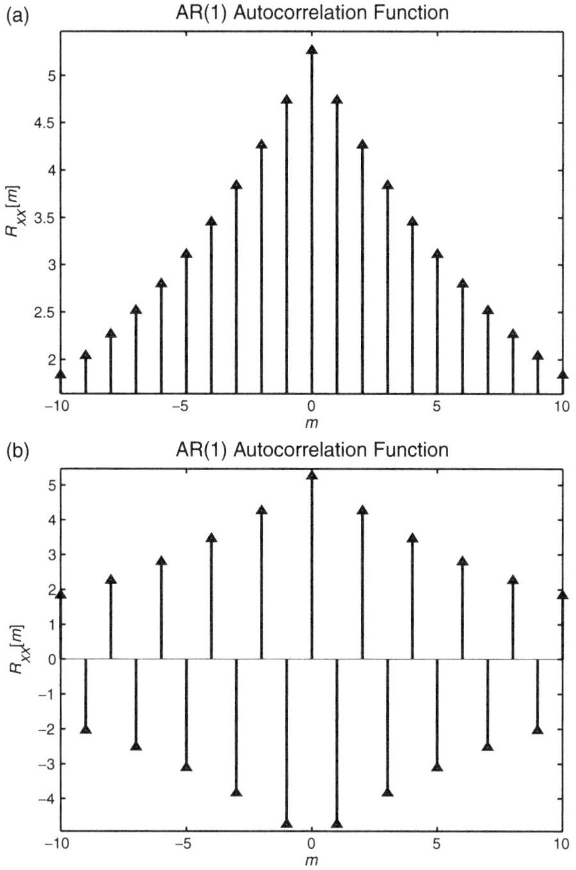

FIGURE 8.30 Autocorrelation function for an AR(1) sequence. (a) $\mathbf{a} = 0.9$. (b) $\mathbf{a} = -0.9$.

for $m > 0$. The second term on the right-hand side is zero because previous samples of $X[k]$ are independent of the white sequence $W[k]$. This leads to the following recursive expression:

$$R_{XX}[m] = \sum_{n=1}^{N} a_n R_{XX}[m-n], \tag{8.281}$$

which can be expressed in matrix form for different values of m as follows:

$$\begin{bmatrix} R_{XX}[1] \\ R_{XX}[2] \\ \vdots \end{bmatrix} = \begin{bmatrix} R_{XX}[0] & R_{XX}[-1] & \cdots & R_{XX}[N-1] \\ R_{XX}[1] & R_{XX}[0] & \cdots & R_{XX}[N-2] \\ \vdots & \vdots & & \vdots \end{bmatrix} \begin{bmatrix} a_1 \\ a_2 \\ \vdots \\ a_N \end{bmatrix}. \tag{8.282}$$

For $m = 0$:

$$\mathcal{E}[X^2[k]] = \sum_{n=1}^{N} a_n \mathcal{E}[X[k-n]X[k]] + \mathcal{E}[W[k]X[k]], \tag{8.283}$$

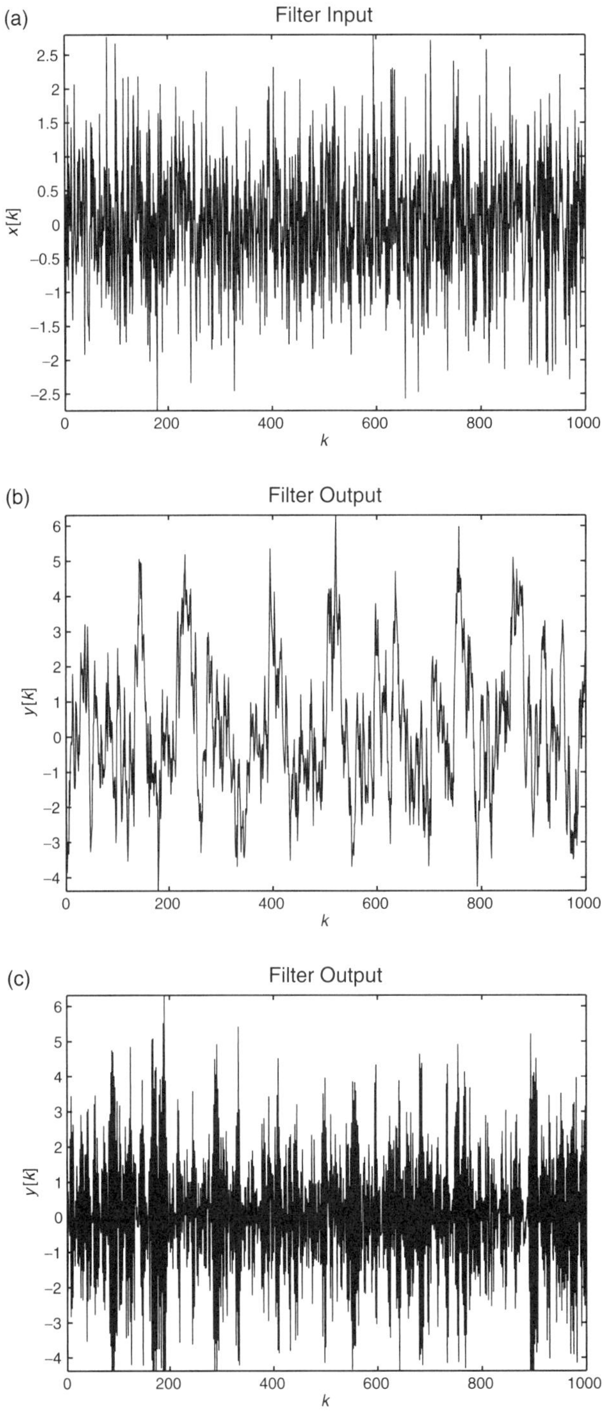

FIGURE 8.31 Example realizations of an AR(1) sequence. (a) White Gaussian input. (b) Output for $\mathbf{a} = 0.9$. (c) Output for $\mathbf{a} = -0.9$.

which gives

$$R_{XX}[0] = \sum_{n=1}^{N} a_n R_{XX}[n] + \sigma_W^2. \tag{8.284}$$

Since the autocorrelation function is an even function, the arguments in (8.282) can be changed such that all are positive, revealing that the matrix in (8.282) is Toeplitz. These equations are known as the *Yule–Walker equations* which are used to solve for the filter coefficients $\{a_n\}$; they can be substituted into (8.284) to estimate the variance σ_W^2. Of course, since $R_{XX}[m]$ is not known, an estimate would be generated from realizations of $X[k]$.

Example 8.14. Suppose the estimated autocorrelation function happens to have the ideal form $\hat{R}_{XX}[m] = (0.9)^{|m|}$, and consider $N = 3$ in the Yule–Walker equations such that

$$\begin{bmatrix} 0.9 \\ 0.81 \\ 0.729 \end{bmatrix} = \begin{bmatrix} 1 & 0.9 & 0.81 \\ 0.9 & 1 & 0.9 \\ 0.81 & 0.9 & 1 \end{bmatrix} \begin{bmatrix} a_1 \\ a_2 \\ a_3 \end{bmatrix}. \tag{8.285}$$

Solving for the coefficients yields $[a_1, a_2, a_3]^T = [0.9, 0, 0]^T$, and from (8.284) the noise variance is $\sigma_W^2 = 1 - 0.81 = 0.19$. The resulting first-order system is not surprising because this particular autocorrelation function has the form in (8.276), and σ_W^2 is related to $R_{XX}[0]$ according to (8.277). Suppose that due to imprecise autocorrelation measurements the Yule–Walker equations are instead

$$\begin{bmatrix} 0.9 \\ 0.8 \\ 0.7 \end{bmatrix} = \begin{bmatrix} 1 & 0.9 & 0.8 \\ 0.9 & 1 & 0.9 \\ 0.8 & 0.9 & 1 \end{bmatrix} \begin{bmatrix} a_1 \\ a_2 \\ a_3 \end{bmatrix}. \tag{8.286}$$

Solving these equations gives $[a_1, a_2, a_3]^T = [0.9444, 0, -0.0556]^T$, and from (8.277) the noise variance is $\sigma_W^2 = 1 - (0.9444)(0.9) + (0.0556)(0.7) \approx 0.1905$. The sensitivity of the estimates to measurement errors can be an important issue, especially if the AR model has a large number of parameters.

8.13.2 Moving-Average Model

An MA model of a random sequence does not have a recursive form; instead, it is a *finite* sum of the weighted white input sequence as follows:

$$X[k] = \sum_{n=0}^{M-1} b_n W[k-n], \tag{8.287}$$

where M is the order. This type of processing is an FIR filter with transfer function

$$H(z) = \sum_{n=0}^{M-1} b_n z^{-n} \triangleq B(z). \tag{8.288}$$

A block diagram of the system is shown in Figure 8.29(b), which has PSD

$$S_{XX}(z) = \sigma_W^2 B(z)B(z^{-1}). \tag{8.289}$$

Because the sequence model is not recursive, the autocorrelation function is easily computed for the general case:

$$\begin{aligned}\mathcal{E}[X[k]X[k+m]] &= \sum_{n=0}^{M-1}\sum_{p=0}^{M-1} b_n b_p \mathcal{E}[W[k-n]W[k+m-p]], \\ &= \sigma_W^2 \sum_{n=0}^{M-1}\sum_{p=0}^{M-1} b_n b_p \delta[m+n-p] = \sigma_W^2 \sum_{n=0}^{M-m-1} b_n b_{m+n} \end{aligned} \tag{8.290}$$

for $m \geq 0$. From symmetry of the autocorrelation function, the result for arbitrary m is

$$R_{XX}[m] = \sigma_W^2 \sum_{n=0}^{M-|m|-1} b_n b_{|m|+n}. \tag{8.291}$$

The sum is actually the deterministic autocorrelation function of the filter parameters $\{b_n\}$. Different values of m cause $b_{|m|+n}$ to be shifted to the left or right relative to b_n. Since there are only M parameters, $R_{XX}[m]$ also has finite duration: it is zero for $|m| > M-1$ as illustrated by the examples in Figure 8.32 where we have

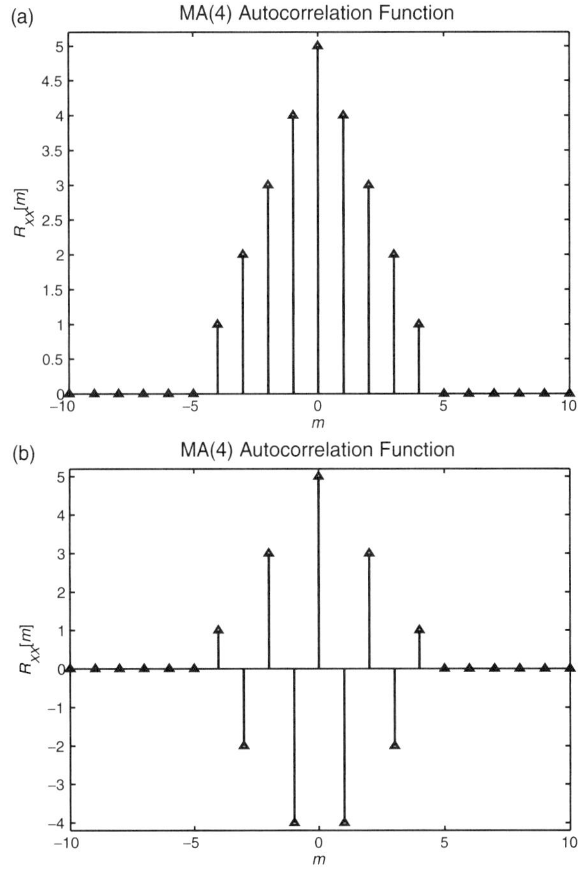

FIGURE 8.32 Example correlation functions for an MA(4) sequence. (a) $\mathbf{b} = [1, 1, 1, 1, 1]^T$. (b) $\mathbf{b} = [1, -1, 1, -1, 1]^T$.

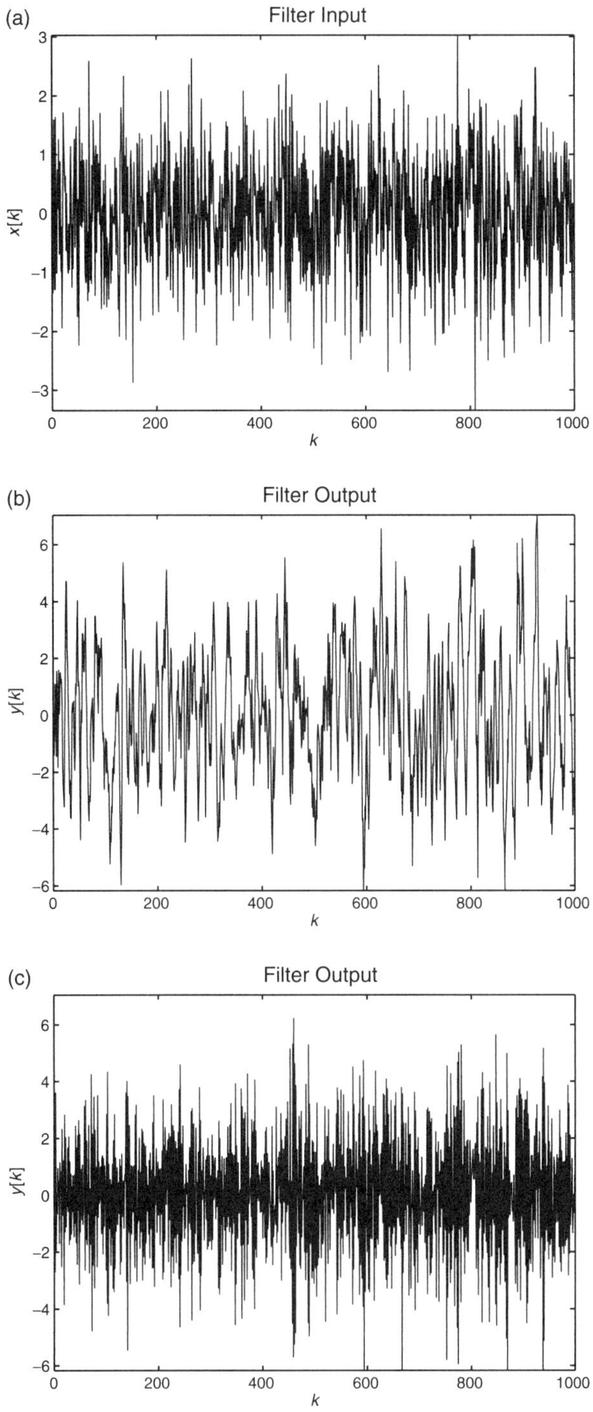

FIGURE 8.33 Example realizations of an MA(4) sequence for different parameter values. (a) White Gaussian input. (b) Output for $\mathbf{b} = [1, 1, 1, 1, 1]^T$. (c) Output for $\mathbf{b} = [1, -1, 1, -1, 1]^T$.

defined the vector $\mathbf{b} \triangleq [b_0, \ldots, b_{M-1}]^T$. Example realizations of an MA sequence are shown in Figure 8.33 for the autocorrelation functions in Figure 8.32 (with $\sigma_W^2 = 1$). The output sequence has noticeable correlation, but not as much as the previous AR model example, which is due to the fact that an AR process has feedback: the autocorrelation function is nonzero for an infinite number of time lags.

Using the coefficient vector $\mathbf{b}$, the random sequence can be expressed as the following inner product:

$$X[k] = \mathbf{b}^T \mathbf{W}[k], \tag{8.292}$$

where $\mathbf{W}[k] \triangleq [W[k], \ldots, W[k-M+1]]^T$. This vector representation is useful for FIR filters and will be seen again in Chapter 12 on adaptive filtering. The autocorrelation function can be written using this form as follows:

$$R_{XX}[m] = \mathcal{E}[X[k]X[k+m]] = \mathbf{b}^T \mathcal{E}[\mathbf{W}[k]\mathbf{W}^T[k+m]]\mathbf{b}, \tag{8.293}$$

where the quantity in the expectation is a square matrix $\in \mathcal{R}^{M\times M}$. Since $W[k]$ is a white sequence, only $M - |m| - 1$ entries are nonzero (as we know from the form of the autocorrelation function derived earlier). This expectation can be written as follows:

$$\mathcal{E}[\mathbf{W}[k]\mathbf{W}^T[k+m]] = \sigma_W^2 \mathbf{U}^{|m|}, \tag{8.294}$$

where $\mathbf{U} \in \mathcal{R}^{M\times M}$ is an upper shift matrix:

$$\mathbf{U} \triangleq \begin{bmatrix} 0 & 1 & 0 & \cdots & 0 \\ 0 & 0 & 1 & \cdots & 0 \\ \vdots & & \ddots & & \vdots \\ 0 & 0 & & \cdots & 1 \\ 0 & & \cdots & 0 & 0 \end{bmatrix}. \tag{8.295}$$

As a result, $\mathbf{U}^m$ has ones along a diagonal that is m elements above the main diagonal, and it is *nilpotent* because $\mathbf{U}^M = \mathbf{0}$. The autocorrelation function is

$$R_{XX}[m] = \sigma_W^2 \mathbf{b}^T \mathbf{U}^{|m|}\mathbf{b}, \tag{8.296}$$

which is another representation for the expression in (8.291). Collecting all the autocorrelation terms in a vector yields the following Yule–Walker-type equations:

$$\begin{bmatrix} R_{XX}[0] \\ R_{XX}[1] \\ \vdots \\ R_{XX}[M-1] \end{bmatrix} = \sigma_W^2 \begin{bmatrix} \mathbf{b}^T\mathbf{b} \\ \mathbf{b}^T\mathbf{U}\mathbf{b} \\ \vdots \\ \mathbf{b}^T\mathbf{U}^{M-1}\mathbf{b} \end{bmatrix}, \tag{8.297}$$

which we see has finite size because the autocorrelation function is zero after time lag $M-1$. Unfortunately, this expression is a nonlinear function of the parameters which is not easily solved in general.

Example 8.15. Consider the simplest nontrivial case of an MA model with $M = 2$ and $\mathbf{b} = [b_0, b_1]^T$ such that

$$\mathbf{b}^T\mathbf{b} = b_0^2 + b_1^2, \quad \mathbf{b}^T\mathbf{U}\mathbf{b} = b_0 b_1. \tag{8.298}$$

Let $\sigma_W^2 = 1$, $R_{XX}[0] = r_0$, and $R_{XX}[1] = r_1$, and substitute $b_0 = r_1/b_1$ into $b_0^2 + b_1^2 = r_0$, yielding

$$b_1^4 - r_0 b_1^2 + r_1^2 = 0, \tag{8.299}$$

which is quadratic in b_1^2. We require $r_0^2 \geq 4r_1^2$ so that b_1^2 is real-valued and $R_{XX}[m]$ is a valid autocorrelation function. For $r_0^2 = 4r_1^2$, we find that $b_1 = \pm\sqrt{r_0/2}$ and $b_0 = \pm r_1\sqrt{2/r_0}$. Thus, even for this trivial example, the solution is not unique due to the quadratic form in (8.297). It is clear that these equations become increasingly difficult to solve for $\mathbf{b}$ with larger values of M, and there will generally be multiple solutions.

8.13.3 Autoregressive Moving-Average Model

An ARMA model is a combination of the AR and MA models as its name implies. The random sequence is written as

$$\begin{aligned} X[k] &= \sum_{n=1}^{N} a_n X[k-n] + \sum_{n=0}^{M-1} b_n W[k-n], \\ &= \mathbf{a}^T \mathbf{X}[k-1] + \mathbf{b}^T \mathbf{W}[k], \end{aligned} \tag{8.300}$$

where vectors $\mathbf{a}$ and $\mathbf{b}$ were previously defined. Its transfer function is

$$H(z) = \frac{\sum_{n=0}^{N} b_n z^{-n}}{1 - \sum_{n=1}^{M-1} a_n z^{-n}} = \frac{B(z)}{1 - A(z)}, \tag{8.301}$$

which results in the following PSD:

$$S_{XX}(z) = \sigma_W^2 \frac{B(z)B(z^{-1})}{[1 - A(z)][1 - A(z^{-1})]}. \tag{8.302}$$

A block diagram of the system is shown in Figure 8.29(c). Example realizations of an ARMA(1,4) process that is a combination of the previous AR(1) and MA(4) examples are shown in Figure 8.34, where we see they have even greater structure because of the increased correlation.

Example 8.16. The PSD for the model used to generate the simulation results in Figure 8.34 is

$$S_{XX}(z) = \sigma_W^2 \frac{B(z)B(z^{-1})}{(1 - az^{-1})(1 - az)}, \tag{8.303}$$

where for $\mathbf{b}^T = [1, 1, 1, 1]$:

$$B(z)B(z^{-1}) = z^{-3} + z^3 + 2(z^{-2} + z^2) + 3(z^{-1} + z) + 4. \tag{8.304}$$

Since the numerator has this relatively simple form, we can use the time-shift property of the bilateral z-transform to determine the individual components of $R_{XX}[m]$. The following z-transform pair is given in Appendix C:

$$\frac{1}{(1 - az^{-1})(1 - az)} \leftrightarrow \frac{|a|^m}{1 - a^2} \tag{8.305}$$

from which we obtain

$$\begin{aligned} R_{XX}[m] = \frac{\sigma_W^2}{1 - a^2} &\left[4a^{|m|} + 3(a^{|m-1|} + a^{|m+1|})\right. \\ &\left. + 2(a^{|m-2|} + a^{|m+2|}) + a^{|m-3|} + a^{|m+3|}\right]. \end{aligned} \tag{8.306}$$

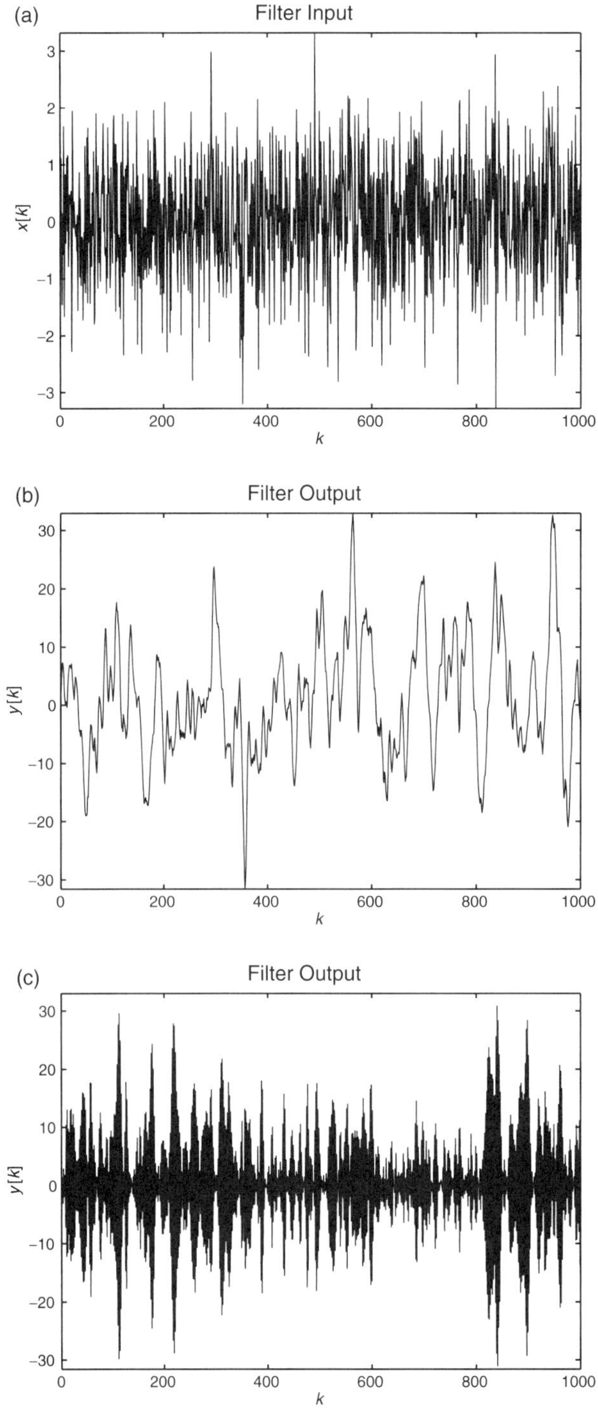

FIGURE 8.34 Example realizations of an ARMA(1,4) sequence for different parameter values. (a) White Gaussian input. (b) Output for $\mathbf{a} = 0.9$ and $\mathbf{b} = [1, 1, 1, 1, 1]^T$. (c) Output for $\mathbf{a} = -0.9$ and $\mathbf{b} = [1, -1, 1, -1, 1]^T$.

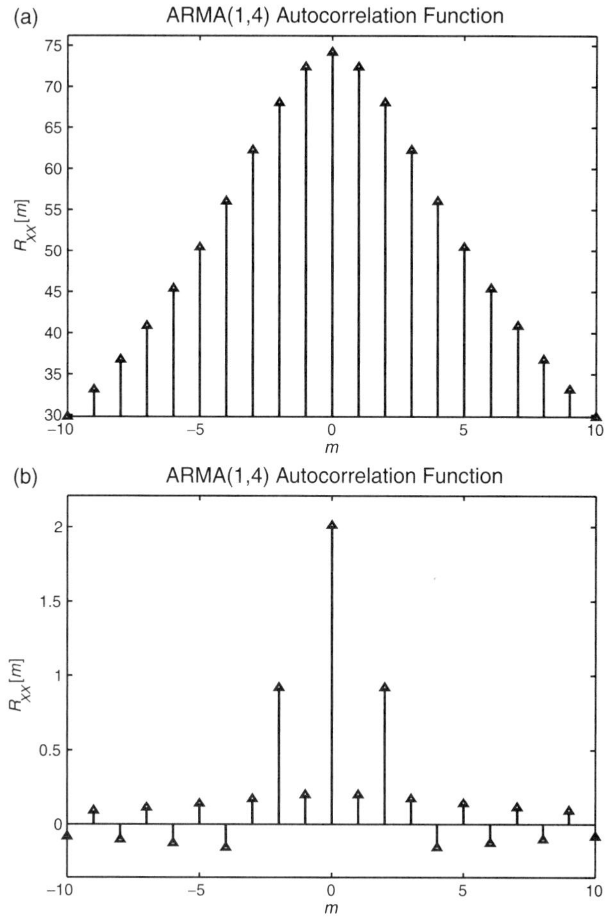

FIGURE 8.35 Example autocorrelation functions for an ARMA(1,4) sequence with $\mathbf{b} = [1, 1, 1, 1, 1]^T$. (a) $\mathbf{a} = 0.9$. (b) $\mathbf{a} = -0.9$.

Examples of this autocorrelation function are plotted in Figure 8.35 for positive and negative values of a (with $\sigma_W^2 = 1$).

The autocorrelation function can be written as follows:

$$\mathcal{E}[X[k]X[k+m]] = \sum_{n=1}^{N} a_n \mathcal{E}[X[k-n]X[k+m]] + \sum_{n=0}^{M-1} b_n \mathcal{E}[W[k-n]X[k+m]] \tag{8.307}$$

so that a recursive form similar to the filter itself is obtained:

$$R_{XX}[m] = \sum_{n=1}^{N} a_n R_{XX}[m+n] + \sum_{n=0}^{M-1} b_n R_{WX}[m+n]. \tag{8.308}$$

Recall that the autocorrelation functions of the input and output are related as follows:

$$R_{WX}[m] = R_{WW}[m] * h[m], \tag{8.309}$$

where $h[m]$ is the impulse-response function of the ARMA transfer function in (8.301). Since $W[k]$ is a white sequence with $R_{WW}[m] = \sigma_W^2 \delta[m]$, the cross-correlation function is proportional to the impulse response of the system:

$$R_{WX}[m] = \sigma_W^2 h[m] \tag{8.310}$$

so that (8.308) becomes

$$R_{XX}[m] = \sum_{n=1}^{N} a_n R_{XX}[m+n] + \sigma_W^2 \sum_{n=0}^{M-1} b_n h[m+n]. \tag{8.311}$$

For the MA model, the first term on the right-hand side is zero, and $h[m+n] = R_{WX}[m+n] = b_{m+n}$ is used in the second term, resulting in a finite-duration autocorrelation function. Solving (8.311) for the general ARMA model is obviously more complicated because of the first term on the right-hand side, and because $h[m]$ depends on $\mathbf{a}$ and $\mathbf{b}$ in the second term, which is a deterministic cross-correlation of b_n and $h[n]$.

Observe that the second sum is not symmetric: for $m < 0$, the impulse response $h[m+n]$ shifts to the left relative to b_n, and since $h(n)$ has infinite length, there are always M terms in the sum. For $m > 0$, the impulse response $h[m+n]$ shifts to the right relative to b_n, and since b_n is zero for $n \geq M$, there is a decreasing number of terms in the sum which is zero when $m \geq M$. Thus, it is possible to compute $\mathbf{a}$ from

$$R_{XX}[m] = \sum_{n=1}^{N} a_n R_{XX}[m+n] \tag{8.312}$$

provided $|m| \geq M$ so that the following Yule–Walker equations can be solved:

$$\begin{bmatrix} R_{XX}[M] \\ R_{XX}[M+1] \\ \vdots \\ R_{XX}[M+N-1] \end{bmatrix} = \begin{bmatrix} R_{XX}[M+1] & \cdots & R_{XX}[M+N] \\ R_{XX}[M+2] & \cdots & R_{XX}[M+N+1] \\ \vdots & & \vdots \\ R_{XX}[M+N] & \cdots & R_{XX}[M+2N] \end{bmatrix} \begin{bmatrix} a_1 \\ a_2 \\ \vdots \\ a_N \end{bmatrix}. \tag{8.313}$$

This expression differs from (8.282) for the AR model in that the time lags of the autocorrelation function must exceed the order of the MA part of the ARMA model. Unfortunately, even though $\mathbf{a}$ can be found relatively easy, it is still difficult to derive $\mathbf{b}$ because it multiplies the impulse response $h[n]$ in (8.311), which in turn depends on $\mathbf{a}$ and $\mathbf{b}$; the problem is more nonlinear than solving for $\mathbf{b}$ of the MA model.

Consider rearranging (8.311) as follows:

$$R_{XX}[m] - \sum_{n=1}^{N} a_n R_{XX}[m+n] = \sigma_W^2 \sum_{n=0}^{M-1} b_n h[m+n], \tag{8.314}$$

which has z-transform:

$$[1 - A(z^{-1})] S_{XX}(z) = \sigma_W^2 B(z) H(z^{-1}). \tag{8.315}$$

Substituting $H(z)$ from (8.301) and rearranging the result gives

$$[1 - A(z)][1 - A(z^{-1})] S_{XX}(z) = \sigma_W^2 B(z) B(z^{-1}), \tag{8.316}$$

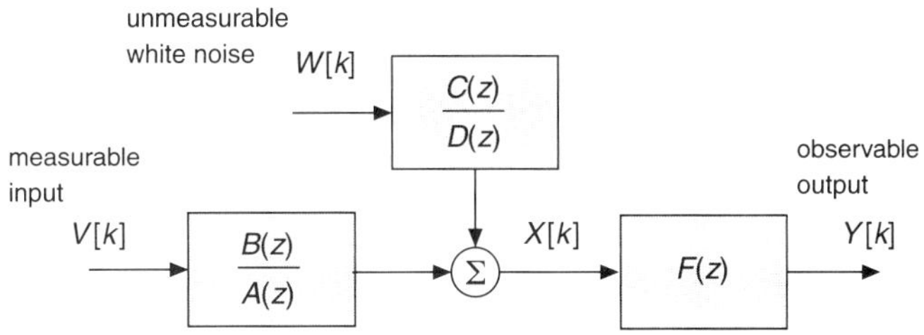

FIGURE 8.36 General system identification model.

which is the same as (8.302). The inverse z-transform of the right-hand side is identical to that in (8.291) for the MA model. Thus, the ARMA model parameters can be generated as follows: (i) estimate $R_{XX}[m]$ and $S_{XX}(z)$ from a time-averaged realization of $X[k]$, (ii) compute the parameters of $A(z)$ using the Yule–Walker equations in (8.313), (iii) take the inverse z-transform of the left-hand side in (8.316), and (iv) solve the right-hand side for b_m. As was the case for the MA model, the last step is nonlinear and there can be multiple solutions.

Finally, we mention another approach to parametric estimation that assumes the random sequence has a specific signal and noise structure for which it is possible to exploit the *eigenstructure* of the autocorrelation matrix. Such *subspace* methods are better suited for detecting narrowband signals in noise rather than for general spectrum estimation. Subspace methods are covered in Chapter 13 where we describe the equivalent problem of separating several narrowband sources that impinge on an *array of antennas* from different directions, corresponding to a system with multiple inputs.

8.14 SYSTEM IDENTIFICATION

Closely related to the signal models in Section 8.13 is *system identification* as depicted in Figure 8.36 where we are interested in estimating an unknown system $B(z)/A(z)$ from measurements of its output $Y[k]$ as well as its input $V[k]$. The input is assumed to be measurable (whereas for the previous signal modeling techniques it is not observable), though we usually assume that $V[k]$ is a white sequence. However, the system identification procedure is not straightforward because there is unmeasurable additive white noise $W[k]$. Also, the noise may be filtered by some transfer function $C(z)/D(z)$ such that the output is nonwhite (with a nonflat PSD). It may be necessary to estimate not only $B(z)/A(z)$ but also $C(z)/D(z)$. Moreover, the output of the summation shown in the figure may be processed further by an unknown filter $F(z)$ such that $X[k]$ is not directly observable. Various algorithms have been developed for identifying the filter components depending on which (if any) of the polynomials are 1. The simplest case occurs when $A(z) = C(z) = D(z) = F(z) = 1$ so that only $B(z)$ needs to be estimated from measurements of $V[k]$ and $X[k]$.

System identification and parametric signal modeling appear to be similar techniques because the goal is to determine filter polynomials that are used to describe the measured signals. In signal modeling, the inverse of the resulting system is used to *synthesize* the measured signal, usually by filtering a white-noise sequence. The filter also provides spectral information. In contrast, the goal in system identification is to generate a model that may then be used for another application, such as in a control system where a model of a complex system is needed for the control mechanism of an actual system.

Assume that we have access to input $V[k]$ and output $Y[k]$ of the system model depicted in Figure 8.36. Note that the actual system is probably not of this form, as it may have nonlinear and time-varying components. However, if the nonlinear parts are not dominant, and if the model parameters are computed and updated sufficiently fast, the filter polynomials may be reasonably accurate. Using mixed time- and z-domain notation, we can write

$$F(z)Y[k] = \frac{B(z)}{A(z)}V[k] + \frac{C(z)}{D(z)}W[k], \tag{8.317}$$

where the polynomials are defined as follows (note that some of these must have a leading 1):

$$A(z) \triangleq 1 + a_1 z^{-1} + \cdots + a_{M_a} z^{-M_a}, \tag{8.318}$$

$$B(z) \triangleq b_0 + b_1 z^{-1} + \cdots + b_{M_b} z^{-M_b}, \tag{8.319}$$

$$C(z) \triangleq c_0 + c_1 z^{-1} + \cdots + c_{M_c} z^{-M_c}, \tag{8.320}$$

$$D(z) \triangleq 1 + d_1 z^{-1} + \cdots + d_{M_d} z^{-M_d}, \tag{8.321}$$

$$F(z) \triangleq 1 + f_1 z^{-1} + \cdots + f_{M_f} z^{-M_f}. \tag{8.322}$$

The mixed notation means the following:

$$F(z)Y[k] = Y[k] + \sum_{m=1}^{M_f} f_m Y[k-m], \tag{8.323}$$

where in this context z^{-1} functions as a *delay operator:* $z^{-1}Y[k] = Y[k-1]$.

The previous model is quite general, and often for most applications, one or more of the polynomials are assumed to be 1. We summarize some special cases.

- *ARMA with exogenous input (ARMAX).* $A(z) = D(z) = 1$:

$$F(z)Y[k] = B(z)V[k] + C(z)W[k]. \tag{8.324}$$

  A special case of the ARMAX model is obtained when $C(z) = 1$:

$$F(z)Y[k] = B(z)V[k] + W[k], \tag{8.325}$$

  which is known as an ARX model.
- *ARMA.* $A(z) = D(z) = 1$, $B(z) = 0$:

$$F(z)Y[k] = C(z)W[k] \Longrightarrow Y[k] = \frac{C(z)}{F(z)}W[k]. \tag{8.326}$$

  Special cases include the MA model ($F(z) = 1$):

$$Y(k) = C(z)W[k], \tag{8.327}$$

  and the AR model ($C(z) = 1$):

$$F(z)Y(k) = W[k] \Longrightarrow Y[k] = \frac{1}{F(z)}W[k]. \tag{8.328}$$

- *Box–Jenkins.* $F(z) = 1$:

$$Y[k] = \frac{B(z)}{A(z)}V[k] + \frac{C(z)}{D(z)}W[k]. \tag{8.329}$$

- *Output error.* $F(z) = 1$ and $C(z) = D(z)$:

$$Y[k] = \frac{B(z)}{A(z)}V[k] + W[k]. \tag{8.330}$$

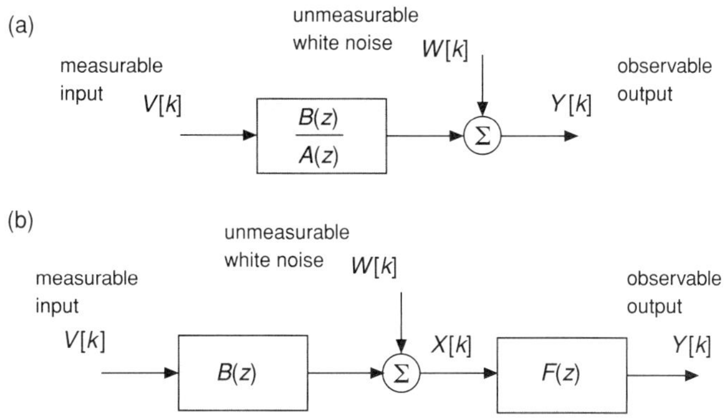

FIGURE 8.37 IIR system identification. (a) Output-error model. (b) Equation-error model.

This corresponds to an IIR filter with additive noise, and when $A(z) = 1$ it is an FIR filter. *Adaptive algorithms* for FIR and IIR output-error filters are discussed in Chapter 12.

- *Equation error*. $A(z) = 1$ and $C(z) = D(z) = 1$:

$$F(z)Y[k] = B(z)V[k] + W[k]. \tag{8.331}$$

An adaptive algorithm for the equation-error filter is also discussed in Chapter 12.

For the output-error and equation-error models shown in Figure 8.37, it is not necessary that the additive noise be a white sequence; thus, we could include $C(z) \neq D(z)$ above, but no attempt is made in the algorithms of Chapter 12 to identify the noise polynomials as does the Box–Jenkins model.

PROBLEMS

Ergodicity

8.1 (a) Derive the necessary and sufficient condition in (8.23) for random sequence $X[k]$ to be ergodic in the mean. (b) Verify the sufficient condition in (8.24).

8.2 Let $X[k]$ have zero mean and autocovariance function $C_{XX}[m] = \alpha^{|m|}$. Determine if both conditions in the first problem are satisfied for this random sequence.

8.3 Suppose that $X(t)$ is ergodic in the mean, and let Y be a random variable that is independent of $X(t)$. Define random process $Z(t) = YX(t)$ and determine if it is also ergodic in the mean. Note that a realization is of the form $z(t) = yx(t)$ such that outcome y remains fixed with time as $x(t)$ varies.

8.4 Random process $X(t)$ has zero mean and the following autocorrelation function:

$$R_{XX}(\tau) = (1 - \tau^2)I_{[-1,1]}(\tau). \tag{8.332}$$

Determine if $X(t)$ is ergodic in the mean.

8.5 Random process $X(t)$ has the following periodic autocovariance function:

$$C_{XX}(\tau) = 2\cos(2\pi f_c \tau) \tag{8.333}$$

for some fixed frequency f_c. Determine if $X(t)$ is ergodic in the mean.

8.6 For $R_{XX}(\tau) = \exp(-|\tau|)$, use the result in Example 8.3 to determine if the zero-mean Gaussian random process $X(t)$ is ergodic in correlation.

Power Spectral Density

8.7 For the cross-PSD, show that (a) $S_{XY}(f) = S_{YX}(-f)$ and (b) the imaginary part is an odd function.

8.8 Consider the following random process:

$$X(t) = a\sin(2\pi f_c t + \Phi), \tag{8.334}$$

where Φ is uniformly distributed on $[-\pi/2, \pi/2]$. Find (a) the autocorrelation function $R_{XX}(\tau)$ and (b) the PSD $S_{XX}(f)$.

8.9 (a) Find the PSD for

$$R_{XX}(\tau) = \exp(-\lambda|\tau|)\cos(2\pi f_c \tau) \tag{8.335}$$

and the average power in the frequency range $f \in [f_c - W, f_c + W]$ for some $W \ll f_c$. (b) Derive an expression for the power spectral *distribution*.

8.10 Repeat Problem 8.9 for

$$R_{XX}(\tau) = \exp(-\tau^2/2\sigma^2) \tag{8.336}$$

and $f \in [-W, W]$, where $W = \sigma$.

8.11 Find the PSD factorization $S_{XX}(s) = S^+_{XX}(s)S^-_{XX}(s)$ for $R_{XX}(\tau)$ in Problem 8.9.

8.12 Let $X(t) = S(t) + W(t)$, where $S(t)$ and $W(t)$ have autocorrelation functions $R_{SS}(\tau) = 5\exp(-|\tau|)$ and $R_{WW}(\tau) = \delta(\tau)$. Assuming that $S(t)$ and $W(t)$ are uncorrelated, find the PSD factorization for $X(t)$.

8.13 Let $X[k]$ be a wide-sense stationary random sequence with zero mean and autocorrelation function $R_{XX}[m] = (N_o/2)\delta[m]$. Demonstrate that in order to have a flat PSD, corresponding to a white-noise sequence, the mean must be zero.

LTI Systems

8.14 An LTI system has the following difference equation:

$$Y[k] = X[k] + (1/2)X[k-1] + (1/4)Y[k-1]. \tag{8.337}$$

Find the PSD $S_{YY}(j\omega)$ if the input has autocorrelation function $R_{XX}[m] = 2\delta[m]$.

8.15 Find the PSD $S_{YY}(f)$ for the following system:

$$Y(t) = \int_0^t X(u)\mathrm{d}u, \tag{8.338}$$

where $R_{XX}(\tau) = 2\delta(\tau)$.

8.16 A linear system has the following transfer function:

$$H(s) = \frac{2}{(s+1)(s+2)}. \tag{8.339}$$

Find the output autocorrelation function $R_{YY}(\tau)$ when $R_{XX}(\tau) = \exp(-2|\tau|)$ for the input.

8.17 Derive the correlation result in (8.152) for an LTI system with wide-sense stationary input $X(t)$.

8.18 Derive the PSD result in (8.156) for an LTI system with wide-sense stationary input $X(t)$.

8.19 An LTI system with input $X(t)$ is represented by the following DE:

$$\frac{\mathrm{d}^2 Y(t)}{\mathrm{d}t^2} + 4\frac{\mathrm{d}Y(t)}{\mathrm{d}t} + 2Y(t) = X(t), \tag{8.340}$$

where $R_{XX}(\tau) = (1/2)\delta(\tau)$. Find expressions for (a) $S_{XY}(f)$ and $R_{XY}(\tau)$, and (b) $S_{YY}(f)$ and $R_{YY}(\tau)$.

8.20 Consider an LTI system with input $X(t)$, output $Y(t)$, and impulse response function

$$h(t) = t\exp(-\lambda t)u(t), \tag{8.341}$$

where $\lambda > 0$. Find expressions for $R_{YX}(\tau)$ and $R_{YY}(\tau)$ when $R_{XX}(\tau) = \delta(\tau)$.

8.21 Suppose that $X_1[k]$ is filtered by $H_1(z)$ to give output $Y_1[k]$, and $X_2[k]$ is filtered by $H_2(z)$ to give output $Y_2[k]$. (a) Find an expression for $S_{Y_1Y_2}(z)$ in terms of $S_{X_1X_2}(z)$. (b) Show that the output processes are orthogonal if the frequency responses of the systems are nonoverlapping (e.g., if they are passband systems).

Bandwidth

8.22 Find (a) the 3-dB bandwidth and (b) the NE bandwidth for the following PSD:

$$S_{XX}(f) = 4(1 - |f|/5)I_{[-5,5]}(f). \tag{8.342}$$

8.23 Repeat Problem 8.22 for

$$S_{XX}(f) = \frac{\sin^2(2\pi f)}{\pi^2 f^2}. \tag{8.343}$$

8.24 Find (a) the RMS bandwidth and (b) the NE bandwidth of the PSD derived in Problem 8.10.

Spectrum Estimation

8.25 Using properties of the DTFT, verify that (8.225) and (8.227) are equivalent expressions for the periodogram.

8.26 Consider the following estimate of the cross-correlation function $R_{XY}[m]$:

$$\hat{R}_{XY}[m] = \frac{1}{N}\sum_{k=0}^{N-|m|-1} x[k]y[k+m]. \tag{8.344}$$

Demonstrate that the expectation of this estimate has a form similar to (8.223) in terms of $R_{YX}[m] = R_{XY}[-m]$.

8.27 Derive an expression for the PSD $S_{XX}(j\omega)$ of the MA sequence in (8.287) with $M = 2$ parameters.

8.28 Consider the following ARMA model:

$$H(z) = \frac{b_0 + b_1 z^{-1}}{1 - a_1 z^{-1}}. \tag{8.345}$$

The zero-mean input $W[k]$ has autocorrelation function $R_{WW}[m] = \sigma_W^2 \delta[m]$. Find expressions for $R_{XX}[k]$ and $S_{XX}(z)$ for the output $X[k]$ in terms of the parameters $\{a_1, b_0, b_1\}$.

8.29 For the general AR model with N parameters, derive an expression for the variance of the output $X[k]$ in terms of $R_{XX}[m]$ when the input $W[k]$ is a white-noise sequence with variance σ_W^2.

Computer Problems

8.30 Use randn to generate 1000 samples of white Gaussian noise $W[k]$ with unit variance, and use filter to generate output $X[k]$ for the following system:

$$H(z) = \frac{2}{1 - (5/6)z^{-1} + (1/6)z^{-2}}. \tag{8.346}$$

Plot example realizations for $W[k]$ and $X[k]$. Process $X[k]$ using periodogram and a rectangular window and then a Bartlett window. Plot the resulting periodograms.

8.31 For the signals in Problem 8.30, derive an estimate of the autocorrelation function $R_{XX}[m]$ using xcorr for lags $\{0, 1, 2, 3, 4\}$. Using MATLAB, estimate the parameters of this system from the Yule–Walker equations for values of N ranging from 2 to 4.

FURTHER READING

The material covered in this chapter can also be found in the references provided at the end of Chapters 6 and 7 on random processes. Additional information about specific topics is available in the following references. Noise: Cohen (2005) and Larson and Shubert (1979). Spectrum estimation: Kay and Marple (1981), Kay (1986), Porat (1994), and Stoica and Moses (1997). Parametric signal models and system identification: Box and Jenkins (1970), Chatfield (1980), Ljung and Söderström (1983), Ljung (1987), and Therrien (1992).

9 Sufficient Statistics and Parameter Estimation

9.1 INTRODUCTION

In this chapter, we are interested in finding a function of a set of random variables that is used to estimate an unknown parameter of the underlying distribution. Such a function is called a *statistic*, and it generally reduces the number of random variables, called *samples*, to a lower dimension without loss of "information" about the parameter. The scenario can be summarized as follows:

- Let $X(t)$ be a random process with fixed probability density function (pdf) $f_X(x)$ that is parameterized by scalar θ. Assume that θ is not observable: it cannot be sampled directly and information about θ is obtained only through samples of $X(t)$.
- The random process is sampled N times, yielding N-independent and identically distributed (iid) random variables $\{X_1, \ldots, X_N\}$. Although the $\{X_n\}$ contain identical information about the unknown θ, the specific outcomes $\{x_n\}$ generally vary.
- The goal after sampling is to combine the $\{X_n\}$ using functions to give "simpler" quantities for estimating the unknown θ. Since N is usually much larger than the number of parameters, combining the $\{X_n\}$ *condenses* the samples to one or more random variables called *sufficient statistics*. These statistics are "sufficient" because information about θ is not lost by the mapping from N variables to a lower dimension.

The $\{X_n\}$ could also be derived from samples of random sequence $X[k]$ or they may just be N samples of a particular random variable without regard to time. For notational convenience, we use a subscript on X_n to indicate that it is a sample (random variable) of either the process $X(t)$, sequence $X[k]$, or random variable X. The samples need not be collected from the process/sequence with uniform spacing in time; we can simply view $\{X_n\}$ as a set of iid random variables with parameter θ.

In order to simplify the equations, we use $\mathbf{X} \triangleq \{X_1, \ldots, X_N\}$ to represent the N samples and $\mathbf{x} \triangleq \{x_1, \ldots, x_N\}$ for the specific outcomes. (Note that this bold notation was also used for random vectors and their outcomes.)

9.2 STATISTICS

Definition: Statistic The *statistic* T is a function of random variables $\mathbf{X}$ which itself is a random variable.

About the notation, the function of $\mathbf{X}$ is *not* random; we could write $T = g(\mathbf{X})$, but this will not be necessary until later in the chapter. For specific outcomes $\mathbf{X} = \mathbf{x}$, the value of the statistic is $T = t$. The symbol T is commonly used in the statistics literature for estimators (so we initially use it in this chapter), and should not be confused with T used for a time duration.

Probability, Random Variables, and Random Processes: Theory and Signal Processing Applications, First Edition. John J. Shynk.

Definition: Estimator An *estimator* of θ is a statistic T that is a function of the samples $\mathbf{X}$ and is chosen to have certain desirable properties.

Example 9.1. The sample mean is probably the most widely used statistic:

$$T_1 = \frac{1}{N}\sum_{n=1}^{N} X_n \triangleq \bar{X}. \tag{9.1}$$

It is an unbiased estimator of the actual mean μ_X. The following statistic

$$T_2 = \sum_{n=1}^{N} X_n \tag{9.2}$$

contains essentially the same information about μ_X as $\bar{X}$, but it is a *biased* estimator. T_1 and T_2 are statistically equivalent because they are related by one-to-one function $g(\cdot)$:

$$T_1 = g(T_2) = \frac{1}{N}T_2, \quad T_2 = g^{-1}(T_1) = NT_1, \tag{9.3}$$

where $g^{-1}(\cdot)$ is the inverse image. Later, we investigate further the idea that statistics are functions or mappings of the samples.

Example 9.2. The sample variance is a statistic:

$$T = \frac{1}{N-1}\sum_{n=1}^{N}(X_n - \bar{X})^2 \triangleq S^2. \tag{9.4}$$

The sum is divided by $N-1$ (and not N which would be the "natural" choice) so that the sample variance is an unbiased estimator of the actual variance σ_X^2.

9.3 SUFFICIENT STATISTICS

Sufficient statistics are useful for parameter-estimation problems. Consider the pdf $f_X(x;\theta)$ with unknown parameter θ, where the notation emphasizes that θ is the parameter to be estimated. For example, θ might be the mean or variance of an iid Gaussian random sequence with parameters $\{\mu, \sigma\}$. Suppose we have N samples $\mathbf{X}$ and let T be a function (mapping) of the samples. Although the original N observations are condensed to sufficient statistic T, it contains the same information *in terms of estimating* θ: it is sufficient for estimation purposes. Thus, if T is available and is sufficient, then knowing the original set of observations $\mathbf{X}$ provides no additional information for estimating θ.

Definition: Sufficient Statistic Statistic T is *sufficient* for estimating parameter θ if the conditional probability $P(\mathbf{X}=\mathbf{x}|T=t)$ is not a function of θ. When the random variables are continuous, T is sufficient for θ if the conditional pdf $f_{\mathbf{X}|T}(\mathbf{x}|t)$ is not a function of θ.

The conditional probability can be rewritten as

$$P(\mathbf{X}=\mathbf{x}|T=t) = \frac{P(\mathbf{X}=\mathbf{x}, T=t)}{P(T=t)} = \frac{P(\mathbf{X}=\mathbf{x})}{P(T=t)}. \tag{9.5}$$

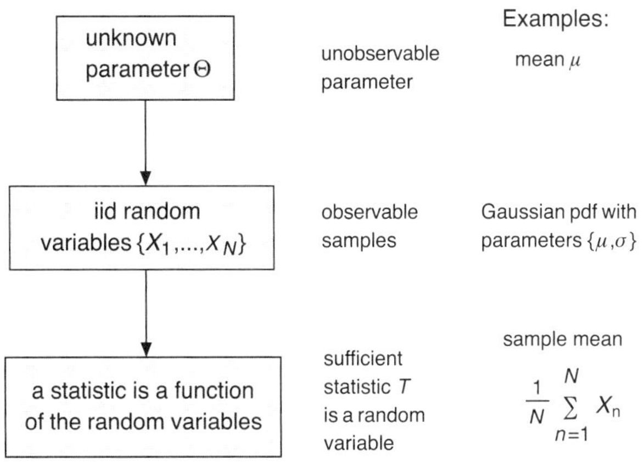

FIGURE 9.1 Summary of a sufficient statistic used for parameter estimation.

The second expression follows because $\{\mathbf{x}, t\}$ is equivalent to $\mathbf{x}$: we can determine t from $\mathbf{x}$. The numerator and denominator in the last expression both depend on θ, although this is not explicitly shown. The numerator is the joint probability of the outcomes $\mathbf{x}$, and the denominator is the probability that the statistic (which is a random variable) has outcome t. The intuition behind the definition of a sufficient statistic is that these two probabilities depend identically on θ, and so functions of θ in the numerator and denominator cancel. A block diagram summarizing a sufficient statistic used for parameter estimation is shown in Figure 9.1.

A sufficient statistic can be multidimensional, even for the scalar parameter θ; later, we define a *minimal* sufficient statistic that is defined by the dimension of the statistic, as well as a *complete* sufficient statistic, which has a more complicated definition.

Example 9.3. Let $\{X_n\}$ be samples of a Bernoulli random variable with parameter p. The sum $T = \sum_{n=1}^{N} X_n$ has a binomial distribution with parameters $\{N, p\}$. From (9.5):

$$\frac{P(\mathbf{X}=\mathbf{x})}{P(T=t)} = \frac{\prod_{n=1}^{N} p^{x_n} q^{1-x_n}}{\binom{N}{t} p^t q^{N-t}} = \frac{p^{\sum_{n=1}^{N} x_n} q^{N-\sum_{n=1}^{N}}}{\binom{N}{t} p^t q^{N-t}} = \frac{1}{\binom{N}{t}}. \tag{9.6}$$

The last result follows because $t = \sum_{n=1}^{N} x_n$. Since the binomial coefficient is not a function of p (or $q \triangleq 1 - p$), T is sufficient for p.

Example 9.4. Let $\{X_n\}$ be exponentially distributed with parameter $\theta = \lambda$ such that their joint pdf is

$$f_{\mathbf{X}}(\mathbf{x}) = \lambda^N \exp\left(-\lambda \sum_{n=1}^{N} x_n\right) \prod_{n=1}^{N} I_{[0,\infty)}(x_n). \tag{9.7}$$

In order to determine if $T = \sum_{n=1}^{N} X_n$ is sufficient, consider the following transformation to an equivalent set of random variables for which it is easier to compute the required conditional pdf: $Y_n = X_n$ for $n = 1, \ldots, N-1$ and $Y_N = \sum_{n=1}^{N} X_n$. This transformation is invertible because $X_N = Y_N - \sum_{n=1}^{N-1} Y_n$. The joint pdf of $\mathbf{Y} \triangleq [Y_1, \ldots, Y_N]^T$ is

$$f_{\mathbf{Y}}(\mathbf{y}) = \lambda^N \exp(-\lambda y_N)\, I_{[0,y_N]}\left(\sum_{n=1}^{N-1} y_n\right) \prod_{n=1}^{N} I_{[0,\infty)}(y_n), \tag{9.8}$$

where the indicator functions ensure that the pdf is nonzero for nonnegative $\{y_n\}$ and for $y_N \geq \sum_{n=1}^{N-1} y_n$. This particular transformation is chosen because Y_N is the statistic under consideration. We emphasize that the joint pdf of $\{Y_n\}$ is *not* the same as that of $\{X_n\}$. However, since the transformation is invertible, showing that Y_N is sufficient for λ also demonstrates that $\sum_{n=1}^{N} X_n$ is sufficient. The conditional pdf is

$$\begin{aligned} f_{Y_1,\ldots,Y_{N-1}|Y_N}(y_1,\ldots,y_{N-1}|y_N) &= \frac{f_{\mathbf{Y}}(\mathbf{y})}{f_{Y_N}(y_N)}, \\ &= \frac{\lambda^N \exp(-\lambda y_N)}{[\lambda^N/\Gamma(N)]y_N^{N-1}\exp(-\lambda y_N)} \\ &\quad \cdot I_{[0,y_N]}\left(\sum_{n=1}^{N-1} y_n\right)\prod_{n=1}^{N} I_{[0,\infty)}(y_n), \end{aligned} \tag{9.9}$$

where the denominator $f_{Y_N}(y_N)$ is the pdf of a gamma random variable with parameters $\{\lambda, r = N\}$. Cancelling terms, we see that the conditional pdf is independent of λ:

$$f_{Y_1,\ldots,Y_{N-1}|Y_N}(y_1,\ldots,y_{N-1}|y_N) = [\Gamma(N)/y_N^{N-1}]I_{[0,y_N]}\left(\sum_{n=1}^{N-1} y_n\right)\prod_{n=1}^{N} I_{[0,\infty)}(y_n), \tag{9.10}$$

and thus $T = Y_N$ is sufficient for λ.

Although the definition of a sufficient statistic is intuitively appealing, it is not always easy to derive the conditional probability: demonstrating that a statistic is sufficient using the definition can be tedious. The following theorem allows one to determine sufficiency by examining only the joint pdf of the samples; there is no need to rewrite it as a conditional pdf.

Theorem 9.1 (Fisher–Neyman factorization). Let $\mathbf{X}$ be samples drawn from the pdf $f_X(x;\theta)$ with unknown parameter θ. Statistic T is sufficient for θ if and only if their joint pdf factors as follows:

$$f_{\mathbf{X}}(\mathbf{x};\theta) = h(\mathbf{x})g(t;\theta), \tag{9.11}$$

where $h(\mathbf{x})$ is a nonnegative function that does not depend on θ, and $g(t;\theta)$ is a nonnegative function that depends on θ and is a function of the outcomes $\mathbf{x}$ only through $T = t$.

Proof. Only if: Assuming that T is a sufficient statistic, we can write

$$f_{\mathbf{X}}(\mathbf{x};\theta) = f_{\mathbf{X},T}(\mathbf{x},t;\theta), \tag{9.12}$$

which follows because T is a function of the observations and provides no information beyond that of the original samples. This can be verified by starting with the joint cumulative distribution function (cdf) and showing that the probability is unchanged by including T, similar to that done in (9.5). Rewriting this expression in terms of a conditional probability gives

$$f_{\mathbf{X}}(\mathbf{x};\theta) = f_{\mathbf{X}|T}(\mathbf{x}|t;\theta)f_T(t;\theta) = h(\mathbf{x})g(t;\theta), \tag{9.13}$$

where $h(\mathbf{x}) \triangleq f_{\mathbf{X}|T}(\mathbf{x}|t;\theta)$ does not depend on θ because of the conditioning on T which we have assumed is sufficient. The other term $g(t;\theta) \triangleq f_T(t;\theta)$ is a function of θ and depends on the samples only through t. *If:* The pdf of the statistic is

$$f_T(t;\theta) = \sum_{\mathbf{x}:T=t}\cdots\sum f(\mathbf{x},t;\theta) = \sum_{\mathbf{x}:T=t}\cdots\sum f(\mathbf{x};\theta), \tag{9.14}$$

which has been written by summing over N variables of a joint pdf. The second expression follows because T is a function of $\mathbf{X}$. Substituting (9.11), $g(t;\theta)$ factors out as follows:

$$f_T(t;\theta) = \sum_{\mathbf{x}:T=t}\cdots\sum h(\mathbf{x})g(t;\theta) = g(t;\theta)\sum_{\mathbf{x}:T=t}\cdots\sum h(\mathbf{x}). \tag{9.15}$$

From the definition of a sufficient statistic, we need to show that the following conditional pdf is independent of θ:

$$\begin{aligned} f_{\mathbf{X}|T}(\mathbf{x}|t;\theta) &= \frac{f_{\mathbf{X},T}(\mathbf{x},t;\theta)}{f_T(t;\theta)} = \frac{f_{\mathbf{X}}(\mathbf{x};\theta)}{f_T(t;\theta)}, \\ &= \frac{h(\mathbf{x})g(t;\theta)}{g(t;\theta)\sum_{\mathbf{x}:T=t}\cdots\sum h(\mathbf{x})} = \frac{h(\mathbf{x})}{\sum_{\mathbf{x}:T=t}\cdots\sum h(\mathbf{x})}. \end{aligned} \tag{9.16}$$

The first expression is the definition of a conditional pdf, the second expression follows from (9.12), and the third expression is obtained by substituting (9.11). Since the terms containing θ cancel, the final result demonstrates that the conditional pdf is not a function of θ, thus completing the proof. □

Example 9.5. From the joint pdf of the exponential distribution in (9.7), we find

$$h(\mathbf{x}) = \prod_{n=1}^{N} I_{[0,\infty)}(x_n), \quad g(t;\lambda) = \lambda^N \exp(-\lambda t), \tag{9.17}$$

and thus $T = \sum_{n=1}^{N} X_n$ is sufficient for λ. Observe that the indicator functions are included in $h(\mathbf{x})$ because they do not depend on λ.

Example 9.6. Indicator functions are particularly useful when the support is finite. Suppose the underlying pdf is uniform on $[0,\theta]$. The joint pdf is

$$f_{\mathbf{X}}(\mathbf{x};\theta) = \frac{1}{\theta^N}\prod_{n=1}^{N} I_{[0,\theta]}(x_n) = \frac{1}{\theta^N} I_{[0,\theta]}(x_{(n)}), \tag{9.18}$$

which depends only on the Nth-*order statistic* (see Chapter 4). If $x_{(N)} \le \theta$, then all other samples are less than θ so that the other $N-1$ indicator functions can be ignored. Thus

$$h(\mathbf{x}) = 1, \quad g(t;\theta) = \frac{1}{\theta^N} I_{[0,\theta]}(t) \tag{9.19}$$

and $T = X_{(N)}$ is sufficient for λ.

Before discussing Example 9.7, we make a general observation about sufficient statistics and order statistics. From the definition of a sufficient statistic, it is clear that the entire sample $\{X_1, \ldots, X_N\}$ is always sufficient.

Of course, this choice of statistic gives no reduction in the number of variables. Likewise, the order statistics $\{X_{(1)}, \ldots, X_{(N)}\}$ are always sufficient. This is easily seen because ordering the product of the N marginal pdfs does not change the joint pdf:

$$\prod_{n=1}^{N} f_{X_n}(x_n) = \prod_{n=1}^{N} f_{X_{(n)}}(x_{(n)}), \tag{9.20}$$

and so the same factorization would be obtained. It can also be shown that the order statistics are sufficient from the definition. The set of all permutations of $\{x_n\}$ has cardinality $N!$ so that

$$P(\mathbf{X} = \mathbf{x} | X_{(1)} = x_{(1)}, \ldots, X_{(N)} = x_{(N)}) = 1/N! \tag{9.21}$$

assuming $\mathbf{x}$ is one of the permutations. Since this conditional pdf is independent of parameter θ (which need not be specified here), the order statistics are sufficient.

Example 9.7. Let $\{X_n\}$ be Gaussian with parameters $\{\theta = \mu, \sigma = 1\}$:

$$f_{\mathbf{X}}(\mathbf{x}; \mu) = \frac{1}{(2\pi)^N} \exp\left(-(1/2)\sum_{n=1}^{N}(x_n^2 - 2\mu x_n + \mu^2)\right), \tag{9.22}$$

where for convenience we have not used indicator functions because the support for each x_n is $\mathcal{R}$. Observe that

$$f_{\mathbf{X}}(\mathbf{x}; \mu) = \frac{1}{(2\pi)^N} \exp\left(-(1/2)\sum_{n=1}^{N} x_n^2\right) \exp\left(\mu \sum_{n=1}^{N} x_n\right) \exp(-N\mu^2/2). \tag{9.23}$$

Thus, the factorization yields

$$h(\mathbf{x}) = \frac{1}{(2\pi)^N} \exp\left(-(1/2)\sum_{n=1}^{N} x_n^2\right), \quad g(t; \mu) = \exp(\mu t)\exp(-N\mu^2/2), \tag{9.24}$$

so that $T = \sum_{n=1}^{N} X_n$ is sufficient for μ. But note that (for N even)

$$\mathbf{T} = \left[\sum_{n=1}^{N/2} X_n, \sum_{n=N/2+1}^{N} X_n\right]^T \tag{9.25}$$

is also sufficient for μ where $\mathbf{T}$ is a vector. One can find other sufficient statistics for μ by splitting these two sums into four sums, and so on. Even though (9.25) and similar multidimensional expressions are sufficient for μ, they are not *minimally* sufficient as discussed later.

The point of the last example is that there are usually many sufficient statistics for the unknown parameter θ. Table 9.1 summarizes several sufficient statistics for the previous Gaussian example with known variance; it also includes a few examples of statistics that are *not* sufficient for μ. The goal of finding a sufficient statistic usually is to reduce the dimension as small as possible while still retaining sufficiency (where no information about θ is lost). The sufficient statistic with the lowest dimension is called a *minimal sufficient statistic*. For the Gaussian distribution in Example 9.7, $T = \sum_{n=1}^{N} X_n$ is obviously minimal sufficient because N samples have been reduced to a single statistic. For a scalar parameter θ, if a single sufficient statistic can be found, then that function is automatically minimal sufficient. The factorization theorem often provides a minimal sufficient statistic, but not always as discussed in Section 9.4. We conclude this section with an example of scalar θ for which it is not possible to obtain a sufficient statistic of dimension one.

TABLE 9.1 Statistics for Gaussian $\{X_n\}$ with Parameters $\{\theta = \mu, \sigma = 1\}$

Statistic **T**	Type	Dimension
Samples: **X**	Sufficient	N
Ordered: $\{X_{(1)}, \ldots, X_{(N)}\}$	Sufficient	N
$\left[\sum_{n=1}^{N/2} X_n, \sum_{n=N/2+1}^{N} X_n\right]^T$	Sufficient	2
Sum: $\sum_{n=1}^{N} X_n$	Minimal sufficient	1
$\{X_1, \ldots, X_{N-1}\}$ (missing a sample)	Not sufficient	$N - 1$
Sum of squares: $\sum_{n=1}^{N} X_n^2$ (missing sign information)	Not sufficient	1

Example 9.8. Consider a uniform distribution with the following joint pdf:

$$\begin{aligned} f_{\mathbf{X}}(\mathbf{x}) &= \prod_{n=1}^{N} I_{[\theta-1/2,\theta+1/2]}(x_n), \\ &= I_{[\theta-1/2,\theta+1/2]}(x_{(1)}) I_{[\theta-1/2,\theta+1/2]}(x_{(N)}) \end{aligned} \tag{9.26}$$

from which $\mathbf{T} = [X_{(1)}, X_{(N)}]^T$ is sufficient. Intuitively, it is apparent that the other order statistics are not needed because we are interested in the *boundaries* of the pdf as depicted in Figure 9.2, for which only $X_{(1)}$ and $X_{(N)}$ are necessary. Thus, even though there is a single parameter θ, the dimension of the minimal sufficient statistic is 2; it is not possible to find a single statistic that is also sufficient. $X_{(1)}$ or $X_{(N)}$ individually are not sufficient for θ, nor is their sum.

9.4 MINIMAL SUFFICIENT STATISTIC

In order to estimate θ, we can generally use any sufficient statistic, but it is more practical to use the sufficient statistic with the lowest dimension. Since all statistics are functions of the samples, a *minimal* sufficient statistic is defined as follows.

Definition: Minimal Sufficient Statistic Sufficient statistic T for parameter θ is *minimal* if it can be expressed as a function of every other sufficient statistic for θ.

Since a function of the N samples $\{X_1, \ldots, X_N\}$ cannot increase the dimension (a function by definition has a unique value for each value of its argument), it is clear that a minimal sufficient statistic must have the lowest dimension possible of all sufficient statistics. If the minimal sufficient statistic is related to another sufficient

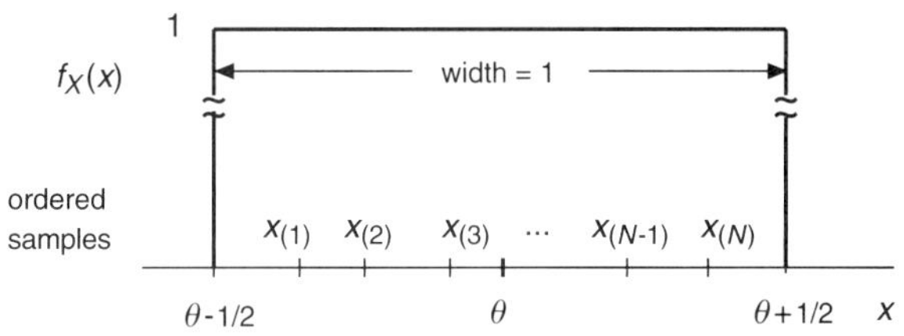

FIGURE 9.2 Example outcomes for the uniform pdf $f_X(x) = I_{[\theta-1/2,\theta+1/2]}(x)$ in Example 9.8 where $\{X_{(1)}, X_{(N)}\}$ are minimal sufficient for θ. (Note the "broken" lines in the vertical direction for a compact figure.)

statistic via a one-to-one function, then both are minimal: they are equivalent. It is important to note that the definition above has two requirements: (i) T must be sufficient and (ii) T is a function of *all* other sufficient statistics for θ.

Example 9.9. For the uniform pdf $f_X(x) = I_{[\theta-1/2,\theta+1/2]}(x)$, the following single statistic is *not* minimal sufficient:

$$T = (1/2)[X_{(1)} + X_{(N)}]. \tag{9.27}$$

Even though T is a function of the sufficient statistics $\{X_{(1)}, X_{(N)}\}$, it is *not* sufficient: some information about θ has been lost. From Figure 9.2, it is intuitive that we might average the minimum and maximum to estimate θ which is in the center of the pdf (θ is the mean). However, from the factorization theorem, we see that it is not possible to combine the remaining two indicator functions of the joint pdf so that $X_{(1)}$ and $X_{(N)}$ sum together. As $N \to \infty$, one could argue that the statistic in (9.27) is a reasonable estimator of θ. In fact, it is unbiased:

$$(1/2)\mathcal{E}[X_{(1)} + X_{(N)}] = \theta \tag{9.28}$$

because it is straightforward to show for this uniform pdf that (see Problem 9.4)

$$\mathcal{E}[X_{(1)}] = \theta - 1/2 + \frac{1}{N+1}, \tag{9.29}$$

$$\mathcal{E}[X_{(N)}] = \theta + 1/2 - \frac{1}{N+1}. \tag{9.30}$$

As $N \to \infty$, the mean of the first-order statistic approaches the left edge of the uniform pdf while that of the Nth-order statistic approaches the right edge. The expectations can never go beyond the edges because $1/(N+1)$ is added in $\mathcal{E}[X_{(1)}]$ and it is subtracted in $\mathcal{E}[X_{(N)}]$. Even though (9.27) appears to be a good estimate for θ, and it is a minimal statistic (of dimension 1), it is not sufficient.

The original samples $\{X_1, \ldots, X_N\}$ can never be minimal sufficient. Since the order statistics are always sufficient, and they are functions of the original samples, they potentially could be minimal: they eliminate the original samples from being minimal. Although not obvious, the order statistics are in fact a reduction of the original samples (even though the dimension is still N) because the original sample indexes are no longer known. It is not possible to express the original samples as a function of the order statistics.

Next, we describe two methods for obtaining a minimal sufficient statistic: one for pdfs with common support and the other when there is no common support. A Gaussian random variable with unknown mean μ has a pdf with common support: the range of samples is the same regardless of the value of θ. This is not the case for a uniform random variable with pdf $f_X(x) = I_{[\theta-1/2,\theta+1/2]}(x)$: a change in the value of θ obviously determines the range of possible outcomes. For the case with common support, assume the unknown parameter takes on a finite number of values.

Theorem 9.2. Let $f_X(x;\theta_m)$ be the pdf of X parameterized by θ_m of which there are $m = 1, \ldots, M$ possible values, and such that this family of pdfs has common support. The following statistic is minimal sufficient:

$$\mathbf{T} = \left[\frac{f_{\mathbf{X}}(\mathbf{x};\theta_2)}{f_{\mathbf{X}}(\mathbf{x};\theta_1)}, \ldots, \frac{f_{\mathbf{X}}(\mathbf{x};\theta_M)}{f_{\mathbf{X}}(\mathbf{x};\theta_1)}\right]^T. \tag{9.31}$$

Proof. For sufficient statistic T, we know from the factorization $f_{\mathbf{X}}(\mathbf{x};\theta) = g(t;\theta)h(\mathbf{x})$ that $g(t;\theta)$ depends on the samples only through t. Because $h(\mathbf{x})$ is not a function of θ, we find that the following ratio depends on the samples only through t for $\theta_1 \neq \theta_2$:

$$\frac{f_{\mathbf{X}}(\mathbf{x};\theta_2)}{f_{\mathbf{X}}(\mathbf{x};\theta_1)} = \frac{g(t;\theta_2)}{g(t;\theta_1)}. \tag{9.32}$$

Since $\mathbf{T}$ contains all possible ratios for the finite number of parameter values and is a function of T, it is necessarily minimal sufficient. □

Usually, one needs to check only one ratio above for one value of θ. The number of ratios required depends on the smoothness of the joint pdf as a function of θ. This procedure extends to countably infinite θ, but is more complicated for uncountable θ because measure theory is involved.

Example 9.10. Consider the exponential pdf with parameter $\theta = \lambda$ which takes on a finite number of values including $\lambda_1 = 1$. From Theorem 9.2, the following ratio is minimal sufficient:

$$\frac{\prod_{n=1}^{N} \lambda \exp(-\lambda x_n) I_{[0,\infty)}(x_n)}{\prod_{n=1}^{N} \exp(-x_n) I_{[0,\infty)}(x_n)} = \frac{\lambda^N \exp\left(-\lambda \sum_{n=1}^{N} x_n\right)}{\exp\left(-\sum_{n=1}^{N} x_n\right)},$$

$$= \lambda^N \exp\left(-(\lambda - 1)\sum_{n=1}^{N} x_n\right), \tag{9.33}$$

which is equivalent to the sample mean $T = \bar{X}$.

Another approach to finding the minimal sufficient statistic for a random variable with common support involves two different sets of N samples, which we state without proof.

Theorem 9.3. Let $\mathbf{X}$ be iid samples of a random variable with pdf $f_X(x;\theta)$, and let $\mathbf{Y}$ be any other set of iid samples from the same distribution. T is minimal sufficient for θ if the ratio

$$\frac{f_{\mathbf{X}}(\mathbf{x};\theta)}{f_{\mathbf{X}}(\mathbf{y};\theta)} = \frac{h(\mathbf{x})g(t_{\mathbf{x}};\theta)}{h(\mathbf{y})g(t_{\mathbf{y}};\theta)} \tag{9.34}$$

is independent of θ if and only if $T_{\mathbf{X}} = T_{\mathbf{Y}}$.

The subscripts on $T_{\mathbf{X}}$ and $T_{\mathbf{Y}}$ denote the samples used in the statistic. Note that the parameters differ in (9.32) whereas the samples differ in (9.34).

Example 9.11. For the conditions of Example 9.10, we have

$$\frac{\lambda^N \exp\left(-\lambda \sum_{n=1}^{N} x_n\right)}{\lambda^N \exp\left(-\lambda \sum_{n=1}^{N} y_n\right)} = \exp\left(-\lambda \sum_{n=1}^{N} x_n + \lambda \sum_{n=1}^{N} y_n\right). \tag{9.35}$$

It is clear that this ratio is independent of λ if and only if the two sums are equal; thus, $T = \sum_{n=1}^{N} X_n$ is minimal sufficient for $\theta = \lambda$.

For random variables without common support, the approach to finding a minimal sufficient statistic is based on the idea of "capturing" the sufficient statistic T from $g(t;\theta)$ as θ is varied. The function $g(t;\theta)$ is plotted

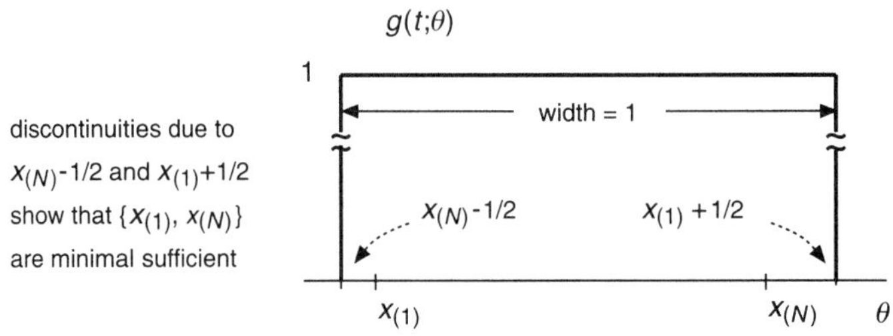

FIGURE 9.3 "Capture" technique for $f_X(x) = I_{[\theta-1/2,\theta+1/2]}(x)$ showing that $\mathbf{T} = [X_{(1)}, X_{(N)}]^T$ is minimal sufficient. (Note the "broken" lines in the vertical direction for a compact figure.)

versus θ and the locations of any *discontinuities* are noted. The minimal sufficient statistic corresponds to the samples where the discontinuities occur.

Example 9.12. For the previous uniform distribution, we repeat the factorization:

$$f_{\mathbf{X}}(\mathbf{x}) = I_{[\theta-1/2,\theta+1/2]}(x_{(1)})I_{[\theta-1/2,\theta+1/2]}(x_{(N)}). \tag{9.36}$$

Observe that the indicator function can be rearranged in terms of θ as follows:

$$I_{[\theta-1/2,\theta+1/2]}(x_{(1)}) = \begin{cases} 1, & \theta - 1/2 \leq x_{(1)} \leq \theta + 1/2 \\ 0, & \text{else.} \end{cases} \tag{9.37}$$

Although it is possible to use the form above to determine minimal sufficiency, it is more easily done when one of the bounds is $\pm\infty$. This is seen from the following property of the indicator function:

$$I_{[\theta-1/2,\theta+1/2]}(x) = I_{[\theta-1/2,\infty)}(x)I_{(-\infty,\theta+1/2]}(x), \tag{9.38}$$

where the closed interval on the left-hand side is the intersection of two semi-closed intervals on the right-hand side. For this problem:

$$I_{[\theta-1/2,\theta+1/2]}(x_{(1)})I_{[\theta-1/2,\theta+1/2]}(x_{(N)}) = I_{[\theta-1/2,\infty)}(x_{(1)})I_{(-\infty,\theta+1/2]}(x_{(N)}), \tag{9.39}$$

from which we find that $g(t;\theta) = 1$ when

$$\theta - 1/2 \leq x_{(1)} \Longrightarrow \theta \leq x_{(1)} + 1/2, \tag{9.40}$$

$$x_{(N)} \leq \theta + 1/2 \Longrightarrow \theta \geq x_{(N)} - 1/2. \tag{9.41}$$

The nonzero region of $g(t;\theta)$ as a function of θ is illustrated in Figure 9.3 with discontinuities at $x_{(1)} + 1/2$ and $x_{(N)} - 1/2$. Thus, $\mathbf{T} = [X_{(1)}, X_{(N)}]^T$ is not only sufficient but also minimal sufficient for θ.

9.5 EXPONENTIAL FAMILIES

In this section, we consider exponential families of random variables for which a sufficient statistic of low dimension is easily found.

Definition: Exponential Family An *exponential family* of random variables with scalar parameter θ has a pdf that can be expressed in the following form:

$$f_X(x) = a(\theta)\exp(b(\theta)c(x))d(x), \tag{9.42}$$

where $a(\theta)$, $b(\theta)$, $c(x)$, and $d(x)$ are known functions. It is called a *family* of random variables because the form above remains the same as θ is varied. For N iid random variables $\{X_n\}$, (9.42) becomes

$$f_{\mathbf{X}}(\mathbf{x}) = a^N(\theta)\exp\left(b(\theta)\sum_{n=1}^{N} c(x_n)\right)\prod_{n=1}^{N} d(x_n). \tag{9.43}$$

For a vector of M parameters $\boldsymbol{\theta} \triangleq [\theta_1, \ldots, \theta_M]^T$ and a single random variable, the pdf is

$$f_X(x) = a(\boldsymbol{\theta})\exp\left(\sum_{m=1}^{M} b_m(\boldsymbol{\theta})c_m(x)\right)d(x), \tag{9.44}$$

and for N iid random variables and M parameters, (9.44) becomes

$$f_{\mathbf{X}}(\mathbf{x}) = a^N(\boldsymbol{\theta})\exp\left(\sum_{m=1}^{M} b_m(\boldsymbol{\theta})\sum_{n=1}^{N} c_m(x_n)\right)\prod_{n=1}^{N} d(x_n). \tag{9.45}$$

An exponential family of random variables should not be confused with the exponential distribution described in Chapter 3 (which itself is an exponential family). As demonstrated below, the pdf of an exponential family can be rearranged (sometimes in a nonobvious way) to obtain one of the forms above containing an exponential. Although a pdf may contain multiple parameters, the exponential form refers to a specific scalar parameter θ or vector of parameters $\boldsymbol{\theta}$ that are of interest; any other parameters of the pdf are assumed to be known (fixed) for estimation purposes. For example, for the Gaussian distribution, we might be interested only in the mean μ and assume that the standard deviation σ is known.

From the factorization theorem,

$$f_{\mathbf{X}}(\mathbf{x}) = h(\mathbf{x})g(t;\theta), \tag{9.46}$$

where $g(t;\theta)$ depends on $\mathbf{x}$ only through the statistic t. Comparing this expression with (9.43), we find

$$g(t;\theta) = a^N(\theta)\exp\left(b(\theta)\sum_{n=1}^{N} c(x_n)\right), \quad h(\mathbf{x}) = \prod_{n=1}^{N} d(x_n), \tag{9.47}$$

and so the sufficient statistic is

$$T = \sum_{n=1}^{N} c(X_n). \tag{9.48}$$

This expression can be obtained for each of the following exponential families; although we show the exponential form only for $N = 1$ samples, the form for N samples is easily determined from (9.43) as summarized later in Table 9.2.

- *Exponential with* $\theta = \lambda$:

$$f_X(x) = \lambda\exp(-\lambda x)I_{[0,\infty)}(x), \tag{9.49}$$

$$\Longrightarrow a(\theta) = \lambda, \quad b(\theta) = -\lambda, \quad c(x) = x, \quad d(x) = I_{[0,\infty)}(x). \tag{9.50}$$

- *Gamma with* $\theta = \lambda$ (known r):

$$f_X(x) = \frac{\lambda^r}{\Gamma(r)} x^{r-1} \exp(-\lambda x) I_{[0,\infty)}(x), \tag{9.51}$$

$$\Longrightarrow a(\theta) = \frac{\lambda^r}{\Gamma(r)}, \quad b(\theta) = -\lambda, \quad c(x) = x, \quad d(x) = x^{r-1} I_{[0,\infty)}(x). \tag{9.52}$$

- *Beta with* $\boldsymbol{\theta} = \{\alpha, \beta\}$:

$$\begin{aligned} f_X(x) &= \frac{1}{B(\alpha, \beta)} x^{\alpha-1} (1-x)^{\beta-1} I_{[0,1]}(x), \\ &= \frac{1}{B(\alpha, \beta)} \left[\exp((\alpha - 1)\ln(x) + (\beta - 1)\ln(1-x)\right] I_{[0,1]}(x), \end{aligned} \tag{9.53}$$

$$\Longrightarrow a(\boldsymbol{\theta}) = \frac{1}{B(\alpha, \beta)}, \quad b_1(\boldsymbol{\theta}) = \alpha - 1, \quad b_2(\boldsymbol{\theta}) = \beta - 1, \tag{9.54}$$

$$c_1(x) = \ln(x), \quad c_2(x) = \ln(1-x), \quad d(x) = I_{[0,1])}(x). \tag{9.55}$$

- *Chi-square with* $\theta = N$:

$$\begin{aligned} f_X(x) &= \frac{x^{N/2-1}}{2^{N/2}\Gamma(N/2)} \exp(-x/2) I_{[0,\infty)}(x), \\ &= \frac{\exp((N/2-1)\ln(x))}{2^{N/2}\Gamma(N/2)} \exp(-x/2) I_{[0,\infty)}(x), \end{aligned} \tag{9.56}$$

$$\Longrightarrow a(\theta) = \frac{2^{-N/2}}{\Gamma(N/2)}, \quad b(\theta) = N/2 - 1, \quad c(x) = \ln(x), \tag{9.57}$$

$$d(x) = \exp(-x/2) I_{[0,\infty)}(x). \tag{9.58}$$

- *Gaussian with* $\boldsymbol{\theta} = \{\mu, \sigma\}$:

$$\begin{aligned} f_X(x) &= \frac{1}{\sqrt{2\pi\sigma^2}} \exp(-(x-\mu)^2/2\sigma^2) I_{(-\infty,\infty)}(x), \\ &= \frac{\exp(-\mu^2/2\sigma^2)}{\sqrt{2\pi\sigma^2}} \exp(-x^2/2\sigma^2 + \mu x/\sigma^2) I_{(-\infty,\infty)}(x), \end{aligned} \tag{9.59}$$

$$\Longrightarrow a(\boldsymbol{\theta}) = \frac{\exp(-\mu^2/2\sigma^2)}{\sqrt{2\pi\sigma^2}}, \quad b_1(\boldsymbol{\theta}) = -\frac{1}{2\sigma^2}, \quad b_2(\boldsymbol{\theta}) = \frac{\mu}{\sigma^2}, \tag{9.60}$$

$$c_1(x) = x^2, \quad c_2(x) = x, \quad d(x) = I_{(-\infty,\infty)}(x). \tag{9.61}$$

- *Rayleigh with* $\theta = \alpha$:

$$f_X(x) = (x/\alpha^2) \exp(-x^2/2\alpha^2) I_{[0,\infty)}(x), \tag{9.62}$$

$$\Longrightarrow a(\theta) = 1/\alpha^2, \quad b(\theta) = -1/2\alpha^2, \quad c(x) = x^2, \quad d(x) = x I_{[0,\infty)}(x). \tag{9.63}$$

- *Laplace with* $\theta = \alpha$ (known $\mu = 0$):

$$f_X(x) = (1/2\alpha)\exp(-|x|/\alpha)I_{(-\infty,\infty)}(x), \tag{9.64}$$

$$\Longrightarrow a(\theta) = 1/2\alpha, \quad b(\theta) = -1/\alpha, \quad c(x) = |x|, \quad d(x) = I_{(-\infty,\infty)}(x). \tag{9.65}$$

- *Poisson with* $\theta = \alpha$:

$$\begin{aligned} p_X[x] &= (1/x!)\exp(-\alpha)\alpha^x I_{\mathcal{Z}^+}[x], \\ &= (1/x!)\exp(-\alpha)\exp(x\ln(\alpha))I_{\mathcal{Z}^+}[x], \end{aligned} \tag{9.66}$$

$$\Longrightarrow a(\theta) = \exp(-\alpha), \quad b(\theta) = \ln(\alpha), \quad c(x) = x, \quad d(x) = (1/x!)I_{\mathcal{Z}^+}[x]. \tag{9.67}$$

- *Geometric with* $\theta = p$:

$$p_X[x] = pq^x I_{\mathcal{Z}^+}[x] = p\exp(x\ln(q))I_{\mathcal{Z}^+}[x], \tag{9.68}$$

$$\Longrightarrow a(\theta) = p, \quad b(\theta) = \ln(q), \quad c(x) = x, \quad d(x) = I_{\mathcal{Z}^+}[x]. \tag{9.69}$$

- *Binomial with* $\theta = p$ (known N):

$$\begin{aligned} p_X[x] &= \binom{N}{x} p^x q^{N-x} I_{\{0,1,\ldots,N\}}[x], \\ &= \binom{N}{x} q^N \exp(x\ln(p/q))I_{\{0,1,\ldots,N\}}[x], \end{aligned} \tag{9.70}$$

$$\Longrightarrow a(\theta) = q^N, \quad b(\theta) = \ln(p/q), \quad c(x) = x, \quad d(x) = \binom{N}{x} I_{\{0,1,\ldots,N\}}[x]. \tag{9.71}$$

Even though the binomial probability mass function (pmf) appears to have two parameters p and q, it is a one-parameter distribution because $q \triangleq 1 - p$. There is a similar constraint for the multinomial distribution described later in Example 9.14, for which we write the M parameters as ratios of $M - 1$ parameters.

- *Bernoulli with* $\theta = p$ (binomial with $N = 1$):

$$p_X[x] = p^x q^{1-x} I_{\{0,1\}}[x] = q\exp(x\ln(p/q))I_{\{0,1\}}[x], \tag{9.72}$$

$$\Longrightarrow a(\theta) = q, \quad b(\theta) = \ln(p/q), \quad c(x) = x, \quad d(x) = I_{\{0,1\}}[x]. \tag{9.73}$$

- *Negative binomial with* $\theta = p$ (known N):

$$p_X[x] = \binom{x+N-1}{x} p^N q^x I_{\mathcal{Z}^+}[x] = \binom{x+N-1}{x} p^N \exp(x\ln(q))I_{\mathcal{Z}^+}[x], \tag{9.74}$$

$$\Longrightarrow a(\theta) = p^N, \quad b(\theta) = \ln(q), \quad c(x) = x, \quad d(x) = \binom{x+N-1}{x} I_{\mathcal{Z}^+}[x]. \tag{9.75}$$

TABLE 9.2 Minimal Sufficient Statistics for Exponential Families

Statistic $\mathbf{T}$	Distribution Family {Parameter $\boldsymbol{\theta}$}
$\sum_{n=1}^{N} X_n$	Exponential $\{\lambda\}$, gamma $\{\lambda$, known $r\}$, Gaussian $\{\mu$, known $\sigma\}$, Poisson $\{\alpha\}$, Bernoulli $\{p\}$, binomial $\{p\}$, geometric $\{p\}$, Negative binomial $\{p$, known $N\}$
$\left[\prod_{n=1}^{N} X_n, \sum_{n=1}^{N} X_n\right]^T$	Beta $\{\alpha, \beta\}$
$\prod_{n=1}^{N} X_n$	Chi-square $\{N\}$
$\left[\sum_{n=1}^{N} X_n, \sum_{n=1}^{N} X_n^2\right]^T$	Gaussian $\{\mu, \sigma\}$
$\sum_{n=1}^{N} \lvert X_n\rvert$	Laplace $\{\alpha$, known $\mu = 0\}$

Since exponential families have a form with terms matching those of the factorization theorem in (9.47), minimal sufficient statistics for each of the previous families are easily identified. They are summarized in Table 9.2.

Example 9.13. The joint pdf for the beta distribution is

$$f_{\mathbf{X}}(\mathbf{x}) = \frac{1}{B^N(\alpha, \beta)} \exp\left((\alpha - 1)\sum_{n=1}^{N} \ln(x_n) + (\beta - 1)\sum_{n=1}^{N} \ln(1 - x_n)\right) \cdot \prod_{n=1}^{N} I_{[0,1]}(x_n). \tag{9.76}$$

Since

$$\sum_{n=1}^{N} \ln(x_n) = \ln\left(\prod_{n=1}^{N} x_n\right), \quad \sum_{n=1}^{N} \ln(1 - x_n) = \ln\left(\prod_{n=1}^{N}(1 - x_n)\right), \tag{9.77}$$

the minimal sufficient statistic for $\boldsymbol{\theta} = [\alpha, \beta]^T$ is

$$\mathbf{T} = \left[\prod_{n=1}^{N} X_n, \prod_{n=1}^{N}(1 - X_n)\right]^T. \tag{9.78}$$

Example 9.14. The multinomial pdf is somewhat more complicated to write in the exponential family form because the parameters are not all independent, so it is presented as a separate example. Let $\{X_1, \ldots, X_M\}$ be M samples with parameters (probabilities) $\{p_1, \ldots, p_M\}$. For known M such that $\sum_{l=1}^{M} x_l = N$, the multinomial pdf is

$$\begin{aligned} f_{X_1,\ldots,X_M}(x_1, \ldots, x_M) &= N! \prod_{m=1}^{M} \frac{p_m^{x_m}}{x_m!} I_{\{0,1,\ldots,N\}}(x_m) I_{\{N\}}\left(\sum_{l=1}^{M} x_l\right), \\ &= N! \exp\left(\sum_{m=1}^{M} x_m \ln(p_m)\right) \\ &\quad \cdot \prod_{m=1}^{M} \frac{1}{x_m!} I_{\{0,1,\ldots,N\}}(x_m) I_{\{N\}}\left(\sum_{l=1}^{M} x_l\right). \end{aligned} \tag{9.79}$$

Since $\sum_{m=1}^{M} p_m = 1$, there are only $M-1$ *unique* parameters; thus, it is convenient to write the parameter vector as $\boldsymbol{\theta} = (p_2/p_1, \ldots, p_M/p_1)$ (recall that the ratio p/q was used for the binomial distribution, which is a special case of the multinomial distribution with $M = 2$). Adding and subtracting $\sum_{m=2}^{M} x_m \ln(p_1)$ to the argument of $\exp(\cdot)$ and substituting the constraint $\sum_{m=1}^{M} x_m = N$ gives

$$\exp\left(N \ln(p_1) + \sum_{m=2}^{M} x_m \ln(p_m/p_1)\right) = \exp\left(-N \ln\left(1 + \sum_{m=2}^{M} p_m/p_1\right)\right) \cdot \exp\left(\sum_{m=2}^{M} x_m \ln(p_m/p_1)\right), \tag{9.80}$$

where we have substituted

$$1/p_1 = \sum_{m=1}^{M} p_m/p_1 = 1 + \sum_{m=2}^{M} p_m/p_1 \tag{9.81}$$

in the first exponential on the right-hand side to write $\ln(p_1)$ in terms of the ratios $\{p_2/p_1, \ldots, p_M/p_1\}$. The exponential family form has components:

$$a(\boldsymbol{\theta}) = N! \exp\left(-N \ln\left(1 + \sum_{m=2}^{M} p_m/p_1\right)\right), \quad b_m(\boldsymbol{\theta}) = \ln(p_m/p_1), \tag{9.82}$$

$$c_m(x_1, \ldots, x_M) = x_m, \;\; d(x_1, \ldots, x_M) = \prod_{m=1}^{M} \frac{1}{x_m!} I_{\{0,1,\ldots,N\}}(x_m) I_{\{N\}}\left(\sum_{l=1}^{M} x_l\right). \tag{9.83}$$

9.6 LOCATION-SCALE FAMILIES

The pdf of a location-scale family of random variables maintains its basic shape when its location or scale parameters are varied.

Definition: Location-Scale Family Let X be a random variable with cdf $F_X(x)$. A *location family* is obtained via the transformation $Y = X + u$ with $u \in \mathcal{R}$. Its distribution has the form

$$F_Y(y) = F_X(y - u). \tag{9.84}$$

A *scale family* is obtained via the transformation $Y = vX$, where $v > 0$. Its distribution has the form

$$F_Y(y) = F_X(y/v). \tag{9.85}$$

Combining the transformations to generate $Y = u + vX$ results in a *location-scale family*, whose distribution has the form

$$F_Y(y) = F_X((y - u)/v) \tag{9.86}$$

TABLE 9.3 Minimal Sufficient Statistics for Location-Scale Families

Statistic **T**	Distribution Family {Parameter $\boldsymbol{\theta}$}
$[X_{(1)}, \ldots, X_{(N)}]^T$	Cauchy $\{c, \text{known } \alpha\}$, Laplace $\{\mu, \text{known } \alpha\}$, logistic $\{\mu, \text{known } \alpha\}$
$X_{(1)}$	Shifted exponential $\{c, \text{known } \lambda\}$, uniform $\{a, \text{known } b\}$
$[X_{(1)}, \sum_{n=1}^{N} X_n]^T$	Shifted exponential $\{c, \lambda\}$
$[X_{(1)}, X_{(N)}]^T$	Uniform $\{a, b\}$
$X_{(N)}$	Uniform $\{b, \text{known } a\}$
$\left[\sum_{n=1}^{N} X_n, \sum_{n=1}^{N} X_n^2\right]^T$	Gaussian $\{\mu, \sigma\}$

and pdf

$$f_Y(y) = (1/v) f_X((y-u)/v), \tag{9.87}$$

where $f_X(x)$ is the pdf of X before transforming to Y.

Minimal sufficient statistics for several location-scale families are provided in Table 9.3. Location-scale families have two parameters and include the following random variables. In each case, we verify the location-scale property by transforming to $Y = u + vX$ and demonstrating that $f_Y(y) = (1/v) f_X((y-u)/v)$ has the same form as $f_X(x)$.

- *Gaussian with* $\boldsymbol{\theta} = \{\mu, \sigma\}$:

$$\begin{aligned} f_Y(y) &= \frac{1}{\sqrt{2\pi v^2 \sigma^2}} \exp(-((y-u)/v - \mu)^2/2\sigma^2) I_{(-\infty,\infty)}(y), \\ &= \frac{1}{\sqrt{2\pi (v\sigma)^2}} \exp(-(y-u-v\mu)^2/2(v\sigma)^2) I_{(-\infty,\infty)}(y), \end{aligned} \tag{9.88}$$

$\Longrightarrow$ Gaussian with parameters $\{u + v\mu, v\sigma\}$.

- *Uniform with* $\boldsymbol{\theta} = \{a, b\}$:

$$f_Y(y) = \frac{1}{v(b-a)} I_{[a,b]}((y-u)/v) = \frac{1}{vb - va} I_{[u+va, u+vb]}(y), \tag{9.89}$$

$\Longrightarrow$ uniform with parameters $\{u + va, u + vb\}$.

- *Cauchy with* $\boldsymbol{\theta} = \{c, \alpha\}$:

$$\begin{aligned} f_Y(y) &= \frac{\alpha/v\pi}{((y-u)/v - c)^2 + \alpha^2} I_{(-\infty,\infty)}(y), \\ &= \frac{v\alpha/\pi}{(y-u-vc)^2 + (v\alpha)^2} I_{(-\infty,\infty)}(y), \end{aligned} \tag{9.90}$$

$\Longrightarrow$ Cauchy with parameters $\{u + vc, v\alpha\}$.

- *Laplace with $\boldsymbol{\theta} = \{\mu, \alpha\}$:*

$$f_Y(y) = (1/2v\alpha)\exp(-|(y-u)/v - \mu|/\alpha)I_{(-\infty,\infty)}(y),$$
$$= (1/2v\alpha)\exp(-|y-u-v\mu|/v\alpha)I_{(-\infty,\infty)}(y), \tag{9.91}$$

$\Longrightarrow$ Laplace with parameters $\{u+v\mu, v\alpha\}$.

- *Logistic with $\boldsymbol{\theta} = \{\mu, \alpha\}$:*

$$f_Y(y) = \frac{\exp(-((y-u)/v-\mu)/\alpha)}{v\alpha(1+\exp(-((y-u)/v-\mu)/\alpha))}I_{(-\infty,\infty)}(y),$$
$$= \frac{\exp(-(y-u-v\mu)/v\alpha)}{v\alpha(1+\exp(-(y-u-v\mu)/v\alpha))}I_{(-\infty,\infty)}(y), \tag{9.92}$$

$\Longrightarrow$ logistic with parameters $\{u+v\mu, v\alpha\}$.

- *Shifted exponential with $\boldsymbol{\theta} = \{c, \lambda\}$:*

$$f_Y(y) = (\lambda/v)\exp(-\lambda((y-u)/v-c))I_{[c,\infty)}((y-u)/v),$$
$$= (\lambda/v)\exp(-(\lambda/v)(y-u-vc))I_{[u+vc,\infty)}(y), \tag{9.93}$$

$\Longrightarrow$ shifted exponential with parameters $\{u+vc, \lambda/v\}$.

- *Discrete uniform with $\boldsymbol{\theta} = \{a, b\}$* and $N \triangleq b-a+1$:

$$p_Y[y] = (1/Nv)I_{\{a,\ldots,b\}}[(y-u)/v] = (1/Nv)I_{\{u+va,\ldots,u+vb\}}[y], \tag{9.94}$$

$\Longrightarrow$ discrete uniform with $N = (u+vb)-(u+va)+1 = v(b-a)+1$ and parameters $\{u+va, u+vb\}$ (assuming that u and v have been chosen such that the parameters are integers).

The Gaussian random variable with $\boldsymbol{\theta} = \{\mu, \sigma\}$ is both an exponential family and a location-scale family. However, this is unusual because most location-scale families cannot be rewritten in the exponential form of (9.42). For example, by generalizing the exponential pdf to include shift parameter c, it is no longer an exponential family.

Example 9.15. Consider N iid samples of the Cauchy distribution with $\alpha = 1$, yielding the joint pdf

$$f_{\mathbf{X}}(\mathbf{x}) = \frac{1}{\pi^N}\prod_{n=1}^{N}\frac{1}{(x_n-c)^2+1}I_{(-\infty,\infty)}(x_n). \tag{9.95}$$

It does not seem possible to simplify this expression to generate a sufficient statistic for $\theta = c$ with dimension lower than N. It turns out that often (but not always), the order statistics are *minimal sufficient* for the location parameter of location-scale families, as we now demonstrate for the Cauchy distribution. Consider another set of samples $\mathbf{y}$ and form the ratio in (9.34):

$$\frac{f_{\mathbf{X}}(\mathbf{x};c)}{f_{\mathbf{X}}(\mathbf{y};c)} = \prod_{n=1}^{N}\frac{(y_n-c)^2+1}{(x_n-c)^2+1} = \prod_{n=1}^{N}\frac{c^2-2cy_n+y_n^2+1}{c^2-2cx_n+x_n^2+1}. \tag{9.96}$$

For each quadratic term in the product, let the numerator have roots $c = y_n \pm j$ and the denominator have roots $c = x_n \pm j$, where $j = \sqrt{-1}$, such that

$$\frac{f_{\mathbf{X}}(\mathbf{x};c)}{f_{\mathbf{X}}(\mathbf{y};c)} = \prod_{n=1}^{N} \frac{(c-y_n-j)(c-y_n+j)}{(c-x_n-j)(c-x_n+j)}. \tag{9.97}$$

Because the product is the same regardless of the order in which the terms are multiplied, the only way that this ratio is independent of c is if the roots of the numerator are a *permutation* of those in the denominator. This means that the numerator and denominator must have the same order statistics, and thus $\mathbf{T} = [X_{(1)}, \ldots, X_{(N)}]^T$ is minimal sufficient for c of the Cauchy distribution. It is shown in Problem 9.8 that the order statistics are likewise minimal sufficient for the Laplace distribution when its mean μ is the parameter to be estimated.

9.7 COMPLETE STATISTIC

Although $\mathbf{T} = [X_{(1)}, X_{(N)}]^T$ is minimal sufficient for θ of the uniform distribution in Example 9.12, and together they contain as much information about θ as the original samples, it turns out they also contain information that is *not* relevant for estimating θ. This irrelevant information is called *ancillary*, and any component of a minimal sufficient statistic that is ancillary is called an ancillary statistic. A minimal sufficient statistic that has no ancillary component is called a *complete* minimal sufficient statistic. We provide two definitions of an ancillary statistic.

Definition: Ancillary Statistic Statistic V is *ancillary* if its distribution is independent of θ.

An ancillary statistic does not contain any information about parameter θ. This is essentially the opposite property of a sufficient statistic which contains the same information about θ as the original samples.

Definition: First-Order Ancillary Statistic Statistic V is *first-order ancillary* if $\mathcal{E}[V]$ is independent of θ.

An ancillary statistic is also first-order ancillary: if its distribution is independent of θ, then its expectation does not depend on θ.

Example 9.16. For the uniform pdf $f_X(x) = I_{[\theta-1/2,\theta+1/2]}(x)$, observe that the statistic $V = X_{(N)} - X_{(1)}$ is first-order ancillary because its mean does not depend on θ:

$$\mathcal{E}[X_{(N)} - X_{(1)}] = 1 - \frac{2}{N+1} = \frac{N-1}{N+1}, \tag{9.98}$$

where (9.29) and (9.30) have been used. We have been able to find a function of the minimal sufficient statistic that is first-order ancillary. Note that (i) V is related to the width of the pdf (which is independent of θ) and (ii) the expectation above asymptotically approaches one.

The next definition refers to a type of statistic that contains no ancillary information about θ.

Definition: Complete Statistic Statistic T is *complete* if for every function $c(\cdot)$:

$$\mathcal{E}[c(T)] = 0, \quad \text{for all } \theta \tag{9.99}$$

implies

$$P(c(T) = 0) = 1, \quad \text{for all } \theta. \tag{9.100}$$

It is possible that a complete statistic contains *no* information about θ. Thus, we are interested in a statistic that is both sufficient and complete, such that it is not missing any information about θ and yet has no irrelevant component. This is provided by the following theorem which states that a *complete sufficient statistic* has no ancillary component.

Theorem 9.4 (Basu). Let T be a complete sufficient statistic for θ. Any ancillary statistic V is independent of T.

Proof. We need to show that $P(V \le v|T = t) = P(V \le v)$. Since V is ancillary, $P(V \le v) = F_V(v)$ does not depend on θ. Define the following conditional probability:

$$g(t) \triangleq P(V \le v|T = t), \tag{9.101}$$

which is a function of t, and for which we can write the expectation:

$$\mathcal{E}[g(T)] = \int_{-\infty}^{\infty} P(V \le v|T = t) f_T(t) \mathrm{d}t = P(V \le v). \tag{9.102}$$

Bringing $F_V(v)$ to the left-hand side gives

$$\mathcal{E}[g(T) - F_V(v)] = 0 \Longrightarrow P(g(T) - F_V(v) = 0) = 1 \Longrightarrow P(g(T) = F_V(v)) = 1 \tag{9.103}$$

since T is complete. The function in the completeness definition in this case is $c(T) = g(T) - F_V(v)$. The last result means that with probability one, $P(V \le v|T = t) = F_V(v)$ for all t and all θ, thus completing the proof. □

Example 9.17. Let $\{X_1, \ldots, X_N\}$ be iid Gaussian random variables with parameters $\{\mu, \sigma = 1\}$. It is shown in Problem 9.10 that the sample mean $T = \bar{X}$ is a complete sufficient statistic for μ. In Problem 4.35, it was shown that $(N - 1)S^2/\sigma^2$ has a chi-square distribution with parameter $N - 1$ where S^2 is the sample variance. Since the distribution of $V = S^2$ does not depend on μ, it is an ancillary statistic, and by Basu's theorem the sample mean and sample variance are independent for a Gaussian distribution.

Example 9.18. Consider a uniform random variable on $[a, \theta]$ with one independent parameter θ and known a such that $\theta > a$. For N iid samples, the joint pdf is

$$f_{\mathbf{X}}(\mathbf{x}) = \frac{1}{(\theta - a)^N} I_{[a,\theta]}(x_{(N)}), \tag{9.104}$$

which shows that $X_{(N)}$ is minimal sufficient for θ because it has dimension one. Moreover, it is a complete sufficient statistic as demonstrated next. From the definition, we first write the following expectation:

$$\mathcal{E}[c(X_{(N)})] = N \int_{-\infty}^{\infty} c(x) F_X^{N-1}(x) f_X(x) \mathrm{d}x, \tag{9.105}$$

where for the uniform pdf:

$$F_X(x) = \frac{1}{\theta - a} \int_{-\infty}^{x} I_{[a,\theta]}(x) \mathrm{d}x = \frac{x - a}{\theta - a} I_{[a,\theta]}(x). \tag{9.106}$$

Substituting the pdf and cdf gives

$$\mathcal{E}[c(X_{(N)})] = \frac{N}{(\theta - a)^N} \int_a^{\theta} c(x)(x - a)^{N-1} \mathrm{d}x. \tag{9.107}$$

Setting this result equal to zero, canceling the constant in front of the integral, and differentiating with respect to θ yields

$$c(\theta)(\theta - a)^{N-1} = 0. \tag{9.108}$$

This must hold for all values of $\theta > a$, which means that the function $c(\theta) = 0$ almost everywhere, that is, with probability one. As a result, $X_{(N)}$ is a complete minimal sufficient statistic for θ for any finite value of a.

Example 9.19. Let $\{X_1, \ldots, X_N\}$ be iid Bernoulli with parameter p:

$$P(\mathbf{X} = \mathbf{x}) = p^{\sum_{n=1}^{N} x_n} q^{1-\sum_{n=1}^{N} x_n} \prod_{n=1}^{N} I_{\{0,1\}}[x_n]. \tag{9.109}$$

From Table 9.2, we know that $T = \sum_{n=1}^{N} X_n$ is minimal sufficient for p. The distribution of T is binomial with parameters $\{N, p\}$ such that

$$\begin{aligned} \mathcal{E}[c(T)] &= \sum_{n=0}^{N} c(n) \binom{N}{n} p^n q^{N-n}, \\ &= c(0)q^N + c(1)Npq^{N-1} + \cdots + c(N)p^N. \end{aligned} \tag{9.110}$$

Because the exponents of p and q differ across all terms in the sum, the only way the above expression can be zero for all p is when $c(n) = 0$ for all n. As a result, T is also complete.

The goal in Section 9.8 is to use completeness of a sufficient statistic to obtain an optimal estimator that is not only unbiased, but also has the lowest variance of all unbiased estimators. It is known as a uniformly minimum-variance unbiased (UMVU) estimator.

Definition: UMVU Estimator The *UMVU estimator* T_{UMVU} of parameter θ based on N iid samples $\{X_1, \ldots, X_N\}$ has the following properties:

- *Unbiased.* $\mathcal{E}[T_{\text{UMVU}}] = \theta$.
- *Minimum variance.* $\text{var}[T_{\text{UMVU}}] \leq \text{var}[T]$ for any other unbiased estimator T of θ.

Before describing how to generate such an estimator using completeness, we first present a technique for generating an estimator with a reduced variance.

9.8 RAO–BLACKWELL THEOREM

The Rao–Blackwell (RB) theorem describes a method for transforming a relatively poor estimator of θ, which is usually straightforward to find, into an improved estimator with lower variance. In order to prove the theorem, we need Jensen's inequality in Appendix F:

$$g\left(\mathcal{E}[X]\right) \leq \mathcal{E}[g(X)], \tag{9.111}$$

where X is a random variable defined on an open interval with cdf $F_X(x)$, and $g(x)$ is a convex function (see Appendix B). Although the Rao–Blackwell theorem applies to any convex function as used in Jensen's inequality, we present the theorem for the variance of an estimator.

Theorem 9.5 (Rao–Blackwell). Let $\{T_1, \ldots, T_P\}$ be a set of statistics that are jointly sufficient for function $c(\theta)$ of parameter θ. Define another statistic T that is an unbiased estimator of $c(\theta)$. Then statistic $T_{\text{RB}} \triangleq \mathcal{E}[T|T_1, \ldots, T_P]$ has the following properties.

- $\mathcal{E}[T_{\text{RB}}]$ is a function of $\{T_1, \ldots, T_P\}$ and thus is sufficient.
- $\mathcal{E}[T_{\text{RB}}] = c(\theta)$ so that T_{RB} is an unbiased estimator of $c(\theta)$.
- $\text{var}[T_{\text{RB}}] \leq \text{var}[T]$ for all θ.
- $\text{var}[T_{\text{RB}}] < \text{var}[T]$ for some θ, except when T_{RB} and T are equal with probability one.

Proof. The first two properties follow from the definition of conditional expectation. The third and fourth properties are derived from Jensen's inequality by letting $g(x) = [x - c(\theta)]^2$ so that

$$(\mathcal{E}[T|T_1 = t_1, \ldots, T_P = t_P] - c(\theta))^2 < \mathcal{E}[(T - c(\theta))^2|T_1 = t_1, \ldots, T_P = t_P]. \tag{9.112}$$

The left-hand side includes T_{RB} written as a function of the specific values $\{t_1, \ldots, t_P\}$, and the right-hand side is the conditional variance of the original unbiased estimator T with fixed conditioning $\{t_1, \ldots, t_P\}$. Taking the expectation of both sides over the conditioning variables gives

$$\begin{aligned}
\mathcal{E}[(\mathcal{E}[T|T_1 = t_1, \ldots, T_P = t_P] - c(\theta))^2] &= \mathcal{E}[(T_{\text{RB}} - c(\theta))^2] = \text{var}[T_{\text{RB}}] \\
&< \mathcal{E}[\mathcal{E}[(T - c(\theta))^2|T_1 = t_1, \ldots, T_P = t_P]], \\
&= \text{var}[T].
\end{aligned} \tag{9.113}$$

The left-hand side of the inequality is the variance of T_{RB}, and the right-hand side reduces to the variance of T because of the nested expectations (see Chapter 5). As discussed in the proof of Jensen's inequality in the appendix, because the variance is strictly convex, we have strict inequality above except when $T = \mathcal{E}[T|T_1, \ldots, T_P] = T_{\text{RB}}$ with probability one. □

Example 9.20. Let $\{X_1, \ldots, X_N\}$ be iid exponential random variables with parameter $\theta = \lambda$. Suppose we are interested in an estimator for

$$c(\lambda) = P(X > x_o) = 1 - F_X(x_o) = \exp(-\lambda x_o) \tag{9.114}$$

with $x_o > 0$. Two components are needed to apply the Rao–Blackwell theorem: (i) a statistic that is sufficient for λ and (ii) an unbiased estimator for $c(\lambda)$ that need not be sufficient. From Example 9.4, we know that $T_1 = \sum_{n=1}^{N} X_n$ is sufficient for λ. Consider the estimator $T = I_{(x_o,\infty)}(X_1)$, which from the properties of the indicator function is an unbiased estimator of $c(\lambda)$:

$$\mathcal{E}[T] = \mathcal{E}[I_{(x_o,\infty)}(X_1)] = P(X_1 > x_o). \tag{9.115}$$

Intuitively, T is a poor estimator (even though it is unbiased) because it depends only on one sample and we know that T_1 based on all N samples is sufficient. In order to find $T_{\text{RB}} = \mathcal{E}[T|T_1]$, the conditional pdf $f_{X_1|T_1}(x_1|t_1)$ is derived by exploiting the fact that $\{X_n\}$ are independent as follows:

$$\begin{aligned}
P(X_1 \leq x_1, T_1 \leq t_1) &= P(X_1 \leq x_1, T_1 - X_1 \leq t_1 - x_1), \\
&= P(X_1 \leq x_1)P(T_1 - X_1 \leq t_1 - x_1), \\
&= P(X_1 \leq x_1)P\left(\sum_{n=2}^{N} X_n \leq t_1 - x_1\right)
\end{aligned} \tag{9.116}$$

with $t_1 \geq x_1$. Such a technique is commonly used when one of the random variables is the *sum* of independent random variables. Differentiating this expression gives the desired joint pdf:

$$f_{X_1,T_1}(x_1,t_1) = f_{X_1}(x_1)I_{[0,\infty)}(x_1)f_{T_2}(t_1 - x_1)I_{[x_1,\infty)}(t_1), \tag{9.117}$$

where $T_2 \triangleq T_1 - X_1$ has an Erlang distribution with parameters $\{\lambda, r = N - 1\}$ because it is the sum of $N - 1$ iid exponential random variables. Thus

$$\begin{aligned} f_{X_1,T_1}(x_1,t_1) &= \lambda \exp(-\lambda x_1)I_{[0,\infty)}(x_1)\frac{\lambda^{N-1}}{(N-2)!}(t_1 - x_1)^{N-2} \\ &\quad \cdot \exp(-\lambda(t_1 - x_1))I_{[x_1,\infty)}(t_1), \\ &= \frac{\lambda^N}{(N-2)!}(t_1 - x_1)^{N-2}\exp(-\lambda t_1)I_{[0,\infty)}(x_1)I_{[x_1,\infty)}(t_1). \end{aligned} \tag{9.118}$$

The marginal pdf of T_1 is obtained by integrating x_1 over $[0, t_1]$:

$$f_{T_1}(t_1) = \frac{\lambda^N}{(N-1)!}t_1^{N-1}\exp(-\lambda t_1)I_{[0,\infty)}(t_1), \tag{9.119}$$

and the ratio $f_{X_1,T_1}(x_1,t_1)/f_{T_1}(t_1)$ gives the conditional pdf:

$$f_{X_1|T_1}(x_1|t_1) = (N-1)[(t_1 - x_1)^{N-2}/t_1^{N-1}]I_{[0,\infty)}(x_1)I_{[x_1,\infty)}(t_1). \tag{9.120}$$

The desired expectation is

$$\begin{aligned} \mathcal{E}\left[I_{(x_o,\infty)}(X_1)|T_1 = t_1\right] &= (N-1)\int_{x_o}^{t_1}[(t_1 - x_1)^{N-2}/t_1^{N-1}]\mathrm{d}x_1, \\ &= \left(\frac{t_1 - x_o}{t_1}\right)^{N-1} I_{[x_o,\infty)}(t_1), \end{aligned} \tag{9.121}$$

which results in the following improved estimator for $c(\lambda) = \exp(-\lambda x_o)$:

$$T_{\mathrm{RB}} = \left(\frac{\sum_{n=1}^N X_n - x_o}{\sum_{n=1}^N X_n}\right)^{N-1} = \left(1 - \frac{x_o}{\sum_{n=1}^N X_n}\right)^{N-1}. \tag{9.122}$$

9.9 LEHMANN–SCHEFFÉ THEOREM

The transformation described in Section 9.8 does *not* necessarily generate the UMVU estimator. It guarantees only that the variance of the improved estimator will not exceed that of the original statistic T. The theorem presented in this section describes how to obtain the UMVU estimator. We first give a result concerning the uniqueness of estimators.

Theorem 9.6. Let T be a complete sufficient statistic for θ. If there exists an unbiased estimator $g(T)$ for $c(\theta)$, then the estimator is unique.

Proof. The theorem is proved by contradiction. Consider another estimator $h(T)$ that is unbiased $\mathcal{E}[h(T)] = c(\theta)$ and differs from $g(T)$ in probability: $P(h(T) - g(T) \neq 0) > 0$. Since both estimators are unbiased:

$$\mathcal{E}[h(T) - g(T)] = 0, \tag{9.123}$$

which implies $h(T) - g(T) = 0$ because T is complete. This contradicts the assumption $P(h(T) - g(T) \neq 0) > 0$, so $h(T)$ must be the same as $g(T)$, which means that $g(T)$ is unique. □

Theorem 9.7 brings together the following properties of an estimator: sufficiency, completeness, unbiasedness, and minimum variance.

Theorem 9.7 (Lehmann–Scheffé). Let T be a complete sufficient statistic for θ. If $\mathcal{E}[g(T)] = c(\theta)$ where $g(\cdot)$ and $c(\cdot)$ are real-valued functions, then $g(T)$ is the UMVU estimator of $c(\theta)$.

Proof. Let $h(\mathbf{X})$ be any unbiased estimator of $c(\theta)$. From Theorem 9.6, we know that

$$\mathcal{E}[h(\mathbf{X})|T] = g(T) \tag{9.124}$$

is the unique unbiased estimator for $c(\theta)$. Moreover, from the Rao–Blackwell theorem the variance of $g(T)$ cannot exceed that of $h(\mathbf{X})$, so $g(T)$ is the UMVU estimator. □

This theorem and its proof suggest two different methods for deriving the UMVU estimator:

- *Unbiasedness*. Find a complete sufficient statistic T for θ and solve $\mathcal{E}[g(T)] = c(\theta)$ for the function $g(\cdot)$. Recall that an estimator is just a particular function of the samples.
- *Conditioning*. Find a complete sufficient statistic T_1 for θ and an unbiased estimator T of $c(\theta)$. Compute the conditional expectation $\mathcal{E}[T|T_1]$ which is also a particular function of the samples.

In both approaches, a complete sufficient statistic is needed for θ. The goal in the first approach is to manipulate the equation for $\mathcal{E}[g(T)]$ to find an expression for $g(\cdot)$. The second approach is essentially the same as that described in the Rao–Blackwell theorem, except that here the sufficient statistic T_1 must be complete; since the conditional expectation is a function of T_1, it provides the unique UMVU estimator due to Theorem 9.6.

Example 9.21. Let $\{X_n\}$ be iid Bernoulli random variables with parameter $\theta = p$ for which $T = \sum_{n=1}^{N} X_n$ is a complete sufficient statistic (see Example 9.19). Using the first approach above, we have

$$(1/N)\mathcal{E}[T] = p, \tag{9.125}$$

and thus $g(T) = \bar{X}$ is the UMVU estimator for p. The second method is more tedious, but is included here for completeness. Note that any of the $\{X_n\}$ can be used as an unbiased estimator since $\mathcal{E}[X_n] = p$. In order to compute the expectation $\mathcal{E}[X_n|T]$, we need the following conditional probability:

$$\begin{aligned} P(X_n = x_n|T = t) &= \frac{P(X_n = x_n, T = t)}{P(T = t)}, \\ &= \frac{P(X_n = x_n)P(T - X_n = t - x_n)}{\binom{N}{t} p^t q^{N-t}}, \end{aligned} \tag{9.126}$$

where the approach in Example 9.20 has been used to simplify the numerator. Thus

$$P(X_n = x_n|T = t) = \frac{p^{x_n}q^{1-x_n}\binom{N-1}{t-x_n}p^{t-x_n}q^{N-1-t+x_n}}{\binom{N}{t}p^t q^{N-t}} = \frac{\binom{N-1}{t-x_n}}{\binom{N}{t}} \tag{9.127}$$

and

$$\mathcal{E}[X_n|T = t] = \frac{1}{\binom{N}{t}}\sum_{x_n=0}^{1} x_n\binom{N-1}{t-x_n} = \frac{\binom{N-1}{t-1}}{\binom{N}{t}} = \frac{t}{N}, \tag{9.128}$$

which gives the same estimator as in the first approach: $\mathcal{E}[X_n|T] = T/N = \bar{X}$.

Example 9.22. Let $\{X_n\}$ be iid random variables with the following uniform distribution:

$$f_X(x) = (1/\theta)I_{[0,\theta]}(x), \tag{9.129}$$

where from Example 9.18 (with $a = 0$), we know that $X_{(N)}$ is a complete sufficient statistic for θ; it has the following pdf:

$$f_{X_{(N)}}(x) = \frac{Nx^{N-1}}{\theta^N}I_{[0,\theta]}(x). \tag{9.130}$$

Using the first method for finding the UMVU estimator for $c(\theta) = \theta$, we evaluate the following expectation:

$$\mathcal{E}\left[g(X_{(N)})\right] = \frac{N}{\theta^N}\int_0^\theta g(x)x^{N-1}\mathrm{d}x. \tag{9.131}$$

In order for this expression to equal θ, the integral must be θ^{N+1}/N. Choosing $g(x) = x$ gives

$$\mathcal{E}\left[g(X_{(N)})\right] = \frac{N}{\theta^N}\int_0^\theta x^N\mathrm{d}x = \frac{N\theta^{N+1}}{(N+1)\theta^N} = \frac{N}{N+1}\theta. \tag{9.132}$$

Thus, the UMVU estimator for θ is

$$g(X_{(N)}) = \frac{N+1}{N}X_{(N)}. \tag{9.133}$$

In the rest of this chapter, we consider several other techniques for estimating parameter θ from a set of N iid samples $\{X_1, \ldots, X_N\}$. These estimators are also based on sufficient statistics, but they are derived by introducing a *risk function* that is used to generate good estimators which may differ from the UMVU estimator. It is not necessary to find a sufficient statistic first because optimization of the risk function leads directly to the estimator. However, a sufficient statistic is usually the end result. We begin with classical Bayes estimation.

9.10 BAYES ESTIMATION

Since an estimator T of θ is a function of the samples $\mathbf{X}$, it is a random variable that is characterized by a distribution which depends on the pdf of the iid samples. In addition to sufficiency for θ, we might be interested in *unbiasedness* and *efficiency* as discussed later in this chapter. We introduce some definitions that are used to describe the quality of an estimator.

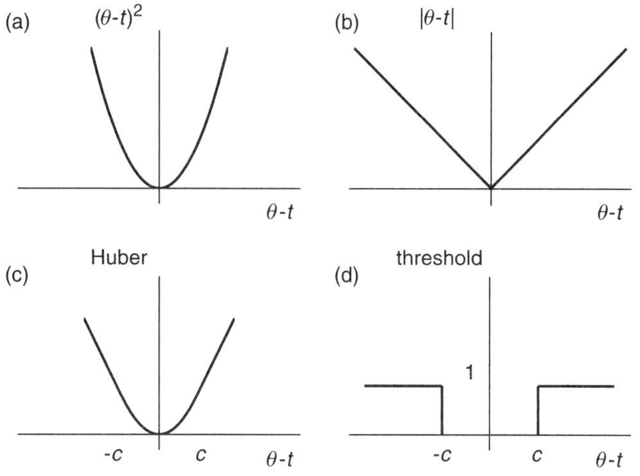

FIGURE 9.4 Loss functions. (a) Squared-error. (b) Absolute-error. (c) Huber. (d) Threshold.

Definition: Loss Function The *loss function* $C(t, \theta)$ maps $T = t$ and θ to a real number.

Various loss functions can be used such as the following examples:

$$\textit{Squared error}: \quad C(t, \theta) = (\theta - t)^2, \tag{9.134}$$

$$\textit{Weighted squared error}: \quad C(t, \theta) = w(\theta)(\theta - t)^2, \tag{9.135}$$

$$\textit{Absolute error}: \quad C(t, \theta) = |\theta - t|, \tag{9.136}$$

$$\textit{Threshold}: \quad C(t, \theta) = \begin{cases} 0, & |\theta - t| \leq c \\ 1, & |\theta - t| > c, \end{cases}$$

$$= u(|\theta - t| - c), \tag{9.137}$$

$$\textit{Huber}: \quad C(t, \theta) = \begin{cases} (\theta - t)^2, & |\theta - t| \leq c \\ 2c|\theta - t| - c^2, & |\theta - t| > c, \end{cases} \tag{9.138}$$

where $w(\theta)$ is a weighting function that depends only on θ, and $c > 0$ is a threshold. Some of these loss functions are illustrated in Figure 9.4. $C(t, \theta)$ is a measure of the level of significance placed on deviations of t from the actual θ. Huber's loss function transitions from the squared-error to the absolute error at threshold c. Often, it is desirable that $C(\cdot)$ be a strictly convex function (see Appendix B) so that the estimator is unique. The squared-error loss function weights the deviations from θ more heavily than the other examples; it is the most commonly used loss function in practice because analyses and derivations are usually more difficult for other loss functions.

Definition: Risk Function The *risk function* for random Θ is the expectation of the loss function:

$$R \triangleq \mathcal{E}[C(T, \Theta)], \tag{9.139}$$

where the expectation is over the joint distribution of Θ and the random variables $\mathbf{X}$ comprising T. It also known as a *cost function* and an *objective function*.

In Bayes estimation, the parameter is assumed to be a random variable, and thus it has the distribution $f_\Theta(\theta)$. The goal is to find statistic T of the samples that minimizes the risk function. Expanding the expectation, R can be written as follows:

$$\begin{aligned}\mathcal{E}[C(T,\Theta)] &= \int_{-\infty}^{\infty}\cdots\int_{-\infty}^{\infty} C(t,\Theta) f_{\Theta,\mathbf{X}}(\theta,\mathbf{x})\mathrm{d}\theta\mathrm{d}\mathbf{x},\\ &= \int_{-\infty}^{\infty}\cdots\int_{-\infty}^{\infty} C(t,\Theta) f_{\mathbf{X}|\Theta}(\mathbf{x}|\theta) f_\Theta(\theta)\mathrm{d}\theta\mathrm{d}\mathbf{x},\end{aligned} \tag{9.140}$$

where in the second expression we have conditioned on Θ. Note that T is the only quantity above that can be varied; the conditional pdf $f_{\mathbf{X}|\Theta}(\mathbf{x}|\theta)$ and the marginal pdf $f_\Theta(\theta)$ are determined by the specific problem. Next, we provide some terminology generally used for two important distributions in estimation.

Definition: Prior and Posterior Distributions The marginal pdf $f_\Theta(\theta)$ of the unknown parameter is called the *prior distribution*. It is also known as the *a priori* distribution. The *posterior distribution* is the conditional pdf $f_{\Theta|X_1,\ldots,X_N}(\theta|x_1,\ldots,x_N)$. It can be viewed as a "measure" of information about the unknown random parameter Θ given samples $\{X_1,\ldots,X_N\}$. It is also known as the *a posteriori* distribution.

Example 9.23. Let the unknown parameter be μ of a Gaussian random variable with known σ. Since the N samples are iid, the conditional pdf is

$$\begin{aligned}f_{\mathbf{X}|\Theta}(\mathbf{x}|\theta) &= \prod_{n=1}^{N} f_{X_n|\Theta}(x_n|\theta),\\ &= \frac{1}{(2\pi\sigma^2)^{N/2}}\exp\left(-\sum_{n=1}^{N}(x_n-\theta)^2/2\sigma^2\right).\end{aligned} \tag{9.141}$$

This is the joint pdf of N-independent Gaussian random variables with known mean θ (because of the conditioning) and known σ (by assumption). In order to continue, the prior distribution of Θ is needed; assume it is uniform as follows:

$$f_\Theta(\theta) = (1/2)I_{[-1,1]}(\theta) \tag{9.142}$$

so that (9.140) becomes

$$\begin{aligned}\mathcal{E}[C(T,\Theta)] = \frac{1}{2(2\pi\sigma^2)^{N/2}}\int_{-1}^{1}\int_{-\infty}^{\infty}\cdots\int_{-\infty}^{\infty} C(t,\theta)\\ \cdot\exp\left(-\sum_{n=1}^{N}(x_n-\theta)^2/2\sigma^2\right)\mathrm{d}\theta\mathrm{d}\mathbf{x}.\end{aligned} \tag{9.143}$$

In order to proceed, a specific loss function $C(t,\theta)$ needs to be chosen. The goal is to find T that minimizes $\mathcal{E}[C(T,\Theta)]$, which necessarily is a function of the samples $\{X_1,\ldots,X_N\}$.

We will return to this example later after examining some specific estimation techniques.

9.11 MEAN-SQUARE-ERROR ESTIMATION

Consider two random variables X and Y that are *correlated* with joint pdf $f_{X,Y}(x, y)$. Assume that X is measurable whereas we do not have direct access to Y. A simple example is

$$X = Y + V, \tag{9.144}$$

where V is not measurable and is independent of Y. Such a model is useful for many problems where an estimate of Y is needed but additive noise V corrupts the samples. From Chapter 4, we know that when Y and V are independent (which is often the case in many problems), the pdf of X is obtained from the following convolution:

$$f_X(x) = \int_{-\infty}^{\infty} f_Y(x - v) f_V(v) \mathrm{d}v. \tag{9.145}$$

In order to estimate Y from X, the joint pdf $f_{X,Y}(x, y)$ is required. In the notation of Section 9.10, the unknown "parameter" is $\Theta = Y$, and there is only one sample X such that $N = 1$. For the simple model above, the joint pdf is

$$f_{X,Y}(x, y) = f_{X|Y}(x|y) f_Y(y). \tag{9.146}$$

Using the techniques in Chapter 4:

$$\begin{aligned} P(X \leq x | Y = y) &= P(Y + V \leq x | Y = y), \\ &= P(V \leq x - y | Y = y) = P(V \leq x - y), \end{aligned} \tag{9.147}$$

where in the last expression we have used the fact that Y and V are independent. Thus

$$f_{X|Y}(x|y) = \frac{\partial}{\partial x} F_V(x - y) = f_V(x - y) \tag{9.148}$$

and

$$f_{X,Y}(x, y) = f_V(x - y) f_Y(y). \tag{9.149}$$

Returning to the general problem of estimating Y from X, the goal in mean-square-error (MSE) estimation is to find the function $h(X) = T$ of the measurable random variable X that minimizes the following risk function:

$$R \triangleq \mathcal{E}[(Y - h(X))^2]. \tag{9.150}$$

For notational convenience, we use ξ_{MSE} to represent R which is the symbol often used in engineering for the MSE. Generally, $h(X)$ is a nonlinear function. Conditions under which it is linear are discussed later, as is the MSE estimation problem when $h(X)$ is constrained to be linear, and is called *linear MSE estimation.* From the results in Chapter 5, the expectation above is written as

$$\xi_{\text{MSE}} = \int_{-\infty}^{\infty} \int_{-\infty}^{\infty} [y - h(x)]^2 f_{X,Y}(x, y) \mathrm{d}x \mathrm{d}y, \tag{9.151}$$

which can be expressed using conditioning as follows:

$$\begin{aligned}\xi_{\text{MSE}} &= \int_{-\infty}^{\infty}\int_{-\infty}^{\infty}[y-h(x)]^2 f_{Y|X}(y|x)f_X(x)\mathrm{d}y\mathrm{d}x,\\ &= \int_{-\infty}^{\infty} f_X(x)\int_{-\infty}^{\infty}[y-h(x)]^2 f_{Y|X}(y|x)\mathrm{d}y\mathrm{d}x. \end{aligned}\tag{9.152}$$

Observe that since $f_X(x) \geq 0$, minimization of ξ_{MSE} is equivalent to minimizing the inner integral for each value x. As a result, the estimation problem reduces to

$$h_{\text{MSE}}(x) = \arg\min_{h(\cdot)} \int_{-\infty}^{\infty}[y-h(x)]^2 f_{Y|X}(y|x)\mathrm{d}y \tag{9.153}$$

for each value x. Differentiating the integral with respect to $h(x)$ and setting the result equal to zero gives

$$-2\int_{-\infty}^{\infty} y f_{Y|X}(y|x)\mathrm{d}y + 2h_{\text{MSE}}(x)\int_{-\infty}^{\infty} f_{Y|X}(y|x)\mathrm{d}y = 0, \tag{9.154}$$

and since the second integral is 1 (because it is a valid pdf with respect to y), we obtain

$$h_{\text{MSE}}(x) = \int_{-\infty}^{\infty} y f_{Y|X}(y|x)\mathrm{d}y = \mathcal{E}[Y|X=x]. \tag{9.155}$$

The MSE estimator is the *conditional mean* of Y given $X = x$. Since this expression holds for all values of X, and because X is random, the MSE estimator can be written as

$$h_{\text{MSE}}(X) = \mathcal{E}[Y|X], \tag{9.156}$$

which itself is a random variable because it is a function of X.

The MSE estimation problem has the following properties.

- *Minimum.* Differentiating (9.154) with respect to $h_{\text{MS}}(x)$ gives

$$\frac{\partial^2 \xi_{\text{MSE}}}{\partial h^2(x)} = 2. \tag{9.157}$$

  This means that ξ_{MSE} is convex so that $h_{\text{MSE}}(x)$ corresponds to the minimum of ξ_{MSE} (not the maximum).
- *Independence.* If X and Y are independent, then $\mathcal{E}[Y|X] = \mathcal{E}[Y]$. Since X provides no information about Y, the MSE estimator reduces to the mean of Y.
- *Minimum MSE.* If $h_{\text{MSE}}(x)$ is substituted into (9.151), the corresponding minimum MSE (MMSE) is

$$\xi_{\text{MSE,min}} = \int_{-\infty}^{\infty} f_X(x)\int_{-\infty}^{\infty}(y-\mathcal{E}[Y|x])^2 f_{Y|X}(y|x)\mathrm{d}y\mathrm{d}x, \tag{9.158}$$

$$= \int_{-\infty}^{\infty} \text{var}(Y|x) f_X(x)\mathrm{d}x, \tag{9.159}$$

  which is the *conditional variance* of Y averaged over all values of X.

- *Bayes' reformulation*. In many problems, it is not easy to compute $f_{Y|X}(y|x)$ directly. In such cases, it is advantageous to rewrite the conditional pdf using Bayes' formula as follows:

$$h_{\text{MSE}}(x) = \frac{\int_{-\infty}^{\infty} y f_{X|Y}(x|y) f_Y(y)\mathrm{d}y}{f_X(x)},$$

$$= \frac{\int_{-\infty}^{\infty} y f_{X|Y}(x|y) f_Y(y)\mathrm{d}y}{\int_{-\infty}^{\infty} f_{X|Y}(x|y) f_Y(y)\mathrm{d}y}. \tag{9.160}$$

This form is usually easier to evaluate because the conditioning is done on the nonmeasurable random variable. In the previous additive noise model, for example, it is straightforward to find $f_{X|Y}(x|y)$ as demonstrated in Example 9.24.

Example 9.24. Let $X = Y + V$ and assume the following pdfs:

$$f_Y(x) = (1/2)[\delta(y-1) + \delta(y+1)], \tag{9.161}$$

$$f_V(v) = \frac{1}{\sqrt{2\pi\sigma_V^2}} \exp(-v^2/2\sigma_V^2), \tag{9.162}$$

so that X is the sum of a symmetric Bernoulli random variable (with outcomes ± 1) and a Gaussian random variable with parameters $\{\mu = 0, \sigma_V\}$. From the result in (9.148), we know that

$$f_{X|Y}(x|y) = f_V(x-y) = \frac{1}{\sqrt{2\pi\sigma_V^2}} \exp(-(x-y)^2/2\sigma_V^2). \tag{9.163}$$

Plots of this conditional pdf for $y = \pm 1$ and $f_X(x)$ are shown in Figure 9.5. A three-dimensional plot of the conditional pdf for a range of values of y is shown in Figure 9.6(a), which will be useful when

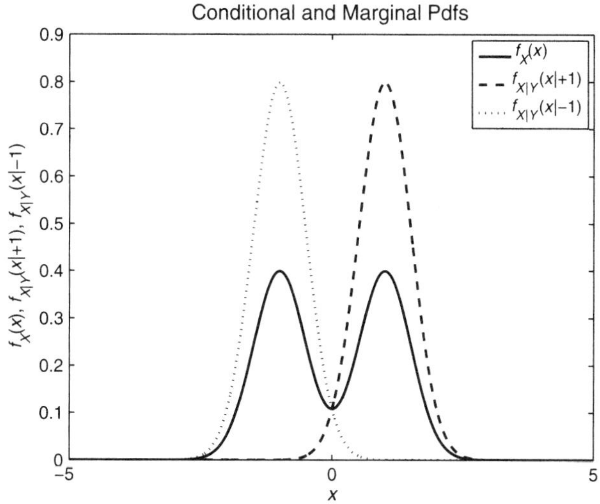

FIGURE 9.5 Conditional and marginal pdfs for Example 9.24.

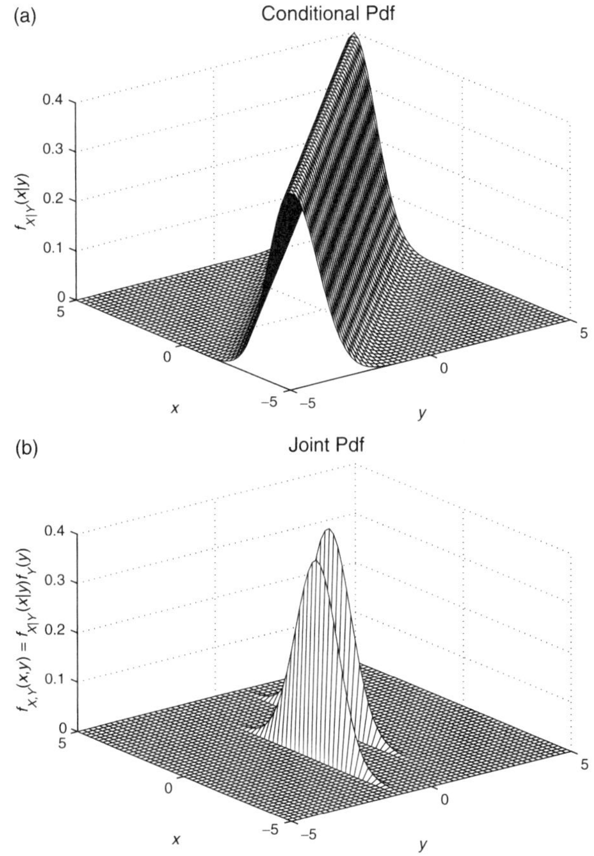

FIGURE 9.6 Pdfs for Example 9.24. (a) $f_{X|Y}(x|y)$ for a range of values of y. (b) $f_{X,Y}(x, y) = f_{X|Y}(x|y)f_Y(y)$ showing that only two "slices" of the plot in part (a) are retained by the Dirac delta functions at $y = \pm 1$.

the joint pdf $f_{X,Y}(x, y)$ is examined. For the numerator of the conditional mean estimator in (9.160), we can write

$$f_{X|Y}(x|y)f_Y(y) = \frac{1}{2\sqrt{2\pi\sigma_V^2}}\left[\exp(-(x-1)^2/2\sigma_V^2)\delta(y-1) + \exp(-(x+1)^2/2\sigma_V^2)\delta(y+1)\right], \quad (9.164)$$

where the sampling property of the Dirac delta function has been used. This expression is the joint pdf $f_{X,Y}(x, y)$ illustrated in Figure 9.6(b). Note in the figure that technically there should be a continuum of Dirac delta functions as x is varied and whose areas are given by the values of the Gaussian components in (9.164).

Integrating the numerator of (9.160) gives

$$\begin{aligned}\int_{-\infty}^{\infty} y f_{X|Y}(x|y)f_Y(y)\mathrm{d}y &= \frac{1}{2\sqrt{2\pi\sigma_V^2}}\left[\exp(-(x-1)^2/2\sigma_V^2) - \exp(-(x+1)^2/2\sigma_V^2)\right], \\ &= \frac{1}{2\sqrt{2\pi\sigma_V^2}}\exp(-(x^2+1)/2\sigma_V^2)\left[\exp(x/\sigma_V^2) - \exp(x/\sigma_V^2)\right], \\ &= \frac{1}{\sqrt{2\pi\sigma_V^2}}\exp(-(x^2+1)/2\sigma_V^2)\sinh(x/\sigma_V^2), \end{aligned} \quad (9.165)$$

where we have used the convolution property of the Dirac delta function. Similarly, the denominator of (9.160) is

$$\begin{aligned}\int_{-\infty}^{\infty} f_{X|Y}(x|y)f_Y(y)\mathrm{d}y &= \frac{1}{2\sqrt{2\pi\sigma_V^2}}\left[\exp(-(x-1)^2/2\sigma_V^2)+\exp(-(x+1)^2/2\sigma_V^2)\right],\\ &= \frac{1}{2\sqrt{2\pi\sigma_V^2}}\exp(-(x^2+1)/2\sigma_V^2)\left[\exp(x/\sigma_V^2)+\exp(-x/\sigma_V^2)\right],\\ &= \frac{1}{\sqrt{2\pi\sigma_V^2}}\exp(-(x^2+1)/2\sigma_V^2)\cosh(x/\sigma_V^2). \end{aligned} \tag{9.166}$$

Thus, the MSE estimator simplifies to

$$h_{\text{MSE}}(X) = \tanh(X/\sigma_V^2). \tag{9.167}$$

Examples of this result are provided in Figure 9.7(a) for different signal-to-noise ratios (SNRs) defined by

$$\text{SNR} \triangleq \sigma_Y^2/\sigma_V^2 = 1/\sigma_V^2. \tag{9.168}$$

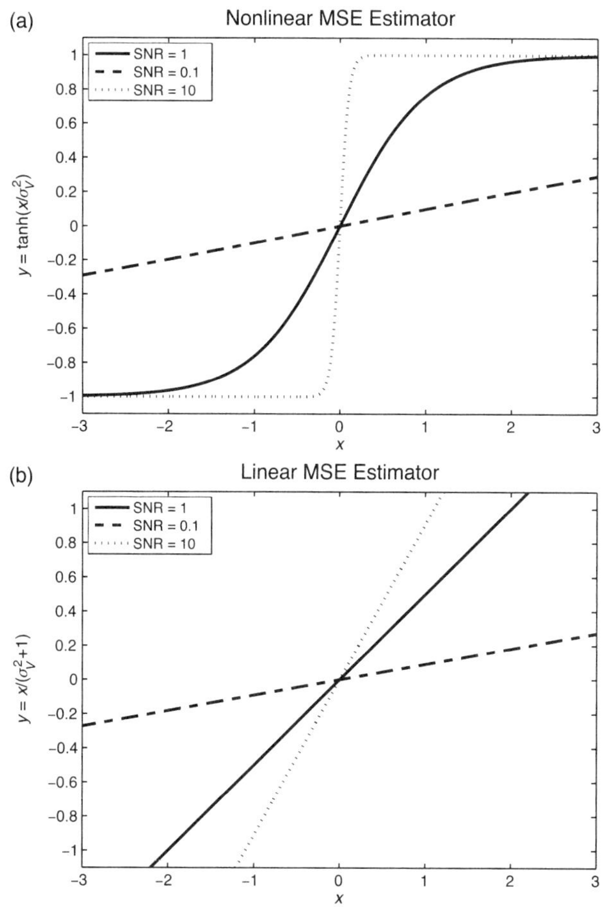

FIGURE 9.7 Examples of MSE estimators for different SNRs. (a) Nonlinear estimator of Example 9.24. (b) Linear estimator of Example 9.25.

Observe in the limit as SNR $\to \infty$, corresponding to $\sigma_V^2 \to 0$, the MSE estimator becomes a signum function:

$$\lim_{\text{SNR}\to\infty} \tanh(X/\sigma_V^2) = \text{sgn}(X). \tag{9.169}$$

In the context of a communication system where Y is the transmitted "signal" and V is additive noise, the MSE estimator can be viewed as a soft-decision detector where its output is a continuum of points on $[-1, 1]$. With increasing SNR, the hyperbolic tangent appears more like the signum function, approaching a hard-decision detector in the limit (with output values ± 1). The signum function is the optimum detector for X in additive white Gaussian noise (see Chapter 10).

Example 9.25. Consider the special case where the estimator in Example 9.24 is restricted to have the following simple linear form:

$$h_{\text{MSE}}(X) = aX, \tag{9.170}$$

where a is the parameter to be estimated. The MSE is

$$\xi_{\text{MSE}} = \mathcal{E}[(Y - aX)^2] \tag{9.171}$$

with minimum achieved when

$$a = \frac{\mathcal{E}[XY]}{\mathcal{E}[X^2]} = \frac{\mathcal{E}[Y^2]}{\mathcal{E}[Y^2] + \mathcal{E}[V^2]} = \frac{1}{1+\sigma_V^2}. \tag{9.172}$$

The *linear* MSE estimator is

$$h_{\text{MSE}}(X) = \frac{X}{\sigma_V^2 + 1}, \tag{9.173}$$

which is plotted in Figure 9.7(b) for different values of the SNR $= 1/\sigma_V^2$. Observe that as the SNR increases, the slope of the line approaches 1. This is intuitively correct because as $\sigma_V^2 \to 0$, the variance of the pdf of X given $Y = \pm 1$ is increasingly smaller about $X = 1$ (or $X = -1$), meaning that with greater probability the correct value of Y is estimated. On the other hand, when σ_V^2 is large, the estimator compensates for this greater uncertainty in X with a smaller slope value; in effect, a wide range of values of X are compressed to estimate Y. This occurs because the conditional pdfs of X about ± 1 have greater variance. It is interesting that for lower SNR values the previous nonlinear estimator approaches the form of this linear estimator as shown in Figure 9.7(a).

In Example 9.25, the estimator is not the conditional-mean estimator because a *constraint* was imposed on the functional form of $h(X)$. In Chapter 11, we consider a generalization of this linear estimator to a linear *filter* operating on a random process.

Example 9.26. Let X and Y be jointly Gaussian with joint pdf

$$\begin{aligned} f_{X,Y}(x, y) = {} & \frac{1}{2\pi\sigma_X\sigma_Y\sqrt{1-\rho^2}} \exp(-[(x-\mu_X)^2/\sigma_X^2 \\ & - 2\rho(x-\mu_X)(y-\mu_Y)/\sigma_X\sigma_Y \\ & + (y-\mu_Y)^2/\sigma_Y^2]/2(1-\rho^2)), \end{aligned} \tag{9.174}$$

where

$$\rho \triangleq \frac{\mathcal{E}[(X-\mu_X)(Y-\mu_Y)]}{\sigma_X\sigma_Y} = \frac{C_{XY}}{\sigma_X\sigma_Y} \tag{9.175}$$

is the correlation coefficient. In order to derive the MSE estimator, expressions for $f_{Y|X}(y|x)$ and $f_X(x)$ are needed. The marginal pdf of X is readily obtained as

$$f_X(x) = \frac{1}{\sqrt{2\pi\sigma_X^2}}\exp(-(x-\mu_X)^2/2\sigma_X^2), \tag{9.176}$$

and the conditional pdf $f_{Y|X}(y|x) = f_{X,Y}(x,y)/f_X(x)$ is

$$\begin{aligned} f_{Y|X}(y|x) = & \frac{1}{\sqrt{2\pi}\sigma_Y\sqrt{1-\rho^2}}\exp(-[(y-\mu_Y)^2/\sigma_Y^2 \\ & - 2\rho(y-\mu_Y)(x-\mu_X)/\sigma_X\sigma_Y \\ & + (x-\mu_X)^2/\sigma_X^2]/2(1-\rho^2) + (x-\mu_X)^2/2\sigma_X^2). \end{aligned} \tag{9.177}$$

As expected, the conditional pdf is also Gaussian, which can be seen by rearranging the exponent as follows:

$$\begin{aligned} f_{Y|X}(y|x) = & \frac{1}{\sqrt{2\pi\sigma_Y^2(1-\rho^2)}}\exp(-[(y-\mu_Y)^2/\sigma_Y^2 \\ & - 2\rho(y-\mu_Y)(x-\mu_X)/\sigma_X\sigma_Y + (x-\mu_X)^2/\sigma_X^2 \\ & - (1-\rho^2)(x-\mu_X)^2/\sigma_X^2]/2(1-\rho^2)), \\ = & \frac{1}{\sqrt{2\pi\sigma_Y^2(1-\rho^2)}}\exp(-[(y-\mu_Y)^2 - 2\rho\sigma_Y(x-\mu_X)(y-\mu_Y)/\sigma_X \\ & + \rho^2\sigma_Y^2(x-\mu_X)^2/\sigma_X^2]/2\sigma_Y^2(1-\rho^2)). \end{aligned} \tag{9.178}$$

This simplifies to

$$f_{Y|X}(y|x) = \frac{1}{\sqrt{2\pi\sigma_Y^2(1-\rho^2)}}\exp(-[y-\mu_Y-\sigma_Y\rho(x-\mu_X)/\sigma_X]^2/2\sigma_Y^2(1-\rho^2)). \tag{9.179}$$

(Note that a similar result was derived in Chapter 4 for zero-mean random variables.) The conditional mean estimator is the mean of (9.179) with random variable X replacing x:

$$h_{\text{MSE}}(X) = \mu_Y + \rho(\sigma_Y/\sigma_X)(X-\mu_X). \tag{9.180}$$

The conditional variance and MMSE is the variance of (9.179):

$$\xi_{\text{MSE,min}} = \sigma_Y^2(1-\rho^2). \tag{9.181}$$

Thus, when the random variables are jointly Gaussian, the MSE estimator is an *affine* function of X. Consider the following special cases. (i) If both random variables have zero mean, then $h_{\text{MSE}}(X) = \rho(\sigma_Y/\sigma_X)X$ is exactly linear. (ii) If the random variables are uncorrelated with $\rho = 0$, then $h_{\text{MS}}(X) = \mu_Y$ is independent of X, and the MMSE is σ_Y^2.

9.12 MEAN-ABSOLUTE-ERROR ESTIMATION

Consider correlated random variables X and Y with joint pdf $f_{X,Y}(x, y)$ of which X is measurable. The mean-absolute-error (MAE) risk function is

$$\xi_{\text{MAE}} = \mathcal{E}[|Y - h(X)|], \tag{9.182}$$

which is minimized by again recognizing that we need only examine the inner integral of the following expression:

$$\begin{aligned}\xi_{\text{MAE}} &= \int_{-\infty}^{\infty}\int_{-\infty}^{\infty} |y - h(x)| f_{X,Y}(x, y)\mathrm{d}x\mathrm{d}y, \\ &= \int_{-\infty}^{\infty} f_X(x) \int_{-\infty}^{\infty} |y - h(x)| f_{Y|X}(y|x)\mathrm{d}x\mathrm{d}y.\end{aligned} \tag{9.183}$$

The absolute value is handled by splitting the integral into the sum of two integrals so that the estimator is expressed as follows:

$$\begin{aligned}h_{\text{MAE}}(x) &= \arg\min_{h(x)} \int_{-\infty}^{\infty} |y - h(x)| f_{Y|X}(y|x)\mathrm{d}y, \\ &= \arg\min_{h(x)} \left[\int_{-\infty}^{h(x)} (h(x) - y) f_{Y|X}(y|x)\mathrm{d}y \right. \\ &\quad \left. + \int_{h(x)}^{\infty} (y - h(x)) f_{Y|X}(y|x)\mathrm{d}y\right].\end{aligned} \tag{9.184}$$

The minimum is achieved by differentiating the integrals with respect to $h(x)$ and applying Leibniz's integral rule (see Appendix E):

$$\frac{\partial}{\partial h(x)} \int_{-\infty}^{h(x)} (h(x) - y) f_{Y|X}(y|x)\mathrm{d}y = \int_{-\infty}^{h(x)} f_{Y|X}(y|x)\mathrm{d}y, \tag{9.185}$$

$$\frac{\partial}{\partial h(x)} \int_{h(x)}^{\infty} (y - h(x)) f_{Y|X}(y|x)\mathrm{d}y = \int_{h(x)}^{\infty} -f_{Y|X}(y|x)\mathrm{d}y, \tag{9.186}$$

where, of course, $f_{Y|X}(y|x)$ does not depend on $h(x)$. Setting the sum of the right-hand sides equal to zero, we recognize that the function minimizing the MAE is the *conditional median*:

$$\int_{-\infty}^{h_{\text{MAE}}(x)} f_{Y|X}(y|x)\mathrm{d}y = \int_{h_{\text{MAE}}(x)}^{\infty} f_{Y|X}(y|x)\mathrm{d}y, \tag{9.187}$$

which means

$$P(Y \leq h_{\text{MAE}}(x)|X = x) = P(Y \geq h_{\text{MAE}}(x)|X = x) = 1/2. \tag{9.188}$$

As was the case for the MSE estimator, it is not always easy to find $f_{Y|X}(y|x)$; the MAE optimality condition can be rewritten using Bayes' formula as follows:

$$\int_{-\infty}^{h_{\text{MAE}}(x)} f_{X|Y}(x|y) f_Y(y)\mathrm{d}y = \int_{h_{\text{MAE}}(x)}^{\infty} f_{X|Y}(x|y) f_Y(y)\mathrm{d}y, \tag{9.189}$$

where $f_X(x)$ has canceled on both sides because the integrals are performed only over y. This result is also obtained by integrating the joint pdf over y:

$$\int_{-\infty}^{h_{\mathrm{MAE}}(x)} f_{X,Y}(x, y)\mathrm{d}y = \int_{h_{\mathrm{MAE}}(x)}^{\infty} f_{X,Y}(x, y)\mathrm{d}y. \tag{9.190}$$

Example 9.27. For the additive noise model in (9.144) where Y is symmetric Bernoulli and V is Gaussian as in Example 9.24, substituting (9.164) into (9.189) yields

$$\begin{aligned}&\int_{-\infty}^{h(x)} [\exp(-(x-1)^2/2\sigma_V^2)\delta(y-1) + \exp(-(x+1)^2/2\sigma_V^2)\delta(y+1)]\mathrm{d}y,\\ &= \int_{h(x)}^{\infty} [\exp(-(x-1)^2/2\sigma_V^2)\delta(y-1) + \exp(-(x+1)^2/2\sigma_V^2)\delta(y+1)]\mathrm{d}y,\end{aligned} \tag{9.191}$$

where common terms have cancelled. The integrands are identical, consisting of two Dirac delta functions located at $y = \pm 1$. It is clear that both integrals are nonzero only when $h(x) \in [-1, 1]$ such that the left-hand side includes the delta function at $y = -1$, and the right-hand side includes the delta function at $y = 1$. Thus, (9.191) yields

$$\exp(-(x+1)^2/2\sigma_V^2) = \exp(-(x-1)^2/2\sigma_V^2), \tag{9.192}$$

such that equality is not achieved in (9.189) except for $x = 0$ (which has probability zero of occurring). The median for discrete random variables is chosen as an integer outcome (see Chapter 5), so (5.187) and (9.191) imply

$$h_{\mathrm{MAE}}(X) = \mathrm{sgn}(X). \tag{9.193}$$

This result is also obtained in Example 9.24 when the SNR $\to \infty$, and it is the maximum likelihood estimator discussed later. This example is considered further in Problem 9.15.

9.13 ORTHOGONALITY CONDITION

Next, we demonstrate an important property of the conditional-mean estimator. Orthogonality was also discussed in Chapter 5 where various properties of conditional expectation were described.

Theorem 9.8. The conditional mean estimator $h_{\mathrm{MSE}}(X)$ has the following orthogonality property for all functions $g(X)$:

$$\mathcal{E}[(Y - h_{\mathrm{MSE}}(X))g(X)] = 0. \tag{9.194}$$

Proof. Conditioning on X in nested expectations gives

$$\begin{aligned}\mathcal{E}[(Y - h_{\mathrm{MSE}}(X))g(X)] &= \mathcal{E}[\mathcal{E}[(Y - h_{\mathrm{MSE}}(X))g(X)|X]],\\ &= \mathcal{E}[\mathcal{E}[(Y - h_{\mathrm{MSE}}(X))|X]g(X)],\\ &= \mathcal{E}[(\mathcal{E}[Y|X] - h_{\mathrm{MSE}}(X))g(X)].\end{aligned} \tag{9.195}$$

In the second line, the conditioning allows us to bring $g(X)$ outside the inner expectation, and in the last line we have used the fact that $\mathcal{E}[h_{\text{MSE}}(X)|X] = h_{\text{MSE}}(X)$ because the estimator is a function of X. Since $\mathcal{E}[Y|X] = h_{\text{MSE}}(X)$, zero is obtained in the last expression, which completes the proof. □

The orthogonality condition can be used to derive the optimal estimator, and thus it is necessary and sufficient. Assuming that (9.194) holds, then

$$\mathcal{E}[h_{\text{MSE}}(X)g(X)] = \mathcal{E}[Yg(X)] \tag{9.196}$$

for all $g(X)$. Conditioning the right-hand side on X in a nested expectation gives

$$\mathcal{E}[h_{\text{MSE}}(X)g(X)] = \mathcal{E}[\mathcal{E}[Yg(X)|X]] = \mathcal{E}[\mathcal{E}[Y|X]g(X)]. \tag{9.197}$$

Since equality must hold for all functions $g(X)$, this can happen only if $h_{\text{MSE}}(X) = \mathcal{E}[Y|X]$. This approach based on the orthogonality condition was also used in Chapter 5 to prove the equivalence of quantities appearing under expectations.

Suppose that instead of X for the MSE estimator we have access only to a function $g(X)$ of the sample. We would like to explore how $\mathcal{E}[Y|X]$ and $\mathcal{E}[Y|g(X)]$ are related. First, condition on X in a nested expectation:

$$\mathcal{E}[Yg(X)] = \mathcal{E}[\mathcal{E}[Yg(X)|X]] = \mathcal{E}[\mathcal{E}[Y|X]g(X)], \tag{9.198}$$

where the outer expectation is over the pdf of X. Next, condition instead on $g(X)$:

$$\mathcal{E}[Yg(X)] = \mathcal{E}[\mathcal{E}[Yg(X)|g(X)]] = \mathcal{E}[\mathcal{E}[Y|g(X)]g(X)], \tag{9.199}$$

where the outer expectation is over the pdf of $g(X)$. We know from Chapter 5 that for $Z = g(X)$:

$$\int_{-\infty}^{\infty} g(x) f_X(x) \mathrm{d}x = \int_{-\infty}^{\infty} z f_Z(z) \mathrm{d}z, \tag{9.200}$$

so the inner expectations of (9.198) and (9.199) must be the same:

$$\mathcal{E}[Y|X] = \mathcal{E}[Y|g(X)]. \tag{9.201}$$

The conditional mean estimator is the same when conditioning on X or on some function $g(X)$. This result is a form of the *smoothing property* of conditional expectation described in Chapter 5.

We can also show that $\mathcal{E}[Y|g(X)]$ is identical to $\mathcal{E}[Y|X]$ using a transformation technique from Chapter 4. Assume that $g(X)$ is monotonic increasing:

$$\begin{aligned} \mathcal{E}[Y|g(X)] &= \int_{-\infty}^{\infty} y f_{Y|g(X)}(y|g(x)) \mathrm{d}y, \\ &= \frac{1}{f_{g(X)}(g(x))} \int_{-\infty}^{\infty} y f_{g(X)|Y}(g(x)|y) f_Y(y) \mathrm{d}y, \end{aligned} \tag{9.202}$$

where Bayes' formula has been used. Substituting the following expression:

$$f_{g(X)|Y}(g(x)|y) = f_{X|Y}(g^{-1}(g(x))|y) \left| \frac{\mathrm{d}x}{\mathrm{d}g(x)} \right| = \frac{1}{g'(x)} f_{X|Y}(x|y) \tag{9.203}$$

yields

$$\mathcal{E}[Y|g(X)] = \frac{1}{f_{g(X)}(g(x))g'(x)} \int_{-\infty}^{\infty} y f_{X|Y}(x|y) f_Y(y) \mathrm{d}y. \tag{9.204}$$

Rearranging this result using Bayes' formula again gives

$$\begin{aligned}\mathcal{E}[Y|g(X)] &= \frac{1}{f_{g(X)}(g(x))g'(x)} \int_{-\infty}^{\infty} y f_{Y|X}(y|x) f_X(x) \mathrm{d}y, \\ &= \frac{f_X(x)}{f_{g(X)}(g(x))g'(x)} \mathcal{E}[Y|X].\end{aligned} \tag{9.205}$$

The term multiplying $\mathcal{E}[Y|X]$ is similar to that in (9.203), except without the conditioning as follows:

$$f_{g(X)}(g(x)) = f_X(x)/g'(x) \Longrightarrow \frac{f_X(x)}{f_{g(X)}(g(x))g'(x)} = 1, \tag{9.206}$$

which shows that $\mathcal{E}[Y|g(X)] = \mathcal{E}[Y|X]$. This result also holds for general transformations $g(X)$, but more work is needed as discussed in Chapter 4 when the function is not monotonic.

9.14 PROPERTIES OF ESTIMATORS

In this section, we describe several important properties of estimators. Since an estimator is a function of the samples (which are random variables), it is also a random variable (or a random vector in the multidimensional case) and thus has a nontrivial distribution. It is important to note that the distribution is generally not fixed; if more samples are included in the estimator, then various features of the distribution of the estimator (such as the variance) could change. Typically, it is desirable that the distribution of the estimator become more concentrated about the parameter θ with increasing N.

Example 9.28. Let $\{X_n\}$ be iid Gaussian random variables with unknown mean μ and known σ. Recall that the sample mean $\bar{X}$ is a minimal sufficient statistic for μ. It is straightforward to show that $\bar{X}$ is also Gaussian with parameters $\{\mu, \sigma/\sqrt{N}\}$. Examples of the pdf are shown in Figure 9.8, which illustrate how the estimator becomes more accurate as N is increased (the width of the pdf decreases).

For the following properties of an estimator, assume the parameter θ (or parameter vector $\boldsymbol{\theta}$) is nonrandom. We use the notation $\hat{\theta}$ for the estimator of θ to denote that it is a random variable (instead of using uppercase T as in previous sections).

9.14.1 Unbiased

The first property describes the location of the estimator relative to θ.

Definition: Unbiased An estimator $\hat{\theta}$ of θ is *unbiased* if $\mathcal{E}[\hat{\theta}] = \theta$.

Unbiasedness is an important attribute of an estimator: it is usually desirable that on average the estimator has the same value as the parameter. However, this is not necessarily the case because it might be better to use a biased estimator that has lower variance, provided the bias is not too large.

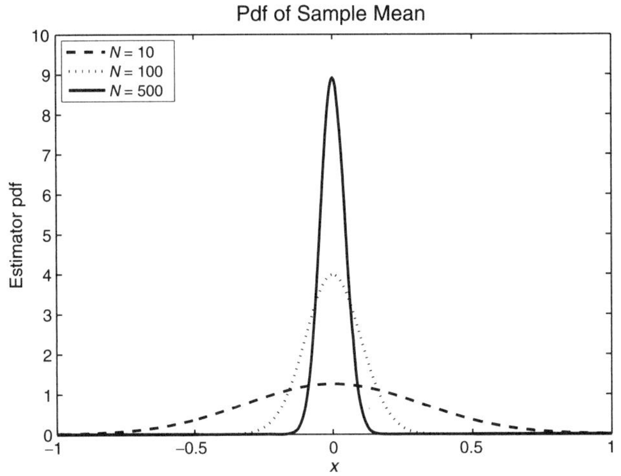

FIGURE 9.8 Pdf of the sample mean $\bar{X}$ in Example 9.28 as the number of samples N is increased.

Definition: Bias *Bias* $B(\hat{\theta})$ of estimator $\hat{\theta}$ is $B(\hat{\theta}) \triangleq \mathcal{E}[\hat{\theta}] - \theta$.

Obviously, when the estimator is unbiased, $B(\hat{\theta}) = 0$.

Example 9.29. For iid $\{X_n\}$ with mean μ_X, $\bar{X}$ is unbiased:

$$\mathcal{E}[\bar{X}] = \frac{1}{N}\sum_{n=1}^{N}\mathcal{E}[X_n] = \frac{1}{N}\sum_{n=1}^{N}\mu_X = \mu_X. \tag{9.207}$$

This result is also evident for the example pdfs in Figure 9.8. Suppose now that we want an unbiased estimator for the variance $\theta = \sigma_X^2$. Consider two cases: (i) μ_X known and (ii) μ_X unknown. For both cases, a sum of the samples is used, but they have different forms:

$$\hat{\theta} = \alpha\sum_{n=1}^{N}(X_n - \mu_X)^2 \quad (\mu_X \text{ known}), \tag{9.208}$$

$$\hat{\theta} = \alpha\sum_{n=1}^{N}(X_n - \bar{X})^2 \quad (\mu_X \text{ unknown}), \tag{9.209}$$

where the sample mean is used in case (ii), and α is a coefficient to be specified so that each estimator is unbiased. For case (i), since the samples are iid,

$$\begin{aligned}\mathcal{E}[\hat{\theta}] &= \alpha\sum_{n=1}^{N}\mathcal{E}[X_n^2 - 2\mu_X X_n + \mu_X^2], \\ &= \alpha N\left[\sigma_X^2 + \mu_X^2 - 2\mu_X^2 + \mu_X^2\right] = \alpha N\sigma_X^2,\end{aligned} \tag{9.210}$$

which shows that $\hat{\theta}$ is unbiased if $\alpha = 1/N$. Case (ii) is more complicated to evaluate because of the sample mean $\bar{X}$:

$$\begin{aligned}\mathcal{E}[\hat{\theta}] &= \alpha \sum_{n=1}^{N} \mathcal{E}[X_n^2 - 2X_n\bar{X} + \bar{X}^2], \\ &= \alpha \left[N(\sigma_X^2 + \mu_X^2) - 2\sum_{n=1}^{N} \mathcal{E}[X_n\bar{X}] + N\mathcal{E}[\bar{X}^2] \right]. \end{aligned} \tag{9.211}$$

Since the samples are iid,

$$\mathcal{E}[X_m X_n] = \begin{cases} \mathcal{E}[X_n^2] = \sigma_X^2 + \mu_X^2, & m = n \\ \mathcal{E}[X_m]\mathcal{E}[X_n] = \mu_X^2, & m \neq n, \end{cases} \tag{9.212}$$

so that

$$\sum_{n=1}^{N} \mathcal{E}[X_n\bar{X}] = (1/N)\left[N(\sigma_X^2 + \mu_X^2) + (N^2 - N)\mu_X^2\right]. \tag{9.213}$$

Likewise

$$\mathcal{E}[\bar{X}^2] = \frac{1}{N^2}\sum_{m=1}^{N}\sum_{n=1}^{N} \mathcal{E}[X_m X_n] = (1/N^2)\left[N(\sigma_X^2 + \mu_X^2) + (N^2 - N)\mu_X^2\right]. \tag{9.214}$$

Combining all terms in (9.211) gives

$$\mathcal{E}[\hat{\theta}] = \alpha\left[N(\sigma_X^2 + \mu_X^2) - (1/N)[N(\sigma_X^2 + \mu_X^2) + (N^2 - N)\mu_X^2]\right] = \alpha(N-1)\sigma_X^2. \tag{9.215}$$

When μ_X is unknown, the estimator for σ_X^2 is unbiased if $\alpha = 1/(N-1)$; this gives the sample variance S^2 that was discussed in previous chapters. Note that a Gaussian assumption was not required; the previous results for the sample mean are general (for iid samples) and do not depend on the specific distribution (assuming the mean exists and the variance is finite).

9.14.2 Consistent

Next, we want to determine if an estimator *converges* in some sense to θ as the number of samples N is increased; such an estimator is called consistent.

Definition: Weakly Consistent Estimator $\hat{\theta}(N)$ is *weakly consistent* if it converges in probability:

$$\lim_{N\to\infty} P(|\hat{\theta} - \theta| \geq \epsilon) = 0 \tag{9.216}$$

for every $\epsilon > 0$.

Although consistency implies that the pdf of $\hat{\theta}$ becomes increasingly concentrated about θ, it does not directly address the variance of the estimator and whether or not it tends to zero. Consistency can also be described in terms of almost sure convergence.

Definition: Strongly Consistent Estimator $\hat{\theta}$ is *strongly consistent* if it converges almost surely:

$$P\left(\lim_{N\to\infty} \hat{\theta} = \theta\right) = 1. \tag{9.217}$$

If an estimator is strongly consistent, then it is also weakly consistent because almost sure convergence implies convergence in probability, as discussed in Chapter 7.

Example 9.30. Consider case (i) of Example 9.29 where $\bar{X}$ is used to estimate the mean μ_X of a Gaussian distribution with known variance σ_X^2. Since $\bar{X}$ has mean μ_X and variance σ_X^2/N:

$$P(|\bar{X} - \mu_X| \geq \epsilon) = P(|\bar{Z}| \geq \epsilon) = 2\int_{\epsilon}^{\infty} \frac{1}{\sqrt{2\pi\sigma_X^2/N}} \exp(-z^2/2(\sigma_X^2/N))\mathrm{d}z, \tag{9.218}$$

where $\bar{Z} \triangleq \bar{X} - \mu_X$ is zero-mean Gaussian with variance σ_X^2/N, and we have used the fact that its pdf is even. Changing variables according to $y \triangleq z/(\sigma_X/\sqrt{N})$ gives

$$P(|\bar{X} - \mu_X| \geq \epsilon) = 2\int_{\sqrt{N}\epsilon/\sigma_X}^{\infty} \frac{1}{\sqrt{2\pi}} \exp(-y^2/2)\mathrm{d}y = 2Q(\sqrt{N}\epsilon/\sigma_X), \tag{9.219}$$

where $Q(\cdot)$ is the Q-function. It is clear that for every $\epsilon > 0$ as $N \to \infty$, the probability above tends to zero so that $\bar{X}$ is weakly consistent. Of course, this result is a restatement of the *weak law of large numbers* discussed in Chapter 7. The probability above is the area under the tails of the pdf in Figure 9.8 for the case of $\mu_X = 0$, and it is clear from the plots that it decreases with increasing N for a fixed ϵ.

We can show that the estimator is weakly consistent also by using Chebyshev's inequality (see Appendix F):

$$P(|\bar{X} - \mu_X| \geq \epsilon) \leq \frac{\sigma_X^2}{N\epsilon^2}, \tag{9.220}$$

For positive ϵ, the right-hand side tends to zero as $N \to \infty$. It turns out that $\bar{X}$ is also strongly consistent corresponding to the *strong law of large numbers* (see Chapter 7), but it is more complicated to derive.

Example 9.31. For case (i) in Example 9.29, the variance of an unbiased estimator for σ_X^2 is derived next assuming that μ_X is known. Since

$$\mathcal{E}[(\hat{\theta} - \theta)^2] = \mathcal{E}[\hat{\theta}^2] - \theta^2 \tag{9.221}$$

and $\hat{\theta}$ is unbiased, only $\mathcal{E}[\hat{\theta}^2]$ needs to be computed. Thus

$$\mathcal{E}[\hat{\theta}^2] = \frac{1}{N^2}\mathcal{E}\left[\left(\sum_{n=1}^{N}(X_n - \mu_X)^2\right)^2\right] = \frac{1}{N^2}\mathcal{E}\left[\left(\sum_{n=1}^{N} Y_n^2\right)^2\right], \tag{9.222}$$

where $Y_n \triangleq X_n - \mu_X$ has zero mean and variance σ_X^2. Since $\{Y_n\}$ are iid, the expectation simplifies to

$$\mathcal{E}\left[\left(\sum_{n=1}^{N} Y_n^2\right)^2\right] = N\mathcal{E}[Y_n^4] + (N^2 - N)\mathcal{E}^2[Y_n^2]. \tag{9.223}$$

To continue, assume that X_n is Gaussian with parameters $\{\mu_X, \sigma_X\}$ so that Y_n is also Gaussian, which allows us to express a fourth-order central moment in terms of the variance (see Chapter 5):

$$\mathcal{E}[Y_n^4] = 3\sigma_X^4. \tag{9.224}$$

Combining all the results above and substituting $\theta^2 = \sigma_X^4$ yields

$$\mathcal{E}[(\hat{\theta} - \theta)^2] = (1/N^2)\left[3N\sigma_X^4 + (N^2 - N)\sigma_X^4\right] - \sigma_X^4 = 2\sigma_X^4/N, \tag{9.225}$$

which tends to zero as $N \to \infty$.

Example 9.32. For Example 9.31 with Gaussian samples, we return to the general estimator of σ_X^2 with known μ_X given by

$$\hat{\theta} = \alpha \sum_{n=1}^{N} (X_n - \mu_X)^2. \tag{9.226}$$

From the result in (9.225), we have the following variance for (9.226):

$$\mathcal{E}[(\hat{\theta} - \theta)^2] = \alpha^2[2N\sigma_X^4 + N^2\sigma_X^4] - \sigma_X^4 = [\alpha^2 N(N+2) - 1]\sigma_X^4. \tag{9.227}$$

In order to have lower variance than the unbiased estimator used in (9.225), we require

$$\alpha^2 N(N+2) - 1 < 2/N \Longrightarrow \alpha < 1/N. \tag{9.228}$$

However, in order to be a valid estimator, it is also necessary that

$$\alpha^2 N(N+2) - 1 > 0 \Longrightarrow \alpha > \frac{1}{\sqrt{N(N+2)}}. \tag{9.229}$$

This last result precludes us from using $\alpha = 1/(N+1)$. The upper and lower bounds on α versus N are shown in Figure 9.9 along with $\alpha = 1/(N + 1/2)$. We see that some advantage is achieved by using a scale factor other than $\alpha = 1/N$ when N is small, but the gain is modest and probably does not justify using a biased estimator for σ_X^2.

9.14.3 Efficient

The efficiency of an estimator is determined by its variance.

Definition: Efficient An unbiased estimator $\hat{\theta}$ is *efficient* if its variance achieves the Cramér–Rao lower bound.

The Cramér–Rao lower bound (CRLB) is an absolute bound on the covariance matrix of $\boldsymbol{\theta}$ that depends on the parameters of the distribution of $\{X_n\}$ and the number of samples N. It is derived in Appendix F:

$$\text{cov}[\hat{\boldsymbol{\theta}}] \geq \mathcal{I}^{-1}(\boldsymbol{\theta}), \tag{9.230}$$

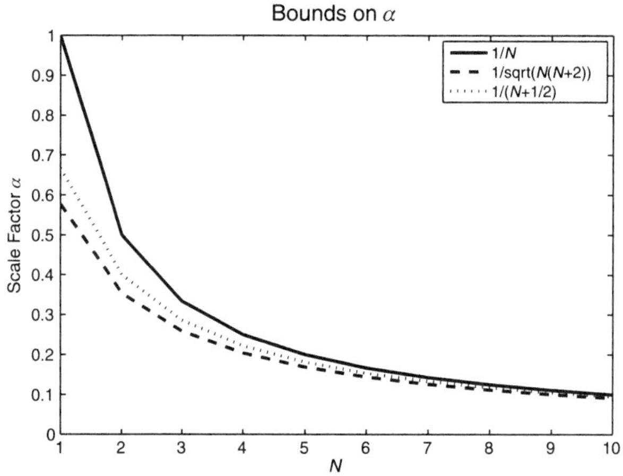

FIGURE 9.9 Values of scale factor α for the sample variance in Example 9.32. (The discrete values have been connected by straight lines for visual clarity.)

where

$$\mathcal{I}(\boldsymbol{\theta}) = \mathcal{E}\left[\left(\frac{\partial}{\partial \boldsymbol{\theta}} \log(f_{\mathbf{X}}(\mathbf{X})\right)\left(\frac{\partial}{\partial \boldsymbol{\theta}} \log(f_{\mathbf{X}}(\mathbf{X})\right)^T\right] \tag{9.231}$$

is the *Fisher information matrix*, $f_{\mathbf{X}}(\mathbf{x})$ is the joint pdf of the samples $\mathbf{X} \triangleq [X_1, \ldots, X_N]^T$, and $\boldsymbol{\theta} \triangleq [\theta_1, \ldots, \theta_M]^T$. Note that the argument of the pdf under the expectation is the random vector $\mathbf{X}$. The inequality above is defined in the positive-semidefinite (PSD) sense as described in Appendix F. Another form of the Fisher information matrix that is often easier to compute is based on the second derivative:

$$\mathcal{I}(\boldsymbol{\theta}) = -\mathcal{E}\left[\frac{\partial^2}{\partial \boldsymbol{\theta}^2} \log(f_{\mathbf{X}}(\mathbf{X}))\right]. \tag{9.232}$$

The logarithm can be any base, but $\ln(f_{\mathbf{X}}(\mathbf{x}))$ is typically used.

Example 9.33. Let $\{X_1, \ldots, X_N\}$ be iid samples of a Poisson random variable with unknown parameter $\theta = \alpha$ and joint pdf

$$f_{\mathbf{X}}(\mathbf{x}) = \frac{\exp(-N\alpha)\alpha^{\sum_{n=1}^N x_n}}{\prod_{n=1}^N x_n!}. \tag{9.233}$$

The logarithm of this expression is

$$\ln(f_{\mathbf{X}}(\mathbf{x})) = -N\alpha + \sum_{n=1}^N x_n \ln(\alpha) - \ln\left(\prod_{n=1}^N x_n!\right). \tag{9.234}$$

Differentiating gives

$$\frac{\partial}{\partial \alpha} \ln(f_{\mathbf{X}}(\mathbf{x})) = -N + \frac{1}{\alpha}\sum_{n=1}^N x_n \tag{9.235}$$

so that the (scalar) Fisher information is

$$\mathcal{I}(\alpha) = \mathcal{E}\left[\left(\frac{\partial}{\partial\alpha}\ln(f_{\mathbf{X}}(\mathbf{X}))\right)^2\right] = \mathcal{E}\left[N^2 - \frac{2N}{\alpha}\sum_{n=1}^{N} X_n + \frac{1}{\alpha^2}\left(\sum_{n=1}^{N} X_n\right)^2\right]. \tag{9.236}$$

Since $\sum_{n=1}^{N} X_n$ is Poisson with parameter $N\alpha$, we have

$$\mathcal{I}(\alpha) = N^2 - \frac{2N}{\alpha}(N\alpha) + \frac{1}{\alpha^2}(N\alpha + N^2\alpha^2) = N/\alpha. \tag{9.237}$$

The CRLB α/N is obtained more directly using the second form of the Fisher information:

$$\mathcal{I}(\alpha) = -\mathcal{E}\left[\frac{\partial^2}{\partial\alpha^2}\log(f_{\mathbf{X}}(\mathbf{X}))\right] = \mathcal{E}\left[\frac{1}{\alpha^2}\sum_{n=1}^{N} X_n\right] = \frac{1}{\alpha^2}(N\alpha) = N/\alpha. \tag{9.238}$$

Consider using the sample mean to estimate α. Since it is unbiased, the variance of $\hat{\alpha}$ is

$$\text{var}[\hat{\alpha}] = \mathcal{E}\left[\left(\bar{X} - \alpha\right)^2\right] = \frac{1}{N^2}\mathcal{E}\left[\left(\sum_{n=1}^{N} X_n - N\alpha\right)^2\right] = \frac{1}{N^2}(N\alpha) = \alpha/N, \tag{9.239}$$

which achieves the CRLB: the sample mean is an efficient estimator for the parameter $\theta = \alpha$ of a Poisson distribution.

9.15 MAXIMUM A POSTERIORI ESTIMATION

Suppose we want to estimate random parameter Θ from N iid samples $\{X_1, \ldots, X_N\}$ assuming that it has known pdf $f_\Theta(\theta)$. Recall that in the context of estimation, $f_\Theta(\theta)$ is called the prior distribution which characterizes Θ before any samples are observed. The posterior distribution $f_{\Theta|X_1,\ldots,X_N}(\theta|x_1, \ldots, x_N)$ characterizes Θ *after* collecting N samples. Using Bayes' formula, the posterior distribution is expressed in terms of the prior distribution as follows:

$$f_{\Theta|\mathbf{X}}(\theta|\mathbf{x}) = \frac{f_{\mathbf{X}|\Theta}(\mathbf{x}|\theta)f_\Theta(\theta)}{f_{\mathbf{X}}(\mathbf{x})}. \tag{9.240}$$

The numerator is the joint pdf of the samples given that $\Theta = \theta$, whereas the denominator is the joint pdf of the samples without any conditioning. Generally, the two terms in the numerator are assumed to be known. The denominator, on the other hand, is not known, but can be derived from the numerator as follows:

$$f_{\mathbf{X}}(\mathbf{x}) = \int_{-\infty}^{\infty} f_{\mathbf{X}|\Theta}(\mathbf{x}|\theta)f_\Theta(\theta)\mathrm{d}\theta. \tag{9.241}$$

The maximum a posteriori (MAP) estimator θ_{MAP} is the value maximizing the posterior distribution on the left-hand side of (9.240). Since the denominator does not depend directly on θ, it does not affect the maximum and can be ignored. The maximization must jointly take into account the two pdfs in the numerator of the right-hand side. Later, we describe the maximum likelihood (ML) estimator which ignores the prior pdf of Θ and instead finds θ_{ML} that maximizes only the conditional pdf on the right-hand side of (9.240) (which is called the *likelihood function*). For the rest of this chapter, a subscript is used to denote the specific type of estimator (instead of the previous notation $\hat{\theta}$).

Definition: MAP Estimator The *MAP estimator* for parameter Θ is the value maximizing the posterior pdf:

$$\theta_{\text{MAP}} = \arg\max_{\theta} f_{\Theta|\mathbf{X}}(\theta|\mathbf{x}). \tag{9.242}$$

Since the denominator of (9.240) does not depend on θ, the MAP criterion can be rewritten as

$$\theta_{\text{MAP}} = \arg\max_{\theta} f_{\mathbf{X}|\Theta}(\mathbf{x}|\theta) f_{\Theta}(\theta) = \arg\max_{\theta} f_{\mathbf{X},\Theta}(\mathbf{x},\theta). \tag{9.243}$$

This latter form is generally preferred because it is often easier to determine the conditional pdf of the samples given Θ. Of course, this expression still requires that we know the prior pdf of Θ, which is necessary in MAP estimation.

Before providing an example, we comment about the notation. In the previous sections on sufficient statistics, $f_{\mathbf{X}}(\mathbf{x};\theta)$ was used to represent the joint pdf of $\{X_n\}$ with parameter θ for which we sought a sufficient statistic. It is equivalent to the notation for the conditional pdf $f_{\mathbf{X}|\Theta}(\mathbf{x}|\theta)$ where, by conditioning on Θ, any randomness of Θ is ignored. However, for the MAP estimator, we need to consider the joint pdf of the samples and the parameter: $f_{\mathbf{X},\Theta}(\mathbf{x},\theta)$.

Example 9.34. Consider a Gaussian random variable with parameters $\{\mu,\sigma\}$ and known σ. Assume that μ is a random variable with the standard Gaussian distribution. For N iid samples $\{X_1,\ldots,X_N\}$, the conditional pdf is

$$f_{\mathbf{X}|\mu}(\mathbf{x}|\mu) = \frac{1}{(2\pi\sigma^2)^{N/2}} \exp\left(-\sum_{n=1}^{N}(x_n-\mu)^2/2\sigma^2\right). \tag{9.244}$$

(Since uppercase μ is identical to the Latin M, we use lowercase μ in the subscript without confusion; it is still a random variable.) Multiplying this expression with the marginal pdf of standard Gaussian μ and integrating gives the joint pdf of the samples:

$$f_{\mathbf{X}}(\mathbf{x}) = \frac{1}{\sqrt{2\pi}(2\pi\sigma^2)^{N/2}} \int_{-\infty}^{\infty} \exp(-\mu^2/2)\exp\left(-\sum_{n=1}^{N}(x_n-\mu)^2/2\sigma^2\right) d\mu. \tag{9.245}$$

Combining the exponentials and expanding the exponent yields

$$\begin{aligned}\mu^2\sigma^2 + \sum_{n=1}^{N}(x_n-\mu)^2 &= \sum_{n=1}^{N}(a\mu^2 - 2\mu x_n + x_n^2),\\ &= aN[\mu^2 - (2\bar{x}/a)\mu + \overline{x^2}/a],\end{aligned} \tag{9.246}$$

where we have ignored the $-1/2\sigma^2$ term for now, $a \triangleq 1+\sigma^2/N$, $\bar{x}$ is a value of the sample mean $\bar{X}$, and $\overline{x^2}$ is a value of the sample second moment:

$$\overline{X^2} \triangleq \frac{1}{N}\sum_{n=1}^{N} X_n^2. \tag{9.247}$$

Since the integration is performed over μ to obtain the joint pdf, we complete the square with respect to μ:

$$\mu^2\sigma^2 + \sum_{n=1}^{N}(x_n-\mu)^2 = aN[(\mu-\bar{x}/a)^2 + \overline{x^2}/a - \bar{x}^2/a^2]. \tag{9.248}$$

Substituting this result into (9.245) gives

$$\begin{aligned} f_{\mathbf{X}}(\mathbf{x}) &= \frac{1}{\sqrt{2\pi}(2\pi\sigma^2)^{N/2}} \exp(-(N/2\sigma^2)(\overline{x^2} - \bar{x}^2/a)) \\ &\quad \cdot \int_{-\infty}^{\infty} \exp(-(aN/2\sigma^2)(\mu_X - \bar{x}/a)^2) \mathrm{d}\mu, \\ &= \frac{\sqrt{\sigma^2/aN}}{(2\pi\sigma^2)^{N/2}} \exp(-(N/2\sigma^2)(\overline{x^2} - \bar{x}^2/a)). \end{aligned} \tag{9.249}$$

The integral equals 1 because the integrand is a Gaussian pdf with mean $\overline{x^2}/a$ and variance σ^2/aN. A plot for $N = 2$ and $\sigma = 1$ is shown in Figure 9.10(a). Observe that it has symmetry about the $x_1 = x_2$ line; this symmetry would extend to higher dimensions because the pdf is a function of the observations through $\bar{x}$ and $\overline{x^2}$ which do not depend on the order of the samples. This pdf result is not important for the MAP estimator because $f_{\mathbf{X}}(\mathbf{x})$ is not a function of μ. We have provided the derivation above to emphasize that the joint pdf is *not* the same as $f_{\mathbf{X}|\Theta}(\mathbf{x}|\theta)$ when Θ is a random variable; it is a complicated expression of $\{x_n\}$, and is not even Gaussian in this example.

Figure 9.10(b) shows the conditional pdf $f_{\mathbf{X}|\mu}(\mathbf{x}|\mu)$ in (9.244) for the specific value $\mu = -1$. Although $f_{\mathbf{X}|\mu}(\mathbf{x}|\mu)$ looks similar to the pdf in Figure 9.10(a), it is actually Gaussian and, because the samples are iid, it has circular symmetry about its center. As μ is varied, the center of the pdf moves along the $x_1 = x_2$ line. In Figure 9.10(c), we show the ratio of these two pdfs for $\mu = -1$ which is used in Bayes' formula in (9.240). As μ is varied, the wave-like shape of this ratio of pdfs moves along the $x_1 = x_2$ line. It is interesting that its maximum occurs along the line defined by $x_2 + \mu = -(x_1 + \mu)$.

We also consider the posterior distribution in (9.240) as $\theta = \mu$ is varied. The joint pdf $f_{\mathbf{X}}(\mathbf{x})$ on the right-hand side does not change with μ; for this example, it depends on the samples only through $\bar{X}$ and $\overline{X^2}$. The prior distribution on the right-hand side of (9.240) is $f_\mu(\mu)$ which is the standard Gaussian pdf. The conditional pdf $f_{\mathbf{X}|\mu}(\mathbf{x}|\mu)$ is the product of N Gaussian pdfs, all with the same mean μ. However, since μ is varied, it is interesting to note that we can view each marginal pdf in the product as a Gaussian distribution for the parameter μ, with each one having a different mean given by x_n. (This type of duality of the samples and the parameter does not exist for most distributions.) As μ is varied, the right-hand side of (9.240) is the product of $N+1$ Gaussian pdfs: one has zero mean and unit variance, and the other N have mean values $\{x_n\}$ and the same variance σ^2. The area of the numerator on the right-hand side is not 1; for specific values of $\{x_n\}$ in $f_{\mathbf{X}|\mu}(\mathbf{x}|\mu)$, the denominator $f_{\mathbf{X}}(\mathbf{x})$ ensures that the left-hand side has unit area (with respect to μ). In order to plot the conditional pdf versus μ, some values for $\{x_n\}$ are needed. Plots of the prior and posterior distributions are shown in Figure 9.11 for $N = 2$ and three different sets of $\{x_1, x_2\}$.

The posterior distribution is, of course, dependent on the particular sample values. Although the mean of random variable μ from the prior distribution is zero, nonzero samples shift the posterior pdf toward those samples. However, as N becomes large, we expect that $\bar{X}$ would approach the mean of μ so that $f_{\mu|\mathbf{X}}(\mu|\mathbf{x})$ is more centered about zero and with a variance that becomes increasingly narrow. This observation follows from the MAP criterion. Since the denominator in (9.240) is ignored (because it does not depend on μ), the MAP criterion becomes

$$\begin{aligned} \mu_{\text{MAP}} &= \arg\max_{\mu} \frac{1}{\sqrt{2\pi}} \exp(-\mu^2/2) \prod_{n=1}^{N} \frac{1}{\sqrt{2\pi\sigma^2}} \exp(-(x_n - \mu)^2/2\sigma^2), \\ &= \arg\max_{\mu} \exp\left(-\mu^2/2 - (1/2\sigma^2)\sum_{n=1}^{N}(x_n - \mu)^2\right), \end{aligned} \tag{9.250}$$

where the constant term containing π has been dropped because it does not affect the maximization. It is often convenient to take the logarithm of an exponential function: since the logarithm is monotonic, the same value of μ is obtained in the maximization. Thus

$$\mu_{\text{MAP}} = \arg\max_{\mu} \left(-\mu^2/2 - (1/2\sigma^2)\sum_{n=1}^{N}(x_n - \mu)^2\right). \tag{9.251}$$

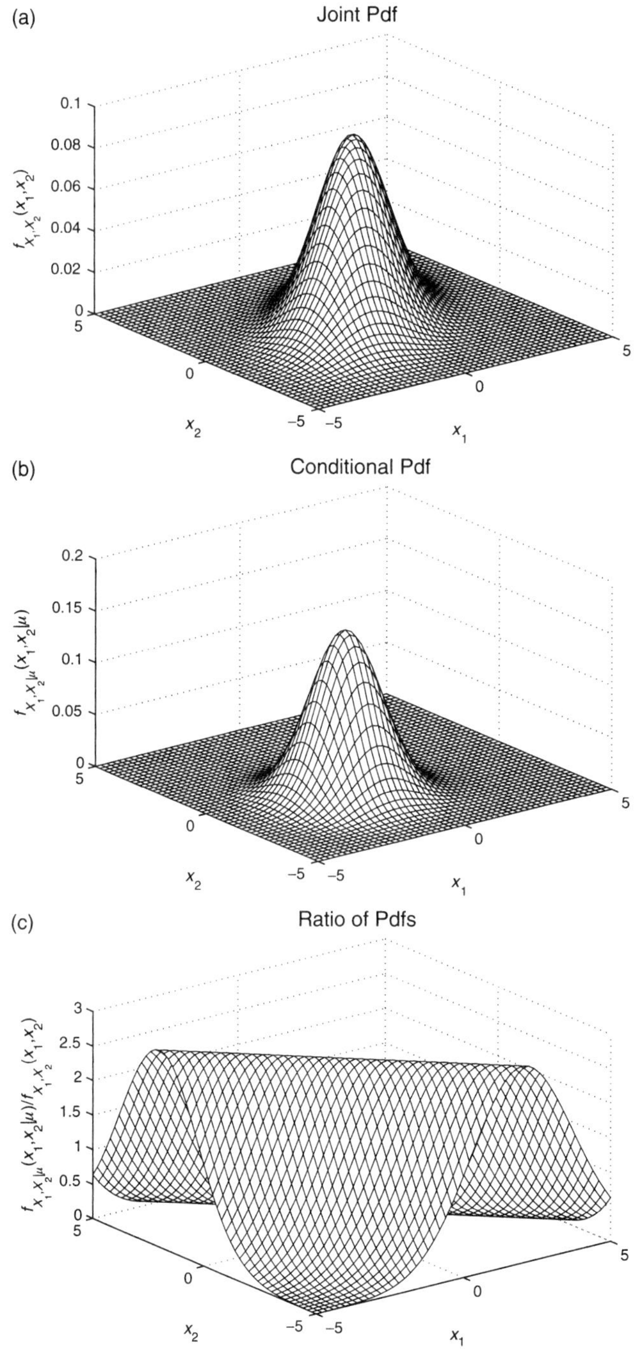

FIGURE 9.10 Pdfs for Example 9.34 with $N = 2$. (a) Joint pdf $f_{\mathbf{X}}(\mathbf{x})$. (b) Conditional pdf $f_{\mathbf{X}|\mu}(\mathbf{x}|\mu)$ for $\mu = -1$. (c) Ratio of pdfs.

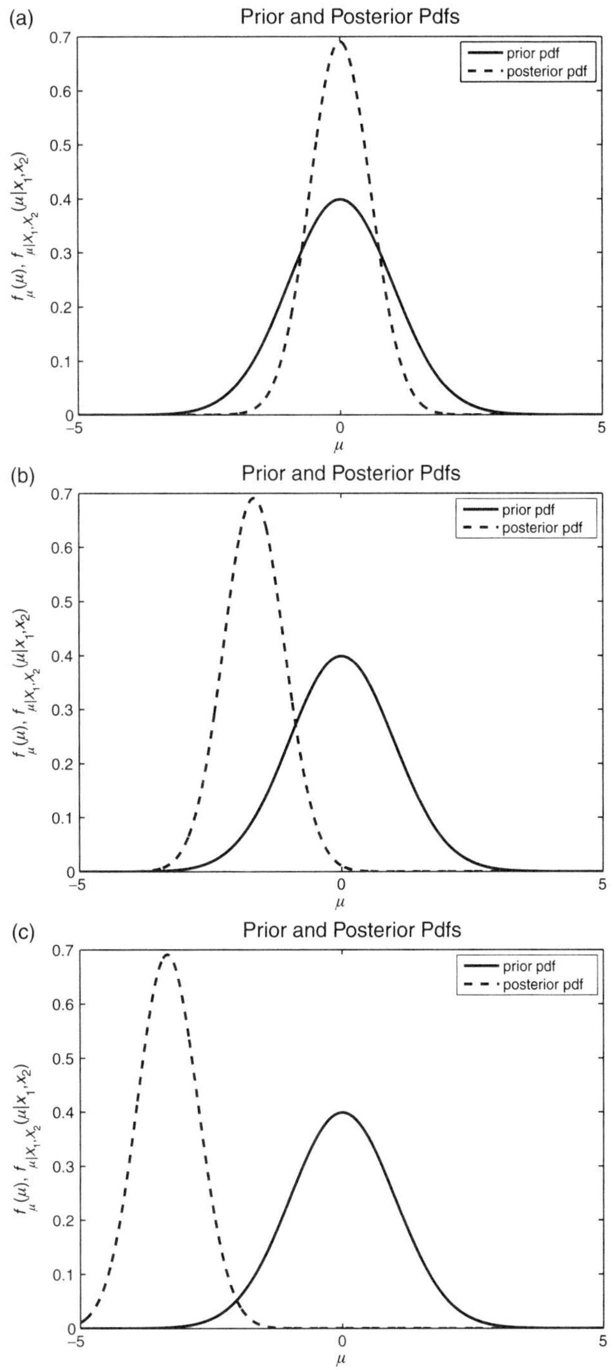

FIGURE 9.11 Prior and posterior pdfs for Example 9.34. (a) $x_1 = x_2 = 0$. (b) $x_1 = 0$ and $x_2 = -5$. (c) $x_1 = x_2 = -5$. The maxima of the posterior pdfs occur at 0, $-5/3$, and $-10/3$, respectively.

Differentiating this expression with respect to μ, setting the result equal to zero, and solving yields

$$\mu_{\text{MAP}} = \frac{1}{\sigma^2 + N} \sum_{n=1}^{N} X_n, \tag{9.252}$$

where we have used samples $\{X_n\}$ in the expression instead of outcomes $\{x_n\}$. Observe that the maximum of the posterior pdf in Figure 9.11 occurs at $(x_1 + x_2)/(\sigma^2 + N) = 0, -5/3$, and $-10/3$ for the specific outcomes used in the three plots. Note that as the variance σ^2 becomes large (for fixed N), the estimator μ_{MAP} approaches zero, which is the mean of the prior distribution. As σ^2 becomes small (meaning that the measurements are tighter about μ of the posterior distribution), μ_{MAP} approaches the sample mean which we see in Section 9.16 is the ML estimator for this example.

Example 9.35. Consider the expression in (9.43) for the exponential family of random variables. For N iid samples, the MAP estimation criterion is

$$\begin{aligned}\theta_{\text{MAP}} &= \arg\max_{\theta} \ln\left(a^N(\theta)\exp\left(b(\theta)\sum_{n=1}^{N} c(x_n)\right)\prod_{n=1}^{N} d(x_n) f_\Theta(\theta)\right),\\ &= \arg\max_{\theta}\left[N\ln(a(\theta)) + b(\theta)\sum_{n=1}^{N} c(x_n) + \ln(f_\Theta(\theta))\right],\end{aligned} \tag{9.253}$$

where the logarithm of (9.243) has been used, and the term containing $\{d(x_n)\}$ has been dropped because it does not affect the maximization. Suppose $\{X_n\}$ have an exponential distribution. We know from Table 9.2 that the sample mean is sufficient for $\theta = \lambda$; in fact, it is the UMVU estimator because it achieves the CRLB. In the expression above, $a(\theta) = -b(\theta) = \lambda$ and $c(x_n) = x_n$. Differentiating with respect to λ and setting the result equal to zero yields an expression for the MAP estimator:

$$N/\lambda_{\text{MAP}} - \sum_{n=1}^{N} X_n + \frac{f'_\Lambda(\lambda_{\text{MAP}})}{f_\Lambda(\lambda_{\text{MAP}})} = 0, \tag{9.254}$$

where $f'_\Lambda(\lambda)$ is the ordinary derivative with respect to λ. Let the prior distribution also be exponential with known parameter λ_o:

$$f_\Lambda(\lambda) = \lambda_o \exp(-\lambda_o\lambda) I_{[0,\infty)}(\lambda). \tag{9.255}$$

Then

$$\frac{f'_\Lambda(\lambda_{\text{MAP}})}{f_\Lambda(\lambda_{\text{MAP}})} = -\lambda_o \Longrightarrow \lambda_{\text{MAP}} = \frac{1}{N}\left(\sum_{n=1}^{N} X_n + \lambda_o\right) = \bar{X} + \lambda_o/N. \tag{9.256}$$

The parameter of the prior distribution causes the estimator to be shifted away from the sample mean (which we know is the UMVU estimator). The prior distribution has less effect with increasing N, and also for small λ_o such that $f_\Lambda(\lambda)$ is relatively flat.

If the prior distribution $f_\Theta(\theta)$ is not included in the MAP criterion, we have the ML criterion described in Section 9.16. It is also evident from (9.253) that when $f_\Theta(\theta)$ has a *uniform distribution*, it has no effect on the MAP estimator, in which case the MAP and ML criteria are the same (this is true in general, not only for an exponential family).

9.16 MAXIMUM LIKELIHOOD ESTIMATION

Consider estimating parameter θ where its prior distribution $f_\Theta(\theta)$ is not known and cannot be used in MAP estimation. For this condition, we can develop an estimator similar to the MAP estimator which does not require any information about the prior distribution.

Definition: Likelihood Function The *likelihood function* associated with N random variables $\{X_1, \ldots, X_N\}$ is the conditional pdf

$$l(\mathbf{x};\theta) \triangleq f_{\mathbf{X}|\Theta}(\mathbf{x}|\theta). \tag{9.257}$$

The *log-likelihood function* is

$$L(\mathbf{x};\theta) \triangleq \log(l(\mathbf{x};\theta)), \tag{9.258}$$

where $\log(\cdot)$ usually is the natural logarithm.

The likelihood function is a function of the unknown parameter θ for specific outcomes of the samples $\{X_1, \ldots, X_N\}$. We are interested in the value of θ for which $l(\mathbf{x};\theta)$ is maximum; it is the "most likely" value for particular outcomes. The log-likelihood function is usually preferred because the logarithm often simplifies the expression so that the maximum is more easily found, as in the case of exponential families. Since the logarithm is a monotonically increasing function, the likelihood and log-likelihood functions are maximized by the same θ_{ML}.

Definition: ML Estimator The *ML estimator* θ_{ML} for parameter θ maximizes the likelihood function $l(\mathbf{x};\theta)$ and the log-likelihood function $L(\mathbf{x};\theta)$.

If the derivative of the likelihood function with respect to θ exists, the ML estimator is derived by solving

$$\frac{\partial}{\partial\theta}L(\mathbf{x};\theta) = 0. \tag{9.259}$$

For an exponential family, we can use the MAP result in (9.253) by ignoring the prior distribution to obtain

$$\frac{N}{a(\theta)}a'(\theta) + b'(\theta)\sum_{n=1}^{N} c(X_n) = 0, \tag{9.260}$$

where the superscripts denote the ordinary derivative with respect to θ. For the exponential distribution in Example 9.35, we find that $a(\theta) = \theta$, $a'(\theta) = 1$, $b'(\theta) = 1$, and $c(X_n) = X_n$ so that $\theta_{\mathrm{ML}} = \bar{X}$. If the derivative is not well defined, then other methods must be used to find θ that maximizes the likelihood function. This is demonstrated later for uniformly distributed samples.

Example 9.36. We would like to estimate the mean μ of Gaussian random variable X with known σ. For N independent samples, the likelihood function is

$$\begin{aligned} l(\mathbf{x};\mu) &= \frac{1}{(2\pi\sigma^2)^{N/2}}\prod_{n=1}^{N}\exp\left(-1/2\sigma^2(x_n-\mu)^2\right), \\ &= \frac{1}{(2\pi\sigma^2)^{N/2}}\exp\left(-1/2\sigma^2\sum_{n=1}^{N}(x_n-\mu)^2\right). \end{aligned} \tag{9.261}$$

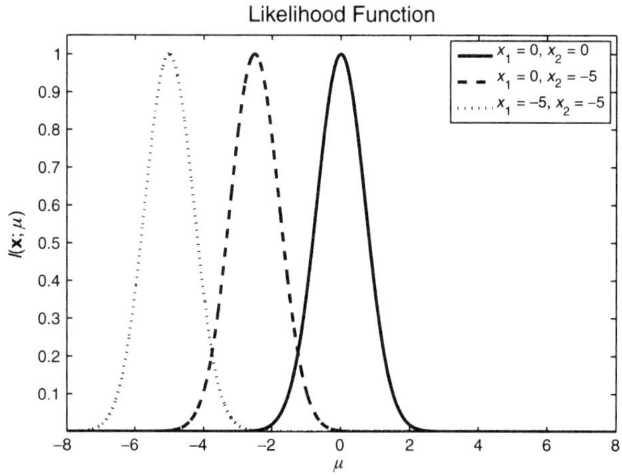

FIGURE 9.12 Likelihood function versus parameter μ for specific values of x_1 and x_2. (The curves have been normalized to have a maximum height of 1.)

Examples of this function for $N = 2$ and the same values of $\{x_1, x_2\}$ used in Figure 9.11 are shown in Figure 9.12. Note that the likelihood function is *not* a pdf with respect to parameter $\theta = \mu$. In order to easily compare the three plots in the figure, they have all been normalized to have a maximum value of 1. Observe that the peak for each plot occurs at the average of the two sample values: 0, -2.5, and -5 for $\{x_1, x_2\} = \{0, 0\}$, $\{0, -5\}$, and $\{-5, -5\}$, respectively. The log-likelihood function is

$$L(\mathbf{x}; \mu) = -(N/2)\ln(2\pi\sigma^2) - \frac{1}{2\sigma^2}\sum_{n=1}^{N}(x_n - \mu)^2. \tag{9.262}$$

Differentiating this expression with respect to μ, setting the result equal to zero, and solving for μ yields the sample mean: $\mu_{\text{ML}} = \bar{X}$. Comparing the plots in Figures 9.11 and 9.12, we see that the results of the MAP estimator (the maximum values of the posterior pdfs) are shifted closer to the origin because of the factor $1/(N + \sigma^2)$ used in the estimator. Asymptotically as $N \to \infty$, the MAP and ML estimators are the same, in which case the influence of the prior distribution of μ on the MAP estimator is diminished because of the large number of samples.

Example 9.37. In order to estimate the variance σ^2 when μ is known in Example 9.36, (9.262) is differentiated with respect to σ^2, yielding

$$\frac{\partial}{\partial\sigma^2}L(\mathbf{x}; \sigma^2) = -\frac{N}{2\sigma^2} + \frac{1}{2\sigma^4}\sum_{n=1}^{N}(x_n - \mu)^2. \tag{9.263}$$

Setting this result equal to zero and solving for σ^2 gives

$$\sigma^2_{\text{ML}} = \frac{1}{N}\sum_{n=1}^{N}(X_n - \mu)^2, \tag{9.264}$$

which is the sample variance with known mean.

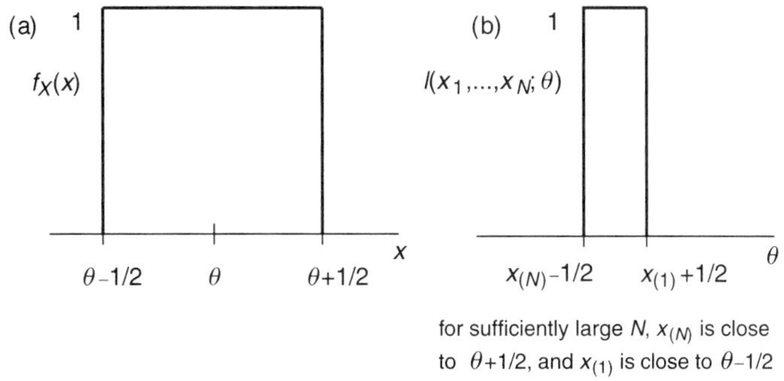

FIGURE 9.13 Plots for Example 9.38. (a) Uniform pdf. (b) Likelihood function (which is not a pdf with respect to parameter θ). Its width is increasingly narrow with increasing N.

Example 9.38. Let $\{X_1, \ldots, X_N\}$ be iid uniform on $[\theta - 1/2, \theta + 1/2]$. The likelihood function is

$$\begin{aligned} l(\mathbf{x};\theta) &= \prod_{n=1}^{N} I_{[\theta-1/2,\theta+1/2]}(x_n), \\ &= I_{[\theta-1/2,\theta+1/2]}(x_{(1)}) I_{[\theta-1/2,\theta+1/2]}(x_{(N)}), \end{aligned} \tag{9.265}$$

where $x_{(1)}$ and $x_{(N)}$ are the first and Nth-order statistics, respectively (the minimum and maximum). By ordering the samples, $l(\mathbf{x};\theta)$ does not depend on $\{x_{(2)}, \ldots, x_{(N-1)}\}$. As discussed in Example 9.8, $\mathbf{T} = [X_{(1)}, X_{(N)}]^T$ is a sufficient statistic for θ; in fact, it is minimal sufficient but not complete. Note that $l(\mathbf{x};\theta)$ cannot be differentiated with respect to θ in order to find the ML estimator; however, if $l(\mathbf{x};\theta)$ is plotted versus θ as shown in Figure 9.13(b), we find that it equals 1 on $[x_{(N)} - 1/2, x_{(1)} + 1/2]$. This is derived from the following:

$$x_{(1)} \geq \theta - 1/2 \Longrightarrow \theta \leq x_{(1)} + 1/2, \tag{9.266}$$

$$x_{(N)} \leq \theta + 1/2 \Longrightarrow \theta \geq x_{(N)} - 1/2. \tag{9.267}$$

Although $x_{(1)}$ and $x_{(N)}$ appear to be interchanged in this result, note again that the plot is versus θ. Moreover, as N is increased, we would expect $x_{(N)}$ to approach $\theta + 1/2$ and $x_{(1)}$ to approach $\theta - 1/2$ so that the rectangle in the figure becomes increasingly narrow. The ML estimate occurs where the likelihood function is maximum, which for this case is not unique: $\theta_{\text{ML}} \in [X_{(N)} - 1/2, X_{(1)} + 1/2]$, but it is defined by the first and Nth-order statistics.

9.17 LIKELIHOOD RATIO TEST

Next, we describe a decision problem that is related to estimation and is based on the likelihood function. In many applications, we are interested in deciding which of two models is most likely to be correct given one or more samples of a random variable. For example, let X be a measurable random variable from which we are to decide if the underlying model has parameter θ_0 or θ_1. Such a scenario can be represented by the following two *hypotheses*:

$$H_0: \quad X = \theta_0 + V, \tag{9.268}$$

$$H_1: \quad X = \theta_1 + V, \tag{9.269}$$

where an additive noise random variable V has been included. H_0 is called the *null hypothesis*, meaning that it is the default model, and the goal is to determine from X if θ_1 is true instead of θ_0. When $\theta_0 = 0$, the null hypothesis means that the sample contains only noise in this example.

Definition: Likelihood Ratio Test The *likelihood ratio test (LRT)* is

$$\Lambda(x) \triangleq \frac{f_{X|\Theta_1}(x|\theta_1)}{f_{X|\Theta_0}(x|\theta_0)} = \frac{l(x;\theta_1)}{l(x;\theta_0)}, \tag{9.270}$$

where the conditional pdfs are the likelihood functions for the two different parameter values. This is also known as the *Neyman–Pearson lemma*.

In order to apply the LRT, the likelihood function for each hypothesis is needed, from which we can write $\Lambda(x)$ as a function of x. The result is compared to some threshold η which allows us to weight one hypothesis more than the other. If $\Lambda(x) > \eta$, we decide that hypothesis H_1 is more likely; otherwise, H_0 is chosen. If $\Lambda(x) = \eta$ (which has zero probability when X is a continuous random variable), the test is not conclusive, and we can choose either hypothesis (for example, by tossing a fair coin).

Example 9.39. Let $\theta_0 = 0$ and $\theta_1 \neq 0$ be nonrandom and assume the noise V in (9.268) is Gaussian with parameters $\{\mu = 0, \sigma\}$. The LRT is

$$\Lambda(x) = \frac{\exp(-(x-\theta_1)^2/2\sigma^2}{\exp(-x^2/2\sigma^2)} = \exp((2\theta_1 x - \theta_1^2)/2\sigma^2) \underset{H_0}{\overset{H_1}{\gtrless}} \eta, \tag{9.271}$$

where all common terms in the numerator and the denominator have cancelled. Taking the logarithm and solving for x gives the test conditions:

$$x \underset{H_0}{\overset{H_1}{\gtrless}} \frac{2\sigma^2 \ln(\eta) + \theta_1^2}{2\theta_1}, \tag{9.272}$$

where $\theta_1 > 0$ is assumed so that the directions of the inequalities are unchanged when dividing by $2\theta_1$. The right-hand side of (9.272) gives the boundary between the two decision regions as illustrated in Figure 9.14 for two values of η. When $\eta = 1$, the two hypotheses are weighted equally, and the decision boundary occurs at the intersection of the two conditional pdfs. When $\eta = 4$ such that more weight is placed on H_0, the boundary shifts to the right. A sample with outcome x to the right of the right-shifted boundary has a lower probability of occurring, meaning that the null hypothesis is more likely to be selected. Alternatively when $\eta > 1$, a larger sample (which is less likely) is needed before deciding against the null hypothesis.

9.18 EXPECTATION–MAXIMIZATION ALGORITHM

The expectation–maximization (EM) algorithm is a recursive technique for finding the ML estimate in situations where there are *unobservable* factors influencing the samples. In conventional ML estimation, the likelihood function $l(\mathbf{x};\theta)$ or, equivalently, the log-likelihood function $L(\mathbf{x};\theta)$ is used to estimate θ from $\{X_1, \ldots, X_N\}$. Suppose the samples are incomplete in the sense that there are other random variables correlated with $\{X_n\}$, but these are not directly measurable. Let those random variables be denoted by $\mathbf{Y} \triangleq [Y_1, \ldots, Y_M]^T$.

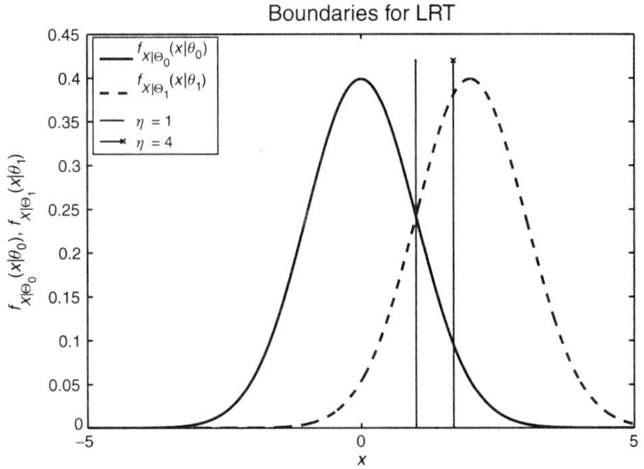

FIGURE 9.14 Conditional pdfs for Example 9.39 showing the LRT boundaries (vertical lines) for different thresholds η.

Consider the joint pdf $f_{\mathbf{X},\mathbf{Y}}(\mathbf{x}, \mathbf{y}; \theta)$ for measurable $\mathbf{X}$ and unmeasurable $\mathbf{Y}$. If $\mathbf{Y}$ is also measurable, then the ML estimator for θ would be the function of $\mathbf{X}$ and $\mathbf{Y}$ that maximizes the joint pdf. However, since only $\mathbf{X}$ is observable, it is possible to estimate θ by maximizing only $f_{\mathbf{X}}(\mathbf{x}; \theta)$, which can be expressed as

$$f_{\mathbf{X}}(\mathbf{x}; \theta) = \int_{-\infty}^{\infty} \cdots \int_{-\infty}^{\infty} f_{\mathbf{X},\mathbf{Y}}(\mathbf{x}, \mathbf{y}; \theta) d\mathbf{y}. \tag{9.273}$$

The influence of the nonmeasurable $\mathbf{Y}$ on the estimate is ignored in conventional ML estimation based on the measurable samples. In doing so, the resulting variance of the estimate cannot be less than that possible if $\mathbf{Y}$ is observable and the joint pdf $f_{\mathbf{X},\mathbf{Y}}(\mathbf{x}, \mathbf{y}; \theta)$ is maximized. The EM algorithm improves upon the maximization of $f_{\mathbf{X}}(\mathbf{x}; \theta)$ by using a two-step procedure where θ is iteratively updated (the M-step), such that in between each update, the distribution of the unobservable $\mathbf{Y}$ is estimated (the E-step). The EM algorithm ensures that the estimate of θ improves (or, at least, does not become worse) due to a lower bound used during the E-step.

As in conventional ML estimation, it is usually more convenient to maximize the log-likelihood function:

$$\begin{aligned}\log(f_{\mathbf{X}}(\mathbf{x}; \theta)) &= \log\left(\int_{-\infty}^{\infty} \cdots \int_{-\infty}^{\infty} f_{\mathbf{X},\mathbf{Y}}(\mathbf{x}, \mathbf{y}; \theta) d\mathbf{y}\right), \\ &= \log\left(\int_{-\infty}^{\infty} \cdots \int_{-\infty}^{\infty} \frac{f_{\mathbf{Y}|\mathbf{X}}(\mathbf{y}|\mathbf{x}; \theta_{\text{EM}}[k]) f_{\mathbf{X},\mathbf{Y}}(\mathbf{x}, \mathbf{y}; \theta)}{f_{\mathbf{Y}|\mathbf{X}}(\mathbf{y}|\mathbf{x}; \theta_{\text{EM}}[k])} d\mathbf{y}\right),\end{aligned} \tag{9.274}$$

where $f_{\mathbf{Y}|\mathbf{X}}(\mathbf{y}|\mathbf{x}; \theta_{\text{EM}}[k])$ based on the EM estimate of θ at iteration k has been substituted into the numerator and denominator. Applying Jensen's inequality from Appendix F to the right-hand side yields

$$\log(f_{\mathbf{X}}(\mathbf{x}; \theta)) \geq \int_{-\infty}^{\infty} \cdots \int_{-\infty}^{\infty} f_{\mathbf{Y}|\mathbf{X}}(\mathbf{y}|\mathbf{x}; \theta_{\text{EM}}[k]) \log\left(\frac{f_{\mathbf{X},\mathbf{Y}}(\mathbf{x}, \mathbf{y}; \theta)}{f_{\mathbf{Y}|\mathbf{X}}(\mathbf{y}|\mathbf{x}; \theta_{\text{EM}}[k])}\right) d\mathbf{y}. \tag{9.275}$$

Thus, instead of maximizing $\log(f_{\mathbf{X}}(\mathbf{x}; \theta))$ (based only on $\mathbf{X}$), θ maximizes the right-hand side of (9.275) which is based on $\mathbf{X}$ and $\mathbf{Y}$. In doing so, we are guaranteed by Jensen's inequality that the right-hand side is a lower bound for the original problem, which is a maximization of the left-hand side of (9.274). Since the logarithm of the ratio in (9.275) gives the difference of two logarithms, the term containing the denominator can be ignored in the maximization because it does not depend on θ. This yields

$$\max_{\theta} \int_{-\infty}^{\infty} \cdots \int_{-\infty}^{\infty} f_{\mathbf{Y}|\mathbf{X}}(\mathbf{y}|\mathbf{x};\theta_{\mathrm{EM}}[k]) \log(f_{\mathbf{X},\mathbf{Y}}(\mathbf{x},\mathbf{y};\theta))d\mathbf{y},$$
$$= \max_{\theta} \mathcal{E}_{\mathbf{Y}|\mathbf{X};\theta_{\mathrm{EM}}[k]}[\log(f_{\mathbf{X},\mathbf{Y}}(\mathbf{x},\mathbf{Y};\theta))], \tag{9.276}$$

where the expectation is performed for the most recent EM estimate $\theta_{\mathrm{EM}}[k]$. This last result describes the EM algorithm: starting with an initial estimate $\theta_{\mathrm{EM}}[0]$, the following two steps are performed for $k = 1, 2, \ldots$:

- *Expectation step.* Compute the following conditional expectation:

$$\xi_{\mathrm{EM}}(\theta, \theta_{\mathrm{EM}}[k]) \triangleq \mathcal{E}_{\mathbf{Y}|\mathbf{X};\theta_{\mathrm{EM}}[k]}[\log(f_{\mathbf{X},\mathbf{Y}}(\mathbf{x},\mathbf{Y};\theta))]. \tag{9.277}$$

- *Maximization step.* Find $\theta_{\mathrm{EM}}[k+1]$ that maximizes the above expectation:

$$\theta_{\mathrm{EM}}[k+1] = \max_{\theta} \xi_{\mathrm{EM}}(\theta, \theta_{\mathrm{EM}}[k]). \tag{9.278}$$

The algorithm is stopped when $\theta_{\mathrm{EM}}[k+1]$ changes relatively little, or after a predetermined number of steps. It is easily extended to a vector of parameters $\boldsymbol{\theta}$.

An important application of the EM algorithm is estimation of the parameters in the mixture models covered in Chapter 4. Consider the following finite mixture:

$$f_X(x) = \sum_{m=1}^{M} a_m f_{Z_m}(x), \tag{9.279}$$

where $a_m \geq 0$ and $\sum_{m=1}^{M} a_m = 1$. Let $\{Z_m\}$ have known distributions and known parameters. In the context of the EM algorithm, the unknown parameters are $\boldsymbol{\theta} = [a_1, \ldots, a_M]^T$, and the samples $\{X_1, \ldots, X_N\}$ are measurable but incomplete. We could also include any unknown parameters of $\{Z_m\}$ in $\boldsymbol{\theta}$; Problem 9.28 describes a Gaussian mixture where the mean values are unknown and also need to be estimated. In order to proceed, the unmeasurable samples must be defined. Let Y_n be a discrete random variable that specifies which Z_m is responsible for the nth observation x_n. It is described by the following joint pdf:

$$f_{X_n,Y_n}(x_n, m; \boldsymbol{\theta}) = a_m f_{Z_m}(x_n), \tag{9.280}$$

which is the mth term of (9.279). The outcomes of Y_n are $\{1, \ldots, M\}$ and correspond to $\{a_1, \ldots, a_M\}$; summing (9.280) over m gives the marginal pdf in (9.279) for $x = x_n$. When one of the random variables is discrete as Y_n is in this example, we can write

$$f_{X_n,Y_n}(x_n, m; \boldsymbol{\theta}) = P(Y_n = m|X_n = x_n; \boldsymbol{\theta}) f_{X_n}(x_n; \boldsymbol{\theta}) \tag{9.281}$$

so that the conditional distribution used in the EM algorithm is

$$P(Y_n = m|X_n = x_n; \boldsymbol{\theta}) = \frac{a_m f_{Z_m}(x_n)}{\sum_{m=1}^{M} a_m f_{Z_m}(x_n)}, \tag{9.282}$$

where (9.279) (with $x = x_n$) and (9.280) have been substituted into (9.281).

The expectation in (9.276) is a set of sums for discrete Y_n:

$$\sum_{m_1=1}^{M} \cdots \sum_{m_N=1}^{M} \prod_{n=1}^{N} P(Y_n = m_n | X_n = x_n; \boldsymbol{\theta}_{\text{EM}}[k]) \log \left(\prod_{l=1}^{N} f_{X_l,Y_l}(x_l, m_l; \boldsymbol{\theta}) \right), \tag{9.283}$$

where a subscript has been included on m to distinguish among the N sums, and the following have been substituted:

$$f_{\mathbf{Y}|\mathbf{X}}(\mathbf{y}|\mathbf{x}; \boldsymbol{\theta}_{\text{EM}}[k]) = \prod_{n=1}^{N} P(Y_n = m_n | X_n = x_n; \boldsymbol{\theta}_{\text{EM}}[k]) \tag{9.284}$$

$$f_{\mathbf{X},\mathbf{Y}}(\mathbf{x}, \mathbf{y}; \boldsymbol{\theta}) = \prod_{l=1}^{N} f_{X_l,Y_l}(x_l, m_l; \boldsymbol{\theta}). \tag{9.285}$$

Since

$$\log \left(\prod_{l=1}^{N} f_{X_l,Y_l}(x_l, m_l; \boldsymbol{\theta}) \right) = \sum_{l=1}^{N} \log(f_{X_l,Y_l}(x_l, m_l; \boldsymbol{\theta})) \tag{9.286}$$

and recognizing for each term in the product over n in (9.283) that $N - 1$ of the sums equal 1 (see Problem 9.21), the expression in (9.283) simplifies to

$$\sum_{l=1}^{N} \sum_{m=1}^{M} P(Y_l = m | X_l = x_l; \boldsymbol{\theta}_{\text{EM}}[k]) \log \left(f_{X_l,Y_l}(x_l, m; \boldsymbol{\theta}) \right) \tag{9.287}$$

such that a subscript on m is no longer needed. The inner sum is an expectation so that the EM cost function for the mixture problem is

$$\xi_{\text{EM}}(\boldsymbol{\theta}, \boldsymbol{\theta}_{\text{EM}}[k]) = \sum_{n=1}^{N} \mathcal{E}_{Y_n|X_n;\boldsymbol{\theta}_{\text{EM}}[k]}[\log(f_{X_n,Y_n}(x_n, \mathbf{Y}_n; \boldsymbol{\theta}))], \tag{9.288}$$

where the subscript has been changed from l to n. Finally, substituting (9.280) and (9.282) gives

$$\begin{aligned} \xi_{\text{EM}}(\boldsymbol{\theta}, \boldsymbol{\theta}_{\text{EM}}[k]) &= \sum_{n=1}^{N} \mathcal{E}_{Y_n|X_n;\boldsymbol{\theta}_{\text{EM}}[k]}[\log(a_m f_{Z_m}(x_n))], \\ &= \sum_{n=1}^{N} \sum_{m=1}^{M} \frac{A_m[k] f_{Z_m}(x_n)}{\sum_{m=1}^{M} A_m[k] f_{Z_m}(x_n)} \log(a_m f_{Z_m}(x_n)), \end{aligned} \tag{9.289}$$

where $\boldsymbol{\theta}_{\text{EM}}[k] = [A_1[k], \ldots, A_M[k]]^T$ is the EM estimate (this is a random vector, so we use uppercase $A_m[k]$).

In order to maximize this expression over $\boldsymbol{\theta}$, we can write $\log(a_m f_{Z_m}(x_n)) = \log(a_m) + \log(f_{Z_m}(x_n))$ and drop the second term because it does not depend on $\boldsymbol{\theta}$. As a result:

$$\xi_{\text{EM}}(\boldsymbol{\theta}, \boldsymbol{\theta}_{\text{EM}}[k]) = \sum_{n=1}^{N} \sum_{m=1}^{M} \frac{A_m[k] f_{Z_m}(x_n)}{\sum_{m=1}^{M} A_m[k] f_{Z_m}(x_n)} \log(a_m) = \sum_{m=1}^{M} B_m[k] \log(a_m), \tag{9.290}$$

where

$$B_m[k] \triangleq \sum_{n=1}^{N} \frac{A_m[k] f_{Z_m}(x_n)}{\sum_{m=1}^{M} A_m[k] f_{Z_m}(x_n)} \tag{9.291}$$

is a random coefficient that depends on the current estimate $\boldsymbol{\theta}_{\text{EM}}[k]$ and the incomplete samples $\{X_n\}$. Since the pdfs $\{f_{Z_m}(x_n)\}$ are known, values of $\{B_m[k]\}$ can be computed with each new set of samples and the updated estimate of $\boldsymbol{\theta}$. This completes the derivation of $\xi_{\text{EM}}(\boldsymbol{\theta}, \boldsymbol{\theta}_{\text{EM}}[k])$ for the E-step of the finite mixture where $\{a_m\}$ are the parameters to be estimated.

The M-step is a maximization of (9.290) which can be done using the method of Lagrange multipliers to constrain the $\{a_m\}$. The modified cost function is

$$\tilde{\xi}_{\text{EM}}(\boldsymbol{\theta}, \boldsymbol{\theta}_{\text{EM}}[k]) = \sum_{m=1}^{M} B_m[k] \log(a_m) + \lambda \left(\sum_{m=1}^{M} a_m - 1 \right), \tag{9.292}$$

where λ is a Lagrange multiplier. Differentiating with respect to a_m and setting the result equal to zero yields the updated estimate:

$$B_m[k]/A_m[k+1] + \lambda = 0 \Longrightarrow A_m[k+1] = -B_m[k]/\lambda. \tag{9.293}$$

Summing over m gives the Lagrange multiplier:

$$\sum_{m=1}^{M} A_m[k+1] = -(1/\lambda) \sum_{m=1}^{M} B_m[k] = 1 \Longrightarrow \lambda = -\sum_{m=1}^{M} B_m[k], \tag{9.294}$$

and thus the maximization for the M-step is

$$A_m[k+1] = \frac{B_m[k]}{\sum_{m=1}^{M} B_m[k]} = B_m[k]/N, \tag{9.295}$$

where the last result follows by summing over (9.291). The resulting parameter estimates turned out to be nonnegative as required in the finite mixture. If this were not the case, we would need to include additional Lagrange multipliers in $\tilde{\xi}_{\text{EM}}(\boldsymbol{\theta}, \boldsymbol{\theta}_{\text{EM}}[k])$ for the nonnegative constraint.

One method of determining when to stop the EM algorithm is to evaluate the log-likelihood function after each update of $\boldsymbol{\theta}_{\text{EM}}[k]$. For the mixture model above, it is given by

$$L(\mathbf{x}; \boldsymbol{\theta}_{\text{EM}}[k]) \log \left(\prod_{n=1}^{N} f_{X_n}(x_n) \right) = \sum_{n=1}^{N} \log \left(\sum_{m=1}^{M} A_m[k] f_{Z_m}(x_n) \right). \tag{9.296}$$

The algorithm would stop updating when the normalized difference achieves

$$D[k] \triangleq (1/N) |L(\mathbf{x}; \boldsymbol{\theta}_{\text{EM}}[k+1]) - L(\mathbf{x}; \boldsymbol{\theta}_{\text{EM}}[k])| < \epsilon \tag{9.297}$$

for some small ϵ. However, because the EM algorithm is nonlinear in general, there is no guarantee that it will converge to the best solution; there is the possibility that it converges to a local maximum.

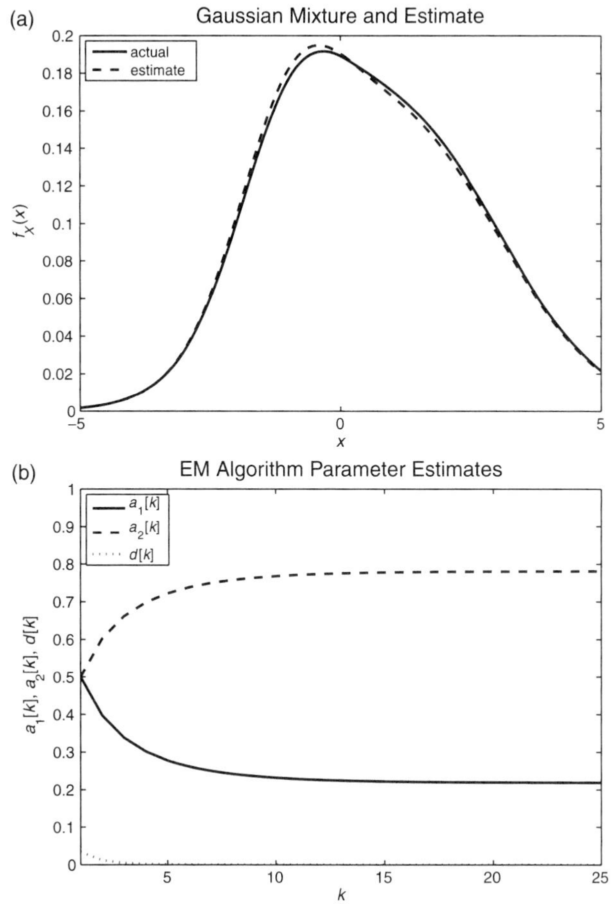

FIGURE 9.15 EM algorithm results for the finite mixture in Example 9.40. (a) Actual and estimated Gaussian mixture. (b) Convergence of parameters $\{A_1[k], A_2[k]\}$ and the log-likelihood performance measure $D[k]$ (which is very close to zero.)

Example 9.40. Consider the following pdf:

$$f_X(x) = (0.2)\frac{1}{\sqrt{2\pi}}\exp(-(x+1)^2/2) + (0.8)\frac{1}{\sqrt{8\pi}}\exp(-(x-1)^2/8), \tag{9.298}$$

which is a finite mixture of two Gaussian pdfs with known parameters $\{\mu_1 = -1, \sigma_1 = 1\}$ and $\{\mu_2 = 1, \sigma_2 = 2\}$, as illustrated in Figure 9.15(a). From $N = 1000$ samples of the mixture, we use the EM algorithm to estimate $\boldsymbol{\theta} = [a_1, a_2]^T = [0.2, 0.8]^T$. The simulation first obtains a sample v from a uniform distribution on [0, 1]. If $v < a_1 = 0.2$, a sample is derived from the first Gaussian pdf; otherwise, a sample is taken from the second Gaussian pdf. This double randomization produces 1000 samples from the Gaussian mixture. Starting with initial condition $\boldsymbol{\theta}_{\text{EM}}[0] = [0.5, 0.5]^T$, the EM algorithm recursively updates $\boldsymbol{\theta}_{\text{EM}}[k]$ using the equations in (9.291) and (9.295). Trajectories (realizations) are shown in Figure 9.15(b) where we see that the EM algorithm converges quickly to the correct weightings. A realization of the difference $D[k]$ is also shown, and the final estimated mixture is plotted in Figure 9.15(a).

In Chapter 12, we examine other types of *adaptive algorithms* that seek to minimize a cost function such as the MSE.

9.19 METHOD OF MOMENTS

The method of moments (MoM) is a parameter-estimation technique that equates sample moments, which depend on *outcomes*, and ensemble moments, which depend on *parameters* of the distribution. From Chapter 5, the mth raw moment of random variable X is

$$\mu_m \triangleq \mathcal{E}[X^m]. \tag{9.299}$$

Let the corresponding sample moment be

$$\overline{X^m} \triangleq \frac{1}{N}\sum_{n=1}^{N} X_n^m. \tag{9.300}$$

For any pdf $f_X(x)$, we know that $\mu_1 = \mu_X$ and $\mu_2 = \sigma_X^2 - \mu_X^2$, where $\{\mu_X, \sigma_X^2\}$ are the ensemble mean and variance of the distribution. For MoM estimation, the following equations are solved for μ_X and σ_X^2:

$$\bar{X} = \mu_1 = \mu_X, \quad \overline{X^2} = \mu_2 = \sigma_X^2 + \mu_X^2, \tag{9.301}$$

which yields the following estimates:

$$\mu_{\text{MoM}} = \bar{X}, \quad \sigma_{\text{MoM}}^2 = \overline{X^2} - \bar{X}^2 = [(N-1)/N]S^2, \tag{9.302}$$

where S^2 is the sample variance. For a Gaussian random variable, μ_X and σ_X^2 appear explicitly in the pdf, and so estimates of the mean and variance are obtained by directly using (9.302). It turns out that the ML estimates are identical to those in (9.302), which is generally not the case for other distributions.

Example 9.41. Consider the gamma random variable with parameters $\{\alpha, r\}$. From Appendix A:

$$\mathcal{E}[X] = r\alpha, \quad \mathcal{E}[X^2] = r(r+1)\alpha^2. \tag{9.303}$$

Equating these with the sample moments and solving for the parameters gives

$$\alpha_{\text{MoM}} = \frac{\overline{X^2} - \bar{X}^2}{\bar{X}}, \quad r_{\text{MoM}} = \frac{\bar{X}^2}{\overline{X^2} - \bar{X}^2}. \tag{9.304}$$

For ML estimation, the likelihood function is

$$l(\mathbf{x}; \alpha, r) = \frac{1}{\alpha^{Nr}\Gamma^N(r)} \exp\left(-\sum_{n=1}^{N} x_n/\alpha\right) \prod_{n=1}^{N} x_n^{r-1} \tag{9.305}$$

and the log-likelihood function is

$$L(\mathbf{x}; \alpha, r) = -N[r\ln(\alpha) + \Gamma(r)] - (1/\alpha)\sum_{n=1}^{N} x_n + (r-1)\sum_{n=1}^{N}\ln(x_n). \tag{9.306}$$

Differentiating this expression with respect to each of the parameters $\{\alpha, r\}$ and setting the results equal to zero, we see that it is not possible to obtain a closed-form expression for an estimate of r, so we focus only on α:

$$\frac{\partial}{\partial\alpha} L(\mathbf{x}; \alpha, r) = -Nr/\alpha + (N/\alpha^2)\bar{X} = 0 \tag{9.307}$$

from which we obtain

$$\alpha_{\text{ML}} = \bar{X}/r. \tag{9.308}$$

This is the ML estimator for *known* r, which is also the result for the MoM estimator of α with known r.

9.20 LEAST-SQUARES ESTIMATION

Assume that information about the parameter vector $\boldsymbol{\theta} \in \mathcal{R}^M$ is obtained via the following measurement model:

$$Y = \mathbf{a}^T\boldsymbol{\theta} + V, \tag{9.309}$$

where Y is a random variable, $\mathbf{a}$ is a known vector, and V is an unobservable additive noise random variable. This is a generalization of the simpler model $Y = X + V$ discussed previously where X is a random variable. If $\boldsymbol{\theta}$ happens to be a random vector, its distribution will *not* be taken into account in least-squares (LS) estimation. Assume there are N iid samples $\mathbf{Y} \triangleq [Y_1, \ldots, Y_N]^T$ such that

$$\mathbf{Y} = \mathbf{A}\boldsymbol{\theta} + \mathbf{V}, \tag{9.310}$$

where $\mathbf{V} \triangleq [V_1, \ldots, V_N]^T$ and

$$\mathbf{A} \triangleq [\mathbf{a}_1, \ldots, \mathbf{a}_N]^T \in \mathcal{R}^{N\times M}. \tag{9.311}$$

Usually, the number of samples far exceeds the number of parameters, that is, $N \gg M$ so that $\mathbf{A}$ is a tall narrow matrix. Let $\hat{\mathbf{Y}} \triangleq \mathbf{A}\hat{\boldsymbol{\theta}}$ be an estimate of the measurements based on the estimated parameters $\hat{\boldsymbol{\theta}}$, and define the estimation error $\tilde{\mathbf{Y}} \triangleq \mathbf{Y} - \hat{\mathbf{Y}}$.

Definition: Weighted LS Estimator The *weighted LS (WLS) estimator* of parameter vector $\boldsymbol{\theta}$ is

$$\boldsymbol{\theta}_{\text{WLS}} = \min_{\boldsymbol{\theta}} \tilde{\mathbf{Y}}^T\mathbf{W}\tilde{\mathbf{Y}}, \tag{9.312}$$

where $\mathbf{W} \in \mathcal{R}^{N\times N}$ is a weighting matrix.

The purpose of the weighting matrix is to place more emphasis on some samples; for example, it is usually advantageous to weight more recent samples if $\boldsymbol{\theta}$ happens to vary with time.

The cost function is

$$\xi_{\text{WLS}} \triangleq \tilde{\mathbf{Y}}^T\mathbf{W}\tilde{\mathbf{Y}} = \mathbf{Y}^T\mathbf{W}\mathbf{Y} - 2\mathbf{Y}^T\mathbf{W}\mathbf{A}\boldsymbol{\theta} + \boldsymbol{\theta}^T\mathbf{A}^T\mathbf{W}\mathbf{A}\boldsymbol{\theta}, \tag{9.313}$$

which is a quadratic function of $\boldsymbol{\theta}$. This is easily minimized by differentiating with respect to $\boldsymbol{\theta}$ and setting the result equal to the zero vector, yielding:

$$\mathbf{A}^T\mathbf{W}\mathbf{A}\boldsymbol{\theta}_{\text{WLS}} = \mathbf{A}^T\mathbf{W}\mathbf{Y}, \tag{9.314}$$

where $\mathbf{A}^T\mathbf{W}\mathbf{A} \in \mathcal{R}^{M\times M}$ and $\mathbf{W}\mathbf{A} \in \mathcal{R}^{N\times M}$. If $\mathbf{A}^T\mathbf{W}\mathbf{A}$ has full rank ($= M$), it is invertible so that the estimator is

$$\boldsymbol{\theta}_{\text{WLS}} = (\mathbf{A}^T\mathbf{W}\mathbf{A})^{-1}\mathbf{A}^T\mathbf{W}\mathbf{Y}. \tag{9.315}$$

Conventional LS is obtained when the weighting matrix is the identity matrix $\mathbf{W} = \mathbf{I}$.

Example 9.42. Consider the simplest case where $\boldsymbol{\theta} \rightarrow \theta$ is a scalar ($M = 1$) and the measurement model is

$$Y = \theta + V, \tag{9.316}$$

where V has zero mean. Assuming N measurements and $\mathbf{W} = \mathbf{I}$, the various quantities are

$$\mathbf{a} = 1, \quad \mathbf{A} = [1, \ldots, 1]^T, \quad \mathbf{A}^T\mathbf{A} = N, \quad \mathbf{A}^T\mathbf{Y} = \sum_{n=1}^{N} Y_n, \tag{9.317}$$

and so the LS estimator is the sample mean: $\theta_{\text{LS}} = \bar{Y}$.

Example 9.43. Next, let the measurement model be slightly more complicated

$$Y = X\theta + V, \tag{9.318}$$

where the parameter is still a scalar, but it is multiplied by a zero-mean random variable X that is uncorrelated with V. Assuming N measurements, $\mathbf{a}_n = X_n$ and

$$\mathbf{A} = [X_1, \ldots, X_N]^T, \quad \mathbf{A}^T\mathbf{A} = \sum_{n=1}^{N} X_n^2, \quad \mathbf{A}^T\mathbf{Y} = \sum_{n=1}^{N} X_n Y_n, \tag{9.319}$$

which give the following LS estimator:

$$\theta_{\text{LS}} = \frac{\sum_{n=1}^{N} X_n Y_n}{\sum_{n=1}^{N} X_n^2} = \frac{1}{N\overline{X^2}} \sum_{n=1}^{N} X_n Y_n. \tag{9.320}$$

This result is the well-known formula obtained in *linear regression* where one wants to determine the best linear fit of N sample pairs $\{X_n, Y_n\}$. The numerator is a deterministic cross-correlation between the two sets of samples, and the denominator is the sample second moment of X. An example of this result for $N = 100$ is shown in Figure 9.16 (labeled one parameter), where $\{X_n\}$ are iid standard Gaussian. The noise V is also iid Gaussian with zero mean, but with smaller variance 0.5. The unknown parameter is $\theta = 2$, and the line in the

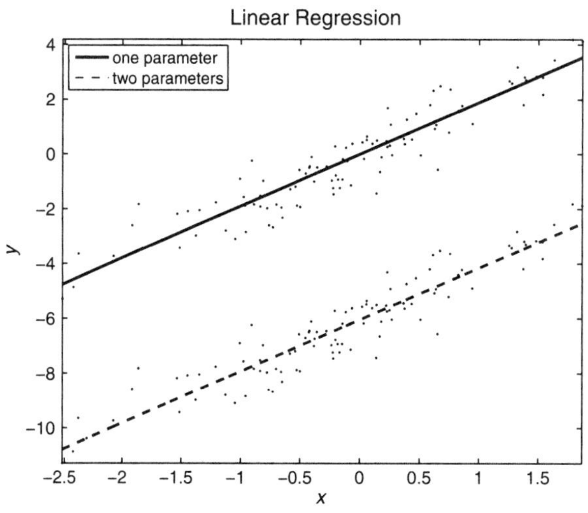

FIGURE 9.16 Linear regression models for Examples 9.43 and 9.44.

figure is $\hat{Y} = \theta_{\text{LS}} X$ which has slope $\theta_{\text{LS}} = 1.9219$, giving an accurate fit of the samples. Observe that the line passes through the origin as it should because an ordinate term was not included in the model, as described in Example 9.44.

Example 9.44. Extending the previous model to have two parameters, we can write

$$Y = \theta_1 + \theta_2 X + V \tag{9.321}$$

such that $\mathbf{a}_n^T = [1, X_n]$ and

$$\mathbf{A} = \begin{bmatrix} 1 & \cdots & 1 \\ X_1 & \cdots & X_N \end{bmatrix}^T, \quad \mathbf{A}^T \mathbf{Y} = \begin{bmatrix} \sum_{n=1}^N Y_n \\ \sum_{n=1}^N X_n Y_n \end{bmatrix} = N \begin{bmatrix} \bar{Y} \\ (1/N) \sum_{n=1}^N X_n Y_n \end{bmatrix}, \tag{9.322}$$

$$\mathbf{A}^T \mathbf{A} = \begin{bmatrix} N & \sum_{n=1}^N X_n \\ \sum_{n=1}^N X_n & \sum_{n=1}^N X_n^2 \end{bmatrix} = N \begin{bmatrix} 1 & \bar{X} \\ \bar{X} & \overline{X^2} \end{bmatrix}. \tag{9.323}$$

The two LS estimates are

$$\theta_{\text{LS},1} = \frac{1}{\Delta} \left[\overline{X^2} \bar{Y} - (1/N) \bar{X} \sum_{n=1}^N X_n Y_n \right], \tag{9.324}$$

$$\theta_{\text{LS},2} = \frac{1}{\Delta} \left[(1/N) \sum_{n=1}^N X_n Y_n - \bar{X} \bar{Y} \right], \tag{9.325}$$

where

$$\Delta \triangleq \overline{X^2} - \bar{X}^2 \tag{9.326}$$

is the determinant of $(1/N)\mathbf{A}^T \mathbf{A}$. The corresponding simulation for this case with $\theta_1 = -6$ and $\theta_2 = 2$ is also shown in Figure 9.16 (labeled two parameter) using the same samples of $\{X_n, Y_n\}$ as in Example 9.43. The line given by

$$\hat{Y} = \theta_{\text{LS},1} + \theta_{\text{LS},2} X \tag{9.327}$$

does not pass through the origin: the estimated values $\theta_{\text{LS},2} = 1.9210$ and $\theta_{\text{LS},1} = -6.0252$ are quite close to the model values. Finally, since the mean of X is zero, we can expect for large N that the second term of the two parameter estimates, as well as the second term of Δ, are approximately zero. Substituting $\bar{X} \approx 0$, the estimates simplify to

$$\theta_{\text{LS},1} \approx \bar{Y}, \quad \theta_{\text{LS},2} \approx \frac{1}{N\overline{X^2}} \sum_{n=1}^N X_n Y_n. \tag{9.328}$$

The slope $\theta_{\text{LS},2}$ is the same as in Example 9.43, and the ordinate $\theta_{\text{LS},1}$ is the sample mean of Y which is the expected result because both X and V have zero mean in the model.

For *multiple linear regression*, the goal is to find parameters $\{\theta_1, \ldots, \theta_M\}$ for the following model:

$$Y = \sum_{m=1}^M \theta_m X_m + V, \tag{9.329}$$

which we see is a generalization of the model in Example 9.44 to include several random variables $\{X_1, \ldots, X_M\}$. For N samples, it is the same as the model presented previously in (9.310) with

$$\mathbf{A} = \begin{bmatrix} X_{11} & \cdots & X_{1M} \\ \vdots & \cdots & \vdots \\ X_{N1} & \cdots & X_{NM} \end{bmatrix}, \tag{9.330}$$

where $\{X_{1m}, \ldots, X_{Nm}\}$ are the N samples of the mth random variable X_m.

Example 9.45. Consider the following matrix containing the Hald data (Hald, 1952) from MATLAB®, which has $N = 13$ measurements of $M = 4$ variables:

$$\mathbf{A}^T = \begin{bmatrix} 7 & 1 & 11 & 11 & 7 & 11 & 3 & 1 & 2 & 21 & 1 & 11 & 10 \\ 26 & 29 & 56 & 31 & 52 & 55 & 71 & 31 & 54 & 47 & 40 & 66 & 68 \\ 6 & 15 & 8 & 8 & 6 & 9 & 17 & 22 & 18 & 4 & 23 & 9 & 8 \\ 60 & 52 & 20 & 47 & 33 & 22 & 6 & 44 & 22 & 26 & 34 & 12 & 12 \end{bmatrix}. \tag{9.331}$$

The output measurements for the Hald data corresponding to $\mathbf{A}$ are given by

$$\mathbf{Y} = [78.5, 74.3, 104.3, 87.6, 95.9, 109.2, 102.7, 72.5, 93.1, 115.9, 83.8, 113.3, 109.4]^T. \tag{9.332}$$

Thus

$$\mathbf{A}^T\mathbf{A} = \begin{bmatrix} 1139 & 4922 & 769 & 2620 \\ 4922 & 33{,}050 & 7201 & 15{,}739 \\ 769 & 7201 & 2293 & 4628 \\ 2620 & 15{,}739 & 4628 & 15{,}062 \end{bmatrix}, \quad \mathbf{A}^T\mathbf{Y} = [10{,}032, 62{,}028, 13{,}982, 34{,}733]^T, \tag{9.333}$$

and $\boldsymbol{\theta}_{\text{LS}} = [2.1930, 1.1533, 0.7585, 0.4863]^T$.

We conclude this section with a brief discussion of *principal component analysis* which is a technique that can be used to generate a reduced-order LS estimate. In the context of linear regression and LS, it is known as *principal component regression (PCR)*. Since the random variables $\{X_1, \ldots, X_M\}$ may not have zero mean, it is necessary that the sample means be subtracted from the elements of $\mathbf{A}$ in order to compute the sample covariance matrix used in PCR. The sample means can be expressed as follows:

$$\bar{\mathbf{X}} = (1/N)\mathbf{A}^T\mathbf{1}, \tag{9.334}$$

where $\mathbf{1} \in \mathcal{R}^N$ is a vector of ones, and the scale factor $1/N$ is included so that $\bar{\mathbf{X}}$ is unbiased. A matrix can then be constructed from $\bar{\mathbf{X}}$ using the Kronecker product (see Appendix G):

$$\mathbf{M} = \mathbf{1} \otimes \bar{\mathbf{X}}^T \tag{9.335}$$

such that the first column of $\mathbf{M}$ is the sample mean for X_1 repeated N times, and likewise for the other columns of $\mathbf{M}$ and X_n. Thus, the sample covariance matrix is

$$\hat{\mathbf{C}}_{\mathbf{XX}} = \frac{1}{N-1}(\mathbf{A} - \mathbf{M})^T(\mathbf{A} - \mathbf{M}), \tag{9.336}$$

where the scale factor $1/(N-1)$ is included so that it is unbiased. The following eigendecomposition is then performed on $\hat{\mathbf{C}}_{\mathbf{XX}}$:

$$\hat{\mathbf{C}}_{\mathbf{XX}} = \mathbf{Q}\boldsymbol{\Lambda}\mathbf{Q}^{-1}, \tag{9.337}$$

where $\boldsymbol{\Lambda}$ contains the eigenvalues of $\hat{\mathbf{C}}_{\mathbf{XX}}$, and the columns of $\mathbf{Q}$ are the corresponding eigenvectors.

Next, we transform the original data to $\mathbf{B} \triangleq \mathbf{AQ}$ and rewrite the model in (9.310) as follows:

$$\mathbf{Y} = \mathbf{AQQ}^{-1}\boldsymbol{\theta} + \mathbf{V} = \mathbf{B}\boldsymbol{\theta}_{\text{PCA}} + \mathbf{V}, \tag{9.338}$$

where $\boldsymbol{\theta}_{\text{PCA}} \triangleq \mathbf{Q}^{-1}\boldsymbol{\theta}$. The LS estimate for this reformulated model is

$$\boldsymbol{\theta}_{\text{PCA,LS}} = (\mathbf{B}^T\mathbf{B})^{-1}\mathbf{B}^T\mathbf{Y}, \tag{9.339}$$

which is equivalent to the original LS solution because

$$\begin{aligned} \mathbf{Q}\boldsymbol{\theta}_{\text{PCA,LS}} &= \mathbf{Q}(\mathbf{Q}^T\mathbf{A}^T\mathbf{AQ})^{-1}\mathbf{Q}^T\mathbf{A}^T\mathbf{Y}, \\ &= (\mathbf{A}^T\mathbf{A})^{-1}\mathbf{A}^T\mathbf{Y} = \boldsymbol{\theta}_{\text{LS.}} \end{aligned} \tag{9.340}$$

From the eigendecomposition, we may decide to retain the eigenvectors associated with the $L < M$ largest eigenvalues of $\boldsymbol{\Lambda}$. Let that matrix be $\mathbf{Q}_{\text{PCA}} \in \mathcal{R}^{M\times L}$ and define $\mathbf{B}_{\text{PCA}} \triangleq \mathbf{AQ}_{\text{PCA}} \in \mathcal{R}^{N\times L}$, resulting in the reduced-order model:

$$\mathbf{Y} = \mathbf{B}_{\text{PCA}}\boldsymbol{\theta}_{\text{PCA}} + \mathbf{V}, \tag{9.341}$$

where now $\boldsymbol{\theta}_{\text{PCA}} \in \mathcal{R}^L$. The LS solution is

$$\boldsymbol{\theta}_{\text{LS,PCA}} = (\mathbf{B}_{\text{PCA}}^T\mathbf{B}_{\text{PCA}})^{-1}\mathbf{B}_{\text{PCA}}^T\mathbf{Y}, \tag{9.342}$$

and the estimate of the measurement vector is $\hat{\mathbf{Y}} = \mathbf{B}_{\text{PCA}}\boldsymbol{\theta}_{\text{LS,PCA}}$. We should mention, however, that the eigenvectors associated with small eigenvalues can be important depending on $\mathbf{A}$ and $\mathbf{Y}$, so it is not always beneficial to ignore such components (Jolliffe, 1982).

Example 9.46. For the Hald data in Example 9.45, the sample means are

$$\bar{\mathbf{x}} = [7.4615, 48.1538, 11.7692, 30.0000]^T, \tag{9.343}$$

and the sample covariance matrix is

$$\hat{\mathbf{C}}_{\mathbf{XX}} = \begin{bmatrix} 34.6026 & 20.9231 & -31.0513 & -24.1667 \\ 20.9231 & 242.1410 & -13.8782 & -253.4167 \\ -31.0513 & -13.8782 & 41.0256 & 3.1667 \\ -24.1667 & -253.4167 & 3.1667 & 280.1667 \end{bmatrix}. \tag{9.344}$$

Performing an eigendecomposition on $\hat{\mathbf{C}}_{\mathbf{XX}}$ gives the following eigenvector matrix:

$$\mathbf{Q} = \begin{bmatrix} -0.0678 & 0.6460 & -0.5673 & 0.5062 \\ -0.6785 & 0.0200 & 0.5440 & 0.4933 \\ 0.0290 & -0.7553 & -0.4036 & 0.5156 \\ 0.7309 & 0.1085 & 0.4684 & 0.4844 \end{bmatrix} \tag{9.345}$$

and eigenvalue matrix $\boldsymbol{\Lambda}$ with diagonal elements $\{517.7969, 67.4964, 12.4054, 0.2372\}$. Since the eigenvalue spread is ≈ 2183, we might ignore the fourth eigenvector with eigenvalue 0.2372. Retaining only the first three columns of $\mathbf{Q}$ for $\mathbf{Q}_{\text{PCA}}$ yields

$$\mathbf{B}_{\text{PCA}}^T\mathbf{B}_{\text{PCA}} = \begin{bmatrix} 7761 & -21 & -4435 \\ -21 & 810 & 60 \\ -4435 & 60 & 12{,}855 \end{bmatrix}, \quad \mathbf{B}_{\text{PCA}}^T\mathbf{Y} = [-16{,}976, 929, 38{,}677]^T, \tag{9.346}$$

and $\boldsymbol{\theta}_{\text{LS,PCA}} = [-0.5830, 0.9227, 2.8032]^T$.

9.21 PROPERTIES OF LS ESTIMATORS

In this section, we present some properties of the weighted LS estimator in (9.315) using matrix $\mathbf{A}$ in (9.311).

9.21.1 Minimum ξ_{WLS}

Substituting the WLS estimator into (9.313) gives

$$\begin{aligned}\xi_{\text{WLS,min}} &= \mathbf{Y}^T\mathbf{W}\mathbf{Y} - 2\mathbf{Y}^T\mathbf{W}\mathbf{A}(\mathbf{A}^T\mathbf{W}\mathbf{A})^{-1}\mathbf{A}^T\mathbf{W}\mathbf{Y} \\ &\quad + \mathbf{Y}^T\mathbf{W}^T\mathbf{A}(\mathbf{A}^T\mathbf{W}\mathbf{A})^{-1}\mathbf{A}^T\mathbf{W}\mathbf{A}(\mathbf{A}^T\mathbf{W}\mathbf{A})^{-1}\mathbf{A}^T\mathbf{W}\mathbf{Y}, \\ &= \mathbf{Y}^T\mathbf{W}\mathbf{Y} - \mathbf{Y}^T\mathbf{W}\mathbf{A}(\mathbf{A}^T\mathbf{W}\mathbf{A})^{-1}\mathbf{A}^T\mathbf{W}\mathbf{Y}, \\ &= \mathbf{Y}^T\mathbf{W}\mathbf{Y} - \mathbf{Y}^T\mathbf{W}\mathbf{A}\boldsymbol{\theta}_{\text{WLS}},\end{aligned} \tag{9.347}$$

where a symmetric weighting $\mathbf{W} = \mathbf{W}^T$ has been assumed. Substituting the LS parameters causes the third term in the expansion to cancel, and the last expression has $\boldsymbol{\theta}_{\text{WLS}}$ reinserted after no further simplification is possible. A plot of the LS cost function ξ_{LS} for Example 9.44 with $\mathbf{W} = \mathbf{I}$ is shown in Figure 9.17(a). It has a convex quadratic form, and the minimum is close to $\boldsymbol{\theta} = [2, -6]^T$. Note that ξ_{WLS} is not an ensemble quantity; it is a function of the samples so that the plot would change with different outcomes for X and Y. However, for large N, the basic shape and location of the minimum are as shown in the figure.

9.21.2 Uniqueness

The LS solution is unique and ξ_{WLS} has a single minimum if $\mathbf{A}$ has full rank, meaning that it has M linearly independent columns. When this occurs (which is "likely" with $N \gg M$), $\mathbf{A}^T\mathbf{A}$ is nonsingular and its inverse exists. Figure 9.17(b) shows an example of the cost function when $\mathbf{A}$ does *not* have full rank; it is no longer quadratic and has an infinity of solutions along the trough in the plot. This example was artificially created by modifying $\mathbf{A}$ in Example 9.44 so that its second column is three times the first column. This causes the "bowl-shaped" plot in Figure 9.17 to collapse as shown. Even though any of the points with minimum ξ_{WLS} in Figure 9.17(b) is a solution, the corresponding $\boldsymbol{\theta}_{\text{LS}}$ is no longer a good estimate of $\boldsymbol{\theta}$, and the resulting line does not properly model the data.

9.21.3 Orthogonality

Rearranging the expression for the WLS parameters gives

$$\mathbf{A}^T\mathbf{W}\mathbf{Y} - \mathbf{A}^T\mathbf{W}\mathbf{A}\boldsymbol{\theta}_{\text{WLS}} = \mathbf{A}^T\mathbf{W}(\mathbf{Y} - \mathbf{A}\boldsymbol{\theta}_{\text{WLS}}) = \mathbf{A}^T\mathbf{W}\tilde{\mathbf{Y}}_{\text{WLS}} = \mathbf{0}, \tag{9.348}$$

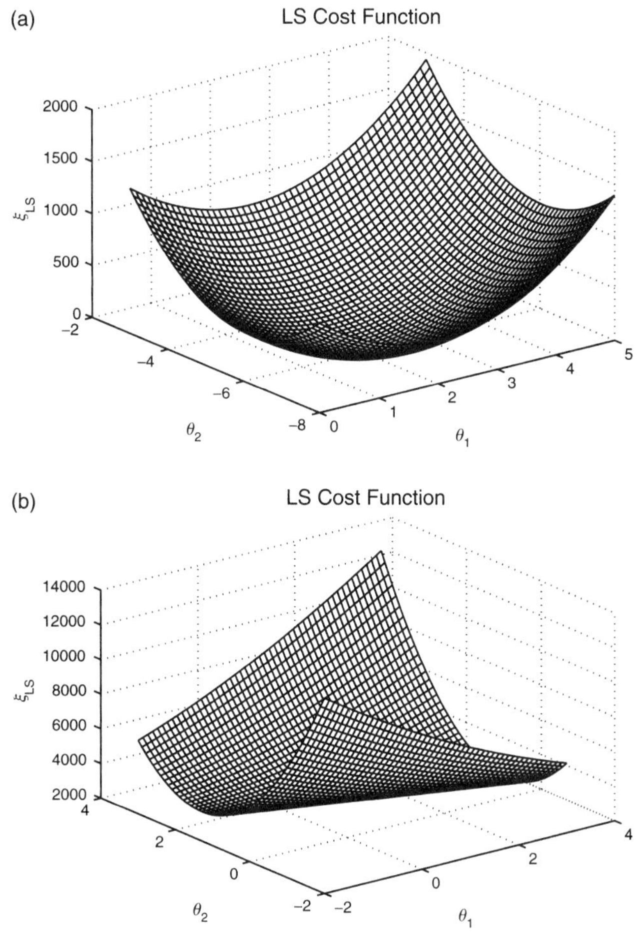

FIGURE 9.17 LS cost function for Example 9.44. (a) Nonsingular $\mathbf{A}^T\mathbf{A}$. (b) Singular $\mathbf{A}^T\mathbf{A}$ such that there is no unique minimum.

where $\tilde{\mathbf{Y}}_{\text{LS}}$ is the estimation error vector evaluated at the WLS solution, and $\mathbf{0} \in \mathcal{R}^M$ is a zero vector. This expression demonstrates that for the WLS parameter vector, the columns of $\mathbf{A}$ (weighted by $\mathbf{W}$) are orthogonal to the estimation error vector. If this expression is multiplied on the left by $\boldsymbol{\theta}_{\text{WLS}}$, then

$$\boldsymbol{\theta}_{\text{WLS}}^T\mathbf{A}^T\mathbf{W}\tilde{\mathbf{Y}}_{\text{WLS}} = \hat{\mathbf{Y}}^T\mathbf{W}\tilde{\mathbf{Y}}_{\text{WLS}} = 0 \tag{9.349}$$

such that the WLS estimator and the estimation error vectors are also orthogonal. This is depicted in Figure 9.18 for $\mathbf{W} = \mathbf{I}$. Generally, $\mathbf{Y}$ is not in the column space (range) of $\mathbf{A}$; otherwise, it would be perfectly estimated. The estimated output $\hat{\mathbf{Y}}_{\text{LS}}$ necessarily lies in the column space of $\mathbf{A}$ because it is obtained as a weighted sum of the columns of $\mathbf{A}$ via $\boldsymbol{\theta}_{\text{LS}}$. Obviously, $\hat{\mathbf{Y}}$ can be any vector in the range of $\mathbf{A}$, but the figure illustrates that the *best* vector is orthogonal to the error $\tilde{\mathbf{Y}}_{\text{LS}}$ because $\tilde{\mathbf{Y}}_{\text{LS}}^T\tilde{\mathbf{Y}}_{\text{LS}}$ (the vector norm squared) is minimum. Figure 9.19 shows a specific example for $N = 2$ samples and $M = 1$ parameter (such that $\mathbf{A}$ is a column vector), demonstrating that the vectors are orthogonal even for a small number of samples; the orthogonality principal is a deterministic property for LS estimation and is not due to large sample averaging.

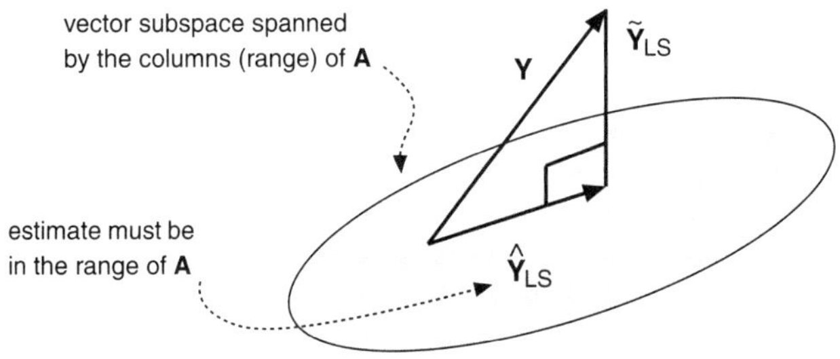

FIGURE 9.18 Vector space illustration of LS orthogonality condition.

9.21.4 Unbiased

For the remaining properties, assume that $\mathbf{Y}$ is generated using the model in (9.310). However, the LS estimate itself does not require that the samples be generated in this manner; $\boldsymbol{\theta}_{\text{LS}}$ is a linear combination of the data and may only be an approximate model for how $\boldsymbol{\theta}$ and Y are actually related.

Observe that $\boldsymbol{\theta}_{\text{LS}}$ is unbiased provided $\mathbf{V}$ has zero mean:

$$\begin{aligned}\mathcal{E}[\boldsymbol{\theta}_{\text{LS}}] &= (\mathbf{A}^T\mathbf{A})^{-1}\mathbf{A}^T\mathcal{E}[\mathbf{Y}],\\ &= (\mathbf{A}^T\mathbf{A})^{-1}\mathbf{A}^T\mathbf{A}\boldsymbol{\theta} + (\mathbf{A}^T\mathbf{A})^{-1}\mathbf{A}^T\mathcal{E}[\mathbf{V}] = \boldsymbol{\theta},\end{aligned} \tag{9.350}$$

where $\mathbf{A}$ has been brought outside the expectation because it is measurable (this could be emphasized by using an expectation that is conditioned on $\mathbf{A}$.)

9.21.5 Covariance Matrix

From the second line above before the expectation is taken, the LS estimator based on the model in (9.310) is

$$\boldsymbol{\theta}_{\text{LS}} = \boldsymbol{\theta} + (\mathbf{A}^T\mathbf{A})^{-1}\mathbf{A}^T\mathbf{V}. \tag{9.351}$$

Subtracting $\boldsymbol{\theta}$ (which is the mean because the estimator is unbiased), we can write the covariance matrix as follows:

$$\begin{aligned}\mathcal{E}[(\boldsymbol{\theta}_{\text{LS}} - \boldsymbol{\theta})(\boldsymbol{\theta}_{\text{LS}} - \boldsymbol{\theta})^T] &= \mathcal{E}\left[(\mathbf{A}^T\mathbf{A})^{-1}\mathbf{A}^T\mathbf{V}\mathbf{V}^T\mathbf{A}(\mathbf{A}^T\mathbf{A})^{-1}\right],\\ &= (\mathbf{A}^T\mathbf{A})^{-1}\mathbf{A}^T\mathcal{E}[\mathbf{V}\mathbf{V}^T]\mathbf{A}(\mathbf{A}^T\mathbf{A})^{-1},\\ &= (\mathbf{A}^T\mathbf{A})^{-1}\mathbf{A}^T\mathbf{R}_{\text{VV}}\mathbf{A}(\mathbf{A}^T\mathbf{A})^{-1}.\end{aligned} \tag{9.352}$$

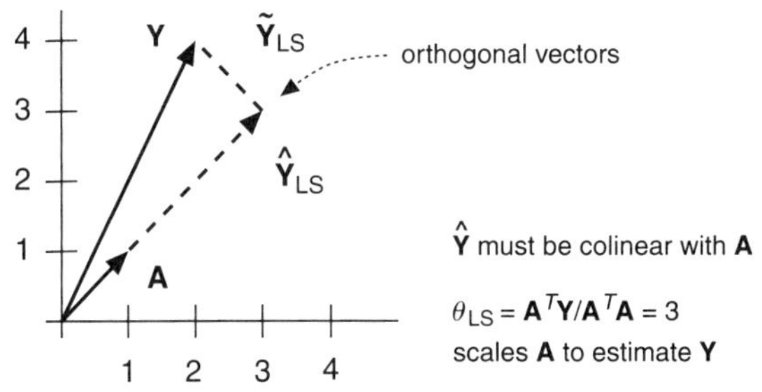

FIGURE 9.19 Numeric example of the LS orthogonality condition in two dimensions.

In order to proceed, assume that $\mathbf{V}$ is uncorrelated such that $\mathbf{R_{VV}} = \sigma_V^2\mathbf{I}$ and

$$\text{cov}[\boldsymbol{\theta}_{\text{LS}}] = \sigma_V^2(\mathbf{A}^T\mathbf{A})^{-1}. \tag{9.353}$$

This result depends not only on the noise variance as expected, but also on the deterministic autocorrelation matrix $\mathbf{A}^T\mathbf{A}$.

9.21.6 Efficient: Achieves CRLB

The LS estimator achieves the CRLB under the assumption that $\mathbf{V}$ is Gaussian and uncorrelated with zero mean. The Fisher information matrix for the LS model is

$$\mathcal{I}(\boldsymbol{\theta}) \triangleq \mathcal{E}\left[\frac{\partial \ln f_{\mathbf{V}}(\mathbf{V})}{\partial\boldsymbol{\theta}}\left(\frac{\partial \ln f_{\mathbf{V}}(\mathbf{V})}{\partial\boldsymbol{\theta}}\right)^T\right], \tag{9.354}$$

where $f_{\mathbf{V}}(\mathbf{v})$ is the joint pdf of $\mathbf{V}$, $\mathcal{I}(\boldsymbol{\theta}) \in \mathcal{R}^{M\times M}$, and M is the number of parameter in $\boldsymbol{\theta}$. For the assumptions above with $\mathbf{R_{VV}} = \sigma_V^2\mathbf{I}$, it is straightforward to evaluate the logarithm:

$$\ln f_{\mathbf{V}}(\mathbf{V}) = \ln\left(\prod_{n=1}^{N} f_{V_n}(V_n)\right) = \ln\left(\frac{1}{(2\pi\sigma_V^2)^{N/2}}\right) - \frac{1}{2\sigma_V^2}\sum_{n=1}^{N} V_n^2, \tag{9.355}$$

where independence has been used in the first line and the Gaussian assumption in the second line. Thus

$$\frac{\partial}{\partial\boldsymbol{\theta}}\ln f_{\mathbf{V}}(\mathbf{V}) = -\frac{1}{2\sigma_V^2}\frac{\partial}{\partial\boldsymbol{\theta}}\mathbf{V}^T\mathbf{V} = -\frac{1}{\sigma_V^2}\left[\frac{\partial\mathbf{V}}{\partial\boldsymbol{\theta}}\right]^T\mathbf{V}, \tag{9.356}$$

where the sum has been replaced by an inner product, and vector differentiation properties in Appendix G have been used. Since $\mathbf{Y} = \mathbf{A}\boldsymbol{\theta} + \mathbf{V}$, the derivative is

$$\frac{\partial\mathbf{V}}{\partial\boldsymbol{\theta}} = \frac{\partial(\mathbf{Y} - \mathbf{A}\boldsymbol{\theta})}{\partial\boldsymbol{\theta}} = -\mathbf{A}^T. \tag{9.357}$$

Substituting this result into the expectation gives

$$\mathcal{I}(\boldsymbol{\theta}) = \frac{1}{\sigma_V^4}\mathcal{E}\left[\left[\frac{\partial\mathbf{V}}{\partial\boldsymbol{\theta}}\right]^T\mathbf{V}\mathbf{V}^T\frac{\partial\mathbf{V}}{\partial\boldsymbol{\theta}}\right] = \frac{1}{\sigma_V^4}\mathbf{A}\mathcal{E}[\mathbf{V}\mathbf{V}^T]\mathbf{A}^T = \frac{1}{\sigma_V^2}\mathbf{A}\mathbf{A}^T \tag{9.358}$$

demonstrating that the covariance matrix of the LS estimator achieves the CRLB bound $\mathcal{I}^{-1}(\boldsymbol{\theta}) = \sigma_V^2(\mathbf{A}\mathbf{A}^T)^{-1}$. The LS estimator for Gaussian white noise $\mathbf{V}$ is the best linear unbiased (BLU) estimator, meaning that it has the smallest covariance of all unbiased linear estimators, as demonstrated next.

9.21.7 BLU Estimator

Let $\hat{\boldsymbol{\theta}}$ be any unbiased estimator of $\boldsymbol{\theta}$ which can be written as a linear combination of $\mathbf{Y}$ as follows:

$$\hat{\boldsymbol{\theta}} = \mathbf{B}\mathbf{Y}, \tag{9.359}$$

where $\mathbf{B} \in \mathcal{R}^{M\times N}$ (which is a wide matrix like $\mathbf{A}^T$). The expectation of this estimator is

$$\mathcal{E}[\hat{\boldsymbol{\theta}}] = \mathbf{B}\mathcal{E}[\mathbf{Y}] = \mathbf{B}\mathbf{A}\boldsymbol{\theta} + \mathbf{B}\mathcal{E}[\mathbf{V}] = \mathbf{B}\mathbf{A}\boldsymbol{\theta} \tag{9.360}$$

so that unbiasedness requires $\mathbf{BA} = \mathbf{I}$, where $\mathbf{I} \in \mathcal{R}^{M \times M}$ is an identity matrix. In general, we cannot solve explicitly for $\mathbf{B}$ because usually $N \gg M$. Since $\boldsymbol{\theta}_{\text{LS}}$ is a vector, the smallest covariance means

$$\text{cov}[\boldsymbol{\theta}_{\text{LS}}] - \text{cov}[\hat{\boldsymbol{\theta}}] \geq \mathbf{0}. \tag{9.361}$$

The inequality indicates that the matrix on the left-hand side is at least PSD: for any nonzero vector $\mathbf{u} \in \mathcal{R}^M$:

$$\mathbf{u}^T \left(\text{cov}[\boldsymbol{\theta}_{\text{LS}}] - \text{cov}[\hat{\boldsymbol{\theta}}] \right) \mathbf{u} \geq 0. \tag{9.362}$$

In order to obtain the covariance matrix for $\hat{\boldsymbol{\theta}}$, note from (9.360) that

$$\hat{\boldsymbol{\theta}} = \boldsymbol{\theta} + \mathbf{BV}, \tag{9.363}$$

and thus

$$\text{cov}[\hat{\boldsymbol{\theta}}] = \mathcal{E}[(\hat{\boldsymbol{\theta}} - \boldsymbol{\theta})(\hat{\boldsymbol{\theta}} - \boldsymbol{\theta})^T] = \mathbf{B}\mathcal{E}[\mathbf{VV}^T]\mathbf{B}^T = \sigma_V^2 \mathbf{BB}^T. \tag{9.364}$$

Subtracting the mean from each estimator, the following error vector is obtained:

$$\begin{aligned} \mathbf{E} &= [(\mathbf{A}^T\mathbf{A})^{-1}\mathbf{A}^T\mathbf{Y} - \boldsymbol{\theta}] - [\mathbf{BY} - \boldsymbol{\theta}], \\ &= (\mathbf{A}^T\mathbf{A})^{-1}\mathbf{A}^T\mathbf{V} - \mathbf{BV}, \end{aligned} \tag{9.365}$$

where the term with $\boldsymbol{\theta}$ has canceled because both estimators are unbiased. The covariance matrix $\mathbf{R}_{\text{EE}} = \mathcal{E}[\mathbf{EE}^T]$ of the error vector is

$$\begin{aligned} \mathbf{R}_{\text{EE}} &= [(\mathbf{A}^T\mathbf{A})^{-1}\mathbf{A}^T - \mathbf{B}]\mathcal{E}[\mathbf{VV}^T][(\mathbf{A}^T\mathbf{A})^{-1}\mathbf{A}^T - \mathbf{B}]^T, \\ &= \sigma_V^2[(\mathbf{A}^T\mathbf{A})^{-1}\mathbf{A}^T((\mathbf{A}^T\mathbf{A})^{-1}\mathbf{A}^T)^T - (\mathbf{A}^T\mathbf{A})^{-1}\mathbf{A}^T\mathbf{B}^T \\ &\quad - \mathbf{B}((\mathbf{A}^T\mathbf{A})^{-1}\mathbf{A}^T)^T + \mathbf{BB}^T], \\ &= \sigma_V^2[\mathbf{BB}^T - (\mathbf{A}^T\mathbf{A})^{-1}], \end{aligned} \tag{9.366}$$

where the first three terms in the second expression combine into a single term because $\mathbf{BA} = \mathbf{A}^T\mathbf{B}^T = \mathbf{I}$ and $\mathbf{B}$ is unbiased. Since $\mathbf{R}_{\text{EE}}$ is a covariance matrix, it is at least PSD so that

$$\mathbf{R}_{\text{EE}} = \text{cov}[\hat{\boldsymbol{\theta}}] - \text{cov}[\boldsymbol{\theta}_{\text{LS}}] \geq \mathbf{0}, \tag{9.367}$$

which demonstrates that the LS estimator has the lowest covariance: it is the BLU estimator when $\mathbf{R}_{\text{VV}} = \sigma_V^2\mathbf{I}$. It turns out that if the noise vector $\mathbf{V}$ is also Gaussian, the LS estimator is the best unbiased estimator among linear and *nonlinear* estimators.

9.22 BEST LINEAR UNBIASED ESTIMATION

In Section 9.21, we assumed $\mathbf{R}_{\text{VV}} = \sigma_V^2\mathbf{I}$ so that the LS estimator is the BLU estimator. When the noise covariance matrix does not have this diagonal form, it is still possible to achieve BLU estimation by using WLS with weighting matrix $\mathbf{W} = \mathbf{R}_{VV}^{-1}$. The resulting estimator is

$$\boldsymbol{\theta}_{\text{WLS}} = (\mathbf{A}^T\mathbf{R}_{\text{VV}}^{-1}\mathbf{A})^{-1}\mathbf{A}^T\mathbf{R}_{\text{VV}}^{-1}\mathbf{Y} = \boldsymbol{\theta}_{\text{BLU}}. \tag{9.368}$$

This result is derived below using a constrained optimization under the following conditions.

- The observation model for $\mathbf{Y}$ is given by (9.310) where $\boldsymbol{\theta} \in \mathcal{R}^M$ is nonrandom but unknown, $\mathbf{A} \in \mathcal{R}^{M\times N}$ is known (measurable), and $\mathbf{V} \in \mathcal{R}^N$ is a random vector with zero mean and covariance matrix $\mathbf{R}_{\mathbf{VV}} \in \mathcal{R}^{N\times N}$.
- The form of the BLU estimator is a linear combination of the measurements:

$$\boldsymbol{\theta}_{\text{BLU}} = \mathbf{BY}, \tag{9.369}$$

where $\mathbf{B} \in \mathcal{R}^{M\times N}$.
- It is unbiased:

$$\mathcal{E}[\boldsymbol{\theta}_{\text{BLU}}] = \mathcal{E}[\mathbf{B}(\mathbf{A}\boldsymbol{\theta} + \mathbf{V})] = \mathbf{BA}\boldsymbol{\theta} + \mathcal{E}[\mathbf{V}] = \mathbf{BA}\boldsymbol{\theta}, \tag{9.370}$$

which requires $\mathbf{BA} = \mathbf{I}$.
- It has minimum covariance $\text{cov}[\boldsymbol{\theta}_{\text{BLU}}]$. The goal is to solve the following constrained minimization:

$$\boldsymbol{\theta}_{\text{BLU}} = \arg\min_{\mathbf{B}} \ \text{cov}[\boldsymbol{\theta}_{\text{BLU}}] \ \text{ s.t. } \ \mathbf{BA} = \mathbf{I}. \tag{9.371}$$

With these assumptions, we proceed to derive the BLU estimator. Since it is unbiased, we can write

$$\boldsymbol{\theta}_{\text{BLU}} = \boldsymbol{\theta} + \mathbf{BV} \tag{9.372}$$

so that

$$\begin{aligned} \text{cov}[\boldsymbol{\theta}_{\text{BLU}}] &= \mathcal{E}\left[(\boldsymbol{\theta}_{\text{BLU}} - \boldsymbol{\theta})(\boldsymbol{\theta}_{\text{BLU}} - \boldsymbol{\theta})^T\right], \\ &= \mathbf{B}\mathcal{E}\left[\mathbf{VV}^T\right]\mathbf{B}^T = \mathbf{BR}_{\mathbf{VV}}\mathbf{B}^T. \end{aligned} \tag{9.373}$$

Minimizing a covariance matrix of the form above is equivalent to minimizing its trace:

$$\min_{\mathbf{B}} \ \text{tr}[\mathbf{BR}_{\mathbf{VV}}\mathbf{B}^T] \ \text{ s.t. } \ \mathbf{BA} = \mathbf{I}, \tag{9.374}$$

so that the cost function for the BLU estimator is

$$\xi_{\text{BLU}} \triangleq \text{tr}[\mathbf{BR}_{\mathbf{VV}}\mathbf{B}^T]. \tag{9.375}$$

Defining $\mathbf{b}_m^T \in \mathcal{R}^M$ to be the mth row of $\mathbf{B}$, each element in the trace is minimized as follows:

$$\min_{\mathbf{b}_m} \left(\mathbf{b}_m^T\mathbf{R}_{\mathbf{VV}}\mathbf{b}_m + [\mathbf{b}_m^T\mathbf{A} - \mathbf{e}_m^T]\boldsymbol{\lambda}_m\right), \tag{9.376}$$

where $\mathbf{e}_m \in \mathcal{R}^M$ is a unit vector with 1 in the mth position, and $\boldsymbol{\lambda}_m \in \mathcal{R}^M$ is a vector of Lagrange multipliers.

Using the differentiation rules in Appendix G:

$$\frac{\partial}{\partial \mathbf{b}_m}\left(\mathbf{b}_m^T\mathbf{R}_{\mathbf{VV}}\mathbf{b}_m + [\mathbf{b}_m^T\mathbf{A} - \mathbf{e}_m^T]\boldsymbol{\lambda}_m\right) = 2\mathbf{R}_{\mathbf{VV}}\mathbf{b}_m + \mathbf{A}\boldsymbol{\lambda}_m, \tag{9.377}$$

which after setting equal to $\mathbf{0}$ and solving gives

$$\mathbf{b}_m = -(1/2)\mathbf{R}_{\mathbf{VV}}^{-1}\mathbf{A}\boldsymbol{\lambda}_m. \tag{9.378}$$

Substituting this result into the constraint yields an expression for the Lagrange multipliers:

$$-(1/2)\mathbf{A}^T\mathbf{R}_{\mathbf{VV}}^{-1}\mathbf{A}\boldsymbol{\lambda}_m = \mathbf{e}_m \Longrightarrow \boldsymbol{\lambda}_m = -2\left(\mathbf{A}^T\mathbf{R}_{\mathbf{VV}}^{-1}\mathbf{A}\right)^{-1}\mathbf{e}_m \tag{9.379}$$

so that (9.378) becomes

$$\mathbf{b}_m = \mathbf{R}_{\mathbf{VV}}^{-1}\mathbf{A}\left(\mathbf{A}^T\mathbf{R}_{\mathbf{VV}}^{-1}\mathbf{A}\right)^{-1}\mathbf{e}_m. \tag{9.380}$$

This expression satisfies the constraint exactly:

$$\mathbf{b}_m^T\mathbf{A} = \mathbf{e}_m^T\left(\mathbf{A}^T\mathbf{R}_{\mathbf{VV}}^{-1}\mathbf{A}\right)^{-1}\mathbf{A}^T\mathbf{R}_{\mathbf{VV}}^{-1}\mathbf{A} = \mathbf{e}_m^T, \tag{9.381}$$

Since $\mathbf{b}_m^T$ is a row of $\mathbf{B}_{\text{BLU}}$, $\mathbf{b}_m$ is a column of $\mathbf{B}_{\text{BLU}}^T$, and we can write

$$\mathbf{B}^T = [\mathbf{b}_1, \ldots, \mathbf{b}_M] = \mathbf{R}_{\mathbf{VV}}^{-1}\mathbf{A}\left(\mathbf{A}^T\mathbf{R}_{\mathbf{VV}}^{-1}\mathbf{A}\right)^{-1}[\mathbf{e}_1, \ldots, \mathbf{e}_M], \tag{9.382}$$

where $[\mathbf{e}_1, \ldots, \mathbf{e}_M] = \mathbf{I} \in \mathcal{R}^{M\times M}$ is an identity matrix. Thus

$$\mathbf{B} = \left(\mathbf{A}^T\mathbf{R}_{\mathbf{VV}}^{-1}\mathbf{A}\right)^{-1}\mathbf{A}^T\mathbf{R}_{\mathbf{VV}}^{-1} \tag{9.383}$$

and the BLU estimator in (9.368) is obtained. The estimated samples are generated using $\hat{\mathbf{Y}} = \mathbf{A}\boldsymbol{\theta}_{\text{BLU}}$.

It is important to note that unlike the LS and WLS estimators, the BLU estimator is based on an underlying model for $\mathbf{Y}$, and the covariance matrix $\mathbf{R}_{\mathbf{VV}}$ of the additive noise is assumed to be known. The cost function ξ_{BLU} for $\boldsymbol{\theta}_{\text{BLU}}$ does not directly involve $\mathbf{Y}$. Minimization of the cost function for LS and WLS is performed over the parameter vector $\boldsymbol{\theta}$, whereas here it is performed over the rows of $\mathbf{B}_{\text{BLU}}$. The samples appear only in the estimate $\boldsymbol{\theta}_{\text{BLU}}$ in (9.369). As a result, the BLU estimator is not generally useful in practice; the LS and WLS estimators are more widely used because an underlying model for $\mathbf{Y}$ is not required. A model was used previously only to derive some properties of the LS and WLS estimators, assuming that it is a valid representation of how the samples are actually generated. Perhaps the most important aspect of the BLU estimator is that the LS and WLS estimators can achieve its performance: they are all the same estimator under certain conditions as summarized in Section 9.23.

Example 9.47. An example of the cost function $\xi_{\text{BLU},1}$ for the first row of $\mathbf{B}$ with $M = 2$, corresponding to $\mathbf{b}_1 = [b_{11}, b_{12}]^T$, and for a small number of samples $N = 2$ is shown in Figure 9.20(a). The noise covariance matrix is

$$\mathbf{R}_{\mathbf{VV}} = \begin{bmatrix} 1 & 0 \\ 0 & 2 \end{bmatrix}. \tag{9.384}$$

Observe that the surface is quadratic and the constraint is a plane intersecting the bowl-shaped surface. Valid BLU estimator solutions occur where the constraint plane intersects the b_{11}–b_{12}-plane at $\xi_{\text{BLU},1} = 0$ in order to satisfy the constraint. The optimal value of the quadratic function is the minimum point of the surface located above where the two planes intersect, denoted by the arrow. In Figure 9.20(b), the constraint plane has been raised by the value of the cost function evaluated at the optimal weights, showing that they intersect at $\boldsymbol{\theta}_{\text{BLU}}$. This type of plot is typical of constrained optimization problems involving a quadratic cost function and linear constraints on the parameters.

Example 9.48. Consider again the single parameter model ($M = 1$) in (9.318), and assume the additive noise vector has a diagonal covariance matrix with nonzero elements $\{\sigma_{V,n}^2\}$. We demonstrate how the LS and BLU estimators differ when the noise variance changes across samples. For this case, the inverted term in (9.383) is

$$\mathbf{A}^T\mathbf{R}_{\mathbf{VV}}^{-1}\mathbf{A} = \sum_{n=1}^{N} X_n^2/\sigma_{V,n}^2 \tag{9.385}$$

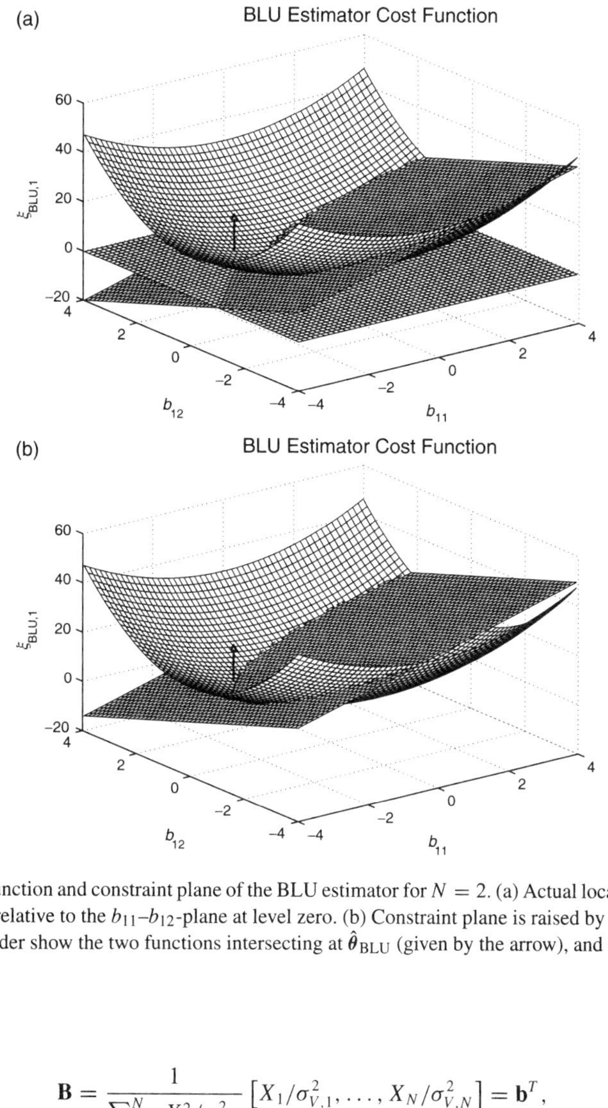

FIGURE 9.20 Cost function and constraint plane of the BLU estimator for $N = 2$. (a) Actual locations of the cost function and the constraint plane relative to the b_{11}–b_{12}-plane at level zero. (b) Constraint plane is raised by the cost function value at the optimal weights in order show the two functions intersecting at $\hat{\boldsymbol{\theta}}_{\text{BLU}}$ (given by the arrow), and with the b_{11}–b_{12}-plane at level zero removed.

so that

$$\mathbf{B} = \frac{1}{\sum_{n=1}^{N} X_n^2/\sigma_{V,n}^2} \left[X_1/\sigma_{V,1}^2, \ldots, X_N/\sigma_{V,N}^2\right] = \mathbf{b}^T, \tag{9.386}$$

which is a row vector for a single parameter. This gives the following BLU estimator:

$$\theta_{\text{BLU}} = \frac{\sum_{n=1}^{N} X_n Y_n/\sigma_{V,n}^2}{\sum_{n=1}^{N} X_n^2/\sigma_{V,n}^2}, \tag{9.387}$$

which reduces to θ_{LS} in (9.320) when the $\{\sigma_{V,n}^2\}$ are all identical. For the general case, samples with greater noise variance are weighted less in the BLU estimator. Example lines generated by the BLU and LS estimators for $M = 1$ are shown in Figure 9.21(a) where the noise variance for $N = 100$ samples is uniform on [0, 10]. The two lines do not coincide because the BLU estimator takes into account the changing noise variance across samples. The actual parameter value is $\theta = 2$, and the two estimates are $\theta_{\text{LS}} = 1.5657$ and $\theta_{\text{BLU}} = 1.9362$, demonstrating that the BLU estimator is more accurate. Of course, as discussed earlier, it is unrealistic that the noise variance for each sample would be known in practice, though some information might be available,

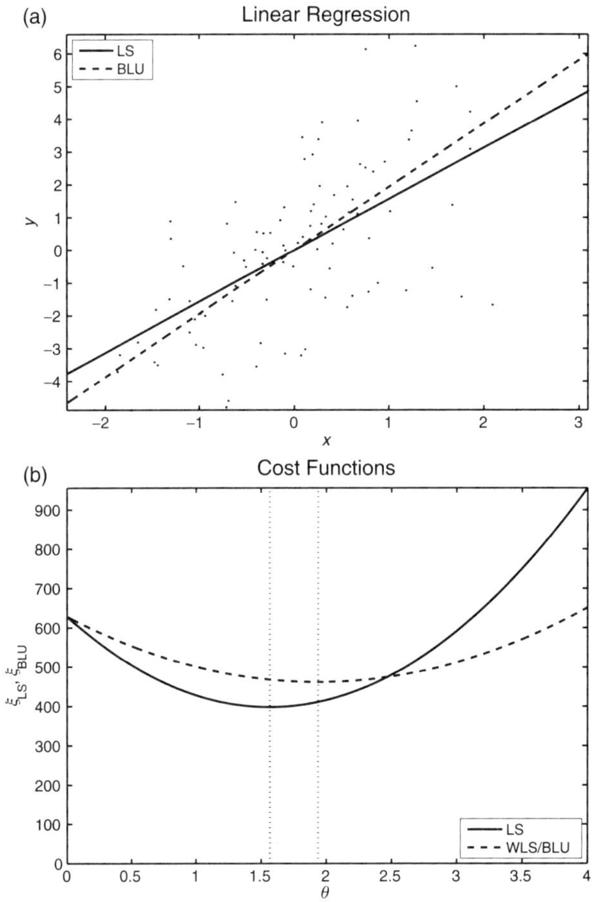

FIGURE 9.21 Estimation results for Example 9.48. (a) Linear regression for the LS and BLU estimators. (b) Cost functions for LS and WLS estimators (with $\mathbf{W} = \mathbf{R}_{\mathbf{VV}}^{-1}$ such that the WLS estimator is equivalent to the BLU estimator).

such as whether or not the noise variance is increasing. Since the estimators are random variables, it is possible for the LS estimator to be more accurate than the BLU estimator for a particular set of samples. The variance of θ_{BLU} is still lower though: for this example, it is 0.0225 versus 0.0564 for θ_{LS}. The LS cost function is plotted in Figure 9.21(b), which is compared to the WLS cost function when weight matrix $\mathbf{W}$ incorporates the changing noise variance (recall that the WLS and BLU estimators are equivalent when $\mathbf{W} = \mathbf{R}_{\mathbf{VV}}^{-1}$). The minimum values for the curves corresponding to the estimators above are indicated by the vertical dotted lines. Although the WLS/BLU curve lies below that for LS, this is not necessarily the case. The relative values of the two cost functions are not necessarily meaningful; their minima and the corresponding estimators are important.

9.23 PROPERTIES OF BLU ESTIMATORS

Although the BLU estimator is derived using a technique that differs from the WLS minimization, the result is in fact a special case.

- *Equivalence of BLU and WLS estimators.* When $\mathbf{W} = \mathbf{R}_{\mathbf{VV}}^{-1}$, the BLU and WLS estimators are identical:

$$\boldsymbol{\theta}_{\mathrm{WLS}}|_{\mathbf{W}=\mathbf{R}_{\mathbf{VV}}^{-1}} = \boldsymbol{\theta}_{\mathrm{BLU}}. \tag{9.388}$$

  When $\mathbf{V}$ is uncorrelated such that $\mathbf{R}_{\mathbf{VV}} = \sigma_V^2\mathbf{I}$, the LS and BLU estimators are equivalent. This is a restatement of the LS property given earlier.
- *Unbiased.* Since the derivation of the BLU estimator includes an unbiasedness constraint, $\boldsymbol{\theta}_{\mathrm{BLU}}$ is obviously unbiased for zero-mean noise:

$$\begin{aligned}\mathcal{E}[\boldsymbol{\theta}_{\mathrm{BLU}}] &= \left(\mathbf{A}^T\mathbf{R}_{\mathbf{VV}}^{-1}\mathbf{A}\right)^{-1}\mathbf{A}^T\mathbf{R}_{\mathbf{VV}}^{-1}\mathcal{E}[\mathbf{Y}],\\ &= \left(\mathbf{A}^T\mathbf{R}_{\mathbf{VV}}^{-1}\mathbf{A}\right)^{-1}\mathbf{A}^T\mathbf{R}_{\mathbf{VV}}^{-1}(\mathbf{A}\boldsymbol{\theta} + \mathcal{E}[\mathbf{V}]) = \boldsymbol{\theta}.\end{aligned} \tag{9.389}$$

- *Covariance matrix.* The covariance matrix of the BLU estimator is

$$\begin{aligned}\mathrm{cov}[\boldsymbol{\theta}_{\mathrm{BLU}}] &= \mathcal{E}[(\boldsymbol{\theta}_{\mathrm{BLU}} - \boldsymbol{\theta})(\boldsymbol{\theta}_{\mathrm{BLU}} - \boldsymbol{\theta})^T],\\ &= \mathcal{E}[\mathbf{B}\mathbf{V}\mathbf{V}^T\mathbf{B}] = \mathbf{B}\mathbf{R}_{\mathbf{VV}}\mathbf{B}^T.\end{aligned} \tag{9.390}$$

  Substituting (9.368) gives

$$\begin{aligned}\mathrm{cov}[\boldsymbol{\theta}_{\mathrm{BLU}}] &= \left(\mathbf{A}^T\mathbf{R}_{\mathbf{VV}}^{-1}\mathbf{A}\right)^{-1}\mathbf{A}^T\mathbf{R}_{\mathbf{VV}}^{-1}\mathbf{R}_{\mathbf{VV}}\mathbf{R}_{\mathbf{VV}}^{-1}\mathbf{A}\left(\mathbf{A}^T\mathbf{R}_{\mathbf{VV}}^{-1}\mathbf{A}\right)^{-1},\\ &= \left(\mathbf{A}^T\mathbf{R}_{\mathbf{VV}}^{-1}\mathbf{A}\right)^{-1}.\end{aligned} \tag{9.391}$$

- *Minimum* ξ_{BLU}. The minimum value of the cost function is the trace of the covariance matrix in (9.391):

$$\xi_{\mathrm{BLU,min}} = \mathrm{tr}\left[\mathbf{B}\mathbf{R}_{\mathbf{VV}}\mathbf{B}^T\right] = \mathrm{tr}\left[\left(\mathbf{A}^T\mathbf{R}_{\mathbf{VV}}^{-1}\mathbf{A}\right)^{-1}\right]. \tag{9.392}$$

  It is interesting to note that the minimum of the cost function for the WLS estimator depends on the samples $\mathbf{Y}$, which is not the case here. This result is due to the fact that the BLU estimator is derived by minimizing its covariance matrix based on an *underlying model* for $\mathbf{Y}$.
- *Efficient.* Finally, we show that the BLU estimator is the most efficient *linear* unbiased estimator; it is the "best" as its name implies.

Theorem 9.9. The BLU estimator $\boldsymbol{\theta}_{\mathrm{BLU}}$ has the minimum variance of all linear unbiased estimators.

Proof. Define the following difference matrix:

$$\mathbf{D} \triangleq \mathrm{cov}[\hat{\boldsymbol{\theta}}] - \mathrm{cov}[\boldsymbol{\theta}_{\mathrm{BLU}}], \tag{9.393}$$

where $\hat{\boldsymbol{\theta}}$ is another linear unbiased estimator: it is of the form $\mathbf{FY}$ with $\mathcal{E}[\mathbf{FY}] = \boldsymbol{\theta}$. The goal is to show that $\mathbf{D}$ is PSD for every nonzero vector $\mathbf{u}$: $\mathbf{u}^T\mathbf{D}\mathbf{u} \geq 0$, and in this sense the covariance of the BLU estimator is the smallest. A similar approach was used to show that the LS estimator is the BLU estimator when $\mathbf{R}_{\mathbf{VV}} = \mathbf{I}$; the result here applies to any noise covariance matrix $\mathbf{R}_{\mathbf{VV}}$. The covariance matrices of the two estimators have the same form:

$$\mathrm{cov}[\boldsymbol{\theta}_{\mathrm{BLU}}] = \mathbf{B}\mathbf{R}_{\mathbf{VV}}\mathbf{B}^T, \quad \mathrm{cov}[\hat{\boldsymbol{\theta}}] = \mathbf{F}\mathbf{R}_{\mathbf{VV}}\mathbf{F}^T. \tag{9.394}$$

When substituting $\mathbf{B}_{\text{BLU}}$, we need to ensure that both estimators are unbiased, which means $\mathbf{BA} = \mathbf{FA} = \mathbf{I}$. Inserting the various quantities and rearranging the result gives

$$\mathbf{D} = (\mathbf{F} - \mathbf{B})\mathbf{R}_{\mathbf{VV}}(\mathbf{F} - \mathbf{B})^T + (\mathbf{F}\mathbf{R}_{\mathbf{VV}}\mathbf{B}^T + \mathbf{B}\mathbf{R}_{\mathbf{VV}}\mathbf{F}^T) - 2\mathbf{B}\mathbf{R}_{\mathbf{VV}}\mathbf{B}^T. \tag{9.395}$$

Substituting (9.368) for $\mathbf{B}$, one of the middle terms becomes

$$\mathbf{B}_{\text{BLU}}\mathbf{R}_{\mathbf{VV}}\mathbf{F}^T = \left(\mathbf{A}^T\mathbf{R}_{\mathbf{VV}}^{-1}\mathbf{A}\right)^{-1}\mathbf{A}^T\mathbf{R}_{\mathbf{VV}}^{-1}\mathbf{R}_{\mathbf{VV}}\mathbf{F}^T = \left(\mathbf{A}^T\mathbf{R}_{\mathbf{VV}}^{-1}\mathbf{A}\right)^{-1}, \tag{9.396}$$

where in the last expression the unbiasedness constraint $\mathbf{A}^T\mathbf{F}^T = (\mathbf{FA})^T = \mathbf{I}$ has been used. The same result is obtained for the other middle term because it is the transpose of $\mathbf{B}\mathbf{R}_{\mathbf{VV}}\mathbf{F}^T$. Substituting $\mathbf{B}_{\text{BLU}}$ into the last term of (9.395) gives the covariance of the BLU estimator in (9.391), and thus four terms cancel, yielding

$$\mathbf{D} = (\mathbf{F} - \mathbf{B}_{\text{BLU}})\mathbf{R}_{\mathbf{VV}}(\mathbf{F} - \mathbf{B}_{\text{BLU}})^T, \tag{9.397}$$

which completes the proof because $\mathbf{D}$ is PSD. □

PROBLEMS

Sufficient Statistics and UMVU Estimation

9.1 For N iid Poisson samples, use the definition of sufficiency to find a sufficient statistic for parameter α.

9.2 For N iid uniform samples with pdf $f_X(x) = I_{[\theta_1,x]}(x)I_{[x,\theta_2]}(x)$, use the factorization theorem to find a sufficient statistic for $\{\theta_1, \theta_2\}$.

9.3 Repeat Problem 9.2 for N iid gamma samples to find a sufficient statistic for $\boldsymbol{\theta} = \{\alpha, r\}$.

9.4 Derive the expectations in (9.30) for the minimum and maximum.

9.5 Find the minimal sufficient statistic from M iid samples of the chi-square distribution with parameter N.

9.6 Use the ratio approach in (9.34) to find the minimal sufficient statistic from N iid samples of the Poisson random variable with parameter α.

9.7 For the exponential random variable with location parameter c:

$$f_X(x) = \lambda \exp(-\lambda(x - c))I_{[c,\infty)}(x) \tag{9.398}$$

show that the minimal sufficient statistic for c (with fixed λ) is $X_{(1)}$ from N iid samples.

9.8 For N iid Laplace samples with parameters $\{\mu, \alpha\}$, find a minimal sufficient statistic for (a) α (with $\mu = 0$ fixed) and then (b) μ (with $\alpha = 1$).

9.9 For N iid uniform samples on $[0, \theta]$, $X_{(N)}$ is minimal sufficient for θ. The joint pdf of $\{X_{(m)}, X_{(n)}\}$ for $1 \leq m \leq n \leq N$ is given in (4.269) of Problem 4.41. Compute $\mathcal{E}[X_{(m)}|X_{(N)}]$ in the Rao–Blackwell theorem and specify the function $c(\theta)$.

9.10 Show that the sample mean $\bar{X}$ is a complete sufficient statistic for μ of the Gaussian distribution with known σ.

9.11 For N iid Gaussian samples with parameters $\{\mu, \sigma = \mu\}$, find a minimal sufficient statistic for μ.

9.12 For N iid half-normal samples with parameter α, find the UMVU estimator for α^2.

9.13 For M iid binomial samples with parameters $\{N, p\}$, find the UMVU estimator for p^2 (assume that $N > 2$ is known).

9.14 The Pitman estimator for location θ is obtained from iid samples as follows:

$$T = \frac{\int_{-\infty}^{\infty} \theta \prod_{n=1}^{N} f_X(x_n - \theta) d\theta}{\int_{-\infty}^{\infty} \prod_{n=1}^{N} f_X(x_n - \theta) d\theta}. \tag{9.399}$$

Find the Pitman estimator for $\theta = \mu$ when X is Gaussian with parameters $\{\mu, \sigma\}$ and σ is known.

MSE Estimation

9.15 For the additive noise problem $X = Y + V$, where Y and V are independent random variables, find the MSE, MAE, and ML estimators of Y when V is Gaussian with parameters $\{\mu = 0, \sigma_V\}$ and Y is symmetric Bernoulli with outcomes $\{\pm a\}$ and parameter p.

9.16 For Problem 9.15, let the MSE estimator be restricted to the following affine form:

$$g_{\text{MSE}}(X) = aX + b. \tag{9.400}$$

Assuming that V has mean $\mu = 1$, find a and b that minimize the MSE.

9.17 Repeat Problem 9.15 when the noise V has a Laplace distribution with parameters $\{\mu = 0, \alpha\}$.

9.18 Let random sequence $X[k]$ for $k \in \mathcal{N}$ have autocorrelation function $R_{XX}[m] = (2/3)^{|m|}$. Consider the following MSE estimator:

$$X_{\text{MSE}}[k] = aX[1] + bX[N] \tag{9.401}$$

for $k \in \{2, \ldots, N-1\}$. Find the optimal parameters $\{a, b\}$ that minimize $\mathcal{E}[(X[k] - X_{\text{MS}}[k])^2]$.

9.19 Suppose that

$$X_{\text{MSE}} = \sum_{m=1}^{M} a_m Y_m = \mathbf{a}^T \mathbf{Y} \tag{9.402}$$

is an estimator for X based on M correlated random variables $\{Y_m\}$, with $\mathbf{a} \triangleq [a_1, \ldots, a_M]^T$ and $\mathbf{Y} \triangleq [Y_1, \ldots, Y_M]^T$. Find an expression for $\mathbf{a}$ that minimizes $\mathcal{E}[(X - X_{\text{MSE}})^2]$.

MAP and ML Estimation

9.20 Find the MAP estimator for the Poisson distribution with parameter α, assuming that α is a standard Gaussian random variable.

9.21 Verify the expression in (9.287).

9.22 Let measurements be generated as follows:

$$Y = X + 2A + A^2, \tag{9.403}$$

where X is exponential with parameter $\lambda = 1$ and A has the following pdf:

$$f_A(a) = (1/4)[3\delta(a-1) + \delta(a-3)]. \tag{9.404}$$

For a single measurement of Y, find (a) the ML estimator for A and (b) the MAP estimator for A.

9.23 Find the ML estimator for θ based on N samples of X which is uniform on $[0, \theta]$.

9.24 Repeat Problem 9.23 when X has a triangular distribution with parameters $\{a = 0, b = 1, c\}$.

9.25 Derive an implicit expression for the ML estimator of θ based on N samples from the Weibull distribution with parameters $\{\theta = r, \alpha = 1\}$.

9.26 Consider the following two hypotheses for the distribution of random variable X. H_0: Gaussian with parameters $\{\mu = 0, \sigma\}$. H_1: Laplace with parameters $\{\mu = 0, \alpha\}$. Use the LRT to find an expression for the test condition for x with $\eta = 1$. The pdfs are compared in Figure 3.20.

9.27 Repeat Problem 9.26 for H_1: Cauchy with parameters $\{c = 0, \alpha\}$.

9.28 Derive the EM algorithm for estimating $\boldsymbol{\theta} = [a_1, \ldots, a_M, \mu_1, \ldots, \mu_M]^T$ of a finite Gaussian mixture with known variances, similar to that leading to (9.289) and (9.295).

Method of Moments

9.29 Find the MoM estimator based on N iid samples from the inverse Gaussian distribution with parameters $\{\mu, \alpha\}$.

9.30 Repeat Problem 9.29 for the negative binomial random variable with parameters $\{N, P\}$.

9.31 Repeat Problem 9.29 for the Laplace random variable with parameters $\{\mu = 1, \alpha\}$.

Cramér–Rao Lower Bound

9.32 (a) Find the CRLB for an unbiased estimator of p for the geometric distribution with support $\mathcal{Z}^+$. (b) Derive the ML estimator for p.

9.33 Find the Fisher information for p of the binomial random variable with parameters $\{p, N\}$.

9.34 Repeat Problem 9.33 for λ of the gamma distribution with parameters $\{\lambda, r\}$.

9.35 Derive the CRLB result in (F.71) of Appendix F for a biased estimator with bias $B(\theta)$.

Least-Squares Estimation

9.36 The LS estimator has the form $\hat{\mathbf{Y}}_{\text{WLS}} = \mathbf{PY}$, where $\mathbf{P} \triangleq \mathbf{A}(\mathbf{A}^T\mathbf{WA})^{-1}\mathbf{A}^T\mathbf{W}$ is a projection matrix because, as shown in Figure 9.18, it projects $\mathbf{Y}$ onto the range of $\mathbf{A}$. Obviously $\tilde{\mathbf{Y}}_{\text{WLS}} = \mathbf{Y} - \hat{\mathbf{Y}}_{\text{WLS}} = (\mathbf{I} - \mathbf{P})\mathbf{Y}$. Demonstrate that $\mathbf{P}$ is an idempotent matrix and it is orthogonal to $\mathbf{I} - \mathbf{P}$.

9.37 Let the LS estimator be constrained as follows:

$$\boldsymbol{\theta}_{\text{LS}} = \min_{\theta} \left[\tilde{\mathbf{Y}}^T\tilde{\mathbf{Y}} + \lambda(\mathbf{c}^T\boldsymbol{\theta} - c)\right], \tag{9.405}$$

where λ is a Lagrange multiplier, and the constraint is $\mathbf{c}^T\boldsymbol{\theta} = c$ for $\mathbf{c} \in \mathcal{R}^M$ and scalar c. Find an expression for $\boldsymbol{\theta}_{\text{LS}}$.

9.38 Suppose the LS model is extended as follows:

$$\mathbf{Y} = \mathbf{A}\boldsymbol{\theta} + \mathbf{b} + \mathbf{V}, \tag{9.406}$$

where $\mathbf{b} \in \mathcal{R}^N$ is a fixed known vector. Formulate the weighted LS problem for this model and find an expression for $\boldsymbol{\theta}_{\text{LS}}$.

Computer Problems

9.39 Use MATLAB to implement the EM algorithm described in this chapter for the finite mixture model, with the E-step and M-step given by (9.291) and (9.295), respectively. Assume that

$$f_X(x) = (0.3)(1/2)\exp(-x/2)u(x) + (0.7)\sqrt{2/\pi}\exp(-x^2/2)u(x), \tag{9.407}$$

which is a mixture of exponential ($\lambda = 1/2$) and half-normal ($\alpha = 1$) distributions. Plot the actual and estimated mixture distributions.

9.40 Consider the following model:

$$Y = \theta_1 + \theta_2 X - \theta_3 X^3 + V, \tag{9.408}$$

where X is uniform on $[0, 1]$ and V is Gaussian with parameters $\{\mu = 0, \sigma = 0.1\}$. Even though X^3 appears in this expression, it is still a linear model with respect to the parameters. Find the WLS estimator with identity weighting $\mathbf{W} = \mathbf{I}$ for the three parameters $\{\theta_1 = 1, \theta_2 = 4, \theta_3 = -2\}$. Use MATLAB to plot $N = 500$ samples of $\{X, Y\}$ and the line generated by the estimate $\boldsymbol{\theta}_{\text{LS}}$, similar to that shown in Figure 9.16.

FURTHER READING

The topics covered in this chapter can be explored further in the following references. Sufficient statistics and estimation: Haykin (2002), Lehmann (1983), Kailath, Sayed, and Hassibi (2000), Kay (1993), Mendel (1995), Poor (1994), Sorenson (1980), and Van Trees (1968). Principal component regression: Jolliffe (1986) and Mosteller and Tukey (1977).

A NOTE ON PART III OF THE BOOK

Part III consists of four chapters on Applications in Signal Processing and Communications. They are located at the Web Site www.wiley.com/go/randomprocesses. The following summarizes their contents:

- Chapter 10: Communication Systems and Information Theory
- Chapter 11: Optimal Filtering
- Chapter 12: Adaptive Filtering
- Chapter 13: Equalization, Beamforming, and Direction Finding

APPENDICES

INTRODUCTION TO APPENDICES

In the following appendices, we provide some background material that the reader might find useful as a review of several topics needed for the various chapters. Of course, most of this information can be found in other textbooks used in undergraduate engineering courses (such as the review of frequency-domain transforms), but it is included here so that the book is reasonably self-contained.

- **Appendix A.** Summaries of all the univariate random variables described in Chapter 3 are included for ease of reference. Each summary provides expressions for the probability density function (pdf), cumulative distribution function (cdf), moments, the characteristic function, and entropy, as well as example plots of each distribution (showing the mean and width $\mu_X \pm \sigma_X$). The random variables are organized alphabetically and with separate sections for continuous and discrete distributions.
- **Appendix B.** Continuity, boundedness, and various properties of functions are covered. We also include specific functions that are used throughout the book and provide example plots to illustrate their behavior. Nearly all of these functions arise because of the wide range of distributions in Chapter 3 and the need to describe the pdf, cdf, moments, and entropy in concise functional forms.
- **Appendix C.** Properties and transform pairs for the following frequency-domain transforms are summarized: (i) Laplace transform, (ii) continuous-time Fourier transform, (iii) z-transform, and (iv) discrete-time Fourier transform. Bilateral and unilateral transforms are discussed. We also include an example of Cauchy's residue theorem for finding the inverse z-transform.
- **Appendix D.** This appendix contains a brief review of the Riemann integral, and describes Riemann–Stieltjes and Lebesgue integrals which are useful for computing expectations. Functional forms of the pdfs for continuous random variables are summarized, which can be useful for computing some integrals. We show how to integrate some "special" functions of two variables such as $\delta(x - y)$ and $u(x - y)$.
- **Appendix E.** Various identities are provided, included trigonometric identities, Stirling's formula, as well as some series expansions and closed-form sums. Pmf sums, which are analogous to the pdf integrals in Appendix D, are also provided for discrete random variables.
- **Appendix F.** Several inequalities are provided in this appendix, which are useful for bounding certain probabilities and expectations. These include the Cauchy–Schwarz and triangle inequalities, the Chernoff bound, Jenson's inequality, and the Cramér–Rao lower bound which is used for parameter estimation in Chapter 9.
- **Appendix G.** Important properties of matrices and vectors are summarized in this appendix. Some matrix decompositions are described, as are the four fundamental subspaces and the matrix inversion lemma. The rules of differentiation with respect to a vector (real and complex) are also provided with examples.

A Summaries of Univariate Parametric Distributions

A.1 NOTATION

The notation below is used in the following summaries of distributions for random variables with one, two, or three parameters:

$f_X(x)$	probability density function (pdf)
$p_X[x]$	probability mass function (pmf)
$F_X(x)$	cumulative distribution function (cdf)
m_e	median
m_o	mode
μ_n	nth raw (noncentral) moment $\mathcal{E}[X^n]$
$\mu_{c,n}$	nth central moment $\mathcal{E}[(X-\mu_X)^n]$
μ_X	mean $\mathcal{E}[X]=\mu_1$
σ_X^2	variance $\mathcal{E}[(X-\mu_X)^2]=\mu_{c,2}$
γ_1	coefficient of skewness $\mu_{c,3}/\sigma_X^3$
γ_2	coefficient of excess kurtosis $\mu_{c,4}/\sigma_X^4-3$
$\Phi_X(\omega)$	characteristic function $\mathcal{E}[\exp(j\omega X)]$ (continuous random variable)
$\Phi_X(j\omega)$	characteristic function $\mathcal{E}[\exp(j\omega X)]$ (discrete random variable)
$H(X)$	entropy (discrete random variable)
$h(X)$	differential entropy (continuous random variable)

For $H(X)$ and $h(X)$, either $\log(x)$ (arbitrary base) or $\ln(x)$ (base e) is used depending on whether or not the entropy expression is more compact for distributions containing an exponential. In order to use the digamma function for some closed forms, $\ln(x)$ is needed. A general guide for the entropy expressions in these tables is as follows: (i) for $\log(x)$, any base can be used in the expression and (ii) for $\ln(x)$, the expression applies only for the natural logarithm and must be converted if another base is desired.

Since a general expression is not easily written for all moments (or central moments) for many of the random variables, only low-order results are provided. The unit-step function $u(x)$ is used to define the support of a continuous random variable, whereas the indicator function is used in Chapter 3. (However, for the triangular distribution the indicator function is used here to write compact expressions.) For discrete random variables, the probability density function (pdf) and cumulative distribution function (cdf) are written as weighted sums of the Dirac delta function $\delta(x)$ and the unit-step function $u(x)$, respectively. As a result, the cdf of a discrete random variable is obtained from its pdf simply by replacing $\delta(x)$ with $u(x)$. The probability mass function (pmf) is also given for completeness, as some readers might prefer that notation.

The following functions and numbers described in Appendix B are used in this appendix:

$B(a,b)$	beta function $=\Gamma(a)\Gamma(b)/\Gamma(a+b)$
$B_y(a,b)$	incomplete beta function

Probability, Random Variables, and Random Processes: Theory and Signal Processing Applications, First Edition. John J. Shynk.

$\Gamma(a)$	gamma function
$\gamma(a, b)$	lower incomplete gamma function
$\Gamma(a, b)$	upper incomplete gamma function
$\psi(x)$	digamma function
$\zeta(x)$	zeta function
$I_{[a,b]}(x)$	indicator function (continuous x)
$I_{\{a,b\}}[x]$	indicator function (discrete x)
$\operatorname{sgn}(x)$	signum function
$Q(x)$	Q-function
$\operatorname{erf}(x)$	error function
$\operatorname{erfi}(x)$	imaginary error function
$I_\alpha(x)$	modified Bessel function (first kind)
$K_\alpha(x)$	modified Bessel function (second kind)
$Q_m(a, b)$	Marcum Q-function
$L_v(x)$	Laguerre polynomial
${}_1F_1(a; b; x)$	confluent hypergeometric function (first kind)
$\Psi(a; b; x)$	confluent hypergeometric function (second kind)
$\lfloor x \rfloor$	floor function
$\lceil x \rceil$	ceiling function
B_n	Bernoulli numbers
γ	Euler–Mascheroni constant (≈ 0.5772)

For notational simplicity, we do not always include brackets or parentheses when dividing by a product of quantities. For example, it should be clear that x^2 is divided by the product $2\sigma^2$ in the expression $\exp(x^2/2\sigma^2)$, even though that would not be the order of operations in most computer programming languages (including MATLAB)®. Finally, some quantities are not included in the summaries for various reasons as follows:

Complicated	$\Longrightarrow$	the expression is complicated and may not be useful (typically, the characteristic function or entropy)
Implicit	$\Longrightarrow$	cannot write an explicit expression (such as the median or mode)
Does not exist	$\Longrightarrow$	moments are undefined or infinite (Cauchy) (also the median for Bernoulli and discrete uniform)
Not unique	$\Longrightarrow$	the mode can take on many values (uniform) (binomial and Poisson can take on two values)

A.2 FURTHER READING

There does not appear to be a single reference containing all of the information in this appendix. The reader can find many of these results in the probability books listed in the bibliography at the end of this book. Perhaps the following books are most useful, especially for moments and other properties of the univariate distributions: Johnson, Kemp, and Kotz (2005), Johnson, Kotz, and Balakrishnan (1994), and Johnson, Kotz, and Balakrishnan (1995). Some additional references with summaries include: Krishnamoorthy (2006), Mood, Graybill, and Boes (1974), Pal, Jin, and Lim (2006), Peebles (1993), Weisstein (2003), Yates and Goodman (2005), and Zwillinger and Kokoska (2000).

Although we do not summarize results here for the multivariate distributions covered in Chapter 4, the reader might find the following references useful: Gumbel (1960), Johnson, Kotz, and Balakrishnan (1997), Kotz, Balakrishnan, and Johnson (2000), and Mardia (1970).

Most of the expressions for differential entropy can be found in Verdugo Lazo and Rathie (1978).

Finally, we provide a few references for specific distributions that are not widely discussed in many textbooks on probability. Half-normal: Leone, Nelson, and Nottingham (1961). Inverse Gaussian: Chhikara and Folks (1989) and Seshadri (1999). Laplace: Kotz, Kozubowski, and Podgórski (2001). Logistic: Balakrishnan, Ed. (1992). Lognormal: Limpert, Stahel, and Abbt (2001). Weibull: Rousu (1973).

A.3 CONTINUOUS RANDOM VARIABLES

A.3.1 Beta (Arcsine for $\alpha = \beta = 1/2$, Power Function for $\beta = 1$)

Parameters $\alpha, \beta > 0$. Support $x \in [0, 1]$.

$$f_X(x) = [1/B(\alpha, \beta)]x^{\alpha-1}(1-x)^{\beta-1}[u(x) - u(x-1)],$$

$$F_X(x) = [B_x(\alpha, \beta)/B(\alpha, \beta)][u(x) - u(x-1)] + u(x-1),$$

$$\mu_X = \alpha/(\alpha + \beta), \quad \sigma_X^2 = \alpha\beta/[(\alpha + \beta + 1)(\alpha + \beta)^2],$$

$$m_e = \text{implicit}, \quad m_o = (\alpha - 1)/(\alpha + \beta - 2) \ \ (\text{for } \alpha, \beta > 1),$$

$$\mu_n = B(n + \alpha, \beta)/B(\alpha, \beta), \quad \gamma_1 = \sqrt{(\alpha + \beta + 1)/\alpha\beta}\,\frac{2(\beta - \alpha)}{(\alpha + \beta + 2)},$$

$$\gamma_2 = 6\frac{\alpha^3 + \alpha^2(1 - 2\beta) - 2\alpha\beta(\beta + 2) + \beta^2(\beta + 1)}{\alpha\beta(\alpha + \beta + 2)(\alpha + \beta + 3)}, \quad \Phi_X(\omega) = {}_1F_1(\alpha; \beta; j\omega),$$

$$h(X) = (\alpha + \beta - 2)\psi(\alpha + \beta) - (\alpha - 1)\psi(\alpha) - (\beta - 1)\psi(\beta) + \ln(B(\alpha, \beta)).$$

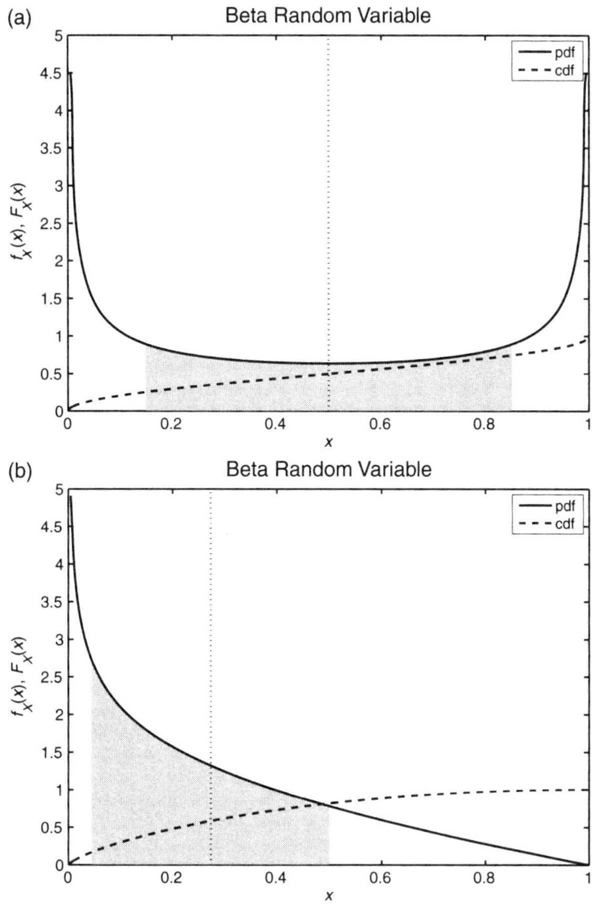

FIGURE A.1 Beta pdf and cdf. (a) $\alpha = \beta = 0.5$. (a) $\alpha = 0.75$ and $\beta = 2$. Dotted lines specify μ_X and shaded regions denote $[\mu_X - \sigma_X, \mu_X + \sigma_X]$.

A.3.2 Cauchy

Parameters $c \in \mathcal{R}$ and $\alpha > 0$. Support $x \in \mathcal{R}$.

$$f_X(x) = (\alpha/\pi)[(x-c)^2 + \alpha^2]^{-1},$$

$$F_X(x) = 1/2 + (1/\pi)\arctan((x-c)/\alpha),$$

$$\mu_X = \text{does not exist}, \quad \sigma_X^2 = \text{does not exist},$$

$$m_e = c, \quad m_o = c,$$

$$\mu_n = \text{do not exist}, \quad \mu_{c,n} = \text{do not exist},$$

$$\gamma_1 = \text{does not exist}, \quad \gamma_2 = \text{does not exist},$$

$$\Phi_X(\omega) = \exp(j\omega c - \alpha|\omega|), \quad h(X) = \log(4\pi\alpha).$$

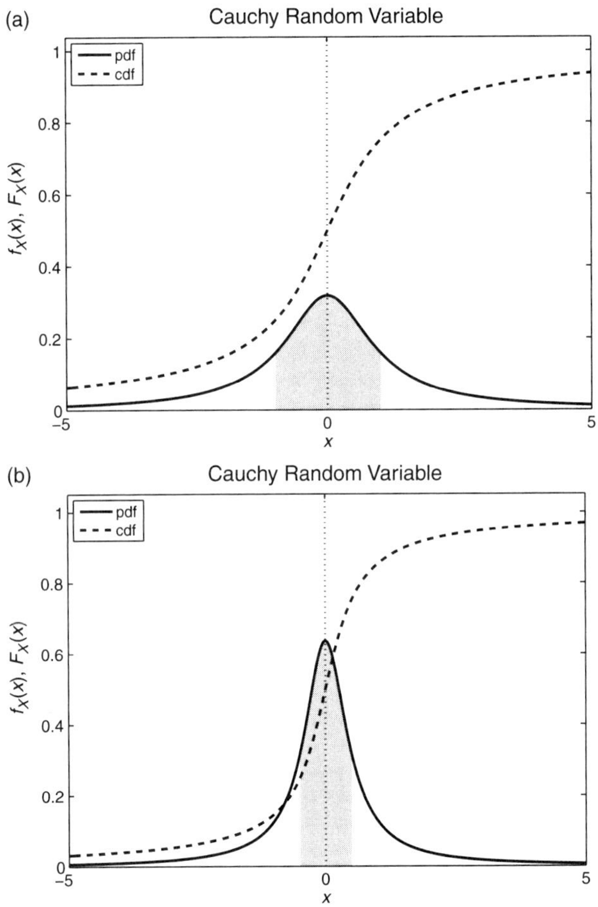

FIGURE A.2 Cauchy pdf and cdf. (a) $c = 0$ and $\alpha = 1$. (b) $c = 0$ and $\alpha = 0.5$. Dotted lines specify m_e and shaded regions denote $\left[F_X^{-1}(1/4), F_X^{-1}(3/4)\right]$ (since μ_X and σ_X do not exist).

A.3.3 Chi

Parameter $N \in \mathcal{N}$. Support $x \in \mathcal{R}^+$.

$$f_X(x) = [x^{N-1}/2^{N/2-1}\Gamma(N/2)]\exp(-x^2/2)u(x),$$

$$F_X(x) = \left[\gamma(N/2, x^2/2)/\,\Gamma(N/2)\right]u(x),$$

$$\mu_X = \sqrt{2}\Gamma((N+1)/2)/\,\Gamma(N/2), \quad \sigma_X^2 = N - 2[\Gamma((N+1)/2)/\,\Gamma(N/2)]^2,$$

$$m_e = \text{implicit}, \quad m_o = \sqrt{N-1}\ (\text{for } N \geq 1),$$

$$\mu_n = 2^{n/2}\Gamma((N+n)/2)/\,\Gamma(N/2), \quad \gamma_1 = \text{complicated}, \quad \gamma_2 = \text{complicated},$$

$$\Phi_X(\omega) = [1/\,\Gamma(N/2)]\sum_{n=0}^{\infty}\left[(j\omega\sqrt{2})^n/n!\right]\Gamma((N+n)/2),$$

$$h(X) = (1/2)[N - (N-1)\psi(N/2) - \ln(2)] + \ln(\Gamma(N/2)).$$

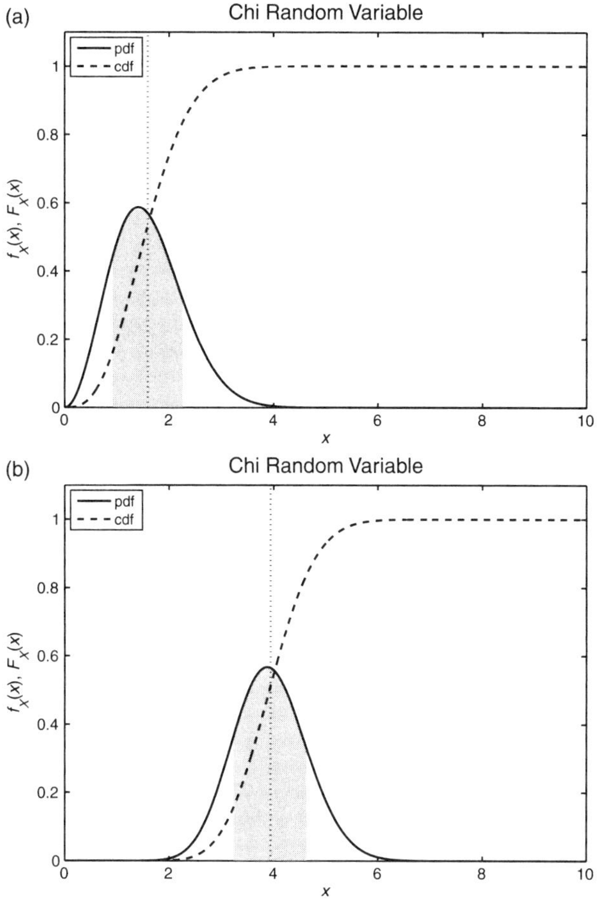

FIGURE A.3 Chi pdf and cdf. (a) $N = 3$. (b) $N = 16$. Dotted lines specify μ_X and shaded regions denote $[\mu_X - \sigma_X, \mu_X + \sigma_X]$.

A.3.4 Chi-Square

Parameter $N \in \mathcal{N}$. Support $x \in \mathcal{R}^+$.

$$f_X(x) = [x^{N/2-1}/2^{N/2}\Gamma(N/2)]\exp(-x/2)u(x),$$

$$F_X(x) = [\gamma(N/2, x/2)/\Gamma(N/2)]\,u(x),$$

$$\mu_X = N, \quad \sigma_X^2 = 2N,$$

$$m_e = \text{implicit}, \quad m_o = \max(0, N-2),$$

$$\mu_n = 2^n\Gamma(n + N/2)/\Gamma(N/2),$$

$$\gamma_1 = \sqrt{8/N}, \quad \gamma_2 = 12/N,$$

$$\Phi_X(\omega) = 1/(1 - 2j\omega)^{N/2},$$

$$h(X) = (1/2)[N + (N-2)\psi(N/2)] + \ln(2\Gamma(N/2)).$$

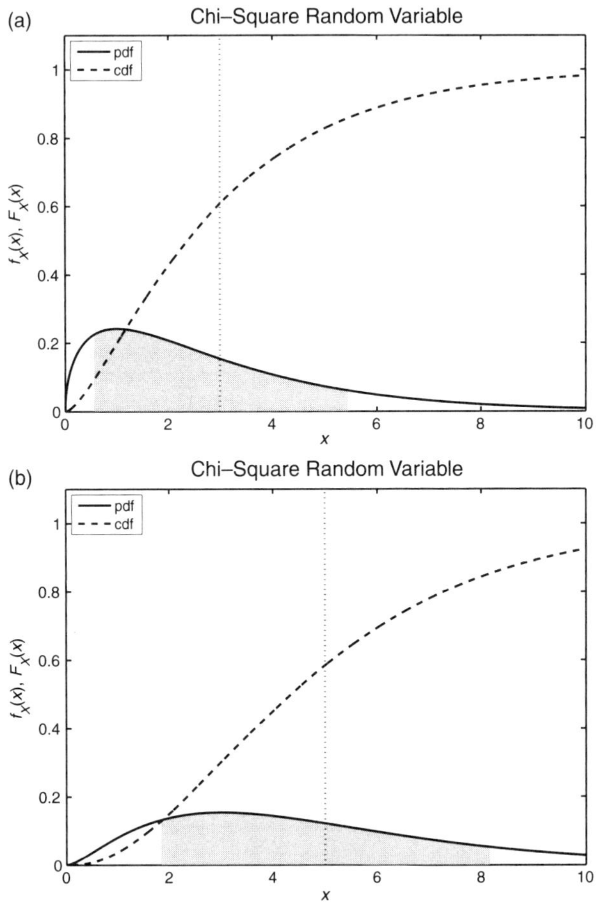

FIGURE A.4 Chi-square pdf and cdf. (a) $N = 3$. (b) $N = 5$. Dotted lines specify μ_X and shaded regions denote $[\mu_X - \sigma_X, \mu_X + \sigma_X]$.

A.3.5 Exponential (Shifted by c)

Parameters $c \in \mathcal{R}$ and $\lambda > 0$. Support $x \in [c, \infty)$.

$$f_X(x) = \lambda \exp(-\lambda(x - c))u(x - c),$$

$$F_X(x) = [1 - \exp(-\lambda(x - c))]u(x - c),$$

$$\mu_X = c + 1/\lambda, \quad \sigma_X^2 = 1/\lambda^2,$$

$$m_e = c + \ln(2)/\lambda, \quad m_o = c,$$

$$\mu_n = \Gamma(n + 1, \lambda c)\exp(\lambda c)/\lambda^n \quad (= n!/\lambda^n \text{ for } c = 0),$$

$$\mu_{c,n} = \Gamma(n + 1, -1)/e\lambda^n, \quad \gamma_1 = 2, \quad \gamma_2 = 6,$$

$$\Phi_X(\omega) = \lambda \exp(j\omega c)/(\lambda - j\omega),$$

$$h(X) = \log(e/\lambda).$$

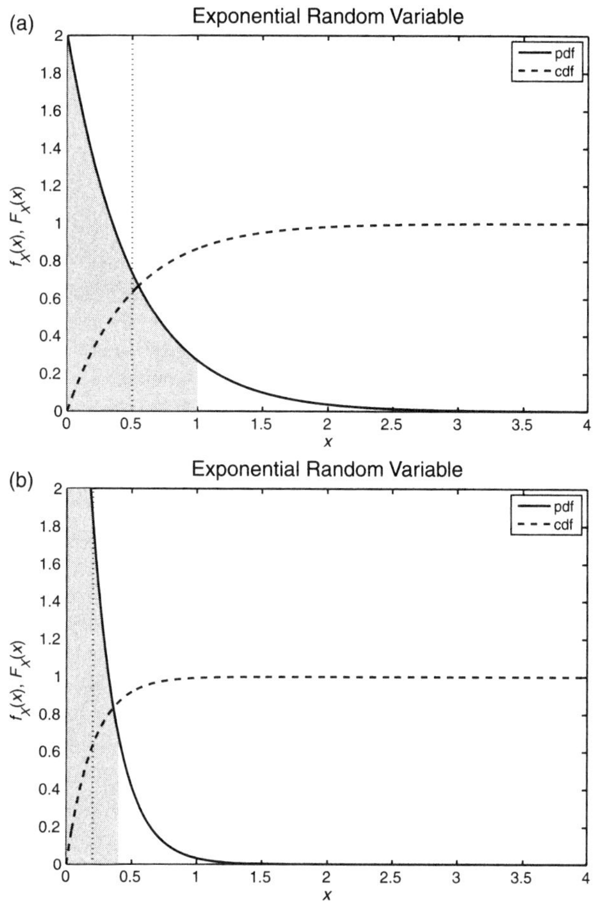

FIGURE A.5 Exponential pdf and cdf. (a) $c = 0$ and $\lambda = 2$. (b) $c = 0$ and $\lambda = 5$. Dotted lines specify μ_X and shaded regions denote $[\mu_X - \sigma_X, \mu_X + \sigma_X]$.

A.3.6 Extreme Value (Type I: Gumbel)

Parameters $c \in \mathcal{R}$ and $\alpha > 0$. Support $x \in \mathcal{R}$.

$$f_X(x) = (1/\alpha)\exp(-(x-c)/\alpha)\exp(-\exp(-(x-c)/\alpha)),$$

$$F_X(x) = \exp(-\exp(-(x-c)/\alpha)),$$

$$\mu_X = c + \alpha\gamma, \quad \sigma_X^2 = \pi^2\alpha^2/6,$$

$$m_e = c - \alpha\ln(\ln(2)), \quad m_o = c,$$

$$\mu_n = \text{complicated},$$

$$\gamma_1 = 12\zeta(3)\sqrt{6}/\pi^3 \approx 1.1395, \quad \gamma_2 = 12/5 = 2.4,$$

$$\Phi_X(\omega) = \Gamma(1 - j\omega\alpha)\exp(j\omega c),$$

$$h(X) = \gamma + \ln(\alpha e).$$

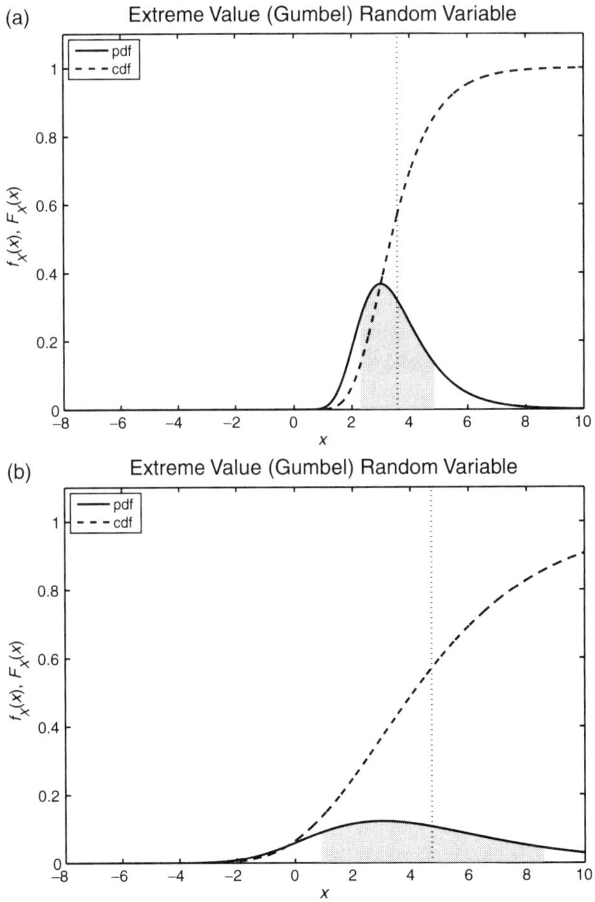

FIGURE A.6 Extreme Value (Gumbel) pdf and cdf. (a) $c = 3$ and $\alpha = 1$. (b) $c = 3$ and $\alpha = 3$. Dotted lines specify μ_X and shaded regions denote $[\mu_X - \sigma_X, \mu_X + \sigma_X]$.

A.3.7 *F*-Distribution

Parameters $m, n \in \mathcal{N}$. Support $x \in \mathcal{R}^+$.

$$f_X(x) = \frac{(m/n)^{m/2} x^{(m-2)/2}}{B(m/2, n/2)[1 + (m/n)x]^{(m+n)/2}} u(x),$$

$$F_X(x) = [B_{mx/(mx+n)}(m/2, n/2)/B(m/2, n/2)]u(x),$$

$$\mu_X = n/(n-2) \text{ (for } n > 2), \quad \sigma_X^2 = \frac{2n^2(m+n-2)}{m(n-2)^2(n-4)} \text{ (for } n > 4), \quad m_e = \text{implicit},$$

$$m_o = n(m-2)/m(n+2) \text{ (for } m \geq 2), \quad \mu_l = (n/m)^l \frac{\Gamma(m/2+l)\Gamma(n/2-l)}{\Gamma(m/2)\Gamma(n/2)} \text{ (for } n > 2l),$$

$$\gamma_1 = \frac{(2m+n-2)\sqrt{8(n-4)}}{(n-6)\sqrt{m(m+n-2)}} \text{ (for } n > 6), \quad \gamma_2 = \frac{12[(n-2)^2(n-4) + m(m+n-2)(5n-22)]}{m(n-6)(n-8)(m+n-2)} \text{(for } n > 8),$$

$$\Phi_X(\omega) = [\Gamma((m+n)/2)/\Gamma(n/2)]\Psi(m/2; 1-n/2; -j\omega n/m),$$

$$h(X) = (1 - m/2)\psi(m/2) - (1 + n/2)\psi(n/2)$$
$$+ [(m+n)/2]\psi((m+n)/2) + \ln((m/n)B(m/2, n/2)).$$

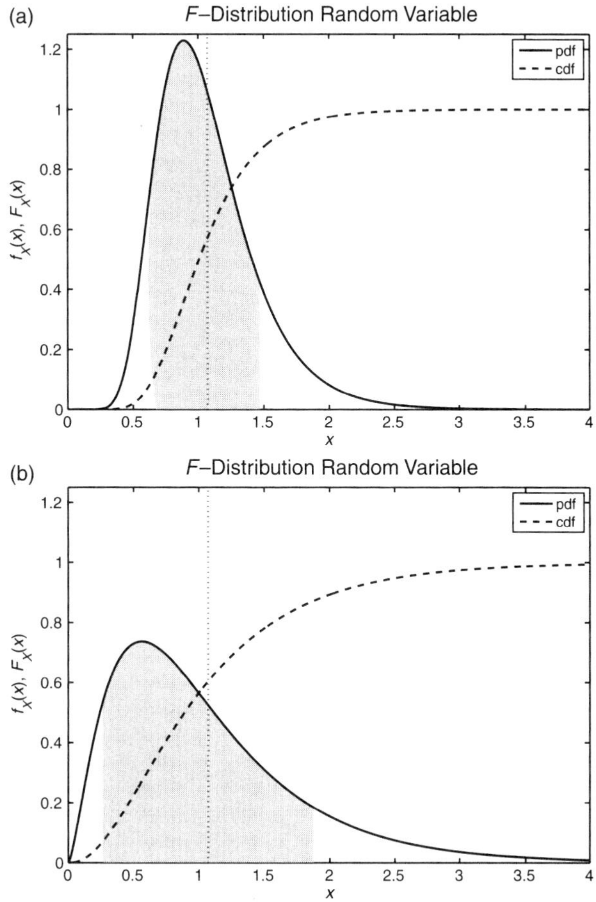

FIGURE A.7 *F*-distribution pdf and cdf. (a) $m = 40$ and $n = 30$. (b) $m = 5$ and $n = 30$. Dotted lines specify μ_X and shaded regions denote $[\mu_X - \sigma_X, \mu_X + \sigma_X]$.

A.3.8 Gamma (Erlang for $r \in \mathcal{N}$ with $\Gamma(r) = (r-1)!$)

Parameters $\lambda, r > 0$. Support $x \in \mathcal{R}^+$.

$$f_X(x) = [x^{r-1}\lambda^r / \Gamma(r)] \exp(-\lambda x) u(x),$$

$$F_X(x) = [\gamma(r, \lambda x) / \Gamma(r)] u(x),$$

$$\mu_X = r/\lambda, \quad \sigma_X^2 = r/\lambda^2,$$

$$m_e = \text{implicit}, \quad m_o = (r-1)/\lambda \text{ (for } r \geq 1),$$

$$\mu_n = \Gamma(r+n)/\lambda^n \Gamma(r),$$

$$\gamma_1 = \sqrt{4/r}, \quad \gamma_2 = 6/r,$$

$$\Phi_X(\omega) = [\lambda/(\lambda - j\omega)]^r,$$

$$h(X) = r + (1-r)\psi(r) + \ln(\Gamma(r)/\lambda).$$

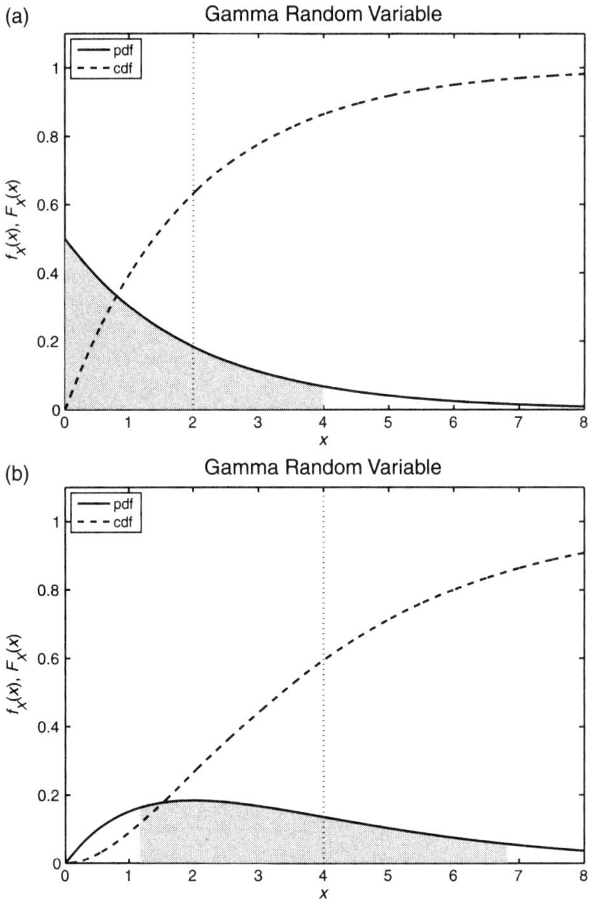

FIGURE A.8 Gamma pdf and cdf. (a) $\lambda = 0.5$ and $r = 1$. (b) $\lambda = 0.5$ and $r = 2$. Dotted lines specify μ_X and shaded regions denote $[\mu_X - \sigma_X, \mu_X + \sigma_X]$.

A.3.9 Gaussian (Normal)

Parameters $\mu \in \mathcal{R}$ and $\sigma > 0$. Support $x \in \mathcal{R}$.

$$f_X(x) = \frac{1}{\sqrt{2\pi\sigma^2}} \exp(-(x-\mu)^2/2\sigma^2),$$

$$F_X(x) = 1 - Q\left(\frac{x-\mu}{\sigma}\right) = (1/2)\left[1 + \text{erf}\left(\frac{x-\mu}{\sqrt{2\sigma^2}}\right)\right],$$

$$\mu_X = \mu, \quad \sigma_X^2 = \sigma^2, \quad m_e = \mu, \quad m_o = \mu, \quad \mu_2 = \sigma^2 + \mu^2, \quad \mu_3 = \mu(\mu^2 + 3\sigma^2),$$

$$\gamma_1 = 0, \quad \gamma_2 = 0, \quad \mu_{c,n} = \begin{cases} 0, & n \text{ odd} \\ \dfrac{\sigma^n n!}{2^{n/2}(n/2)!}, & n \text{ even}, \end{cases}$$

$$\Phi_X(\omega) = \exp(j\omega\mu - \sigma^2\omega^2/2), \quad h(X) = (1/2)\log(2\pi\sigma^2 e).$$

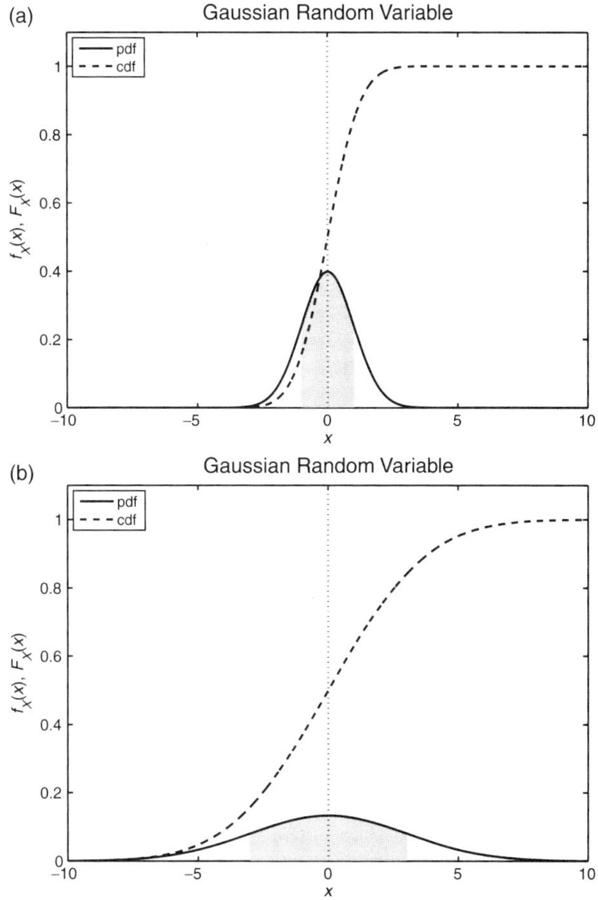

FIGURE A.9 Gaussian pdf and cdf. (a) $\mu = 0$ and $\sigma = 1$. (b) $\mu = 0$ and $\sigma = 3$. Dotted lines specify μ_X and shaded regions denote $[\mu_X - \sigma_X, \mu_X + \sigma_X]$.

A.3.10 Half-Normal (Folded Normal)

Parameter $\alpha > 0$. Support $x \in \mathcal{R}^+$.

$$f_X(x) = \sqrt{2/\pi\alpha^2}\exp(-x^2/2\alpha^2)u(x),$$

$$F_X(x) = \mathrm{erf}(x/\alpha\sqrt{2})u(x),$$

$$\mu_X = \alpha\sqrt{2/\pi}, \quad \sigma_X^2 = \alpha^2(\pi - 2)/\pi, \quad m_e = \text{implicit}, \quad m_o = 0,$$

$$\mu_n = (2^{n/2}\alpha^n/\sqrt{\pi})\Gamma((n+1)/2),$$

$$\gamma_1 = \sqrt{2}(4-\pi)/(\pi-2)^{3/2} \approx 0.9953,$$

$$\gamma_2 = 8(\pi-3)/(\pi-2)^2 \approx 0.8692,$$

$$\Phi_X(\omega) = \left[1 + j\,\mathrm{sgn}(\omega)\mathrm{erfi}(\alpha|\omega|/\sqrt{2})\right]\exp(-\omega^2\alpha^2/2),$$

$$h(X) = (1/2)\log(\pi\alpha^2 e/2).$$

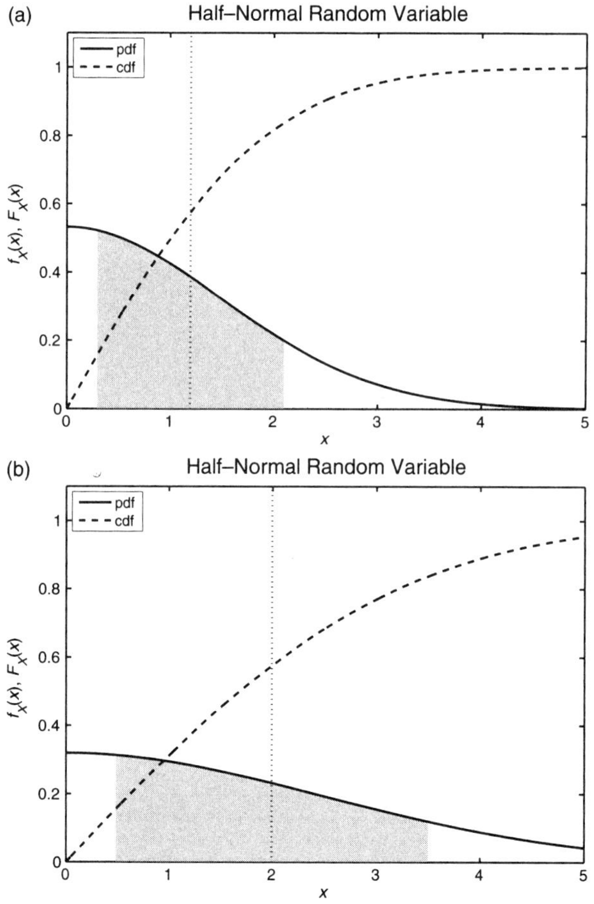

FIGURE A.10 Half-normal pdf and cdf. (a) $\alpha = 1.5$. (b) $\alpha = 2.5$. Dotted lines specify μ_X and shaded regions denote $[\mu_X - \sigma_X, \mu_X + \sigma_X]$.

A.3.11 Inverse Gaussian (Wald)

Parameters $\mu, \alpha > 0$. Support $x \in \mathcal{R}^+$.

$$f_X(x) = \sqrt{\frac{\alpha}{2\pi x^3}} \exp(-\alpha(x-\mu)^2/2\mu^2 x)u(x),$$

$$F_X(x) = \left(1 - Q\left((x/\mu - 1)\sqrt{\alpha/x}\right) + \exp(2\alpha/\mu)\left[Q\left((x/\mu + 1)\sqrt{\alpha/x}\right)\right]\right)u(x),$$

$$\mu_X = \mu, \quad \sigma_X^2 = \mu^3/\alpha, \quad m_e = \text{implicit}, \quad m_o = \mu\sqrt{1 + (3\mu/2\alpha)^2} - 3\mu^2/2\alpha,$$

$$\mu_2 = \mu^2(1 + \mu/\alpha), \quad \mu_3 = \mu^3(1 + 3\mu/\alpha + 3\mu^2/\alpha^2), \quad \mu_{c,3} = 3\mu^5/\alpha^2,$$

$$\gamma_1 = 3\sqrt{\mu/\alpha}, \quad \gamma_2 = 15\mu/\alpha,$$

$$\Phi_X(\omega) = \exp\left((\alpha/\mu)\left[1 - \sqrt{1 - 2j\omega\mu^2/\alpha}\right]\right), \quad h(X) = \text{complicated}.$$

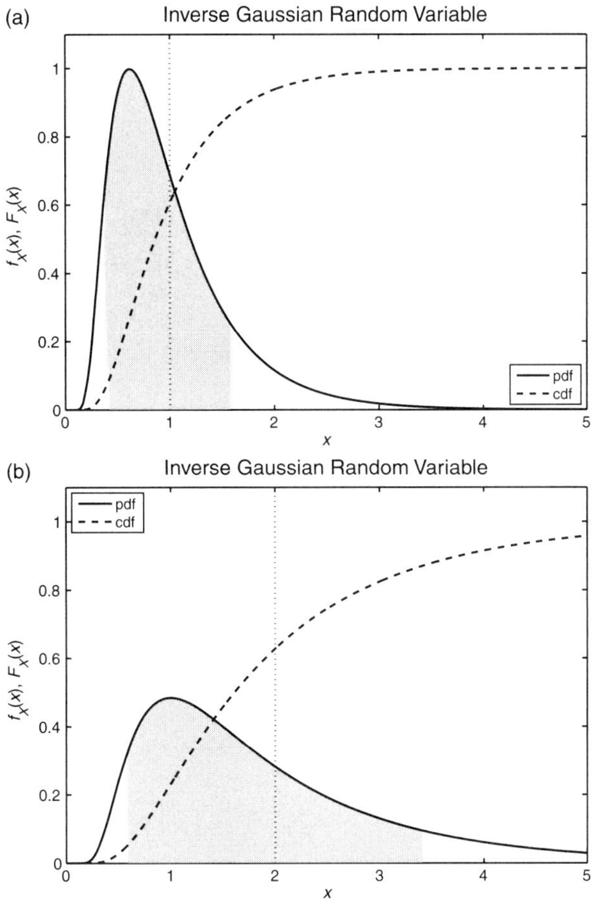

FIGURE A.11 Inverse Gaussian pdf and cdf. (a) $\mu = 1$ and $\alpha = 3$. (b) $\mu = 2$ and $\alpha = 4$. Dotted lines specify μ_X and shaded regions denote $[\mu_X - \sigma_X, \mu_X + \sigma_X]$.

A.3.12 Laplace (Double-Sided Exponential)

Parameters $\mu \in \mathcal{R}$ and $\alpha > 0$. Support $x \in \mathcal{R}$.

$$f_X(x) = (1/2\alpha)\exp(-|x - \mu|/\alpha),$$

$$F_X(x) = (1/2)(1 + \text{sgn}(x - \mu)[1 - \exp(-|x - \mu|/\alpha)]),$$

$$\mu_X = \mu, \quad \sigma_X^2 = 2\alpha^2, \quad m_e = \mu, \quad m_o = \mu,$$

$$\mu_{c,n} = \begin{cases} 0, & n \text{ odd} \\ \alpha^n n!, & n \text{ even}, \end{cases}$$

$$\gamma_1 = 0, \quad \gamma_2 = 3,$$

$$\Phi_X(\omega) = \exp(j\omega\mu)/(1 + \alpha^2\omega^2),$$

$$h(X) = \log(2\alpha e).$$

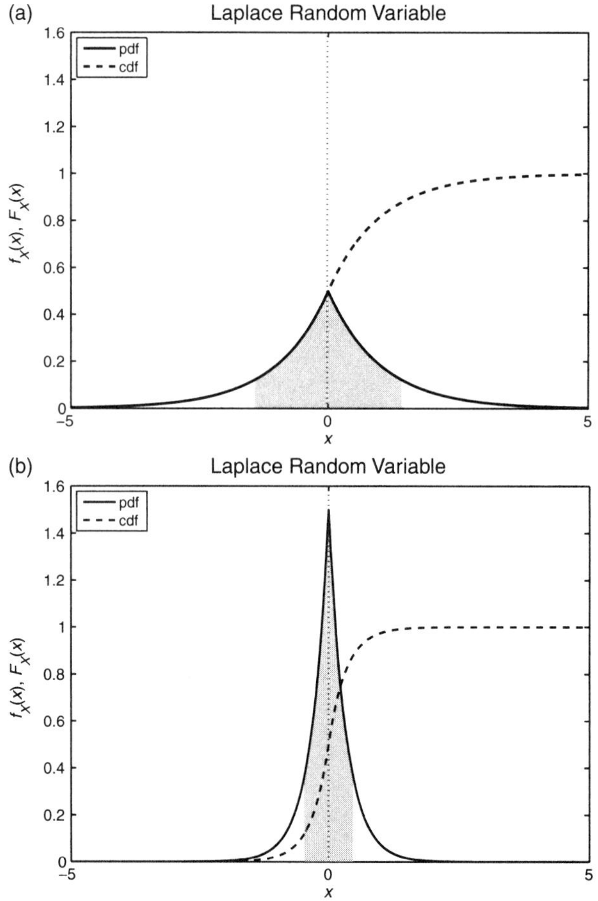

FIGURE A.12 Laplace pdf and cdf. (a) $\mu = 0$ and $\alpha = 1$. (b) $\mu = 0$ and $\alpha = 1/3$. Dotted lines specify μ_X and shaded regions denote $[\mu_X - \sigma_X, \mu_X + \sigma_X]$.

A.3.13 Logistic (Sigmoid for $\{\mu = 0, \alpha = 1\}$)

Parameters $\mu \in \mathcal{R}$ and $\alpha > 0$. Support $x \in \mathcal{R}$.

$$f_X(x) = (1/\alpha)\exp(-(x-\mu)/\alpha)\left[1+\exp(-(x-\mu)/\alpha)\right]^{-2} = (1/4\alpha)\text{sech}^2((x-\mu)/2\alpha),$$

$$F_X(x) = [1+\exp(-(x-\mu)/\alpha)]^{-1} = (1/2)[1+\tanh((x-\mu)/2\alpha)],$$

$$\mu_X = \mu, \quad \sigma_X^2 = \pi^2\alpha^2/3,$$

$$m_e = \mu, \quad m_o = \mu,$$

$$\mu_{c,n} = (\alpha\pi)^n(2^n-2)|B_n|,$$

$$\gamma_1 = 0, \quad \gamma_2 = 6/5,$$

$$\Phi_X(\omega) = \omega\pi\alpha\exp(j\omega\mu)/\sinh(\omega\pi\alpha)$$

$$h(X) = 2 + \ln(\alpha).$$

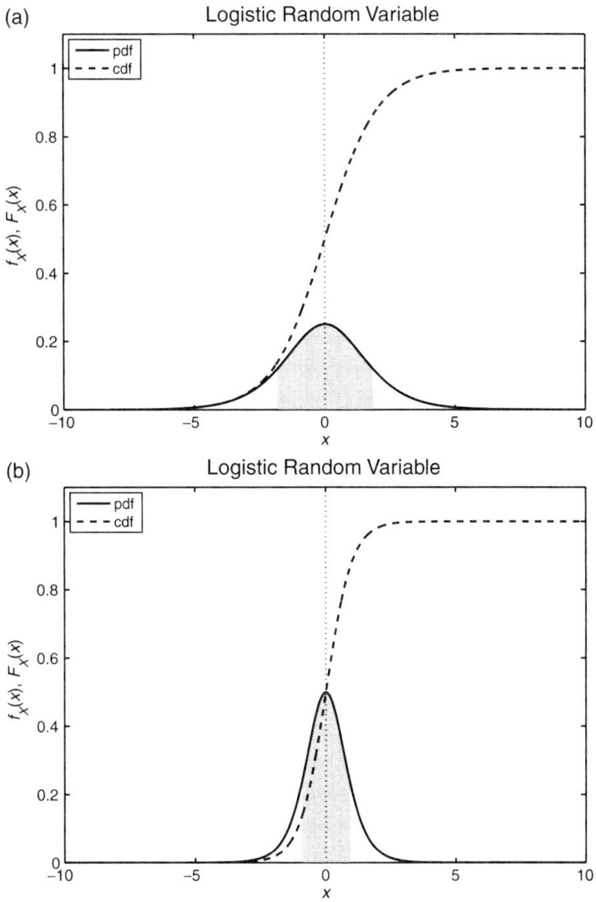

FIGURE A.13 Logistic pdf and cdf. (a) $\mu = 0$ and $\alpha = 1$. (b) $\mu = 0$ and $\alpha = 0.5$. Dotted lines specify μ_X and shaded regions denote $[\mu_X - \sigma_X, \mu_X + \sigma_X]$.

A.3.14 Log-Normal

Parameters $\mu \in \mathcal{R}$ and $\sigma > 0$. Support $x \in \mathcal{R}^+$.

$$f_X(x) = \frac{1}{x\sqrt{2\pi\sigma^2}} \exp(-(\ln(x) - \mu)^2/2\sigma^2)u(x),$$

$$F_X(x) = \left[1 - Q\left(\frac{\ln(x) - \mu}{\sigma}\right)\right]u(x) = (1/2)\left[1 + \text{erf}\left(\frac{\ln(x) - \mu}{\sqrt{2\sigma^2}}\right)\right]u(x),$$

$$\mu_X = \exp(\mu + \sigma^2/2), \quad \sigma_X^2 = \exp(2\mu + 2\sigma^2) - \exp(2\mu + \sigma^2),$$

$$m_e = \exp(\mu), \quad m_o = \exp(\mu - \sigma^2), \quad \mu_n = \exp(n\mu + n^2\sigma^2/2),$$

$$\gamma_1 = [\exp(\sigma^2) + 2]\sqrt{\exp(\sigma^2) - 1}, \quad \gamma_2 = \exp(4\sigma^2) + 2\exp(3\sigma^2) + 3\exp(2\sigma^2) - 6,$$

$$\Phi_X(\omega) = \text{complicated}, \quad h(X) = \mu + (1/2)\ln(2\pi\sigma^2 e).$$

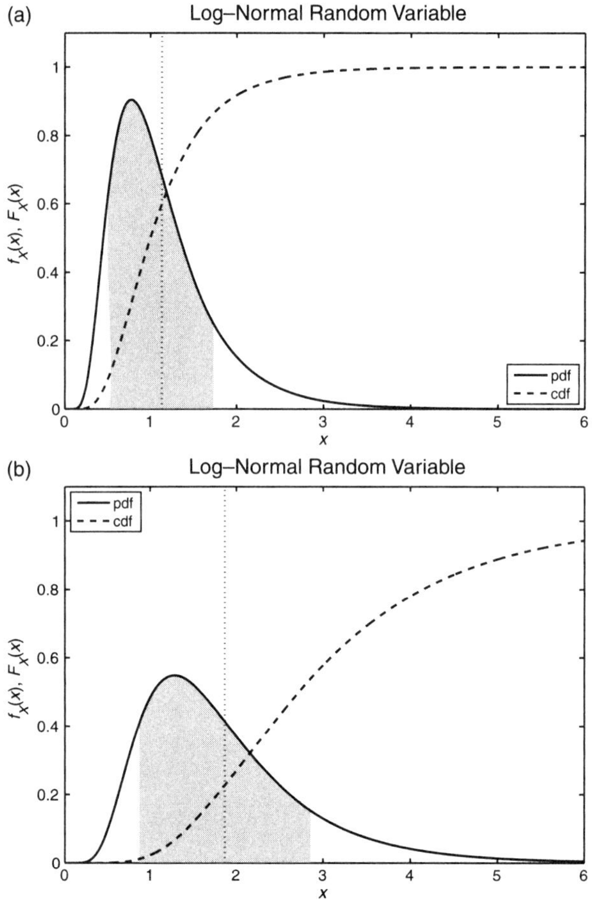

FIGURE A.14 Log-normal pdf and cdf. (a) $\mu = 0$ and $\sigma = 0.5$. (b) $\mu = 0.5$ and $\sigma = 0.5$. Dotted lines specify μ_X and shaded regions denote $[\mu_X - \sigma_X, \mu_X + \sigma_X]$.

A.3.15 Maxwell–Boltzmann

Parameter $\alpha > 0$. Support $x \in \mathcal{R}^+$.

$$f_X(x) = (1/\alpha^3)\sqrt{2/\pi}\,x^2 \exp(-x^2/2\alpha^2)u(x),$$

$$F_X(x) = \left[\mathrm{erf}(x/\alpha\sqrt{2}) - \sqrt{2/\pi}(x/\alpha)\exp(-x^2/2\alpha^2)\right]u(x),$$

$$\mu_X = 2\alpha\sqrt{2/\pi}, \quad \sigma_X^2 = \alpha^2(3\pi - 8)/\pi,$$

$$m_e = \text{implicit}, \quad m_o = \alpha\sqrt{2}, \quad \mu_n = \left[2^{(n+2)/2}\alpha^n/\sqrt{\pi}\right]\Gamma((n+3)/2),$$

$$\gamma_1 = 2\sqrt{2}(5\pi - 16)/(3\pi - 8)^{3/2} \approx -0.4857,$$

$$\gamma_2 = 4(3\pi^2 - 40\pi + 96)/(3\pi - 8)^2 \approx -0.1082,$$

$$\Phi_X(\omega) = j\omega\alpha\sqrt{2/\pi} + (1 - \omega^2\alpha^2)[1 + j\,\mathrm{sgn}(\omega)\mathrm{erfi}(\alpha|\omega|/\sqrt{2})]\exp(-\omega^2\alpha^2/2),$$

$$h(X) = \gamma + (1/2)[\ln(2\pi\alpha^2) - 1].$$

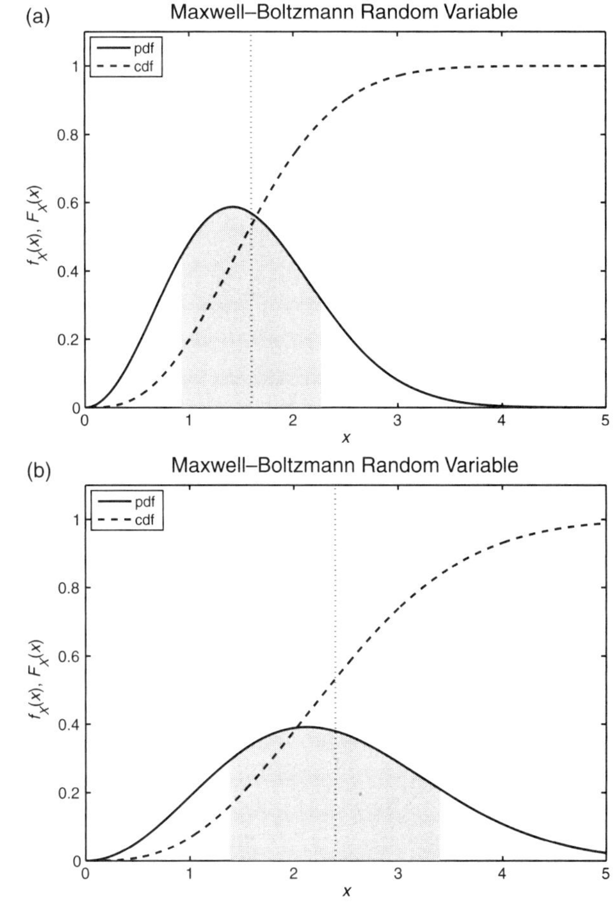

FIGURE A.15 Maxwell–Boltzmann pdf and cdf. (a) $\alpha = 1$. (b) $\alpha = 1.5$. Dotted lines specify μ_X and shaded regions denote $[\mu_X - \sigma_X, \mu_X + \sigma_X]$.

A.3.16 Pareto

Parameters $r, \alpha > 0$. Support $x \in [\alpha, \infty)$.

$$f_X(x) = [r\alpha^r/x^{r+1}]u(x-\alpha),$$

$$F_X(x) = [1-(\alpha/x)^r]u(x-\alpha),$$

$$\mu_X = r\alpha/(r-1) \text{ (for } r>1), \quad \sigma_X^2 = r\alpha^2/[(r-1)^2(r-2)] \text{ (for } r>2),$$

$$m_e = \alpha 2^{1/r}, \quad m_o = \alpha, \quad \mu_n = r\alpha^n/(r-n) \text{ (for } r>n),$$

$$\gamma_1 = [2(r+1)/(r-3)]\sqrt{(r-2)/r} \text{ (for } r>3), \quad \gamma_2 = \frac{6(r^3+r^2-6r-2)}{r(r-3)(r-4)} \text{ (for } r>4),$$

$$\Phi_X(\omega) = r(-j\omega\alpha)^r\Gamma(-r,-j\omega\alpha),$$

$$h(X) = (r+1)/r + \ln(\alpha/r).$$

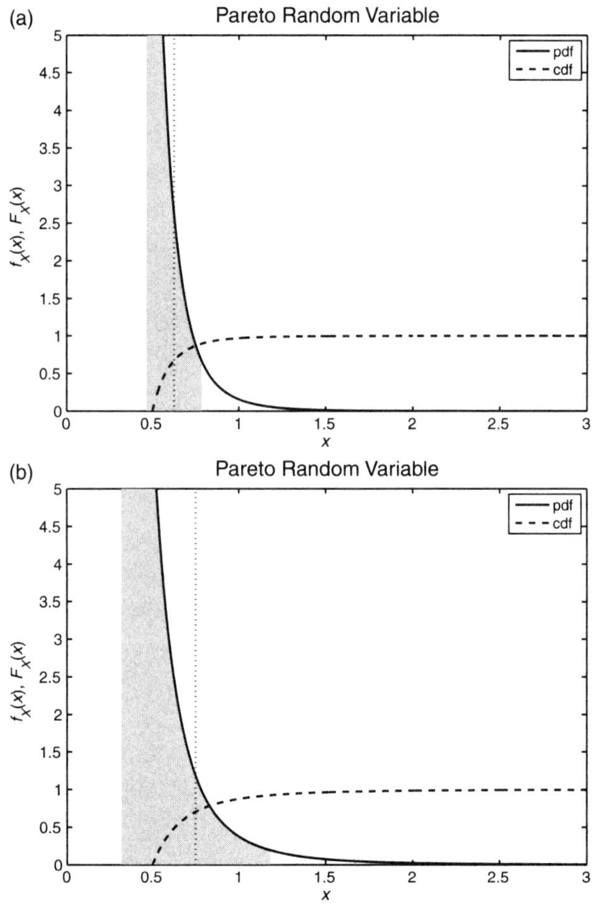

FIGURE A.16 Pareto pdf and cdf. (a) $\alpha = 0.5$ and $r = 5$. (b) $\alpha = 0.5$ and $r = 3$. Dotted lines specify μ_X and shaded regions denote $[\mu_X - \sigma_X, \mu_X + \sigma_X]$.

A.3.17 Rayleigh

Parameter $\alpha > 0$. Support $x \in \mathcal{R}^+$.

$$f_X(x) = (x/\alpha^2)\exp(-x^2/2\alpha^2)u(x),$$

$$F_X(x) = [1 - \exp(-x^2/2\alpha^2)]u(x),$$

$$\mu_X = \alpha\sqrt{\pi/2}, \quad \sigma_X^2 = (4-\pi)\alpha^2/2, \quad m_e = \alpha\sqrt{\ln(4)}, \quad m_o = \alpha,$$

$$\mu_n = 2^{n/2}\alpha^n\Gamma(n/2+1), \quad \mu_{c,n} = \sqrt{\pi/2}(\pi-3)\alpha^3,$$

$$\gamma_1 = 2(\pi-3)\sqrt{\pi}/(4-\pi)^{3/2} \approx 0.6311, \quad \gamma_2 = (24\pi - 6\pi^2 - 16)/(4-\pi)^2 \approx 0.2451,$$

$$\Phi_X(\omega) = 1 - \omega\alpha\sqrt{\pi/2}\exp(-\alpha^2\omega^2/2)[\text{erfi}(\omega\alpha/\sqrt{2}) - j],$$

$$h(X) = \gamma/2 + 1 + \ln(\alpha/\sqrt{2}).$$

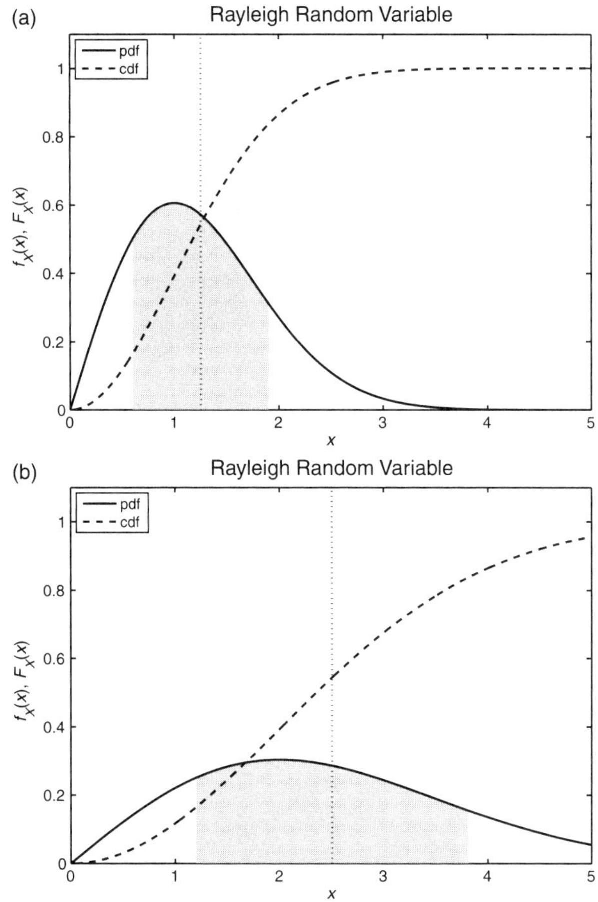

FIGURE A.17 Rayleigh pdf and cdf. (a) $\alpha = 1$. (b) $\alpha = 2$. Dotted lines specify μ_X and shaded regions denote $[\mu_X - \sigma_X, \mu_X + \sigma_X]$.

A.3.18 Rice

Parameters $c, \alpha > 0$. Support $x \in \mathcal{R}^+$.

$$f_X(x) = (x/\alpha^2)\exp(-(x^2+c^2)/2\alpha^2)I_0(cx/\alpha^2)u(x),$$

$$F_X(x) = [1 - Q_1(c/\alpha, x/\alpha)]u(x),$$

$$\mu_X = \alpha\sqrt{\pi/2}L_{1/2}(-c^2/2\alpha^2), \quad \sigma_X^2 = c^2 + 2\alpha^2 - (\pi\alpha^2/2)L_{1/2}^2(-c^2/2\alpha^2),$$

$$m_e = \text{implicit}, \quad m_o = \text{implicit}, \quad \gamma_1 = \text{complicated}, \quad \gamma_2 = \text{complicated},$$

$$\mu_n = (2\alpha^2)^{n/2}\Gamma(n/2+1)L_{n/2}(-c^2/2\alpha),$$

$$\Phi_X(\omega) = \textstyle\sum_{n=0}^{\infty}[(j\omega\alpha\sqrt{2})^n/n!]\Gamma(n/2+1)L_{n/2}(-c^2/2\alpha),$$

$$h(X) = \text{complicated}.$$

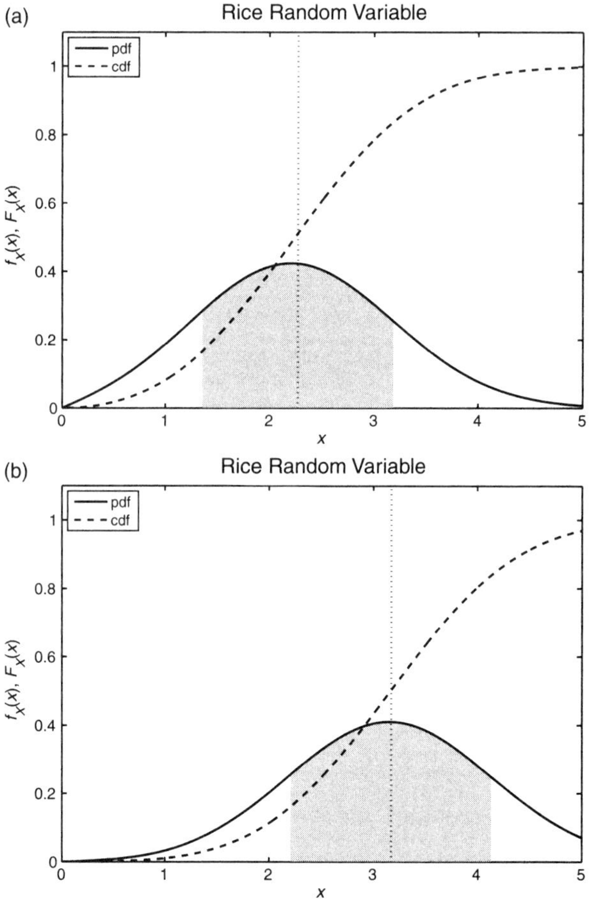

FIGURE A.18 Rice pdf and cdf. (a) $c = 2$ and $\alpha = 1$. (a) $c = 3$ and $\alpha = 1$. Dotted lines specify μ_X and shaded regions denote $[\mu_X - \sigma_X, \mu_X + \sigma_X]$.

A.3.19 Student's *t* Distribution

Parameter $r > 0$. Support $x \in \mathcal{R}$.

$$f_X(x) = (1 + x^2/r)^{-(r+1)/2}/\sqrt{r}B(1/2, r/2),$$

$$F_X(x) = B_{(x+\sqrt{x^2+r})/2\sqrt{x^2+r}}(r/2, r/2)/B(r/2, r/2),$$

$$\mu_X = 0 \text{ (for } r > 1), \quad \sigma_X^2 = r/(r-2) \text{ (for } r > 2), \quad m_e = 0, \quad m_o = 0,$$

$$\mu_{c,n} = \begin{cases} 0, & n \text{ odd} \\ r^{n/2}B((n+1)/2, (r-n)/2)/B(1/2, k/2), & n \text{ even} \end{cases} \text{ (for } r > n),$$

$$\gamma_1 = 0 \text{ (for } r > 3), \quad \gamma_2 = 6/(r-4) \text{ (for } r > 4),$$

$$\Phi_X(\omega) = K_{r/2}(\sqrt{r}|\omega|)\sqrt{(\sqrt{r}|\omega|)^r}\,/\,\Gamma(r/2)2^{r/2-1},$$

$$h(X) = [\psi((r+1)/2) - \psi(r/2)](r+1)/2 + \ln(\sqrt{r}B(1/2, r/2)).$$

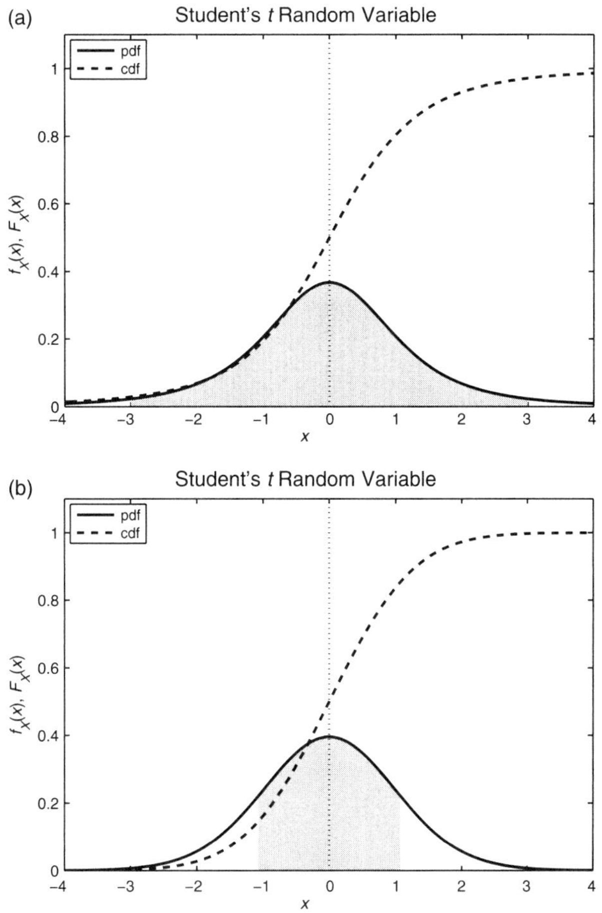

FIGURE A.19 Student's t pdf and cdf. (a) $r = 3$. (b) $r = 30$. Dotted lines specify μ_X and shaded regions denote $[\mu_X - \sigma_X, \mu_X + \sigma_X]$.

A.3.20 Triangular

Parameters $a, b, c \in \mathcal{R}$ (with $a < c < b$). Support $x \in [a, b]$.

$$f_X(x) = [2/(b-a)]\left[\frac{x-a}{c-a}I_{[a,c]}(x) + \frac{b-x}{b-c}I_{(c,b]}(x)\right],$$

$$F_X(x) = [1/(b-a)]\left[\frac{(x-a)^2}{c-a}I_{[a,c]}(x) + \left(b-a-\frac{(b-x)^2}{b-c}\right)I_{(c,b]}(x)\right] + I_{(b,\infty)}(x),$$

$$\mu_X = (a+b+c)/3, \quad \sigma_X^2 = (a^2+b^2+c^2-ab-ac-bc)/18,$$

$$m_o = c, \quad m_e = \begin{cases} b - \sqrt{(b-a)(b-c)/2}, & c \le (a+b)/2 \\ a + \sqrt{(b-a)(c-a)/2}, & c > (a+b)/2, \end{cases}$$

$$\mu_2 = (a^2+b^2+c^2+ab+ac+bc)/6, \quad \gamma_1 = \mu_{c,3}/\sigma_X^3, \quad \gamma_2 = -3/5 = -0.6,$$

$$\mu_{c,3} = -(a+b-2c)(a+c-2b)(b+c-2a)/270, \quad h(X) = 1/2 + \log((b-a)/2),$$

$$\Phi_X(\omega) = 2\frac{(b-a)\exp(j\omega c) - (b-c)\exp(j\omega a) - (c-a)\exp(j\omega b)}{\omega^2(b-a)(b-c)(c-a)}.$$

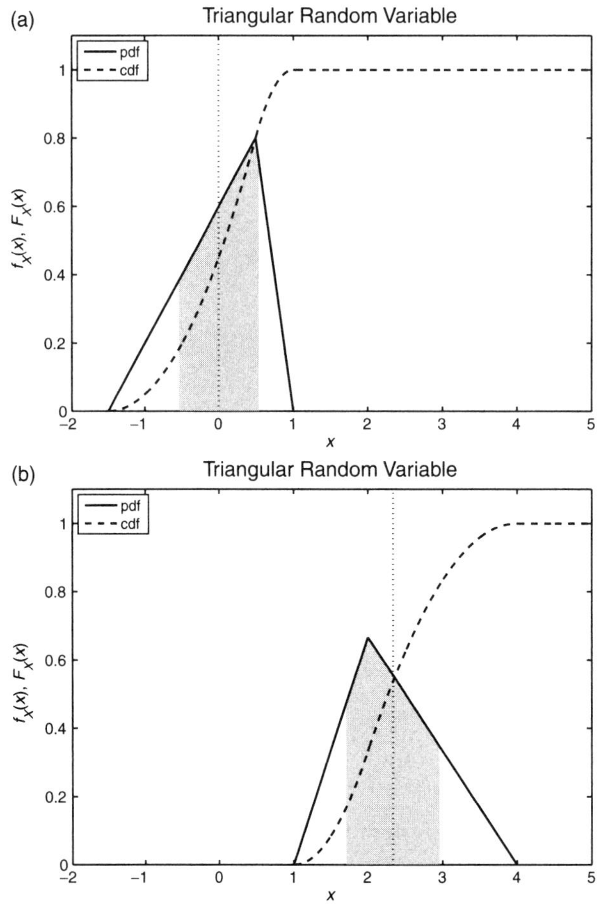

FIGURE A.20 Triangular pdf and cdf. (a) $a = -1.5, b = 1$, and $c = 0.5$. (a) $a = 1, b = 4$, and $c = 2$. Dotted lines specify μ_X and shaded regions denote $[\mu_X - \sigma_X, \mu_X + \sigma_X]$.

A.3.21 Uniform (Continuous)

Parameters $a, b \in \mathcal{R}$ (with $a < b$). Support $x \in [a, b]$.

$$f_X(x) = [1/(b-a)][u(x-a) - u(x-b)],$$

$$F_X(x) = [(x-a)/(b-a)]I_{[a,b]}(x) + I_{(b,\infty)}(x),$$

$$\mu_X = (a+b)/2, \quad \sigma_X^2 = (b-a)^2/12,$$

$$m_e = (a+b)/2, \quad m_o \in [a, b] \text{ (not unique)},$$

$$\mu_n = [b^{n+1} - a^{n+1}]/[(n+1)(b-a)], \quad \mu_{c,n} = \begin{cases} 0, & n \text{ odd} \\ (b-a)^n/[(n+1)2^n], & n \text{ even}, \end{cases}$$

$$\gamma_1 = 0, \quad \gamma_2 = -6/5 = -1.2,$$

$$\Phi_X(\omega) = [\exp(j\omega b) - \exp(j\omega a)]/[j\omega(b-a)], \quad h(X) = \log(b-a).$$

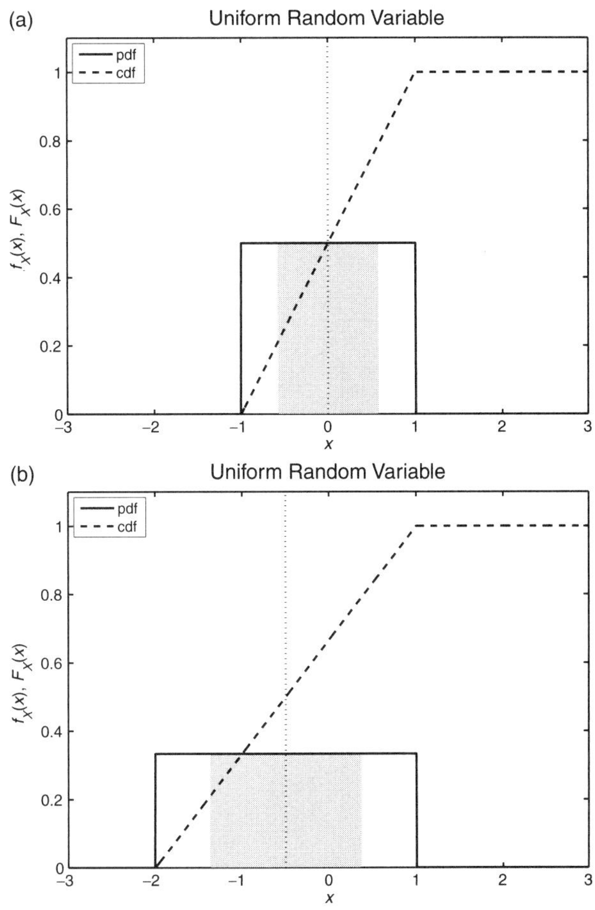

FIGURE A.21 Uniform pdf and cdf. (a) $a = -1$ and $b = 1$. (a) $a = -2$ and $b = 1$. Dotted lines specify μ_X and shaded regions denote $[\mu_X - \sigma_X, \mu_X + \sigma_X]$.

A.3.22 Weibull

Parameters $r, \alpha > 0$. Support $x \in \mathcal{R}^+$.

$$f_X(x) = (r/\alpha)(x/\alpha)^{r-1} \exp(-(x/\alpha)^r)u(x),$$

$$F_X(x) = [1 - \exp(-(x/\alpha)^r)]u(x),$$

$$\mu_X = \alpha\Gamma(1 + 1/r), \quad \sigma_X^2 = \alpha^2 \left[\Gamma(1 + 2/r) - \Gamma^2(1 + 1/r)\right],$$

$$m_e = \alpha[\ln(2)]^{1/r}, \quad m_o = \alpha[(r - 1)/r]^{1/r} \text{ (for } r > 1), \quad \mu_n = \alpha^n\Gamma(1 + n/r),$$

$$\gamma_1 = \frac{\Gamma(1 + 3/r) - 3\Gamma(1 + 1/r)\Gamma(1 + 2/r) + 2\Gamma^3(1 + 1/r)}{[\Gamma(1 + 2/r) - \Gamma^2(1 + 1/r)]^{3/2}},$$

$$\gamma_2 = \frac{\Gamma(1 + 4/r) - 4\Gamma(1 + 1/r)\Gamma(1 + 3/r) - 3\Gamma^2(1 + 2/r) + 12\Gamma^2(1 + 1/r)\Gamma(1 + 2/r) - 6\Gamma^4(1 + 1/r)}{[\Gamma(1 + 2/r) - \Gamma^2(1 + 1/r)]^2},$$

$$\Phi_X(\omega) = \sum_{n=0}^{\infty}[(j\omega\alpha)^n/n!]\Gamma(1 + n/r), \quad h(X) = 1 + (r - 1)\gamma/r + \ln(\alpha/r).$$

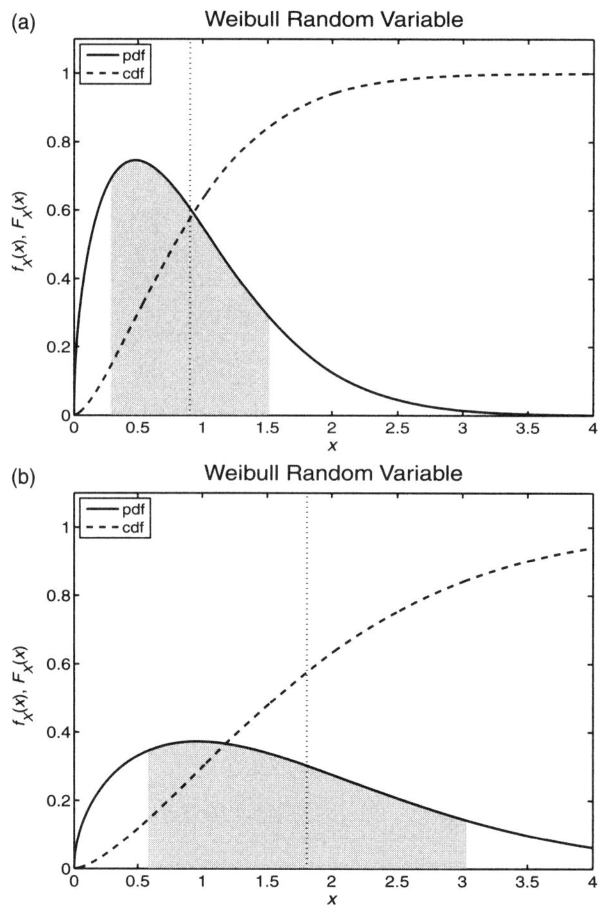

FIGURE A.22 Weibull pdf and cdf. (a) $\alpha = 1$ and $r = 1.5$. (b) $\alpha = 2$ and $r = 1.5$. Dotted lines specify μ_X and shaded regions denote $[\mu_X - \sigma_X, \mu_X + \sigma_X]$.

A.4 DISCRETE RANDOM VARIABLES

A.4.1 Bernoulli (with Support {0, 1})

Parameter $p \in [0, 1]$, $q \triangleq 1 - p$. Support $x \in \{0, 1\}$.

$$f_X(x) = q\delta(x) + p\delta(x-1), \quad p_X[x] = p^x q^{1-x} I_{\{0,1\}}[x],$$

$$F_X(x) = qu(x) + pu(x-1),$$

$$\mu_X = p, \quad \sigma_X^2 = pq, \quad \mu_n = p, \quad \mu_{c,3} = pq(1-2p), \quad \mu_{c,4} = pq(1-3pq),$$

$$m_e = \text{does not exist}, \quad m_o = \begin{cases} 0, & p \leq q \\ 1, & p \geq q, \end{cases}$$

$$\gamma_1 = (1-2p)/\sqrt{pq}, \quad \gamma_2 = 1/pq - 6,$$

$$\Phi_X(j\omega) = q + p\exp(j\omega), \quad H(X) = -p\log(p) - q\log(q).$$

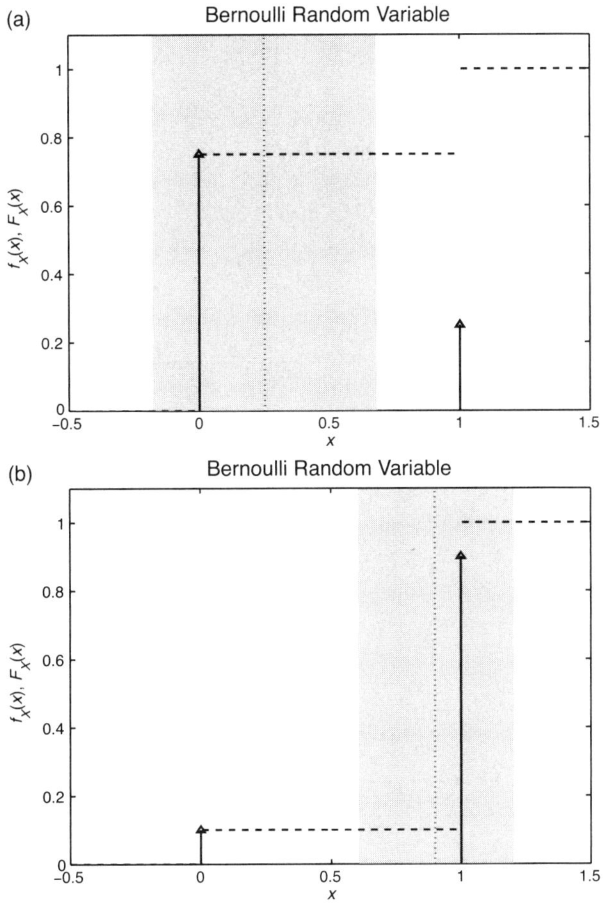

FIGURE A.23 Bernoulli pdf and cdf. (a) $p = 0.25$. (b) $p = 0.9$. Dotted lines specify μ_X and shaded regions denote $[\mu_X - \sigma_X, \mu_X + \sigma_X]$.

A.4.2 Bernoulli (Symmetric with Support {−1, 1})

Parameter $p \in [0, 1]$, $q \triangleq 1 - p$. Support $x \in \{-1, 1\}$.

$$f_X(x) = q\delta(x+1) + p\delta(x-1), \quad p_X[x] = p^{(1+x)/2} q^{(1-x)/2} I_{\{-1,1\}}[x],$$

$$F_X(x) = qu(x+1) + pu(x-1), \quad \mu_X = p - q, \quad \sigma_X^2 = 1 - (p-q)^2,$$

$$\mu_{c,3} = 2(p-q)[(p-q)^2 - 1], \quad \mu_{c,4} = 1 + 2(p-q)^2 - 3(p-q)^4,$$

$$\mu_n = \begin{cases} p - q, & n \text{ odd} \\ 1, & n \text{ even}, \end{cases} \quad m_e = \text{does not exist}, \quad m_o = \begin{cases} -1, & p \le q \\ 1, & p \ge q, \end{cases}$$

$$\gamma_1 = -2(p-q)/\sqrt{1-(p-q)^2}, \quad \gamma_2 = -2[1 - 4(p-q)^2 + 3(p-q)^4]/[1 - 2(p-q)^2 + (p-q)^4],$$

$$\Phi_X(j\omega) = q\exp(-j\omega) + p\exp(j\omega), \quad H(X) = -p\log(p) - q\log(q).$$

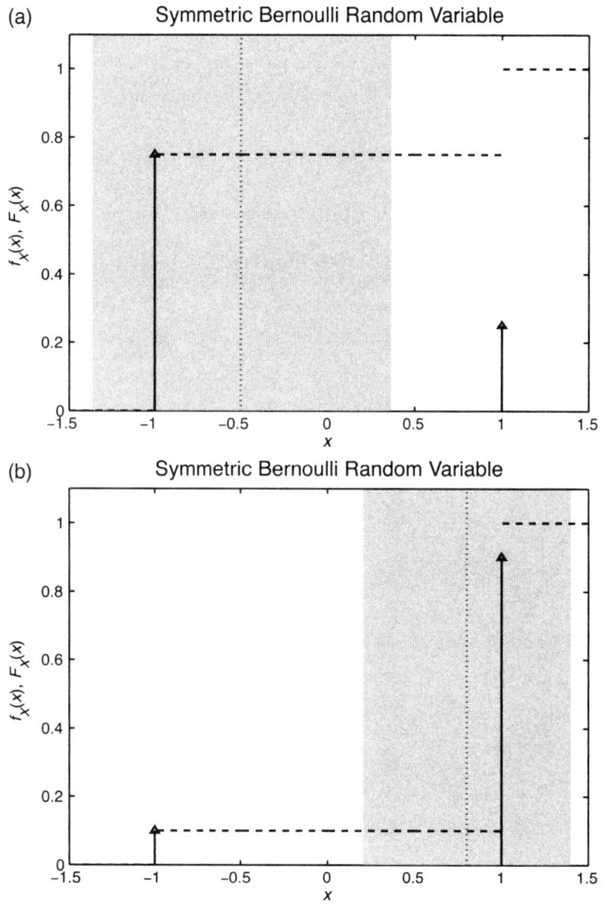

FIGURE A.24 Symmetric Bernoulli pdf and cdf. (a) $p = 0.25$. (b) $p = 0.9$. Dotted lines specify μ_X and shaded regions denote $[\mu_X - \sigma_X, \mu_X + \sigma_X]$.

A.4.3 Binomial

Parameters $p \in [0, 1]$, $q \triangleq 1 - p$, and $N \in \mathcal{N}$. Support $x \in \{0, \ldots, N\}$.

$$f_X(x) = \sum_{n=0}^{N} \binom{N}{n} p^n q^{N-n} \delta(x - n), \quad p_X[x] = \binom{N}{x} p^x q^{N-x} I_{\{0,\ldots,N\}}[x],$$

$$F_X(x) = \sum_{n=0}^{N} \binom{N}{n} p^n q^{N-n} u(x - n) = \sum_{n=0}^{\lfloor x \rfloor} \binom{N}{n} p^n q^{N-n},$$

$$\mu_X = Np, \quad \sigma_X^2 = Npq, \quad \mu_2 = Np(q + Np), \quad m_e = \lfloor Np \rfloor \text{ or } \lceil Np \rceil,$$

$$m_o = \begin{cases} \lfloor (N+1)p \rfloor, & \text{noninteger } (N+1)p \\ (N+1)p \text{ and } (N+1)p - 1, & \text{integer } (N+1)p, \end{cases}$$

$$\mu_{c,3} = Npq(q - p), \quad \gamma_1 = (1 - 2p)/\sqrt{Npq}, \quad \gamma_2 = (1 - 6pq)/Npq,$$

$$\Phi_X(j\omega) = [q + p\exp(j\omega)]^N, \quad H(X) = \text{complicated}.$$

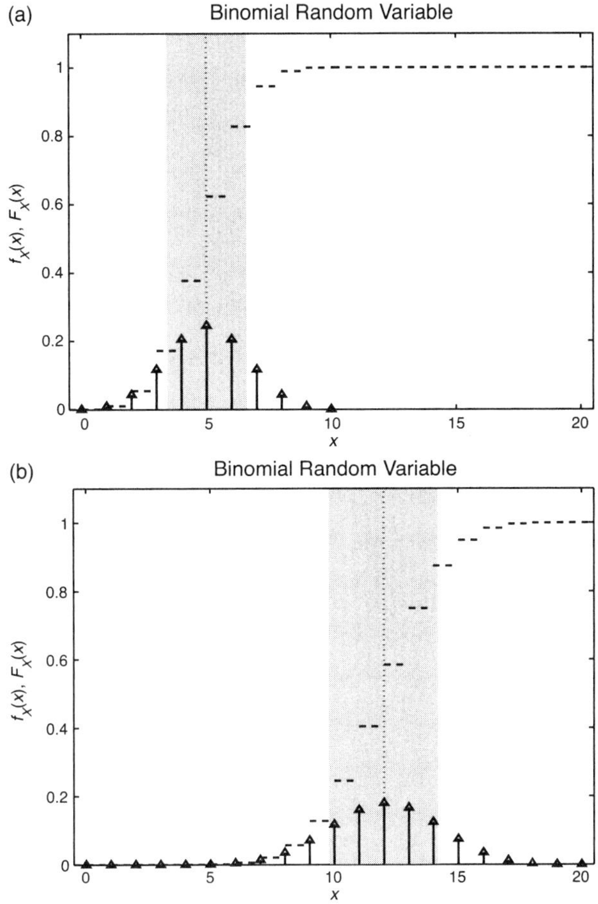

FIGURE A.25 Binomial pdf and cdf. (a) $N = 10$ and $p = 0.5$. (a) $N = 20$ and $p = 0.6$. Dotted lines specify μ_X and shaded regions denote $[\mu_X - \sigma_X, \mu_X + \sigma_X]$.

A.4.4 Geometric (with Support $\mathcal{Z}^+$)

Parameter $p \in [0, 1]$, $q \triangleq 1 - p$. Support $x \in \mathcal{Z}^+$.

$$f_X(x) = \sum_{n=0}^{\infty} pq^n \delta(x - n), \quad p_X[x] = pq^x I_{\mathcal{Z}^+}[x],$$

$$F_X(x) = \sum_{n=0}^{\infty} pq^n u(x - n) = (1 - q^{x+1}) I_{\mathcal{Z}^+}[x],$$

$$\mu_X = q/p, \quad \sigma_X^2 = q/p^2, \quad m_e = \lceil -1/\log_2(q) \rceil - 1, \quad m_o = 0,$$

$$\mu_2 = q(2-p)/p^2, \quad \mu_3 = q(6q + p^2)/p^3, \quad \mu_{c,3} = q(2-p)/p^3, \quad \mu_{c,4} = q(9q + p^2)/p^4,$$

$$\gamma_1 = (2-p)/\sqrt{q}, \quad \gamma_2 = p^2/q + 6,$$

$$\Phi_X(j\omega) = p[1 - q\exp(j\omega)]^{-1}, \quad H(X) = -\log(p) - (q/p)\log(q).$$

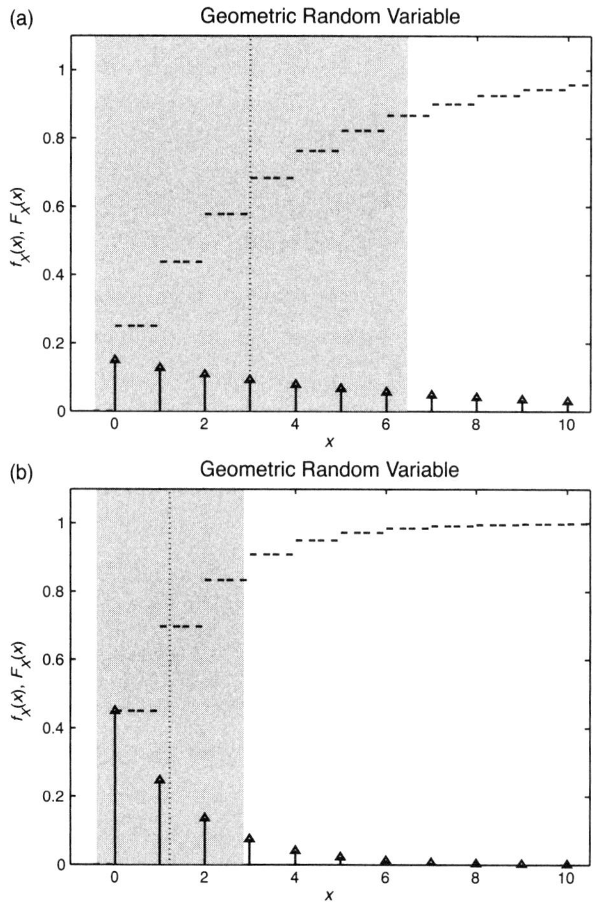

FIGURE A.26 Geometric pdf and cdf. (a) $p = 0.25$. (b) $p = 0.45$. Dotted lines specify μ_X and shaded regions denote $[\mu_X - \sigma_X, \mu_X + \sigma_X]$.

A.4.5 Geometric (Shifted with Support $\mathcal{N}$)

$$\text{Parameter } p \in [0,1],\ q \triangleq 1-p. \text{ Support } x \in \mathcal{N}.$$

$$f_X(x) = \sum_{n=1}^{\infty} pq^{n-1}\delta(x-n), \quad p_X[x] = pq^{x-1}I_{\mathcal{N}}[x],$$

$$F_X(x) = \sum_{n=1}^{\infty} pq^{n-1}u(x-n) = (1-q^x)I_{\mathcal{N}}[x],$$

$$\mu_X = 1/p, \quad \sigma_X^2 = q/p^2, \quad m_e = \lceil -1/\log_2(q) \rceil, \quad m_o = 1,$$

$$\mu_2 = (2-p)/p^2, \quad \mu_3 = (6q+p^2)/p^3, \quad \mu_{c,3} = q(2-p)/p^3, \quad \mu_{c,4} = q(9q+p^2)/p^4,$$

$$\gamma_1 = (2-p)/\sqrt{q}, \quad \gamma_2 = p^2/q + 6,$$

$$\Phi_X(j\omega) = p\exp(j\omega)[1-q\exp(j\omega)]^{-1}, \quad H(X) = -\log(p) - (q/p)\log(q).$$

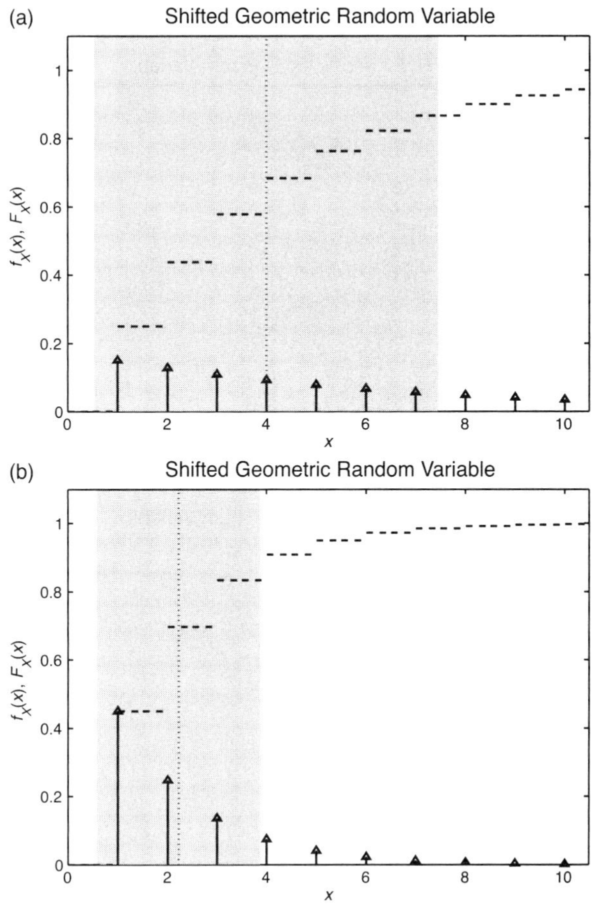

FIGURE A.27 Shifted geometric pdf and cdf. (a) $p = 0.25$. (b) $p = 0.45$. Dotted lines specify μ_X and shaded regions denote $[\mu_X - \sigma_X, \mu_X + \sigma_X]$.

A.4.6 Hypergeometric

Parameters $N \in \mathcal{N}$, $M \in \{0, \ldots, N\}$, and $n \in \{1, \ldots, N\}$. Support $x \in \{x_L \triangleq \max(0, n+M-N), \ldots, \min(M, n) \triangleq x_H\}$.

$$f_X(x) = \sum_{m=x_L}^{x_H} \frac{\binom{M}{m}\binom{N-M}{n-m}}{\binom{N}{n}} \delta(x-m), \quad p_X[x] = \frac{\binom{M}{x}\binom{N-M}{n-x}}{\binom{N}{n}} I_{\{x_L,\ldots,x_H\}}[x],$$

$$F_X(x) = \sum_{m=x_L}^{x_H} \frac{\binom{M}{m}\binom{N-M}{n-m}}{\binom{N}{n}} u(x-m) = \sum_{m=x_L}^{\lfloor x \rfloor} \frac{\binom{M}{m}\binom{N-M}{n-m}}{\binom{N}{n}},$$

$$\mu_X = nM/N, \quad \sigma_X^2 = nM(N-M)(N-n)/N^2(N-1), \quad m_e = \text{implicit},$$

$$\mu_{c,3} = nM(N-M)(N-2M)(N-n)(N-2n)/N^3(N-1)(N-2),$$

$$m_o = \lfloor (n+1)(M+1)/(N+2) \rfloor, \quad \gamma_2 = \text{complicated}, \quad \Phi_X(j\omega) = \text{complicated},$$

$$\gamma_1 = (N-2M)(N-2n)\sqrt{N-1}/(N-2)\sqrt{nM(N-M)(N-n)}, \quad H(X) = \text{complicated}.$$

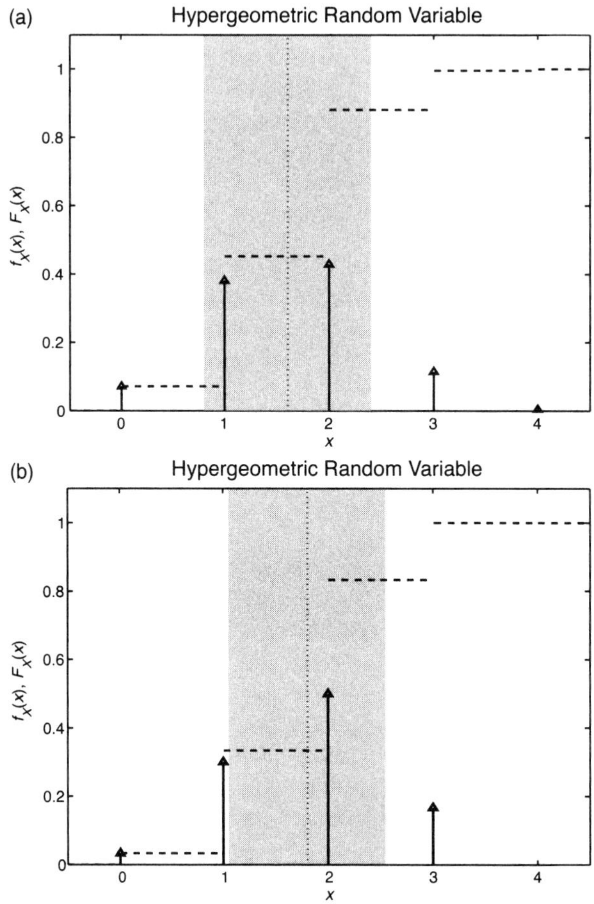

FIGURE A.28 Hypergeometric pdf and cdf. (a) $N = 10$ and $M = n = 4$. (b) $N = 10$, $M = 6$, and $n = 3$. Dotted lines specify μ_X and shaded regions denote $[\mu_X - \sigma_X, \mu_X + \sigma_X]$.

A.4.7 Logarithmic (Log-Series)

Parameter $p \in [0, 1]$, $q \triangleq 1 - p$. Support $x \in \mathcal{N}$.

$$f_X(x) = [-1/\ln(q)] \sum_{n=1}^{\infty} (p^n/n)\delta(x - n), \quad p_X[x] = [-p^x/x \ln(q)] I_{\mathcal{N}}[x],$$

$$F_X(x) = [-1/\ln(q)] \sum_{n=1}^{\infty} (p^n/n) u(x - n) = [-1/\ln(q)] \sum_{n=1}^{\lfloor x \rfloor} p^n/n,$$

$$\mu_X = -p/q \ln(q), \quad \sigma_X^2 = -p[p + \ln(q)]/q^2 \ln^2(q),$$

$$m_e = \text{implicit}, \quad m_o = 1, \quad \mu_2 = -p/q^2 \ln(q), \quad \mu_3 = -p(p+1)/q^3 \ln(q),$$

$$\gamma_1 = [2p^2 + 3p\ln(q) + (p+1)\ln^2(q)]/[p + \ln(q)]\sqrt{-p\ln(q) - p^2},$$

$$\gamma_2 = -[3p^3 + 6p^2 \ln(q) + 4p(p+1)\ln^2(q) + (p^2 + 4p + 1)\ln^3(q)]/p[p + \ln(q)]^2,$$

$$\Phi_X(j\omega) = \ln(1 - p\exp(j\omega))/\ln(q),$$

$$H(X) = (p/q)\ln(p/q) - \ln(-1/\ln(q)) - [1/\ln(q)] \sum_{n=1}^{\infty} \ln(n) p^n/n.$$

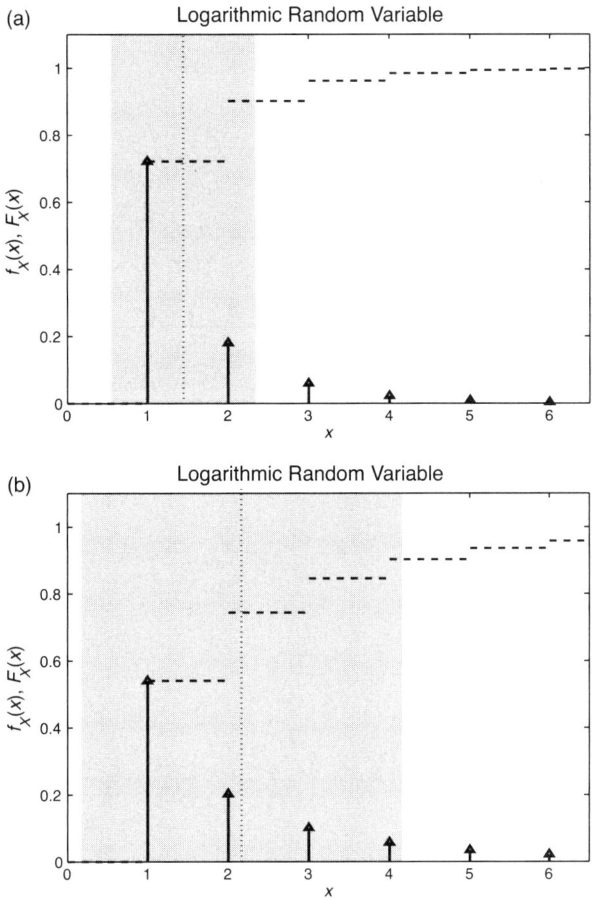

FIGURE A.29 Logarithmic pdf and cdf. (a) $p = 0.5$. (b) $p = 0.75$. Dotted lines specify μ_X and shaded regions denote $[\mu_X - \sigma_X, \mu_X + \sigma_X]$.

A.4.8 Negative Binomial (Pascal)

Parameters $p \in [0, 1]$, $q \triangleq 1 - p$, and $N \in \mathcal{N}$. Support $x \in \mathcal{Z}^+$.

$$f_X(x) = \sum_{n=0}^{\infty} \binom{n+N-1}{n} p^N q^n \delta(x-n), \quad p_X[x] = \binom{x+N-1}{x} p^N q^x I_{\mathcal{Z}^+}[x],$$

$$F_X(x) = \sum_{n=0}^{\infty} \binom{n+N-1}{n} p^N q^n u(x-n) = \sum_{n=0}^{\lfloor x \rfloor} \binom{n+N-1}{n} p^N q^n,$$

$$\mu_X = Nq/p, \quad \sigma_X^2 = Nq/p^2,$$

$$m_e = \text{implicit}, \quad m_o = 0 \text{ or } \lfloor (N-1)q/p \rfloor \text{ (for } N > 1), \quad \mu_{c,3} = Nq(1+q)/p^3,$$

$$\mu_{c,4} = 3N^2q^2/p^4 + Nq(p^2 + 6q)/p^4, \quad \gamma_1 = (2-p)/\sqrt{Nq}, \quad \gamma_2 = p^2/Nq + 6/N,$$

$$\Phi_X(j\omega) = p^N/[1 - q\exp(j\omega)]^N, \quad H(X) = \text{complicated}.$$

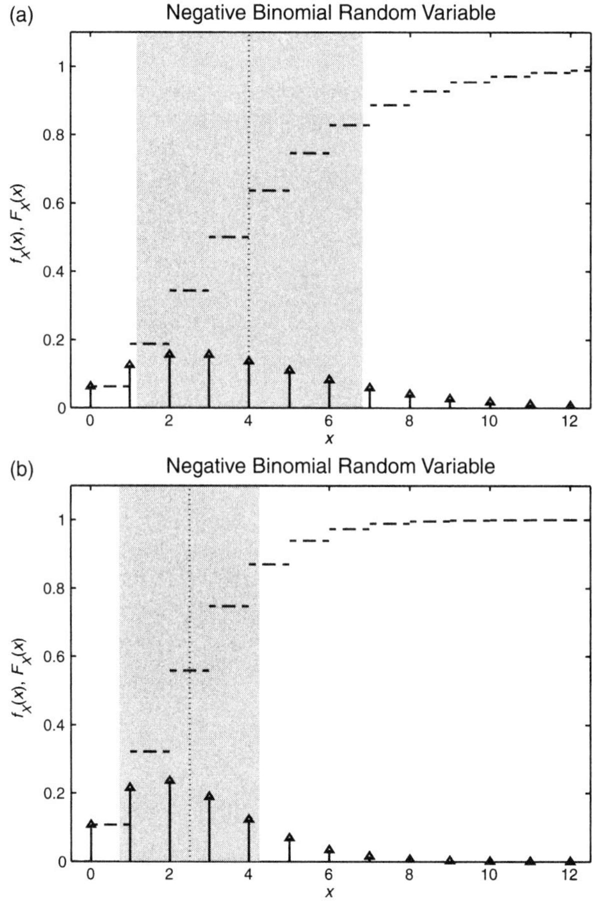

FIGURE A.30 Negative binomial pdf and cdf. (a) $N = 4$ and $p = 0.5$. (b) $N = 10$ and $p = 0.8$. Dotted lines specify μ_X and shaded regions denote $[\mu_X - \sigma_X, \mu_X + \sigma_X]$.

A.4.9 Poisson

Parameter $\alpha > 0$. Support $x \in \mathcal{Z}^+$.

$$f_X(x) = \sum_{n=0}^{\infty} \exp(-\alpha)(\alpha^n/n!)\delta(x-n), \quad p_X[x] = \exp(-\alpha)(\alpha^x/x!)I_{\mathcal{Z}^+}[x],$$

$$F_X(x) = \sum_{n=0}^{\infty} \exp(-\alpha)(\alpha^n/n!)u(x-n) = \sum_{n=0}^{\lfloor x \rfloor} \exp(-\alpha)(\alpha^n/n!),$$

$$\mu_X = \alpha, \quad \sigma_X^2 = \alpha, \quad m_e = \text{implicit}, \quad m_o = \begin{cases} \alpha \text{ and } \alpha - 1, & \alpha \in \mathcal{N} \\ \lfloor \alpha \rfloor, & \text{else,} \end{cases}$$

$$\mu_{c,3} = \alpha, \quad \mu_{c,4} = 3\alpha^2 + \alpha, \quad \mu_{c,5} = 10\alpha^2 + \alpha, \quad \gamma_1 = 1/\sqrt{\alpha}, \quad \gamma_2 = 1/\alpha,$$

$$\Phi_X(j\omega) = \exp(\alpha[\exp(j\omega) - 1]), \quad H(X) = \exp(-\alpha)\sum_{n=0}^{\infty} \alpha^n \log(n!)/n! + \alpha[1 - \log(\alpha)].$$

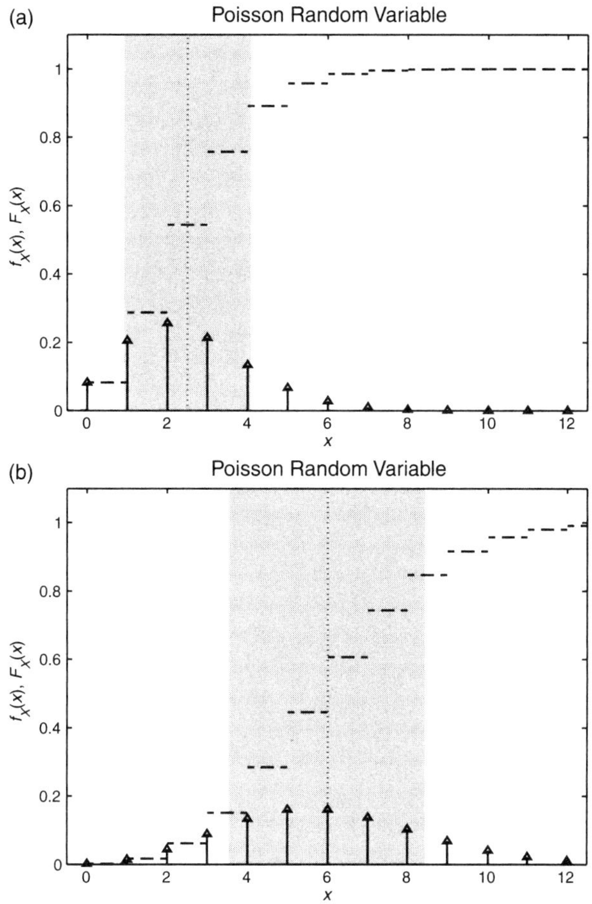

FIGURE A.31 Poisson pdf and cdf. (a) $\alpha = 2.5$. (a) $\alpha = 6$. Dotted lines specify μ_X and shaded regions denote $[\mu_X - \sigma_X, \mu_X + \sigma_X]$.

A.4.10 Uniform (Discrete)

Parameters $a, b \in \mathcal{Z}$ (with $a < b$) and $N \triangleq b - a + 1$. Support $x \in \{a, \ldots, b\}$.

$$f_X(x) = (1/N)\sum_{n=a}^{b} \delta(x-n), \quad p_X[x] = (1/N) I_{\{a,\ldots,b\}}[x],$$

$$F_X(x) = (1/N)\sum_{n=a}^{b} u(x-n) = (1/N)(\lfloor x \rfloor - a) I_{[a,b]}(x) + I_{(b,\infty)}(x),$$

$$\mu_X = (a+b)/2, \quad \sigma_X^2 = (N^2-1)/12, \quad m_e = \begin{cases} \lfloor (a+b+1)/2 \rfloor, & (a+b) \text{ even } (N \text{ odd}) \\ \text{does not exist}, & (a+b) \text{ odd } (N \text{ even}), \end{cases}$$

$$m_o \in \{a, \ldots, b\} \text{ (not unique)}, \quad \mu_{c,3} = 0, \quad \mu_{c,4} = (N^2-1)(3N^2-7)/240,$$

$$\gamma_1 = 0, \quad \gamma_2 = -(6/5)(N^2+1)/(N^2-1),$$

$$\Phi_X(j\omega) = (1/N)\exp(j\omega a)[1-\exp(j\omega N)]/[1-\exp(j\omega)], \quad H(X) = \log(N).$$

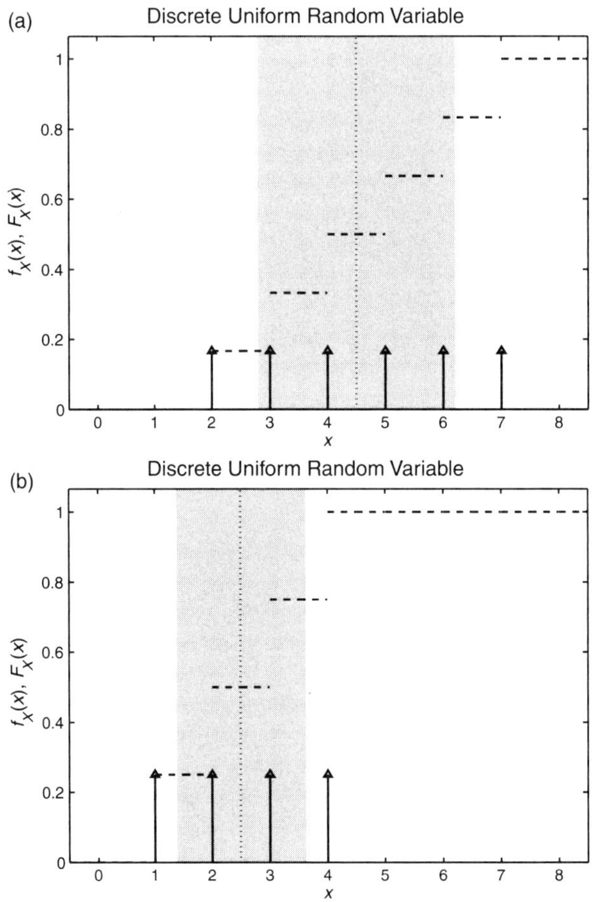

FIGURE A.32 Discrete uniform pdf and cdf. (a) $a = 2$ and $b = 7$. (b) $a = 1$ and $b = 4$. Dotted lines specify μ_X and shaded regions denote $[\mu_X - \sigma_X, \mu_X + \sigma_X]$.

A.4.11 Zeta (Zipf)

Parameter $\alpha \in (1, \infty)$. Support $x \in \mathcal{N}$.

$$f_X(x) = [1/\zeta(\alpha)] \sum_{n=1}^{\infty} (1/n^{\alpha})\delta(x - n), \quad p_X[x] = [1/\zeta(\alpha)x^{\alpha}] I_{\mathcal{N}}[x],$$

$$F_X(x) = [1/\zeta(\alpha)] \sum_{n=1}^{\infty} (1/n^{\alpha}) u(x - n) = [1/\zeta(\alpha)] \sum_{n=1}^{\lfloor x \rfloor} 1/n^{\alpha},$$

$$\mu_X = \zeta(\alpha - 1)/\zeta(\alpha) \text{ (for } \alpha > 2), \quad \sigma_X^2 = [\zeta(\alpha)\zeta(\alpha - 2) - \zeta^2(\alpha - 1)]/\zeta^2(\alpha) \text{ (for } \alpha > 3),$$

$$m_e = \text{implicit}, \quad m_o = 1, \quad \mu_n = \zeta(\alpha - n)/\zeta(\alpha) \text{ (for } \alpha > n + 1),$$

$$\gamma_1 = \text{complicated}, \quad \gamma_2 = \text{complicated}, \quad \Phi_X(j\omega) = [1/\zeta(\alpha)] \sum_{n=1}^{\infty} \exp(j\omega n)/n^{\alpha},$$

$$H(X) = [1/\zeta(\alpha)] \sum_{n=1}^{\infty} \log(n^{\alpha}\zeta(\alpha))/n^{\alpha}.$$

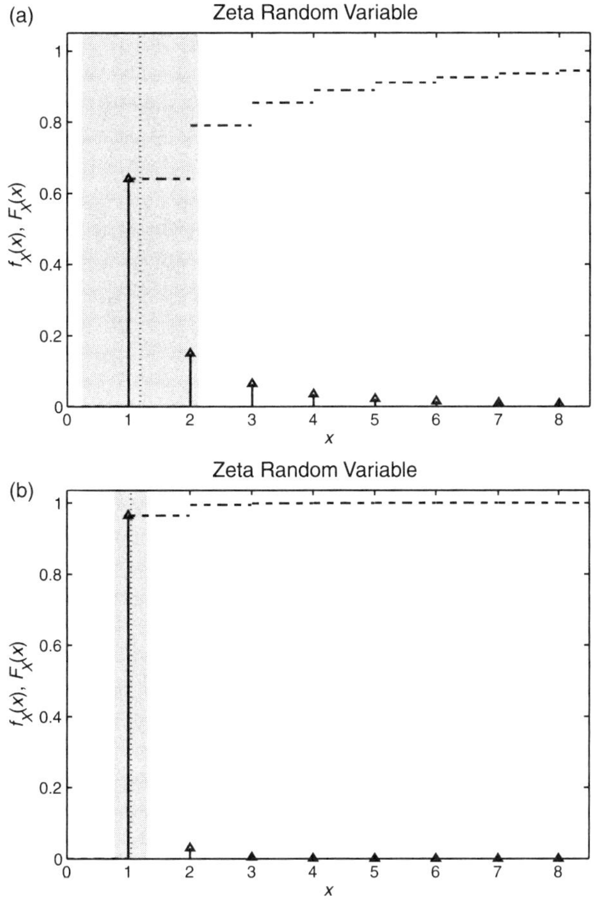

FIGURE A.33 Zeta pdf and cdf. (a) $\alpha = 3.5$. (b) $\alpha = 5$. Dotted lines specify μ_X and shaded regions denote $[\mu_X - \sigma_X, \mu_X + \sigma_X]$.

B Functions and Properties

In this appendix, we review some basic properties of functions and then summarize specific functions and the notation used in several chapters of this book.

B.1 CONTINUITY AND BOUNDED VARIATION

Definition: Function *Function* $y = g(x)$ is a mapping from x to y.

This definition implies that for a given x, the mapping to y is *unique*. The *inverse image* $x = g^{-1}(y)$ may not be unique; if it is unique, then $g(x)$ is one-to-one.

Definition: Continuous Function Function $g(x)$ is *continuous* at x_o if for every $\delta > 0$, there exists $\epsilon > 0$ such that

$$|x - x_o| < \epsilon \Longrightarrow |g(x) - g(x_o)| < \delta. \tag{B.1}$$

We can also write for a continuous function $g(x)$ at x_o that

$$\lim_{\epsilon \to 0} |g(x_o + \epsilon) - g(x_o)| = 0, \tag{B.2}$$

where in this case ϵ is positive or negative: the limit must hold approaching x_o from the right and the left.

Thus, we can simply write that $g(x)$ is continuous at x_o if

$$\lim_{x \to x_o} g(x) = g(x_o) \tag{B.3}$$

for finite $g(x_o)$. The definition in (B.1) applies to a specific point x_o, meaning that a particular pair $\{\epsilon, \delta\}$ apply to that point. This can be emphasized by writing δ as a function of x_o and ϵ: $\delta(x_o, \epsilon)$ (not to be confused with the Dirac delta function defined later). For some other point x_1, a different $\delta(x_1, \epsilon)$ may be needed.

Example B.1. Function $g(x) = \sin(2\pi/x)$ is not continuous at $x = 0$. As $x \to 0$, the function oscillates between -1 and 1; the limit at $x = 0$ does not exist as shown in Figure B.1(a). There is no ϵ satisfying the continuity condition in (B.1) for every δ.

Example B.2. Consider $g(x) = x^2 \sin(2\pi/x)$, where x^2 "compresses" $\sin(1/x)$ near the origin, so that $g(0) = 0$ as shown in Figure B.1(b). Since $|x^2 \sin(2\pi/x)| \leq x^2$, we can choose $\delta = \epsilon^2$, so that for every $\delta > 0$, the continuity condition holds: $|x - 0| < \epsilon \Longrightarrow |x^2 \sin(2\pi/x) - 0| \leq x^2 = \epsilon^2$ which is $\delta(x = 0, \epsilon)$.

Probability, Random Variables, and Random Processes: Theory and Signal Processing Applications, First Edition. John J. Shynk.

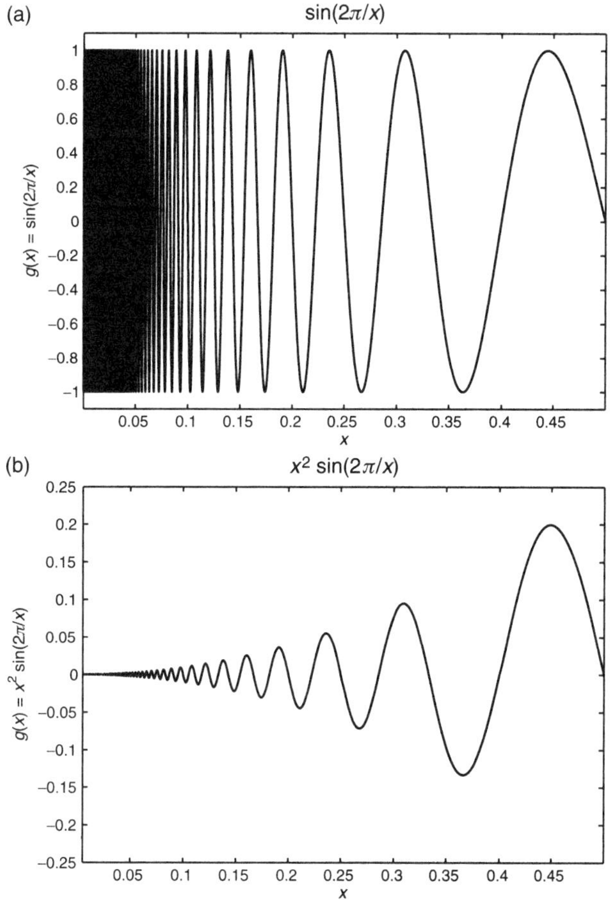

FIGURE B.1 Continuity of functions. (a) $g(x) = \sin(2\pi/x)$ is not continuous at $x = 0$. (b) $g(x) = x^2 \sin(2\pi/x)$ is continuous at $x = 0$.

If (B.1) holds only for positive or negative ϵ (not both), then $g(x)$ is *right continuous* or *left continuous*, respectively.

Definition: Right- and Left-Continuous Functions Function $g(x)$ is *right continuous* if for every $\delta > 0$ there exists $\epsilon > 0$ such that

$$x_o < x < x_o + \epsilon \Longrightarrow |g(x) - g(x_o)| < \delta. \tag{B.4}$$

Likewise, for a *left-continuous* function:

$$x_o - \epsilon < x < x_o \Longrightarrow |g(x) - g(x_o)| < \delta. \tag{B.5}$$

Obviously, a continuous function is both left and right continuous at x_o. An example of a right-continuous function is shown in Figure B.2.

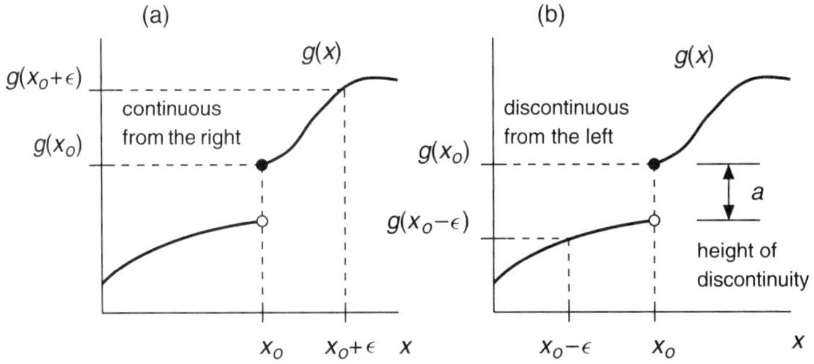

FIGURE B.2 Right-continuous function at x_o. (a) The solid circle defines $g(x_o)$, so that from the right $g(x+\epsilon) \to g(x_o)$ as $\epsilon \to 0$. (b) From the left, $g(x_o - \epsilon)$ approaches $g(x_o) - a$, where $a > 0$ is the height of the discontinuity. The open circle indicates that the function is not left continuous: $g(x_o - \epsilon)$ "jumps" to $g(x_o)$ when $\epsilon \to 0$.

Example B.3. The cumulative distribution function (cdf) of a discrete random variable is defined to be right continuous; an example for the binomial random variable is shown in Figure A.25 of the distribution summaries in Appendix A. Consider the discontinuity of the cdf at $x = 1$, where the probability mass is $\binom{N}{1}pq^{N-1}$. It is clear that for every $\delta > 0$, we can find $\epsilon > 0$ such that $1 < x < 1+\epsilon \Longrightarrow |F_X(x) - F_X(1)| < \delta$. In fact, $|F_X(x) - F_X(1)| = 0$ for $1 < x < 2$. On the other hand, $\epsilon > 0$ cannot be chosen such that $x = 1 - \epsilon$ is arbitrarily close to one and $|F_X(1) - F_X(x)| < \delta$ for every δ because of the "jump" at $x = 1$.

Example B.4. Consider the quadratic function

$$g(x) = ax^2 + bx + c \tag{B.6}$$

with finite real coefficients $\{a, b, c\}$. Let $|x - x_o| < \epsilon$ for $\epsilon > 0$. Then

$$\begin{aligned} |g(x) - g(x_o)| &= |ax^2 + bx + c - ax_o^2 - bx_o - c| = |a(x - x_o)(x + x_o) + b(x - x_o)|, \\ &= |x - x_o||a(x + x_o) + b| < \epsilon|a(x + x_o) + b| = \delta(x_o, \epsilon), \end{aligned} \tag{B.7}$$

where in the last expression $|x - x_o| < \epsilon$ has been substituted, demonstrating that $g(x)$ is continuous at $x_o \in \mathcal{R}$. We can also use the alternative description in (B.2):

$$\begin{aligned} |g(x_o + \epsilon) - g(x_o)| &= |a(x_o + \epsilon)^2 + b(x_o + \epsilon) + c - ax_o^2 - bx_o - c|, \\ &= |a(2x_o\epsilon + \epsilon^2) + b\epsilon| = |\epsilon||a(2x_o + \epsilon) + b|, \end{aligned} \tag{B.8}$$

which tends to zero as $\epsilon \to 0$. In fact, it can be shown that all polynomial functions with finite coefficients are continuous.

Example B.5. Consider the exponential function

$$g(x) = \exp(-\lambda x) \tag{B.9}$$

with $\lambda > 0$ for which

$$\begin{aligned} |g(x_o + \epsilon) - g(x_o)| &= |\exp(-\lambda(x_o + \epsilon)) - \exp(-\lambda x_o)|, \\ &= \exp(-\lambda x_o)|\exp(-\lambda\epsilon) - 1|. \end{aligned} \tag{B.10}$$

This expression tends to zero as $\epsilon \to 0$, so that $\exp(-\lambda x)$ is continuous at $x_o \in \mathcal{R}$.

Definition: Right and Left Limits The *right limit* b of function $g(x)$ at x_o is

$$\lim_{x \to x_o^+} g(x) = b \tag{B.11}$$

and the *left limit* a is

$$\lim_{x \to x_o^-} g(x) = a. \tag{B.12}$$

For a continuous function at x_o (i.e., right and left continuous), $a = b$.

Example B.6. The floor function $\lfloor x \rfloor$ (defined later) is the largest integer not exceeding x. It has the following right and left limits for integer $x_o = n$:

$$\lim_{x \to n^+} \lfloor x \rfloor = n, \qquad \lim_{x \to n^-} \lfloor x \rfloor = n - 1. \tag{B.13}$$

It is an example of a right-continuous function. The ceiling function $\lceil x \rceil$ (also defined later) is the smallest integer not less than x. It has the following right and left limits for integer $x_o = n$:

$$\lim_{x \to n^+} \lceil x \rceil = n + 1, \qquad \lim_{x \to n^-} \lceil x \rceil = n, \tag{B.14}$$

which means it is continuous from the left.

A stronger type of continuity allows for the same δ to apply to *any* point in the closed interval $[a, b]$ for $b > a$.

Definition: Uniformly Continuous Function $g(x)$ is *uniformly continuous* on $[a, b]$ if for every $\delta > 0$ there exists $\epsilon > 0$ such that

$$|x_1 - x_2| < \epsilon \quad \Longrightarrow \quad |g(x_1) - g(x_2)| < \delta \tag{B.15}$$

for every $x_1, x_2 \in [a, b]$. A function that does not satisfy this definition at one or more points on $[a, b]$ is *discontinuous*.

This definition implies that there exists a single pair $\{\epsilon, \delta\}$ in the first definition of continuity in (B.1) for every $x_o \in [a, b]$; thus, $\delta = \delta(\epsilon)$ holds *uniformly* on the interval, and is a function only of ϵ and not any particular $x_o \in [a, b]$. If a function is uniformly continuous on $[a, b]$, then it is continuous at any $x_o \in [a, b]$. Although this definition also applies to open and semi-open intervals, it can be shown that if $g(x)$ is continuous at every point in the *closed* interval $[a, b]$, then it is uniformly continuous.

Example B.7. The exponential function $g(x) = \exp(-\lambda x)$ with $\lambda > 0$ is uniformly continuous on $\mathcal{R}^+$ as demonstrated below. For $|x_2 - x_1| < \epsilon$:

$$\begin{aligned} |g(x_2) - g(x_1)| &= |\exp(-\lambda x_2) - \exp(-\lambda x_1)| = \exp(-\lambda x_1)|\exp(-\lambda(x_2 - x_1)) - 1| \\ &< \exp(-\lambda x_1)|\exp(\lambda \epsilon) - 1| < |\exp(\lambda \epsilon) - 1| = \delta(\epsilon), \end{aligned} \tag{B.16}$$

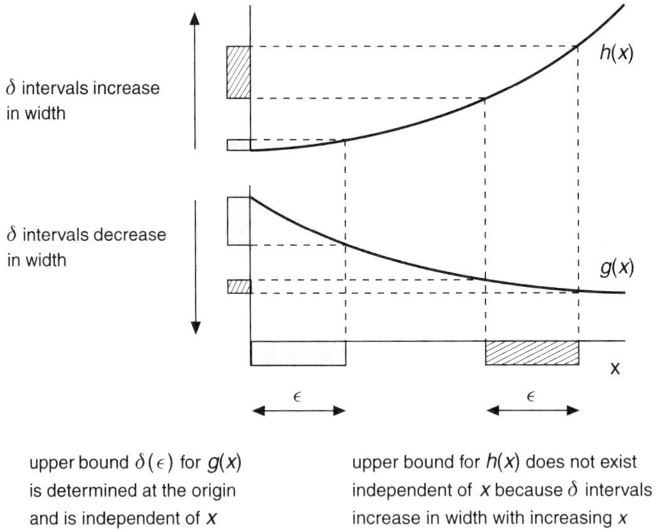

FIGURE B.3 Example of uniformly continuous function $g(x)$ and discontinuous $h(x)$, both on $\mathcal{R}^+$. (The vertical axis has been split at 1 for ease of viewing the two sets of intervals.)

where in the last inequality we have used the fact that $\exp(-\lambda x_1)$ is greatest at $x_1 = 0$. For every $\delta(\epsilon) > 0$, we can specify an $\epsilon > 0$ such that the inequality holds for every $x_1, x_2 \in \mathcal{R}^+$. This is demonstrated in Figure B.3. Note that $g(x) = \exp(-\lambda x)$ is *not* uniformly continuous for the entire real line $\mathcal{R}$.

Example B.8. Next, consider $h(x) = \exp(\lambda x)$ with $\lambda > 0$. It is straightforward to show that $h(x)$ is continuous for any $x_o \in \mathcal{R}^+$, but is *not* uniformly continuous. Similar to (B.16) for $|x_2 - x_1| < \epsilon$:

$$\begin{aligned} |h(x_2) - h(x_1)| &= |\exp(\lambda x_2) - \exp(\lambda x_1)|, \\ &= \exp(\lambda x_1)|\exp(\lambda(x_2 - x_1)) - 1| \\ &< \exp(\lambda x_1)|\exp(\lambda\epsilon) - 1| = \delta(x_1, \epsilon). \end{aligned} \tag{B.17}$$

The right-hand side cannot be written independently of x_1; Figure B.3 demonstrates that $h(x)$ is not uniformly continuous on $\mathcal{R}^+$.

In general, a function is not uniformly continuous if there exists $x_1, x_2 \in [a, b]$ such that $g(x_1)$ and $g(x_2)$ both approach infinity for small ε (because $\infty - \infty$ is not defined).

Example B.9. Function $g(x) = 1/x$ is uniformly continuous on $[1, \infty)$. Observe that $|1/x_1 - 1/x_2| = |(x_2 - x_1)/x_1x_2| \leq |x_2 - x_1|$ since $x_1, x_2 \geq 1$. Thus, we can choose $\delta = \epsilon$ (independently of x) and the uniform continuity condition holds for every $\{x_1, x_2\}$: $|x_1 - x_2| < \epsilon \Longrightarrow |1/x_1 - 1/x_2| < \epsilon = \delta$.

Example B.10. The unit-step function and the signum function covered later are both discontinuous at $x = 0$, whereas the absolute value function $f(x) = |x|$ is uniformly continuous on $[-1, 1]$ even though its first derivative changes sign at $x = 0$; there are no "jumps" in $|x|$ for any x. The signum function and the absolute

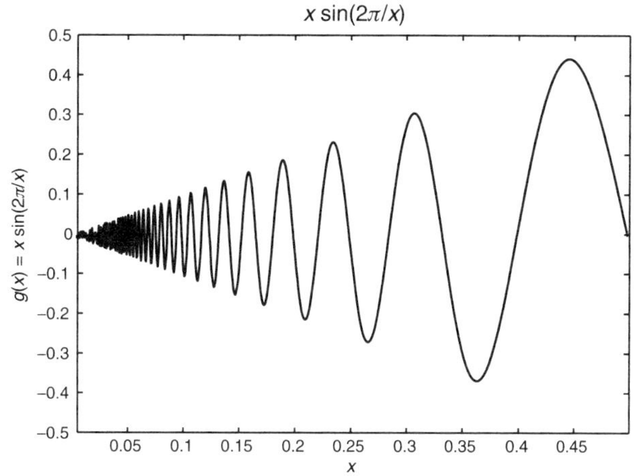

FIGURE B.4 Function $g(x) = x\sin(2\pi/x)$ is not absolutely continuous on $[0, b]$ for $b > 0$.

value function are related as follows:

$$\frac{\mathrm{d}}{\mathrm{d}x}|x| = \mathrm{sgn}(x) \tag{B.18}$$

from which it is clear that the *derivative* of $|x|$ is not continuous at $x = 0$.

There is yet another type of continuity that involves *sequences of points* on $[a, b]$.

Definition: Absolutely Continuous Function $g(x)$ is *absolutely continuous* on $[a, b]$ if for all $\delta > 0$ and finite $N \in \mathcal{N}$ there exists $\epsilon > 0$ such that

$$\sum_{n=1}^{N} |x_{1n} - x_{2n}| < \epsilon \Longrightarrow \sum_{n=1}^{N} |g(x_{1n}) - g(x_{2n})| < \delta, \tag{B.19}$$

where the subintervals $\{[x_{1n}, x_{2n}]\}$ of $[a, b]$ for $n = 1, \ldots, N$, are mutually disjoint.

This type of continuity requires that δ depend only on ϵ for any finite set of subintervals. Note that if the set $A \triangleq \cup_{n=1}^{N}[x_{1n}, x_{2n}]$ is defined, then using the Lebesgue measure from Chapter 2, the left-hand side of (B.19) is equivalent to $L(A) < \epsilon$.

Example B.11. Function $g(x) = x^2\sin(2\pi/x)$ shown previously in Figure B.1(b) is absolutely continuous on $[0, b]$ for $b > 0$. Function $g(x) = x\sin(2\pi/x)$ is not absolutely continuous on $[0, b]$, as shown in Figure B.4.

Definition: Bounded Variation Function $g(x)$ has *bounded variation* on $[a, b]$ if there exists finite $M > 0$ such that

$$\sum_{n=1}^{N} |g(x_n) - g(x_{n-1})| \leq M \tag{B.20}$$

for every set of ordered points $\{x_0 \leq x_1 \leq \cdots \leq x_N\} \in [a, b]$ and each finite $N \in \mathcal{N}$.

Note that $a = x_0$ and $b = x_N$ unless the interval is open or semi-open. It can be shown that if $g(x)$ is absolutely continuous on $[a, b]$, then it has bounded variation on that interval.

Example B.12. Functions $\sin(2\pi/x)$ and $x\sin(2\pi/x)$ shown previously in Figures B.1(a) and B.4, respectively, do not have bounded variation on $[0, b]$. Regardless of how much the x-axis is expanded near $x = 0$, corresponding to finer partitions in (B.20), we would continue to see rapid variation; no finite M exists for (B.20) on $[0, b]$. On the other hand, function $x^n\sin(2\pi/x)$ has bounded variation on $[0, b]$ provided $n > 1$; the plot for $n = 2$ is shown in Figure B.1(b).

This brief review of continuity is useful when we consider continuous random variables in Chapter 3 and discuss the various types of continuity for a random process in Chapter 7. In the rest of this appendix, we define specific functions and describe some of their properties.

B.2 SUPREMUM AND INFIMUM

Definition: Supremum The *supremum* of set $E \in \mathcal{R}$ is the least upper bound: it is the smallest real number that is $\geq$ all elements in E.

The supremum is useful when working with intervals because

$$\sup[a, b) = \sup[a, b] = b. \tag{B.21}$$

The maximum is *not* defined for $E = [a, b)$ because it must be an element of the set; b is not in this set, and for any $b - \epsilon \in E$ for $\epsilon > 0$, we can always find an element of E that is greater. The maximum is also called the *greatest element*. If a set has a greatest element, then the supremum and maximum are identical.

Definition: Infimum The *infimum* of set $E \in \mathcal{R}$ is the greatest lower bound: it is the largest real number that is $\leq$ all elements in E.

Thus

$$\inf(a, b] = \inf[a, b] = a. \tag{B.22}$$

The minimum of a set is called the *least element* and is not defined for semi-open intervals such as $(a, b]$. If a set has a least element, then the infimum and minimum are identical.

B.3 ORDER NOTATION

Order notation is used to represent the behavior of a function $g(x)$ as $x \to \infty$.

Definition: Big O Notation Function $g(x)$ is of *order* $h(x)$ for large x if

$$\lim_{x\to\infty} \frac{g(x)}{h(x)} = x_o \tag{B.23}$$

for some nonzero $x_o \in \mathcal{R}$. It is written using *big O* notation as $g(x) = O(h(x))$.

This means that the two functions change at a similar rate.

Definition: Little O Notation Function $g(x)$ is of *order zero* relative to $h(x)$ for large x if

$$\lim_{x\to\infty} \frac{g(x)}{h(x)} = 0. \tag{B.24}$$

It is written using *little o* notation as $g(x) = o(h(x))$.

This means that $h(x)$ increases faster than $g(x)$.

Example B.13. For the polynomial $g(x) = 2x^3 + x + 1$, we can state the following:

$$g(x) = O(x^3), \quad g(x) = o(x^n) \text{ for } n > 3. \tag{B.25}$$

From this simple example, we see that $O(\cdot)$ describes the "type" of function. Thus, $O(1) \Longrightarrow$ a constant, $O(x) \Longrightarrow$ a linear function, and $O(x^2) \Longrightarrow$ a quadratic function. Other common cases are factorial $O(x!)$, logarithmic $O(\log(x))$, and exponential $O(a^x)$ for $a > 1$. Little o notation, on the other hand, is typically used to designate high-order terms of a function that tend to zero in the limit (and thus can be ignored). For the function above, if we are interested only in the cubic term for large x, it could be written as $g(x) = 2x^3 + o(x^2)$.

B.4 FLOOR AND CEILING FUNCTIONS

Definition: Floor Function The *floor function* for $x \in \mathcal{R}$ gives the largest integer $\leq x$:

$$\text{floor}(x) = \lfloor x \rfloor \triangleq \max\{n \in \mathcal{Z} \text{ s.t. } n \leq x\}. \tag{B.26}$$

Definition: Ceiling Function The *ceiling function* for $x \in \mathcal{R}$ gives the smallest integer $\geq x$:

$$\text{ceil}(x) = \lceil x \rceil \triangleq \min\{n \in \mathcal{Z} \text{ s.t. } n \geq x\}. \tag{B.27}$$

B.5 CONVEX AND CONCAVE FUNCTIONS

Definition: Convex and Concave Function $g(x)$ is *convex* if for any two values $x_1, x_2 \in \mathcal{R}$ and for $\alpha \in [0, 1]$:

$$g(\alpha x_1 + (1-\alpha)x_2) \leq \alpha g(x_1) + (1-\alpha)g(x_2). \tag{B.28}$$

The function is *strictly convex* if for $x_1 \neq x_2$, this expression has strict inequality. Function $h(x)$ is *concave* if $-h(x)$ is convex. Likewise, the function is strictly concave if $-h(x)$ is strictly convex for $x_1 \neq x_2$.

Examples of convex and concave functions are illustrated in Figure B.5. Convex functions are used in Jensen's inequality (see Appendix F).

B.6 EVEN AND ODD FUNCTIONS

Definition: Even and Odd *Even function* $g_E(x)$ and *odd function* $g_O(x)$ are defined by having the following properties about $x = 0$:

$$g_E(x) = g_E(-x), \quad g_O(x) = -g_O(-x). \tag{B.29}$$

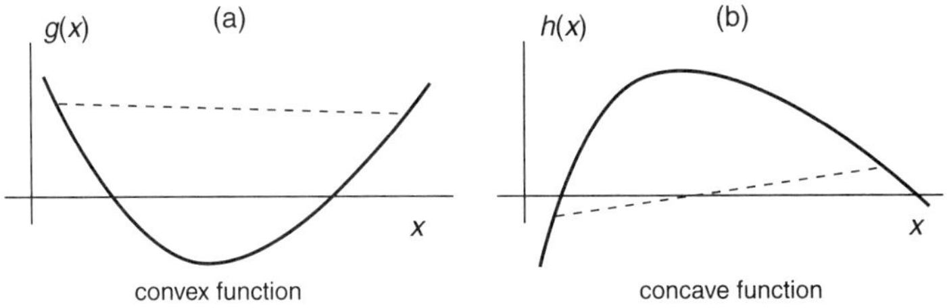

FIGURE B.5 Examples of (a) convex $g(x)$ and (b) concave $h(x)$.

Any function $g(x)$ can be decomposed as follows:

$$g(x) = g_E(x) + g_O(x), \tag{B.30}$$

where

$$g_E(x) = [g(x) + g(-x)]/2, \quad g_O(x) = [g(x) - g(-x)]/2. \tag{B.31}$$

An example of this decomposition is shown in Figure B.6. The only type of function that is even and odd, satisfying both equations in (B.31), is a constant function.

Several properties of even and odd functions are provided in Table B.1 without proof. It is easy to see that for finite a:

$$\int_{-a}^{a} g_O(x)\mathrm{d}x = 0, \tag{B.32}$$

which is useful when calculating moments of a distribution. When an integral of the following form is obtained

$$\int_{-a}^{a} g_O(x) g_E(x)\mathrm{d}x, \tag{B.33}$$

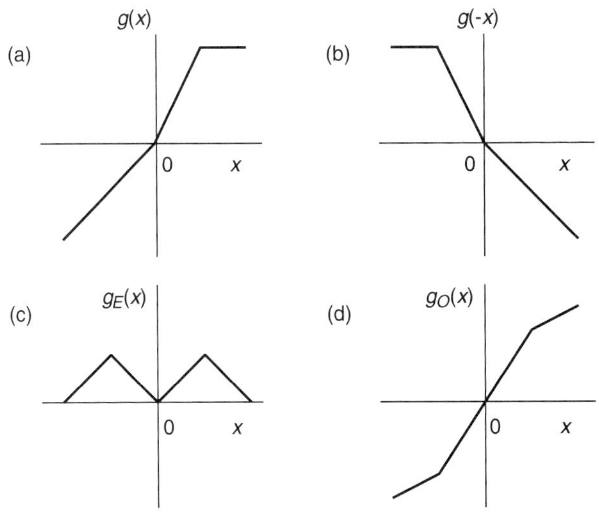

FIGURE B.6 Example even and odd parts of a piecewise linear function. (a) $g(x)$. (b) $g(-x)$. (c) $g_E(x)$. (d) $g_O(x)$.

TABLE B.1 Properties of Even and Odd Functions

Function	Property
$g_{1E}(x)g_{2E}(x)$	Even
$g_{1O}(x)g_{2O}(x)$	Even
$g_E(t)g_O(x)$	Odd
$g_{1E}(x)+g_{2E}(x)$	Even
$g_{1O}(x)+g_{2O}(x)$	Odd
$g_E(t)+g_O(x)$	Neither
$\mathrm{d}g_E(x)/\mathrm{d}x$	Odd
$\mathrm{d}g_O(x)/\mathrm{d}x$	Even

we know it is zero. For example, when calculating odd raw moments (see Chapter 5), the expectation is zero whenever $f_X(x)$ is an even probability density function (pdf) (such as for a zero-mean Gaussian random variable). The following property is also convenient when computing expectations:

$$\int_{-a}^{a} g_E(x)\mathrm{d}x = 2\int_{0}^{a} g_E(x)\mathrm{d}x. \tag{B.34}$$

For example, if $g_E(x) = |x|$ and the pdf $f_X(x)$ is also even, then we can write

$$\int_{-a}^{a} |x| f_X(x)\mathrm{d}x = 2\int_{0}^{a} x f_X(x)\mathrm{d}x, \tag{B.35}$$

which allows us to remove the absolute value. These last results may not hold for $a = \infty$ if the integral is not defined, as is the case for the mean of the Cauchy random variable.

B.7 SIGNUM FUNCTION

Definition: Signum Function The *signum function* is

$$\operatorname{sgn}(x) \triangleq \begin{cases} -1, & x < 0 \\ 0, & x = 0 \\ 1, & x > 0, \end{cases} \tag{B.36}$$

although usually in practice sgn(0) = 1.

The signum function can also be expressed as

$$\operatorname{sgn}(x) = \frac{\mathrm{d}}{\mathrm{d}x}|x| = x/|x|, \tag{B.37}$$

where $x \neq 0$ in the last expression, and its derivative can be written in terms of the Dirac delta function as follows:

$$\frac{\mathrm{d}}{\mathrm{d}x}\operatorname{sgn}(x) = 2\delta(x). \tag{B.38}$$

For complex-valued x:

$$\operatorname{sgn}(x) \triangleq (x_R + jx_I)/\sqrt{x_R^2 + x_I^2} = x/|x|, \tag{B.39}$$

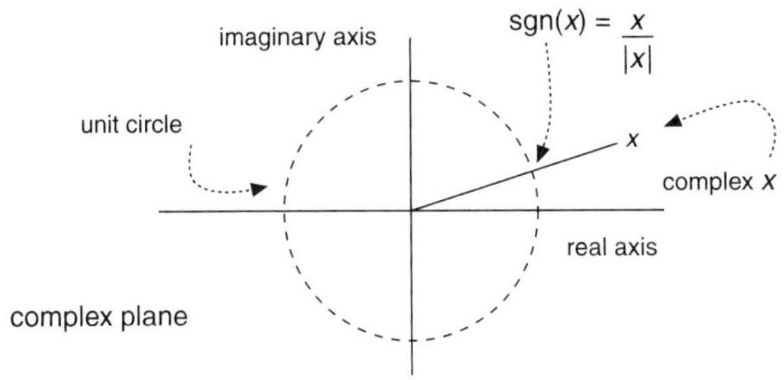

FIGURE B.7 Signum function for a complex argument: sgn(x) is the point on the unit circle closest to x.

where $\{x_R, x_I\}$ are the real and imaginary parts of x. Note that $\text{sgn}(x) \neq \text{sgn}(x_R) + j\text{sgn}(x_I)$ (even though the right-hand side is the maximum likelihood (ML) detector for quadrature phase shift keying (QPSK) signals in an additive white Gaussian noise (AWGN) channel: see Chapter 10). Instead, the complex signum function is the point on the unit circle of the complex plane that is closest to x, as depicted in Figure B.7. Complex sgn(x) could be any point on the unit circle as x is varied, whereas $\text{sgn}(x_R) + j\text{sgn}(x_I)$ is one of only four points: $\pm 1 \pm j$.

B.8 DIRAC DELTA FUNCTION

A *generalized function* is used to describe mappings that technically are not functions. Perhaps the most well-known generalized function is the Dirac delta function.

Definition: Dirac Delta Function The *Dirac delta function* is

$$\delta(x) \triangleq \begin{cases} 0, & x \neq 0 \\ \infty, & x = 0 \end{cases} \quad \text{s.t.} \quad \int_{-\infty}^{\infty} \delta(x)\text{d}x = 1. \tag{B.40}$$

However, this definition is not very useful; for example, it is not apparent how to multiply another function $g(x)$ with $\delta(x)$. It turns out that we can *define* such a multiplication as follows:

$$g(x)\delta(x - x_o) \triangleq g(x_o)\delta(x - x_o), \tag{B.41}$$

where the right-hand side is still a delta function at the same location but now has area $g(x_o)$. This is the so-called *sampling property* of $\delta(x)$. The Dirac delta function is a convenient description of an impulse in physical systems. In particular, we are interested in how $\delta(x)$ behaves when filtered (via convolution) and when using integral transforms (such as the Fourier transform). From the sampling property:

$$\begin{aligned} \int_{-\infty}^{\infty} g(x)\delta(x - x_o)\text{d}x &= \int_{-\infty}^{\infty} g(x_o)\delta(x - x_o)\text{d}x, \\ &= g(x_o)\int_{-\infty}^{\infty} \delta(x - x_o)\text{d}x = g(x_o). \end{aligned} \tag{B.42}$$

This result is the *sifting property* of $\delta(x)$ because it "sifts" out the value of the function $g(x)$ (with which it is multiplied) at the location of the delta function.

TABLE B.2 Properties of the Dirac Delta Function

Property	Expression
Scaling	$\delta(\alpha x) = (1/\|\alpha\|)\delta(x)\ (\alpha \neq 0)$
Sampling	$g(x)\delta(x - x_o) = g(x_o)\delta(x - x_o)$
Sifting	$\int_{-\infty}^{\infty} g(x)\delta(x - x_o)\mathrm{d}x = g(x_o)$
Convolution	$\int_{-\infty}^{\infty} g(x)\delta(x_o - x)\mathrm{d}x = g(x_o)$
Integral	$\int_{-\infty}^{x} \delta(y)\mathrm{d}y = u(x)$
Derivative	$\delta(x) = \mathrm{d}u(x)/\mathrm{d}x$
Composition	$\delta(g(x)) = \delta(x - x_o)/\|g'(x_o)\|$
Laplace transform	$\int_{-\infty}^{\infty} \delta(x - x_o)\exp(-sx)\mathrm{d}x = \exp(-x_o s)$

We can also view (and define) the Dirac delta function as the limiting result of a unit-area rectangle as its width tends to zero and its height tends to infinity:

$$\delta(x) \triangleq \lim_{a\to 0}(1/a)\mathrm{rect}(x/a). \tag{B.43}$$

It can likewise be expressed in terms of a zero-mean Gaussian pdf as the standard deviation σ tends to zero:

$$\delta(x) = \lim_{\sigma\to 0} \frac{1}{\sqrt{2\pi\sigma^2}} \exp(-x^2/2\sigma^2), \tag{B.44}$$

which also has unit area for any σ. Later, it is defined in terms of the sinc function as illustrated in Figure B.9. Table B.2 summarizes some useful properties of the Dirac delta function. The convolution and sifting expressions are identical because the delta function is symmetric about its argument. The Dirac delta function is useful for defining pdfs for discrete random variables which can be integrated (e.g., to find moments).

Definition: Dirac Delta Function of Two Variables The *two-dimensional Dirac delta function* is

$$\delta(x, y) = \delta(x)\delta(y) \triangleq \begin{cases} \infty, & x = y = 0 \\ 0, & \text{else,} \end{cases} \tag{B.45}$$

such that

$$\int_{-\infty}^{\infty}\int_{-\infty}^{\infty} \delta(x, y)\mathrm{d}x\mathrm{d}y = 1. \tag{B.46}$$

B.9 KRONECKER DELTA FUNCTION

Definition: Kronecker Delta Function The *Kronecker delta function* is

$$\delta[k] \triangleq \begin{cases} 1, & k = 0 \\ 0, & k \neq 0 \end{cases} \tag{B.47}$$

for $k \in \mathcal{Z}$.

Brackets are used so that the Kronecker delta function is not confused with the Dirac delta function. (The symbol δ_k is also used for the Kronecker delta function; however, it is not used here because it resembles the

Dirac measure discussed in Chapter 2.) It has the following basic properties:

$$\delta[k-n] = \begin{cases} 1, & k = n \\ 0, & k \neq n \end{cases} \quad \text{(shift)}, \tag{B.48}$$

$$\sum_{n=-\infty}^{\infty} \alpha_n \delta[k-n] = \alpha_k \quad \text{(sifting)} \tag{B.49}$$

for $k, n \in \mathcal{Z}$. The Kronecker delta function is convenient for defining probability mass functions (pmfs) and also for representing discrete-time signals and systems.

Definition: Kronecker Delta Function of Two Variables The *two-dimensional Kronecker delta function* is

$$\delta[x, y] = \delta[x]\delta[y] \triangleq \begin{cases} 1, & x = y = 0 \\ 0, & \text{else.} \end{cases} \tag{B.50}$$

B.10 UNIT-STEP FUNCTIONS

Definition: Unit-Step Function The *unit-step function* is

$$u(x) \triangleq \begin{cases} 1, & x \geq 0 \\ 0, & x < 0. \end{cases} \tag{B.51}$$

It is also known as the Heaviside step function, and the symbol $H(x)$ is sometimes used.

The unit-step function can also be defined in terms of the signum function:

$$u(x) \triangleq [1 + \text{sgn}(x)]/2 = \begin{cases} 0, & x < 0 \\ 1/2, & x = 0 \\ 1, & x > 0. \end{cases} \tag{B.52}$$

Since $x = 0$ has Lebesgue measure zero, it is generally more convenient to use the definition in (B.51), which is the form mostly used in this book, especially when $u(x)$ defines the support of a random variable. The advantage of (B.52) is that it is consistent with the definition of the signum function, which has no ambiguity about the value at $x = 0$ (though, as mentioned previously $\text{sgn}(0) = 1$ is often used in practice). The unit-step function is obtained from the Dirac delta function as follows:

$$u(x) = \int_{-\infty}^{x} \delta(y) dy. \tag{B.53}$$

Definition: Discrete Unit-Step Function The *discrete unit-step function* is

$$u[k] \triangleq \begin{cases} 1, & k \in \mathcal{Z}^+ \\ 0, & \text{else.} \end{cases} \tag{B.54}$$

It can be expressed in terms of the Kronecker delta function as follows:

$$u[k] = \sum_{n=0}^{\infty} \delta[k-n]. \tag{B.55}$$

B.11 RECTANGLE FUNCTIONS

Definition: Rectangle Function The *rectangle function* is

$$\mathrm{rect}(x) \triangleq \begin{cases} 1, & |x| \leq 1/2 \\ 0, & |x| > 1/2. \end{cases} \tag{B.56}$$

The symbol $\Pi(x)$ is sometimes used.

It can be written in terms of the unit-step function as follows:

$$\mathrm{rect}(x) = u(x + 1/2) - u(x - 1/2). \tag{B.57}$$

We have arbitrarily defined $\mathrm{rect}(\pm 1/2) \triangleq 1$ since the Lebesgue measure of $\pm 1/2$ is zero, although it is possible to assign zero to $x = \pm 1/2$. Another definition assigns 1/2 to $x = \pm 1/2$:

$$\mathrm{rect}(x) \triangleq \begin{cases} 0, & |x| > 1/2 \\ 1/2, & |x| = 1/2 \\ 1, & |x| < 1/2. \end{cases} \tag{B.58}$$

This last form is consistent with the definition of the signum function, so that $\mathrm{sgn}(x)$ can be written in terms of $\mathrm{rect}(x)$ by appropriate scaling and shifting. The rectangle function based on the first definition is often used to specify the support of a function (see the triangle function below).

Definition: Discrete Rectangle Function The *discrete rectangle function* is

$$\mathrm{rect}[x] \triangleq \begin{cases} 1, & |x| = 0, 1, \ldots, N-1 \\ 0, & \text{else}, \end{cases} \tag{B.59}$$

where $N \in \mathcal{Z}$.

Note that in digital filter design, this function is called a rectangular *window*.

B.12 TRIANGLE AND RAMP FUNCTIONS

Definition: Triangle Function The *triangle function* is

$$\mathrm{tri}(x) \triangleq (1 - |x|)\mathrm{rect}(x/2) = \begin{cases} 1 - |x|, & |x| \leq 1 \\ 0, & |x| > 1. \end{cases} \tag{B.60}$$

The symbol $\Lambda(x)$ is sometimes used.

The rectangle function is used above to specify the support of $\mathrm{tri}(x)$, by appropriately truncating $1 - |x|$. The triangle function can be derived by convolving two rectangle functions:

$$\mathrm{tri}(x) = \int_{-\infty}^{\infty} \mathrm{rect}(x - v)\mathrm{rect}(v)dv. \tag{B.61}$$

Definition: Ramp Function The *ramp function* is

$$R(x) \triangleq xu(x) = \begin{cases} x, & x \geq 0 \\ 0, & x < 0, \end{cases} \tag{B.62}$$

The unit-step function in the second expression defines the support of $R(x)$ by truncating the linear function $g(x) = x$. The ramp function is obtained by integrating the unit-step function:

$$R(x) = \int_{-\infty}^{x} u(v)\mathrm{d}v. \tag{B.63}$$

B.13 INDICATOR FUNCTIONS

Definition: Indicator Function The *indicator function* for set E is

$$I_E(x) \triangleq \begin{cases} 1, & x \in E \\ 0, & \text{else}. \end{cases} \tag{B.64}$$

When $E = [a, b]$ for $a < b$, the indicator function specifies a closed interval:

$$I_{[a,b]}(x) = \begin{cases} 1, & a \leq x \leq b \\ 0, & \text{else}. \end{cases} \tag{B.65}$$

Open and semi-open intervals are similarly given by $I_{[a,b)}(x)$, $I_{(a,b]}(x)$, and $I_{(a,b)}(x)$. It is possible for $a = -\infty$ or $b = \infty$ or both, in which case we might use $I_{\mathcal{R}}(x)$. The indicator function can be written in terms of the unit-step function as follows:

$$I_{[0,\infty)}(x) = u(x), \quad I_{[a,\infty)}(x) = u(x - a), \quad I_{[a,b)}(x) = u(x - a) - u(x - b). \tag{B.66}$$

It can also be used for countable sets $E \in \mathcal{Z}$ such as

$$I_{\{0,1,\ldots\}}[x] = I_{\mathcal{Z}^+}[x] = \begin{cases} 1, & x = 0, 1, \ldots \\ 0, & \text{else}, \end{cases} \tag{B.67}$$

where brackets have been used to emphasize that x is discrete. The indicator function is convenient for specifying the support of a pdf (for the exponential random variable):

$$f_X(x) = \lambda \exp(-\lambda x) I_{[0,\infty)}(x) \tag{B.68}$$

or a pmf (for the Poisson random variable):

$$p_X[x] = \exp(-\alpha)(\alpha^x/x!) I_{\mathcal{Z}^+}[x]. \tag{B.69}$$

The indicator function for set complement is

$$\begin{aligned} I_{E^c}(x) &= \begin{cases} 0, & x \in E \\ 1, & \text{else}, \end{cases} \\ &= 1 - I_E(x), \end{aligned} \tag{B.70}$$

and for set intersection is

$$I_{E\cap F}(x) = \begin{cases} 1, & x \in E \cap F \\ 0, & \text{else}, \end{cases}$$
$$= I_E(x) I_F(x). \tag{B.71}$$

For the union of disjoint sets E and F, we can write $I_{E\cup F}(x) = I_E(x) + I_F(x)$. However, if the sets are not disjoint, a term must be subtracted as follows:

$$I_{E\cup F}(x) = I_E(x) + I_F(x) - I_{E\cap F}(x),$$
$$= I_E(x) + I_F(x) - I_E(x) I_F(x). \tag{B.72}$$

Note the similarity of this expression and that for probability in (2.49) for overlapping events, that is, $P(E \cup F) = P(E) + P(F) - P(EF)$. In fact, the expectation property for the indicator function described in Chapter 5 yields

$$\mathcal{E}[I_{E\cup F}(x)] = P(x \in E \cup F),$$
$$= P(x \in E) + P(x \in F) - P(x \in EF). \tag{B.73}$$

B.14 SINC FUNCTION

Definition: Sinc Function The *sinc function* is

$$\text{sinc}(x) \triangleq \frac{\sin(\pi x)}{\pi x}. \tag{B.74}$$

Note that π is not shown explicitly in the notation sinc(x). The sinc function is plotted in Figure B.8, demonstrating that the zero crossings occur for $x \in \mathcal{Z}$ except zero. There is a form of this function used primarily in mathematics that does not include π, given by

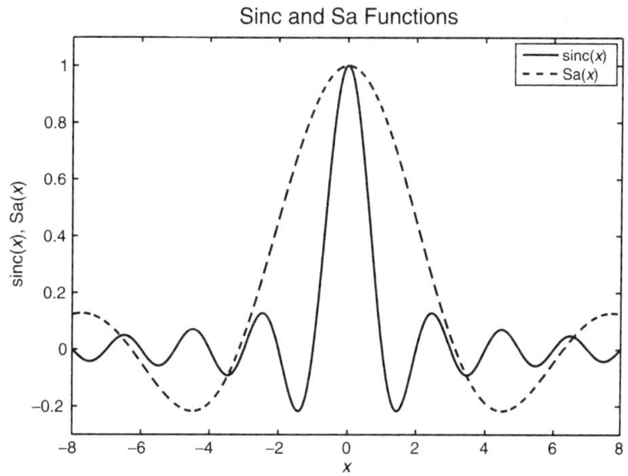

FIGURE B.8 Sinc and Sa functions.

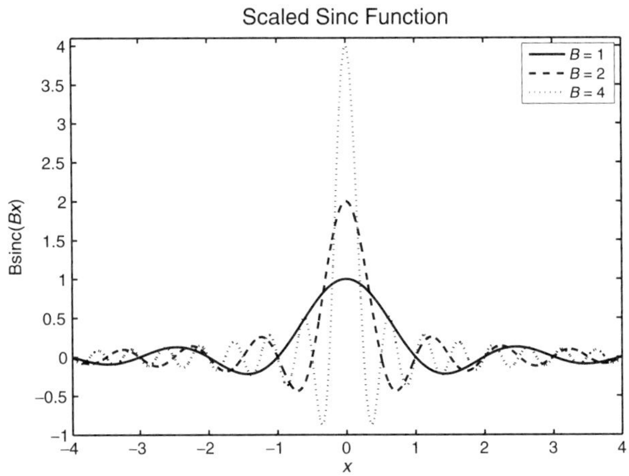

FIGURE B.9 $B\text{sinc}(Bx)$ for various values of B.

$$\text{Sa}(x) \triangleq \frac{\sin(x)}{x}, \tag{B.75}$$

which is also plotted in Figure B.8. It is also called the unnormalized sinc function, and its zero crossings occur at integer multiples of π except zero.

The sinc function is related to the gamma function (described in Section B.15) as follows:

$$\text{sinc}(x) = [\Gamma(1+x)\Gamma(1-x)]^{-1}. \tag{B.76}$$

The sinc function and the rectangle function are a Fourier transform pair:

$$\text{sinc}(x) = \int_{-\infty}^{\infty} \text{rect}(f)\exp(j2\pi f x)\text{d}f, \tag{B.77}$$

$$\text{rect}(f) = \int_{-\infty}^{\infty} \text{sinc}(x)\exp(-j2\pi f x)\text{d}x, \tag{B.78}$$

which shows that it has unit area:

$$\int_{-\infty}^{\infty} \text{sinc}(x)\text{d}x = \text{rect}(0) = 1. \tag{B.79}$$

The sinc function can also be used to define the Dirac delta function as follows:

$$\lim_{B\to\infty} B\ \text{sinc}(Bx) = \delta(x), \tag{B.80}$$

where the scaling B ensures that the function has unit area. Examples are shown in Figure B.9.

B.15 LOGARITHM FUNCTIONS

The logarithm function is used in the definitions of entropy and mutual information (see Chapter 10). It also appears in the logarithm random variable and is used for some transformations of random variables. The natural

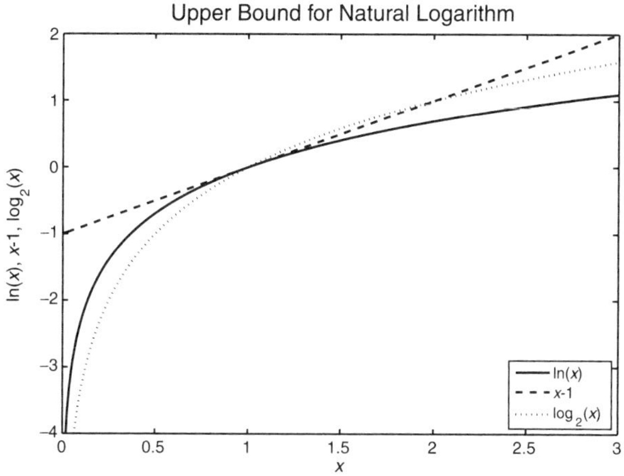

FIGURE B.10 Upper bound for $\ln(x)$.

logarithm is plotted in Figure B.10 where we see that it is a monotonically increasing function. It is shown in Problem 10.14 that the natural logarithm is bounded as follows:

$$\ln(x) \leq x - 1, \tag{B.81}$$

which is also demonstrated in the figure. This bound is useful for proving various properties in information theory; an example is provided in Problem 10.19 to show that the mutual information is nonnegative. The upper bound is reached only when $x = 1$ such that $\ln(1) = 0$. Note that the bound generally does not apply to other logarithm bases; for example, it is shown in the figure that $\log_2(x)$ exceeds $x - 1$ for an interval just above $x = 1$. We can convert between different bases using the formula:

$$\log_b(x) = \log_b(a)\log_a(x) = \log_a(x)/\log_a(b). \tag{B.82}$$

Thus, if an upper bound is needed for $\log_2(x)$, the following can be used:

$$\log_2(x) = \log_2(e)\ln(x) \leq 1.4427(x - 1). \tag{B.83}$$

Because both the slope and the ordinate of $x - 1$ are changed by the scale factor 1.4427, $\ln(x)$ exceeds the upper bound in (B.83) for an interval just below $x = 1$ (this upper bound is not shown in the figure).

B.16 GAMMA FUNCTIONS

Definition: Gamma Function The *gamma function* is the following definite integral:

$$\Gamma(a) \triangleq \int_0^\infty x^{a-1}\exp(-x)\mathrm{d}x, \tag{B.84}$$

where $a \in \mathcal{C}$ (though we only consider real values in this book).

It has the following property: $\Gamma(a+1) = a\Gamma(a)$, and for $a = n \in \mathcal{N}$:

$$\begin{aligned}\Gamma(n) &= (n-1)\Gamma(n-1), \\ &= (n-1)(n-2)\cdots 2\Gamma(1) = (n-1)!. \end{aligned} \tag{B.85}$$

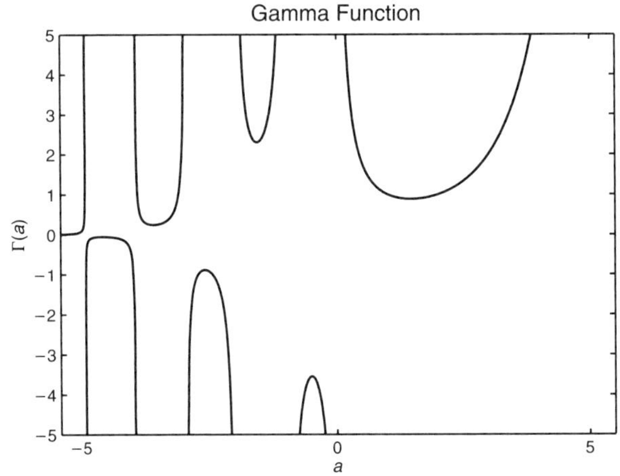

FIGURE B.11 Gamma function $\Gamma(a)$.

For noninteger numbers a, the gamma function can be viewed as an extension of the factorial function. A plot of $\Gamma(a)$ is provided in Figure B.11. Observe that for negative integers, the function alternates between $\pm\infty$, and for positive integers it is the factorial function (with argument shifted down by one), as given in (B.85). It can also be shown that for $n \in \mathcal{N}$:

$$\Gamma(1/2 + n) = [(2n)!/4^n n!]\sqrt{\pi}, \quad \Gamma(1/2 - n) = [(-4)^n n!/(2n)!]\sqrt{\pi} \tag{B.86}$$

with $\Gamma(1/2) = \sqrt{\pi}$ and $\Gamma(-1/2) = -2\sqrt{\pi}$. The Euler–Masheroni constant (described later) is related to the gamma function as follows:

$$\gamma = -\frac{\mathrm{d}}{\mathrm{d}x}\Gamma(x)\bigg|_{x=1}. \tag{B.87}$$

For ease of reference, we summarize in Table B.3 several values of the gamma function. (Recall from Figure B.11 that the gamma function is unbounded ($\pm\infty$) for negative integers.)

If the upper limit in the integral of the gamma function is changed to a finite number $0 < b < \infty$, then the result is called the *upper incomplete gamma function*:

$$\Gamma(a, b) \triangleq \int_b^\infty x^{a-1} \exp(-x)\mathrm{d}x \tag{B.88}$$

for $a, b > 0$. Obviously, $\Gamma(a, 0) \equiv \Gamma(a)$. The *lower incomplete gamma function* is

$$\gamma(a, b) \triangleq \int_0^b x^{a-1} \exp(-x)\mathrm{d}x \tag{B.89}$$

TABLE B.3 Gamma Function Values

n	$\Gamma(n)$	n	$\Gamma(n)$	n	$\Gamma(n)$
1	1	1/2	$\sqrt{\pi}$	$-1/2$	$-2\sqrt{\pi}$
2	1	3/2	$\sqrt{\pi}/2$	$-3/2$	$4\sqrt{\pi}/3$
3	2	5/2	$3\sqrt{\pi}/4$	$-5/2$	$-8\sqrt{\pi}/15 \approx 0.5333\sqrt{\pi}$
4	6	7/2	$15\sqrt{\pi}/8 = 1.875\sqrt{\pi}$	$-7/2$	$16\sqrt{\pi}/105 \approx 0.1523\sqrt{\pi}$
5	24	9/2	$105\sqrt{\pi}/16 = 6.5625\sqrt{\pi}$	$-9/2$	$-32\sqrt{\pi}/945 \approx 0.0339\sqrt{\pi}$

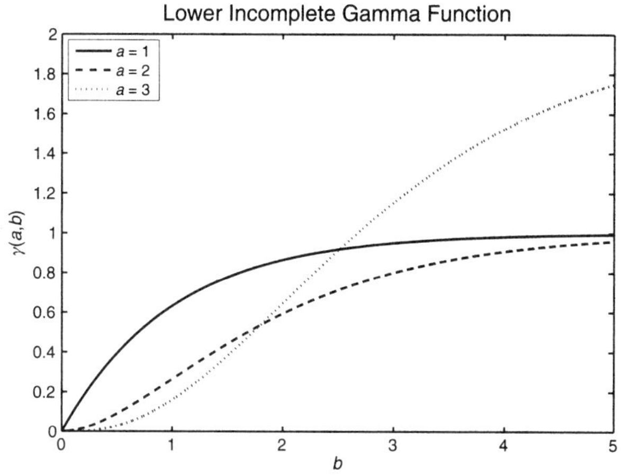

FIGURE B.12 Lower incomplete gamma function $\gamma(a, b)$.

with $\gamma(a, \infty) \equiv \Gamma(a)$. The incomplete gamma functions are described by the following recursions:

$$\Gamma(a, b) = (a-1)\Gamma(a-1, b) + b^{a-1}\exp(-b), \tag{B.90}$$

$$\gamma(a, b) = (a-1)\gamma(a-1, b) - b^{a-1}\exp(-b). \tag{B.91}$$

Note also that $\Gamma(a) = \Gamma(a, b) + \gamma(a, b)$. The lower incomplete gamma function is useful for representing the cdf of some random variables, such as the chi-square distribution. Examples are plotted in Figure B.12. It also has the following series expansion:

$$\gamma(a, b) = b^a \exp(-b)(a-1)! \sum_{n=0}^{\infty} \frac{b^n}{(a+n)!}. \tag{B.92}$$

Definition: Digamma Function The *digamma function* is

$$\psi(x) \triangleq \frac{\mathrm{d}}{\mathrm{d}x} \ln(\Gamma(x)) = \Gamma'(x)/\Gamma(x) \tag{B.93}$$

for $x \in \mathcal{R}^+$ where $\Gamma(x)$ is the gamma function and $\Gamma'(x)$ is its ordinary derivative.

It is used to write the differential entropy $h(X)$ for some random variables (see Appendix A), and is plotted in Figure B.13.

B.17 BETA FUNCTIONS

Definition: Beta Function The *beta function* is

$$B(a, b) \triangleq \int_0^1 x^{a-1}(1-x)^{b-1}\mathrm{d}x = \frac{\Gamma(a)\Gamma(b)}{\Gamma(a+b)} \tag{B.94}$$

for $a, b > 0$.

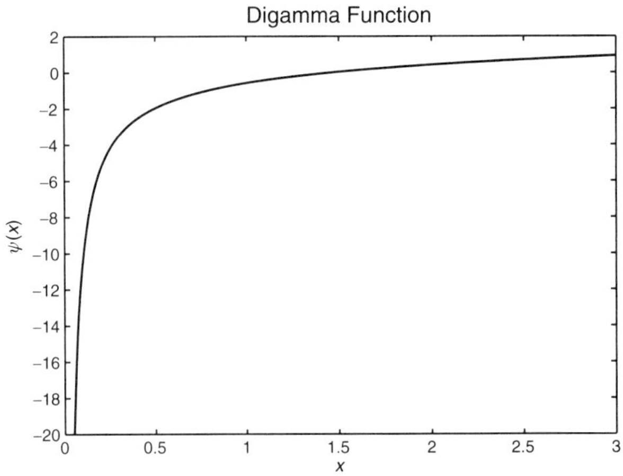

FIGURE B.13 Digamma function $\psi(x)$.

The beta function is symmetric with respect to its parameters: $B(a, b) = B(b, a)$, as illustrated in Figure B.14. The *incomplete beta function* is obtained by replacing the definite integral with an indefinite integral as follows:

$$B_y(a, b) \triangleq \int_0^y x^{a-1}(1 - x)^{b-1}\mathrm{d}x. \tag{B.95}$$

Obviously, $B_1(a, b) = B(a, b)$. Sometimes it is notationally convenient to use the *regularized* incomplete beta function:

$$I_y(a, b) \triangleq B_y(a, b)/B(a, b), \tag{B.96}$$

which is a normalized form (this notation should not be confused with the indicator function).

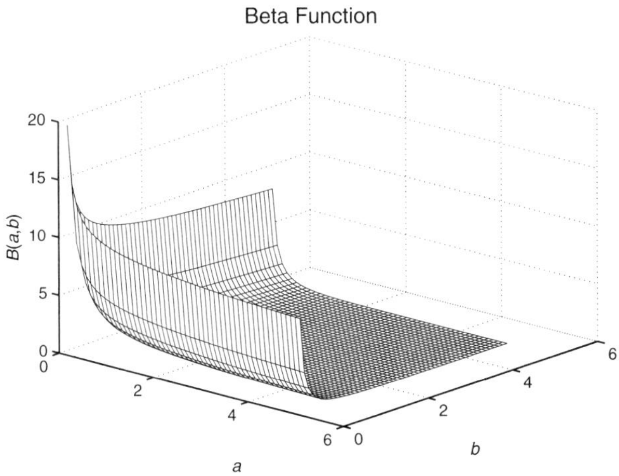

FIGURE B.14 Beta function $B(a, b)$.

B.18 BESSEL FUNCTIONS

Definition: Bessel Functions *Bessel functions of the first kind* can be defined using a Maclaurin series (which is specified about $x = 0$):

$$J_\alpha(x) = \sum_{n=0}^{\infty} \frac{(-1)^n}{\Gamma(n+\alpha+1)n!}(x/2)^{2n+\alpha} \tag{B.97}$$

where $\alpha \in \mathcal{R}$ and $\Gamma(n)$ is the gamma function. *Modified Bessel functions of the first kind* are generated from $J_\alpha(x)$ by imposing a purely imaginary argument with $j = \sqrt{-1}$ as follows:

$$I_\alpha(x) \triangleq j^{-\alpha} J_\alpha(jx), \tag{B.98}$$

where $I_\alpha(x)$ is a real-valued function of x (not to be confused with the indicator function). The series expansion of $I_\alpha(x)$ is identical to that of $J_\alpha(x)$ except for the $(-1)^n$ term in the numerator:

$$I_\alpha(x) = \sum_{n=0}^{\infty} \frac{1}{\Gamma(n+\alpha+1)n!}(x/2)^{2n+\alpha}. \tag{B.99}$$

Modified Bessel functions of the second kind are given by

$$K_\alpha(x) \triangleq (\pi/2)\frac{I_{-\alpha}(x) - I_\alpha(x)}{\sin(\alpha\pi)}. \tag{B.100}$$

Examples of the various Bessel functions are shown in Figure B.15.

B.19 *Q*-FUNCTION AND ERROR FUNCTIONS

Definition: *Q*-Function The *Q-function* is defined as an area under the standard Gaussian pdf:

$$Q(x) \triangleq \frac{1}{\sqrt{2\pi}} \int_x^{\infty} \exp(-v^2/2)dv \tag{B.101}$$

for $x \in \mathcal{R}$ with $Q(-\infty) = 1$, $Q(\infty) = 0$, $Q(0) = 0.5$, and

$$Q(-x) = 1 - Q(x). \tag{B.102}$$

The Q-function is used to express a definite integral of a Gaussian pdf for which no closed form exists. The cdf of the standard Gaussian random variable can be written as

$$F_X(x) = \frac{1}{\sqrt{2\pi}} \int_{-\infty}^{x} \exp(-v^2/2)dv = 1 - Q(x). \tag{B.103}$$

A plot of the Q-function is shown in Figure B.16(a). (Note that the Q-function is *not* a special case of the Marcum Q-function described later.) Since a closed form does not exist for $Q(x)$, some tabulated results are provided in Table B.4. However, it is often convenient to have a closed-form *approximation* for $Q(x)$. It is relatively straightforward to show that for $x \in \mathcal{R}^+$, $Q(x)$ is upper bounded as follows:

$$Q(x) \leq (1/2)\exp(-x^2). \tag{B.104}$$

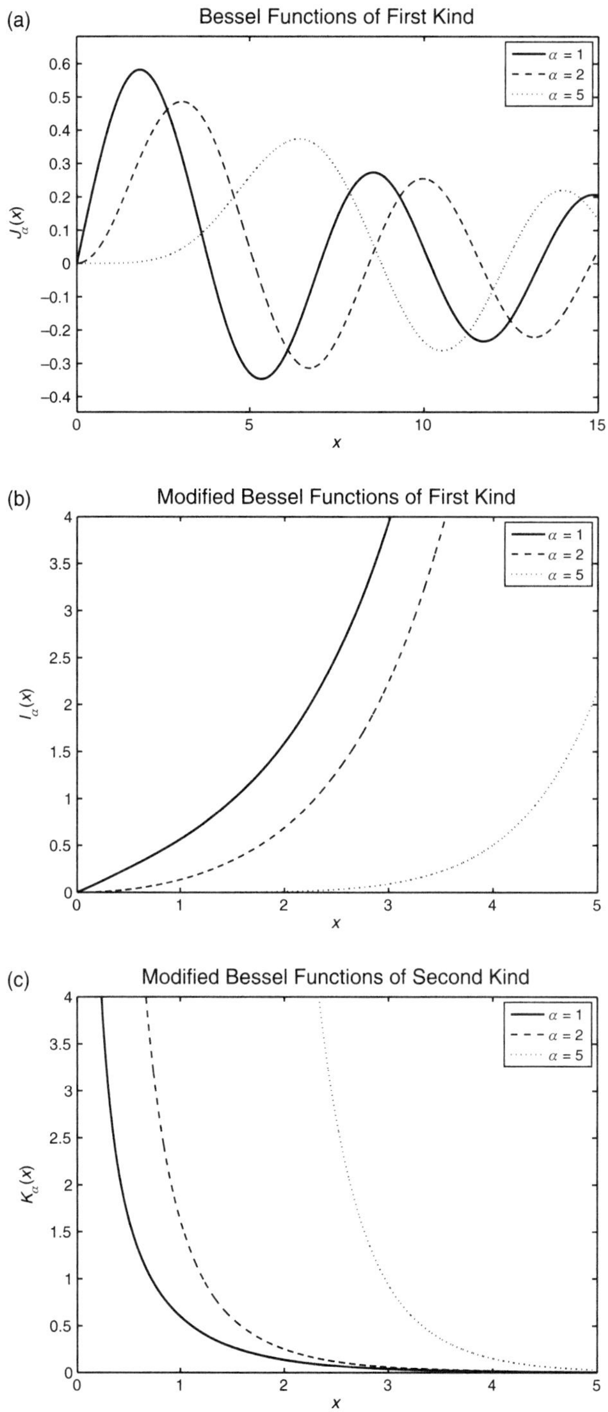

FIGURE B.15 Bessel functions. (a) Bessel functions of the first kind $J_\alpha(x)$. (b) Modified Bessel functions of the first kind $I_\alpha(x)$. (c) Modified Bessel functions of the second kind $K_\alpha(x)$.

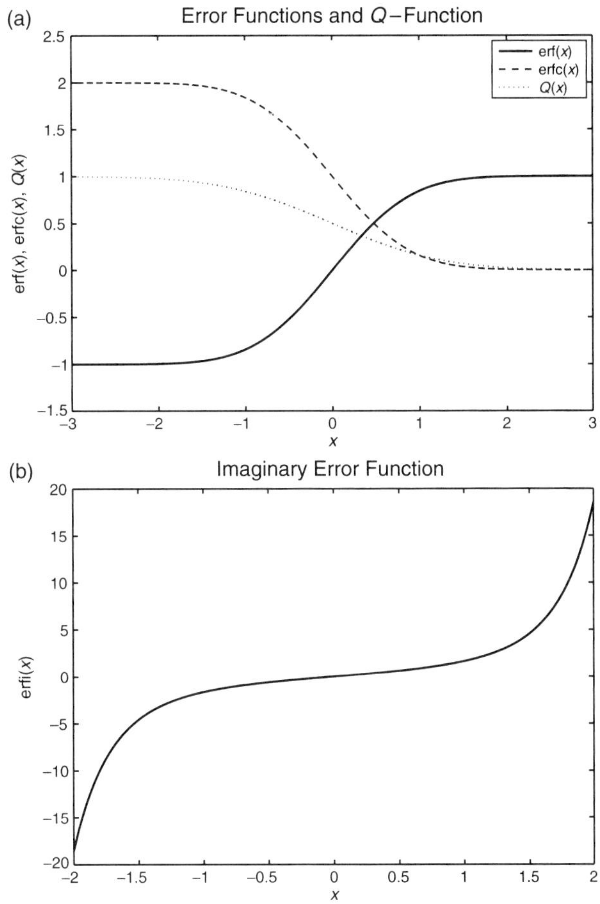

FIGURE B.16 Functions based on the Gaussian pdf. (a) Error functions and Q-function. (b) Imaginary error function.

Definition: Error Function The *error function* is

$$\operatorname{erf}(x) \triangleq \frac{2}{\sqrt{\pi}} \int_0^x \exp(-v^2) \mathrm{d}v \tag{B.105}$$

for $x \in \mathcal{R}^+$ with $\operatorname{erf}(0) = 0$, $\operatorname{erf}(\infty) = 1$, and

$$\operatorname{erf}(-x) = -\operatorname{erf}(x) \tag{B.106}$$

The error function is another tabulated function used to express the area under the standard Gaussian pdf; it is also plotted in Figure B.16(a).

Definition: Complementary Error Function The *complementary error function* is defined for $x \in \mathcal{R}$ as

$$\operatorname{erfc}(x) \triangleq 1 - \operatorname{erf}(x), \tag{B.107}$$

TABLE B.4 $Q(x)$ and Erfc(x)

x	$Q(x)$	erfc(x)	x	$Q(x)$	erfc(x)
0.0	0.5000000	1.0000000	1.8	0.0359303	0.0109095
0.1	0.4601722	0.8875371	1.9	0.0287166	0.0072096
0.2	0.4207440	0.7772974	2.0	0.0227501	0.0046777
0.3	0.3820886	0.6713732	2.1	0.0178644	0.0029795
0.4	0.3445783	0.5716076	2.2	0.0139034	0.0018628
0.5	0.3085375	0.4795001	2.3	0.0107241	0.0011432
0.6	0.2742531	0.3961439	2.4	0.0081975	0.0006885
0.7	0.2419637	0.3221988	2.5	0.0062097	0.0004070
0.8	0.2118554	0.2578990	2.6	0.0046612	0.0002360
0.9	0.0184060	0.2030918	2.7	0.0034670	0.0001343
1.0	0.1586553	0.1572992	2.8	0.0025555	0.0000750
1.1	0.1356660	0.1197949	2.9	0.0018658	0.0000411
1.2	0.1150697	0.0896860	3.0	0.0013499	0.0000221
1.3	0.0968005	0.0659921	3.1	0.0009676	0.0000116
1.4	0.0807567	0.0477149	3.2	0.0006871	0.0000060
1.5	0.0668072	0.0338949	3.3	0.0004834	0.0000031
1.6	0.0547993	0.0236516	3.4	0.0003369	0.0000015
1.7	0.0445655	0.0162095	3.5	0.0002326	0.0000007

and thus

$$\text{erfc}(x) = \frac{2}{\sqrt{\pi}} \int_x^{\infty} \exp(-v^2) dv \tag{B.108}$$

with $\text{erfc}(\infty) = 0$, $\text{erfc}(0) = 1$, $\text{erfc}(-\infty) = 2$, and

$$\text{erfc}(-x) = 2 - \text{erfc}(x). \tag{B.109}$$

It is plotted in Figure B.16(a) and some numerical values are provided in Table B.4. Observe from (B.101) and (B.108) that $Q(x)$ and erfc(x) are related as follows:

$$Q(x) = (1/2)\text{erfc}(x/\sqrt{2}), \quad \text{erfc}(x) = 2Q(\sqrt{2}x). \tag{B.110}$$

Thus

$$Q(x) = (1/2)\left[1 - \text{erf}(x/\sqrt{2})\right] \tag{B.111}$$

and the Gaussian cdf is

$$F_X(x) = (1/2)\left[1 + \text{erf}(x/\sqrt{2})\right]. \tag{B.112}$$

Definition: Imaginary Error Function The *imaginary error function* is

$$\text{erfi}(x) \triangleq -j\text{erf}(jx) \tag{B.113}$$

for $x \in \mathcal{R}$ and $j = \sqrt{-1}$.

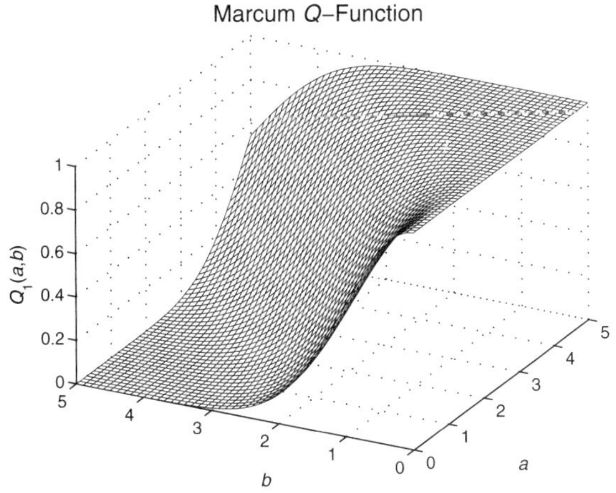

FIGURE B.17 Marcum Q-function $Q_m(a, b)$ for $m = 1$.

Note that $\text{erfi}(x)$ is a real-valued function of x as shown in Figure B.16(b). It is used in the characteristic function of the Rayleigh random variable.

B.20 MARCUM Q-FUNCTION

Definition: Marcum Q-Function The *generalized Marcum Q-function* is

$$Q_m(a, b) \triangleq \frac{1}{a^{m-1}} \int_b^\infty x^m \exp(-(x^2 + a^2)/2) I_{m-1}(ax) \mathrm{d}x,$$

$$= \exp(-(a^2 + b^2)/2) \sum_{k=1-m}^{\infty} (a/b)^k I_k(ab), \tag{B.114}$$

where $m \in \mathcal{Z}^+$, $\{a, b\} > 0$, and $I_m(x)$ is the mth-order modified Bessel function of the first kind.

It is used in the cdf of the Rice random variable; an example is plotted in Figure B.17 for $m = 1$.

B.21 ZETA FUNCTION

Definition: Riemann Zeta Function The *Riemann zeta function* is

$$\zeta(s) \triangleq \sum_{n=1}^{\infty} \frac{1}{n^s}, \tag{B.115}$$

where $s \triangleq \sigma + j\omega$ is a complex variable.

This series converges provided $\text{Re}(s) = \sigma > 1$. Figure B.18 shows a plot of $\zeta(s)$ for real values of s, which converges to zero for $\sigma < 0$, converges to 1 for $\sigma > 0$, and has a pole at $s = 1$. The zeta function is used in the pmf of the zeta random variable with real-valued $s \to \alpha$, which is a scale parameter of the distribution.

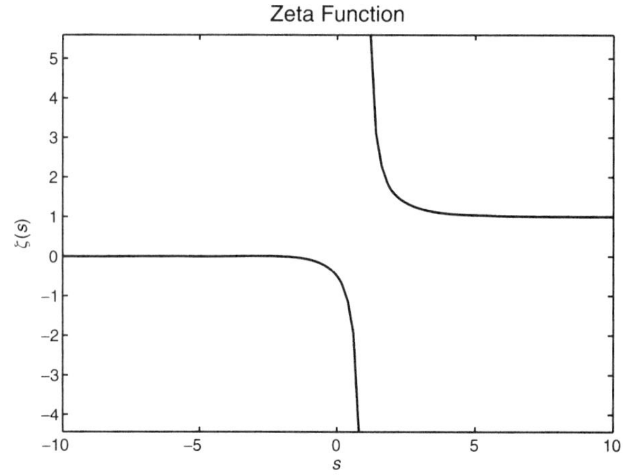

FIGURE B.18 Zeta function $\zeta(s)$ for real-valued s.

B.22 RISING AND FALLING FACTORIALS

Definition: Rising and Falling Factorials The *rising factorial* is

$$x^{(n)} \triangleq x(x+1)\cdots(x+n-1) \tag{B.116}$$

for $x \in \mathcal{R}$ and $n \in \mathcal{N}$. The *falling factorial* is

$$(x)_n \triangleq x(x-1)\cdots(x-n+1). \tag{B.117}$$

Except for $x^{(0)} = (x)_0 = 1$, these functions yield polynomials in x, such as

$$(x)_3 = x^3 - 3x^2 + 2x, \quad x^{(3)} = x^3 + 3x^2 + 2x. \tag{B.118}$$

They are related as follows:

$$x^{(n)} = (-1)^n(-x)_n, \quad x^{(n)} = (x+n-1)_n \tag{B.119}$$

and have the following properties:

$$x^{(n)} = \frac{\Gamma(x+n)}{\Gamma(x)}, \quad (x)_n = \frac{\Gamma(x+1)}{\Gamma(x-n+1)}, \quad 1^{(n)} = n!, \quad (n)_n = n!, \tag{B.120}$$

$$\frac{x^{(n)}}{n!} = \binom{x+n-1}{n}, \quad \frac{(x)_n}{n!} = \binom{x}{n}, \tag{B.121}$$

where $\Gamma(x)$ is the gamma function and $\binom{x}{n}$ is a binomial coefficient. When $x \in \mathcal{N}$:

$$x^{(n)} = \frac{(x+n-1)!}{(x-1)!}, \quad (x)_n = \frac{x!}{(x-n)!} \tag{B.122}$$

B.23 LAGUERRE POLYNOMIALS

Definition: Laguerre Polynomials *Laguerre polynomials* are defined by the following recursion:

$$L_{v+1}(x) \triangleq [(2v + 1 - x)L_v(x) - vL_{v-1}(x)]/(v + 1) \tag{B.123}$$

with initial conditions $L_0(x) = 1$ and $L_1(x) = 1 - x$.

They correspond to solutions of the following differential equation as v is varied:

$$x\frac{d^2y}{dx^2} + (1 - x)\frac{dy}{dx} + vy = 0. \tag{B.124}$$

An explicit expression for the Laguerre polynomials is

$$L_v(x) = \sum_{n=0}^{v}(-1)^n \binom{v}{v - n} x^n/n!. \tag{B.125}$$

Figure B.19 shows plots of the first three polynomials:

$$L_1(x) = 1 - x, \tag{B.126}$$

$$L_2(x) = x^2/2 - 2x + 1, \tag{B.127}$$

$$L_3(x) = -x^3/6 + 3x^2/2 - 3x + 1. \tag{B.128}$$

Also included is $L_{1/2}(x)$ which is used for the mean and variance of the Rice random variable. It can be written in terms of the modified Bessel functions of the first kind as follows:

$$L_{1/2}(x) = \exp(x/2)\,[(1 - x)I_0(-x/2) - xI_1(-x/2)]\,. \tag{B.129}$$

The Laguerre polynomials, including fractional forms $L_{n/2}(x)$, can be expressed in terms of the confluent hypergeometric function described in Section B.24.

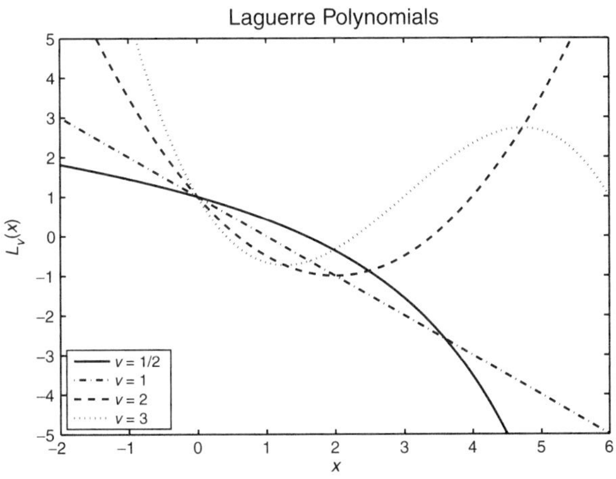

FIGURE B.19 Laguerre polynomials $L_v(x)$.

B.24 HYPERGEOMETRIC FUNCTIONS

Definition: Confluent Hypergeometric Function The *confluent hypergeometric function of the first kind* is the following infinite series:

$$ {}_1F_1(a;b;x) \triangleq \sum_{n=0}^{\infty} \frac{a^{(n)}x^n}{b^{(n)}n!}, \tag{B.130}$$

where $\{a, b\}$ are parameters and $\{a^{(n)}, b^{(n)}\}$ are rising factorials (see Section B.22). For $a, b > 0$, it is represented by the following integral:

$$ {}_1F_1(a;b;x) = \frac{\Gamma(b)}{\Gamma(b-a)\Gamma(a)} \int_0^1 v^{a-1}(1-v)^{b-a-1} \exp(xv)\mathrm{d}v, \tag{B.131}$$

where $\Gamma(\cdot)$ is the gamma function.

It is a specific case of the generalized hypergeometric function ${}_pF_q(a;b;x)$ (which we do not define because the general form is not used in this book). ${}_1F_1(a;b;x)$ is one solution of the following second-order differential equation:

$$ x\frac{\mathrm{d}^2y}{\mathrm{d}x^2} + (b-x)\frac{\mathrm{d}y}{\mathrm{d}x} - ay = 0. \tag{B.132}$$

Comparing this equation with (B.124), we see that the Laguerre polynomials are a special case as follows:

$$ L_v(x) = {}_1F_1(-v; 1, ; x). \tag{B.133}$$

The Laguerre polynomial used to describe the moments of the Rice distribution is given by $L_{n/2}(x) = {}_1F_1(-n/2; 1; x)$. The other solution to (B.132) is called the *confluent hypergeometric function of the second kind*, which has the following notation and integral representation:

$$ \Psi(a;b;x) = \frac{1}{\Gamma(a)} \int_0^{\infty} v^{a-1}(1+v)^{b-a-1} \exp(-xv)\mathrm{d}v. \tag{B.134}$$

Several other functions are special cases of the confluent hypergeometric function of the first kind. These include the error function:

$$ \mathrm{erf}(x) = (2x/\sqrt{\pi})\, {}_1F_1(1/2; 3/2; -x^2) \tag{B.135}$$

and the lower incomplete gamma function:

$$ \gamma(a, b) = (b^a/a)\, {}_1F_1(a; a+1; -x). \tag{B.136}$$

Both of these can be verified from the integral in (B.131), which usually requires a change of variables to obtain the familiar form of a particular function, as demonstrated next.

Example B.14. For the error function, we can write

$$\begin{aligned}\text{erf}(x) &= (2x/\sqrt{\pi})\frac{\Gamma(3/2)}{\Gamma(1)\Gamma(1/2)}\int_0^1 v^{-1/2}(1-v)^0\exp(-vx^2)\mathrm{d}v,\\ &= (1/\sqrt{\pi})\int_0^1 (x/\sqrt{v})\exp(-vx^2)\mathrm{d}v. \end{aligned}\tag{B.137}$$

Changing variables to $z^2 = vx^2$ such that $\mathrm{d}v = (2/x)\sqrt{v}\mathrm{d}z$ with limits of integration $\{0, x\}$, we obtain

$$(1/\sqrt{\pi})\int_0^1 (x/\sqrt{v})\exp(-vx^2)\mathrm{d}v = (2/\sqrt{\pi})\int_0^x \exp(-z^2)\mathrm{d}z, \tag{B.138}$$

which is the desired result.

B.25 BERNOULLI NUMBERS

Definition: Bernoulli Numbers The *Bernoulli numbers* $\{B_n\}$ are rational numbers specified by the coefficients in the following series:

$$x/[\exp(x) - 1] = \sum_{n=0}^{\infty} B_n x^n/n!. \tag{B.139}$$

They include: $B_0 = 1, B_1 = -1/2, B_2 = 1/6, B_3 = 0, B_4 = -1/30, B_5 = 0, B_6 = 1/42, B_7 = 0, B_8 = -1/30$, $B_9 = 0$, $B_{10} = 5/66$, and $B_{20} = -174{,}611/330$. Note that $B_n = 0$ if the subscript can be expressed as $n = 2m + 1$ for $m \in \mathcal{N}$, that is, for all odd n (except $n = 1$). The Bernoulli numbers appear in expressions for the central moments of the logistic random variable.

B.26 HARMONIC NUMBERS

Definition: Harmonic Numbers The *generalized harmonic numbers* $\{H_n(v)\}$ are defined by the following finite sum:

$$H_n(v) = \sum_{m=1}^{n} \frac{1}{m^v} \tag{B.140}$$

for $v \in (1, \infty)$ and $x \in \mathcal{N}$.

The Riemann zeta function is obtained when $n \to \infty$. The ordinary harmonic numbers are given by $H_n \triangleq H_n(1)$, of which the first 10 values are $H_1 = 1$, $H_2 = 3/2$, $H_3 = 11/6$, $H_4 = 25/12$, $H_5 = 137/60$, $H_6 = 441/180$, $H_7 = 3267/1260$, $H_8 = 6849/2520$, $H_9 = 64{,}161/22{,}680$, and $H_{10} = 66{,}429/22{,}680$. The harmonic numbers are related to the Bernoulli numbers as follows:

$$H_n = \ln(n) + \gamma + (1/2n) - (1/2)\sum_{m=1}^{\infty} B_{2m} n^{-2m}/m, \tag{B.141}$$

where γ is the Euler–Mascheroni constant described in Section B.27. From this expansion, we find that $H_n \approx \ln(n) + \gamma$ for large n.

B.27 EULER–MASCHERONI CONSTANT

Definition: Euler–Mascheroni Constant The *Euler–Mascheroni constant* is defined by the following limit:

$$\gamma \triangleq \lim_{n\to\infty}\left[\sum_{m=1}^{n}(1/m) - \ln(n)\right] \approx 0.5772. \tag{B.142}$$

It appears in the differential entropy expression $h(X)$ for some random variables, and in the Laplace transform of $x(t) = \ln(t) \Longrightarrow X(s) = -[\ln(s) + \gamma]/s$.

B.28 DIRICHLET FUNCTION

Definition: Dirichlet Function The *Dirichlet function* is

$$D(x) \triangleq I_{\mathcal{Q}}[x] = \begin{cases} 1, & x \in \mathcal{Q} \\ 0, & x \in \mathcal{R} - \mathcal{Q} \end{cases} \tag{B.143}$$

for $x \in \mathcal{R}$ where $I_{\mathcal{Q}}[x]$ is the discrete indicator function.

It is an example of a function that is not Riemann integrable and is discontinuous everywhere. However, the Dirichlet function is Lebesgue integrable; for example, its Lebesgue integral on [0, 1] is zero.

FURTHER READING

These functions can be found in many mathematical handbooks, as well as several of the references listed at the end of the chapters. These include the following: Beyer (1976), Bracewell (1978), Gradshteyn and Ryzhik (1980), Kreyszig (1979), Sokolnikoff and Redheffer (1966), Thomas (1968), Weisstein (2003), and Zwillinger and Kokoska (2000).

C Frequency-Domain Transforms and Properties

In this appendix, we provide definitions of frequency-domain transforms for continuous- and discrete-time functions, summarize important properties, and include tables of several transform pairs.

C.1 LAPLACE TRANSFORM

Definition: Laplace Transform $X(s)$ The *Laplace transform* of $x(t)$ for $t \in \mathcal{R}$ is

$$X(s) \triangleq \int_{-\infty}^{\infty} x(t)\exp(-st)\mathrm{d}t \qquad \text{(bilateral)}, \tag{C.1}$$

where $s \triangleq \sigma + j\omega \in \mathcal{C}$ is a complex variable with units of s^{-1}. The real part σ is neper frequency (nepers/second), and the imaginary part ω is radian frequency (radians/s). The region of convergence (ROC) specifies the range of σ for which $X(s)$ exists.

Equation (C.1) is the *bilateral* Laplace transform. For the *unilateral* Laplace transform, the integral has a finite lower limit, usually zero:

$$X(s) \triangleq \int_{0}^{\infty} x(t)\exp(-st)\mathrm{d}t \qquad \text{(unilateral)}. \tag{C.2}$$

The unilateral and bilateral Laplace transforms are identical when $x(t)$ is nonzero only for $t \geq 0$, which can be emphasized by writing $x(t)u(t)$, where $u(t)$ is the unit-step function. The inverse Laplace transform is

$$x(t) = \int_{\sigma-j\infty}^{\sigma+j\infty} X(s)\exp(st)\mathrm{d}s, \tag{C.3}$$

where the integration is performed along a line parallel to the $j\omega$ axis, and σ lies in the ROC. However, when $X(s)$ is a rational function, it is generally easier to perform a partial fraction expansion (PFE) and use a table of transform pairs to find $x(t)$. Figure C.1 shows the s-plane and summarizes the ROC for each of the four basic types of functions.

Several properties of the unilateral Laplace transform are summarized in Table C.1, and some transform pairs are given in Table C.2. Recall that the convolution of two continuous-time functions is

$$x(t) * h(t) \triangleq \int_{-\infty}^{\infty} x(\tau)h(t-\tau)\mathrm{d}\tau = \int_{-\infty}^{\infty} h(\tau)x(t-\tau)\mathrm{d}\tau, \tag{C.4}$$

Probability, Random Variables, and Random Processes: Theory and Signal Processing Applications, First Edition. John J. Shynk.

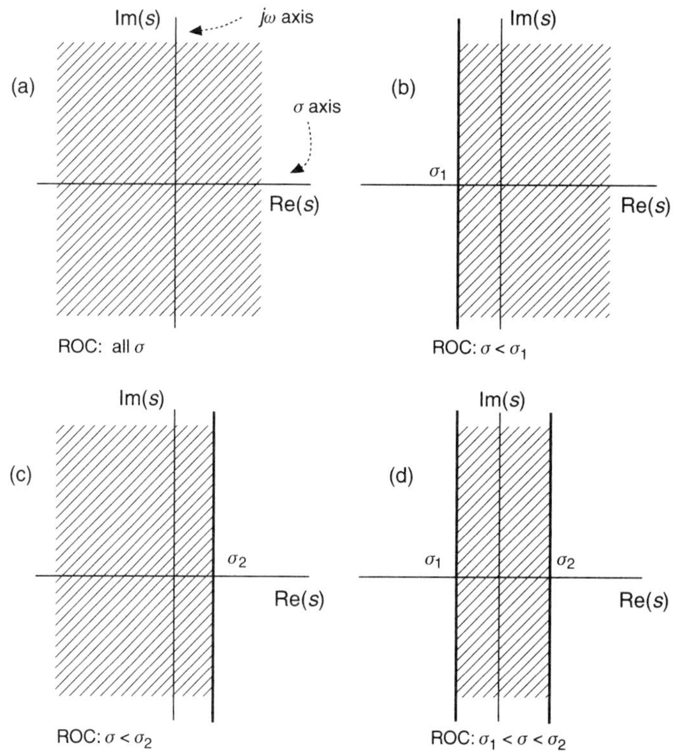

FIGURE C.1 s-plane and ROCs for $X(s)$. (a) Finite-duration function. (b) Right-sided function. (c) Left-sided function. (d) Two-sided function. A function is bounded (stable) if the ROC includes the imaginary ($j\omega$) axis, which means $\sigma_1 < 0$ and $\sigma_2 > 0$. The Laplace transform does not exist if $\sigma_1 > \sigma_2$.

which is a symmetric operation: $x(t) * h(t) \rightarrow X(s)H(s) = H(s)X(s) \rightarrow h(t) * x(t)$. Their cross-correlation is

$$x(t) \star h(t) \triangleq \int_{-\infty}^{\infty} x(\tau)h(\tau + t)\mathrm{d}\tau = \int_{-\infty}^{\infty} x(\tau - t)h(\tau)\mathrm{d}\tau, \tag{C.5}$$

which, unlike convolution, is *not* a symmetric operation: $x(t) \star h(t) \rightarrow X(-s)H(s)$. Note that some books define cross-correlation with the arguments interchanged in (C.5). We use this definition because it is similar to the ensemble cross-correlation function in Chapter 6.

TABLE C.1 Properties of the Laplace Transform

Property	Function	Laplace Transform
Linearity	$a_1x_1(t) + a_2x_2(t)$	$a_1X_1(s) + a_2X_2(s)$
Time shift	$x(t - t_o)$	$\exp(-t_os)X(s)$
Time scaling	$x(\alpha t)$	$(1/\|\alpha\|)X(s/\alpha)$
Frequency shift	$\exp(s_ot)x(t)$	$X(s - s_o)$
Derivative	$\mathrm{d}^nx(t)/\mathrm{d}t^n$	$s^nX(s)$
Integral	$\int_{-\infty}^{t} x(\tau)\mathrm{d}\tau$	$(1/s)X(s)$
Convolution	$x(t) * h(t)$	$X(s)H(s)$
Cross-correlation	$x(t) \star h(t)$	$X(-s)H(s)$
Multiplication	$x(t)h(t)$	$(1/2\pi j)\int_{\sigma-j\infty}^{\sigma+j\infty} X(v)H(s - v)\mathrm{d}v$

TABLE C.2 Laplace Transform Pairs

Time-Domain $x(t)$	Laplace Transform $X(s)$	ROC ($\sigma = \mathrm{Re}(s)$)
$\delta(t)$	1	$\sigma \in \mathcal{R}$
$t^n \exp(-at)u(t)$	$n!/(s+a)^{n+1}$	$\sigma > -a \ (n \in \mathcal{Z})$
$t^n \exp(-at)u(-t)$	$-n!/(s+a)^{n+1}$	$\sigma < -a \ (n \in \mathcal{Z})$
$\exp(-a\|t\|)$	$2a/(a^2 - s^2)$	$-a < \sigma < a$
$\cos(\omega_o t)u(t)$	$s/(s^2 + \omega_o^2)$	$\sigma > 0$
$\sin(\omega_o t)u(t)$	$\omega_o/(s^2 + \omega_o^2)$	$\sigma > 0$
$\cosh(at)u(t)$	$s/(s^2 - a^2)$	$\sigma > \|a\|$
$\sinh(at)u(t)$	$a/(s^2 - a^2)$	$\sigma > \|a\|$
$\exp(-at)\cos(\omega_o t)u(t)$	$(s+a)/[(s+a)^2 + \omega_o^2]$	$\sigma > -a$
$\exp(-at)\sin(\omega_o t)u(t)$	$\omega_o/[(s+a)^2 + \omega_o^2]$	$\sigma > -a$

In this book, we use the unilateral Laplace transform primarily as a means for solving linear differential equations with constant coefficients, which represent causal linear time-invariant systems (such as linear circuits). For such causal applications, the derivative property in Table C.1 is modified as follows:

$$\frac{\mathrm{d}^n x(t)}{\mathrm{d}t^n} \rightarrow s^n X(s) - \sum_{m=0}^{n-1} s^{n-m-1} x^{(m)}(0^-), \tag{C.6}$$

where $x^{(m)}(t)$ is the mth ordinary derivative of $x(t)$, and it is understood that $x^{(m)}(0^-)$ is evaluated at $t = 0^-$ after differentiation. The notation 0^- indicates that the initial condition is defined before any discontinuities that might exist at $t = 0$ (such as a voltage source switched on at that time). The Laplace transform is also used for power spectral densities (PSDs), in which case the bilateral form is needed because the autocorrelation function is two-sided (it is an even function).

Example C.1. Consider the autocorrelation function:

$$R_{XX}(\tau) = 5\exp(-2|\tau|). \tag{C.7}$$

Its Laplace transform is the following PSD:

$$\begin{aligned} S_{XX}(s) &= 5\int_{-\infty}^{0} \exp(2\tau)\exp(-s\tau)\mathrm{d}\tau + 5\int_{0}^{\infty} \exp(-2\tau)\exp(-s\tau)\mathrm{d}\tau, \\ &= \frac{5}{2-s} + \frac{5}{s+2} = \frac{20}{4-s^2}, \end{aligned} \tag{C.8}$$

which matches the result in Table C.2. The first integral converges for $\mathrm{Re}(2-s) > 0 \Longrightarrow \sigma < 2$, and the second integral converges for $\mathrm{Re}(s+2) > 0 \Longrightarrow \sigma > -2$, yielding the ROC $-2 < \sigma < 2$. It includes the imaginary axis, as expected, because $R_{XX}(\tau)$ is a bounded function.

Example C.2. Next, we find the inverse Laplace transform of the following transfer function for three different ROCs: (i) $\sigma > 3$, (ii) $\sigma < -3$, and (iii) $-3 < \sigma < 3$:

$$H(s) = 12\frac{(s-1)}{s^2 - 9} = \frac{a}{s-3} + \frac{b}{s+3}. \tag{C.9}$$

The ROCs are defined by the following intersections on the s-plane: (i) $\{\sigma > -3\} \cap \{\sigma > 3\}$, (ii) $\{\sigma < -3\} \cap \{\sigma < 3\}$, and (iii) $\{\sigma > -3\} \cap \{\sigma < 3\}$. The numerator coefficients of (C.9) are obtained in the same way for all three cases:

$$a = H(s)(s-3)|_{s=3} = 24/6 = 4, \tag{C.10}$$

$$b = H(s)(s+3)|_{s=-3} = 48/6 = 8, \tag{C.11}$$

yielding

$$H(s) = \frac{4}{s-3} + \frac{8}{s+3}. \tag{C.12}$$

The ROC determines the specific inverse Laplace transform of each term. For case (i), $h(t)$ is a right-sided function:

$$h(t) = [4\exp(3t) + 8\exp(-3t)]u(t), \tag{C.13}$$

and for case (ii), it is a left-sided function:

$$h(t) = -[4\exp(3t) + 8\exp(-3t)]u(-t). \tag{C.14}$$

From Figure C.1(d), we see that only case (iii) yields a stable $h(t)$ because the ROC includes the imaginary axis. It is a two-sided function:

$$h(t) = -4\exp(3t)u(-t) + 8\exp(-3t)u(t). \tag{C.15}$$

All three functions are plotted in Figure C.2. Observe the slight drop in the right-sided function due to the bounded term of $h(t)$, which quickly becomes dominated by the unbounded term. For the two-sided function, the value is 4 at $t = 0$ where there is a discontinuity.

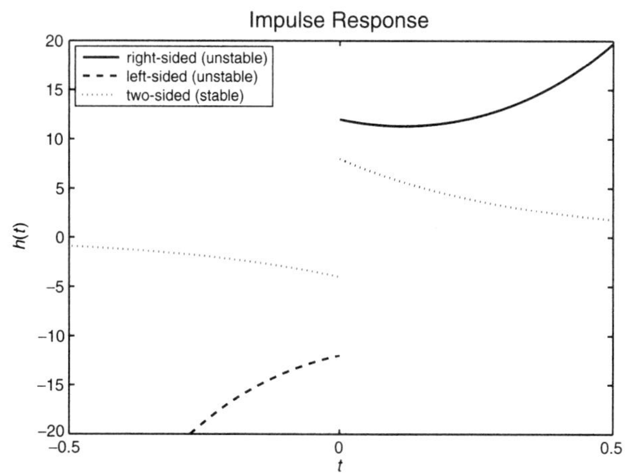

FIGURE C.2 Impulse response functions for Example C.2.

C.2 CONTINUOUS-TIME FOURIER TRANSFORM

Definition: Fourier Transform $X(\omega)$ The *continuous-time Fourier transform* of $x(t)$ for $t \in \mathcal{R}$ is

$$X(\omega) \triangleq \int_{-\infty}^{\infty} x(t)\exp(-j\omega t)\mathrm{d}t, \tag{C.16}$$

where ω is radian frequency.

It is obtained from the bilateral Laplace transform $X(s)$ by substituting $s = j\omega$ (with real part $\sigma = 0$), assuming the ROC includes the imaginary axis (as is shown for the four cases in Figure C.1). The inverse Fourier transform is

$$x(t) = \frac{1}{2\pi}\int_{-\infty}^{\infty} X(\omega)\exp(j\omega t)\mathrm{d}\omega. \tag{C.17}$$

Since radian frequency is related to ordinary frequency as $\omega = 2\pi f$, the Fourier transform can also be expressed using the following symmetric transform pair:

$$X(f) \triangleq \int_{-\infty}^{\infty} x(t)\exp(-j2\pi f t)\mathrm{d}t, \tag{C.18}$$

$$x(t) = \int_{-\infty}^{\infty} X(f)\exp(j2\pi f t)\mathrm{d}f. \tag{C.19}$$

The units of f are s^{-1} denoted by hertz (Hz); cycles/s is also used. Several properties of the Fourier transform are summarized in Table C.3, and some Fourier transform pairs are provided in Table C.4 for radian frequency and ordinary frequency. In this book, we are primarily interested in using the Fourier transform for PSDs and characteristic functions.

Example C.3. The Fourier transform of the autocorrelation function in Example C.1 is obtained from (C.8) by substituting $s = j\omega$:

$$S_{XX}(\omega) = \frac{20}{(j\omega + 2)(2 - j\omega)} = \frac{20}{\omega^2 + 4}, \tag{C.20}$$

which is consistent with the result in Table C.4.

TABLE C.3 Properties of the Fourier Transform

Property	Function	Fourier Transform (ω)	Fourier Transform (f)
Linearity	$a_1x_1(t) + a_2x_2(t)$	$a_1X_1(\omega) + a_2X_2(\omega)$	$a_1X_1(f) + a_2X_2(f)$
Time shift	$x(t - t_o)$	$\exp(-j\omega t_o)X(\omega)$	$\exp(-j2\pi f t_o)X(f)$
Time scaling	$x(\alpha t)$	$(1/\lvert\alpha\rvert)X(\omega/\alpha)$	$(1/\lvert\alpha\rvert)X(f/\alpha)$
Frequency shift	$\exp(j\omega_o t)x(t)$	$X(\omega - \omega_o)$	$X(f - f_o)$
Derivative	$\mathrm{d}^n x(t)/\mathrm{d}t^n$	$(j\omega)^n X(\omega)$	$(j2\pi f)^n X(f)$
Integral	$\int_{-\infty}^{t} x(\tau)\mathrm{d}\tau$	$(1/j\omega)X(\omega) + \pi X(0)\delta(\omega)$	$(1/j2\pi f)X(f) + (1/2)X(0)\delta(f)$
Convolution	$x(t) * h(t)$	$X(\omega)H(\omega)$	$X(f)H(f)$
Cross-correlation	$x(t) \star h(t)$	$X(-\omega)H(\omega)$	$X(-f)H(f)$
Autocorrelation	$x(t) \star x(t)$	$\lvert X(\omega)\rvert^2$	$\lvert X(f)\rvert^2$
Multiplication	$x(t)h(t)$	$(1/2\pi)\int_{-\infty}^{\infty} X(v)H(\omega - v)\mathrm{d}v$	$\int_{-\infty}^{\infty} X(v)H(f - v)\mathrm{d}v$
Area	$x(0)$	$(1/2\pi)\int_{-\infty}^{\infty} X(\omega)\mathrm{d}\omega$	$\int_{-\infty}^{\infty} X(f)\mathrm{d}f$
Parseval	$\int_{-\infty}^{\infty} x^2(t)\mathrm{d}t$	$(1/2\pi)\int_{-\infty}^{\infty} \lvert X(\omega)\rvert^2\mathrm{d}\omega$	$\int_{-\infty}^{\infty} \lvert X(f)\rvert^2\mathrm{d}f$

TABLE C.4 Fourier Transform Pairs

Time-Domain $x(t)$	Fourier Transform $X(\omega)$	Fourier Transform $X(f)$
$\delta(t)$	1	1
1	$2\pi\delta(\omega)$	$\delta(f)$
$u(t)$	$\pi\delta(\omega) + 1/j\omega$	$(1/2)\delta(f) + 1/j2\pi f$
$\text{sgn}(t)$	$2/j\omega$	$1/j\pi f$
$1/t$	$-j\pi\,\text{sgn}(\omega)$	$-j\pi\,\text{sgn}(f)$
$\cos(\omega_o t)$	$\pi[\delta(\omega - \omega_o) + \delta(\omega + \omega_o)]$	$(1/2)[\delta(f - \omega_o/2\pi) + \delta(f + \omega_o/2\pi)]$
$\sin(\omega_o t)$	$j\pi[\delta(\omega + \omega_o) - \delta(\omega - \omega_o)]$	$(j/2)[\delta(f + \omega_o/2\pi) - \delta(f - \omega_o/2\pi)]$
$\exp(-at)u(t)$	$1/(j\omega + a)$ (for $a > 0$)	$1/(j2\pi f + a)$ (for $a > 0$)
$\exp(-a\|t\|)$	$2a/(\omega^2 + a^2)$ (for $a > 0$)	$2a/(4\pi^2 f^2 + a^2)$ (for $a > 0$)
$\exp(-at^2)$	$\sqrt{\pi/a}\exp(\omega^2/4a)$ (for $a > 0$)	$\sqrt{\pi/a}\exp(-\pi^2 f^2/a)$ (for $a > 0$)
$\text{rect}(t)$	$\text{sinc}(\omega/2\pi)$	$\text{sinc}(f)$
$\text{sinc}(t)$	$\text{rect}(\omega/2\pi)$	$\text{rect}(f)$
$\text{sinc}^2(t)$	$\text{tri}(\omega/2\pi)$	$\text{tri}(f)$
$\text{tri}(t)$	$\text{sinc}^2(\omega/2\pi)$	$\text{sinc}^2(f)$
$\sum_{m=-\infty}^{\infty}\delta(t - mT)$	$(2\pi/T)\sum_{m=-\infty}^{\infty}\delta(\omega - 2\pi m/T)$	$(1/T)\sum_{m=-\infty}^{\infty}\delta(f - m/T)$

Since real-valued functions are assumed in Table C.3, $H(-\omega) = H^*(\omega)$, where the superscript denotes complex conjugation. Thus, for the cross-correlation property, $x(t) \star h(t) \to X^*(\omega)H(\omega)$. When $h(t) = x(t)$, the autocorrelation property is obtained: $x(t) \star x(t) \to |X(\omega)|^2$, which leads to Parseval's result in the table. The energy of $x(t)$, given by $\int_{-\infty}^{\infty} x^2(t)\mathrm{d}t$, is obtained as the area under the magnitude squared of its Fourier transform, which means that $|X(\omega)|^2$ is an *energy density*. As a result, $x(t) \star x(t) \leftrightarrow |X(\omega)|^2$ is the deterministic version of the ensemble autocorrelation function of random process $X(t)$, whose Fourier transform is a PSD: $R_{XX}(\tau) \leftrightarrow S_{XX}(\omega)$.

Example C.4. The characteristic function for a continuous random variable is

$$\Phi_X(\omega) = \mathcal{E}[\exp(j\omega X)] = \int_{-\infty}^{\infty} f_X(x)\exp(j\omega x)\mathrm{d}x, \tag{C.21}$$

which is the Fourier transform of the probability density function (pdf) $f_X(x)$, but with $\omega \to -\omega$. As a result, all the properties in Table C.3 apply to the characteristic function, and the transform pairs in Table C.4 can be used provided ω is replaced with $-\omega$ (or f with $-f$). For example, for the exponential random variable with pdf $f_X(x) = \lambda\exp(-\lambda x)u(x)$, we have from Table C.4:

$$\Phi_X(\omega) = \lambda/(\lambda - j\omega). \tag{C.22}$$

However, note that not many pdfs are among the "standard" functions in the table. Characteristic functions for the continuous random variables mentioned in this book are provided in Appendix A.

C.3 *z*-TRANSFORM

Definition: *z*-Transform $X(z)$ The *z-transform* of $x[k]$ for $k \in \mathcal{Z}$ is

$$X(z) \triangleq \sum_{k=-\infty}^{\infty} x[k]z^{-k} \quad \text{(bilateral)}, \tag{C.23}$$

where $z \triangleq r\exp(j\omega)$ is a complex variable. The sum converges to $X(z)$ for some ROC on the z-plane as depicted in Figure C.3 for the four basic types of sequences.

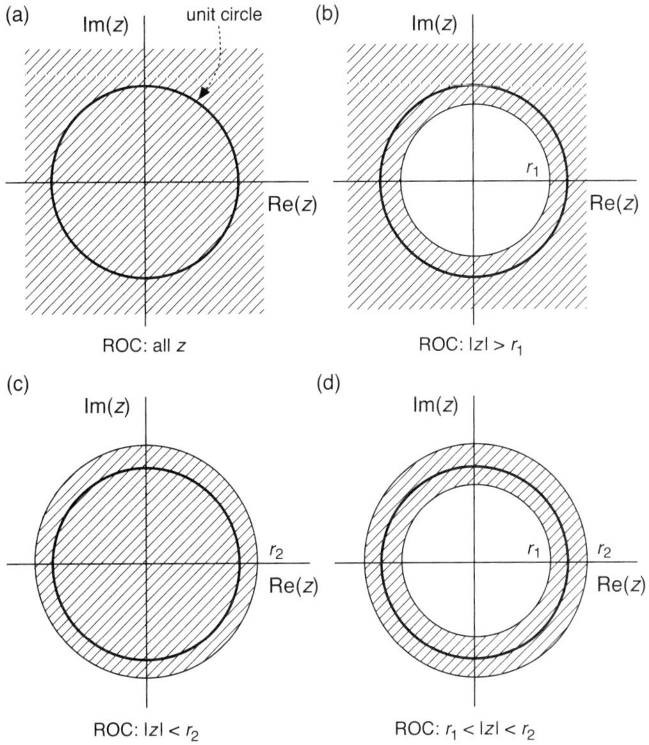

FIGURE C.3 z-plane and ROCs for $X(z)$. The unit circle is defined by $|z| = 1$. (a) Finite-duration sequence. (b) Right-sided sequence. (c) Left-sided sequence. (d) Two-sided sequence. A sequence is bounded (stable) if the ROC includes the unit circle, which means $r_1 < 1$ and $r_2 > 1$. The z-transform does not exist if $r_1 > r_2$.

The mapping from the s-plane to the z-plane when a continuous-time signal $x(t)$ is uniformly sampled to generate the discrete-time sequence $x[k]$ is described in Chapter 1. For the *unilateral* Laplace transform, the integral has a finite lower limit, usually zero:

$$X(z) \triangleq \sum_{k=0}^{\infty} x[k]z^{-k} \qquad \text{(unilateral).} \tag{C.24}$$

The unilateral and bilateral z-transforms are identical when $x[k]$ is nonzero only for $k \geq 0$, which can be emphasized by writing $x[k]u[k]$, where $u[k]$ is the discrete unit-step function.

Example C.5. Consider the following two-sided discrete-time sequence:

$$x[k] = \alpha\beta^{|k|}, \tag{C.25}$$

where α and $|\beta| < 1$ are real-valued. The bilateral z-transform is

$$\begin{aligned} X(z) &= \alpha \sum_{k=-\infty}^{-1} (\beta z)^{-k} + \alpha \sum_{k=0}^{\infty} (\beta z^{-1})^k, \\ &= \alpha \sum_{m=1}^{\infty} (\beta z)^m + \frac{\alpha}{1 - \beta z^{-1}}, \end{aligned} \tag{C.26}$$

where $m = -k$ has been substituted in the first sum. The closed-form series result in Appendix E was used for the second sum, which converges when $|\beta z^{-1}| < 1 \Longrightarrow |z| > |\beta|$. The first sum is handled by adding and subtracting the term α for $m = 0$ as follows:

$$\begin{aligned} X(z) &= \alpha \sum_{m=0}^{\infty} (\beta z)^m - \alpha + \frac{\alpha}{1 - \beta z^{-1}}, \\ &= \frac{\alpha}{1 - \beta z} - \alpha + \frac{\alpha}{1 - \beta z^{-1}}, \end{aligned} \tag{C.27}$$

which converges when $|\beta z| < 1 \Longrightarrow |z| < 1/|\beta|$. The overall ROC is $|\beta| < |z| < 1/|\beta|$ which requires $|\beta| < 1$ as we have assumed. (If $|\beta| > 1$, there is no ROC because $|z| < 1/|\beta|$ and $|z| > |\beta|$ do not overlap.) The ROC includes the unit circle $|z| = 1$, which is necessary for a bounded sequence. Combining terms, we finally have

$$\begin{aligned} X(z) &= \frac{\alpha(1 - \beta z^{-1}) - \alpha(1 - \beta z^{-1})(1 - \beta z) + \alpha(1 - \beta z)}{(1 - \beta z^{-1})(1 - \beta z)}, \\ &= \frac{\alpha(1 - \beta^2)}{(1 - \beta z^{-1})(1 - \beta z)}. \end{aligned} \tag{C.28}$$

This is a useful result for computing PSDs (see Chapter 8). In particular, for the following PSD:

$$S_{XX}(z) = \frac{b}{(1 - az^{-1})(1 - az)}, \tag{C.29}$$

we can immediately write the corresponding autocorrelation function:

$$R_{XX}[m] = \frac{ba^{|m|}}{1 - a^2}. \tag{C.30}$$

This type of PSD arises often in Wiener filtering (see Chapter 11).

Example C.6. Consider the following z-transform of the right-sided (causal) sequence $h[k] = a^k u[k]$ where a is real valued:

$$H(z) = \frac{1}{1 - az^{-1}}, \tag{C.31}$$

which has ROC $z > |a|$. The sequence is bounded if $|a| < 1$, so that the ROC includes the unit circle. The magnitude squared of $H(z)$ is

$$|H(z)|^2 = H(z)H^*(z) = \frac{zz^*}{(z - a)(z^* - a)}. \tag{C.32}$$

(This expression is *not* the same as $H(z)H(z^{-1})$ which is the output PSD of a linear system with a unit-variance, white-noise input.) Defining $z \triangleq c + jd$, we obtain the following real-valued function:

$$|H(z)|^2 = \frac{c^2 + d^2}{c^2 + d^2 + a^2 - 2ac} = \frac{c^2 + d^2}{(c - a)^2 + d^2}. \tag{C.33}$$

The form in (C.32) is not usually discussed because it is a function of two variables (c and d) as shown in the last expression (it can also be written in terms of the radius r and frequency ω). Instead, the magnitude squared $|H(j\omega)|^2$ of the discrete-time Fourier transform (DTFT) is examined because it is a function of the single variable ω (it is defined in Section C.3). The form in (C.32) is used here because we want to view the pole (at $z = a = 0.9$) and zero (at $z = 0$) on the three-dimensional plot shown in Figure C.4(a). (A log scale is used for ease of viewing. The resolution of the axes was chosen such that c is close to $\text{Re}(z) = 0$ and $\text{Re}(z) = 0.9$, but not exactly. Of course, for c equal to these two values, $10 \log |H(z)|^2 \rightarrow -\infty$ and ∞, respectively, and cannot

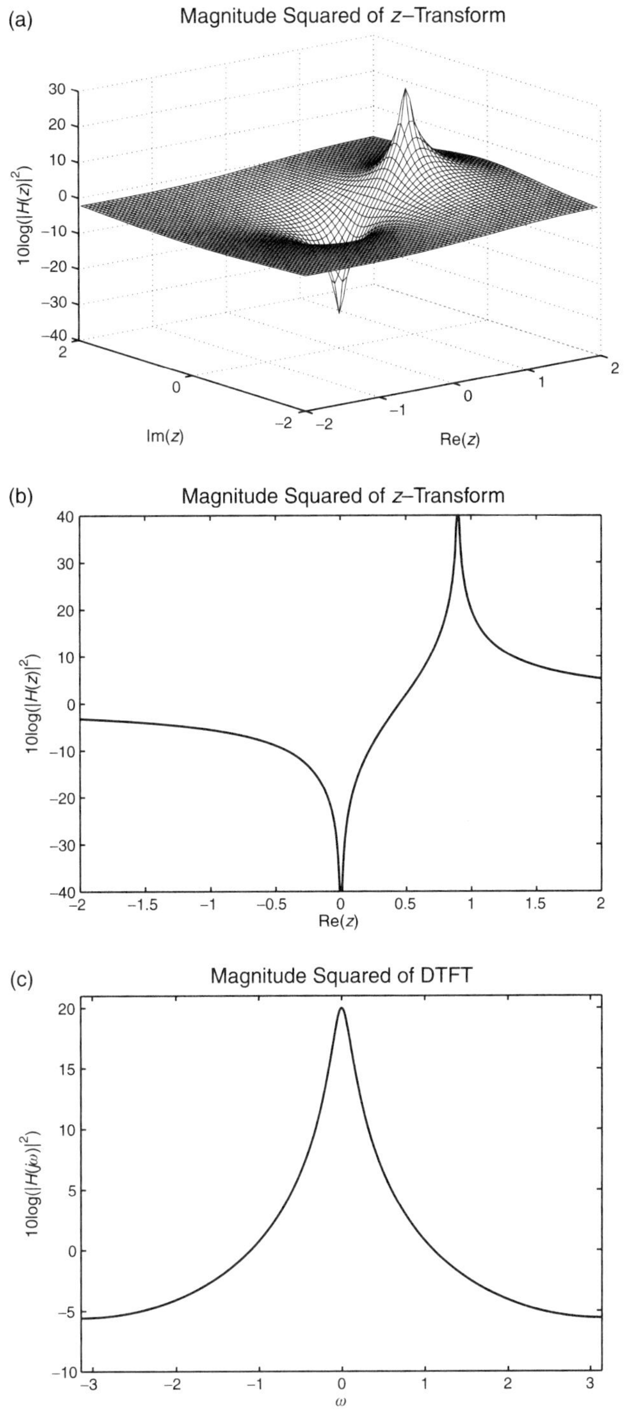

FIGURE C.4 z-transform of the first-order system in Example C.6 with one real zero (at $z = 0$) and one real pole (at $z = a = 0.9$). (a) $10 \log |H(z)|^2$ on the z-plane. (b) $10 \log |H(z)|^2$ on the real axis (versus c of $z = c + jd$). (c) $10 \log |H(j\omega)|^2$ on the unit circle (versus ω of $z = \exp(j\omega)$ with $r = 1$). $H(j\omega)$ is the DTFT.

TABLE C.5 Properties of the z-Transform

Property	Function	z-Transform
Linearity	$a_1x_1[k] + a_2x_2[k]$	$a_1X_1(z) + a_2X_2(z)$
Time shift	$x[k - k_o]$	$z^{-k_o}X(z)$
Time reversal	$x[-k]$	$X(z^{-1})$
Geometric weighting	$a^kx[k]$	$X(za^{-1})$
Time weighting	$kx[k]$	$-z\mathrm{d}X(z)/\mathrm{d}z$
Convolution	$x[k] * h[k]$	$X(z)H(z)$
Cross-correlation	$x[k] \star h[k]$	$X(z^{-1})H(z)$
Multiplication	$x[k]h[k]$	$(1/2\pi j)\oint_C X(v)H(zv^{-1})v^{-1}\mathrm{d}v$
Parseval	$\sum_{k=-\infty}^{\infty} x[k]h[k]$	$(1/2\pi j)\oint_C X(v)H(v^{-1})v^{-1}\mathrm{d}v$

be plotted.) The function in Figure C.4(b) is $|H(z)|^2$ evaluated along the real axis $c = \mathrm{Re}(z)$ with $d = 0$, which is a "slice" through the previous plot. The magnitude squared of the DTFT $H(j\omega)$ is shown in Figure C.4(c); it is obtained by evaluating the z-transform on the unit circle: $z = \exp(j\omega)$ (radius $r = 1$). The pole at $z = 0.9$ causes the frequency response to have the characteristic of a lowpass filter.

Several properties of the z-transform are summarized in Table C.5, and some transform pairs are provided in Table C.6. Note that sine and cosine are the usual continuous-time functions, but they are evaluated in the table only at integer multiples of ω_o. The inverse z-transform is given by the following integral:

$$x[k] = \frac{1}{2\pi j}\oint_C X(z)z^{k-1}\mathrm{d}z, \tag{C.34}$$

where C denotes a closed contour performed counterclockwise in the ROC. When $X(z)$ is a rational function, it is often easier to evaluate this integral using a PFE, or instead Cauchy's residue theorem which we provide without proof.

Theorem C.1. (Cauchy residue). Assume that N_1 poles $\{p_n\}$ are inside the contour C and N_2 poles $\{q_n\}$ are outside the contour. The following two expressions can be used to obtain the inverse z-transform:

$$x[k] = \sum_{n=1}^{N_1} \mathrm{Res}[X(z)z^{k-1}, p_n], \tag{C.35}$$

$$x[k] = -\sum_{n=1}^{N_2} \mathrm{Res}[X(z)z^{k-1}, q_n] - \mathrm{Res}[X(z)z^{k-1}, \infty], \tag{C.36}$$

TABLE C.6 z-Transform Pairs

Time-Domain $x[k]$	z-Transform $X(z)$	ROC ($\lvert z\rvert = r$)
$\delta[k]$	1	$\lvert z\rvert \in \mathcal{R}$
$a^ku[k]$	$1/(1 - az^{-1})$	$\lvert z\rvert > \lvert a\rvert$
$a^ku[-k-1]$	$-1/(1 - az^{-1})$	$\lvert z\rvert < \lvert a\rvert$
$ka^ku[k]$	$az^{-1}/(1 - az^{-1})^2$	$\lvert z\rvert > \lvert a\rvert$
$ka^ku[-k-1]$	$-az^{-1}/(1 - az^{-1})^2$	$\lvert z\rvert < \lvert a\rvert$
$k^2a^ku[k]$	$az^{-1}(1 + az^{-1})/(1 - az^{-1})^3$	$\lvert z\rvert > \lvert a\rvert$
$k^2a^ku[-k-1]$	$-az^{-1}(1 + az^{-1})/(1 - az^{-1})^3$	$\lvert z\rvert < \lvert a\rvert$
$a^k\cos(\omega_ok)u[k]$	$[1 - a\cos(\omega_o)z^{-1}]/[1 - 2a\cos(\omega_o)z^{-1} + a^2z^{-2}]$	$\lvert z\rvert > \lvert a\rvert$
$a^k\sin(\omega_ok)u[k]$	$a\sin(\omega_o)z^{-1}/[1 - 2a\cos(\omega_o)z^{-1} + a^2z^{-2}]$	$\lvert z\rvert > \lvert a\rvert$

where Res$[\cdot, p]$ is the residue of the argument evaluated at the pole p. For a pole of order N at p_n, the residue is

$$\text{Res}[X(z)z^{k-1}, p_n] = \frac{1}{(N-1)!} \lim_{z \to p_n} \frac{\mathrm{d}^{N-1}}{\mathrm{d}z^{N-1}} \left\{ (z-p_n)^N X(z) z^{k-1} \right\}, \tag{C.37}$$

which simplifies to the following expression for a first-order pole at p_n:

$$\text{Res}[X(z)z^{k-1}, p_n] = \lim_{z \to p_n} (z - p_n) X(z) z^{k-1}. \tag{C.38}$$

The residue for a first-order zero at infinity is

$$\text{Res}[X(z)z^{k-1}, \infty] = -\lim_{z \to \infty} z X(z) z^{k-1}, \tag{C.39}$$

and the residue for a higher order zero at infinity is zero.

Note that a zero at $z = \infty$ is equivalent to a pole at $z = 0$, as illustrated in Example C.7.

Example C.7. Consider again the PSD in (C.29) where $a < 1$ is real valued and the ROC is $a < |z| < 1/a$ as depicted in Figure C.4(d) with $r_1 = a$ and $r_2 = 1/a$. The pole at a is associated with the right side of the autocorrelation function $R_{XX}[m]$, and the pole at $1/a$ is associated with the left side. Using Cauchy's residue theorem, note that for some values of m it is easier to compute the residue at a, while for other values it is preferable to compute the residue at $1/a$. First, rearrange the PSD (scaled by z^{m-1}) in "standard form" as follows:

$$S_{XX}(z)z^{m-1} = -\frac{(b/a)z^m}{(z-a)(z-1/a)}. \tag{C.40}$$

The number of poles due to z^m in the numerator changes with m, which in turn determines which residue is easier to evaluate:

$$m : \begin{cases} = 0 & \Longrightarrow \quad \text{no additional poles} & \Longrightarrow \quad \text{evaluate residue at } a \text{ or } 1/a \\ > 0 & \Longrightarrow \quad \text{additional poles at } \infty & \Longrightarrow \quad \text{evaluate residue at } a \\ < 0 & \Longrightarrow \quad \text{additional poles at } 0 & \Longrightarrow \quad \text{evaluate residue at } 1/a. \end{cases} \tag{C.41}$$

Since $m > 0$ does not introduce any additional poles inside the unit circle, (C.35) can be used to evaluate the residue for the single pole at a. If (C.35) is used for $m < 0$, we would need to evaluate the residues for the additional poles inside the unit circle (at $z = 0$). Instead, (C.36) should be used, which generally requires that we also evaluate the residue for the first-order zero at infinity. Thus

$$\begin{aligned} R_{XX}[m] &= \begin{cases} \text{Res}\left[S_{XX}(z)z^{m-1}, a\right], & m \geq 0 \\ -\text{Res}\left[S_{XX}(z)z^{m-1}, 1/a\right] & \\ -\text{Res}\left[S_{XX}(z)z^{m-1}, \infty\right], & m < 0, \end{cases} \\ &= -\frac{(b/a)a^m}{a-(1/a)}u[m] + \frac{(b/a)a^{-m}}{(1/a)-a}u[-m-1] = \frac{ba^{|m|}}{1-a^2}, \end{aligned} \tag{C.42}$$

which is the result in (C.30). Note that the residue at infinity for $m = -1$ happens to be zero for this example.

C.4 DISCRETE-TIME FOURIER TRANSFORM

Definition: Discrete-Time Fourier Transform $X(j\omega)$ The DTFT of $x[k]$ for $k \in \mathcal{Z}$ is

$$X(j\omega) \triangleq \sum_{k=-\infty}^{\infty} x[k] \exp(-j\omega k), \tag{C.43}$$

where j is included in the argument of $X(j\omega)$ to distinguish the DTFT from the continuous-time Fourier transform $X(\omega)$ (the notation $X(e^{j\omega})$ is also used for the DTFT).

The DTFT is *not* the same as the discrete Fourier transform (DFT) which is a function of discrete frequency. It is well known that the DFT is a useful transform because of the computational advantages of a fast Fourier transform (FFT) implementation. We do not directly use the DFT for any of the topics covered in this book, so a discussion of it is not included here, except to mention that the DFT can be derived from the DTFT by uniformly sampling ω on the unit circle.

The DTFT is obtained from the bilateral z-transform by evaluating $X(z)$ on the unit circle where $z = \exp(j\omega)$ (with $r = 1$), which implies that the ROC of $X(z)$ includes $|z| = 1$. The inverse DTFT is

$$x[k] = \frac{1}{2\pi} \int_{-\pi}^{\pi} X(j\omega) \exp(j\omega k) \mathrm{d}\omega. \tag{C.44}$$

It is straightforward to show that this expression is equivalent to the inverse z-transform when the contour of integration is performed along the unit circle with limits $[-\pi, \pi]$, by substituting $z = \exp(j\omega)$ and $\mathrm{d}z = j\exp(j\omega)\mathrm{d}\omega$ into (C.34). Several properties of the DTFT are summarized in Table C.7, and some transform pairs are given in Table C.8. Note that rect$[x]$ is the discrete rectangle function defined in Appendix B; it is nonzero for $k \in \{-(N-1), \ldots, (N-1)\}$. sinc$(Wk)$ is the usual continuous-time function, but it is evaluated in the table only for integer multiples of W. Its DTFT is the continuous rectangle function. Like the Fourier transform for continuous functions, we are primarily interested in the DTFT for PSDs and characteristic functions.

Example C.8. The characteristic function for a discrete random variable is

$$\Phi_X(j\omega) = \mathcal{E}[\exp(j\omega X)] = \sum_{x=-\infty}^{\infty} p_X[x] \exp(j\omega x), \tag{C.45}$$

TABLE C.7 Properties of the DTFT

Property	Function	DTFT
Linearity	$a_1 x_1[k] + a_2 x_2[k]$	$a_1 X_1(j\omega) + a_2 X_2(j\omega)$
Time shift	$x[k - k_o]$	$\exp(-j\omega k_o) X(j\omega)$
Time reversal	$x[-k]$	$X(-j\omega)$
Geometric weighting	$a^k x[k]$	$X(j\omega/a)$
Time weighting	$kx[k]$	$j\mathrm{d}X(j\omega)/\mathrm{d}\omega$
Convolution	$x[k] * h[k]$	$X(j\omega)H(j\omega)$
Cross-correlation	$x[k] \star h[k]$	$X(-j\omega)H(j\omega)$
Autocorrelation	$x[k] \star x[k]$	$\lvert X(j\omega)\rvert^2$
Multiplication	$x[k]h[k]$	$(1/2\pi)\int_{-\pi}^{\pi} X(jv)H(j(\omega - v))\mathrm{d}v$
Area	$x[0]$	$(1/2\pi)\int_{-\pi}^{\pi} X(j\omega)\mathrm{d}\omega$
Parseval	$\sum_{k=-\infty}^{\infty} x^2[k]$	$(1/2\pi)\int_{-\pi}^{\pi} \lvert X(j\omega)\rvert^2 \mathrm{d}\omega$

TABLE C.8 DTFT Pairs

Time-Domain $x[k]$	DTFT $X(j\omega)$
$\delta[k]$	1
1	$2\pi \sum_{m=-\infty}^{\infty} \delta(\omega + 2\pi m)$
$u[k]$	$1/[1 - \exp(-j\omega)] + \pi \sum_{m=-\infty}^{\infty} \delta(\omega + 2\pi m)$
$a^k u[k]$	$1/[1 - a\exp(-j\omega)]$ (for $\lvert a\rvert < 1$)
$(k+1)a^k u[k]$	$1/[1 - a\exp(-j\omega)]^2$
$\cos(\omega_o k)$	$\pi \sum_{m=-\infty}^{\infty} [\delta(\omega - \omega_o + 2\pi m) + \delta(\omega + \omega_o + 2\pi m)]$
$\sin(\omega_o k)$	$j\pi \sum_{m=-\infty}^{\infty} [\delta(\omega + \omega_o + 2\pi m) - \delta(\omega - \omega_o + 2\pi m)]$
$\text{rect}[k]$	$\sin((N - 1/2)\omega)/\sin(\omega/2)$
$\text{sinc}(Wk)$	$(1/W)\text{rect}(\omega/2\pi W)$ (for $0 < W < 1$)
$\text{sinc}^2(Wk)$	$(1/W)\text{tri}(\omega/2\pi W)$ (for $0 < W < 1/2$)
$\sum_{n=-\infty}^{\infty} \delta[k - nN]$	$(1/N) \sum_{m=-\infty}^{\infty} \delta(\omega/2\pi - n/N)$

which is the DTFT of the probability mass function (pmf) $p_X[x]$, except that $\omega \to -\omega$. Thus, all the properties in Table C.7 can be used for the characteristic function. Note that Table C.8 contains DTFTs only for two pmfs, corresponding to the geometric and discrete uniform random variables. For the geometric random variable with pmf $p_X[x] = pq^x I_{Z^+}[x]$, where $I_{Z^+}[x]$ is the discrete indicator function, the fourth transform in the table gives

$$\Phi_X(j\omega) = p/[1 - q\exp(j\omega)], \tag{C.46}$$

where we have replaced ω with $-\omega$. Characteristic functions for the discrete random variables covered in this book are summarized in Appendix A.

FURTHER READING

The frequency-domain transforms summarized in this appendix can be studied further in Bracewell (1978), Lathi (1965), Lathi (2000), Mitra (2010), Oppenheim and Schafer (2009), Oppenheim, Willsky, and Nawab (1996), Stearns and Hush (1990), and Tretter (1976).

D Integration and Integrals

In this appendix, we provide a description of different types of integration for computing expectations of random variables. The Riemann integral is briefly reviewed, so that the more general (and less familiar) Riemann–Stieltjes integral can be described. We will not attempt to be as rigorous as most textbooks on measure theory and integration. Instead, for the probability space $\{\mathcal{R}, \mathcal{B}(\mathcal{R}), P_X\}$, we would like to provide a basic understanding of the meaning of the following integral:

$$\int_{-\infty}^{\infty} g(x)\mathrm{d}F_X(x), \tag{D.1}$$

where $F_X(x)$ is the cumulative distribution function (cdf) of X, and $g(x)$ is the function to be integrated. We also briefly describe the Lebesgue integral, which is more general than the Riemann integral and can be used for functions that are not Riemann integrable. For the abstract probability space $\{\Omega, \mathcal{F}, P\}$, it is generally written as

$$\int_{\Omega} g(\omega)\mathrm{d}P(\omega), \tag{D.2}$$

where $\omega \in \Omega$ are elements of the sample space.

This material on integration is not necessary for understanding the definition of expectation and the various moments described in Chapter 5. However, it does provide some insight into integration for discrete and mixed random variables, and the roles of the Dirac delta function and the unit-step function used in the probability density function (pdf) and cdf, respectively.

D.1 REVIEW OF RIEMANN INTEGRAL

The Riemann integral is covered in introductory calculus courses, and is the integral used in Chapters 3 and 4 on random variables. Consider the function $y = g(x)$ depicted in Figure D.1 which is continuous and bounded on $[a, b]$. Note that in general an integrable function need not be continuous. Differentiable functions form a subset of continuous functions, and continuous functions form a subset of integrable functions. Later we consider integrals of functions that are not continuous.

Example D.1. The rectangle function can be written as $\text{rect}(x) = u(x + 1/2) - u(x - 1/2)$ where $u(x)$ is the unit-step function. It is clear that $\text{rect}(x)$ is not differentiable or continuous at $x = \pm 1/2$, but is integrable on $[-1/2, 1/2]$: it has unit area. The triangle function $\text{tri}(x) = (1 - |x|)\text{rect}(x/2)$ is continuous and integrable on $[-1, 1]$, but is not differentiable at $x = 0$.

Observe in Figure D.1 that $[a, b]$ on the x-axis has been partitioned into a collection of subintervals $\{[x_n, x_{n+1}]\}$ for $n = 1, \ldots, N$. Let α_n be a point in the nth subinterval $[x_n, x_{n+1}]$. The *mesh* of the partition is

Probability, Random Variables, and Random Processes: Theory and Signal Processing Applications, First Edition. John J. Shynk.

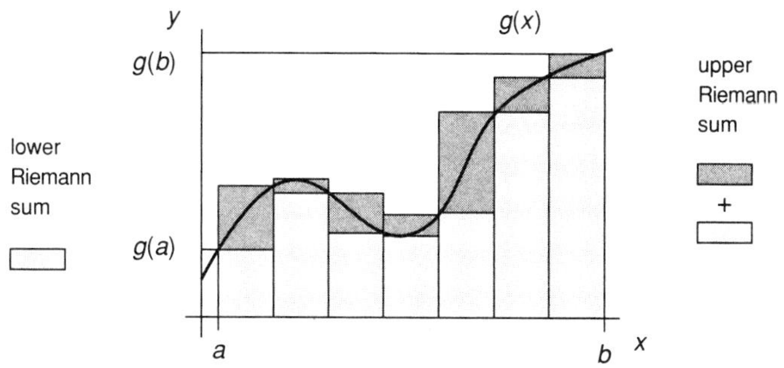

FIGURE D.1 Function $y = g(x)$ partitioned on the x-axis into subintervals on $[a, b]$. Light-shaded rectangles comprise the lower Riemann sum. Light-plus-dark-shaded rectangles comprise the upper Riemann sum.

the subinterval with the largest width $\Delta_n \triangleq x_{n+1} - x_n$ (though often the subintervals have the same width as in this example). The area under the curve is approximated by the following *Riemann sum*:

$$R(N) = \sum_{n=1}^{N} g(\alpha_n)(x_{n+1} - x_n) = \sum_{n=1}^{N} g(\alpha_n)\Delta_n, \tag{D.3}$$

which is a sum of the areas of N rectangles; the nth rectangle has width Δ_n, height $g(\alpha_n)$, and area $g(\alpha_n)\Delta_n$. In the figure, the results are shown for two different sums depending on which value of the function is used for the height of a rectangle. The *lower Riemann sum* is obtained by choosing α_n to be the infimum m_n of $g(x)$ on $[x_n, x_{n+1}]$:

$$R_L(N) = \sum_{n=1}^{N} m_n \Delta_n. \tag{D.4}$$

This results in the lightly shaded rectangles. The lower Riemann sum is necessarily upper bounded by $g(x)$ for all values of x on $[a, b]$. The *upper Riemann sum* is obtained by choosing instead that α_n be the supremum M_n of $g(x)$ on $[x_n, x_{n+1}]$:

$$R_U(N) = \sum_{n=1}^{N} M_n \Delta_n. \tag{D.5}$$

This causes the rectangles of the lower sum to be extended by the darker rectangles, so that the upper Riemann sum is lower bounded by $g(x)$ for all values of x on $[a, b]$. Because infimum and supremum are used in these definitions, the subintervals on the x-axis can be semi-open and the same results are obtained for a continuous function.

The Riemann integral R is derived by simultaneously increasing the number of intervals $N \to \infty$ and decreasing their widths $\Delta_n \to 0$. It can be defined as follows: there exists $\epsilon > 0$ for every $\delta > 0$ such that

$$\max(\Delta_n) < \epsilon \Longrightarrow \left| \sum_{n=1}^{N} g(\alpha_n)\Delta_n - R \right| < \delta. \tag{D.6}$$

Also, the Riemann integral on $[a, b]$ exists if $R_L(N) \to R_U(N)$ as $N \to \infty$. It is obvious that with an increasing number of rectangles with decreasing widths, the sum of their areas will approach the area under the continuous function $g(x)$ on $[a, b]$. We denote the Riemann integral as follows:

$$R = \int_a^b g(x)\mathrm{d}x = G(b) - G(a), \tag{D.7}$$

where $G(x)$ is the antiderivative of the *integrand* $g(x)$ such that $G'(x) = g(x)$. The last expression follows from the fundamental theorem of calculus.

In the previous discussion of sums, we considered a nonnegative function. If the function has negative parts, then the integral for each positive and negative part can be computed separately; the results of the negative parts are subtracted from those of the positive parts to generate the entire integral: $R = R_+ - R_-$.

Example D.2. Consider the linear function $g(x) = x$ on $[a, b]$ for which we know

$$\int_a^b x\mathrm{d}x = (x^2/2)\big|_a^b = (b^2 - a^2)/2. \tag{D.8}$$

For the Riemann sum, the interval can be partitioned such that $x_{n+1} = a + n(b-a)/N$ and $\Delta_n = n(b-a)/N - (n-1)(b-a)/N = (b-a)/N$. Choosing $\alpha_n = x_{n+1}$ in (D.3) (the largest value of $g(x) = x$ in the nth subinterval) gives the upper Riemann sum:

$$\begin{aligned} R_U(N) &= \sum_{n=1}^{N} [a + n(b-a)/N](b-a)/N, \\ &= a(b-a) + \frac{(b-a)^2}{N^2}\sum_{n=1}^{N} n, \\ &= ab - a^2 + \frac{(b-a)^2}{N^2}(N + N^2)/2, \end{aligned} \tag{D.9}$$

where a closed-form expression has been substituted for the finite series from Appendix E. Rearranging this expression and taking the limit gives R:

$$\lim_{N\to\infty} R_U(N) = \lim_{N\to\infty} \frac{(b-a)^2}{2N} + (b^2 - a^2)/2 = (b^2 - a^2)/2, \tag{D.10}$$

which is the result in (D.8). For completeness, we also include the lower Riemann sum where $\alpha_n = x_n$ is used (the smallest value of $g(x) = x$ in the nth subinterval):

$$R_L(N) = \sum_{n=1}^{N} [a + (n-1)(b-a)/N](b-a)/N, \tag{D.11}$$

which contains $n - 1$ instead of n as above. Thus, we need only subtract a term from the final expression in (D.10) as follows:

$$R_L(N) = \frac{(b-a)^2}{2N} + (b^2 - a^2)/2 - \frac{(b-a)^2}{N}, \tag{D.12}$$

which also gives $(b^2 - a^2)/2$ as $N \to \infty$.

D.2 RIEMANN–STIELTJES INTEGRAL

The Riemann–Stieltjes integral is a generalization of the Riemann integral where the subintervals $\{\Delta_n = x_{n+1} - x_n\}$ of the Riemann sum in (D.3) are replaced as follows:

$$RS(N) = \sum_{n=1}^{N} g(\alpha_n)[h(x_{n+1}) - h(x_n)], \tag{D.13}$$

where $h(x)$ is a *nondecreasing* function. In this way, different *weightings* can be assigned to the subintervals in the partition on the x-axis. This integration is performed with respect to the real function $h(x)$; the original function $g(x)$ is still called the integrand, and $h(x)$ is known as the *integrator*. The Riemann–Stieltjes integral RS is defined in a manner similar to the Riemann integral R as follows: there exists $\epsilon > 0$ for every $\delta > 0$ such that

$$\max(\Delta_n^h) < \epsilon \Rightarrow \left| \sum_{n=1}^{N} g(\alpha_n)\Delta_n^h - RS \right| < \delta, \tag{D.14}$$

where $\Delta_n^h \triangleq h(x_{n+1}) - h(x_n)$. The following notation is used for the Riemann–Stieltjes integral on $[a, b]$:

$$RS = \int_a^b g(x)\mathrm{d}h(x). \tag{D.15}$$

If the function has negative parts, then the integral for each positive and negative part is computed separately, yielding the overall integral: $RS = RS_+ - RS_-$.

The integrator $h(x)$ need not be differentiable or continuous; in fact, the Riemann–Stieltjes integral is useful for computing expectations of *discrete* and *mixed* random variables whose cdfs are not continuous. If X is a random variable with cdf $F_X(x)$ (which by definition is nondecreasing), then setting $h(x) = F_X(x)$ gives the *expectation* (discussed in Chapter 5):

$$\mathcal{E}[g(X)] = \int_a^b g(x)\mathrm{d}F_X(x). \tag{D.16}$$

Thus, we find that (D.1) mentioned in the introduction is the Riemann–Stieltjes integral of $g(x)$ with weighting given by the cdf $F_X(x)$.

Example D.3. Consider the Bernoulli random variable X with parameter p which has the following cdf:

$$F_X(x) = \sum_{n=0}^{1} p^n q^{1-n} u(x - n), \tag{D.17}$$

where $u(x)$ is the unit-step function and $q \triangleq 1 - p$. Suppose we want to compute the following Riemann–Stieltjes integral:

$$\mathcal{E}[X] = \int_{-\infty}^{\infty} x\mathrm{d}F_X(x), \tag{D.18}$$

which is the expectation (mean) of X. Figure D.2 shows a plot of the integrand $g(x) = x$ and the integrator $h(x) = F_X(x)$ for $p = 1/2$. The Bernoulli cdf has two discontinuities from the left at $x = \{0, 1\}$, but is defined to be continuous from the right (indicated by the solid circles). A partition of the x-axis is shown with $x_1 = 0$,

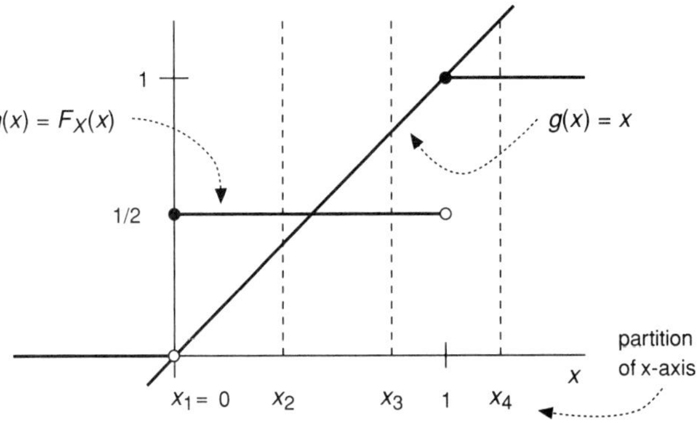

FIGURE D.2 Integrand $g(x) = x$ and integrator $h(x) = F_X(x)$ for the Riemann–Stieltjes integral in Example D.3.

$x_2 = 2/5$, $x_3 = 4/5$, and $x_4 = 6/5$. For this course partition with $N = 3$, the lower Riemann sum is

$$R_L(3) = (0)(x_2 - x_1) + (2/5)(x_3 - x_2) + (4/5)(x_4 - x_3) = 12/25, \tag{D.19}$$

which is the area of the rectangles on $[0, 6/5]$ approximating the integral of $g(x) = x$. By increasing N and using a finer partition, the Riemann sum approaches the actual area of $(6/5)^2/2 = 18/25$ in that interval.

For the lower Riemann–Stieltjes sum, the weighting $F_X(x_{n+1}) - F_X(x_n)$ is used instead of $x_{n+1} - x_n$ for the rectangles:

$$\begin{aligned} RS_L(3) &= (0)[F_X(x_2) - F_X(x_1)] + (2/5)[F_X(x_3) - F_X(x_2)] \\ &\quad + (4/5)[F_X(x_4) - F_X(x_3)], \\ &= (0)(1/2 - 1/2) + (2/5)(1 - 1) + (4/5)(1 - 1/2) = 2/5, \end{aligned} \tag{D.20}$$

where only subinterval $[x_3, x_4]$ contributes to the sum. It is clear that with a finer partition only the subinterval including $x = 1$ will contribute to the weighted sum, and in the limit gives the value $1 \cdot (1 - 1/2) = 1/2$, which is the mean of the Bernoulli random variable: $\mathcal{E}[X] = p = 1/2$.

Example D.4. From Example D.3, we recognize how to compute the Riemann–Stieltjes integral for any discrete random variable. Consider the binomial random variable with

$$F_X(x) = \sum_{n=0}^{N} \binom{N}{n} p^n q^{N-n} u(x - n). \tag{D.21}$$

As the partition on the x-axis becomes finer with increasing N, the Riemann–Stieltjes sum "captures" only those values of $g(x)$ at the discontinuities of the staircase function $F_X(x)$, specifically where $F_X(x_{n+1}) - F_X(x_n) \neq 0$ as $x_n \to x_{n+1}$. Thus, for the binomial cdf:

$$\begin{aligned} \mathcal{E}[g(X)] &= \int_{-\infty}^{\infty} g(x) \mathrm{d}F_X(x) = \sum_{n=0}^{N} \binom{N}{n} p^n q^{N-n} \int_{-\infty}^{\infty} g(x) \mathrm{d}u(x - n), \\ &= \sum_{n=0}^{N} \binom{N}{n} p^n q^{N-n} g(n) = \sum_{n=0}^{N} p_X[n] g(n), \end{aligned} \tag{D.22}$$

where $p_X[n]$ is the probability mass function (pmf) of X.

In Examples D.3 and D.4, the integrals (expectations) are computed without using the Dirac delta function to represent the pdf of a discrete random variable. As mentioned previously, the Dirac delta function is often used in basic courses on probability because it facilitates computing expectations using Riemann integration. For example, we can write for the binomial random variable

$$\mathrm{d}F_X(x) = \sum_{n=0}^{N} \binom{N}{n} p^n q^{N-n} \delta(x-n)\mathrm{d}x, \tag{D.23}$$

so that

$$\begin{aligned} \mathcal{E}[g(X)] &= \int_{-\infty}^{\infty} g(x)\mathrm{d}F_X(x) = \sum_{n=0}^{N} \binom{N}{n} p^n q^{N-n} \int_{-\infty}^{\infty} g(x)\delta(x-n)\mathrm{d}x, \\ &= \sum_{n=0}^{N} \binom{N}{n} p^n q^{N-n} g(n) = \sum_{n=0}^{N} p_X[n]g(n), \end{aligned} \tag{D.24}$$

where the sifting property of the Dirac delta function has been used. However, as shown by the previous examples, the Riemann–Stieltjes integral directly handles cdfs with discontinuities.

If $h(x)$ is differentiable everywhere on $[a, b]$ with ordinary derivative $h'(x)$, then

$$h'(x) = \frac{\mathrm{d}h(x)}{\mathrm{d}x} \Rightarrow \mathrm{d}h(x) = h'(x)\mathrm{d}x \tag{D.25}$$

and

$$RS = \int_a^b g(x)h'(x)\mathrm{d}x. \tag{D.26}$$

If $h(x) = F_X(x)$ is a cdf, then $f_X(x) = F'_X(x)$ and

$$RS = \int_a^b g(x)f_X(x)\mathrm{d}x, \tag{D.27}$$

which is the form of expectation more commonly known. If $F_X(x)$ is mixed with continuous $F_{X,c}(x)$ and discrete $F_{X,d}(x)$ components:

$$F_X(x) = F_{X,c}(x) + F_{X,d}(x) \tag{D.28}$$

then expectations are computed as follows:

$$\mathcal{E}[g(X)] = \int_{-\infty}^{\infty} g(x)\mathrm{d}F_X(x) = \int_{-\infty}^{\infty} g(x)f_{X,c}(x)\mathrm{d}x + \sum_{n=-\infty}^{\infty} g(n)p_{X,d}[n], \tag{D.29}$$

where $f_{X,c}(x)$ is the pdf of the continuous part of the random variable, and $p_{X,d}[n]$ is the pmf of the discrete part.

Example D.5. Consider a mixed random variable with

$$F_X(x) = (1/2)[1-\exp(-\lambda x)]I_{[0,\infty)}(x) + (1/2)\sum_{n=0}^{\infty} pq^n u(x-n), \tag{D.30}$$

which is actually a *finite mixture* of exponential and geometric cdfs (see Chapter 4). The mean of this random variable is

$$
\begin{aligned}
\mathcal{E}[X] &= \int_0^{\infty} x \mathrm{d}F_X(x) = (1/2)\int_0^{\infty} x\lambda \exp(-\lambda x)\mathrm{d}x + (1/2)\sum_{n=0}^{\infty} npq^n, \\
&= (1/2)[1/\lambda + (q/p)],
\end{aligned} \tag{D.31}
$$

which is the average of the mean values of the exponential and geometric random variables.

The Riemann and Riemann–Stieltjes integrals are well defined for bounded functions where the x-axis can be partitioned into a set of nonoverlapping subintervals. Recall that the length of the closed interval $[a, b]$ is given by the Lebesgue measure of that interval: $L([a, b]) = b - a$. For functions that are not continuous, such as the Dirichlet function defined by

$$
g(x) \triangleq \begin{cases} 1, & x \in \mathcal{Q} \\ 0, & x \in \mathcal{R} - \mathcal{Q}, \end{cases} \tag{D.32}
$$

where $\mathcal{Q}$ is the set of rational numbers, partitioning the domain into subintervals is not useful for computing the integral. To handle this type of situation, we can generalize the length of a subinterval on the x-axis to a *measure* of a *set* (instead of the Lebesgue measure of a subinterval). This leads to the Lebesgue integral.

D.3 LEBESGUE INTEGRAL

For function $y = g(x)$, a particular value of x is mapped only to one value of y; this, of course, is the defining property of a function. If the function is not one-to-one, then many values of x could map to the same y, such that the inverse image $g^{-1}(y)$ is one-to-many. This distinction is needed when initially describing the Lebesgue integral, which is obtained by partitioning the y-axis instead of the x-axis (as is done for the Riemann integral). Partitioning the y-axis allows integrals to be computed for functions that are not Riemann integrable, where the x-axis cannot be partitioned in a useful manner.

Consider $y = g(x)$ shown in Figure D.3(a) where the y-axis in the interval $[g(a), g(b)]$ has been partitioned into subintervals $\{[y_n, y_{n+1})\}$. Observe that this yields horizontal rectangles whose ends are constrained by the function. We would like to compute for each y_n the area of all vertical rectangles formed by subintervals on the x-axis with height given by y_n. For the example in Figure D.3(b), we see there is one vertical rectangle with height $g(a)$ (denoted by the diagonal lines). There are three vertical rectangles (with the darkest shading) whose heights are given by the next y value in the partition, and then two rectangles for the next y value. Multiple rectangles occur because the inverse image from y to x is not one-to-one. Each of the other values in the partition of the y-axis corresponds to a single vertical rectangle (with the lightest shading). "Missing" vertical rectangles in the middle of the function would be included with a finer partition of the y-axis.

As was done for the Riemann integral, we approximate the Lebesgue integral by a sum of the areas of the rectangles. The *Lebesgue sum* is

$$
LB(N) = \sum_{n=1}^{N} y_n L(\{x \in [a, b] : y_n \leq g(x) < y_{n+1}\}), \tag{D.33}
$$

where $L(\cdot)$ is the Lebesgue measure such that $L(\{x \in [a, b] : y_n \leq g(x) < y_{n-1}\})$ gives the overall width of *all* rectangles with the *same* height. With increasingly finer partitions of the y-axis as $N \to \infty$, $LB(N)$ approaches the area under the function on $[a, b]$. Obviously, if $g(x)$ is continuous and bounded, then the Riemann integral and the Lebesgue integral are identical: this is evident because we have just rearranged how the rectangles are organized in the sum approximations. For the Riemann sum, they are defined relative to a partition of the x-axis

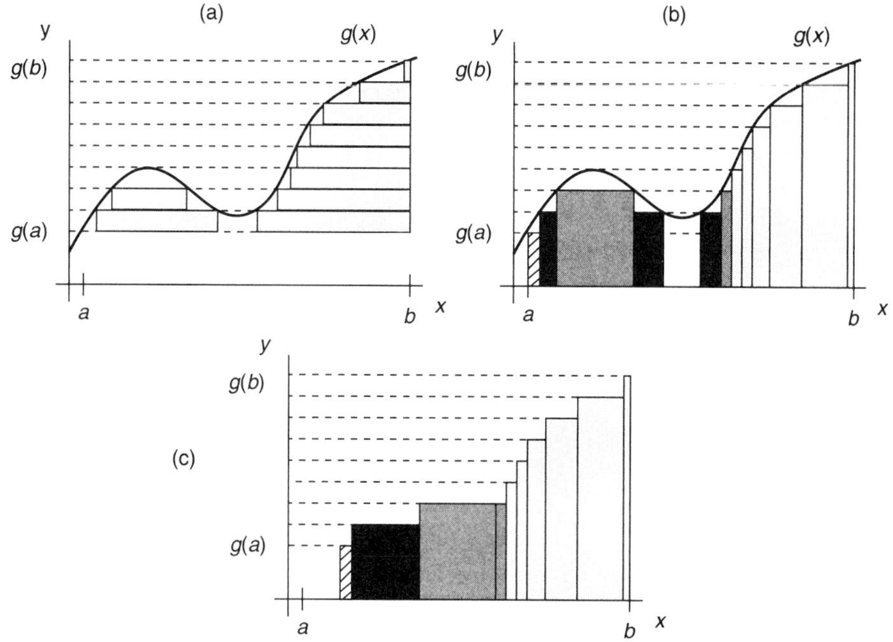

FIGURE D.3 Lebesgue sum. (a) Function $y = g(x)$ on $[a, b]$ partitioned on the y-axis into subintervals $\{[y_n, y_{n+1})\}$, yielding horizontal rectangles. (b) Vertical rectangles defined by the endpoints of the horizontal rectangles. Those with similar shadings indicate subintervals on the x-axis that have the same rectangular height. (c) Vertical rectangles arranged according to increasing y.

into disjoint subintervals, whereas for the Lebesgue sum the rectangles are defined relative to a partition of the y-axis.

The Lebesgue integral can handle functions that are not continuous and where it is not possible to usefully partition the x-axis into subintervals. Moreover, measures other than the Lebesgue measure can be used when constructing the Lebesgue sum. Toward this end, note that because the order of terms in the sum is irrelevant for the final value of the integral, consider rearranging the rectangles with increasing y as depicted in Figure D.3(c). Of course, the rectangles no longer follow the function $g(x)$ (which is not shown in this plot), but obviously the cumulative area is the same as that in Figure D.3(b).

Consider now a function $g(x)$ that is not continuous such as the Dirichlet function mentioned earlier. It is clear that by partitioning the x-axis into subintervals, we cannot construct meaningful rectangles to compute the Riemann integral of $g(x)$. However, it is easy to compute the Lebesgue sum for this function because the y-axis is partitioned and $g(x)$ takes on only two values: $\{0, 1\}$. As was done in Figure D.3(c), rearrange the order of the numbers on the x-axis so that the irrational numbers are grouped first, followed by the rational numbers. This is depicted "pictorially" in Figure D.4 for finite interval $x \in [a, b]$ where the partition on the y-axis requires only $N = 2$ subintervals for the Dirichlet function: $[0, 1)$ and $[1, \infty)$. The Lebesgue sum is written as follows:

$$\begin{aligned} LB(N) &= \sum_{n=1}^{N} y_n L(\{x \in [a, b] : y_n \leq g(x) < y_{n+1}\}), \\ &= (0)L(\{x \in [a, b] : 0 \leq g(x) < 1\}) \\ &\quad + (1)L(\{x \in [a, b] : 1 \leq g(x) < \infty\}) = L(\mathcal{Q}). \end{aligned} \tag{D.34}$$

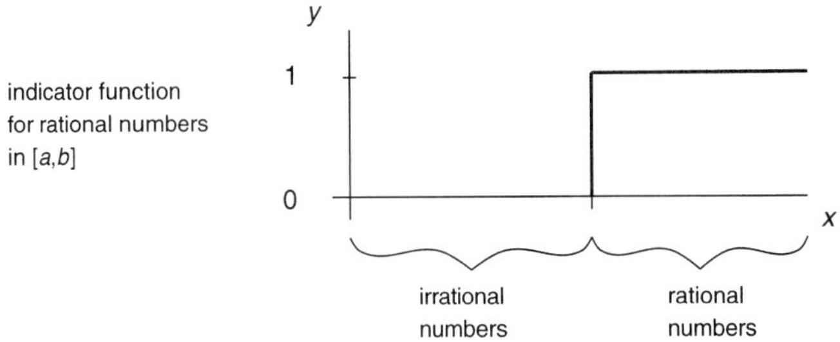

FIGURE D.4 Pictorial example of the real numbers in a finite interval rearranged so that all irrational numbers are grouped together in increasing order, and likewise for all rational numbers.

Since $L(\mathcal{Q}) = 0$ (countable sets have zero Lebesgue measure), the Lebesgue integral for the Dirichlet function is zero. The Riemann integral does not exist for this function. By grouping the rational and irrational numbers in this manner and partitioning the y-axis, we have computed the integral by generalizing the notion of size from the length of a subinterval to the Lebesgue measure of a set. Moreover, other measures can be used, as demonstrated next for measure μ.

Definition: Simple Function A *simple function* is a weighted linear combination of indicator functions of events $\{E_n\}$ with weights given by values on the y-axis:

$$g(x) = \sum_{n=1}^{N} y_n I_{E_n}(x). \tag{D.35}$$

It is a mapping from the measurable sets $\{E_n\}$ to $\mathcal{R}$.

Example D.6. The Dirichlet function is a simple function. There are only two sets: (i) the rational numbers $E_1 = \mathcal{Q}$ and (ii) the irrational numbers $E_2 = \mathcal{R} - \mathcal{Q}$. Thus, for some interval $[a, b] \in \mathcal{R}$:

$$g(x) = y_1 I_{\mathcal{Q}}(x) + y_2 I_{\mathcal{R}-\mathcal{Q}}(x) = (1)I_{\mathcal{Q}}(x) + (0)I_{\mathcal{R}-\mathcal{Q}}(x) = I_{\mathcal{Q}}(x), \tag{D.36}$$

which is defined for all $x \in [a, b]$.

It is important to note that the sum in (D.35) is defined by specific values $\{y_n\}$ which can be viewed as a partition of the y-axis. There is no partitioning of the x-axis into subintervals; we are concerned only with sets $\{E_n\}$ and the corresponding values of x in those sets, without regard to subintervals on the x-axis. This is evident for the Dirichlet function where $x \in \mathcal{Q}$ on $[a, b]$ are countable and spread unevenly throughout this interval. Analogous to (D.33), the Lebesgue sum of $g(x)$ in (D.35) with respect to μ is

$$LB(N) = \sum_{n=1}^{N} y_n \mu(E_n), \tag{D.37}$$

which is more general than the Riemann sum: (i) sets $\{E_n\}$ have replaced subintervals and (ii) the length of a subinterval has been replaced by the measure μ on a set.

The Lebesgue sum in (D.37) can be extended to any *measurable* function $g(x)$ that is nonnegative as follows. Consider a staircase approximation of $g(x)$ obtained by partitioning the y-axis into subintervals of equal length:

$$g_n(x) = \begin{cases} m/2^n, & m/2^n \leq g(x) < (m+1)/2^n, \ 0 \leq m \leq 2^{2n} - 1 \\ 2^n, & g(x) \geq 2^n, \end{cases} \tag{D.38}$$

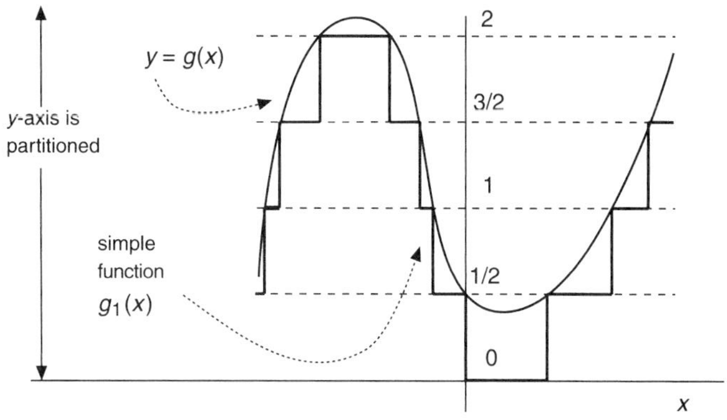

FIGURE D.5 "Staircase" approximation of $g(x)$ by simple function $g_1(x)$.

where $m, n \in \mathcal{Z}$. This type of approximation is similar to *quantization* in signal processing, which is a technique used to convert continuous waveforms to digital signals (this is briefly discussed in Chapter 10 on communications). For $n = 1$, the subintervals on the y-axis are $[0, 1/2)$, $[1/2, 1)$, $[1, 3/2)$, $[3/2, 2)$, $[2, \infty)$, and the corresponding staircase approximation of $g(x)$ is shown in Figure D.5. Although the x-axis is divided into subintervals, this is due to a partition of the y-axis and the fact that $g(x)$ is a continuous function of x. The number of distinct y values in the simple function $g_1(x)$ is five for the example in the figure.

Increasing n by one, the range of the partition increases by a factor of two and the width of each subinterval on the y-axis decreases by a factor of two. Thus, in the limit we have

$$\lim_{n\to\infty} \sup |g_n(x) - g(x)| = 0, \tag{D.39}$$

where the supremum is over all simple functions $g_n(x)$ not exceeding $g(x)$ (as demonstrated by the example in Figure D.5 for $n = 1$). The Lebesgue sum for the nth simple function is

$$LB(N_n) = \sum_{m=0}^{2^{2n}-1} (m/2^n)\mu(E_{m,n}) = \sum_{m=0}^{N_n-1} y_{m,n}\mu(E_{m,n}), \tag{D.40}$$

where $N_n \triangleq 2^{2n}$ is the number of terms in the sum, $y_{m,n} \triangleq m/2^n$ is a value on the y-axis, and $E_{m,n}$ is the set of x with function value $y_{m,n}$. The Lebesgue integral LB of $g(x)$ is obtained in the limit as the number of steps approaches infinity and the width of the subintervals on the y-axis approaches zero; it has the following notation:

$$LB = \int_a^b g(x)\mathrm{d}\mu(x). \tag{D.41}$$

If $g(x)$ is negative anywhere on $[a, b]$, it can be separated into positive and negative parts (as was done for the Riemann integral), and the Lebesgue integral of the negative part is subtracted from that of the positive part: $LB = LB_+ - LB_-$. Finally, for the abstract probability space $\{\Omega, \mathcal{F}, P\}$, the expectation of function $g(\omega)$ is the following Lebesgue integral:

$$LB = \int_\Omega g(\omega)\mathrm{d}P(\omega), \tag{D.42}$$

where ω are elements of Ω. Thus, (D.2) mentioned in the introduction is the Lebesgue integral of $g(\omega)$ with measure given by the probability measure P, and it corresponds to an expectation for an abstract sample space.

Example D.7. Consider the experiment of tossing two fair coins with sample space $\Omega = \{HH, TT, HT, TH\}$. Let the atoms of the σ-field $\mathcal{F}$ be $\{\{HH, TT\}, TH, HT\}$ such that

$$\mathcal{F} = \{\phi, \Omega, \{HH, TT\}, TH, HT, \{TH, HT\}, \{HH, TT, TH\}, \{HH, TT, HT\}\}. \tag{D.43}$$

Recall that the probability measure P is defined for events in $\mathcal{F}$, not the outcomes of Ω, unless the atoms are the same as the outcomes. The probabilities of the atoms must sum to one because (i) they are disjoint and (ii) their union must be Ω. Thus, we know for fair coins that

$$\begin{aligned} P(\{HH, TT\}) = 1/2, \quad P(TH) = P(HT) = 1/4, \\ P(\{HH, TT, TH\}) = P(\{HH, TT, HT\}) = 3/4, \\ P(\Omega) = P(\{HH, TT, TH, HT\}) = 1, \quad P(\phi) = 0. \end{aligned} \tag{D.44}$$

For this particular σ-field, outcomes $\{HH\}$ and $\{TT\}$ have been excluded as events, so they are not assigned probabilities individually; (D.44) gives the probabilities of all eight events in $\mathcal{F}$ which form a subset of the power set $\mathcal{P}(\Omega)$. (If we had used the power set, then of course $P(HH) = P(TT) = 1/4$.) Suppose we are interested in the integral of the following function:

$$g(\omega) = \begin{cases} 1, & \omega = HH \\ 0, & \omega = TT, HT, TH, \end{cases} \tag{D.45}$$

which is the discrete indicator function: $g(\omega) = I_{\{HH\}}[\omega]$. Since a function is defined for the outcomes of Ω (and not the events in $\mathcal{F}$), $g(\omega)$ is *not* a random variable, even though it maps outcomes to $\mathcal{R}$, because it is not measurable: the inverse images are $g^{-1}(1) = HH \notin \mathcal{F}$ and $g^{-1}(0) = \{TT, HT, TH\} \notin \mathcal{F}$. Since the Lebesgue integral assumes a measurable function, consider instead the following function which is a random variable:

$$X(\omega) = I_{\{HH,TT\}}[\omega]. \tag{D.46}$$

Note that $X^{-1}(1) = \{HH, TT\} \in \mathcal{F}$ (it is an atom of $\mathcal{F}$) and $X^{-1}(0) = \{HT, TH\} \in \mathcal{F}$ (it is the union of two atoms of $\mathcal{F}$). The Lebesgue integral for this case is

$$\begin{aligned} \int_\Omega X(\omega)\mathrm{d}P(\omega) &= (1)P(\{\omega \in \Omega : X(\omega) = 1\}) + (0)P(\{\omega \in \Omega : X(\omega) = 0\}), \\ &= (1)P(\{HH, TT\}) + (0)P(\{HT, TH\}), \\ &= (1)P(\{HH, TT\}) + (0)[P(HT) + P(TH)], \\ &= (1)(1/2) + (0)(1/4 + 1/4) = 1/2. \end{aligned} \tag{D.47}$$

We recognize that $X(\omega)$ is the Bernoulli random variable in Example D.3 where its expectation was computed using a Riemann–Stieltjes integral. The result above is the corresponding Lebesgue integral over Ω in the abstract probability space $\{\Omega, \mathcal{F}, P\}$, which of course is the same as in Example D.3.

D.4 PDF INTEGRALS

The various pdfs covered in Chapter 3 and summarized in Appendix A are functions whose shapes are useful models for the random outcomes in various applications. Most of the pdfs include an exponential term as shown in Table D.1 where we have collected them into three groups (except the continuous uniform and triangular pdfs). The pdfs in the first group include an exponential function of x^2 (or higher, as in the case of the Weibull

TABLE D.1 Pdf Functional Forms for Continuous Random Variables

Random Variable	Functional Form	Support	Area
Chi	$x^{N-1}\exp(-x^2/2)$	$\mathcal{R}^+$	$\Gamma(N/2)2^{N/2-1}$
Gaussian	$\exp(-(x-\mu)^2/2\sigma^2)$	$\mathcal{R}$	$\sqrt{2\pi\sigma^2}$
Half-normal	$\exp(-x^2/2\alpha^2)$	$\mathcal{R}^+$	$\alpha\sqrt{\pi/2}$
Inverse Gaussian	$(1/x^{3/2})\exp(-\alpha(x-\mu)^2/2\mu^2x)$	$\mathcal{R}^+$	$\sqrt{2\pi/\alpha}$
Log-normal	$x^{-1}\exp(-(\ln(x)-\mu)^2/2\sigma^2)$	$\mathcal{R}^+$	$\sqrt{2\pi\sigma^2}$
Maxwell–Boltzmann	$x^2\exp(-x^2/2\alpha^2)$	$\mathcal{R}^+$	$\alpha^3\sqrt{\pi/2}$
Rayleigh	$x\exp(-x^2/2\alpha^2)$	$\mathcal{R}^+$	α^2
Rice	$x\exp(-(x^2+c^2)/2\alpha^2)I_0(cx/\alpha^2)$	$\mathcal{R}^+$	α^2
Weibull	$x^{r-1}\exp(-(x/\alpha)^r)$	$\mathcal{R}^+$	α^r/r
Chi-square	$x^{N/2-1}\exp(-x/2)$	$\mathcal{R}^+$	$\Gamma(N/2)2^{N/2}$
Exponential	$\exp(-\lambda x)$	$\mathcal{R}^+$	$1/\lambda$
Extreme value	$\exp(-(x-c)/\alpha)$ $\cdot\exp(-\exp(-(x-c)/\alpha))$	$\mathcal{R}$	α
Gamma	$x^{r-1}\exp(-\lambda x)$	$\mathcal{R}^+$	$\Gamma(r)/\lambda^r$
Laplace	$\exp(-\lvert x-\mu\rvert/\alpha)$	$\mathcal{R}$	2α
Logistic	$\exp(-(x-\mu)/\alpha)$ $\cdot[1+\exp(-(x-\mu))]^{-2}$	$\mathcal{R}$	α
Beta	$x^{\alpha-1}(1-x)^{\beta-1}$	$[0,1]$	$B(\alpha,\beta)$
Cauchy	$[(x-c)^2+\alpha^2]^{-1}$	$\mathcal{R}$	π/α
F-distribution	$x^{(m-2)/2}/[1+(m/n)x]^{(m+n)/2}$	$\mathcal{R}^+$	$B(m/2,n/2)(n/m)^{m/2}$
Pareto	x^{-r-1}	$[\alpha,\infty)$	$1/r\alpha^r$
Student's t	$1/(1+x^2/r)^{(r+1)/2}$	$\mathcal{R}$	$\sqrt{r}B(1/2,r/2)$

distribution). The second group also includes an exponential function but of x, and the logistic and extreme value pdfs have two exponential functions. The third group of pdfs do not include any exponential functions, and depend on x with various exponents.

The area under each function and the support are also included in the table. The inverse of each term in the last column is the weighting applied to the functional form so that the pdf has unit area. It is the *functional form* that defines the pdf, and the weighting is just a constant for specific values of the pdf parameters. For example, consider the gamma random variable with pdf

$$f_X(x) = \frac{\lambda^r}{\Gamma(r)}x^{r-1}\exp(-\lambda x)u(x), \tag{D.48}$$

where

$$\Gamma(r) \triangleq \int_0^\infty v^{r-1}\exp(-v)\mathrm{d}v \tag{D.49}$$

is the gamma function (see Appendix B). The integrand of the definite integral that defines $\Gamma(r)$ resembles the functional form of the pdf. Thus, $\Gamma(r)$ has a role similar to that of the Q-function: it is a representation for an integral that has no closed form. Likewise for the beta function:

$$B(\alpha,\beta) \triangleq \int_0^1 v^{\alpha-1}(1-v)^{\beta-1}\mathrm{d}v \tag{D.50}$$

whose integrand is exactly the functional form of the beta random variable shown in the table. $B(\alpha,\beta)$ is just notation for the definite integral above which does not have a closed form in terms of α and β.

With this functional view, it is possible to exploit the fact that the area of a pdf is one in order to evaluate a definite integral whose integrand resembles the functional form of a particular pdf. This is done by rearranging the integrand (by a change of variables or completing the square) so that it resembles the functional form of a known pdf. (A similar approach is used in Appendix E for sums of pmfs.)

Example D.8. Consider the following integral:

$$\int_{-\infty}^{\infty} \exp(-x^2)\mathrm{d}x = (1/\sqrt{2}) \int_{-\infty}^{\infty} \exp(-y^2/2)\mathrm{d}y \tag{D.51}$$

whose integrand resembles the functional form of the Gaussian pdf. Thus, with $\mu = 0$ and $\sigma = 1$ we have

$$\int_{-\infty}^{\infty} \exp(-x^2)\mathrm{d}x = \sqrt{\pi}. \tag{D.52}$$

A similar approach can be used to obtain the following definite integrals:

$$\int_{-\infty}^{\infty} \exp(-(x-\mu)^2/\alpha^2)\mathrm{d}x = \alpha\sqrt{\pi}, \tag{D.53}$$

$$\int_{-\infty}^{\infty} \exp(-ax^2 + bx + c)\mathrm{d}x = \sqrt{\pi/a}\exp(c + b^2/4a), \tag{D.54}$$

where we completed the square in the exponent to obtain the last expression.

Example D.9. Suppose we are interested in the moments of the exponential random variable:

$$\mathcal{E}[X^n] = \int_0^{\infty} \lambda x^n \exp(-\lambda x)\mathrm{d}x. \tag{D.55}$$

From the functional form of the Gamma random variable in the table with $r - 1 = n$, we have

$$\mathcal{E}[X^n] = \lambda\Gamma(n+1)/\lambda^{n+1} = \Gamma(n+1)/\lambda^n, \tag{D.56}$$

which is the expression in Appendix A.

Note that one must be careful when using this approach to derive definite integrals: the limits of integration must match the support in the table. For example, the Maxwell–Boltzmann result in the table with support $\mathcal{R}^+$ cannot be used directly for the second moment of the Gaussian random variable which has support $\mathcal{R}$.

D.5 INDEFINITE AND DEFINITE INTEGRALS

Several integrals that are useful for some of the distributions are provided below:

- *Cauchy:*

$$\int \frac{\mathrm{d}x}{a^2 + b^2x^2} = (1/ab)\tan^{-1}(bx/a), \tag{D.57}$$

$$\int \frac{x\mathrm{d}x}{a^2+x^2} = (1/2)\ln(a^2+x^2), \tag{D.58}$$

$$\int \frac{x^2\mathrm{d}x}{a^2+x^2} = x - a\tan^{-1}(x/a). \tag{D.59}$$

- *Chi-square, exponential, gamma, and Laplace:*

$$\int \exp(ax)\mathrm{d}x = (1/a)\exp(ax), \tag{D.60}$$

$$\int x\exp(ax)\mathrm{d}x = (1/a)(x-1/a)\exp(ax), \tag{D.61}$$

$$\int x^2\exp(ax)\mathrm{d}x = (1/a)(x^3-2x/a+2/a^2)\exp(ax), \tag{D.62}$$

$$\int x^m\exp(ax)\mathrm{d}x = (1/a)x^m\exp(ax) - (m/a)\int x^{m-1}\exp(ax)\mathrm{d}x, \tag{D.63}$$

$$\int_0^\infty x^m\exp(-ax)\mathrm{d}x = \frac{\Gamma(m+1)}{a^{m+1}}, \quad (\text{for } a>0, n>-1). \tag{D.64}$$

- *Chi, Gaussian, half-normal, Maxwell–Boltzmann, and Rayleigh:*

$$\int_0^\infty x^{2m}\exp(-ax^2)\mathrm{d}x = \frac{1\cdot 3\cdot 5\cdots(2m-1)}{2^{m+1}a^m}\sqrt{\pi/a}, \tag{D.65}$$

$$\int_0^\infty x^{2m+1}\exp(-ax^2)\mathrm{d}x = \frac{m!}{2a^{m+1}}, \quad (\text{for } a>0). \tag{D.66}$$

- *Pareto:*

$$\int \frac{x\mathrm{d}x}{(bx)^m} = -\frac{1/b^2}{(m-2)(bx)^{m-2}}, \quad (m\neq 1,2). \tag{D.67}$$

- *Student's t:*

$$\int \frac{\mathrm{d}x}{(a^2+x^2)^m} = \frac{1}{2a^2(m-1)}\left[\frac{x}{(a^2+x^2)^{m-1}} + (2m-3)\int\frac{\mathrm{d}x}{(a^2+x^2)^{m-1}}\right], \tag{D.68}$$

$$\int \frac{x\mathrm{d}x}{(a^2+x^2)^{m+1}} = -\frac{1}{2m(a^2+x^2)^m}. \tag{D.69}$$

- *Weibull:*

$$\int_0^\infty x^m\exp(-ax^n)\mathrm{d}x = \frac{\Gamma((m+1)/n)}{na^{(m+1)/n}}, \quad (\text{for } a>0, m>-1, n>0). \tag{D.70}$$

D.6 INTEGRAL FORMULAS

For convenience, we provide expressions for Leibniz's integral rule:

$$\frac{\partial}{\partial v}\int_{a(v)}^{b(v)} g(u,v)\mathrm{d}u = \int_{a(v)}^{b(v)} \frac{\partial}{\partial v} g(u,v)\mathrm{d}u + g(b(v),v)\frac{\partial}{\partial v}b(v) - g(a(v),v)\frac{\partial}{\partial v}a(v) \tag{D.71}$$

and for integration by parts:

$$\int_{x_1}^{x_2} g(x)\mathrm{d}h(x) = g(x_2)h(x_2) - g(x_1)h(x_1) - \int_{x_1}^{x_2} h(x)\mathrm{d}g(x), \tag{D.72}$$

which is written using Riemann–Stieltjes integrals. If the functions are differentiable everywhere on $[x_1, x_2]$, then $\mathrm{d}h(x) = h'(x)\mathrm{d}x$ and $\mathrm{d}g(x) = g'(x)\mathrm{d}x$, yielding

$$\int_{x_1}^{x_2} g(x)h'(x)\mathrm{d}x = g(x_2)h(x_2) - g(x_1)h(x_1) - \int_{x_1}^{x_2} h(x)g'(x)\mathrm{d}x. \tag{D.73}$$

For an indefinite integral and suppressing the variable x, (D.72) can be written as

$$\int g\mathrm{d}h = gh - \int h\mathrm{d}g, \tag{D.74}$$

which is the familiar form covered in calculus courses.

D.7 DOUBLE INTEGRALS OF SPECIAL FUNCTIONS

Finally, we summarize some double integrals of a few special functions of two variables. They include $\delta(x-y)$, $u(x-y)$, $\min(x,y)$, $\max(x,y)$, and $|x-y|$. Some of these are useful in Chapter 7 on mean-square calculus where the autocorrelation function $R_{XX}(t_1, t_2)$ for nonstationary random process $X(t)$ is integrated. All the results depend on which upper limit is larger.

- *Dirac delta function* $\delta(y-x)$. This function is nonzero on the x–y-plane for all $x = y$. When it is integrated over y, unit-step functions are obtained along the line $x = y$, giving the appearance of a three-dimensional step as shown in Figure D.6(a):

$$\int_0^y \delta(y-x)\mathrm{d}y = u(y-x). \tag{D.75}$$

  The second integral gives

$$\int_0^x u(y-x)\mathrm{d}x = \int_0^{\min(x,y)} \mathrm{d}x = \min(x,y), \tag{D.76}$$

  which is shown in Figure D.6(b).
- *Unit-step function* $u(y-x)$. Integrating over x gives the expression in (D.76). Integrating over y gives

$$\int_0^y \min(x,y)\mathrm{d}y = \int_0^x y\mathrm{d}y + \int_x^y x\mathrm{d}y = (yx - x^2/2)u(y-x), \tag{D.77}$$

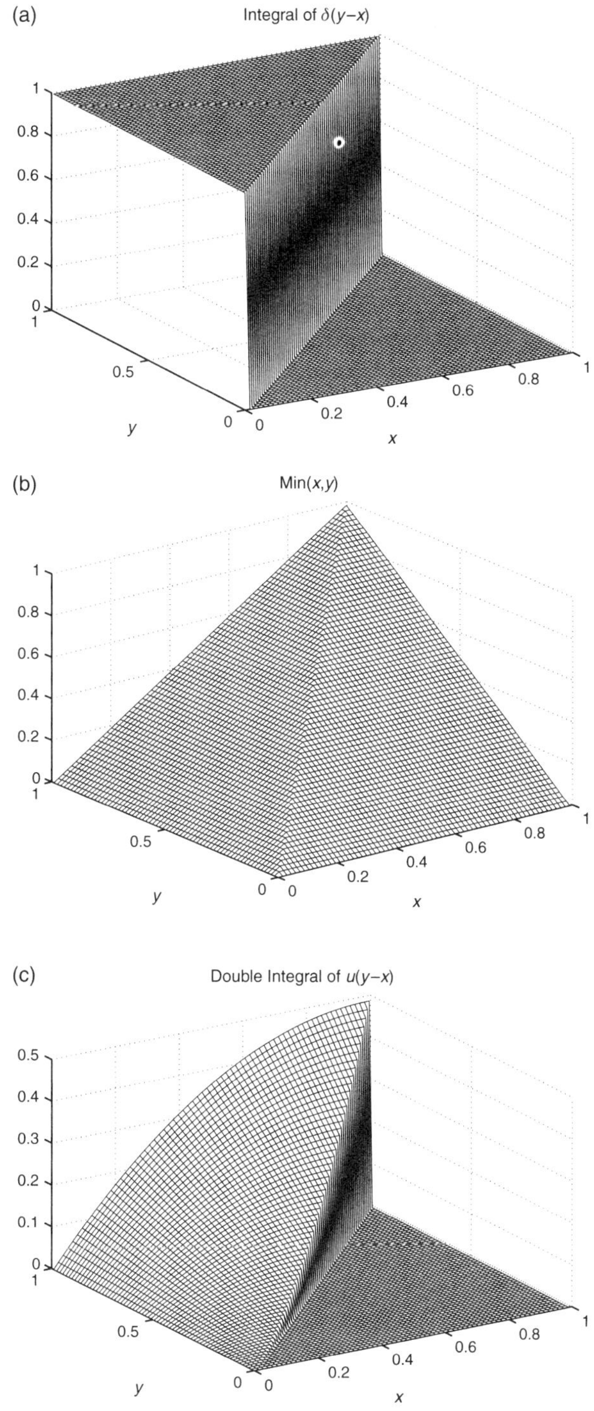

FIGURE D.6 Integrals of $\delta(y-x)$ and $u(y-x)$. (a) $\int_0^y \delta(y-x)\mathrm{d}y$. (b) $\int_0^x \int_0^y \delta(y-x)\mathrm{d}y\mathrm{d}x$. This is also $\int_0^x u(y-x)\mathrm{d}x$ and corresponds to $\min(x, y)$. (c) $\int_0^y \int_0^x u(y-x)\mathrm{d}x\mathrm{d}y$. (Note scale change from (b) to (c).)

where it is understood that $y \geq x$ since we started with $u(y - x)$ in (D.75). This quadratic shape is shown in Figure D.6(c). The surface is not symmetric about $y = x$ because the unit-step function is likewise not symmetric. It has a maximum of 1/2 as x and y approach 1. The double integral of $u(x - y)$ requires that we interchange x and y in (D.77).

- *Minimum function* $\min(x, y)$. Integrating over y for $y \geq x$ gives the expression in (D.77). Integrating over x gives

$$\int_0^x (yx - x^2/2)\mathrm{d}x = (yx^2/2 - x^3/6)u(y - x). \quad \text{(D.78)}$$

Thus, in general

$$\int_0^x \int_0^y \min(x, y)\mathrm{d}y\mathrm{d}x = \max(x, y)\min^2(x, y)/2 - \min^3(x, y)/6, \quad \text{(D.79)}$$

which is plotted in Figure D.7(a). For $x = y = 1$, (D.79) is 1/3 in the plot.

- *Maximum function* $\max(x, y)$. A plot of the maximum function is provided in Figure D.7(b). Integrating over y for $y \geq x$ gives

$$\int_0^y \max(x, y)\mathrm{d}y = \int_0^x x\mathrm{d}y + \int_x^y y\mathrm{d}y = (y^2/2 + x^2/2)u(y - x) \quad \text{(D.80)}$$

and integrating over x yields

$$\int_0^x (y^2/2 + x^2/2)\mathrm{d}x = (xy^2/2 + x^3/6)u(y - x). \quad \text{(D.81)}$$

Thus, in general

$$\int_0^x \int_0^y \max(x, y)\mathrm{d}y\mathrm{d}x = \min(x, y)\max^2(x, y)/2 + \min^3(x, y)/6. \quad \text{(D.82)}$$

This result is shown in Figure D.7(c) which we see increases more rapidly than the integration result in (D.79) for $\min(x, y)$. For $x = y = 1$, (D.82) is 2/3 in the plot.

- *Absolute value function* $|y - x|$. A plot of the absolute value function is provided in Figure D.8(a). Integrating over y for $y \geq x$ gives

$$\int_0^y |y - x|\mathrm{d}y = \int_0^x (x - y)\mathrm{d}y + \int_x^y (y - x)\mathrm{d}y = (x^2 + y^2/2 - xy)u(y - x) \quad \text{(D.83)}$$

and integrating over x yields

$$\int_0^x (x^2 + y^2/2 - xy)\mathrm{d}x = (x^3/3 + xy^2/2 - x^2y/2)u(y - x). \quad \text{(D.84)}$$

Thus, in general

$$\int_0^x \int_0^y |y - x|\mathrm{d}y\mathrm{d}x = \min^3(x, y)/3 + \min(x, y)\max^2(x, y)/2 - \min^2(x, y)\max(x, y)/2, \quad \text{(D.85)}$$

which is shown in Figure D.8(b). For $x = y = 1$, (D.85) is 1/3 in the plot.

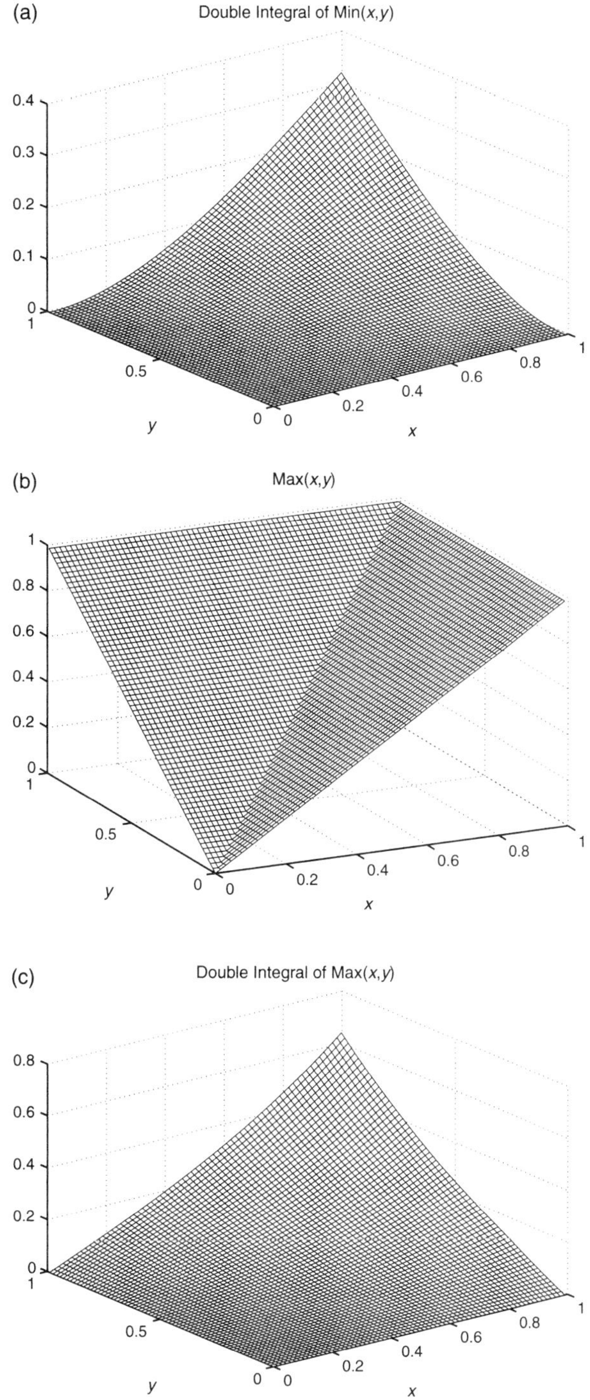

FIGURE D.7 Integrals of $\min(x, y)$ and $\max(x, y)$. (a) $\int_0^x \int_0^y \min(x, y)\mathrm{d}y\mathrm{d}x$. (b) $\max(x, y)$. (c) $\int_0^x \int_0^y \max(x, y)\mathrm{d}y\mathrm{d}x$. (Note scale change from (a) to (c).)

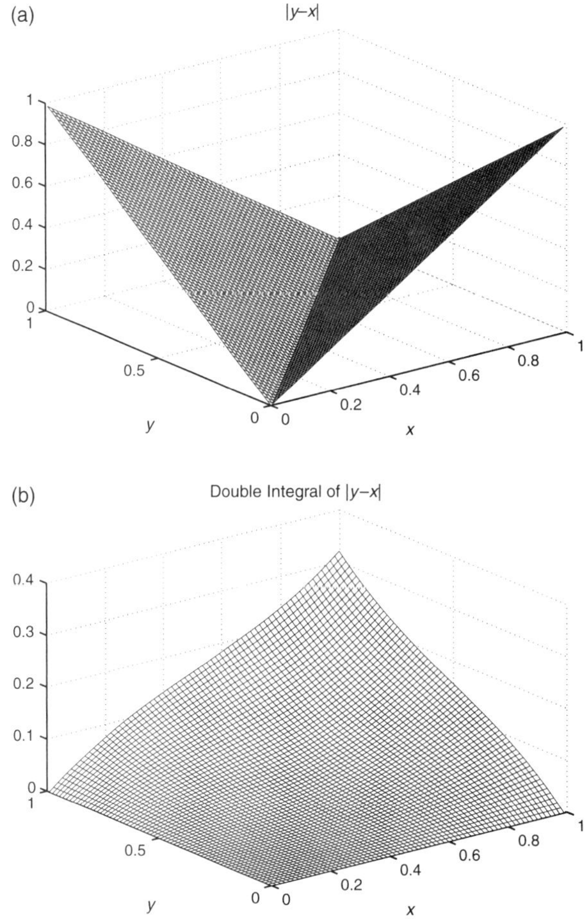

FIGURE D.8 Integrals of $|y-x|$. (a) $|y-x|$. (b) $\int_0^x \int_0^y |y-x| \mathrm{d}y \mathrm{d}x$. (Note scale change from (a) to (b).)

FURTHER READING

The following references provide additional material on the different types of integration: Athreya and Lahiri (2006), Billingsley (1986), Richardson (2009), Ross (1980), Sokolnikoff and Redheffer (1966), and Thomas, Jr. (1968).

E Identities and Infinite Series

In this appendix, we include several identities that are used in various chapters, along with some results for series that are useful for discrete random variables. Most results are provided without proof.

E.1 ZERO AND INFINITY

- *Addition and subtraction:*

$$0+0=0, \qquad 0-0=0, \qquad 0+\infty=\infty, \qquad \infty+\infty=\infty, \tag{E.1}$$

$$0-\infty=-\infty, \qquad \infty-0=\infty, \qquad \infty-\infty=\text{undefined}. \tag{E.2}$$

  This last result occurs when calculating the mean of the Cauchy distribution, which is why it does not exist.

- *Multiplication and division:*

$$\infty\cdot\infty=\infty, \qquad 0\cdot 0=0, \qquad 0\cdot\infty=\text{undefined}, \tag{E.3}$$

$$0/0=\text{undefined}, \qquad \infty/\infty=\text{undefined}, \qquad 0/\infty=0, \qquad \infty/0=\infty. \tag{E.4}$$

- *Exponents:*

$$0^0=\text{undefined}, \qquad 0^\infty=0, \qquad \infty^0=\text{undefined}, \qquad \infty^\infty=\infty. \tag{E.5}$$

E.2 MINIMUM AND MAXIMUM

The following identities are useful when minimum and maximum are included in functions to be integrated or differentiated:

$$\min(x_1,x_2)=(1/2)(x_1+x_2-|x_1-x_2|), \tag{E.6}$$

$$\max(x_1,x_2)=(1/2)(x_1+x_2+|x_1-x_2|). \tag{E.7}$$

Probability, Random Variables, and Random Processes: Theory and Signal Processing Applications, First Edition. John J. Shynk.

E.3 TRIGONOMETRIC IDENTITIES

- *Basic identities:*

$$\cos(x)\cos(y) = (1/2)[\cos(x-y) + \cos(x+y)], \tag{E.8}$$

$$\sin(x)\sin(y) = (1/2)[\cos(x-y) - \cos(x+y)], \tag{E.9}$$

$$\sin(x)\cos(y) = (1/2)[\sin(x-y) + \sin(x+y)], \tag{E.10}$$

$$\cos(x \pm y) = \cos(x)\cos(y) \mp \sin(x)\cos(y), \tag{E.11}$$

$$\sin(x \pm y) = \sin(x)\cos(y) \pm \cos(x)\sin(y), \tag{E.12}$$

$$\cos(x \pm \pi/2) = \mp\sin(x), \tag{E.13}$$

$$\sin(x \pm \pi/2) = \pm\cos(x). \tag{E.14}$$

- *Euler's formulas:*

$$\cos(x) = (1/2)[\exp(jx) + \exp(-jx)], \qquad \sin(x) = (1/2)[\exp(jx) - \exp(-jx)], \tag{E.15}$$

which give Euler's identity: $\exp(j\pi) = -1$.

- *Quadrature and polar forms:*

$$r\cos(x+\phi) = r\cos(\phi)\cos(x) - r\sin(\phi)\sin(x) \tag{E.16}$$

$$\triangleq a\cos(x) - b\sin(x), \tag{E.17}$$

$$a = r\cos(\phi), \qquad b = r\sin(\phi), \qquad r = \sqrt{a^2+b^2}, \qquad \phi = \arctan(b/a). \tag{E.18}$$

- *Hyperbolic functions:*

$$\sinh(x) = (1/2)[\exp(x) - \exp(-x)] = \sum_{m=0}^{\infty} \frac{x^{2m+1}}{(2m+1)!},$$

$$\cosh(x) = (1/2)[\exp(x) + \exp(-x)] = \sum_{m=0}^{\infty} \frac{x^{2m}}{(2m)!},$$

$$\tanh(x) = \frac{\sinh(x)}{\cosh(x)}, \tag{E.19}$$

$$\cosh^2(x) - \sinh^2(x) = 1, \qquad \cosh(x) \pm \sinh(x) = \exp(\pm x). \tag{E.20}$$

E.4 STIRLING'S FORMULA

The expression for factorial is, of course, given by

$$n! = \prod_{m=0}^{n-1}(n-m). \tag{E.21}$$

For large n, it can be approximated by Stirling's formula:

$$n! \approx (n/e)^n\sqrt{2\pi n}. \tag{E.22}$$

TABLE E.1 Factorials and Stirling's Approximation

n	$n!$	Stirling's	n	$n!$	Stirling's
0	1	0	8	40,320	39,902.4
1	1	0.9	9	362,880	359,536.9
2	2	1.9	10	3,628,800	3,598,695.6
3	6	5.8	11	39,916,800	39,615,625.1
4	24	23.5	12	479,001,600	475,687,486.5
5	120	118.0	13	6,227,020,800	6,187,239,475.2
6	720	710.1	14	87,178,291,200	86,661,001,740.6
7	5040	4980.4	15	1,307,674,368,000	1,300,430,722,199.5

For convenience when computing probabilities for combinatoric problems (involving discrete random variables), we include some factorials in Table E.1. The approximations provided by Stirling's formula (rounded to one decimal place) are also shown, which we see are quite accurate even for relatively small values of n.

E.5 TAYLOR SERIES

The Taylor series of function $g(x)$ is a representation given by an infinite sum of derivatives at x_o. The general formula is

$$g(x) = \sum_{n=0}^{\infty} \frac{g^{(n)}(x_o)}{n!}(x - x_o)^n, \tag{E.23}$$

where $g^{(n)}(x_o)$ is the nth ordinary derivative of $g(x)$ evaluated at $x = x_o$. The Maclaurin series is a special case of a Taylor series evaluated at $x_o = 0$.

E.6 SERIES EXPANSIONS AND CLOSED-FORM SUMS

- *Exponential:*

$$\exp(x) = \sum_{n=0}^{\infty} \frac{x^n}{n!}, \qquad x \in \mathcal{R}. \tag{E.24}$$

We also include the useful result:

$$\lim_{n\to\infty} (1 + x/n)^n = \exp(x). \tag{E.25}$$

- *Trigonometric:*

$$\cos(x) = \sum_{n=0}^{\infty} (-1)^n \frac{x^{2n}}{(2n)!}, \qquad \sin(x) = \sum_{n=0}^{\infty} (-1)^n \frac{x^{2n+1}}{(2n+1)!}, \qquad x \in \mathcal{R}. \tag{E.26}$$

- *Natural logarithm:*

$$\ln(x + 1) = \sum_{n=0}^{\infty} \frac{(-1)^n}{(n+1)} x^{n+1}, \qquad |x| < 1. \tag{E.27}$$

- *Sum of powers:*

$$\sum_{n=0}^{\infty} x^n = \frac{1}{1-x}, \qquad |x| < 1, \tag{E.28}$$

$$\sum_{n=m}^{\infty} x^n = \frac{x^m}{1-x}, \qquad |x| < 1, \tag{E.29}$$

$$\sum_{n=1}^{\infty} nx^n = \frac{x}{(1-x)^2}, \qquad |x| < 1, \tag{E.30}$$

$$\sum_{n=1}^{\infty} n^2x^n = \frac{x(1+x)}{(1-x)^3}, \qquad |x| < 1, \tag{E.31}$$

$$\sum_{n=1}^{\infty} n^3x^n = \frac{x(1+4x+x^2)}{(1-x)^4}, \qquad |x| < 1. \tag{E.32}$$

- *Finite sums:*

$$\sum_{n=0}^{N} x^n = \frac{1-x^{N+1}}{1-x}, \qquad x \neq 1, \tag{E.33}$$

$$\sum_{n=1}^{N} nx^n = \frac{x}{(1-x)^2}[1-(N+1)x^N + Nx^{N+1}], \qquad x \neq 1, \tag{E.34}$$

$$\sum_{n=1}^{N} n = (1/2)N(N+1), \tag{E.35}$$

$$\sum_{n=1}^{N} n^2 = (1/6)N(N+1)(2N+1), \tag{E.36}$$

$$\sum_{n=1}^{N} n^3 = (1/4)N^2(N+1)^2. \tag{E.37}$$

L'Hôpital's rule can be used to evaluate the closed-form results in (E.33) and (E.34) for $x = 1$, though it easy to see that the first sum is $N + 1$, and the second sum is (E.35).

- *Binomial theorem:*

$$(x+y)^n = \sum_{m=0}^{n} \binom{n}{m} x^{n-m} y^m, \qquad x, y \in \mathcal{R}, \quad n \in \mathcal{Z}^+. \tag{E.38}$$

Special case for $x = y = 1$:

$$\sum_{m=1}^{n} \binom{n}{m} = 2^n. \tag{E.39}$$

For convenience, we provide some binomial coefficients in Table E.2.

TABLE E.2 Binomial Coefficients

n	$\binom{n}{1}$	$\binom{n}{2}$	$\binom{n}{3}$	$\binom{n}{4}$	$\binom{n}{5}$	$\binom{n}{6}$	$\binom{n}{7}$	$\binom{n}{8}$	$\binom{n}{9}$
1	1								
2	2	1							
3	3	3	1						
4	4	6	4	1					
5	5	10	10	5	1				
6	6	15	20	15	6	1			
7	7	21	35	35	21	7	1		
8	8	28	56	70	56	28	8	1	
9	9	36	84	126	126	84	36	9	1
10	10	45	120	210	252	210	120	45	10

- *Generalized binomial theorem:*

$$(x+y)^n = \sum_{m=0}^{\infty} \binom{n}{m} x^{n-m} y^m, \qquad x, y \in \mathcal{R}, |x| > |y|, \quad n \in \mathcal{Z}^+, \tag{E.40}$$

where the upper limit has been extended to infinity. The following are special cases for $|x| < 1$:

$$(1-x)^n = \sum_{m=0}^{\infty} \binom{n}{m} x^m, \qquad (1+x)^n = \sum_{m=0}^{\infty} \binom{n}{m} (-x)^m. \tag{E.41}$$

- *Multinomial theorem:*

$$\left(\sum_{k=1}^{m} x_k\right)^n = \sum_{p_1} \cdots \sum_{p_m} \binom{m}{p_1, \ldots, p_m} x_1^{p_1} \cdots x_m^{p_m} \tag{E.42}$$

such that $p_1 + \cdots + p_m = n$.

- *Harmonic:*

$$\sum_{n=1}^{\infty} \frac{1}{n} = \infty. \tag{E.43}$$

This is a special (diverging) case of the p-series.

- *p-series:*

$$\sum_{n=1}^{\infty} \frac{1}{n^p} = \zeta(p), \qquad p > 1, \tag{E.44}$$

which is the zeta function in Appendix B. For $p = 2$:

$$\sum_{n=1}^{\infty} \frac{1}{n^2} = \pi^2/6. \tag{E.45}$$

E.7 VANDERMONDE'S IDENTITY

Theorem E.1 (Vandermonde). For $m, n, k \in \mathcal{Z}^+$:

$$\binom{m+n}{k} = \sum_{p=0}^{k} \binom{m}{p}\binom{n}{k-p}. \tag{E.46}$$

Proof. Consider the following form of the binomial theorem:

$$\sum_{k=0}^{m+n} \binom{m+n}{k} x^k = (x+1)^{m+n}, \tag{E.47}$$

which is a polynomial in x of degree $m+n$. Expressing the right-hand side above as a product and applying the binomial theorem twice gives

$$\sum_{k=0}^{m+n} \binom{m+n}{k} x^k = (x+1)^m (x+1)^n = \sum_{r=0}^{m} \binom{m}{r} x^r \sum_{q=0}^{n} \binom{n}{q} x^q. \tag{E.48}$$

Collecting terms with like-exponents of x, it is straightforward to rearrange this expression as follows:

$$\sum_{r=0}^{m} \binom{m}{r} x^r \sum_{q=0}^{n} \binom{n}{q} x^q = \sum_{k=0}^{m+n} \sum_{p=0}^{k} \binom{m}{p}\binom{n}{k-p} x^k, \tag{E.49}$$

which gives

$$\sum_{k=0}^{m+n} \binom{m+n}{k} x^k = \sum_{k=0}^{m+n} \sum_{p=0}^{k} \binom{m}{p}\binom{n}{k-p} x^k. \tag{E.50}$$

Note that binomial coefficients where the bottom number exceeds the top number are defined to be zero. Both sides have sums over k that provide the exponents of x in a polynomial of degree $m+n$. Equating coefficients of x^k on both sides gives the result in (E.46), which completes the proof. □

Example E.1. This proof is verified by the following simple case. Consider the product $(x+1)^2(x+1)$ in (E.48) where $m = 2$ and $n = 1$. The last part of the proof in (E.49) is the following expression:

$$\sum_{k=0}^{3} \binom{3}{k} x^k = \sum_{k=0}^{3} \sum_{p=0}^{k} \binom{2}{p}\binom{1}{k-p} x^k. \tag{E.51}$$

The left-hand side is

$$\sum_{k=0}^{3} \binom{3}{k} x^k = \binom{3}{0} x^0 + \binom{3}{1} x^1 + \binom{3}{2} x^2 + \binom{3}{3} x^3 = 1 + 3x + 3x^2 + x^3 \tag{E.52}$$

and the right-hand side is

$$\begin{aligned}\sum_{k=0}^{3}\sum_{p=0}^{k}\binom{2}{p}\binom{1}{k-p}x^k &= \binom{2}{0}\binom{1}{0}x^0 + \left[\binom{2}{0}\binom{1}{1} + \binom{2}{1}\binom{1}{0}\right]x^1 \\ &\quad + \left[\binom{2}{0}\binom{1}{2} + \binom{2}{1}\binom{1}{1} + \binom{2}{2}\binom{1}{0}\right]x^2 \\ &\quad + \left[\binom{2}{0}\binom{1}{3} + \binom{2}{1}\binom{1}{2} + \binom{2}{2}\binom{1}{1} + \binom{2}{3}\binom{1}{0}\right]x^3. \end{aligned} \tag{E.53}$$

Eliminating four binomial coefficients that equal zero yields

$$\begin{aligned}\sum_{k=0}^{3}\sum_{p=0}^{k}\binom{2}{p}\binom{1}{k-p}x^k &= \binom{2}{0}\binom{1}{0}x^0 + \left[\binom{2}{0}\binom{1}{1} + \binom{2}{1}\binom{1}{0}\right]x^1 \\ &\quad + \left[\binom{2}{1}\binom{1}{1} + \binom{2}{2}\binom{1}{0}\right]x^2 + \binom{2}{2}\binom{1}{1}x^3. \end{aligned} \tag{E.54}$$

Thus

$$\sum_{k=0}^{3}\sum_{p=0}^{k}\binom{2}{p}\binom{1}{k-p}x^k = 1 + (1+2)x + (2+1)x^2 + x^3, \tag{E.55}$$

which is the same as (E.52). The first sum on the right-hand side of (E.49) determines the initial set of coefficients for all four powers of x (including exponent zero), and the second sum adds to those coefficients the amounts needed to match the left-hand side. Note also from the right-hand side of (E.48):

$$\sum_{r=0}^{2}\binom{2}{r}x^r \sum_{q=0}^{1}\binom{1}{q}x^q = (1+2x+x^2)(1+x) = 1+3x+3x^2+x^3, \tag{E.56}$$

which is the same as (E.55).

E.8 PMF SUMS AND FUNCTIONAL FORMS

Each probability mass function (pmf) covered in Chapter 3 is either a finite sum or an infinite series whose sum is 1. The elements of a pmf have a functional form as shown in Table E.3 for the various random variables (excluding the discrete uniform pmf), where constants that are independent of x have been dropped. Without these weightings, the functional forms sum to the values in the last column of the table. The first five functions are all based on Bernoulli experiments with probability of success p.

It is the *functional form* that defines the pmf, whereas the inverse of the quantity in the last column of the table is just a weighting included to *normalize* the function so that the pmf sums to 1. For the Poisson random variable, for example, the functional form in the second column of the table is a series for the exponential function. Thus, the inverse of the sum given by $1/\exp(\alpha)$ weights the functional form to give a valid pmf. Likewise for the hypergeometric random variable, the functional form must be scaled by $1/\binom{N}{n}$ to obtain a valid pmf. These pmf sums can be used to find closed-form expressions for various functions of discrete random variables, usually by rearranging the equation so that a recognizable sum is obtained. (A similar approach is used in Appendix D for integrals of probability density functions (pdfs).)

TABLE E.3 Pmf Functional Forms for Discrete Random Variables

Random Variable	Functional Form	Support	Sum
Bernoulli	$(p/q)^x$	$\{0, 1\}$	$1/q$
Bernoulli (symmetric)	$(\sqrt{p/q})^x$	$\{-1, 1\}$	$1/\sqrt{pq}$
Binomial	$\binom{N}{x}(p/q)^x$	$\{0, \ldots, N\}$	$1/q^N$
Geometric	q^x	$\mathcal{Z}^+$	$1/p$
Geometric (shifted)	q^x	$\mathcal{N}$	q/p
Hypergeometric	$\binom{M}{x}\binom{N-M}{n-x}$	$\{0, \ldots, N\}$	$\binom{N}{n}$
Logarithmic	$-p^x/x$	$\mathcal{N}$	$\ln(q)$
Negative binomial	$\binom{x+N-1}{x}q^x$	$\mathcal{Z}^+$	$1/p^N$
Poisson	$\alpha^x/x!$	$\mathcal{Z}^+$	$\exp(\alpha)$
Zeta	$1/x^\alpha$	$\mathcal{N}$	$\zeta(\alpha)$

E.9 COMPLETING THE SQUARE

Consider the quadratic equation

$$g(x) = ax^2 + bx + c, \tag{E.57}$$

which we want to write in the form

$$g(x) = a(x + d_1)^2 + d_2. \tag{E.58}$$

It is easily shown that

$$d_1 = b/2a, \qquad d_2 = c - b^2/4a, \tag{E.59}$$

which is verified by factoring a in (E.57), adding and subtracting $b^2/4a^2$, and rearranging the expression as follows:

$$\begin{aligned} g(x) &= a[x^2 + (b/a)x] + c, \\ &= a[x^2 + (b/a)x + b^2/4a^2 - b^2/4a^2] + c, \\ &= a[x^2 + (b/a)x + b^2/4a^2] + c - b^2/4a, \end{aligned} \tag{E.60}$$

which has the desired squared term:

$$g(x) = a(x + b/2a)^2 + c - b^2/4a. \tag{E.61}$$

This technique is frequently used to rearrange the exponent of a Gaussian pdf.

Example E.2. When integrating a function over the entire range of variable x, often it is convenient to rewrite the integrand as the product of two parts: one in terms of x which has the form of a known pdf, and the other which is independent of x. For example, it may be possible to write

$$\int_{-\infty}^{\infty} g(x)\mathrm{d}x = \int_{-\infty}^{\infty} cf_X(x)\mathrm{d}x = c\int_{-\infty}^{\infty} f_X(x)\mathrm{d}x = c, \tag{E.62}$$

where $f_X(x)$ is a valid pdf, and c is a constant or a function of some variable other than x. The parameters of $f_X(x)$ may also be a function of another variable, as illustrated next. Suppose we want to integrate the following function over x:

$$g(x, y) = \frac{1}{\sqrt{2\pi\sigma^2}} \exp(-(x^2 - 2xy)/2\sigma^2) \tag{E.63}$$

yielding a function of a single variable. Completing the square gives

$$\begin{aligned} g(y) &= \int_{-\infty}^{\infty} \frac{1}{\sqrt{2\pi\sigma^2}} \exp(-[(x-y)^2 - y^2]/2\sigma^2)\mathrm{d}x, \\ &= \exp(-y^2/2\sigma^2)\frac{1}{\sqrt{2\pi\sigma^2}} \int_{-\infty}^{\infty} \exp(-(x-y)^2/2\sigma^2)\mathrm{d}x. \end{aligned} \tag{E.64}$$

Since the last integral is performed over a Gaussian pdf with parameters $\{\mu = y, \sigma\}$, the final result is

$$g(y) = \exp(-y^2/2\sigma^2). \tag{E.65}$$

E.10 SUMMATION BY PARTS

Summation by parts is a technique for evaluating a finite sum of sequences that is analogous to integration by parts used for continuous functions.

Lemma E.1. For sequences $x[k]$ and $y[k]$:

$$\sum_{k=k_1}^{k_2} x[k](y[k+1] - y[k]) = x[k_2]y[k_2+1] - x[k_1-1]y[k_1] - \sum_{k=k_1}^{k_2} y[k](x[k] - x[k-1]). \tag{E.66}$$

Proof. This expression is verified by bringing the sums to the left-hand side and combining them to cancel the $x[k]y[k]$ terms:

$$\sum_{k=k_1}^{k_2} (x[k]y[k+1] - y[k]x[k-1]) = x[k_2]y[k_2+1] - y[k_1]x[k_1-1]. \tag{E.67}$$

The right-hand side follows because all other terms in the sum cancel. □

Note that other forms are possible, such as

$$\sum_{k=k_1}^{k_2} x[k-1](y[k] - y[k-1]) = x[k_2]y[k_2] - x[k_1-1]y[k_1-1] - \sum_{k=k_1}^{k_2} y[k](x[k] - x[k-1]). \tag{E.68}$$

Defining the difference sequences $\Delta y[k] \triangleq y[k] - y[k-1]$ and $\Delta x[k] \triangleq x[k] - x[k-1]$, (E.68) can be written as

$$\sum_{k=k_1}^{k_2} x[k-1]\Delta y[k] = x[k_2]y[k_2] - x[k_1-1]y[k_1-1] - \sum_{k=k_1}^{k_2} y[k]\Delta x[k]. \tag{E.69}$$

This form resembles integration by parts:

$$\int_{t_1}^{t_2} x(t)\mathrm{d}y(t) = x(t_2)y(t_2) - x(t_1)y(t_1) - \int_{t_1}^{t_2} y(t)\mathrm{d}x(t), \tag{E.70}$$

which has been written using Riemann–Stieltjes integrals (see Appendix D).

Example E.3. Summation by parts can be used to verify the finite sum formula in (E.34). Letting $x[k] = k$ and $y[k] = x^k$ gives

$$\begin{aligned} \sum_{k=1}^{N} k(x^{k+1} - x^k) &= (Nx^{N+1} - 0 \cdot x^1) - \sum_{k=1}^{N} x^k[k - (k-1)], \\ &= Nx^{N+1} - \sum_{k=1}^{N} x^k. \end{aligned} \tag{E.71}$$

The sum on the right-hand side is simpler to evaluate because k does not multiply x. Factoring $x - 1$ on the left-hand side and using the finite sum in (E.33) for the right-hand side yields

$$\begin{aligned} (x-1)\sum_{k=1}^{N} kx^k &= Nx^{N+1} - \frac{1 - x^{N+1}}{1-x} + 1, \\ &= \frac{x}{1-x}\left[(N+1)x^N - Nx^{N+1} - 1\right]. \end{aligned} \tag{E.72}$$

Dividing both sides by $x - 1$ and multiplying the numerator and denominator by -1 gives the result in (E.34).

FURTHER READING

Additional information about the material in this appendix can be found in the following references: Beyer (1976), Gradshteyn and Ryzhik (1980), Kreyszig (1979), Sokolnikoff and Redheffer (1966), and Weisstein (2003).

F Inequalities and Bounds for Expectations

In this appendix, we describe several inequalities that provide bounds for various expectations. Some results are provided with proofs.

F.1 CAUCHY–SCHWARZ AND HÖLDER INEQUALITIES

Theorem F.1 (Cauchy–Schwarz inequality). For random variables $\{X, Y\}$ and functions $\{g(X), h(Y)\}$:

$$|\mathcal{E}[g(X)h(Y)]| \leq \sqrt{\mathcal{E}[g^2(X)]}\sqrt{\mathcal{E}[h^2(Y)]}. \tag{F.1}$$

Equality is achieved if and only if $P(g(X) = ah(Y)) = 1$ for some nonrandom a.

Proof. Consider the following mean-square error (MSE):

$$\mathcal{E}[(g(X) - ah(Y))^2] = \mathcal{E}[g^2(X)] + a^2\mathcal{E}[h^2(Y)] - 2a\mathcal{E}[g(X)h(Y)], \tag{F.2}$$

where $a \in \mathcal{R}$ is a constant. Differentiating this expression with respect to a and setting the result equal to zero, the expectation is minimum when

$$a_o = \frac{\mathcal{E}[g(X)h(Y)]}{\mathcal{E}[h^2(Y)]}. \tag{F.3}$$

Parameter a_o is the scalar Wiener solution for this particular MSE (see Chapter 11), and the product $a_o h(Y)$ is the minimum MSE estimator for $g(X)$ given measurement $h(Y)$ (see Chapter 9). Substituting a_o into (F.2) gives

$$\mathcal{E}[g^2(X)] + \frac{(\mathcal{E}[g(X)h(Y)])^2}{\mathcal{E}[h^2(Y)]} - 2\frac{(\mathcal{E}[g(X)h(Y)])^2}{\mathcal{E}[h^2(Y)]} \geq 0 \tag{F.4}$$

and

$$\mathcal{E}[g^2(X)] \geq \frac{(\mathcal{E}[g(X)h(Y)])^2}{\mathcal{E}[h^2(Y)]}. \tag{F.5}$$

Rearranging this expression, taking the square root of each side, and retaining the magnitude $|\mathcal{E}[g(X)h(Y)]|$ completes the proof. In order to have equality, let the functions be proportional $g(X) = ah(Y)$ with real-valued constant a; the left-hand side of (F.1) becomes

$$|\mathcal{E}[g(X)h(Y)]| = a\mathcal{E}[h^2(Y)] \tag{F.6}$$

Probability, Random Variables, and Random Processes: Theory and Signal Processing Applications, First Edition. John J. Shynk.

and the right-hand side is

$$\sqrt{\mathcal{E}[g^2(X)]\mathcal{E}[h^2(Y)]} = \sqrt{a^2(\mathcal{E}[h^2(Y)])^2} = a\mathcal{E}[h^2(Y)], \tag{F.7}$$

thus achieving equality. If the MSE in (F.2) is zero, then $P(g(X) = ah(Y)) = 1$. □

Example F.1. Consider the specific functions $g(X) = (X - \mu_X)/\sigma_X$ and $h(Y) = (Y - \mu_Y)/\sigma_Y$, where $\{\mu_X, \mu_Y\}$ and $\{\sigma_X^2, \sigma_Y^2\}$ are the means and variances of random variables $\{X, Y\}$. Then

$$\mathcal{E}[g(X)h(Y)] = \frac{C_{XY}}{\sigma_X \sigma_Y} \triangleq \rho_{XY} \tag{F.8}$$

is the correlation coefficient discussed in Chapter 5. It immediately follows from the Cauchy–Schwarz inequality that

$$|\rho_{XY}| \leq \sqrt{\mathcal{E}[g^2(X)]\mathcal{E}[h^2(Y)]} = 1. \tag{F.9}$$

The Cauchy–Schwarz inequality is a special case of Hölder's inequality, which we state without proof.

Theorem F.2 (Hölder's inequality). For random variables $\{X, Y\}$ and functions $\{g(X), h(Y)\}$:

$$\mathcal{E}[|g(X)h(Y)|] \leq (\mathcal{E}[|g(X)|^p])^{1/p}(\mathcal{E}[|h(Y)|^q])^{1/q}, \tag{F.10}$$

where $p, q \geq 1$ and $1/p + 1/q = 1$. Equality is achieved if and only if $P(|g(X)|^p = a|h(Y)|^q) = 1$ for some nonrandom a.

Note that the absolute value on the left-hand side is inside the expectation. For $p = q = 2$:

$$\mathcal{E}[|g(X)h(Y)|] \leq \sqrt{\mathcal{E}[|g(X)|^2]}\sqrt{\mathcal{E}[|h(Y)|^2]}, \tag{F.11}$$

which is another form of the Cauchy–Schwarz inequality since $|\mathcal{E}[g(X)h(Y)]| \leq \mathcal{E}[|g(X)h(Y)|]$.

F.2 TRIANGLE AND MINKOWSKI INEQUALITIES

Theorem F.3 (Triangle inequality). For random variables $\{X, Y\}$:

$$\sqrt{\mathcal{E}[(X+Y)^2]} \leq \sqrt{\mathcal{E}[X^2]} + \sqrt{\mathcal{E}[Y^2]}. \tag{F.12}$$

Proof. Expanding the left-hand side under the square root gives

$$\begin{aligned} \mathcal{E}[(X+Y)^2] &\leq \mathcal{E}[|X+Y|(|X|+|Y|)] = \mathcal{E}[|X+Y||X|] + \mathcal{E}[|X+Y||Y|], \\ &\leq \sqrt{\mathcal{E}[(X+Y)^2]\mathcal{E}[X^2]} + \sqrt{\mathcal{E}[(X+Y)^2]\mathcal{E}[Y^2]}, \end{aligned} \tag{F.13}$$

where the Cauchy–Schwarz inequality has been applied to both terms on the right-hand side. Dividing both sides by $\sqrt{\mathcal{E}[(X+Y)^2]}$ completes the proof. □

The triangle inequality is a special case of Minkowski's inequality, which we state without proof.

Theorem F.4 (Minkowski's inequality). For random variables $\{X, Y\}$ and $p \geq 1$:

$$(\mathcal{E}[|X+Y|^p])^{1/p} \leq (\mathcal{E}[|X|^p])^{1/p} + (\mathcal{E}[|Y|^p])^{1/p}. \tag{F.14}$$

Another form of the triangle inequality is obtained from (F.14) with $p = 1$:

$$\mathcal{E}[|X+Y|] \leq \mathcal{E}[|X|] + \mathcal{E}[|Y|]. \tag{F.15}$$

F.3 BIENAYMÉ, CHEBYSHEV, AND MARKOV INEQUALITIES

Theorem F.5 (Bienaymé's inequality). For random variable X and nonrandom c:

$$P(|X-c| \geq \delta) \leq (1/\delta^n)\mathcal{E}[|X-c|^n] \tag{F.16}$$

for every $\delta > 0$ and $n \in \mathcal{Z}^+$.

Proof. Define the translated random variable $Y \triangleq |X-c|^n$ with probability density function (pdf) $f_Y(y)$. Since Y and $f_Y(y)$ are nonnegative:

$$\mathcal{E}[|X-c|^n] = \int_0^\infty y f_Y(y)\mathrm{d}y \geq \int_{\delta^n}^\infty y f_Y(y)\mathrm{d}y, \tag{F.17}$$

where the lower limit has been increased from zero. Using the smallest value of y in the integrand and factoring it from the integral gives

$$\begin{aligned}\mathcal{E}[|X-c|^n] &\geq \delta^n \int_{\delta^n}^\infty f_Y(y)\mathrm{d}y, \\ &= \delta^n P(|X-c|^n \geq \delta^n) = \delta^n P(|X-c| \geq \delta).\end{aligned} \tag{F.18}$$

Rearranging the last expression completes the proof. □

Theorem F.6 (Chebyshev's inequality). For random variable X with mean μ_X and variance σ_X^2:

$$P(|X-\mu_X| \geq \delta) \leq \sigma_X^2/\delta^2 \tag{F.19}$$

for every $\delta > 0$.

Proof. Chebyshev's inequality is a special case of Bienaymé's inequality with $n = 2$ and $c = \mu_X$. □

This inequality is a precise way of saying that "most" values of X are close to the mean of the underlying pdf. The result is intuitive because there is a connection between (i) the probability that a random variable exceeds some distance from its mean and (ii) the width (variance) of the pdf. Figure F.1 illustrates this connection for a Gaussian random variable.

Theorem F.7. For nonnegative function $g(x)$, a generalized form of Bienaymé's inequality is

$$P(g(X) \geq \delta) \leq \mathcal{E}[g(X)]/\delta \tag{F.20}$$

for every $\delta > 0$.

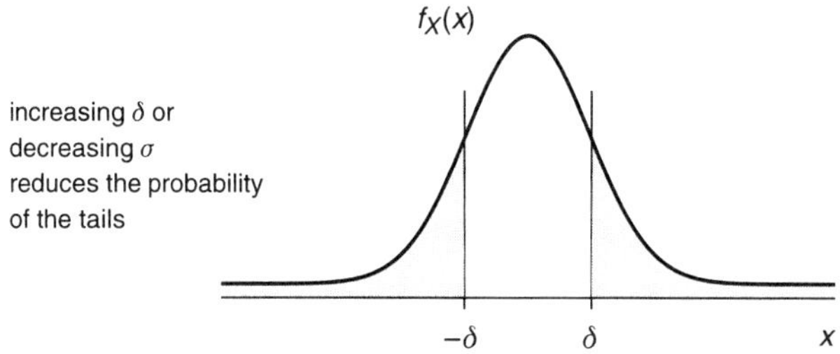

FIGURE F.1 Chebyshev's inequality relates the probability of the tails of a distribution to their distance from the mean, shown for a Gaussian random variable with parameters $\{\mu = 0, \sigma = \sqrt{2}\}$.

Proof. From the expectation of a function:

$$\begin{aligned} \mathcal{E}[g(X)] &= \int_{-\infty}^{\infty} g(x) f_X(x) \mathrm{d}x, \\ &= \int_{-\infty}^{g(x)<\delta} g(x) f_X(x) \mathrm{d}x + \int_{g(x)\geq\delta}^{\infty} g(x) f_X(x) \mathrm{d}x. \end{aligned} \tag{F.21}$$

Both integrals are nonnegative because $g(x)$ and $f_X(x)$ are nonnegative. Thus

$$\begin{aligned} \mathcal{E}[g(X)] &\geq \int_{g(x)\geq\delta}^{\infty} g(x) f_X(x) \mathrm{d}x, \\ &\geq \int_{g(x)\geq\delta}^{\infty} \delta f_X(x) \mathrm{d}x = \delta P(g(X) \geq \delta), \end{aligned} \tag{F.22}$$

where $g(x)$ has been replaced with its smallest value δ. Rearranging the last expression completes the proof. □

It is important to note that $g(x)$ must be a nonnegative function in the previous inequality.

Theorem F.8 (Markov's inequality). For random variable X:

$$P(|X| \geq \delta) \leq \mathcal{E}[|X|]/\delta, \tag{F.23}$$

and for nonnegative random variable Y:

$$P(Y \geq \delta) \leq \mathcal{E}[Y]/\delta \tag{F.24}$$

both for every $\delta > 0$.

Proof. Markov's inequality is another special case of Bienaymé's inequality with $n = 1$ and $c = 0$. □

A generalization of Markov's inequality is

$$P(|X| \geq \delta) \leq \mathcal{E}[|X|^n]/\delta^n \tag{F.25}$$

for $n \in \mathcal{Z}^+$.

F.4 CHERNOFF'S INEQUALITY

Theorem F.9 (Chernoff's inequality). For random variable X and every $\delta > 0$:

$$P(X \geq \delta) \leq \exp(-t\delta)\mathcal{E}[\exp(tX)], \tag{F.26}$$

where $\mathcal{E}[\exp(tX)] \triangleq m_X(t)$ is the moment generating function of X and $t \geq 0$.

Proof. From the definition of the moment generating function:

$$\begin{aligned} m_X(t) &= \int_{-\infty}^{\delta} \exp(tx) f_X(x)\mathrm{d}x + \int_{\delta}^{\infty} \exp(tx) f_X(x)\mathrm{d}x, \\ &\geq \int_{\delta}^{\infty} \exp(tx) f_X(x)\mathrm{d}x, \\ &\geq \exp(t\delta) \int_{\delta}^{\infty} f_X(x)\mathrm{d}x = \exp(t\delta)P(X \geq \delta). \end{aligned} \tag{F.27}$$

The first inequality is obtained because $\exp(tx) \geq 0$, and the second inequality follows because the smallest value $\exp(t\delta)$ has replaced $\exp(tx)$ in the integrand. Rearranging the final expression completes the proof. □

Since Chernoff's bound must hold for all $t \geq 0$, the tightest bound is

$$\begin{aligned} P(X \geq \delta) &\leq \min_{t \geq 0} \left(\exp(-t\delta) m_X(t)\right), \\ &= \min_{t \geq 0} \left(\exp(-t\delta + \ln(m_X(t)))\right), \end{aligned} \tag{F.28}$$

which depends on δ and the parameters of the specific moment generating function.

Example F.2. Consider the exponential random variable with parameter λ for which

$$P(X \geq \delta) = \lambda \int_{\delta}^{\infty} \exp(-\lambda x)\mathrm{d}x = \exp(-\lambda\delta). \tag{F.29}$$

Since X is nonnegative, Markov's inequality with $\mathcal{E}[X] = 1/\lambda$ becomes

$$P(X \geq \delta) \leq 1/\lambda\delta. \tag{F.30}$$

Chernoff's inequality is

$$P(X \geq \delta) \leq \min_{t \geq 0} \left(\exp(-t\delta + \ln(\lambda) - \ln(\lambda - t))\right), \tag{F.31}$$

where $m_X(t) = \lambda/(\lambda - t)$ for the exponential random variable. Differentiating the expression in the exponent with respect to t and setting the result equal to zero gives the "optimal" t_o:

$$-\delta + 1/(\lambda - t_o) = 0 \Longrightarrow t_o = \lambda - 1/\delta. \tag{F.32}$$

Note that $t_o \geq 0$ only when $\delta \geq 1/\lambda$; for $\delta < 1/\lambda$, $t_o = 0$ is used in the bound.

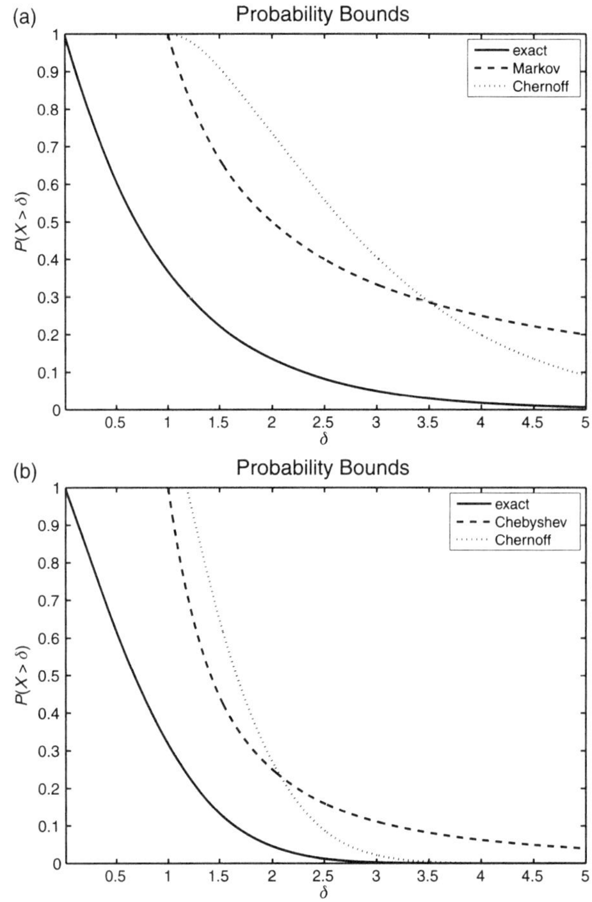

FIGURE F.2 Probability bounds for Example F.2. (a) Exponential distribution with $\lambda = 1$. (b) Standard Gaussian distribution with $\{\mu = 0, \sigma = 1\}$.

Substituting these results into Chernoff's inequality yields

$$P(X \geq \delta) \leq \begin{cases} \lambda\delta \exp(1 - \lambda\delta), & \delta \geq 1/\lambda \\ 1, & 0 \leq \delta < 1/\lambda. \end{cases} \tag{F.33}$$

The bounds and the exact probability are plotted in Figure F.2(a) versus δ for $\lambda = 1$. Observe that the Markov bound is not restricted to be less than 1, and the Chernoff bound is tighter for larger values of δ.
Results for the standard Gaussian random variable are shown in Figure F.2(b). The corresponding equations are (using Chebyshev instead of Markov)

$$P(|X| \geq \delta) = 2Q(\delta) \quad \text{(exact)}, \tag{F.34}$$

$$P(|X| \geq \delta) \leq 1/\delta^2 \quad \text{(Chebyshev)}, \tag{F.35}$$

$$P(|X| \geq \delta) \leq 2 \min_{t \geq 0}(\exp(-t\delta + t^2/2)) \quad \text{(Chernoff)}, \tag{F.36}$$

where $Q(\cdot)$ is the Q-function. Note that a factor of two is included in the exact result and the Chernoff bound because the probability involves $|X|$, so that both tails for this symmetric distribution are considered. The last equation is minimum when $t = \delta$, so that the Chernoff bound used in the plot is

$$P(|X| \geq \delta) \leq 2\exp(-\delta^2/2). \tag{F.37}$$

F.5 JENSEN'S INEQUALITY

Theorem F.10 (Jensen's inequality). Let X be a random variable defined on an open interval with cdf $F_X(x)$, and let $g(x)$ be a convex function. Then

$$g\left(\mathcal{E}[X]\right) \leq \mathcal{E}[g(X)]. \tag{F.38}$$

Proof. From the definition of a convex function (see Appendix B), there exists a line $c(x) = a(x - x_o) + g(x_o)$ such that $c(x) \leq g(x)$ for all x in the open interval. As a result for random variable X:

$$\mathcal{E}[c(X)] \leq \mathcal{E}[g(X)]. \tag{F.39}$$

Letting $x_o = \mathcal{E}[X]$ in $c(X)$ gives

$$\mathcal{E}[c(X)] = a(\mathcal{E}[X] - \mathcal{E}[X]) + g(\mathcal{E}[X]) = g(\mathcal{E}[X]). \tag{F.40}$$

Substituting this result into (F.39) completes the proof. If $g(x)$ is strictly convex, then (F.38) has strict inequality except when $g(X) = \mathcal{E}[g(X)]$ with probability one. □

If $g(x)$ is instead a concave function, the inequality in (F.38) is reversed. The following example gives a result in information theory (see Chapter 10) that is also useful for the expectation-maximization algorithm described in Chapter 9.

Example F.3. Consider random variable X with pdf $f_X(x)$ and let $g(x)$ be a nonnegative function (which may or may not be a pdf). Define random variable $Y \triangleq g(X)/f_X(X)$. Although $f_X(x)$ is the pdf of X, it is used here as a *function* of random variable X. Recall from Chapter 5 that it is not necessary to find $f_Y(y)$ in order to compute $\mathcal{E}[Y]$; instead, we can write

$$\mathcal{E}[Y] = \int_{-\infty}^{\infty} y(x) f_X(x) \mathrm{d}x. \tag{F.41}$$

Substituting $y(x) = g(x)/f_X(x)$, this expression simplifies to

$$\mathcal{E}[g(X)/f_X(X)] = \int_{-\infty}^{\infty} g(x) \mathrm{d}x, \tag{F.42}$$

which is just the area under $g(x)$. Consider next the convex function $\log(Y)$ such that

$$\mathcal{E}[\log(g(X)/f_X(X))] = \int_{-\infty}^{\infty} \log(g(x)/f_X(x)) f_X(x) \mathrm{d}x. \tag{F.43}$$

Jensen's inequality yields the following lower bound:

$$\mathcal{E}[\log(g(X)/f_X(X))] \geq \log(\mathcal{E}[g(X)/f_X(X)]) = \log\left(\int_{-\infty}^{\infty} g(x)\mathrm{d}x\right), \tag{F.44}$$

where (F.42) has been used. If $g(x)$ happens to be a pdf, then (F.42) equals 1 and the bound becomes

$$\mathcal{E}[\log(g(X)/f_X(X))] \geq 0. \tag{F.45}$$

F.6 CRAMÉR–RAO INEQUALITY

The Cramér–Rao inequality provides a lower bound on the variance of an estimator; it is also called the Cramér-Rao lower bound (CRLB). The CRLB is useful as a performance measure for biased and unbiased estimators of the parameters of a distribution, as described in Chapter 9. We provide a derivation for an unbiased estimator, and state the results for a biased estimator and then for a vector of parameters.

Theorem F.11 (Cramér–Rao inequality). Let T be an unbiased estimator for the function $g(\theta)$ of scalar parameter θ, where $\{X_1, \ldots, X_N\}$ are iid samples with pdf $f_X(x;\theta)$. Let $T = t$ be a specific value of the estimator for outcomes $\{x_1, \ldots, x_N\}$, and assume the following regularity conditions:

- Differentiation with respect to the parameter:

$$\frac{\partial}{\partial\theta}\log(f_X(x;\theta)) \text{ exists for all } x \text{ and } \theta. \tag{F.46}$$

- Differentiation and integration exchange I:

$$\frac{\partial}{\partial\theta}\int_{-\infty}^{\infty}\cdots\int_{-\infty}^{\infty}\prod_{n=1}^{N} f(x_n;\theta)dx_n = \int_{-\infty}^{\infty}\cdots\int_{-\infty}^{\infty}\frac{\partial}{\partial\theta}\prod_{n=1}^{N} f(x_n;\theta)dx_n. \tag{F.47}$$

- Differentiation and integration exchange II:

$$\frac{\partial}{\partial\theta}\int_{-\infty}^{\infty}\cdots\int_{-\infty}^{\infty}\prod_{n=1}^{N} tf(x_n;\theta)dx_n = \int_{-\infty}^{\infty}\cdots\int_{-\infty}^{\infty} t\frac{\partial}{\partial\theta}\prod_{n=1}^{N} f(x_n;\theta)dx_n. \tag{F.48}$$

- Bounded positive expectation:

$$0 < \mathcal{E}\left[\left(\frac{\partial}{\partial\theta}\log(f_X(X;\theta))\right)^2\right] < \infty. \tag{F.49}$$

The variance of an unbiased estimator T of θ is bounded from below according to

$$\text{var}[T] \geq \frac{\left(\frac{\partial}{\partial\theta}g(\theta)\right)^2}{N\mathcal{E}\left[\left(\frac{\partial}{\partial\theta}\log(f_X(X;\theta))\right)^2\right]}. \tag{F.50}$$

Note that random variable X appears in the argument of the pdf under the expectation.

Proof. Since T is an unbiased estimator of $g(\theta)$, we can write

$$g(\theta) = \int_{-\infty}^{\infty}\cdots\int_{-\infty}^{\infty} t\prod_{n=1}^{N} f(x_n;\theta)\mathrm{d}x_n. \tag{F.51}$$

Differentiating gives

$$\frac{\partial}{\partial\theta}g(\theta) = \int_{-\infty}^{\infty}\cdots\int_{-\infty}^{\infty} t\frac{\partial}{\partial\theta}\prod_{n=1}^{N} f(x_n;\theta)\mathrm{d}x_n. \tag{F.52}$$

Since

$$\int_{-\infty}^{\infty}\cdots\int_{-\infty}^{\infty}\prod_{n=1}^{N} f(x_n;\theta)\mathrm{d}x_n = 1, \tag{F.53}$$

(F.52) can be augmented as follows:

$$\begin{aligned}\frac{\partial}{\partial\theta}g(\theta) &= \int_{-\infty}^{\infty}\cdots\int_{-\infty}^{\infty} t\frac{\partial}{\partial\theta}\prod_{k=n}^{N} f(x_n;\theta)\mathrm{d}x_n \\ &\quad - g(\theta)\frac{\partial}{\partial\theta}\int_{-\infty}^{\infty}\cdots\int_{-\infty}^{\infty}\prod_{n=1}^{N} f(x_n;\theta)\mathrm{d}x_n,\end{aligned} \tag{F.54}$$

where only zero has been subtracted because it is the derivative of the constant in (F.53). Thus

$$\frac{\partial}{\partial\theta}g(\theta) = \int_{-\infty}^{\infty}\cdots\int_{-\infty}^{\infty}[t - g(\theta)]\frac{\partial}{\partial\theta}\prod_{n=1}^{N} f(x_n;\theta)\mathrm{d}x_n. \tag{F.55}$$

For arbitrary function $c(\theta)$, we have the following useful identity:

$$\frac{\partial}{\partial\theta}\log(c(\theta)) = \frac{1}{c(\theta)}\frac{\partial}{\partial\theta}c(\theta) \Longrightarrow \frac{\partial}{\partial\theta}c(\theta) = c(\theta)\frac{\partial}{\partial\theta}\log(c(\theta)). \tag{F.56}$$

Using this property in (F.55) yields

$$\begin{aligned}\frac{\partial}{\partial\theta}g(\theta) &= \int_{-\infty}^{\infty}\cdots\int_{-\infty}^{\infty}[t - g(\theta)]\frac{\partial}{\partial\theta}\log\left(\prod_{n=1}^{N} f(x_n;\theta)\right)\prod_{n=1}^{N} f(x_n;\theta)\mathrm{d}x_n, \\ &= \mathcal{E}\left[[T - g(\theta)]\frac{\partial}{\partial\theta}\log\left(\prod_{k=n}^{N} f(X_n;\theta)\right)\right].\end{aligned} \tag{F.57}$$

From the Cauchy–Schwarz inequality:

$$\left(\frac{\partial}{\partial\theta}g(\theta)\right)^2 \leq \mathcal{E}\left[(T - g(\theta))^2\right]\mathcal{E}\left[\left(\frac{\partial}{\partial\theta}\log\left(\prod_{n=1}^{N} f(X_n;\theta)\right)\right)^2\right], \tag{F.58}$$

where the first expectation on the right-hand side is the variance of the unbiased estimator T of $g(\theta)$. Since the samples $\{X_1, \ldots, X_N\}$ are iid:

$$\mathcal{E}\left[\left(\frac{\partial}{\partial\theta}\log\left(\prod_{n=1}^{N} f(X_n;\theta)\right)\right)^2\right] = N\mathcal{E}\left[\left(\frac{\partial}{\partial\theta}\log(f(X;\theta))\right)^2\right]. \tag{F.59}$$

Substituting this expression into (F.58) and rearranging the result completes the proof. □

It is important to note that the CRLB does *not* depend on the particular unbiased estimator T.

Definition: Fisher Information The denominator of (F.50) is known as the *Fisher information*:

$$\mathcal{I}(\theta) \triangleq N\mathcal{E}\left[\left(\frac{\partial}{\partial\theta}\log(f_X(X;\theta))\right)^2\right]. \tag{F.60}$$

The base of the logarithm is not specified above, but most often base e is used.

Theorem F.12. If the previous regularity conditions extend to the second derivative, the Fisher information can also be expressed as follows:

$$\mathcal{I}(\theta) = -N\mathcal{E}\left[\frac{\partial^2}{\partial\theta^2}\log(f_X(X;\theta))\right]. \tag{F.61}$$

Proof. Consider the following derivative:

$$N\frac{\partial}{\partial\theta}\int_{-\infty}^{\infty} f(x;\theta)\mathrm{d}x = 0, \tag{F.62}$$

which is zero because the integration is over a valid pdf. Exchanging integration and differentiation and using the identity in (F.56) with $c(\theta) = f(x;\theta)$ gives

$$N\mathcal{E}\left[\frac{\partial}{\partial\theta}\log(f(X;\theta))\right] = 0. \tag{F.63}$$

Differentiating again and using the chain rule yields

$$N\mathcal{E}\left[\frac{\partial^2}{\partial\theta^2}\log(f(X;\theta))\right] + N\mathcal{E}\left[\left(\frac{\partial}{\partial\theta}\log(f(X;\theta))\right)^2\right] = 0. \tag{F.64}$$

The second term is the Fisher information in (F.60), thus completing the proof. □

Example F.4. For the exponential random variable with parameter λ, let the sample mean with N samples be an estimator for $g(\lambda) = 1/\lambda$. The Fisher information is

$$\mathcal{I}(\lambda) = -N\mathcal{E}\left[\frac{\partial^2}{\partial\lambda^2}\ln(\lambda\exp(-\lambda X))\right] = N/\lambda^2, \tag{F.65}$$

and since

$$\frac{\partial}{\partial\lambda}g(\lambda) = -1/\lambda^2 \tag{F.66}$$

the CRLB is

$$\mathrm{var}[T] \geq \frac{1}{N\lambda^2}. \tag{F.67}$$

The variance of the sample mean is

$$\mathcal{E}[(\bar{X} - 1/\lambda)^2] = \mathcal{E}[\bar{X}^2] - (2/\lambda)\mathcal{E}[\bar{X}] + 1/\lambda^2 = \mathcal{E}[\bar{X}^2] - 1/\lambda^2, \tag{F.68}$$

where $\mathcal{E}[\bar{X}] = 1/\lambda$ has been substituted. Since the samples are iid,

$$\begin{aligned} \mathcal{E}[\bar{X}^2] &= \frac{1}{N^2}\left(N\mathcal{E}[X^2] + (N^2 - N)(\mathcal{E}[X])^2\right), \\ &= \frac{1}{N}\left[2/\lambda^2 + (N-1)/\lambda^2\right] = \frac{N+1}{N\lambda^2}. \end{aligned} \tag{F.69}$$

Combining this result with (F.68), we see that the sample mean achieves the CRLB in (F.67); $T = \bar{X}$ is the uniformly minimum-variance unbiased (UMVU) estimator of $g(\lambda)$ (see Chapter 9).

Theorem F.13. For biased estimator T of $g(\theta)$ with bias $B(\theta)$ such that

$$\mathcal{E}[T] = g(\theta) + B(\theta), \tag{F.70}$$

the CRLB is

$$\text{var}[T] \geq \frac{\left[\dfrac{\partial}{\partial\theta}(g(\theta) + B(\theta))\right]^2}{N\mathcal{E}\left[\left(\dfrac{\partial}{\partial\theta}\log(f_X(X;\theta))\right)^2\right]}. \tag{F.71}$$

Proof. See Problem 9.35. □

Finally, consider a vector of M parameters $\boldsymbol{\theta} \triangleq [\theta_1, \ldots, \theta_M]^T$ and let $\mathbf{g}(\boldsymbol{\theta}) \triangleq [g_1(\boldsymbol{\theta}), \ldots, g_K(\boldsymbol{\theta})]^T$ be a vector of functions of the parameters. Define the following vector of K estimators for the components of $\mathbf{g}(\boldsymbol{\theta})$: $\mathbf{T} \triangleq [T_1, \ldots, T_K]^T$, and assume it is unbiased: $\mathcal{E}[\mathbf{T}] = \mathbf{g}(\boldsymbol{\theta})$. Let $\mathbf{J} \in \mathcal{R}^{K\times M}$ be the following Jacobian matrix (see Appendix G):

$$\mathbf{J} \triangleq \frac{\partial}{\partial\boldsymbol{\theta}}\mathbf{g}(\boldsymbol{\theta}) = \begin{bmatrix} \dfrac{\partial}{\partial\theta_1}g_1(\boldsymbol{\theta}) & \cdots & \dfrac{\partial}{\partial\theta_M}g_1(\boldsymbol{\theta}) \\ \vdots & & \vdots \\ \dfrac{\partial}{\partial\theta_1}g_K(\boldsymbol{\theta}) & \cdots & \dfrac{\partial}{\partial\theta_M}g_K(\boldsymbol{\theta}) \end{bmatrix}. \tag{F.72}$$

Theorem F.14. For unbiased estimator $\mathbf{T}$ of parameter vector $\mathbf{g}(\boldsymbol{\theta})$:

$$\text{cov}[\mathbf{T}] \geq \mathbf{J}\mathcal{I}^{-1}(\boldsymbol{\theta})\mathbf{J}^T, \tag{F.73}$$

where

$$\mathcal{I}(\boldsymbol{\theta}) = \mathcal{E}\left[\frac{\partial}{\partial\boldsymbol{\theta}}\log(f(X;\boldsymbol{\theta}))\left(\frac{\partial}{\partial\boldsymbol{\theta}}\log(f(X;\boldsymbol{\theta}))\right)^T\right], \tag{F.74}$$

$$= -\mathcal{E}\left[\frac{\partial^2}{\partial\boldsymbol{\theta}^2}\log(f(X;\boldsymbol{\theta}))\right] \tag{F.75}$$

is the Fisher information matrix $\in \mathcal{R}^{M\times M}$.

The inequality in (F.73) is defined in the positive semidefinite (PSD) sense, meaning that

$$\mathbf{a}^T \operatorname{cov}[\mathbf{T}]\mathbf{a} \geq \mathbf{a}^T \mathbf{J}\boldsymbol{\mathcal{I}}^{-1}(\boldsymbol{\theta})\mathbf{J}^T \mathbf{a} \tag{F.76}$$

for every nonzero vector **a**. Equivalently, the matrix obtained by subtracting the right-hand side from the left-hand side in (F.73) is PSD.

FURTHER READING

The inequalities in this appendix are discussed in some of the references at the end of Chapters 5, 9, and 10. These include Lehmann (1983), McEliece (1977), Mendel (1995), Mood, Graybill, and Boes (1974), and Papoulis (1965).

G Matrix and Vector Properties

In this final appendix, some important properties of matrices and vectors are summarized. We focus on quantities with real-valued elements, but also provide some results for complex matrices and vectors.

G.1 BASIC PROPERTIES

Consider matrix $\mathbf{A} \in \mathcal{R}^{M \times N}$ with N columns $\{\mathbf{a}_n\}$ and M rows $\{\tilde{\mathbf{a}}_m^T\}$ as follows:

$$\mathbf{A} = [\mathbf{a}_1, \ldots, \mathbf{a}_N] = \begin{bmatrix} \tilde{\mathbf{a}}_1^T \\ \vdots \\ \tilde{\mathbf{a}}_M^T \end{bmatrix}, \tag{G.1}$$

where the superscript ^{T} denotes matrix/vector transpose. The scalar elements of $\mathbf{A}$ are denoted by $\{a_{mn}\}$. All vectors in this book are defined to be *column* vectors; row vectors are obtained by using T.

Definition: Linearly Dependent The columns of $\mathbf{A}$ are *linearly dependent* if there exist nonzero $\{x_n\}$ such that

$$\sum_{n=1}^{N} x_n \mathbf{a}_n = \mathbf{A}\mathbf{x} = \mathbf{0}, \tag{G.2}$$

where $\mathbf{x} \triangleq [x_1, \ldots, x_N]^T$ and $\mathbf{0} \in \mathcal{R}^M$ is a column vector of zeros. Otherwise, they are *linearly independent.*

A similar definition applies to the rows of $\mathbf{A}$.

Definition: Rank The *rank* r of matrix $\mathbf{A}$ is the number of linearly independent columns, which is also the number of linearly independent rows.

Obviously $r \leq \min\{M, N\}$. Assume for the next set of definitions that $\mathbf{A} \in \mathcal{R}^{N \times N}$ is a square matrix with $M = N$.

Definition: Symmetric Matrix $\mathbf{A}$ is *symmetric* if $\mathbf{A} = \mathbf{A}^T$ and is *skew symmetric* if $\mathbf{A}^T = -\mathbf{A}$.

For matrices of appropriate dimensions:

$$(\mathbf{A}_1\mathbf{A}_2)^T = \mathbf{A}_2^T\mathbf{A}_1^T. \tag{G.3}$$

Probability, Random Variables, and Random Processes: Theory and Signal Processing Applications, First Edition. John J. Shynk.

Definition: Nonsingular Matrix $\mathbf{A}$ is *nonsingular* if it has an inverse $\mathbf{A}^{-1}$ such that

$$\mathbf{A}\mathbf{A}^{-1} = \mathbf{A}^{-1}\mathbf{A} = \mathbf{I}, \tag{G.4}$$

where $\mathbf{I} \in \mathcal{R}^{N\times N}$ is the identity matrix. If $\mathbf{A}$ does not have an inverse, it is *singular*.

Nonsingular matrix $\mathbf{A}$ has the following properties:

- $(\mathbf{A}^T)^{-1} = (\mathbf{A}^{-1})^T \triangleq \mathbf{A}^{-T}$.
- $\mathbf{A}\mathbf{x} = \mathbf{y}$ has a unique solution given by $\mathbf{x} = \mathbf{A}^{-1}\mathbf{y}$.
- The determinant is nonzero: $\det(\mathbf{A}) \neq 0$.
- It has full rank $r = N$, and so the columns are linearly independent, and likewise for the rows.
- For nonsingular matrices of appropriate dimensions:

$$(\mathbf{A}_1\mathbf{A}_2)^{-1} = \mathbf{A}_2^{-1}\mathbf{A}_1^{-1}. \tag{G.5}$$

Definition: Positive Definite Matrix $\mathbf{A}$ is *positive definite* (PD) if $\mathbf{x}^T\mathbf{A}\mathbf{x} > 0$ for all nonzero $\mathbf{x} \in \mathcal{R}^N$. Likewise, matrix $\mathbf{A}$ is *positive semidefinite* (PSD), *negative definite*, or *negative semidefinite* if $\mathbf{x}^T\mathbf{A}\mathbf{x} \geq 0$, $\mathbf{x}^T\mathbf{A}\mathbf{x} < 0$, or $\mathbf{x}^T\mathbf{A}\mathbf{x} \leq 0$, respectively, for all nonzero $\mathbf{x} \in \mathcal{R}^N$.

Definition: Toeplitz and Circulant Matrix $\mathbf{A}$ is *Toeplitz* if the elements of each descending diagonal (from left to right) are identical. Toeplitz matrix $\mathbf{A}$ is *circulant* if each row is obtained by circularly shifting the previous row to the right by one element.

A Toeplitz matrix has at most N unique elements: the elements of any row or column specify the entire matrix.

Definition: Idempotent Matrix $\mathbf{A}$ is *idempotent* when $\mathbf{A}^2 = \mathbf{A}$.

Definition: Orthogonal Matrix $\mathbf{A}$ is *orthogonal* if $\mathbf{A}^T\mathbf{A} = \mathbf{A}\mathbf{A}^T = \mathbf{I}$ such that $\mathbf{A}^{-1} = \mathbf{A}^T$.

Definition: Nilpotent Matrix $\mathbf{A}$ is *nilpotent* when $\mathbf{A}^K = \mathbf{0}$ for some $K \in \mathcal{N}$.

All triangular matrices with zeros on the main diagonal are nilpotent. However, this is not a requirement; it is possible for all elements of a nilpotent matrix to be nonzero, but they must have some pattern in order for terms to eventually cancel when the matrix is raised to power K.

Example G.1. The autocorrelation matrix of random sequence vector $\mathbf{X}[k]$ is defined as the following expectation:

$$\mathbf{R}_{\mathbf{XX}} \triangleq \mathcal{E}[\mathbf{X}[k]\mathbf{X}^T[k]], \tag{G.6}$$

where $\mathbf{X}[k] \triangleq [X[k], \ldots, X[k-N+1]]^T$ contains N samples of the sequence. It is a symmetric matrix:

$$\mathbf{R}_{\mathbf{XX}}^T = \mathcal{E}^T[\mathbf{X}[k]\mathbf{X}^T[k]] = \mathcal{E}[\mathbf{X}[k]\mathbf{X}^T[k]] = \mathbf{R}_{\mathbf{XX}}, \tag{G.7}$$

and its eigenvalues are *real-valued*. The autocorrelation matrix is also PSD since for every nonzero vector $\mathbf{v}$:

$$\mathbf{v}^T\mathbf{R}_{\mathbf{XX}}\mathbf{v} = \mathcal{E}[\mathbf{v}^T\mathbf{X}[k]\mathbf{X}^T[k]\mathbf{v}] = \mathcal{E}[(\mathbf{v}^T\mathbf{X}(k))^2] \geq 0, \tag{G.8}$$

and its eigenvalues are *nonnegative*. Observe that $\mathbf{R}_{\mathrm{XX}}$ is also Toeplitz:

$$\mathbf{R}_{\mathrm{XX}} = \begin{bmatrix} \mathcal{E}[X^2[k]] & \mathcal{E}[X[k]X[k-1]] & \cdots & \mathcal{E}[X[k]X[k-N+1]] \\ \mathcal{E}[X[k-1])X[k]] & \mathcal{E}[X^2[k-1]] & & \\ \vdots & & \ddots & \vdots \\ \mathcal{E}[X[k-N+1]X[k] & & \cdots & \mathcal{E}[X^2[k-N+1]] \end{bmatrix},$$

$$= \begin{bmatrix} R_{XX}[0] & R_{XX}[1] & \cdots & R_{XX}[N-1] \\ R_{XX}[1] & R_{XX}[0] & & \\ \vdots & & \ddots & \vdots \\ R_{XX}[N-1] & & \cdots & R_{XX}[0] \end{bmatrix}, \tag{G.9}$$

where $R_{XX}[m] \triangleq \mathcal{E}[X[k]X[k+m]] = R_{XX}[-m]$. This type of matrix appears in the optimal filtering and adaptive filtering applications discussed in Chapters 11 and 12.

Suppose that $\mathbf{A} \in \mathcal{C}^{N \times N}$ is a complex-valued square matrix.

Definition: Hermitian Matrix $\mathbf{A}$ is *Hermitian* if $\mathbf{A} = \mathbf{A}^H$ and is *skew Hermitian* if $\mathbf{A}^H = -\mathbf{A}$.

The superscript ^{H} denotes complex conjugate transpose (also called Hermitian transpose). For matrices of appropriate dimensions:

$$(\mathbf{A}_1\mathbf{A}_2)^H = \mathbf{A}_1^H\mathbf{A}_2^H. \tag{G.10}$$

Definition: Unitary Matrix $\mathbf{A}$ is *unitary* if $\mathbf{A}^H\mathbf{A} = \mathbf{A}\mathbf{A}^H = \mathbf{I}$ and thus $\mathbf{A}^{-1} = \mathbf{A}^H$.

For complex-valued matrices, Hermitian and unitary are the definitions corresponding to symmetric and orthogonal, respectively, for real-valued matrices.

G.2 FOUR FUNDAMENTAL SUBSPACES

Consider again the case where matrix $\mathbf{A} \in \mathcal{R}^{M \times N}$ is real-valued and has rank r.

Definition: Column Space The *column space* of matrix $\mathbf{A}$ has dimension r and consists of all linear combinations of its columns. It is also called the *range* of $\mathbf{A}$.

In the following expression, $\mathbf{y} \in \mathcal{R}^M$ lies in the column space of $\mathbf{A}$:

$$\mathbf{y} = \mathbf{A}\mathbf{x} = \sum_{n=1}^{N} x_n\mathbf{a}_n, \tag{G.11}$$

where $\mathbf{x} \triangleq [x_1, \ldots, x_N]^T$ and $\{\mathbf{a}_n\}$ are the columns of $\mathbf{A}$.

Definition: Null Space The *null space* of matrix $\mathbf{A}$ has dimension $N - r$ and consists of all vectors $\mathbf{x}$ such that

$$\mathbf{A}\mathbf{x} = \mathbf{0}, \tag{G.12}$$

where $\mathbf{0} \in \mathcal{R}^M$. The null space is also known as the *kernel* of $\mathbf{A}$, and its dimension is called the *nullity*.

Definition: Row Space The *row space* of matrix $\mathbf{A}$ has dimension r and consists of all linear combinations of the rows.

In the following expression, $\mathbf{y} \in \mathcal{R}^N$ lies in the row space of $\mathbf{A}$:

$$\mathbf{y} = \mathbf{A}^T\mathbf{x} = \sum_{m=1}^{M} x_m \tilde{\mathbf{a}}_m, \tag{G.13}$$

where $\mathbf{x} \triangleq [x_1, \ldots, x_M]^T$ and $\{\tilde{\mathbf{a}}_m^T\}$ are the rows of $\mathbf{A}$. The row space of $\mathbf{A}$ is the column space of $\mathbf{A}^T$.

Definition: Left Null Space The *left null space* of matrix $\mathbf{A}$ has dimension $M - r$ and consists of all vectors $\mathbf{x}$ such that

$$\mathbf{A}^T\mathbf{x} = \mathbf{0}, \tag{G.14}$$

where $\mathbf{0} \in \mathcal{R}^N$.

The left null space of $\mathbf{A}$ is the null space of $\mathbf{A}^T$.

G.3 EIGENDECOMPOSITION

In this section, we consider a particular type of matrix decomposition known as an *eigendecomposition* that is a useful representation for studying the properties of a linear system. Other types of matrix decompositions are summarized in Section G.4. Since an eigendecomposition yields a diagonal matrix, it is also referred to as the *diagonal form* of $\mathbf{A}$.

Definition: Eigenvalues and Eigenvectors An *eigenvalue* λ of square matrix $\mathbf{A} \in \mathcal{R}^{N \times N}$ satisfies the following system of equations:

$$\mathbf{A}\mathbf{q} = \lambda\mathbf{q} \tag{G.15}$$

for nonzero vector $\mathbf{q}$ called an *eigenvector*.

Observe that (G.15) can be rewritten as $(\mathbf{A} - \lambda\mathbf{I})\mathbf{q} = \mathbf{0}$, showing that $\mathbf{q}$ lies in the null space of $\mathbf{A} - \lambda\mathbf{I}$. The eigenvalues are the specific λs such that this null space is nonempty, which means $\mathbf{A} - \lambda\mathbf{I}$ is a singular matrix. Since the determinant of a singular matrix is zero, we can find the eigenvalues by solving the following *characteristic equation*:

$$\det(\mathbf{A} - \lambda\mathbf{I}) = 0. \tag{G.16}$$

It is straightforward to show that (G.16) yields a polynomial in λ of degree N:

$$\lambda^N + \alpha_{N-1}\lambda^{N-1} + \cdots + \alpha_1\lambda + \alpha_0 = 0, \tag{G.17}$$

where the coefficients $\{\alpha_n\}$ are determined by the elements of $\mathbf{A}$. There are N solutions to this equation for λ; however, they may not be distinct, some may be zero, and they could be complex valued. In the following discussion, the eigenvalues are assumed to be real valued (as they would be for a symmetric matrix).

If λ_n is a solution to (G.16), the corresponding eigenvector $\mathbf{q}_n$ is obtained by solving (G.15):

$$(\mathbf{A} - \lambda_n \mathbf{I})\mathbf{q}_n = \mathbf{0}. \tag{G.18}$$

Since λ_n is an eigenvalue, $\mathbf{A} - \lambda_n \mathbf{I}$ is singular, its null space is nonempty, and $\mathbf{q}_n$ is any vector in that null space. The eigenvectors are not unique: if $\mathbf{q}_n$ is an eigenvector, then so is $c\mathbf{q}_n$ for any nonzero $c \in \mathcal{R}$, which follows from (G.15). Often it is convenient that $\mathbf{q}_n$ be scaled to have unit norm: $\mathbf{q}_n^T \mathbf{q}_n \triangleq ||\mathbf{q}_n||^2 = 1$.

Theorem G.1 (Eigendecomposition). If square matrix $\mathbf{A} \in \mathcal{R}^{N\times N}$ has full rank $r = N$ such that there are N linearly independent eigenvectors, it can be decomposed as follows:

$$\mathbf{A} = \mathbf{Q}\mathbf{\Lambda}\mathbf{Q}^{-1}, \tag{G.19}$$

where diagonal matrix $\mathbf{\Lambda} \in \mathcal{R}^{N\times N}$ contains the eigenvalues, and the columns of matrix $\mathbf{Q} \in \mathcal{R}^{N\times N}$ are the eigenvectors. If $\mathbf{A}$ is a symmetric matrix, then $\mathbf{Q}^{-1} = \mathbf{Q}^T$.

Proof. Define the eigenvector matrix

$$\mathbf{Q} \triangleq [\mathbf{q}_1, \ldots, \mathbf{q}_N], \tag{G.20}$$

and the diagonal eigenvalue matrix

$$\mathbf{\Lambda} \triangleq \begin{bmatrix} \lambda_1 & 0 & \cdots & 0 \\ 0 & \lambda_2 & & \vdots \\ \vdots & & \ddots & 0 \\ 0 & \cdots & 0 & \lambda_N \end{bmatrix}. \tag{G.21}$$

From the result in (G.15), the eigenvectors can be arranged columnwise:

$$\mathbf{A}[\mathbf{q}_1, \ldots, \mathbf{q}_N] = [\lambda_1 \mathbf{q}_1, \ldots, \lambda_N \mathbf{q}_N], \tag{G.22}$$

which can be written as

$$\mathbf{A}\mathbf{Q} = \mathbf{Q}\mathbf{\Lambda}. \tag{G.23}$$

Since the eigenvectors are linearly independent, $\mathbf{Q}^{-1}$ exists and the result in (G.19) is obtained. When $\mathbf{A}$ is a symmetric matrix, $\mathbf{A}^T = (\mathbf{Q}\mathbf{\Lambda}\mathbf{Q}^{-1})^T = \mathbf{Q}^{-T}\mathbf{\Lambda}\mathbf{Q}^T$ implies that $\mathbf{Q}^{-1} = \mathbf{Q}^T$, which completes the proof. □

The previous definition of eigenvectors is the one most commonly used, though technically they are called *right eigenvectors*. We can likewise write an eigenvalue equation for the *left eigenvectors* as follows:

$$\mathbf{v}^T \mathbf{A} = \lambda \mathbf{v}^T. \tag{G.24}$$

It turns out that the eigenvalues are the same as before, but generally the left and right eigenvectors are not the same. Arranging the left eigenvectors row-wise gives

$$\begin{bmatrix} \mathbf{v}_1^T \\ \vdots \\ \mathbf{v}_N^T \end{bmatrix} \mathbf{A} = \begin{bmatrix} \lambda_1 \mathbf{v}_1^T \\ \vdots \\ \lambda_N \mathbf{v}_N^T \end{bmatrix} \Longrightarrow \mathbf{V}^T \mathbf{A} = \mathbf{\Lambda} \mathbf{V}^T \Longrightarrow \mathbf{A} = \mathbf{V}^{-T} \mathbf{\Lambda} \mathbf{V}^T, \tag{G.25}$$

where $\mathbf{\Lambda}$ is the previously defined matrix of eigenvalues, and $\mathbf{V} \triangleq [\mathbf{v}_1, \ldots, \mathbf{v}_N]$. The last result assumes the left eigenvectors are linearly independent. The rows of $\mathbf{V}^T$ are the left eigenvectors of $\mathbf{A}$. If $\mathbf{A}$ happens to be a symmetric matrix, then the left and right eigenvectors are the transpose of each other such that $(\mathbf{V}^T)^T = \mathbf{Q} \Longrightarrow \mathbf{V} = \mathbf{Q}$ and

$$\mathbf{A}^T = \mathbf{V} \mathbf{\Lambda} \mathbf{V}^{-1} = \mathbf{Q} \mathbf{\Lambda} \mathbf{Q}^{-1} = \mathbf{Q} \mathbf{\Lambda} \mathbf{Q}^T = \mathbf{A}. \tag{G.26}$$

A singular value decomposition (SVD) is a generalization of an eigendecomposition to rectangular matrices, which we present in the following theorem without proof.

Theorem G.2 (SVD). Rectangular matrix $\mathbf{A} \in \mathcal{R}^{M \times N}$ can be decomposed as follows:

$$\mathbf{A} = \mathbf{U} \mathbf{\Sigma} \mathbf{V}^T, \tag{G.27}$$

where $\mathbf{\Sigma} \in \mathcal{R}^{M \times N}$ is a diagonal matrix with nonnegative diagonal elements called *singular values*, $\mathbf{U} \in \mathcal{R}^{M \times M}$ is an orthogonal matrix whose columns are called left singular vectors, and $\mathbf{V} \in \mathcal{R}^{N \times N}$ is another orthogonal matrix whose columns are called right singular vectors.

G.4 LU, LDU, AND CHOLESKY DECOMPOSITIONS

Before describing the Cholesky decomposition for a PD matrix, we summarize other matrix decompositions, given without proofs.

Definition: Leading Principal Submatrix The nth *leading principal submatrix* $\mathbf{A}_n \in \mathcal{R}^{n \times n}$ is a square matrix generated from matrix $\mathbf{A} \in \mathcal{R}^{M \times N}$ by retaining only the first n rows and first n columns.

Of course, the previous definition assumes $n \leq \min(M, N)$.

Theorem G.3 (LU decomposition). If all leading principal submatrices $\mathbf{A}_n$ for $n \leq \min(M - 1, N)$ of $\mathbf{A} \in \mathcal{R}^{M \times N}$ are nonsingular, then

$$\mathbf{A} = \mathbf{L} \mathbf{U} \tag{G.28}$$

where $\mathbf{L} \in \mathcal{R}^{M \times M}$ is a lower triangular matrix with ones along the main diagonal, and $\mathbf{U} \in \mathcal{R}^{M \times N}$ is an upper triangular matrix.

Note that $\mathbf{L}$ is always square whereas $\mathbf{U}$ is rectangular if $\mathbf{A}$ is not square. A lower (upper) triangular matrix with ones along the main diagonal is called a lower (upper) *unit triangular matrix*.

Theorem G.4 (LDU decomposition). If all leading principal submatrices $\mathbf{A}_n$ of square matrix $\mathbf{A} \in \mathcal{R}^{N\times N}$ are nonsingular, then

$$\mathbf{A} = \mathbf{LDU}, \tag{G.29}$$

where $\mathbf{L} \in \mathcal{R}^{N\times N}$ is a lower unit triangular matrix, $\mathbf{U} \in \mathcal{R}^{N\times N}$ is an upper unit triangular matrix, and $\mathbf{D} \in \mathcal{R}^{N\times N}$ is a diagonal matrix.

If $\mathbf{A}$ is PD, the diagonal elements of $\mathbf{D}$ are all positive.

Theorem G.5 (Cholesky decomposition). If $\mathbf{A} \in \mathcal{R}^{N\times N}$ is a symmetric PD matrix, then

$$\mathbf{A} = \mathbf{BB}^T, \tag{G.30}$$

where $\mathbf{B} \in \mathcal{R}^{N\times N}$ is a lower triangular matrix with positive diagonal elements.

Since the diagonal elements of $\mathbf{D}$ are positive in the LDU decomposition, it is clear that $\mathbf{B} = \mathbf{LD}^{1/2}$, where $\mathbf{D}^{1/2}$ is the square root of $\mathbf{D}$, that is, $\mathbf{D} = (\mathbf{D}^{1/2})^2$.

Example G.2. Consider the symmetric PD matrix

$$\mathbf{A} = \begin{bmatrix} 2 & 1 & 1/2 \\ 1 & 2 & 1 \\ 1/2 & 1 & 2 \end{bmatrix}, \tag{G.31}$$

which has the following LU decomposition:

$$\mathbf{L} = \begin{bmatrix} 1 & 0 & 0 \\ 1/2 & 1 & 0 \\ 1/4 & 1/2 & 1 \end{bmatrix}, \quad \mathbf{U} = \begin{bmatrix} 2 & 1 & 1/2 \\ 0 & 3/2 & 3/4 \\ 0 & 0 & 3/2 \end{bmatrix}. \tag{G.32}$$

Its LDU decomposition has diagonal matrix

$$\mathbf{D} = \begin{bmatrix} 2 & 0 & 0 \\ 0 & 3/2 & 0 \\ 0 & 0 & 3/2 \end{bmatrix}, \tag{G.33}$$

and the Cholesky decomposition has lower triangular matrix

$$\mathbf{B} = \begin{bmatrix} 1.4142 & 0 & 0 \\ 0.7071 & 1.2247 & 0 \\ 0.3536 & 0.6124 & 1.2247 \end{bmatrix}. \tag{G.34}$$

For completeness, we also include the eigendecomposition of $\mathbf{A}$:

$$\mathbf{Q} = \begin{bmatrix} -0.4544 & -0.7071 & 0.5418 \\ 0.7662 & 0 & 0.6426 \\ -0.4544 & 0.7071 & 0.5418 \end{bmatrix}, \quad \mathbf{\Lambda} = \begin{bmatrix} 0.8139 & 0 & 0 \\ 0 & 1.5 & 0 \\ 0 & 0 & 3.6861 \end{bmatrix}. \tag{G.35}$$

G.5 JACOBIAN MATRIX AND THE JACOBIAN

Let $y_m = g_m(x_1, \ldots, x_N)$ be M real-valued functions from $\{x_1, \ldots, x_N\}$ to $\{y_1, \ldots, y_M\}$.

Definition: Jacobian Matrix The *Jacobian matrix* $\mathbf{J}(x_1, \ldots, x_N) \in \mathcal{R}^{M\times N}$ contains the following partial derivatives:

$$\mathbf{J}(x_1, \ldots, x_N) \triangleq \begin{bmatrix} \partial y_1/\partial x_1 & \cdots & \partial y_1/\partial x_N \\ \vdots & \ddots & \vdots \\ \partial y_M/\partial x_1 & \cdots & \partial y_M/\partial x_N \end{bmatrix}, \tag{G.36}$$

where the argument of $\mathbf{J}(x_1, \ldots, x_N)$ specifies the independent variables, and thus the elements of the matrix are functions of $\{x_1, \ldots, x_N\}$.

Assume that $\mathbf{J}(x_1, \ldots, x_N) \in \mathcal{R}^{N\times N}$ is a square matrix.

Definition: Jacobian Determinant The *Jacobian determinant* (also simply called the *Jacobian*) of square matrix $\mathbf{J}(x_1, \ldots, x_N)$ is the following determinant:

$$J(x_1, \ldots, x_N) \triangleq \det(\mathbf{J}(x_1, \ldots, x_N)) = |\mathbf{J}(x_1, \ldots, x_N)|. \tag{G.37}$$

Consider the case of $M = N = 2$ such that $y_1 = g_1(x_1, x_2)$ and $y_2 = g_2(x_1, x_2)$. The following two Jacobians for the same set of variables can be defined:

$$J(x_1, x_2) = \begin{vmatrix} \partial y_1/\partial x_1 & \partial y_1/\partial x_2 \\ \partial y_2/\partial x_1 & \partial y_2/\partial x_2 \end{vmatrix}, \quad J(y_1, y_2) = \begin{vmatrix} \partial x_1/\partial y_1 & \partial x_1/\partial y_2 \\ \partial x_2/\partial y_1 & \partial x_2/\partial y_2 \end{vmatrix}, \tag{G.38}$$

where again the arguments specify the independent variables. It is straightforward to verify the following property:

$$J(x_1, x_2)J(y_1, y_2) = 1. \tag{G.39}$$

The absolute value of the Jacobian $|J|$ is used when a transformation of variables is performed for an integrand, usually to simplify the integration. Let $x_1 = h_1(y_1, y_2)$ and $x_2 = h_2(y_1, y_2)$ be the inverse transformations. Then the following integral of function $c(x_1, x_2)$ can be evaluated in two different ways:

$$\begin{aligned} \iint c(x_1, x_2)\mathrm{d}x_1\mathrm{d}x_2 &= \iint c(h_1(y_1, y_2), h_2(y_1, y_2))|J(y_1, y_2)|\mathrm{d}y_1\mathrm{d}y_2, \\ &= \iint c(h_1(y_1, y_2), h_2(y_1, y_2))|J(x_1, x_2)|^{-1}\mathrm{d}y_1\mathrm{d}y_2, \end{aligned} \tag{G.40}$$

which follow from (G.39). Note that the limits of integration on the right-hand side must be defined for $\{y_1, y_2\}$, and the inverse transformations have been substituted into $c(x_1, x_2)$. The second equation is perhaps more familiar where the Jacobian is based on the transformations $g_1(x_1, x_2)$ and $g_2(x_1, x_2)$, which are usually given in the problem statement. However, the first equation may be simpler to evaluate because sometimes the inverse transformations $h_1(y_1, y_2)$ and $h_2(y_1, y_2)$ are more easily differentiated.

The Jacobian is useful for finding the joint pdf of a transformed set of random variables, as covered in Chapter 4. In Example G.3, the Jacobian is used to evaluate a definite integral.

Example G.3. One of the more widely used transformations is from rectangular to polar coordinates:

$$x = r\cos(\phi), \quad y = r\sin(\phi) \tag{G.41}$$

for which

$$\begin{aligned} J(r,\phi) &= \begin{vmatrix} \partial x/\partial r & \partial x/\partial\phi \\ \partial y/\partial r & \partial y/\partial\phi \end{vmatrix} = \begin{vmatrix} \cos(\phi) & -r\sin(\phi) \\ \sin(\phi) & r\cos(\phi) \end{vmatrix}, \\ &= r\cos^2(\phi) + r\sin^2(\phi) = r. \end{aligned} \tag{G.42}$$

The inverse transformations are

$$x^2 + y^2 = r^2\cos^2(\phi) + r^2\sin^2(\phi) \Longrightarrow r = \sqrt{x^2+y^2}, \tag{G.43}$$

$$y/x = r\sin(\phi)/r\cos(\phi) \Longrightarrow \phi = \tan^{-1}(y/x). \tag{G.44}$$

Although the Jacobian in (G.42) was straightforward to evaluate, we include the other Jacobian for completeness:

$$\begin{aligned} J(x,y) &= \begin{vmatrix} \partial r/\partial x & \partial r/\partial y \\ \partial\phi/\partial x & \partial\phi/\partial y \end{vmatrix} = \begin{vmatrix} x/\sqrt{x^2+y^2} & y/\sqrt{x^2+y^2} \\ -y/(x^2+y^2) & x/(x^2+y^2) \end{vmatrix}, \\ &= 1/\sqrt{x^2+y^2} = 1/r = J^{-1}(r,\phi). \end{aligned} \tag{G.45}$$

Suppose we want to evaluate the following definite integral:

$$\int_0^\alpha \int_0^{\sqrt{\alpha^2-y^2}} (x^2+y^2)\mathrm{d}x\mathrm{d}y. \tag{G.46}$$

The following substitutions are used for the transformation above:

$$x^2+y^2 = r^2, \quad \mathrm{d}x\mathrm{d}y = J(r,\phi)\mathrm{d}r\mathrm{d}\phi = r\mathrm{d}r\mathrm{d}\phi. \tag{G.47}$$

The lower limits of r and ϕ are both zero, and the upper limits are

$$\begin{aligned} r &= \sqrt{x^2+y^2}\Big|_{x=\sqrt{\alpha^2-y^2}} = \sqrt{\alpha^2-y^2+y^2} = \alpha, \\ \phi &= \tan^{-1}(y/\sqrt{\alpha^2-y^2})\Big|_{y=\alpha} = \pi/2. \end{aligned} \tag{G.48}$$

Thus

$$\int_0^\alpha \int_0^{\sqrt{\alpha^2-y^2}} (x^2+y^2)\mathrm{d}x\mathrm{d}y = \int_0^{\pi/2}\int_0^\alpha r^3\mathrm{d}r\mathrm{d}\phi = \alpha^4\pi/8. \tag{G.49}$$

G.6 KRONECKER AND SCHUR PRODUCTS

Definition: Kronecker Product The *Kronecker product* of matrices $\mathbf{A} \in \mathcal{R}^{M\times N}$ and $\mathbf{B} \in \mathcal{R}^{P\times Q}$ is a size $MP \times NQ$ block matrix:

$$\mathbf{A} \otimes \mathbf{B} \triangleq \begin{bmatrix} a_{11}\mathbf{B} & \cdots & a_{1N}\mathbf{B} \\ \vdots & \ddots & \vdots \\ a_{M1}\mathbf{B} & \cdots & a_{MN}\mathbf{B} \end{bmatrix}, \tag{G.50}$$

where the upper-left block is

$$a_{11}\mathbf{B} = \begin{bmatrix} a_{11}b_{11} & \cdots & a_{11}b_{1Q} \\ \vdots & \ddots & \vdots \\ a_{11}b_{P1} & \cdots & a_{11}b_{PQ} \end{bmatrix}, \tag{G.51}$$

and likewise for the other blocks.

Although the Kronecker product is associative, it is not commutative: $\mathbf{A} \otimes \mathbf{B} \neq \mathbf{B} \otimes \mathbf{A}$. Note also that $\mathbf{A}$ and $\mathbf{B}$ can be vectors: if $\mathbf{A} \rightarrow \mathbf{a} \in \mathcal{R}^M$ is a column vector and $\mathbf{B} \rightarrow \mathbf{b}^T \in \mathcal{R}^N$ is a row vector, then $\mathbf{A} \otimes \mathbf{B} \in \mathcal{R}^{M\times N}$ is a matrix. For this case, the Kronecker product reduces to the standard vector *outer product* $\mathbf{ab}^T$.

Definition: Schur Product For matrices $\mathbf{A} \in \mathcal{R}^{M\times N}$ and $\mathbf{B} \in \mathcal{R}^{M\times N}$, the *Schur product* is obtained by a multiplication of their entries element-wise:

$$\mathbf{A} \circ \mathbf{B} \triangleq \begin{bmatrix} a_{11}b_{11} & \cdots & a_{1N}b_{1N} \\ \vdots & \ddots & \vdots \\ a_{M1}b_{M1} & \cdots & a_{MN}b_{MN} \end{bmatrix}, \tag{G.52}$$

such that $\mathbf{A} \circ \mathbf{B} \in \mathcal{R}^{M\times N}$. It is also known as the *Hadamard product*.

The Schur product is associative and commutative.

G.7 PROPERTIES OF TRACE AND DETERMINANT

Definition: Trace The *trace* of square matrix $\mathbf{A} \in \mathcal{R}^{N\times N}$ is

$$\text{tr}[\mathbf{A}] \triangleq \sum_{n=1}^{N} a_{nn}, \tag{G.53}$$

where $\{a_{nn}\}$ are the diagonal elements of $\mathbf{A}$.

The trace has the following properties:

- *Eigenvalues* $\{\lambda_n\}$:

$$\text{tr}[\mathbf{A}] = \sum_{n=1}^{N} \lambda_n, \quad \text{tr}\left[\mathbf{A}^K\right] = \sum_{n=1}^{N} \lambda_n^K. \tag{G.54}$$

- *Addition and multiplication by scalars* $\{c, d\}$:

$$\text{tr}[c\mathbf{A} + d\mathbf{B}] = c\text{tr}[\mathbf{A}] + d\text{tr}[\mathbf{B}], \tag{G.55}$$

- *Associative pairwise*: $\text{tr}[\mathbf{AB}] = \text{tr}[\mathbf{BA}]$ and

$$\text{tr}[\mathbf{ABC}] = \text{tr}[\mathbf{BCA}] = \text{tr}[\mathbf{CAB}], \tag{G.56}$$

$$\neq \text{tr}[\mathbf{CBA}] \text{ or } \text{tr}[\mathbf{ACB}] \text{ or } \text{tr}[\mathbf{BAC}]. \tag{G.57}$$

- *Kronecker product*:

$$\text{tr}[\mathbf{A} \otimes \mathbf{B}] = \text{tr}[\mathbf{A}]\text{tr}[\mathbf{B}]. \tag{G.58}$$

- *Similarity invariant*:

$$\text{tr}[\mathbf{P}^{-1}\mathbf{AP}] = \text{tr}[\mathbf{A}] \tag{G.59}$$

  for invertible matrix $\mathbf{P}$.
- *Bounds for PSD matrices* $\mathbf{A}$ and $\mathbf{B}$:

$$0 \leq \text{tr}\left[(\mathbf{AB})^K\right] \leq (\text{tr}[\mathbf{A}]\text{tr}[\mathbf{B}])^K. \tag{G.60}$$

The determinant of square matrix $\mathbf{A} \in \mathcal{R}^{N \times N}$ has the following properties:

- *Transpose*: $\det(\mathbf{A}) = \det(\mathbf{A}^T)$.
- *Inverse*: $\det(\mathbf{A}^{-1}) = 1/\det(\mathbf{A})$.
- *Multiplication by a constant* c: $\det(c\mathbf{A}) = c^N \det(\mathbf{A})$.
- *Distributive*: $\det(\mathbf{AB}) = \det(\mathbf{A})\det(\mathbf{B})$ for $\mathbf{B} \in \mathcal{R}^{N \times N}$.

G.8 MATRIX INVERSION LEMMA

The following property of matrices, which is known as the Sherman–Morrison–Woodbury formula, is useful for deriving the recursive least-squares (RLS) algorithm in Chapter 11.

Lemma G.1 (Matrix inversion). For square nonsingular matrices $\mathbf{B}$ and $\mathbf{D}$:

$$(\mathbf{B} + \mathbf{UDV}^T)^{-1} = \mathbf{B}^{-1} - \mathbf{B}^{-1}\mathbf{U}(\mathbf{D}^{-1} + \mathbf{V}^T\mathbf{B}^{-1}\mathbf{U})^{-1}\mathbf{V}^T\mathbf{B}^{-1}, \tag{G.61}$$

where matrices $\mathbf{U}$ and $\mathbf{V}$ are not necessarily square.

Proof. See Problem 11.17. □

When $\mathbf{U} \rightarrow \mathbf{u}$ and $\mathbf{V} \rightarrow \mathbf{v}$ are column vectors, the matrix inversion lemma simplifies to the Sherman–Morrison formula:

$$(\mathbf{B} + d\mathbf{uv}^T)^{-1} = \mathbf{B}^{-1} - \mathbf{B}^{-1}\mathbf{uv}^T\mathbf{B}^{-1}/(d^{-1} + \mathbf{v}^T\mathbf{B}^{-1}\mathbf{u}). \tag{G.62}$$

Since $\mathbf{D} \rightarrow d$ is necessarily a scalar, this case is considerably simpler because of the division by scalar $d^{-1} + \mathbf{v}^T\mathbf{B}^{-1}\mathbf{u}$. This is the form used to derive the RLS algorithm.

G.9 CAUCHY–SCHWARZ INEQUALITY

The Cauchy–Schwarz inequality for vectors of the same size is

$$(\mathbf{x}^T\mathbf{y})^2 \leq \mathbf{x}^T\mathbf{x}\mathbf{y}^T\mathbf{y} = ||\mathbf{x}||^2||\mathbf{y}||^2 \tag{G.63}$$

with equality if and only if $\mathbf{x} = c\mathbf{y}$ for some nonzero $c \in \mathcal{R}$. An equivalent inequality is

$$|\mathbf{x}^T\mathbf{y}| \leq ||\mathbf{x}||\;||\mathbf{y}||. \tag{G.64}$$

The Cauchy–Schwarz inequality for *expectations* of random variables is given in Appendix F.

G.10 DIFFERENTIATION

Let $\mathbf{A}$ be an arbitrary nonzero matrix, and $\mathbf{x}$ and $\mathbf{v}$ be column vectors with appropriate dimensions. The following results are useful when differentiating matrix and vector expressions with respect to a column vector:

$$\frac{\partial}{\partial\mathbf{x}}\mathbf{x} = \mathbf{I}, \quad \frac{\partial}{\partial\mathbf{x}}\mathbf{v}^T\mathbf{x} = \frac{\partial}{\partial\mathbf{x}}\mathbf{x}^T\mathbf{v} = \mathbf{v}, \quad \frac{\partial}{\partial\mathbf{x}}\mathbf{x}^T\mathbf{x} = 2\mathbf{x}, \tag{G.65}$$

$$\frac{\partial}{\partial\mathbf{x}}\mathbf{x}^T\mathbf{A}\mathbf{x} = (\mathbf{A}+\mathbf{A}^T)\mathbf{x} = 2\mathbf{A}\mathbf{x} \text{ (for symmetric } \mathbf{A}), \tag{G.66}$$

$$\frac{\partial}{\partial\mathbf{x}}\mathbf{x}^T\mathbf{A}\mathbf{v} = \mathbf{A}\mathbf{v}, \quad \frac{\partial}{\partial\mathbf{v}}\mathbf{x}^T\mathbf{A}\mathbf{v} = \mathbf{A}^T\mathbf{x}, \tag{G.67}$$

$$\frac{\partial}{\partial\mathbf{x}}\mathbf{A}\mathbf{x} = \mathbf{A}^T, \quad \frac{\partial}{\partial\mathbf{x}}\mathbf{x}^T\mathbf{A} = \mathbf{A}. \tag{G.68}$$

Example G.4. Consider the following system of linear equations:

$$\mathbf{A}\mathbf{x} = \mathbf{y}. \tag{G.69}$$

If square matrix $\mathbf{A} \in \mathcal{R}^{N\times N}$ has full rank $r = N$, then it has an inverse and the solution is $\mathbf{x} = \mathbf{A}^{-1}\mathbf{y}$. However, if $\mathbf{A} \in \mathcal{R}^{M\times N}$ is rectangular such that $M > N$, the system of equations is overdetermined ($\mathbf{A}$ is a tall narrow matrix if $M \gg N$). For this case, we seek to minimize the following LS cost function:

$$\begin{aligned}\xi_{\text{LS}} &\triangleq ||\mathbf{A}\mathbf{x}-\mathbf{y}||^2 = (\mathbf{A}\mathbf{x}-\mathbf{y})^T(\mathbf{A}\mathbf{x}-\mathbf{y}),\\ &= \mathbf{x}^T\mathbf{A}^T\mathbf{A}\mathbf{x} - \mathbf{x}^T\mathbf{A}^T\mathbf{y} - \mathbf{y}^T\mathbf{A}\mathbf{x} + \mathbf{y}^T\mathbf{y}.\end{aligned} \tag{G.70}$$

Differentiating ξ_{LS} with respect to $\mathbf{x}$ and setting the result equal to zero gives

$$2\mathbf{A}^T\mathbf{A}\mathbf{x}_{\text{LS}} - 2\mathbf{A}^T\mathbf{y} = \mathbf{0}. \tag{G.71}$$

The LS solution is

$$\mathbf{x}_{\text{LS}} = (\mathbf{A}^T\mathbf{A})^{-1}\mathbf{A}\mathbf{y}, \tag{G.72}$$

where $\mathbf{A}^T\mathbf{A} \in \mathcal{R}^{N\times N}$ is square and has full rank. LS techniques are covered in Chapter 9. The quantity

$$\mathbf{A}^+ \triangleq (\mathbf{A}^T\mathbf{A})^{-1}\mathbf{A} \tag{G.73}$$

is known as the *Moore–Penrose pseudoinverse*.

Example G.5. Consider the following constrained optimization problem:

$$\mathbf{x}_o = \arg\max_{\mathbf{x}} \left[\mathbf{x}^T \mathbf{A}\mathbf{x} + \lambda(\mathbf{x}^T \mathbf{c} - c) \right], \tag{G.74}$$

where λ is a Lagrange multiplier (not to be confused with an eigenvalue) and $\mathbf{x}^T \mathbf{c} = c$ is a linear constraint on $\mathbf{x}$ with scalar c. Let $\mathbf{A} \in \mathcal{R}^{N \times N}$ such that $\mathbf{x}, \mathbf{c} \in \mathcal{R}^N$. Differentiating this expression with respect to $\mathbf{x}$ and λ yields

$$2\mathbf{A}\mathbf{x}_o + \lambda_o \mathbf{c} = \mathbf{0}, \quad \mathbf{x}_o^T \mathbf{c} - c = 0. \tag{G.75}$$

(Note that differentiation with respect to λ is not usually required because the result is just the constraint.) Rearrange the first equation as follows:

$$\mathbf{x}_o = -(\lambda_o/2)\mathbf{A}^{-1}\mathbf{c}. \tag{G.76}$$

Premultiplying this expression by $\mathbf{c}^T$ and applying the constraint gives

$$c = \mathbf{c}^T \mathbf{x}_o = -(\lambda_o/2)\mathbf{c}^T \mathbf{A}^{-1}\mathbf{c} \Longrightarrow \lambda_o = -2c/\mathbf{c}^T \mathbf{A}^{-1}\mathbf{c}. \tag{G.77}$$

Substituting this scalar quantity into the first expression of (G.75) yields the final result:

$$\mathbf{x}_o = \frac{c\mathbf{A}^{-1}\mathbf{c}}{\mathbf{c}^T \mathbf{A}^{-1}\mathbf{c}}. \tag{G.78}$$

The scalar c in the constraint controls the norm of $\mathbf{x}_o$:

$$||\mathbf{x}_o||^2 = \frac{c^2 \mathbf{c}^T \mathbf{A}^{-1}\mathbf{c}}{(\mathbf{c}^T \mathbf{A}^{-1}\mathbf{c})^2} = \frac{c^2}{\mathbf{c}^T \mathbf{A}^{-1}\mathbf{c}}. \tag{G.79}$$

G.11 COMPLEX DIFFERENTIATION

For complex-valued matrix $\mathbf{A} \in \mathcal{C}^{M \times N}$ and complex column vectors $\mathbf{x}$ and $\mathbf{v}$ with appropriate dimensions:

$$\frac{\partial}{\partial \mathbf{x}}\mathbf{x} = \mathbf{I}, \quad \frac{\partial}{\partial \mathbf{x}^*}\mathbf{x} = \mathbf{0}, \tag{G.80}$$

where the superscript * denotes complex conjugation. The second expression is called the *conjugate derivative*. The derivatives used in this book involving complex quantities, such as the gradient described in Chapter 13 on beamforming, are based on the conjugate derivative. Some derivative results for complex-valued $\mathbf{A}$, $\mathbf{x}$, and $\mathbf{v}$ are given by the following:

$$\frac{\partial}{\partial \mathbf{x}^*}\mathbf{v}^H \mathbf{x} = \mathbf{0}, \quad \frac{\partial}{\partial \mathbf{x}^*}\mathbf{x}^H \mathbf{v} = \mathbf{v}, \quad \frac{\partial}{\partial \mathbf{x}^*}\mathbf{x}^H \mathbf{x} = \mathbf{x}, \tag{G.81}$$

$$\frac{\partial}{\partial \mathbf{x}^*}\mathbf{x}^H \mathbf{A}\mathbf{x} = \mathbf{A}\mathbf{x}, \quad \frac{\partial}{\partial \mathbf{x}^*}\mathbf{x}^H \mathbf{A}\mathbf{v} = \mathbf{A}\mathbf{v}, \quad \frac{\partial}{\partial \mathbf{x}^*}\mathbf{v}^H \mathbf{A}\mathbf{x} = \mathbf{0}, \tag{G.82}$$

$$\frac{\partial}{\partial \mathbf{x}^*}\mathbf{A}\mathbf{x} = \mathbf{0}, \quad \frac{\partial}{\partial \mathbf{x}^*}\mathbf{x}^H \mathbf{A} = \mathbf{A}. \tag{G.83}$$

Example G.6. Consider again the optimization problem in Example G.5 but with complex quantities:

$$\begin{aligned} \mathbf{x}_o &= \arg\max_{\mathbf{x}}[\mathbf{x}^H \mathbf{A}\mathbf{x} + \lambda^*(\mathbf{x}^H \mathbf{c} - c) + \lambda(\mathbf{c}^H \mathbf{x} - c^*)], \\ &= \arg\max_{\mathbf{x}}[\mathbf{x}^H \mathbf{A}\mathbf{x} + 2\mathrm{Re}(\lambda^*(\mathbf{x}^H \mathbf{c} - c))], \end{aligned} \tag{G.84}$$

where λ is a *complex-valued* Lagrange multiplier. Differentiating this expression with respect to $\mathbf{x}^*$ and λ^*, and using the rules above yields

$$\mathbf{A}\mathbf{x}_o + \lambda_o^*\mathbf{c} = \mathbf{0}, \quad \mathbf{c}^H\mathbf{x}_o - c = 0. \tag{G.85}$$

The remaining steps in the derivation are identical to those in Example G.6:

$$\mathbf{x}_o = -\lambda_o^*\mathbf{A}^{-1}\mathbf{c} \Longrightarrow \mathbf{c}^H\mathbf{x}_o = -\lambda_o^*\mathbf{c}^H\mathbf{A}^{-1}\mathbf{c} \tag{G.86}$$

and

$$\lambda_o^* = -c^*/\mathbf{c}^H\mathbf{A}^{-1}\mathbf{c} \Longrightarrow \mathbf{x}_o = \frac{c^*\mathbf{A}^{-1}\mathbf{c}}{\mathbf{c}^H\mathbf{A}^{-1}\mathbf{c}}, \tag{G.87}$$

which has a form similar to real-valued $\mathbf{x}_o$.

FURTHER READING

The topics in this appendix can be explored further in the following references: Golub and Van Loan (1983), Haykin (1991), Kreyszig (1979), Laub (2005), Leon (1990), Schilling and Lee (1988), and Strang (1976).

GLOSSARY

SUMMARY OF NOTATION

X	uppercase letter: random variable
x	lowercase letter: outcome of random variable X
$\|X\|$	absolute value of random variable X
$X_{(n)}$	nth-order statistic
$\hat{X}$	estimator of X
$\tilde{X}$	error random variable $X - \hat{X}$
$\mathbf{X}$	bold uppercase letter: random vector
$\mathbf{x}$	bold lowercase letter: outcome of random vector $\mathbf{X}$
$X[k]$	uppercase function of k: random sequence
$x[k]$	lowercase function of k: realization of $X[k]$
$X(t)$	uppercase function of t: random process
$x(t)$	lowercase function of t: realization of $X(t)$
$X'(t)$	derivative random process
$\mathbf{X}[k]$	bold uppercase function of k: random vector sequence
$\mathbf{x}[k]$	bold lowercase function of k: realization of $\mathbf{X}[k]$
$\mathbf{X}(t)$	bold uppercase function of t: random vector process
$\mathbf{x}(t)$	bold lowercase function of t: realization of $\mathbf{X}(t)$
$X(t) * Y(t)$, $X[k] * Y[k]$	convolution (continuous, discrete)
$X(t) \star Y(t)$, $X[k] \star Y[k]$	deterministic cross-correlation (continuous, discrete)
$X(s)$	Laplace transform
$X(z)$	z-transform
$X(\omega)$	Fourier transform or random variable mapping
$X(j\omega)$	discrete-time Fourier transform
$\mathbf{X}(\omega)$	random vector mapping
$X^{-1}(B)$, $\mathbf{X}^{-1}(B)$	inverse image of event B
$\mathbf{a}$	column vector
$\|\|\mathbf{a}\|\|^2$	squared norm $\mathbf{a}^T\mathbf{a}$
$\mathbf{a}^*$	complex conjugate
$\mathbf{a}^T$	row vector
$\mathbf{A}$	nonrandom matrix
$\mathbf{A}^T$	matrix transpose
$\mathbf{A}^H$	matrix conjugate transpose
$\mathbf{A}^{-1}$	matrix inverse
$\mathbf{A}^{-T}$	matrix inverse and transpose
$\mathbf{A}^{-H}$	matrix inverse and conjugate transpose
$\mathbf{A}^+$	Moore-Penrose pseudoinverse
$\mathbf{A} \circ \mathbf{B}$	Schur product

Probability, Random Variables, and Random Processes: Theory and Signal Processing Applications, First Edition. John J. Shynk.

$\mathbf{A} \otimes \mathbf{B}$	Kronecker product
$\|\mathbf{A}\|$	determinant of square matrix $\mathbf{A}$
$A \cup B$, $A + B$	union of sets
$A \cap B$, AB	intersection of sets
$A - B$, $A \setminus B$	difference of sets
$A \oplus B$	exclusive or of sets
A^c, $\bar{A}$	complement of set
$A \times B$	Cartesian product of sets A and B
$\|A\|$	cardinality of set A
$[a, b]$, $[a, b)$, $(a, b]$, (a, b)	interval subset of real numbers $\mathcal{R}$
$\{a, \ldots, b\}$	finite subset of integers $\mathcal{Z}$
$g'(x)$	ordinary derivative
$g^{(n)}(x)$	nth ordinary derivative
$g_E(x)$	even function
$g_O(x)$	odd function
$\binom{n}{x}$	binomial coefficient
$\binom{n}{x_1, x_2, \ldots, x_m}$	multinomial coefficient
$x!$	factorial
$x^{(n)}$	rising factorial
$(x)_n$	falling factorial

GENERAL SYMBOLS

$\mathbf{0}$	matrix or column vector of zeros
$\mathbf{1}$	column vector of ones
$\beth_n$	nth beth number (cardinality of infinite sets)
∞, $-\infty$	infinity symbols (not real numbers)
$\arg(s)$	angle of complex s
$\arg\min_\theta g(x;\theta)$	θ minimizing $g(x;\theta)$
$\arg\max_\theta g(x;\theta)$	θ maximizing $g(x;\theta)$
B_n	Bernoulli number
$B(a, b)$	beta function
$B_y(a, b)$	incomplete beta function
$B_{3\text{dB}}$	3-dB bandwidth
B_{NE}	noise-equivalent bandwidth
B_{RMS}	root-mean-square bandwidth
$B_m[k]$	mth-order backward prediction error
$B(\theta)$	bias of an estimator for parameter θ or beampattern of an antenna array
$\text{ceil}(x)$	ceiling function
$\text{cov}[X, Y]$	covariance of X and Y
$\text{cov}[\mathbf{X}]$	autocovariance matrix of $\mathbf{X}$
c	general location parameter
$c(t)$	cumulant generating function
c_X	coefficient of variation
C	Cantor set, channel capacity per symbol, or contour of integration
$\{C_m\}$	sequence of sets leading to the Cantor set
C_r	channel capacity rate
$C(A)$	counting measure of set A
$C(t, \theta)$	loss function
C_{XX}, C_{XY}	autocovariance, cross-covariance
$C_{XX}(t_1, t_2)$, $C_{XX}[m, n]$	autocovariance function (continuous, discrete)

$C_{XY}(t_1, t_2), C_{XY}[m, n]$	cross-covariance function (continuous, discrete)
$\mathbf{C_{XX}}, \mathbf{C_{XY}}$	autocovariance matrix, cross-covariance matrix
$\tilde{\mathbf{C}}_{\mathbf{XX}}$	$\mathbf{C_{XX}}$ with ones along the main diagonal
$\det(\mathbf{A})$	determinant of square matrix **A**
d	distance between antennas of a uniform linear array
$D(x)$	Dirichlet function
$D(p_X \| q_X)$	Kullback–Leibler divergence
$\mathbf{D}$	diagonal square matrix
e	base of natural logarithm: 2.71828182845904523536...
$\mathrm{erf}(x)$	error function
$\mathrm{erfc}(x)$	complementary error function
$\mathrm{erfi}(x)$	imaginary error function
$\exp(x)$	exponential function (instead of e^x)
$\mathbf{e}_m$	mth unit vector
E	energy or a set
E_b	energy per bit
$E[k]$	error signal
$\mathrm{floor}(x)$	floor function
f	frequency in Hertz
f_s	sample frequency in Hertz
$f_X(x)$	probability density function (pdf)
$f_{X,Y}(x, y)$	joint pdf
$f_{X\|Y}(x\|y)$	conditional pdf
F	a set
$F_X(x)$	cumulative distribution function (cdf)
$F^{-1}(x)$	inverse cdf
$F_{X,Y}(x, y)$	joint cdf
$F_{X\|Y}(x\|y)$	conditional cdf
$F_C(x)$	Cantor function
$\{F_{C_m}(x)\}$	sequence of functions leading to the Cantor function
$F_m[k]$	mth-order forward prediction error
${}_1F_1(a; b; x)$	confluent hypergeometric function of the first kind
$\{g_n(t)\}$	Gram–Schmidt basis functions
$G_X(t)$	probability generating function
$\mathbf{G}[k]$	least-squares gain vector or Kalman filter gain matrix
H_0, H_n	null hypothesis, nth hypothesis
$h(X)$	differential entropy of continuous X
$H(X)$	entropy of discrete X
$H(Y\|X)$	conditional entropy
$H(\omega), H(f), H(s)$	transfer function (continuous time)
$H(j\omega), H(z)$	transfer function (discrete time)
$H_n(v)$	harmonic number
$\{H, T\}$	heads, tails of a coin toss
$H_T(f), H_R(f)$	transmit, receive filters
$\mathbf{H}[k]$	Hessian matrix
$\inf(a, b)$	infimum
$i(t)$	circuit current
$I(x_n)$	self information of x_n
$I(x_n; y_m)$	mutual information of x_n and y_m
$I(X; Y)$	average mutual information
$I(X; Y\|Z)$	conditional mutual information
$I_{[a,b]}(x)$	indicator function for continuous x

$I_{\{a,b\}}[x]$	indicator function for discrete x
$I_\alpha(x)$	modified Bessel function of the first kind
$I_y(a, b)$	regularized incomplete beta function
I_s	current source
$\text{Im}(s)$	imaginary part of s
$\mathbf{I}$	identity matrix
j	$\sqrt{-1}$
$J_\alpha(x)$	Bessel function of the first kind
J	Jacobian
$\mathbf{J}$	Jacobian matrix
k	discrete time
$K_\alpha(x)$	Bessel function of the second kind
$l(\mathbf{x})$	likelihood function
$L(\mathbf{x})$	log-likelihood function
$\ln(x)$	natural logarithm function
$\log_a(x)$	logarithm function, base a
$L_v(x)$	Laguerre polynomial
L^m	L^m norm
$L(A)$	Lebesgue measure of set A
$L[k]$	likelihood ratio of a random sequence
$LB(N)$	Lebesgue sum
$\mathbf{L}$	lower triangular or lower shift matrix
$\min\{x, y\}$	minimum function
$\max\{x, y\}$	maximum function
m, n	parameters of F-distribution
m_e	median
m_o	mode
$m_X(t)$	moment generating function
$M_X(t)$	central moment generating function
N	number of variables or degrees of freedom
$N_o/2$	white-noise power spectral density
$o(x)$	little o notation for "of order"
$O(x)$	big O notation for "of order zero"
p, q	probability of "success" p and "failure" $q \triangleq 1 - p$
$p_X[x]$	probability mass function (pmf)
$p_{X,Y}[x, y]$	joint pmf
$p_{X\|Y}[x\|y]$	conditional pmf
P	power of $x(t)$ or pitch of speech
$P(E)$	probability of event E
$P(E\|F)$	conditional probability
$P_X(x)$	$P(X = x)$ (not a pmf)
P_{XX}, P_{XY}	power of $X(t)$, cross-power of $X(t)$ and $Y(t)$
P_e	average probability of error
$P_{ac}(X \leq x)$	absolutely continuous cdf
$P_s(X \leq x)$	singular cdf
$P(\theta)$	function for estimating angles of arrival
$\tilde{p}_n^{(k)}$	probability that a Markov chain returns to state n
p_{mn}	conditional probability for a Markov chain
$\mathbf{p}$	state probabilities of a Markov chain
$\mathbf{P}$	state-transition matrix of a one-step Markov chain with elements $\{p_{mn}\}$
$\mathbf{P}^{(r)}$	state-transition matrix of an r-step Markov chain with elements $\{p_{mn}^{(r)}\}$
q_m	mth quantile

$q_X[x]$	pmf used in Kullback–Leibler divergence
$\mathbf{q}$	eigenvector
$\mathbf{Q}$	matrix of eigenvectors
$Q(x)$	Q-function
$Q_m(a, b)$	Marcum Q-function
r	general shape parameter, degrees of freedom, radius, or magnitude of a complex quantity
r_0, r_1, r_2	gain parameters for blind equalization algorithms
$\{R, L, C\}$	resistor R, inductor L, capacitor C
R_n	nth decision region for maximum-likelihood signal detection
R	Riemann integral or risk function
$R(N)$, $R_L(N)$, $R_U(N)$	Riemann sum, lower and upper Riemann sums
RS	Riemann-Stieltjes integral
$RS(N)$	Riemann-Stieltjes sum
$R(x)$	ramp function
R_{XX}, R_{XY}	autocorrelation, cross-correlation
$R_{XX}(t_1, t_2)$, $R_{XX}[m, n]$	autocorrelation function (continuous, discrete)
$R_{XY}(t_1, t_2)$, $R_{XY}[m, n]$	cross-correlation function (continuous, discrete)
$\bar{R}_{XX}$	time average of random process $X(t)X(t + \tau)$
$\bar{\bar{R}}_{XX}$	time average of realization $x(t)x(t + \tau)$
$\text{rank}[\mathbf{A}]$	rank of matrix $\mathbf{A}$
$\text{rect}(x)$, $\text{rect}[x]$	rectangle function (continuous, discrete)
$\text{Re}(s)$	real part of s
$\text{Res}[X(z), p]$	residue of $X(z)$ evaluated at p
s	complex s-domain variable: $s = \sigma + j\omega$
$\text{sgn}(x)$	signum (or sign) function
$\text{sinc}(x)$	sinc function
$\text{stdev}[X]$	standard deviation of X
$\text{sup}(a, b)$	supremum
$\text{Sa}(x)$	Sa function
$S_{XX}(s)$, $S_{XX}(z)$	power spectral density (s, z domains)
$S^+_{XX}(z)$, $S^-_{XX}(z)$	spectral factorization: poles/zeros inside, outside the unit circle
$S^+_{XX}(s)$, $S^-_{XX}(s)$	spectral factorization: poles/zeros on the left-half, right-half of the s-plane
$S_{XX}(\omega)$, $S_{XX}(f)$, $S_{XX}(j\omega)$	power spectral density (frequency domain)
$S^D_{XX}(s)$, $S^D_{XX}(z)$	power spectral distribution (s, z domains)
$S^D_{XX}(\omega)$, $S^D_{XX}(f)$, $S^D_{XX}(j\omega)$	power spectral distribution (frequency domain)
S^2	sample variance
t	continuous time or variable of moment and cumulant generating functions
$\text{tr}[\mathbf{A}]$	trace of matrix $\mathbf{A}$
$\text{tri}(x)$	triangle function
T	statistic/estimator of a parameter or time duration
$\mathbf{T}$	statistic/estimator of a vector of parameters
T_d	time duration of a pulse
T_o	time interval between pulses
T_s	sample period
$\text{var}[X]$	variance of X
$u(t)$	unit-step (Heaviside) function
$u[k]$	discrete unit-step (Heaviside) function
$\mathbf{U}$	upper triangular or upper shift matrix
$\text{var}[X]$	variance of X
V	Vitali set
W	one-sided bandwidth

$\mathbf{W}$	weighting matrix
$X_{\text{DM}}[k]$	Doob-Meyer predictable part of $X[k]$
$X_{\text{MS}}[k]$	mean-square predictable part of $X[k]$
$\tilde{X}(t)$	baseband signal
$X_I(t), X_Q(t)$	in-phase and quadrature baseband components
$X_q(t)$	quantized signal
$\bar{X}$	sample mean
$\overline{X^n}$	sample nth moment
z	complex z-domain variable: $z = r\exp(j\omega)$
z^{-1}	delay operator: $z^{-1}x[k] = x[k-1]$

GREEK SYMBOLS

α	pdf scale parameter or algorithm step-size parameter
α_c	$1 - \alpha_c =$ confidence level (confidence interval)
α_s	significance level (hypothesis testing)
$\alpha_{n,m}$	gain factor of constant modulus array
β	location, scale, shape parameter of beta distribution, "leakage" factor of least-mean-square algorithm, or roll-off factor of raised-cosine pulse
$\beta_{n,m}$	constant modulus array shift factor
γ	Euler–Mascheroni constant: 0.57721566490153286060 . . .
γ_1	skewness
γ_2	kurtosis (excess)
$\gamma(a, b)$	lower incomplete gamma function
Γ_m	mth-order reflection coefficient
$\Gamma(x)$	gamma function
$\Gamma(a, b)$	upper incomplete gamma function
δ	small positive quantity
$\delta(x), \delta(x-y), \delta(x, y)$	Dirac delta function
$\delta[m], \delta[m-n]$	Kronecker delta function
$\delta_\omega(A)$	Dirac measure of set A
Δ	matrix determinant or quantization interval
Δ_{cd}	number of concordant pairs minus number of discordant pairs of samples
Δt	time interval t/k
Δ_N	time interval $(t - t_o)/N$
$\Delta X[k]$	difference sequence $X[k] - X[k-1]$
∇	gradient
$\boldsymbol{\nabla}[k]$	gradient vector
ϵ	small positive quantity
ζ	element of random process sample space $\mathcal{S}$
$\zeta(s)$	Riemann zeta function
η	"leakage" factor of least-mean-square algorithm, threshold of likelihood ratio test, or parameter of normalized least-mean-square algorithm
θ	parameter or an angle
$\boldsymbol{\theta}$	parameter vector
κ_n	nth cumulant
λ	eigenvalue, wavelength, Lagrange multiplier, pdf location/scale parameter, or forgetting factor
$\Lambda(x)$	triangle function, likelihood ratio test
$\boldsymbol{\lambda}$	vector of Lagrange multipliers
$\boldsymbol{\Lambda}$	square diagonal matrix

$\mu(A)$	measure of set A
μ	mean or parameter of μ-law quantizer
μ_n	nth raw moment
$\mu_{c,n}$	nth central moment
$\tilde{\mu}_{c,n}$	nth standardized central moment
$\bar{\mu}_X$	time average of random process $X(t)$
$\bar{\bar{\mu}}_X$	time average of realization $x(t)$
ξ	cost function
π	3.14159265358979323846...
$\{\pi(1), \ldots, \pi(N)\}$	permutation of $\{1, \ldots, N\}$
$\prod_n$	product over index n
ρ	correlation coefficient
$\bar{\rho}$	sample correlation coefficient
σ	standard deviation or real part of complex s
σ^2	variance
$\sigma(\mathcal{G})$	sigma-field of set $\mathcal{G}$
$\mathbf{\Sigma}$	diagonal rectangular matrix
Σ_n	summation over index n
τ	time lag $t_2 - t_1$ or time constant
τ_K	Kendall's rank correlation coefficient
τ_S	Spearman's rank correlation coefficient
ϕ	phase of a complex quantity or the null set
$\{\phi_n(t)\}$	Karhunen–Loève expansion basis functions
$\Phi_X(\omega)$	characteristic function (continuous pdf)
$\Phi_X(j\omega)$	characteristic function (discrete pmf)
χ_N	chi random variable (N degrees of freedom)
χ_N^2	chi-square random variable (N degrees of freedom)
$\Psi(a; bx)$	confluent hypergeometric function of the second kind
ω	element of Ω or radian frequency (radians/s)
ω_s	sample frequency (radians/s)
Ω	sample space for a random variable

CALLIGRAPHIC SYMBOLS

$\mathcal{A}$	set of atoms of a field
$\mathcal{B}(\mathcal{R})$	Borel σ-field
$\mathcal{C}$	complex numbers
$\mathcal{E}[X]$	expectation of X
$\mathcal{E}[X\|Y]$	conditional expectation
$\mathcal{E}_Y[\cdot]$	expectation over the pdf of Y
$\mathcal{F}$	field or σ-field
$\mathcal{G}$	generator of a σ-field
$\mathcal{I}(\theta)$	Fisher information
$\mathcal{I}(\boldsymbol{\theta})$	Fisher information matrix
$\mathcal{M}$	misadjustment of the least-mean-square algorithm
$\mathcal{N}$	natural numbers $\{1, 2, \ldots\}$
$\mathcal{P}(E)$	power set of E
$\mathcal{Q}$	rational numbers
$\mathcal{R}$	real numbers $(-\infty, \infty)$
$\mathcal{R}^+$	nonnegative real numbers $[0, \infty)$
$\mathcal{R}^-$	nonpositive real numbers $(-\infty, 0]$

$\mathcal{R}^N$	size $N \times 1$ vector of real numbers
$\mathcal{R}^{M \times N}$	size $M \times N$ matrix of real numbers
$\mathcal{S}$	sample space for a random process
$\mathcal{T}$	set of time instants
$\mathcal{Z}$	integers $\{\ldots, -2, -1, 0, 1, 2, \ldots\}$
$\mathcal{Z}^+$	nonnegative integers $\{0, 1, 2, \ldots\}$
$\mathcal{Z}^-$	nonpositive integers $\{\ldots, -1, 0\}$

MATHEMATICAL SYMBOLS

$\rightarrow$	frequency-domain transform or replaced with
$\xrightarrow{a.s.}$	almost sure convergence
$\xrightarrow{d}$	convergence in distribution
$\xrightarrow{L^m}$	convergence in mean
$\xrightarrow{m.s.}$	mean-square convergence
$\xrightarrow{p}$	convergence in probability
$\xrightarrow{s}$	sure convergence
$\sim$	distributed according to or similar to
$\Longrightarrow$	implies
$\triangleq$	defined as
$\equiv$	equivalent to
$\gg$	much greater than
$\ll$	much less than
$\subset$	subset
$\subseteq$	subset with equality
$\in$	element of
$\notin$	not an element of

ABBREVIATIONS

a.s.	almost surely
cdf	cumulative distribution function
icdf	inverse of the cdf
iid	independent and identically distributed
isi	independent and stationary increment
pdf	probability density function
plim	convergence in probability
pmf	probability mass function
s.	surely
s.t.	subject to
w.p.1	with probability one
A	ampere
AM	amplitude modulation
AOA	angle of arrival
AR	autoregressive
ARMA	autoregressive moving-average
ASK	amplitude shift keying
AWGN	additive white Gaussian noise
BER	bit error rate

BLU	best linear unbiased
BPSK	binary phase shift keying
BSC	binary symmetric channel
CF	characteristic function
CGF	cumulant generating function
CLT	central limit theorem
CM	constant modulus
CMA	constant modulus algorithm
CMGF	central moment generating function
CRLB	Cramér–Rao lower bound
dB	decibel
DD	decision directed
DE	differential equation or difference equation
DF	direction finding
DFT	discrete Fourier transform
DM	Doob-Meyer
DTFT	discrete-time Fourier transform
EM	expectation maximization
F	Farenheit
FFT	fast Fourier transform
FIR	finite impulse response
FM	frequency modulation
FMGF	factorial moment generating function
FR	forced response
FSK	frequency shift keying
FSM	finite-state machine
GS	Gram–Schmidt
Hz	Hertz
IIR	infinite impulse response
ISI	intersymbol interference
KL	Kullback–Leibler
KLE	Karhunen–Loève expansion
LD	Levinson–Durbin
LMF	least-mean fourth
LMS	least-mean square
LoS	line of sight
LPC	linear predictive coding
LPF	lowpass filter
LRT	likelihood ratio test
LS	least squares
LTI	linear time invariant
MA	moving average
MAE	mean absolute error
MAP	maximum a posteriori
MF	matched filter
MGF	moment generating function
MISO	multiple-input single-output
ML	maximum likelihood
MMSE	minimum mean-square error
MoM	method of moments
MQ	mean quartic
MQE	mean-quartic error

MS	mean square
MSE	mean-square error
MVDR	minimum-variance distortionless response
MUSIC	multiple signal classification
MVUE	minimum-variance unbiased estimate
NE	noise equivalent
NLMS	normalized least-mean-square algorithm
NM	Newton's method
NR	natural response
NS	noise subspace
PAM	pulse amplitude modulation
PCA	principal component analysis
PCM	pulse code modulation
PCR	principal component regression
PD	positive definite
PFE	partial fraction expansion
PGF	probability generating function
PLL	phase-locked loop
PRN	pseudorandom noise or pseudorandom number
PSD	power spectral density or positive semidefinite
PSK	phase-shift keying
QAM	quadrature amplitude modulation
QPSK	quadrature phase-shift keying
RB	Rao–Blackwell
RC	raised cosine
RLC	circuit: resistor R, inductor L, capacitor C
RLS	recursive least squares
RMS	root mean-square
ROC	region of convergence
SD	steepest descent
SINR	signal-to-interference-plus-noise ratio
SISO	single-input single-output
SNR	signal-to-noise ratio
SOI	signal of interest
SQNR	signal-to-quantization-noise ratio
SRRC	square-root raised cosine
TDL	tapped delay line
ULA	uniform linear array
UMVU	uniformly minimum variance unbiased
V	volt
V/UV	voiced/unvoiced
WLS	weighted least squares
ZF	zero forcing

REFERENCES

K. Abed-Meraim, W. Qiu, and Y. Hua, "Blind system identification," *Proceedings of the IEEE*, vol. 85, no. 8, pp. 1310–1322, August 1997.

S. T. Alexander, *Adaptive Signal Processing: Theory and Applications*, Springer-Verlag, New York, 1986.

B. D. O. Anderson and J. B. Moore, *Optimal Filtering*, Prentice-Hall, Englewood Cliffs, NJ, 1979.

S. P. Applebaum and D. J. Chapman, "Adaptive arrays with main beam constraints," *IEEE Transactions on Antennas and Propagation*, vol. AP-24, no. 5, pp. 650–662, September 1976.

R. B. Ash, *Basic Probability Theory*, John Wiley & Sons, New York, 1970.

R. B. Ash, *Real Analysis and Probability*, Academic Press, New York, 1972.

K. B. Athreya and S. N. Lahiri, *Measure Theory and Probability Theory*, Springer, New York, 2006.

C. Au and J. Tam, "Transforming variables using the Dirac generalized function," *The American Statistician*, vol. 53, no. 3, pp. 270–272, August 1999.

A. V. Balakrishnan, *Introduction to Random Processes in Engineering*, John Wiley & Sons, New York, 1995.

N. Balakrishnan, Ed., *Handbook of the Logistic Distribution*, Marcel Dekker, New York, 1992.

M. G. Bellanger, *Adaptive Digital Filters: Signal Processing and Communications*, 2nd edition, Marcel Dekker, New York, 2001.

S. Bellini, "Bussgang techniques for blind deconvolution and equalization," in *Blind Deconvolution*, S. Haykin, Ed., Prentice-Hall, Englewood Cliffs, NJ, pp. 8–59, 1994.

A. Benveniste and M. Goursat, "Blind equalizers," *IEEE Transactions on Communications*, vol. COM-32, no. 8, pp. 871–883, August 1984.

J. C. M. Bermudez and N. J. Bershad, "Transient and tracking performance analysis of the quantized LMS algorithm for time-varying system identification," *IEEE Transactions on Signal Processing*, vol. 44, no. 8, pp. 1990–1997, August 1996.

N. J. Bershad, "Analysis of the normalized LMS algorithm with Gaussian inputs," *IEEE Transactions on Acoustics, Speech, and Signal Processing*, vol. ASSP-34, no. 4, pp. 793–806, August 1986.

D. Bertsekas and J. N. Tsitsiklis, *Introduction to Probability*, Athena Scientific, Belmont, MA, 2002.

W. H. Beyer, *Standard Mathematical Tables*, 24th edition, CRC Press, Cleveland, OH, 1976.

P. Billingsley, *Probability and Measure*, 2nd edition, John Wiley & Sons, New York, 1986.

R. E. Blahut, *Principles and Practice of Information Theory*, Addison-Wesley, Reading, MA, 1987.

G. E. P. Box, G. M. Jenkins, and G. C. Reinsel, *Time Series Analysis: Forecasting, and Control*, 4th edition, John Wiley & Sons, Hoboken, NJ, 2008.

S. Boyd and L. Vandenberghe, *Convex Optimization*, Cambridge University Press, Cambridge, 2004.

R. N. Bracewell, *The Fourier Transform and Its Applications*, McGraw-Hill, New York, 1978.

C. Breining, P. Dreiseitel, E. Hänsler, A. Mader, B. Nitsch, H. Puder, T. Schertler, G. Schmidt, and J. Tip, "Acoustic echo control: An application of very-high-order adaptive filters," *IEEE Signal Processing Magazine*, vol. 16, no. 4, pp. 42–69, July 1999.

Y. Bresler and A. Macovski, "Exact maximum likelihood parameter estimation of superimposed exponential signals in noise," *IEEE Transactions on Acoustics, Speech, and Signal Processing*, vol. ASSP-34, no. 5, pp. 1081–1089, October 1986.

S. Brown and Z. Vranesic, *Fundamentals of Digital Logic with Verilog Design*, 2nd edition, McGraw-Hill, New York, 2008.

Probability, Random Variables, and Random Processes: Theory and Signal Processing Applications, First Edition. John J. Shynk.

K. M. Buckley, "Spatial/spectral filtering with linearly constrained minimum variance beamformers," *IEEE Transactions on Acoustics, Speech, and Signal Processing*, vol. ASSP-35, no. 3, pp. 249–266, March 1987.

J. A. Cadzow, "Spectral estimation: An overdetermined rational model equation approach," *Proceedings of the IEEE*, vol. 70, no. 9, pp. 907–939, September 1982.

J. A. Cadzow, *Foundations of Digital Signal Processing and Data Analysis*, Macmillan, New York, 1987.

R. E. Cagley, J. J. Shynk, and R. P. Gooch, "SINR enhancement in an IS-95 dowlink SIC receiver," *Wireless Personal Communications*, vol. 41, no. 4, pp. 563–580, June 2007.

M. Capiński and T. Zastawniak, *Probability Through Problems*, Springer, New York, 2001.

J. Capon, "High-resolution frequency-wavenumber spectrum analysis," *Proceedings of the IEEE*, vol. 57, no. 8, pp. 1408–1418, August 1969.

G. F. Carrier and C. E. Pearson, *Partial Differential Equations: Theory and Techniques*, Academic Press, New York, 1976.

J. A. Chambers, O. Tanrikulu, and A. G. Constantinides, "Least mean mixed-norm adaptive filtering," *Electronics Letters*, vol. 30, no. 19, pp. 1574–1575, September 1994.

S. Chandrasekaran, G. H. Golub, M. Gu, and A. H. Sayed, "Parameter estimation in the presence of bounded data uncertainties," *SIAM Journal on Matrix Analysis and Applications*, vol. 19, no. 1, pp. 235–252, January 1998.

L. F. Chaparro, *Signals and Systems Using MATLAB®*, Academic Press, Burlington, MA, 2011.

C. Chatfield, *The Analysis of Time Series: An Introduction*, 2nd edition, Chapman and Hall, New York, 1980.

C. Chatterjee, Z. Kang, and V. P. Roychowdjury, "Algorithms for accelerated convergence of adaptive PCA," *IEEE Transactions on Neural Networks*, vol. 11, no. 2, pp. 338–355, March 2000.

R. S. Chhikara and J. L. Folks, *The Inverse Gaussian Distribution: Theory, Methodology, and Applications*, Marcel Dekker, New York, 1989.

D. G. Childers, *Probability and Random Processes: Using MATLAB with Applications in Continuous and Discrete Time Systems*, Irwin/McGraw-Hill, Boston, 1997.

J. M. Cioffi, "Limited-precision effects in adaptive filtering," *IEEE Transactions on Circuits and Systems*, vol. CAS-34, no. 7, pp. 821–833, July 1987.

L. Cohen, "The history of noise," *IEEE Signal Processing Magazine*, vol. 22, no. 6, pp. 20–45, November 2005.

R. T. Compton, Jr., *Adaptive Antennas: Concepts and Performance*, John Wiley & Sons, Englewood Cliffs, NJ, 1988.

F. Conway and J. Siegelman, *Dark Hero of the Information Age: In Search of Norbert Wiener, the Father of Cybernetics*, Basic Books, New York, 2005.

T. M. Cover and J. A. Thomas, *Elements of Information Theory*, 2nd edition, John Wiley & Sons, Hoboken, NJ, 2006.

H. Cramér, *Mathematical Methods of Statistics*, Princeton University Press, Princeton, NJ, 1946.

W. B. Davenport, Jr., and W. L. Root, *An Introduction to the Theory of Random Signals and Noise*, McGraw-Hill, New York, 1958.

W. B. Davenport, Jr., *Probability and Random Processes: An Introduction for Applied Scientists and Engineers*, McGraw-Hill, New York, 1970.

H. A. David and H. N. Nagaraja, *Order Statistics*, 3rd edition, John Wiley & Sons, Hoboken, NJ, 2003.

R. de Neufville and J. H. Stafford, *Systems Analysis for Engineers and Managers*, McGraw-Hill, New York, 1971.

M. H. DeGroot and M. J. Schervish, *Probability and Statistics*, 4th edition, Addison-Wesley, Reading, MA, 2011.

Z. Ding, C. R. Johnson, Jr., and R. A. Kennedy, "Global convergence issues with linear blind adaptive equalizers," in *Blind Deconvolution*, S. Haykin, Ed., Prentice-Hall, Englewood Cliffs, NJ, pp. 60–120, 1994.

P. S. R. Diniz, *Adaptive Filtering: Algorithms and Practical Implementation*, 3rd edition, Springer, New York, 2008.

P. M. Djurić, "Spatial spectrum estimation," in *Handbook on Array Processing and Sensor Networks*, S. Haykin and K. J. R. Liu, Eds., John Wiley & Sons, Hoboken, NJ, 2009.

J. L. Doob, *Stochastic Processes*, John Wiley & Sons, New York, 1953.

J. L. Doob, *Measure Theory*, Springer-Verlag, New York, 1993.

R. C. Dorf, *Modern Control Systems*, 2nd edition, Addison-Wesley, Reading, MA, 1974.

E. R. Dougherty, *Probability and Statistics for the Engineering, Computing, and Physical Sciences*, Prentice-Hall, Englewood Cliffs, NJ, 1990.

A. El Gamal and Y.-H. Kim, *Network Information Theory*, Cambridge University Press, Cambridge, 2012.

H. Fan, "A structural view of asymptotic convergence speed of adaptive IIR filtering algorithms: Part I—Infinite precision implementation," *IEEE Transactions on Signal Processing*, vol. 41, no. 4, pp. 1493–1517, April 1993.

B. Farhang-Boroujeny, *Adaptive Filters: Theory and Applications*, John Wiley & Sons, New York, 1998.

G. Farmelo, *The Strangest Man: The Hidden Life of Paul Dirac, Mystic of the Atom*, Basic Books, New York, 2009.

W. Feller, *An Introduction to Probability Theory and Its Applications*, vol. I, 3rd edition, John Wiley & Sons, New York, 1968.

W. Feller, *An Introduction to Probability Theory and Its Applications*, vol. II, John Wiley & Sons, New York, 1971.

A. Feuer and E. Weinstein, "Convergence analysis of LMS filters with uncorrelated Gaussian data," *IEEE Transactions on Acoustics, Speech, and Signal Processing*, vol. ASSP-33, no. 1, pp. 222–230, February 1985.

T. L. Fine, *Probability and Probabilistic Reasoning for Electrical Engineering*, Prentice-Hall, Upper Saddle River, NJ, 2005.

T. E. Forster, *Set Theory with a Universal Set: Exploring an Untyped Universe*, 2nd edition, Clarendon Press, Oxford, 1995.

G. F. Franklin and J. D. Powell, *Digital Control of Dynamic Systems*, Addison-Wesley, Reading, MA, 1980.

B. Friedlander, "Lattice filters for adaptive processing," *Proceedings of the IEEE*, vol. 70, no. 8, pp. 829–867, August 1982.

O. L. Frost, III, "An algorithm for linearly constrained adaptive array processing," *Proceedings of the IEEE*, vol. 60, no. 8, pp. 926–935, August 1972.

W. F. Gabriel, "Adaptive arrays: An introduction," *Proceedings of the IEEE*, vol. 64, no. 2, pp. 239–272, February 1976.

W. F. Gabriel, "Spectral analysis and adaptive array superresolution techniques," *Proceedings of the IEEE*, vol. 68, no. 6, pp. 654–666, June 1980.

R. Gagliardi, *Introduction to Communications Engineering*, John Wiley & Sons, New York, 1978.

R. G. Gallager, *Information Theory and Reliable Communication*, John Wiley & Sons, New York, 1968.

W. A. Gardner, *Introduction to Random Processes with Applications to Signals & Systems*, 2nd edition, McGraw-Hill, New York, 1990.

A. Gelb, Ed., *Applied Optimal Estimation*, MIT Press, Cambridge, 1974.

A. Gersho, "Adaptive filtering with binary reinforcement," *IEEE Transactions on Information Theory*, vol. IT-30, no. 2, pp. 191–199, March 1984.

G. B. Giannakis, Y. Hua, P. Stoica, and L. Tong, Eds., *Signal Processing Advances in Wireless and Mobile Communications, Volume 1: Trends in Channel Estimation and Equalization*, Prentice-Hall, Upper Saddle River, NJ, 2000.

J. D. Gibson, "Adaptive prediction in speech differential encoding systems," *Proceedings of the IEEE*, vol. 68, no. 4, pp. 488–525, April 1980.

J. D. Gibson, Ed., *The Communications Handbook*, CRC Press, Boca Raton, FL, 1997.

J. D. Gibson, T. Berger, T. Lookabaugh, D. Lindbergh, and R. L. Baker, *Digital Compression for Multimedia: Principles and Standards*, Morgan Kaufmann, San Francisco, 1998.

W. J. Gilbert, *Modern Algebra with Applications*, John Wiley & Sons, New York, 1976.

R. M. Glorioso, *Engineering Cybernetics*, Prentice-Hall, Englewood Cliffs, NJ, 1975.

J. R. Glover, Jr., "Adaptive noise cancelling applied to sinusoidal interferences," *IEEE Transactions on Acoustics, Speech, and Signal Processing*, vol. ASSP-25, no. 6, pp. 484–491, December 1977.

D. N. Godard, "Self-recovering equalization and carrier tracking in two-dimensional data communication systems," *IEEE Transactions on Communications*, vol. COM-28, no. 11, pp. 1867–1875, November 1980.

M. A. Golberg, *An Introduction to Probability Theory with Statistical Applications*, Plenum Press, New York, 1984.

A. Goldsmith, *Wireless Communications*, Cambridge University Press, Cambridge, 2005.

G. H. Golub and C. F. Van Loan, *Matrix Computations*, Johns Hopkins University Press, Baltimore, MD, 1983.

R. P. Gooch and J. D. Lundell, "The CM array: An adaptive beamformer for constant modulus signals," in *Proceedings of the IEEE International Conference on Acoustics, Speech, and Signal Processing*, Tokyo, Japan, pp. 2523–2526, April 1986.

S. E. Goodman and S. T. Hedetniemi, *Introduction to the Design and Analysis of Algorithms*, McGraw-Hill, New York, 1977.

G. C. Goodwin and K. S. Sin, *Adaptive Filtering, Prediction, and Control*, Prentice-Hall, Englewood Cliffs, NJ, 1984.

I. S. Gradshteyn and I. M. Ryzhik, *Table of Integrals, Series, and Products*, Academic Press, New York, 1980.

R. M. Gray, *Probability, Random Processes, and Ergodic Properties*, Springer-Verlag, New York, 1987.

R. M. Gray, *Entropy and Information*, Springer-Verlag, New York, 1990.

R. M. Gray and L. D. Davisson, *An Introduction to Statistical Signal Processing*, Cambridge University Press, Cambridge, 2004.

R. M. Gray and J. G. Goodman, *Fourier Transforms: An Introduction for Engineers*, Springer, New York, 1995.

M. S. Grewal and A. P. Andrews, *Kalman Filtering: Theory and Practice*, 3rd edition, John Wiley & Sons, Hoboken, NJ, 2008.

L. J. Griffiths, "A simple adaptive algorithm for real-time processing in antenna arrays," *Proceedings of the IEEE*, vol. 57, no. 10, pp. 1696–1704, October 1969.

L. J. Griffiths and C. W. Jim, "An alternative approach to linearly constrained adaptive beamforming," *IEEE Transactions on Antennas and Propagation*, vol. AP-30, no. 1, pp. 27–34, January 1982.

V. S. Groza and S. Shelley, *Precalculus Mathematics*, Holt, Rinehart, and Winston, New York, 1972.

J. A. Gubner, *Probability and Random Processes for Electrical and Computer Engineers*, Cambridge University Press, Cambridge, 2006.

E. J. Gumbel, "Bivariate exponential distributions," *Journal of the American Statistical Society*, vol. 55, no. 292, pp. 698–707, December 1960.

B. Hajek, *Notes for ECE 534: An Exploration of Random Processes for Engineers*, unpublished notes: http://www.ifp.illinois.edu/~hajek/Papers/randomprocJan09.pdf, 2009.

A. Hald, *Statistical Theory with Engineering Applications*, John Wiley & Sons, New York, 1952.

R. W. Hamming, *Digital Filters*, 2nd edition, Prentice-Hall, Englewood Cliffs, NJ, 1983.

B. Hassibi, A. H. Sayed, and T. Kailath, "$\mathcal{H}^\infty$ optimality of the LMS algorithm," *IEEE Transactions on Signal Processing*, vol. 44, no. 2, pp. 267–280, February 1996.

D. Hatzinakos and C. L. Nikias, "Blind equalization using a tricepstrum-based algorithm," *IEEE Transactions on Communications*, vol. 39, no. 5, pp. 669–682, May 1991.

S. Haykin, *Adaptive Filter Theory*, 2nd edition, Prentice-Hall, Englewood Cliffs, NJ, 1991.

S. Haykin *Advances in Spectrum Analysis and Array Processing*, vol. II, Prentice-Hall, Englewood Cliffs, NJ, 1991.

S. Haykin, Ed., *Blind Deconvolution*, Prentice-Hall, Englewood Cliffs, NJ, 1994.

S. Haykin, *Adaptive Filter Theory*, 4th edition, Prentice-Hall, Upper Saddle River, NJ, 2002.

S. Haykin, *Communication Systems*, fourth edition, John Wiley & Sons, New York, 2001.

S. Haykin and M. Moher, *Introduction to Analog and Digital Communications*, 2nd edition, John Wiley & Sons, Hoboken, NJ, 2007.

M. L. Honig and D. G. Messerschmitt, *Adaptive Filters: Structures, Algorithms, and Applications*, Kluwer Academic Publishers, Boston, 1984.

C. W. Helstrom, *Probability and Stochastic Processes for Engineers*, 2nd edition, Macmillan, New York, 1991.

C. W. Helstrom, *Elements of Signal Detection and Estimation*, Prentice-Hall, Englewood Cliffs, NJ, 1995.

F. S. Hillier and G. J. Lieberman, *Operations Research*, 2nd edition, Holden-Day, San Francisco, 1974.

P. G. Hoel, S. C. Port, and C. J. Stone, *Introduction to Stochastic Processes*, Houghton Mifflin, Boston, 1972.

M. L. Honig, S. L. Miller, M. J. Shensa, and L. B. Milstein, "Performance of adaptive linear interference suppression in the presence of dynamic fading," *IEEE Transactions on Communications*, vol. 49, no. 4, pp. 635–645, April 2001.

L. L. Horowitz and K. D. Senne, "Performance advantage of complex LMS for controlling narrow-band adaptive arrays," *IEEE Transactions on Circuits and Systems*, vol. CAS-28, no. 6, pp. 562–576, June 1981.

Y. Hu and P. C. Loizou, "Subjective comparison and evaluation of speech enhancement algorithms," *Speech Communication*, vol. 49, no. 7, pp. 588–601, July 2007.

S. Hwang and J. J. Shynk, "Blind GPS receiver with a modified despreader for interference suppression," *IEEE Transactions on Aerospace and Electronic Systems*, vol. 42, no. 2, pp. 503–513, April 2006.

O. C. Ibe, *Fundamentals of Applied Probability and Random Processes*, Elsevier Academic Press, London, 2005.

R. A. Iltis, J. J. Shynk, and K. Giridhar, "Bayesian algorithms for blind equalization using parallel adaptive filtering," *IEEE Transactions on Communications*, vol. 42, no. 2/3/4, pp. 1017–1032, February/March/April 1994.

V. K. Ingle and J. G. Proakis, *Digital Signal Processing Using MATLAB®*, 3rd edition, Cengage Learning, Stamford, CT, 2011.

Y. Isukapalli and B. D. Rao, "An analytically tractable approximation for the Gaussian Q-function," *IEEE Communications Letters*, vol. 12, no. 9, pp. 669–671, September 2008.

N. K. Jablon, "Steady state analysis of the generalized sidelobe canceller by adaptive noise cancelling techniques," *IEEE Transactions on Antennas and Propagation*, vol. AP-34, no. 3, pp. 330–337, March 1986.

W. C. Jakes, Ed., *Microwave Mobile Communications*, Wiley-IEEE Press, New York, 1994.

N. S. Jayant and P. Noll, *Digital Coding of Waveforms: Principles and Applications to Speech and Video*, Prentice-Hall, Englewood Cliffs, NJ, 1984.

C. R. Johnson, Jr., "Adaptive IIR filtering: Current results and open issues," *IEEE Transactions on Information Theory*, vol. IT-30, no. 2, pp. 237–250, March 1984.

C. R. Johnson, Jr., *Lectures on Adaptive Parameter Estimation*, Prentice-Hall, Englewood Cliffs, NJ, 1988.

C. R. Johnson, Jr., P. Schniter, T. J. Endres, J. D. Behm, D. R. Brown, and R. A. Casas, "Blind equalization using the constant modulus criterion: A review," *Proceedings of the IEEE*, vol. 86, no. 10, pp. 1927–1950, October 1998.

N. L. Johnson, S. Kotz, and N. Balakrishnan, *Continuous Univariate Distributions*, vol. 1, 2nd edition, John Wiley & Sons, New York, 1994.

N. L. Johnson, S. Kotz, and N. Balakrishnan, *Continuous Univariate Distributions*, vol. 2, 2nd edition, John Wiley & Sons, New York, 1995.

N. L. Johnson, S. Kotz, and N. Balakrishnan, *Discrete Multivariate Distribution*, John Wiley & Sons, New York, 1997.

N. L. Johnson, A. W. Kemp, and S. Kotz, *Univariate Discrete Distributions*, 3rd edition, John Wiley & Sons, Hoboken, NJ, 2005.

I. T. Jolliffe, "A note on the use of principal components in regression," *Journal of the Royal Statistical Society. Series C (Applied Statistics)*, vol. 31, no. 3, pp. 300–303, 1982.

I. T. Jolliffe, *Principal Component Analysis*, Springer-Verlag, New York, 1986.

T. Kailath, "A view of three decades of linear filtering theory," *IEEE Transactions on Information Theory*, vol. IT-20, no. 2, pp. 146–181, March 1974.

T. Kailath, *Lectures on Linear Least-Squares Estimation*, Springer-Verlag, New York, 1976.

T. Kailath, *Linear Systems*, Prentice-Hall, Englewood Cliffs, NJ, 1980.

T. Kailath, A. H. Sayed, and B. Hassibi, *Linear Estimation*, Prentice-Hall, Upper Saddle River, NJ, 2000.

R. E. Kalman, "A new approach to linear filtering and prediction problems," *Transactions of the ASME–Journal of Basic Engineering*, vol. 82, series D, pp. 34–45, March 1960.

R. E. Kalman and R. S. Bucy, "New results in linear filtering an prediction theory," *Transactions of the ASME–Journal of Basic Engineering*, vol. 83, series D, pp. 95–108, March 1961.

S. Karlin and H. M. Taylor, *A First Course in Stochastic Processes*, 2nd edition, Academic Press, New York, 1975.

S. M. Kay and S. L. Marple, "Spectrum analysis—A modern perspective," *Proceedings of the IEEE*, vol. 69, no. 11, pp. 1380–1419, November 1981.

S. M. Kay, *Modern Spectral Estimation: Theory and Applications*, Prentice-Hall, Englewood Cliffs, NJ, 1988.

S. M. Kay, *Fundamentals of Statistical Signal Processing, Volume I: Estimation Theory*, Prentice-Hall, Englewood Cliffs, NJ, 1993.

S. M. Kay, *Fundamentals of Statistical Signal Processing, Volume 2: Detection Theory*, Prentice-Hall, Englewood Cliffs, NJ, 1998.

A. V. Keerthi, A. Mathur, and J. J. Shynk, "Misadjustment and tracking analysis of the constant modulus array," *IEEE Transactions on Signal Processing* vol. 46, no. 1, pp. 51–58, January 1998.

J. F. C. Kingman and S. J. Taylor, *Introduction to Measure and Probability*, Cambridge University Press, Cambridge, 1966.

B. Kolman, *Elementary Linear Algebra*, 2nd edition, Macmillan, New York, 1977.

A. N. Kolmogorov, *Foundations of the Theory of Probability*, 2nd edition, Chelsea, New York, 1960.

B. Kosko, Ed., *Neural Networks for Signal Processing*, Prentice-Hall, Englewood Cliffs, NJ, 1992.

S. Kotz, N. Balakrishnan, and N. L. Johnson, *Continuous Multivariate Distributions: Models and Applications*, vol. 1, John Wiley & Sons, New York, 2000.

S. Kotz, T. J. Kozubowski, and K. Podgórski, *The Laplace Distribution and Generalizations: A Revisit with Applications to Communications, Economics, Engineering, and Finance*, Birkhäuser, Boston, 2001.

E. Kreyszig, *Advanced Engineering Mathematics*, 4th edition, John Wiley & Sons, New York, 1979.

H. Krim and M. Viberg, "Two decades of array signal processing research: The parametric approach," *IEEE Signal Processing Magazine*, vol. 13, no. 4, pp. 67–94, July 1996.

K. Krishnamoorthy, *Handbook of Statistical Distributions with Applications*, Chapman & Hall/CRC Press, Boca Raton, FL, 2006.

V. Krishnan, *Probability and Random Processes*, John Wiley & Sons, Hoboken, NJ, 2006.

M. Krstić, I. Kanellakopoulos, and P. Kokotović, *Nonlinear and Adaptive Control Design*, John Wiley & Sons, New York, 1995.

A. Kuh, "Comparison of tracking algorithms for single layer threshold networks in the presence of random drift," *IEEE Transactions on Signal Processing*, vol. 45, no. 3, pp. 640–649, March 1997.

S. Kullback and R. A. Leibler, "On information and sufficiency," *Annals of Mathematical Statistics*, vol. 22, no. 1, pp. 79–86, March 1951.

R. Kumaresan and D. W. Tufts, "Estimating the angles of arrival of multiple plane waves," *IEEE Transactions on Aerospace and Electronic Systems*, vol. AES-19, no. 1, pp. 134–139, January 1983.

M. Kunt, *Digital Signal Processing*, Artech House, Norwood, MA, 1987.

K.-C. Lai and J. J. Shynk, "Performance evaluation of a generalized linear SIC for DS/CDMA signals," *IEEE Transactions on Signal Processing*, vol. 51, no. 6, pp. 1604–1614, June 2003.

I. D. Landau, *Adaptive Control: The Model Reference Approach*, Marcel Dekker, New York, 1979.

H. J. Larson and B. O. Shubert, *Probabilistic Models in Engineering Sciences: Random Variables and Stochastic Processes*, vol. I, John Wiley & Sons, New York, 1979.

H. J. Larson and B. O. Shubert, *Probabilistic Models in Engineering Sciences: Random Noise, Signals and Dynamic Systems*, vol. II, John Wiley & Sons, New York, 1979.

B. P. Lathi, *Signals, Systems, and Communication*, John Wiley & Sons, New York, 1965.

B. P. Lathi, *Signal Processing & Linear Systems*, Oxford University Press, New York, 2000.

B. P. Lathi and Z. Ding, *Modern Digital and Analog Communication Systems*, 4th edition, Oxford University Press, New York, 2009.

A. J. Laub, *Matrix Analysis for Scientists and Engineers*, Society for Industrial and Applied Mathematics, Philadelphia, 2005.

W. C. Y. Lee, *Mobile Cellular Telecommunications Systems*, McGraw-Hill, New York, 1989.

E. A. Lee and D. G. Messerschmitt, *Digital Communication*, 2nd edition, Kluwer Academic Publishers, Boston, 1994.

E. L. Lehmann, *Testing Statistical Hypotheses*, John Wiley & Sons, New York, 1959.

E. L. Lehmann, *Theory of Point Estimation*, John Wiley & Sons, New York, 1983.

S. J. Leon, *Linear Algebra with Applications*, 3rd edition, Macmillan, New York, 1990.

F. C. Leone, L. S. Nelson, and R. B. Nottingham, "The folded normal distribution," *Technometrics*, vol. 3, no. 4, pp. 543–550, November 1961.

A. Leon-Garcia, *Probability, Statistics, and Random Processes for Electrical Engineers*, Prentice-Hall, Upper Saddle River, NJ, 2008.

H. Lev-Ari, T. Kailath, and J. M. Cioffi, "Least-squares adaptive lattice and transversal filters: A unified geometric theory," *IEEE Transactions on Information Theory*, vol. IT-30, no. 2, pp. 222–236, March 1984.

B. C. Levy, *Principles of Signal Detection and Parameter Estimation*, Springer, New York, 2008.

Y. Li and Z. Ding, "Convergence analysis of finite length blind adaptive equalizers," *IEEE Transactions on Signal Processing*, vol. 43, no. 9, pp. 2120–2129, September 1995.

M. Liebling, T. Blu, and M. Unser, "Fresnelets: New multiresolution wavelet bases for digital holography," *IEEE Transactions on Image Processing*, vol. 12, no. 1, pp. 29–43, January 2003.

E. Limpert, W. A. Stahel, and M. Abbt, "Log-normal distributions across the sciences: Keys and clues," *BioScience*, vol. 51, no. 5, pp. 341–352, May 2001.

R. E. Lippmann, "An introduction to computing with neural nets," *IEEE ASSP Magazine*, vol. 4, no. 2, pp. 4–22, April 1987.

L. Ljung, *System Identification: Theory for the User*, Prentice-Hall, Englewood Cliffs, NJ, 1987.

L. Ljung and T. Söderström, *Theory and Practice of Recursive Identification*, MIT Press, Cambridge, MA, 1983.

R. Lucky, "Automatic equalization for digital communications," *Bell System Technical Journal*, vol. 44, no. 4, pp. 547–588, April 1965.

L. C. Ludeman, *Random Processes: Filtering, Estimation, and Detection*, John Wiley & Sons, Hoboken, NJ, 2003.

D. G. Luenberger, *Linear and Nonlinear Programming*, 2nd edition, Addison-Wesley, Reading, MA, 1984.

J. Lundell and B. Widrow, "Applications of the constant modulus adaptive beamformer to constant and non-constant modulus signals," in *Proceedings of the Twenty-First Asilomar Conference on Signals, Systems, and Computers*, Pacific Grove, CA, pp. 432–436, November 1987.

D. J. C. MacKay, *Information Theory, Inference, and Learning Algorithms*, Cambridge University Press, Cambridge, 2003.

U. Madhow, *Fundamentals of Digital Communication*, Cambridge University Press, Cambridge, 2008.

V. Madisetti, Ed., *The Digital Signal Processing Handbook*, 2nd edition, CRC Press, Boca Raton, FL, 2009.

J. Makhoul, "Linear prediction: A tutorial rcview," *Proceedings of the IEEE*, vol. 63, no. 4, pp. 561–580, April 1975.

B. S. Manjunath, P. Salembier, and T. Sikora, Eds., *Introduction to MPEG-7: Mulimedia Content Description Interface*, John Wiley & Sons, New York, 2002.

D. G. Manolakis, V. K. Ingle, and S. M. Kogon, *Statistical and Adaptive Signal Processing*, McGraw-Hill, New York, 2000.

K. V. Mardia, *Families of Bivariate Distributions*, Griffin, London, 1970.

J. D. Markel and A. H. Gray, Jr., *Linear Prediction of Speech*, Springer-Verlag, New York, 1976.

G. Marsaglia, "Ratios of normal variables," *Journal of Statistical Software*, vol. 16, no. 4, pp. 1–10, May 2006.

V. J. Mathews, "Performance analysis of adaptive filters equipped with the dual sign algorithm," *IEEE Transactions on Signal Processing*, vol. 39, no. 1, pp. 85–91, January 1991.

J. H. McClellan, R. W. Schafer, and M. A. Yoder, *DSP First: A Multimedia Approach*, Prentice-Hall, Upper Saddle River, NJ, 1998.

R. J. McEliece, *The Theory of Information and Coding: A Mathematical Framework for Communication*, Addison-Wesley, Reading, MA, 1977.

E. L. McMahon, "An extension of Price's theorem," *IEEE Transactions on Information Theory*, vol. 10, no. 2, p. 168, April 1964.

E. L. Melnick and A. Tenenbein, "The misspecifications of the normal distribution," *The American Statistician*, vol. 36, no. 4, pp. 372–373, November 1982.

Members of the Technical Staff, Bell Laboratories, *Transmission Systems for Communications*, 4th edition, Bell Telephone Laboratories/Western Electric, Winston-Salem, NC, 1970.

J. M. Mendel, *Discrete Techniques of Parameter Estimation: The Equation Error Formulation*, Marcel Dekker, New York, 1973.

J. M. Mendel, "Tutorial on higher-order statistics (spectra) in signal processing and systems theory: Theoretical results and some applications," *Proceedings of the IEEE*, vol. 79, no. 3, pp. 278–305, March 1991.

J. M. Mendel, *Lessons in Estimation Theory for Signal Processing, Communications, and Control*, Prentice-Hall, Englewood Cliffs, NJ, 1995.

D. Middleton, *An Introduction to Statistical Communication Theory*, Wiley-IEEE Press, New York, 1996.

I. Miller and J. E. Freund, *Probability and Statistics for Engineers*, 2nd edition, Prentice-Hall, Englewood Cliffs, NJ, 1977.

T. M. Mills, *Problems in Probability*, World Scientific Publishing Company, River Edge, NJ, 2001.

M. L. Minsky and S. A. Papert, *Perceptrons: An Introduction to Computational Geometry*, expanded edition, MIT Press, Cambridge, MA, 1988.

S. K. Mitra and J. F. Kaiser, Eds., *Handbook for Digital Signal Processing*, John Wiley & Sons, New York, 1993.

S. K. Mitra, *Digital Signal Processing: A Computer-Based Approach*, 4th edition, McGraw-Hill, New York, 2010.

N. Mohanty, *Random Signals, Estimation, and Identification: Analysis and Applications*, Van Nostrand Reinhold, New York, 1986.

R. A. Monzingo and T. W. Miller, *Introduction to Adaptive Arrays*, John Wiley & Sons, New York, 1980.

A. M. Mood, F. A. Graybill, and D. C. Boes, *Introduction to the Theory of Statistics*, 3rd edition, McGraw-Hill, New York, 1974.

T. K. Moon, "The expectation-maximization algorithm," *IEEE Signal Processing Magazine*, vol. 13, no. 6, pp. 47–60, November 1996.

T. K. Moon and W. C. Stirling, *Mathematical Methods and Algorithms for Signal Processing*, Prentice-Hall, Upper Saddle, NJ, 2000.

F. Mosteller and J. W. Tukey, *Data Analysis and Regression: A Second Course in Statistics*, Addison-Wesley, Reading, MA, 1977.

B. Mulgrew and C. F. N. Cowan, *Adaptive Filters and Equalisers*, Kluwer Academic Publishers, Boston, 1988.

H. T. Nagle, Jr., B. D. Carroll, and J. D. Irwin, *An Introduction to Computer Logic*, Prentice-Hall, Englewood Cliffs, NJ, 1975.

M. Nayeri, H. Fan, and W. K. Jenkins, "Some Characteristics of error surfaces for insufficient order adaptive IIR filters," *IEEE Transactions on Acoustics, Speech, and Signal Processing*, vol. 38, no. 6, pp. 1222–1227, July 1990.

C. L. Nikias and A. P. Petropulu, *Higher-Order Spectra Analysis: A Nonlinear Signal Processing Framework*, Prentice-Hall, Upper Saddle River, NJ, 1993.

J. W. Nilsson and S. A Riedel, *Electric Circuits*, 7th edition, Pearson Prentice-Hall, Upper Saddle River, NJ, 2005.

J. J. O'Connor and E. F. Robertson, *The MacTutor History of Mathematics Archive*, http://www-history.mcs.st-andrews.ac.uk/index.html, 2011.

T. Ogunfunmi and M. Narasimha, *Principles of Speech Coding*, CRC Press, Boca Raton, FL, 2010.

P. Olofsson, *Probability, Statistics, and Stochastic Processes*, John Wiley & Sons, Hoboken, NJ, 2005.

A. V. Oppenheim and R. W. Schafer, *Digital Signal Processing*, Prentice-Hall, Englewood Cliffs, NJ 1975.

A. V. Oppenheim, A. S. Willsky, and S. H. Nawab, *Signals and Systems*, 2nd edition, Prentice-Hall, Englewood Cliffs, NJ, 1996.

A. V. Oppenheim and R. W. Schafer, *Discrete-Time Signal Processing*, 3rd edition, Prentice-Hall, Upper Saddle River, NJ, 2009.

N. Pal, C. Jin, and W. K. Lim, *Handbook of Exponential and Related Distributions for Engineers and Scientists*, Chapman & Hall/CRC Press, Boca Raton, FL, 2006.

A. Papoulis, *Probability, Random Variables, and Stochastic Processes*, McGraw-Hill, New York, 1965.

A. Papoulis, *Probability & Statistics*, Prentice-Hall, Englewood, Cliffs, NJ, 1990.

E. Parzen, *Modern Probability Theory and Its Applications*, John Wiley & Sons, New York, 1960.

E. Parzen, *Stochastic Processes*, Holden-Day, San Francisco, 1962.

R. F. Pawula, "A modified version of Price's theorem," *IEEE Transactions on Information Theory*, vol. IT-13, no. 2, pp. 285–288, April 1967.

P. Z. Peebles, Jr., *Probability, Random Variables, and Signal Principles*, 4th edition, McGraw-Hill, New York, 2001.

M. R. Petraglia, R. G. Alves, and P. S. R. Diniz, "New structures for adaptive filtering in subbands with critical sampling," *IEEE Transactions on Signal Processing*, vol. 48, no. 12, pp. 3316–3327, December 2000.

G. Picchi and G. Prati, "Blind equalization and carrier recovery using a "stop-and-go" decision-directed algorithm," *IEEE Transactions on Communications*, vol. COM-35, no. 9, pp. 877–887, September 1987.

S. U. Pillai, T. Suel, and S. Cha, "The Perron-Frobenius theorem," *IEEE Signal Processing Magazine*, vol. 22, no. 2, pp. 62–75, March 2005.

K. C. Pohlmann, *Principles of Digital Audio*, 4th edition, McGraw-Hill, New York, 2000.

C. Polprasert and J. A. Ritcey, "A Nakagami fading phase difference distribution and its impact on BER performance," *IEEE Transactions on Wireless Communications*, vol. 7, no. 7, pp. 2805–2813, July 2008.

H. V. Poor, *An Introduction to Signal Detection and Estimation*, 2nd edition, Springer, New York, 1994.

H. V. Poor and O. Hadjiliadis, *Quickest Detection*, Cambridge University Press, Cambridge, 2008.

B. Porat, *Digital Processing of Random Signals*, Prentice-Hall, Englewood Cliffs, NJ, 1994.

R. Price, "A useful theorem for nonlinear devices having Gaussian inputs," *IRE Transactions on Information Theory*, vol. 4, no. 2, pp. 69–72, June 1958.

J. G. Proakis, *Digital Communications*, fourth edition, McGraw-Hill, New York, 2001.

J. G. Proakis and M. Salehi, *Communication Systems Engineering*, Prentice-Hall, Upper Saddle River, NJ, 2002.

P. E. Protter, *Stochastic Integration and Differential Equations*, 2nd edition, Springer, New York, 2004.

M. B. Pursley, *Random Processes in Linear Systems*, Prentice-Hall, Upper Saddle River, NJ, 2002.

M. B. Pursley, *Introduction to Digital Communications*, Pearson Prentice-Hall, Upper Saddle River, NJ, 2005.

S. U. H. Qureshi, "Adaptive equalization," *Proceedings of the IEEE*, vol. 73, no. 9, pp. 1349–1387, September 1985.

L. R. Rabiner and B. Gold, *Theory and Application of Digital Signal Processing*, Prentice-Hall, Englewood Cliffs, NJ, 1975.

L. R. Rabiner and R. W. Schafer, *Digital Processing of Speech Signals*, Prentice-Hall, Englewood Cliffs, NJ, 1978.

H. Raiffa, *Decision Analysis: Introductory Lectures on Choices under Uncertainty*, Addison-Wesley, Reading, MA, 1970.

T. S. Rappaport, *Wireless Communications*, Prentice-Hall, Upper Saddle River, NJ, 1996.

I. S. Reed, "On a moment theorem for complex Gaussian processes" *IRE Transactions on Information Theory*, vol. 8, no. 3, pp. 194–195, April 1962.

P. A. Regalia, *Adaptive IIR Filtering in Signal Processing and Control*, Marcel Dekker, New York, 1995.

S. O. Rice, "Mathematical analysis of random noise," *Bell System Technical Journal*, vol. 24, no. 1, pp. 46–156, January 1945.

L. F. Richardson, *Measure and Integration: A Concise Introduction to Real Analysis*, John Wiley & Sons, Hoboken, NJ, 2009.

E. A. Robinson, *Probability Theory and Applications*, International Human Development Corporation, Boston, 1985.

K. Rose and S. L. Regunathan, "Toward optimality in scalable predictive coding," *IEEE Transactions on Image Processing*, vol. 10, no. 7, pp. 965–976, July 2001.

K. A. Ross, *Elementary Analysis: The Theory of Calculus*, Springer-Verlag, New York, 1980.

S. Ross, *A First Course in Probability*, MacMillan, New York, 1976.

S. M. Ross, *Stochastic Processes*, John Wiley & Sons, New York, 1983.

S. A. Ross, *Introduction to Ordinary Differential Equations*, 2nd edition, John Wiley & Sons, New York, 1977.

D. W. Rousu, "Weibull skewness and kurtosis as a function of the shape parameter," *Technometrics*, vol. 15, no. 4, pp. 927–930, November 1973.

S. Roy, "Subspace blind adaptive detection for multiuser CDMA," *IEEE Transactions on Communications*, vol. 48, no. 1, pp. 169–175, January 2000.

M. Rupp and A. H. Sayed, "On the convergence of blind adaptive equalizers for constant modulus signals," *IEEE Transactions on Communications*, vol. 48, no. 5, pp. 795–803, May 2000.

D. E. Rumelhart, J. L. McClelland, and the PDP Research Group, *Parallel Distributed Processing: Explorations in the Microstructure of Cognition, Volume 1: Foundations*, MIT Press, Cambridge, MA, 1986.

J. Salz, Optimum mean-square decision feedback equalization," *Bell System Technical Journal*, vol. 52, no. 8, pp. 1341–1373, October 1973.

Y. Sato, "A method of self-recovering equalization for multilevel amplitude-modulation systems," *IEEE Transactions on Communications*, vol. COM-23, no. 6, pp. 679–682, June 1975.

A. H. Sayed, *Fundamentals of Adaptive Filtering*, John Wiley & Sons, Hoboken, NJ, 2003.

A. H. Sayed, *Adaptive Filters*, John Wiley & Sons, Hoboken, NJ, 2008.

L. L. Scharf, *Statistical Signal Processing*, Prentice-Hall, Englewood Cliffs, NJ, 1990.

R. J. Schilling and H. Lee, *Engineering Analysis: A Vector Space Approach*, John Wiley & Sons, New York, 1988.

R. O. Schmidt, "Multiple emitter location and signal parameter estimation," *IEEE Transactions on Antennas and Propagation*, vol. AP-34, no. 3, pp. 276–281, March 1986.

T. A. Schonhoff and A. A. Giordano, *Detection and Estimation Theory and Its Applications*, Prentice-Hall, Upper Saddle River, NJ, 2006.

M. Schwartz, *Information, Transmission, Modulation, and Noise*, 3rd edition, McGraw-Hill, New York, 1980.

V. Seshadri, *The Inverse Gaussian Distribution: Statistical Theory and Applications*, Springer-Verlag, New York, 1999.

W. A. Sethares, "Adaptive algorithms with nonlinear data and error functions," *IEEE Transactions on Signal Processing*, vol. 40, no. 9, pp. 2199–2206, September 1992.

O. Shalvi and E. Weinstein, "New criteria for blind deconvolution of nonminimum phase systems (channels)," *IEEE Transactions on Information Theory*, vol. 36, no. 2, pp. 312–321, March 1990.

K. S. Shanmugan and A. M. Breipohl, *Random Signals: Detection, Estimation, and Data Analysis*, John Wiley & Sons, New York, 1988.

C. E. Shannon, "A mathematical theory of communication," *Bell System Technical Journal*, vol. 27, no. 3, pp. 379–423, July 1948.

C. E. Shannon, "A mathematical theory of communication," *Bell System Technical Journal*, vol. 27, no. 4, pp. 623–656, October 1948.

C. E. Shannon and W. Weaver, *The Mathematical Theory of Communication*, University of Illinois Press, Urbana, IL, 1963.

A. N. Shiryayev, *Probability*, Springer-Verlag, New York, 1984.

J. J. Shynk, "Adaptive IIR Filtering," *IEEE ASSP Magazine*, vol. 6, no. 2, pp. 4–21, April 1989.

J. J. Shynk and C. K. Chan, "Performance surfaces of the constant modulus algorithm based on a conditional Gaussian model," *IEEE Transactions on Signal Processing*, vol. 41, no. 5, pp. 1965–1969, May 1993.

J. J. Shynk and R. P. Gooch, "The constant modulus array for cochannel signal copy and direction finding" *IEEE Transactions on Signal Processing*, vol. 44, no. 3, pp. 652–660, March 1996.

M. K. Simon, *Probability Distributions Involving Gaussian Random Variables: A Handbook for Engineers, Scientists, and Mathematicians*, Springer, New York, 2002.

D. T. M. Slock, "On the convergence behavior of the LMS and the normalized LMS algorithms," *IEEE Transactions on Signal Processing*, vol. 41, no. 9, pp. 2811–2825, September 1993.

I. S. Sokolnikoff and R. M. Redheffer, *Mathematics of Physics and Modern Engineering*, McGraw-Hill, New York, 1966.

V. Solo and X. Kong, *Adaptive Signal Processing Algorithms*, Prentice-Hall, Englewood Cliffs, NJ, 1995.

H. W. Sorenson, *Parameter Estimation: Principles and Problems*, Marcel Dekker, New York, 1980.

M. D. Srinath, P. K. Rajasekaran, and R. Viswanathan, *Introduction to Statistical Signal Processing with Applications*, Prentice-Hall, Upper Saddle River, NJ, 1996.

H. Stark and J. W. Woods, *Probability and Random Processes with Applications to Signal Processing*, 3rd edition, Prentice-Hall, Upper Saddle River, NJ, 2002.

S. D. Stearns and D. R. Hush, *Digital Signal Analysis*, Prentice-Hall, Englewood Cliffs, NJ, 1990.

P. Stoica and R. Moses, *Introduction to Spectral Analysis*, Prentice-Hall, Upper Saddle River, NJ, 1997.

P. Stoica and A. Nehorai, "MUSIC, maximum likelihood, and Cramér-Rao bound," *IEEE Transactions on Acoustics, Speech, and Signal Processing*, vol. 37, no. 5, pp. 720–741, May 1989.

G. Strang, *Linear Algebra and Its Applications*, Academic Press, New York, 1976.

P. Strobach, *Linear Prediction Theory: A Mathematical Basis for Adaptive Systems*, Springer-Verlag, New York, 1990.

B. L. Sturm and J. J. Shynk, "Sparse approximation and the pursuit of meaningful signal models with interference adaptation," *IEEE Transactions on Audio, Speech, and Language Processing*, vol. 18, no. 3, pp. 461–472, March 2010.

A. L. Swindlehurst and T. Kailath, "A performance analysis of subspace-based methods in the presence of model errors, part I: The MUSIC algorithm," *IEEE Transactions on Signal Processing*, vol. 40, no. 7, pp. 1758–1774, July 1992.

M. E. Tarter and V. A. Clark, "Properties of the median and other order statistics of logistic variates," *The Annals of Mathematical Statistics*, vol. 36, no. 6, pp. 1779–1786, December 1965.

H. M. Taylor and S. Karlin, *An Introduction to Stochastic Modeling*, Academic Press, New York, 1984.

C. W. Therrien, *Discrete Random Signals and Statistical Signal Processing*, Prentice-Hall, Englewood Cliffs, NJ, 1992.

C. W. Therrien and M. Tummala, *Probability and Random Processes for Electrical and Computer Engineers*, 2nd edition, CRC Press, Boca Raton, FL, 2012.

R. E. Thomas, A. J. Rosa, and G. J. Toussaint, *The Analysis and Design of Linear Circuits*, 6th edition, John Wiley & Sons, Hoboken, NJ, 2009.

G. B. Thomas, Jr., *Calculus and Analytic Geometry, Part One: Functions of One Variable and Analytic Geometry*, 4th edition, Addison-Wesley, Reading, MA, 1968.

G. B. Thomas, Jr., *Calculus and Analytic Geometry, Part Two: Vectors and Functions of Several Variables*, 4th edition, Addison-Wesley, Reading, MA, 1968.

L. Tong, G. Xu, and T. Kailath, "Blind identification and equalization based on second-order statistics: A time-domain approach," *IEEE Transactions on Information Theory*, vol. 40, no. 2, pp. 340–349, March 1994.

J. R. Treichler and B. G. Agee, "A new approach to multipath correction of constant modulus signals," *IEEE Transactions on Acoustics, Speech, and Signal Processing*, vol. ASSP-31, no. 2, pp. 459–472, April 1983.

J. R. Treichler, C. R. Johnson, Jr., and M. G. Larimore, *Theory and Design of Adaptive Filters*, Prentice-Hall, Upper Saddle River, NJ, 2001.

S. A. Tretter, *Introduction to Discrete-Time Signal Processing*, John Wiley & Sons, New York, 1976.

J. K. Tugnait, "On fractionally spaced blind adaptive equalization under symbol timing offsets using Godard and related equalizers," *IEEE Transactions on Signal Processing*, vol. 44, no. 7, pp. 1817–1821, July 1996.

P. P. Vaidyanathan, *Multirate Systems and Filter Banks*, Prentice-Hall, Englewood Cliffs, NJ, 1993.

H. L. Van Trees, *Detection, Estimation, and Modulation Theory, Part I: Detection, Estimation, and Linear Modulation Theory*, John Wiley & Sons, New York, 1968.

H. L. Van Trees, *Optimum Array Processing, Part IV of Detection Estimation, and Modulation Theory*, John Wiley & Sons, New York, 2002.

B. D. Van Veen and K. M. Buckley, "Beamforming: A versatile approach to spatial filtering," *IEEE ASSP Magazine*, pp. 4–24, April 1988.

V. Venkataraman and J. J. Shynk, "A multistage hybrid constant modulus array with constrained adaptation for correlated sources," *IEEE Transactions on Signal Processing*, vol. 55, no. 6, pp. 2509–2519, June 2007.

S. Verdú, *Multiuser Detection*, Cambridge University Press, Cambridge, 1998.

A. C. G. Verdugo Lazo and P. N. Rathie, "On the entropy of continuous probability distributions," *IEEE Transactions on Information Theory*, vol. IT-24, no. 1, pp. 120–122, January 1978.

Y. Viniotis, *Probability and Random Processes for Electrical Engineers*, McGraw-Hill, New York, 1998.

E. Walach and B. Widrow, "The least mean fourth (LMF) adaptive algorithm and its family," *IEEE Transactions on Information Theory*, vol. IT-30, no. 2, pp. 275–283, March 1984.

E. W. Weisstein, *CRC Concise Encyclopedia of Mathematics*, 2nd edition, Chapman & Hall/CRC Press, Boca Raton, FL, 2003.

B. Widrow, K. M. Duvall, R. P. Gooch, and W. C. Newman, "Signal cancellation phenomena in adaptive antennas: Causes and cures," *IEEE Transactions on Antennas and Propagation*, vol. AP-30, no. 3, pp. 469–478, May 1982.

B. Widrow and S. D. Stearns, *Adaptive Signal Processing*, Prentice-Hall, Englewood Cliffs, NJ 1985.

B. Widrow and E. Walach, *Adaptive Inverse Control*, Prentice-Hall, Upper Saddle River, NJ, 1996.

B. Widrow and M. A. Lehr, "30 years of adaptive neural networks: Perceptron, madaline, and backpropagation," *Proceedings of the IEEE*, vol. 78, no. 9, pp. 1415–1442, September 1990.

B. Widrow and I. Kollár, *Quantization Noise: Roundoff Error in Digital Computation, Signal Processing, Control, and Communications*, Cambridge University Press, Cambridge, 2008.

N. Wiener, *Extrapolation, Interpolation, and Smoothing of Stationary Time Series with Engineering Applications*, MIT Press, Cambridge, MA, 1949.

N. Wiener, *Time Series*, MIT Press, Cambridge, MA, 1964.

D. Williams, *Probability with Martingales*, Cambridge University Press, Cambridge, 1991.

R. H. Williams, *Electrical Engineering Probability*, West Publishing, St. Paul, MN, 1991.

G. A. Williamson, "Implementations of adaptive IIR filters with lowest complexity," *IEEE Transactions on Circuits and Systems—II: Analog and Digital Signal Processing*, vol. 44, no. 8, pp. 673–678, August 1997.

E. Wong, *Stochastic Processes in Information and Dynamical Systems*, Robert E. Krieger Publishing Company, Huntington, NY, 1971.

E. Wong and B. Hajek, *Stochastic Processes in Engineering Systems*, Springer-Verlag, New York, 1985.

J. M. Wozencraft and I. M. Jacobs, *Principles of Communication Engineering*, John Wiley & Sons, New York, 1965.

Z. Xu, P. Liu, and X. Wang, "Blind multiuser detection: From MOE to subspace methods," *IEEE Transactions on Signal Processing*, vol. 52, no. 2, pp. 510–524, February 2004.

R. D. Yates and D. J. Goodman, *Probability and Stochastic Processes: A Friendly Introduction for Electrical and Computer Engineers*, John Wiley & Sons, New York, 2005.

M.-H. Yeon and J. J. Shynk, "A stochastic analysis of an iterative semi-blind beamformer for TDMA systems," *Signal Processing*, vol. 89, no. 7, pp. 1347–1358, July 2009.

J. R. Zeidler, "Performance analysis of LMS adaptive prediction filters," *Proceedings of the IEEE*, vol. 78, no. 12, pp. 1781–1806, December 1990.

D. Zwillinger and S. Kokoska, *CRC Standard Probability and Statistics Tables and Formulae*, Chapman & Hall/CRC Press, Boca Raton, FL, 2000.

INDEX

Italicized entries (page numbers 769 and higher) are to be found at www.wiley.com/go/randomprocesses.

Probability, Random Variables, and Random Processes: Theory and Signal Processing Applications, First Edition. John J. Shynk.